Fast Track
OBJECTIVE
ARITHMETIC

COMPLETELY **REVISED EDITION**

Fast Track
OBJECTIVE
ARITHMETIC

RBI & SBI & IBPS PO/Clerk, SSC (10+2)/CGL/CPO/Multitasking, LIC AAO/ADO, CDS I & II, IAS CSAT, States CSAT, CMAT, MAT & Other Management Entrances, Hotel Management, Railways, Paramilitary Forces, State Police Recruitments & All Other Entrances, Recruitments and Aptitude Tests

Rajesh Verma

✹ **arihant**
ARIHANT PUBLICATIONS (INDIA) LIMITED

COMPLETELY **REVISED EDITION**

❋arihant
Arihant Publications (India) Ltd.
All Rights Reserved

❋ © Publisher
No part of this publication may be re-produced, stored in a retrieval system or distributed in any form or by any means, electronic, mechanical, photocopying, recording, scanning, web or otherwise without the written permission of the publisher. Arihant has obtained all the information in this book from the sources believed to be reliable and true. However, Arihant or its editors or authors or illustrators don't take any responsibility for the absolute accuracy of any information published and the damages or loss suffered there upon.

All disputes subject to Meerut (UP) jurisdiction only.

❋ Administrative & Production Offices
Regd. Office
'Ramchhaya' 4577/15, Agarwal Road, Darya Ganj, New Delhi -110002
Tele: 011- 47630600, 43518550; Fax: 011- 23280316

Head Office
Kalindi, TP Nagar, Meerut (UP) - 250002
Tele: 0121-2401479, 2512970, 4004199; Fax: 0121-2401648

❋ Sales & Support Offices
Agra, Ahmedabad, Bengaluru, Bhubaneswar, Bareilly, Chennai, Delhi, Guwahati, Hyderabad, Jaipur, Jhansi, Kolkata, Lucknow, Meerut, Nagpur & Pune

❋ ISBN : 978-93-12149-83-6

❋ Price : ₹ 465.00

PO No. : TXT-59-T045386-2-22

PUBLISHED BY ARIHANT PUBLICATIONS (INDIA) LTD.

For further information about the books published by Arihant log on to www.arihantbooks.com or email to info@arihantbooks.com

COMPLETELY REVISED EDITION

Fast Track
OBJECTIVE ARITHMETIC

TODAY, THERE IS A PLETHORA OF BOOKS AVAILABLE IN THE MARKET ON OBJECTIVE ARITHMETIC WHICH SEEMS TO BE COMPLETE IN THEIR WAY, BUT STILL THEY ARE UNABLE TO FULLY SATISFY THE NEEDS OF THE ASPIRANTS.

LET US KNOW SOME OF THE REASONS

Lack of Understanding the Basic Concepts
Mostly, students face a competitive examination on the base of their knowledge about mathematical rules, formulae and concepts. Inspite of having the knowledge, he lacks behind when he faces questions in the examination. Does he realise this inability? Yes, he does but feels confused and blocked when he is unable to solve them and is left with a sense of grudge that he could solve it. The only reason behind this problem is the understanding of basic concepts. If he would have been clear with them, he could solve any of the questions because as a matter of fact, every question is based on a particular concept which is just twisted in the examinations to judge the overall ability of a student.

Inappropriate Use of Short Tricks
This is the second biggest problem in front of the aspirants. The number of questions asked in the competitive examination is much more than the time assigned for them. This leads the aspirants to use shortcut methods. Although, these methods prove to be beneficial in some cases, but due to time management problems, he gets bound to use these methods irrationally and inappropriately. As a result, he jumbles between all the shortcuts which lead to wrong answers which could have been solved, if he knew when and where to apply the shortcut methods.

Inability to Distinguish Between the Applications of Formulae
We all are aware of the amount of stress and pressure, a competitive examination creates on the mindset of an aspirant. Succumbed to such pressure, an aspirant is unable to decide the appropriate formula to be applied in a particular step. During the crisis of time, such confusion adds to the problems and squeezes in more time and results to an unsatisfactory score.

Keeping in mind all kinds of problems faced by an aspirant in a competitive examination, we have developed this book with profound interest in a stepwise method to encounter all your queries and worries. This book named 'FAST TRACK ARITHMETIC' is worthy to fulfill your expectations and will help you as loyal guide throughout.

OUTSTANDING QUALITIES OF FAST TRACK OBJECTIVE ARITHMETIC

Use of Fundamental Formulae and Method

In this book, all the fundamental formulae and methods have been presented in such a striking yet friendly and systematic manner that just going through them once will give you an effective grasp. They have been presented in such a manner that they will never let you get confused between fast track technique and basic method.

Appropriate Shortcut Methods

An important feature of this book is its shortcut methods or tricks given with the name of 'Fast Track Formulae or Techniques'. Each technique is given with its basic or fundamental method, so that a student can use these tricks according to their desire and save their precious time in exams.

Division of Exercises According to the Difficulty Level

Based on the standard and level of difficulty of various questions, the exercises are divided into two parts i.e. 'Base Level Questions' for relatively easier questions and 'Higher Skill Level Questions' for difficult questions. 'Multi Concept Questions' which requires a use of different concepts in a single question have also been incorporated with important chapters.

Special Emphasise on Geometry, Trigonometry and Mensuration

Now-a-days, questions from geometry, trigonometry and mensuration are asked in large numbers in different exams. So, a large variety and number of questions are provided for these chapters.

Completely Updated with Questions from Recent Exams

This book is incorporated with the questions from all the recent competitive exams, held in year 2017.

This book is a brain child of Mr. Deepesh Jain (Director), Arihant Publications (India) Limited. The entire project has been managed and supervised by Mr. Mahendra Singh Rawat and Mr. Keshav Mohan (Publishing Managers), Arihant Publications (India) Limited. Tarun Sharma (Project Head) and Jai Prakash Chamola have given their best and sincere efforts for the completion and final presentations of the book.

Aas Mohammad Malik and Mazher Chaudhary are to be complemented for very apt designing to the book cover. Ravindra Kumar, Amit Bansal, Sandeep Saini and Mayank Saini have given their expertise in page layout of the book. Also, contribution of Deepak Sharma, Harvinder Singh and Monika Agarwal for this book is very special and is worthy of great applause.

Reader's recommendations will be highly treasured.

With best compliments
Rajesh Verma

CONTENTS

1. **Number System**   1-18
   Numerals · Face and Place Values of the Digits in a Number · Types of Numbers · Operations on Numbers · Divisibility Tests · Unit's Digit of an Expression

2. **Series and Progressions**   19-36
   Series · Types of Series · Types of Questions Asked on Number Series · Progressions

3. **HCF and LCM**   37-60
   Factors · Multiples · Least Common Multiple (LCM) · Highest Common Factor (HCF) · Method to Calculate LCM and HCF of Fractions · Fast Track Techniques to Solve the Questions · Method to Solve Questions Based on Bells

4. **Simple and Decimal Fractions**   61-78
   Simple Fraction · Comparison of Simple Fractions · Operations on Simple Fractions · Decimal Fraction · Operations on Decimal Fractions · Fast Track Formulae to Solve the Questions

5. **Square Root and Cube Root**   79-98
   Square · Square Root · Properties of Squares and Square Roots · Cube · Cube Root · Properties of Cubes and Cube Roots · Fast Track Formulae to Solve the Questions · Fast Track Techniques to Solve the Questions

6. **Indices and Surds**   99-113
   Indices · Surds · Operations on Surds

7. **Simplification**   114-132
   VBODMAS Rule · Some Basic Formulae

8. **Approximation**   133-141
   Basic Rules to Solve the Problems Based on Approximation

9. **Word Problems Based on Numbers**   142-151
   Types of Word Problems Based on Numbers

10. **Average**   152-174
    Properties of Average · Important Formulae Related to Average of Numbers · Fast Track Techniques to Solve the Questions

11. **Percentage** — 175-200
   Fast Track Techniques to Solve the Questions

12. **Profit and Loss** — 201-229
   Fast Track Techniques to Solve the Questions

13. **Discount** — 230-244
   Marked Price (List Price) · Successive Discount · Fast Track Techniques to Solve the Questions

14. **Simple Interest** — 245-267
   Simple Interest (SI) · Instalments · Fast Track Techniques to Solve the Questions

15. **Compound Interest** — 268-285
   Basic Formulae Related to Compound Interest · Instalments · Fast Track Techniques to Solve the Questions

16. **True Discount and Banker's Discount** — 286-294
   True Discount · Banker's Discount · Fast Track Formulae to Solve the Questions

17. **Ratio and Proportion** — 295-316
   Ratio · Comparison of Ratios · Proportion · Fast Track Techniques to Solve the Questions

18. **Mixture or Alligation** — 317-329
   Mixture · Rule of Mixture or Alligation · Fast Track Techniques to Solve the Questions

19. **Partnership** — 330-345
   Types of Partnership · Types of Partners · Fast Track Techniques to Solve the Questions

20. **Unitary Method** — 346-355
   Direct Proportion · Indirect Proportion

21. **Problems Based on Ages** — 356-365
   Important Rules for Problems Based on Ages · Fast Track Techniques to Solve the Questions

22. **Work and Time** — 366-392
   Basic Rules Related to Work and Time · Fast Track Techniques to Solve the Questions

23. **Work and Wages** — 393-402
   Some Important Points · Fast Track Formulae to Solve the Questions

24. **Pipes and Cisterns** — 403-420
   Important Facts Related to Pipes and Cisterns · Fast Track Techniques to Solve the Questions

25. **Speed, Time and Distance**   421-445
   Basic Formulae Related to Speed, Time and Distance · Fast Track Techniques to Solve the Questions

26. **Problems Based on Trains**   446-467
   Basic Rules Related to Problems Based on Trains · Fast Track Techniques to Solve the Questions

27. **Boats and Streams**   468-482
   Fast Track Techniques to Solve the Questions

28. **Races and Games of Skill**   483-492
   Important Terms Related to Races and Games of Skill · Some Facts about Race · Fast Track Techniques to Solve the Questions

29. **Clock and Calendar**   493-507
   Clock · Fast Track Techniques to Solve the Questions · Calender · Day Gain/Loss

30. **Linear Equations**   508-518
   Linear Equations in One Variable · Linear Equations in Two Variables · Consistency for the System of Linear Equations

31. **Quadratic Equations**   519-533
   Methods of Solving Quadratic Equations · Important Points Related to Quadratic Equations · Fast Track Formulae to Solve the Questions

32. **Permutations and Combinations**   534-547
   Factorial · Permutation · Combination · Fundamental Principles of Counting · Fast Track Formulae to Solve the Questions

33. **Probability**   548-564
   Important Terms Related to Probability · Event · Rules/Theorems Related to Probability · Conditional Probability · Types of Questions

34. **Area and Perimeter**   565-606
   Area · Perimeter · Triangle · Types of Triangles · Properties of Triangles · Quadrilateral · Types of Quadrilaterals · Regular Polygon · Circle · Semi-circle · Circular Ring · Fast Track Techniques to Solve the Questions

35. **Volume and Surface Area**   607-643
   Volume · Surface Area · Cube · Cuboid · Room · Box · Right Circular Cylinder · Solid Cylinder · Hollow Cylinder · Right

Circular Cone · Frustum of Right Circular Cone · Sphere · Hollow Sphere or Spherical Shell · Hemisphere · Prism · Pyramid · Fast Track Techniques to Solve the Questions

36. **Geometry**   644-686
Basic Definitions Related to Geometry · Angle · Triangle · Important Terms Related to Triangles · Quadrilateral · Cyclic Quadrilateral · Polygons · Circle

37. **Coordinate Geometry**   687-700
Rectangular Coordinate Axes · Quadrants · Basic Points Related to Straight Lines

38. **Trigonometry**   701-731
Trigonometric Ratios · Trigonometric Identities · Sign of Trigonometric Functions · Trigonometric Ratios of Compound Angles

39. **Height and Distance**   732-744
Line of Sight · Horizontal Line · Angle of Elevation · Angle of Depression

40. **Set Theory**   745-755
Sets · Representation of Sets · Types of Sets · Subset · Power Set · Universal Set · Venn Diagram · Operations on Sets · Important Results Based on Sets

41. **Statistics**   756-769
Collection of Data · Presentation of Data · Frequency Distribution · Cumulative Frequency · Mean or Arithmetic Mean · Properties of Mean · Median · Mode · Relation between Mean, Median and Mode

42. **Data Table**   770-788

43. **Pie Chart**   789-801

44. **Bar Chart**   802-816

45. **Line Graph**   817-829

46. **Mixed Graph**   830-841

47. **Data Sufficiency**   842-852

- Fast Track Practice Sets (1-5)   853-870

Chapter 01

Number System

A system in which we study different types of numbers, their relationship and rules govern in them is called **number system**.
In the Hindu-Arabic system, we use the symbols 0, 1, 2, 3, 4, 5, 6, 7, 8 and 9. These symbols are called **digits**. Out of these ten digits, 0 is called an **insignificant digit**, whereas the others are called **significant digits**.

Numerals

A mathematical symbol representing a number in a systematic manner is called a numeral represented by a set of digits.

How to Write a Number?

To write a number, we put digits from right to left at the places designated as unit's, ten's, hundred's, thousand's, ten thousand's, lakh's, ten lakh's, crore's, ten crore's.
Let us see how the number 308761436 *is denoted*

Ten crore's	Crore's	Ten lakh's	Lakh's	Ten thousand's	Thousand's	Hundred's	Ten's	Unit's
10^8	10^7	10^6	10^5	10^4	10^3	10^2	10^1	10^0
3	0	8	7	6	1	4	3	6

It is read as 'Thirty crore eighty seven lakh sixty one thousand four hundred thirty six'.

Face and Place Values of the Digits in a Number

Face Value

In a numeral, the face value of a digit is the value of the digit itself irrespective of its place in the numeral.
For example In the numeral 486729, the face value of 8 is 8, the face value of 7 is 7, the face value of 6 is 6, the face value of 4 is 4 and so on.

Place (Local) Value

In a numeral, the place value of a digit changes according to the change of its place.

In a number,
Place value of unit's digit = (Digit at one's place) $\times 10^0$
Place value of ten's digit = (Digit at ten's place) $\times 10^1$
Place value of hundred's digit = (Digit at hundred's place) $\times 10^2$ and so on.
The place value of number is also called the local value of the number.
For example In the number 28397,
Place value of 8 = Place value of thousand's digit
= (Digit at thousand's place) $\times 10^3 = 8 \times 10^3 = 8000$

Types of Numbers

There are various types of numbers as follow

1. Natural Numbers

Natural numbers are counting numbers and these are denoted by N,
i.e. $N = \{1, 2, 3, ...\}$.
* All natural numbers are positive.
* Zero is not a natural number, therefore 1 is the smallest natural number.

2. Whole Numbers

All natural numbers and zero form the set of whole numbers and these are denoted by W,
i.e. $W = \{0, 1, 2, 3, ...\}$.
* Zero is the smallest whole number.
* Whole numbers are also called as non-negative integers.

3. Integers

Whole numbers and negative numbers form the set of integers and these are denoted by I,
i.e. $I = \{..., -4, -3, -2, -1, 0, 1, 2, 3, 4, ...\}$.
Integers are of following two types

(i) **Positive Integers** Natural numbers are called as positive integers and these are denoted by I^+,
i.e. $I^+ = \{1, 2, 3, 4, ...\}$

(ii) **Negative Integers** Negative of natural numbers are called as negative integers and these are denoted by I^-,
i.e. $I^- = \{-1, -2, -3, -4, ...\}$.
* '0' is neither +ve nor –ve integer.

4. Even Numbers

A counting number, which is divisible by 2, is called an even number.
For example 2, 4, 6, 8, 10, 12, ... etc.
* The unit's place of every even number will be 0, 2, 4, 6 or 8.

5. Odd Numbers

A counting number, which is not divisible by 2, is known as an odd number.
For example 1, 3, 5, 7, 9, 11, 13, 15, 17, 19, ... etc.
* The unit's place of every odd number will be 1, 3, 5, 7 or 9.

6. Prime Numbers

A counting number is called a prime number when it is exactly divisible by only 1 and itself.

For example   2, 3, 5, 7, 11, 13, ... etc.

* 2 is the only even number which is prime.
* A prime number is always greater than 1.
* 1 is not a prime number, therefore the lowest odd prime number is 3.
* Every prime number greater than 3 can be represented by $6n + 1$, where n is integer.

How to test a number is prime or not?

If P = Given number, then
(i) find the whole number x such that $x > \sqrt{P}$.
(ii) take all the prime numbers less than or equal to x.
(iii) if none of these divides P exactly, then P is prime; otherwise P is non-prime.

For example   Let $P = 193$, clearly $14 > \sqrt{193}$
Prime numbers upto 14 are 2, 3, 5, 7, 11, 13.
No one of these divides 193 exactly.
Hence, 193 is a prime number.

---

7. Composite Numbers

Composite numbers are non-prime natural numbers. They must have atleast one factor apart from 1 and itself.

For example   4, 6, 8, 9, etc.

* Composite numbers can be both odd and even.
* 1 is neither a prime number nor a composite number.

8. Coprimes

Two natural numbers are said to be coprimes, if their common divisor is 1.

For example   (7, 9), (15, 16), etc.

* Coprime numbers may or may not be prime.
* Every pair of consecutive numbers is coprime.

9. Rational Numbers

A number that can be expressed in the form of p/q, is called a rational number, where p, q are integers and $q \neq 0$.

For example $\frac{3}{5}, \frac{7}{9}, \frac{8}{9}, \frac{13}{15}$, etc.

10. Irrational Numbers

The numbers that cannot be expressed in the form of p/q, are called irrational numbers, where p, q are integers and $q \neq 0$.

For example $\sqrt{2}, \sqrt{3}, \sqrt{7}, \sqrt{11}$, etc.

* π is an irrational number as $22/7$ is not the actual value of π but it is its nearest value.
* Non-periodic infinite decimal fractions are called irrational numbers.

11. Real Numbers

Real numbers include both rational and irrational numbers. They are denoted by R.

For example $\frac{7}{9}, \sqrt{2}, \sqrt{5}, \pi, \frac{8}{9}$, etc.

Operations on Numbers

There are four operations on numbers as given below

Addition

When two or more numbers are combined together, then it is called addition. Addition is denoted by '+' sign.

For example $24 + 23 + 26 = 73$

Subtraction

When one or more numbers are taken out from a larger number, then it is called subtraction. Subtraction is denoted by '−' sign.
For example $100 - 4 - 13 = 100 - 17 = 83$

Multiplication

Multiplication is repeated addition. When 'a' is multiplied by 'b', then 'a' is added 'b' times or 'b' is added 'a' times. Multiplication is denoted by '×'.
Let us see the following operation of multiplication
If $a = 2$ and $b = 4$, then $2 \times 4 = 8$ or $(2 + 2 + 2 + 2) = 8$
Here, 'a' is added 'b' times or in other words, 2 is added 4 times.
Similarly, $4 \times 2 = 8$ or $(4 + 4) = 8$
In this case, 'b' is added 'a' times or in other words, 4 is added 2 times.

Division

Division is repeated subtraction. If D and d are two numbers, then D/d is called the operation of division, where D is the **dividend** and d is the **divisor**. A number which tells how many times a divisor (d) exists in dividend (D) is called the **quotient** (Q).

If dividend (D) is not a multiple of divisor (d), then D is not exactly divisible by d and in this case, **remainder** (R) is obtained.
Let us see the following operation of division

Let $D = 17$ and $d = 3$, then $\dfrac{D}{d} = \dfrac{17}{3} = 5\dfrac{2}{3}$

Here, 5 = Quotient (Q), 3 = Divisor (d) and 2 = Remainder (R)
We see that, $(3 \times 5) + 2 = 17$
Hence, we can write a formula

$$\text{Dividend} = (\text{Divisor} \times \text{Quotient}) + \text{Remainder}$$

Divisibility Tests

There are following rules to test the divisibility by different numbers

Divisibility by 2 When the last digit of a number is either 0 or even, then the number is divisible by 2.
For example 12, 86, 472, 520, 1000 etc., are divisible by 2.

Divisibility by 3 When the sum of the digits of a number is divisible by 3, then the number is divisible by 3.
For example
(i) 1233 $1 + 2 + 3 + 3 = 9$, which is divisible by 3, so 1233 must be divisible by 3.
(ii) 156 $1 + 5 + 6 = 12$, which is divisible by 3, so 156 must be divisible by 3.

Divisibility by 4 When the number made by last two digits of a number is divisible by 4, then that particular number is divisible by 4. Apart from this, the number having two or more zeros at the end, is also divisible by 4.

For example
(i) 6428 is divisible by 4 as the number made by its last two digits, i.e. 28 is divisible by 4.
(ii) The numbers 4300, 153000, 9530000 etc., are divisible by 4 as they have two or more zeros at the end.

Divisibility by 5 — Numbers having 0 or 5 at the end are divisible by 5.
For example 45, 4350, 135, 14850 etc., are divisible by 5 as they have 0 or 5 at the end.

Divisibility by 6 — When a number is divisible by both 3 and 2, then that particular number is divisible by 6 also.
For example 18, 36, 720, 1440 etc., are divisible by 6 as they are divisible by both 3 and 2.

Divisibility by 7 — A number is divisible by 7 when the difference between twice the digit at ones place and the number formed by other digits is either zero or a multiple of 7.
For example 658 is divisible by 7, because
$$65 - 2 \times 8 = 65 - 16 = 49.$$
As 49 is divisible by 7, so the number 658 is also divisible by 7.

Divisibility by 8 — When the number made by last three digits of a number is divisible by 8, then the number is also divisible by 8. Apart from this, if the last three or more digits of a number are zero, then the number is divisible by 8.
For example
(i) 2256   As 256 (the last three digits of 2256) is divisible by 8, therefore 2256 is also divisible by 8.
(ii) 4362000   As 4362000 has three zeros at the end, therefore it will definitely be divisible by 8.

Divisibility by 9 — When the sum of all the digits of a number is divisible by 9, then the number is also divisible by 9.
For example
(i) 936819 $9 + 3 + 6 + 8 + 1 + 9 = 36$, which is divisible by 9. Therefore, 936819 is also divisible by 9.
(ii) 4356 $4 + 3 + 5 + 6 = 18$, which is divisible by 9. Therefore, 4356 is also divisible by 9.

Divisibility by 10 — When a number ends with zero, then it is divisible by 10.
For example 20, 40, 150, 123450, 478970 etc., are divisible by 10, as they all end with zero.

Divisibility by 11 — When the sum of digits at odd and even places are equal or differ by a number divisible by 11, then the number is also divisible by 11.
For example
(i) 2865423   Let us see
Sum of digits at odd places $(A) = 2 + 6 + 4 + 3 = 15$
Sum of digits at even places $(B) = 8 + 5 + 2 = 15$
∵ $A = B$
Hence, 2865423 is divisible by 11.

(ii) 217382   Let us see
Sum of digits at odd places $(A) = 2 + 7 + 8 = 17$
Sum of digits at even places $(B) = 1 + 3 + 2 = 6$
∵ $A - B = 17 - 6 = 11$
Hence, 217382 is divisible by 11.

Divisibility by 12 A number which is divisible by both 4 and 3, is also divisible by 12.
For example 2244 is divisible by both 3 and 4. Therefore, it is divisible by 12 also.

Divisibility by 14 A number which is divisible by both 7 and 2, is also divisible by 14.
For example 1232 is divisible by both 7 and 2. Therefore, it is divisible by 14 also.

Divisibility by 15 A number which is divisible by both 5 and 3, is also divisible by 15.
For example 1275 is divisible by both 5 and 3. Therefore, it is divisible by 15 also.

Divisibility by 16 A number is divisible by 16 when the number made by its last 4 digits is divisible by 16.
For example 126304 is divisible by 16 as the number made by its last 4 digits, i.e. 6304 is divisible by 16.

Divisibility by 18 A number is divisible by 18 when it is even and divisible by 9.
For example 936198 is divisible by 18 as it is even and divisible by 9.

Divisibility by 25 A number is divisible by 25 when its last 2 digits are either zero or divisible by 25.
For example 500, 1275, 13550 are divisible by 25 as last 2 digits of these numbers are either zero or divisible by 25.

Divisibility by 125 A number is divisible by 125 when the number made by its last 3 digits is divisible by 125.
For example 630125 is divisible by 125 as the number made by its last 3 digits is divisible by 125.

To Find a Number Completely Divisible by Given Number

Consider a number x, which when divided by d, gives a quotient q and leaves a remainder r. Then,

$$d \overline{\smash{)}x} (q$$
$$\phantom{d \overline{)x} (} r$$

To find the number which is completely divisible by d such that remainder r is zero, follow the example given below.

Ex. 1 Find the number, which on (i) addition (ii) subtraction from the number 5029 is completely divisible by 17.

Sol. Dividing 5029 by 17, we find remainder = 14
(i) Required number which on adding the given number is completely divisible by 17 = Divisor − Remainder = 17 − 14 = 3.
(ii) Required number on subtraction of which the given number is completely divisible by 17 = Remainder = 14.

$$17 \overline{\smash{)}5029} (295$$
$$\underline{34}$$
$$162$$
$$\underline{153}$$
$$99$$
$$\underline{85}$$
$$14$$

Ex. 2 What least number must be added to 1056, so that the sum is completely divisible by 23? [IB 2016]

Sol. Dividing 1056 by 23, we find remainder = 21
∴ Least number that must be added = 23 – 21 = 2

```
    23)1056(45
        92
        ---
        136
        115
        ---
         21
```

Unit's Digit of an Expression

Given expression can be of following two types

1. When Number is Given in the Form of Product of Numbers

To find the unit's digit in the product of two or more numbers, we take unit's digit of every numbers and then multiply them. Then, the unit's digit of the resultant product is the unit's digit of the product of original numbers.

For example $207 \times 781 \times 39 \times 94$
Taking unit's digit of every number and then multiplying them
$$= 7 \times 1 \times 9 \times 4 = 7 \times 36$$
Again, taking unit's digits and then multiplying $= 7 \times 6 = 42$
∴ Unit's digit for $207 \times 781 \times 39 \times 94$ is 2.

2. When Number is Given in the Form of Index

Suppose that the number is of the form a^b. Then, following cases arise

Case I If b is a multiple of 4.
(i) If a is an even number, i.e. 2, 4, 6 or 8, then unit's digit is 6.
(ii) If a is an odd number, i.e. 1, 3, 7 or 9, then unit's digit is 1.

For example
(i) Unit's digit in $(4137)^{756}$ is 1 as 756 divisible by 4 and 4137 is odd.
(ii) Unit's digit in $(2138)^{392}$ is 6 as 392 is divisible by 4 and 2138 is even.

Case II If b is not a multiple of 4.
Let r be the remainder when b is divided by 4, i.e. $b = 4q + r$, then unit's place digit of a^b is equal to unit's place digit of a^r.

For example Consider 7^{105}. Here, 105 is not divisible by 4, so when 105 is divided by 4, we get remainder as 1.
∴ Unit's digit in 7^{105} = Unit's digit in $7^1 = 7$

Case III If the unit's digit of a is 0, 1, 5 or 6, then the resultant unit's digit of a^b remains same.

For example (i) Unit's digit of $(576)^{1151} = 6$ (ii) Unit's digit of $(155)^{120} = 5$
(iii) Unit's digit of $(191)^{19} = 1$ (iv) Unit's digit of $(900)^{51} = 0$

Case IV If unit's digit of a is 9 and the power of a is even, then unit's digit will be 1 and if the power of a is odd, then unit's digit will be 9.

For example

(i) $(539)^{140}$ Since, the power is even and unit's digit is 9.
   ∴ Unit's digit in $(539)^{140} = 1$

(ii) $(539)^{141}$ Since, the power is odd and unit's digit is 9.
   ∴ Unit's digit in $(539)^{141} = 9$

Ex. 3 What is the unit's digit of 7^{139}? [CDS 2016 (II)]

Sol. Here, 139 is not a multiple of 4
∴ $139 = 4 \times 34 + 3$
Hence, unit's digit in 7^{139} = unit's digit in $7^3 = 3$

Ex. 4 What is the units digit of $6^{15} - 7^4 - 9^3$?

Sol. Unit's digit in 6^{15} = Unit's digit in $6^{4 \times 3 + 3}$ = Unit's digit in $6^3 = 6$
Unit's digit in $7^4 = 1$ [∵ 4 is a multiple of 4 and 7 is odd]
Unit's digit in $9^3 = 9$
∴ Unit's digit in $6^{15} - 7^4 - 9^3 = 6 - 1 - 9 = 5 - 9 = 6$

✦ We are not subtracting 9 from 5 instead. We take carry over from the ten's place and then subtract 9 from 15. Hence, unit's digit comes out to be 6.

Some Important Facts

✦ Square of every even number is an even number while square of every odd number is an odd number.
✦ A number obtained by squaring a number does not have 2, 3, 7 or 8 at its unit place.
✦ Sum of first n natural numbers = $\dfrac{n(n+1)}{2}$
✦ Sum of first n odd numbers = n^2
✦ Sum of first n even numbers = $n(n+1)$
✦ Sum of squares of first n natural numbers = $\dfrac{n(n+1)(2n+1)}{6}$
✦ Sum of cubes of first n natural numbers = $\left[\dfrac{n(n+1)}{2}\right]^2$
✦ There are 15 prime numbers between 1 and 50, and 10 prime numbers between 50 and 100.
✦ If p divides q and r, then p divides their sum and difference also.
For example 4 divides 12 and 20, then $20 + 12 = 32$ and $20 - 12 = 8$ are also divisible by 4.
✦ For any natural number n, $(n^3 - n)$ is divisible by 6.
✦ The product of three consecutive natural numbers is always divisible by 6.
✦ $(x^m - a^m)$ is divisible by $(x - a)$ for all values of m.
✦ $(x^m - a^m)$ is divisible by $(x + a)$ for even values of m.
✦ $(x^m + a^m)$ is divisible by $(x + a)$ for odd values of m.
✦ Number of prime factors of $a^p\ b^q\ c^r\ d^s$ is $p + q + r + s$, where a, b, c and d are prime numbers.

Ex. 5 Find the sum of first 12 multiples of 9.

Sol. Since, first 12 multiples of 9 are 9, 18, 27, ..., 108, i.e. 9(1, 2, 3, ..., 12).
Here, $n = 12$
∴ Required sum = $9\left[\dfrac{n(n+1)}{2}\right] = 9\left[\dfrac{12(12+1)}{2}\right] = \dfrac{9 \times 12 \times 13}{2} = 702$

Ex. 6 Find the sum of first 37 odd numbers.

Sol. Here, $n = 37$
∴ Required sum = $37^2 = 1369$ [∵ sum of first n odd numbers = n^2]

Multi Concept Questions

1. If n is any odd number greater than 1, then $n(n^2 - 1)$ is always divisible by
   - (a) 96
   - (b) 48
   - (c) 24
   - (d) None of these

   ➥ (c) On solving the question by taking two odd numbers greater than 1, i.e. 3 and 5, then
   $$n(n^2 - 1) = 3(9 - 1) = 24 \quad \text{[for } n = 3\text{]}$$
   and $\quad n(n^2 - 1) = 5(25 - 1) = 120 \quad \text{[for } n = 5\text{]}$
   By using the option, we find that both the numbers are divisible by 24.

2. $7^{6n} - 6^{6n}$, where n is an integer greater than 0, is divisible by
   - (a) 13
   - (b) 127
   - (c) 556
   - (d) None of these

   ➥ (b) $7^{6n} - 6^{6n} = 7^6 - 6^6$ [for $n = 1$]
   $$= (7^3)^2 - (6^3)^2$$
   $$= (7^3 - 6^3)(7^3 + 6^3) \quad [\because a^2 - b^2 = (a + b)(a - b)]$$
   $$= (343 - 216)(343 + 216)$$
   $$= 127 \times 559$$
   Hence, $7^{6n} - 6^{6n}$ is divisible by 127.

3. Find the remainder of $\dfrac{17^{18^{19^{20 \cdots \infty}}}}{8}$.
   - (a) 2
   - (b) 3
   - (c) 4
   - (d) 1

   ➥ (d) $\dfrac{17^{18^{19^{20 \cdots \infty}}}}{8} = \text{Remainder} \left\{ \dfrac{(8 \times 2 + 1)^{18^{19^{20 \cdots \infty}}}}{8} \right\}$

   $= \text{Remainder} \left\{ \dfrac{(1)^{18^{19^{20 \cdots \infty}}}}{8} \right\} = \text{Remainder} \left\{ \dfrac{1}{8} \right\}$

   $\therefore$ Remainder = 1

Fast Track Practice

Exercise 1 Base Level Questions

1. Find the sum of the face values of 9 and 6 in 907364. [Hotel Mgmt. 2007]
   (a) 15 (b) 20 (c) 9 (d) 18

2. Find the sum of place and face values of 8 in 43836. [Hotel Mgmt. 2008]
   (a) 88 (b) 808 (c) 880 (d) 888

3. The sum of place values of 2 in 2424 is [CTET 2012]
   (a) 4 (b) 220 (c) 2002 (d) 2020

4. How many numbers are there between 99 and 1000 such that the digit 8 occupies the units place? [UPSC CSAT 2017]
   (a) 64 (b) 80 (c) 90 (d) 104

5. The product of any number and the 1st whole number is
   (a) 0 (b) 2 (c) 1 (d) –1
   (e) None of these

6. Which of the following is a prime number?
   (a) 35 (b) 53 (c) 88 (d) 90
   (e) None of these

7. A rational number is expressed as ......, where p and q are integers and $q \neq 0$.
   (a) pq (b) $p+q$ (c) $p-q$ (d) p/q
   (e) None of these

8. The number of all prime numbers less than 40 is
   (a) 15 (b) 18 (c) 17 (d) 12
   (e) None of these

9. The number of prime numbers, which are less than 100, is [CDS 2017 (I)]
   (a) 24 (b) 25 (c) 26 (d) 27

10. The pair of numbers which are relatively prime to each other, is [CDS 2012]
    (a) (68, 85) (b) (65, 91)
    (c) (92, 85) (d) (102, 153)

11. 2/3 is a rational number, whereas $\sqrt{2}/\sqrt{3}$ is [CL 2013]
    (a) also a rational number
    (b) an irrational number
    (c) not a number
    (d) a natural periodic number

12. If m and n are distinct natural numbers, then which of the following is/are integer(s)?
    I. $\dfrac{m}{n} + \dfrac{n}{m}$
    II. $mn\left(\dfrac{m}{n} + \dfrac{n}{m}\right)(m^2 + n^2)^{-1}$
    III. $\dfrac{mn}{m^2 + n^2}$
    Select the correct answer using the code given below. [CDS 2016 (I)]
    (a) I and II (b) Only II
    (c) II and III (d) Only III

13. When 121012 is divided by 12, the remainder is [CTET 2012]
    (a) 0 (b) 2 (c) 3 (d) 4

14. Find the quotient when 445 is divided by 5.
    (a) 78 (b) 48 (c) 79 (d) 89
    (e) None of these

15. Find the dividend when divisor is 13, quotient is 30 and remainder is 12.
    (a) 402 (b) 543 (c) 436 (d) 455
    (e) None of these

16. What is the remainder in the expression $29\dfrac{18}{26}$?
    (a) 29 (b) 26 (c) 18 (d) 0
    (e) None of these

17. Find the dividend from the expression $41\dfrac{4}{19}$.
    (a) 783 (b) 800 (c) 893 (d) 387
    (e) None of these

18. In a division sum, the divisor is ten times the quotient and five times the remainder. If the remainder is 46, then find the dividend.
    (a) 5388 (b) 5343 (c) 5336 (d) 5391
    (e) None of these

19. A number when divided by 627 leaves a remainder 43. By dividing the same number by 19, the remainder will be [SBI Clerk (Pre) 2016]
    (a) 24 (b) 43 (c) 13 (d) 5
    (e) 7

Number System / 11

20. A number when divided by 361 gives a remainder 47. If the same number is divided by 19, then the remainder obtained is [SSC CGL (Mains) 2015]
    (a) 8   (b) 1   (c) 3   (d) 9

21. The number 58129745812974 is divisible by [CDS 2012]
    (a) 11   (b) 9
    (c) 4   (d) None of these

22. What least number must be added to 1057 to get a number exactly divisible by 23?
    (a) 1   (b) 3   (c) 2   (d) 4
    (e) None of these

23. The smallest number, which should be added to 756896 so as to obtain a multiple of 11, is [SSC CGL (Mains) 2016]
    (a) 1   (b) 2   (c) 3   (d) 5

24. The digit in the unit's place of the product $81 \times 82 \times 83 \times 84 \times \ldots \times 99$ is [CDS 2015 (I)]
    (a) 0   (b) 4   (c) 6   (d) 8

25. Find the unit's digit in the product of (4326×5321). [Hotel Mgmt. 2010]
    (a) 6   (b) 8   (c) 1   (d) 3

26. The unit's digit in the product $(2467)^{153} \times (341)^{72}$ is [SSC CGL (Mains) 2015]
    (a) 9   (b) 3   (c) 1   (d) 7

27. The sum of the natural numbers which are divisors of 100, is
    (a) 116   (b) 117   (c) 21   (d) 217

28. Find the sum of the cubes of first 15 natural numbers.
    (a) 15400   (b) 14400
    (c) 16800   (d) 13300

29. Find the sum of first 37 odd numbers. [Hotel Mgmt. 2010]
    (a) 1369   (b) 1295   (c) 1388   (d) 1875

30. What is the difference between the sum of the cubes and that of squares of first ten natural numbers? [CDS 2016 (II)]
    (a) 2200   (b) 2640
    (c) 3820   (d) 4130

Exercise ❷ Higher Skill Level Questions

1. The sum of first 15 multiples of 8 is [CLAT 2013]
   (a) 960   (b) 660
   (c) 1200   (d) 1060

2. The seven-digit number $876p37q$ is divisible by 225. The values of p and q can be respectively [CDS 2015 (II)]
   (a) 9, 0   (b) 0, 0
   (c) 0, 9   (d) 9, 5

3. In a question on division with zero remainder, a candidate took 12 as divisor instead of 21. The quotient obtained by him was 35. Find the correct quotient.
   (a) 10   (b) 12
   (c) 20   (d) 15
   (e) None of these

4. How many numbers between −11 and 11 are multiples of 2 or 3? [CDS 2012]
   (a) 11   (b) 14
   (c) 15   (d) None of these

5. The sum of four consecutive even numbers is 107 more than the sum of three consecutive odd numbers. If the sum of smallest odd number and the smallest even number is 55, then what is the smallest even number? [SBI Clerk (Pre) 2016]
   (a) 36   (b) 40   (c) 32   (d) 38
   (e) 34

6. The sum of the four consecutive even numbers is 284. What would be the smallest number? [Bank PO 2010]
   (a) 72   (b) 74   (c) 68   (d) 66
   (e) None of these

7. When 1/7 of a number is subtracted from the number itself, it gives the same value as the sum of all the angles of a triangle. What is the number? [Bank PO 2010]
   (a) 224   (b) 210
   (c) 140   (d) 350
   (e) 187

8. The sum of the digits of a two-digit number is 14 and the difference between the two digits of the number is 2. What is the product of the two digits of the two-digit number? [Bank Clerk 2009]
   (a) 56   (b) 48   (c) 45
   (d) Cannot be determined
   (e) None of the above

12 / Fast Track Objective Arithmetic

9. There are certain 2-digit numbers. The difference between the number and the one obtained on reversing it is always 27. How many such maximum 2-digit numbers are there? **[UPSC CSAT 2017]**
   (a) 3 (b) 4
   (c) 5 (d) None of these

10. A number X is successively divided by 11, 4 and 3 leaving remainder 6, 2 and 1, respectively. X when divided successively by 3, 4 and 11 leaves remainder **[CMAT 2015]**
    (a) 6, 0, 0 (b) 0, 1, 2
    (c) 0, 0, 6 (d) 1, 2, 6

11. A number when divided by a divisor leaves a remainder of 24. When twice the original number is divided by the same divisor, the remainder is 11. What is the value of the divisor? **[IB ACIO 2013]**
    (a) 13 (b) 59 (c) 35 (d) 37

12. What are the last two digits of 7^{2008}? **[CLAT 2015]**
    (a) 01 (b) 21 (c) 61 (d) 71

13. A 2-digit number is reversed. The larger of the two numbers is divided by the smaller one. What is the largest possible remainder? **[UPSC CSAT 2017]**
    (a) 95 (b) 27 (c) 36 (d) 45

14. Consider the following statements
    I. To obtain prime numbers less than 121, we have to reject all the multiples of 2, 3, 5 and 7.
    II. Every composite number less than 121 is divisible by a prime number less than 11.
    Which of the statement(s) given above is/are correct? **[CDS 2013]**
    (a) Only I (b) Only II
    (c) Both I and II (d) Neither I nor II

15. Consider the following statements
    I. 7710312401 is divisible by 11.
    II. 173 is a prime number.
    Which of the statement(s) given above is/are correct? **[CDS 2013]**
    (a) Only I (b) Only II
    (c) Both I and II (d) Neither I nor II

16. The digit in the unit's place of the resulting number of the expression $(234)^{100} + (234)^{101}$, is **[CDS 2015 (II)]**
    (a) 6 (b) 4 (c) 2 (d) 0

17. Consider the following statements
    I. Of two consecutive integers, one is even.

II. Square of an odd integer is of the form $8n + 1$.
Which of the above statements is/are correct? **[CDS 2017 (I)]**
(a) Only I (b) Only II
(c) Both I and II (d) Neither I nor II

18. A number when divided by 7 leaves a remainder 3 and the resulting quotient, when divided by 11 leaves a remainder 6. If the same number when divided by 11 leaves a remainder m and the resulting quotient when divided by 7 leaves a remainder n. What are the values of m and n, respectively? **[CDS 2015 (II)]**
    (a) 1 and 4 (b) 4 and 1
    (c) 3 and 6 (d) 6 and 3

19. If a mobile has 5 digits password, then how many different passwords can be set, so that it must be divisible by 4? **[NIFT UG 2015]**
    (a) 25000 (b) $10^5 - 10^2$
    (c) $5! - 4!$ (d) 24000

20. A common factor of $(41^{43} + 43^{43})$ and $(41^{41} + 43^{41})$ is
    (a) $(43 - 41)$ (b) $(41^{41} + 43^{41})$
    (c) $(41^{43} + 43^{43})$ (d) $(41 + 43)$
    (e) None of these

21. What is the remainder when 4^{1000} is divisible by 7? **[CDS 2014]**
    (a) 1 (b) 2
    (c) 4 (d) None of these

22. The remainder when $9^{19} + 6$ is divided by 8, is **[SSC CGL (Mains) 2012]**
    (a) 2 (b) 3 (c) 5 (d) 7

23. What will be the remainder when 19^{100} is divided by 20? **[SSC CGL (Mains) 2012]**
    (a) 19 (b) 20 (c) 3 (d) 1

24. $7^{10} - 5^{10}$ is divisible by **[CDS 2016 (I)]**
    (a) 5 (b) 7 (c) 10 (d) 11

25. If $x = (16^3 + 17^3 + 18^3 + 19^3)$, then x divided by 70 leaves a remainder of **[CLAT 2015]**
    (a) 0 (b) 1 (c) 69 (d) 35

26. What is the remainder, when $13^5 + 14^5 + 15^5 + 16^5$ is divided by 29? **[CDS 2016 (II)]**
    (a) 8 (b) 5 (c) 3 (d) 0

27. $2^{122} + 4^{62} + 8^{42} + 4^{64} + 2^{130}$ is divisible by which one of the following integers? **[CDS 2016 (II)]**
    (a) 3 (b) 5 (c) 7 (d) 11

Number System / 13

28. It is given that $(2^{32} + 1)$ is exactly divisible by a certain number. Which of the following is also definitely divisible by the same number?
    (a) $(2^{16} + 1)$ (b) $(2^{16} - 1)$
    (c) 7×2^{13} (d) $(2^{96} + 1)$
    (e) None of these

29. $19^5 + 21^5$ is divisible by   [CDS 2013]
    (a) only 10
    (b) only 20
    (c) Both 10 and 20
    (d) Neither 10 nor 20

30. The number $(6x^2 + 6x)$ for natural number x is always divisible by
    (a) 6 and 12   (b) only 12
    (c) only 6   (d) only 3
    (e) None of these

31. If a is a natural number, then the largest number dividing $(a^3 - a)$ is
    (a) 4   (b) 5   (c) 6   (d) 7
    (e) None of these

32. The largest natural number which divides every natural number of the form $(n^3 - n)(n - 2)$, where n is a natural number greater than 2, is
       [CDS 2015 (II)]
    (a) 6   (b) 12   (c) 24   (d) 48

33. If N, $(N + 2)$ and $(N + 4)$ are prime numbers, then the number of possible solutions for N are   [CDS 2013]
    (a) 1   (b) 2
    (c) 3   (d) None of these

34. The smallest positive prime (say p) such that $2^p - 1$ is not a prime, is
       [CDS 2013]
    (a) 5   (b) 11   (c) 17   (d) 29

35. If b is the largest square divisor of c and a^2 divides c, then which one of the following is correct, where a, b and c are integers?   [CDS 2013]
    (a) b divides a
    (b) a does not divide b
    (c) a divides b
    (d) a and b are coprime

36. $(N^{p-1} - 1)$ is a multiple of p, if N is prime to p and p is a   [CDS 2017 (I)]
    (a) prime number
    (b) rational number
    (c) real number
    (d) composite number

37. If n is a whole number greater than 1, then $n^2(n^2 - 1)$ is always divisible by
       [CDS 2014]
    (a) 12   (b) 24   (c) 48   (d) 60

38. Certain 3-digit numbers have the following characteristics
    I. All the three digits are different.
    II. The number is divisible by 7.
    III. The number on reversing the digits is also divisible by 7.
    How many such 3-digit numbers are there?   [UPSC CSAT 2017]
    (a) 2   (b) 4   (c) 6   (d) 8

39. What is the remainder when the number $(4444)^{4444}$ is divided by 9?
       [CDS 2017 (I)]
    (a) 4   (b) 6   (c) 7   (d) 8

40. If k is a positive integer, then every square integer is of the form   [CDS 2013]
    (a) only $4k$ (b) $4k$ or $4k + 3$
    (c) $4k + 1$ or $4k + 3$ (d) $4k$ or $4k + 1$

41. Every prime number of the form $3k + 1$ can be represented in the form $6m + 1$ (where k, m are integers), when
    (a) k is odd   [CDS 2013]
    (b) k is even
    (c) k can be both odd and even
    (d) No such form is possible

42. Consider the following statements in respect of positive odd integers x and y.
    I. $x^2 + y^2$ is even integer.
    II. $x^2 + y^2$ is divisible by 4.
    Which of the above statement(s) is/are correct?   [CDS 2016 (II)]
    (a) Only I   (b) Only II
    (c) Both I and II   (d) Neither I nor II

43. Which of the following is correct in respect of the number 1729? [CDS 2016 (II)]
    (a) It cannot be written as the sum of the cubes of two positive integers
    (b) It can be written as the sum of the cubes of two positive integers in one way only
    (c) It can be written as the sum of the cubes of two positive integers in two ways only
    (d) It can be written as the sum of the cubes of two positive integers in three ways only

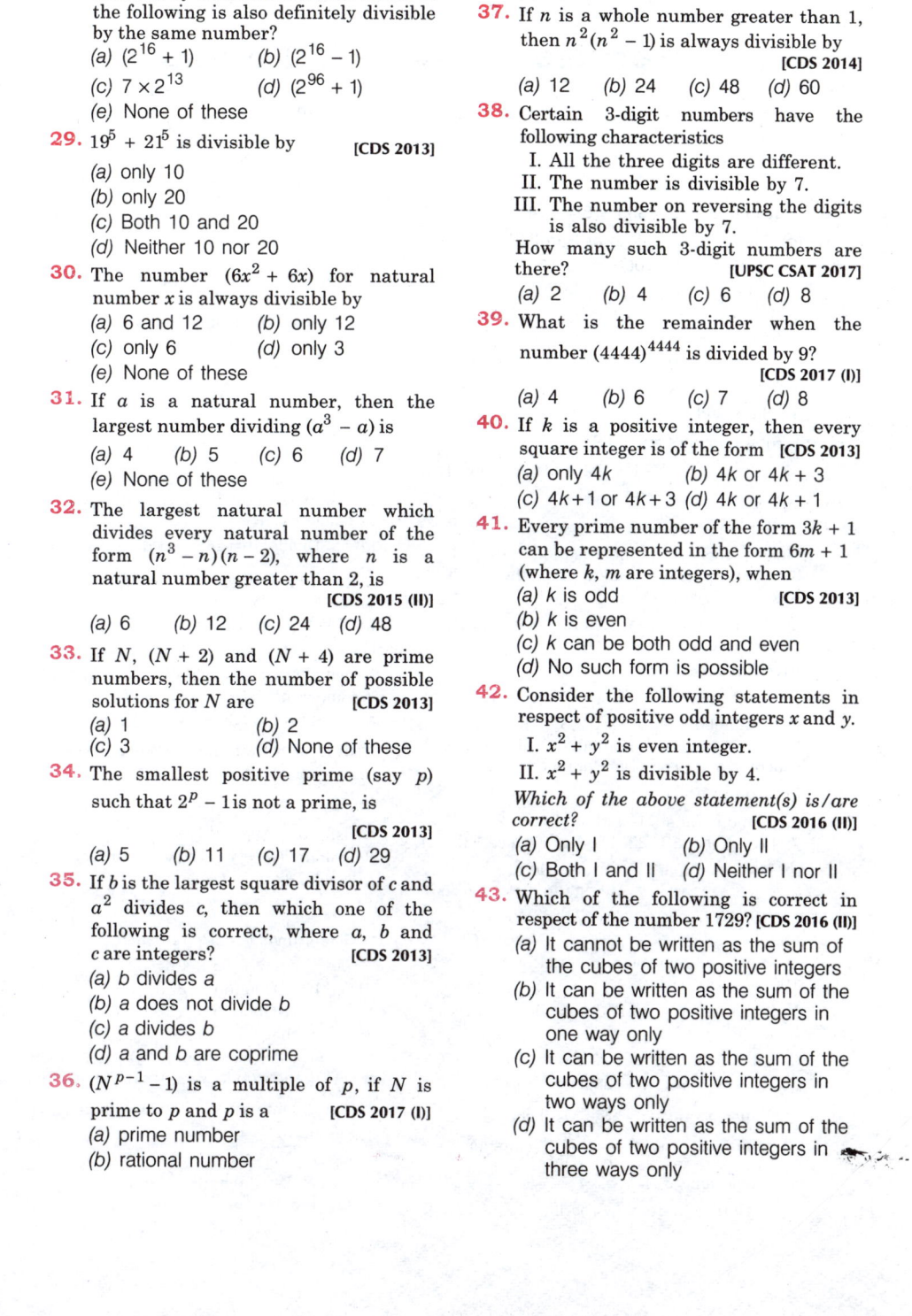

Answer with Solutions

Exercise 1 Base Level Questions

1. (a) The face value is the value of digit itself.
   So, required sum = 9 + 6 = 15

2. (b) Place value of 8 = 800
   and face value of 8 = 8
   ∴ Required sum = 800 + 8 = 808

3. (d) Sum of place values of 2 in 2424
   = 2 × 1000 + 2 × 10
   = 2000 + 20 = 2020

4. (c) Between 99 and 1000, all numbers are 3-digit numbers.
   From 101-200, there are 10 such numbers.
   Similarly, from 201-300, there are 10 such numbers and so on.
   ∴ Total numbers = 10 × 9
   [as there are 9 groups]
   = 90

5. (a) 1st whole number = 0
   Clearly, when any number is multiplied with 0 (the 1st whole number), then the result is 0.

6. (b) 53 has only two factors itself and 1. Hence, it is a prime number.
   Alternate Method
   Here, $8 > \sqrt{53}$
   Prime numbers upto 8 are 2,3,5 and 7. None of these prime numbers divides 53 exactly. So, 53 is a prime number.

7. (d) A proven fact.

8. (d) Prime numbers less than 40 are 2, 3, 5, 7, 11, 13, 17, 19, 23, 29, 31, 37, i.e. 12.

9. (b) Prime numbers less than 100 are 2, 3, 5, 7, 11, 13, 17, 19, 23, 29, 31, 37, 41, 43, 47, 53, 59, 61, 67, 71, 73, 79, 83, 89, 97.
   ∴ There are 25 prime numbers less than 100.

10. (c) 92 and 85 are coprime numbers, because their HCF is 1.

11. (b) $\sqrt{2}$ is irrational and $\sqrt{3}$ is also irrational. So, $\sqrt{2}/\sqrt{3}$ is also irrational.

12. (b) I. If m and n are distinct natural numbers, then $\dfrac{m}{n} + \dfrac{n}{m}$ is integer if and only if $m = n$.
    Hence, Statement I is incorrect.

    II. $mn\left(\dfrac{m}{n} + \dfrac{n}{m}\right)(m^2 + n^2)^{-1}$
    $= mn\left(\dfrac{m^2 + n^2}{mn}\right)\dfrac{1}{m^2 + n^2} = 1$
    Hence, Statement II is correct for all values of m and n.

    III. Now, $\dfrac{mn}{m^2 + n^2}$ is fraction.
    Hence, Statement III is incorrect.

13. (d) 12)121012(10084
       12
       ―――
       101
        96
       ―――
        52
        48
        ――
         4
    Hence, when 121012 is divided by 12, then remainder is 4.

14. (d) ∵ Dividend = 445 and divisor = 5
    ∴ Required quotient = $\dfrac{445}{5}$ = 89

15. (a) Given, divisor $(d) = 13$,
    Quotient $(Q) = 30$,
    Remainder $(R) = 12$ and dividend $(D) = ?$
    We know that, $(D) = d \times Q + R$
    ∴ $D = 13 \times 30 + 12 = 390 + 12 = 402$

16. (c) It is clear from the given expression that remainder is 18.

17. (a) Given, quotient $(Q) = 41$,
    Divisor $(d) = 19$,
    Remainder $(R) = 4$ and dividend $(D) = ?$
    ∴ Dividend $(D) = d \times Q + R = 19 \times 41 + 4$
        = 779 + 4 = 783

18. (c) Given, divisor = 5 × remainder
        = 5 × 46 = 230
    Also, 10 × Quotient = 230
    ∴   Quotient = 23
    We know that,
    Dividend = (Divisor × Quotient) + Remainder
    = (230 × 23) + 46 = 5290 + 46 = 5336

19. (a) Let the number be x and quotient be q.
    Then, $x = 627q + 43$
    ⇒ $x = (33 \times 19) q + 19 + 24$
       = 19(33 q + 1) + 24
    ∴ Required remainder = 24

Number System / 15

20. (d) Let the number $= N = 361k + 47$
When $k = 1$, then $N = 408$
$\therefore \quad \dfrac{N}{19} = \dfrac{408}{19} = 9$ (remainder)

Alternate Method
Let the required number be N.
Then, $N = 361k + 47$
$\Rightarrow \quad N = 361k + 38 + 9 = 19(19k + 2) + 9$
$\therefore$ When N is divided by 19, then remainder is 9.

21. (a) We know that a number is divisible by 11 when the difference between the sum of its digits at even places and sum of digits at odd places is either 0 or the difference is divisible by 11.
Given number is 58129745812974.
$\therefore$ Sum of digits at odd places
$= 4 + 9 + 1 + 5 + 7 + 2 + 8 = 36$
and sum of digits at even places
$= 7 + 2 + 8 + 4 + 9 + 1 + 5 = 36$
$\therefore$ Required difference $= 36 - 36 = 0$
Hence, the number is divisible by 11.

22. (a)
```
   23) 1057 (45
       92
       ---
       137
       115
       ---
        22
```
$\therefore$ Number to be added $= 23 - 22 = 1$

23. (c)
```
   11) 756896 (68808
       66
       --
        96
        88
        --
         88
         88
         --
         0096
           88
           --
            8
```
Here, we see that remainder is 8, which is not divisible by 11. If we add 3 in it, then it would become 11 and it is divisible by 11, i.e. number will be multiple of 11.
$\therefore$ Required smallest number is 3.

24. (a) Product of unit's digits
$= 1 \times 2 \times 3 \times 4 \times 5 \times 6 \times 7 \times 8 \times 9 \times 0... = 0$
$\therefore$ Required digit in the unit's place is 0.

25. (a) Product of unit's digit $= 6 \times 1 = 6$
$\therefore$ Required digit $= 6$

26. (d) Unit's digit of $(2467)^{153} \times (341)^{72}$
$=$ Unit's digit of $7^{153} \times 1^{72}$
$= 7^{(4 \times 38)+1} \times 1$
$= 7^1 \times 1 = 7$

27. (d) Here, divisors of 100 are
1, 2, 4, 5, 10, 20, 25, 50, 100
$\therefore$ Sum $= 1 + 2 + 4 + 5 + 10 + 20 + 25 + 50 + 100 = 217$

28. (b) We know that,
Sum of the cubes of first n natural numbers
$= \left[\dfrac{n(n+1)}{2}\right]^2$
Given, $n = 15$
$\therefore$ Required sum $= \left[\dfrac{15(15+1)}{2}\right]^2$
$= \left[\dfrac{15 \times 16}{2}\right]^2$
$= (15 \times 8)^2 = (120)^2 = 14400$

29. (a) We know that,
Sum of first n odd numbers $= n^2$
Given, $n = 37$
$\therefore$ Required sum $= (37)^2$
$= 37 \times 37 = 1369$

30. (b) Sum of cubes of first 10 natural numbers
$= \left\{\dfrac{10(10+1)}{2}\right\}^2 = 3025$
Sum of squares of first 10 natural numbers
$= \dfrac{10(10+1)(2 \times 10 + 1)}{6} = 385$
$\therefore$ Required difference $= 3025 - 385 = 2640$

Exercise 2 Higher Skill Level Questions

1. (a) First 15 multiples of 8 are
8, 16, 24, ...,120
i.e. 8 (1, 2, 3, 4, ..., 15).
$\therefore$ Required sum $= 8\left[\dfrac{n(n+1)}{2}\right]$
$= 8\left[\dfrac{15(15+1)}{2}\right]$ [here, $n = 15$]
$= 8 \times 15 \times 8 = 960$

2. (d) Seven-digit number $876p37q$ is divisible by 225, if this number is divisible by 9 and 5. This number is divisible by 9, then sum of its digits is divisible by 9.
Now, sum of digits
$= 8 + 7 + 6 + p + 3 + 7 + q$
$= 31 + p + q$
$\therefore \quad p + q = 5$ or $p + q = 14$
q must be 0, 5 as $876p37q$ is divisible by 5.
If $q = 5$, then $p = 0$ or 9.

16 / Fast Track Objective Arithmetic

3. (c) Number $= 35 \times 12 = 420$

$\therefore$ Required correct quotient $= \dfrac{420}{21} = 20$

4. (c) Following are the numbers between -11 and 11 which are multiples of 2 or 3.
$-10, -9, -8, -6, -4, -3, -2, 0, 2, 3, 4, 6, 8, 9, 10$

$\therefore$ The number of multiples of 2 or 3, between -11 and 11 is 15.

5. (d) Let the four consecutive even numbers be $x, x+2, x+4$ and $x+6$, respectively.
Again, let the three consecutive odd numbers be $y, y+2$ and $y+4$, respectively. Then,
$x + x + 2 + x + 4 + x + 6$
$\qquad\qquad = 107 + y + y + 2 + y + 4$
$\Rightarrow \quad 4x + 12 = 107 + 3y + 6$
$\Rightarrow \quad 4x - 3y = 101$ ...(i)
Also, $\quad x + y = 55$ ...(ii)
On solving Eqs. (i) and (ii), we get
Smallest even number $= 38$.

6. (c) Let the four consecutive even numbers be $x, x+2, x+4$ and $x+6$.
According to the question,
$x + x + 2 + x + 4 + x + 6 = 284$
$\Rightarrow 4x + 12 = 284 \Rightarrow 4x = 284 - 12 = 272$
$\therefore \quad x = \dfrac{272}{4} = 68$

7. (b) Let the number be x.
According to the question,
$x - \dfrac{x}{7} = 180 \Rightarrow \dfrac{6x}{7} = 180$
$\therefore \quad x = \dfrac{180 \times 7}{6} = 210$

8. (b) Let the ten's digit be x and unit's digit be y. Then, two-digit number $= 10x + y$
[where, $x > y$]
According to the question,
$\qquad x + y = 14$ ...(i)
and $\qquad x - y = 2$ ...(ii)
On solving Eqs. (i) and (ii), we get
$\qquad x = 8$ and $y = 6$
$\therefore$ Required product $= 8 \times 6 = 48$

9. (d) Let the numbers be of the form $10x + y$.
Then, $10x + y - (10y + x) = 9x - 9y$
$\qquad\qquad\qquad\qquad = 9(x - y)$
Now, difference is 27 when $x - y = 3$.
So, required numbers are 14, 25, 36, 47, 58, 69, 41, 52, 63, 74, 85 and 96.

10. (d) When the number X is divided by 3, the remainder will be 1, when X is divided by 4, the remainder will be 2 and when the number X is divided by 11, the remainder will be 6.

11. (d) Let the divisor be x and quotient be y.
Then, number $= xy + 24$

Twice the number $= 2xy + 48$
Now, $2xy$ is completely divisible by x.
On dividing 48 by x, remainder is 11.
$\therefore \qquad x = 48 - 11 = 37$

12. (a) Given, $(7)^{2008} = (7)^{4 \times 502} = (24\underline{01})^{502}$
Hence, the last two digits will be 01.

13. (d) Consider the numbers 19, 29, 39, 49 and 59. Reverse the numbers and find the largest possible remainder which comes out to be 45 using number 49.

Note We have taken these numbers because after 59, the remainder starts decreasing as number gets increasing.

14. (c) I. As 121 is the square of 11. So, to obtain prime numbers less than 121, we reject all the multiples of prime numbers less than 11, i.e. 2, 3, 5 and 7.

II. Every composite number less than 121 is divisible by a prime number less than 11, i.e. 2, 3, 5 or 7.

Hence, both the given statements are correct.

15. (c) I. Any number in order to get completely divisible by 11 must have the difference between the sum of even place digits and odd place digits equal to 0 or the multiple of 11. In 7710312401, difference between sum of even place digits and the sum of odd place digit 0.

So, it is divisible by 11.

II. To check divisibility of 173, we can divide the number by all the prime numbers from 2 to 13. It is not divisible by 2, 3, 5, 7, 11 and 13.

So, it is a prime number.

Hence, both the given statements are correct.

16. (d) Unit's digit in $(234)^{100} + (234)^{101}$
$=$ Unit's digit in $(4^{100} + 4^{101})$
$=$ Unit's digit in $(4^{4 \times 25} + 4^{4 \times 25 + 1})$
$=$ Unit's digit in $(6 + 4)$
$= 0$

17. (c) I. Any integer number is either even or odd. Let first number be odd, i.e. $2n + 1$, then next consecutive number will be $2n + 2 = 2(n + 1) = 2m$, i.e. even. If we take first number as even, then next number will be $2n + 1$, i.e. odd.

So, one number is always even.

II. $1^2 = 1 = 8 \times 0 + 1$
$3^2 = 9 = 8 \times 1 + 1$
$5^2 = 25 = 8 \times 3 + 1$ and so on

$\therefore$ Square of any odd integer is always of the form $8n + 1$.

Number System / 17

18. (a) Let the number be y.
Then, $y = 7q + 3$ and $q = 11p + 6$
∴ $y = 7(11p + 6) + 3$
⇒ $y = 77p + 45 = 11(7p + 4) + 1$
When y is divided by 11, then remainder is $1 (= m)$ and quotient is $7p + 4$ and when quotient is divided by 7, then remainder is $4 (= n)$.

19. (a) Since, the password is to be divisible by 4, so the number made by last two digits must be divisible by 4. There are 25 possible cases in which the number made by last two digits is divisible by 4. The numbers are 00, 44, 04, 08, 88,12,..., 96. At the remaining three places, there can be all the ten digits.
Thus, required number of passwords
$= 10 \times 10 \times 10 \times 25$
$= 25000$

20. (d) We know that if m is odd, then $(x^m + a^m)$ is divisible by $(x + a)$.
∴ Each one is divisible by $(41 + 43)$.
∴ Common factor is $(41 + 43)$.

21. (c) Remainder of $\dfrac{4^{1000}}{7} = \dfrac{(4^2)^{500}}{7}$
$= \dfrac{(16)^{500}}{7} = \dfrac{2^{500}}{7}$
$= \dfrac{2^2 \times (2^3)^{166}}{7}$
$= \dfrac{4 \times (8)^{166}}{7} = 4$

22. (d) Required remainder $= 9^{19} + 6$
$= (1)^{19} + 6 = 7$
[∵ $8 = 9 - 1$, so replaced by 1]
Now, $\dfrac{(9)^{19} + 6}{8} = \dfrac{1^{19} + 6}{8} = \dfrac{1+6}{8} = \dfrac{7}{8}$
∴ Remainder $= 7$

23. (d) $\dfrac{19^{100}}{20} = \dfrac{(20-1)^{100}}{20} = \dfrac{(-1)^{100}}{20} = \dfrac{1}{20}$
∴ Remainder $= 1$

24. (d) $7^{10} - 5^{10} = (7^5)^2 - (5^5)^2$
$= (7^5 + 5^5)(7^5 - 5^5)$
[∵ $a^2 - b^2 = (a-b)(a+b)$]
$= (16807 + 3125)(7^5 - 5^5)$
$= 19932 \times (7^5 - 5^5)$
As sum of digits at odd places
$= 1 + 9 + 2 = 12$
and sum of digits at even places
$= 9 + 3 = 12$
∴ Required difference $= 12 - 12 = 0$
Hence, $7^{10} - 5^{10}$ is divisible by 11.

25. (a) $x = (16^3 + 17^3 + 18^3 + 19^3)$
$= (4096 + 4913 + 5832 + 6859)$
$= 21700$
According to the question,
When x is divided by 70, then
$\dfrac{21700}{70} = 310$ [∵ $x = 21700$]
So, there is no remainder. Hence, answer is '0'.

26. (d) $13^5 + 16^5$ is divisible by $13 + 16 = 29$.
[as 5 is odd]
Also, $14^5 + 15^5$ is divisible by $14 + 15 = 29$.
[as 5 is odd]
∴ Remainder of $\left(\dfrac{13^5 + 14^5 + 15^5 + 16^5}{29}\right)$
$=$ Remainder of $\left(\dfrac{13^5 + 16^5}{29}\right)$
$+$ Remainder $\left(\dfrac{14^5 + 15^5}{29}\right)$
$= 0 + 0 = 0$

27. (d) We have,
$2^{122} + 4^{62} + 8^{42} + 4^{64} + 2^{130}$
$= 2^{122} + 2^{124} + 2^{126} + 2^{128} + 2^{130}$
$= 2^{122}(1 + 2^2 + 2^4 + 2^6 + 2^8)$
$= 2^{122}(1 + 4 + 16 + 64 + 256)$
$= 2^{122}(341)$
$= 2^{122} \times 31 \times 11$
which is divisible by 11.

28. (d) Let $2^{32} = x$ and $(2^{32} + 1) = (x + 1)$ be divisible by a number n.
Then, $(2^{96} + 1) = (x^3 + 1)$
$= (x + 1)(x^2 - x + 1)$
which is clearly divisible by n as $(x + 1)$ is divisible by n.

29. (c) We can check divisibility of $19^5 + 21^5$ by 10 by adding the unit's digits of 9^5 and 1^5 which is equal to $9 + 1 = 10$.
So, it must be divisible by 10.
Now, for divisibility by 20, we add 19 and 21 which is equal to 40. So, it is clear that it is also divisible by 20.
So, $19^5 + 21^5$ is divisible by both 10 and 20.

30. (a) Clearly, $(6x^2 + 6x) = 6x(x + 1)$ is divisible by 6 and 12 as $x(x + 1)$ is even.

31. (c) $(2^3 - 2) = 6$ is the largest natural number that divides $(a^3 - a)$ for every number a.

32. (c) Let $x = (n^3 - n)(n - 2)$, where $n > 2$.
On taking $n = 3$, we get
$x = (3^3 - 3)(3 - 2)$
$= (27 - 3)(1) = 24$
[which is divisible by 6, 12 and 24]
On taking $n = 4$, we get
$x = (4^3 - 4)(4 - 2)$
$= (64 - 4) \times 2 = 120$
[which is again divisible by 6, 12 and 24]
Now, on taking $n = 5$, we get
$x = (5^3 - 5)(5 - 2) = (125 - 5) \times 3$
$= 120 \times 3 = 360$
[which is again divisible by 6, 12 and 24]
Hence, 24 is the largest natural number.

33. (c) When N is a prime number, then there is only one possible case that N, $(N + 2)$ and $(N + 4)$ are prime numbers.
When $N = 3$, then N, $(N + 2)$, $(N + 4) = 3, 5, 7$ all are primes.

34. (b) Taking $p = 5$,
$2^p - 1 = 2^5 - 1 = 31$, which is prime.
Taking $p = 11$,
$2^p - 1 = 2^{11} - 1 = 2047$
Since, 2047 is divisible by 23, so it is not prime. Thus, required least positive prime number is 11.

35. (c) Since, b is largest square divisor of c.
So, $c = bx$
[where, x is not a whole square number]
Also, a^2 divides c.
So, a^2 will divide bx or a will divide b.
[since, it cannot divide x as it is not a whole square]

36. (a) According to Fermat's theorem, if p is a prime number and N is prime to p, then $N^{p-1} - 1$ is divisible by p.

37. (a) If n is greater than 1, then $n^2(n^2 - 1)$ is always divisible by 12.
Illustration 1 Put $n = 2$, then
$n^2(n^2 - 1) = (2)^2(2^2 - 1)$
$= 4 \times 3 = 12$
Illustration 2 Now, put $n = 3$, then
$n^2(n^2 - 1) = (3)^2(3^2 - 1) = 9 \times 8 = 72$

38. (b) There are four numbers, i.e. 168, 861, 952, 259.

39. (c) 4444 mod 9 = 7
[here, a mod b means the remainder when a is divided by b]
$\Rightarrow (4444^2)$ mod 9 = 4
$\Rightarrow (4444^3)$ mod 9 = 1
$\Rightarrow (4444^4)$ mod 9 = 7
and so on.
It follows a certain pattern which can be summarised as
If (4444^k) mod 9 = y.
Then, if (k mod 3) = 1, y = 7
if (k mod 3) = 2, y = 4
if (k mod 3) = 0, y = 1
As, $k = 4444$ and 4444 mod 3 = 1, hence $y = 7$
$\therefore (4444^{4444})$ mod 9 = 7
Hence, the remainder is 7.

40. (d) If k is a positive integer, then every square integer is of the form $4k$ or $4k + 1$, as every square number is either a multiple of 4 or exceeds multiple of 4 by unity.

41. (b) Every prime number of the form $3k + 1$ can be represented in the form $6m + 1$ only, when k is even.

42. (a) We have, x and y both are positive odd integers.
Let $x = 2m + 1$ and $y = 2n + 1$, when $m, n \in N$
$\therefore x^2 + y^2 = (2m + 1)^2 + (2n + 1)^2$
$= 4m^2 + 4m + 1 + 4n^2 + 4n + 1$
$= 4(m^2 + n^2 + m + n) + 2$
$= 2[2(m^2 + n^2 + m + n) + 1]$
$= 2k$
Hence, $x^2 + y^2$ is even integer, but $x^2 + y^2$ is not divisible by 4.

43. (c) We have,
$1729 = 1728 + 1 = (12)^3 + (1)^3$
or $1729 = 1000 + 729$
$= (10)^3 + (9)^3$
Thus, 1729 can be written as sum of the cubes of two positive integers in two ways only.

Chapter 2

Series and Progressions

Series

A number series is a sequence of numbers written from left to right in a certain pattern. To solve the questions on series, we have to detect/find the pattern that is followed in the series between the consecutive terms, so that the wrong/missing term can be find out.

Types of Series

There can be following types of series

1. Prime Number Series

The number which is divisible by 1 and itself, is called a prime number. The series formed by using prime number is called prime number series.

Ex. 1 Find out the next term in the series

7, 11, 13, 17, 19, ...

Sol. Given series is a consecutive prime number series. Therefore, the next term will be 23.

Ex. 2 Find out the next term in the series 3, 7, 17, 31, ...

Sol. Here, every next prime number takes place skipping one, two, three and four prime numbers, respectively. Hence, the required answer will be 53. Let us see

③ 5 ⑦ 11 13 ⑰ 19 23 29 ㉛ 37 41 43 47 ㊽
    Skipped Skipped   Skipped   Skipped

2. Addition Series

The series in which next term is obtained by adding a specific number to the previous term, is known as addition series. Addition series are increasing order series and difference between consecutive term is equal.

◆ Number which is added to consecutive terms, can be fixed or variable.

Ex. 3 Find the missing term in the series 6, ?, 21, 33, 48.

Sol. Series pattern:

$$6 \xrightarrow{+6} \boxed{12} \xrightarrow{+9} 21 \xrightarrow{+12} 33 \xrightarrow{+15} 48$$

∴ Missing term = 6 + 6 = 12

3. Difference Series

Difference series is decreasing order series in which next term is obtained by subtracting a fixed/specific number from the previous term.

◆ Number which is subtracted from consecutive terms, can be fixed or variable.

Ex. 4 Find the missing term in the series 125, 80, 45, ?, 5.

Sol. Series pattern:

$$125 \xrightarrow{-45} 80 \xrightarrow{-35} 45 \xrightarrow{-25} \boxed{20} \xrightarrow{-15} 5$$

∴ Missing term = 45 − 25 = 20

4. Multiplication Series

When each term of a series is obtained by multiplying a number with the previous term, then the series is called a multiplication series.

◆ Number which is multiplied to consecutive terms, can be fixed or variable.

Ex. 5 Find the missing term in the series 4, 12, 36, ?, 324, 972.

Sol. Series pattern:

$$4 \xrightarrow{\times 3} 12 \xrightarrow{\times 3} 36 \xrightarrow{\times 3} \boxed{108} \xrightarrow{\times 3} 324 \xrightarrow{\times 3} 972$$

∴ Missing term = 36 × 3 = 108

Ex. 6 Find the missing term in the series 12, 24, 72, 144, 432, ?.

Sol. Series pattern:

$$12 \xrightarrow{\times 2} 24 \xrightarrow{\times 3} 72 \xrightarrow{\times 2} 144 \xrightarrow{\times 3} 432 \xrightarrow{\times 2} \boxed{864}$$

∴ Missing term = 432 × 2 = 864

5. Division Series

When the next term of a series is obtained by dividing the previous term by a number, then the series is called a division series.

◆ Number which divides consecutive terms, can be fixed or variable.

Series and Progressions / 21

Ex. 7 Find out the missing term in the series 10080, 1440, 240, ..., 12, 4.

Sol. Series pattern: $\dfrac{10080}{7} = 1440, \dfrac{1440}{6} = 240, \dfrac{240}{5} = \mathbf{48}, \dfrac{48}{4} = 12, \dfrac{12}{3} = 4$

Hence, the missing term is 48.

Ex. 8 Find out the missing term in the series 432, 216, 72, ..., 12.

Sol. Series pattern: $\dfrac{432}{2} = 216, \ \dfrac{216}{3} = 72, \ \dfrac{72}{2} = 36, \ \dfrac{36}{3} = 12$

Hence, the missing term is 36.

6. n^2 Series

When a number is multiplied with itself, then it is known as **square of a number** and the series formed by square of numbers is called n^2 series.

Ex. 9 Find out the missing term in the series ?, 9, 25, 49, 81, 121.

Sol. The given series is squares of consecutive odd numbers. Let us see
$$1^2 = \mathbf{1}, 3^2 = 9, 5^2 = 25, 7^2 = 49, 9^2 = 81, 11^2 = 121$$
Hence, the missing term is 1.

Ex. 10 Find out the missing term in the series 1, 4, 9, 16, 25, ..., 49.

Sol. The given series is squares of consecutive natural numbers. Let us see
$$1^2 = 1, 2^2 = 4, 3^2 = 9, 4^2 = 16, 5^2 = 25, 6^2 = \mathbf{36}, 7^2 = 49$$
Hence, the missing term is 36.

7. $(n^2 + 1)$ Series

In a series, if each term is a sum of a square term and 1, then this series is called $(n^2 + 1)$ series.

Ex. 11 Find out the missing term in the series 10, 17, 26, 37, ..., 65.

Sol. Series pattern:
$$3^2 + 1 = 10, 4^2 + 1 = 17, 5^2 + 1 = 26, 6^2 + 1 = 37, 7^2 + 1 = \mathbf{50}, 8^2 + 1 = 65$$
Hence, the missing term is 50.

Ex. 12 Find out the missing term in the series 122, 145, 170, 197, ..., 257.

Sol. Series pattern:
$$11^2 + 1 = 122, 12^2 + 1 = 145, 13^2 + 1 = 170,$$
$$14^2 + 1 = 197, 15^2 + 1 = \mathbf{226}, 16^2 + 1 = 257$$
Hence, the missing term is 226.

8. $(n^2 - 1)$ Series

In a series, if each term is obtained by subtracting 1 from square of a number, then this series is called $(n^2 - 1)$ series.

Ex. 13 Find out the missing term in the series 0, 3, 8, 15, 24, ..., 48.

Sol. Series pattern:
$$1^2 - 1 = 0, \ 2^2 - 1 = 3, \ 3^2 - 1 = 8, \ 4^2 - 1 = 15,$$
$$5^2 - 1 = 24, \ 6^2 - 1 = \mathbf{35}, \ 7^2 - 1 = 48$$
Hence, the missing term is 35.

Ex. 14 Find out the missing term in the series 224, 195, 168, ... , 120.

Sol. Series pattern : $15^2 - 1, 14^2 - 1, 13^2 - 1, 12^2 - 1, 11^2 - 1$

Hence, the missing term is $12^2 - 1$, i.e. 143.

9. ($n^2 + n$) Series

The series in which each term is a sum of a number and square of that number, is called ($n^2 + n$) series.

Ex. 15 Find out the missing term in the series 12, 20, 30, 42, ..., 72.

Sol. Series pattern : $3^2 + 3, 4^2 + 4, 5^2 + 5, 6^2 + 6, 7^2 + 7, 8^2 + 8$

Hence, the missing term is $7^2 + 7$, i.e. 56.

Ex. 16 Find out the missing term in the series 420 , 930, 1640, ..., 3660.

Sol. Series pattern : $20^2 + 20, 30^2 + 30, 40^2 + 40, 50^2 + 50, 60^2 + 60$

Hence, the missing term is $50^2 + 50$, i.e. 2550.

10. ($n^2 - n$) Series

The series in which each term is obtained by subtracting a number from square of that number, is called ($n^2 - n$) series.

Ex. 17 Find out the missing term in the series 42, 30, ..., 12, 6.

Sol. Series pattern : $7^2 - 7, 6^2 - 6, 5^2 - 5, 4^2 - 4, 3^2 - 3$

Hence, the missing term is $5^2 - 5$, i.e. 20.

Ex. 18 Find out the missing term in the series 210, 240, 272, 306, ..., 380.

Sol. Series pattern : $15^2 - 15, 16^2 - 16, 17^2 - 17, 18^2 - 18, 19^2 - 19, 20^2 - 20$

Hence, the missing term is $19^2 - 19$, i.e. 342.

11. n^3 Series

If a number is multiplied two times with itself, then the resulting number is called the cube of a number and series which consist of cube of different numbers following a specified sequence, is called n^3 series.

Ex. 19 Find out the missing term in the series 1, 8, 27, ..., 125, 216.

Sol. Series pattern : $1^3, 2^3, 3^3, 4^3, 5^3, 6^3$

Hence, the missing term is 4^3, i.e. 64.

Ex. 20 Find out the missing term in the series 1000, 8000, 27000, 64000, 125000, ...

Sol. Series pattern : $10^3, 20^3, 30^3, 40^3, 50^3, 60^3$

Hence, the missing term is 60^3, i.e. 216000.

12. ($n^3 + 1$) Series

The series in which each term is sum of a cube of a number and 1, is called ($n^3 + 1$) series.

Ex. 21 Find out the missing term in the series 126, 217, 344, ..., 730.

Sol. Series pattern : $5^3 + 1, 6^3 + 1, 7^3 + 1, 8^3 + 1, 9^3 + 1$

Hence, the missing term is $8^3 + 1$, i.e. 513.

Series and Progressions / 23

Ex. 22 Find out the missing term in the series 1001, 8001, ..., 64001, 125001, 216001.

Sol. Series pattern : $10^3 + 1, 20^3 + 1, 30^3 + 1, 40^3 + 1, 50^3 + 1, 60^3 + 1$

Hence, the missing term $30^3 + 1$, i.e. 27001.

13. $(n^3 - 1)$ Series

The series in which each term is obtained by subtracting 1 from the cube of a number, is called $(n^3 - 1)$ series.

Ex. 23 Find out the missing term in the series 0, 7, 26, 63, 124, 215, ...

Sol. Series pattern : $1^3 - 1, 2^3 - 1, 3^3 - 1, 4^3 - 1, 5^3 - 1, 6^3 - 1, 7^3 - 1$

Hence, the missing term is $7^3 - 1$, i.e. 342.

Ex. 24 Find out the missing term in the series ..., 7999, 26999, 63999, 124999.

Sol. Series pattern : $10^3 - 1, 20^3 - 1, 30^3 - 1, 40^3 - 1, 50^3 - 1$

Hence, the missing term is $10^3 - 1$, i.e. 999.

14. $(n^3 + n)$ Series

When each term of a series is a sum of a number and its cube, then the series is called $(n^3 + n)$ series.

Ex. 25 Find out the missing term in the series 2, 10, 30, ..., 130, 222.

Sol. Series pattern : $1^3 + 1, 2^3 + 2, 3^3 + 3, 4^3 + 4, 5^3 + 5, 6^3 + 6$

Hence, the missing term is $4^3 + 4$, i.e. 68.

Ex. 26 Find out the missing term in the series ..., 8020, 27030, 64040.

Sol. Series pattern : $10^3 + 10, 20^3 + 20, 30^3 + 30, 40^3 + 40$

Hence, the missing term is $10^3 + 10$, i.e. 1010.

15. $(n^3 - n)$ Series

When each term of a series is obtained by subtracting a number from its cube, then series is called $(n^3 - n)$ series.

Ex. 27 Find out the missing term in the series 0, 6, 24, 60, 120, ...

Sol. Series pattern : $1^3 - 1, 2^3 - 2, 3^3 - 3, 4^3 - 4, 5^3 - 5, 6^3 - 6$

Hence, the missing term is $6^3 - 6$, i.e. 210.

Ex. 28 Find out the missing term in the series ..., 7980, 26970, 63960, 124950.

Sol. Series pattern : $10^3 - 10, 20^3 - 20, 30^3 - 30, 40^3 - 40, 50^3 - 50$

Hence, the missing term is $10^3 - 10$, i.e. 990.

16. Alternating Series

In alternating series, the successive terms increase and decrease alternately.
- Alternating series is a combination of two different series.
- Two different operations are performed on successive terms alternately.

Ex. 29 Find the next term in the series 15, 14, 19, 11, 23, 8, ...

Sol. Series pattern:

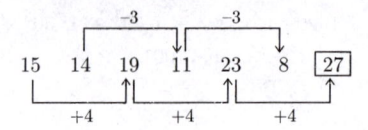

```
         -3        -3
    ┌────┐    ┌────┐
 15  14  19  11  23   8  [27]
    └────┘ └────┘ └────┘ ↑
     +4     +4     +4
```

Hence, the missing term = 23 + 4 = 27.

Ex. 30 Find the next term in the series 50, 200, 100, 100, 200, 50, 400, ...

Sol. Series pattern:

```
        ÷2        ÷2        ÷2
    ┌────┐    ┌────┐    ┌────┐
 50  200  100  100  200  50  400  [25]
    └────┘ └────┘ └────┘
      ×2     ×2     ×2
```

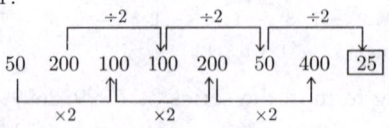

Hence, the missing term = $\dfrac{50}{2}$ = 25.

Types of Questions Asked on Number Series

There are mainly three types of questions which are asked from this chapter which are as follows

Type ❶ To Find the Missing Term

In these type of questions, a series is given in which one of the term is missing and it is required to find the missing term by detecting the pattern of series.

Ex. 31 What should be the next term in the following series?
12, 24, 72, 144, ?

Sol. Series pattern : $12 \times 2 = 24$; $24 \times 3 = 72$;
$72 \times 2 = 144$; $144 \times 3 = \boxed{432}$

Hence, the next term is 432.

Ex. 32 What should come in place of question mark in the series given below?
7, 8, 18, ?, 232, 1165

Sol. Series pattern :

$7 \times 1 + 1 = 8,$ $57 \times 4 + 4 = 232,$
$8 \times 2 + 2 = 18,$ $232 \times 5 + 5 = 1165,$
$18 \times 3 + 3 = \boxed{57}$

Clearly, 57 should come in place of question mark.

Type ❷ To Find the Wrong Term

In these type of questions, a series is given in which one of the terms does not follow the pattern of series and we have to find that wrong term.

Series and Progressions / 25

Ex. 33 Find the wrong number in the series 850, 600, 550, 500, 475, 462.5, 456.25.

Sol. Series pattern :
$$850 - 200 = \boxed{600}\ 650$$
$$650 - 100 = 550$$
$$550 - 50 = 500$$
$$500 - 25 = 475$$
$$475 - 12.5 = 462.5$$
$$462.5 - 6.25 = 456.25$$

Hence, 600 is the wrong number and will be replaced by 650.

Ex. 34 Find the wrong number in the series 84, 83, 79, 70, 52, 29.

Sol. Series pattern : $84 - 1^2 = 83$
$$83 - 2^2 = 79$$
$$79 - 3^2 = 70$$
$$70 - 4^2 = \boxed{52}\ 54$$
$$54 - 5^2 = 29$$

Hence, the wrong number is 52.

Type ③ To Find the Missing Series Based on a Given Series

Two number series are given in this type of questions. The pattern of the second series is same as the pattern of the first series and on the basis of the first series, we have to find the unknown terms of second series.

Ex. 35. There are two number series given below. First number series is arranged in a particular way and second series is based on first series. On the basis of this, which number will come at the place of D?

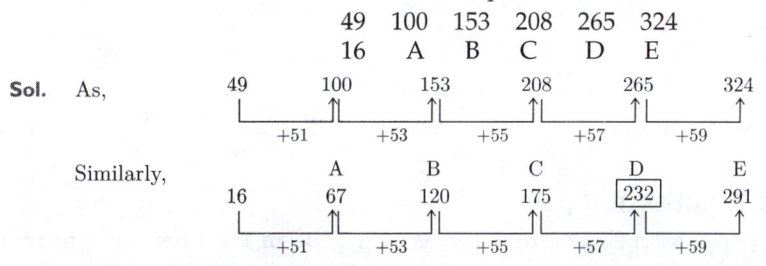

$$\begin{array}{cccccc} 49 & 100 & 153 & 208 & 265 & 324 \\ 16 & A & B & C & D & E \end{array}$$

Sol. As,

49 100 153 208 265 324
   +51   +53   +55   +57   +59

Similarly,

      A   B   C   D   E
16   67 120 175 232 291
   +51   +53   +55   +57   +59

Hence, from the above it is clear that the value at the place of D is 232.

Progressions

There are three types of progressions, which are as follow

Arithmetic Progression (AP)

The progression of the form $a, a + d, a + 2d, a + 3d,...$ is known as an **arithmetic progression** with first term a and common difference d. We have,

(i) nth term, $T_n = a + (n-1)d$

(ii) Sum of n terms of AP, $S_n = \dfrac{n}{2}[2a + (n-1)d]$

(iii) Sum of n terms, $S_n = \dfrac{n}{2}(a + l)$, where, l = last term

Ex. 36 In series 359, 365, 371, ..., what will be the 10th term?

Sol. The given series is in the form of AP, since common difference, i.e. d is same.
Here, $n = 10, a = 359$ and $d = 6$.
∴ 10th term $= a + (n − 1) d = 359 + (10 − 1) 6 = 359 + (9 \times 6) = 413$

Ex. 37 What will be the sum of 6 terms of the series $7 + 14 + 21 + 28 + ...$?

Sol. Here, $a = 7, d = 7$ and $n = 6$.
∴ $S_6 = \dfrac{n}{2} [2a + (n − 1)d] = \dfrac{6}{2} [2 \times 7 + (6 − 1) \times 7] = 3[14 + 35] = 3 \times 49 = 147$

Geometric Progression (GP)

The progression of the form $a, ar, ar^2, ar^3, ...$ is known as a **geometric progression** with first term a and common ratio r. We have,

(i) nth term of GP, $T_n = ar^{n-1}$

(ii) Sum of n terms of GP, $S_n = \dfrac{a(1-r^n)}{(1-r)}$, where $r < 1$

(iii) Sum of n terms of GP, $S_n = \dfrac{a(r^n - 1)}{r - 1}$, where $r > 1$

(iv) Sum of an infinite GP, $S_\infty = \dfrac{a}{1-r}$

Ex. 38 In the series 7, 14, 28, ..., what will be the 10th term?

Sol. The given series is in the form of GP, since the common ratio, i.e. r is same.
Here, $n = 10, a = 7$ and $r = 2$.
∴ 10th term $= ar^{n-1} = 7(2)^{(10-1)} = 7 \times 2^9 = 3584$

Ex. 39 What will be the sum of first 10 terms of the series $1 + 5 + 5^2 + 5^3 + ...$?

Sol. Given series is $1 + 5 + 5^2 + 5^3 + ...$
Here, $n = 10, a = 1$ and $r = 5 (r > 1)$
∴ Sum of 10 terms, $S_{10} = \dfrac{a(r^n - 1)}{r - 1} = \dfrac{1(5^{10} - 1)}{5 - 1} = \dfrac{5^{10} - 1}{4} = \dfrac{9765624}{4} = 2441406$

Harmonic Progression (HP)

A series of quantities $a_1, a_2, a_3, a_4, ..., a_n$ is said to be in HP when their reciprocals $\dfrac{1}{a_1}, \dfrac{1}{a_2}, \dfrac{1}{a_3}, ..., \dfrac{1}{a_n}$ are in AP.

* The converse is also true.
* Simply, there is no general formula for the sum of any number of quantities in HP. All questions in HP are generally solved by inverting the terms and making use of the properties of the corresponding AP.

Ex. 40 Find the 10th term of HP $1, \dfrac{1}{3}, \dfrac{1}{5}, \dfrac{1}{7}, \dfrac{1}{9}, \dfrac{1}{11}, ...$.

Sol. Since, $\dfrac{1}{1}, \dfrac{1}{3}, \dfrac{1}{5}, \dfrac{1}{7}, ...$ are in HP. So, $1, 3, 5, 7, ...$ are in AP.
∴ $T_{10} = a + (n − 1) d = 1 + (10 − 1) 2 = 1 + 9 \times 2 = 19$
Since, 10th term of the AP is 19. Therefore, 10th term of HP is $\dfrac{1}{19}$.

Fast Track Practice

Exercise 1 Base Level Questions

Directions (Q. Nos. 1-31) *Find the missing term in each of the following series.*

1. 2, 15, 41, 80, 132, ?
   (a) 197   (b) 150   (c) 178   (d) 180
   (e) None of these

2. 3, 8, 18, 38, 78, ?   [RRB Clerk (Pre) 2017]
   (a) 158   (b) 154   (c) 150   (d) 162
   (e) 166

3. 5, 28, 47, 64, 77, ?   [RRB PO (Pre) 2017]
   (a) 84   (b) 86
   (c) 89   (d) 88
   (e) None of these

4. 160, 151, 133, 106, ?, 25   [LIC AAO 2016]
   (a) 70   (b) 32   (c) 55   (d) 40
   (e) 25

5. 89, 88, 85, 78, 63, ?   [RRB PO (Pre) 2017]
   (a) 30   (b) 34   (c) 36   (d) 32
   (e) None of these

6. 3, 6, 12, 24, 48, 96, ?   [IBPS Clerk 2011]
   (a) 192   (b) 182   (c) 186   (d) 198
   (e) None of these

7. 165, 195, 255, 285, ?, 435
   (a) 340   (b) 341   (c) 345   (d) 401
   (e) None of these

8. 5, ?, 15, 75, 525, 4725, 51975   [IBPS Clerk 2011]
   (a) 5   (b) 10   (c) 8   (d) 6
   (e) None of these

9. 6, 3, 3, 6, 24, ?   [RRB Clerk (Pre) 2017]
   (a) 184   (b) 186   (c) 188   (d) 190
   (e) 192

10. 16, ?, 8, 16, 64, 512   [LIC AAO 2016]
    (a) 12   (b) 10   (c) 32   (d) 8
    (e) 9

11. 10000, 2000, 400, 80, 16, 3.2, ?   [IBPS Clerk 2011]
    (a) 0.38   (b) 0.45   (c) 0.64   (d) 0.54
    (e) None of these

12. 840, ?, 420, 140, 35, 7   [Bank PO 2007]
    (a) 408   (b) 840
    (c) 480   (d) 804
    (e) None of these

13. 12, 16, 32, 68, 132, ?   [NICL AO (Pre) 2017]
    (a) 196   (b) 232   (c) 276   (d) 213
    (e) None of these

14. 104, 300, 469, 613, 734, ?
    [NICL AO (Pre) 2017]
    (a) 982   (b) 715   (c) 834   (d) 755
    (e) None of these

15. 41, 40, 36, ?, 11
    (a) 35   (b) 27   (c) 29   (d) 30
    (e) None of these

16. 5, 9, 25, 89, ?, 1369   [IBPS Clerk (Pre) 2016]
    (a) 343   (b) 355   (c) 349   (d) 341
    (e) 345

17. 400, 375, 424, 343, 464, ?   [LIC AAO 2016]
    (a) 251   (b) 385   (c) 295   (d) 371
    (e) 562

18. 135, 134, ?, 99, 35
    (a) 126   (b) 115   (c) 85   (d) 111
    (e) None of these

19. 2, 9, 28, 65, 126, ?   [SSC (10+2) 2010]
    (a) 195   (b) 199
    (c) 208   (d) 217

20. 6, 13, 32, ?, 130, 221
    (a) 75   (b) 69   (c) 100   (d) 85
    (e) None of these

21. 7, 26, 63, 124, 215, 342, ?
    [SSC CGL (Mains) 2012]
    (a) 481   (b) 511   (c) 391   (d) 421

22. 15, 365, 587, 717, 785, 815, ?
    [SBI Clerk (Mains) 2016]
    (a) 825   (b) 835   (c) 828   (d) 832
    (e) 838

23. 2, 6, 14, 30, ?, 126   [SSC (10+2) 2009]
    (a) 62   (b) 63   (c) 73   (d) 95

24. 17, 9, 10, 16.5, ?, 90   [SBI PO (Pre) 2017]
    (a) 44   (b) 35   (c) 48   (d) 38
    (e) 33

25. 14, 33, 104, ?, 2110   [MCA 2009]
    (a) 421   (b) 433   (c) 372   (d) 840

26. 4, 18, ?, 100, 180, 294   [IB ACIO 2013]
    (a) 32   (b) 36   (c) 48   (d) 40

28 / Fast Track Objective Arithmetic

27. 2, 10, 42, 170, ?, 2730, 10922
　　　　　　　　　　　　　[IBPS Clerk 2011]
　(a) 588　(b) 568　(c) 596　(d) 682
　(e) None of these

28. 23, 42.2, 80.6, 157.4, 311, ? [IDBI SO 2012]
　(a) 618.2　(b) 623.6　(c) 624.2　(d) 616.6
　(e) None of these

29. 190, 94, 46, 22, ?, 4　[RRB PO (Pre) 2017]
　(a) 12　(b) 14　(c) 10　(d) 8
　(e) None of these

30. 6, 17, 50, 149,?, 1337 [IBPS Clerk (Pre) 2016]
　(a) 454　(b) 446　(c) 442　(d) 452
　(e) 432

31. 33, 40, 29, 42, 25, ?　[SBI PO (Pre) 2017]
　(a) 40　(b) 44　(c) 52　(d) 48
　(e) 46

32. The next term in the sequence
　　3, 12, 39, 120, ... is
　(a) 363　(b) 352　(c) 345　(d) 350

Directions (Q. Nos. 33-37) *Find the wrong number in each of the following series.*

33. 7, 9, 16, 25, 41, 68, 107, 173
　　　　　　　　　　　　　[Bank Clerk 2008]
　(a) 107　(b) 16　(c) 41　(d) 68
　(e) 25

34. 14, 19, 29, 40, 44, 51, 59, 73
　　　　　　　　　　　　　　[SSC CGL 2012]
　(a) 59　(b) 51　(c) 44　(d) 29

35. 4, 2, 3.5, 7.5, 26.25, 118.125
　　　　　　　　　　　　　[Bank Clerk 2009]
　(a) 118.125　　(b) 26.25
　(c) 3.5　　　　(d) 2
　(e) 7.5

36. 2, 11, 38, 197, 1172, 8227, 65806
　(a) 11　(b) 38　(c) 197　(d) 1172
　(e) None of these

37. 2, 9, 28, 65, 126, 216, 344
　　　　　　　　　　　　　[SSC (10+2) 2012]
　(a) 9　(b) 65　(c) 216　(d) 28

38. The 10th term of the sequence
　　$\dfrac{-7}{6}, \dfrac{1}{3}, \dfrac{11}{6}, \dfrac{10}{3}, \cdots$ is
　(a) $\dfrac{35}{3}$　(b) $\dfrac{37}{3}$　(c) $\dfrac{27}{6}$　(d) $\dfrac{40}{3}$

39. The sum of n terms of the series
　　$1 + \dfrac{3}{2} + 2 + \dfrac{5}{2} + 3 + \cdots$ is
　(a) $\dfrac{n(n+3)}{2}$　　(b) $\dfrac{(n+1)(n+2)}{4}$
　(c) $\dfrac{n(n+1)}{2}$　　(d) $\dfrac{n(n+3)}{4}$

40. The first and last terms of an arithmetic progression are 33 and -57. What is the sum of the series, if it has 16 terms?
　　　　　　　　　　　　　[SSC (10+2) 2017]
　(a) -135　(b) -192　(c) -207　(d) -165

41. The middle term(s) of the series
　　$2 + 4 + 6 + \ldots + 198$ is　[SSC (10+2) 2013]
　(a) 98　(b) 96　(c) 94　(d) 100

42. Find the sum of all positive multiples of 3 less than 50.　[SSC CGL (Mains) 2014]
　(a) 400　(b) 404　(c) 408　(d) 412

43. If p, q and r are in GP, then which is true among the following?　[SSC CGL 2013]
　(a) $q = \dfrac{p+r}{2}$　　(b) $p^2 = qr$
　(c) $q = \sqrt{pr}$　　(d) $\dfrac{p}{r} = \dfrac{r}{q}$

44. If 4th and 7th terms of a HP are $\dfrac{1}{2}$ and $\dfrac{2}{7}$ respectively, then the first term is
　(a) 2　　(b) 3
　(c) 5　　(d) 7

Exercise 2 *Higher Skill Level Questions*

Directions (Q. Nos. 1-32) *What should come in place of the question mark in each of the following number series?*

1. 142, 70, 34, 16, ?, 2.5 [SBI Clerk (Pre) 2016]
　(a) 5　(b) 7　(c) 3　(d) 12
　(e) 8

2. 849, 282, 93, 30, 9, ?　[IBPS SO 2016]
　(a) 1　(b) 3　(c) 4　(d) 6
　(e) 2

3. 36, 154, 232, 278, 300, ?　[IDBI SO 2012]
　(a) 306　(b) 313　(c) 308　(d) 307
　(e) None of these

4. 656, 432, 320, 264, 236, ?　[Bank PO 2010]
　(a) 222　(b) 229　(c) 232　(d) 223
　(e) None of these

5. 18, 18.8, 20.4, 23.6, 30, ?　[IBPS SO 2016]
　(a) 44.4　(b) 43.5　(c) 49.2　(d) 49.6
　(e) 42.8

Series and Progressions / 29

6. 5, 9, 33, 72, 121, ?   [SBI PO (Pre) 2017]
   (a) 169   (b) 163   (c) 175   (d) 184
   (e) Other than those given as options

7. 85, 53, 33, 22, 17, ?   [SBI PO (Pre) 2016]
   (a) 5   (b) 9   (c) 10   (d) 8
   (e) 15

8. 456.5, 407, 368.5, 341, 324.5, ?
   [IDBI SO 2012]
   (a) 321   (b) 319   (c) 317   (d) 323
   (e) None of these

9. 6, 5, 6, 10, ?   [IBPS SO 2016]
   (a) 25   (b) 26   (c) 19   (d) 29
   (e) 40

10. 24, 536, 487, 703, 678, ?   [IDBI SO 2012]
   (a) 768   (b) 748   (c) 764   (d) 742
   (e) None of these

11. 17, 9, 10, 16.5, 35, ?   [IBPS SO 2016]
   (a) 85   (b) 70   (c) 92.5   (d) 90
   (e) 84.5

12. 6, 4, 5, 11, ?, 189   [IBPS SO 2016]
   (a) 82   (b) 39   (c) 44   (d) 65
   (e) 96

13. 14, 6, 5, 6.5, 12, ?   [SBI PO 2015]
   (a) 29   (b) 27   (c) 23   (d) 33
   (e) 35

14. 6, 7, 16, 51, ?, 1045   [IBPS SO 2016]
   (a) 257   (b) 194   (c) 200   (d) 350
   (e) 208

15. 7, 6, 10, 27, ?, 515   [SBI PO (Pre) 2017]
   (a) 112   (b) 104   (c) 114   (d) 96
   (e) 108

16. 600, 125, 30, ?, 7.2, 6.44, 6.288
   (a) 6   (b) 10
   (c) 15   (d) 11
   (e) None of these

17. 9, 62, ?, 1854, 7415, 22244 [IDBI SO 2010]
   (a) 433   (b) 309   (c) 406   (d) 371
   (e) None of these

18. 274, 136, 66, 30, 11, ?   [IBPS SO 2016]
   (a) 5.5   (b) 0.5   (c) 5   (d) 1
   (e) 2

19. 6, 4, 5, 11, 39, ?   [RRB PO (Pre) 2017]
   (a) 159   (b) 169   (c) 189   (d) 198
   (e) None of these

20. 179, 180, 172, 199, 135, ?   [SBI PO 2015]
   (a) 236   (b) 272   (c) 240   (d) 256
   (e) 260

21. 10, 6, 8, 15, 34, ?   [SBI PO (Pre) 2016]
   (a) 95   (b) 80   (c) 90   (d) 75
   (e) 85

22. 7, 4, 5, 9, ?, 52.5, 160.5   [Bank PO 2010]
   (a) 32   (b) 16   (c) 14   (d) 20
   (e) None of these

23. 12, 13, 20, 39, 82, ?   [IBPS SO 2016]
   (a) 259   (b) 232   (c) 210   (d) 198
   (e) 173

24. 61, 82, 124, 187, ?, 376   [SBI PO 2015]
   (a) 271   (b) 263   (c) 257   (d) 287
   (e) 249

25. 12, 15, 36, ?, 480, 2415, 14508
   [Bank SO 2010]
   (a) 115   (b) 109   (c) 117   (d) 121
   (e) None of these

26. 18, 43, 204, 1145, 8190, ?
   (a) 73915   (b) 73925
   (c) 73935   (d) 73945
   (e) 73955

27. 1, 2, 5, 16, 65, ?   [LIC AAO 2016]
   (a) 312   (b) 294   (c) 326   (d) 482
   (e) 257

28. 12, 93, 730, 5097, 30570, ?
   (a) 152835   (b) 152837
   (c) 152839   (d) 152841
   (e) 152833

29. 13, 35, 57, 79, 911, ?
   (a) 1113   (b) 1123   (c) 1114   (d) 1124
   (e) None of these

30. 8, 288, 968, 2048, 3528, 5408, ?
   [SBI Clerk (Mains) 2016]
   (a) 7288   (b) 7388   (c) 7488   (d) 7688
   (e) 7588

31. 50, 60, 75, 97.5, ?, 184.275, 267.19875
   [Bank SO 2010]
   (a) 120.50   (b) 130.50
   (c) 131.625   (d) 124.25
   (e) None of these

32. 90, 91, 98, 115, 149, ?   [IBPS SO 2016]
   (a) 274   (b) 240   (c) 209   (d) 252
   (e) 250

Directions (Q. Nos. 33-35) *Find the wrong number in each of the following series.*

33. 12, 6, 7.5, 12.75, 27.5, 71.25
   [Bank PO 2011]
   (a) 6   (b) 7.5   (c) 27.5   (d) 12
   (e) None of these

34. 16, 24, 37, 54, 81,121.5 [Hotel Mgmt. 2011]
   (a) 24   (b) 54   (c) 37   (d) 121.5

35. 12, 12, 18, 48, 180, 1080 [Bank Clerk 2009]
   (a) 180   (b) 12   (c) 18   (d) 48
   (e) None of these

30 / Fast Track Objective Arithmetic

Directions (Q. Nos. 36-42) *There are two series given below, of which one is complete and follows a certain pattern. Based on the pattern of complete series, you have to find the different answering terms of series.*

36. 1275 1307 1371 1467 1595 1755
    972   A   B   C   D   E

   Which number will come at the place of D?
   [IBPS PO 2013]
   (a) 1292   (b) 550   (c) 500   (d) 462.5
   (e) None of these

37. 6   9   18   45   135   472.5
    20 A   B   C   D   E

   Which number will come at the place of C?
   [IBPS PO 2013]
   (a) 30   (b) 1125   (c) 375   (d) 150
   (e) None of these

38. 3   4   10   33   136   685
    7   A   B   C   D   E

   Which number will come at the place of E?
   [IBPS PO 2013]
   (a) 57   (b) 1165   (c) 18   (d) 1398
   (e) None of these

39. 4   23   113   449   1343   2681
    7   A   B   C   D   E

   Which number will come at the place of E?
   [IBPS PO 2013]
   (a) 4793   (b) 4782   (c) 4841   (d) 4932
   (e) None of these

40. 4   16   48   120   272   584
    124   A   B   C   D   E

   Which number will come from the following at place of B?
   (a) 256   (b) 528   (c) 1080   (d) 4424

41. 2   9   57   337   1681   6721
    7   A   B   C   D   E

   Which number will come from the following at place of E?
   [IBPS PO 2013]
   (a) 673   (b) 3361   (c) 13441
   (d) 17   (e) None of these

42. 5   9   25   91   414   2282.5
    3   A   B   C   D   E   [IBPS PO 2015]

   What will come in place of C?
   (a) 63.25   (b) 63.75
   (c) 64.25   (d) 64.75
   (e) None of these

43. The 5th and 9th terms of an arithmetic progression are 7 and 13, respectively. What is the 15th term?
    [SSC CGL (Pre) 2017]
    (a) 22   (b) 21   (c) 55   (d) 59

44. In an arithmetic progression, if 17 is the 3rd term, –25 is the 17th term, then –1 is which term?   [SSC (10 + 2) 2017]
    (a) 10   (b) 11   (c) 9   (d) 12

45. What is the sum of the following series?
    – 64, – 66, – 68, …, – 100   [XAT 2015]
    (a) –1458 (b) –1558 (c) –1568 (d) –1664
    (e) None of these

46. Consider the following statements in respect of the expression $S_n = \dfrac{n(n+1)}{2}$, where n is an integer.
    I. There are exactly two values of n for which $S_n = 861$.
    II. $S_n = S_{-(n+1)}$ and hence for any integer m, we have two values of n for which $S_n = m$.

    Which of the above statement(s) is/are correct?   [CDS 2016 (I)]
    (a) Only I   (b) Only II
    (c) Both I and II   (d) Neither I nor II

47. If A, G and H are the arithmetic, geometric and harmonic means between a and b respectively, then which one of the following relations is correct?
    [CDS 2015 (I)]
    (a) G is the geometric mean between A and H
    (b) A is the arithmetic mean between G and H
    (c) H is the harmonic mean between A and G
    (d) None of the above

48. If each term of GP is positive and is the sum of two preceding terms, then the common ratio of GP is   [MAT 2015]
    (a) $\dfrac{(\sqrt{5}-1)}{2}$ (b) $\dfrac{(1-\sqrt{5})}{2}$
    (c) $\dfrac{(\sqrt{5}+1)}{2}$ (d) $\dfrac{(\sqrt{3}+1)}{2}$

49. The value of $\dfrac{1}{1\times 4} + \dfrac{1}{4\times 7} + \dfrac{1}{7\times 10} + \ldots + \dfrac{1}{16\times 19}$ is
    [CDS 2015 (II)]
    (a) $\dfrac{5}{19}$ (b) $\dfrac{6}{19}$ (c) $\dfrac{8}{19}$ (d) $\dfrac{9}{19}$

50. The sum of first 47 terms of the series $\dfrac{1}{4} + \dfrac{1}{5} - \dfrac{1}{6} - \dfrac{1}{4} + \dfrac{1}{5} + \dfrac{1}{6} + \dfrac{1}{4} - \dfrac{1}{5} - \dfrac{1}{6} + \ldots$ is
    [CDS 2015 (II)]
    (a) 0   (b) $-\dfrac{1}{6}$ (c) $\dfrac{1}{6}$ (d) $\dfrac{9}{20}$

Answer with Solutions

Exercise 1 Base Level Questions

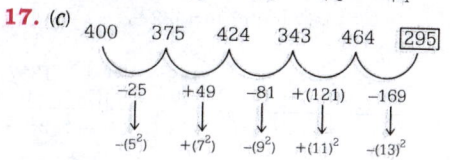

1. (a) Series pattern:
$+13, +26, +39, +52, +65$
∴ Missing term $= 132 + 65 = 197$

2. (a) The pattern of the series is
3, 8, 18, 38, 78, [158]
$+5, +10, +20, +40, +80$

3. (d) The pattern of the series is
5, 28, 47, 64, 77, [88]
$+23, +19, +17, +13, +11$
Adding prime number $= 77 + 11 = 88$

4. (a) 160, 151, 133, 106, [70], 25
$-9, -18, -27, -36, -45$

5. (d) The pattern of the series is
89, 88, 85, 78, 63, [32]
$-1, -3, -7, -15, -31$
$-2, -4, -8, -16$

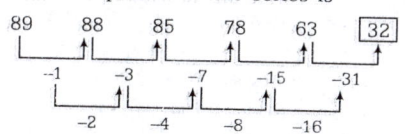

6. (a)
3, 6, 12, 24, 48, 96, [192]
$\times 2, \times 2, \times 2, \times 2, \times 2, \times 2$
∴ Missing term = 192

7. (c) Series pattern:
$(15 \times 11), (15 \times 13), (15 \times 17), (15 \times 19),$
$(15 \times 23), (15 \times 29)$
[multiplication of 15 with prime numbers]
∴ Missing term $= (15 \times 23) = 345$

8. (a)
5, [5], 15, 75, 525, 4725, 51975
$\times 1, \times 3, \times 5, \times 7, \times 9, \times 11$
∴ Missing term = 5

9. (e) The pattern of the series is
6, 3, 3, 6, 24, [192]
$\times 0.5, \times 1, \times 2, \times 4, \times 8$

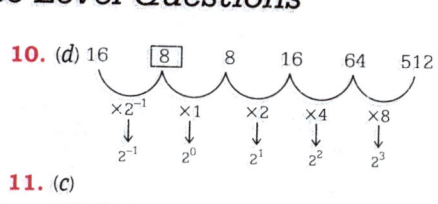

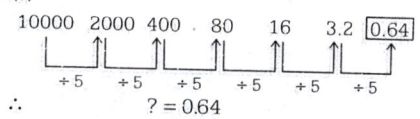

10. (d) 16, [8], 8, 16, 64, 512
$\times 2^{-1}, \times 1, \times 2, \times 4, \times 8$
$2^{-1}, 2^0, 2^1, 2^2, 2^3$

11. (c)
10000, 2000, 400, 80, 16, 3.2, [0.64]
$\div 5, \div 5, \div 5, \div 5, \div 5, \div 5$
∴ ? $= 0.64$

12. (b) Series pattern:
$\div 1, \div 2, \div 3, \div 4, \div 5$
∴ Missing term $= \dfrac{840}{1} = 840$

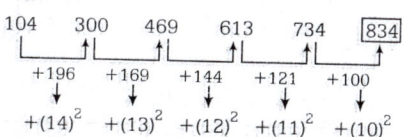

13. (b)
12, 16, 32, 68, 132, [232]
$+2^2, +4^2, +6^2, +8^2, +10^2$

14. (c)
104, 300, 469, 613, 734, [834]
$+196, +169, +144, +121, +100$
$+(14)^2, +(13)^2, +(12)^2, +(11)^2, +(10)^2$

15. (b) Series pattern:
$-(1)^2, -(2)^2, -(3)^2, -(4)^2$
∴ Missing term $= 36 - 3^2 = 36 - 9 = 27$

16. (e) 5, 9, 25, 89, [345], 1369
$+4^1, +4^2, +4^3, +4^4, +4^5$

17. (c) 400, 375, 424, 343, 464, [295]
$-25, +49, -81, +(121), -169$
$-(5^2), +(7^2), -(9^2), +(11)^2, -(13)^2$

18. (a) Series pattern:
$-(1)^3, -(2)^3, -(3)^3, -(4)^3$
∴ Missing term
$= 134 - 2^3 = 134 - 8 = 126$

19. (d) Series pattern:
$1^3 + 1, 2^3 + 1, 3^3 + 1, 4^3 + 1, 5^3 + 1, 6^3 + 1$
∴ Missing term $= 6^3 + 1 = 217$

32 / Fast Track Objective Arithmetic

20. (b) Series pattern :
$1^3 + 5, 2^3 + 5, 3^3 + 5, 4^3 + 5, 5^3 + 5, 6^3 + 5$
$\therefore$ Missing term $= 4^3 + 5 = 69$

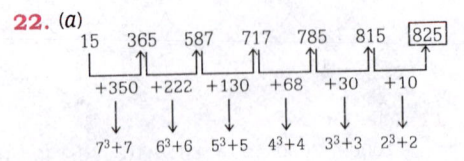

21. (b) 7, 26, 63, 124, 215, 342, [511]
$2^3-1, 3^3-1, 4^3-1, 5^3-1, 6^3-1, 7^3-1, 8^3-1$

22. (a) 15, 365, 587, 717, 785, 815, [825]
+350, +222, +130, +68, +30, +10
$7^3+7, 6^3+6, 5^3+5, 4^3+4, 3^3+3, 2^3+2$

23. (a) Series pattern :
(2 × Previous number) + 2
$\therefore$ Missing term $= (2 \times 30) + 2 = 62$

24. (b) 17, 9, 10, 16.5, [35], 90
×0.5+0.5, ×1+1, ×1.5+1.5, ×2+2, ×2.5+2.5

25. (a) Series pattern :
× 2 + 5, × 3 + 5, × 4 + 5, × 5 + 5
$\therefore$ Missing term $= 104 \times 4 + 5$
$= 416 + 5 = 421$

26. (c) 4, 18, [48], 100, 180, 294
$2^2 \times 1, 3^2 \times 2, 4^2 \times 3, 5^2 \times 4, 6^2 \times 5, 7^2 \times 6$

27. (d) 2, 10, 42, 170, [682], 2730, 10922
×4+2, ×4+2, ×4+2, ×4+2, ×4+2, ×4+2
$\therefore$ Missing term $= 682$

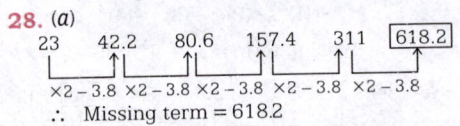

28. (a) 23, 42.2, 80.6, 157.4, 311, [618.2]
×2−3.8, ×2−3.8, ×2−3.8, ×2−3.8, ×2−3.8
$\therefore$ Missing term $= 618.2$

29. (c) The pattern of the series is, $\div 2-1, \div 2-1$
$\therefore$ Missing term $= (22 \div 2) - 1 = \boxed{10}$

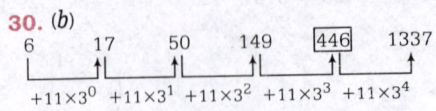

30. (b) 6, 17, 50, 149, [446], 1337
+11×3⁰, +11×3¹, +11×3², +11×3³, +11×3⁴

31. (b) 33, 40, 29, 42, 25, [44]
−4, +2, −4, +2

32. (a) $3^1; 3^1 + 3^2 = 12;$
$3^1 + 3^2 + 3^3 = 39;$
$3^1 + 3^2 + 3^3 + 3^4 = 120;$
and $3^1 + 3^2 + 3^3 + 3^4 + 3^5 = \boxed{363}$

33. (d) Series pattern:
$7 + 9 = 16$
$9 + 16 = 25$
$16 + 25 = 41$
$25 + 41 = \boxed{68}$ 66
$41 + 66 = 107$
and $66 + 107 = 173$
Hence, 68 is the wrong number and will be replaced by 66.

34. (b) 14, 19, 29, 40, 44, [52]/51, 59, 73
+(1+4), +(1+9), +(2+9), +(4+0), +(4+4), +(5+2), +(5+9)
Hence, 51 is the wrong number and will be replaced by 52.

35. (c) Series pattern : $4 \times 0.5 = 2$
$2 \times 1.5 = \boxed{3.5}$ 3
$3 \times 2.5 = 7.5$
$7.5 \times 3.5 = 26.25$
$26.25 \times 4.5 = 118.125$
Hence, 3.5 is the wrong number and will be replaced by 3.

36. (d) Series pattern:
$2 \times 3 + 5 = 11$
$11 \times 4 - 6 = 38$
$38 \times 5 + 7 = 197$
$197 \times 6 - 8 = \boxed{1172}$ 1174
$1174 \times 7 + 9 = 8227$
and $8227 \times 8 - 10 = 65806$
Hence, 1172 is the wrong number and will be replaced by 1174.

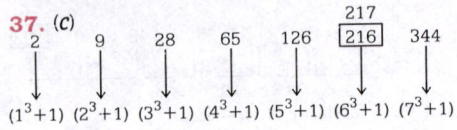

37. (c) 2, 9, 28, 65, 126, [216]/217, 344
$(1^3+1), (2^3+1), (3^3+1), (4^3+1), (5^3+1), (6^3+1), (7^3+1)$
Hence, 216 is the wrong number and will be replaced by 217.

38. (b) Given series is
$\dfrac{-7}{6}, \dfrac{1}{3}, \dfrac{11}{6}, \dfrac{10}{3}, \ldots$
$+\dfrac{3}{2}, +\dfrac{3}{2}, +\dfrac{3}{2}$
which is an AP with $d = \dfrac{3}{2}$ and $a = \dfrac{-7}{6}$.
$\therefore T_{10} = a + (n-1)d$
$= \dfrac{-7}{6} + (10-1)\dfrac{3}{2} = \dfrac{-7}{6} + \dfrac{9 \times 3}{2}$
$= \dfrac{-7+81}{6} = \dfrac{74}{6} = \dfrac{37}{3}$

39. (d) Here, $a = 1, d = \dfrac{1}{2}$ and $n = n$

Series and Progressions / 33

$\therefore S_n = \dfrac{n}{2}[2 \times a + (n-1)d]$

$= \dfrac{n}{2}\left[2 \times 1 + (n-1) \times \dfrac{1}{2}\right]$

$= \dfrac{n}{2}\left[2 + \dfrac{n}{2} - \dfrac{1}{2}\right]$

$= \dfrac{n}{2}\left[\dfrac{n}{2} + \dfrac{3}{2}\right] = \dfrac{n(n+3)}{4}$

40. (b) We know that,
$S_n = \dfrac{n}{2}(a+l)$
Here, $n = 16, a = 33$ and $l = -57$
$\therefore S_n = \dfrac{16}{2}(33-57)$
$= 8(-24) = -192$

41. (d) Given series is an arithmetic progression series, where
$a = 2, T_n = 198, d = 2$
and number of terms $= n$.
$\therefore \quad T_n = a + (n-1)d$
$\Rightarrow \quad 198 = 2 + (n-1)2$
$\Rightarrow \quad 2(n-1) = 198 - 2 = 196$
$\Rightarrow \quad n - 1 = \dfrac{196}{2} = 98 \Rightarrow n = 99$
Middle term $= \dfrac{n+1}{2} = \dfrac{99+1}{2}$
$= 50\text{th term}$
$\therefore \quad T_{50} = 2 + (50-1)2 = 2 + 98 = 100$

42. (c) Multiples of 3 upto 50 are
$3, 6, 9, 12, \ldots, 42, 45, 48.$
The above terms form an AP consisting 16 terms.
Here, $n = 16, a = 3, l = 48$
$\therefore$ Sum of AP $= \dfrac{n}{2}(a+l) = \dfrac{16}{2}(3+48)$
$= 8 \times 51 = 408$

43. (c) Since, p, q and r are in geometric progression.
$\therefore \qquad q^2 = pr$
Then, $\qquad q = \sqrt{pr}$

44. (a) $\because$ 4th term in HP $= \dfrac{1}{2}$
$\therefore$ 4th term in AP $= 2$
and 7th term in HP $= \dfrac{2}{7}$
$\therefore$ 7th term in AP $= \dfrac{7}{2}$
Now, $T_4 = a + 3d = 2$ ...(i)
and $T_7 = a + 6d = \dfrac{7}{2}$ ...(ii)
From Eqs. (i) and (ii), we get $d = \dfrac{1}{2}$
$\therefore \qquad a = \dfrac{1}{2}$
which is the first term of the AP.
Hence, the first term of HP is 2.

Exercise 2 Higher Skill Level Questions

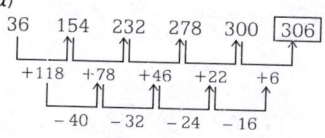

1. (b) 142　70　34　16　[7]　2.5
　　　÷2−1　÷2−1　÷2−1　÷2−1　÷2−1

2. (e)
849　282　93　30　9　[2]
÷3−1　÷3−1　÷3−1　÷3−1　÷3−1

3. (a)
36　154　232　278　300　[306]
+118　+78　+46　+22　+6
−40　−32　−24　−16

4. (a) Series pattern :
$656 - 224 = 432$
$432 - 112 = 320$
$320 - 56 = 264$
$264 - 28 = 236$
and　$236 - 14 = \boxed{222}$

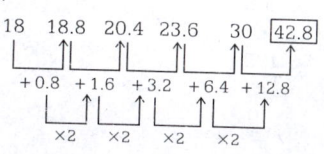

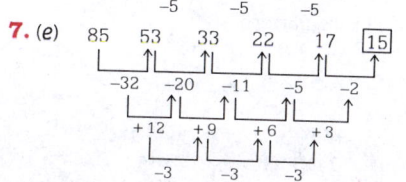

5. (e)
18　18.8　20.4　23.6　30　[42.8]
+0.8　+1.6　+3.2　+6.4　+12.8
×2　×2　×2　×2

6. (c)
5　9　33　72　121　[175]
+4　+24　+39　+49　+54
+20　+15　+10　+5
−5　−5　−5

7. (e) 85　53　33　22　17　[15]
−32　−20　−11　−5　−2
+12　+9　+6　+3
−3　−3　−3

34 / Fast Track Objective Arithmetic

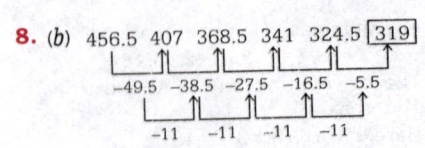

8. (b) 456.5 407 368.5 341 324.5 319
  −49.5 −38.5 −27.5 −16.5 −5.5
  −11 −11 −11 −11

9. (c) 6 5 6 10 19
  -1^2 $+1^2$ $+2^2$ $+3^2$

10. (d) 24 536 487 703 678 742
  +512 −49 +216 −25 +64
  $+(8^3)$ $-(7^2)$ $+(6^3)$ $-(5^2)$ $+(4^3)$

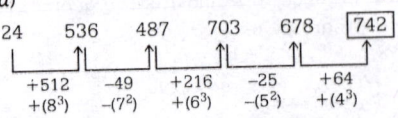

11. (d) 17 9 10 16.5 35 90
  ×0.5 +0.5 ×1+1 ×1.5+1.5 ×2+2 ×2.5+2.5

12. (b) 6 4 5 11 39 189
  ×1−2 ×2−3 ×3−4 ×4−5 ×5−6

13. (a) 14 6 5 6.5 12 29
  ×0.5−1 ×1−1 ×1.5−1 ×2−1 ×2.5−1

14. (e) 6 7 16 51 208 1045
  ×1+1 ×2+2 ×3+3 ×4+4 ×5+5

15. (b) 7 6 10 27 104 515
  ×1−1 ×2−2 ×3−3 ×4−4 ×5−5

16. (d) Series pattern :
  $\frac{600}{5} + 5 = 125$
  $\frac{125}{5} + 5 = 30$
  $\frac{30}{5} + 5 = 11$
  $\frac{11}{5} + 5 = 7.2$
  $\frac{7.2}{5} + 5 = 6.44$
  $\frac{6.44}{5} + 5 = 6.288$

17. (d) Series pattern :
  $9 \times 7 - 1 = 62$
  $62 \times 6 - 1 = 371$
  $371 \times 5 - 1 = 1854$
  $1854 \times 4 - 1 = 7415$
  and $7415 \times 3 - 1 = 22244$

18. (b) $274 \times \frac{1}{2} - 1 = 136$
  $136 \times \frac{1}{2} - 2 = 66$
  $66 \times \frac{1}{2} - 3 = 30$
  $30 \times \frac{1}{2} - 4 = 11$
  $11 \times \frac{1}{2} - 5 = 0.5$

19. (c) The pattern of the series is
  $(6 \times 1) - 2 = 4$
  $(4 \times 2) - 3 = 5$
  $(5 \times 3) - 4 = 11$
  $(11 \times 4) - 5 = 39$
  $(39 \times 5) - 6 = 189$

20. (e) 179 180 172 199 135 260
  $+1^3$ -2^3 $+3^3$ -4^3 $+5^3$

21. (c) 10 6 8 15 34 90
  $\times\frac{1}{2}+1$ ×1+2 $\times\frac{3}{2}+3$ ×2+4 $\times\frac{5}{2}+5$

22. (d) Series pattern :
  $7 \times 0.5 + 0.5 = 4$
  $4 \times 1 + 1 = 5$
  $5 \times 1.5 + 1.5 = 9$
  $9 \times 2 + 2 = 20$
  $20 \times 2.5 + 2.5 = 52.5$
  and $52.5 \times 3 + 3 = 160.5$

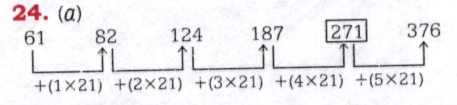

23. (e) 12 13 20 39 82 173
  ×2−11 ×2−6 ×2−1 ×2−(−4) ×2−(−9)

24. (a) 61 82 124 187 271 376
  +(1×21) +(2×21) +(3×21) +(4×21) +(5×21)

25. (c) Series pattern :
  $12 \times 1 + 3 \times 1 = 15$
  $15 \times 2 + 3 \times 2 = 36$
  $36 \times 3 + 3 \times 3 = 117$
  $117 \times 4 + 3 \times 4 = 480$
  $480 \times 5 + 3 \times 5 = 2415$
  and $2415 \times 6 + 3 \times 6 = 14508$

26. (c) 18 43 204 1145 8190 73935
  ×1+25 ×3+75 ×5+125 ×7+175 ×9+225

Series and Progressions / 35

27. (c) Series pattern :
$0 \times 0 + 1 = 1;\ 1 \times 1 + 1 = 2;$
$2 \times 2 + 1 = 5;\ 5 \times 3 + 1 = 16;$
$16 \times 4 + 1 = 65$
and $65 \times 5 + 1 = \boxed{326}$

28. (c)
12   93   730   5097   30570 $\boxed{152839}$
  ×9−15 ×8−14 ×7−13 ×6−12 ×5−11

29. (a) The elements of the given series are the numbers formed by joining together consecutive odd numbers in order, 1 and 3, 3 and 5, 5 and 7, 7 and 9, 9 and 11, ...
∴ Missing term = Number formed by joining 11 and 13 = 1113

30. (d)
8   288   968   2048
↓   ↓   ↓   ↓
2×4 12×24 22×44 32×64
3528 5408 $\boxed{7688}$
↓   ↓   ↓
42×84 52×104 62×124

31. (c) Series pattern :
$50 \times 1.2 = 60;\ 60 \times 1.25 = 75;$
$75 \times 1.3 = 97.5;\ 97.5 \times 1.35 = \boxed{131.625}$
$131.625 \times 1.4 = 184.275$
and $184.275 \times 1.45 = 267.19875$

32. (c) Series pattern :
$90 + (1^2) = 91$
$91 + (1^2 + 2^2 + 2) = 98$
$98 + (1^2 + 2^2 + 3^2 + 3) = 115$
$115 + (1^2 + 2^2 + 3^2 + 4) = 149$
$149 + (1^2 + 2^2 + 3^2 + 4^2 + 5^2 + 5) = \boxed{209}$

33. (a) Series pattern :
$12 \times 0.5 + 0.5 = \boxed{6.5}$
$6.5 \times 1 + 1 = 7.5$
$7.5 \times 1.5 + 1.5 = 12.75$
$12.75 \times 2 + 2 = 27.5$
and $27.5 \times 2.5 + 2.5 = 71.25$

34. (c) Series pattern :
$16 \times \dfrac{3}{2} = 8 \times 3 = 24$
$24 \times \dfrac{3}{2} = 12 \times 3 = \boxed{36}$
$36 \times \dfrac{3}{2} = 18 \times 3 = 54$
$54 \times \dfrac{3}{2} = 27 \times 3 = 81$
and $81 \times \dfrac{3}{2} = 40.5 \times 3 = 121.5$

35. (d) Series pattern :
$12 \times 1 = 12;\ 12 \times 1.5 = 18;$
$18 \times 2.5 = \boxed{45}$
$45 \times 4 = 180$ and $180 \times 6 = 1080$

36. (a) Series pattern:
$+32, +64, +96, +128, ...$
On this pattern, starting from 972, 1292 will come at place of D.

37. (d) Series pattern :
$\times 1.5, \times 2, \times 2.5, \times 3, \times 3.5, \times 4, ...$
On this pattern, starting from 20, 150 will come at the place of C.

38. (b) Series pattern :
$\times 1 + 1, \times 2 + 2, \times 3 + 3, \times 4 + 4, ...$
On this pattern, starting from 7, 1165 will come at the place of E.

39. (c) Series pattern :
$\times 6 - 1, \times 5 - 2, \times 4 - 3, \times 3 - 4, ...$
On this pattern, starting from 7, 4841 will come at place of E.

40. (b)
4   16   48   120   272   584
  +4×2 +8×2 +12×2 +16×2 +20×2

On this pattern, starting from 124, 528 will come at the place of B.

41. (e) Series pattern :
$\times 8 - 7, \times 7 - 6, \times 6 - 5, \times 5 - 4, ...$
On this pattern, starting from 3, 40321 will come at the place of E.

42. (d) Series pattern :
$\times 1.5 + 1.5, \times 2.5 + 2.5, \times 3.5 + 3.5,$
$\times 4.5 + 4.5, ...$
On this pattern, starting from 3, 64.75 will come in place of C.

43. (a) Given, fifth term of an AP $= T_5 = 7$
and ninth term of an AP $= T_9 = 13$
We know that,
$T_n = a + (n-1)d$
∴ $T_5 = a + (5-1)d$
⇒ $7 = a + 4d$
and $T_9 = a + (9-1)d$
⇒ $13 = a + 8d$ ...(ii)
Now, solving Eqs. (i) and (ii), we get
$a = 1$ and $d = \dfrac{3}{2}$
∴ $T_{15} = 1 + (15-1)\dfrac{3}{2}$
$= 1 + 14 \times \dfrac{3}{2} = 1 + 21 = 22$
Hence, 15th term of that AP = 22.

44. (c) $T_3 = 17 = a + 2d$ ...(i)
$T_{17} = a + 16d = -25$ ...(ii)
From Eqs. (i) and (ii), we get
$d = -3$
$\therefore \quad a = 23$
Let nth term of the given AP be -1, then $T_n = -1$
$a + (n-1)d = -1$
$\Rightarrow 23 + (n-1)(-3) = -1$
$\Rightarrow 23 - 3n + 3 = -1$
$\Rightarrow -3n = -1 - 26$
$\Rightarrow -3n = -27 \Rightarrow n = 9$

45. (b) The given series is
$-64, -66, -68, ..., -100$
which is an AP.
Here, first term $(a) = -64$
Common difference $(b) = -66 - (-64)$
$= -66 + 64 = -2$
and last term $(l) = -100$
We know that, $l = a + (n-1)d$
$\Rightarrow -100 = -64 + (n-1)(-2)$
$\Rightarrow -100 = -64 - 2n + 2$
$\Rightarrow 2n = 38 \Rightarrow n = 19$
$\therefore$ Sum of n terms of an AP,
$S_n = \frac{n}{2}[2a + (n-1)d]$
$= \frac{19}{2}[2(-64) + (19-1)(-2)]$
$= \frac{19}{2}(-128 - 36) = -1558$

46. (a) I. $S_n = \frac{n(n+1)}{2} = 861$
$\Rightarrow n^2 + n - 861 \times 2 = 0$
$\Rightarrow (n+42)(n-41) = 0$
$\Rightarrow n = -42, 41$
Hence, Statement I is correct.
II. Given, $S_n = S_{-(n+1)}$.
If $S_n = m$, then we have two values of n if and only if m is positive integer.
Hence, Statement II is incorrect.

47. (a) Given, A, G and H are the arithmetic, geometric and harmonic means between a and b, respectively.
$\therefore \quad A = \frac{a+b}{2}$ ...(i)
$G = \sqrt{ab}$ ...(ii)

and $H = \frac{2ab}{a+b}$ ...(iii)
On multiplying Eqs. (i) and (iii), we get
$AH = \frac{a+b}{2} \times \frac{2ab}{a+b} = ab = (\sqrt{ab})^2$
$\Rightarrow AH = G^2$ [from Eq. (ii)]

48. (c) Let the first term be a and common ratio be r, then
GP series is $a, ar, ar^2, ar^3, ...$
According to the question,
$ar^2 = a + ar$
$\Rightarrow ar^2 = a(1+r)$
$\Rightarrow r^2 - r - 1 = 0$
$\Rightarrow r = \frac{1 \pm \sqrt{(-1)^2 - 4(1)(-1)}}{2 \times 1}$
$\left[\because x = \frac{-b \pm \sqrt{b^2 - 4ac}}{2a} \right]$
$= \frac{1 \pm \sqrt{1+4}}{2} = \frac{1 \pm \sqrt{5}}{2}$
Since, the terms are positive, hence
$r = (1 + \sqrt{5})/2$.

49. (b) $\frac{1}{1 \times 4} + \frac{1}{4 \times 7} + \frac{1}{7 \times 10} + \cdots + \frac{1}{16 \times 19}$
$= \sum_{n=1}^{6} \frac{1}{(3n-2)(3n+1)}$
$= \sum_{n=1}^{6} \frac{1}{3}\left(\frac{1}{(3n-2)} - \frac{1}{(3n+1)} \right)$
$= \frac{1}{3}\left[\left(\frac{1}{1} - \frac{1}{4}\right) + \left(\frac{1}{4} - \frac{1}{7}\right) + \cdots + \left(\frac{1}{16} - \frac{1}{19}\right) \right]$
$= \frac{1}{3}\left(1 - \frac{1}{19}\right) = \frac{1}{3} \times \frac{18}{19} = \frac{6}{19}$

50. (b) Given series is
$\frac{1}{4} + \frac{1}{5} - \frac{1}{6} - \frac{1}{4} + \frac{1}{5} + \frac{1}{6} + \frac{1}{4} - \frac{1}{5} - \frac{1}{6}$
$+ \cdots 47$ terms
It is clear that sum of first 6 terms is zero.
Similarly, sum of 42 terms is zero.
Now, sum of 43, 44, 45, 46, 47 terms are
$\frac{1}{4} + \frac{1}{5} - \frac{1}{6} - \frac{1}{4} - \frac{1}{5} = \frac{-1}{6}$

Chapter 3

HCF and LCM

Factors

If a number x divides another number y exactly (without leaving any remainder), then x is a factor of y and y is a multiple of x.

or

Factors are the set of numbers which exactly divides the given number.

Multiples

Set of numbers, which are exactly divisible by the given number, are multiples of the given number.

For example If the number is 8, then {8, 4, 2, 1} is the set of factors, while {8, 16, 24, 32, ...} is the set of multiples of 8.

- Factors of a number are always less than or equal to the given number.
- Multiples of a number are always more than or equal to the given number.
- 1 is the factor of every number.
- Every number is a factor and multiple of itself.

Common Multiple

A common multiple of two or more numbers is a number which is completely divisible (without leaving remainder) by each of them.

For example We can obtain common multiples of 3, 5 and 10 as follows

Multiples of 3 = {3, 6, 9, 12, 15, 18, 21, 24, 27, **30**, 33, ...}
Multiples of 5 = {5, 10, 15, 20, 25, **30**, 35, ...}
Multiples of 10 = {10, 20, **30**, 40, 50, 60, ...}
∴ Common multiples of 3, 5 and 10
= {30, 60, 90, 120, ...}

Least Common Multiple (LCM)

The LCM of two or more given numbers is the least number which is exactly divisible by each of them.

For example We can obtain LCM of 4 and 12 as follows

Multiples of 4 = 4, 8, **12**, 16, 20, **24**, 28, 32, **36**, ...
Multiples of 12 = **12**, **24**, **36**, 48, 60, 72, ...
Common multiples of 4 and 12 = 12, 24, 36, ...
∴   LCM of 4 and 12 = 12

Methods to Calculate LCM

There are two methods to find the LCM of two or more numbers which are explained below

1. Prime Factorisation Method

Following are the steps to obtain LCM through prime factorisation method

Step I   Resolve the given numbers into the product of their prime factors.

Step II Find the product of all the prime factors (with highest powers) that occur in the given numbers.

Step III This product of all the prime factors (with highest powers) is the required LCM.

Ex. 1 Find the LCM of 8, 12 and 15.

Sol.

```
2 | 8      2 | 12     3 | 15
2 | 4      2 | 6      5 | 5
2 | 2      3 | 3          1
    1          1
```

Factors of $8 = 2 \times 2 \times 2 = 2^3$
Factors of $12 = 2 \times 2 \times 3 = 2^2 \times 3^1$
and factors of $15 = 3 \times 5 = 3^1 \times 5^1$

Here, the prime factors that occur in the given numbers are 2, 3 and 5 and their highest powers are 2^3, 3^1 and 5^1.

∴ Required LCM = $2^3 \times 3^1 \times 5^1 = 8 \times 3 \times 5 = 120$

2. Division Method

Following are the steps to obtain LCM through division method

Step I   Write down the given numbers in a row, separating them by commas.

Step II Divide them by a prime number, which exactly divides atleast two of the given numbers.

Step III Write down the quotients and the undivided numbers in a line below the 1st.

Step IV Repeat the process until you get a line of numbers which are prime to one another.

Step V   The product of all divisors and the numbers in the last line will be the required LCM.

Ex. 2 What will be the LCM of 15, 24, 32 and 45?

Sol. LCM of 15, 24, 32 and 45 is calculated as

2	15,	24,	32,	45
2	15,	12,	16,	45
2	15,	6,	8,	45
3	15,	3,	4,	45
5	5,	1,	4,	15
	1,	1,	4,	3

∴ Required LCM = $2 \times 2 \times 2 \times 3 \times 5 \times 4 \times 3 = 1440$

Common Factor

A common factor of two or more numbers is that particular number which divides each of them exactly.

For example We can obtain common factors of 12, 48, 54 and 63 as follows
Factors of 12 = 12, 6, 4, **3**, 2, **1**
Factors of 48 = 48, 24, 16, 12, 8, 6, 4, **3**, 2, **1**
Factors of 54 = 54, 27, 18, 9, 6, **3**, 2, **1**
and factors of 63 = 63, 21, 9, 7, **3**, **1**
∴ Common factors of 12, 48, 54 and 63 = 3

Highest Common Factor (HCF)

HCF of two or more numbers is the greatest number which divides each of them exactly.
For example 6 is the HCF of 12 and 18 as there is no number greater than 6 that divides both 12 and 18. Similarly, 3 is the HCF of 6 and 9.

✦ HCF is also known as Highest Common Divisor (HCD) and Greatest Common Measure (GCM).

Methods to Calculate HCF

There are two methods to calculate the HCF of two or more numbers which are explained below

1. Prime Factorisation Method

Following are the steps for calculating HCF through prime factorisation method
Step I Resolve the given numbers into product of their prime factors.
Step II Find the product of all the prime factors (with least power) common to all the numbers.
Step III The product of common prime factors (with the least powers) gives HCF.

Ex. 3 Find the HCF of 24, 30 and 42.

Sol. Resolving 24, 30 and 42 into their prime factors.

2	24
2	12
2	6
3	3
	1

2	30
3	15
5	5
	1

2	42
3	21
7	7
	1

40 / Fast Track Objective Arithmetic

∴ Factors of 24 = 2 × 2 × 2 × 3 = ($2^3 \times 3^1$)
Factors of 30 = 2 × 3 × 5 = ($2^1 \times 3^1 \times 5^1$)
and factors of 42 = 2 × 3 × 7 = ($2^1 \times 3^1 \times 7^1$)
∴ The product of common prime factors with the least powers = $2^1 \times 3^1 = 6$
Hence, HCF of 24, 30 and 42 = 6

2. Division Method

Following are the steps to obtain HCF through division method for two numbers

Step I To find the HCF of two given numbers, divide the largest number by the smaller one.

Step II Divide the divisor of step I by the remainder obtained in step I.

Step III Repeat step II till the remainder becomes zero. The last divisor is the required HCF.

- To calculate the HCF of more than two numbers, calculate the HCF of first two numbers, then take the third number and HCF of first two numbers and calculate their HCF and so on. The resulting HCF will be the required HCF of numbers.

Ex. 4 Find the HCF of 26 and 455.

Sol.
$$\begin{array}{r} 26\overline{)455}(17 \\ 26 \\ \hline 195 \\ 182 \\ \hline 13\overline{)26}(2 \\ 26 \\ \hline \times \end{array}$$

∴ Required HCF = 13

Ex. 5 What will be the HCF of 1785, 1995 and 3381?

Sol. In the 1st step, use any two of three numbers.
HCF of 1785 and 1995

$$\begin{array}{r} 1785\overline{)1995}(1 \\ 1785 \\ \hline 210\overline{)1785}(8 \\ 1680 \\ \hline 105\overline{)210}(2 \\ 210 \\ \hline \times \end{array}$$

∴ HCF for 1785 and 1995 = 105

In the 2nd step, use HCF 105 and the 3rd given number 3381.
HCF of 105 and 3381

$$\begin{array}{r} 105\overline{)3381}(32 \\ 315 \\ \hline 231 \\ 210 \\ \hline 21\overline{)105}(5 \\ 105 \\ \hline \times \end{array}$$

∴ Required HCF = 21
Hence, HCF of 1785, 1995 and 3381 is 21.

HCF and LCM / 41

> **MIND IT!** To find the HCF of given numbers, we can divide the numbers by their lowest possible difference. If these numbers are divisible by this difference, then this difference itself is the HCF of the given numbers, otherwise any other factor of this difference will be its HCF.

Ex. 6 Find the HCF of 30, 42 and 135.

Sol. We can notice that the difference between 30 and 42 is less than difference between 135 and 42 or 135 and 30.

Difference between 30 and 42 is 12, but 12 does not divide 30, 42 and 135 completely.

∴ Factors of 12 = 12, 6, 4, 3, 2, 1

Clearly, 3 is the highest factor, which divides all the three numbers completely.

Hence, 3 is the HCF of 30, 42 and 135.

Method to Calculate LCM and HCF of Fractions

The LCM and HCF of fractions can be obtained by the following formulae

(i) LCM of fractions = $\dfrac{\text{LCM of numerators}}{\text{HCF of denominators}}$

(ii) HCF of fractions = $\dfrac{\text{HCF of numerators}}{\text{LCM of denominators}}$

- All the fractions must be in their lowest terms. If they are not in their lowest terms, then conversion in the lowest form is required before finding the HCF or LCM.
- The required HCF of two or more fractions is the highest fraction which exactly divides each of the fractions.
- The required LCM of two or more fractions is the least fraction/integer which is exactly divisible by each of them.
- The HCF of numbers of fractions is always a fraction but this is not true in case of LCM.

Ex. 7 Calculate the LCM of $\dfrac{72}{250}, \dfrac{126}{75}$ and $\dfrac{162}{165}$.

Sol. Here, $\dfrac{72}{250} = \dfrac{36}{125}, \dfrac{126}{75} = \dfrac{42}{25}$ and $\dfrac{162}{165} = \dfrac{54}{55}$ [converting the fractions to lowest terms]

According to the formula,

Required LCM = $\dfrac{\text{LCM of 36, 42 and 54}}{\text{HCF of 125, 25 and 55}} = \dfrac{756}{5} = 151\dfrac{1}{5}$

Ex. 8 Find the HCF of $\dfrac{36}{51}$ and $3\dfrac{9}{17}$.

Sol. Here, $\dfrac{36}{51} = \dfrac{12}{17}$ and $3\dfrac{9}{17} = \dfrac{60}{17}$ [converting the fractions to lowest terms]

Now, we have to find the HCF of $\dfrac{12}{17}$ and $\dfrac{60}{17}$.

According to the formula,

HCF of fractions = $\dfrac{\text{HCF of numerators}}{\text{LCM of denominators}}$

= $\dfrac{\text{HCF of 12 and 60}}{\text{LCM of 17 and 17}} = \dfrac{12}{17}$

- The LCM and HCF of decimals can also be obtained from the above formulae (by converting the decimal into fraction).

Ex. 9 Find the LCM of 0.6, 9.6 and 0.12.

Sol. Here, the given numbers are equivalent to $\frac{6}{10}, \frac{96}{10}, \frac{12}{100}$, i.e. $\frac{3}{5}, \frac{48}{5}, \frac{3}{25}$.

Now, we have to find the LCM of $\frac{3}{5}, \frac{48}{5}, \frac{3}{25}$.

According to the formula,

$$\text{LCM of fractions} = \frac{\text{LCM of numerators}}{\text{HCF of denominators}}$$

$$= \frac{\text{LCM of 3, 48, 3}}{\text{HCF of 5, 5, 25}} = \frac{48}{5} = 9.6$$

Fast Track Techniques to solve the QUESTIONS

Technique 1 Product of two numbers = HCF of numbers × LCM of numbers

Ex. 10 The LCM of two numbers is 2079 and their HCF is 27. If the first number is 189, then find the second number.

Sol. Here, LCM = 2079, HCF = 27, first number = 189, second number = ?

According to the formula,

Product of two numbers = HCF × LCM

$\Rightarrow$ 189 × Second number = 27 × 2079

$\therefore$ Second number = $\frac{27 \times 2079}{189}$ = 297

Technique 2 The greatest number which divides the numbers x, y and z, leaving remainders a, b and c respectively, is given by

HCF of $(x - a), (y - b), (z - c)$.

♦ The above technique is also true for more than three numbers.

Ex. 11 Find the greatest number which divides 29, 60 and 103 leaving remainders 5, 12 and 7, respectively.

Sol. Given that, $x = 29, y = 60, z = 103$,

$a = 5, b = 12$ and $c = 7$

Now, according to the formula,

Required number = HCF of $[(29 - 5), (60 - 12), (103 - 7)]$

= HCF of 24, 48, 96

HCF and LCM / 43

Now,

```
2 | 24      2 | 48      2 | 96
2 | 12      2 | 24      2 | 48
2 | 6       2 | 12      2 | 24
    3       2 | 6       2 | 12
                3       2 | 6
                            3
```

∴ Factors of $24 = 2 \times 2 \times 2 \times 3 = 2^3 \times 3^1$
Factors of $48 = 2 \times 2 \times 2 \times 2 \times 3 = 2^4 \times 3^1$
Factors of $96 = 2 \times 2 \times 2 \times 2 \times 2 \times 3 = 2^5 \times 3^1$
∴ Required HCF of 24, 48 and 96 $= 2^3 \times 3^1 = 8 \times 3 = 24$
Hence, 24 is the required number.

Technique ③ The least number which when divided by x, y and z leaves the remainders a, b and c respectively, is given by [LCM of (x, y, z)] − k, where $k = (x - a) = (y - b) = (z - c)$.

✦ The above technique is also true for more than three numbers.

Ex. 12 Find the least number which when divided by 24, 32 and 36 leaves the remainders 19, 27 and 31, respectively.

Sol. Given that, $x = 24, y = 32, z = 36, a = 19, b = 27$ and $c = 31$
Then, $24 - 19 = 5, 32 - 27 = 5, 36 - 31 = 5$
∴ $\quad k = 5$
According to the formula,
Required number = (LCM of 24, 32 and 36) − 5
Now, LCM of 24, 32 and 36 $= 2 \times 2 \times 2 \times 3 \times 4 \times 3 = 288$
∴ Required number $= 288 - 5 = 283$

```
2 | 24,  32,  36
2 | 12,  16,  18
2 |  6,   8,   9
3 |  3,   4,   9
  |  1,   4,   3
```

Technique ④ The least number which when divided by x, y and z leaves the same remainder k in each case, is given by
[LCM of $(x, y, z) + k$].

✦ The above technique is also true for more than three number.

Ex. 13 Find the least number which when divided by 24, 30 and 54 leaves 5 as remainder in each case.

Sol. Given that, $x = 24, y = 30, z = 54$ and $k = 5$
According to the formula,
Required number = [LCM of (24, 30 and 54)] + 5
Now, LCM of 24, 30, 54

```
2 | 24,  30,  54
3 | 12,  15,  27
  |  4,   5,   9
```

∴ LCM $= 2 \times 3 \times 4 \times 5 \times 9 = 1080$
∴ Required number $= 1080 + 5 = 1085$

Technique 5 The greatest number that will divide $x, y, z, \ldots$ leaving the same remainder in each case, is given by

[HCF of $|x-y|, |y-z|, |z-x|, \ldots$].

Ex. 14 What is the greatest number that will divide 99, 123 and 183 leaving the same remainder in each case? Also, find the common remainder.

Sol. Given that, $x = 99, y = 123$ and $z = 183$.

According to the formula,

Required number = HCF of $(|x-y|, |y-z|, |z-x|)$

Now, $|x - y| = |99 - 123| = 24$

$|y - z| = |123 - 183| = 60$

and $|z - x| = |183 - 99| = 84$

Therefore,

Factors of $24 = 2^3 \times 3^1$

Factors of $60 = 2^2 \times 3^1 \times 5^1$

Factors of $84 = 2^2 \times 3^1 \times 7^1$

$\therefore$ Required number = HCF of 24, 60 and 84

$= 2^2 \times 3^1 = 4 \times 3 = 12$

Common remainder $= \dfrac{99}{12} = 8\dfrac{3}{12}, \dfrac{123}{12} = 10\dfrac{3}{12}, \dfrac{183}{12} = 15\dfrac{3}{12}$

Hence, the required common remainder is 3.

Technique 6 When the HCF of each pair of n given numbers is a and their LCM is b, then the product of these numbers is given by

$(a)^{n-1} \times b$ or $(HCF)^{n-1} \times LCM$.

Ex. 15 There are five numbers. HCF of each possible pair is 4 and LCM of all the five numbers is 27720. What will be the product of all the five numbers?

Sol. Given, HCF $= 4$, LCM $= 27720$ and $n = 5$.

According to the formula,

Required product $= (HCF)^{n-1} \times LCM = (4)^{5-1} \times 27720 = (4)^4 \times 27720$

$= 256 \times 27720 = 7096320$

Some Useful Results

(i) If a is the HCF of two parts of b, then b must be divisible by a. This rule has been derived from the proven fact that, if two numbers are divisible by a certain number, then their sum is also divisible by that number.

(ii) If the two numbers are prime to each other (coprimes), then their HCF should be equal to 1. Conversely, if their HCF is equal to 1, the numbers are prime to each other.

(iii) LCM of numbers which are prime to each other, is the product of numbers itself.

(iv) If k is the HCF of p and q, then
  (a) k is the HCF of p and $p + q$ also.
  (b) k is the HCF of p and $q - p$ also.

(v) LCM of numbers is always divisible by the HCF of given numbers, i.e. if x is the LCM of a, b, c, d, e and y is the HCF, then x is completely divisible by y.

HCF and LCM / 45

Technique 7 The greatest *n*-digit number which when divided by *x*, *y* and *z* leaves

(i) no remainder, then
   Required number = Greatest *n*-digit number − R

(ii) remainder *k*, then
   Required number = [Greatest *n*-digit number − R] + k

where, R is the remainder obtained when *n*-digit greatest number is divided by the LCM of *x*, *y* and *z*.

Ex. 16 Find the largest possible number of five digits which is exactly divisible by 32, 36 and 40.

Sol. Given numbers are 32, 36 and 40.
i.e. $x = 32$, $y = 36$ and $z = 40$
LCM of 32, 36 and 40 = 1440
Greatest 5-digit number = 99999

4	32,	36,	40
4	8,	9,	10
2	2,	9,	10
	1,	9,	5

$$1440 \overline{)99999} (69$$
$$\underline{8640}$$
$$13599$$
$$\underline{12960}$$
$$639 = R$$

∴ Required number = Greatest 5-digit number − R = 99999 − 639 = 99360

Ex. 17 Find the greatest 4-digit number such that when divided by 16, 24 and 36, leaves remainder 4 in each case.

Sol. Given numbers are 16, 24 and 36.
i.e. $x = 16$, $y = 24$, $z = 36$ and $k = 4$
∴ LCM of 16, 24 and 36 = $2 \times 2 \times 2 \times 2 \times 3 \times 3 = 144$
Greatest 4-digit number = 9999
On dividing 9999 by 144,

2	16,	24,	36
2	8,	12,	18
2	4,	6,	9
3	2,	3,	9
	2,	1,	3

$$144 \overline{)9999} (69$$
$$\underline{864}$$
$$1359$$
$$\underline{1296}$$
$$63 = R$$

∴ Required number = (Greatest 4-digit number − R) + k = (9999 − 63) + 4 = 9940

Technique 8 The smallest *n*-digit number which when divided by *x*, *y* and *z* leaves

(i) no remainder, then
   Required number = [Smallest *n*-digit number + (L − R)]

(ii) remainder *k*, then
   Required number = [Smallest *n*-digit number + (L − R)] + k

where, R is the remainder obtained when *n*-digit smallest number is divided by LCM of *x*, *y*, *z* and L is the LCM of *x*, *y*, *z*.

46 / Fast Track Objective Arithmetic

Ex. 18 Find the smallest 3-digit number, which is exactly divisible by 3, 4 and 5.

Sol. Here, $x = 3, y = 4$ and $z = 5$

$\therefore$ LCM of 3, 4 and 5 $= 3 \times 4 \times 5 = 60$, i.e. $L = 60$

and smallest 3-digit number $= 100$

Now, dividing 100 by 60,

$$60) \overline{100} (1$$
$$\underline{60}$$
$$\overline{40} = R$$

$\therefore$ Required number $=$ [Smallest n-digit number $+ (L - R)$] $= [100 + (60 - 40)] = 100 + 20 = 120$

Ex. 19 Find the least possible 5-digit number, which when divided by 10, 12, 16 and 18 leaves remainder 27.

Sol. Here, $x = 10, y = 12, z = 16, w = 18$ and $k = 27$

First of all, we will find the LCM of 10, 12, 16 and 18.

2	10, 12, 16, 18
2	5, 6, 8, 9
3	5, 3, 4, 9
	5, 1, 4, 3

$\therefore$ LCM $= 2 \times 2 \times 3 \times 5 \times 4 \times 3 = 720$, i.e. $L = 720$

We know that,

Smallest 5-digit number $= 10000$

On dividing 10000 by 720,

$$720) \overline{10000} (13$$
$$\underline{720}$$
$$\overline{2800}$$
$$\underline{2160}$$
$$\overline{640} = R$$

$\therefore$ Required number $=$ [Smallest 5-digit number $+ (L - R)] + k$

$= [10000 + (720 - 640)] + 27 = (10000 + 80) + 27 = 10107$

Ex. 20 Is it possible to divide 1394 into 2 parts such that their HCF may be 34?

Sol. If 34 is the HCF of two parts of 1394, then 1394 must be divisible by 34.

Let us see $\dfrac{1394}{34} = 41$

Hence, it is possible to divide 1394 into 2 parts such that their HCF may be 34.

Ex. 21 Are 6 and 11 coprimes?

Sol. Clearly, HCF of 6 and 11 is equal to 1. Therefore, 6 and 11 are coprimes or prime to each other.

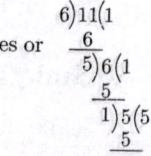

$$6) \overline{11} (1$$
$$\underline{6}$$
$$5) \overline{6} (1$$
$$\underline{5}$$
$$1) \overline{5} (5$$
$$\underline{5}$$
$$\times$$

Method to Solve Questions Based on Bells

To solve such questions, following steps are used

Step I Find the LCM of given time intervals.

Step II Obtained LCM is added to the initial time and result of this addition will be our answer (the next time when bells ring together).

✦ Before calculating LCM, make sure that all numbers are in the same unit. This method is also applicable to runner running round a circular track or traffic light glowing at different intervals.

HCF and LCM / 47

Ex. 22 Seven bells ring at intervals of 2, 3, 4, 6, 8, 9 and 12 min, respectively. They started ringing simultaneously at 5 : 00 in the morning. What will be the next time when they all ring simultaneously?

Sol. LCM of 2, 3, 4, 6, 8, 9 and 12

$$\begin{array}{r|l} 2 & 2, 3, 4, 6, 8, 9, 12 \\ \hline 2 & 1, 3, 2, 3, 4, 9, 6 \\ \hline 3 & 1, 3, 1, 3, 2, 9, 3 \\ \hline & 1, 1, 1, 1, 2, 3, 1 \end{array}$$

LCM = 2 × 2 × 2 × 3 × 3 = 72 min = 1 h 12 min

∴ Required time = (5 + 1 : 12) O'clock = 6 : 12 O'clock in the morning.

Ex. 23 Six bells ring at intervals of 2, 4, 6, 8, 10 and 12 s, respectively. They started ringing simultaneously. How many times, will they ring together in 30 min?

Sol. ∵ LCM of 2, 4, 6, 8, 10 and 12 = 120

i.e. bells will ring together after every 120 s or 2 min.

∴ Required number of times = $\left(\dfrac{30}{2} + 1\right)$ = 16

[here, 1 is added because bells started ringing together]

Ex. 24 Three rings complete 60, 36 and 24 revolutions in a minute. They start from a certain point in their circumference downwards. By what time they come together again in the same position?

Sol. Time taken by each ring in one revolution are $\dfrac{60}{60}$ s, $\dfrac{60}{36}$ s and $\dfrac{60}{24}$ s, i.e. 1s, $\dfrac{5}{3}$ s and $\dfrac{5}{2}$ s, respectively.

∴ Required time = LCM of 1, $\dfrac{5}{3}$, $\dfrac{5}{2}$ = 5 s

Ex. 25 The traffic lights at three different road crossings change after every 48 s, 72 s and 108 s, respectively. If they all change simultaneously at 8 : 20 : 00 h, when will they again change simultaneously?

Sol. Let us first calculate LCM of 48, 72 and 108.

$$\begin{array}{r|l} 2 & 48, 72, 108 \\ \hline 2 & 24, 36, 54 \\ \hline 2 & 12, 18, 27 \\ \hline 3 & 6, \ 9, 27 \\ \hline 3 & 2, \ 3, \ 9 \\ \hline & 2, \ 1, \ 3 \end{array}$$

∴ LCM of 48, 72, 108 = 2 × 2 × 2 × 3 × 3 × 2 × 3 = 432 s

Thus, the three lights will change after 432 s, i.e. 7 min 12 s in the second time.

Hence, the three lights will change simultaneously at the next time

= 8 : 20 : 00 + 0 : 7 : 12 = 8 : 27 : 12.

Multi Concept Questions

1. What is the quotient, when LCM is divided by the HCF of geometric progression with first term a and common ratio r?
   (a) r^{n+1} (b) r^{n-2} (c) r^{n-1} (d) r^{n+2}

   ➥ (c) Given that, first term = a and common ratio = r
   ∴ Required GP = $a, ar, ar^2, ar^3, ..., ar^{n-1}$
   HCF of GP = a and LCM of GP = ar^{n-1}
   According to the question, $\dfrac{\text{LCM of GP}}{\text{HCF of GP}} = \dfrac{ar^{n-1}}{a} = r^{n-1}$

2. What is the LCM of $x^2 + 2x - 8$, $x^3 - 4x^2 + 4x$ and $x^2 + 4x$?
   (a) $x(x+4)(x-2)^2$ (b) $x(x+4)(x-2)$ (c) $x(x+4)(x+2)^2$ (d) $x(x+4)^2(x-2)$

   ➥ (a) $x^2 + 2x - 8 = x^2 + 4x - 2x - 8$
   $= x(x+4) - 2(x+4) = (x-2)(x+4)$
   $x^3 - 4x^2 + 4x = x^3 - 2x^2 - 2x^2 + 4x$
   $= x^2(x-2) - 2x(x-2) = (x^2 - 2x)(x-2)$
   $= x(x-2)(x-2)$
   and $x^2 + 4x = x(x+4)$
   Now, LCM of $(x^2 + 2x - 8)$, $(x^3 - 4x^2 + 4x)$ and $(x^2 + 4x)$
   $= x(x-2)(x+4)(x-2) = x(x+4)(x-2)^2$

3. What is the HCF of $36(3x^4 + 5x^3 - 2x^2)$, $9(6x^3 + 4x^2 - 2x)$ and $54(27x^4 - x)$?
   [CDS 2012]
   (a) $9x(x+1)$ (b) $9x(3x-1)$ (c) $18x(3x-1)$ (d) $18x(x+1)$

   ➥ (c) Let $P(x) = 36(3x^4 + 5x^3 - 2x^2)$
   $= 36x^2(3x^2 + 5x - 2)$
   $= 36x^2\{3x^2 + 6x - x - 2\}$
   $= 36x^2\{3x(x+2) - 1(x+2)\}$
   $= 2 \times 2 \times 3 \times 3 \times x \times x \times (x+2)(3x-1)$
   $Q(x) = 9(6x^3 + 4x^2 - 2x)$
   $= 9x(6x^2 + 4x - 2)$
   $= 18x(3x^2 + 2x - 1)$
   $= 18x\{3x^2 + 3x - x - 1\}$
   $= 18x\{3x(x+1) - 1(x+1)\}$
   $= 2 \times 3 \times 3 \times x \times (3x-1)(x+1)$
   and $R(x) = 54(27x^4 - x) = 54x(27x^3 - 1)$
   $= 2 \times 3 \times 3 \times 3 \times x \times (3x-1)(9x^2 + 3x + 1)$ [∵ $a^3 - b^3 = (a-b)(a^2 + b^2 + ab)$]
   So, HCF of $[P(x), Q(x), R(x)] = 2 \times 3 \times 3 \times x(3x-1) = 18x(3x-1)$

Fast Track Practice

Exercise 1 Base Level Questions

1. If three numbers are $2a$, $5a$ and $7a$, then what will be their LCM? [Bank Clerk 2011]
   (a) $70a$ (b) $65a$ (c) $75a$ (d) $70a^3$
   (e) None of these

2. Find the LCM of 8, 15, 24 and 72.
   (a) 350 (b) 360 (c) 720 (d) 735
   (e) None of these

3. Find the LCM of $(2^3 \times 3 \times 5^2 \times 7)$ $(2^4 \times 3^2 \times 5 \times 7^2 \times 11)$ and $(2 \times 3^3 \times 5^4)$
   (a) $2^4 \times 3^3 \times 5^4$ [RRB 2008]
   (b) $2 \times 3 \times 7 \times 5 \times 11$
   (c) $2^4 \times 3^3 \times 5^4 \times 7^2 \times 11$
   (d) $2^4 \times 3^4 \times 5^4 \times 7$

4. If the HCF of a and b are 12, where a, b are positive integers and $a > b > 12$, then what will be the values of a and b? [RRB 2012]
   (a) 12, 24 (b) 24, 12 (c) 24, 36 (d) 36, 24

5. What will be the HCF of $(2 \times 3 \times 7 \times 9)$, $(2 \times 3 \times 9 \times 11)$ and $(2 \times 3 \times 4 \times 5)$? [SSC CGL 2008]
   (a) $2 \times 3 \times 7$ (b) $2 \times 3 \times 9$
   (c) 2×3 (d) $2 \times 7 \times 9 \times 11$

6. Find the HCF of 144, 180 and 192. [SSC CPO 2015]
   (a) 540 (b) 432 (c) 36 (d) 12

7. If the product of two coprime numbers is 117, then their LCM is [SSC CGL 2013]
   (a) 9 (b) 13 (c) 39 (d) 117

8. In a store, there are 345 L mustard oil, 120 L sunflower oil and 225 L soyabean oil. What will be the capacity of the largest container to measure the above three types of oil?
   (a) 8 L (b) 20 L (c) 23 L (d) 15 L
   (e) None of these

9. If HCF of two numbers is 8, then which of the following can never be their LCM? [RBI Clerk 2007]
   (a) 24 (b) 48 (c) 56 (d) 60
   (e) None of these

10. Find the LCM of $\frac{1}{3}, \frac{2}{9}, \frac{5}{6}$ and $\frac{4}{27}$. [RRB 2007]
    (a) $\frac{1}{54}$ (b) $\frac{10}{27}$ (c) $\frac{20}{3}$ (d) $\frac{3}{20}$

11. Find the HCF of $\frac{1}{2}, \frac{3}{4}$ and $\frac{4}{5}$.
    (a) $\frac{1}{20}$ (b) $\frac{1}{40}$ (c) 20 (d) 15
    (e) None of these

12. Which of the following will be the LCM of 0.25, 0.1 and 0.125?
    (a) 0.25 (b) 0.005 (c) 0.05 (d) 0.5
    (e) None of these

13. Find the LCM of 2.5, 1.2, 20 and 7.5.
    (a) 60 (b) 65 (c) 70 (d) 50
    (e) None of these

14. If a number is exactly divisible by 11 and 13, then the number must be [Hotel Mgmt. 2008]
    (a) divisible by (11 + 13)
    (b) divisible by (13 − 11)
    (c) divisible by (11 × 13)
    (d) divisible by (13 ÷ 11)

15. The product of HCF and LCM of 18 and 15 is [CDS 2012]
    (a) 120 (b) 150 (c) 175 (d) 270

16. The LCM of two numbers is 2376 while their HCF is 33. If one of the numbers is 297, then the another number is [CDS 2013]
    (a) 216 (b) 264 (c) 642 (d) 792

17. The HCF and LCM of two numbers are 13 and 1989, respectively. If one of the numbers is 117, then determine the another. [DMRC CRA 2012]
    (a) 121 (b) 131 (c) 221 (d) 231

18. The HCF of two expressions p and q is 1. What is the reciprocal of their LCM?
    (a) $p + q$ (b) $p - q$
    (c) pq (d) $(pq)^{-1}$

19. The HCF and LCM of two natural numbers are 12 and 72, respectively. What is the difference between the two numbers, if one of the number is 24? [CDS 2012]
    (a) 12 (b) 18 (c) 21 (d) 24

20. The ratio of two numbers is 3 : 4 and their HCF is 4. What will be their LCM? [Hotel Mgmt. 2007]
    (a) 12 (b) 16 (c) 24 (d) 48

21. If the ratio of two numbers is 5 : 6 and their LCM is 480, then their HCF is [SSC Multitasking 2013]
    (a) 20   (b) 16   (c) 6   (d) 5

22. The product of two whole numbers is 1500 and their HCF is 10. Find the LCM. [Bank Clerk 2008]
    (a) 15000   (b) 150   (c) 1500   (d) 15
    (e) None of these

23. Find the greatest number that divides 130, 305 and 245 leaving remainders 6, 9 and 17, respectively. [RBI Clerk 2008]
    (a) 4   (b) 5   (c) 14   (d) 24
    (e) None of these

24. What will be the greatest number that divides 1023 and 750 leaving remainders 3 and 2, respectively?
    (a) 68   (b) 65   (c) 78   (d) 19
    (e) None of these

25. What is the greatest number that divides 13850 and 17030 leaves a remainder 17? [CDS 2012]
    (a) 477   (b) 159   (c) 107   (d) 87

26. What will be the greatest number that divides 1356, 1868 and 2764 leaving 12 as remainder in each case? [Delhi Police 2007]
    (a) 64   (b) 124   (c) 156   (d) 260

27. What is the greatest four-digit number which when divided by 10, 15, 21 and 28 leaves remainders 4, 9, 15 and 22, respectively? [LIC ADO 2008]
    (a) 9654   (b) 9666   (c) 9664   (d) 9864
    (e) None of these

28. What will be the least number which when divided by 12, 21 and 35 leaves 6 as remainder in each case? [UP Police 2007]
    (a) 426   (b) 326   (c) 536   (d) 436

29. Find the least number which when divided by 16, 18 and 20 leaves a remainder 4 in each case, but is completely divisible by 7.
    (a) 2884   (b) 2256   (c) 865   (d) 3332
    (e) None of these

30. Find the least number which when divided by 12, 16 and 18 leaves 5 as remainder in each case. [RBI Clerk 2009]
    (a) 139   (b) 144   (c) 149   (d) 154
    (e) None of these

31. Find the largest number which divides 1305, 4665 and 6905 leaving same remainder in each case. Also, find the common remainder. [CBI Clerk 2009]
    (a) 1210, 158   (b) 1120, 158
    (c) 1120, 185   (d) 1210, 185
    (e) None of these

32. There are four numbers. The HCF of each pair is 5 and the LCM of all the four numbers is 2310. What is the product of four numbers?
    (a) 288750   (b) 288570
    (c) 828570   (d) 288650

33. Find the greatest number of three digits which when divided by 6, 9 and 12 leaves 3 as remainder in each case. [CBI 2008; BOI 2007]
    (a) 975   (b) 996   (c) 903   (d) 939
    (e) None of these

34. Find the greatest 4-digit number, which when divided by 12, 18, 21 and 28, leaves a remainder 3 in each case.
    (a) 9831   (b) 9913   (c) 9940   (d) 9911

35. The least number of four digits which is divisible by each one of the numbers 12, 18, 21 and 28, is
    (a) 1008   (b) 1006   (c) 1090   (d) 1080

36. Three electronic devices make a beep after every 48 s, 72 s and 108 s, respectively. They beeped together at 10 am. The time when they will next make a beep together at the earliest is [SSC CGL (Mains) 2016]
    (a) 10 : 07 : 12   (b) 10 : 07 : 24
    (c) 10 : 07 : 36   (d) 10 : 07 : 48

37. Five bells begin to toll together at intervals of 9 s, 6 s, 4 s, 10 s and 8 s, respectively. How many times will they toll together in the span of 1 h (excluding the toll at the start)? [Bank Clerk 2007]
    (a) 5   (b) 8   (c) 10
    (d) Cannot be determined
    (e) None of these

38. Monica, Veronica and Rachat begin to jog around a circular stadium. They complete their revolutions in 42 s, 56 s and 63 s, respectively. After how many seconds will they be together at the starting point? [Bank Clerk 2008]
    (a) 366   (b) 252   (c) 504
    (d) Cannot be determined
    (e) None of these

39. A General can draw up his soldiers in the rows of 10, 15 or 18 soldiers and he can also draw them up in the form of a solid square. Find the least number of soldiers with the General. [SSC CGL 2007]
    (a) 100   (b) 3600   (c) 900   (d) 90

40. The least number which should be added to 2497, so that the sum is exactly divisible by 5, 6, 4 and 3, is
    (a) 3   (b) 13   (c) 23   (d) 33

HCF and LCM / 51

41. What is the least number which when increased by 9, is divisible by each one of 24, 32, 36 and 54?
    (a) 855 (b) 890 (c) 756 (d) 895
    (e) None of these

42. The HCF of three numbers is 23. If they are in the ratio of 1 : 2 : 3, then find the numbers.
    (a) 69, 15, 22   (b) 23, 46, 69
    (c) 25, 31, 41   (d) 23, 21, 35
    (e) None of these

43. Three numbers are in the ratio of 3 : 4 : 5 and their LCM is 1200. Find the HCF of the numbers.
    (a) 40   (b) 30   (c) 80   (d) 20
    (e) None of these

44. The HCF and LCM of two numbers m and n are respectively 6 and 210. If $m + n = 72$, then $\frac{1}{m} + \frac{1}{n}$ is equal to

    (a) $\frac{1}{35}$ (b) $\frac{3}{35}$ (c) $\frac{5}{37}$ (d) $\frac{2}{35}$
    (e) None of these

45. The LCM of two numbers is 48. The numbers are in the ratio of 2 : 3. Find the sum of the numbers. **[SSC (10+2) 2011]**
    (a) 28   (b) 32   (c) 40   (d) 64

46. Four numbers are in the ratio of $10 : 12 : 15 : 18$. If their HCF is 3, then find their LCM.
    (a) 420   (b) 540
    (c) 620   (d) 680
    (e) None of these

47. What is the least number which is exactly divisible by 8, 9, 12, 15 and and is also a perfect square?
    (a) 3600   (b) 7200
    (c) 5200   (d) 6500
    (e) None of these

Exercise 2 *Higher Skill Level Questions*

1. The least number which when divided by 4, 6, 8 and 9 leaves zero remainder in each case and when divided by 13 leaves a remainder of 7, is **[SSC CGL (Pre) 2016]**
   (a) 144   (b) 72   (c) 36   (d) 85

2. A is a set of those positive integers such that when these are divided by 2, 3, 4, 5 and 6 leaves the remainders 1, 2, 3, 4 and 5, respectively. How many integers between 0 and 100 belong to the set A? **[CDS 2016 (II)]**
   (a) No integer   (b) One
   (c) Two   (d) Three

3. There are two numbers p and q such that their HCF is 1. Which of the following statements are correct?
   I. Both p and q may be prime.
   II. One number may be prime and the other composite.
   III. Both the numbers may be composite.
   Select the correct answer using the code given below. **[CDS 2016 (II)]**
   (a) I and II   (b) II and III
   (c) I and III   (d) I, II and III

4. Two pipes of lengths 1.5 m and 1.2 m are to be cut into equal pieces without leaving any extra length of pipes. The greatest length of the pipe pieces of same size, which can be cut from these two lengths, will be **[SSC CGL (Mains) 2016]**
   (a) 0.13 m (b) 0.4 m (c) 0.3 m (d) 0.41 m

5. The difference of two numbers is 1/9 of their sum and their sum is 45. Find the LCM. **[SSC CGL 2007]**
   (a) 225   (b) 100   (c) 150   (d) 200

6. The sum of two numbers is 1056 and their HCF is 66, find the number of such pairs.
   (a) 6   (b) 2   (c) 4   (d) 8
   (e) None of these

7. The sum of HCF and LCM of two numbers is 403 and their LCM is 12 times their HCF. If one number is 93, then find the other number. **[MBA 2007]**
   (a) 115   (b) 122   (c) 124   (d) 138

8. The LCM of two numbers is 495 and their HCF is 5. If sum of the numbers is 100, then find the difference of the numbers. **[Hotel Mgmt. 2008]**
   (a) 10   (b) 46   (c) 70   (d) 90

9. The LCM of two numbers is 20 times of their HCF and (LCM + HCF) = 2520. If one number is 480, then what will be the triple of another number?
   (a) 1200 (b) 1500 (c) 2100 (d) 1800
   (e) None of these

10. Find the side of the largest possible square slabs which can be paved on the floor of a room 2 m 50 cm long and 1 m 50 cm broad. Also, find the number of such slabs to pave the floor.
    [LIC AAO 2007]

(a) 25, 20 (b) 30, 15
(c) 50, 15 (d) 55, 10
(e) None of these

11. What is the sum of digits of the least multiple of 13, which when divided by 6, 8 and 12, leaves 5, 7 and 11 respectively, as the remainders? [CDS 2015 (II)]
(a) 5 (b) 6
(c) 7 (d) 8

12. What is the LCM of $x^3 + 8$, $x^2 + 5x + 6$ and $x^3 + 4x^2 + 4x$? [CDS 2017 (I)]
(a) $x(x + 2)^2(x + 3)(x^2 - 2x + 4)$
(b) $x(x - 2)^2(x - 3)(x^2 + 2x + 4)$
(c) $(x + 2)^2(x + 3)(x^2 - 2x + 4)$
(d) $(x - 2)^2(x - 3)(x^2 - 2x + 4)$

13. If $(x + 1)$ is the HCF of $Ax^2 + Bx + C$ and $Bx^2 + Ax + C$, where $A \neq B$, then the value of C is [CDS 2015 (II)]
(a) A (b) B
(c) A − B (d) 0

14. If a and b are positive integers, then what is the value of $\text{HCF}\left(\dfrac{a}{\text{HCF}(a, b)}, \dfrac{b}{\text{HCF}(a, b)}\right)$? [CDS 2014]
(a) a (b) b (c) 1 (d) $\dfrac{a}{\text{HCF}(a, b)}$

15. What is the HCF of $8(x^5 - x^3 + x)$ and $28(x^6 + 1)$? [CDS 2014]
(a) $4(x^4 - x^2 + 1)$ (b) $x^3 - x + 4x^2$
(c) $x^3 - x + 3x^2$ (d) None of these

16. The HCF of $(x^4 - y^4)$ and $(x^6 - y^6)$ is [CDS 2013]
(a) $x^2 - y^2$ (b) $x - y$
(c) $x^3 - y^3$ (d) $x^4 - y^4$

17. The HCF of $(x^3 - x^2 - 2x)$ and $(x^3 + x^2)$ is [CDS 2013]
(a) $x^3 - x^2 - 2x$ (b) $x^2 + x$
(c) $x^4 - x^3 - 2x^2$ (d) $x - 2$

18. What is the HCF of $a^2b^4 + 2a^2b^2$ and $(ab)^7 - 4a^2b^9$? [CDS 2013]
(a) ab (b) a^2b^3
(c) a^2b^2 (d) a^3b^2

19. For any integer n, what is HCF $(22n + 7, 33n + 10)$ equal to? [CDS 2014]
(a) 0 (b) 1 (c) 11 (d) 2

20. For any integers a and b with HCF $(a, b) = 1$, what is HCF $(a + b, a − b)$ equal to? [CDS 2013]
(a) It is always 1 (b) It is always 2
(c) Either 1 or 2 (d) None of these

21. The sum and difference of two expressions are $5x^2 - x - 4$ and $x^2 + 9x - 10$, respectively. The HCF of the two expressions will be [CDS 2016]
(a) $(x + 1)$ (b) $(x - 1)$
(c) $(3x + 7)$ (d) $(2x - 3)$

22. Consider the following statements in respect of natural numbers a, b and c
I. LCM $(ab, ac) = a$ LCM (b, c)
II. HCF $(ab, ac) = a$ HCF (b, c)
III. HCF $(a, b) <$ LCM (a, b)
IV. HCF (a, b) divides LCM (a, b).
Which of the above statements are correct? [CDS 2016 (I)]
(a) I and II (b) III and IV
(c) I, II and IV (d) I, II, III and IV

23. Consider the following statements
I. If $a = bc$ with HCF $(b, c) = 1$, then HCF $(c, bd) =$ HCF (c, d).
II. If $a = bc$ with HCF $(b, c) = 1$, then LCM $(a, d) =$ LCM (c, bd).
Which of the above statements is/are correct? [CDS 2017 (I)]
(a) Only I (b) Only II
(c) Both I and II (d) Neither I nor II

24. The HCF and LCM of two polynomials are $(x + y)$ and $(3x^5 + 5x^4y + 2x^3y^2 - 3x^2y^3 - 5xy^4 - 2y^5)$, respectively. If one of the polynomials is $(x^2 - y^2)$, then the other polynomial is [CDS 2015 (I)]
(a) $3x^4 - 8x^3y + 10x^2y^2 + 7xy^3 - 2y^4$
(b) $3x^4 - 8x^3y - 10x^2y^2 + 7xy^3 + 2y^4$
(c) $3x^4 + 8x^3y + 10x^2y^2 + 7xy^3 + 2y^4$
(d) $3x^4 + 8x^3y - 10x^2y^2 + 7xy^3 + 2y^4$

25. An ascending series of numbers satisfies the following conditions
I. When divided by 3, 4, 5 and 6, the numbers leave a remainder of 2.
II. When divided by 11, the numbers leave no remainder.
The 6th number in this series will be [XAT 2015]
(a) 242 (b) 2882
(c) 3542 (d) 4202
(e) None of these

Answer with Solutions

Exercise 1 Base Level Questions

1. (a) Required LCM = $a \times 2 \times 5 \times 7 = 70a$

2. (b) By prime factorisation method,
Factors of $8 = 2 \times 2 \times 2 = 2^3$
Factors of $15 = 3 \times 5 = 3^1 \times 5^1$
Factors of $24 = 2 \times 2 \times 2 \times 3 = 2^3 \times 3^1$
and factors of $72 = 2 \times 2 \times 2 \times 3 \times 3 = 2^3 \times 3^2$
Here, the prime factors that occur in the given numbers are 2, 3 and 5 and their highest powers are 2^3, 3^2 and 5^1.
$\therefore$ LCM of 8, 15, 24 and $72 = 2^3 \times 3^2 \times 5^1$
$= 8 \times 9 \times 5 = 360$

Alternate Method
By division method,

2	8,	15,	24,	72
2	4,	15,	12,	36
2	2,	15,	6,	18
3	1,	15,	3,	9
	1,	5,	1,	3

$\therefore$ LCM of 8, 15, 24 and 72
$= 2 \times 2 \times 2 \times 3 \times 3 \times 5 = 360$

3. (c) Given prime factors are
$(2^3 \times 3 \times 5^2 \times 7); (2^4 \times 3^2 \times 5 \times 7^2 \times 11);$
$(2 \times 3^3 \times 5^4)$
$\therefore$ Required LCM = Product of common prime factors having highest powers
$= 2^4 \times 3^3 \times 5^4 \times 7^2 \times 11$

4. (d) From option (d), we can say that the HCF of 36 and 24 is 12 and it also satisfies the given condition $a > b > 12$.
$\therefore a = 36$ and $b = 24$

5. (c) Given factors are
$(2 \times 3 \times 7 \times 9); (2 \times 3 \times 9 \times 11); (2 \times 3 \times 4 \times 5)$
$\therefore$ Required HCF = Product of common prime factors having least powers
$= 2 \times 3$

6. (d) In the 1st step, we take the two numbers 144 and 180.
HCF of 144 and 180

$$144\overline{)180}(1$$
$$\underline{144}$$
$$36\overline{)144}(4$$
$$\underline{144}$$
$$\times$$

$\therefore$ HCF of 144 and 180 is 36.

In the 2nd step, we take 36 and the third given number 192.
HCF of 36 and 192

$$36\overline{)192}(5$$
$$\underline{180}$$
$$12\overline{)36}(3$$
$$\underline{36}$$
$$\times$$

Hence, HCF of 144, 180 and 192 is 12.

7. (d) LCM of two coprimes is equal to their product.

8. (d) Required capacity = HCF of 345 L, 120 L and 225 L
Firstly, find the HCF of any two numbers, i.e. 120 and 225.

$$120\overline{)225}(1$$
$$\underline{120}$$
$$105\overline{)120}(1$$
$$\underline{105}$$
$$15\overline{)105}(7$$
$$\underline{105}$$
$$\times$$

Here, HCF of 120 and 225 = 15
Now, find the HCF of 15 and 345.

$$15\overline{)345}(23$$
$$\underline{30}$$
$$45$$
$$\underline{45}$$
$$\times$$

$\therefore$ HCF of 345, 120 and 225 = 15
Hence, required capacity of container to measure the oil is 15 L.

9. (d) We know that, LCM of two numbers must be the multiple of their HCF. In the given options, 60 is not a multiple of 8 and hence 60 cannot be the LCM of the numbers.

10. (c) We know that,
LCM of fractions = $\dfrac{\text{LCM of numerators}}{\text{HCF of denominators}}$
$\therefore$ Required LCM = $\dfrac{\text{LCM of 1, 2, 5 and 4}}{\text{HCF 3, 9, 6 and 27}}$
$= \dfrac{20}{3}$

11. (a) Required HCF = $\dfrac{\text{HCF of numerators}}{\text{LCM of denominators}}$
$= \dfrac{\text{HCF of 1, 3 and 4}}{\text{LCM of 2, 4 and 5}} = \dfrac{1}{20}$

12. (d) Required LCM = (LCM of 250, 100 and 125) $\times$ 0.001

Now, LCM of 250, 100 and 125

$$\begin{array}{c|ccc} 2 & 250, & 100, & 125 \\ \hline 5 & 125, & 50, & 125 \\ \hline 5 & 25, & 10, & 25 \\ \hline 5 & 5, & 2, & 5 \\ \hline & 1, & 2, & 1 \end{array}$$

∴ LCM of 250, 100 and 125
$= 2 \times 2 \times 5 \times 5 \times 5 = 500$
∴ Required LCM $= 500 \times 0.001 = 0.5$

Alternate Method
The given numbers are equivalent to
$\dfrac{25}{100}, \dfrac{1}{10}, \dfrac{125}{1000}$, i.e. $\dfrac{1}{4}, \dfrac{1}{10}$ and $\dfrac{1}{8}$.

Now, LCM of fractions
$= \dfrac{\text{LCM of numerators}}{\text{HCF of denominators}}$
$= \dfrac{\text{LCM of 1, 1, 1}}{\text{HCF of 4, 10, 8}} = \dfrac{1}{2} = 0.5$

13. (a) Required LCM = (LCM of 25, 12, 200 and 75) × 0.1
LCM of 25, 12, 200, 75

$$\begin{array}{c|cccc} 2 & 25, & 12, & 200, & 75 \\ \hline 2 & 25, & 6, & 100, & 75 \\ \hline 3 & 25, & 3, & 50, & 75 \\ \hline 5 & 25, & 1, & 50, & 25 \\ \hline 5 & 5, & 1, & 10, & 5 \\ \hline & 1, & 1, & 2, & 1 \end{array}$$

LCM $= 2 \times 2 \times 2 \times 3 \times 5 \times 5 = 600$
∴ Required LCM $= 600 \times 0.1 = 60$

Alternate Method
The given numbers are equivalent to
$\dfrac{25}{10}, \dfrac{12}{10}, \dfrac{20}{1}, \dfrac{75}{10}$ i.e. $\dfrac{5}{2}, \dfrac{6}{5}, \dfrac{20}{1}, \dfrac{15}{2}$.

Now, we have to find the LCM of $\dfrac{5}{2}, \dfrac{6}{5}, \dfrac{20}{1}$ and $\dfrac{15}{2}$.

According to the formula,
LCM of fractions $= \dfrac{\text{LCM of numerators}}{\text{HCF of denominators}}$

∴ Required LCM $= \dfrac{\text{LCM of 5, 6, 20 and 15}}{\text{HCF of 2, 5, 1 and 2}}$
$= 60$

14. (c) LCM of 11 and 13 will be (11 × 13). Hence, if a number is exactly divisible by 11 and 13, then the same number must be exactly divisible by their LCM or by (11 × 13).

15. (d) ∵ HCF × LCM = Product of numbers
[by Technique 1]
∴ HCF × LCM = 18 × 15 = 270

16. (b) Given, LCM of two numbers = 2376
HCF of two numbers = 33
and one of the number = 297
According to the formula,
HCF of two numbers × LCM of two numbers
= First number × Second number
[by Technique 1]
∴ Second number $= \dfrac{33 \times 2376}{297} = 264$

17. (c) Let the another number be x.
Given, LCM of two numbers = 1989
HCF of two numbers = 13
and first number = 117
According to the formula,
Product of LCM and HCF
= Product of two numbers
[by Technique 1]
∴ $1989 \times 13 = 117 \times x$
⇒ $x = \dfrac{1989 \times 13}{117} = 221$

18. (d) We have,
HCF of two expressions p and $q = 1$
We know that,
LCM × HCF = Product of numbers
[by Technique 1]
⇒ LCM $\times 1 = p \times q$
⇒ LCM $= pq$
∴ Reciprocal of LCM $= \dfrac{1}{pq} = (pq)^{-1}$
$\left[\because \text{reciprocal of } a = \dfrac{1}{a} \right]$

19. (a) Given, HCF of two numbers = 12
LCM of two numbers = 72
and first number = 24
We know that,
First number × Second number = HCF × LCM
[by Technique 1]
∴ Second number $= \dfrac{\text{LCM} \times \text{HCF}}{\text{First number}}$
$= \dfrac{72 \times 12}{24} = 36$
Now, difference between the two numbers
$= 36 - 24 = 12$

20. (d) Let two numbers be $3m$ and $4m$,
$m =$ HCF.
Given, HCF $= m = 4$
We know that,
LCM $= \dfrac{\text{Product of two numbers}}{\text{HCF}}$
[by Technique 1]
$= \dfrac{3m \times 4m}{m} = 12m = 12 \times 4 = 48$

HCF and LCM / 55

21. (b) Let the numbers be $5x$ and $6x$.
Now, HCF of these two numbers is x.
We know that,
LCM × HCF = Product of two numbers
[by Technique 1]
$\Rightarrow \quad 480 \times x = 5x \times 6x$
$\Rightarrow \quad 480x = 30x^2$
$\Rightarrow \quad x = 16$

22. (b) Given, product of two numbers = 1500
and   HCF = 10
According to the formula,
Product of two numbers = HCF × LCM
[by Technique 1]
$\Rightarrow \quad 1500 = 10 \times$ LCM
$\therefore \quad$ LCM $= \dfrac{1500}{10} = 150$

23. (a) Given that, $x = 130, y = 305, z = 245$,
$a = 6, b = 9$ and $c = 17$.
According to the formula,
Required greatest number
= HCF of $[(x - a), (y - b), (z - c)]$
[by Technique 2]
= HCF of $[(130 - 6), (305 - 9), (245 - 17)]$
= HCF of (124, 296, 228) = 4

24. (a) Given that,
$x = 1023, y = 750, a = 3$ and $b = 2$
$\therefore$ Required number
= HCF of $[(x - a), (y - b)]$
[by Technique 2]
= HCF of $[(1023 - 3), (750 - 2)]$
= HCF of 1020 and 748
Now, find the HCF of 1020 and 748.

$$748\overline{)1020}(1$$
$$\underline{748}$$
$$272\overline{)748}(2$$
$$\underline{544}$$
$$204\overline{)272}(1$$
$$\underline{204}$$
$$68\overline{)204}(3$$
$$\underline{204}$$
$$\times$$

$\therefore$ Required greatest number = 68

25. (b) Given, $x = 13850, y = 17030, a = 17$
and $b = 17$.
Now, according to the formula,
Required greatest number
= HCF of $[(x - a), (y - b)]$
[by Technique 2]
= HCF of $(13850 - 17)$ and $(17030 - 17)$,
i.e. 13833 and 17013.

$$13833\overline{)17013}(1$$
$$\underline{13833}$$
$$3180\overline{)13833}(4$$
$$\underline{12720}$$
$$1113\overline{)3180}(2$$
$$\underline{2226}$$
$$954\overline{)1113}(1$$
$$\underline{954}$$
$$159\overline{)954}(6$$
$$\underline{954}$$
$$\times$$

Hence, the greatest number is 159.

26. (a) Given that, $x = 1356, y = 1868, z = 2764$
and $\quad a = b = c = 12$
According to the formula,
Required greatest number
= HCF of $[(x - a), (y - b), (z - c)]$
[by Technique 2]
$\therefore$ Required number = HCF of $[(1356 - 12),$
$(1868 - 12), (2764 - 12)]$
= HCF of (1344, 1856 and 2752)
Firstly, find the HCF of 1344 and 1856.

$$1344\overline{)1856}(1$$
$$\underline{1344}$$
$$512\overline{)1344}(2$$
$$\underline{1024}$$
$$320\overline{)512}(1$$
$$\underline{320}$$
$$192\overline{)320}(1$$
$$\underline{192}$$
$$128\overline{)192}(1$$
$$\underline{128}$$
$$64\overline{)128}(2$$
$$\underline{128}$$
$$\times$$

Here, HCF of 1344 and 1856 = 64
Now, find the HCF of 64 and third number 2752.

$$64\overline{)2752}(43$$
$$\underline{256}$$
$$192$$
$$\underline{192}$$
$$\times$$

$\therefore$ HCF of 1344, 1856 and 2752 = 64
Hence, the required greatest number is 64.

27. (a) LCM of 10, 15, 21 and 28

2	10,	15,	21,	28
3	5,	15,	21,	14
5	5,	5,	7,	14
7	1,	1,	7,	14
	1,	1,	1,	2

$\therefore$ LCM $= 2 \times 2 \times 3 \times 5 \times 7 = 420$
Greatest number of 4-digits = 9999

Now, $\dfrac{9999}{420} = 23\dfrac{339}{420}$

∴ Remainder = 339
∴ 4-digit number divisible by 10, 15, 21 and 28 = 9999 − 339 = 9660
Here, $a = 4, b = 9, c = 15$ and $d = 22$
∴ $k = 10 − 4 = 15 − 9 = 21 − 15$
$= 28 − 22 = 6$
∴ Required number = (9660 − k)
[by Technique 3]
$= (9660 − 6) = 9654$

28. (a) Given, $x = 12, y = 21, z = 35$ and $k = 6$
∴ Required least number
= LCM of (12, 21, 35) + k   [by Technique 4]
LCM of 12, 21, 35 is

3	12, 21, 35
7	4, 7, 35
	4, 1, 5

Here, LCM = $3 \times 4 \times 5 \times 7 = 420$
∴ Required least number
$= 420 + 6 = 426$

29. (a) Given, $x = 16, y = 18, z = 20$ and $k = 4$.
∴ Required least number
= LCM of (16, 18, 20) + k   [by Technique 4]
Now, LCM of 16, 18 and 20 is

2	16, 18, 20
2	8, 9, 10
	4, 9, 5

Here, LCM = $2 \times 2 \times 4 \times 5 \times 9 = 720$
∴ Required number = $720m + 4$
where, m is a natural number.
Now, $(720m + 4)$ will be a multiple of 7.
∵ Smallest value of $m = 4$
∴ Required number = $720 \times 4 + 4 = 2884$

30. (c) Given, $x = 12, y = 16, z = 18$ and $k = 5$.
According to the formula,
Required number = LCM of $(x, y$ and $z) + k$
[by Technique 4]
= LCM of (12, 16, 18) + 5
LCM of 12, 16 and 18 is

2	12, 16, 18
2	6, 8, 9
3	3, 4, 9
	1, 4, 3

Here, LCM = $2 \times 2 \times 3 \times 3 \times 4 = 144$
∴ Required number = $144 + 5 = 149$

31. (c) Given, $x = 1305, y = 4665$ and $z = 6905$.
Then,
$|x − y| = |1305 − 4665| = 3360$
$|y − z| = |4665 − 6905| = 2240$
and $|z − x| = |6905 − 1305| = 5600$

∴ Required number
= HCF of 3360, 2240 and 5600
[by Technique 5]
Firstly, find the HCF of any two numbers i.e. 3360 and 2240.

$2240\overline{)3360}(1$
$\underline{2240}$
$1120\overline{)2240}(2$
$\underline{2240}$
$\times$

Here, HCF of 3360 and 2240 = 1120
Now, find the HCF of 1120 and 5600.

$1120\overline{)5600}(5$
$\underline{5600}$
$\times$

∴ HCF of 3360, 2240 and 5600 = 1120
Here, $\dfrac{1305}{1120} = 1\dfrac{185}{1120}, \dfrac{4665}{1120} = 4\dfrac{185}{1120}$
and $\dfrac{6905}{1120} = 6\dfrac{185}{1120}$.
Hence, the common remainder is 185.

32. (a) Given, HCF = 5, LCM = 2310
and $n = 4$
∴ Required product = $(HCF)^{n−1} \times LCM$
[by Technique 6]
$= (5)^{4−1} \times 2310 = 5^3 \times 2310$
$= 125 \times 2310 = 288750$

33. (a) ∵ Greatest number of 3-digits = 999
LCM of 6, 9 and 12

2	6, 9, 12
2	3, 9, 6
3	3, 9, 3
3	1, 3, 1
	1, 1, 1

∴ Required LCM = $2 \times 2 \times 3 \times 3 = 36$
Now, $\dfrac{999}{36} = 27\dfrac{27}{36}$
∴ Remainder, $R = 27$ and $k = 3$.
Hence, required number
= (Greatest 3-digit number −R) + k
[by Technique 7]
$= (999 − 27 + 3) = 975$

34. (a) Given numbers are 12, 18, 21, 28 and remainder, $k = 3$.
∵ LCM of 12, 18, 21 and 28 = 252
and greatest 4-digit number = 9999
On dividing 9999 by 252,

$252\overline{)9999}(39$
$\underline{756}$
2439
$\underline{2268}$
$171 = R$

∴ Required number
= [Greatest 4-digit number − R] + k
　　　　　　　　　　[by Technique 7]
= (9999 − R) + k
= (9999 − 171) + 3
= 9828 + 3 = 9831

35. (a) LCM of (12, 18, 21 and 28) = 252,
i.e. L = 252
∵ Smallest 4-digit number = 1000
On dividing 1000 by 252,

$$252\overline{)1000}(3$$
$$\underline{756}$$
$$244 = R$$

∴ Required number
= [Smallest 4-digit number + (L − R)]
　　　　　　　　　　[by Technique 8]
= 1000 + (252 − 244) = 1008

36. (a) First we have to find out the LCM of 48, 72 and 108 s.
∴ LCM of 48, 72 and 108 s

2	48, 72, 108
2	24, 36, 54
2	12, 18, 27
2	6, 9, 27
3	3, 9, 27
3	1, 3, 9
3	1, 1, 3
	1, 1, 1

∴ LCM = 2 × 2 × 2 × 2 × 3 × 3 × 3 s
= 432 s = $\dfrac{432}{60}$ min
= 7 min 12 s
∴ Time of beep together
= 10 : 00 + 07 : 12
= 10 : 07 : 12

37. (c) The bells will toll together after the time (in seconds) which is equal to the LCM of 9, 6, 4, 10 and 8.
LCM of 9, 6, 4, 10 and 8 is

2	9, 6, 4, 10, 8
2	9, 3, 2, 5, 4
3	9, 3, 1, 5, 2
	3, 1, 1, 5, 2

∴ LCM = 2 × 2 × 2 × 3 × 3 × 5 = 360
In one hour, the rings will toll together $\dfrac{3600}{360}$, i.e. 10 times.

38. (c) Required time = LCM of 42, 56 and 63 s
LCM of 42, 56, 63 is

2	42, 56, 63
3	21, 28, 63
7	7, 28, 21
	1, 4, 3

∴ Required time = 2 × 3 × 3 × 4 × 7 = 504 s

39. (c) LCM of 10, 15 and 18 is

2	10, 15, 18
3	5, 15, 9
5	5, 5, 3
	1, 1, 3

LCM = 2 × 3 × 3 × 5 = 90
To make it perfect square, we multiply it with 2 × 5 = 10
∴ Required number of soldiers
= 90 × 10 = 900

40. (c) LCM of 5, 6, 4 and 3 = 60
On dividing 2497 by 60, the remainder is 37.
∴ Least number to be added = 60 − 37 = 23

41. (a) LCM of 24, 32, 36, 54

2	24, 32, 36, 54
2	12, 16, 18, 27
2	6, 8, 9, 27
3	3, 4, 9, 27
3	1, 4, 3, 9
	1, 4, 1, 3

LCM = 2 × 2 × 2 × 3 × 3 × 3 × 4 = 864
∴ Required least number
= LCM of (24, 32, 36, 54) − 9
= 864 − 9 = 855

42. (b) Let the numbers be x, $2x$ and $3x$ in which x = HCF.
Given that, HCF = x = 23
So, the numbers are 23, 46 and 69.

43. (d) Let the numbers be $3x$, $4x$ and $5x$ in which x = HCF and LCM = $60x$
Given, 　　LCM = 1200
Now, 　　$60x = 1200$
∴ 　　　　$x = 20$

44. (d) Given, HCF = 6 and LCM = 210
Now, $m \times n = 6 \times 210 = 1260$
and $m + n = 72$
∴ $\dfrac{1}{m} + \dfrac{1}{n} = \dfrac{m+n}{mn} = \dfrac{72}{1260} = \dfrac{4}{70} = \dfrac{2}{35}$

45. (c) Let two numbers be $2x$ and $3x$.
∴ HCF = x and LCM = $x \times 2 \times 3 = 6x$

According to the question,
$$6x = 48$$
$$\Rightarrow x = 8$$
$\therefore$ Required sum $= (2x + 3x) = 5x$
$$= 5 \times 8 = 40$$

46. (b) Let the numbers be $10x, 12x, 15x$ and $18x$, respectively.
Then, LCM $= 180x$ and HCF $= x$
But given that, HCF $= x = 3$
Hence, required LCM $= 180 \times 3 = 540$

47. (a) Required number = Multiple of LCM of 8, 9, 12, 15 and 18.

Now, LCM of 8, 9, 12, 15 and 18

2	8,	9,	12,	15,	18
2	4,	9,	6,	15,	9
3	2,	9,	3,	15,	9
3	2,	3,	1,	5,	3
	2,	1,	1,	5,	1

Here, LCM $= 2 \times 2 \times 2 \times 3 \times 3 \times 5 = 360$
$\therefore$ The factors make it clear that to make a perfect square, 360 must be multiplied by (2×5).
$\therefore$ Required number $= 360 \times 2 \times 5 = 3600$

Exercise 2 Higher Skill Level Questions

1. (b) LCM of 4, 6, 8 and 9

2	4, 6, 8, 9
2	2, 3, 4, 9
3	1, 3, 2, 9
	1, 1, 2, 3

$\therefore$ LCM $= 2 \times 2 \times 3 \times 2 \times 3 = 72$
$\therefore$ Required number $= 72$

2. (b) Let $p = 2, q = 3, r = 4, s = 5$ and $t = 6$ and remainders
$a = 1, b = 2, c = 3, d = 4, e = 5$
Now, $2 - 1 = 1; 3 - 2 = 1; 4 - 3 = 1;$
$5 - 4 = 1; 6 - 5 = 1$
$\therefore k = 1$
Number that is divisible by 2, 3, 4, 5 and 6 leaving the remainders 1, 2, 3, 4 and 5
$=$ LCM of (2, 3, 4, 5, 6) $- k$
$= 60 - 1 = 59$ [by Technique 3]
Hence, there is only one integer between 0 and 100 which satisfies the given condition.

3. (d) Let two prime numbers be 2 and 3, then their HCF $= 1$.
Hence, Statement I is true.
Let $p = 5$ and $q = 6$, then their HCF is also 1.
Hence, Statement II is also true.
Let $p = 8$ and $q = 9$.
p and q both are composite numbers and their HCF is also 1.
Hence, Statement III is also true.

4. (c) To find the greatest length of same pipe pieces, we have to find the HCF of 1.5 m and 1.2 m.
$\therefore$ Required length of pipe pieces = HCF of 1.5 m and 1.2 m

$$1.2 \overline{)1.5}(1$$
$$\underline{1.2}$$
$$0.3\overline{)1.2}(4$$
$$\underline{1.2}$$
$$\times$$

$\therefore$ Required length of pipe pieces $= 0.3$ m

5. (b) Let the numbers be x and y.
According to the question,
$$x + y = 45 \quad \ldots(i)$$
Difference of two numbers
$$= \frac{1}{9} \times \text{Sum of two numbers}$$
$\Rightarrow \quad x - y = 5 \quad \ldots(ii)$
On adding Eqs. (i) and (ii), we get
$$x + y = 45$$
$$x - y = 5$$
$$\overline{2x = 50} \Rightarrow x = 25$$
From Eq. (i),
$$x + y = 45$$
$\Rightarrow \quad y = 45 - x$
$\Rightarrow \quad y = 45 - 25 = 20$
Now, LCM of 25 and 20

5	25, 20
	5, 4

$\therefore$ Required LCM $= 4 \times 5 \times 5 = 100$

6. (c) Let the numbers be $66a$ and $66b$, where a and b are coprimes.
According to the question,
$$66a + 66b = 1056$$
$\Rightarrow \quad 66(a + b) = 1056$
$\Rightarrow \quad (a + b) = \dfrac{1056}{66} = 16$
$\therefore$ Possible values of a and b are
$(a = 1, b = 15), (a = 3, b = 13)$
$(a = 5, b = 11), (a = 7, b = 9)$
$\therefore$ Numbers are
$(66 \times 1, 66 \times 15), (66 \times 3, 66 \times 13),$
$(66 \times 5, 66 \times 11), (66 \times 7, 66 \times 9).$
Hence, the possible number of pairs is 4.

7. (c) Let LCM $= m$ and HCF $= n$.
According to the question,
$$m = 12n \quad \ldots(i)$$
and $\quad m + n = 403 \quad \ldots(ii)$
$\Rightarrow \quad 12n + n = 403 \quad$ [from Eq. (i)]

HCF and LCM / 59

$\Rightarrow \quad 13n = 403$

$\therefore \quad n = \dfrac{403}{13} = 31$

Now, $\quad m = 12n = 12 \times 31 = 372$

Given, first number = 93

Let another number be k.

Product of two numbers = HCF × LCM

[by Technique 1]

$\therefore \ 93 \times k = 372 \times 31 \Rightarrow k = \dfrac{372 \times 31}{93} = 124$

8. (a) Given, LCM = 495 and HCF = 5

Let first number = x

and second number = y

∵ Product of two numbers = HCF × LCM

[by Technique 1]

$\therefore \quad xy = 495 \times 5 \Rightarrow xy = 2475$

But given that, $x + y = 100$

We know that, $(x - y)^2 = (x + y)^2 - 4xy$

$= (100)^2 - 4 \times 2475$

$= 10000 - 9900 = 100$

$\therefore \quad (x - y) = \sqrt{100} = 10$

9. (d) Let HCF be x.

According to the question, LCM = $20x$

Given that, HCF + LCM = 2520

$\Rightarrow \quad x + 20x = 2520$

$\Rightarrow \quad x = \dfrac{2520}{21} = 120$

Now, LCM = $20x = 20 \times 120 = 2400$

We know that,

First number × Second number

= HCF × LCM

[by Technique 1]

$\Rightarrow$ Second number = $\dfrac{\text{LCM} \times \text{HCF}}{\text{First number}}$

$= \dfrac{120 \times 2400}{480} = 600$

$\therefore$ Required answer = $600 \times 3 = 1800$

10. (c) HCF of 250 cm and 150 cm

$$\begin{array}{r}150\overline{)250}(1\\150\\\hline 100\overline{)150}(1\\100\\\hline 50\overline{)100}(2\\100\\\hline \times\end{array}$$

$\therefore$ HCF = 50

Now, number of slabs = $\dfrac{\text{Area of the floor}}{\text{Area of the slab}}$

$= \dfrac{250 \times 150}{50 \times 50} = 15$

11. (d) Here, $6 - 5 = 1$, $8 - 7 = 1$, $12 - 11 = 1$

and LCM of 6, 8 and 12 = 24

$\therefore$ Required number = $24k - 1$

$= 24 \times 6 - 1 = 144 - 1 = 143$

So, sum of digits = $1 + 4 + 3 = 8$

12. (a) We have, $x^3 + 8 = (x)^3 + (2)^3$

$= (x + 2)\{(x)^2 - (x)(2) + (2)^2\}$

$= (x + 2)(x^2 - 2x + 4)$

$x^2 + 5x + 6 = x^2 + 2x + 3x + 6$

$= x(x + 2) + 3(x + 2) = (x + 2)(x + 3)$

and $\quad x^3 + 4x^2 + 4x = x(x^2 + 4x + 4)$

$= x(x + 2)^2$

$\therefore$ LCM = $x(x + 2)^2 (x + 3)(x^2 - 2x + 4)$

13. (c) Since, $(x + 1)$ is the HCF of

$Ax^2 + Bx + C$ and $Bx^2 + Ax + C$.

$\therefore \ A(-1)^2 + B(-1) + C = 0$ [put $x + 1 = 0$]

$\Rightarrow \quad A - B + C = 0 \Rightarrow C = B - A$

and $\quad B(-1)^2 + A(-1) + C = 0$

$\Rightarrow \quad B - A + C = 0 \Rightarrow C = A - B$

Since, $A \ne B$, hence $C = A - B$

14. (c) HCF $\left(\dfrac{a}{\text{HCF}(a, b)}, \dfrac{b}{\text{HCF}(a, b)}\right)$ is always

equal to 1 because when a and b are divided by their HCF, then resulting number are coprime.

Illustration 1 Let the two positive integers be $a = 24$ and $b = 36$.

$\therefore \quad$ HCF $\left(\dfrac{24}{\text{HCF}(24, 36)}, \dfrac{36}{\text{HCF}(24, 36)}\right)$

$= $ HCF $\left(\dfrac{24}{12}, \dfrac{36}{12}\right) = $ HCF $(2, 3) = 1$

Illustration 2 Let the two positive integers be $a = 13$ and $b = 17$.

$\therefore \quad$ HCF $\left(\dfrac{13}{\text{HCF}(13, 17)}, \dfrac{17}{\text{HCF}(13, 17)}\right)$

$= $ HCF $\left(\dfrac{13}{1}, \dfrac{17}{1}\right) = 1$

15. (a) Let $p(x) = 8(x^5 - x^3 + x)$

$= 4 \times 2 \times x (x^4 - x^2 + 1)$

and $\quad q(x) = 28(x^6 + 1)$

$= 7 \times 4 \ [(x^2)^3 + (1)^3]$

$= 4 \times 7 \times (x^2 + 1)(x^4 - x^2 + 1)$

$[\because a^3 + b^3 = (a + b)(a^2 + b^2 - ab)]$

$\therefore$ HCF of $p(x)$ and $q(x) = 4(x^4 - x^2 + 1)$

16. (a) Let $f(x) = (x^4 - y^4) = (x^2 - y^2)(x^2 + y^2)$

$[\because a^2 - b^2 = (a + b)(a - b)]$

$= (x - y)(x + y)(x^2 + y^2)$

and $\ g(x) = (x^6 - y^6) = (x^3)^2 - (y^3)^2$

$= (x^3 + y^3)(x^3 - y^3)$

$$= (x+y)(x^2 - xy + y^2)$$
$$(x-y)(x^2 + xy + y^2)$$
$$[\because a^3 - b^3 = (a-b)(a^2 + b^2 + ab)$$
$$\text{and } a^3 + b^3 = (a+b)(a^2 + b^2 - ab)]$$
$$= (x-y)(x+y)(x^2 - xy + y^2)(x^2 + xy + y^2)$$
$$\therefore \text{ HCF of }[f(x), g(x)] = (x-y)(x+y)$$
$$= x^2 - y^2$$

17. (b) Let $f(x) = x^3 - x^2 - 2x = x(x^2 - x - 2)$
$$= x(x^2 - 2x + x - 2)$$
$$= x\{x(x-2) + 1(x-2)\}$$
$$= x(x+1)(x-2)$$
and $g(x) = x^3 + x^2$
$$= x^2(x+1) = x \cdot x(x+1)$$
Now, HCF of $f(x)$ and $g(x)$
$$= x(x+1) = x^2 + x$$

18. (c) $a^2b^4 + 2a^2b^2 = a^2b^2(b^2 + 2)$ ...(i)
and $(ab)^7 - 4a^2b^9 = a^7b^7 - 4a^2b^9$
$$= a^2b^2(a^5b^5 - 4b^7) \quad ...(ii)$$
From Eqs. (i) and (ii), we get HCF $= a^2b^2$

19. (b) HCF of $(22n + 7, 33n + 10)$ is always 1.
Illustration
For $n = 1$, HCF $(29, 43) \Rightarrow $ HCF $= 1$
For $n = 2$, HCF $(51, 76) \Rightarrow $ HCF $= 1$
For $n = 3$, HCF $(73, 109) \Rightarrow $ HCF $= 1$

20. (c) HCF $(a+b, a-b)$ is either 1 or 2.
Illustration 1 Let $a = 9$ and $b = 8$
$\therefore \quad $ HCF $(8+9, 9-8) = $ HCF $(17, 1) = 1$
Illustration 2 Let $a = 23$ and $b = 17$
HCF $(17 + 23, 23 - 17)$
HCF $(40, 6) = 2$
Hence, HCF $(a+b, a-b)$ can either be 1 or 2.

21. (b) Let $f(x) + g(x) = 5x^2 - x - 4$ ...(i)
and $f(x) - g(x) = x^2 + 9x - 10$ ...(ii)
On solving Eqs. (i) and (ii), we get
$$f(x) = 3x^2 + 4x - 7 = (x-1)(3x+7)$$
and $g(x) = 2x^2 - 5x + 3$
$$= (x-1)(2x-3)$$
Hence, the required HCF is $(x-1)$.

22. (d) Given, a, b and c are natural numbers.
I. LCM of $(ab, ac) = $ Least multiple of (ab, ac)
and $a \times $ LCM of (b, c)
$= a \times $ Least multiple of (b, c)
Hence, Statement I is correct.
II. HCF of $(ab, ac) = $ Common factor of (ab, ac)
and $a \times $ HCF of (b, c)
$= a \times $ Common factor of (b, c)
Hence, Statement II is correct.

III. We know that, HCF is always less than LCM.
Hence, Statement III is correct.
IV. HCF (a, b) divides LCM (a, b) because a common factor between a, b always divides $(a \times b)$.
Hence, Statement IV is correct.

23. (c) I. If $a = bc$ with HCF $(b, c) = 1$
$\Rightarrow b$ and c are coprime numbers.
$\therefore \quad $ HCF $(c, bd) = $ HCF (c, d)
which is the correct.
II. If $a = bc$ with HCF $(b, c) = 1$
$\Rightarrow b$ and c are coprime numbers.
$\therefore \quad $ LCM $(b, c) = bc$
Now, LCM $(a, d) = $ LCM (bc, d)
$\therefore $ LCM $(a, d) = $ LCM (c, bd)
which is the correct.

24. (c) Given, HCF $= (x+y)$
LCM $= 3x^5 + 5x^4y + 2x^3y^2 - 3x^2y^3$
$$-5xy^4 - 2y^5$$
$$= 3x^5 - 3x^2y^3 + 5x^4y - 5xy^4 + 2x^3y^2 - 2y^5$$
$$= 3x^2(x^3 - y^3) + 5xy(x^3 - y^3) + 2y^2(x^3 - y^3)$$
$$= (3x^2 + 5xy + 2y^2)(x^3 - y^3)$$
and first polynomial $= x^2 - y^2$
$$= (x-y)(x+y)$$
We know that,
First polynomial $\times$ Second polynomial
$= $ HCF $\times$ LCM [by Technique 1]
$\therefore $ Second polynomial
$$= \frac{(x+y)(x^3 - y^3)(3x^2 + 5xy + 2y^2)}{(x-y)(x+y)}$$
$$= \frac{(x-y)(x^2 + y^2 + xy)(3x^2 + 5xy + 2y^2)}{(x-y)}$$
$$= (x^2 + y^2 + xy)(3x^2 + 5xy + 2y^2)$$
$$= 3x^4 + 5x^3y + 2x^2y^2 + 3x^2y^2 + 5xy^3 + 2y^4$$
$$+ 3x^3y + 5x^2y^2 + 2xy^3$$
$$= 3x^4 + 8x^3y + 10x^2y^2 + 7xy^3 + 2y^4$$

25. (c) LCM of $(3, 4, 5$ and $6) = 60$
Remainder required $= 2$
$\therefore \quad $ Number must be of the form $60n + 2$.
The required number should also be completely divisible by 11. Thus, the first number of the series will be 242 and the whole series will be like 242, 902, 1562, 2222, 2882, 3542 having a common difference of 660.
Therefore, the 6th number in this series will be 3542.

Chapter 4

Simple and Decimal Fractions

A number which can be represented in p/q form, where $q \neq 0$, and p, q are positive integers, is called a **fraction**. Here, p is called the numerator and q is called the denominator.
For example 3/5 is a fraction, where 3 is called numerator and 5 is called denominator.

or

When a unit is divided into any number of equal parts, then these parts are termed as a fraction of the unit.
For example If 1 is to be divided into two equal parts, then 1 is divided by 2 and is represented as $\frac{1}{2}$.

Simple Fraction

The fraction, which has denominator other than power of 10, is called simple fraction.
For example $\frac{3}{7}, \frac{5}{11}, \frac{7}{9}$, etc.

✦ Simple fraction is also known as **vulgar fraction**.

Types of Simple Fractions

There are following types of fractions

1. **Proper Fraction** When the numerator of a fraction is less than its denominator, then fraction is called proper fraction.
   For example $\frac{1}{2}, \frac{15}{17}, \frac{21}{43}$, etc.

2. **Improper Fraction** When the numerator of a fraction is greater than its denominator, then fraction is called improper fraction.
   For example $\frac{17}{13}, \frac{18}{14}, \frac{45}{19}$, etc.

3. **Like and Unlike Fractions** The fractions whose denominators are same, is called like fractions, whereas fractions whose denominators are different, is called unlike fractions.

   For example $\dfrac{2}{11}, \dfrac{3}{11}, \dfrac{6}{11}$ etc., are like fractions

   and $\dfrac{2}{3}, \dfrac{8}{7}, \dfrac{6}{4}$ etc., are unlike fractions.

4. **Equivalent Fractions** The fractions whose values are same, is called equivalent fractions.

   For example $\dfrac{2}{3} = \dfrac{4}{6} = \dfrac{6}{9}$.

5. **Compound Fraction** A fraction, in which numerator or denominator or both are in fraction, is called compound fraction.

   For example $\dfrac{1}{7/9}, \dfrac{11/9}{13}, \dfrac{1/4}{7/13}$, etc.

6. **Inverse Fraction** If we inverse the numerator and the denominator of a fraction, then the resultant fraction will be the inverse fraction of the original fraction.

   For example If fraction $= \dfrac{3}{8}$, then its inverse fraction $= \dfrac{8}{3}$.

7. **Mixed Fraction** The fraction, which is the combination of integer and fraction, is called mixed fraction.

   For example $3\dfrac{2}{5}, 7\dfrac{1}{9}$, etc.

8. **Continuous Fraction** It has no certain definition but only say that a fraction contains additional fractions in its denominators, is called continuous fraction.

   For example (i) $2 + \dfrac{1}{2 + \dfrac{2}{5 + \dfrac{2}{3}}}$ (ii) $6 + \dfrac{1}{1 + \dfrac{1}{\dfrac{1}{2} + \dfrac{1}{2}}}$

   ✦ To simplify a continuous fraction, start from bottom and work upwards.

Comparison of Simple Fractions

Following are some techniques to compare the fractions.

1. **Cross-multiplication Method**

   If $\dfrac{a}{b}$ and $\dfrac{c}{d}$ are two fractions, then

   (i) If $ad > bc$, then $\dfrac{a}{b} > \dfrac{c}{d}$ (ii) If $ad < bc$, then $\dfrac{a}{b} < \dfrac{c}{d}$

   (iii) If $ad = bc$, then $\dfrac{a}{b} = \dfrac{c}{d}$

Simple and Decimal Fractions / 63

Ex. 1 (i) Between $\dfrac{4}{7}$ and $\dfrac{3}{8}$, which fraction is bigger?

(ii) Which one of the fraction is largest among $\dfrac{2}{3}, \dfrac{3}{4}, \dfrac{4}{3}, \dfrac{5}{4}$?

Sol. (i) Given, $\dfrac{4}{7}$ and $\dfrac{3}{8}$ $\because 4 \times 8 > 7 \times 3$ $\therefore \dfrac{4}{7} > \dfrac{3}{8}$

(ii) First find the largest among the two fractions

$\dfrac{2}{3}, \dfrac{3}{4}$ $\because 3 \times 3 > 4 \times 2$ $\therefore \dfrac{3}{4} > \dfrac{2}{3}$

$\dfrac{4}{3}, \dfrac{5}{4}$ $\because 4 \times 4 > 5 \times 3$ $\therefore \dfrac{4}{3} > \dfrac{5}{4}$

Now, taking the two largest fractions, find which one is largest

$\dfrac{3}{4}, \dfrac{4}{3}$ $\because 4 \times 4 > 3 \times 3$ $\therefore \dfrac{4}{3} > \dfrac{3}{4}$

Hence, $\dfrac{4}{3}$ is the largest fraction.

2. By Changing Fractions in Decimal Form

To compare two or more fractions, first convert fractions into decimal form and then compare.

Ex. 2 Between $\dfrac{1}{7}$ and $\dfrac{2}{9}$, which fraction is bigger?

Sol. Here, $\dfrac{1}{7} = 0.14$ and $\dfrac{2}{9} = 0.22$

It is clear that $0.22 > 0.14 \Rightarrow \dfrac{2}{9} > \dfrac{1}{7}$

Hence, the bigger fraction is $\dfrac{2}{9}$.

3. By Equating Denominators of Given Fractions

For comparison of fractions, take LCM of the denominators of all fractions, so that the denominators of all fractions are same. Now, the fraction having largest numerator is the largest fraction.

Ex. 3 Arrange the fractions $\dfrac{3}{5}, \dfrac{7}{9}, \dfrac{11}{13}$ in decreasing order.

Sol. $\because$ LCM of 5, 9 and 13 = $5 \times 9 \times 13 = 585$

$\therefore \dfrac{3}{5} = \dfrac{3 \times 117}{5 \times 117} = \dfrac{351}{585}$; $\dfrac{7}{9} = \dfrac{7 \times 65}{9 \times 65} = \dfrac{455}{585}$ and $\dfrac{11}{13} = \dfrac{11 \times 45}{13 \times 45} = \dfrac{495}{585}$

Now, the fraction having largest numerator will be largest.

$\therefore$ Decreasing order will be $\dfrac{495}{585}, \dfrac{455}{585}, \dfrac{351}{585}$.

Hence, the decreasing order is $\dfrac{11}{13}, \dfrac{7}{9}, \dfrac{3}{5}$.

4. By Equating Numerators of Given Fractions

For comparison of fractions, take LCM of the numerator of all fractions, so that numerators of all the fractions are same. Now, the fraction having smallest denominator will be largest.

Ex. 4 Which fraction is largest among $\dfrac{3}{13}, \dfrac{2}{15}, \dfrac{4}{17}$?

Sol. $\because$ LCM of 2, 3 and 4 $= 2 \times 2 \times 3 = 12$

$\therefore \quad \dfrac{3}{13} = \dfrac{3 \times 4}{13 \times 4} = \dfrac{12}{52}; \quad \dfrac{2}{15} = \dfrac{2 \times 6}{15 \times 6} = \dfrac{12}{90}$

and $\quad \dfrac{4}{17} = \dfrac{3 \times 4}{3 \times 17} = \dfrac{12}{51}$

Now, the fraction having smallest denominator will be largest.

Hence, $\dfrac{4}{17}$ is the largest number.

Operations on Simple Fractions

There are various operations on simple fractions as given below

1. Addition of Simple Fractions

(i) **When Denominators are Same** If denominators of fractions are same, then numerators of fractions are added and their addition is divided by denominator.

For example $\dfrac{1}{4} + \dfrac{2}{4} = (1+2)\dfrac{1}{4} = \dfrac{3}{4}$

(ii) **When Denominators are Different** If denominators of fractions are not same, then make their denominators equal (by taking their LCM) and then add their numerators.

For example $\dfrac{1}{2} + \dfrac{1}{3} + \dfrac{1}{4} = \dfrac{(1 \times 6) + (1 \times 4) + (1 \times 3)}{12} = \dfrac{6+4+3}{12} = \dfrac{13}{12}$

2. Subtraction of Simple Fractions

(i) **When Denominators are Same** If denominators of fractions are same, then numerators of fractions are subtracted and their subtraction is divided by the denominator.

For example $\dfrac{3}{4} - \dfrac{1}{4} = (3-1)\dfrac{1}{4} = \dfrac{2}{4} = \dfrac{1}{2}$

(ii) **When Denominators are Different** If denominators of fractions are not same, then make their denominators equal and then subtract their numerators.

For example $\dfrac{2}{3} - \dfrac{1}{2} = \dfrac{(2 \times 2) - (3 \times 1)}{6} = \dfrac{4-3}{6} = \dfrac{1}{6}$

3. Multiplication of Simple Fractions

(i) To multiply two or more simple fractions, multiply their numerators and denominators.

For example $\dfrac{1}{2} \times \dfrac{3}{4} = \dfrac{(1 \times 3)}{(2 \times 4)} = \dfrac{3}{8}$

(ii) If fractions are given in mixed form, first convert them into improper fraction and then multiply.

For example $2\dfrac{4}{5} \times 1\dfrac{8}{3} = \dfrac{14}{5} \times \dfrac{11}{3} = \dfrac{154}{15}$ $\left[\text{here}, a\dfrac{b}{c} = \dfrac{a \times c + b}{c}\right]$

4. Division of Simple Fractions

To divide two fractions, first fraction is multiplied by the inverse of second fraction.

For example $\dfrac{2}{3} \div \dfrac{3}{5} = \dfrac{2}{3} \times \dfrac{5}{3} = \dfrac{10}{9}$

Decimal Fraction

If the fraction has denominator in the powers of 10, then fraction is called decimal fraction.

For example (i) 10th part of unit $= \dfrac{1}{10} = 0.1$ (ii) 10th part of 6 $= \dfrac{6}{10} = 0.6$

To Convert a Decimal Number into a Vulgar Fraction Firstly, place 1 in the denominator under the decimal point. After removing the decimal point, place as many zeroes after 1 as the number of digits after the decimal point. Then, it becomes decimal fraction. Finally, reduce the fraction to its lowest terms which is a vulgar fraction.

For example (i) $0.23 = \dfrac{23}{100}$ (ii) $0.0035 = \dfrac{35}{10000} = \dfrac{7}{2000}$

- Placing zeroes to the right of a decimal fraction does not make any change in value. Hence, 0.5, 0.50, 0.500 and 0.5000 are equal.
- If the numerator and denominator of a fraction have same number of decimal places, then each of the decimal points can be removed.

Recurring Decimal Fractions

The decimal fraction, in which one or more decimal digits are repeated again and again, is called recurring decimal fraction. To represent these fractions, a line is drawn on the digits which are repeated.

For example (i) $\dfrac{2}{3} = 0.6666\ldots = 0.\overline{6}$ (ii) $\dfrac{22}{7} = 3.142857142857 = 3.\overline{142857}$

There are two types of recurring decimal fractions

1. **Pure Recurring Decimal Fraction** When all the digits in a decimal fraction are repeated after the decimal point, then the decimal fraction is called pure recurring decimal fraction.

   For example $0.\overline{5}, 0.\overline{489}$, etc.

   To Convert Pure Recurring Decimal Fractions into Simple Fractions (Vulgar Fractions) Firstly, write down the repeated digits only once in numerator and then place as many nines in the denominator as the number of digits repeating.

   For example (i) $0.\overline{3} = \dfrac{3}{9} = \dfrac{1}{3}$

   Since, there is only 1 repeated digit.
   Therefore, only single 9 is placed in denominator.

   (ii) $0.\overline{57} = \dfrac{57}{99} = \dfrac{19}{33}$

   Since, there are only 2 repeated digits.
   Therefore, two 9's are placed in denominator.

2. **Mixed Recurring Decimal Fraction** A decimal fraction in which some digits are repeated and some are not repeated after decimal, is called mixed recurring decimal fraction.

   For example $3.2\overline{23}, 0.12\overline{36}$, etc.

To Convert Mixed Recurring Decimal Fractions into Simple Fractions In the numerator, take the difference between the number formed by all the digits after decimal point (repeated digits will be taken only once) and the number formed by non-repeating digits. In the denominator, place as many nines as there are repeating digits and after nine, put as many zeroes as the number of non-repeating digits.

For example (i) $0.3\bar{6} = \dfrac{(36-3)}{90} = \dfrac{33}{90} = \dfrac{11}{30}$

(ii) $0.4\overline{267} = \dfrac{(4267-42)}{9900} = \dfrac{4225}{9900} = \dfrac{169}{396}$

Operations on Decimal Fractions

There are various operations on decimal fractions as given below

1. Addition and Subtraction of Decimal Fractions

To add or subtract decimal fractions, the given numbers are written under each other such that the decimal points lie in one column and the numbers so arranged can now be added or subtracted as per the conventional method of addition and subtraction.

Ex. 5 (i) $353.5 + 2.32 + 43.23 = ?$ (ii) $1000 - 132.23 = ?$

Sol. (i) 353.50
       2.32
     + 43.23
     ───────
       399.05

(ii) 1000.00
    − 132.23
    ─────────
       867.77

Ex. 6 $8.3\bar{1} + 0.\bar{6} + 0.00\bar{2} = ?$

Sol. First, convert the decimal into fraction and then add.

$\therefore 8 + \dfrac{31-3}{90} + \dfrac{6}{9} + \dfrac{2}{900} = \dfrac{7200 + 280 + 600 + 2}{900}$

$= \dfrac{8082}{900} = 8\dfrac{882}{900} = 8 + \dfrac{979-97}{900} = 8.9\overline{79}$

2. Multiplication of Two or More Decimal Fractions

Given fractions are multiplied without considering the decimal points and then in the product, decimal point is marked from the right hand side to as many places of decimal as the sum of the numbers of decimal places in the multiplier and the multiplicand together.

Ex. 7 (i) $4.3 \times 0.13 = ?$ (ii) $0.\overline{936} \times 0.\overline{12} = ?$

Sol. (i) $43 \times 13 = 559$

$\because$ Sum of the decimal places $= (1 + 2) = 3$

$\therefore$ Required product $= 0.559$

(ii) $0.\overline{936} \times 0.\overline{12} = \dfrac{936}{999} \times \dfrac{12}{99} = \dfrac{104 \times 4}{333 \times 11} = \dfrac{416}{3663} = 0.113$

3. Multiplication of Decimal Fraction by an Integer

Given integer is multiplied by the fraction without considering the decimal point and then in the product, decimal is marked as many places before as that in the given decimal fraction.

Ex. 8 Find the value of the following.
    (i) 19.72×4 (ii) 0.0745×10 (iii) 3.52×14

Sol. (i) 19.72×4
Multiplying without taking decimal point into consideration, $1972 \times 4 = 7888$
So, $19.72 \times 4 = 78.88$
Since, in the given decimal fraction, decimal point is two places before. So, in the product, decimal point will also be put two places before.
Similarly, (ii) $0.0745 \times 10 = 0.7450 = 0.745$ (iii) $3.52 \times 14 = 49.28$

4. Dividing a Decimal Fraction by an Integer

Do simple division, i.e. divide the given decimal number without considering the decimal point and place the decimal point as many places of decimal as in the dividend.

Ex. 9 Divide the following.
    (i) $0.81 \div 9$ (ii) $1.2875 \div 25$ (iii) $0.00049 \div 7$

Sol. (i) $\dfrac{81}{9} = 9 \Rightarrow \dfrac{0.81}{9} = 0.09$ [two places of decimal]

(ii) $\dfrac{12875}{25} = 515 \Rightarrow \dfrac{1.2875}{25} = 0.0515$ [four places of decimal]

(iii) $\dfrac{49}{7} = 7 \Rightarrow \dfrac{0.00049}{7} = 0.00007$ [five places of decimal]

5. Division of Decimal Fractions

In such divisions, dividend and divisor both are multiplied first by a suitable multiple of 10 to convert divisor into a whole number and then above mentioned rule of division is followed.

Ex. 10 Divide the following.
    (i) $42 \div 0.007$ (ii) $0.00048 \div 0.8$

Sol. (i) $\dfrac{42}{0.007} = \dfrac{42}{0.007} \times \dfrac{1000}{1000} = \dfrac{42000}{7} = 6000$

(ii) $\dfrac{0.00048}{0.8} = \dfrac{0.00048}{0.8} \times \dfrac{10}{10} = \dfrac{0.0048}{8} = 0.0006$

Important Facts Related to Simple and Decimal Fractions

(i) In a fraction, if numerator is equal to denominator, then the value of fraction equal to 1.
(ii) If the numerator of a fraction is always non-zero and denominator is zero, then the value of fraction is infinity (∞).
(iii) If the numerator of a fraction is zero and denominator is not equal to zero, then the value of fraction is zero.
(iv) If the numerator or denominator of any fraction is either multiplied or divided by same number, then the value of fraction remains unchanged.
(v) If the numerator and denominator have no common factor other than 1, then the fraction is said to be in its lowest form.

Fast Track Formulae to solve the QUESTIONS

Formula 1

To represent any fraction in simplified form, divide its numerator and denominator by their HCF.

Ex. 11 Write $\dfrac{27}{81}$ in simplified form.

Sol. Here, $27 = 3 \times 3 \times 3$ and $81 = 3 \times 3 \times 3 \times 3$

Hence, $\dfrac{27}{81}$ can be simplified as $\dfrac{27/27}{81/27}$, i.e. $\dfrac{1}{3}$. [∵ HCF of 27 and 81 = $3 \times 3 \times 3 = 27$]

Formula 2

If in the given fractions, the difference between numerator and denominator are same, then fraction having larger numerator is the largest and fraction having smaller numerator is the smallest.

Ex. 12 Arrange the given fractions in increasing order, $\dfrac{4}{5}, \dfrac{5}{6}, \dfrac{6}{7}$.

Sol. Since, all the fractions have difference in numerator and denominator are same. Hence, the increasing order is $\dfrac{4}{5}, \dfrac{5}{6}, \dfrac{6}{7}$.

Formula 3

If in the given fractions, the numerators are increasing by a definite value and the denominator is also increasing by a definite value but the value of denominator is greater than numerator, then the fraction having smaller numerator will be the smallest fraction and the fraction having larger numerator will be the largest fraction.

Ex. 13 Which of the following fractions is largest?

$$\dfrac{2}{5}, \dfrac{5}{11}, \dfrac{8}{17}, \dfrac{11}{23}$$

Sol. Since, in the given fractions, numerator value is increasing by 3 and denominator value is increasing by 6 and $6 > 3$. Then, the fraction having larger numerator will be the larger fraction.

Hence, $\dfrac{11}{23}$ is largest among the given fractions.

Simple and Decimal Fractions / 69

▶ Formula 4

If any number is divided by a/b instead of multiplying by $\dfrac{a}{b}$, then the obtained value will be x greater than original value and the given number will be $\dfrac{abx}{b^2 - a^2}$.

Ex. 14 Arun was to find 6/7 of a fraction. Instead of multiplying, he divided the fraction by 6/7 and the result obtained was 13/70 greater than original value. Find the fraction given to Arun?

Sol. Let the fraction be x.
Then, according to the question,

$$\frac{x}{6/7} = \frac{13}{70} + \frac{6x}{7}$$

$$\Rightarrow \frac{7x}{6} - \frac{6x}{7} = \frac{13}{70}$$

$$\Rightarrow \frac{49x - 36x}{42} = \frac{13}{70}$$

$$\Rightarrow \frac{13x}{42} = \frac{13}{70}$$

$$\Rightarrow x = \frac{42}{13} \times \frac{13}{70} = \frac{3}{5}$$

Fast Track Method

Given, $a = 6, b = 7$ and $x = 13/70$

$$\therefore \text{Required fraction} = \frac{abx}{b^2 - a^2}$$

$$= \frac{6 \times 7 \times \dfrac{13}{70}}{7^2 - 6^2}$$

$$= \frac{6 \times 13}{10 \times 13} = \frac{3}{5}$$

Hence, the fraction given to Arun is 3/5.

Fast Track Practice

Exercise 1 Base Level Questions

1. $\dfrac{19999}{21111} = ?$
   (a) 0.947   (b) 0.749
   (c) 0.497   (d) 0.794
   (e) 0.974

2. $33 + 371 \div 7 = ?$ [Bank Clerk 2011]
   (a) 89   (b) 85
   (c) 86   (d) 84
   (e) None of these

3. Find the sum $\dfrac{3}{10} + \dfrac{5}{100} + \dfrac{8}{1000}$ in decimal form.
   (a) 0.853   (b) 0.358
   (c) 3.58   (d) 8.35
   (e) None of these

4. Find the value of $\dfrac{1}{3} + \dfrac{1}{15} + \dfrac{1}{35} + \dfrac{1}{63} + \dfrac{1}{99}$.
   (a) $\dfrac{10}{11}$ (b) $\dfrac{5}{11}$
   (c) $\dfrac{9}{11}$ (d) $\dfrac{7}{11}$

5. $1 + \dfrac{1}{2} + \dfrac{1}{4} + \dfrac{1}{7} + \dfrac{1}{14} + \dfrac{1}{28} = ?$
   (a) 2   (b) 5   (c) 4   (d) 6
   (e) None of these

6. $1\dfrac{3}{5} + 1\dfrac{8}{9} + 2\dfrac{4}{5} = ?$ [Bank Clerk 2011]
   (a) $6\dfrac{19}{45}$ (b) $6\dfrac{16}{45}$
   (c) $6\dfrac{17}{45}$ (d) $6\dfrac{13}{45}$
   (e) None of these

7. If $\dfrac{1}{1\dfrac{1}{4}} + \dfrac{1}{6\dfrac{2}{3}} - \dfrac{1}{x} + \dfrac{1}{10} = \dfrac{11}{12}$, then find the value of x.
   (a) $\dfrac{15}{3}$ (b) $\dfrac{20}{15}$ (c) $\dfrac{15}{2}$ (d) $\dfrac{12}{13}$

8. $\dfrac{16}{23} \times \dfrac{47}{288} \times \dfrac{92}{141} = ?$ [Bank Clerk 2011]
   (a) $\dfrac{4}{27}$ (b) $\dfrac{2}{27}$ (c) $\dfrac{2}{29}$ (d) $\dfrac{3}{28}$
   (e) None of these

9. $\dfrac{3}{4}$ of $\dfrac{5}{6}$ of $\dfrac{7}{10}$ of $1664 = ?$ [Bank Clerk 2009]
   (a) 648   (b) 762
   (c) 612   (d) 728
   (e) None of these

10. $8\dfrac{1}{3} \div 10\dfrac{5}{6} = ?$
    (a) $\dfrac{5}{6}$ (b) $\dfrac{2}{3}$
    (c) $\dfrac{10}{13}$ (d) $\dfrac{11}{13}$
    (e) None of these

11. If $1\dfrac{2}{3} \div \dfrac{2}{7} \times \dfrac{x}{7} = 1\dfrac{1}{4} \times \dfrac{2}{3} \div \dfrac{1}{6}$, then find the value of x. [SSC CGL 2011]
    (a) 0.006   (b) 1/6
    (c) 0.6   (d) 6

12. If $x + \dfrac{1}{1 + \dfrac{1}{2 + \dfrac{1}{3}}} = 2$, then what is x equal to? [CDS 2016 (II)]
    (a) $\dfrac{7}{10}$ (b) $\dfrac{13}{10}$
    (c) $\dfrac{11}{10}$ (d) $\dfrac{17}{10}$

13. The value of $4 - \dfrac{5}{1 + \dfrac{1}{3 + \dfrac{1}{2 + \dfrac{1}{4}}}}$ is [SSC CGL 2015]
    (a) $\dfrac{1}{8}$ (b) $\dfrac{1}{32}$ (c) $\dfrac{1}{64}$ (d) $\dfrac{1}{16}$

14. Arrange $\dfrac{7}{12}, \dfrac{2}{3}$ and $\dfrac{3}{8}$ in the ascending order.
    (a) $\dfrac{3}{8} < \dfrac{7}{12} < \dfrac{2}{3}$ (b) $\dfrac{2}{3} < \dfrac{7}{12} < \dfrac{3}{8}$
    (c) $\dfrac{7}{12} < \dfrac{2}{3} < \dfrac{3}{8}$ (d) $\dfrac{3}{8} < \dfrac{2}{3} < \dfrac{7}{12}$
    (e) None of these

Simple and Decimal Fractions / 71

15. Which one among the following is the largest? [CDS 2017 (I)]
(a) $\frac{7}{9}$
(b) $\frac{11}{14}$
(c) $\frac{3}{4}$
(d) $\frac{10}{13}$

16. Out of the fractions $\frac{4}{7}, \frac{5}{13}, \frac{6}{11}, \frac{3}{5}$ and $\frac{2}{3}$, which is the second smallest fraction? [SSC CGL 2010]
(a) $\frac{4}{7}$
(b) $\frac{5}{13}$
(c) $\frac{6}{11}$
(d) $\frac{3}{5}$

17. Out of the fractions $\frac{5}{7}, \frac{7}{13}, \frac{4}{7}, \frac{4}{15}$ and $\frac{9}{14}$, which is the third highest?
(a) $\frac{5}{7}$
(b) $\frac{7}{13}$
(c) $\frac{4}{7}$
(d) $\frac{4}{15}$
(e) $\frac{9}{14}$

18. Out of the fractions $\frac{5}{7}, \frac{4}{9}, \frac{6}{11}, \frac{2}{5}$ and $\frac{3}{4}$, what is the difference between the largest and the smallest fractions? [IBPS Clerk 2011]
(a) $\frac{6}{13}$
(b) $\frac{11}{18}$
(c) $\frac{7}{18}$
(d) $\frac{11}{20}$
(e) None of these

19. If the fractions $\frac{19}{21}, \frac{21}{25}, \frac{25}{29}, \frac{29}{31}$ and $\frac{31}{37}$ are arranged in ascending order of their values, then which one will be the 2nd? [Bank Clerk 2009]
(a) $\frac{19}{21}$
(b) $\frac{21}{25}$
(c) $\frac{25}{29}$
(d) $\frac{29}{31}$
(e) None of these

20. When 0.252525... is converted into a fraction, then find the result. [RRB 2009]
(a) $\frac{25}{99}$
(b) $\frac{25}{90}$
(c) $\frac{25}{999}$
(d) $\frac{25}{9999}$

21. 0.4777... is equal to
(a) $\frac{477}{100000}$
(b) $\frac{477}{100}$
(c) $\frac{437}{100}$
(d) $\frac{43}{90}$
(e) None of these

22. Find the value of $9.46\overline{7}$ in a vulgar fraction.
(a) $9\frac{421}{900}$
(b) $9\frac{422}{900}$
(c) $9\frac{435}{900}$
(d) $9\frac{437}{900}$
(e) None of these

23. Express $0.8\overline{268}$ as a vulgar fraction.
(a) $\frac{3093}{4950}$
(b) $\frac{3043}{4850}$
(c) $\frac{4093}{4950}$
(d) $\frac{3039}{4950}$
(e) None of these

24. Representation of $0.2\overline{341}$ in the form p/q, where p and q are integers, $q \neq 0$, is [CDS 2013]
(a) $\frac{781}{3330}$
(b) $\frac{1171}{4995}$
(c) $\frac{2341}{9990}$
(d) $\frac{2339}{9990}$

25. $1088.88 + 1800.08 + 1880.80 = ?$ [Bank PO 2010]
(a) 8790.86
(b) 8890.86
(c) 5588.80
(d) 4769.76
(e) None of these

26. $6435.9 + 7546.4 + 1203.5 = ?$ [Bank PO 2010]
(a) 15188.5
(b) 15185.8
(c) 15155.5
(d) 15815.8
(e) None of these

27. $\frac{1212}{0.5} = 6.06 \times ?$ [SSC (10+2) 2007]
(a) 4.04
(b) 400
(c) 0.4
(d) 0.44

28. If $35 \times 35 = 1225$, then 3.5×3.5 equals [IB PA 2016, 15]
(a) 122.5
(b) 12.25
(c) 1225
(d) 1121.5

29. $(39.3 \times 53.4) + (26.7 \times 5.9) = ?$ [Bank Clerk 2011]
(a) 2520.15
(b) 2256.15
(c) 2562.15
(d) 2652.15
(e) None of these

30. $783 \div 9 \div 0.75 = ?$ [Bank Clerk 2009]
(a) 130
(b) 124
(c) 118
(d) 116
(e) None of these

31. Find the value of $2 \times \left\{ \frac{3.6 \times 0.48 \times 2.50}{0.12 \times 0.09 \times 0.5} \right\}$.
(a) 800
(b) 500
(c) 900
(d) 1600
(e) None of these

72 / Fast Track Objective Arithmetic

32. The value of $(0.\overline{63} + 0.\overline{37})$ is [CDS 2015 (II)]

(a) 1 (b) $\dfrac{100}{91}$

(c) $\dfrac{100}{99}$ (d) $\dfrac{1000}{999}$

33. The value of $0.\overline{3} + 0.\overline{6} + 0.\overline{7} + 0.\overline{8}$ in fraction will be

(a) $2\dfrac{3}{10}$ (b) $2\dfrac{2}{3}$

(c) $20.\overline{35}$ (d) $5\dfrac{3}{10}$

34. The simplification $(0.\overline{63} + 0.\overline{37} + 0.\overline{80})$ yields the result [SSC CGL (Pre) 2016]

(a) $1.\overline{80}$ (b) $1.\overline{81}$

(c) $1.\overline{79}$ (d) 1.80

35. Find the value of $27 \times 1.\overline{2} \times 5.5262 \times 0.\overline{6}$.

(a) $121.\overline{57}$ (b) $121.\overline{75}$

(c) $121.7\overline{5}$ (d) None of these

36. $(0.\overline{142857} \div 0.\overline{285714}) = ?$

(a) $\dfrac{1}{2}$ (b) $\dfrac{1}{3}$

(c) 2 (d) 10

(e) None of these

37. Which one of the following is true? [SSC (10+2) 2015]

(a) $0.5 < \dfrac{2}{3} < \dfrac{3}{4} < \left(\dfrac{16}{25}\right)^{0.5}$

(b) $\dfrac{7}{24} > \dfrac{1}{3} > \dfrac{3}{8} > \dfrac{5}{12}$

(c) $\dfrac{1}{2} > \dfrac{2}{3} > \dfrac{3}{4} > \dfrac{4}{5}$

(d) $0 > \dfrac{7}{17} > \dfrac{3}{7} > \dfrac{3}{5}$

38. Sachin was to find $\dfrac{4}{5}$ of a fraction. Instead of multiplying, he divided the fraction by $\dfrac{4}{5}$ and the result obtained was $\dfrac{9}{70}$ greater than original value. Find the fraction given to Sachin.

(a) $\dfrac{2}{7}$ (b) $\dfrac{3}{7}$

(c) $\dfrac{3}{5}$ (d) $\dfrac{4}{5}$

39. If the fraction a/b is positive, then which of the following must be true? [SSC Multitasking 2014]

(a) $a > 0$ (b) $b > 0$

(c) $ab > 0$ (d) $a + b > 0$

Exercise 2 Higher Skill Level Questions

1. The pair of rational numbers that lies between $\dfrac{1}{4}$ and $\dfrac{3}{4}$ is [CDS 2014 (II)]

(a) $\dfrac{262}{1000}, \dfrac{752}{1000}$ (b) $\dfrac{24}{100}, \dfrac{74}{100}$

(c) $\dfrac{9}{40}, \dfrac{31}{40}$ (d) $\dfrac{252}{1000}, \dfrac{748}{1000}$

2. If $1.5x = 0.04y$, then find the value of $\left(\dfrac{y-x}{y+x}\right)$. [SSC CPO 2007]

(a) $\dfrac{730}{77}$ (b) $\dfrac{73}{77}$

(c) $\dfrac{73}{770}$ (d) $\dfrac{703}{77}$

3. If $0.764y = 1.236x$, then what is the value of $\left(\dfrac{y-x}{y+x}\right)$? [CDS 2012]

(a) 0.764 (b) 0.236

(c) 2 (d) 0.472

4. $\dfrac{0.04}{0.03}$ of $\dfrac{\left(3\dfrac{1}{3} - 2\dfrac{1}{2}\right) \div \dfrac{1}{2} \text{ of } 1\dfrac{1}{4}}{\dfrac{1}{3} + \dfrac{1}{5} \text{ of } \dfrac{1}{9}} = ?$ [SSC CPO 2011]

(a) 1 (b) 5

(c) $\dfrac{1}{5}$ (d) $\dfrac{1}{2}$

5. Find the value of

$999\dfrac{1}{7} + 999\dfrac{2}{7} + 999\dfrac{3}{7} + 999\dfrac{4}{7} + 999\dfrac{5}{7} + 999\dfrac{6}{7}$.

(a) 5997 (b) 5979

(c) 5994 (d) 2997

6. What is the value of $0.007 + 0.\overline{7} + 17.\overline{83} + 310.02\overline{02}$? [CDS 2012]

(a) 327.86638 (b) 328.644

(c) 327.86683 (d) $327.866\overline{8}$

Simple and Decimal Fractions / 73

7. If $\dfrac{37}{13} = 2 + \dfrac{1}{x + \dfrac{1}{y + \dfrac{1}{z}}}$, where x, y and z are natural numbers, then what is z equal to? [CDS 2015 (I)]
(a) 1
(b) 2
(c) 3
(d) Cannot be determined

8. If $\dfrac{61}{19} = 3 + \dfrac{1}{x + \dfrac{1}{y + \dfrac{1}{z}}}$, where x, y and z are natural numbers, then what is z equal to? [CDS 2016 (I)]
(a) 1 (b) 2 (c) 3 (d) 4

9. What number must be subtracted from both the numerator and the denominator of the fraction $\dfrac{27}{35}$ so that it becomes $\dfrac{2}{3}$? [CDS 2017 (I)]
(a) 6 (b) 8 (c) 9 (d) 11

10. If a fraction is multiplied by itself and then divided by the reciprocal of the same fraction, the result is $18\dfrac{26}{27}$, then find the fraction.
(a) $\dfrac{8}{27}$
(b) $1\dfrac{1}{3}$
(c) $2\dfrac{2}{3}$
(d) $3\dfrac{2}{3}$
(e) None of these

11. 1/8 part of a pencil is black and 1/2 part of the remaining is white. If the remaining part is blue and length of this blue part is $3\dfrac{1}{2}$ cm, then find the length of the pencil.
(a) 6 cm   (b) 7 cm
(c) 8 cm   (d) 9 cm
(e) None of these

12. If the numerator of a fraction is increased by 200% and the denominator of the fraction is increased by 150%, then the resultant fraction is 9/35. What is the original fraction? [SBI Clerk 2011]
(a) $\dfrac{3}{10}$ (b) $\dfrac{2}{15}$ (c) $\dfrac{3}{16}$ (d) $\dfrac{2}{7}$
(e) None of these

13. The numerator of a fraction is 4 less than its denominator. If the numerator is decreased by 2 and the denominator is increased by 1, then the denominator becomes eight times the numerator, then find the fraction. [SSC CGL 2013]
(a) $\dfrac{3}{7}$ (b) $\dfrac{4}{8}$ (c) $\dfrac{2}{7}$ (d) $\dfrac{3}{8}$

14. Sum of three fractions is $2\dfrac{11}{24}$. If the greatest fraction is divided by the smallest fraction, the result is 7/6, which is greater than the middle fraction by 1/3. Find all the three fractions.
(a) $\dfrac{3}{5}, \dfrac{4}{7}, \dfrac{2}{3}$
(b) $\dfrac{7}{8}, \dfrac{5}{6}, \dfrac{3}{4}$
(c) $\dfrac{7}{9}, \dfrac{2}{3}, \dfrac{3}{5}$
(d) $\dfrac{7}{8}, \dfrac{7}{9}, \dfrac{7}{10}$
(e) None of these

15. A school group charters three identical buses and occupies 4/5 of the seats. After 1/4 of the passengers leave, the remaining passengers use only two of the buses. The fraction of the seats on the two buses that are now occupied, is [SSC CGL (Mains) 2014]
(a) $\dfrac{7}{10}$
(b) $\dfrac{9}{10}$
(c) $\dfrac{8}{9}$
(d) $\dfrac{7}{9}$

16. If p is a prime number other than 2 or 5. One would like to express the vulgar fraction $1/p$ in the form of a recurring decimal. Then, the decimal will be [CDS 2015 (I)]
(a) a pure recurring decimal and its period will be necessarily $(p-1)$
(b) a mixed recurring decimal and its period will be necessarily $(p-1)$
(c) a pure recurring decimal and its period will be some factor of $(p-1)$
(d) a mixed recurring decimal and its period will be some factor of $(p-1)$

17. Ranu's monthly salary is four-fifth of Ali's monthly salary. Ranu and Ali save one-fourth and two-fifth amount from their respective monthly salary. If the difference between the amount saved by Ranu and that saved by Ali is ₹ 7000, what is Ranu's monthly salary? [IBPS Clerk (Pre) 2016]
(a) ₹ 30000   (b) ₹ 35600
(c) ₹ 35000   (d) ₹ 28000
(e) ₹ 21000

Answer with Solutions

Exercise 1 Base Level Questions

1. (a) $\dfrac{19999}{21111} = 0.947$

2. (c) $33 + 371 \div 7 = 33 + \dfrac{371}{7} = 33 + 53 = 86$

3. (b) $\dfrac{3}{10} + \dfrac{5}{100} + \dfrac{8}{1000}$
$= 0.3 + 0.05 + 0.008 = 0.358$

4. (b) $\dfrac{1}{3} + \dfrac{1}{15} + \dfrac{1}{35} + \dfrac{1}{63} + \dfrac{1}{99}$

$= \left(\dfrac{1}{3} + \dfrac{1}{15}\right) + \dfrac{1}{35} + \dfrac{1}{63} + \dfrac{1}{99}$

$= \left(\dfrac{6}{15} + \dfrac{1}{35}\right) + \dfrac{1}{63} + \dfrac{1}{99}$

$= \dfrac{45}{105} + \dfrac{1}{63} + \dfrac{1}{99} = \left(\dfrac{3}{7} + \dfrac{1}{63}\right) + \dfrac{1}{99}$

$= \dfrac{28}{63} + \dfrac{1}{99} = \dfrac{4}{9} + \dfrac{1}{99}$

$= \dfrac{44 + 1}{99} = \dfrac{45}{99} = \dfrac{5}{11}$

5. (a) $1 + \dfrac{1}{2} + \dfrac{1}{4} + \dfrac{1}{7} + \dfrac{1}{14} + \dfrac{1}{28}$

$= \dfrac{28 + 14 + 7 + 4 + 2 + 1}{28} = \dfrac{56}{28} = 2$

6. (d) $1\dfrac{3}{5} + 1\dfrac{8}{9} + 2\dfrac{4}{5}$

$= (1 + 1 + 2) + \left(\dfrac{3}{5} + \dfrac{8}{9} + \dfrac{4}{5}\right)$

$= 4 + \dfrac{27 + 40 + 36}{45}$

$= 4 + \dfrac{103}{45} = 4 + 2\dfrac{13}{45} = 6\dfrac{13}{45}$

Alternate Method

$1\dfrac{3}{5} + 1\dfrac{8}{9} + 2\dfrac{4}{5}$

$= \dfrac{8}{5} + \dfrac{17}{9} + \dfrac{14}{5}$

$= \dfrac{8 \times 9 + 17 \times 5 + 14 \times 9}{45}$

$= \dfrac{72 + 85 + 126}{45} = \dfrac{283}{45} = 6\dfrac{13}{45}$

7. (c) $\dfrac{1}{1\dfrac{1}{4}} + \dfrac{1}{6\dfrac{2}{3}} - \dfrac{1}{x} + \dfrac{1}{10} = \dfrac{11}{12}$

$\Rightarrow \dfrac{4}{5} + \dfrac{3}{20} - \dfrac{1}{x} + \dfrac{1}{10} = \dfrac{11}{12}$

$\Rightarrow \dfrac{1}{x} = \dfrac{4}{5} + \dfrac{3}{20} + \dfrac{1}{10} - \dfrac{11}{12}$

$\Rightarrow \dfrac{1}{x} = \dfrac{48 + 9 + 6 - 55}{60}$

$\Rightarrow \dfrac{1}{x} = \dfrac{63 - 55}{60} = \dfrac{8}{60} = \dfrac{2}{15}$

$\Rightarrow \dfrac{1}{x} = \dfrac{2}{15} \Rightarrow x = \dfrac{15}{2}$

8. (b) $\dfrac{16}{23} \times \dfrac{47}{288} \times \dfrac{92}{141} = \dfrac{4}{18 \times 3} = \dfrac{2}{9 \times 3} = \dfrac{2}{27}$

9. (d) $? = \dfrac{3}{4} \times \dfrac{5}{6} \times \dfrac{7}{10} \times 1664 = 728$

10. (c) $? = \dfrac{25}{3} \div \dfrac{65}{6} = \dfrac{25}{3} \times \dfrac{6}{65} = \dfrac{10}{13}$

11. (d) $1\dfrac{2}{3} \div \dfrac{2}{7} \times \dfrac{x}{7} = 1\dfrac{1}{4} \times \dfrac{2}{3} \div \dfrac{1}{6}$

$\Rightarrow \dfrac{5}{3} \times \dfrac{7}{2} \times \dfrac{x}{7} = \dfrac{5}{4} \times \dfrac{2}{3} \times 6$

$\Rightarrow \dfrac{5x}{3 \times 2} = 5 \Rightarrow x = \dfrac{5 \times 6}{5}$

$\therefore x = 6$

12. (b) Given, $x + \dfrac{1}{1 + \dfrac{1}{2 + \dfrac{1}{3}}} = 2$

$\Rightarrow x + \dfrac{1}{1 + \dfrac{3}{7}} = 2$

$\Rightarrow x + \dfrac{7}{10} = 2$

$\therefore x = 2 - \dfrac{7}{10} = \dfrac{13}{10}$

13. (a) $4 - \dfrac{5}{1 + \dfrac{1}{3 + \dfrac{1}{2 + \dfrac{1}{4}}}} = 4 - \dfrac{5}{1 + \dfrac{1}{3 + \dfrac{4}{9}}}$

$= 4 - \dfrac{5}{1 + \dfrac{9}{31}}$

$= 4 - \dfrac{5 \times 31}{40} = \dfrac{32 - 31}{8} = \dfrac{1}{8}$

Simple and Decimal Fractions / 75

14. (a) $\dfrac{7}{12} = 0.583, \dfrac{2}{3} = 0.666$ and $\dfrac{3}{8} = 0.375$

Hence, the ascending order is
$$\dfrac{3}{8} < \dfrac{7}{12} < \dfrac{2}{3}.$$

Alternate Method
LCM of 12, 3 and 8

$$\begin{array}{r|lll} 2 & 12, & 3, & 8 \\ \hline 2 & 6, & 3, & 4 \\ \hline 3 & 3, & 3, & 2 \\ \hline & 1, & 1, & 2 \end{array}$$

∴ LCM = 2 × 2 × 3 × 2 = 24

∴ $\dfrac{7}{12} = \dfrac{7 \times 2}{12 \times 2} = \dfrac{14}{24}; \dfrac{2}{3} = \dfrac{2 \times 8}{3 \times 8} = \dfrac{16}{24}$

and $\dfrac{3}{8} = \dfrac{3 \times 3}{8 \times 3} = \dfrac{9}{24}$

Now, the fraction having smallest numerator will be smallest.

∴ Ascending order is
$\dfrac{9}{24} < \dfrac{14}{24} < \dfrac{16}{24}$ i.e. $\dfrac{3}{8} < \dfrac{7}{12} < \dfrac{2}{3}$

15. (b) We have, $\dfrac{7}{9}, \dfrac{11}{14}, \dfrac{3}{4}, \dfrac{10}{13}$

L.C.M. of (9, 14, 4, 13) = 3276
Now, make the denominator of all the fractions equal to 3276. We get
$\dfrac{2548}{3276}, \dfrac{2574}{3276}, \dfrac{2457}{3276}, \dfrac{2520}{3276}$

∴ Largest number = $\dfrac{2574}{3276} = \dfrac{11}{14}$

16. (c) $\dfrac{4}{7} = 0.57, \dfrac{5}{13} = 0.38,$
$\dfrac{6}{11} = 0.54, \dfrac{3}{5} = 0.6, \dfrac{2}{3} = 0.67$

Clearly, the second smallest fraction is $\dfrac{6}{11}$.

17. (c) $\dfrac{5}{7} = 0.71, \dfrac{7}{13} = 0.54, \dfrac{4}{7} = 0.57,$
$\dfrac{4}{15} = 0.27, \dfrac{9}{14} = 0.64$

Clearly, $\dfrac{4}{7}$ is the third highest fraction.

18. (e) $\dfrac{5}{7} = 0.71, \dfrac{4}{9} = 0.44, \dfrac{6}{11} = 0.54, \dfrac{2}{5} = 0.40$

and $\dfrac{3}{4} = 0.75$

Here, the largest fraction = $\dfrac{3}{4}$

and the smallest fraction = $\dfrac{2}{5}$

So, required difference
$= \dfrac{3}{4} - \dfrac{2}{5} = \dfrac{15-8}{20} = \dfrac{7}{20}$

19. (b) $\dfrac{19}{21} = 0.904, \dfrac{21}{25} = 0.84,$
$\dfrac{25}{29} = 0.86, \dfrac{29}{31} = 0.93,$
$\dfrac{31}{37} = 0.837$

0.837 < 0.84 < 0.86 < 0.904 < 0.93

Clearly, $\dfrac{21}{25}$ will be on second number.

20. (a) $0.252525... = 0.\overline{25} = \dfrac{25}{99}$

21. (d) $0.4777... = 0.4\overline{7} = \dfrac{47-4}{90} = \dfrac{43}{90}$

22. (a) $9.46\overline{7} = 9 + \dfrac{(467-46)}{900}$
$= 9 + \dfrac{421}{900} = 9\dfrac{421}{900}$

23. (c) $0.82\overline{68} = \dfrac{8268-82}{9900} = \dfrac{8186}{9900} = \dfrac{4093}{4950}$

24. (d) $0.2\overline{341} = \dfrac{2341-2}{9990} = \dfrac{2339}{9990}$

25. (d)
```
   1088.88
   1800.08
 + 1880.80
 ─────────
   4769.76
```

26. (b)
```
   6435.9
   7546.4
 + 1203.5
 ─────────
  15185.8
```

27. (b) $? = \dfrac{1212}{0.5 \times 6.06} = \dfrac{1212}{5 \times 606} \times 10 \times 100$
$= \dfrac{1212}{3030} \times 1000 = \dfrac{1212 \times 100}{303} = 400$

28. (b) ∵ 35 × 35 = 1225 [given]
On dividing the both sides by 100, we get
$\dfrac{35 \times 35}{100} = \dfrac{1225}{100}$
⇒ 3.5 × 3.5 = 12.25

29. (b) (39.3 × 53.4) + (26.7 × 5.9)
= 2098.62 + 157.53 = 2256.15

30. (d) $? = \dfrac{783}{9 \times 0.75} = 116$

31. (d) $2 \times \left\{\dfrac{3.6 \times 0.48 \times 2.50}{0.12 \times 0.09 \times 0.5}\right\}$
$= 2 \times \left\{\dfrac{36 \times 48 \times 250}{12 \times 9 \times 5}\right\}$
$= 2 \times 4 \times 4 \times 50 = 1600$

32. (c) $0.\overline{63} + 0.\overline{37} = \dfrac{63}{99} + \dfrac{37}{99} = \dfrac{100}{99}$

76 / Fast Track Objective Arithmetic

33. (b) $0.\overline{3} + 0.\overline{6} + 0.\overline{7} + 0.\overline{8}$
$= \frac{3}{9} + \frac{6}{9} + \frac{7}{9} + \frac{8}{9}$
$= \frac{24}{9} = \frac{8}{3} = 2\frac{2}{3}$

34. (b) Given expression
$= 0.\overline{63} + 0.\overline{37} + 0.\overline{80}$
$= \frac{63}{99} + \frac{37}{99} + \frac{80}{99}$
$= \frac{63 + 37 + 80}{99} = \frac{180}{99}$
$= 1\frac{81}{99} = 1.\overline{81}$

35. (d) $27 \times 1.\overline{2} \times 5.5\overline{262} \times 0.\overline{6}$
$= 27 \times \frac{12-1}{9} \times \frac{55262-5526}{9000} \times \frac{6}{9}$
$= 27 \times \frac{11}{9} \times \frac{49736}{9000} \times \frac{6}{9}$
$= \frac{1094192}{9000} = 121.576888... = 121.57\overline{68}$

36. (a) $\frac{0.\overline{142857}}{0.\overline{285714}} = \frac{\frac{142857}{999999}}{\frac{285714}{999999}} = \frac{142857}{285714} = \frac{1}{2}$

37. (a) By taking option (a),
$\left(\frac{16}{25}\right)^{0.5} = \left(\frac{16}{25}\right)^{1/2} = \frac{4}{5}$
Given, $0.5 < \frac{2}{3} < \frac{3}{4} < \frac{4}{5}$

$\Rightarrow \frac{1}{2} < \frac{2}{3} < \frac{3}{4} < \frac{4}{5}$

∵ LCM of (2, 3, 4 and 5) = 60
∴ $\frac{1}{2} \times \frac{30}{30} < \frac{2}{3} \times \frac{20}{20} < \frac{3}{4} \times \frac{15}{15} < \frac{4}{5} \times \frac{12}{12}$

$\Rightarrow \frac{30}{60} < \frac{40}{60} < \frac{45}{60} < \frac{48}{60}$

So, the order of the fractions given in option (a) is correct.

Alternate Method
In option (a),
$\left(\frac{16}{25}\right)^{0.25} = \left(\frac{16}{25}\right)^{1/2} = \frac{4}{5}$
Given, $0.5 < \frac{2}{3} < \frac{3}{4} < \frac{4}{5}$

$\Rightarrow \frac{1}{2} < \frac{2}{3} < \frac{3}{4} < \frac{4}{5}$

Hence, the above optioin is true.
[by Formula 2]

38. (a) Given, $a = 4$, $b = 5$ and $x = \frac{9}{70}$

∴ Required fraction $= \frac{abx}{b^2 - a^2}$
[by Formula 4]
$= \frac{4 \times 5 \times \frac{9}{70}}{5^2 - 4^2}$
$= \frac{4 \times 9}{14 \times 9} = \frac{2}{7}$

39. (c) If the fraction $\frac{a}{b}$ is positive, then $ab > 0$ must be true.

Exercise 2 Higher Skill Level Questions

1. (d) Here, $\frac{1}{4} = 0.25$ and $\frac{3}{4} = 0.75$
Only option (d) with $\frac{252}{1000} = 0.252$
and $\frac{748}{1000} = 0.748$ lies between 0.25 and 0.75.

2. (b) Given, $1.5x = 0.04y$
$\Rightarrow \frac{y}{x} = \frac{1.5}{0.04} = \frac{1.50}{0.04}$
$= \frac{150}{4} = \frac{75}{2}$
∴ $\left(\frac{y-x}{y+x}\right) = \left(\frac{\frac{y}{x} - 1}{\frac{y}{x} + 1}\right) = \frac{\left(\frac{75}{2} - 1\right)}{\left(\frac{75}{2} + 1\right)} = \frac{73}{77}$

3. (b) Given, $\frac{y}{x} = \frac{1.236}{0.764} = \frac{309}{191}$

∴ $\left(\frac{y-x}{y+x}\right) = \left(\frac{\frac{y}{x} - 1}{\frac{y}{x} + 1}\right) = \left(\frac{\frac{309}{191} - 1}{\frac{309}{191} + 1}\right) = \left(\frac{\frac{118}{191}}{\frac{500}{191}}\right)$
$= \frac{118}{500} = 0.236$

4. (b) Given expression
$= \frac{0.04}{0.03} \times \frac{\left(3\frac{1}{3} - 2\frac{1}{2}\right) \div \frac{1}{2} \times 1\frac{1}{4}}{\frac{1}{3} + \frac{1}{5} \times \frac{1}{9}}$
$= \frac{4}{3} \times \frac{\left(\frac{10}{3} - \frac{5}{2}\right) \div \frac{1}{2} \text{ of } \frac{5}{4}}{\frac{1}{3} + \frac{1}{45}}$

Simple and Decimal Fractions / 77

$$= \frac{4}{3} \times \frac{\left(\frac{20-15}{6}\right) \div \frac{5}{8}}{\left(\frac{15+1}{45}\right)}$$

$$= \frac{4}{3} \times \frac{\frac{5}{6} \times \frac{8}{5}}{\frac{16}{45}} = \frac{4}{3} \times \frac{5}{6} \times \frac{8}{5} \times \frac{45}{16} = 5$$

5. (a) $999\frac{1}{7} + 999\frac{2}{7} + 999\frac{3}{7} + 999\frac{4}{7}$
$\qquad\qquad\qquad\qquad + 999\frac{5}{7} + 999\frac{6}{7}$
$= (999 + 999 + 999 + 999 + 999 + 999)$
$\qquad + \left(\frac{1}{7} + \frac{2}{7} + \frac{3}{7} + \frac{4}{7} + \frac{5}{7} + \frac{6}{7}\right)$
$= 6 \times 999 + \frac{21}{7} = 6(1000-1) + \frac{21}{7}$
$= 6000 - 6 + 3 = 5997$

6. (b) $0.007 + \frac{7}{9} + 17\frac{83}{99} + 310\frac{(202-20)}{9000}$
$= 0.007 + \frac{7}{9} + \frac{1766}{99} + 310\frac{182}{9000}$
$= 0.007 + \frac{7}{9} + \frac{1766}{99} + 310.022$
$= 0.007 + 0.777 + 17.838 + 310.022$
$= 328.644$

7. (d) RHS $= 2 + \cfrac{1}{x + \cfrac{1}{y + \cfrac{1}{z}}} = 2 + \cfrac{1}{x + \cfrac{1}{\frac{yz+1}{z}}}$

$= 2 + \cfrac{1}{x + \cfrac{z}{yz+1}}$

$= 2 + \cfrac{1}{\frac{xyz+x+z}{yz+1}} = 2 + \frac{yz+1}{(xyz+x+z)}$

We can check that no natural numbers satisfy the equation.
So, we cannot determine the value of z.

8. (c) $\because \quad \frac{61}{19} = 3 + \cfrac{1}{x + \cfrac{1}{y + \frac{1}{z}}}$

$\Rightarrow \quad \frac{61}{19} - 3 = \cfrac{1}{x + \cfrac{1}{\frac{yz+1}{z}}}$

$\Rightarrow \quad \frac{4}{19} = \cfrac{1}{x + \cfrac{z}{yz+1}}$

$\Rightarrow \quad \frac{4}{19} = \frac{yz+1}{xyz+x+z}$

$\Rightarrow \quad (4xy + 4 - 19y)z = 19 - 4x$
$\Rightarrow \quad z = \frac{19 - 4x}{4xy + 4 - 19y}$

As x, y and z are natural numbers.
Again, let $x = 4$ and $y = 1$
Then, $z = \frac{19 - 4 \times 4}{4 \times 4 \times 1 + 4 - 19 \times 1}$
$= \frac{19 - 16}{20 - 19} = 3$

which is a natural number.

9. (d) Let the number be x.
Then, $\frac{27 - x}{35 - x} = \frac{2}{3}$
$\Rightarrow \quad 3(27 - x) = 2(35 - x)$
$\Rightarrow \quad 81 - 3x = 70 - 2x$
$\Rightarrow \quad 11 = x$

Hence, the required number is 11.

10. (c) Let the required fraction be x.
According to the question,
$\frac{x \times x}{\frac{1}{x}} = 18\frac{26}{27}$

$\Rightarrow \quad x^3 = \frac{512}{27} = \left(\frac{8}{3}\right)^3$

$\therefore \quad x = \frac{8}{3} = 2\frac{2}{3}$

11. (c) Let total length of pencil be x cm.
Then, black part $= \frac{x}{8}$
Remaining part $= x - \frac{x}{8} = \frac{7x}{8}$
White part $= \frac{1}{2}\left(\frac{7x}{8}\right) = \frac{7x}{16}$
Remaining part $= x - \left(\frac{x}{8} + \frac{7x}{16}\right) = \frac{7x}{16}$

$\therefore$ Length of blue part $= \frac{7x}{16}$
According to the question,
$\frac{7x}{16} = 3\frac{1}{2} \Rightarrow \frac{7x}{16} = \frac{7}{2}$
$\therefore \quad x = 8$ cm

12. (e) Let the original fraction be $\frac{x}{y}$.
Then, numerator is increased by 200%.
$\therefore$ Numerator $= x + 200\%$ of x
$= x + \frac{200x}{100}$
$= \frac{100x + 200x}{100} = \frac{300x}{100}$

and denominator of the fraction is increased by 150%.

Denominator $= y + \dfrac{150y}{100}$

$= \dfrac{100y + 150y}{100} = \dfrac{250y}{100}$

Then, according to the question,

$\dfrac{300x/100}{250y/100} = \dfrac{9}{35}$

$\dfrac{300x}{250y} = \dfrac{9}{35}$

$\therefore \dfrac{x}{y} = \dfrac{9}{35} \times \dfrac{250}{300} = \dfrac{3}{14}$

13. (a) Let denominator of fraction be x.
Then, numerator $= (x - 4)$
$\therefore$ Fraction $= \dfrac{(x-4)}{x}$

Now, according to the question,
$8\,[(x - 4) - 2] = (x + 1)$

$\Rightarrow (x - 4) - 2 = \dfrac{(x+1)}{8}$

$\Rightarrow x - 6 = \dfrac{x+1}{8}$

$\Rightarrow 8x - 48 = x + 1$
$\Rightarrow 8x - x = 48 + 1$
$\Rightarrow 7x = 49 \Rightarrow x = \dfrac{49}{7}$
$\Rightarrow x = 7$

$\therefore$ Fraction $= \dfrac{7-4}{7} = \dfrac{3}{7}$

14. (b) Let the greatest, middle and smallest fractions be x, y and z respectively in decreasing order.
According to the question,

$\dfrac{\text{Greatest fraction}}{\text{Smallest fraction}} = \dfrac{7}{6}$

$\Rightarrow \dfrac{x}{z} = \dfrac{7}{6} \Rightarrow x = \dfrac{7z}{6}$ ...(i)

and $y = \dfrac{7}{6} - \dfrac{1}{3}$

$= \dfrac{7-2}{6} = \dfrac{5}{6}$ ...(ii)

Now, $x + y + z = 2\dfrac{11}{24}$ ...(iii)

On putting the values of x and y from Eqs. (i) and (ii) in Eq. (iii), we get

$\dfrac{7}{6}z + \dfrac{5}{6} + z = \dfrac{59}{24}$

$\Rightarrow \dfrac{7z + 5 + 6z}{6} = \dfrac{59}{24}$

$\Rightarrow \dfrac{13z + 5}{6} = \dfrac{59}{24}$

$\Rightarrow 13z = \dfrac{59}{4} - 5 = \dfrac{39}{4}$

$\Rightarrow z = \dfrac{39}{4 \times 13}$

$\therefore z = \dfrac{3}{4}$

On putting the value of z in Eq. (i), we get

$x = \dfrac{7}{6}z = \dfrac{7}{6} \times \dfrac{3}{4} = \dfrac{7}{8}$

15. (b) Let the number of seats in each bus be x.
$\therefore$ Total number of seats in all the three buses $= 3x$
According to the question,
Number of seats occupied in the bus

$= \dfrac{4}{5} \times 3x = \dfrac{12x}{5}$

and number of passenger leave the bus

$= \dfrac{1}{4} \times \dfrac{12x}{5} = \dfrac{12x}{20}$

Number of seats occupied now after $\dfrac{1}{4}$ passengers leave $= \dfrac{12x}{5} - \dfrac{12x}{20}$

$= \dfrac{36x}{20} = \dfrac{9x}{5}$

$\therefore$ Fraction of the seats on the two buses

that are now occupied $= \dfrac{\frac{9x}{5}}{2x} = \dfrac{9}{10}$

16. (c) Value of p may be 3, 7, 11 and 13.

$\therefore \dfrac{1}{3} = 0.\overline{3}$, Period $= 1$

Here, $p - 1 = 2$ and 1 is a factor of 2.

$\therefore \dfrac{1}{7} = 0.\overline{142857}$, Period $= 6$

Here, $p - 1 = 6$ and 6 is a factor of 6.

$\therefore \dfrac{1}{11} = 0.\overline{09}$, Period $= 2$

Here, $p - 1 = 10$ and 2 is a factor of 10.

17. (d) Let Ranu's montly salary be ₹ x and Ali's monthly salary be ₹ y.
Then, according to the question,

$x = \dfrac{4}{5}y \Rightarrow y = \dfrac{5x}{4}$

and $\dfrac{2}{5}y - \dfrac{1}{4}x = 7000$

$\Rightarrow 8y - 5x = 7000 \times 4 \times 5$
$\Rightarrow 8y - 5x = 140000$

$\Rightarrow 8\left(\dfrac{5x}{4}\right) - 5x = 140000 \quad \left[\because y = \dfrac{5x}{4}\right]$

$\Rightarrow 10x - 5x = 140000$
$\Rightarrow 5x = 140000 \Rightarrow x = ₹\,28000$

Hence, Ranu's monthly salary is ₹ 28000.

Chapter 5

Square Root and Cube Root

Square

If a number is multiplied with itself, then the result of this multiplication is called the **square** of that number.

For example
(i) Square of $6 = 6 \times 6 = 36$
(ii) Square of $12 = 12 \times 12 = 144$
(iii) Square of $100 = 100 \times 100 = 10000$

Methods to Find the Square

Different methods to calculate the square of a number are as follow

1. Multiplication Method

In this method, the square of any 2-digit number can be calculated by the following steps

Step I Square the unit's digit.
{If the square has two digits, then write ten's digit as carry.}

Step II $2 \times$ Ten's digit $\times$ Unit's digit $+$ Carry

Step III (Ten's digit)^2 + Carry from Step II

Step IV Now, arrange the numbers starting with Step III number, then Step II and finally Step I number at unit's place.

Ex. 1 Find the square of 74.

Sol. **Step I** $(4)^2 = 16$ {Carry = 1}
Step II $2 \times 7 \times 4 + 1 = 57$ {Carry = 5}
Step III $(7)^2 + 5 = 49 + 5 = 54$
Step IV $(74)^2 = 5476$

2. Algebraic Method

To calculate the square by algebraic method, following two formulae are used

(i) $(a + b)^2 = a^2 + b^2 + 2ab$ (ii) $(a - b)^2 = a^2 + b^2 - 2ab$

Ex. 2 Find the square of 34.

Sol. $(34)^2 = (30 + 4)^2 = (30)^2 + (4)^2 + 2 \times 30 \times 4$
$= 900 + 16 + 240 = 1156$

Square of Decimal Numbers

To find the square of any decimal number, write the square of the number ignoring the decimal and then place the decimal after twice the place of the decimal in the original number starting from unit's place.

Ex. 3 Find the square of 3.5.

Sol. $\because (35)^2 = 1225$

Here, the decimal is after one-digit in 3.5. Hence, the decimal will be placed after twice the place of decimal in the original number.

$\therefore \quad (3.5)^2 = 12.25$

Square Root

The square root of a number is that number, the square of which is equal to the given number. There are two square roots of a number, positive and negative. It is denoted by the sign '$\sqrt{}$'.

For example 49 has two square roots 7 and -7, because $(7)^2 = 49$ and $(-7)^2 = 49$. Hence, we can write $\sqrt{49} = \pm 7$.

Methods to Find the Square Root

Different methods to calculate the square root of a number are as follow

1. Prime Factorisation Method

Following steps are used in this method

Step I Express the given number as the product of prime factors.
Step II Arrange the factors in pairs of same prime numbers.
Step III Take the product of these prime factors taking one out of every pair of the same primes. This product gives us the square root of the given number.

Ex. 4 Find the square root of 1089.

Sol. $\because$ Prime factors of $1089 = 11 \times 11 \times 3 \times 3$

Also, $\sqrt{1089} = \sqrt{11 \times 11 \times 3 \times 3}$

Now, taking one number from each pair and multiplying them, we get

$\sqrt{1089} = 11 \times 3 = 33$

11	1089
11	99
3	9
3	3
	1

Square Root and Cube Root / 81

Ex. 5 Find the square root of 1024.

Sol. ∵ Prime factors of $1024 = 2 \times 2 \times 2 \times 2 \times 2 \times 2 \times 2 \times 2 \times 2 \times 2$

Also, $\sqrt{1024} = \sqrt{2 \times 2 \times 2 \times 2 \times 2 \times 2 \times 2 \times 2 \times 2 \times 2}$

Now, taking one number from each pair and multiplying them, we get

$\sqrt{1024} = 2 \times 2 \times 2 \times 2 \times 2$
$= 32$

```
2 | 1024
2 |  512
2 |  256
2 |  128
2 |   64
2 |   32
2 |   16
2 |    8
2 |    4
2 |    2
  |    1
```

2. Division Method

If it is not easy to evaluate square root using prime factorisation method, then we use division method.

The steps of this method can be easily understood with the help of following examples.

Ex. 6 Find the square root of 18769.

Sol. Step I In the given number, mark off the digits in pairs starting from the unit digit. Each pair and the remaining one digit (if any) is called a period.

Step II Choose a number whose square is less than or equal to 1. Here, $1^2 = 1$, on subtracting, we get 0 (zero) as remainder.

Step III Bring down the next period, i.e. 87. Now, the trial divisor is $1 \times 2 = 2$ and trial dividend is 87. So, we take 23 as divisor and put 3 as quotient. The remainder is 18 now.

Step IV Bring down the next period, which is 69. Now, trial divisor is $13 \times 2 = 26$ and trial dividend is 1869. So, we take 267 as dividend and 7 as quotient. The remainder is 0.

Step V The process (processes like III and IV) goes on till all the periods (pairs) come to an end and we get remainder as 0 (zero) now.

Hence, the required square root = 137

```
          | 137
     1 | 1 87 69
       | 1
     23|   87
       |   69
    267|  1869
       |  1869
       |     ×
```

Ex. 7 What is the square root of 151321?

Sol. Required square root = 389

```
         | 389
     3 | 15 13 21
       |  9
    68 |  613
       |  544
   769 | 6921
       | 6921
       |    ×
```

Properties of Squares and Square Roots

(i) The difference of squares of two consecutive numbers will always be equal to the sum of the number, i.e. $(a^2 - b^2) = (a + b)(a - b)$. Here, $a > b$ and (a, b) being consecutive implies $(a - b) = 1$.

For example If $a = 12$ and $b = 11$, then $(12^2 - 11^2) = (12 + 11)(12 - 11) = 23$

(ii) The square of any number always ends with 0, 1, 4, 5, 6 or 9 but will never end with 2, 3, 7 or 8.
(iii) If the square of any number ends with 1, then its square root will end with 1 or 9.
(iv) If the square of any number ends with 4, then its square root will end with 2 or 8.
(v) If the square of any number ends with 5, then its square root will end with 5.
(vi) If the square of any number ends with 6, then its square root will end with 4 or 6.
(vii) If the square of any number ends with 9, then its square root will end with 3 or 7.
(viii) Square root of negative number is imaginary.

Some Important Relations

- $\sqrt{x} = y \Rightarrow x = y^2$
- $\sqrt{a^3 \times b^3} = ab\sqrt{ab}$
- $\sqrt{a^n \times b^m} = a^{n/2} \times b^{m/2}$
- $\dfrac{\sqrt{x}}{\sqrt{y}} = \sqrt{\dfrac{x}{y}}$
- $\sqrt{a^2 \times b^2} = ab$
- $\sqrt{a^4 \times b^4 \times c^4} = a^2 b^2 c^2$
- $\sqrt{x} \times \sqrt{y} = \sqrt{xy}$

Square Root of Decimal Numbers

If in a given decimal number, the number of digits after decimal are not even, then we put 0 (zero) at the extreme right, so that there are even number of digits after the decimal point. Now, periods are marked in the same way as in previous explanation starting from right hand side before the decimal point and from the left hand after the decimal digit.

Ex. 8 Find the square root of 147.1369.

Sol. Here, 147.1369 contains even digits after decimal, so there is no need to add zero after the last digit.
Now, periods are marked at
$\overline{147}.\overline{1369}$
After marking the periods, division method is used to find the square root.

	12.13
1	147.1369
	1
22	47
	44
241	313
	241
2423	7269
	7269
	×

∴ Required square root = 12.13

Ex. 9 Find the square root of 149.597361.

Sol. Here, 149.597361 contains even number of digits after decimal, so there is no need to add zero after the last digit.

	12.231
1	149.597361
	1
22	49
	44
242	559
	484
2443	7573
	7329
24461	24461
	24461
	×

∴ Required square root = 12.231

Square Root of a Fraction

To find square root of a fraction, we have to find the square root of numerator and denominator, separately.

Ex. 10 $\sqrt{\dfrac{2704}{81}} = ?$

Sol. $\sqrt{\dfrac{2704}{81}} = \dfrac{\sqrt{2704}}{\sqrt{81}} = \dfrac{52}{9}$

	52
5	2704
	25
102	204
	204
	×

Square Root and Cube Root / 83

Ex. 11 Find the square root of $\dfrac{461}{8}$.

Sol. $\sqrt{\dfrac{461}{8}} = \sqrt{\dfrac{461 \times 2}{8 \times 2}} = \sqrt{\dfrac{922}{16}} = \dfrac{\sqrt{922}}{\sqrt{16}} = \dfrac{30.3644}{4} = 7.5911$ (approx.)

✦ Sometimes, numerator and denominator are not a complete square. In these types of cases, it is better to convert the given fraction into decimal fraction to find the square root.

Ex. 12 $\sqrt{\dfrac{9261}{8400}} = ?$

Sol. $\sqrt{\dfrac{9261}{8400}} = \sqrt{1.1025} = 1.05$

```
           | 1.05
     1     | 1.1025
           | 1
    205    | 1025
           | 1025
           |   ×
```

Cube

If a number is multiplied two times with itself, then the result of this multiplication is called the **cube** of that number.
For example
(i) Cube of $6 = 6 \times 6 \times 6 = 216$ (ii) Cube of $8 = 8 \times 8 \times 8 = 512$

Method to Find the Cube
Different methods to calculate the cube of a number are as follow

Algebraic Method
To calculate the cube by this method, following two formulae are used
(i) $(a + b)^3 = a^3 + 3ab(a + b) + b^3$
(ii) $(a - b)^3 = a^3 - 3ab(a - b) - b^3$

For example Cube of $16 = (16)^3 = (10 + 6)^3$
$= (10)^3 + 3 \times 10 \times 6 (10 + 6) + (6)^3$
$= 1000 + 2880 + 216 = 4096$

Alternate Method
Following steps used in this method
Step I The answer consists of 4 parts each of which has to be calculated separately.
Step II First write down the cube of ten's digit to the extreme left. Write the next two terms to the right of it by creating GP (Geometric Progression) having common ratio which is equal to $\dfrac{\text{Unit's digit}}{\text{Ten's digit}}$ and the fourth number will be cube of unit's digit.
Step III Write the double of 2nd and 3rd number below them.
Step IV Now, add the number with numbers written below it and write the unit's place digit of the sum obtained in a straight line and remaining number is carried forward to the next number.

Ex. 13 Find the cube of 35.
Sol. Here, unit's digit is 5 and ten's digit is 3.
Step I Write the cube of ten's digit at extreme left.
i.e. $(3)^3 = 27$

Step II Now, the next two terms on the right will be in a GP of common ratio equals to $\dfrac{\text{unit's digit}}{\text{ten's digit}}$.

So, the next two terms will be $27 \times \dfrac{5}{3} = 45$ and $45 \times \dfrac{5}{3} = 75$

and last term will be cube of unit's digit, i.e. $(5)^3 = 125$.

So, they are arranged as 27 45 75 125.

Step III Twice the second and third terms are written under them and are added.

So,   ⑮ ㉓ ⑫ Carry

```
   27  45  75  125
 +      90  150
  ─────────────────
   42   8   7   5
```

∴ $(35)^3 = 42875$

Cube Root

The cube root of a given number is the number whose cube is the given number. The cube root is denoted by the sign '$\sqrt[3]{}$'. Cube root of a positive integer is always positive.

For example (i) $\sqrt[3]{8} = \sqrt[3]{2 \times 2 \times 2} = 2$ (ii) $\sqrt[3]{512} = \sqrt[3]{8 \times 8 \times 8} = 8$

Method to Find the Cube Root
There is only one method to calculate the cube root of a number which is as follows

Prime Factorisation Method
Following steps are used in this method

Step I Express the given number as the product of prime factors.

Step II Arrange the factors in a group of three of same prime numbers.

Step III Take the product of these prime factors picking one out of every group (group of three) of the same primes. This product gives us the cube root of given number.

Ex. 14 Find the cube root of 9261.

Sol. Prime factors of $9261 = (3 \times 3 \times 3) \times (7 \times 7 \times 7)$

$\Rightarrow \sqrt[3]{9261} = \sqrt[3]{3 \times 3 \times 3 \times 7 \times 7 \times 7}$

Now, taking one number from each group of three, we get

$\sqrt[3]{9261} = 3 \times 7 = 21$

3	9261
3	3087
3	1029
7	343
7	49
7	7
	1

Properties of Cubes and Cube Roots

(i) If the cube of a number is of 2 or 3 digits, then its cube root will be of 1 digit.

(ii) If the cube of a number is of 4, 5 or 6 digits, then its cube root will be of 2 digits.

(iii) If the cube of a number have 0, 1, 2, 3, 4, 5, 6, 7, 8, 9 in its unit's place, then its cube root will have 0, 1, 8, 7, 4, 5, 6, 3, 2 or 9 in its unit's place, respectively.

(iv) There are only three numbers whose cube is equal to the number,

i.e. $(0)^3 = 0;$ $(1)^3 = 1;$ $(-1)^3 = -1.$

Square Root and Cube Root / 85

- If $\frac{p}{q}$ is a fraction, then $\sqrt[3]{\frac{p}{q}} = \frac{\sqrt[3]{p}}{\sqrt[3]{q}}$, where p and q are integers.
- If p is an integer, then $\sqrt[3]{-p} = -\sqrt[3]{p}$.

Ex. 15 Find the value of $\sqrt[3]{\frac{0.000729}{0.085184}}$.

Sol. $\sqrt[3]{\frac{0.000729}{0.085184}} = \sqrt[3]{\frac{729}{85184}} = \sqrt[3]{\frac{9 \times 9 \times 9}{44 \times 44 \times 44}} = \frac{9}{44}$

Ex. 16 Find the cube root of -5832.

Sol. $\sqrt[3]{(-5832)} = -\sqrt[3]{5832} = -\sqrt[3]{18 \times 18 \times 18} = -18$

Ex. 17 Find the cube root of -17576.

Sol. $\sqrt[3]{(-17576)} = -\sqrt[3]{17576} = -\sqrt[3]{26 \times 26 \times 26} = -26$

Ex. 18 $\sqrt[3]{17.576} \times 15 = ?$ [SBI IT 2016]

Sol. $\because \sqrt[3]{17.576} \times 15 = ?$

$\therefore \quad ? = 2.6 \times 5 = 39$

Fast Track Formulae
to solve the QUESTIONS

▶ **Formula 1**

If in a given number, the total number of digits are n and if n is even, then square root of that number will have $n/2$ digits and if n is odd, then square root of that number will have $\frac{n+1}{2}$ digits.

Ex. 19 How many digits are there in square root of 1838736?

Sol. Since, the total number of digits are 7, which is odd number.

$\therefore$ Number of digits $= \frac{7+1}{2} = 4$ digits

▶ **Formula 2**

If any number has 5 in its unit's place, then its square can be calculated as

$$(\underline{A}\cdot\underline{5})^2 = \underline{A \times (A+1)} \cdot \underline{25}$$

- Student should understand the dot (·) signifies just a separation between numbers.

Ex. 20 Find the square of 125.

Sol. $(125)^2 = 12\,(12+1) \cdot 25 = 15625$

Fast Track Techniques to solve the QUESTIONS

Technique 1
(i) To find the largest *n*-digit number which is a perfect square, firstly calculate the square root of the largest *n*-digit number by division method and then subtract the remainder from the largest *n*-digit number.

(ii) To find the smallest *n*-digit number which is a perfect square, firstly calculate the square root of smallest *n*-digit numbers by division method and then take the next nearest square number of *n* digits.

Ex. 21 Find the 5-digit number which is a perfect square.

Sol. Largest 5-digit number = 99999

```
        | 3 16
    3   | 9 99 99
        | 9
    61  | 99
        | 61
   626  | 3899
        | 3756
        | 143
```

Clearly, the largest 5-digit number which is a perfect square, is less by 143.

∴ Number = 99999 − 143 = 99856

Ex. 22 Find the least number of 6 digits which is a perfect square.

Sol. ∵ Least 6-digit number = 100000.

```
         | 316.2
     3   | 10 00 00 . 00
         | 9
     61  | 100
         | 61
    626  | 3900
         | 3756
   6322  | 14400
         | 12644
         | 1756
```

Since, $(316)^2 < 100000 < (317)^2$

∴ Least 6-digit number which is a perfect square = $(317)^2 = 100489$

Square Root and Cube Root / 87

Technique 2 To find the value of $\sqrt{x+\sqrt{x+\sqrt{x+...}}}$, factorise x. If $x = m(m+1)$, then $(m+1)$ is your answer.

Ex. 23 Find the value of $\sqrt{6+\sqrt{6+\sqrt{6+...}}}$.

Sol. We know that, $6 = 2 \times 3 = 2(2+1)$

∴ $\sqrt{6+\sqrt{6+\sqrt{6+...}}} = 3$

Technique 3 To find the value of $\sqrt{x-\sqrt{x-\sqrt{x-...}}}$, factorise x. If $x = m(m+1)$, then m is your answer.

Ex. 24 Find the value of $\sqrt{12-\sqrt{12-\sqrt{12-...}}}$.

Sol. We know that, $12 = 3 \times 4 = 3(3+1)$

∴ $\sqrt{12-\sqrt{12-\sqrt{12-...}}} = 3$

Technique 4 $\sqrt{x\sqrt{x\sqrt{x\sqrt{x...\infty}}}} = x$

Ex. 25 Find the value of $\sqrt{7\sqrt{7\sqrt{7\sqrt{...\infty}}}}$.

Sol. As the expression continues to infinity, so $\sqrt{7\sqrt{7\sqrt{7...\infty}}} = 7$.

Technique 5 $\sqrt{x\sqrt{x\sqrt{x...n}}}$ times $= x^{\left(\frac{2^n-1}{2^n}\right)}$, where n is the number of times x is repeated.

Ex. 26 Find the value of $\sqrt{7\sqrt{7\sqrt{7}}}$.

Sol. Here, $n = 3$

∴ $\sqrt{7\sqrt{7\sqrt{7}}} = 7^{\left(\frac{2^3-1}{2^3}\right)} = 7^{7/8}$

Multi Concept Questions

1. If $2 * 3 = \sqrt{13}$ and $3 * 4 = 5$, then the value of $5 * 12$ is
   (a) 17 (b) $\sqrt{29}$ (c) 21 (d) 13

   ↪ (d) By observing, we see
   $$2 * 3 = \sqrt{(2)^2 + (3)^2} = \sqrt{4 + 9} = \sqrt{13}$$
   and $3 * 4 = \sqrt{(3)^2 + (4)^2} = \sqrt{9 + 16} = 5$
   Hence, value of $5 * 12 = \sqrt{(5)^2 + (12)^2}$
   $$= \sqrt{25 + 144} = \sqrt{169} = 13$$

2. Which of the following equations are equivalent?

   I. $\left(\dfrac{1}{2}M + \dfrac{2}{3}N\right)^2$

   II. $\dfrac{4}{9}N^2 + \dfrac{1}{4}M^2 + \dfrac{2}{3}MN$

   III. $\left(\dfrac{M}{2} + \dfrac{2}{3}N\right)\left(\dfrac{1}{2}M - \dfrac{2}{3}N\right)$

   IV. $\dfrac{1}{4}\left(M + \dfrac{4}{3}N\right)^2$

   (a) II and III (b) I and IV (c) I and II (d) I, II and IV

   ↪ (d) Simplifying all the above equations,

   I. $\left(\dfrac{1}{2}M + \dfrac{2}{3}N\right)^2 = \dfrac{1}{4}M^2 + \dfrac{4}{9}N^2 + \dfrac{2}{3}MN$

   II. $\dfrac{4}{9}N^2 + \dfrac{1}{4}M^2 + \dfrac{2}{3}MN$

   III. $\left(\dfrac{M}{2} + \dfrac{2}{3}N\right)\left(\dfrac{1}{2}M - \dfrac{2}{3}N\right) = \dfrac{1}{4}M^2 + \dfrac{1}{3}MN - \dfrac{1}{3}MN - \dfrac{4}{9}N^2 = \dfrac{1}{4}M^2 - \dfrac{4}{9}N^2$

   IV. $\dfrac{1}{4}\left(M + \dfrac{4}{3}N\right)^2 = \dfrac{1}{4}\left(M^2 + \dfrac{16}{9}N^2 + \dfrac{8}{3}MN\right) = \dfrac{1}{4}M^2 + \dfrac{4}{9}N^2 + \dfrac{2}{3}MN$

   From the above four solutions, we find that I, II and IV are equivalent.

3. If $\left(\dfrac{x}{y}\right) = \left(\dfrac{z}{w}\right)$, then what is $(xy + zw)^2$ equal to?

   (a) $(x^2 + z^2)(y^2 + w^2)$
   (b) $x^2y^2 + z^2w^2$
   (c) $x^2w^2 + y^2z^2$
   (d) $(x^2 + w^2)(y^2 + z^2)$

   ↪ (a) Here, $\dfrac{x}{y} = \dfrac{z}{w} = k$ (say) $\Rightarrow x = yk$ and $z = wk$

   Now, $(xy + zw)^2 = (y^2k + w^2k)^2$
   $$= k^2(y^2 + w^2)^2$$
   $$= (y^2 + w^2)(k^2y^2 + k^2w^2)$$
   $$= (y^2 + w^2)(x^2 + z^2)$$

Fast Track Practice

Exercise 1 — Base Level Questions

1. Find the square root of 144.
   (a) 15 (b) 16
   (c) 12 (d) 13
   (e) None of these

2. $\sqrt{1089} + \sqrt{289} = \sqrt{?}$ [Bank Clerk 2007]
   (a) 625 (b) 50
   (c) 2500 (d) 1378
   (e) None of these

3. What will be the square root of 7921? [SSC CGL 2008]
   (a) 89 (b) 87
   (c) 37 (d) 47

4. $\sqrt{?} + 19 = \sqrt{1225}$ [SBI Clerk (Pre) 2016]
   (a) 15 (b) 225
   (c) 16 (d) 256
   (e) 289

5. $\sqrt{?} + 43 = \sqrt{19881}$ [Bank Clerk 2009]
   (a) 9604 (b) 7744
   (c) 9216 (d) 8464
   (e) None of these

6. $22^2 + \sqrt{?} = 516$ [Bank Clerk 2009]
   (a) 1028 (b) 1024
   (c) 1124 (d) 1128
   (e) None of these

7. $\sqrt{?} + 28 = \sqrt{1681}$ [Bank Clerk 2009]
   (a) 13 (b) 225
   (c) 216 (d) 169
   (e) None of these

8. $\sqrt{15 \times 163 \div 5 - 89} = ?$ [Bank Clerk 2011]
   (a) 15 (b) 25
   (c) 10 (d) 20
   (e) None of these

9. $\sqrt{8 \times 7 - ? + 208 \div 16} = 8$ [Bank Clerk 2009]
   (a) 5 (b) 7
   (c) 4 (d) 3
   (e) None of these

10. $\sqrt{15^2 + 11 \times 3^2} = ?$ [Bank Clerk 2007]
    (a) 18.25 (b) 19
    (c) 18 (d) 19.5
    (e) None of these

11. $\sqrt{8^2 \times 7 \times 5^2 - 175} = ?$ [SBI Clerk (Mains) 2016]
    (a) 105 (b) 95 (c) 115 (d) 125
    (e) 135

12. $140\sqrt{?} + 315 = 1015$ [Hotel Mgmt. 2008]
    (a) 16 (b) 25
    (c) 36 (d) 5

13. $\sqrt{176 + \sqrt{2401}} = ?$ [Hotel Mgmt. 2007]
    (a) 14 (b) 15
    (c) 18 (d) 24

14. $\sqrt{?} + 136 = \dfrac{5}{8}$ of 320 [Bank Clerk 2008]
    (a) 1936 (b) 4624
    (c) 4196 (d) 4096
    (e) None of these

15. $\sqrt{?} + \dfrac{3}{5}$ of $80 = 60 \times \dfrac{1}{2} \times 8$
    (a) 36864 (b) 46864
    (c) 56864 (d) 66864
    (e) None of these

16. If $\sqrt{18 \times 14 \times x} = 168$, then x is equal to
    (a) 113 (b) 112
    (c) 115 (d) 117
    (e) None of these

17. If $\sqrt{24} = 4.899$, then find the value of $\sqrt{\dfrac{8}{3}}$. [SSC (10+2) 2009]
    (a) 1.633 (b) 1.333
    (c) 2.666 (d) 0.544

18. Find the value of $\sqrt{128} + \sqrt{160} - \sqrt{21}$. [LIC ADO 2015]
    (a) 16.39 (b) 19.39
    (c) 15.56 (d) 18.35
    (e) None of these

19. The square root of $0.\overline{4}$ is
    (a) $0.\overline{6}$ (b) $0.\overline{7}$
    (c) $0.\overline{8}$ (d) $0.\overline{9}$
    (e) None of these

20. Evaluate $\sqrt{0.9}$ upto three places of decimal.
    (a) 0.948 (b) 0.984
    (c) 0.988 (d) 938
    (e) None of these

21. If $0.14 \div x^2 = 14$, then x is equal to
(a) 0.01 (b) 10 (c) 100 (d) 0.1
(e) None of these

22. Find the square root of $105\dfrac{4}{64}$.
[RBI Clerk 2008]
(a) $15\dfrac{1}{4}$ (b) $15\dfrac{12}{4}$
(c) $10\dfrac{1}{4}$ (d) $6\dfrac{2}{4}$
(e) None of these

23. $\sqrt{\dfrac{324}{81}} + \sqrt{\dfrac{324}{81}} = ?$
(a) 4 (b) 12 (c) 6 (d) 18
(e) None of these

24. The square root of $\left(\dfrac{1}{4}\right) \times \left(\dfrac{1}{49}\right) \div \left(\dfrac{25}{121}\right)$ is
[SSC (10+2) 2012]
(a) $\dfrac{11}{5}$ (b) $\dfrac{11}{70}$
(c) $\dfrac{7}{11}$ (d) $\dfrac{11}{7}$

25. $\sqrt{\dfrac{160 \times 8}{5}} \times 8 = ?$
[SBI Clerk (Pre) 2016]
(a) 112 (b) 164
(c) 128 (d) 116
(e) 136

26. $\dfrac{\sqrt{24} + \sqrt{216}}{\sqrt{96}} = ?$
[SSC (10+2) 2009]
(a) $2\sqrt{6}$ (b) 2
(c) $6\sqrt{2}$ (d) $\dfrac{2}{\sqrt{6}}$

27. If $\sqrt{\dfrac{x}{169}} = \dfrac{18}{13}$, then x is equal to
[LIC AAO 2008]
(a) 108 (b) 324
(c) 2916 (d) 4800
(e) None of these

28. If $\sqrt{1 + \dfrac{x}{144}} = \dfrac{13}{12}$, then x equal to
[SSC CGL (Pre) 2016]
(a) 1 (b) 13
(c) 27 (d) 25

29. $30^2 - 117 = 24.5^2 + ?$ [SBI Clerk (Mains) 2016]
(a) 127.70 (b) 127.65
(c) 182.75 (d) 172.05
(e) 165.28

30. The value of $\sqrt{400} + \sqrt{0.04} - \sqrt{0.000004}$ is
[SSC CPO 2013]
(a) 20.22 (b) 20.198
(c) 20.188 (d) 20.022

31. Evaluate $\sqrt{16} + \sqrt{\dfrac{9.5 \times 0.0085 \times 18.9}{0.021 \times 0.0017 \times 1.9}}$.
(a) 154 (b) 158 (c) 160 (d) 169
(e) None of these

32. Evaluate $\sqrt{\dfrac{0.289}{0.00121}} + \sqrt{\dfrac{64}{16}}$.
(a) $\dfrac{192}{11}$ (b) $\dfrac{170}{11}$ (c) $\dfrac{182}{11}$ (d) $\dfrac{172}{11}$
(e) None of these

33. If $\dfrac{\sqrt{1296}}{x} = \dfrac{x}{2.25}$, then find the value of x.
[Bank Clerk 2008]
(a) 6 (b) 8 (c) 7 (d) 9
(e) None of these

34. If $x = 2 + \sqrt{2}$ and $y = 2 - \sqrt{2}$, then find the value of $(x^2 + y^2)$.
(a) 12 (b) 14
(c) 6 (d) 18
(e) None of these

35. The value of $\dfrac{\sqrt{180} \times \sqrt{18} \times \sqrt{5} \times \sqrt{3}}{\sqrt{96} \times 5}$ is
[SBI Associate Clerk 2015]
(a) $5\dfrac{1}{2}$ (b) $2\dfrac{3}{4}$
(c) $1\dfrac{1}{2}$ (d) $4\dfrac{1}{2}$
(e) $5\dfrac{2}{3}$

36. Find the square root of $\dfrac{0.204 \times 42}{0.07 \times 3.4}$.
[SSC (10+2) 2007]
(a) 5 (b) 6
(c) 3 (d) 9

37. If $\sqrt{3} = 1.732$, then find the value of $4\left(\sqrt{192} - \dfrac{1}{2}\sqrt{48} - \sqrt{75}\right)$ correct to three places of decimal.
(a) 6.928 (b) 5.928
(c) 2.732 (d) 3.732
(e) None of these

38. What is the value of $\sqrt[3]{4\dfrac{12}{125}}$? [CDS 2017 (I)]
(a) $1\dfrac{3}{5}$ (b) $1\dfrac{2}{5}$
(c) $1\dfrac{4}{5}$ (d) $2\dfrac{2}{5}$

Square Root and Cube Root / 91

39. $(64)^2 \div \sqrt[3]{32768} = ?$ [Bank Clerk 2008]
(a) 128   (b) 132   (c) 142   (d) 104
(e) None of these

40. $\sqrt[3]{1331} \times \sqrt[3]{216} + \sqrt[3]{729} + \sqrt[3]{64}$ is equal to [RBI Clerk 2009]
(a) 13.62   (b) 79
(c) 14.82   (d) 90.88
(e) None of these

41. What is the least number to be multiplied with 294 to make it a perfect square? [Bank Clerk 2007]
(a) 2   (b) 3   (c) 6   (d) 24
(e) None of these

42. What is the least number to be added to 8200 to make it a perfect square? [Bank Clerk 2009]
(a) 81   (b) 100   (c) 264   (d) 154
(e) None of these

43. Find the difference between 777 and its nearest perfect square number.
(a) 4   (b) 7   (c) 27   (d) 28
(e) None of these

44. If $(46)^2$ is subtracted from the square of a number, the answer so obtained is 485. What is the number? [SBI 2012]
(a) 49   (b) 51   (c) 56   (d) 53
(e) None of these

45. If $(\sqrt{a} + \sqrt{b}) = 17$ and $(\sqrt{a} - \sqrt{b}) = 1$, then find $\sqrt{ab}$. [LIC ADO 2008]
(a) 17   (b) 18   (c) 72   (d) 21
(e) None of these

46. What least number should be subtracted from 6860, so that 19 is the cube root of the result from this subtraction? [LIC ADO 2009]
(a) 1   (b) 2
(c) 3   (d) 5
(e) None of these

47. $\sqrt{72 + \sqrt{72 + \sqrt{72 + \ldots}}}$ is equal to [SSC CGL (Mains) 2014]
(a) 8   (b) 12
(c) 9   (d) 18

48. Find the value of $\sqrt{20 - \sqrt{20 - \sqrt{20 - \ldots}}}$.
(a) 2   (b) 3
(c) 4   (d) 5

49. Find the value of $\sqrt{9\sqrt{9\sqrt{9\ldots\infty}}}$.
(a) 81   (b) 3
(c) 9   (d) ∞

50. If $x = \sqrt{8 + \sqrt{8 + \sqrt{8 + \ldots}}}$ and $y = \sqrt{8 - \sqrt{8 - \sqrt{8 - \ldots}}}$, then
(a) $x + y = 1$ (b) $x + y + 1 = 0$
(c) $x - y = 1$ (d) $x - y + 1 = 0$

51. Find the value of $\sqrt{6\sqrt{6\sqrt{6\sqrt{6\sqrt{6}}}}}$.
(a) 6   (b) $6^{31/32}$
(c) $6^{31/34}$ (d) 3

Exercise 2   *Higher Skill Level Questions*

1. A General of an Army wants to create a formation of square from 36562 army men. After arrangement, he found some army men remained unused. Then, the number of such army men remained unused was [SSC CGL (Mains) 2016]
(a) 36   (b) 65
(c) 81   (d) 97

2. Each student of class 10 contributed some money for a picnic. The money contributed by each student was equal to the square of the total number of students. If the total collected amount was ₹ 29791, then find the total number of students. [UP Police 2008]
(a) 15   (b) 27
(c) 31   (d) 34

3. A gardener has 1000 plants. He wants to plant them in such a way that the number of rows and the number of columns remains the same. What is the minimum number of plants that he needs more for this purpose?
(a) 14   (b) 24   (c) 32   (d) 34

4. A toy factory manufactured a batch of electronic toys. If the toys were packed in boxes of 115 each, 13 boxes would not be filled completely. If the toys were packed in boxes of 65 each, 22 such boxes would not be enough to pack all of them. Coincidentally, in the end, the toys were packed in n boxes containing n toys each, without any remainder. The total number of toys was
(a) 1424   (b) 1434   (c) 1444   (d) 1454

92 / Fast Track Objective Arithmetic

5. The least number which is a perfect square and has 7936 as one of its factors, is equal to [LIC ADO 2007]
   (a) 12.008 (b) 246016
   (c) 61504 (d) 240616
   (e) None of these

6. The smallest whole number that is to be multiplied with 59535 to make a perfect square number is x. The sum of digits of x is [SSC CPO 2015]
   (a) 9 (b) 5 (c) 6 (d) 7

7. $\dfrac{\sqrt{10+\sqrt{25+\sqrt{108+\sqrt{154+\sqrt{225}}}}}}{\sqrt[3]{8}} = ?$ [SSC CPO 2015]
   (a) 4 (b) 8
   (c) 2 (d) 1/2

8. If $a = \dfrac{2+\sqrt{3}}{2-\sqrt{3}}$ and $b = \dfrac{2-\sqrt{3}}{2+\sqrt{3}}$, then the value of $(a^2 + b^2 + ab)$ is [SSC CGL 2016]
   (a) 195 (b) 200
   (c) 175 (d) 185

9. $\dfrac{6^2 + 7^2 + 8^2 + 9^2 + 10^2}{\sqrt{7+4\sqrt{3}} - \sqrt{4+2\sqrt{3}}}$ is equal to [SSC CGL (Mains) 2015]
   (a) 366 (b) 355
   (c) 305 (d) 330

10. If $x = \sqrt{a\sqrt[3]{b\sqrt{a\sqrt[3]{b}\ldots\infty}}}$, then the value of x is [SSC CGL 2015]
    (a) $\sqrt[5]{ab^3}$ (b) $\sqrt[5]{a^5b}$
    (c) $\sqrt[3]{a^3b}$ (d) $\sqrt[5]{a^3b}$

11. If function $(x + 809436 \times 809438)$ is a perfect square, then the value of x is [SSC CGL (Mains) 2015]
    (a) 0 (b) 1 (c) 2 (d) 2^{16}

12. Find the value of
    $\sqrt{\dfrac{(0.03)^2 + (0.21)^2 + (0.065)^2}{(0.003)^2 + (0.021)^2 + (0.0065)^2}} \div 2$.
    (a) 2.8 (b) 5 (c) 10^2 (d) 5^2
    (e) None of these

13. The sum of two numbers, when multiplied with each of the numbers separately, then the results of the above multiplications are 2418 and 3666, respectively. Find the difference between the numbers. [UP Police 2007]
    (a) 16 (b) 21
    (c) 19 (d) 23

14. What is the square root of $\dfrac{(0.35)^2 + 0.70 + 1}{2.25} + 0.19$? [CDS 2017 (I)]
    (a) 1 (b) 2
    (c) 3 (d) 4

15. The positive square root of $(\sqrt{48} - \sqrt{45})$ is [SSC CGL 2011]
    (a) $\dfrac{\sqrt[4]{3}}{\sqrt{2}}(\sqrt{5} - \sqrt{3})$
    (b) $\dfrac{\sqrt[4]{3}}{2}(\sqrt{5} - \sqrt{3})$
    (c) $\dfrac{\sqrt{2}}{\sqrt[4]{3}}(\sqrt{5} - \sqrt{3})$
    (d) $\dfrac{\sqrt{2}}{\sqrt[4]{3}}(\sqrt{5} + \sqrt{3})$

16. What is $\sqrt{4+\sqrt{4-\sqrt{4+\sqrt{4-\ldots}}}}$ equal to? [CDS 2015 (II)]
    (a) 3 (b) $\dfrac{\sqrt{13}-1}{2}$
    (c) $\dfrac{\sqrt{13}+1}{2}$ (d) 0

17. The square root of $\dfrac{(0.75)^3}{1-0.75} + [0.75 + (0.75)^2 + 1]$ is [CDS 2015 (I)]
    (a) 1 (b) 2
    (c) 3 (d) 4

18. What should be added to $x(x+a)(x+2a)(x+3a)$, so that the sum is a perfect square? [CDS 2014]
    (a) a^2 (b) a^4
    (c) a^3 (d) None of these

Answer with Solutions

Exercise 1 Base Level Questions

1. (c) Prime factors of 144

2	144
2	72
2	36
2	18
3	9
3	3
	1

∴ $144 = 2 \times 2 \times 2 \times 2 \times 3 \times 3$

⇒ $\sqrt{144} = \sqrt{2 \times 2 \times 2 \times 2 \times 3 \times 3}$

Now, taking one number from each pair and multiplying them, we get

$\sqrt{144} = 2 \times 2 \times 3 = 12$

2. (c) By prime factorisation method,

3	1089
3	363
11	121
11	11
	1

17	289
17	17
	1

$\sqrt{1089} = \sqrt{3 \times 3 \times 11 \times 11}$
$= 3 \times 11 = 33$

$\sqrt{289} = \sqrt{17 \times 17} = 17$

∴ $\sqrt{1089} + \sqrt{289} = 33 + 17 = 50$

Here, $\sqrt{?} = 50$

∴ $? = 50 \times 50 = 2500$

3. (a) By division method,

```
        8 9
      ┌──────
    8 │ 79 21
      │ 64
      ├──────
  169 │ 1521
      │ 1521
      ├──────
      │   ×
```

∴ Required square root = 89

4. (d) $\sqrt{?} + 19 = \sqrt{1225}$
⇒ $\sqrt{?} = 35 - 19 = 16$
⇒ $? = (16)^2 = 256$
⇒ $? = 256$

5. (a) $\sqrt{?} = \sqrt{19881} - 43 = 141 - 43 = 98$
⇒ $? = (98)^2 = 9604$

6. (b) $22^2 + \sqrt{?} = 516$
⇒ $\sqrt{?} = 516 - 484 = 32$
∴ $? = (32)^2 = 32 \times 32 = 1024$

7. (d) $\sqrt{?} = \sqrt{1681} - 28 = 41 - 28 = 13$
∴ $? = 13^2 = 13 \times 13 = 169$

8. (d) $\sqrt{15 \times \dfrac{163}{5} - 89} = \sqrt{3 \times 163 - 89}$
$= \sqrt{489 - 89} = \sqrt{400} = 20$

9. (a) $\sqrt{8 \times 7 - ? + \dfrac{208}{16}} = 8$
⇒ $56 - ? + 13 = 64$
[∵ squaring on both sides]
⇒ $69 - ? = 64$
∴ $? = 69 - 64 = 5$

10. (c) $\sqrt{(15)^2 + 11 \times (3)^2} = \sqrt{225 + 99} = \sqrt{324}$

By prime factorisation method,

2	324
2	162
9	81
9	9
	1

$\sqrt{324} = \sqrt{2 \times 2 \times 9 \times 9} = 2 \times 9 = 18$

11. (a) $\sqrt{8^2 \times 7 \times 5^2 - 175} = ?$
∴ $? = \sqrt{8^2 \times 7 \times 5^2 - 5^2 \times 7}$
$= 5\sqrt{8^2 \times 7 - 7} = 5\sqrt{7 \times 63}$
$= 5 \times 3 \times 7 = 105$

12. (b) $140\sqrt{?} + 315 = 1015$
⇒ $140\sqrt{?} = 1015 - 315$
⇒ $140\sqrt{?} = 700 \Rightarrow \sqrt{?} = \dfrac{700}{140} = 5$
∴ $? = (5)^2 = 5 \times 5 = 25$

13. (b) $\sqrt{2401} = \sqrt{7 \times 7 \times 7 \times 7} = 7 \times 7 = 49$
∴ $\sqrt{176 + \sqrt{2401}} = \sqrt{176 + 49} = \sqrt{225}$
$= \sqrt{3 \times 3 \times 5 \times 5} = 3 \times 5 = 15$

14. (d) Given, $\sqrt{?} + 136 = 320 \times \dfrac{5}{8}$
⇒ $\sqrt{?} + 136 = 200 \Rightarrow \sqrt{?} = 200 - 136 = 64$
∴ $? = (64)^2 = 64 \times 64 = 4096$

15. (a) $\sqrt{?} + \dfrac{3}{5} \times 80 = 60 \times \dfrac{1}{2} \times 8$

$\Rightarrow \quad \sqrt{?} + 3 \times 16 = 60 \times 4$

$\Rightarrow \quad \sqrt{?} = 240 - 48 = 192$

$\therefore \quad ? = 192 \times 192 = 36864$

16. (b) $\sqrt{18 \times 14 \times x} = 168$

On squaring both sides, we get

$18 \times 14 \times x = 168 \times 168$

$\therefore \quad x = \dfrac{168 \times 168}{18 \times 14} = 28 \times 4 = 112$

17. (a) $\sqrt{\dfrac{8}{3}} = \sqrt{\dfrac{8 \times 3}{3 \times 3}} = \sqrt{\dfrac{24}{3 \times 3}} = \dfrac{\sqrt{24}}{3} = \dfrac{4.899}{3}$

$[\because \sqrt{24} = 4.899]$

$\therefore \quad \sqrt{\dfrac{8}{3}} = 1.633$

18. (e)

```
      | 11.31              |     | 12.64
    1 | 1 28. 00 00        |   1 | 1 60. 00 00
   ×1 | 1                  |  ×1 | 1
   ---+-----               |  ---+-----
   21 | 28                 |  22 | 60
   × 1| 21                 |  × 2| 44
   ---+-----               |  ---+-----
  223 | 700                | 246 | 1600
  × 3 | 669                | × 6 | 1476
   ---+-----               |  ---+-----
 2261 | 3100               |2524 | 12400
  × 1 | 2261               | × 4 | 10096
   ---+-----               |  ---+-----
      | 839                |     | 2304
```

```
           | 4.58
         4 | 21.00 00
        × 4| 16
        ---+-----
         85| 500
        × 5| 425
        ---+-----
        908| 7500
        × 8| 7264
        ---+-----
           | 236
```

$\therefore \sqrt{128} + \sqrt{160} - \sqrt{21}$

$= 11.31 + 12.64 - 4.58$

$= 23.95 - 4.58 = 19.37$

19. (a) $\sqrt{0.4} = \sqrt{\dfrac{4}{9}} = \dfrac{2}{3} = 0.666... = 0.\overline{6}$

20. (a)

```
             | 0.948
           9 | 0.90 00 00
             | 81
         ----+-----
         184 | 900
             | 736
         ----+-----
        1888 | 16400
             | 15104
         ----+-----
             | 1296
```

$\therefore \quad \sqrt{0.9} = 0.948$

21. (d) $\dfrac{0.14}{x^2} = 14 \Rightarrow x^2 = \dfrac{0.14}{14} = \dfrac{1}{100}$

$\therefore \quad x = \sqrt{\dfrac{1}{100}} = \dfrac{1}{10} = 0.1$

22. (c) $\sqrt{105\dfrac{4}{64}} = \sqrt{\dfrac{6724}{64}}$

$= \dfrac{82}{8} = \dfrac{41}{4} = 10\dfrac{1}{4}$

23. (a) $\sqrt{\dfrac{324}{81}} + \sqrt{\dfrac{324}{81}} = 2 \times \sqrt{\dfrac{324}{81}}$

By prime factorisation method,

```
 2 | 324
 2 | 162
 3 | 81
 3 | 27
 3 | 9
 3 | 3
   | 1
```

$2 \times \sqrt{\dfrac{324}{81}} = 2 \times \sqrt{\dfrac{2 \times 2 \times 3 \times 3 \times 3 \times 3}{3 \times 3 \times 3 \times 3}}$

$= 2 \times \dfrac{2 \times 3 \times 3}{3 \times 3} = 2 \times 2 = 4$

24. (b) Square root of $\dfrac{1}{4} \times \dfrac{1}{49} \div \left(\dfrac{25}{121}\right)$

$= \sqrt{\dfrac{1}{4} \times \dfrac{1}{49} \times \dfrac{121}{25}} = \dfrac{11}{2 \times 7 \times 5} = \dfrac{11}{70}$

25. (c) $? = \sqrt{\dfrac{160 \times 8}{5}} \times 8$

$\Rightarrow \quad ? = \sqrt{32 \times 8} \times 8$

$\Rightarrow \quad ? = \sqrt{4 \times 8 \times 8} \times 8$

$\Rightarrow \quad ? = 2 \times 8 \times 8 \Rightarrow ? = 128$

26. (b) $\dfrac{\sqrt{24} + \sqrt{216}}{\sqrt{96}}$

$= \dfrac{\sqrt{2 \times 2 \times 2 \times 3} + \sqrt{3 \times 3 \times 3 \times 2 \times 2 \times 2}}{\sqrt{2 \times 2 \times 2 \times 2 \times 2 \times 3}}$

$= \dfrac{2\sqrt{6} + 6\sqrt{6}}{4\sqrt{6}} = \dfrac{8\sqrt{6}}{4\sqrt{6}} = 2$

27. (b) Given, $\sqrt{\dfrac{x}{169}} = \dfrac{18}{13}$

On squaring both sides, we get

$\dfrac{x}{169} = \dfrac{18}{13} \times \dfrac{18}{13}$

$\therefore \quad x = \dfrac{18 \times 18 \times 169}{13 \times 13} = 18 \times 18 = 324$

Square Root and Cube Root / 95

28. (d) Given, $\sqrt{1 + \dfrac{x}{144}} = \dfrac{13}{12}$ ...(i)

On squaring both sides of Eq. (i), we get

$1 + \dfrac{x}{144} = \dfrac{169}{144}$

$\Rightarrow \dfrac{x}{144} = \dfrac{169}{144} - 1 \Rightarrow \dfrac{x}{144} = \dfrac{25}{144}$

$\therefore \quad x = 25$

29. (c) $30^2 - 117 = 24.5^2 + ?$

$\Rightarrow \quad 900 - 117 = 24.5^2 + ?$

$\Rightarrow \quad 783 - (24.5)^2 = ?$

$\Rightarrow \quad 783 - 600.25 = ? \Rightarrow ? = 182.75$

30. (b) $\sqrt{400} + \sqrt{0.04} - \sqrt{0.000004}$

$= 20 + 0.2 - 0.002 = 20.2 - 0.002 = 20.198$

31. (a) $\sqrt{16} + \sqrt{\dfrac{9.5 \times 0.0085 \times 18.9}{0.021 \times 0.0017 \times 1.9}}$

$= \sqrt{16} + \sqrt{\dfrac{95 \times 85 \times 18900}{21 \times 17 \times 19}}$

$= 4 + \sqrt{5 \times 5 \times 900}$

$= 4 + 5 \times 30 = 4 + 150 = 154$

32. (a) $\sqrt{\dfrac{0.289}{0.00121}} + \sqrt{\dfrac{64}{16}} = \sqrt{\dfrac{28900}{121}} + \sqrt{\dfrac{64}{16}}$

$= \sqrt{\dfrac{17 \times 17 \times 10 \times 10}{11 \times 11}} + \sqrt{4}$

$= \dfrac{170}{11} + 2 = \dfrac{170 + 22}{11} = \dfrac{192}{11}$

33. (d) $\dfrac{\sqrt{1296}}{x} = \dfrac{x}{2.25}$

$\Rightarrow x^2 = 2.25 \times \sqrt{1296}$

$= 2.25 \times \sqrt{2 \times 2 \times 2 \times 2 \times 3 \times 3 \times 3 \times 3}$

$= 2.25 \times 2 \times 2 \times 3 \times 3 = 2.25 \times 36$

$\therefore \quad x = \sqrt{2.25 \times 36} = 1.5 \times 6 = 9$

34. (a) $x^2 + y^2 = (2 + \sqrt{2})^2 + (2 - \sqrt{2})^2$

$= (4 + 2 + 4\sqrt{2}) + (4 + 2 - 4\sqrt{2}) = 12$

35. (d) $\dfrac{\sqrt{180} \times \sqrt{18} \times \sqrt{5} \times \sqrt{3}}{\sqrt{96} \times 5}$

$= \sqrt{\dfrac{180 \times 18 \times 5 \times 3}{96 \times 25}} = \sqrt{20.25} = 4.5 = 4\dfrac{1}{2}$

36. (b) $\because \dfrac{0.204 \times 42}{0.07 \times 3.4} = \dfrac{204 \times 42}{7 \times 34} = 6 \times 6 = 36$

$\therefore \sqrt{\dfrac{0.204 \times 42}{0.07 \times 3.4}} = \sqrt{36} = \sqrt{6 \times 6} = 6$

37. (a) $4\left(\sqrt{192} - \dfrac{1}{2}\sqrt{48} - \sqrt{75}\right)$

$= 4\left(\sqrt{64 \times 3} - \dfrac{1}{2}\sqrt{16 \times 3} - \sqrt{25 \times 3}\right)$

$= 4\left(8\sqrt{3} - \dfrac{1}{2} \times 4\sqrt{3} - 5\sqrt{3}\right)$

$= 4(3\sqrt{3} - 2\sqrt{3}) = 4 \times \sqrt{3}$

$= 4 \times 1.732 = 6.928$

38. (a) $\sqrt[3]{4\dfrac{12}{125}} = \sqrt[3]{\dfrac{4 \times 125 + 12}{125}} = \sqrt[3]{\dfrac{500 + 12}{125}}$

$= \sqrt[3]{\dfrac{512}{125}} = \sqrt[3]{\left(\dfrac{8}{5}\right)^3}$

$= \left[\left(\dfrac{8}{5}\right)^3\right]^{\frac{1}{3}} = \left(\dfrac{8}{5}\right)^{3 \times \frac{1}{3}} = \dfrac{8}{5} = 1\dfrac{3}{5}$

39. (a) $\because (64)^2 \div \sqrt[3]{32768} = ?$

$\therefore \quad ? = (64)^2 \div \sqrt[3]{32 \times 32 \times 32}$

$= 64 \times 64 \div 32$

$= \dfrac{64 \times 64}{32} = 64 \times 2 = 128$

40. (b) $\sqrt[3]{1331} \times \sqrt[3]{216} + \sqrt[3]{729} + \sqrt[3]{64}$

$= \sqrt[3]{11 \times 11 \times 11} \times \sqrt[3]{2 \times 2 \times 2 \times 3 \times 3 \times 3}$

$+ \sqrt[3]{3 \times 3 \times 3 \times 3 \times 3 \times 3}$

$+ \sqrt[3]{2 \times 2 \times 2 \times 2 \times 2 \times 2}$

$= 11 \times 6 + 9 + 4 = 66 + 9 + 4 = 79$

41. (c) Given number = 294

2	294
3	147
7	49
7	7
	1

$294 = 2 \times 3 \times 7 \times 7 = 2^1 \times 3^1 \times 7^2$

Hence, to make '294' a perfect square, we have to multiply it with $2 \times 3 = 6$.

42. (a) Given number = 8200

First, we find the nearest square values of given number.

i.e. $(90)^2 = 8100$

and $(91)^2 = 8281$

$\because (90)^2 < 8200 < (91)^2$

Hence, required number = $8281 - 8200 = 81$

43. (b) $28^2 = 784 > 777 > 729 = 27^2$

$[\because 28^2$ is the nearest$]$

$\therefore$ Required difference = $784 - 777 = 7$

44. (b) Let the number be x.

According to the question,

$x^2 - (46)^2 = 485$

$\Rightarrow \quad x^2 - 2116 = 485$

$\Rightarrow \quad x^2 = 2116 + 485 = 2601$

$\therefore \quad x = \sqrt{2601} = 51$

96 / Fast Track Objective Arithmetic

45. (c) We know that, $(a+b)^2 - (a-b)^2 = 4ab$
$\therefore \quad 4\sqrt{ab} = (\sqrt{a}+\sqrt{b})^2 - (\sqrt{a}-\sqrt{b})^2$
$= 17^2 - 1^2 = 289 - 1 = 288$
$\therefore \quad \sqrt{ab} = \dfrac{288}{4} = 72$

46. (a) Given number = 6860
$(19)^3 = 6859$
$\because \quad (19)^3 < 6860$
Hence, required number = 6860 − 6859 = 1

47. (c) The factors of 72 with difference of 1 are 9 and 8. $\quad [\because 72 = 8 \times 9]$
Here, the larger number is 9.
[by Technique 2]
$\therefore \sqrt{72+\sqrt{72+\sqrt{72+\ldots}}} = 9$

48. (c) We know that, $20 = 4 \times 5 = 4(4+1)$
$\therefore \sqrt{20-\sqrt{20-\sqrt{20-\ldots}}} = 4$
[by Technique 3]

49. (c) Since, $\sqrt{x\sqrt{x\sqrt{x\ldots\infty}}} = x$
$\therefore \sqrt{9\sqrt{9\sqrt{9\ldots\infty}}} = 9$ [by Technique 4]

50. (c) Given, $x = \sqrt{8 + \sqrt{8 + \sqrt{8 + \ldots}}}$
$\Rightarrow \quad x^2 = 8 + \sqrt{8 + \sqrt{8 + \sqrt{8 + \ldots}}}$
$\therefore \quad x^2 = 8 + x$
Similarly, $y^2 = 8 - y$
$\therefore \quad x^2 - y^2 = (x+8) - (8-y)$
$(x+y)(x-y) = (x+y)$
$\Rightarrow \quad x - y = 1$
✦ Here, x and y cannot be splitted in the form of $n(n+1)$, so we cannot apply shortcut method.

51. (b) $\sqrt{6\sqrt{6\sqrt{6\sqrt{6\sqrt{6}}}}} = 6^{\frac{2^5-1}{2^5}}$
$= 6^{\left(\frac{32-1}{32}\right)} = 6^{\frac{32-1}{32}} = 6^{\frac{31}{32}}$ [by Technique 5]

Exercise 2 Higher Skill Level Questions

1. (c) First, we have to find out the square root of 36562.

```
           1 9 1
      ┌─────────────
   1  │ 3 65 62
   1  │ 1
  ────┼─────
  29  │ 2 65
   9  │ 2 61
  ────┼─────
  381 │   4 62
      │   3 81
      ────────
           81
```
Here, the remainder is 81.
Hence, the unused army men = 81

2. (c) Given, total cost amount = 29791
According to the question,
Money contributed by each student
= Square of the total number of students
Hence, total number of students
$= \sqrt[3]{29791}$
$= \sqrt[3]{31 \times 31 \times 31} = 31$

3. (b) Let the number of rows and columns be m. Then, total plants should be $m \times m$.
Now, 1000 is not a square of any number.
Let $\quad m = 30$
Then, $m \times m = 30 \times 30 = 900$
which is less than total plants.
Now, let $\quad m = 32$

Then, $m \times m = 32 \times 32 = 1024$
which is greater than 1000.
So, minimum number of plants needed
= 1024 − 1000 = 24 plants

4. (c) According to the question, total number of toys is a perfect square number because the toys were packed in n boxes containing n toys each, without any remainder.
From all the options, the only perfect square root is $\sqrt{1444} = 38$
Hence, option (c) is the correct answer.

5. (b)

2	7936
2	3968
2	1984
2	992
2	496
2	248
2	124
2	62
	31

$\therefore 7936 = 2\times2\times2\times2\times2\times2\times2\times2\times31$
To make it a perfect square, it must be multiplied by 31.
$\therefore$ Required number = 7936 × 31
= 246016

Square Root and Cube Root / 97

6. (c)

3	59535
3	19845
3	6615
3	2205
3	735
5	245
7	49
7	7
	1

$59535 = 3 \times 3 \times 3 \times 3 \times 3 \times 5 \times 7 \times 7$

Here, 5 and 3 are to be multiplied with 59535 to make it a perfect square number.

∴ Required sum = 6

7. (c) $? = \dfrac{\sqrt{10 + \sqrt{25 + \sqrt{108 + \sqrt{154 + \sqrt{225}}}}}}{\sqrt[3]{8}}$

$= \dfrac{\sqrt{10 + \sqrt{25 + \sqrt{108 + \sqrt{154 + 15}}}}}{2}$

$= \dfrac{\sqrt{10 + \sqrt{25 + \sqrt{108 + \sqrt{169}}}}}{2}$

$= \dfrac{\sqrt{10 + \sqrt{25 + \sqrt{108 + 13}}}}{2}$

$= \dfrac{\sqrt{10 + \sqrt{25 + \sqrt{121}}}}{2}$

$= \dfrac{\sqrt{10 + \sqrt{25 + 11}}}{2}$

$= \dfrac{\sqrt{10 + \sqrt{36}}}{2}$

$= \dfrac{\sqrt{10 + 6}}{2} = \dfrac{\sqrt{16}}{2} = \dfrac{4}{2} = 2$

8. (a) Given, $a = \dfrac{2 + \sqrt{3}}{2 - \sqrt{3}}$ and $b = \dfrac{2 - \sqrt{3}}{2 + \sqrt{3}}$

We have, $a^2 + b^2 + ab = (a+b)^2 - ab$

$= \left(\dfrac{2+\sqrt{3}}{2-\sqrt{3}} + \dfrac{2-\sqrt{3}}{2+\sqrt{3}}\right)^2$

$\quad - \left(\dfrac{2+\sqrt{3}}{2-\sqrt{3}} \times \dfrac{2-\sqrt{3}}{2+\sqrt{3}}\right)$

$= \left[\dfrac{2(4+3)}{4-3}\right]^2 - 1$

$= (14)^2 - 1 = 195$

9. (d) Here, adding and subtracting $1^2 + 2^2 + 3^2 + 4^2 + 5^2$ in numerator, we get

$= \dfrac{\begin{bmatrix}1^2+2^2+3^2+4^2+5^2+6^2+7^2+8^2+9^2 \\ +10^2 - (1^2+2^2+3^2+4^2+5^2)\end{bmatrix}}{\sqrt{(2)^2 + (\sqrt{3})^2 + 4\sqrt{3}} - \sqrt{(\sqrt{3})^2 + (1)^2 + 2\sqrt{3}}}$

$= \dfrac{\dfrac{10 \times 11 \times 21}{6} - \dfrac{5 \times 6 \times 11}{6}}{\sqrt{(2+\sqrt{3})^2} - \sqrt{(\sqrt{3}+1)^2}}$

$= \dfrac{\dfrac{10 \times 11 \times 21}{6} - \dfrac{5 \times 6 \times 11}{6}}{(2+\sqrt{3}) - (\sqrt{3}+1)}$

$\left[\because 1^2 + 2^2 + 3^2 + \ldots + n^2 = \dfrac{n(n+1)(2n+1)}{6}\right]$

$= \dfrac{385 - 55}{1} = 330$

10. (d) Given, $x = \sqrt{a\sqrt[3]{b}\sqrt{a\sqrt[3]{b}\ldots\infty}}$ ...(i)

On squaring both sides, we get

$x^2 = a\sqrt[3]{b}\sqrt{a\sqrt[3]{b}\ldots\infty}$

On cubing both sides, we get

$x^6 = a^3 b\sqrt{a\sqrt[3]{b}\ldots\infty}$

$\Rightarrow \quad x^6 = a^3 bx$ [from Eq. (i)]

$\Rightarrow \quad x^6 - a^3 bx = 0$

$\Rightarrow \quad x(x^5 - a^3 b) = 0$

$\Rightarrow \quad x = 0, \; x^5 = a^3 b$

$\Rightarrow \quad x = \sqrt[5]{a^3 b}$

11. (b) Let $N^2 = x + 809436 \times 809438$

$\Rightarrow N^2 = x + (809437 - 1)(809437 + 1)$

$\Rightarrow N^2 = x + (809437)^2 - 1$

$[\because (a-b)(a+b) = a^2 - b^2]$

Hence, N is a perfect square number, when the value of x is 1.

12. (b) Given expression

$= \sqrt{\dfrac{(0.03)^2 + (0.21)^2 + (0.065)^2}{\left(\dfrac{0.03}{10}\right)^2 + \left(\dfrac{0.21}{10}\right)^2 + \left(\dfrac{0.065}{10}\right)^2}} \div 2$

$= \sqrt{\dfrac{100\,[(0.03)^2 + (0.21)^2 + (0.065)^2]}{(0.03)^2 + (0.21)^2 + (0.065)^2}} \div 2$

$= \sqrt{100} \div 2 = 10 \div 2 = 5$

13. (a) Let the numbers be x and y.

Now, according to the question,

$x(x+y) = 2418$ ...(i)

and $y(x+y) = 3666$ ...(ii)

On adding Eqs. (i) and (ii), we get
$$x^2 + xy + yx + y^2 = 6084$$
$$\Rightarrow x^2 + 2xy + y^2 = 6084$$
$$\Rightarrow (x+y)^2 = 6084$$
$$\therefore x+y = \sqrt{6084} = 78$$

On subtracting Eq. (ii) from Eq. (i), we get
$$x^2 + xy - yx - y^2 = -1248$$
$$\Rightarrow x^2 - y^2 = -1248$$
$$\Rightarrow (x+y)(x-y) = -1248$$
$$\Rightarrow 78(x-y) = -1248$$
$$\Rightarrow x - y = -\frac{1248}{78} = -16$$
$$\therefore (y - x) = 16$$

14. (a) Let $y = \dfrac{(0.35)^2 + 0.70 + 1}{2.25} + 0.197$

$$= \frac{(0.35)^2 + 2 \times 0.35 \times 1 + (1)^2}{(1.5)^2} + 0.197$$

$$= \frac{(0.35 + 1)^2}{(1.5)^2} + 0.197$$

$$= \left(\frac{1.35}{1.5}\right)^2 + 0.197$$

$$= (0.9)^2 + 0.197$$

$$= 0.81 + 0.197 = 1.007 \approx 1$$

$$\therefore \sqrt{y} = \sqrt{1} = 1$$

15. (a) $\sqrt{48} - \sqrt{45} = \sqrt{3}(\sqrt{16} - \sqrt{15})$

[multiplying numerator and denominator by 2]

$$= \frac{2\sqrt{3}(\sqrt{16} - \sqrt{15})}{2}$$

$$= \frac{\sqrt{3}(\sqrt{64} - 2\sqrt{15})}{2}$$

$$= \frac{\sqrt{3}}{2}(8 - 2 \times \sqrt{5} \times \sqrt{3})$$

$$= \frac{\sqrt{3}}{2}(5 + 3 - 2 \times \sqrt{5} \times \sqrt{3})$$

$$= \frac{\sqrt{3}}{2}((\sqrt{5})^2 + (\sqrt{3})^2 - 2 \times \sqrt{5} \times \sqrt{3})$$

$$= \frac{\sqrt{3}}{2}(\sqrt{5} - \sqrt{3})^2$$

$\therefore$ Required square root $= \dfrac{\sqrt[4]{3}}{\sqrt{2}}(\sqrt{5} - \sqrt{3})$

16. (c) Let $x = \sqrt{4 + \sqrt{4 - \sqrt{4 + ...}}}$

$\therefore \quad x = \sqrt{4 + \sqrt{4-x}}$
$\Rightarrow \quad x^2 = 4 + \sqrt{4-x}$
$\Rightarrow \quad (x^2 - 4)^2 = 4 - x$
$\Rightarrow \quad (x^2 - 4)^2 + x = 4 \quad ...(i)$

From option (c), $x = \dfrac{\sqrt{13} + 1}{2}$

$\therefore$ From Eq. (i),

$$\text{LHS} = \left[\left(\frac{\sqrt{13}+1}{2}\right)^2 - 4\right]^2 + \frac{\sqrt{13}+1}{2}$$

$$= \left[\frac{13 + 1 + 2\sqrt{13} - 16}{4}\right]^2 + \frac{\sqrt{13}+1}{2}$$

$$= \left(\frac{2\sqrt{13} - 2}{4}\right)^2 + \left(\frac{\sqrt{13}+1}{2}\right)$$

$$= \left(\frac{\sqrt{13}-1}{2}\right)^2 + \left(\frac{\sqrt{13}+1}{2}\right)$$

$$= \frac{13 + 1 - 2\sqrt{13}}{4} + \frac{\sqrt{13}+1}{2}$$

$$= \frac{14 - 2\sqrt{13}}{4} + \frac{\sqrt{13}+1}{2}$$

$$= \frac{14 - 2\sqrt{13} + 2\sqrt{13} + 2}{4}$$

$$= \frac{16}{4} = 4 = \text{RHS}$$

17. (b) $\sqrt{\dfrac{(0.75)^3}{1 - 0.75} + [0.75 + (0.75)^2 + 1]}$

$$= \sqrt{\frac{(0.75)^3}{0.25} + [1.75 + (0.75)^2]}$$

$$= \sqrt{1.6875 + [1.75 + 0.5625]}$$

$$= \sqrt{1.6875 + 2.3125}$$

$$= \sqrt{4} = 2$$

Hence, the required square root is 2.

18. (b) $x(x + a)(x + 2a)(x + 3a)$

$$= (x^2 + 3ax)(x^2 + 3ax + 2a^2)$$

$$= (x^2 + 3ax - a^2 + a^2)(x^2 + 3ax + 2a^2)$$

$$= [(x^2 + 3ax + a^2) - a^2]$$
$$[(x^2 + 3ax + a^2) + a^2]$$

$$= (x^2 + 3ax + a^2)^2 - a^4$$

$[\because (a+b)(a-b) = a^2 - b^2]$

$\therefore$ For making it a perfect square, we need to add a^4.

Chapter 6

Indices and Surds

Indices

When a number 'P' is multiplied by itself 'n' times, then the product is called nth power of 'P' and is written as P^n. Here, P is called the **base** and n is known as the **index** of the power.

Therefore, P^n is the exponential expression. P^n is read as 'P raised to the power n' or 'P to the power n'.

Laws of Indices

Let P, Q be two real numbers and m, n be two positive integers, then

(i) $P^m \times P^n = P^{m+n}$ (ii) $\dfrac{P^m}{P^n} = P^{m-n}$

(iii) $(P^m)^n = P^{mn}$ (iv) $(PQ)^n = P^n \times Q^n$

(v) $\left(\dfrac{P}{Q}\right)^n = \dfrac{P^n}{Q^n}$ (vi) $P^0 = 1$

(vii) $P^{-n} = \dfrac{1}{P^n}$

Ex. 1 Simplify $(256)^{3/4}$.

Sol. Given expression $= (256)^{3/4} = (4^4)^{3/4}$
$= (4)^{4 \times 3/4} = 4^3 = 64$

Ex. 2 Simplify $(1024)^{-3/5}$.

Sol. Given expression $= (1024)^{-3/5} = (4^5)^{-3/5} = (4)^{5 \times \left(\frac{-3}{5}\right)}$
$= 4^{-3} = \dfrac{1}{4^3} = \dfrac{1}{64}$

Ex. 3 Simplify $(a^4 b^{5/3})^{-3/4}$.

Sol. Given expression $= (a^4 b^{5/3})^{-3/4} = (a^4)^{-3/4} (b^{5/3})^{-3/4}$
$= (a)^{4 \times \left(\frac{-3}{4}\right)} (b)^{\frac{5}{3} \times \left(\frac{-3}{4}\right)} = a^{-3} b^{-5/4}$

Surds

When root of a non-negative rational number (i.e. quantities of type $\sqrt[n]{a}$, a being a rational number) does not provide an exact solution, then this root is called a surd.
For example $\sqrt{2}$, $\sqrt{5}$, $3\sqrt{8}$, $a + \sqrt{b}$, etc.

* All surds are irrational numbers.
* All irrational numbers are not surds.

Order of Surds

Let P be a rational number and m be a positive integer such that $P^{1/m} = \sqrt[m]{P}$ is irrational. Then, $\sqrt[m]{P}$ is called a surd of mth order and P is called the **radicand**.
Here, $7^{1/2} = \sqrt{7}$ = Surd of 2nd order and $6^{1/5} = \sqrt[5]{6}$ = Surd of 5th order

* Surds of order 2 are known as **quadratic surds**.
* Surds of order 3 are known as **cubic surds**.

Laws of Surds

Let P, Q be two positive rational numbers and m, n be two positive integers, then

(i) $\sqrt[m]{P} = P^{1/m}$

(ii) $\sqrt[m]{PQ} = \sqrt[m]{P} \times \sqrt[m]{Q}$

(iii) $\sqrt[m]{\dfrac{P}{Q}} = \dfrac{\sqrt[m]{P}}{\sqrt[m]{Q}}$

(iv) $(\sqrt[m]{P})^m = P$

(iv) $(\sqrt[m]{P})^n = (P^{1/m})^n = P^{n/m} = \sqrt[m]{P^n}$

Types of Surds

There are four types of surds, which are as follow

1. Pure Surds

Those surds which do not have factor other than 1, are known as pure surds.
For example $\sqrt{2}$, $\sqrt[3]{5}$.

2. Mixed Surds

Those surds which have factor other than 1, are known as mixed surds.
For example $3\sqrt{5}$, $4\sqrt{6}$, $7^2\sqrt[3]{5}$.

3. Like Surds

When the radicands of two surds are same, then those are known as like surds.
For example $4\sqrt{5}$ and $(6^3)\sqrt{5}$.

4. Unlike Surds

When radicands are different, then they are called unlike surds.
For example $3\sqrt{5}$ and $\sqrt{3}$.

Indices and Surds / 101

Properties of Surds

(i) A quadratic surd cannot be equal to the sum and difference of a rational number and a quadratic surd.
For example $a + b \neq \sqrt{c}$ or $\sqrt{a} - b \neq \sqrt{c}$

(ii) If $a + \sqrt{b} = c + \sqrt{d}$ or $a - \sqrt{b} = c - \sqrt{d}$, then $a = c$ and $b = d$.

(iii) If $a + \sqrt{b} = c + \sqrt{d}$, then $a - \sqrt{b} = c - \sqrt{d}$ and vice-versa.

(iv) If $\sqrt{a + \sqrt{b}} = \sqrt{c} + \sqrt{d}$, then $\sqrt{a - \sqrt{b}} = \sqrt{c} - \sqrt{d}$ and vice-versa.

MIND IT ! If x is a positive real number and a, b, c are real numbers, then

(i) $\left(\dfrac{x^b}{x^c}\right)^a \cdot \left(\dfrac{x^c}{x^a}\right)^b \cdot \left(\dfrac{x^a}{x^b}\right)^c = 1$

(ii) $\left(\dfrac{x^a}{x^b}\right)^{a+b} \cdot \left(\dfrac{x^b}{x^c}\right)^{b+c} \cdot \left(\dfrac{x^c}{x^a}\right)^{c+a} = 1$

(iii) $\left(\dfrac{x^a}{x^b}\right)^{(a^2 + b^2 + ab)} \cdot \left(\dfrac{x^b}{x^c}\right)^{(b^2 + c^2 + bc)} \cdot \left(\dfrac{x^c}{x^a}\right)^{(c^2 + a^2 + ca)} = 1$

Operations on Surds

There are various operations on surds as given below

1. Addition and Subtraction of Surds

Only like surds can be added or subtracted. Therefore, to add or subtract two or more surds, first simplify them and then add or subtract like surds.

$\sqrt{a} + \sqrt{b} \neq \sqrt{a+b}$ $\sqrt{a} - \sqrt{b} \neq \sqrt{a-b}$ $x\sqrt{a} + x\sqrt{b} \neq x\sqrt{a+b}$

Ex. 4 Find the value of $\sqrt{80} + 3\sqrt{245} - \sqrt{125}$.

Sol. $\because \sqrt{80} = \sqrt{16 \times 5} = 4\sqrt{5}$, $3\sqrt{245} = 3\sqrt{49 \times 5} = 21\sqrt{5}$ and $\sqrt{125} = \sqrt{25 \times 5} = 5\sqrt{5}$

$\therefore \sqrt{80} + 3\sqrt{245} - \sqrt{125} = 4\sqrt{5} + 21\sqrt{5} - 5\sqrt{5} = 20\sqrt{5}$

2. Multiplication and Division of Surds

To multiply or divide the surds, we make the denominators of the powers equal to each other. Then, multiply or divide as usual.

Ex. 5 Find the product of $\sqrt{5}$, $\sqrt[6]{6}$ and $\sqrt[3]{4}$.

Sol. $\because$ LCM of 2, 6 and 3 = 6

$\therefore \sqrt{5} = 5^{1/2} = 5^{3/6} = (125)^{1/6}$; $\sqrt[6]{6} = (6)^{1/6}$ and $\sqrt[3]{4} = 4^{1/3} = 4^{2/6} = (16)^{1/6}$

$\therefore$ Required product $= (125 \times 6 \times 16)^{1/6} = (12000)^{1/6}$

Ex. 6 Divide $12 \times 4^{1/3}$ by $3\sqrt{2}$.

Sol. $\dfrac{12 \times 4^{1/3}}{3 \times 2^{1/2}} = \dfrac{4 \times 4^{2/6}}{1 \times 2^{3/6}} = \dfrac{4 \times (16)^{1/6}}{(8)^{1/6}} = 4\left(\dfrac{16}{8}\right)^{1/6} = 4(2)^{1/6} = 4\sqrt[6]{2}$

Comparison of Surds

Suppose the given surds are $p^{1/a}, q^{1/b}, r^{1/c}$. First of all, take the LCM of a, b, c and then use it to make the denominator of the powers same. Then, we can easily find the required order.

Ex. 7 Arrange $\sqrt[4]{3}, \sqrt[3]{2}$ and $\sqrt[6]{5}$ in the decreasing order.

Sol. Here, terms $\sqrt[4]{3}, \sqrt[3]{2}$ and $\sqrt[6]{5}$ are written as $(3)^{1/4}, (2)^{1/3}$ and $(5)^{1/6}$.
So, $a = 4, b = 3$ and $c = 6$
Now, LCM of 4, 3 and 6 = 12
Then, we will make the denominator of the power of every coefficient equal, i.e. 12.
Also, $\sqrt[4]{3} = (3)^{1/4} = 3^{3/12} = \sqrt[12]{27}$; $\sqrt[3]{2} = (2)^{1/3} = 2^{4/12} = \sqrt[12]{16}$
and $\sqrt[6]{5} = (5)^{1/6} = 5^{2/12} = \sqrt[12]{25}$
Thus, $\sqrt[12]{27} > \sqrt[12]{25} > \sqrt[12]{16} \Rightarrow \sqrt[4]{3} > \sqrt[6]{5} > \sqrt[3]{2}$

Ex. 8 Which of the following surds is greatest?
$\sqrt[3]{2}, \sqrt{3}, \sqrt[4]{5}, \sqrt[6]{7}$

Sol. $\because \sqrt[3]{2} = (2)^{1/3}, \sqrt{3} = (3)^{1/2}, \sqrt[4]{5} = (5)^{1/4}$ and $\sqrt[6]{7} = (7)^{1/6}$
Here, LCM of 3, 2, 4 and 6 is 12.
$\therefore \quad (2)^{1/3} = (2)^{4/12} = (2^4)^{1/12} = (16)^{1/12},$
$(3)^{1/2} = (3)^{6/12} = (3^6)^{1/12} = (729)^{1/12},$
$(5)^{1/4} = (5)^{3/12} = (5^3)^{1/12} = (125)^{1/12}$
and $(7)^{1/6} = (7)^{2/12} = (7^2)^{1/12} = (49)^{1/12}$
Hence, $\sqrt{3}$ is the greatest among all the given surds.

Rationalisation of Surds

The method of obtaining a rational number from a surd by multiplying it with another surd is known as rationalisation of surds. Both the surds are known as rationalising factor of each other.

✦ If any number is in the form of $\dfrac{1}{\sqrt{a} \pm \sqrt{b}}$, then to rationalise its denominator, the numerator and denominator should be multiplied by $\sqrt{a} \mp \sqrt{b}$.

Ex. 9 Find the fraction equivalent to $\dfrac{5}{2-\sqrt{3}}$, such that denominator of the fraction is not irrational.

Sol. $\dfrac{5}{2-\sqrt{3}} = \dfrac{5}{2-\sqrt{3}} \times \dfrac{2+\sqrt{3}}{2+\sqrt{3}} = \dfrac{5(2+\sqrt{3})}{(2)^2 - (\sqrt{3})^2} = 5(2+\sqrt{3})$

Ex. 10 Simplify $\dfrac{\sqrt{5}+\sqrt{3}}{\sqrt{5}-\sqrt{3}}$.

Sol. $\dfrac{\sqrt{5}+\sqrt{3}}{\sqrt{5}-\sqrt{3}} = \dfrac{\sqrt{5}+\sqrt{3}}{\sqrt{5}-\sqrt{3}} \times \dfrac{\sqrt{5}+\sqrt{3}}{\sqrt{5}+\sqrt{3}} = \dfrac{5+3+2\sqrt{15}}{5-3} = \dfrac{8+2\sqrt{15}}{2} = 4+\sqrt{15}$

Multi Concept Questions

1. If $xyz = 1$, then find the value of $\dfrac{1}{(1 + x + y^{-1})} + \dfrac{1}{(1 + y + z^{-1})} + \dfrac{1}{(1 + z + x^{-1})}$.

(a) 1 (b) 2 (c) 3 (d) 4

↪ (a) Given that, $xyz = 1$

Then, $\dfrac{1}{(1 + x + y^{-1})} + \dfrac{1}{(1 + y + z^{-1})} + \dfrac{1}{(1 + z + x^{-1})}$

$= \dfrac{1}{\left(1 + x + \dfrac{1}{y}\right)} + \dfrac{1}{\left(1 + y + \dfrac{1}{z}\right)} + \dfrac{1}{\left(1 + z + \dfrac{1}{x}\right)}$ $\quad \left[\because a^{-1} = \dfrac{1}{a}\right]$

$= \dfrac{y}{y + xy + 1} + \dfrac{z}{z + yz + 1} + \dfrac{x}{x + xz + 1}$

$= \dfrac{y}{y + xy + 1} + \dfrac{z}{z + \dfrac{1}{x} + 1} + \dfrac{x}{x + \dfrac{1}{y} + 1}$ $\quad \left[\because yz = \dfrac{1}{x} \text{ and } xz = \dfrac{1}{y}\right]$

$= \dfrac{y}{1 + xy + y} + \dfrac{xz}{xz + 1 + x} + \dfrac{xy}{xy + 1 + y}$

$= \dfrac{y}{1 + xy + y} + \dfrac{1/y}{\dfrac{1}{y} + 1 + x} + \dfrac{xy}{1 + xy + y}$

$= \dfrac{y}{1 + xy + y} + \dfrac{1}{1 + xy + y} + \dfrac{xy}{1 + xy + y}$

$= \dfrac{1 + xy + y}{1 + xy + y} = 1$

2. If $a^x = b^y = c^z$ and $abc = 1$, then what is the value of $xy + yz + zx$?

(a) 1 (b) 3 (c) 0 (d) 5

↪ (c) Here, $a^x = b^y = c^z$

Let $a^x = b^y = c^z = K$, then $a = K^{1/x}$, $b = K^{1/y}$ and $c = K^{1/z}$

$\Rightarrow \quad abc = K^{\left(\dfrac{1}{x} + \dfrac{1}{y} + \dfrac{1}{z}\right)}$

$\because \quad abc = 1$ [given]

$\Rightarrow \quad K^{\left(\dfrac{1}{x} + \dfrac{1}{y} + \dfrac{1}{z}\right)} = 1$

$\Rightarrow \quad K^{\left(\dfrac{1}{x} + \dfrac{1}{y} + \dfrac{1}{z}\right)} = K^0$

On comparing, we get

$\dfrac{1}{x} + \dfrac{1}{y} + \dfrac{1}{z} = 0$ or $xy + yz + zx = 0$

104 / Fast Track Objective Arithmetic

3. If $a + b + c = 0$, then what is the value of
$$\frac{1}{(x^a + x^{-b} + 1)} + \frac{1}{(x^b + x^{-c} + 1)} + \frac{1}{(x^c + x^{-a} + 1)}?$$

(a) 1   (b) 2   (c) 3   (d) 4

↳ (a) $\dfrac{1}{(x^a + x^{-b} + 1)} + \dfrac{1}{(x^b + x^{-c} + 1)} + \dfrac{1}{(x^c + x^{-a} + 1)}$

$= \dfrac{1}{x^a + \dfrac{1}{x^b} + 1} + \dfrac{1}{x^b + \dfrac{1}{x^c} + 1} + \dfrac{1}{x^c + \dfrac{1}{x^a} + 1}$

$= \dfrac{x^b}{x^{a+b} + 1 + x^b} + \dfrac{x^c}{x^{b+c} + 1 + x^c} + \dfrac{x^a}{x^{a+c} + 1 + x^a}$

$= \dfrac{x^b}{x^{a+b} + x^b + 1} + \dfrac{x^c}{x^{-a} + 1 + x^c} + \dfrac{x^a}{x^{-b} + 1 + x^a}$

$[\because a + b + c = 0 \Rightarrow a + c = -b \text{ and } b + c = -a]$

$= \dfrac{x^b}{x^{a+b} + x^b + 1} + \dfrac{x^{a+c}}{1 + x^a + x^{a+c}} + \dfrac{x^{a+b}}{1 + x^b + x^{a+b}}$

$= \dfrac{x^b}{x^{a+b} + x^b + 1} + \dfrac{x^{-b} \cdot x^b}{1 + x^b + x^{a+b}} + \dfrac{x^{a+b}}{1 + x^b + x^{a+b}}$

$= \dfrac{x^b + 1 + x^{a+b}}{x^{a+b} + x^b + 1} = 1$ $[\because x^{-b} x^b = x^0 = 1]$

4. Find the value of the expression $\dfrac{4^n \times 20^{m-1} \times 12^{m-n} \times 15^{m+n-2}}{16^m \times 5^{2m+n} \times 9^{m-1}}$.

(a) $\dfrac{1}{200}$ (b) $\dfrac{1}{500}$ (c) $\dfrac{1}{700}$ (d) $\dfrac{1}{900}$

↳ (b) $\dfrac{4^n \times 20^{m-1} \times 12^{m-n} \times 15^{m+n-2}}{16^m \times 5^{2m+n} \times 9^{m-1}}$

$= \dfrac{(2)^{2n} \times (2^2)^{m-1} \times 5^{m-1} \times (2^2)^{m-n} \times (3)^{m-n} \times (3)^{m+n-2} \times (5)^{m+n-2}}{(2^4)^m \times 5^{2m+n} \times (3^2)^{m-1}}$

$= \dfrac{(2)^{2n+2m-2+2m-2n} \times 5^{m-1+m+n-2} \times 3^{m-n+m+n-2}}{2^{4m} \times 5^{2m+n} \times 3^{2m-2}}$

$= \dfrac{2^{4m-2} \times 5^{2m+n-3} \times 3^{2m-2}}{2^{4m} \times 5^{2m+n} \times 3^{2m-2}}$

$= 2^{4m-2-4m} \times 5^{2m+n-3-2m-n} \times 3^{2m-2-2m+2}$

$= 2^{-2} \times 5^{-3} \times 3^0$

$= \dfrac{1}{2^2} \times \dfrac{1}{5^3} = \dfrac{1}{4} \times \dfrac{1}{125} = \dfrac{1}{500}$

Fast Track Practice

Exercise 1 Base Level Questions

1. $\dfrac{[(12)^{-2}]^2}{[(12)^2]^{-2}} = ?$ [Bank Clerk 2009]
   (a) 12 (b) 4.8 (c) $\dfrac{12}{144}$ (d) 1
   (e) None of these

2. Find the value of $(10)^{200} \div (10)^{196}$. [MBA 2008]
   (a) 10000 (b) 1000 (c) 100 (d) 100000

3. Evaluate $(0.00032)^{2/5}$.
   (a) $\dfrac{1}{625}$ (b) $\dfrac{1}{225}$ (c) $\dfrac{1}{125}$ (d) $\dfrac{1}{25}$
   (e) None of these

4. $\left[\left\{\left(-\dfrac{1}{2}\right)^2\right\}^{-2}\right]^{-1}$ is equal to
   (a) $\dfrac{1}{16}$ (b) 16 (c) $-\dfrac{1}{16}$ (d) -16

5. If $289 = 17^{x/5}$, then $x = ?$ [Bank PO 2010]
   (a) 16 (b) 8 (c) 10 (d) 2/5
   (e) None of these

6. If $5^a = 3125$, then the value of 5^{a-3} is [CGPSC 2016]
   (a) 625 (b) 25 (c) 5 (d) 225
   (e) None of these

7. If $a^{2x+2} = 1$, where a is a positive real number other than 1, then $x = ?$ [SSC CGL 2007]
   (a) -2 (b) -1 (c) 0 (d) 1

8. If $\{(2^4)^{1/2}\}^? = 256$, then find the value of '?'. [Bank PO 2007]
   (a) 1 (b) 2 (c) 4 (d) 8
   (e) None of these

9. $(16)^9 \div (16)^4 \times 16^3 = (16)^?$
   (a) 6.75 (b) 8 (c) 10 (d) 12
   (e) None of these

10. Solve $112 \times 5^4 = ?$. [IB PA 2016]
    (a) 67000 (b) 70000 (c) 76500 (d) 77200

11. What is the value of x in the following equation? [CLAT 2015]
    $x^{0.4} \div 16 = 32 \div x^{2.6}$
    (a) 8 (b) 9 (c) 6 (d) 7

12. $17^{3.5} \times 17^{7.3} \div 17^{4.2} = 17^?$ [Bank Clerk 2010]
    (a) 8.4 (b) 8 (c) 6.6 (d) 6.4
    (e) None of these

13. If $\left(\dfrac{1}{5}\right)^{3a} = 0.008$, then find the value of $(0.25)^a$.
    (a) 20.5 (b) 22.5 (c) 0.25 (d) 6.25
    (e) None of these

14. If $\left(\dfrac{p}{q}\right)^{n-1} = \left(\dfrac{q}{p}\right)^{n-3}$, then the value of n is
    (a) $\dfrac{1}{2}$ (b) $\dfrac{7}{2}$ (c) 1 (d) 2

15. $[p^{(b-c)}]^{b+c} \cdot [p^{(c-a)}]^{c+a} \cdot [p^{(a-b)}]^{(a+b)} = ?$
    (a) 0 (b) p^{abc} (c) 1 (d) p^{a+b+c}

16. If $a^x = b$, $b^y = c$ and $xyz = 1$, then what is the value of c^z? [CDS 2012]
    (a) a (b) b (c) ab (d) a/b

17. If $16 \times 8^{n+2} = 2^m$, then m is equal to [CDS 2013]
    (a) $n + 8$ (b) $2n + 10$
    (c) $3n + 2$ (d) $3n + 10$

18. $\sqrt[L]{M}$ is a surd of order ......, where M is a rational number, L is a positive integer and $\sqrt[L]{M}$ is irrational.
    (a) L (b) M (c) 2 (d) 4

19. $\left[\left(\sqrt[5]{x^{-3/5}}\right)^{-5/3}\right]^5$ is equal to
    (a) x^5 (b) x^{-5} (c) x (d) $\dfrac{1}{x}$

20. If $P = 124$, then $\sqrt[3]{P(P^2 + 3P + 3) + 1} = ?$
    (a) 5 (b) 7 (c) 123 (d) 125

21. The expression $(\sqrt{2})^{\sqrt{2}^{\sqrt{2}}}$ gives [CDS 2013]
    (a) a natural number
    (b) a integer and not a natural number
    (c) a rational number but not an integer
    (d) a real number but not a rational number

106 / Fast Track Objective Arithmetic

22. If m and n are natural numbers, then $\sqrt[m]{n}$ is [CDS 2013]
(a) always irrational
(b) irrational unless n is the mth power of an integer
(c) irrational unless m is the nth power of an integer
(d) irrational unless m and n are coprime

23. If $\sqrt{10 + \sqrt[3]{x}} = 4$, then what is the value of x? [CDS 2012]
(a) 150   (b) 216   (c) 316   (d) 450

24. $81^{2.5} \times 9^{4.5} \div 3^{4.8} = 9^?$ [Bank Clerk 2009]
(a) 7.1   (b) 9.4   (c) 4.7   (d) 4.5
(e) None of these

25. The value of '?' in the expression $7^{8.9} \div (343)^{1.7} \times (49)^{4.8} = 7^?$ is
(a) 13.4   (b) 12.8   (c) 11.4   (d) 9.6
(e) None of these

26. $(42 \times 229) \div (9261)^{1/3} = ?$
(a) 448   (b) 452   (c) 456   (d) 458
(e) None of these

27. If $2^x + 2^{x+1} = 48$, then the value of x^x is [SNAP 2016]
(a) 4   (b) 64   (c) 256   (d) 16

28. Find the quotient when $(a^{-1} - 1)$ is divided by $(a - 1)$.
(a) $\frac{2}{a}$ (b) $2a$ (c) $\frac{a}{2}$ (d) $-\frac{1}{a}$

29. Simplify $\frac{6a^{-2}bc^{-3}}{4ab^{-3}c^2} \div \frac{5a^{-3}b^2c^{-1}}{3ab^{-2}c^3}$.
(a) $\frac{9}{10}ac$ (b) $\frac{9}{10}ac^{-1}$
(c) $\frac{9}{10}ac^2$ (d) $\frac{9}{10}ac^{-3}$
(e) None of these

30. If $2^{x-1} + 2^{x+1} = 320$, then find the value of x.
(a) 6   (b) 8   (c) 7   (d) 5

31. If $3^x - 3^{x-1} = 18$, then x^x is equal to
(a) 3   (b) 8
(c) 27   (d) 216

32. Consider the following in respect of the numbers $\sqrt{2}, \sqrt[3]{3}$ and $\sqrt[6]{6}$.
I. $\sqrt[6]{6}$ is the greatest number.
II. $\sqrt{2}$ is the smallest number.
Which of the above statement(s) is/are correct? [CDS 2014]
(a) Only I   (b) Only II
(c) Both I and II   (d) Neither I nor II

33. Which one is greatest out of $\sqrt{2}, \sqrt[6]{3}, \sqrt[3]{4}$ and $\sqrt[4]{5}$?
(a) $\sqrt[3]{4}$ (b) $\sqrt[4]{5}$
(c) $\sqrt{2}$ (d) $\sqrt[6]{3}$
(e) None of these

34. The greatest number among $3^{50}, 4^{40}, 5^{30}$ and 6^{20} is [SSC CGL (Mains) 2015]
(a) 6^{20} (b) 5^{30}
(c) 3^{50} (d) 4^{40}

35. If $a = \frac{\sqrt{3} - \sqrt{2}}{\sqrt{3} + \sqrt{2}}$, then $a + \frac{1}{a}$ is equal to [Delhi Police SI 2009]
(a) 4   (b) 6   (c) 9   (d) 10

36. If $\frac{\sqrt{7} - \sqrt{5}}{\sqrt{7} + \sqrt{5}} = a + b\sqrt{35}$, then the value of $(a - b)$ is
(a) 5   (b) 6
(c) 8   (d) None of these

37. If $a = 2 + \sqrt{3}$, then what is the value of $(a^2 + a^{-2})$? [CDS 2012]
(a) 12   (b) 14   (c) 16   (d) 18

38. If $x = 3 + 2\sqrt{2}$, then what will be the value of $x^2 + \frac{1}{x^2}$? [SNAP 2016]
(a) 35   (b) 32   (c) 36   (d) 34

Exercise 2 Higher Skill Level Questions

1. If $2x^{1/3} + 2x^{-1/3} = 5$, then $x^{1/3}$ is equal to [Bank Clerk 2008]
(a) 1 or -1 (b) 2 or $\frac{1}{2}$
(c) 8 or $\frac{1}{8}$ (d) 3 or $\frac{1}{3}$
(e) None of these

2. If $\left(\frac{p^{-1}q^2}{p^3q^{-2}}\right)^{1/3} \div \left(\frac{p^5q^{-3}}{p^{-2}q^3}\right)^{1/3} = p^a q^b$, then the value of $a + b$, where p and q are different positive primes, is [SSC CGL (Mains) 2015]
(a) 0   (b) 1
(c) 2   (d) None of these

Indices and Surds / 107

3. If $\dfrac{2^{n+4} - 2 \cdot 2^n}{2 \cdot 2^{n+3}} + 2^{-3} = x$, then the value of x is
   (a) $-2^{n+1} + \dfrac{1}{8}$ (b) 1
   (c) 2^{n+1} (d) $\dfrac{n}{8} - 2^n$

4. The sum of $\dfrac{1}{\sqrt{2}+1} + \dfrac{1}{\sqrt{3}+\sqrt{2}} + \dfrac{1}{\sqrt{4}+\sqrt{3}}$
   $+ \ldots + \dfrac{1}{\sqrt{100}+\sqrt{99}}$ is [SSC CPO 2016]
   (a) 9 (b) 10
   (c) 11 (d) 12

5. $\dfrac{3\sqrt{2}}{\sqrt{3}+\sqrt{6}} - \dfrac{4\sqrt{3}}{\sqrt{6}+\sqrt{2}} + \dfrac{\sqrt{6}}{\sqrt{3}+\sqrt{2}}$ is equal to [MPPSC 2015]
   (a) 4 (b) 0
   (c) $\sqrt{2}$ (d) $3\sqrt{6}$

6. The value of $3 + \dfrac{1}{\sqrt{3}} + \dfrac{1}{3+\sqrt{3}} + \dfrac{1}{\sqrt{3}-3}$ is
   (a) $3 + \sqrt{3}$ (b) 3
   (c) 1 (d) 0

7. If $3^{x+y} = 81$ and $81^{x-y} = 3$, then the value of x is [SSC CGL 2016]
   (a) 42 (b) 15/8
   (c) 17/8 (d) 39

8. Find the value of $m - n$, if
   $\dfrac{9^n \times 3^2 \times \left(3^{\tfrac{-n}{2}}\right)^{-2} - (27)^n}{3^{3m} \times 2^3} = \dfrac{1}{27}$.
   (a) 1 (b) -2
   (c) -1 (d) 2

9. If $a = \dfrac{\sqrt{3}}{2}$, then $\sqrt{1+a} + \sqrt{1-a} = ?$ [SSC CGL 2007]
   (a) $(2-\sqrt{3})$ (b) $(2+\sqrt{3})$
   (c) $\dfrac{\sqrt{3}}{2}$ (d) $\sqrt{3}$

10. If $m = 7 - 4\sqrt{3}$, then $\left(\sqrt{m} + \dfrac{1}{\sqrt{m}}\right) = ?$
    (a) 8 (b) 3
    (c) 4 (d) 9
    (e) None of these

11. If $2^p + 3^q = 17$ and $2^{p+2} - 3^{q+1} = 5$, then find the values of p and q.
    (a) $-2, 3$ (b) $2, -3$ (c) $3, 2$ (d) $2, 3$

12. Find the value of $\left(\dfrac{a^p}{a^q}\right)^{p+q-r} \times \left(\dfrac{a^r}{a^p}\right)^{r+p-q} \times \left(\dfrac{a^q}{a^r}\right)^{q+r-p}$.
    (a) a^{pqr} (b) a^{p+q+r}
    (c) $a^{pq+qr+pr}$ (d) 1

13. The values of x which satisfy the equation $5^{1+x} + 5^{1-x} = 26$ are [CDS 2017 (I)]
    (a) $-1, 1$ (b) $0, 1$
    (c) $1, 2$ (d) $-1, 0$

14. If $x = \dfrac{\sqrt{13}+\sqrt{11}}{\sqrt{13}-\sqrt{11}}$ and $y = \dfrac{1}{x}$, then the value of $3x^2 - 5xy + 3y^2$ is [SSC CGL (Mains) 2015]
    (a) 1771 (b) 1177
    (c) 1717 (d) 1171

15. If $x = \dfrac{\sqrt{a+2b}+\sqrt{a-2b}}{\sqrt{a+2b}-\sqrt{a-2b}}$, then $bx^2 - ax + b$ is equal to (given that, $b \neq 0$) [CDS 2016 (I)]
    (a) 0 (b) 1
    (c) ab (d) $2ab$

16. $\dfrac{6^2 + 7^2 + 8^2 + 9^2 + 10^2}{\sqrt{7+4\sqrt{3}} - \sqrt{4+2\sqrt{3}}}$ is equal to [CDS 2016 (II)]
    (a) 366 (b) 355
    (c) 305 (d) 330

17. If $x^2 = y + z$, $y^2 = z + x$ and $z^2 = x + y$, then what is the value of $\dfrac{1}{x+1} + \dfrac{1}{y+1} + \dfrac{1}{z+1}$?
    (a) -1 (b) 1 (c) 2 (d) 4 [CDS 2016 (II)]

18. If $2^a = 3^b = 6^{-c}$, then $\dfrac{1}{a} + \dfrac{1}{b} + \dfrac{1}{c} = ?$
    (a) $\dfrac{7}{32}$ (b) 0 (c) $\dfrac{7}{16}$ (d) $\dfrac{7}{48}$

19. If $9^x 3^y = 2187$ and $2^{3x} 2^{2y} - 4^{xy} = 0$, then what can be the value of $(x + y)$? [CDS 2017 (I)]
    (a) 1 (b) 3 (c) 5 (d) 7

20. If $x = t^{\tfrac{1}{t-1}}$ and $y = t^{\tfrac{t}{t-1}}$, $t > 0, t \neq 1$, then what is the relation between x and y? [CDS 2017 (I)]
    (a) $y^x = x^{1/y}$ (b) $x^{1/y} = y^{1/x}$
    (c) $x^y = y^x$ (d) $x^y = y^{1/x}$

Answer with Solutions

Exercise 1 Base Level Questions

1. (d) $? = \dfrac{[(12)^{-2}]^2}{[(12)^2]^{-2}} = \dfrac{(12)^{-4}}{(12)^{-4}} = 1$

2. (a) $(10)^{200} \div (10)^{196} = (10)^{200-196}$
$= 10^4 = 10000$

3. (d) $(0.00032)^{2/5} = \left(\dfrac{32}{100000}\right)^{2/5}$
$= \left(\dfrac{2^5}{10^5}\right)^{2/5} = \left\{\left(\dfrac{2}{10}\right)^5\right\}^{2/5}$
$= \left(\dfrac{2}{10}\right)^{5 \times \frac{2}{5}} = \left(\dfrac{1}{5}\right)^2 = \dfrac{1}{25}$

4. (a) $\left[\left\{\left(-\dfrac{1}{2}\right)^2\right\}^{-2}\right]^{-1} = \left[\left\{\dfrac{1}{4}\right\}^{-2}\right]^{-1} = \left[\dfrac{1}{(1/4)^2}\right]^{-1}$
$= \left[\dfrac{1}{1/16}\right]^{-1} = (16)^{-1} = \dfrac{1}{16}$

5. (c) Given that, $289 = 17^{x/5}$
$\Rightarrow \quad 17^2 = 17^{x/5}$
On comparing, we get
$\dfrac{x}{5} = 2 \Rightarrow x = 10$

6. (b) Given, $5^a = 3125$
$\Rightarrow \quad 5^a = 5^5$
On comparing, we get
$a = 5$
$\therefore \quad 5^{a-3} = 5^{5-3} = 5^2 = 25$

7. (b) Given that, $a^{2x+2} = 1 \Rightarrow a^{2x+2} = a^0$
$\Rightarrow \quad 2x + 2 = 0$
$\Rightarrow \quad x = \dfrac{-2}{2} = -1$

8. (c) Given that, $(2^{4 \times \frac{1}{2}})^? = 256 \Rightarrow (2^2)^? = 2^8$
$\Rightarrow \quad 2 \times ? = 8 \Rightarrow ? = \dfrac{8}{2} = 4$

9. (b) Given that, $(16)^9 \div (16)^4 \times (16)^3 = (16)^?$
$\Rightarrow \quad (16)^? = \dfrac{(16)^9 \times (16)^3}{(16)^4}$
$\Rightarrow \quad (16)^? = (16)^{9+3-4}$
$\Rightarrow \quad ? = 12 - 4 = 8$

10. (b) $112 \times 5^4 = (100 + 10 + 2) \times 625$
$= 62500 + 6250 + 1250$
$= 70000$

11. (a) $\because \quad \dfrac{x^{0.4}}{16} = \dfrac{32}{x^{2.6}}$
$\Rightarrow x^{0.4} \times x^{2.6} = 16 \times 32$
$\Rightarrow \quad x^3 = 512 \Rightarrow x^3 = (8)^3$
On comparing the power of both sides, we get $x = 8$

12. (c) $17^{3.5} \times 17^{7.3} \div 17^{4.2} = 17^?$
$\Rightarrow \quad 17^{3.5+7.3-4.2} = 17^?$
$\Rightarrow \quad 17^{6.6} = 17^?$
$\therefore \quad ? = 6.6$

13. (c) $\left(\dfrac{1}{5}\right)^{3a} = 0.008 = \dfrac{8}{1000} = \dfrac{1}{125} = \left(\dfrac{1}{5}\right)^3$
$\Rightarrow \quad 3a = 3 \Rightarrow a = 1$
$\therefore \quad (0.25)^a = (0.25)^1 = 0.25$

14. (d) $\left(\dfrac{p}{q}\right)^{n-1} = \left(\dfrac{q}{p}\right)^{n-3}$
$\Rightarrow \quad \left(\dfrac{p}{q}\right)^{n-1} = \left(\dfrac{p}{q}\right)^{3-n}$
$\Rightarrow \quad n - 1 = 3 - n$
$\Rightarrow \quad 2n = 4 \Rightarrow n = 2$

15. (c) $? = [p^{(b-c)}]^{b+c} \cdot [p^{(c-a)}]^{c+a} \cdot [p^{(a-b)}]^{(a+b)}$
$= p^{b^2-c^2} \cdot p^{c^2-a^2} \cdot p^{a^2-b^2}$
$[\because (a-b)(a+b) = a^2 - b^2]$
$= p^{b^2-c^2+c^2-a^2+a^2-b^2} = p^0 = 1$

16. (a) Given, $xyz = 1$, $a^x = b$, $b^y = c$
Now, $\quad b = a^x$
$\Rightarrow \quad b^y = a^{xy}$
$\Rightarrow \quad b^{yz} = a^{xyz} \Rightarrow c^z = a$

17. (d) Given that, $16 \times 8^{n+2} = 2^m$
$\Rightarrow \quad (2)^4 \times 2^{3(n+2)} = 2^m$
$\Rightarrow \quad (2)^{(4+3n+6)} = 2^m$
$\Rightarrow \quad 2^{(3n+10)} = 2^m$
On comparing, we get
$m = 3n + 10$

Indices and Surds / 109

18. (a) $\sqrt[L]{M} = M^{1/L}$
$\Rightarrow$ Surd of Lth order.

19. (c) $\left[\left(\sqrt[5]{x^{-3/5}}\right)^{-\frac{5}{3}}\right]^5 = \left[\left(x^{-\frac{3}{5} \times \frac{1}{5}}\right)^{-\frac{5}{3}}\right]^5$
$= x^{-\frac{3}{5} \times \frac{1}{5} \times \frac{-5}{3} \times 5} = x$

20. (d) $\sqrt[3]{P(P^2 + 3P + 3) + 1}$
$= \sqrt[3]{P^3 + 3P^2 + 3P + 1}$
$= \sqrt[3]{(P+1)^3} = P + 1$
$[\because (a+b)^3 = a^3 + b^3 + 3ab(a+b)]$
$\because P = 124 \Rightarrow P + 1 = 125$

21. (d) Given expression $= (\sqrt{2})^{\sqrt{2}^{\sqrt{2}}}$
$= (\sqrt{2})^{(2)^{\frac{\sqrt{2}}{2}}} = (\sqrt{2})^{(2)^{\frac{1}{\sqrt{2}}}}$
$= (2)^{\frac{1}{2} \times 2^{\frac{1}{\sqrt{2}}}} = 2^{\left(\frac{2^{\frac{1}{\sqrt{2}}}}{2}\right)} = (2)^{(2)^{\left(\frac{1}{\sqrt{2}} - 1\right)}}$
which denotes a real number but not a rational number.

22. (b) If m and n are natural numbers, then $\sqrt[m]{n}$ is irrational unless n is mth power of an integer.

23. (b) Given, $\sqrt{10 + \sqrt[3]{x}} = 4$
On squaring both sides, we get
$10 + \sqrt[3]{x} = 16 \Rightarrow \sqrt[3]{x} = 6$
Now, cubing both sides, we get
$x = (6)^3 = 216$

24. (a) $81^{2.5} \times 9^{4.5} \div 3^{4.8} = 9^?$
$\Rightarrow \dfrac{81^{2.5} \times 9^{4.5}}{3^{4.8}} = 9^? \Rightarrow \dfrac{(3^4)^{2.5} \times (3^2)^{4.5}}{3^{4.8}} = 9^?$
$\Rightarrow \dfrac{3^{10} \times 3^9}{3^{4.8}} = 9^? \Rightarrow 3^{(10+9-4.8)} = 9^?$
$\Rightarrow 9^? = 3^{14.2} \Rightarrow (3)^{2 \times ?} = 3^{14.2}$
$\Rightarrow 2 \times ? = 14.2 \Rightarrow ? = \dfrac{14.2}{2} = 7.1$

25. (a) $7^? = 7^{8.9} \div (343)^{1.7} \times (49)^{4.8}$
$= \dfrac{7^{8.9} \times (7^2)^{4.8}}{(7^3)^{1.7}}$
$= \dfrac{7^{8.9+9.6}}{7^{5.1}} = \dfrac{7^{18.5}}{7^{5.1}}$
$= 7^{18.5 - 5.1} = 7^{13.4}$
$\therefore ? = 13.4$

26. (d) Given that, $(42 \times 229) \div (9261)^{1/3} = ?$
$\therefore ? = \dfrac{42 \times 229}{(21^3)^{1/3}} = \dfrac{42 \times 229}{21} = 458$
$[\because (9261)^{1/3} = (21 \times 21 \times 21)^{1/3} = 21^{3 \times 1/3} = 21]$

27. (c) $2^x + 2^{x+1} = 48 \Rightarrow 2^x + 2^x \times 2 = 48$
$\Rightarrow 2^x(1 + 2) = 48 \Rightarrow 2^x \times 3 = 48$
$\Rightarrow 2^x = 16 \Rightarrow 2^x = 2^4$
On comparing the power of both sides, we get
$x = 4$
$\therefore x^x = 4^4 = 256$

28. (d) $\dfrac{a^{-1} - 1}{a - 1} = \dfrac{\frac{1}{a} - 1}{a - 1} = \dfrac{(1-a)}{a} \times \dfrac{1}{(a-1)} = -\dfrac{1}{a}$
$\therefore$ Required quotient $= -\dfrac{1}{a}$

29. (b) $\dfrac{6a^{-2}bc^{-3}}{4ab^{-3}c^2} \div \dfrac{5a^{-3}b^2c^{-1}}{3ab^{-2}c^3}$
$= \dfrac{6a^{-2}bc^{-3}}{4ab^{-3}c^2} \times \dfrac{3ab^{-2}c^3}{5a^{-3}b^2c^{-1}}$
$= \dfrac{18a^{-2+1}b^{1-2}c^{-3+3}}{20a^{1-3}b^{-3+2}c^{2-1}}$
$= \dfrac{9}{10} \cdot \dfrac{a^{-1}b^{-1}c^0}{a^{-2}b^{-1}c^1} = \dfrac{9}{10} a^{-1+2} b^{-1+1} c^{0-1}$
$= \dfrac{9}{10} a^1 b^0 c^{-1} = \dfrac{9}{10} ac^{-1}$ $[\because b^0 = 1]$

30. (c) Given, $2^{x-1} + 2^{x+1} = 320$
$\Rightarrow 2^x(2^{-1} + 2) = 320$
$\Rightarrow 2^x\left(\dfrac{1}{2} + 2\right) = 320 \Rightarrow 2^x \times \dfrac{5}{2} = 320$
$\Rightarrow 2^x = 64 \times 2 = 2^7 \Rightarrow x = 7$

31. (c) $\because 3^x - 3^{x-1} = 18$
$\Rightarrow 3^{x-1}(3 - 1) = 18$
$\Rightarrow 3^{x-1} = 9 = 3^2$
$\Rightarrow x - 1 = 2 \Rightarrow x = 3$
$\therefore x^x = (3)^3 = 27$

32. (d) $\sqrt{2}, \sqrt[3]{3}, \sqrt[6]{6}$; LCM of 2, 3 and 6 = 12
Now, $\sqrt{2} = (2)^{1/2} = (2)^{6/12} = \sqrt[12]{2^6} = \sqrt[12]{64}$
$\sqrt[3]{3} = (3)^{1/3} = (3)^{4/12} = \sqrt[12]{3^4} = \sqrt[12]{81}$
$\sqrt[6]{6} = (6)^{1/6} = (6)^{2/12} = \sqrt[12]{6^2} = \sqrt[12]{36}$
$\therefore \sqrt[12]{36} < \sqrt[12]{64} < \sqrt[12]{81}$
So, neither I nor II are correct.

33. (a) Given surds are in the form $2^{1/2}, 3^{1/6}, 4^{1/3}$ and $5^{1/4}$.

LCM of (2, 6, 3, 4) = 12
$2^{1/2} = (2^6)^{1/12} = (64)^{1/12}$
$3^{1/6} = (3^2)^{1/12} = (9)^{1/12}$
$4^{1/3} = (4^4)^{1/12} = (256)^{1/12}$
$5^{1/4} = (5^3)^{1/12} = (125)^{1/12}$
$\therefore \quad 64 < 9^{1/12} < 125 < 256^{1/12}$
Clearly, greatest surd = $(256)^{1/12} = \sqrt[3]{4}$

34. (d) To compare the given numbers, we can write them as
$3^{50} = (3^5)^{10} = (243)^{10}$
$4^{40} = (4^4)^{10} = (256)^{10}$
$5^{30} = (5^3)^{10} = (125)^{10}$
$6^{20} = (6^2)^{10} = (36)^{10}$
So, the greatest number is $(4)^{40}$.

35. (d) $a + \dfrac{1}{a} = \dfrac{\sqrt{3}-\sqrt{2}}{\sqrt{3}+\sqrt{2}} + \dfrac{1}{\dfrac{\sqrt{3}-\sqrt{2}}{\sqrt{3}+\sqrt{2}}}$

$= \dfrac{\sqrt{3}-\sqrt{2}}{\sqrt{3}+\sqrt{2}} + \dfrac{\sqrt{3}+\sqrt{2}}{\sqrt{3}-\sqrt{2}}$

$= \dfrac{(\sqrt{3}-\sqrt{2})^2 + (\sqrt{3}+\sqrt{2})^2}{(\sqrt{3}+\sqrt{2})(\sqrt{3}-\sqrt{2})}$

$= \dfrac{\begin{bmatrix}(\sqrt{3})^2 + (\sqrt{2})^2 - 2\sqrt{3}\times\sqrt{2} + (\sqrt{3})^2 \\ + (\sqrt{2})^2 + 2\sqrt{3}\times\sqrt{2}\end{bmatrix}}{(\sqrt{3})^2 - (\sqrt{2})^2}$

$\begin{bmatrix}\because (a+b)^2 = a^2 + b^2 + 2ab, \\ (a-b)^2 = a^2 + b^2 - 2ab \\ \text{and } a^2 - b^2 = (a+b)(a-b)\end{bmatrix}$

$= \dfrac{3+2+3+2}{3-2} = 10$

36. (d) $\dfrac{\sqrt{7}-\sqrt{5}}{\sqrt{7}+\sqrt{5}} = a + b\sqrt{35}$

$\therefore \dfrac{\sqrt{7}-\sqrt{5}}{\sqrt{7}+\sqrt{5}} = \dfrac{(\sqrt{7}-\sqrt{5})\times(\sqrt{7}-\sqrt{5})}{(\sqrt{7}+\sqrt{5})\times(\sqrt{7}-\sqrt{5})}$

$= \dfrac{(\sqrt{7})^2 + (\sqrt{5})^2 - 2\sqrt{7\times 5}}{(\sqrt{7})^2 - (\sqrt{5})^2}$

$[\because (a-b)^2 = a^2 + b^2 - 2ab]$

$= \dfrac{7+5-2\sqrt{35}}{7-5} = \dfrac{12 - 2\sqrt{35}}{2} = 6 - \sqrt{35}$

On comparing with $a + b\sqrt{35}$, we get
$a = 6$ and $b = -1$
$\therefore \quad a - b = 6 - (-1) = 7$

37. (b) Given that, $a = 2 + \sqrt{3}$
Then, $\dfrac{1}{a} = \dfrac{1}{2+\sqrt{3}}$

$= \dfrac{(2-\sqrt{3})}{(2+\sqrt{3})(2-\sqrt{3})} = 2 - \sqrt{3}$

Now, $a^2 + a^{-2} = \left(a + \dfrac{1}{a}\right)^2 - 2$

$= (2 + \sqrt{3} + 2 - \sqrt{3})^2 - 2$
$= (4)^2 - 2 = 16 - 2 = 14$

38. (d) Given, $x = 3 + 2\sqrt{2}$
$\therefore \quad x^2 = (3 + 2\sqrt{2})^2 = 9 + 8 + 12\sqrt{2}$
$= 17 + 12\sqrt{2}$

$\therefore \quad \dfrac{1}{x^2} = \dfrac{1}{17 + 12\sqrt{2}}$

$= \dfrac{1}{17 + 12\sqrt{2}} \times \dfrac{17 - 12\sqrt{2}}{17 - 12\sqrt{2}}$

$= \dfrac{17 - 12\sqrt{2}}{289 - 288} = 17 - 12\sqrt{2}$

$\therefore \quad x^2 + 1/x^2 = 17 + 12\sqrt{2} + 17 - 12\sqrt{2}$
$= 34$

Exercise 2 *Higher Skill Level Questions*

1. (b) Given that, $2x^{1/3} + 2x^{-1/3} = 5$
Let $x^{1/3} = m$, then $2m + \dfrac{2}{m} = 5$
$\Rightarrow \quad 2m^2 - 5m + 2 = 0$
$\Rightarrow \quad (2m-1)(m-2) = 0$
$\therefore \quad m = \dfrac{1}{2}$ or $m = 2 \Rightarrow x^{1/3} = 2$ or $\dfrac{1}{2}$

2. (d) $\left(\dfrac{p^{-1}q^2}{p^3q^{-2}}\right)^{\frac{1}{3}} \div \left(\dfrac{p^5q^{-3}}{p^{-2}q^3}\right)^{\frac{1}{3}} = p^a q^b$

$\Rightarrow \quad (p^{-4}q^4)^{1/3} \div (p^7 q^{-6})^{1/3} = p^a q^b$

$\Rightarrow \quad p^{\frac{-4}{3}} q^{\frac{4}{3}} \div p^{\frac{7}{3}} q^{\frac{-6}{3}} = p^a q^b$

$\Rightarrow \quad p^{\left(-\frac{4}{3}-\frac{7}{3}\right)} q^{\left(\frac{4}{3}+\frac{6}{3}\right)} = p^a q^b$

$\Rightarrow \quad p^{\frac{-11}{3}} q^{\frac{10}{3}} = p^a q^b$

$\therefore \quad a = \dfrac{-11}{3}$ and $b = \dfrac{10}{3}$

Now, $a + b = \dfrac{-11}{3} + \dfrac{10}{3} = \dfrac{-1}{3}$

Indices and Surds / 111

3. (b) $x = \dfrac{2^{n+4} - 2 \cdot 2^n}{2 \cdot 2^{n+3}} + 2^{-3}$

$= \dfrac{2^{n+4} - 2^{n+1}}{2^{n+4}} + 2^{-3} = \dfrac{2^{n+1}(2^3 - 1)}{2^{n+1} \cdot 2^3} + \dfrac{1}{2^3}$

$= \dfrac{8-1}{2^3} + \dfrac{1}{2^3} = \dfrac{7}{8} + \dfrac{1}{8} = 1$

4. (a) $\dfrac{1}{\sqrt{2}+1} + \dfrac{1}{\sqrt{3}+\sqrt{2}} + \dfrac{1}{\sqrt{4}+\sqrt{3}}$

$+ \ldots + \dfrac{1}{\sqrt{100}+\sqrt{99}}$

$= \dfrac{\sqrt{2}-1}{2-1} + \dfrac{\sqrt{3}-\sqrt{2}}{3-2} + \dfrac{\sqrt{4}-\sqrt{3}}{4-3}$

$+ \ldots + \dfrac{\sqrt{100}-\sqrt{99}}{100-99}$

[by rationalisation]

$= \sqrt{2} - 1 + \sqrt{3} - \sqrt{2} + \sqrt{4} - \sqrt{3}$

$+ \ldots + \sqrt{100} - \sqrt{99}$

$= \sqrt{100} - 1 = 10 - 1 = 9$

5. (b) $\dfrac{3\sqrt{2}}{\sqrt{3}+\sqrt{6}} - \dfrac{4\sqrt{3}}{\sqrt{6}+\sqrt{2}} + \dfrac{\sqrt{6}}{\sqrt{3}+\sqrt{2}}$

$= \dfrac{3\sqrt{2}(\sqrt{6}-\sqrt{3})}{(\sqrt{6}+\sqrt{3})(\sqrt{6}-\sqrt{3})}$

$- \dfrac{4\sqrt{3}(\sqrt{6}-\sqrt{2})}{(\sqrt{6}+\sqrt{2})(\sqrt{6}-\sqrt{2})}$

$+ \dfrac{\sqrt{6}(\sqrt{3}-\sqrt{2})}{(\sqrt{3}+\sqrt{2})(\sqrt{3}-\sqrt{2})}$ [by rationalisation]

$= \dfrac{3\sqrt{2}(\sqrt{6}-\sqrt{3})}{(\sqrt{6})^2 - (\sqrt{3})^2} - \dfrac{4\sqrt{3}(\sqrt{6}-\sqrt{2})}{(\sqrt{6})^2 - (\sqrt{2})^2}$

$+ \dfrac{\sqrt{6}(\sqrt{3}-\sqrt{2})}{(\sqrt{3})^2 - (\sqrt{2})^2}$

$[\because (a+b)(a-b) = a^2 - b^2]$

$= \dfrac{(3\sqrt{12} - 3\sqrt{6})}{3} - \dfrac{(4\sqrt{18} - 4\sqrt{6})}{4}$

$+ \dfrac{\sqrt{18} - \sqrt{12}}{1}$

$= (\sqrt{12} - \sqrt{6}) - (\sqrt{18} - \sqrt{6}) + (\sqrt{18} - \sqrt{12})$

$= \sqrt{12} - \sqrt{6} - \sqrt{18} + \sqrt{6} + \sqrt{18} - \sqrt{12} = 0$

6. (b) $3 + \dfrac{1}{\sqrt{3}} + \dfrac{1}{3+\sqrt{3}} + \dfrac{1}{\sqrt{3}-3}$

$= 3 + \dfrac{1}{\sqrt{3}} + \dfrac{(3-\sqrt{3})}{(3+\sqrt{3})(3-\sqrt{3})}$

$+ \dfrac{(\sqrt{3}+3)}{(\sqrt{3}-3)(\sqrt{3}+3)}$

$= 3 + \dfrac{1}{\sqrt{3}} + \dfrac{3-\sqrt{3}}{9-3} + \dfrac{\sqrt{3}+3}{3-9}$

$= 3 + \dfrac{1}{\sqrt{3}} + \dfrac{3-\sqrt{3}}{6} - \dfrac{\sqrt{3}+3}{6}$

$= 3 + \dfrac{1}{\sqrt{3}} + \dfrac{3 - \sqrt{3} - \sqrt{3} - 3}{6}$

$= 3 + \dfrac{1}{\sqrt{3}} - \dfrac{1}{\sqrt{3}} = 3$

7. (c) Given, $3^{x+y} = 81$

$\Rightarrow \quad 3^{x+y} = 3^4$

$\Rightarrow \quad x + y = 4$ ...(i)

and $\quad 81^{x-y} = 3$

$\Rightarrow \quad (3^4)^{x-y} = 3 \Rightarrow 3^{4x-4y} = 3^1$

$\Rightarrow \quad 4x - 4y = 1$ ...(ii)

On multiplying by 4 in Eq. (i) and then adding Eqs. (i) and (ii), we get

$4x + 4y = 16$
$4x - 4y = 1$
$\overline{\quad 8x = 17 \quad}$

$\Rightarrow \quad x = \dfrac{17}{8}$

8. (a) $\dfrac{9^n \times 3^2 \times (3^{-n/2})^{-2} - (27)^n}{3^{3m} \times 2^3} = \dfrac{1}{27}$

$\Rightarrow \dfrac{3^{2n} \times 3^2 \times 3^n - 3^{3n}}{3^{3m} \times 2^3} = \dfrac{1}{27}$

$\Rightarrow \dfrac{3^{3n+2} - 3^{3n}}{3^{3m} \times 8} = \dfrac{1}{27} \Rightarrow \dfrac{3^{3n}(3^2 - 1)}{3^{3m} \times 8} = \dfrac{1}{27}$

$\Rightarrow \dfrac{3^{3n} \times 8}{3^{3m} \times 8} = \dfrac{1}{27} \Rightarrow (3^3)^{n-m} = 3^{-3}$

$\Rightarrow \quad n - m = -1 \text{ or } m - n = 1$

9. (d) $(\sqrt{1+a} + \sqrt{1-a})^2$

$= (1+a) + (1-a) + 2\sqrt{1-a^2}$

$= 2(1 + \sqrt{1-a^2}) = 2\left(1 + \sqrt{1 - \dfrac{3}{4}}\right)$

$= 2\left(1 + \dfrac{1}{2}\right) = 2 \times \dfrac{3}{2} = 3$

$\therefore \quad (\sqrt{1+a} + \sqrt{1-a}) = \sqrt{3}$

10. (c) Given, $m = 7 - 4\sqrt{3}$

$\therefore \quad \dfrac{1}{m} = \dfrac{1}{7-4\sqrt{3}} \times \dfrac{7+4\sqrt{3}}{7+4\sqrt{3}} = \dfrac{7+4\sqrt{3}}{49-48}$

$= 7 + 4\sqrt{3}$

Now, $\quad m + \dfrac{1}{m} = 7 - 4\sqrt{3} + 7 + 4\sqrt{3} = 14$

$\Rightarrow \quad m + \dfrac{1}{m} + 2 = 14 + 2 = 16$

Now, $\left(\sqrt{m} + \dfrac{1}{\sqrt{m}}\right)^2 = m + \dfrac{1}{m} + 2$

$\Rightarrow \left(\sqrt{m} + \dfrac{1}{\sqrt{m}}\right)^2 = 4^2 \Rightarrow \left(\sqrt{m} + \dfrac{1}{\sqrt{m}}\right) = 4$

11. (c) Given, $2^p + 3^q = 17$ ...(i)
and $2^{p+2} - 3^{q+1} = 5$ or $4 \cdot 2^p - 3 \cdot 3^q = 5$ ...(ii)
On multiplying Eq. (i) by 3 and adding it with Eq. (ii), we get

$$3 \cdot 2^p + 3 \cdot 3^q = 51$$
$$\underline{4 \cdot 2^p - 3 \cdot 3^q = 5}$$
$$7 \cdot 2^p = 56$$

$\Rightarrow 2^p = 8 = 2^3 \Rightarrow p = 3$
On putting the value of p in Eq. (i), we get
$2^3 + 3^q = 17 \Rightarrow 3^q = 9 \Rightarrow q = 2$
$\therefore \quad p = 3$ and $q = 2$

12. (d) $\left(\dfrac{a^p}{a^q}\right)^{(p+q-r)} \times \left(\dfrac{a^r}{a^p}\right)^{(r+p-q)}$
$\qquad \times \left(\dfrac{a^q}{a^r}\right)^{(q+r-p)}$
$= (a^{p-q})^{p+q-r} \times (a^{r-p})^{r+p-q}$
$\qquad \times (a^{q-r})^{q+r-p}$
$= a^{p^2 - q^2 - rp + rq} \times a^{r^2 - p^2 - qr + pq}$
$\qquad \times a^{q^2 - r^2 - pq + pr}$
$= a^{p^2 - q^2 - rp + rq + r^2 - p^2 - qr + pq}$
$\qquad {}^{+ q^2 - r^2 - pq + pr} = a^0 = 1$

13. (a) We have, $5^{1+x} + 5^{1-x} = 26$
$\Rightarrow \quad 5 \cdot 5^x + 5 \cdot 5^{-x} = 26$
$\Rightarrow \quad 5 \cdot 5^x + \dfrac{5}{5^x} = 26$
Let $\quad 5^x = y$
$\therefore \quad 5y + \dfrac{5}{y} = 26$
$\Rightarrow \quad 5y^2 - 26y + 5 = 0$
$\Rightarrow \quad 5y^2 - 25y - y + 5 = 0$
$\Rightarrow \quad 5y(y-5) - 1(y-5) = 0$
$\Rightarrow \quad (y-5)(5y-1) = 0$
$\Rightarrow \quad y = 5, \dfrac{1}{5}$
$\Rightarrow \quad 5^x = 5$ or 5^{-1}
$\Rightarrow \quad x = 1$ or -1

14. (c) Given, $x = \dfrac{\sqrt{13} + \sqrt{11}}{\sqrt{13} - \sqrt{11}}$
and $y = \dfrac{\sqrt{13} - \sqrt{11}}{\sqrt{13} + \sqrt{11}}$
$\therefore x + y = \dfrac{\sqrt{13} + \sqrt{11}}{\sqrt{13} - \sqrt{11}} + \dfrac{\sqrt{13} - \sqrt{11}}{\sqrt{13} + \sqrt{11}}$
$= \dfrac{(\sqrt{13} + \sqrt{11})^2 + (\sqrt{13} - \sqrt{11})^2}{(\sqrt{13})^2 - (\sqrt{11})^2}$
$= \dfrac{2[(\sqrt{13})^2 + (\sqrt{11})^2]}{13 - 11} = 13 + 11 = 24$
$[\because (a+b)^2 + (a-b)^2 = 2(a^2 + b^2)]$
and $xy = \dfrac{\sqrt{13} + \sqrt{11}}{\sqrt{13} - \sqrt{11}} \times \dfrac{\sqrt{13} - \sqrt{11}}{\sqrt{13} + \sqrt{11}} = 1$
$\therefore 3x^2 - 5xy + 3y^2 = 3(x+y)^2 - 11xy$
$= 3(24)^2 - 11 = 1717$

15. (a) Given, $x = \dfrac{\sqrt{a+2b} + \sqrt{a-2b}}{\sqrt{a+2b} - \sqrt{a-2b}}$
By rationalising, we get
$x = \dfrac{\left[\begin{array}{c}(\sqrt{a+2b})^2 + (\sqrt{a-2b})^2 \\ + 2\sqrt{(a+2b)(a-2b)}\end{array}\right]}{(\sqrt{a+2b})^2 - (\sqrt{a-2b})^2}$
$\Rightarrow x = \dfrac{a + 2b + a - 2b + 2\sqrt{a^2 - 4b^2}}{a + 2b - a + 2b}$
$\Rightarrow x = \dfrac{2a + 2\sqrt{a^2 - 4b^2}}{4b}$
$\Rightarrow x = \dfrac{a + \sqrt{a^2 - 4b^2}}{2b}$
$\therefore \quad bx^2 - ax + b$
$= b\left(\dfrac{a + \sqrt{a^2 - 4b^2}}{2b}\right)^2 - a\left(\dfrac{a + \sqrt{a^2 - 4b^2}}{2b}\right) + b$
$= b\left(\dfrac{a^2 + a^2 - 4b^2 + 2a\sqrt{a^2 - 4b^2}}{4b^2}\right)$
$\qquad - \left(\dfrac{a^2 + a\sqrt{a^2 - 4b^2}}{2b}\right) + b$
$= \dfrac{\left[\begin{array}{c}2a^2 - 4b^2 + 2a\sqrt{a^2 - 4b^2} \\ - 2a^2 - 2a\sqrt{a^2 - 4b^2} + 4b^2\end{array}\right]}{4b}$
$= 0$

16. (d) $\dfrac{6^2 + 7^2 + 8^2 + 9^2 + 10^2}{\sqrt{7 + 4\sqrt{3}} - \sqrt{4 + 2\sqrt{3}}}$
$= \dfrac{36 + 49 + 64 + 81 + 100}{\left[\begin{array}{c}\sqrt{2^2 + (\sqrt{3})^2 + 2 \cdot 2 \cdot \sqrt{3}} \\ - \sqrt{(\sqrt{3})^2 + 1^2 + 2 \cdot 1 \cdot \sqrt{3}}\end{array}\right]}$
$= \dfrac{330}{\sqrt{(2 + \sqrt{3})^2} - \sqrt{(\sqrt{3} + 1)^2}}$
$[\because (a+b)^2 = a^2 + b^2 + 2ab]$
$= \dfrac{330}{2 + \sqrt{3} - \sqrt{3} - 1} = 330$

17. (*b*) Given, $x^2 = y + z \Rightarrow x = \dfrac{y+z}{x}$ ...(i)

$y^2 = z + x \Rightarrow y = \dfrac{z+x}{y}$ ...(ii)

and $z^2 = x + y \Rightarrow z = \dfrac{x+y}{z}$ ...(iii)

Now, $\dfrac{1}{x+1} + \dfrac{1}{y+1} + \dfrac{1}{z+1}$

$= \dfrac{1}{\dfrac{y+z}{x}+1} + \dfrac{1}{\dfrac{z+x}{y}+1} + \dfrac{1}{\dfrac{x+y}{z}+1}$

$= \dfrac{x}{x+y+z} + \dfrac{y}{x+y+z} + \dfrac{z}{x+y+z}$

$= \dfrac{x+y+z}{x+y+z} = 1$

18. (*b*) Let $2^a = 3^b = 6^{-c} = K$

$\therefore 2 = K^{1/a}, 3 = K^{1/b}, 6 = K^{-1/c}$

We know that, $2 \times 3 = 6$

$\therefore K^{1/a} \times K^{1/b} = K^{-1/c}$

$\Rightarrow K^{\left(\frac{1}{a}+\frac{1}{b}\right)} = K^{\frac{-1}{c}}$

On comparing, we get

$\dfrac{1}{a} + \dfrac{1}{b} = \dfrac{-1}{c}$

$\Rightarrow \dfrac{1}{a} + \dfrac{1}{b} + \dfrac{1}{c} = 0$

19. (*c*) We have,

$9^x \cdot 3^y = 2187$

$\Rightarrow (3^2)^x \cdot 3^y = 2187$

$\Rightarrow 3^{2x+y} = 3^7$

$\Rightarrow 2x + y = 7$ ...(i)

Again,

$2^{3x} \cdot 2^{2y} - 4^{xy} = 0$

$\Rightarrow 2^{3x} \cdot 2^{2y} = 4^{xy}$

$\Rightarrow 2^{3x+2y} = (2^2)^{xy}$

$\Rightarrow 3x + 2y = 2xy$ ...(ii)

From Eqs. (i) and (ii)

$3x + 2(7 - 2x) = 2x(7 - 2x)$

$\Rightarrow 3x + 14 - 4x = 14x - 4x^2$

$\Rightarrow 4x^2 - 15x + 14 = 0$

$\Rightarrow (x-2)(4x-7) = 0$

$\Rightarrow x = 2, \dfrac{7}{4}$

$\therefore y = 3, \dfrac{7}{2}$

$\therefore x + y = 5$ or $\dfrac{21}{4}$

20. (*c*) We have,

$x = t^{\frac{1}{t-1}}$ and $y = t^{\frac{t}{t-1}}, t > 0, t \neq 1$

Now, $y = t^{\left(\frac{1}{t-1}\right)t} = x^t$ ...(i)

Again $\dfrac{y}{x} = \dfrac{t^{\frac{t}{t-1}}}{t^{\frac{1}{t-1}}} = t^{\frac{t}{t-1}-\frac{1}{t-1}} = t^{\frac{t-1}{t-1}} = t$

$\Rightarrow \dfrac{y}{x} = t$ ...(ii)

From Eqs. (i) and (ii), we get $y = x^{y/x}$

$\Rightarrow y^x = x^y$

Chapter 7

Simplification

Simplification is a process of reducing a complex arithmetical expression into a simple expression. A simple technique for arranging the expression in the proper sequence is chronology involving 'VBODMAS' rule which is explained below.

VBODMAS Rule

To simplify arithmetic expressions, which involve various operations, like brackets, multiplication, addition, etc., a particular sequence of the operations has to be followed.

The operations have to be carried out in the order, in which they appear in the word **VBODMAS**, where different letters of the word stand for following operations.

$$V = \text{Vinculum or Bar '—'}$$
$$B = \text{Brackets}$$
$$O = \text{Of}$$
$$D = \text{Division}$$
$$M = \text{Multiplication}$$
$$A = \text{Addition}$$
$$S = \text{Subtraction}$$

Order of above mentioned operations is same as the order of letters in the 'VBODMAS' from left to right as

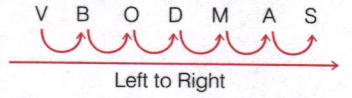

V B O D M A S
Left to Right

Order of removing the brackets

First	Small brackets (Circular brackets) '( )'
Second	Middle brackets (Curly brackets) '{ }'
Third	Square brackets (Big brackets) '[ ]'

+ Order of the letter which is used in this rule is always fixed and absence of any operation or more than one operation does not change the order of the rule.

Simplification / 115

🔴 MIND IT! Absolute Value of a Real Number

If m is a real number, then its absolute value is defined as

$$|m| = \begin{cases} m, & \text{if } m > 0 \\ -m, & \text{if } m < 0 \end{cases}$$

For example $|3| = 3$ and $|-3| = -(-3) = 3$

Ex. 1 Simplify $4 - [6 - \{12 - (10 - \overline{8 - 6})\}]$.

Sol. Given expression $= 4 - [6 - \{12 - (10 - (\overline{8-6}))\}]$
$= 4 - [6 - \{12 - (10 - 2)\}]$ Solve vinculum
$= 4 - [6 - \{12 - 8\}]$ Solve ()
$= 4 - [6 - 4]$ Solve { }
$= 4 - 2 = 2$ Solve [ ]

Ex. 2 Simplify $\left(9.6 \times 3.6 \div 7.2 + 10.8 \text{ of } \dfrac{1}{18} - \dfrac{1}{10}\right)$.

Sol. Given expression $= 9.6 \times 3.6 \div 7.2 + 10.8 \text{ of } \dfrac{1}{18} - \dfrac{1}{10}$

$= 9.6 \times 3.6 \div 7.2 + 0.6 - \dfrac{1}{10}$ Solve 'of'

$= 9.6 \times \dfrac{3.6}{7.2} + 0.6 - 0.1$ Solve '÷'

$= 4.8 + 0.6 - 0.1$ Solve '×'
$= 5.4 - 0.1$ Solve '+'
$= 5.3$ Solve '−'

Ex. 3 Simplify $n - [n - (m + n) - \{n - (n - \overline{m - n})\} + 2m]$.

Sol. Given expression $= n - [n - (m + n) - \{n - (n - \overline{m - n})\} + 2m]$
$= n - [n - (m + n) - \{n - (n - m + n)\} + 2m]$
$= n - [n - (m + n) - \{n - (2n - m)\} + 2m]$
$= n - [n - m - n - \{n - 2n + m\} + 2m]$
$= n - [-m - \{-n + m\} + 2m]$
$= n - [-m + n - m + 2m]$
$= n - [n] = n - n = 0$

Ex. 4 Simplify $\dfrac{\dfrac{7}{2} \div \dfrac{5}{2} \times \dfrac{3}{2}}{\dfrac{7}{2} \div \dfrac{5}{2} \text{ of } \dfrac{3}{2}} \div \dfrac{15}{14}$.

Sol. Given expression $= \dfrac{\dfrac{7}{2} \div \dfrac{5}{2} \times \dfrac{3}{2}}{\dfrac{7}{2} \div \dfrac{5}{2} \text{ of } \dfrac{3}{2}} \div \dfrac{15}{14} = \dfrac{\dfrac{7}{2} \div \dfrac{5}{2} \times \dfrac{3}{2}}{\dfrac{7}{2} \div \dfrac{15}{4}}$

$= \dfrac{\dfrac{7}{2} \times \dfrac{2}{5} \times \dfrac{3}{2}}{\dfrac{7}{2} \times \dfrac{4}{15}} \div \dfrac{15}{14} = \dfrac{\dfrac{21}{10}}{\dfrac{14}{15}} \div \dfrac{15}{14}$

$= \dfrac{21}{10} \times \dfrac{15}{14} \times \dfrac{14}{15} = \dfrac{21}{10} = 2\dfrac{1}{10}$

Some Basic Formulae

Following formulae are useful in various operations of simplification

(i) $(a+b)^2 = a^2 + 2ab + b^2$

(ii) $(a-b)^2 = a^2 - 2ab + b^2$

(iii) $(a+b)^2 + (a-b)^2 = 2(a^2 + b^2)$

(iv) $(a+b)^2 - (a-b)^2 = 4ab$

(v) $a^2 - b^2 = (a+b)(a-b)$

(vi) $(a+b)^3 = a^3 + b^3 + 3ab(a+b)$

(vii) $(a-b)^3 = a^3 - b^3 - 3ab(a-b)$

(viii) $a^3 + b^3 = (a+b)(a^2 - ab + b^2)$

(ix) $a^3 - b^3 = (a-b)(a^2 + ab + b^2)$

(x) $\dfrac{a^3 + b^3 + c^3 - 3abc}{(a^2 + b^2 + c^2 - ab - bc - ca)} = (a+b+c)$

Or $a^3 + b^3 + c^3 - 3abc = (a+b+c)(a^2 + b^2 + c^2 - ab - bc - ca)$

$= (a+b+c) \dfrac{1}{2}(2a^2 + 2b^2 + 2c^2 - 2ab - 2bc - 2ca)$

$= \dfrac{(a+b+c)}{2}[(a-b)^2 + (b-c)^2 + (c-a)^2]$

If $a+b+c = 0$, then $a^3 + b^3 + c^3 = 3abc$

(xi) $(a+b+c)^2 = (a^2 + b^2 + c^2) + 2(ab + bc + ca)$

(xii) $\left(a + \dfrac{1}{a}\right)^2 = \left(a^2 + \dfrac{1}{a^2}\right) + 2 = \left(a - \dfrac{1}{a}\right)^2 + 4$

(xiii) $\left(a - \dfrac{1}{a}\right)^2 = \left(a^2 + \dfrac{1}{a^2}\right) - 2 = \left(a + \dfrac{1}{a}\right)^2 - 4$

(xiv) $\left(a + \dfrac{1}{a}\right)^3 = \left(a^3 + \dfrac{1}{a^3}\right) + 3\left(a + \dfrac{1}{a}\right)$

(xv) $\left(a - \dfrac{1}{a}\right)^3 = \left(a^3 - \dfrac{1}{a^3}\right) - 3\left(a - \dfrac{1}{a}\right)$

Ex. 5 Solve $\dfrac{(5.9)^3 + (1.8)^3 + (4.8)^3 - 3 \times 5.9 \times 1.8 \times 4.8}{(5.9)^2 + (1.8)^2 + (4.8)^2 - 5.9 \times 1.8 - 1.8 \times 4.8 - 4.8 \times 5.9}$.

Sol. We know that,

$\dfrac{a^3 + b^3 + c^3 - 3abc}{a^2 + b^2 + c^2 - ab - bc - ca} = (a+b+c)$

Here, $a = 5.9$, $b = 1.8$ and $c = 4.8$

$\therefore \dfrac{(5.9)^3 + (1.8)^3 + (4.8)^3 - 3 \times 5.9 \times 1.8 \times 4.8}{(5.9)^2 + (1.8)^2 + (4.8)^2 - 5.9 \times 1.8 - 1.8 \times 4.8 - 4.8 \times 5.9}$

$= (5.9 + 1.8 + 4.8) = 12.5$

Ex. 6 Solve $\dfrac{(9.8)^3 - (6.8)^3}{(9.8)^2 + 9.8 \times 6.8 + (6.8)^2}$.

Sol. We know that, $a^3 - b^3 = (a-b)(a^2 + ab + b^2)$

$\therefore \dfrac{a^3 - b^3}{a^2 + ab + b^2} = (a-b)$

Simplification / 117

Here, $a = 9.8$ and $b = 6.8$

$\therefore \quad \dfrac{(9.8)^3 - (6.8)^3}{(9.8)^2 + 9.8 \times 6.8 + (6.8)^2} = (9.8 - 6.8) = 3$

Ex. 7 Solve $\dfrac{(835 + 378)^2 + (835 - 378)^2}{835 \times 835 + 378 \times 378}$.

Sol. We know that, $\dfrac{(a + b)^2 + (a - b)^2}{a^2 + b^2} = \dfrac{2(a^2 + b^2)}{(a^2 + b^2)} = 2$

$\therefore \quad \dfrac{(835 + 378)^2 + (835 - 378)^2}{835 \times 835 + 378 \times 378} = 2$

Ex. 8 If $x = 3 + \sqrt{8}$, then find the value of $x^2 + \dfrac{1}{x^2}$.

Sol. Given, $x = 3 + \sqrt{8}$

$\therefore \quad \dfrac{1}{x} = \dfrac{1}{3 + \sqrt{8}} \times \dfrac{3 - \sqrt{8}}{3 - \sqrt{8}}$ [rationalising the denominator]

$= \dfrac{3 - \sqrt{8}}{(3)^2 - (\sqrt{8})^2} = \dfrac{3 - \sqrt{8}}{9 - 8} = 3 - \sqrt{8}$

$\therefore x^2 + \dfrac{1}{x^2} = \left(x + \dfrac{1}{x}\right)^2 - 2 \qquad \left[\because \left(a + \dfrac{1}{a}\right)^2 = \left(a^2 + \dfrac{1}{a^2}\right) + 2\right]$

$= (3 + \sqrt{8} + 3 - \sqrt{8})^2 - 2 = 36 - 2 = 34$

Ex. 9 If $\left(a + \dfrac{1}{a}\right) = 4\sqrt{2}$, then what is the value of $(a^6 + a^{-6})$?

Sol. Given, $\left(a + \dfrac{1}{a}\right) = 4\sqrt{2}$

On squaring both the sides, we get

$\left(a + \dfrac{1}{a}\right)^2 = (4\sqrt{2})^2 \Rightarrow a^2 + \dfrac{1}{a^2} + 2 = 32 \qquad [\because (a + b)^2 = a^2 + b^2 + 2ab]$

$\Rightarrow \qquad a^2 + \dfrac{1}{a^2} = 32 - 2 = 30$

Now, taking cube on both sides, we get

$\left(a^2 + \dfrac{1}{a^2}\right)^3 = (30)^3 \Rightarrow a^6 + \dfrac{1}{a^6} + 3 \cdot a^2 \cdot \dfrac{1}{a^2}\left(a^2 + \dfrac{1}{a^2}\right) = 27000$

$\qquad [\because (a + b)^3 = a^3 + b^3 + 3ab(a + b)]$

$\Rightarrow \quad a^6 + \dfrac{1}{a^6} + 3(30) = 27000 \Rightarrow a^6 + \dfrac{1}{a^6} = 27000 - 90 = 26910$

$\Rightarrow \qquad a^6 + a^{-6} = 26910$

Ex. 10 If $x + \dfrac{1}{x} = 2$, then find the value of $x^{2013} + \dfrac{1}{x^{2014}}$.

Sol. Given, $x + \dfrac{1}{x} = 2 \Rightarrow x^2 - 2x + 1 = 0$

$\Rightarrow \qquad (x - 1)^2 = 0 \qquad [\because (a - b)^2 = a^2 + b^2 - 2ab]$

$\Rightarrow \qquad x - 1 = 0 \Rightarrow x = 1$

$\therefore \qquad x^{2013} + \dfrac{1}{x^{2014}} = 1 + 1 = 2$

Multi Concept Questions

1. If $x + y + z = 0$, then the value of $\dfrac{x^2y^2 + y^2z^2 + z^2x^2}{x^4 + y^4 + z^4}$ is

   (a) 0 (b) $\dfrac{1}{2}$ (c) 1 (d) 2

   ↪ (b) ∵ $x + y + z = 0$ [given]

   On squaring both sides, we get
   $$(x + y + z)^2 = 0$$
   ⇒ $x^2 + y^2 + z^2 + 2(xy + yz + zx) = 0$
   ⇒ $x^2 + y^2 + z^2 = -2(xy + yz + zx)$

   Again, squaring both sides, we get
   $$(x^2 + y^2 + z^2)^2 = 4(xy + yz + zx)^2$$
   ⇒ $x^4 + y^4 + z^4 + 2(x^2y^2 + y^2z^2 + z^2x^2)$
   $= 4[x^2y^2 + y^2z^2 + z^2x^2 + \underbrace{2xyz(x + y + z)}_{0}]$ [∵ $x + y + z = 0$, given]

   ⇒ $x^4 + y^4 + z^4 = 2(x^2y^2 + y^2z^2 + z^2x^2)$

   ∴ $\dfrac{x^2y^2 + y^2z^2 + z^2x^2}{x^4 + y^4 + z^4} = \dfrac{1}{2}$

2. If $x(x + y + z) = 9$, $y(x + y + z) = 16$ and $z(x + y + z) = 144$, then the value of x will be

   (a) $\dfrac{9}{5}$ (b) $\dfrac{9}{7}$ (c) $\dfrac{9}{13}$ (d) $\dfrac{16}{13}$

   ↪ (c) Given,
   $x(x + y + z) = 9$ ...(i)
   $y(x + y + z) = 16$ ...(ii)
   and $z(x + y + z) = 144$ ...(iii)

   On adding Eqs. (i), (ii) and (iii), we get
   $x(x + y + z) + y(x + y + z) + z(x + y + z) = 9 + 16 + 144$
   ⇒ $(x + y + z)(x + y + z) = 169$
   ⇒ $(x + y + z)^2 = 169$
   ⇒ $x + y + z = 13$ ...(iv)

   On putting the value of $(x + y + z)$ from Eq. (iv) in Eq. (i), we get
   $x(13) = 9$
   ∴ $x = \dfrac{9}{13}$

Fast Track Practice

Exercise 1 Base Level Questions

1. Simplify [0.9 − {2.3 − 3.2 − (7.1 − 5.4 − 3.5)}]. **[FCI 2015]**
   (a) 0.18 (b) 1.8 (c) 0 (d) 2.6

2. Simplify $6 - [9 - \{18 - (15 - \overline{12 - 9})\}]$.
   (a) 1 (b) 4 (c) 5 (d) 3
   (e) None of these

3. $(154 \times 2.5 \div 0.5) \div ? = 192.5$ **[IBPS Clerk (Pre) 2016]**
   (a) 6 (b) 8 (c) 2 (d) 4
   (e) 18

4. $\dfrac{\frac{1}{5} + \left(999 + \frac{494}{495}\right) \times 99}{4} = ?$
   (a) 25000 (b) 24225
   (c) 24800 (d) 24750
   (e) None of these

5. $\sqrt{? + 483} \div 6 = 125^{1/3}$ **[IBPS Clerk (Pre) 2016]**
   (a) 429 (b) 411 (c) 423 (d) 413
   (e) 417

6. $\dfrac{5136}{(523 + 333) \text{ of } \frac{3}{4}} + 459 = ?$ **[SBI PO 2015]**
   (a) 520 (b) 541 (c) 513 (d) 493
   (e) 467

7. $56 - 742 / 53 / 2 = ?^2$ **[SBI PO 2015]**
   (a) 3 (b) 4 (c) 6 (d) 7
   (e) 2

8. $9^2 \times \sqrt[4]{1296} - 254 = (? \times 9 + 151)$ **[SBI PO 2015]**
   (a) 14 (b) 9 (c) 6 (d) 12
   (e) 20

9. $(4 \times 4 \times 4 \times 4 \times 4 \times 4)^5 \times (4 \times 4 \times 4)^8 \div (4)^3$
   $= (64)^?$ **[IBPS Clerk 2011]**
   (a) 17 (b) 10 (c) 16 (d) 11
   (e) None of these

10. 60% of $725 = 174 \times ?$ **[IBPS Clerk (Pre) 2016]**
    (a) 0.2 (b) 0.9 (c) 2.5 (d) 0.8
    (e) 0.3

11. $? - (0.6)^2 = (0.7)^2 \div (0.35)$ **[IBPS Clerk (Pre) 2016]**
    (a) 1.54 (b) 1.32 (c) 1.92 (d) 1.62
    (e) 1.76

12. $(0.49)^4 \times (0.343)^4 \div (0.2401)^4$
    $= (70 \div 100)^{?+3}$ **[IDBI SO 2012]**
    (a) 3 (b) 1 (c) 4 (d) 7
    (e) None of these

13. $\dfrac{5 - \left[\frac{3}{4} + \left\{2\frac{1}{2} - \left(\frac{1}{2} + \overline{\frac{1}{6} - \frac{1}{7}}\right)\right\}\right]}{2} = ?$
    (a) $1\frac{23}{168}$ (b) $2\frac{23}{168}$ (c) $3\frac{23}{168}$ (d) $4\frac{23}{168}$
    (e) None of these

14. Simplify
    $\dfrac{\left[3\frac{1}{4} \div \left\{1\frac{1}{4} - 0.5 \left(2\frac{1}{2} - \overline{\frac{1}{4} - \frac{1}{6}}\right)\right\}\right]}{4 \times \frac{1}{12}}$.
    (a) 245 (b) 233 (c) 234 (d) 299
    (e) None of these

15. If $0.764y = 1.236x$, then what is the value of $\left(\dfrac{y-x}{y+x}\right)$?
    (a) 0.764 (b) 0.236 (c) 2 (d) 0.472

Directions (Q. Nos. 16-19) *What value come in the place of question mark (?) in the following questions?*

16. $8^{12} \div 16^2$ of $32^3 \times \sqrt{256} = 2^?$ **[IBPS Clerk (Mains) 2016]**
    (a) 17 (b) 18 (c) 19 (d) 20
    (e) 15

17. $\sqrt{?} \div 8 \times 31 = 127.5 - 3.5$ **[IBPS Clerk (Pre) 2016]**
    (a) 1158 (b) 784 (c) 578 (d) 1024
    (e) 484

18. $486 \div ? \times 7392 \div 66 = 1008$ **[SBI Clerk (Mains) 2016]**
    (a) 54 (b) 55 (c) 52 (d) 53
    (e) 51

19. 17.8% of $? = 427.2 \times 8.4\%$ of 135 **[SBI Clerk (Mains) 2016]**
    (a) 21734 (b) 24378
    (c) 27216 (d) 28120
    (e) 25315

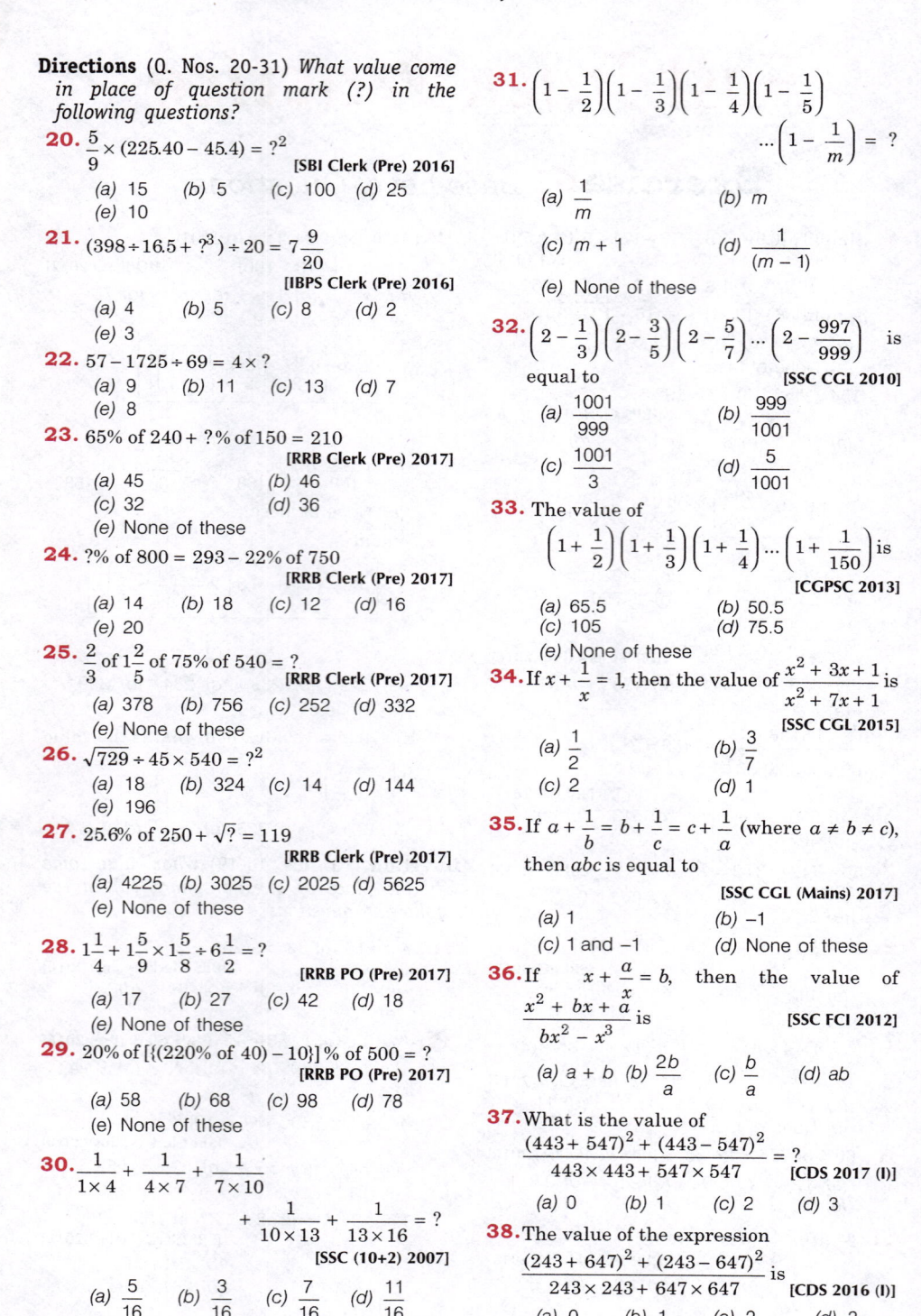

Directions (Q. Nos. 20-31) *What value come in place of question mark (?) in the following questions?*

20. $\frac{5}{9} \times (225.40 - 45.4) = ?^2$
[SBI Clerk (Pre) 2016]
(a) 15 (b) 5 (c) 100 (d) 25
(e) 10

21. $(398 \div 16.5 + ?^3) \div 20 = 7\frac{9}{20}$
[IBPS Clerk (Pre) 2016]
(a) 4 (b) 5 (c) 8 (d) 2
(e) 3

22. $57 - 1725 \div 69 = 4 \times ?$
(a) 9 (b) 11 (c) 13 (d) 7
(e) 8

23. 65% of 240 + ?% of 150 = 210
[RRB Clerk (Pre) 2017]
(a) 45 (b) 46
(c) 32 (d) 36
(e) None of these

24. ?% of 800 = 293 – 22% of 750
[RRB Clerk (Pre) 2017]
(a) 14 (b) 18 (c) 12 (d) 16
(e) 20

25. $\frac{2}{3}$ of $1\frac{2}{5}$ of 75% of 540 = ?
[RRB Clerk (Pre) 2017]
(a) 378 (b) 756 (c) 252 (d) 332
(e) None of these

26. $\sqrt{729} \div 45 \times 540 = ?^2$
(a) 18 (b) 324 (c) 14 (d) 144
(e) 196

27. 25.6% of 250 + $\sqrt{?}$ = 119
[RRB Clerk (Pre) 2017]
(a) 4225 (b) 3025 (c) 2025 (d) 5625
(e) None of these

28. $1\frac{1}{4} + 1\frac{5}{9} \times 1\frac{5}{8} \div 6\frac{1}{2} = ?$
[RRB PO (Pre) 2017]
(a) 17 (b) 27 (c) 42 (d) 18
(e) None of these

29. 20% of [{(220% of 40) – 10}] % of 500 = ?
[RRB PO (Pre) 2017]
(a) 58 (b) 68 (c) 98 (d) 78
(e) None of these

30. $\frac{1}{1 \times 4} + \frac{1}{4 \times 7} + \frac{1}{7 \times 10}$
$+ \frac{1}{10 \times 13} + \frac{1}{13 \times 16} = ?$
[SSC (10+2) 2007]
(a) $\frac{5}{16}$ (b) $\frac{3}{16}$ (c) $\frac{7}{16}$ (d) $\frac{11}{16}$

31. $\left(1 - \frac{1}{2}\right)\left(1 - \frac{1}{3}\right)\left(1 - \frac{1}{4}\right)\left(1 - \frac{1}{5}\right)$
$\ldots \left(1 - \frac{1}{m}\right) = ?$
(a) $\frac{1}{m}$ (b) m
(c) $m + 1$ (d) $\frac{1}{(m-1)}$
(e) None of these

32. $\left(2 - \frac{1}{3}\right)\left(2 - \frac{3}{5}\right)\left(2 - \frac{5}{7}\right)\ldots\left(2 - \frac{997}{999}\right)$ is equal to
[SSC CGL 2010]
(a) $\frac{1001}{999}$ (b) $\frac{999}{1001}$
(c) $\frac{1001}{3}$ (d) $\frac{5}{1001}$

33. The value of
$\left(1 + \frac{1}{2}\right)\left(1 + \frac{1}{3}\right)\left(1 + \frac{1}{4}\right)\ldots\left(1 + \frac{1}{150}\right)$ is
[CGPSC 2013]
(a) 65.5 (b) 50.5
(c) 105 (d) 75.5
(e) None of these

34. If $x + \frac{1}{x} = 1$, then the value of $\frac{x^2 + 3x + 1}{x^2 + 7x + 1}$ is
[SSC CGL 2015]
(a) $\frac{1}{2}$ (b) $\frac{3}{7}$
(c) 2 (d) 1

35. If $a + \frac{1}{b} = b + \frac{1}{c} = c + \frac{1}{a}$ (where $a \neq b \neq c$), then abc is equal to
[SSC CGL (Mains) 2017]
(a) 1 (b) –1
(c) 1 and –1 (d) None of these

36. If $x + \frac{a}{x} = b$, then the value of $\frac{x^2 + bx + a}{bx^2 - x^3}$ is
[SSC FCI 2012]
(a) $a + b$ (b) $\frac{2b}{a}$ (c) $\frac{b}{a}$ (d) ab

37. What is the value of
$\frac{(443 + 547)^2 + (443 - 547)^2}{443 \times 443 + 547 \times 547} = ?$
[CDS 2017 (I)]
(a) 0 (b) 1 (c) 2 (d) 3

38. The value of the expression
$\frac{(243 + 647)^2 + (243 - 647)^2}{243 \times 243 + 647 \times 647}$ is
[CDS 2016 (I)]
(a) 0 (b) 1 (c) 2 (d) 3

Simplification / 121

39. Find the value of $a^3 + b^3 + c^3 - 3abc$, when $a = 225$, $b = 226$ and $c = 227$.
 [SSC CGL 2012]
(a) 2304 (b) 2430
(c) 2034 (d) 2340

40. The value of $\dfrac{(0.96)^3 - (0.1)^3}{(0.96)^2 + 0.096 + 0.01}$ is
 [SSC (10+2) 2012]
(a) 0.86 (b) 1.06 (c) 0.95 (d) 0.97

41. $\dfrac{\begin{bmatrix}0.5 \times 0.5 \times 0.5 + 0.2 \times 0.2 \times 0.2 \\ + 0.3 \times 0.3 \times 0.3 - 3 \times 0.5 \times 0.3 \times 0.2\end{bmatrix}}{\begin{bmatrix}0.5 \times 0.5 + 0.2 \times 0.2 + 0.3 \times 0.3 \\ - 0.5 \times 0.2 - 0.2 \times 0.3 - 0.5 \times 0.3\end{bmatrix}} = ?$
 [SSC CGL 2010]
(a) 1 (b) 0.6
(c) 0.4 (d) 0.03

42. $\dfrac{\begin{bmatrix}(2.247)^3 + (1.730)^3 + (1.023)^3 \\ - 3 \times 2.247 \times 1.730 \times 1.023\end{bmatrix}}{\begin{bmatrix}(2.247)^2 + (1.730)^2 + (1.023)^2 - 2.247 \\ \times 1.730 - 1.730 \times 1.023 - 2.247 \times 1.023\end{bmatrix}} = ?$
(a) 1.730 (b) 4
(c) 5 (d) 5.247

43. $\dfrac{8.73 \times 8.73 \times 8.73 + 4.27 \times 4.27 \times 4.27}{8.73 \times 8.73 - 8.73 \times 4.27 + 4.27 \times 4.27}$
is equal to [SSC CGL 2012]
(a) 11
(b) 13
(c) 11/7
(d) None of the above

44. If $a^2 + b^2 = 234$ and $ab = 108$, then find the value of $\dfrac{a+b}{a-b}$.
(a) 10 (b) 8
(c) 5 (d) 4
(e) None of these

45. If $a^2 + 1 = a$, then the value of $a^{12} + a^6 + 1$ is [SSC CGL 2013]
(a) 2 (b) 3
(c) –3 (d) 1

46. If $a + b + c = 0$, then the value of $\left(\dfrac{a+b}{c} + \dfrac{b+c}{a} + \dfrac{c+a}{b}\right)$ $\left(\dfrac{a}{b+c} + \dfrac{b}{c+a} + \dfrac{c}{a+b}\right)$ is
 [SSC CGL 2013]
(a) 9 (b) 0
(c) 8 (d) – 3

47. If $x = \dfrac{\sqrt{2}+1}{\sqrt{2}-1}$ and $x - y = 4\sqrt{2}$, then the value of $(x^2 + y^2)$ is
 [SSC CPO 2013]
(a) 34 (b) 38 (c) 30 (d) 32

48. If $a = (\sqrt{2} - 1)^{1/3}$, then the value of $(a - a^{-1})^3 + 3(a - a^{-1})$ is
 [SSC (10+2) 2012]
(a) –2 (b) 2 (c) $2\sqrt{2}$ (d) $\sqrt{2}$

49. If $x = \dfrac{\sqrt{3}+1}{\sqrt{3}-1}$ and $y = \dfrac{\sqrt{3}-1}{\sqrt{3}+1}$, then the value of $\dfrac{x^2}{y} + \dfrac{y^2}{x}$ is [SSC (10+2) 2012]
(a) 52 (b) 76 (c) 4 (d) 64

50. If $p + q = 10$ and $pq = 5$, then the numerical value of $\dfrac{p}{q} + \dfrac{q}{p}$ will be
 [SSC (10+2) 2012]
(a) 22 (b) 18 (c) 16 (d) 20

51. If $a + b = 8$ and $ab = 15$, then what is the value of $a^3 + b^3$? [SSC CGL (Pre) 2017]
(a) 98 (b) 152
(c) 124 (d) 260

52. If $x + y + z = 1$, $xy + yz + zx = -1$ and $xyz = -1$, then $x^3 + y^3 + z^3$ is equal to
 [SSC CGL 2016]
(a) –2 (b) –1
(c) 0 (d) 1

53. If $a^2 + b^2 + c^2 = 250$ and $ab + bc + ca = 3$, then $a + b + c$ is equal to [SSC CGL 2016]
(a) 16 (b) –16
(c) ±16 (d) None of these

54. If $a + b + c = 14$ and $a^2 + b^2 + c^2 = 96$, then $(ab + bc + ca) = ?$
(a) 51 (b) 55
(c) 50 (d) 65
(e) None of these

55. If $x = 11$, then the value of $x^5 - 12x^4 + 12x^3 - 12x^2 + 12x - 1$ is
 [SSC CGL (Mains) 2016]
(a) 11 (b) 10
(c) 12 (d) – 10

56. If $a = \dfrac{1}{a-5}$ $(a > 0)$, then the value of $a + \dfrac{1}{a}$ is [SSC CGL (Pre) 2016]
(a) $\sqrt{29}$ (b) $-\sqrt{27}$
(c) $-\sqrt{29}$ (d) $\sqrt{27}$

Exercise 2 Higher Skill Level Questions

1. Simplify $\dfrac{\sqrt[4]{0.0625} + \sqrt[3]{0.008} + \sqrt{0.09} - 1}{\sqrt[3]{62.5}\,\sqrt[5]{32}}$.
   [SSC (10+2) 2015]
   (a) 1.25 (b) 1/5
   (c) 0   (d) 2.4

2. If $x = 2 + 2^{2/3} + 2^{1/3}$, then what is the value of $x^3 - 6x^2 + 6x$? [CDS 2017 (I)]
   (a) 3 (b) 2
   (c) 1 (d) 0

3. If $x^2 + \dfrac{1}{x^2} = 98\,(x>0)$, then the value of $x^3 + \dfrac{1}{x^3}$ is
   [SSC CGL (Mains) 2016]
   (a) 970   (b) 1030
   (c) –970 (d) –1030

4. If $x = p + \dfrac{1}{p}$ and $y = p - \dfrac{1}{p}$, then the value of $x^4 - 2x^2y^2 + y^4$ is
   [SSC (10+2) 2014]
   (a) 24 (b) 4
   (c) 16 (d) 8

5. If $3(a^2 + b^2 + c^2) = (a+b+c)^2$, then the relation between a, b and c is
   [SSC CGL 2015]
   (a) $a \neq b \neq c$ (b) $a = b = c$
   (c) $a \neq b = c$ (d) $a = b \neq c$

6. If $\dfrac{x}{xa + yb + zc} = \dfrac{y}{ya + zb + xc} = \dfrac{z}{za + xb + yc}$ and $x + y + z \neq 0$, then each ratio is
   [SSC (10+2) 2014]
   (a) $\dfrac{1}{a-b-c}$ (b) $\dfrac{1}{a+b-c}$
   (c) $\dfrac{1}{a-b+c}$ (d) $\dfrac{1}{a+b+c}$

7. If $x + \dfrac{1}{x} = 2$, then the value of $x^7 + \dfrac{1}{x^5}$ is
   [SSC CPO 2015]
   (a) 2^5 (b) 2^{12}
   (c) 2   (d) 2^7

8. If $x + \dfrac{1}{x} = 3$, then the value of $\dfrac{3x^2 - 4x + 3}{x^2 - x + 1}$ is
   [SSC (10+2) 2014]
   (a) $\dfrac{4}{3}$ (b) $\dfrac{3}{2}$ (c) $\dfrac{5}{2}$ (d) $\dfrac{5}{3}$

9. If $x = 2^{1/3} + 2^{-1/3}$, then the value of $2x^3 - 6x - 5$ is [CDS 2016 (I)]
   (a) 0 (b) 1 (c) 2 (d) 3

10. If $x = \dfrac{\sqrt{3}}{2}$, then the value of $\dfrac{1+x}{1+\sqrt{1+x}} + \dfrac{1-x}{1-\sqrt{1-x}}$ is
    [SSC CGL 2012]
    (a) 0 (b) 1
    (c) $\dfrac{\sqrt{3}}{2}$ (d) $\sqrt{3}$

11. If $x^2 + y^2 + z^2 = xy + yz + zx$, then the value of $\dfrac{3x^4 + 7y^4 + 5z^4}{5x^2y^2 + 7y^2z^2 + 3z^2x^2}$ is
    [SSC CGL 2015]
    (a) –1 (b) 2 (c) 1 (d) 0

12. If $x = 3 + 2\sqrt{2}$, then $\dfrac{x^6 + x^4 + x^2 + 1}{x^3}$ is equal to
    [SSC (10+2) 2014]
    (a) 216 (b) 192
    (c) 198 (d) 204

13. If $a - \dfrac{1}{a-3} = 5$, then the value of $(a-3)^3 - \dfrac{1}{(a-3)^3}$ is
    [SSC CGL 2015]
    (a) 7 (b) 2 (c) 5 (d) 14

14. If $x = a^{1/2} + a^{-1/2},\, y = a^{1/2} - a^{-1/2}$, then the value of $(x^4 - x^2y^2 - 1) + (y^4 - x^2y^2 + 1)$ is
    [SSC CGL 2015]
    (a) 16 (b) 12 (c) 13 (d) 14

15. If $x^2 + \dfrac{1}{x^2} = \dfrac{17}{4}$, then what is $\left(x^3 - \dfrac{1}{x^3}\right)$ equal to?
    [CDS 2014 (I)]
    (a) $\dfrac{75}{16}$ (b) $\dfrac{63}{8}$
    (c) $\dfrac{95}{8}$ (d) None of these

16. If $x - \sqrt{3} - \sqrt{2} = 0$ and $y - \sqrt{3} + \sqrt{2} = 0$, then the value of $(x^3 - 20\sqrt{2}) - (y^3 + 2\sqrt{2})$ is
    [SSC CGL 2015]
    (a) 1 (b) 3
    (c) 0 (d) 2

Simplification / 123

17. If $\sqrt{\frac{x}{y}} = \frac{24}{5} + \sqrt{\frac{y}{x}}$ and $x + y = 26$, then what is the value of xy? [CDS 2017 (I)]
(a) 5 (b) 15 (c) 25 (d) 30

18. If $\sqrt{\frac{x}{y}} = \frac{10}{3} - \sqrt{\frac{y}{x}}$ and $x - y = 8$, then the value of xy is equal to [CDS 2016 (I)]
(a) 36 (b) 24 (c) 16 (d) 9

19. If $x + \frac{1}{x} = 3$, then $x^5 + \frac{1}{x^5}$ is equal to
[SSC CPO 2013]
(a) 123 (b) 83 (c) 92 (d) 112

20. If $a^3 = 335 + b^3$ and $a = 5 + b$, then what is the value of $a + b$ (given that $a > 0$ and $b > 0$)? [CDS 2017 (I)]
(a) 7 (b) 9
(c) 16 (d) 49

21. If $a^3 = 117 + b^3$ and $a = 3 + b$, then the value of $a + b$ is (given that, $a > 0$ and $b > 0$) [CDS 2016 (I)]
(a) 7 (b) 9 (c) 11 (d) 13

22. If $a - b = 4$ and $a^2 + b^2 = 40$, where a and b are positive integers, then $a^3 + b^6$ is equal to [CDS 2015 (I)]
(a) 264 (b) 280 (c) 300 (d) 324

23. $\frac{(m-n)^3 + (n-r)^3 + (r-m)^3}{6(m-n)(n-r)(r-m)} = ?$
(a) $\frac{1}{2}$ (b) $\frac{1}{3}$ (c) $\frac{1}{5}$ (d) $\frac{1}{6}$
(e) None of these

24. If $ax + by = 1$ and $bx + ay = \frac{2ab}{a^2 + b^2}$, then $(x^2 + y^2)(a^2 + b^2)$ is equal to
[SSC CGL (Mains) 2016]
(a) 1 (b) 2 (c) 0.5 (d) 0

Answer with Solutions

Exercise 1 Base Level Questions

1. (c) $[0.9 - \{2.3 - 3.2 - (7.1 - 5.4 - 3.5)\}]$
$= [0.9 - \{2.3 - 3.2 - (7.1 - 8.9)\}]$
$= [0.9 - \{2.3 - 3.2 - (-1.8)\}]$
$= [0.9 - \{2.3 - 3.2 + 1.8\}]$
$= [0.9 - \{0.9\}] = 0$

2. (d) Given expression
$= 6 - [9 - \{18 - (15 - \overline{12 - 9})\}]$
$= 6 - [9 - \{18 - (15 - 3)\}]$
$= 6 - [9 - \{18 - 12\}]$
$= 6 - [9 - 6] = 6 - 3 = 3$

3. (d) $(154 \times 2.5 \div 0.5) \div ? = 192.5$
$\Rightarrow \left(154 \times \frac{2.5}{0.5}\right) \div x = 192.5$
$\Rightarrow \frac{(154 \times 5)}{x} = 192.5$
$\Rightarrow x = \frac{154 \times 5}{192.5} \Rightarrow x = 4$

4. (d) Given expression
$= \frac{\frac{1}{5} + \left(999 + \frac{494}{495}\right) \times 99}{4}$
$= \frac{\frac{1}{5} + 999 \times 99 + \frac{494}{495} \times 99}{4}$

$= \frac{\frac{1}{5} + (1000 - 1) \times 99 + \frac{494}{5}}{4}$
$= \frac{\frac{495}{5} + (99000 - 99)}{4}$
$= \frac{99 + 99000 - 99}{4}$
$= \frac{99000}{4} = 24750$

5. (e) $\sqrt{? + 483} \div 6 = 125^{1/3}$
$\Rightarrow \frac{\sqrt{x + 483}}{6} = 5$
$\Rightarrow \sqrt{x + 483} = 30 \Rightarrow x + 483 = (30)^2$
$\Rightarrow x = 900 - 483 = 417$

6. (e) $\frac{5136}{(523 + 333) \text{ of } \frac{3}{4}} + 459 = ?$
$\Rightarrow ? = \frac{5136}{856 \times \frac{3}{4}} + 459$
$= \frac{5136}{214 \times 3} + 459$
$= 8 + 459 = 467$

124 / Fast Track Objective Arithmetic

7. (d) $56 - 742/53/2 = ?^2$
$\Rightarrow \quad 56 - 7 = ?^2$
$\Rightarrow \quad 49 = ?^2 \Rightarrow ? = \sqrt{49} = 7$

8. (b) $9^2 \times \sqrt[4]{1296} - 254 = (? \times 9 + 151)$
$\Rightarrow \quad 81 \times 6 - 254 = ? \times 9 + 151$
$\Rightarrow \quad 486 - 254 = ? \times 9 + 151$
$\Rightarrow \quad 232 - 151 = ? \times 9$
$\Rightarrow \quad ? = \dfrac{81}{9} = 9$

9. (a) $(4 \times 4 \times 4 \times 4 \times 4 \times 4)^5$
$\times (4 \times 4 \times 4)^8 \div (4)^3 = (64)^?$
$\Rightarrow (4^6)^5 \times (4^3)^8 \times \dfrac{1}{(4)^3} = (4^3)^?$
$\Rightarrow \dfrac{(4)^{30} \times (4)^{24}}{(4)^3} = (4)^{3 \times ?}$
$\Rightarrow 4^{30+24-3} = 4^{3 \times ?}$
$\Rightarrow 4^{51} = 4^{3 \times ?}$
$\Rightarrow 3 \times ? = 51$
$\therefore \quad ? = \dfrac{51}{3} = 17$

10. (c) 60% of $725 = 174 \times ?$
$\Rightarrow \dfrac{60}{100} \times 725 = 174 \times x \Rightarrow \dfrac{60}{100} \times \dfrac{725}{174} = x$
$\Rightarrow x = 2.5$

11. (e) $? - (0.6)^2 = (0.7)^2 \div 0.35$
$\Rightarrow x - (0.6)^2 = \dfrac{(0.7)^2}{0.35} \Rightarrow x = \dfrac{0.49}{0.35} + (0.36)$
$\Rightarrow \quad x = 1.76$

12. (b) $(70 \div 100)^{?+3}$
$= (0.49)^4 \times (0.343)^4 \div (0.2401)^4$
$\Rightarrow (0.7)^{?+3} = \left(\dfrac{0.49 \times 0.343}{0.2401}\right)^4$
$\Rightarrow (0.7)^{?+3} = (0.7)^4$
On comparing the exponents from both sides, we get
$? + 3 = 4$
$\therefore \quad ? = 4 - 3 = 1$

13. (a) $? = \dfrac{5 - \left[\dfrac{3}{4} + \left\{\dfrac{5}{2} - \left(\dfrac{1}{2} + \dfrac{1}{6} - \dfrac{1}{7}\right)\right\}\right]}{2}$

$= \dfrac{5 - \left[\dfrac{3}{4} + \left\{\dfrac{5}{2} - \left(\dfrac{1}{2} + \dfrac{7-6}{42}\right)\right\}\right]}{2}$

$= \dfrac{5 - \left[\dfrac{3}{4} + \left\{\dfrac{5}{2} - \left(\dfrac{1}{2} + \dfrac{1}{42}\right)\right\}\right]}{2}$

$= \dfrac{5 - \left[\dfrac{3}{4} + \left\{\dfrac{5}{2} - \left(\dfrac{21+1}{42}\right)\right\}\right]}{2}$

$= \dfrac{5 - \left[\dfrac{3}{4} + \left\{\dfrac{5}{2} - \dfrac{22}{42}\right\}\right]}{2}$

$= \dfrac{5 - \left[\dfrac{3}{4} + \left\{\dfrac{105 - 22}{42}\right\}\right]}{2}$

$= \dfrac{5 - \left[\dfrac{3}{4} + \dfrac{83}{42}\right]}{2} = \dfrac{5 - \left[\dfrac{63 + 166}{84}\right]}{2}$

$= \dfrac{5 - \dfrac{229}{84}}{2} = \dfrac{\dfrac{420 - 229}{84}}{2} = \dfrac{\dfrac{191}{84}}{2}$

$= \dfrac{191}{84 \times 2} = \dfrac{191}{168} = 1\dfrac{23}{168}$

14. (c) Given expression
$= \dfrac{\left[\dfrac{13}{4} \div \left\{\dfrac{5}{4} - \dfrac{1}{2}\left(\dfrac{5}{2} - \dfrac{3-2}{12}\right)\right\}\right]}{4 \times \dfrac{1}{12}}$

$= \dfrac{\left[\dfrac{13}{4} \div \left\{\dfrac{5}{4} - \dfrac{1}{2}\left(\dfrac{5}{2} - \dfrac{1}{12}\right)\right\}\right]}{\dfrac{1}{3}}$

$= \dfrac{\left[\dfrac{13}{4} \div \left\{\dfrac{5}{4} - \dfrac{1}{2}\left(\dfrac{30-1}{12}\right)\right\}\right]}{\dfrac{1}{3}}$

$= \dfrac{\left[\dfrac{13}{4} \div \left\{\dfrac{5}{4} - \dfrac{29}{24}\right\}\right]}{\dfrac{1}{3}}$

$= \dfrac{\left[\dfrac{13}{4} \div \left\{\dfrac{30-29}{24}\right\}\right]}{\dfrac{1}{3}} = \dfrac{\left[\dfrac{13}{4} \div \dfrac{1}{24}\right]}{\dfrac{1}{3}}$

$= \dfrac{13}{4} \times 24 \times 3 = 13 \times 18 = 234$

15. (b) Given, $0.764 y = 1.236 x$
$\Rightarrow \quad \dfrac{y}{x} = \dfrac{1.236}{0.764}$

Now, $\dfrac{y - x}{y + x} = \dfrac{\dfrac{y}{x} - 1}{\dfrac{y}{x} + 1} = \dfrac{\dfrac{1.236}{0.764} - 1}{\dfrac{1.236}{0.764} + 1}$

$= \dfrac{1.236 - 0.764}{1.236 + 0.764} = \dfrac{0.472}{2.000} = 0.236$

Simplification / 125

16.(a) $8^{12} \div 16^2$ of $32^3 \times \sqrt{256} = 2^?$
$\Rightarrow (2^3)^{12} \div (2^4)^2$ of $(2^5)^3 \times 16 = 2^?$
$\Rightarrow 2^{36} \div 2^8$ of $2^{15} \times 2^4 = 2^? \Rightarrow 2^{17} = 2^?$
$\Rightarrow \qquad ? = 17$

17.(d) $\sqrt{?} \div 8 \times 31 = 127.5 - 3.5$
$\Rightarrow \dfrac{\sqrt{x}}{8} \times 31 = 124 \Rightarrow \sqrt{x} = \dfrac{124 \times 8}{31}$
$\Rightarrow \sqrt{x} = 32$
$\Rightarrow x = (32)^2 = 1024$

18.(a) $486 \div ? \times 7392 \div 66 = 1008$
$\Rightarrow \dfrac{486}{?} \times \dfrac{7392}{66} = 1008$
$\therefore ? = \dfrac{486 \times 7392}{1008 \times 66} = 54$

19.(c) 17.8% of $? = 427.2 \times 8.4\%$ of 135
$\Rightarrow \dfrac{17.8}{100} \times ? = 427.2 \times \dfrac{8.4}{100} \times 135$
$\Rightarrow ? = \dfrac{427.2 \times 8.4 \times 135 \times 100}{100 \times 17.8}$
$\Rightarrow ? = 27216$

20.(e) $?^2 = \dfrac{5}{9} \times (225.40 - 45.40)$
$\Rightarrow ?^2 = \dfrac{5}{9} \times 180 \Rightarrow ?^2 = 5 \times 20$
$\therefore ?^2 = 100 \Rightarrow ? = \sqrt{100} = 10$

21. (b) $(398 \div 16.5 + ?^3) \div 20 = 7\dfrac{9}{20}$
$\Rightarrow \left(\dfrac{398}{16.5} + x^3\right) \times \dfrac{1}{20} = \dfrac{149}{20}$
$\Rightarrow 24.12 + x^3 = 149$
$\Rightarrow x^3 = 124.87$
$\Rightarrow x \approx 5$

22.(e) $57 - 1725 \div 69 = 4 \times ?$
$\Rightarrow 57 - \dfrac{1725}{69} = 4 \times ?$
$\Rightarrow 57 - 25 = 4 \times ?$
$\Rightarrow 4 \times ? = 32$
$\therefore ? = \dfrac{32}{4} = 8$

23. (d) $\dfrac{65}{100}$ of $240 + \dfrac{?}{100}$ of $150 = 210$
$\Rightarrow 156 + 1.5 \times ? = 210$
$\therefore ? = \dfrac{210 - 156}{1.5} = 36$

24. (d) $\dfrac{800 \times ?}{100} = 293 - \dfrac{750 \times 22}{100}$
$\Rightarrow 8 \times ? = 293 - 165 = 128$
$\Rightarrow ? = \dfrac{128}{8} = 16$

25. (a) $? = \dfrac{2}{3}$ of $\dfrac{7}{5}$ of $\dfrac{75}{100}$ of $540 = 7 \times 54 = 378$

26.(a) $\sqrt{729} \div 45 \times 540 = ?^2$
$\Rightarrow ?^2 = 27 \div 45 \times 540$
$\Rightarrow ?^2 = \dfrac{27}{45} \times 540 = \dfrac{3}{5} \times 540$
$\Rightarrow ?^2 = 3 \times 108 \Rightarrow ?^2 = 324$
$\Rightarrow ? = \sqrt{324} \Rightarrow ? = 18$

27.(b) $250 \times \dfrac{25.6}{100} + \sqrt{?} = 119$
$\Rightarrow 64 + \sqrt{?} = 119$
$\Rightarrow \sqrt{?} = 119 - 64 = 55$
$\Rightarrow ? = 55 \times 55 = 3025$

28.(e) $? = 1\dfrac{1}{4} + 1\dfrac{5}{9} \times 1\dfrac{5}{8} \div 6\dfrac{1}{2} = \dfrac{5}{4} + \dfrac{14}{9} \times \dfrac{13}{8} \div \dfrac{13}{2}$
$= \dfrac{5}{4} + \dfrac{14}{9} \times \dfrac{13}{8} \times \dfrac{2}{13}$
$= \dfrac{5}{4} + \dfrac{7}{18} = \dfrac{45 + 14}{36} = \dfrac{59}{36} = 1\dfrac{23}{36}$

29.(d) $\dfrac{20}{100} \times \left[\left\{\left(\dfrac{220}{100} \times 40\right) - 10\right\}\right]\%$ of $500 = ?$
$\Rightarrow \dfrac{1}{5} \times [\{88 - 10\}]\%$ of $500 = ?$
$\Rightarrow \dfrac{1}{5} \times \dfrac{78}{100} \times 500 = ?$
$\therefore ? = 78$

30.(a) Given expression can be written as
$\dfrac{1}{3}\left[\left(1 - \dfrac{1}{4}\right) + \left(\dfrac{1}{4} - \dfrac{1}{7}\right) + \left(\dfrac{1}{7} - \dfrac{1}{10}\right)\right.$
$\left. + \left(\dfrac{1}{10} - \dfrac{1}{13}\right) + \left(\dfrac{1}{13} - \dfrac{1}{16}\right)\right]$
$= \dfrac{1}{3}\left[1 - \dfrac{1}{16}\right] = \dfrac{1}{3}\left(\dfrac{16-1}{16}\right) = \dfrac{1}{3} \times \dfrac{15}{16} = \dfrac{5}{16}$

31.(a) Given expression
$= \left(1 - \dfrac{1}{2}\right)\left(1 - \dfrac{1}{3}\right)\left(1 - \dfrac{1}{4}\right)\left(1 - \dfrac{1}{5}\right)$
$\qquad \ldots \left(1 - \dfrac{1}{m-1}\right)\left(1 - \dfrac{1}{m}\right)$
$= \dfrac{1}{2} \times \dfrac{2}{3} \times \dfrac{3}{4} \times \dfrac{4}{5} \times \ldots \times \dfrac{(m-2)}{(m-1)}$
$\qquad \times \dfrac{(m-1)}{m} = \dfrac{1}{m}$

32.(c) Given expression
$= \left(2 - \dfrac{1}{3}\right)\left(2 - \dfrac{3}{5}\right)\left(2 - \dfrac{5}{7}\right) \ldots \left(2 - \dfrac{997}{999}\right)$
$= \dfrac{5}{3} \times \dfrac{7}{5} \times \dfrac{9}{7} \times \ldots \times \dfrac{1001}{999} = \dfrac{1001}{3}$

33. (d) $\left(1+\dfrac{1}{2}\right)\left(1+\dfrac{1}{3}\right)\left(1+\dfrac{1}{4}\right)...\left(1+\dfrac{1}{150}\right)$

$= \dfrac{3}{2} \times \dfrac{4}{3} \times \dfrac{5}{4} \times ... \times \dfrac{151}{150} = \dfrac{151}{2} = 75.5$

34. (a) Given, $x + \dfrac{1}{x} = 1$ ...(i)

$\therefore \dfrac{x^2 + 3x + 1}{x^2 + 7x + 1} = \dfrac{x\left(x+\dfrac{1}{x}\right) + 3x}{x\left(x+\dfrac{1}{x}\right) + 7x}$

$= \dfrac{x + 3x}{x + 7x} = \dfrac{4x}{8x} = \dfrac{1}{2}$ [from Eq. (i)]

35. (c) $a + \dfrac{1}{b} = b + \dfrac{1}{c}$

$\Rightarrow a - b = \dfrac{1}{c} - \dfrac{1}{b} = \dfrac{b-c}{cb}$ ...(i)

Similarly, $b - c = \dfrac{1}{a} - \dfrac{1}{c} = \dfrac{c-a}{ac}$ ...(ii)

and $c - a = \dfrac{1}{b} - \dfrac{1}{a} = \dfrac{a-b}{ab}$ ...(iii)

From Eqs. (i), (ii) and (iii), we get

$a - b = \dfrac{b-c}{cb} = \dfrac{c-a}{abc^2} = \dfrac{a-b}{a^2b^2c^2}$

$\Rightarrow a^2 b^2 c^2 = 1$

$\Rightarrow abc = \pm 1$

36. (b) $\because x + \dfrac{a}{x} = b \Rightarrow \dfrac{x^2 + a}{x} = b$

$\Rightarrow x^2 + a = bx$ ...(i)

Now, $\dfrac{x^2 + bx + a}{bx^2 - x^3} = \dfrac{(x^2 + a) + bx}{bx^2 - x^3}$

$= \dfrac{2bx}{bx^2 - x^3} = \dfrac{2b}{bx - x^2}$

$= \dfrac{2b}{a}$ $\begin{bmatrix} \because x^2 + a = bx \\ \Rightarrow bx - x^2 = a \end{bmatrix}$

37. (c) $\dfrac{(443 + 547)^2 + (443 - 547)^2}{443 \times 443 + 547 \times 547}$

$= \dfrac{\begin{bmatrix}(443)^2 + (547)^2 + 2 \times 443 \times 547 \\ +(443)^2 + (547)^2 - 2 \times 443 \times 547\end{bmatrix}}{(443)^2 + (547)^2}$

$= \dfrac{2[(443)^2 + (547)^2]}{[(443)^2 + (547)^2]} = 2$

38. (c) $\dfrac{(243 + 647)^2 + (243 - 647)^2}{243 \times 243 + 647 \times 647}$

[$\because (a+b)^2 + (a-b)^2 = 2(a^2 + b^2)$]

$= \dfrac{2[(243)^2 + (647)^2]}{[(243)^2 + (647)^2]} = 2$

39. (c) We know that,

$a^3 + b^3 + c^3 - 3abc$

$= (a + b + c)\dfrac{1}{2}[(a-b)^2 + (b-c)^2 + (c-a)^2]$

$= (225 + 226 + 227)\dfrac{1}{2}[1 + 1 + 4]$

$= 678 \times 3 = 2034$

40. (a) $\dfrac{(0.96)^3 - (0.1)^3}{(0.96)^2 + 0.096 + 0.01}$

$\because (0.96)^3 - (0.1)^3 = (0.96 - 0.1)$

$[(0.96)^2 + 0.96 \times 0.1 + (0.1)^2]$

$[\because a^3 - b^3 = (a-b)(a^2 + ab + b^2)]$

$\therefore \dfrac{(0.96 - 0.1)[(0.96)^2 + 0.096 + 0.01]}{(0.96)^2 + 0.096 + 0.01}$

$= 0.96 - 0.1 = 0.86$

41. (a) Given expression

$= \dfrac{a^3 + b^3 + c^3 - 3abc}{a^2 + b^2 + c^2 - ab - bc - ca}$

$= a + b + c = 0.5 + 0.2 + 0.3 = 1$

where, $a = 0.5, b = 0.2, c = 0.3$

42. (c) We know that, $a^3 + b^3 + c^3 - 3abc$

$= (a+b+c)(a^2 + b^2 + c^2 - ab - bc - ca)$

$\Rightarrow (a + b + c) = \left(\dfrac{a^3 + b^3 + c^3 - 3abc}{a^2 + b^2 + c^2 - ab - bc - ca}\right)$ ...(i)

Here, $a = 2.247, b = 1.730, c = 1.023$

$\therefore \dfrac{\begin{bmatrix}(2.247)^3 + (1.730)^3 + (1.023)^3 \\ - 3 \times 2.247 \times 1.730 \times 1.023\end{bmatrix}}{\begin{bmatrix}(2.247)^2 + (1.730)^2 + (1.023)^2 - (2.247 \\ \times 1.730) - (1.730 \times 1.023) - (2.247 \times 1.023)\end{bmatrix}}$

$= (2.247 + 1.730 + 1.023)$ [from Eq. (i)]

$= 5.000 = 5$

43. (b) Let $8.73 = a$ and $4.27 = b$

Given expression $= \dfrac{a^3 + b^3}{a^2 - ab + b^2}$

$= \dfrac{(a+b)(a^2 - ab + b^2)}{(a^2 - ab + b^2)} = (a + b)$

$= 8.73 + 4.27 = 13$

44. (c) Given, $a^2 + b^2 = 234$ and $ab = 108$

$\therefore (a+b)^2 = a^2 + b^2 + 2ab$

$= 234 + 2 \times 108 = 450$

and $(a-b)^2 = a^2 + b^2 - 2ab$

$= 234 - 2 \times 108 = 18$

Simplification / 127

$\therefore \dfrac{(a+b)^2}{(a-b)^2} = \dfrac{450}{18} = 25$

$\Rightarrow \left(\dfrac{a+b}{a-b}\right)^2 = 25$

$\Rightarrow \dfrac{a+b}{a-b} = \sqrt{25} = 5$

45. (b) $\because a^2 + 1 = a \Rightarrow a + \dfrac{1}{a} = 1$

On squaring both sides, we get

$a^2 + \dfrac{1}{a^2} + 2 = 1 \Rightarrow a^2 + \dfrac{1}{a^2} = -1$

On cubing both sides, we get

$\left(a^2 + \dfrac{1}{a^2}\right)^3 = (-1)^3$

$\Rightarrow a^6 + \dfrac{1}{a^6} + 3 \cdot a^2 \cdot \dfrac{1}{a^2}\left(a^2 + \dfrac{1}{a^2}\right) = -1$

$\Rightarrow a^6 + \dfrac{1}{a^6} + 3 \cdot (-1) = -1$

$\Rightarrow a^6 + \dfrac{1}{a^6} + 1 = 3$

$\Rightarrow a^6 + \dfrac{1}{a^6} = 2$ ...(i)

For Eq. (i) to be true, a can take value of -1 or $+1$ only.

$\therefore a^{12} + a^6 + 1 = (\pm 1)^{12} + (\pm 1)^6 + 1 = 3$

46. (a) $\because a + b + c = 0$

$\therefore \quad a + b = -c$ ...(i)
$\quad a + c = -b$ ...(ii)
$\quad b + c = -a$ ...(iii)

Now, $\left(\dfrac{a+b}{c} + \dfrac{b+c}{a} + \dfrac{c+a}{b}\right)$

$\left(\dfrac{a}{b+c} + \dfrac{b}{c+a} + \dfrac{c}{a+b}\right)$

$= \left[\dfrac{(-c)}{c} + \dfrac{(-a)}{a} + \dfrac{(-b)}{b}\right]\left[\dfrac{a}{-a} + \dfrac{b}{-b} + \dfrac{c}{-c}\right]$

[put the values of $a+b$, $b+c$ and $c+a$ from Eqs. (i), (ii) and (iii)]

$= [(-1) + (-1) + (-1)][(-1) + (-1) + (-1)]$

$= (-3) \times (-3) = 9$

47. (a) $x = \dfrac{\sqrt{2}+1}{\sqrt{2}-1}$

$\Rightarrow x = \dfrac{\sqrt{2}+1}{\sqrt{2}-1} \times \dfrac{\sqrt{2}+1}{\sqrt{2}+1}$

[by rationalising]

$\Rightarrow x = \dfrac{2 + 1 + 2\sqrt{2}}{1}$

$\Rightarrow x = 3 + 2\sqrt{2}$ ...(i)

and $x - y = 4\sqrt{2}$

$\Rightarrow y = x - 4\sqrt{2}$
$= 3 + 2\sqrt{2} - 4\sqrt{2}$ [from Eq. (i)]
$= 3 - 2\sqrt{2}$

Now, $x^2 + y^2 = (3 + 2\sqrt{2})^2 + (3 - 2\sqrt{2})^2$
$= 9 + 8 + 12\sqrt{2} + 9 + 8 - 12\sqrt{2} = 34$

48. (a) Given, $a = (\sqrt{2} - 1)^{1/3}$

$\therefore a^3 = \sqrt{2} - 1$

Now, $(a - a^{-1})^3 + 3(a - a^{-1})$

$= \left(a - \dfrac{1}{a}\right)^3 + 3\left(a - \dfrac{1}{a}\right)$

$= a^3 - \dfrac{1}{a^3} - 3\left(a - \dfrac{1}{a}\right) + 3\left(a - \dfrac{1}{a}\right)$

$= \sqrt{2} - 1 - \dfrac{1}{\sqrt{2} - 1}$

$= (\sqrt{2} - 1) - \dfrac{1}{(\sqrt{2} - 1)} \times \dfrac{(\sqrt{2} + 1)}{(\sqrt{2} + 1)}$

[by rationalising]

$= \sqrt{2} - 1 - (\sqrt{2} + 1) = -2$

49. (a) Given, $x = \dfrac{\sqrt{3}+1}{\sqrt{3}-1}$

and $y = \dfrac{\sqrt{3}-1}{\sqrt{3}+1}$

Now, $x + y = \dfrac{(\sqrt{3}+1)^2 + (\sqrt{3}-1)^2}{(\sqrt{3}-1)(\sqrt{3}+1)}$

$= \dfrac{3 + 1 + 2\sqrt{3} + 3 + 1 - 2\sqrt{3}}{3 - 1}$

$= \dfrac{8}{2} = 4$

and $xy = \dfrac{\sqrt{3}+1}{\sqrt{3}-1} \times \dfrac{\sqrt{3}-1}{\sqrt{3}+1} = 1$

$\therefore \dfrac{x^2}{y} + \dfrac{y^2}{x} = \dfrac{x^3 + y^3}{xy}$

$= \dfrac{(x+y)^3 - 3xy(x+y)}{xy}$

$= \dfrac{(4)^3 - 3 \times 1 \times 4}{1} = \dfrac{64 - 12}{1} = 52$

50. (b) Given, $p + q = 10$ and $pq = 5$

Now, $\dfrac{p}{q} + \dfrac{q}{p} = \dfrac{p^2 + q^2}{pq} = \dfrac{(p+q)^2 - 2pq}{pq}$

$= \dfrac{(10)^2 - 2 \times 5}{5}$

$= \dfrac{100 - 10}{5} = \dfrac{90}{5} = 18$

51. (b) Given, $a + b = 8, ab = 15$
We know that,
$$(a + b)^3 = a^3 + b^3 + 3ab(a + b)$$
$\Rightarrow \quad (8)^3 = a^3 + b^3 + 3 \times 15 \times (8)$
$\Rightarrow \quad 512 = a^3 + b^3 + 360$
$\Rightarrow \quad a^3 + b^3 = 512 - 360$
$\Rightarrow \quad a^3 + b^3 = 152$

52. (d) $x^3 + y^3 + z^3 - 3xyz$
$= (x + y + z)(x^2 + y^2 + z^2 - xy - yz - zx)$
$\therefore x^3 + y^3 + z^3 - 3(-1) = (1)[x^2 + y^2 + z^2 - (-1)]$
$\Rightarrow x^3 + y^3 + z^3 + 3 = x^2 + y^2 + z^2 + 1$
Also, $(x + y + z)^2 = x^2 + y^2 + z^2 + 2(xy + yz + zx)$
$1 = x^2 + y^2 + z^2 - 2$
$\Rightarrow x^2 + y^2 + z^2 = 3$
$\therefore x^3 + y^3 + z^3 = 3 + 1 - 3 = 1$

53. (c) $(a + b + c)^2 = a^2 + b^2 + c^2 + 2(ab + bc + ca)$
$\Rightarrow (a + b + c)^2 = 250 + 2 \times 3 = 256$
$[\because a^2 + b^2 + c^2 = 250$ and $ab + bc + ca = 3]$
$\therefore \quad a + b + c = \sqrt{256} = \pm 16$

54. (c) We know that,
$(a + b + c)^2 = (a^2 + b^2 + c^2)$
$\qquad + 2(ab + bc + ca)$
$\Rightarrow \quad 196 = 96 + 2(ab + bc + ca)$
$\Rightarrow \quad 2(ab + bc + ca) = 196 - 96 = 100$

$\therefore \quad (ab + bc + ca) = \dfrac{100}{2} = 50$

55. (b) Given, $x = 11$
$\therefore x^5 - 12x^4 + 12x^3 - 12x^2 + 12x - 1$
$= x^4(x - 12) + 12x^2(x - 1) + 12x - 1$
$= x^4(11 - 12) + 12x^2(11 - 1) + 12 \times 11 - 1$
$= x^4(-1) - 12x^2(10) + 132 - 1$
$= -x^4 + 120x^2 + 131$
$= x^2(-x^2 + 120) + 131$
$= (11)^2[-(11)^2 + 120] + 131$
$= 121(-121 + 120) + 131$
$= 121(-1) + 131$
$= -121 + 131 = 10$

56. (a) Given, $a = \dfrac{1}{a-5} \Rightarrow \dfrac{1}{a} = a - 5$
$\Rightarrow \quad a - \dfrac{1}{a} = 5$
On squaring both sides, we get
$a^2 + \dfrac{1}{a^2} - 2 = 25$
$\Rightarrow \quad a^2 + \dfrac{1}{a^2} = 27$
$\therefore \quad \left(a + \dfrac{1}{a}\right)^2 = a^2 + \dfrac{1}{a^2} + 2$
$= 27 + 2 = 29$
$\therefore \quad a + \dfrac{1}{a} = \sqrt{29}$

Exercise 2 *Higher Skill Level Questions*

1. (c) $\dfrac{\sqrt[4]{0.0625} + \sqrt[3]{0.008} + \sqrt{0.09} - 1}{\sqrt[3]{62.5} \sqrt[5]{32}}$

$= \dfrac{\sqrt[4]{(0.5)^4} + \sqrt[3]{(0.2)^3} + \sqrt{(0.3)^2} - 1}{\sqrt[3]{62.5} \sqrt[5]{2^5}}$

$= \dfrac{0.5 + 0.2 + 0.3 - 1}{\sqrt[3]{62.5 \times 2}} = \dfrac{1 - 1}{\sqrt[3]{125}} = \dfrac{0}{5} = 0$

2. (b) We have,
$\qquad x = 2 + 2^{2/3} + 2^{1/3}$ ...(i)
$\Rightarrow \quad x - 2 = 2^{1/3}(2^{1/3} + 1)$
On cubing both the sides, we get
$\qquad (x - 2)^3 = 2(2^{1/3} + 1)^3$
$\Rightarrow x^3 - 3(x^2)(2) + 3(x)(2)^2 - 8$
$= 2[(2^{1/3})^3 + 3(2^{1/3})^2(1) + 3(2^{1/3})(1) + (1)^3)]$
$\Rightarrow x^3 - 6x^2 + 12x - 8 = 2$
$\qquad [2 + 3 \cdot 2^{2/3} + 3 \cdot 2^{1/3} + 1]$

$\Rightarrow x^3 - 6x^2 + 12x - 8 = 2[3 + 3 \cdot 2^{2/3} + 3 \cdot 2^{1/3}]$
$\Rightarrow x^3 - 6x^2 + 12x - 8 = 6[1 + 2^{2/3} + 2^{1/3}]$
$\Rightarrow x^3 - 6x^2 + 12x - 8 = 6[1 + x - 2]$
$\qquad$ [from Eq. (i)]
$\Rightarrow x^3 - 6x^2 + 12x - 8 = 6(x - 1)$
$\Rightarrow x^3 - 6x^2 + 12x - 8 = 6x - 6$
$\Rightarrow \quad x^3 - 6x^2 + 6x = 2$

3. (a) Given, $x^2 + \dfrac{1}{x^2} = 98$
$\Rightarrow \quad x^2 + \dfrac{1}{x^2} + 2 = 98 + 2 = 100$
$\Rightarrow \quad \left[(x)^2 + \left(\dfrac{1}{x}\right)^2 + 2 \cdot x \cdot \dfrac{1}{x}\right] = 100$
$\Rightarrow \quad \left(x + \dfrac{1}{x}\right)^2 = (10)^2 \Rightarrow \left(x + \dfrac{1}{x}\right) = 10$

Simplification / 129

Now, cubing on both sides, we get
$$x^3 + \frac{1}{x^3} + 3 \cdot x \cdot \frac{1}{x}\left(x + \frac{1}{x}\right) = 1000$$
$$\Rightarrow x^3 + \frac{1}{x^3} + 3 \times 10 = 1000$$
$$\Rightarrow \left(x^3 + \frac{1}{x^3}\right) = 1000 - 30$$
$$\Rightarrow \left(x^3 + \frac{1}{x^3}\right) = 970$$

4. (c) Given, $x = p + \frac{1}{p}$ and $y = p - \frac{1}{p}$

$\therefore \quad x^4 - 2x^2y^2 + y^4 = [(x^2 - y^2)]^2$
$= [(x+y)(x-y)]^2$
$= \left[\left(p + \frac{1}{p} + p - \frac{1}{p}\right)\left(p + \frac{1}{p} - p + \frac{1}{p}\right)\right]^2$

[put the values of x and y]

$= \left[(2p)\left(\frac{2}{p}\right)\right]^2 = 4 \times 4 = 16$

5. (b) Given, $3(a^2 + b^2 + c^2) = (a + b + c)^2$
$\Rightarrow 3a^2 + 3b^2 + 3c^2 = a^2 + b^2 + c^2 + 2ab + 2bc + 2ca$
$\Rightarrow 2a^2 + 2b^2 + 2c^2 - 2ab - 2bc - 2ca = 0$
$\Rightarrow (a-b)^2 + (b-c)^2 + (c-a)^2 = 0$
$\therefore \quad a = b = c$

6. (d) $\dfrac{x+y+z}{xa+yb+zc+ya+zb+xc+za+xb+yc}$

[adding all the ratios]
$= \dfrac{x+y+z}{x(a+b+c) + y(a+b+c) + z(a+b+c)}$
$= \dfrac{x+y+z}{(x+y+z)(a+b+c)} = \dfrac{1}{a+b+c}$

7. (c) Given, $x + \dfrac{1}{x} = 2 \Rightarrow x^2 + 1 = 2x$
$\Rightarrow x^2 - 2x + 1 = 0 \Rightarrow (x-1)^2 = 0$
[$\because (a-b)^2 = a^2 + b^2 - 2ab$]
$\Rightarrow x - 1 = 0 \Rightarrow x = 1$
$\therefore \quad x^7 + \dfrac{1}{x^5} = (1)^7 + \dfrac{1}{(1)^5}$
$= 1 + \dfrac{1}{1} = 1 + 1 = 2$

8. (c) Given, $\left(x + \dfrac{1}{x}\right) = 3$

$\therefore \dfrac{3x^2 - 4x + 3}{x^2 - x + 1} = \dfrac{x\left(3x - 4 + \dfrac{3}{x}\right)}{x\left(x - 1 + \dfrac{1}{x}\right)} = \dfrac{3x + \dfrac{3}{x} - 4}{x + \dfrac{1}{x} - 1}$

$= \dfrac{3\left(x + \dfrac{1}{x}\right) - 4}{\left(x + \dfrac{1}{x}\right) - 1} = \dfrac{3 \times 3 - 4}{3 - 1}$
$= \dfrac{9-4}{2} = \dfrac{5}{2}$

9. (a) Given, $x = 2^{1/3} + 2^{-1/3}$
$\therefore 2x^3 - 6x - 5 = 2(2^{1/3} + 2^{-1/3})^3 - 6(2^{1/3} + 2^{-1/3}) - 5$
$= 2[2 + 2^{-1} + 3(2^{1/3} + 2^{-1/3})] - 6(2^{1/3} + 2^{-1/3}) - 5$
$= 4 + 2 \times \dfrac{1}{2} + 6(2^{1/3} + 2^{-1/3}) - 6(2^{1/3} + 2^{-1/3}) - 5$
$= 4 + 1 - 5 = 0$

10. (b) Given expression
$= \dfrac{1+x}{1+\sqrt{1+x}} + \dfrac{1-x}{1-\sqrt{1-x}}$
$= \dfrac{1+\dfrac{\sqrt{3}}{2}}{1+\sqrt{1+\dfrac{\sqrt{3}}{2}}} + \dfrac{1-\dfrac{\sqrt{3}}{2}}{1-\sqrt{1-\dfrac{\sqrt{3}}{2}}}$

$\left[\text{put } x = \dfrac{\sqrt{3}}{2}\right]$

$= \dfrac{2+\sqrt{3}}{\sqrt{2}(\sqrt{2}+\sqrt{2+\sqrt{3}})} + \dfrac{2-\sqrt{3}}{\sqrt{2}(\sqrt{2}-\sqrt{2-\sqrt{3}})}$

$= \dfrac{2+\sqrt{3}}{2+\sqrt{4+2\sqrt{3}}} + \dfrac{2-\sqrt{3}}{2-\sqrt{4-2\sqrt{3}}}$

$= \dfrac{2+\sqrt{3}}{2+\sqrt{(1)^2 + (\sqrt{3})^2 + 2\sqrt{3}}}$
$\quad + \dfrac{2-\sqrt{3}}{2-\sqrt{(1)^2 + (\sqrt{3})^2 - 2\sqrt{3}}}$

$= \dfrac{2+\sqrt{3}}{2+\sqrt{(1+\sqrt{3})^2}} + \dfrac{2-\sqrt{3}}{2-\sqrt{(\sqrt{3}-1)^2}}$

$= \dfrac{2+\sqrt{3}}{2+(1+\sqrt{3})} + \dfrac{2-\sqrt{3}}{2-(\sqrt{3}-1)}$

$= \dfrac{2+\sqrt{3}}{3+\sqrt{3}} + \dfrac{2-\sqrt{3}}{3-\sqrt{3}}$

$= \dfrac{(2+\sqrt{3})(3-\sqrt{3}) + (2-\sqrt{3})(3+\sqrt{3})}{(3+\sqrt{3})(3-\sqrt{3})}$

$= \dfrac{6 + 3\sqrt{3} - 2\sqrt{3} - 3 + 6 - 3\sqrt{3} + 2\sqrt{3} - 3}{(3+\sqrt{3})(3-\sqrt{3})}$

$= \dfrac{6}{9 + 3\sqrt{3} - 3\sqrt{3} - 3} = \dfrac{6}{6} = 1$

11. (c) Given, $x^2 + y^2 + z^2 = xy + yz + zx$

$\therefore \quad x^2 + y^2 + z^2 - xy - yz - zx = 0$

$\Rightarrow \dfrac{1}{2}[(x-y)^2 + (y-z)^2 + (z-x)^2] = 0$

$\therefore \qquad x = y = z$

Now, $\dfrac{3x^4 + 7y^4 + 5z^4}{5x^2y^2 + 7y^2z^2 + 3z^2x^2} = \dfrac{3+7+5}{5+7+3}$

$\qquad\qquad = \dfrac{15}{15} = 1$

12. (d) Given, $x = 3 + 2\sqrt{2} = 3 + \sqrt{8}$

$\therefore \dfrac{1}{x} = \dfrac{1}{3+\sqrt{8}} \times \dfrac{3-\sqrt{8}}{3-\sqrt{8}} = \dfrac{3-\sqrt{8}}{9-8} = (3-\sqrt{8})$

$\Rightarrow x + \dfrac{1}{x} = 3 + \sqrt{8} + 3 - \sqrt{8} = 6$

$\Rightarrow \qquad x + \dfrac{1}{x} = 6 \qquad\qquad \text{...(i)}$

On cubing both sides, we get

$\left(x + \dfrac{1}{x}\right)^3 = (6)^3$

$\Rightarrow \quad x^3 + \dfrac{1}{x^3} + 3 \cdot x \cdot \dfrac{1}{x}\left(x + \dfrac{1}{x}\right) = 216$

$\Rightarrow \quad x^3 + \dfrac{1}{x^3} + 3 \times 6 = 216$

$\Rightarrow \quad x^3 + \dfrac{1}{x^3} = 216 - 18 = 198 \qquad \text{...(ii)}$

Now, $\quad \dfrac{x^6 + x^4 + x^2 + 1}{x^3}$

$= \dfrac{x^6}{x^3} + \dfrac{x^4}{x^3} + \dfrac{x^2}{x^3} + \dfrac{1}{x^3}$

$= x^3 + x + \dfrac{1}{x} + \dfrac{1}{x^3}$

$= \left(x^3 + \dfrac{1}{x^3}\right) + \left(x + \dfrac{1}{x}\right)$

$= 198 + 6 = 204$

[from Eqs. (i) and (ii)]

13. (d) Given, $a - \dfrac{1}{a-3} = 5 \qquad \text{...(i)}$

Let $\qquad a - 3 = x$

Then, $\qquad a = x + 3$

From Eq. (i), $x + 3 - \dfrac{1}{x} = 5 \Rightarrow x - \dfrac{1}{x} = 2 \quad \text{...(ii)}$

On cubing both sides, we get

$\left(x - \dfrac{1}{x}\right)^3 = (2)^3$

$\Rightarrow \quad x^3 - \dfrac{1}{x^3} - 3 \cdot x \cdot \dfrac{1}{x}\left(x - \dfrac{1}{x}\right) = 8$

$\Rightarrow \quad x^3 - \dfrac{1}{x^3} - 3 \times (2) = 8 \quad$ [from Eq. (ii)]

$\Rightarrow \quad x^3 - \dfrac{1}{x^3} = 14 \Rightarrow (a-3)^3 - \dfrac{1}{(a-3)^3} = 14$

14. (a) Given, $x = a^{1/2} + a^{-1/2}$

$\Rightarrow \qquad x = a^{1/2} + \dfrac{1}{a^{1/2}} = \dfrac{a+1}{a^{1/2}}$

On squaring both sides, we get

$x^2 = \left(\dfrac{a+1}{a^{1/2}}\right)^2 = \dfrac{a^2 + 1 + 2a}{a} = a + \dfrac{1}{a} + 2$

$\Rightarrow \qquad x^2 = a + \dfrac{1}{a} + 2 \qquad \text{...(i)}$

Similarly, $\quad y^2 = a + \dfrac{1}{a} - 2 \qquad \text{...(ii)}$

Now, $(x^4 - x^2y^2 - 1) + (y^4 - x^2y^2 + 1)$

$= x^4 - x^2y^2 - 1 + y^4 - x^2y^2 + 1$

$= x^4 + y^4 - 2x^2y^2 = (x^2 - y^2)^2$

$= \left(a + \dfrac{1}{a} + 2 - a - \dfrac{1}{a} + 2\right)^2$

[from Eqs. (i) and (ii)]

$= (4)^2 = 16$

15. (b) Given, $x^2 + \dfrac{1}{x^2} = \dfrac{17}{4}$

$\Rightarrow \quad x^2 + \dfrac{1}{x^2} + 2 - 2 = \dfrac{17}{4} \Rightarrow \left(x - \dfrac{1}{x}\right)^2 + 2 = \dfrac{17}{4}$

$\Rightarrow \quad \left(x - \dfrac{1}{x}\right)^2 = \dfrac{17}{4} - 2 \Rightarrow \left(x - \dfrac{1}{x}\right)^2 = \dfrac{9}{4}$

$\Rightarrow \quad \left(x - \dfrac{1}{x}\right) = \dfrac{3}{2} \qquad \text{...(i)}$

On cubing both sides, we get

$\left(x - \dfrac{1}{x}\right)^3 = \left(\dfrac{3}{2}\right)^3$

$\Rightarrow \quad x^3 - \dfrac{1}{x^3} - 3 \cdot \dfrac{1}{x} \cdot x \left(x - \dfrac{1}{x}\right) = \dfrac{27}{8}$

$\Rightarrow \quad x^3 - \dfrac{1}{x^3} = \dfrac{27}{8} + 3\left(\dfrac{3}{2}\right) \quad$ [from Eq. (i)]

$\Rightarrow \quad x^3 - \dfrac{1}{x^3} = \dfrac{27}{8} + \dfrac{9}{2} \Rightarrow x^3 - \dfrac{1}{x^3} = \dfrac{63}{8}$

16. (c) Given, $x = \sqrt{3} + \sqrt{2}$ and $y = \sqrt{3} - \sqrt{2}$

$\therefore (x^3 - 20\sqrt{2}) - (y^3 + 2\sqrt{2})$

$= [(\sqrt{3} + \sqrt{2})^3 - 20\sqrt{2}] - [(\sqrt{3} - \sqrt{2})^3 + 2\sqrt{2}]$

$= (3\sqrt{3} + 2\sqrt{2} + 9\sqrt{2} + 6\sqrt{3} - 20\sqrt{2})$

$\qquad - (3\sqrt{3} - 2\sqrt{2} - 9\sqrt{2} + 6\sqrt{3} + 2\sqrt{2})$

$= (9\sqrt{3} - 9\sqrt{2}) - (9\sqrt{3} - 9\sqrt{2})$

$= 9\sqrt{3} - 9\sqrt{2} - 9\sqrt{3} + 9\sqrt{2} = 0$

17. (c) We have,

$\sqrt{\dfrac{x}{y}} = \dfrac{24}{5} + \sqrt{\dfrac{y}{x}}$

Simplification / 131

Let $\sqrt{\dfrac{x}{y}} = z$

$\therefore \quad z = \dfrac{24}{5} + \dfrac{1}{z} \Rightarrow z = \dfrac{24z + 5}{5z}$

$\Rightarrow \quad 5z^2 - 24z - 5 = 0$

$\Rightarrow \quad 5z^2 - 25z + z - 5 = 0$

$\Rightarrow \quad 5z(z-5) + 1(z-5) = 0$

$\Rightarrow \quad (z-5)(5z+1) = 0$

$\Rightarrow \quad z = 5 \text{ or } -\dfrac{1}{5} \Rightarrow \sqrt{\dfrac{x}{y}} = 5 \text{ or } -\dfrac{1}{5}$

$\Rightarrow \quad \sqrt{\dfrac{x}{y}} = 5 \qquad \left[\because \sqrt{\dfrac{x}{y}} \neq -\dfrac{1}{5}\right]$

$\Rightarrow \quad x = 25y \qquad \dots(i)$

Again, $x + y = 26$

$\Rightarrow \quad 25y + y = 26$ [from Eq. (i)]

$\Rightarrow \quad 26y = 26$

$\Rightarrow \quad y = 1$

$\therefore \quad x = 25$

$\therefore \quad xy = 25 \times 1 = 25$

18. (d) We have, $x - y = 8$ ...(i)

and $\sqrt{\dfrac{x}{y}} = \dfrac{10}{3} - \sqrt{\dfrac{y}{x}} \Rightarrow \sqrt{\dfrac{x}{y}} + \sqrt{\dfrac{y}{x}} = \dfrac{10}{3}$

$\Rightarrow \quad \dfrac{(\sqrt{x})^2 + (\sqrt{y})^2}{\sqrt{xy}} = \dfrac{10}{3}$

$\Rightarrow \quad x + y = \dfrac{10}{3}\sqrt{xy}$

On squaring both sides, we get

$(x + y)^2 = \dfrac{100}{9} xy$

$\Rightarrow \quad (x - y)^2 + 4xy = \dfrac{100}{9} xy$

$\qquad [\because (x+y)^2 = (x-y)^2 + 4xy]$

$\Rightarrow \quad (8)^2 = \left(\dfrac{100}{9} - 4\right) xy$ [from Eq. (i)]

$\Rightarrow \quad 64 = \dfrac{64}{9} xy$

$\Rightarrow \quad xy = 9$

19. (a) Given, $x + \dfrac{1}{x} = 3$ ...(i)

On squaring both sides, we get

$\left(x + \dfrac{1}{x}\right)^2 = (3)^2$

$\Rightarrow \quad x^2 + \dfrac{1}{x^2} + 2 = 9 \Rightarrow x^2 + \dfrac{1}{x^2} = 7$ ...(ii)

Again, $x + \dfrac{1}{x} = 3$

On cubing both sides, we get

$x^3 + \dfrac{1}{x^3} + 3 \cdot x \cdot \dfrac{1}{x}\left(x + \dfrac{1}{x}\right) = 27$

$\Rightarrow \quad x^3 + \dfrac{1}{x^3} + 9 = 27$

$\Rightarrow \quad x^3 + \dfrac{1}{x^3} = 18$ ...(iii)

On multiplying Eqs. (ii) and (iii), we get

$\left(x^2 + \dfrac{1}{x^2}\right) \times \left(x^3 + \dfrac{1}{x^3}\right) = 18 \times 7$

$\Rightarrow \quad x^5 + x + \dfrac{1}{x} + \dfrac{1}{x^5} = 126$

$\Rightarrow \quad \left(x^5 + \dfrac{1}{x^5}\right) + \left(x + \dfrac{1}{x}\right) = 126$

$\Rightarrow \quad x^5 + \dfrac{1}{x^5} + 3 = 126$

$\Rightarrow \quad x^5 + \dfrac{1}{x^5} = 126 - 3 = 123$

$\therefore \quad x^5 + \dfrac{1}{x^5} = 123$

20. (b) We have, $a^3 = 335 + b^3$

$\Rightarrow \quad a^3 - b^3 = 335$ ...(i)

and $a = 5 + b$

$\Rightarrow \quad a - b = 5$ ...(ii)

Now, $(a - b)^3 = a^3 - b^3 - 3ab(a-b)$

$\Rightarrow \quad (5)^3 = 335 - 3ab(5)$

[from Eqs. (i) and (ii)]

$\Rightarrow \quad 125 = 335 - 15 ab \Rightarrow ab = 14$

Again, $(a+b)^2 = (a-b)^2 + 4ab = (5)^2 + 4(14)$

$= 25 + 56$

$(a + b)^2 = 81$

$\therefore \quad a + b = 9$

21. (a) Given, $a^3 - b^3 = 117$ ...(i)

and $a - b = 3$ ...(ii)

On cubing both sides of Eq. (ii), we get

$(a - b)^3 = (3)^3 \Rightarrow a^3 - b^3 - 3ab(a - b) = 27$

$\Rightarrow \quad 117 - 3ab(3) = 27 \Rightarrow 117 - 9ab = 27$

[from Eqs. (i) and (ii)]

$\Rightarrow \quad 9ab = 90 \Rightarrow ab = 10$

Now, $(a+b)^2 = (a-b)^2 + 4ab$

$\Rightarrow \quad (a+b)^2 = (3)^2 + 4(10)$

$\Rightarrow \quad (a+b)^2 = 9 + 40 = 49$

$\Rightarrow \quad a + b = 7 \qquad [\because a > 0 \text{ and } b > 0]$

22. (b) Given, $a - b = 4$ ...(i)

On squaring both sides, we get

$(a - b)^2 = (4)^2 \Rightarrow a^2 + b^2 - 2ab = 16$

$\Rightarrow \quad 40 - 2ab = 16 \qquad [\because a^2 + b^2 = 40, \text{ given}]$

$\Rightarrow \quad 2ab = 40 - 16 \Rightarrow 2ab = 24 \Rightarrow ab = 12$

$\therefore \quad a+b = \sqrt{a^2+b^2+2ab} = \sqrt{40+2\times 12}$
$\qquad = \sqrt{40+24} = \sqrt{64}$
$\Rightarrow \quad a+b=8 \qquad \ldots(\text{ii})$

On adding Eqs. (i) and (ii), we get
$\qquad 2a=12 \Rightarrow a=6$

On putting the value of a in Eq. (i), we get
$\qquad 6-b=4 \Rightarrow b=2$

Now, $a^3+b^6 = 6^3+2^6 = 2^3 \times 3^3 + 2^6$
$\qquad = 2^3(3^3+2^3)$
$\qquad = 8(27+8) = 8\times 35 = 280$

Hence, the value of a^3+b^6 is 280.

23. (a) Let $(m-n)=a$, $(n-r)=b$, $(r-m)=c$
Then, $\qquad a+b+c=0$
$\therefore \qquad a^3+b^3+c^3 = 3abc$
Now, $\qquad \dfrac{a^3+b^3+c^3}{6abc} = \dfrac{3abc}{6abc} = \dfrac{1}{2}$

24. (a) Given, $ax+by=1 \qquad \ldots(\text{i})$
and $\qquad bx+ay = \dfrac{2ab}{a^2+b^2} \qquad \ldots(\text{ii})$

On multiplying Eq. (i) by a and Eq. (ii) by b and then subtract, we get

$\qquad a(ax+by=1)$
$\qquad b\left(bx+ay = \dfrac{2ab}{a^2+b^2}\right)$
$\qquad \overline{}$
$\qquad a^2x - b^2x = a - \dfrac{2ab}{a^2+b^2}$

$\Rightarrow \quad x = \dfrac{a^3+ab^2-2ab^2}{(a^2+b^2)(a^2-b^2)}$
$\qquad = \dfrac{a^3-ab^2}{(a^2+b^2)(a^2-b^2)} = \dfrac{a}{a^2+b^2}$

$\therefore \quad y = \dfrac{1-ax}{b} = \dfrac{1-\dfrac{a^2}{a^2+b^2}}{b}$
$\qquad = \dfrac{a^2+b^2-a^2}{b(a^2+b^2)} = \dfrac{b}{a^2+b^2}$

$\therefore (x^2+y^2)(a^2+b^2)$
$\qquad = \left[\dfrac{a^2}{(a^2+b^2)^2} + \dfrac{b^2}{(a^2+b^2)^2}\right](a^2+b^2)$
$\qquad = \dfrac{(a^2+b^2)(a^2+b^2)}{(a^2+b^2)^2} = 1$

Chapter 8

Approximation

In mathematical expression, which includes division and multiplication of decimal values of large number, it becomes quite complicated to solve them. So, to reduce this complexity, we use **approximation** method.

In approximation method, we need not calculate the exact value of an expression, but we calculate the nearest (round off) values. When we use approximation method, then final result obtained is not equal to the exact result, but it is very close to the final result (either a little less or little more).

Basic Rules to Solve the Problems Based on Approximation

Rule ❶

To solve the complex mathematical expression, take the nearest value of numbers given in the expression.

For example 199.03 is approximated to 200 and 94.6% is approximated to 95% etc.

Ex. 1 89% of (599.88 ÷ 30 × 400) + 50 = ?

Sol. ∵ 89% of (599.88 ÷ 30 × 400) + 50 = ? [given]

∴ ? ≈ $\frac{90}{100}$ × (600 ÷ 30 × 400) + 50

= $\frac{9}{10}$ × 8000 + 50 = 7200 + 50 = 7250

Rule ❷

To multiply large numbers, we can take the approximate value (round off) of numbers by increasing one number and decreasing the other accordingly, so that the calculation is eased.

For example 589 × 231 is approximated to 590 × 230.

Ex. 2 393 × 197 + 5600 × $\frac{5}{4}$ + 8211.80 = ?

Sol. ? = 393 × 197 + 5600 × $\frac{5}{4}$ + 8211.80

≈ 390 × 200 + 5600 × $\frac{5}{4}$ + 8200

= 78000 + 7000 + 8200 = 93200

Rule ❸

When we divide two large decimal numbers, we can increase or decrease both numbers accordingly.
For example 7987.26 ÷ 38.69 is approximated as 8000 ÷ 40.

Ex. 3 (9615.36 + 1247.18) ÷ (2435.72 + 1937.92) = ?

Sol. ? = (9615.36 + 1247.18) ÷ (2435.72 + 1937.92)
= 10862.54 ÷ 4373.64 ≈ 10860 ÷ 4370
= 2.48 ≈ 2.5

Rule ❹

To find the percentage of any number, we can use the following shortcut methods
- To calculate 10% of any number, we simply put a decimal after a digit from the right end.
- To calculate 1% of any number, we simply put a decimal after two digits from the right end.
- To calculate 25% of any number, we simply divide the number by 4.

Ex. 4 24% of 3580 + 799.99 ÷ $\dfrac{1000}{25}$ = ?

Sol. ? = 24% of 3580 + 799.99 ÷ $\dfrac{1000}{25}$
≈ 25% of 3600 + 800 ÷ 40
= 900 + 20 = 920

Ex. 5 10% of 1350 + ? = 365

Sol. 10% of 1350 + ? = 365
⇒ 135 + ? = 365
∴ ? = 365 − 135 = 230

Fast Track Practice

Exercise 1 Base Level Questions

Directions (Q. Nos. 1-65) *What approximate value should come in place of the question mark (?) in the following questions? (Note that you are not expected to calculate the exact value).*

1. 12.49% of 839.859 = ? [RBI Clerk 2008]
   (a) 95   (b) 100   (c) 115   (d) 105
   (e) 90

2. 124.35% of 8096 = ? [OBC PO 2004]
   (a) 2000   (b) 10000   (c) 1000   (d) 12000
   (e) None of these

3. 5454 ÷ 54 ÷ 5 = ? [IBPS Clerk 2011]
   (a) 15   (b) 25   (c) 30   (d) 20
   (e) None of these

4. 99.999 ÷ 0.99 ÷ 0.00991234 = ? [RBI 2003]
   (a) 100   (b) 10000   (c) 1000   (d) 100000
   (e) 99999

5. $2\frac{3}{5} \times \frac{15}{26} \times 283.75 = ?$ [UBI PO 2005]
   (a) 440   (b) 435   (c) 410   (d) 425
   (e) 400

6. 23.003 × 22.998 + 100.010 = ? [IBPS Clerk 2011]
   (a) 630   (b) 550   (c) 700   (d) 720
   (e) 510

7. 16.003 × 27.998 − 209.010 = ? [IBPS Clerk 2011]
   (a) 150   (b) 200   (c) 75   (d) 240
   (e) 110

8. 19.003 × 22.998 − 280.010 = ? [IBPS Clerk 2011]
   (a) 220   (b) 110   (c) 160   (d) 90
   (e) None of these

9. 26.003 × 37.998 − 309.010 = ? [SBI Clerk (Mains) 2016]
   (a) 685   (b) 695   (c) 680   (d) 679
   (e) 675

10. 42.8 × 13.5 × 16.2 × ? = 2340.09 [SBI Clerk (Mains) 2016]
    (a) 0.15   (b) 0.25   (c) 0.5   (d) 0.75
    (e) 1

11. 815.002 + 29.98 − 53.998 + (3.01)² = ? [SBI PO (Pre) 2016]
    (a) 800   (b) 880   (c) 840   (d) 900
    (e) 750

12. (9.95)² × (2.01)³ = 2 × (?)² [Dena Bank Clerk 2005]
    (a) 12   (b) 15   (c) 25   (d) 20
    (e) 16

13. 96.894 + 33.002 + 15.02 × 7.99 = ? [SBI PO (Pre) 2016]
    (a) 180   (b) 123   (c) 140   (d) 250
    (e) 170

14. 425 ÷ 16.95 × ? = 225 [OBC PO 2004]
    (a) 11   (b) 0.8   (c) 9   (d) 19
    (e) 0.9

15. $36.0001 ÷ 5.9998 \times \sqrt{?} = 108.0005$ [Andhra Bank PO 2003]
    (a) 18   (b) 16   (c) 256   (d) 316
    (e) 325

16. 26.003 − 154.001 ÷ 6.995 = ? [Corporation Bank 2004]
    (a) 4   (b) 18   (c) 9   (d) 10
    (e) 14

17. ?% of $218 = 3\frac{1}{3} \times 3.3121$ [IBPS PO 2011]
    (a) 2.5   (b) 5   (c) 10   (d) 15
    (e) None of these

18. ?% of (6274 ÷ 6.14) = 646.51 − 241.58 [IBPS Clerk (Mains) 2017]
    (a) 40   (b) 45   (c) 39   (d) 65
    (e) 74

19. 10.56 × ? ÷ 3 = 576.73 × 2.02 [IBPS Clerk (Mains) 2017]
    (a) 328   (b) 295   (c) 412   (d) 381
    (e) 235

20. $768 ÷ 11.82 \times \sqrt{168} − 42 = ?$ [IBPS Clerk (Mains) 2017]
    (a) 705   (b) 910   (c) 980   (d) 790
    (e) 722

21. ?² × 124.03 ÷ (19.07 × 5.98) = 131 [IBPS Clerk (Mains) 2017]
    (a) 20   (b) 21   (c) 16   (d) 11
    (e) 9

22. (11.92)² + (16.01)² = ?² × (3.85)² [SBI PO 2015]
    (a) 15   (b) 2   (c) 4   (d) 5
    (e) None of these

23. ?² − 137.99 ÷ 6 = 21.99 × 23.01 [SBI PO (Pre) 2017]
    (a) 23   (b) 50   (c) 42   (d) 29
    (e) 35

24. ?% of 400.02 + (12.93)² = 285
　　　　　　　　　　　　　　　[SBI PO (Pre) 2017]
　(a) 18　(b) 15　(c) 24　(d) 34
　(e) 29

25. (3327.99 − 27.93) ÷ ? = 110 × 5.99
　　　　　　　　　　　　　　　[SBI PO (Pre) 2017]
　(a) 9　(b) 1　(c) 19　(d) 15
　(e) 5

26. 5520 ÷ 12.01 + √226 × 5.99 = ?
　　　　　　　　　　　　　　　[SBI PO (Pre) 2017]
　(a) 350　(b) 550　(c) 500　(d) 450
　(e) 250

27. (21 + 99) × (30 − 19.02) = ?
　　　　　　　　　　　　　　　[Indian Bank PO 2003]
　(a) 3581　(b) 1311　(c) 1290　(d) 1600
　(e) None of these

28. 160.01 + 40 ÷ (16.5 ÷ 33) = ?
　　　　　　　　　　　　　　　[SBI PO (Pre) 2017]
　(a) 310　(b) 290　(c) 250　(d) 350
　(e) 240

29. ?% of (5284.89 ÷ 7.08) = 986.01 − 533.06
　　　　　　　　　　　　　　　[RRB PO (Pre) 2017]
　(a) 42　(b) 39　(c) 74　(d) 65
　(e) 60

30. 421 × 0.9 + 130 × 101 + 10000 = ?
　　　　　　　　　　　　　　　[IBPS Clerk 2012]
　(a) 33500　　　(b) 23500
　(c) 225000　　(d) 24500
　(e) None of these

31. 30.9 × 3000 − 10.1 × 1100 + 8298 − 4302 = ?
　　　　　　　　　　　　　　　[IBPS Clerk 2012]
　(a) 80000　　　(b) 90000
　(c) 105000　　(d) 85000
　(e) None of these

32. (42.11 × 5.006) − √17 × 15.08 = ?
　　　　　　　　　　　　　　　[SBI PO (Pre) 2016]
　(a) 250　　(b) 150
　(c) 45　　(d) 200
　(e) 125

33. $\frac{3}{10}$ of 111 = ? ÷ (1.8 × 0.499)
　　　　　　　　　　　　　　　[IBPS SO 2016]
　(a) 100　　(b) 10
　(c) 3　　　(d) 30
　(e) 60

34. (1041.84 + ?) ÷ 3.02 = 1816.25 ÷ 4.01
　　　　　　　　　　　　　　　[RRB PO (Pre) 2017]
　(a) 442　(b) 337　(c) 385　(d) 268
　(e) 320

35. 69.3% of 445.12 ÷ 14.06 = 623.08 ÷ ?
　　　　　　　　　　　　　　　[RRB PO (Pre) 2017]
　(a) 28　(b) 19　(c) 21　(d) 33
　(e) 37

36. 9876 ÷ 24.96 + 215.005 − ? = 309.99
　　　　　　　　　　　　　　　[SBI Associates PO 2002]
　(a) 395　(b) 295　(c) 300　(d) 315
　(e) 51

37. ?² + 114.09 − 24.06 × 5.14 = 163.19
　　　　　　　　　　　　　　　[RRB PO (Pre) 2017]
　(a) 7　(b) 13　(c) 11　(d) 15
　(e) 19

38. 768.16 ÷ 11.87 × √257 − 58.05 = ?
　　　　　　　　　　　　　　　[RRB PO (Pre) 2017]
　(a) 1033　(b) 1175　(c) 966　(d) 880
　(e) 975

39. 724 ÷ 25 × 31.05 + 101 = ?
　　　　　　　　　　　　　　　[United Bank of India PO 2005]
　(a) 900　(b) 1000　(c) 950　(d) 1050
　(e) 1010

40. 2375.85 ÷ 18.01 − 4.525 × 8.05 = ?
　　　　　　　　　　　　　　　[Corporation Bank PO 2004]
　(a) 103　(b) 96　(c) 88　(d) 90
　(e) 112

41. 2508 ÷ 15.02 + ? × 11 = 200
　　　　　　　　　　　　　　　[Corporation Bank PO 2004]
　(a) 13　(b) 8　(c) 3　(d) 4
　(e) 6

42. 6.39 × 128.948 + 5.215 ÷ 12.189 + 25.056 = ?
　　　　　　　　　　　　　　　[IBPS PO 2011]
　(a) 800　(b) 900　(c) 850　(d) 950
　(e) None of these

43. 1559.999 ÷ 24.001 + 11.005 × 6.999 = ?
　　　　　　　　　　　　　　　[RBI 2003]
　(a) 137　(b) 132　(c) 152　(d) 149
　(e) 142

44. 98743 ÷ 198 = 800 − ?　[IBPS Clerk 2012]
　(a) 200　(b) 250　(c) 300　(d) 350
　(e) None of these

45. 685.005 ÷ 5 − √? = 16.99 × 6.01
　　　　　　　　　　　　　　　[Indian Bank PO 2002]
　(a) 625　(b) 1225　(c) 1156　(d) 841
　(e) None of these

46. (24.96)² / (34.11 + 20.05) + 67.96 + 89.11 = ?
　　　　　　　　　　　　　　　[IBPS IT 2015]
　(a) 884　(b) 546　(c) 252　(d) 424
　(e) 170

47. √? = (1248.28 + 51.7) ÷ 99.9 − 7.98
　　　　　　　　　　　　　　　[IBPS SO 2016]
　(a) 49　(b) 81　(c) 84　(d) 16
　(e) 25

48. 63.5% of 8924.19 + 22% of 5324.42 = ?
　　　　　　　　　　　　　　　[SBI IT 2016]
　(a) 6278　(b) 6128　(c) 6228　(d) 5624
　(e) 6817

49. 105.27% of 1200.11 + 11.80% of 2360.85 = 21.99% of ? + 1420.99
　　　　　　　　　　　　　　　[IBPS IT 2015]
　(a) 500　(b) 240　(c) 310　(d) 550
　(e) 960

Approximation / 137

50. 0.98% of 7824 + 4842 ÷ 119.46 − ? = 78
 [IBPS IT 2015]
 (a) 30 (b) 60 (c) 40 (d) 50
 (e) 70

51. 134% of 3894 + 38.94% of 134 = ?
 [Indian Bank PO 2003]
 (a) 5000 (b) 5300 (c) 5500 (d) 5270
 (e) 4900

52. 35% of 121 + 85% of 230.25 = ?
 [Corporation Bank PO 2004]
 (a) 225 (b) 230 (c) 240 (d) 245
 (e) 228

53. 22.5% of 1350 + 135% of 225 = ?
 [UBI PO 2005]
 (a) 570 (b) 610 (c) 670 (d) 630
 (e) 590

54. 31% of 1508 + 26% of 2018 = ?
 [IBPS PO 2011]
 (a) 1500 (b) 2000 (c) 1000 (d) 1200
 (e) None of these

55. 39.8% of 400 + ?% of 350 = 230
 [Indian Bank PO 2004]
 (a) 15 (b) 25 (c) 18 (d) 24
 (e) 20

56. $(15.96)^2$ + 75% of 285 = ?
 [Syndicate Bank PO 2004]
 (a) 435 (b) 485 (c) 440 (d) 420
 (e) 470

57. $0.0005 \times 76 \times (10)^3$ + ? = 170% of 29.98
 [UBI PO 2005]
 (a) 12 (b) 16 (c) 18 (d) 15
 (e) 20

58. 161% of 3578 + 139.85 + $\frac{5}{8}$ of 161.56 = ?
 (a) 6000 (b) 6500 (c) 6700 (d) 5500

59. 68.67% of 369.87 ÷ 15.01 = 221 ÷ ?
 [IBPS Clerk (Mains) 2017]
 (a) 16 (b) 18 (c) 13 (d) 21
 (e) 10

60. (19.97% of 781) + ? + (30% of 87) = 252
 [SBI PO 2015]
 (a) 40 (b) 50 (c) 25 (d) 70
 (e) None of these

61. $(3 \times 9.2) \div \frac{2}{9}$ of 62 = ? [IBPS SO 2013]
 (a) 40 (b) 8 (c) 24 (d) 16
 (e) 2

62. $\sqrt{144}$ − (113.79 + 65.89) ÷ ? = 6
 [IBPS SO 2016]
 (a) 15 (b) 90 (c) 50 (d) 60
 (e) 30

63. $59.998 \times \sqrt{256.002} \div 9.98 - 38$ = ?
 [IBPS SO 2016]
 (a) 40 (b) 93 (c) 60 (d) 58
 (e) 49

64. ?% of 750.11 × 34.90 + 6.995 = 30000
 [SBI PO (Pre) 2016]
 (a) 100 (b) 80 (c) 35 (d) 75
 (e) 60

65. (24.99% of 399.995) ÷ ? = $(125\%$ of $4.111)^2$
 [IBPS SO 2016]
 (a) 80 (b) 4 (c) 60 (d) 16
 (e) 40

Exercise 2 *Higher Skill Level Questions*

Directions (Q. Nos. 1-8) *What approximate value should come in place of the question mark (?) in the following questions?*

1. $[(41.99)^2 - (18.04)^2] - ? = (13.11)^2 - 138.99$
 [IBPS IT 2015]
 (a) 4004 (b) 1200 (c) 1720 (d) 8432
 (e) 1410

2. (24.99% of 900.911) ÷ (10.30% of 25.011) = $?^4$
 [IBPS SO 2016]
 (a) 9 (b) 30 (c) 27 (d) 3
 (e) 6

3. $(348)^{1/3} \times (14.001) \times (27.998)^2 \div (1.997)^3 = 2^? \times 7^4$
 [IBPS SO 2016]
 (a) 4 (b) 6 (c) 5 (d) 3
 (e) 2

4. $(64.01)^2 \times (65)^{1/3} \times (25.99)^2 \div [2^{11} \times (12.97)^2] = 2^?$
 [IBPS SO 2016]
 (a) 4 (b) 5 (c) 2 (d) 3
 (e) 6

5. $\sqrt{1024.002} \div 3.996 \div 9.98 \div 2.9 = ?$
 [IBPS SO 2016]
 (a) 0.3 (b) 9
 (c) 30   (d) 90
 (e) 80

6. $(95)^{3.7} \div (95)^{0.9989} = (95)^?$
 [Canara Bank PO 2003]
 (a) 1.9 (b) 3
 (c) 2.99 (d) 3.6
 (e) 2.7

7. $(8)^{0.75} \times (63.96)^{0.25} \div (64)^{-1} = (8)^?$
 [SBI IT 2016]
 (a) 2.25 (b) 0.30
 (c) 3.25 (d) 3.50
 (e) 3.75

8. $\sqrt{2025.11} \times \sqrt{256.04} + \sqrt{399.95} \times \sqrt{?}$
 = 33.98 × 40.11   [IBPS IT 2015]
 (a) 1682 (b) 1024
 (c) 1582 (d) 678
 (e) 1884

Answer with Solutions

Exercise 1 Base Level Questions

1. (b) ? = 12.49% of 839.859 ≈ 12% of 840
$$\approx \frac{840 \times 12}{100} = 100.80 \approx 100$$

2. (b) ? = 124.35% of 8096
$$\approx \frac{124}{100} \times 8096 = 10039.04 \approx 10000$$

3. (d) ? = 5454 ÷ 54 ÷ 5
$$= \left(\frac{5454}{54}\right) \div 5 = \frac{101}{5} = 20.2 \approx 20$$

4. (b) ? = 99.999 ÷ 0.99 ÷ 0.00991234
≈ 100 ÷ 1 ÷ 0.01 = 10000

5. (d) ? = $2\frac{3}{5} \times \frac{15}{26} \times 283.75$
$$\approx \frac{13}{5} \times \frac{15}{26} \times 284 = 426 \approx 425$$

6. (a) 23.003 × 22.998 + 100.010 = ?
⇒ ? ≈ 23 × 23 + 100   [by Rule 2]
⇒ ? = 529 + 100 = 629 ≈ 630

7. (d) ? = 16.003 × 27.998 − 209.010
≈ 16 × 28 − 210 = 448 − 210
= 238 ≈ 240

8. (c) ? = 19.003 × 22.998 − 280.010
≈ 19 × 23 − 280 = 437 − 280 = 157 ≈ 160

9. (d) ? = 26.003 × 37.998 − 309.010
≈ 26 × 38 − 309 = 988 − 309 = 679

10. (b) 42.8 × 13.5 × 16.2 × ? = 2340.09
$$\Rightarrow ? = \frac{2340.09}{42.8 \times 13.5 \times 16.2}$$
$$\approx \frac{2340}{43 \times 14 \times 16} = 0.25$$

11. (a) ? = 815.002 + 29.98 − 53.998 + (3.01)²
≈ 815 + 30 − 54 + (3)²
= 845 − 54 + 9 = 800

12. (d) (9.95)² × (2.01)³ = 2 × (?)²
⇒ 2 × (?)² ≈ (10)² × (2)³ = 100 × 8 = 800
⇒ (?)² = 400 ⇒ ? = 20

13. (d) ? = 96.894 + 33.002 + 15.02 × 7.99
≈ 97 + 33 + 15 × 8 = 130 + 120 = 250

14. (c) 425 ÷ 16.95 × ? = 225
$$\Rightarrow ? \approx \frac{225 \times 17}{425} = 9$$

15. (e) 36.0001 ÷ 5.9998 × $\sqrt{?}$ = 108.0005
⇒ 36 ÷ 6 × $\sqrt{?}$ ≈ 108 ⇒ 6 × $\sqrt{?}$ = 108
∴ $\sqrt{?}$ = 18 ⇒ ? = 324 ≈ 325

16. (a) ? = 26.003 − 154.001 ÷ 6.995
≈ 26 − 154 ÷ 7 = 26 − 22 = 4

17. (b) ?% of 218 = $3\frac{1}{3} \times 3.3121$
$$\Rightarrow \frac{218 \times ?}{100} \approx \frac{10}{3} \times 3$$
$$\Rightarrow ? = 10 \times \frac{100}{218}$$
∴ ? = 4.5 ≈ 5

18. (c) ?% of (6274 ÷ 6.14) = 646.51 − 241.58
⇒ x% of 6274 ÷ 6 ≈ 646 − 242
$$\Rightarrow \frac{6274 \times x}{100} \div 6 = 404$$
$$\Rightarrow \frac{6274 \times x}{100 \times 6} = 404$$
$$\Rightarrow x = \frac{404 \times 100 \times 6}{6274}$$
$$\Rightarrow x = \frac{242400}{6274} = 38.63$$
⇒ x = 38.63 ⇒ x ≈ 39

19. (a) 10.56 × ? ÷ 3 = 576.73 × 2.02
$$\Rightarrow \frac{10.56 \times x}{3} \approx 577 \times 2$$
$$\Rightarrow x = \frac{3 \times 577 \times 2}{10.56} \Rightarrow x = \frac{3462}{10.56} = 327.84$$
⇒ x = 327.84 ⇒ x ≈ 328

20. (d) 768 ÷ 11.82 × $\sqrt{168}$ − 42 = ?
⇒ ? ≈ 768 ÷ 12 × $\sqrt{169}$ − 42
$$= \frac{768}{12} \times 13 - 42$$
= 64 × 13 − 42 = 832 − 42
= 790

21. (d) ?² × 124.03 ÷ (19.07 × 5.98) = 131
⇒ ?² × 124 ÷ (19 × 6) ≈ 131
⇒ ?² × 124 ÷ 114 = 131
$$\Rightarrow \frac{?^2 \times 124}{114} = 131 \Rightarrow ?^2 = \frac{114 \times 131}{124}$$
$$\Rightarrow ?^2 = \frac{14934}{124} = 120.43$$
⇒ ? = $\sqrt{120.43} \approx \sqrt{121} = 11$

Approximation / 139

22. (d) $(11.92)^2 + (16.01)^2 = (?)^2 \times (3.85)^2$
$\Rightarrow (12)^2 + (16)^2 \approx (?)^2 \times (4)^2$
$\Rightarrow 144 + 256 = (?)^2 \times 16$
$\Rightarrow 400 = (?)^2 \times 16$
$\Rightarrow ?^2 = \frac{400}{16} = 25 \Rightarrow ? = 5$

23. (a) $?^2 - 137.99 \div 6 = 21.99 \times 23.01$
$\Rightarrow ?^2 - \frac{138}{6} \approx 22 \times 23$
$\Rightarrow ?^2 - 23 = 506 \Rightarrow ?^2 = 506 + 23$
$\Rightarrow ?^2 = 529 \Rightarrow ? = 23$

24. (e) ?% of $400.02 + (12.93)^2 = 285$
$\Rightarrow \frac{?}{100} \times 400 + (13)^2 \approx 285$
$\Rightarrow ? \times 4 + 169 = 285$
$\Rightarrow ? \times 4 = 285 - 169 \Rightarrow ? \times 4 = 116$
$\therefore ? = \frac{116}{4} = 29$

25. (e) $(3327.99 - 27.93) \div ? = 110 \times 5.99$
$\Rightarrow \frac{(3328 - 28)}{?} \approx 110 \times 6$
$\therefore ? = \frac{3300}{110 \times 6} = \frac{30}{6} = 5$

26. (b) $? = 5520 \div 12.01 + \sqrt{226} \times 5.99$
$\Rightarrow ? \approx \frac{5520}{12} + \sqrt{225} \times 6$
$\therefore ? = 460 + 90 = 550$

27. (e) $? = (21 + 99) \times (30 - 19.02)$
$= (21 + 99) \times (30 - 19)$
$= 120 \times 11 = 1320$

28. (e) $? = 160.01 + 40 \div (16.5 \div 33)$
$\Rightarrow ? \approx 160 + 40 \div \left(\frac{16.5}{33}\right)$
$\Rightarrow ? \approx 160 + 40 \div \left(\frac{1}{2}\right)$
$\Rightarrow ? = 160 + 40 \times \frac{2}{1}$
$\Rightarrow ? = 160 + 80 = 240$

29. (e) $\frac{?}{100}$ of $(5285 \div 7) \approx 986 - 533$
$\Rightarrow \frac{?}{100} \times 755 = 453 \Rightarrow ? = 60$

30. (b) $? = 421 \times 0.9 + 130 \times 101 + 10000$
$\approx 421 + 13000 + 10000$
$= 23421 \approx 23500$

31. (e) $? = 30.9 \times 3000 - 10.1 \times 1100 + 8298 - 4302$
$\approx 31 \times 3000 - 10 \times 1100 + 8298 - 4302$
$= 93000 - 11000 + 8298 - 4302$
$= 101298 - 15302 = 85996 \approx 86000$

32. (b) $? = (42.11 \times 5.006) - \sqrt{17} \times 15.08$
$\approx (42 \times 5) - \sqrt{16} \times 15$
$= 210 - 4 \times 15 = 210 - 60 = 150$

33. (d) $\frac{3}{10}$ of $111 = ? \div (1.8 \times 0.499)$
$\Rightarrow 110 \times \frac{3}{10} \approx ? \div (2 \times 0.5)$
$\Rightarrow 33 = ? \div 1 \Rightarrow 33 = ?$
$\Rightarrow ? = 33 \approx 30$

34. (e) $(1041.84 + ?) \div 3.02 = 1816.25 \div 4.01$
$\Rightarrow (1042 + ?) \div 3 \approx 1816 \div 4$
$\Rightarrow \frac{1042 + ?}{3} = \frac{1816}{4}$
$\Rightarrow 1042 + ? = 454 \times 3$
$\Rightarrow ? = 1362 - 1042 = 320$

35. (a) $\frac{70}{100} \times \frac{445}{14} \approx \frac{623}{?} \Rightarrow ? = 7 \times 4 = 28$

36. (c) $? = \frac{9876}{24.96} + 215.005 - 309.99$
$\approx \frac{9875}{25} + 215 - 310$
$= 395 + 215 - 310 = 300$

37. (b) $?^2 + 114 - 24 \times 5 \approx 163$
$\Rightarrow ?^2 = 163 + 6 \Rightarrow ? = 13$

38. (c) $? = 64 \times 16 - 58 \Rightarrow ? = 966$

39. (b) $724 \div 25 \times 31.05 + 101 = ?$
$\Rightarrow ? \approx 725 \div 25 \times 31 + 101$
$= 29 \times 31 + 101 = 899 + 101 = 1000$

40. (b) $? = 2375.85 \div 18.01 - 4.525 \times 8.05$
$\approx 2376 \div 18 - 4.5 \times 8 = 132 - 36 = 96$

41. (c) $2508 \div 15.02 + ? \times 11 \approx 200$
$\Rightarrow \frac{2508}{15} + ? \times 11 \approx 200$
$\therefore ? = (200 - 167.2) \times \frac{1}{11}$
$\approx (200 - 167) \times \frac{1}{11} = 33 \times \frac{1}{11} = 3$

42. (c) $? = 6.39 \times 128.948 + 5.215 \div 12.189 + 25.056$
$\approx 6.4 \times 129 + 5.2 \div 12 + 25$
$= 825.6 + 0.4333 + 25 = 851.033 \approx 850$

43. (e) $1559.999 \div 24.001 + 11.005 \times 6.999 = ?$
$\Rightarrow ? \approx 1560 \div 24 + 11 \times 7 = 65 + 77 = 142$

44. (c) $98743 \div 198 = 800 - ?$
$\Rightarrow ? \approx 800 - 98800 \div 200$
$\Rightarrow ? = 800 - 494 = 306 \approx 300$

45. (b) $685.005 \div 5 - \sqrt{?} = 16.99 \times 6.01$
$\Rightarrow \sqrt{?} \approx \frac{685}{5} - 17 \times 6$

$\Rightarrow$ $\sqrt{?} = 137 - 102 = 35$

$\therefore$ $? = (35)^2 = 1225$

46. (e) $(24.96)^2 / (34.11 + 20.05) + 67.96 + 89.11 = ?$

$\Rightarrow ? = \dfrac{(25)^2}{54} + 68 + 89$

$\Rightarrow ? = \dfrac{625}{54} + 68 + 89$

$\Rightarrow ? = 11.5 + 68 + 89 = 168.5 \approx 170$

47. (e) $\sqrt{?} = (1248.28 + 51.7) \div 99.9 - 7.98$

$\Rightarrow \sqrt{?} \approx (1248 + 52) \div 100 - 8$

$\Rightarrow \sqrt{?} = 1300 \div 100 - 8 = 13 - 8$

$\Rightarrow \sqrt{?} = 5 \Rightarrow (\sqrt{?})^2 = (5)^2$

$\Rightarrow ? = 25$

48. (e) $\dfrac{63.5 \times 8924.19}{100} + \dfrac{22 \times 5324.42}{100} = ?$

$\Rightarrow 5666.8 + 1171.3 = ?$

$\Rightarrow ? = 6838.1 \approx 6817$

49. (d) 105.27% of $1200.11 + 11.80\%$ of $2360.85 = 21.99\%$ of $? + 1420.99$

$\Rightarrow 105\%$ of $1200 + 12\%$ of $2360 \approx 22\%$ of $? + 1421$

$\Rightarrow \dfrac{105}{100} \times 1200 + \dfrac{12}{100} \times 2360$
$\qquad = 0.22 \times ? + 1421$

$\Rightarrow 1260 + 283.2 = 0.22 \times ? + 1421$

$\Rightarrow 0.22 \times ? = 122.2$

$\Rightarrow ? = \dfrac{122.2}{0.22} = 555.45 \approx 550$

50. (c) 0.98% of $7824 + 4842 \div 119.46 - ? = 78$

$\Rightarrow 1\%$ of $7824 + 4842 \div 120 - 78 = ?$

$\Rightarrow ? = 78.24 + 40.35 - 78 = 40.59 \approx 40$

51. (d) 134% of $3894 + 38.94\%$ of $134 = ?$

$\Rightarrow ? = \dfrac{3894 \times 134}{100} + \dfrac{38.94 \times 134}{100}$

$= 5217.96 + 52.1796$
$= 5270.1396 \approx 5270$

52. (c) $? = 35\%$ of $121 + 85\%$ of 230.25

$\approx \dfrac{121 \times 35}{100} + \dfrac{230 \times 85}{100}$

$= 42.35 + 195.5 = 237.85 \approx 240$

53. (b) $? = 22.5\%$ of $1350 + 135\%$ of 225

$= \dfrac{1350 \times 22.5}{100} + \dfrac{225 \times 135}{100}$

$= 303.75 + 303.75 = 607.5 \approx 610$

54. (c) $? = 31\%$ of $1508 + 26\%$ of 2018

$= \dfrac{31}{100} \times 1508 + \dfrac{26}{100} \times 2018$

$= 31 \times 15.08 + 26 \times 20.18$

$\approx 31 \times 15 + 26 \times 20$

$= 465 + 520 = 985 \approx 1000$

55. (e) 39.8% of $400 + ?\%$ of $350 = 230$

$\Rightarrow 40 \times \dfrac{400}{100} + \dfrac{350 \times ?}{100} \approx 230$

$\Rightarrow ? = (230 - 160) \times \dfrac{100}{350} = 20$

56. (e) $? = (15.96)^2 + 75\%$ of 285

$\approx (16)^2 + 213.75 \approx 256 + 214 = 470$

57. (a) $0.0005 \times 76 \times (10)^3 + ? = 170\%$ of 29.98

$\Rightarrow 0.038 \times 1000 + ? \approx \dfrac{170 \times 30}{100}$

$\therefore ? = 51 - 38 = 13 \approx 12$

58. (a) $? = 161\%$ of $3578 + 139.85 + \dfrac{5}{8}$ of 161.56

$\approx \dfrac{161}{100} \times 3580 + 140 + \dfrac{5}{8} \times 160$

$= 5763.8 + 140 + 100 = 6003.8 \approx 6000$

59. (c) 68.67% of $369.87 \div 15.01 = 221 \div ?$

$\Rightarrow 69\%$ of $370 \div 15 \approx 221 \div x$

$\Rightarrow 370 \times \dfrac{69}{100} \div 15 = \dfrac{221}{x}$

$\Rightarrow \dfrac{255.3}{15} = \dfrac{221}{x}$

$\Rightarrow x \approx \dfrac{15 \times 221}{255.3} = \dfrac{3315}{255.3}$

$\Rightarrow x = 12.984 \approx 13$

60. (d) $(19.97\%$ of $781) + ? + (30\%$ of $87) = 252$

$\Rightarrow (20\%$ of $780) + ? + (30\%$ of $87) \approx 252$

$\Rightarrow 156 + ? + 26 = 252$

$\Rightarrow 182 + ? = 252$

$\Rightarrow ? = 252 - 182$

$\therefore ? = 70$

61. (e) $(3 \times 9.2) \div 62 \times \dfrac{2}{9} = ?$

$\Rightarrow ? \approx 3 \times 10 \div 63 \times \dfrac{2}{9}$

$\Rightarrow ? = 30 \div 63 \times \dfrac{2}{9} = 30 \div 14$

$\Rightarrow ? = 2.14 \approx 2$

62. (e) $\sqrt{144} - (113.79 + 65.89) \div ? = 6$

$\Rightarrow \sqrt{144} - (114 + 66) \div ? \approx 6 \Rightarrow 12 - \dfrac{180}{?} = 6$

Let $? = x$, then

$12 - \dfrac{180}{x} = 6 \Rightarrow 12 - 6 = \dfrac{180}{x}$

$\Rightarrow 6 = \dfrac{180}{x}$

$\Rightarrow 6x = 180 \Rightarrow x = 30$

63. (d) $59.998 \times \sqrt{256.002} \div 9.98 - 38 = ?$

$\Rightarrow 60 \times \sqrt{256} \div 10 - 38 \approx ?$

$\Rightarrow ? = \dfrac{60 \times 16}{10} - 38$

$\Rightarrow ? = 96 - 38 = 58$

Approximation / 141

64. (a) ?% of 750.11 × 34.90 + 6.995 = 30000

$\Rightarrow \dfrac{?}{100} \times 750 \times 35 + 7 \approx 30000$

$\Rightarrow \dfrac{?}{2} \times 75 \times 7 = 30000 - 7$

$\Rightarrow ? = \dfrac{29993 \times 2}{75 \times 7} = \dfrac{59986}{525}$

$\Rightarrow ? = 114.25 \approx 100$

65. (b) (24.99% of 399.995) ÷ ?
$= (125\% \text{ of } 4.111)^2$

$\Rightarrow (25\% \text{ of } 400) \div ? \approx (125\% \text{ of } 4)^2$

$\Rightarrow 400 \times \dfrac{25}{100} \div ? = \left(4 \times \dfrac{125}{100}\right)^2$

$\Rightarrow 100 \div ? = (5)^2 \Rightarrow \dfrac{100}{?} = 25$

$\therefore ? = \dfrac{100}{25} = 4$

Exercise 2 *Higher Skill Level Questions*

1. (e) $[(41.99)^2 - (18.04)^2] - ? = (13.11)^2 - 138.99$
$\Rightarrow (42^2 - 18^2) - ? \approx (13)^2 - 139$
$\Rightarrow \{(42 + 18)(42 - 18)\} - ? = 169 - 139$
$\Rightarrow \{60 \times 24\} - ? = 30$
$\Rightarrow 1440 - ? = 30 \Rightarrow ? = 1410$

2. (d) (24.99% of 900.911)
$\div (10.30\% \text{ of } 25.011) = ?^4$

$\Rightarrow (25\% \text{ of } 900) \div (10\% \text{ of } 25) \approx ?^4$

$\Rightarrow \left(\dfrac{900 \times 25}{100}\right) \div \left(\dfrac{25 \times 10}{100}\right) = ?^4$

$\Rightarrow \dfrac{900 \times 25}{100} \times \dfrac{100}{25 \times 10} = ?^4$

$\Rightarrow ?^4 = 90 \approx 81 \Rightarrow ?^4 = (3)^4 \Rightarrow ? = 3$

3. (e) $(348)^{1/3} \times (14.001) \times (27.998)^2 \div (1.997)^3$
$= 2^? \times 7^4$

$\Rightarrow (343)^{1/3} \times 14 \times (28)^2 \div (2)^3 \approx 2^? \times 7^4$

$\Rightarrow \dfrac{(7^3)^{1/3} \times 14 \times 28 \times 28}{8} = 2^? \times 7^4$

$\Rightarrow \dfrac{7 \times 14 \times 28 \times 28}{8 \times 7^4} = 2^?$

$\Rightarrow \dfrac{7 \times 7 \times 2 \times 4 \times 7 \times 4 \times 7}{8 \times 7^4} = 2^?$

$\Rightarrow \dfrac{7^4 \times 8 \times 4}{8 \times 7^4} = 2^?$

$\Rightarrow 2^? = 4 \Rightarrow 2^? = (2)^2 \Rightarrow ? = 2$

4. (b) $(64.01)^2 \times (65)^{1/3} \times (25.99)^2$
$\div \{2^{11} \times (12.97)^2\} = 2^?$

$\Rightarrow \dfrac{(64)^2 \times (64)^{1/3} \times (26)^2}{(2)^{11} \times (13)^2} \approx 2^?$

$\Rightarrow 2^? = \dfrac{(64)^{7/3} \times (2)^2 \times (13)^2}{(2)^{11} \times 13^2}$

$\Rightarrow 2^? = \dfrac{(2^6)^{7/3} \times (2)^2}{(2)^{11}}$

$\Rightarrow 2^? = (2)^{14} \times (2)^2 \times (2)^{-11}$

$\Rightarrow 2^? = (2)^{14+2-11} = (2)^5$

$\Rightarrow ? = 5$

5. (a) $\sqrt{1024.002} \div 3.996 \div 9.98 \div 2.9 = ?$

$\Rightarrow ? \approx \sqrt{1024} \div 4 \div 10 \div 3$

$\Rightarrow ? = 32 \div 4 \div 10 \div 3$

$= \dfrac{32}{4 \times 10 \times 3} = 0.266 \approx 0.3$

6. (e) $(95)^{3.7} \div (95)^{0.9989} = (95)^?$

$\Rightarrow 95^{(3.7 - 0.9989)} = (95)^?$

$\Rightarrow (95)^{2.7011} = (95)^?$

$\therefore ? = 2.7011 \approx 2.7$

7. (c) $(8)^{0.75} \times (63.96)^{0.25} \div (64)^{-1} = (8)^?$

$\Rightarrow (8)^{3/4} \times (8^2)^{1/4} \div (8^2)^{-1} \approx (8)^?$

$\Rightarrow 8^{3/4} \times 8^{1/2} \times 8^2 = 8^?$

$\Rightarrow 8^{13/4} = 8^? \Rightarrow ? = \dfrac{13}{4} = 3.25$

8. (b) $\sqrt{2025.11} \times \sqrt{256.04} + \sqrt{399.95} \times \sqrt{?}$
$= 33.98 \times 40.11$

$\Rightarrow \sqrt{2025} \times \sqrt{256} + \sqrt{400} \times \sqrt{?} \approx 34 \times 40$

$\Rightarrow 45 \times 16 + 20 \times \sqrt{?} = 34 \times 40$

$\Rightarrow 720 + 20 \times \sqrt{?} = 1360$

$\Rightarrow 20 \times \sqrt{?} = 1360 - 720$

$\Rightarrow 20 \times \sqrt{?} = 640$

$\Rightarrow \sqrt{?} = \dfrac{640}{20} = 32$

$\therefore ? = (32)^2 = 1024$

Chapter 9

Word Problems Based on Numbers

Numbers play an important role in our day-to-day life. Puzzles based on these numbers are known as **word problems**. To solve such problems, you have to extract the information correctly and form equations based on given information.

The equations formed can be of single variable, multi variables, linear, quadratic etc., depending on the type of problem asked.

Types of Word Problems Based on Numbers

There are basically following types of questions that are asked in word problems on numbers.

Type ❶ Based on Operation with Numbers

These types of questions includes the operations, like subtraction, addition, multiplication, division of number with other number, calculation of average of consecutive numbers, calculation of parts of a number, operation on even or odd numbers, calculation of sum or difference of reciprocal of numbers, etc.

Important Points to be Remembered

- Consecutive natural numbers can be assumed as
  $x, x+1, x+2, \ldots$
- Consecutive numbers can be assumed as
  $x-2, x-1, x, x+1, x+2, \ldots$
- Consecutive even/odd numbers can be assumed
  $x-3, x-1, x+1, x+3, \ldots$
- If sum of two numbers is given as S, then take one number as x and other as $(S-x)$.
- If difference of two numbers is d, then take one number as x and other as $(x+d)$ or $(x-d)$.

Word Problems Based on Numbers / 143

Ex. 1 Three-fourth of two-third of a number is 782. What is three-fifth of one-fourth of the same number?

Sol. Let the number be x.

Then, according to the question,
$$\frac{3}{4} \text{ of } \frac{2}{3} \text{ of } x = 782$$
$$\Rightarrow \quad x \times \frac{3}{4} \times \frac{2}{3} = 782$$
$$\therefore \quad x = \frac{782 \times 4 \times 3}{2 \times 3} = 1564 \quad \text{[by cross-multiplication]}$$

Now, $\frac{3}{5} \times \frac{1}{4} \times 1564 = 234.6$

Ex. 2 The sum of three consecutive odd natural number is 87. Find the smallest number.

Sol. Let three consecutive odd natural numbers be $x, (x+2)$ and $(x+4)$.

Then, according to the question,
$$x + (x+2) + (x+4) = 87$$
$$\Rightarrow \quad 3x + 6 = 87 \quad \text{[by separating]}$$
$$\Rightarrow \quad x = \frac{87-6}{3} = \frac{81}{3} = 27$$

Hence, the smallest number is 27.

Ex. 3 One-fourth of a number exceeds its one-seventh by 24. What is the number?

Sol. Let the required number be x.

Then, according to the question,
$$x \times \frac{1}{4} - x \times \frac{1}{7} = 24$$
$$\Rightarrow \quad \frac{x}{4} - \frac{x}{7} = 24$$
$$\Rightarrow \quad \frac{7x - 4x}{28} = 24 \quad \text{[taking LCM]}$$
$$\Rightarrow \quad \frac{3x}{28} = 24$$
$$\therefore \quad x = \frac{28 \times 24}{3} = 224 \quad \text{[by cross-multiplication]}$$

Ex. 4 The sum and product of two numbers are 12 and 35, respectively. What is the sum of their reciprocals?

Sol. Let two numbers be x and y.

Then, according to the question,
$$x + y = 12 \quad \ldots(i)$$
and $\quad x \times y = 35$

We know that, $(x-y)^2 = (x+y)^2 - 4xy$

On putting $x + y = 12$ and $xy = 35$ in the above equation, we get
$$(x-y)^2 = (12)^2 - 4 \times 35 = 144 - 140 = 4$$

Now, taking square root on both sides, we get
$$x - y = \sqrt{4} = 2 \quad \ldots(ii)$$

On adding Eqs. (i) and (ii), we get
$$x + y = 12$$
$$x - y = 2$$
$$\overline{2x = 14}$$
$$\therefore x = 7$$
On putting the value of x in Eq. (i), we get
$$7 + y = 12$$
$$\Rightarrow y = 5$$
$$\therefore x = 7 \text{ and } y = 5$$
Now, sum of reciprocals of the two numbers $= \dfrac{1}{x} + \dfrac{1}{y} = \dfrac{1}{7} + \dfrac{1}{5} = \dfrac{5+7}{35} = \dfrac{12}{35}$

Alternate Method
Let two numbers be x and y.
According to the question, $x + y = 12$ and $xy = 35$
Now, sum of reciprocals of the two numbers $= \dfrac{1}{x} + \dfrac{1}{y} = \dfrac{x+y}{xy}$

On putting $x + y = 12$ and $xy = 35$ in the above formula, we get
Required number $= \dfrac{12}{35}$

Type 2 Based on Formation of Number with Digits

These types of questions include formation of a number with digits and its difference with reciprocal of the same number, calculation of a number, if a number is added or subtracted to it, so that the digits get reversed etc.

Important Points to be Remembered

+ A two-digit number with x as unit digit and y as ten's digit is formed as $(10y + x)$ and if the digits are reversed, then number is represented as $(10x + y)$.
+ A three-digit number with x as unit digit, y as ten's digit and z as hundred's digit is formed as $(100z + 10y + x)$.

Ex. 5 A number consists of two digits whose sum is 8. If 18 is subtracted from the number, the digits interchange their places, then what is the number?

Sol. Let the unit's digit be y and ten's digit be x, then the number $= 10x + y$
When digits are interchanged, then the number $= 10y + x$
According to the question,
$$x + y = 8 \quad \ldots(i)$$
and $(10x + y) - 18 = 10y + x$
$\Rightarrow 10x + y - 10y - x = 18 \Rightarrow 9x - 9y = 18$
$\therefore x - y = 2 \quad \ldots(ii)$
On adding Eqs. (i) and (ii), we get
$$x + y = 8$$
$$x - y = 2$$
$$\overline{2x = 10}$$
$$\therefore x = 5$$

On putting the value of x in Eq. (i), we get
$$5 + y = 8$$
$$\Rightarrow y = 3$$
$\therefore$ Required number $= 10x + y = 10 \times 5 + 3 = 53$

Type 3 *Based on Calculation of Head and Feet of Animals*

If a group of animals having either two feet (like ducks, hens, etc.) or four feet (like horses, cows, etc.) is there and total number of heads in the group is H and number of feets of these animals is L, then

(i) Number of animals with four feet $= \dfrac{L - 2H}{2}$

(ii) Number of animals with two feet = Total number of heads − Total number of four feeted animals

Ex. 6 In a park, there are some cows and some ducks. If total number of heads in the park is 68 and number of their legs together is 198, then find the total number of ducks in the park.

Sol. Let the number of cows in the park be x and the number of ducks in the park be y.
Then, according to the question,
$$\text{Total number of heads} = \text{Total number of cows and ducks}$$
So, $\quad x + y = 68 \quad \text{...(i)}$
Total number of legs $= (4 \times \text{Number of cows}) + (2 \times \text{Number of ducks})$
$\Rightarrow \quad 4x + 2y = 198 \quad$ [$\because$ a cow has 4 legs and a duck has 2 legs]
$\Rightarrow \quad 2x + y = 99 \quad$ [dividing both sides by 2] ...(ii)
On subtracting Eq. (i) from Eq. (ii), we get
$$2x + y - x - y = 99 - 68$$
$\Rightarrow \quad x = 31$
On putting the value of x in Eq. (i), we get
$$31 + y = 68$$
$\therefore \quad y = 37$
Hence, the total number of ducks is 37.

Fast Track Method

Here, $L = 198$ and $H = 68$
$\therefore \quad$ Number of cows $= \dfrac{L - 2H}{2} = \dfrac{198 - 2 \times 68}{2}$
$= \dfrac{198 - 136}{2} = \dfrac{62}{2} = 31$

Now, total number of ducks = Total number of heads − Total number of cows
$= 68 - 31 = 37$

Fast Track Practice

1. If a number is multiplied by three-fourth of itself, the value thus obtained is 10800. What is that number? **[Bank Clerk 2011]**
   (a) 210 (b) 180 (c) 120 (d) 160
   (e) 140

2. If the sum of two numbers, one of which is 2/5 times the another is 50, then the numbers are **[SSC CGL 2015]**
   (a) $\frac{240}{7}, \frac{110}{7}$ (b) $\frac{250}{7}, \frac{100}{7}$
   (c) $\frac{115}{7}, \frac{235}{7}$ (d) $\frac{150}{7}, \frac{200}{7}$

3. If $\frac{4}{5}$th of a number exceeds its $\frac{3}{4}$th by 8, then the number is **[SSC CGL (Pre) 2016]**
   (a) 130 (b) 120 (c) 160 (d) 150

4. One-fifth of half of a number is 20. Then, 20% of that number is **[SSC (10+2) 2015]**
   (a) 60 (b) 20 (c) 40 (d) 80

5. 20% of a number when added to 20 becomes the number itself, then the number is **[CDS 2015 (II)]**
   (a) 20 (b) 25
   (c) 50 (d) 80

6. The sum of a non-zero number and 4 times its reciprocal is $\frac{17}{2}$. What is the number? **[SSC CGL (Pre) 2017]**
   (a) 8 (b) 12
   (c) 16 (d) 4

7. 16 times X is equal to 5 times Y. If 8 is subtracted from Y it is 10 more than 2 times X. What is the sum of X and Y? **[SBI Clerk 2015]**
   (a) 78 (b) 39
   (c) 48 (d) 92
   (e) 63

8. The difference between the squares of two consecutive odd integers is always divisible by **[CDS 2015 (I)]**
   (a) 3 (b) 7
   (c) 8 (d) 16

9. The sum of five consecutive even numbers is equal to 250. What is the sum of the largest and the smallest numbers? **[OBC Clerk 2011]**
   (a) 98 (b) 96
   (c) 102 (d) 38
   (e) None of these

10. Out of three given numbers, the first number is twice the second and thrice the third. If the average of these three numbers is 154, then what is the difference between the first and the third numbers? **[Bank Clerk 2011]**
    (a) 126 (b) 42 (c) 168 (d) 52
    (e) None of these

11. Twice the difference between two numbers is equal to their sum. If one number is 15, then find the another number. **[RRB (Non-Tech) 2016]**
    (a) 15 (b) 10 (c) 5 (d) 20

12. A number consists of two digits whose sum is 10. If the digits of the number are reversed, then the number is decreased by 36.
    I. The number is divisible by a composite number.
    II. The number is a multiple of a prime number.
    Which of the following is/are correct? **[CDS 2013]**
    (a) Only I (b) Only II
    (c) Both I and II (d) Neither I nor II

13. The difference between a two-digit number and the number obtained by interchanging the two digits of the number is 18. The sum of the two digits of the number is 12. What is the product of the digits of two-digit number? **[IBPS Clerk 2012]**
    (a) 35 (b) 27 (c) 32
    (d) Cannot be determined
    (e) None of these

14. In a two-digit positive number, the unit digit is equal to the square of ten's digit. The difference between the original number and the number formed by interchanging the digits is 54. What is 40% of the original number? **[IBPS PO 2011]**
    (a) 64 (b) 73 (c) 84
    (d) Cannot be determined
    (e) None of these

15. On Children's Day, sweets were to be equally distributed amongst 300 children. But on that particular day, 50 children remained absent, hence each child got one extra sweet. How many sweets were distributed? **[SSC CGL 2013]**
    (a) 1450 (b) 1700
    (c) 1500 (d) 1650

Word Problems Based on Numbers / 147

16. If a two-digit number is k times the sum of its digits. If the number formed by interchanging the digits is m times the sum of the digits, then the value of m is
   [CDS 2016 (II)]
   (a) $9 - k$ (b) $10 - k$
   (c) $11 - k$ (d) $k - 1$

17. In a group of buffaloes and ducks, the number of legs are 24 more than twice the number of heads. What is the number of buffaloes in the group?
   (a) 6 (b) 8
   (c) 10 (d) 12

18. A man has some hens and cows. If the number of heads is 48 and the number of feet is 140, then the number of hens will be
   (a) 22 (b) 23
   (c) 24 (d) 26

19. Two baskets together have 640 oranges. If 1/5th of the oranges in the first basket be taken to the second basket, the number of oranges in both the baskets becomes equal, then the number of oranges in the first basket is
   [SSC CGL (Mains) 2017]
   (a) 800 (b) 600
   (c) 400 (d) 300

20. In an examination, a student was asked to divide a certain number by 8. By mistake, he multiplied it by 8 and got the answer 2016 more than the correct answer. What was the number?
   [CDS 2016 (II)]
   (a) 252 (b) 256 (c) 258 (d) 260

21. The sum of the squares of two positive integers is 208. If the square of the larger number is 18 times the smaller number, then what is the difference of the larger and smaller number?
   [CDS 2016 (II)]
   (a) 2 (b) 3 (c) 4 (d) 6

22. Twenty one times of a positive number is less than its square by 100. The value of the positive number is
   [SSC CGL (Mains) 2016]
   (a) 25 (b) 26 (c) 42 (d) 41

23. Ram left $\frac{1}{3}$ of his property to his widow and $\frac{3}{5}$ of the remainder to his daughter. He gave the rest to his son who received ₹ 6400. How much was his original property worth?
   [SSC (10+2) 2014]
   (a) ₹ 16000 (b) ₹ 32000
   (c) ₹ 24000 (d) ₹ 1600

24. The product of two numbers is 48. If one number equals "the number of wings of a bird plus 2 times the number of fingers on your hand divided by the number of wheels of a tricycle", then the other number is
   [SSC CGL (Mains) 2016]
   (a) 9 (b) 10 (c) 12 (d) 18

25. X, Y and Z had taken a dinner together. The cost of the meal of Z was 20% more than that of Y and the cost of the meal of X was 5/6 as much as the cost of the meal of Z. If Y paid ₹ 100, then what was the total amount that all the three of them had paid?
   [CDS 2013]
   (a) ₹ 285 (b) ₹ 300
   (c) ₹ 355 (d) None of these

26. Find the maximum number of trees which can be planted 20 m apart on the two sides of a straight road 1760 m long.
   [SSC CGL 2013]
   (a) 174 (b) 176
   (c) 180 (d) 178

27. In a three-digit number, the digit in the unit's place is four times the digit in the hundred's place. If the digit in the unit's place and the ten's place are interchanged, the new number so formed is 18 more than the original number. If the digit in the hundred's place is one-third of the digit in the ten's place, then what is 25% of the original number?
   [LIC AAO 2013]
   (a) 67 (b) 84 (c) 137
   (d) Cannot be determined
   (e) None of these

28. There are two examination halls P and Q. If 10 students shifted from P to Q, then the number of students will be equal in both the examination halls. If 20 students shifted from Q to P, then the students of P would be doubled to the students of Q. The numbers of students would be in P and Q, respectively are
   [RRB 2012]
   (a) 60, 40 (b) 70, 50
   (c) 80, 60 (d) 100, 80

29. There are 200 questions in a 3 h examination. Among 200 questions, 50 are from Mathematics, 100 are from GK and 50 are from Science. Ram spent twice as much time on each Mathematics question as for each other question. How many minutes did he spend on Mathematics questions?
   (a) 36 (b) 72
   (c) 100 (d) 60

30. A three-digit number has digits in strictly descending order and divisible by 10. By changing the places of the digits, a new three-digit number is constructed in such a way that the new number is also divisible by 10. The difference between the original number and the new number is divisible by 40. How many numbers will satisfy all these conditions? [XAT 2015]
(a) 5 (b) 6 (c) 7 (d) 8
(e) None of these

31. A tin of oil was $\frac{4}{5}$ full. When 6 bottles of oil were taken out from this tin and 4 bottles of oil were poured into it, it was $\frac{3}{4}$ full. Oil of how many bottles can the tin contain? (all bottles are of equal volume) [CDS 2015 (II)]
(a) 35 (b) 40
(c) 45 (d) 50

Answer with Solutions

1. (c) Let the number be x.
According to the question,
$$x \times \left(x \times \frac{3}{4}\right) = 10800$$
$$\Rightarrow \frac{3x^2}{4} = 10800$$
$$\Rightarrow x^2 = \frac{10800 \times 4}{3} = 14400$$
On taking square root both sides, we get
$$x = \sqrt{14400} = 120$$

2. (b) Let the first number be x.
Then, another number $= \frac{2}{5}x$
According to the question, $x + \frac{2}{5}x = 50$
$$\Rightarrow \frac{7x}{5} = 50 \Rightarrow x = \frac{250}{7}$$
∴ Another number $= \frac{2}{5}x = \frac{2}{5} \times \frac{250}{7} = \frac{100}{7}$
Hence, the numbers are $\left(\frac{250}{7}, \frac{100}{7}\right)$.

3. (c) Let the number be x.
According to the question,
$$x \times \frac{4}{5} - \frac{3}{4} \times x = 8 \Rightarrow \frac{16x - 15x}{20} = 8$$
$$\Rightarrow x = 8 \times 20 = 160$$
Hence, the required number is 160.

4. (c) Let the number be x.
According to the question,
$$\frac{1}{5}\left(\frac{1}{2}x\right) = 20 \Rightarrow \frac{1}{10}x = 20 \Rightarrow x = 200$$
∴ 20% of $x = \frac{20}{100} \times 200 = 40$

5. (b) Let the number be x.
Then, $20 + 20\%$ of $x = x$
$$\Rightarrow 20 + \frac{20}{100}x = x$$

$$\Rightarrow x\left(1 - \frac{20}{100}\right) = 20$$
$$\Rightarrow x \times \frac{80}{100} = 20$$
∴ $x = 25$

6. (a) Let the number be x.
Then, according to the question,
$$x + \frac{4}{x} = \frac{17}{2} \Rightarrow x + \frac{4}{x} - \frac{17}{2} = 0$$
$$\Rightarrow \frac{2x^2 + 8 - 17x}{2x} = 0$$
$$\Rightarrow 2x^2 - 17x + 8 = 0$$
$$\Rightarrow 2x^2 - 16x - x + 8 = 0$$
$$\Rightarrow 2x(x - 8) - 1(x - 8) = 0$$
$$\Rightarrow (x - 8)(2x - 1) = 0$$
$$\Rightarrow x = 8, \frac{1}{2}$$
But number cannot be $\frac{1}{2}$.
Hence, the required number is 8.

7. (e) Given, $16X = 5Y$
$$\Rightarrow 16X - 5Y = 0 \qquad \ldots(i)$$
and $Y - 8 = 2X + 10$
$$\Rightarrow 2X - Y = -18 \qquad \ldots(ii)$$
On solving Eqs. (i) and (ii), we get
$$X = 15, Y = 48$$
∴ Sum of X and $Y = 15 + 48 = 63$

8. (c) Let two consecutive odd integers be $(2x - 1)$ and $(2x + 1)$, respectively.
Then, $(2x + 1)^2 - (2x - 1)^2$
$$= (4x^2 + 1 + 4x) - (4x^2 + 1 - 4x)$$
$$\left[\begin{array}{l}\because (a + b)^2 = a^2 + b^2 + 2ab \text{ and}\\ (a - b)^2 = a^2 + b^2 - 2ab\end{array}\right]$$
$$= 4x^2 + 1 + 4x - 4x^2 - 1 + 4x$$
$$= 8x$$
Hence, the difference between the squares of two consecutive odd integers is always divisible by 8.

Word Problems Based on Numbers / 149

9. (e) Let five consecutive even numbers be $2x, 2x + 2, 2x + 4, 2x + 6$ and $2x + 8$, respectively.
Now, according to the question,
$$2x + 2x + 2 + 2x + 4 + 2x + 6 + 2x + 8 = 250$$
$\Rightarrow \quad 10x + 20 = 250$
$\Rightarrow \quad 10x = 250 - 20$
$\Rightarrow \quad 10x = 230$
$\Rightarrow \quad x = 23$
∴ Required sum = Smallest number + Largest number
$= 2x + 2x + 8$
$= 4x + 8 = 4 \times 23 + 8 = 100$

10. (c) Let the third number be x.
Then, the first number = $3x$
and second number = $\dfrac{3x}{2}$
According to the question,
$$\dfrac{x + 3x + \dfrac{3x}{2}}{3} = 154$$
$\Rightarrow \quad \dfrac{2x + 6x + 3x}{6} = 154$
$\Rightarrow \quad x = \dfrac{154 \times 6}{11} = 84$
∴ Required difference = $3x - x = 2x$
$= 2 \times 84 = 168$

11. (c) Let the another number be x.
Given, first number = 15
Now, according to the question,
$2(15 - x) = (15 + x)$
$\Rightarrow \quad 30 - 2x = 15 + x$
$\Rightarrow \quad x + 2x = 30 - 15$
$\Rightarrow \quad 3x = 15$
∴ $\quad x = 5$
Hence, the another number is 5.

12. (b) Let the unit's digit and ten's digit be y and x, respectively.
∴ Number = $10x + y$
When digits are interchanged, then the number = $10y + x$
Now, according to the question,
$x + y = 10$ ...(i)
and $10y + x + 36 = 10x + y$
$\Rightarrow \quad 9x - 9y = 36$
$\Rightarrow \quad x - y = 4$ ...(ii)
On adding Eqs. (i) and (ii), we get
$2x = 14 \Rightarrow x = 7$
On putting the value of x in Eq. (i), we get
$7 + y = 10 \Rightarrow y = 3$
So, $\quad x = 7$ and $y = 3$
∴ Required number = $10x + y$
$= 10 \times 7 + 3 = 73$

So, the number is a multiple of a prime number.

13. (a) Let the unit's digit be y and ten's digit be x.
Then, the number = $10x + y$
When interchanging the place, the number is $10y + x$.
According to the question,
$(10x + y) - (10y + x) = 18$
$\Rightarrow \quad 10x + y - 10y - x = 18$
$\Rightarrow \quad 9x - 9y = 18$
∴ $\quad x - y = 2$ ...(i)
and $\quad x + y = 12$ ...(ii)
On adding Eqs. (i) and (ii), we get
$x - y = 2$
$x + y = 12$
$\overline{2x = 14}$
$\Rightarrow \quad x = 7$
So, $\quad x = 7$ and $y = 5$
∴ Product = $xy = 7 \times 5 = 35$

14. (e) Let ten's digit be x and unit's digit be x^2.
Original number = $10x + x^2$
New number = $10x^2 + x$
According to the question,
$(10x^2 + x) - (10x + x^2) = 54$
$\Rightarrow \quad 10x^2 + x - 10x - x^2 = 54$
$\Rightarrow \quad 9x^2 - 9x = 54$
$\Rightarrow \quad 9(x^2 - x) = 54$
$\Rightarrow \quad x^2 - x - 6 = 0$
$\Rightarrow \quad x^2 - 3x + 2x - 6 = 0$
$\Rightarrow \quad x(x - 3) + 2(x - 3) = 0$
$\Rightarrow \quad (x - 3)(x + 2) = 0$
$\Rightarrow \quad x = 3, -2$
∴ Ten's digit = $x = 3$
and unit's digit = $x^2 = 3^2 = 9$
Original number = 39
∴ Required number = $39 \times \dfrac{40}{100} = 15.6$

15. (c) Let the total number of sweets be x.
According to the question,
$\dfrac{x}{250} - \dfrac{x}{300} = 1$
$\Rightarrow \quad \dfrac{6x - 5x}{1500} = 1$
$\Rightarrow \quad x = 1500$
∴ Required number of sweets = 1500

16. (c) Let the unit's place number be y and ten's place number be x.
Then, the original number = $10x + y$
Now, after interchanging the digits,
New number = $10y + x$

and sum of digits $= x + y$
According to the question,
$$10x + y = k(x + y) \quad \ldots(i)$$
and
$$10y + x = m(x + y) \quad \ldots(ii)$$
On adding Eqs. (i) and (ii), we get
$$11(x + y) = (k + m)(x + y)$$
$$\Rightarrow k + m = 11 \Rightarrow m = 11 - k$$

17. (d) Let the number of buffaloes be x and the number of ducks be y.
Then, according to the question,
$$4x + 2y = 2(x + y) + 24$$
$$\Rightarrow 2x = 24 \Rightarrow x = 12$$
∴ Number of buffaloes is 12.

Fast Track Method
∵ L = Number of total legs
and H = Number of heads
Here, $L = 24 + 2H$
[according to the question]
Number of buffaloes $= \dfrac{L - 2H}{2}$ [by Type 3]
$= \dfrac{L - (L - 24)}{2}$ $[\because 2H = L - 24]$
$= \dfrac{L - L + 24}{2} = 12$

18. (d) Let the number of hens be x and number of cows be y.
According to the question,
$$x + y = 48 \quad \ldots(i)$$
and
$$2x + 4y = 140$$
$$\Rightarrow x + 2y = 70 \quad \ldots(ii)$$
On solving Eqs. (i) and (ii), we get
$$x = 26 \text{ and } y = 22$$
Hence, the number of hens is 26.

Fast Track Method
Here, number of heads, $H = 48$
and number of legs, $L = 140$
∴ Total number of cows $= \dfrac{L - 2H}{2}$
[by Type 3]
$= \dfrac{140 - 2 \times 48}{2}$
$= \dfrac{140 - 96}{2} = 22$
∴ Number of hens $= 48 - 22 = 26$

19. (c) Let the number of oranges in first basket $= x$ and number of oranges in second basket $= y$
Now, $x + y = 640$ ...(i)
Also, $\dfrac{4x}{5} = y + \dfrac{x}{5}$
∴ $\dfrac{3x}{5} = y$
$\Rightarrow \dfrac{x}{y} = \dfrac{5}{3}$ ...(ii)

From Eqs. (i) and (ii), we get
$$x = 400 \text{ and } y = 240$$
∴ Number of oranges in first basket = 400.

20. (b) Let x be the required number.
Then, according to the question,
$$8x = \dfrac{x}{8} + 2016$$
$$\Rightarrow 8x - \dfrac{x}{8} = 2016 \Rightarrow \dfrac{63x}{8} = 2016$$
$$\Rightarrow x = \dfrac{2016 \times 8}{63} = 256$$

21. (c) Let two positive integers be x and y.
According to the question,
$$x^2 + y^2 = 208 \quad \ldots(i)$$
and
$$y^2 = 18x \quad \ldots(ii)$$
On putting $y^2 = 18x$ in Eq. (i), we get
$$x^2 + 18x = 208$$
$$\Rightarrow x^2 + 18x - 208 = 0$$
$$\Rightarrow x^2 + 26x - 8x - 208 = 0$$
$$\Rightarrow x(x + 26) - 8(x + 26) = 0$$
$$\Rightarrow (x + 26)(x - 8) = 0$$
$$\Rightarrow x + 26 = 0 \text{ or } x - 8 = 0$$
$$\Rightarrow x = -26 \text{ or } x = 8$$
But x is positive.
∴ $x = 8$
On putting $x = 8$ in Eq. (ii), we get
$$y^2 = 18x \Rightarrow y^2 = 18 \times 8$$
$$\Rightarrow y = \pm\sqrt{16 \times 9} \Rightarrow y = 4 \times 3$$
$$\Rightarrow y = 12 \quad [\because y > 0]$$
∴ Difference between larger and smaller number $= 12 - 8 = 4$.

22. (a) Let positive number $= x$
According to the question,
$$x \times 21 = x^2 - 100$$
$$\Rightarrow x^2 - 21x - 100 = 0$$
$$\Rightarrow x^2 - 25x + 4x - 100 = 0$$
$$\Rightarrow x(x - 25) + 4(x - 25) = 0$$
$$\Rightarrow (x - 25)(x + 4) = 0$$
$$\Rightarrow x = 25, -4$$
Since, the number is positive.
∴ $x = 25$

23. (c) Let total property of Ram be x.
Then, share of his wife $= \dfrac{x}{3}$
and share of his daughter
$= \left(x - \dfrac{x}{3}\right) \times \dfrac{3}{5} = \dfrac{2x}{3} \times \dfrac{3}{5} = \dfrac{2x}{5}$
Now, share of his son $= x - \dfrac{x}{3} - \dfrac{2x}{5}$
$= \dfrac{15x - 5x - 6x}{15} = \dfrac{4x}{15}$

Word Problems Based on Numbers / 151

Given, share of son = 6400

$\Rightarrow \quad \dfrac{4x}{15} = 6400 \Rightarrow 4x = 6400 \times 15$

$\Rightarrow \quad x = \dfrac{6400 \times 15}{4}$

$\therefore \quad x = 1600 \times 15 = ₹ 24000$

24. (c) Let first number = x
and second number = y
According to the question,
$$xy = 48 \qquad \ldots(i)$$
and $\quad x = \dfrac{2 + 5 \times 2}{3} = \dfrac{2 + 10}{3}$

$\Rightarrow \quad x = \dfrac{12}{3}$

$\Rightarrow \quad x = 4$

Now, putting the value of x in Eq. (i), we get
$4 \times y = 48$

$\Rightarrow \quad y = \dfrac{48}{4} = 12$

Hence, the second number is 12.

25. (d) Given, cost of the meal of Y = ₹ 100

Now, according to the question,
Cost of the meal of Z
= 20% more than that of Y
$= \left(100 + \dfrac{20}{100} \times 100\right)$
$= (100 + 20) = ₹ 120$

and cost of the meal of X
$= \dfrac{5}{6}$ as much as cost of the meal of Z
$= \dfrac{5}{6} \times 120 = ₹ 100$

$\therefore$ Total amount that all the three of them has paid = 100 + 120 + 100 = ₹ 320

26. (d) Number of trees that can be planted on one side of road $= \dfrac{1760}{20} + 1$
$= 88 + 1 = 89$

$\therefore$ Trees on the both sides = $2 \times 89 = 178$

27. (a) Let hundred's digit be x.
Then, unit's digit = $4x$ and ten's digit = $3x$
Number = $100x + 30x + 4x = 134x$
Again, hundred's digit = x
Ten's digit = $4x$ and unit's digit = $3x$
Number = $100x + 40x + 3x = 143x$
According to the question,
$143x - 134x = 18$
$\Rightarrow \quad 9x = 18$
$\therefore \quad x = 2$
Original number = $134x = 134 \times 2 = 268$

$\therefore 25\%$ of original number = $268 \times \dfrac{25}{100} = 67$

28. (d) Let number of students in examination halls P and Q be x and y, respectively.
Then, according to the first condition,
$x - 10 = y + 10$
$\Rightarrow \quad x - y = 20 \qquad \ldots(i)$
and according to the second condition,
$x + 20 = 2(y - 20)$
$\Rightarrow \quad x + 20 = 2y - 40$
$\Rightarrow \quad x - 2y = -60 \qquad \ldots(ii)$
On subtracting Eq. (ii) from Eq. (i), we get
$-y + 2y = 20 + 60$
$\Rightarrow \quad y = 80$
On putting the value of y in Eq. (i), we get
$x - 80 = 20 \Rightarrow x = 100$
Hence, number of students in examination halls P and Q are 100 and 80, respectively.

29. (b) Let Ram spends x min on each Mathematics question.
According to the question,
$50x + 100 \times \dfrac{x}{2} + 50 \times \dfrac{x}{2} = 3 \times 60$
$\Rightarrow \quad x(50 + 50 + 25) = 180$
$\Rightarrow \quad x = \dfrac{180}{125}$
$\therefore$ Required time = $50 \times \dfrac{180}{125} = 72$ min

30. (b) Let us assume the three-digit number be $100x + 10y + 0$ (as it is divisible by 10). The new number will be $100y + 10x + 0$ (as this should also be divisible by 10). Then, the difference between them is $90(x - y)$.
If $90(x - y)$ is divisible by 40, then 9 $(x - y)$ should be divisible by 4.
In this three-digit number, digits are in strictly descending order. Thus, x or y cannot be 0.
$\therefore$ Possible combinations of (x, y) can be $(9, 1)$, $(9, 5)$, $(8, 4)$, $(7, 3)$, $(6, 2)$ and $(5, 1)$, i.e. 6.

31. (b) Let a tin of oil contain x bottles.
According to the question,
$\dfrac{4}{5} - \dfrac{6}{x} + \dfrac{4}{x} = \dfrac{3}{4}$
$\Rightarrow \quad \dfrac{4}{5} - \dfrac{3}{4} = \dfrac{6}{x} - \dfrac{4}{x}$
$\Rightarrow \quad \dfrac{1}{20} = \dfrac{2}{x}$
$\Rightarrow \quad x = 40$
So, the tin contains 40 bottles.

Chapter 10

Average

An **average** or **an arithmetic mean** of given data is the sum of the given observations divided by number of observations.

For example If we have to find out the average of 10, 15, 25 and 30, then the required average will be

$$= \frac{10 + 15 + 25 + 30}{4} = \frac{80}{4} = 20$$

Similarly, average of 30, 10 and 50

$$= \frac{30 + 10 + 50}{3} = \frac{90}{3} = 30$$

Therefore, we can write the formula

$$\text{Average } (A) = \frac{\text{Sum of observations } (S)}{\text{Number of observations } (N)}$$

Properties of Average

(i) Average of a given data is less than the greatest observation and greater than the smallest observation of the given data.

For example Average of 3, 7, 9 and 13

$$= \frac{3 + 7 + 9 + 13}{4} = \frac{32}{4} = 8$$

Clearly, 8 is less than 13 and greater than 3.

(ii) If the observations of given data are equal, then the average will also be the same as observations.

For example Average of 6, 6, 6 and 6 will be 6 because

$$\frac{6 + 6 + 6 + 6}{4} = \frac{24}{4} = 6$$

(iii) If 0 (zero) is one of the observations of a given data, then that 0 (zero) will also be included while calculating average.

For example Average of 3, 6 and 0 is 3 because

$$\frac{3 + 6 + 0}{3} = \frac{9}{3} = 3$$

✦ If all the numbers get increased by a, then their average is also increased by a.
✦ If all the numbers get decreased by a, then their average is also decreased by a.
✦ If all the numbers are multiplied by a, then their average is also multiplied by a.
✦ If all the numbers are divided by a, then their average is also divided by a.

Ex. 1 Find out the average of 308, 125, 45, 120 and 102.

Sol. Required average = $\dfrac{\text{Sum of given observations}}{\text{Number of observations}}$

$= \dfrac{308 + 125 + 45 + 120 + 102}{5} = \dfrac{700}{5} = 140$

Ex. 2 If the weight of A is 60 kg, weight of B is 45 kg and weight of C is 54 kg, then what is the average weight of three persons?

Sol. Required average = $\dfrac{\text{Sum of observations}}{\text{Number of observations}}$

$= \dfrac{60 + 45 + 54}{3} = \dfrac{159}{3} = 53 \text{ kg}$

Ex. 3 The average expenditure of Chandan in four days is ₹ 90. If his expenditures for the first three days are ₹ 100, ₹ 125 and ₹ 85 respectively, then what is the expenditure of Chandan for the fourth day?

Sol. Let the expenditure for fourth day = ₹ x

Then, average expenditure = $\dfrac{\text{Sum of the expenditures of four days}}{4}$

∴ $90 = \dfrac{100 + 125 + 85 + x}{4} \Rightarrow 310 + x = 360 \Rightarrow x = 360 - 310 = ₹ 50$

Important Formulae Related to Average of Numbers

(i) Average of first n natural numbers $= \left(\dfrac{n+1}{2}\right)$

(ii) Average of first n even numbers $= (n+1)$

(iii) Average of first n odd numbers $= n$

(iv) Average of consecutive numbers or average of numbers in AP

$= \dfrac{\text{First number + Last number}}{2}$

(v) Average of 1 to n odd numbers $= \dfrac{\text{Last odd number} + 1}{2}$

(vi) Average of 1 to n even numbers $= \dfrac{\text{Last even number} + 2}{2}$

(vii) Average of squares of first n natural numbers $= \dfrac{(n+1)(2n+1)}{6}$

(viii) Average of the cubes of first n natural numbers $= \dfrac{n(n+1)^2}{4}$

(ix) Average of first n multiples of any number $= \dfrac{\text{Number} \times (n+1)}{2}$

What will be the average of numbers from 1 to 51?

Sol. According to the formula,

Average of first n natural numbers $= \left(\dfrac{n+1}{2}\right)$

Here, $n = 51$

∴ Required average $= \dfrac{51+1}{2} = \dfrac{52}{2} = 26$

Ex. 5 Find out the average of 2, 4, 6, 8, 10, 12 and 14.

Sol. As we know, average of first n even numbers $= (n+1)$

Here, $n = 7$

∴ Required average $= (7+1) = 8$

Ex. 6 Calculate the average of 1, 3, 5, 7, 9, 11, 13, 15 and 17.

Sol. As we know, average of first n odd numbers $= n$

Here, $n = 9$

∴ Required average $= 9$

Ex. 7 What will be the average of 3, 4, ..., 51, 52, 53?

Sol. As we know, average of consecutive numbers $= \dfrac{\text{First number} + \text{Last number}}{2}$

Here, first number $= 3$ and last number $= 53$

∴ Required average $= \dfrac{3+53}{2} = \dfrac{56}{2} = 28$

Ex. 8 Find out the average of 4, 7, 10, 13, ..., 28, 31.

Sol. Here, the difference between any two numbers written in continuous sequence is 3. So, this is a series of consecutive numbers.

As we know, average of numbers in AP $= \dfrac{\text{First number} + \text{Last number}}{2}$

Here, first number $= 4$ and last number $= 31$

∴ Required average $= \dfrac{4+31}{2} = \dfrac{35}{2} = 17.5$

Ex. 9 Find the average of all the odd numbers and average of all the even numbers from 1 to 45.

According to the formula,

Average of 1 to n odd numbers $= \dfrac{\text{Last odd number} + 1}{2}$

Here, last odd number $= 45$

∴ Average of 1 to 45 odd numbers $= \dfrac{45+1}{2} = \dfrac{46}{2} = 23$

Again, according to the formula,

Average of 1 to n even numbers $= \dfrac{\text{Last even number} + 2}{2}$

Here, last even number $= 44$

∴ Average of 1 to 44 even numbers $= \dfrac{44+2}{2} = \dfrac{46}{2} = 23$

Average / 155

Ex. 10 Calculate the average of the squares of natural numbers from 1 to 25.

Sol. According to the formula,

Average of squares of first n natural numbers = $\dfrac{(n+1)(2n+1)}{6}$

Here, $n = 25$

∴ Required average = $\dfrac{(25+1)(2 \times 25+1)}{6} = \dfrac{26 \times 51}{6} = \dfrac{1326}{6} = 221$

Ex. 11 Calculate the average of the cubes of first 9 natural numbers.

Sol. According to the formula,

Average of cubes of first n natural numbers = $\dfrac{n(n+1)^2}{4}$

Here, $n = 9$

∴ Required average = $\dfrac{9(9+1)^2}{4} = \dfrac{9 \times 100}{4} = 9 \times 25 = 225$

Ex. 12 What will be the average of first 9 multiples of 5?

Sol. According to the formula,

Average of first n multiples of a number = $\dfrac{\text{Number} \times (n+1)}{2}$

Here, $n = 9$ and number = 5

∴ Required average = $\dfrac{5 \times (9+1)}{2} = \dfrac{50}{2} = 25$

Fast Track Techniques to solve the QUESTIONS

Technique 1 If the average of n_1 observations is a_1, the average of n_2 observations is a_2 and so on, then

Average of all the observations = $\dfrac{n_1 a_1 + n_2 a_2 + \ldots}{n_1 + n_2 + \ldots}$

Ex. 13 There are 30 boys and 60 girls in a class. If the average age of boys is 12 yr and average age of girls is 10 yr, then find out the average age of the whole class.

Sol. Here, $n_1 = 30, n_2 = 60, a_1 = 12$ and $a_2 = 10$

∴ Average age of the whole class = $\dfrac{n_1 a_1 + n_2 a_2}{n_1 + n_2} = \dfrac{30 \times 12 + 60 \times 10}{30 + 60}$

$= \dfrac{360 + 600}{90} = \dfrac{960}{90} = 10.66$ yr

Ex. 14 Pinky bought 20 books at the rate of ₹ 10 each, 45 pens at the rate of ₹ 5 each and 15 pencils at the rate of ₹ 3 each. Calculate the average price of all the stationary goods.

Sol. Here, $n_1 = 20, n_2 = 45, n_3 = 15, a_1 = 10, a_2 = 5$ and $a_3 = 3$

∴ Average price of all the stationary goods $= \dfrac{n_1 a_1 + n_2 a_2 + n_3 a_3}{n_1 + n_2 + n_3}$

$= \dfrac{20 \times 10 + 45 \times 5 + 15 \times 3}{20 + 45 + 15} = \dfrac{200 + 225 + 45}{80} = \dfrac{470}{80} = ₹ 5.875$

Technique 2

If the average of m observations is a and the average of n observations taken out of m is b, then

Average of rest of the observations $= \dfrac{ma - nb}{m - n}$

Ex. 15 A man bought 20 cows in ₹ 200000. If the average cost of 12 cows is ₹ 12500, then what will be the average cost of remaining cows?

Sol. Average cost of 12 cows = ₹ 12500

Total cost of 12 cows = 12×12500 = ₹ 150000

Cost of remaining 8 cows = $200000 - 150000$ = ₹ 50000

∴ Average of remaining 8 cows = $\dfrac{50000}{8}$ = ₹ 6250

Fast Track Method

Average cost of 20 cows = $\dfrac{200000}{20}$ = ₹ 10000

Here, $m = 20, n = 12, a = 10000, b = 12500$

∴ Average cost of remaining $(20 - 12)$ cows $= \dfrac{20 \times 10000 - 12 \times 12500}{20 - 12}$

$= \dfrac{200000 - 150000}{8} = \dfrac{50000}{8} = ₹ 6250$

Technique 3

(i) If average of n observations is a but the average becomes b when one observation is eliminated, then value of eliminated observation $= n(a - b) + b$

(ii) If average of n observations is a but the average becomes b when a new observation is added, then value of added observation $= n(b - a) + b$

Ex. In a cricket team, the average age of 11 players and the coach is 18 yr. If the age of the coach is not considered, then the average decreases by 1 yr. Find out the age of the coach.

Sol. Total age of 11 players and the coach = $12 \times 18 = 216$ yr

Total age of 11 players = $11 \times 17 = 187$ yr

∴ Age of the coach = $216 - 187 = 29$ yr

Fast Track Method

Here, $n = 11 + 1 = 12$, initial average, $a = 18$ yr and last average, $b = 18 - 1 = 17$ yr

∴ Age of the coach = $n(a - b) + b = 12(18 - 17) + 17 = 12 + 17 = 29$ yr

Ex. 17 The average runs scored by a batsman in 20 innings is 32. After 21st innings, the average runs becomes 34. How much runs does the batsman score in his 21st inning?

Sol. Runs scored in 20 innings = 20 × 32 = 640
Runs scored in 21 innings = 21 × 34 = 714
∴ Runs scored in the 21st inning = 714 − 640 = 74

Fast Track Method
Here, $n = 20$, initial average, $a = 32$ and last average, $b = 34$
∴ Runs scored in 21st inning = $n(b - a) + b$ = 20 (34 − 32) + 34
$$= 20 \times 2 + 34 = 74$$

Technique 4 We have n observations out of which some observations ($a_1, a_2, a_3, ...$) are replaced by some other new observations and in this way, if the average increases or decreases by b, then
Value of new observations = $a \pm nb$, where $a = a_1 + a_2 + a_3 + ...$

✦ In this formula, the signs of '+' and '−' depend upon the increment or decrement in the average.

Ex. 18 The average weight of 3 women is increased by 4 kg, when one of them whose weight is 100 kg, is replaced by another woman. What is the weight of the new woman?

Sol. Total weight increased = 4 × 3 = 12 kg
∴ Weight of new woman = 100 + 12 = 112 kg

Fast Track Method
Here, $n = 3$, $a = 100$ kg and $b = 4$ kg
∴ Weight of new woman = $a + nb$ = 100 + 3 × 4 = 112 kg
[here, '+' sign has been taken as average increases in this case]

Ex. 19 The average age of 25 boys in a class decreases by 6 months, when a new boy takes the place of a 20 yr old boy. Find out the age of the new boy.

Sol. The average age is decreased by 6 months, i.e. $\frac{1}{2}$ yr.
Total age decreased = $25 \times \frac{1}{2}$ = 12.5 yr
∴ Age of new boy = Age of boy replaced − Total age decreased = 20 − 12.5 = 7.5 yr

Fast Track Method
Here, $n = 25$, $a = 20$, $b = 6$ months = $\frac{6}{12}$ = 0.5 yr
∴ Age of the new boy = $a - nb$ = 20 − 25 × 0.5 = 20 − 12.5 = 7.5 yr
[here, '−' sign has been taken as average decreases in this case]

Technique 5 If the average of n students in a class is a, where average of passed students is x and average of failed students is y, then
Number of students passed
$$= \frac{\text{Total students (Total average − Average of failed students)}}{\text{Average of passed students − Average of failed students}} = \frac{n(a-y)}{(x-y)}$$

Ex. 20 In a class, there are 75 students and their average marks in the annual examination is 35. If the average marks of passed students is 55 and average marks of failed students is 30, then find out the number of students who failed.

158 / Fast Track Objective Arithmetic

Sol. Let number of failed students = x
Then, number of passed students = $75 - x$
According to the question, $75 \times 35 = 30x + 55(75 - x)$
$\Rightarrow \quad 15 \times 35 = 6x + 11(75 - x) \Rightarrow 525 = 6x + 825 - 11x \Rightarrow 5x = 300$
$\therefore \quad x = 60$

Fast Track Method

Here, $n = 75$, $a = 35$, $x = 55$ and $y = 30$
$\because$ Number of students who passed $= \dfrac{n(a - y)}{x - y} = \dfrac{75(35 - 30)}{55 - 30} = \dfrac{75 \times 5}{25} = 15$
$\therefore$ Number of students who failed = $75 - 15 = 60$

Technique 6 If the average of total components in a group is a, where average of n components (1st part) is b and average of remaining components (2nd part) is c, then

Number of remaining components (2nd part) $= \dfrac{n(a - b)}{(c - a)}$

Ex. 21 The average salary of the entire staff in an office is ₹ 200 per day. The average salary of officers is ₹ 550 and that of non-officers is ₹ 120. If the number of officers is 16, then find the number of non-officers in the office.

Sol. Let number of non-officers be x.
Then, $120x + 550 \times 16 = 200(16 + x)$
$\Rightarrow \quad 12x + 55 \times 16 = 20(16 + x) \Rightarrow 3x + 55 \times 4 = 5(16 + x)$
$\Rightarrow \quad 3x + 220 = 80 + 5x \Rightarrow 5x - 3x = 220 - 80 \Rightarrow 2x = 140$
$\therefore \quad x = \dfrac{140}{2} = 70$

Fast Track Method

Here, $n = 16$, $a = 200$, $b = 550$ and $c = 120$
According to the formula,
Number of non-officers $= \dfrac{n(a - b)}{c - a} = \dfrac{16(200 - 550)}{120 - 200} = \dfrac{16 \times 35}{8} = 70$

MIND IT! Average speed is defined as total distance travelled divided by total time taken,
i.e. Average speed $= \dfrac{\text{Total distance travelled}}{\text{Total time taken}}$

Technique 7 If a person covers a certain distance at a speed of A km/h and again covers the same distance at a speed of B km/h, then the average speed during the whole journey will be $\dfrac{2AB}{A + B}$.

Ex. 22 A person covers a certain distance by car at a speed of 25 km/h and comes back at a speed of 40 km/h. What is his average speed during his travel?

Sol. Here, $A = 25$ km/h and $B = 40$ km/h
$\therefore$ Required average speed $= \dfrac{2AB}{A + B} = \dfrac{2 \times 25 \times 40}{25 + 40} = \dfrac{2000}{65} = 30.76$ km/h

Fast Track Practice

Exercise 1 Base Level Questions

1. Find the average of the following set of scores.
253, 124, 255, 534, 836, 375, 101, 443, 760 **[IBPS Clerk 2011]**
(a) 427   (b) 413   (c) 141   (d) 490
(e) None of these

2. The mean temperature of Monday to Wednesday was 37°C and of Tuesday to Thursday was 34°C. If the temperature on Thursday was 4/5 that of Monday, then the temperature of Thursday was **[SSC FCI 2012]**
(a) 35.5°C   (b) 34°C
(c) 36.5°C   (d) 36°C

3. Find the average of all prime numbers between 60 and 90. **[SSC (10+2) 2012]**
(a) 72   (b) 74.7
(c) 74   (d) 73.6

4. If $47a + 47b = 5452$, then what is the average of a and b? **[SNAP 2012]**
(a) 116   (b) 23.5
(c) 96   (d) 58

5. The average of the test scores of a class of 'm' students is 70 and that of 'n' students is 91. When the scores of both the classes are combined, the average is 80. What is n/m?
(a) 11/10   (b) 13/10
(c) 10/13   (d) 10/11

6. The average of ten numbers is 7. If every number is multiplied by 12, then what will be the average of new numbers? **[SSC (10+2) 2009]**
(a) 7   (b) 9
(c) 82   (d) 84

7. The mean marks obtained by seven students in a group is 226. If the marks obtained by six of them are 340, 180, 260, 56, 275 and 307 respectively, then find the marks obtained by the seventh student.
(a) 164
(b) 226
(c) 340
(d) Cannot be determined

8. The average of four positive integers is 73.5. The highest integer is 108 and the lowest integer is 29. The difference between the remaining two integers is 15. Which of the following is the smaller of the remaining two integers? **[Bank Clerk 2009]**
(a) 80   (b) 86   (c) 73
(d) Cannot be determined
(e) None of the above

9. The average of 13 results is 60. If the average of first 7 results is 59 and that of last 7 results is 61, then what will be the seventh result?
(a) 90   (b) 50
(c) 75   (d) 60
(e) None of these

10. The average of nine numbers is 50. The average of the first five numbers is 54 and that of the last three numbers is 52. Then, the sixth number is
(a) 34   (b) 24
(c) 44   (d) 30

11. The average marks of 30 students are 45 but after checking, there are two mistakes found. After adjustment, if a student got 45 more marks and other student got 15 less marks, then what will be the adjusted average? **[SSC CGL 2012]**
(a) 45   (b) 44
(c) 47   (d) 46

12. In an exam, the average marks obtained by John in English, Maths, Hindi and Drawing were 50. His average marks in Maths, Science, Social Studies and Craft were 70. If the average marks in all seven subjects is 58, his score in Maths was **[SSC CGL (Mains) 2016]**
(a) 50   (b) 52
(c) 60   (d) 74

13. Five years ago, the average age of P and Q was 15 yr. Now, average age of P, Q and R is 20 yr. What would be the age of R after 10 yr?
(a) 35 yr   (b) 40 yr
(c) 30 yr   (d) 50 yr

14. A batsman scores 80 runs in his sixth innings and thus increases his average by 5. What is his average after six innings? **[SSC CGL (Mains) 2016]**
    (a) 50   (b) 55   (c) 60   (d) 65

15. The average of 12 observations is 8, later it was observed that one observation 10 is wrongly written as 13. The correct average of observations is **[IBPS SO 2016]**
    (a) 7   (b) 17.5
    (c) 7.75   (d) 8
    (e) 5

16. A man was assigned to find the average age of a class of 13 students. By mistake, he included the 35 yr old teacher as well and hence the average went up by 2 yr. Find the actual average age of the class.
    [SBI Clerk (Mains) 2016]
    (a) 8 yr   (b) 7 yr
    (c) 15 yr   (d) 11 yr
    (e) 9 yr

17. A man spends an average of ₹ 1694.70 per month for the first 7 month and ₹ 1810.50 per month for the next 5 months. His monthly salary, if he saves ₹ 3084.60 during the whole year, is
    (a) ₹ 2400   (b) ₹ 3000
    (c) ₹ 1000   (d) ₹ 2000

18. A batsman has a certain average of runs for 12 innings. In the 13th inning, he scores 96 runs thereby increasing his average by 5 runs. What will be his average after 13th inning?
    [SSC CGL (Mains) 2016]
    (a) 28   (b) 32
    (c) 36   (d) 42

19. The average runs of a cricketer in a tournament, in which he played 14 matches, are 47. His average runs in the first seven matches are 57 and that in the last five matches are 44. If the runs made by him in 8th match are 15, then how many runs did he make in 9th match? **[SBI Clerk (Pre) 2016]**
    (a) 24   (b) 32
    (c) 26   (d) 22
    (e) 28

20. A cricketer has a certain average of 10 innings. In the eleventh inning, he scored 108 runs, thereby increasing his average by 6 runs. What is his new average? **[CDS 2016 (II)]**
    (a) 42   (b) 47
    (c) 48   (d) 60

21. A cricketer has a certain average for 10 innings. In the eleventh inning, he scored 216 runs, thereby increasing his average by 12 runs. Find out his new average. **[Bank Clerk 2011]**
    (a) 96   (b) 84   (c) 97   (d) 87
    (e) None of these

22. The average age of students of a class is 15.8 yr. The average age of boys in the class is 16.4 yr and that of the girls is 15.4 yr. Find out the ratio of the number of boys to the number of girls in the class. **[SSC CGL 2011]**
    (a) 3 : 1   (b) 5 : 2
    (c) 2 : 3   (d) 3 : 7

23. In a certain examination, the average marks of an examinee is 64 per paper. If he had obtained 18 more marks for his Mathematics paper and 4 more marks for his English paper, his average per paper would have been 66. How many papers were there in the examination?
    [Bank PO 2010]
    (a) 11   (b) 13
    (c) 9   (d) 15
    (e) None of these

24. Calculate the average of the cubes of first five natural numbers. **[LIC ADO 2007]**
    (a) 55   (b) 65
    (c) 45   (d) 35
    (e) None of these

25. The average of four consecutive numbers A, B, C and D is 49.5. What is the product of B and D?
    (a) 2499   (b) 2352
    (c) 2450   (d) 2550
    (e) None of these

26. The average of four consecutive even numbers is 27. What is the greatest number? **[SSC CGL 2010]**
    (a) 28   (b) 26
    (c) 32   (d) 30

27. The average of 5 consecutive odd positive integers is 9. The least one among them is
    [SSC Multitasking 2014]
    (a) 5   (b) 3
    (c) 1   (d) 7

28. Eight consecutive numbers are given. If the average of the two numbers that appear in the middle is 6, then the sum of the eight given numbers is
    [SSC (10+2) 2012]
    (a) 36   (b) 48
    (c) 54   (d) 64

29. A factory buys 8 machines, 3 Machine A, 2 Machine B and rest Machine C. Prices of the machines are ₹ 100000, ₹ 80000, and ₹ 45000 respectively. Calculate the average cost of these machines.
[SSC (10+2) 2017]
(a) 74375 (b) 75000
(c) 75625 (d) 72875

30. A class is divided into two sections A and B. Passing average of 20 students of section A is 80% and passing average of 30 students of section B is 70%. What is the passing average of both the sections?
[SSC CGL 2009]
(a) 72% (b) 74%
(c) 75% (d) 77%

31. The numbers of boys and girls are x and y respectively. Ages of a boy and a girl are 'a' yr and 'b' yr respectively. The average age (in years) of all boys and girls is [SSC CGL (Pre) 2016]
(a) $\dfrac{x+y}{bx+ay}$ (b) $\dfrac{bx+ay}{x+y}$
(c) $\dfrac{ax+by}{x+y}$ (d) $\dfrac{x+y}{ax+by}$

32. The average weight of 19 men is 74 kg and the average weight of 38 women is 63 kg. What is the average weight (rounded off to the nearest integer) of all the men and the women together?
[SBI Clerk 2012]
(a) 59 kg (b) 65 kg
(c) 69 kg (d) 67 kg
(e) 71 kg

33. Roshan bought 5 pants at ₹ 25 each, 10 shirts at ₹ 50 each and 15 ties at ₹ 35 each. Find the average price of all the articles.
(a) ₹ 38.33 (b) ₹ 45
(c) ₹ 60 (d) ₹ 45.33
(e) None of these

34. The average height of the basketball team A is 5 ft 11 inch and that of B is 6 ft 2 inch. There are 20 players in team A and 18 players in team B. The overall average height is
(a) 72.42 inch (b) 72 inch
(c) 70.22 inch (d) 70 inch

35. The average age of a class is 35 yr. 6 new students with an average age of 33 yr joined in that class, thereby decreasing the average by half year. The original strength of the class was [SSC (10+2) 2012]
(a) 14 (b) 18 (c) 16 (d) 20

36. The average age of a group of workers is 42 yr. If eight new workers of average age 36 yr join the group, the average age of the group becomes 40 yr. How many workers were there in the group initially? [CMAT 2015]
(a) 12 (b) 15
(c) 18 (d) 16

37. The mean weight of 150 students in a class is 60 kg. The mean weight of boys is 70 kg and that of girls is 55 kg. What is the number of boys in the class?
[CDS 2012]
(a) 50 (b) 60
(c) 75 (d) 100

38. The average age of 30 girls is 13 yr. The average of first 18 girls is 15 yr. Find out the average age of remaining 12 girls.
[Hotel Mgmt. 2009]
(a) 12 yr (b) 10 yr
(c) 16 yr (d) 10.5 yr

39. The average weight of a class of 15 boys and 10 girls is 38.4 kg. If the average weight of the boys is 40 kg, then what is the average weight of the girls? [CDS 2013]
(a) 36.5 kg (b) 35 kg
(c) 36 kg (d) 35.6 kg

40. In a class of 39 students, there are 26 girls. The average weight of these girls is 42 kg and average weight of the full class is 48 kg. What is the average weight of the boys of the class? [SSC CGL (Pre) 2017]
(a) 54 kg (b) 66 kg
(c) 60 kg (d) 62 kg

41. The average age of 5 boys is 16 yr, of which that of 4 boys is 16 yr 3 months. The age of the 5th boy is
(a) 15 yr (b) 15 yr 6 months
(c) 15 yr 4 months (d) 15 yr 2 months

42. A businessman purchased 30 TV sets in ₹ 300000. If the average cost of 15 TV sets is ₹ 9000, then find out the average cost of remaining TV sets.
(a) ₹ 20000 (b) ₹ 12000
(c) ₹ 11000 (d) ₹ 10000
(e) None of these

43. In a class of 100 students, there are 70 boys whose average marks in a subject are 75. If the average marks of the complete class is 72, then what is the average marks of the girls? [CDS 2016 (II)]
(a) 64 (b) 65
(c) 68 (d) 74

44. The average salary of all the workers in a workshop is ₹ 8000. The average salary of 7 technicians is ₹ 12000 and the average salary of the rest is ₹ 6000. The total number of workers in the workshop is [SSC (10+2) 2015]
(a) 20 (b) 21 (c) 22 (d) 23

45. The average monthly income of 4 earning members of a family is ₹ 7350. One member passes away and the average monthly income becomes ₹ 6500. What was the monthly income of the person, who is no more? [SSC CGL 2007]
(a) ₹ 6928 (b) ₹ 8200
(c) ₹ 9900 (d) ₹ 13850

46. The average age of 14 girls and their teacher's age is 15 yr. If teacher's age is excluded, then the average reduced by 1. What is the teacher's age? [SSC CGL 2013]
(a) 29 yr (b) 35 yr (c) 32 yr (d) 30 yr

47. The average weight of 21 boys was recorded as 64 kg. If the weight of the teacher was added, the average increased by 1 kg. What was the teacher's weight? [Bank Clerk 2011]
(a) 86 kg (b) 64 kg
(c) 72 kg (d) 98 kg
(e) None of these

48. The mean weight of 34 students of a school is 42 kg. If the weight of the teacher be included, the mean rises by 400 g. Find the weight (in kg) of the teacher.
(a) 66 (b) 56 (c) 55 (d) 57

49. The average age of 4 members of a family is 25 yr. If head of the family is included in this group, then average age increases by 20%. Find out the age of the head. [Delhi Police 2008]
(a) 45 yr (b) 50 yr
(c) 55 yr (d) 60 yr

50. The average age of a committee of 11 persons increases by 2 yr when 3 men of 32 yr, 34 yr and 33 yr are replaced by 3 women. What will be the average age of those 3 women? [Bank Clerk 2009]
(a) 40 yr (b) $41\frac{1}{3}$ yr
(c) 41 yr (d) $40\frac{1}{3}$ yr
(e) None of these

51. There are 50 students in a class. One of them weighing 50 kg goes away and a new student joins. By this, the average weight of the class increases by $\frac{1}{2}$ kg. The weight of the new student is [SSC CGL 2016]
(a) 70 kg (b) 72 kg
(c) 75 kg (d) 76 kg

52. The average age of 30 women decreases by 3 months, if a new person Priyanka is included in place of a 25 yr old woman. Calculate the age of Priyanka.
(a) 17.5 yr (b) 20 yr
(c) 30 yr (d) 22 yr
(e) None of these

53. The average marks of 120 students are 35. If the average of passed students was 39 and failed students was 15, then find the number of students who have passed. [SSC CGL 2012]
(a) 80 (b) 100 (c) 120 (d) 140

54. The average marks obtained by 120 students was 30. If the average of the passed candidates was 40 and that of the failed candidates was 10, then what is 25% of the candidates who passed the examination?
(a) 30 (b) 20 (c) 80 (d) 60

55. The average salary per head of all workers of an institution is ₹ 60. The average salary per head of 12 officers is ₹ 400. The average salary per head of the rest is ₹ 56. Then, the total number of workers in the institution is
(a) 1030 (b) 1032
(c) 1062 (d) 1060

56. A car travels a certain distance from town A to town B at the speed of 42 km/h and from town B to town A at a speed of 48 km/h. What is the average speed of the car? [Bank Clerk 2008]
(a) 45 km/h (b) 46 km/h
(c) 44 km/h (d) 48 km/h
(e) None of these

57. A motorist travels to a place 150 km away at an average speed of 50 km/h and returns at 30 km/h. What is the average speed for the whole journey? [CDS 2016 (II)]
(a) 35 km/h (b) 37 km/h
(c) 37.5 km/h (d) 40 km/h

58. In the afternoon, a student read 100 pages at the rate of 60 pages/h. In the evening, when she was tired, she read 100 more pages at the rate of 40 pages/h. What was her average rate of reading, in pages per hour?
(a) 48 (b) 50 (c) 60 (d) 70

Average / 163

Exercise 2 — Higher Skill Level Questions

1. If the algebraic sum of deviations of 20 observations measured from 23 is 70, mean of these observations would be [SNAP 2012]
   (a) 21.5   (b) 22   (c) 24.5   (d) 26.5

2. The average of 39 numbers is 17. Each number is multiplied by 12 and then 16 is added to each number. At last, each number thus formed is divided by 11. What is the average of new set of numbers thus formed? [CMAT 2015]
   (a) 19   (b) 39
   (c) 20   (d) 17

3. The average age of two boys and their father is greater than the average age of those two boys and their mother by 3 yr. The average age of the four is 19 yr. If the average age of the two boys is $5\frac{1}{2}$ yr, then find the age of the father and mother. [SSC Multitasking 2014]
   (a) 37 yr and 28 yr   (b) 47 yr and 38 yr
   (c) 50 yr and 41 yr   (d) 35 yr and 32 yr

4. Let a, b, c, d, e, f, g be consecutive even numbers and j, k, l, m, n be consecutive odd numbers. What is the average of all the numbers? [CDS 2017 (I)]
   (a) $\dfrac{3(a+n)}{2}$
   (b) $\dfrac{(5l+7d)}{4}$
   (c) $\dfrac{(a+b+m+n)}{4}$
   (d) None of the above

5. The average runs scored by a cricketer in 42 innings, is 30. The difference between his maximum and minimum scores in an inning is 100. If these two innings are not taken into consideration, then the average score of remaining 40 innings is 28. Calculate the maximum runs scored by him in an inning? [SSC SAS 2010]
   (a) 125   (b) 120   (c) 110   (d) 100

6. The average age of 5 sisters is 20 yr. If the age of the youngest sister be 4 yr, then what was the average age of the group of sisters at the birth of the youngest sister?
   (a) 25 yr   (b) 15 yr   (c) 18 yr   (d) 20 yr
   (e) None of these

7. The average of m numbers is n^4 and the average of n numbers is m^4. The average of $(m+n)$ numbers is [CDS 2015 (II)]
   (a) mn
   (b) $m^2 + n^2$
   (c) $mn(m^2 + n^2)$
   (d) $mn(m^2 + n^2 - mn)$

8. The average age of 4 members of a family is 20 yr. If youngest member is 4 yr old, then what was the average age of family at the time of the birth of youngest member?
   (a) 27 yr   (b) $23\frac{1}{3}$ yr
   (c) $21\frac{1}{3}$ yr   (d) 22 yr
   (e) None of these

9. If the average ages of Rakesh and Mohan is 15, average ages of Mohan and Ramesh is 12 and the average ages of Rakesh and Ramesh is 13, then the age of Mohan is [CGPSC 2013]
   (a) 16 yr   (b) 13 yr   (c) 14 yr   (d) 12 yr
   (e) None of these

10. Nine friends have a dinner in a hotel. Eight of them spent ₹ 12 each on their meals and the ninth spent ₹ 16 more than the average expenditure of all the nine. Find out the total money spent by them. [Hotel Mgmt. 2010]
    (a) ₹ 126   (b) ₹ 135
    (c) ₹ 111   (d) ₹ 141

11. A person bought some oranges worth ₹ 36 from each of the five markets at ₹ 1, ₹ 1.50, ₹ 1.80, ₹ 2 and ₹ 2.25 per orange, respectively. What is the average price of an orange? [SSC FCI]
    (a) ₹ 1.91   (b) ₹ 2.00
    (c) ₹ 1.58   (d) ₹ 1.80

12. The average weight of 3 men A, B and C is 84 kg. Another man D joins the group and the average now becomes 80 kg. If another man E whose weight is 3 kg more than that of D, replaces A, then the average weight of B, C, D and E becomes 79 kg. What is the weight of A? [SSC CGL (Mains) 2016]
    (a) 70 kg   (b) 72 kg
    (c) 75 kg   (d) 80 kg

13. The arithmetic mean of the scores of a group of students in a test was 52. The brightest 20% of them secured a mean score of 80 and the dullest 25%, a mean score of 31. The mean score of remaining 55% is
    (a) 54.6%   (b) 45%
    (c) 50%   (d) 51.4%

14. The average of a, b and c is 11; average of c, d and e is 17; average of e and f is 22 and average of e and c is 17. Find out the average of a, b, c, d, e and f. [CBI Clerk 2009]
    (a) $15\frac{2}{3}$ (b) $18\frac{1}{2}$
    (c) $21\frac{1}{3}$ (d) $16\frac{1}{2}$
    (e) None of these

15. There are five boxes in a cargo hold. The weight of the first box is 200 kg and the weight of the second box is 20% higher than the weight of the third box, whose weight is 25% higher than the first box's weight. The fourth box at 350 kg is 30% lighter than the fifth box. Find the difference in the average weight of the four heaviest boxes and the four lightest boxes. [MAT 2015]
    (a) 51.5 kg   (b) 75 kg
    (c) 37.5 kg   (d) 112.5 kg

16. There are six consecutive odd numbers. The difference between the square of the average of last three numbers and the first three numbers is 288. What is the last odd number? [SBI PO (Pre) 2016]
    (a) 31   (b) 27   (c) 29   (d) 25
    (e) 33

17. The average weight of first 11 persons among 12 persons is 95 kg. The weight of 12th person is 33 kg more than the average weight of all the 12 persons. The weight of the 12th person is [SSC CGL (Pre) 2015]
    (a) 128 kg   (b) 97.45 kg
    (c) 128.75 kg (d) 131 kg

18. There are 3 consecutive odd numbers and 3 consecutive even numbers. The smallest even number is 9 more than largest odd number. If the square of average of all the 3 given odd number is 507 less than the square of the average of all the 3 given even number, what is the smallest odd number?
    [RRB PO (Pre) 2017]
    (a) 11   (b) 13   (c) 17   (d) 19
    (e) 9

19. Given the set of n numbers, n > 1, of which one is 1 − (1/n) and all the other are 1. The arithmetic mean of the n numbers is
    (a) 1
    (b) $n - \frac{1}{n}$
    (c) $n - \frac{1}{n^2}$
    (d) $1 - \frac{1}{n^2}$

20. There are three positive numbers, one-third of average of all the three numbers is 8 less than the value of the highest number. The average of the lowest and the second lowest numbers is 8. Which is the highest number? [LIC ADO 2015]
    (a) 11   (b) 14
    (c) 10   (d) 13
    (e) None of these

21. The average weight of 5 men is decreased by 3 kg when one of them weighing 150 kg, is replaced by another person. This new person is again replaced by another person whose weight is 30 kg lower than the person he replaced. What is overall change in the average due to this dual change? [MAT 2015]
    (a) 6 kg   (b) 9 kg
    (c) 12 kg   (d) 15 kg

22. Three Science classes A, B and C take a Life Science test. The average score of class A is 83. The average score of class B is 76. The average score of class C is 85. The average score of classes A and B is 79 and average score of classes B and C is 81. Then, the average score of classes A, B and C is [NICL AO (Pre) 2017]
    (a) 80   (b) 80.5
    (c) 81   (d) 81.5
    (e) 82

23. The average weight of the students in four sections A, B, C and D is 60 kg. The average weight of the students of A, B, C and D individually are 45 kg, 50 kg, 72 kg and 80 kg, respectively. If the average weight of the students of section A and B together is 48 kg and that of B and C together is 60 kg, then what is the ratio of the number of students in sections A and D? [SSC CPO 2013]
    (a) 12 : 7   (b) 4 : 3
    (c) 3 : 2   (d) 8 : 5

Answer with Solutions

Exercise 1 Base Level Questions

1. (e) Required average = $\dfrac{\text{Sum of all scores}}{\text{Number of scores}}$

$= \dfrac{\begin{bmatrix} 253 + 124 + 255 + 534 + 836 \\ + 375 + 101 + 443 + 760 \end{bmatrix}}{9}$

$= \dfrac{3681}{9} = 409$

2. (d) Temperature of (Mon + Tue + Wed)
$= 37 \times 3 = 111°C$
Temperature of (Tue + Wed + Thu)
$= 34 \times 3 = 102°C$
Temperature of (Mon − Thu)
$= 111°C − 102°C = 9°C$
Temperature of $\left[\text{Mon} - \dfrac{4}{5}(\text{Mon})\right] = 9$
Temperature of Monday
$= 9 \times 5 = 45°C$
∴ Temperature of Thursday
$= 45 \times \dfrac{4}{5} = 36°\,C$

3. (b) Average of all prime numbers between 60 and 90
$= \dfrac{61 + 67 + 71 + 73 + 79 + 83 + 89}{7}$
$= \dfrac{523}{7} = 74.7$

4. (d) Given, $47(a + b) = 5452$
$\Rightarrow \quad a + b = \dfrac{5452}{47} = 116$
∴ Average value $= \dfrac{a+b}{2} = \dfrac{116}{2} = 58$

5. (d) According to the question,
$70m + 91n = 80(m + n)$
$\Rightarrow \quad 70m + 91n = 80m + 80n$
$\Rightarrow \quad 10m = 11n$
$\Rightarrow \quad \dfrac{n}{m} = \dfrac{10}{11}$

6. (d) Here, initial average = 7
As we know that, if all the numbers are multiplied by a certain number, then their average must also be multiplied by that number.
∴ New average $= 7 \times 12 = 84$

7. (a) Sum of marks of 7 students
= Mean × Number of students
$= 226 \times 7 = 1582$

Now, total marks of 6 students
$= (340 + 180 + 260 + 56 + 275 + 307)$
$= 1418$
∴ Marks obtained by 7th student
= Total marks of 7th students
− Total marks of 6 students
$= 1582 − 1418 = 164$

8. (e) Let one integer be x.
∴ Another integer $= x + 15$
According to the question,
$29 + x + x + 15 + 108 = 4 \times 73.5$
$\Rightarrow \quad 2x + 152 = 294$
$\Rightarrow \quad 2x = 294 − 152 = 142$
$\Rightarrow \quad x = \dfrac{142}{2} = 71$

9. (d) According to the fundamental formula,
$A = \dfrac{S}{N}$
[here, A = average, S = sum, N = number]
From the question, $60 = \dfrac{S}{13}$
∴ $S = 60 \times 13 = 780$
Sum of first seven results $= 59 \times 7 = 413$
Sum of last seven results $= 61 \times 7 = 427$
∴ 7th result = Sum of first seven results
+ Sum of last seven results
− Sum of all the results
$= (413 + 427 − 780)$
$= (840 − 780) = 60$
∵ Sum of first seven results when added to the sum of last seven results, then repetition of 7th result takes place.

10. (b) Required number
$= 50 \times 9 − (54 \times 5 + 3 \times 52)$
$= 450 − (270 + 156)$
$= 450 − 426 = 24$

11. (b) Total marks of all students of the class
$= 30 \times 45 = 1350$
∴ New average after adjustment
$= \dfrac{1350 − 45 + 15}{30} = \dfrac{1320}{30} = 44$

12. (d) Total marks obtained by John in English, Maths, Hindi and Drawing
$= 50 \times 4 = 200$
Total marks obtained by John in Maths, Science, Social Studies and Craft $= 70 \times 4$
$= 280$

Total marks obtained by John in all seven subjects = $58 \times 7 = 406$
∴ John's marks in Maths = $200 + 280 - 406$
$= 480 - 406 = 74$

13. (c) According to the question,
$\left(\dfrac{P+Q}{2}\right) - 5 = 15 \Rightarrow \dfrac{P+Q}{2} = 20$
$\Rightarrow \qquad P + Q = 40 \qquad ...(i)$
and $\dfrac{P+Q+R}{3} = 20$
$\Rightarrow \qquad P + Q + R = 60 \qquad ...(ii)$
On solving Eqs. (i) and (ii), we get $R = 20$ yr
∴ Age of R after 10 yr from now
$= 20 + 10 = 30$ yr

14. (b) Let the average of 5 innings be x.
Score in sixth inning = 80
∴ Total of 5 innings = $5x$
According to the question,
$\dfrac{5x + 80}{6} = x + 5$
$\Rightarrow \qquad 5x + 80 = 6x + 30$
$\Rightarrow \qquad x = 80 - 30 = 50$
∴ His average after six innings
$= 50 + 5 = 55$

15. (c) ∵ Sum of 12 observations
$= 12 \times 8 = 96$
∴ Correct sum of 12 observations
$= 96 - 13 + 10 = 93$
Hence, correct average of 12 observations
$= \dfrac{93}{12} = 7.75$

16. (b) Let the average age of 13 students be x.
Then, sum of ages of 13 students = $13x$
As per question,
$13x + 35 = 14(x + 2)$
$\Rightarrow \qquad 13x + 35 = 14x + 28$
$\Rightarrow \qquad x = 7$
Hence, actual average age of the class is 7 yr.

17. (d) Let the monthly salary of a man be ₹ x.
Then, annual salary = ₹ $12x$
According to the question,
Annual spending of man
$= 7 \times 1694.70 + 5 \times 1810.50 = ₹\ 20915.40$
∴ His monthly spending
$= \dfrac{20915.40}{12} = ₹\ 1742.95$
and monthly saving = $\dfrac{3084.60}{12} = ₹\ 257.05$
∴ His monthly salary = $1742.95 + 257.05$
$= ₹\ 2000$

18. (c) Let average runs for 12 innings = x
Then, total runs for 12 innings = $12x$
In the 13th inning he scored = 96
Now, according to the question,
$x + 5 = \dfrac{12x + 96}{13}$
$\Rightarrow \qquad 13x + 65 = 12x + 96$
$\Rightarrow \qquad 13x - 12x = 96 - 65$
$\Rightarrow \qquad x = 31$
∴ Average runs after 13th inning
$= 31 + 5 = 36$

19. (a) Total runs in first seven matches
$= 7 \times 57 = 399$
Total runs in last five matches
$= 5 \times 44 = 220$
Total runs in 8th match = 15
Now, according to the question,
$\dfrac{1}{14}$ [Runs in {first seven matches
$+$ 8th match $+$ 9th match
$+$ last five matches}] = 47
$\Rightarrow \dfrac{1}{14}[399 + 15 + 9\text{th match} + 220] = 47$
$\Rightarrow$ 9th match $+ 634 = 47 \times 14 = 658$
$\Rightarrow$ 9th match $= 658 - 634$
$\Rightarrow$ 9th match $= 24$
Hence, the cricketer made 24 runs in his 9th match.

20. (c) Let the average run of cricketer in 10 innings be x.
∴ Total runs in 10 innings = $10x$
Total runs in 11 innings = $10x + 108$
According to the question,
$\dfrac{10x + 108}{11} = x + 6$
$\Rightarrow \qquad 10x + 108 = 11x + 66$
∴ $\qquad x = 42$
Hence, new average $= 42 + 6 = 48$

21. (a) Let average after 10 innings = x
According to the question,
$10x + 216 = 11(x + 12)$
$\Rightarrow \qquad x = 216 - 132 = 84$
∴ New average $= 84 + 12 = 96$

22. (c) Let the number of boys be x and the number of girls be y.
Then, according to the question,
$16.4 \times x + 15.4 \times y = 15.8(x + y)$
$\Rightarrow \qquad (16.4 - 15.8)x = (15.8 - 15.4)y$
$\Rightarrow \qquad 0.6x = 0.4y \Rightarrow \dfrac{x}{y} = \dfrac{2}{3}$

23. (a) Let the number of papers be x.
Then, according to the question,
$64x + 18 + 4 = 66x$
$\Rightarrow \qquad 2x = 22 \Rightarrow x = \dfrac{22}{2} = 11$

Average / 167

24. (c) As per the formula,
Average of the cubes of first n natural numbers $= \dfrac{n(n+1)^2}{4}$

Here, $n = 5$

∴ Required average $= \dfrac{5(5+1)^2}{4}$

$= \dfrac{5 \times 36}{4} = 5 \times 9 = 45$

25. (a) Let the numbers A, B, C and D be $x, (x+1), (x+2)$ and $(x+3)$, respectively.

Average of consecutive numbers
$= \dfrac{\text{First number } + \text{ Last number}}{2}$

$\Rightarrow 49.5 = \dfrac{x + (x+3)}{2}$

$\Rightarrow 99 = 2x + 3$

$\Rightarrow 2x = 96 \Rightarrow x = 48$

∴ Required product $= B \times D$
$= (x+1)(x+3)$
$= (48+1)(48+3)$
$= 49 \times 51 = 2499$

26. (d) Let four consecutive even numbers be $x, (x+2), (x+4)$ and $(x+6)$, respectively.

Clearly, this is a series of consecutive even numbers.

According to the formula,
Average of consecutive numbers
$= \dfrac{\text{First number } + \text{ Last number}}{2}$

$\Rightarrow 27 = \dfrac{x + (x+6)}{2}$

$\Rightarrow x + 3 = 27$

∴ $x = 24$

∴ Greatest number = 4th consecutive even number
$= (x+6) = 24 + 6 = 30$

27. (a) Let the 5 consecutive odd positive integers be $x+1, x+3, x+5, x+7, x+9$.

∵ Average of consecutive numbers
$= \dfrac{\text{First number } + \text{ Last number}}{2}$

$\Rightarrow 9 = \dfrac{x+1+x+9}{2}$

$\Rightarrow 9 = \dfrac{2x+10}{2} \Rightarrow 18 = 2x + 10$

$\Rightarrow 8 = 2x$

$\Rightarrow x = 4$

Now, least one $= x + 1 = 4 + 1 = 5$

28. (b) Let eight consecutive numbers be $x, x+1, x+2, x+3, x+4, x+5, x+6$ and $x+7$, respectively.

According to the question,
$\dfrac{x+3+x+4}{2} = 6$

$\Rightarrow \dfrac{2x+7}{2} = 6 \Rightarrow 2x + 7 = 12$

$\Rightarrow 2x = 5 \Rightarrow x = \dfrac{5}{2}$

∴ Sum of all eight numbers
$= x + x + 1 + x + 2 + x + 3 + x + 4$
$+ x + 5 + x + 6 + x + 7 = 8x + 28$

$= 8 \times \dfrac{5}{2} + 28 = 20 + 28 = 48$

29. (a) Average cost of all the machines
$= \dfrac{₹(3 \times 100000 + 2 \times 80000 + 3 \times 45000)}{8}$

$= ₹\dfrac{595000}{8} = ₹74375$

30. (b) Here, $n_1 = 20$, $n_2 = 30$, $a_1 = 80\%$
and $a_2 = 70\%$

∴ Total passing average $= \dfrac{n_1 a_1 + n_2 a_2}{n_1 + n_2}$

[by Technique 1]

$= \dfrac{20 \times 80 + 30 \times 70}{20 + 30}$

$= \dfrac{1600 + 2100}{50} = 74\%$

31. (c) Here, number of boys $= x$
and number of girls $= y$
Age of a boy $= a$ yr
and age of a girl $= b$ yr

∴ Average age of all boys and girls
$= \dfrac{x \times a + y \times b}{x + y}$ [by Technique 1]

$= \dfrac{ax + by}{x + y}$

32. (d) Here, $n_1 = 19$, $a_1 = 74$, $n_2 = 38$, $a_2 = 63$

∴ Total average weight $= \dfrac{n_1 a_1 + n_2 a_2}{n_1 + n_2}$

[by Technique 1]

$= \dfrac{19 \times 74 + 38 \times 63}{19 + 38} = \dfrac{3800}{57} \approx 67$ kg

33. (a) Here, $n_1 = 5$, $n_2 = 10$, $n_3 = 15$, $a_1 = 25$, $a_2 = 50$ and $a_3 = 35$

According to the formula,

Average price $= \dfrac{n_1 a_1 + n_2 a_2 + n_3 a_3}{n_1 + n_2 + n_3}$

[by Technique 1]

$= \dfrac{5 \times 25 + 10 \times 50 + 15 \times 35}{5 + 10 + 15}$

$= \dfrac{25 + 100 + 105}{1 + 2 + 3} = \dfrac{230}{6} = ₹38.33$

34. (a) Here, $n_1 = 20, a_1 = 5$ ft 11 inch
$= 5 \times 12 + 11 = 71$ inch
and $n_2 = 18, a_2 = 6$ ft 2 inch
$= 6 \times 12 + 2 = 74$ inch

∴ Overall average height $= \dfrac{n_1 a_1 + n_2 a_2}{n_1 + n_2}$

[by Technique 1]

$= \dfrac{(20)(71) + (18)(74)}{20 + 18}$

$= \dfrac{1420 + 1332}{38} = 72.42$ inch

35. (b) Let the number of students in the class be x.
According to the question,
$\dfrac{35x + 6 \times 33}{x + 6} = 35 - \dfrac{1}{2}$

⇒ $35x + 198 = 34.5(x + 6)$
⇒ $35x + 198 = 34.5x + 207$
⇒ $0.5x = 207 - 198$
∴ $x = \dfrac{9}{0.5} = 18$

Fast Track Method
Here, $a_1 = 35, a_2 = 33, n_2 = 6, n_1 = ?$

∴ Average $= \dfrac{n_1 a_1 + n_2 a_2}{n_1 + n_2}$ [by Technique 1]

⇒ $34.5 = \dfrac{n_1 \times 35 + 6 \times 33}{n_1 + 6}$

⇒ $(34.5)(n_1 + 6) = 35 n_1 + 198$
⇒ $34.5 n_1 + 207 = 35 n_1 + 198$
⇒ $0.5 n_1 = 9 \Rightarrow n_1 = 18$

36. (d) Let there be x workers present initially.
∵ Average age of x workers $= 42$ yr [given]
∴ Total age of x workers $= 42x$ yr
Average age of 8 new workers $= 36$ yr
Total age of 8 new workers $= 36 \times 8$
$= 288$ yr
Now, according to the question,
$\dfrac{42x + 288}{x + 8} = 40$

⇒ $42x + 288 = 40(x + 8)$
⇒ $42x - 40x = 40 \times 8 - 288$
⇒ $2x = 32$
∴ $x = \dfrac{32}{2} = 16$

Hence, initially there were 16 workers.

Fast Track Method
Here, $a_1 = 42, a_2 = 36, n_2 = 8, n_1 = ?$
∴ Average $= \dfrac{n_1 a_1 + n_2 a_2}{n_1 + n_2}$ [by Technique 1]

⇒ $40 = \dfrac{42 n_1 + 8 \times 36}{n_1 + 8}$

⇒ $40 n_1 + 320 = 42 n_1 + 288$

⇒ $2 n_1 = 32$
⇒ $n_1 = 16$
Hence, initially there were 16 workers.

37. (a) Let n_1 and n_2 be the number of boys and girls, respectively.
Given that, total number of students
$= 150 = (n_1 + n_2)$ and mean weight of 150 students $= 60$ kg.
Also, the mean weight of boys, $a_1 = 70$ kg and the mean weight of girls, $a_2 = 55$ kg.
Now, $n_1 + n_2 = 150$ ...(i)
and $11 n_1 + 11 n_2 = 1650$ ...(ii)

∴ Average weight $= \dfrac{n_1 a_1 + n_2 a_2}{n_1 + n_2}$

[by Technique 1]

⇒ $60 = \dfrac{70 n_1 + 55 n_2}{150}$

⇒ $70 n_1 + 55 n_2 = 9000$
⇒ $14 n_1 + 11 n_2 = 1800$ ...(iii)
On subtracting Eq. (ii) from Eq. (iii), we get
$3 n_1 = 150 \Rightarrow n_1 = 50$
Hence, the required number of boys is 50.

38. (b) Total age of 30 girls $= 30 \times 13 = 390$ yr
Total age of 18 girls $= 18 \times 15 = 270$ yr
∴ Age of remaining 12 girls
$= 390 - 270 = 120$ yr

∴ Required average $= \dfrac{120}{12} = 10$ yr

Fast Track Method
Here, $m = 30, n = 18, a = 13, b = 15$

∴ Average of remaining 12 girls $= \dfrac{ma - nb}{m - n}$

[by Technique 2]

$= \dfrac{30 \times 13 - 18 \times 15}{30 - 18}$

$= \dfrac{390 - 270}{12} = \dfrac{120}{12} = 10$ yr

39. (c) Here, $m = 15 + 10 = 25$
$a = 38.4, n = 15, b = 40$

∴ Average weight of girls $= \dfrac{ma - nb}{m - n}$

[by Technique 2]

$= \dfrac{25 \times 38.4 - 15 \times 40}{25 - 15}$

$= \dfrac{960 - 600}{10} = \dfrac{360}{10} = 36$ kg

40. (c) Total number of students in the class $= 39$
Total number of girls in the class $= 26$
∴ Total number of boys in the class
$= 39 - 26 = 13$
Average weight of girls $= 42$ kg
∴ Total weight of girls $= 26 \times 42$ kg

Average weight of full class = 48 kg
∴ Total weight of full class = 39 × 48 kg
Now, average weight of boys
$$= \frac{39 \times 48 - 26 \times 42}{13}$$
$$= \frac{13(3 \times 48 - 2 \times 42)}{13}$$
$$= 3 \times 48 - 2 \times 42 = 144 - 84 = 60 \text{ kg}$$

Fast Track Method
Here, $m = 39, n = 26, a = 48, b = 42$
∴ Average weight of boys $= \frac{ma - nb}{m - n}$
[by Technique 2]
$$= \frac{39 \times 48 - 26 \times 42}{39 - 26}$$
$$= \frac{1872 - 1092}{13} = \frac{780}{13} = 60 \text{ kg}$$

41. (a) Average age of 5 boys = 16 yr
Sum of age of 5 boys = 16 × 5 = 80 yr
∴ Average age of 4 boys = 16 yr 3 months
$$= 16 \text{ yr} + \frac{3}{12} \text{ yr} = 16 \text{ yr} + \frac{1}{4} \text{ yr}$$
Sum of ages of 4 boys $= 4 \times \left(16 + \frac{1}{4}\right)$ yr
$$= 64 + 1 = 65 \text{ yr}$$
∴ Age of 5th boy = (80 − 65) yr = 15 yr

Fast Track Method
Here, $m = 5, a = 16, n = 4, b = 16\frac{3}{12} = 16\frac{1}{4}$
∴ Age of 5th boy $= \frac{ma - nb}{m - n}$
[by Technique 2]
$$= \frac{5 \times 16 - 4 \times 16\frac{1}{4}}{1} = 80 - 65 = 15 \text{ yr}$$

42. (c) Total cost of 30 TV sets = ₹ 300000
Total cost of 15 TV sets = 9000 × 15
$$= 135000$$
Total cost of remaining 15 TV sets
$$= 300000 - 135000 = 165000$$
∴ Average cost of remaining TV sets
$$= \frac{165000}{15} = ₹ 11000$$

Fast Track Method
Here, $m = 30, n = 15, a = 10000, b = 9000$
∴ Average of remaining TV sets $= \frac{ma - nb}{m - n}$
[by Technique 2]
Average cost of remaining (30 − 15) TV sets
$$= \frac{30 \times 10000 - 15 \times 9000}{30 - 15}$$
$$= \frac{300000 - 135000}{15} = \frac{165000}{15}$$
$$= ₹ 11000$$

43. (b) We have, 100 students
Number of boys = 70
and number of girls = 30
Average marks of boys = 75
∴ Total marks of boys = 75 × 70 = 5250
Average marks of class = 72
Total marks of students = 72 × 100 = 7200
Let the average marks of girls be x.
Total marks of girls = 30x
∴ Total marks of students in class
= Total marks of boys + Total marks of girls
⇒ 7200 = 5250 + 30x
⇒ 30x = 7200 − 5250
∴ $x = \frac{1950}{30} = 65$

Fast Track Method
Here, $m = 100, n = 70, a = 72$ and $b = 75$
∴ Average of girls $= \frac{ma - nb}{m - n}$
[by Technique 2]
$$= \frac{100 \times 72 - 70 \times 75}{30}$$
$$= \frac{7200 - 5250}{30} = 65$$

44. (b) Let total number of workers in workshop be x.
Then, total salary of all workers = 8000x
According to the question,
$$8000x = 7 \times 12000 + (x - 7) \times 6000$$
⇒ $8000x = 84000 + 6000x - 42000$
⇒ $8000x - 6000x = 84000 - 42000$
⇒ $2000x = 42000$
∴ $x = 21$

Fast Track Method
Here, $a = 8000, n = 7, b = 12000$ and $m = ?$
∴ Average salary of remaining workers
$$= \frac{ma - nb}{m - n}$$ [by Technique 2]
⇒ $6000 = \frac{m \times 8000 - 7 \times 12000}{m - 7}$
⇒ $6000m - 42000 = 8000m - 84000$
⇒ $2000m = 42000 \Rightarrow m = 21$

45. (c) Monthly income of 4 persons
$$= 7350 \times 4 = ₹ 29400$$
Monthly income of 3 persons (excluding the dead person) = 6500 × 3 = ₹ 19500
∴ Monthly income of dead person
$$= 29400 - 19500 = ₹ 9900$$

Fast Track Method
Here, $n = 4$
Initial average, $a = ₹ 7350$
Last average, $b = ₹ 6500$
∴ Required income $= n(a - b) + b$
[by Technique 3(i)]

Monthly income of the person, who is no more = 4(7350 − 6500) + 6500
= 4 × 850 + 6500
= 3400 + 6500 = ₹ 9900

46. (a) Sum of 14 girl's and teacher's age
= 15 × 15 = 225
Excluding teacher's age, sum of 14 girls age
= 14 × 14 = 196
∴ Teacher's age = 225 − 196 = 29 yr
Fast Track Method
Here, $n = 15, a = 15, b = 14$
∴ Teacher's age = $n(a − b) + b$
[by Technique 3 (i)]
= 15 (15 − 14) + 14 = 15 + 14 = 29 yr

47. (a) Here, $n = 21, a = 64$ and $b = 65$
Weight of teacher = $n(b − a) + b$
[by Technique 3(ii)]
= 21(65 − 64) + 65 = 21 + 65 = 86 kg

48. (b) Here, $n = 34, a = 42$ and $b = 42.4$
Weight of the teacher
= $n(b − a) + b$ = 34(42.4 − 42) + 42.4
[by Technique 3 (ii)]
= 13.6 + 42.4 = 56.0 kg

49. (b) Total age of 4 members = 25 × 4 = 100 yr
Total age of 5 members including the head
= $5 \times \left(25 + 25 \times \dfrac{20}{100}\right) = 5 \times 30 = 150$ yr
∴ Age of the head = 150 − 100 = 50 yr
Fast Track Method
Here, $n = 4$, initial average, $a = 25$
and last average, $b = 25 + 20\% \times 25$
$= 25 + \dfrac{20 \times 25}{100} = 30$
∴ Age of the head = $n(b − a) + b$
[by Technique 3 (ii)]
= 4(30 − 25) + 30
= 4 × 5 + 30 = 50 yr

50. (d) Here, $n = 11, a = 32 + 33 + 34 = 99$ yr, $b = 2$ yr
∴ Age of 3 women = $a + nb$
[by Technique 4]
= 99 + 11 × 2 = 99 + 22 = 121 yr
∴ Average age of 3 women = $\dfrac{121}{3} = 40\dfrac{1}{3}$ yr
[here, '+' sign $(a + nb)$ has been taken as average increases in this case]

51. (c) Total weight increased = $\dfrac{1}{2} \times 50 = 25$ kg
∴ Weight of the new man
= 50 + 25 = 75 kg
Fast Track Method
Here, $n = 50, a = 50$ and $b = \dfrac{1}{2}$

∴ Weight of new man = $a + nb$
[by Technique 4]
= $50 + 50 \times \dfrac{1}{2} = 75$ kg

52. (a) Here, $a = 25, n = 30, b = \dfrac{3}{12} = \dfrac{1}{4}$
∴ Age of Priyanka = $a − nb$
[by Technique 4]
= $25 − 30 \times \dfrac{1}{4}$
= 25 − 7.5
= 17.5 yr

53. (b) Here, $n = 120, a = 35, x = 39$ and $y = 15$
∴ Number of students passed = $\dfrac{n(a − y)}{(x − y)}$
[by Technique 5]
= $\dfrac{120(35 − 15)}{(39 − 15)} = \dfrac{120 \times 20}{24} = 100$

54. (b) Let the number of passed students be x.
∴ Total marks = 120 × 30
According to the question,
$40x + (120 − x) \times 10 = 120 \times 30$
⇒ $3600 = 40x + 1200 − 10x$
⇒ $30x = 2400$
∴ $x = \dfrac{2400}{30} = 80$
∴ 25% of 80 = $\dfrac{80}{4} = 20$ students
Fast Track Method
Here, $n = 120, a = 30, x = 40$ and $y = 10$
∴ Number of passed students
= $\dfrac{n(a − y)}{x − y}$ [by Technique 5]
= $\dfrac{120(30 − 10)}{40 − 10}$
= 4(30 − 10) = 80
∴ 25% of 80 = $\dfrac{80}{4} = 20$ students

55. (b) Let the total number of workers be x.
According to the question,
$60x = 12 \times 400 + 56(x − 12)$
⇒ $60x − 56x = 4800 − 672$
⇒ $4x = 4128 \Rightarrow x = 1032$
Fast Track Method
Here, $a = 60, n = 12, b = 400$ and $c = 56$
∴ Number of remaining workers
= $\dfrac{n(a − b)}{c − a}$ [by Technique 6]
= $\dfrac{12(60 − 400)}{56 − 60} = \dfrac{12 \times 340}{4} = 1020$
Hence, total number of workers
= 1020 + 12 = 1032

56. (e) If two equal distances are covered at different speeds at A km/h and B km/h respectively, then
Average speed during the whole journey
$$= \frac{2AB}{A+B} \quad \text{[by Technique 7]}$$
$$= \frac{2 \times 42 \times 48}{42 + 48} = \frac{2 \times 42 \times 48}{90}$$
$$= 44.8 \text{ km/h}$$

57. (c) Total distance travelled by motorist
$$= 150 + 150 = 300 \text{ km}$$
Total time taken = Time taken going to a place + Time taken on returning
$$= \frac{150}{50} + \frac{150}{30} = 8 \text{ h}$$
$$\therefore \text{Average speed} = \frac{\text{Total distance}}{\text{Total time taken}}$$
$$= \frac{300}{8} = 37.5 \text{ km/h}$$

Fast Track Method
Here, $A = 50$ km/h, $B = 30$ km/h
$$\therefore \text{Average speed} = \frac{2AB}{A+B} \quad \text{[by Technique 7]}$$
$$= \frac{2 \times 50 \times 30}{50 + 30}$$
$$= \frac{2 \times 50 \times 30}{80}$$
$$= 37.5 \text{ km/h}$$

58. (a) Required average rate of reading
$$= \frac{2AB}{A+B} \quad \text{[by Technique 7]}$$
$$= \frac{2 \times 60 \times 40}{60 + 40}$$
$$= \frac{2 \times 60 \times 40}{100}$$
$$= 48 \text{ pages/h}$$

Exercise 2 Higher Skill Level Questions

1. (d) Mean $= \dfrac{\text{Sum of observations}}{\text{Number of observations}}$
$\because$ Sum of observations = 70 [given]
and number of observations = 20
$\therefore \quad$ Mean $= \dfrac{70}{20} = 3.5$
But, here observation starts from 23.
$\therefore$ Mean $= 23 + 3.5 = 26.5$

2. (c) Given, average of 39 numbers = 17
$\therefore$ Sum of 39 numbers = $17 \times 39 = 663$
New sum when each number multiplied by $12 = 663 \times 12 = 7956$
Sum of all 39 numbers when 16 is added to each = $7956 + 39 \times 16 = 8580$
Now, sum of all numbers when each is divided by $11 = \dfrac{8580}{11} = 780$
$\therefore$ Average of new set of numbers
$$= \frac{\text{Sum of numbers}}{\text{Total number of numbers}} = \frac{780}{39} = 20$$

3. (a) Let the ages of mother, father and boys be M, F, B_1 and B_2, respectively.
Total age of four members = $19 \times 4 = 76$ yr
Given, $\dfrac{B_1 + B_2}{2} = \dfrac{11}{2}$
$\Rightarrow B_1 + B_2 = 11$ and $M + F + B_1 + B_2 = 76$
$\Rightarrow \quad M + F = 76 - 11$
$\Rightarrow \quad M + F = 65 \quad$ ...(i)
According to the question,
$$\frac{B_1 + B_2 + F}{3} = \frac{B_1 + B_2 + M}{3} + 3$$

$\Rightarrow B_1 + B_2 + F = B_1 + B_2 + M + 9$
$\Rightarrow F = M + 9 \Rightarrow F - M = 9 \quad$ ...(ii)
From Eqs. (i) and (ii), we get
$F = 37$ yr and $M = 28$ yr

4. (d) Let $a = x$ (even), then
$b = x + 2$
$c = x + 4$
$d = x + 6$
$e = x + 8$
$f = x + 10$
$g = x + 12$
$\therefore \quad a + b + c + d + e + f + g = 7x + 42$
$= 7(x + 6) = 7d$
Again, let $j = y$ (odd), then
$k = y + 2$
$l = y + 4$
$m = y + 6$
$n = y + 8$
$\therefore j + k + l + m + n = 5y + 20$
$= 5(y + 4) = 5l$
$\therefore$ Average of all the numbers $= \dfrac{7d + 5l}{12}$

5. (b) Let the minimum score = x
Then, maximum score = $x + 100$
$\therefore \quad x + (x + 100) = 30 \times 42 - 40 \times 28$
$\Rightarrow \quad 2x + 100 = 1260 - 1120 = 140$
$\Rightarrow \quad 2x = 140 - 100 = 40$
$\Rightarrow \quad x = 20$
Hence, maximum score $= x + 100$
$= 20 + 100 = 120$ runs

6. (d) Total age of 5 sisters = $20 \times 5 = 100$ yr
4 yr ago, total sum of ages = $100 - (5 \times 4)$
$\qquad = 100 - 20 = 80$ yr
But at that time (4 yr ago), there were 4 sisters in the group.
∴ Average age at that time (4 yr ago)
$\qquad = \dfrac{80}{4} = 20$ yr

7. (d) Let m numbers be $a_1, a_2, \ldots, a_m$ and n numbers be $b_1, b_2, \ldots, b_n$.
According to the question,
$$\dfrac{a_1 + a_2 + \ldots + a_m}{m} = n^4$$
and $\dfrac{b_1 + b_2 + \ldots + b_n}{n} = m^4$
Now, average of $(m + n)$ numbers
$= \dfrac{a_1 + a_2 + \ldots + a_m + b_1 + b_2 + \ldots + b_n}{m + n}$
$= \dfrac{mn^4 + nm^4}{m + n} = \dfrac{mn(m^3 + n^3)}{m + n}$
$= mn \cdot \dfrac{(m + n)(m^2 + n^2 - mn)}{m + n}$
$= mn(m^2 + n^2 - mn)$

Fast Track Method
Here, $n_1 = m, n_2 = n, a_1 = n^4$ and $a_2 = m^4$
∴ Required average $= \dfrac{n_1 a_1 + n_2 a_2}{n_1 + n_2}$
[by Technique 1]
$= \dfrac{mn^4 + nm^4}{m + n} = \dfrac{mn(m^3 + n^3)}{m + n}$
$= mn(m^2 + n^2 - mn)$

8. (c) Total age of all members of the family (at present) be 20×4, i.e. 80 yr.
4 yr ago, total age of all members
$= 80 - (4 \times 4) = 80 - 16 = 64$ yr
As 4 yr ago, there were only 3 members in the family.
∴ Average age of the family at the time of the birth of the youngest member
$= \dfrac{64}{3} = 21\dfrac{1}{3}$ yr

9. (c) Let ages of Rakesh, Mohan and Ramesh be R, M and r, respectively.
Then, $R + M = 15 \times 2 = 30$ ...(i)
$M + r = 12 \times 2 = 24$ ...(ii)
and $r + R = 13 \times 2 = 26$ ...(iii)
On adding Eqs. (i), (ii) and (iii), we get
$2(R + M + r) = 30 + 24 + 26 = 80$
$\Rightarrow R + M + r = 40$ ...(iv)

On subtracting Eq. (iii) from Eq. (iv), we get
$M = 40 - 26 = 14$
Hence, the age of Mohan is 14 yr.

10. (a) Let the average expenditure of 9 persons be x.
According to the question,
$12 \times 8 + (x + 16) = 9x \Rightarrow 8x = 112$
$\Rightarrow x = \dfrac{112}{8} = 14$
∴ Total money spent $= 9x = 9 \times 14 = ₹ 126$

11. (c) Number of oranges bought by the person
$= \dfrac{36}{1} + \dfrac{36}{1.50} + \dfrac{36}{1.80} + \dfrac{36}{2} + \dfrac{36}{2.25}$
$= 36 + 24 + 20 + 18 + 16 = 114$
Total expenditure $= 36 \times 5 = ₹ 180$
Thus, average price of each orange
$= ₹ \dfrac{180}{114} = ₹ 1.58$

12. (c) Here, $A + B + C = 84 \times 3 = 252$ ...(i)
Similarly,
$A + B + C + D = 80 \times 4 = 320$ ...(ii)
From Eqs. (i) and (ii), we get $D = 68$
$E = D + 3 = 68 + 3$
$\Rightarrow E = 71$
Now, $B + C + D + E = 79 \times 4 = 316$
$\Rightarrow B + C + 68 + 71 = 316$
$\Rightarrow B + C + 139 = 316$
$\Rightarrow B + C = 316 - 139$
$\Rightarrow B + C = 177$ ...(iii)
Now, from Eqs. (i) and (iii), we get
$A + B + C = 252$
$\Rightarrow A + 177 = 252$
$\Rightarrow A = 252 - 177 = 75$ kg
Hence, the weight of A is 75 kg.

13. (d) Let the total number of students be 100.
∵ Mean score = 52
Let the mean of remaining students be x.
According to the question,
$20 \times 80 + 25 \times 31 + (100 - 20 - 25) \times x$
$\qquad = 100 \times 52$
$\Rightarrow 1600 + 775 + 55 \times x = 5200$
$\Rightarrow 55x = 5200 - 1600 - 775$
∴ $x = \dfrac{2825}{55} = 51.36 \approx 51.4\%$
So, the mean score of remaining 55% is 51.4%.

14. (a) According to the question,
$a + b + c = 11 \times 3 = 33$ ...(i)
$c + d + e = 17 \times 3 = 51$ ...(ii)
$e + f = 22 \times 2 = 44$ ...(iii)
and $e + c = 17 \times 2 = 34$ ...(iv)
From Eqs. (ii) and (iv), we get
$34 + d = 51$ or $d = 17$ ...(v)

Now, adding Eqs. (i), (iii) and (v), we get
$a + b + c + d + e + f = 33 + 44 + 17 = 94$
$\therefore$ Average of a, b, c, d, e and f
$= \dfrac{94}{6} = \dfrac{47}{3} = 15\dfrac{2}{3}$

15. *(b)* Weight of 1st box = 200 kg
Weight of 2nd box $= \dfrac{250 \times 120}{100} = 300$ kg
Weight of 3rd box $= \dfrac{200 \times 125}{100} = 250$ kg
Weight of 4th box = 350 kg
Weight of 5th box $= \dfrac{350 \times 100}{70} = 500$ kg
$\therefore$ Difference between average weight of the four heaviest boxes and the four lightest boxes
$= \left(\dfrac{500 + 350 + 300 + 250}{4} - \dfrac{200 + 300 + 250 + 350}{4} \right)$
$= \dfrac{1}{4}(1400 - 1100) = \dfrac{1}{4} \times 300 = 75$ kg

16. *(c)* Let the six odd numbers be $x, (x + 2), (x + 4), (x + 6), (x + 8)$ and $(x + 10)$.
Average of first three numbers
$= \dfrac{x + x + 2 + x + 4}{3}$
$= \dfrac{3x + 6}{3} = (x + 2)$
Average of last three numbers
$= \dfrac{(x + 6) + (x + 8) + (x + 10)}{3}$
$= \dfrac{3x + 24}{3} = (x + 8)$
Now, according to the question,
$(x + 8)^2 - (x + 2)^2 = 288$
$\Rightarrow x^2 + 64 + 16x - (x^2 + 4 + 4x) = 288$
$\Rightarrow x^2 + 16x + 64 - x^2 - 4x - 4 = 288$
$\Rightarrow 12x + 60 = 288$
$\Rightarrow 12x = 288 - 60$
$\Rightarrow 12x = 228$
$\Rightarrow x = 19$
$\therefore$ Last odd number $= x + 10$
$= (19 + 10) = 29$

17. *(d)* $\because$ Sum of weight of 11 persons
$= 11 \times 95 = 1045$ kg
Now, let the average weight of 12 persons be x.
Then, weight of 12th person $= x + 33$

According to the question,
Average weight of 12 persons
$= \dfrac{\text{Sum of weight of 11 persons} + \text{Weight of 12th person}}{12}$
$\Rightarrow \dfrac{1045 + x + 33}{12} = x$
$\Rightarrow 1078 + x = 12x \Rightarrow 11x = 1078$
$\Rightarrow x = \dfrac{1078}{11} = 98$ kg
$\therefore$ Weight of 12th person $= 98 + 33$
$= 131$ kg

18. *(a)* Let the consecutive odd numbers be $x - 2, x$ and $x + 2$ and consecutive even numbers be $y - 2, y, y + 2$.
Then, $y - 2 = 9 + x + 2$
$\Rightarrow y - x = 13$ ...(i)
and $(x)^2 + 507 = (y)^2$
$y^2 - x^2 = 507$
$\Rightarrow (x + y)(y - x) = 507$
$\Rightarrow (x + y) = \dfrac{507}{13}$
$\Rightarrow x + y = 39$ ...(ii)
On solving Eqs. (i) and (ii), we get
$y = 26$ and $x = 13$.
So, smallest odd numbers $= x - 2$
$= 13 - 2 = 11$

19. *(d)* Sum of numbers
$= \left(1 - \dfrac{1}{n}\right) + 1 + 1 + 1 + (n - 1)$ times
$= 1 - \dfrac{1}{n} + (n - 1)$
$= n - \dfrac{1}{n}$
$\therefore$ Arithmetic mean of n numbers
$= \dfrac{n - \dfrac{1}{n}}{n} = \dfrac{n^2 - 1}{n^2} = 1 - \dfrac{1}{n^2}$

20. *(a)* Let the average of all the three numbers be x.
Then, highest number $= \left(\dfrac{x}{3} + 8\right)$
Total sum of 3 numbers $= 3x$
Sum of lowest and second lowest number $= 2 \times 8 = 16$
According to the question,
$3x = \left(\dfrac{x}{3} + 8\right) + 16$
$\Rightarrow 3x - \dfrac{x}{3} = 24 \Rightarrow \dfrac{8x}{3} = 24$
$\Rightarrow x = 9$
$\therefore$ Highest number $= \dfrac{9}{3} + 8 = 11$

21. *(b)* Let the weights of five persons be x_1, x_2, x_3, x_4, x_5 and average weight of five persons be x kg.

$$\therefore \quad \frac{x_1 + x_2 + x_3 + x_4 + x_5}{5} = x$$

$$\Rightarrow \quad x_1 + x_2 + x_3 + x_4 + x_5 = 5x$$

Now, one of them whose weight is 150 kg, is replaced by other person.

So, $x_1 + x_2 + x_3 + x_4 + 150 = 5x$

$\Rightarrow \quad x_1 + x_2 + x_3 + x_4 = 5x - 150$ ...(i)

Again, let 150 be replaced by person z, then average decrease by 3.

$$\Rightarrow \quad \frac{x_1 + x_2 + x_3 + x_4 + z}{5} = x - 3$$

Now, putting the value from Eq. (i), we get

$$\frac{5x - 150 + z}{5} = x - 3$$

$\Rightarrow \quad 5x - 150 + z = 5x - 15$

$\therefore \quad z = -15 + 150$

$= 135$ kg   [new person]

Again, replaced by new person whose weight is lower than 30 kg that the person he replaced.

So, next new person weight
$= 135 - 30 = 105$ kg

Final change in weight
$= 150 - 105 = 45$ kg

$\therefore$ Overall change in average due to this dual change $= \dfrac{45}{5} = 9$ kg

Fast Track Method
Here, $a = 150$, $n = 5$ and $b = 3$
$\therefore$ Weight of new person $= a - nb$
[by Technique 4]
$= 150 - 5 \times 3 = 150 - 15 = 135$

Again, 135 kg person is replaced by new person.

$\therefore$ Weight of next new person
$= 135 - 30 = 105$

$\therefore$ Overall change in average
$= \dfrac{150 - 105}{5} = 9$ kg

22. *(d)* We have,
Average score of class $A = 83$
Average score of class $B = 76$
Average score of class $C = 85$
Let n_1, n_2 and n_3 students are in class A, B and C, respectively.
$\therefore$ Total scores of class $A = 83 n_1$
   Total scores of class $B = 76 n_2$
and total scores of class $C = 85 n_3$
Average score of class A and B is

$$\frac{83n_1 + 76n_2}{n_1 + n_2} = 79$$

$\Rightarrow \quad 4n_1 = 3n_2$

$\Rightarrow \quad n_1 = \dfrac{3}{4} n_2$ ...(i)

Similarly, average score of class B and C is

$$\frac{76n_2 + 85n_3}{n_2 + n_3} = 81$$

$\Rightarrow \quad 5n_2 = 4n_3 \Rightarrow n_3 = \dfrac{5}{4} n_2$ ...(ii)

$\therefore$ Average scores of A, B and C

$$= \frac{83n_1 + 76n_2 + 85n_3}{n_1 + n_2 + n_3}$$

$$= \frac{83 \times \dfrac{3}{4} + 76 + 85 \times \dfrac{5}{4}}{\dfrac{3}{4} + 1 + \dfrac{5}{4}}$$

[from Eqs. (i) and (ii)]

$$= \frac{249 + 304 + 425}{12} = 81.5$$

23. *(b)* Let number of students in the sections A, B, C and D be a, b, c and d, respectively.
Then, total weight of students of section A
$= 45a$
Total weight of students of section $B = 50b$
Total weight of students of section $C = 72c$
Total weight of students of section $D = 80d$
According to the question,
Average weight of students of sections A and $B = 48$ kg

$$\therefore \quad \frac{45a + 50b}{a + b} = 48$$

$\Rightarrow \quad 45a + 50b = 48a + 48b$

$\Rightarrow \quad 3a = 2b$

$\Rightarrow \quad 15a = 10b$ ...(i)

and average weight of students of sections B and $C = 60$ kg

$\Rightarrow \quad 50b + 72c = 60(b + c)$

$\Rightarrow \quad 10b = 12c$ ...(ii)

Now, average weight of students of A, B, C and $D = 60$ kg

$\therefore \quad 45a + 50b + 72c + 80d$
$\quad\quad\quad = 60(a + b + c + d)$

$\Rightarrow \quad 15a + 10b - 12c - 20d = 0$

$\Rightarrow \quad 15a = 20d$ [using Eq. (ii)]

$\Rightarrow \quad a : d = 4 : 3$

Chapter 11

Percentage

The term per cent means *'for every hundred'*. It can be defined as follows

"A per cent is a fraction whose denominator is 100 and the numerator of the fraction is the **rate per cent**." Per cent is denoted by the sign '%'.

Formula to Calculate Per cent

If we have to find y% of x, then

$$y\% \text{ of } x = \frac{y}{100} \times x$$

Some Quick Results

5% of a number = $\frac{\text{Number}}{20}$,   10% of a number = $\frac{\text{Number}}{10}$

$12\frac{1}{2}$% of a number = $\frac{\text{Number}}{8}$,   20% of a number = $\frac{\text{Number}}{5}$

25% of a number = $\frac{\text{Number}}{4}$,   50% of a number = $\frac{\text{Number}}{2}$

Ex. 1 20% of 300 = ?

Sol. According to the formula,

$$20\% \text{ of } 300 = 300 \times \frac{20}{100} = 3 \times 20 = 60$$

Alternate Method

Here, number = 300

∴ 20% of number = $\frac{\text{Number}}{5} = \frac{300}{5} = 60$

Ex. 2 If 30% of a = 60, then find the value of a.

Sol. According to the formula,

$$30\% \text{ of } a = 60 \Rightarrow \frac{30a}{100} = 60$$

∴ $$a = \frac{60 \times 100}{30} = 200$$

Conversion of Per cent into Fraction

Expressing per cent ($x\%$) into fraction,

$$\text{Required fraction} = \frac{x}{100}$$

Ex. 3 Express 25% in fraction.

Sol. $25\% = \frac{25}{100} = \frac{1}{4}$

Ex. 4 Express 84% in fraction.

Sol. $84\% = \frac{84}{100} = \frac{21}{25}$

Conversion of Fraction into Percentage

Expressing a fraction $\left(\frac{x}{y}\right)$ into per cent,

$$\text{Required percentage} = \left(\frac{x}{y} \times 100\right)\%$$

Ex. 5 Convert $\frac{3}{8}$ into per cent.

Sol. Required percentage = $\left(\frac{3}{8} \times 100\right)\% = 37.5\%$

Ex. 6 Express $2\frac{1}{4}$ in per cent.

Sol. Required percentage = $\left(2\frac{1}{4} \times 100\right)\% = \left(\frac{9}{4} \times 100\right)\% = 225\%$

Conversion of Per cent into Decimal

Expressing per cent ($x\%$) in decimal,

$$\text{Required decimal} = \frac{x}{100} = 0.0x$$

Ex. 7 Express 9% in decimal.

Sol. Required decimal = $\frac{9}{100} = 0.09$

Ex. 8 Express 18% in decimal.

Sol. Required decimal = $\frac{18}{100} = 0.18$

Expressing One Quantity as a Per cent with Respect to Other

To express a quantity as a per cent with respect to other quantity, following formula is used

$$\left[\frac{\text{The quantity to be expressed in per cent}}{\text{2nd quantity (in respect of which the per cent has to be obtained)}} \times 100\right]\%$$

✦ To apply this formula, both the quantities must be in same metric unit.

Percentage / 177

Ex. 9 60 kg is what per cent of 240 kg?

Sol. According to the formula,

Required percentage

$$= \left[\frac{\text{The quantity to be expressed in per cent}}{\text{2nd quantity (in respect of which the per cent has to be obtained)}} \times 100\right]\%$$

$$= \frac{60}{240} \times 100\% = \frac{100}{4}\% = 25\%$$

Ex. 10 10 g is what per cent of 1 kg?

Sol. Here, units of both the quantities are different. Hence, first, we will make both the units same.

1 quantity = 10 g ; 2 quantity = 1 kg = 1000 g

Now, both the quantities are in same unit.

According to the formula,

Required per cent = $\frac{10}{1000} \times 100\% = 1\%$

Fast Track Techniques to solve the QUESTIONS

Technique ① If $x\%$ of A is equal to $y\%$ of B, then $z\%$ of $A = \left(\frac{yz}{x}\right)\%$ of B.

Ex. 11 If 10% of A is equal to 12% of B, then 15% of A is equal to what per cent of B?

Sol. Given, 10% of A = 12% of B

$\Rightarrow \quad \frac{10}{100} \times A = \frac{12}{100} \times B$

$\Rightarrow \quad A = \frac{6}{5} B$ ...(i)

Now, let 15% of $A = x\%$ of B

$\Rightarrow \quad \frac{15}{100} \times A = \frac{x}{100} \times B$

$\Rightarrow \quad \frac{15}{100} \times \frac{6}{5} \times B = \frac{x}{100} \times B$ [from Eq. (i)]

$\therefore \quad x = 3 \times 6 = 18\%$

Fast Track Method

Given that, $x = 10$, $y = 12$ and $z = 15$

$\therefore \quad 15\%$ of $A = \left(\frac{12 \times 15}{10}\right)\%$ of $B = 18\%$ of B

178 / Fast Track Objective Arithmetic

Technique 2 When a number x is increased or decreased by $y\%$, then the new number will be $\dfrac{100 \pm y}{100} \times x$.

+ (i) '+' sign is used in case of increasing. (ii) '–' sign is used in case of decreasing.

Ex. 12 The monthly income of a person is ₹ 8000. If his income is increased by 20%, then what will be his new monthly income?

Sol. Monthly income of a person = ₹ 8000
Increment in income = $8000 \times \dfrac{20}{100}$ = ₹ 1600
New income = 8000 + 1600 = ₹ 9600

Fast Track Method
Here, x = ₹ 8000 and y = 20%
According to the formula,
New income = $\dfrac{100 + 20}{100} \times 8000$ ['+' sign is used for increase in income]
= $\dfrac{120}{100} \times 8000$ = ₹ 9600

Ex. 13 The price of a computer is ₹ 20000. What will be the price of computer after reduction of 25%?

Sol. Here, x = ₹ 20000 and y = 25%
According to the formula,
New price = $\dfrac{100 - y}{100} \times x = \dfrac{100 - 25}{100} \times 20000$ ['–' sign is used for decrease in price]
= $\dfrac{75}{100} \times 20000 = 75 \times 200$ = ₹ 15000

Technique 3
(i) If x is $a\%$ more than y, then y is $\left(\dfrac{a}{100 + a} \times 100\right)\%$ less than x.

(ii) If x is $a\%$ less than y, then y is $\left(\dfrac{a}{100 - a} \times 100\right)\%$ more than x.

Ex. 14 If income of Ravi is 20% more than that of Ram, then income of Ram is how much per cent less than that of Ravi?

Sol. Let Ram's income be ₹ 100.
Then, Ravi's income = ₹ 120
∴ Required percentage = $\dfrac{120 - 100}{120} \times 100\% = \dfrac{20}{120} \times 100\% = 16\dfrac{2}{3}\%$

Fast Track Method
Here, a = 20%
According to the formula,
Required percentage = $\left(\dfrac{a}{100 + a} \times 100\right)\% = \left(\dfrac{20}{100 + 20} \times 100\right)\% = \dfrac{50}{3}\% = 16\dfrac{2}{3}\%$

Alternate Method
Ravi's income = 20% more than that of Ram = $\dfrac{1}{5}$ more than that of Ram

i.e. 1 out of 5 part is more than Ravi have, hence we can say Ravi's income is 6, if Ram's income is 5 and 1 part is less in the salary of Ram as compared to Ravi.

$$\therefore \text{Less\%} = \frac{\text{Part which is less in salary of Ram}}{\text{Ravi's salary}} \times 100$$

$$= \frac{1}{6} \times 100 = 16\frac{2}{3}\%$$

Ex. 15 If in an examination, the marks obtained by Preeti is 20% less than that of Vandana, then marks obtained by Vandana is how much per cent more than marks obtained by Preeti?

Sol. Here, $a = 20\%$

According to the formula,

$$\text{Required percentage} = \left(\frac{a}{100-a} \times 100\right)\% = \left(\frac{20}{100-20} \times 100\right)\%$$

$$= \left(\frac{20}{80} \times 100\right)\% = 25\%$$

Technique 4 If the value of a number is first increased by *a*% and later decreased by *a*%, then the net effect is always a decrease which is equal to *a*% of *a* and is written as $\frac{a^2}{100}\%$ or $\left(\frac{a}{10}\right)^2\%$.

Ex. 16 The salary of a worker is first increased by 5% and then it is decreased by 5%. What is the change in his salary?

Sol. Let the initial salary of the worker be ₹ 100.

Firstly, the salary of worker is increased by 5%.

So, increased salary = 105% of 100 = $\frac{105 \times 100}{100}$ = ₹ 105

Now, the salary is reduced by 5% after the increase.

$\therefore$ Reduced salary = 95% of 105 = $\frac{95 \times 105}{100}$ = 99.75

$\therefore$ Required change is a decrease, i.e. 100 − 99.75 = 0.25

So, required percentage decrease in salary = $\frac{0.25 \times 100}{100}\%$ = 0.25%

Fast Track Method

Here, $a = 5\%$

We know that, the salary of worker is decreasing.

According to the formula,

$$\text{Decreased percentage} = \frac{a^2}{100}\% = \frac{5^2}{100}\% = \frac{25}{100}\% = 0.25\%$$

Technique 5 When the value of an object is first changed (increased or decreased) by *a*% and then changed (increased or decreased) by *b*%, then

$$\text{Net effect} = \left[\pm a \pm b + \frac{(\pm a)(\pm b)}{100}\right]\%$$

Net effect is an increase or a decrease according to the +ve or −ve sign, respectively of the final result.

180 / Fast Track Objective Arithmetic

◈ MIND IT! 1. Signs of *a* and *b* depend on increment and decrement of the quantity. '+' sign is used for increment and '−' sign for decrement.

2. The above formula can also be used to find net change in the product of two numbers, if they are increased or decreased by *a*% and *b*%.

Ex. 17 The price of an article is first increased by 20% and later on the price were decreased by 25% due to reduction in sales. Find the net percentage change in final price of the article.

Sol. Let the original price of article be ₹ 100. Then, after the first change i.e. increase of 20%,

Price of the article $= \dfrac{100 + 20}{100} \times 100 = ₹ 120$

Again, the price were decreased by 25%, so the reduced price

$= \dfrac{100 - 25}{100} \times 120 = \dfrac{75 \times 120}{100} = ₹ 90$

∴ Required decrease $= 100 - 90 = ₹ 10$ [∵ final price < initial price]

Net percentage change $= \dfrac{\text{Final price} - \text{Initial price}}{\text{Initial price}} \times 100 = \dfrac{-10}{100} \times 100 = -10\%$

Fast Track Method

Here, $a = 20\%, b = 25\%$

Required change $= \left[(\pm a) + (\pm b) + \dfrac{(\pm a)(\pm b)}{100}\right]\% = \left[20 - 25 + \dfrac{20 \times (-25)}{100}\right]\%$

[+ ve sign for increase and − ve sign for decrease]

$= [-5 - 5]\% = -10\%$

∴ Net percentage change is a decrease of 10% because final result is negative.

Technique 6 If the price of a commodity increases or decreases by *a*%, then the decrease or increase in consumption, so as not to increase or decrease the expenditure is equal to $\left(\dfrac{a}{100 \pm a}\right) \times 100\%$.

Ex. 18 If the price of a commodity is raised by 40%, by how much per cent must a householder reduce his consumption of that commodity, so as not to increase his expenditure?

Sol. Let initial price be ₹ 100.

Price after increase $= 140\%$ of $100 = \dfrac{140 \times 100}{100} = ₹ 140$

∴ Required reduction $= \left(\dfrac{140 - 100}{140} \times 100\right)\% = \dfrac{40}{140} \times 100\% = 28\dfrac{4}{7}\%$

Fast Track Method

Here, $a = 40\%$

According to the formula,

Reduction in consumption $= \left(\dfrac{40}{100 + 40} \times 100\right)\%$

$= \left(\dfrac{40}{140} \times 100\right)\% = \dfrac{200}{7}\% = 28\dfrac{4}{7}\%$

Percentage / 181

Ex. 19 If the price of milk falls down by 20%, by how much per cent must a householder increase its consumption, so as not to decrease his expenditure on this item?

Sol. Here, $a = 20\%$
According to the formula,
$$\text{Increase in consumption} = \left(\frac{20}{100-20} \times 100\right)\% = \left(\frac{20}{80} \times 100\right)\% = 25\%$$

Technique 7 The passing marks in an examination is $P\%$. If a candidate scores R marks and fails by F marks, then
$$\text{Maximum marks, } M = \frac{100(R+F)}{P}$$

Ex. 20 A student has to score 30% marks to get throughout. If he gets 30 marks and fails by 30 marks, then find the maximum marks set for the examination.

Sol. Let the maximum marks be x.
Now, according to the question,
$$30\% \text{ of } x = 30 + 30$$
$$\Rightarrow \frac{30x}{100} = 30 + 30$$
$$\Rightarrow \frac{30x}{100} = 60 \Rightarrow x = 200$$

Fast Track Method
Here, $P = 30$, $R = 30$ and $F = 30$
According to the formula,
$$\text{Maximum marks} = \frac{100(R+F)}{P} = \frac{100(30+30)}{30} = \frac{100 \times 60}{30} = 200$$

Technique 8 A candidate scores $x\%$ marks in an examination and fails by a marks, while another candidate who scores $y\%$ marks, gets b marks more than the minimum required passing marks. The maximum marks for the examination is given as
$$M = \frac{100(a+b)}{y-x} = \frac{\text{Sum of scores}}{\text{Difference in marks}} \times 100$$

Ex. 21 A candidate scores 25% and fails by 60 marks, while another candidate who scores 50% marks, gets 40 marks more than the minimum required marks to pass the examination. Find the maximum marks for the examination.

Sol. Let the maximum marks be x.
Now, according to the question,
Marks scored by first candidate + 60 = Marks scored by second candidate − 40
$$\Rightarrow 25\% \text{ of } x + 60 = 50\% \text{ of } x - 40$$
$$\Rightarrow \frac{25x}{100} + 60 = \frac{50x}{100} - 40$$
$$\Rightarrow \frac{50x}{100} - \frac{25x}{100} = 60 + 40 \Rightarrow \frac{25x}{100} = 100$$
$$\therefore x = 400$$

Fast Track Method

Here, $x = 25$, $y = 50$, $a = 60$ and $b = 40$

According to the formula,
$$M = \frac{100(a+b)}{y-x} = \frac{100(60+40)}{50-25}$$
$$= \frac{100 \times 100}{25} = 400$$

Technique 9 Suppose in an examination, $x\%$ of total number of students failed in subject A and $y\%$ of total number of students failed in subject B and $z\%$ failed in both the subjects. Then,

(i) Percentage of students who passed in both the subjects = $[100 - (x + y - z)]\%$

(ii) Percentage of students who failed in either subject = $(x + y - z)\%$

Ex. 22 In an examination, 20% of total number of students failed in History, 15% of total number of students failed in Hindi and 5% of total number of students failed in both. Find the percentage of students who passed in both the subjects. Also, find the percentage of students who failed in either subject.

Sol. Let total number of students be 100.
Then, students failed in History only = $20 - 5 = 15$
and students failed in Hindi only = $15 - 5 = 10$
$\therefore$ Total number of failed students = $15 + 10 + 5 = 30$
$\therefore$ Number of students passed in both subjects = $100 - 30 = 70$
$\therefore$ Required percentage = 70%
Total number of students who failed either in Hindi or in History
$= 20 + 10 = 30$
$\therefore$ Required percentage = 30%

Fast Track Method

Here, $x = 20$, $y = 15$ and $z = 5$
According to the formula,
Required percentage of students who passed in both subjects
$= [100 - (x + y - z)]\%$
$= [100 - (20 + 15 - 5)]\%$
$= (100 - 30)\% = 70\%$
Again, required percentage of students who failed either in Hindi or in History
$= (x + y - z)\% = (20 + 15 - 5) = 30\%$

Ex. 23 In an examination, 26% students failed in Mathematics and 30% students failed in Physics. If 5% students failed in both the subjects, then find the percentage of students who failed in either subject.

Sol. Here, $x = 26$, $y = 30$ and $z = 5$
$\therefore$ Required percentage = $(x + y - z)\%$
$= (26 + 30 - 5)\% = 51\%$

Percentage / 183

Technique 10 If due to r% decrease in the price of an item, a person can buy A kg more in ₹ x, then

Actual price of that item = $₹\dfrac{rx}{(100-r)A}$ per kg

Ex. 24 If due to 10% decrease in the price of sugar, Ram can buy 5 kg more sugar in ₹ 100, then find the actual price of sugar.

Sol. Let the actual price of sugar be ₹ y per kg.

Amount of sugar bought in ₹ 100 = $\dfrac{100}{y}$ ...(i)

Price of sugar after 10% decrease = 90% of $y = \dfrac{9}{10}y$

Now, amount of sugar bought in ₹ 100 = $\dfrac{1000}{9y}$ ...(ii)

According to the question, $\dfrac{1000}{9y} - \dfrac{100}{y} = 5$

$\Rightarrow \dfrac{1000 - 900}{9y} = 5 \Rightarrow y = \dfrac{100}{9 \times 5}$

$\therefore y = \dfrac{20}{9} = ₹ 2\dfrac{2}{9}$

Fast Track Method

Here, $r = 10\%$, $x = ₹ 100$ and $A = 5$ kg

$\therefore$ Actual price of sugar = $\dfrac{r \times x}{(100-r)A} = \dfrac{10 \times 100}{(100-10) \times 5} = \dfrac{1000}{450} = \dfrac{20}{9} = ₹ 2\dfrac{2}{9}$

Technique 11 If two candidates contested in an election and one candidate got x% of total votes casted and still lose by y votes, then

Total number of votes casted = $\dfrac{100 \times y}{100 - 2x}$.

Ex. 25 In an election contested by two candidates, one candidate got 30% of total votes and still lost by 500 votes, then find the total number of votes casted.

Sol. Let total number of votes casted be x.

Number of votes got by first candidate = $\dfrac{30}{100}x$

Number of votes got by second candidate = $\dfrac{70}{100}x$

According to the question,

Difference in votes = 500

$\Rightarrow \dfrac{70}{100}x - \dfrac{30}{100}x = 500$

$\Rightarrow \dfrac{40}{100}x = 500 \Rightarrow x = \dfrac{500 \times 100}{40} = 1250$

Fast Track Method

Here, $x = 30$ and $y = 500$

$\therefore$ Total number of votes = $\dfrac{100 \times y}{100 - 2x} = \dfrac{100 \times 500}{100 - 2 \times 30} = \dfrac{100 \times 500}{40} = 1250$

184 / Fast Track Objective Arithmetic

Technique 12   If the population of a town is P and it increases (or decreases) at the rate of R% per annum, then

(i) Population after n yr $= P\left(1 \pm \dfrac{R}{100}\right)^n$

(ii) Population, n yr ago $= \dfrac{P}{\left(1 \pm \dfrac{R}{100}\right)^n}$

✦ Use '+' sign for increment and '–' sign for decrement.

Ex. 26 The population of a town is 352800. If it increases at the rate of 5% per annum, then what will be its population 2 yr hence. Also, find the population 2 yr ago.

Sol. Given that, $P = 352800$, $R = 5\%$ and $n = 2$
According to the formula,

Population after 2 yr $= P\left(1 + \dfrac{R}{100}\right)^n = 352800 \times \left(1 + \dfrac{5}{100}\right)^2$

$= 352800 \times \left(\dfrac{100+5}{100}\right)^2 = 352800 \times \left(\dfrac{21}{20} \times \dfrac{21}{20}\right) = 388962$

Population 2 yr ago $= \dfrac{P}{\left(1 + \dfrac{R}{100}\right)^n} = \dfrac{352800}{\left(1 + \dfrac{5}{100}\right)^2} = 352800 \times \dfrac{20}{21} \times \dfrac{20}{21} = 320000$

✦ Students may also solve problems through simple increasing-decreasing method of percentage, because 5% increase for 2 yr means population after 2 yr, which is equal to $352800 \times \dfrac{105}{100} \times \dfrac{105}{100}$. In case, when population decrease at the rate of 5%, then population after 2 yr is equal to $352800 \times \dfrac{95}{100} \times \dfrac{95}{100}$.

Technique 13   If the present population of a city is P and there is a increment or decrement of $R_1\%$, $R_2\%$ and $R_3\%$ in first, second and third year respectively, then

Population of city after 3 yr $= P\left(1 \pm \dfrac{R_1}{100}\right)\left(1 \pm \dfrac{R_2}{100}\right)\left(1 \pm \dfrac{R_3}{100}\right)$

Ex. 27 Population of a city in 2004 was 1000000. If in 2005, there is an increment of 15%, in 2006, there is a decrement of 35% and in 2007, there is an increment of 45%, then find the population of city at the end of year 2007.

Sol. Given that, $P = 1000000$, $R_1 = 15\%$, $R_2 = 35\%$ (decrease) and $R_3 = 45\%$
Population of city at the end of year 2007

$= P\left(1 + \dfrac{R_1}{100}\right)\left(1 - \dfrac{R_2}{100}\right)\left(1 + \dfrac{R_3}{100}\right)$

$= 1000000 \left(1 + \dfrac{15}{100}\right)\left(1 - \dfrac{35}{100}\right)\left(1 + \dfrac{45}{100}\right)$

$= 1000000 \times \dfrac{115}{100} \times \dfrac{65}{100} \times \dfrac{145}{100} = 1083875$

Fast Track Practice

Exercise 1 Base Level Questions

1. Express 0.6% as a fraction.
   (a) $\frac{3}{500}$ (b) $\frac{4}{125}$ (c) $\frac{6}{197}$ (d) $\frac{6}{100}$

2. Half of 1 per cent written as a decimal is
   (a) 0.005 (b) 0.05 (c) 0.02 (d) 0.2

3. What is 5% of 50% of 500? [CDS 2012]
   (a) 12.5 (b) 25 (c) 1.25 (d) 6.25

4. 89% of ? + 365 = 1075.22 [Bank Clerk 2010]
   (a) 798 (b) 897 (c) 898 (d) 752
   (e) None of these

5. 27 is 3.6% of? [SSC (10 + 2) 2017]
   (a) 750 (b) 75
   (c) 1500 (d) 1875

6. 12 of 26% of $\frac{5}{78}$ of 38% of $\frac{7}{152}$ of 10000 = ?
   [Hotel Mgmt. 2010]
   (a) 38 (b) 35 (c) 41 (d) 52

7. 25% of what amount of money is equal to $12\frac{1}{2}$% of ₹ 180? [SSC FCI 2012]
   (a) ₹ 120 (b) ₹ 75 (c) ₹ 80 (d) ₹ 90

8. 30% of a 3-digit number is 190.8. What will be 125% of that number?
   [Bank Clerk 2011]
   (a) 759 (b) 785 (c) 795 (d) 779
   (e) None of these

9. A person's salary has increased from ₹ 7200 to ₹ 8100. What is the percentage increase in his salary? [CDS 2013]
   (a) 25% (b) 18%
   (c) $16\frac{2}{3}$% (d) $12\frac{1}{2}$%

10. There are 1240 employees in an organisation out of which, 25% are promoted. How many such employees are there who got promotion? [Bank Clerk 2009]
    (a) 398 (b) 345 (c) 310 (d) 372
    (e) None of these

11. The difference between 78% of a number and 59% of the same number is 323. What is 62% of that number?
    [Bank Clerk 2009]
    (a) 1054 (b) 1178 (c) 1037 (d) 1159
    (e) None of these

12. A saves 20% of his monthly salary. If his monthly expenditure is ₹ 6000, then his monthly savings is [SSC (10+2) 2012]
    (a) ₹ 1200 (b) ₹ 4800
    (c) ₹ 1500 (d) ₹ 1800

13. A number increased by $137\frac{1}{2}$% and the increment is 33. The number is
    [SSC CGL 2013]
    (a) 27 (b) 22 (c) 24 (d) 25

14. When 35 is subtracted from a number, it reduces to its 80%. What is $\frac{4}{5}$ th of that number?
    (a) 70 (b) 90 (c) 120 (d) 140
    (e) None of these

15. A man losses 20% of his money. After spending 25% of the remainder, he has ₹ 480 left. What is the amount of money he originally had? [CDS 2012]
    (a) ₹ 600 (b) ₹ 720
    (c) ₹ 800 (d) ₹ 840

16. Two numbers are respectively 20% and 50% more than a 3rd number. What is the percentage of 2nd with respect to 1st? [SSC CGL 2011]
    (a) 125% (b) 90% (c) 80% (d) 75%

17. In a school, 10% of boys are equal to the one-fourth of the girls. What is the ratio of boys and girls in that school? [RRB 2012]
    (a) 3 : 2 (b) 5 : 2 (c) 2 : 1 (d) 4 : 3

18. 48% of the 1st number is 60% of the 2nd number. What is the ratio of the 1st number to the 2nd number?
    [Bank Clerk 2008]
    (a) 4 : 7
    (b) 3 : 4
    (c) 5 : 4
    (d) Cannot be determined
    (e) None of the above

19. The sum of 15% of a positive number and 20% of the same number is 126. What is one-third of that number? [IBPS Clerk 2011]
    (a) 360 (b) 1080 (c) 120 (d) 40
    (e) None of these

20. In a test, A scored 10% more than B and B scored 5% more than C. If C scored 300 marks out of 400, then A's marks are [SSC CGL 2012]
    (a) 310 (b) 325
    (c) 350 (d) 360

21. In an office, 40% of the staff is female. 40% of the female and 60% of the male voted for me. The percentage of votes I got was [SSC CPO 2011]
    (a) 24% (b) 42% (c) 50% (d) 52%

22. Mathew scored 42 marks in Biology, 51 marks in Chemistry, 58 marks in Mathematics, 35 marks in Physics and 48 marks in English. The maximum marks, a student can score in each subject, are 60. How much overall percentage did Mathew get in this exam? [SBI Clerk 2012]
    (a) 76% (b) 82% (c) 68% (d) 78%
    (e) None of these

3. In a class of 40 students and 5 teachers, each student got sweets that are 15% of the total number of students and each teacher got sweets that are 20% of the total number of students. How many sweets were there? [Bank Clerk 2008]
    (a) 280 (b) 240
    (c) 320 (d) 360
    (e) None of these

24. A water pipe is cut into two pieces. The longer piece is 70% of the length of the pipe. By how much percentage is the longer piece longer than the shorter piece? [CDS 2014]
    (a) 140% (b) $\frac{400}{3}$%
    (c) 40% (d) None of these

25. The prices of two articles are as 3 : 4. If the price of the first article is increased by 10% and that of the second by ₹ 4, original ratio remains the same. The original price of the second article is [SSC CPO 2013]
    (a) ₹ 40 (b) ₹ 10
    (c) ₹ 30 (d) ₹ 35

26. In a particular constituency, 75% of voters cast their votes, out of which 2% were rejected. The winning candidate received 75% of the valid votes and bagged a total of 9261 votes. The total number of voters in the constituency is [SSC (10+2) 2012]
    (a) 14500 (b) 18900
    (c) 16800 (d) 24000

27. Last year, there were 610 boys in a school. The number decreased by 20% this year. How many girls are there in the school, if the number of girls is 175% of the total number of boys in the school this year? [SBI Clerk 2012]
    (a) 854 (b) 848 (c) 798 (d) 782
    (e) None of these

28. If increasing 20 by P percentage gives the same result as decreasing 60 by P percentage, what is P percentage of 70? [IB Assistant 2013]
    (a) 50 (b) 140 (c) 14 (d) 35

29. In a class X of 30 students, 24 passed in first class; in another class Y of 35 students, 28 passed in first class. In which class was the percentage of students getting first class more? [FCI 2015]
    (a) Class X has more percentage of students getting first class
    (b) Class Y has more percentage of students getting first class
    (c) Both classes have equal percentage of students getting first class
    (d) None of the above

30. A man spend $7\frac{1}{2}$% of his money and after spending 75% of the remaining, he had ₹ 370 left. How much money did he have? [SSC CGL (Mains) 2016]
    (a) 1200 (b) 1600
    (c) 1500 (d) 1400

31. In order to pass in an exam, a student is required to get 780 marks out of the aggregate marks. Sonu got 728 marks and was declared failed by 5%. What are the maximum aggregate marks a student can get in the examination?
    (a) 1040 (b) 1100
    (c) 1000 (d) 960
    (e) None of these

32. The salary of a person is increased by 10% of his original salary. But he received the same amount even after increment. What is the percentage of his salary he did not receive? [CDS 2016 (I)]
    (a) 11% (b) 10% (c) $\frac{100}{11}$% (d) $\frac{90}{11}$%

33. Priya got 9 marks more in History than what she got in Geography. Her History marks are 56% of the sum of her History and Geography marks. What are her Geography marks? [SSC CGL (Pre) 2017]
    (a) 42 (b) 65 (c) 53 (d) 33

Percentage / 187

34. A box has 100 blue balls, 50 red balls and 50 black balls. 25% of blue balls and 50% of red balls are taken away. Then, percentage of black balls at present is **[SSC CGL 2013]**
    (a) 25%   (b) $33\frac{1}{3}$%   (c) 40%   (d) 50%

35. A jogger desires to run a certain course in $\frac{1}{4}$ less time than he usually takes. By what per cent must be increase his average running speed to accomplish the goal? **[SSC CPO 2013]**
    (a) 50%   (b) 20%   (c) 25%   (d) $33\frac{1}{3}$%

36. From 2008 to 2009, the sales of a book decreased by 80%. If the sales in 2010 was the same as in 2008, by what per cent did it increase from 2009 to 2010? **[SSC (10+2) 2012]**
    (a) 80%   (b) 100%   (c) 120%   (d) 400%

37. If 15% of A is equal to 20% of B, then 25% of A is equal to what per cent of B?
    (a) 30%   (b) $33\frac{1}{3}$%   (c) 35%   (d) 25%

38. The monthly income of a person is ₹ 5000. If his income is increased by 30%, then what is his monthly income now?
    (a) ₹ 7000   (b) ₹ 5500
    (c) ₹ 4500   (d) ₹ 6500
    (e) None of these

39. The price of a certain article is ₹ 15000. But due to slump in the market, its price decreases by 8%. Find the new price of the article.
    (a) ₹ 14000   (b) ₹ 13800
    (c) ₹ 16500   (d) ₹ 12600
    (e) None of these

40. If A's salary is 25% higher than B's salary, then how much per cent is B's salary lower than A's salary? **[RRB Clerk (Pre) 2017]**
    (a) 15%   (b) 20%   (c) 25%   (d) $33\frac{1}{3}$%
    (e) None of these

41. A's salary is 20% less than B's salary. Then, B's salary is more than A's salary by **[SSC CGL 2013]**
    (a) $33\frac{1}{2}$%   (b) $16\frac{2}{3}$%   (c) 20%   (d) 25%

42. In an examination, the marks obtained by Shantanu is 40% less than the marks obtained by Kamal, then marks obtained by Kamal is how much per cent more than the marks obtained by Shantanu? **[Bank Clerk 2009]**
    (a) $55\frac{2}{3}$%   (b) $44\frac{3}{5}$%   (c) $33\frac{1}{3}$%   (d) $66\frac{2}{3}$%
    (e) None of these

43. Income of Suman is first increased by 7% and then it is decreased by 7%. What is the change in her income? **[Hotel Mgmt. 2008]**
    (a) 0.49% (increase)
    (b) 0.39% (decrease)
    (c) 0.39% (increase)
    (d) 0.49% (decrease)

44. Because of scarcity of rainfall, the price of a land decreases by 12% and its production also decreases by 4%. What is the total effect on revenue? **[LIC ADO 2008]**
    (a) Loss of 16%   (b) Gain of 15%
    (c) Loss of 15.48%   (d) Gain of 15.48%
    (e) Loss of 15.52%

45. The marked price of brand A watches is 15% higher than its original price. Due to increase in demand, the price is further increased by 10%. How much profit will be obtained in selling the watches?
    (a) 25%   (b) 35%   (c) 26.5%   (d) 27%
    (e) None of these

46. If length and breadth of a rectangle became half and double respectively, then what will be the resultant area?
    (a) 25%   (b) 55%   (c) 75%   (d) 80%
    (e) None of these

47. A student was asked to measure the length and breadth of a rectangle. By mistake, he measured the length 20% less and the breadth 10% more. If its original area is 200 sq cm, then find the area after this measurement?
    (a) 176 sq cm   (b) 206 sq cm
    (c) 226 sq cm   (d) 316 sq cm
    (e) None of these

48. If the price of petrol is increased by 20%, by what percentage should the consumption be decreased by the consumer, if the expenditure on petrol remains unchanged? **[CGPSC 2013]**
    (a) $16\frac{2}{3}$%   (b) $6\frac{2}{3}$%   (c) 8%   (d) 15%
    (e) None of these

49. The price of sugar is increased by 25%. If a family wants to keep its expenses on sugar unaltered, then the family will have to reduce the consumption of sugar by **[SSC CGL 2010]**
    (a) 20%   (b) 21%   (c) 22%   (d) 25%

188 / Fast Track Objective Arithmetic

50. A student has to score 40% marks to get throughout. If he gets 40 marks and fails by 40 marks, then find the maximum marks set for the examination.
    (a) 200 (b) 250 (c) 300 (d) 150
    (e) None of these

51. A candidate scoring x% marks in an examination and fails by a marks, while another candidate who scores y% marks, gets b marks more than the minimum required pass marks. What is the maximum marks for the examination? [CDS 2016 (II)]
    (a) $\dfrac{100(a+b)}{x-y}$ (b) $\dfrac{100(a-b)}{x+y}$
    (c) $\dfrac{100(a+b)}{y-x}$ (d) $\dfrac{100(a-b)}{x-y}$

52. In an examination, 34% of the students failed in Mathematics and 42% failed in English. If 20% of the students failed in both the subjects, then what is the percentage of students who passed in both the subjects? [SBI Clerk (Mains) 2016]
    (a) 40% (b) 42% (c) 44% (d) 46%
    (e) 48%

53. There are only two candidates contesting the election, a person who got 35% votes lost by 600 votes. Assuming that there were no invalid votes, then the total number of votes casted are
    (a) 2000 (b) 1500 (c) 8000 (d) 1000

54. The population of a city is 250000. It is increasing at the rate of 2% every year. The growth in the population after 2 yr is [CLAT 2013]
    (a) 2500 (b) 10000
    (c) 252000 (d) 10100

55. The population of a state in the year was 93771. If the rate of increase was 8% and 15% respectively from the previous two years before 2013, what was the population in the year 2011? [NICL AO 2015]
    (a) 55650 (b) 49550
    (c) 75500 (d) 84500
    (e) 65200

56. During the first year, the population of a village is increased by 5% and the second year, it is diminished by 5%. At the end of the second year, its population was 47880. What was the population at the beginning of the first year? [SSC Multitasking 2014]
    (a) 45500 (b) 48000
    (c) 43500 (d) 53000

57. The population of a particular area 'A' of a city is 5000. It increases by 10% in first year, decreases by 20% in the second year because of some reason. In the third year, the population increases by 30%. What will be the population of area 'A' at the end of third year? [RBI Officer Grade 2015]
    (a) 5225 (b) 5720
    (c) 4895 (d) 5560
    (e) 5407

Exercise 2 Higher Skill Level Questions

1. A line of length 1.5 m was measured as 1.55 m by mistake. What will be the value of error per cent? [SSC CGL (Mains) 2016]
   (a) 0.05% (b) 3.33% (c) 3.67% (d) 0.5%

2. A boy found the answer for the question "Subtract the sum of 1/4 and 1/5 from unity and express the answer in decimals" as 0.45. The percentage of error in his answer was [SSC CGL (Mains) 2016]
   (a) (100/11)% (b) 50%
   (c) 10% (d) (200/11)%

3. If a% of $a + b$% of $b = 2$% of ab, then what per cent of a is b? [CDS 2017 (I)]
   (a) 50% (b) 75% (c) 100%
   (d) Cannot be determined

4. Mohan gave 25% of a certain amount of money to Ram. From the money Ram received, he spent 20% on buying books and 35% on buying a watch. After the mentioned expenses, Ram has ₹ 2700 remaining. How much did Mohan have initially? [IBPS SO 2016]
   (a) ₹ 16000 (b) ₹ 15000
   (c) ₹ 24000 (d) ₹ 27000
   (e) ₹ 20000

5. Sonali invests 15% of her monthly salary in insurance policies. She spends 55% of her monthly salary in shopping and on household expenses. She saves the remaining amount at ₹ 12750. What is Sonali's monthly income? [SNAP 2012]
   (a) ₹ 42500 (b) ₹ 38800
   (c) ₹ 40000 (d) ₹ 35500

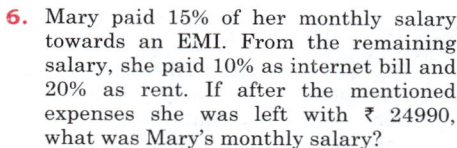

Percentage / 189

6. Mary paid 15% of her monthly salary towards an EMI. From the remaining salary, she paid 10% as internet bill and 20% as rent. If after the mentioned expenses she was left with ₹ 24990, what was Mary's monthly salary?
   [SBI Clerk (Pre) 2016]
   (a) ₹ 45000   (b) ₹ 48000
   (c) ₹ 42000   (d) ₹ 36000
   (e) ₹ 40000

7. Sohan spends 23% of an amount of money on an insurance policy, 33% on food, 19% on children's education and 16% on recreation. He deposits the remaining amount of ₹ 504 in bank. How much total amount does he spend on food and insurance policy together?
   [Bank Clerk 2010]
   (a) ₹ 3200   (b) ₹ 3126
   (c) ₹ 3136   (d) ₹ 3048
   (e) None of these

8. Two students appeared for an examination. One of them secured 10 marks more than the other and his marks were 75% of the sum of their marks. The marks obtained by them are
   [SSC (10 + 2) 2017]
   (a) 52 and 42   (b) 68 and 58
   (c) 63 and 53   (d) 15 and 5

9. If the mean age of combined group of boys and girls is 18 yr and the mean of age of boys is 20 and that of girls is 16, then what is the percentage of boys in the group?
   [CDS 2016 (II)]
   (a) 60   (b) 50   (c) 45   (d) 40

10. A person could save 10% of his income. But 2 yr later, when his income increased by 20%, he could save the same amount only as before. By how much percentage has his expenditure increased?   [CDS 2015 (I)]
    (a) $22\frac{2}{9}\%$ (b) $23\frac{1}{3}\%$
    (c) $24\frac{2}{9}\%$ (d) $25\frac{2}{9}\%$

11. Out of the total population, 40% were females. Out of the total number of males, 75% were literates and the remaining 225 were illiterates. If 70% of the total number of females are literates, how many females were illiterate?
    [LIC AAO 2016]
    (a) 200   (b) 210   (c) 240   (d) 150
    (e) 180

12. Ram sells his goods 25% cheaper than Shyam and 25% dearer than Hari. How much percentage is Hari's goods cheaper than Shyam?   [SSC CPO 2011]
    (a) 25%   (b) $33\frac{1}{3}\%$
    (c) 40%   (d) 50%

13. Due to an increase of 30% in the price of eggs, 6 eggs less are available for ₹ 7.80. The present rate of eggs per dozen is
    (a) ₹ 5.50   (b) ₹ 4.68
    (c) ₹ 6.49   (d) ₹ 3.58
    (e) None of these

14. In an examination out of 480 students, 85% of the girls and 70% of the boys passed. How many boys appeared in the examination, if total pass percentage was 75%?   [SNAP 2012]
    (a) 370   (b) 340   (c) 320   (d) 360

15. Due to a 25% increase in the price of rice per kilogram, a person is able to purchase 20 kg less for ₹ 400. What is the increased price of rice per kilogram?
    [IB ACIO 2012]
    (a) ₹ 5   (b) ₹ 6
    (c) ₹ 10   (d) ₹ 4

16. 1 L of water is added to 5 L of alcohol and water solution containing 40% alcohol strength. The strength of alcohol in the new solution will be
    [SSC CGL 2007]
    (a) 30%   (b) $33\frac{1}{3}\%$
    (c) $33\frac{2}{3}\%$ (d) 33%

17. Wheat is now being sold at ₹ 27 per kg. During last month, its cost was ₹ 24 per kg. Find by how much per cent a family reduces its consumption, so as to keep the expenditure fixed?
    (a) 10.2%   (b) 12.1%
    (c) 12.3%   (d) 11.1%

18. Candidates in a competitive examination consisted of 60% men and 40% women. 70% men and 75% women cleared the qualifying test and entered the final test where 80% men and 70% women were successful. Which of the following statements is correct?   [UPSC CSAT 2015]
    (a) Success rate is higher for women
    (b) Overall success rate is below 50%
    (c) More men cleared the examination than women
    (d) Both (a) and (b) are correct

19. In a city, 12% of households earn less than ₹30000 per year, 6% households earn more than ₹200000 per year, 22% households earn more than ₹100000 per year and 990 households earn between ₹30000 and ₹100000 per year. How many households earn between ₹100000 and ₹200000 per year? [UPSC CSAT 2017]
    (a) 250   (b) 240   (c) 230   (d) 225

20. Sia gave one-fourth of the money she had with her to her brother. Her brother, from the money he received from Sia, spent one-eight on his business and one-fourth on tuition fees. After the mentioned expenses, Sia's brother had ₹125 remaining with him. How much money did Sia have initially? [IBPS SO 2016]
    (a) ₹ 750   (b) ₹ 800   (c) ₹ 780   (d) ₹ 790
    (e) ₹ 770

21. In March 2014, Rashmi paid EMI which was 30% of her monthly salary. The remaining salary she spent on shopping of groceries and clothes in the respecting ratio of 4 : 3. She spent 15000 on shopping of clothes. If in April 2014, her salary increasing by 12%, what was her salary in April? [IBPS SO 2016]
    (a) ₹ 48000   (b) ₹ 50000
    (c) ₹ 56000   (d) ₹ 66000
    (e) ₹ 55000

22. A student was awarded certain marks in an examination. However, after re-evaluation, his marks were reduced by 40% of the marks that were originally awarded to him, so that the new score now became 96. How many marks did the student lose after re-evaluation?
    (a) 64   (b) 68   (c) 63   (d) 56

23. In a competitive exam, there were 5 sections. 10% of the total number of students cleared the cut-off in all the sections and 5% cleared none of the sections. From the remaining candidates, 30% cleared only section I, 20% cleared only section 2, 10% cleared only section 3 and remaining 1020 candidates cleared only section 4. How many students appeared in the competitive exam? [SNAP 2016]
    (a) 2550   (b) 2800   (c) 3000   (d) 3200

24. 5/9th part of the population in a village are males. If 30% of the males are married, the percentage of unmarried females in the total population is [CDS 2017 (I)]

(a) $20\frac{2}{9}$ (b) $27\frac{2}{9}$ (c) $27\frac{7}{9}$ (d) $29\frac{2}{9}$

25. The price of ghee is increased by 32%. Therefore, a family reduces its consumption, so that the increment in price of ghee is only 10%. If consumption of ghee is 10 kg before the increment, then what is the consumption now?
    (a) $8\frac{1}{3}$ kg   (b) $8\frac{3}{4}$ kg
    (c) $8\frac{1}{2}$ kg   (d) 9 kg

26. Out of her monthly salary, Ridhi spends 34% on various expenses. From the remaining, she gives one-sixth to her brother, two-third to her sister and the remaining she keeps as savings. If the difference between the amounts she gave to her sister and her brother was ₹ 10560, what was Ridhi's savings? [SBI PO (Pre) 2017]
    (a) ₹ 3740   (b) ₹ 3520
    (c) ₹ 4230   (d) ₹ 3230
    (e) Other than those given as options

27. In a village, 70% registered voters cast their votes in the election. Only two candidates (A and B) are contesting the election. A won the election by 400 votes. Had A received 12.5% less votes, the result would have been different. How many registered voters are there in the village? [RBI Officer Grade 2015]
    (a) 7200   (b) 4500   (c) 8000   (d) 4250
    (e) 8572

28. P = (40% of A) + (65% of B) and Q = (50% of A) + (50% of B), where A is greater than B. In this context, which of the following statements is correct? [UPSC CSAT 2017]
    (a) P is greater than Q
    (b) Q is greater than P
    (c) P is equal to Q
    (d) None of the above can be concluded with certainty

29. Suri gave 25% of her monthly salary to her mother. From the remaining salary, she paid 15% towards rent and 25%, she kept aside for her monthly expenses. The remaining amount she kept in bank account. The sum of the amount she kept in bank and that she gave to her mother was ₹ 42000. What was her monthly salary? [SBI PO (Pre) 2016]
    (a) ₹ 50000   (b) ₹ 60000
    (c) ₹ 65000   (d) ₹ 64000
    (e) ₹ 72000

Percentage / 191

30. In 2003, the total population of a village was 4800, out of which 40% were females. In 2004, the total population increased by 10% as compared to the previous year. If the number of females remained the same in 2004, what was the percentage increase in the number of males in 2004 as compared to 2003?
[IBPS Clerk (Pre) 2016]

(a) $18\frac{1}{5}\%$ (b) 20%

(c) 15%   (d) $16\frac{2}{3}\%$

(e) $12\frac{1}{3}\%$

31. Three candidates A, B, C participated in an election. A gets 40% of the votes more than B. C gets 20% votes more that B. A also overtakes C by 4000 votes. If 90% voters voted and no invalid or illegal votes were cast, then what will be the number of voters in the voting list? [SNAP 2016]

(a) 72000   (b) 80000
(c) 70000   (d) 78500

32. Puneet distributed a sum of money among his wife, two sons and one daughter and kept some money for himself. 20% of the total money that he had, he gave to his wife and kept 22% of it for himself. 60% of the remaining money he distributed among his two sons and gave the remaining to his daughter. If the daughter got ₹ 2940 more than the money he kept for himself, what was the total money that he distributed among his sons?
[NICL AO 2015]

(a) ₹ 87178   (b) ₹ 87108
(c) ₹ 85260   (d) ₹ 86800
(e) ₹ 86786

Answer with Solutions

Exercise 1   Base Level Questions

1. (a) $0.6\% = \frac{0.6}{100} = \frac{6}{1000} = \frac{3}{500}$

2. (a) Half of 1 per cent
$= \frac{1}{2}\% = \left(\frac{1}{2} \times \frac{1}{100}\right) = \frac{0.5}{100} = 0.005$

3. (a) 5% of 50% of 500
$= \frac{5}{100} \times \frac{50}{100} \times 500 = 12.5$

4. (a) 89% of ? + 365 = 1075.22
$\Rightarrow \frac{89}{100} \times ? = 1075.22 - 365$
$\therefore ? = \frac{71022}{89} = 798$

5. (a) Let the number be x.
Then, according to the question,
$x \times \frac{3.6}{100} = 27$
$\Rightarrow x = \frac{27 \times 100}{3.6} = \frac{27 \times 1000}{36}$
$\therefore x = \frac{3 \times 1000}{4} = 3 \times 250 = 750$

6. (b) $12 \times \frac{26}{100} \times \frac{5}{78} \times \frac{38}{100} \times \frac{7}{152} \times 10000$
$= 12 \times \frac{5}{3} \times \frac{7}{4} = 35$

7. (d) Let required amount of money be x.
Then, $x \times 25\% = 180 \times 12\frac{1}{2}\%$
$\therefore x = \frac{180 \times 12.5}{25} = ₹ 90$

8. (c) Let x be the number.
Then, 30% of $x = 190.8$
$\Rightarrow x = \frac{190.8 \times 100}{30} = 636$
$\therefore$ 125% of $636 = \frac{125}{100} \times 636 = 795$

9. (d) Percentage increase in salary
$= \frac{8100 - 7200}{7200} \times 100$
$= \frac{900}{7200} \times 100 = 12.5\% = 12\frac{1}{2}\%$

10. (c) Required number $= \frac{1240 \times 25}{100} = 310$

11. (a) Let the number be x.
According to the question,
$(78 - 59)\%$ of $x = 323$
$\Rightarrow \frac{19 \times x}{100} = 323$
$\Rightarrow x = \frac{323 \times 100}{19} = 1700$
$\therefore$ 62% of $1700 = \frac{62 \times 1700}{100} = 1054$

12. (c) Let A's monthly income be ₹ x.
According to the question,
Monthly expenditure of A = ₹ 6000
$x \times (100 - 20)\% = 6000$
$\Rightarrow \quad x \times \dfrac{80}{100} = 6000$
$\Rightarrow \quad x = \dfrac{6000 \times 100}{80} = 7500$
∴ Monthly savings of A = 20% of 7500
$= \dfrac{7500 \times 20}{100}$ = ₹ 1500

13. (c) Let the number be x.
Then, 137.5% of x = 33
∴ $x = \dfrac{33 \times 100}{137.5} = 24$

14. (d) Let the number be x.
Then, $x - 35 = \dfrac{80}{100} x$
$\Rightarrow \quad x - \dfrac{4}{5} x = 35$
$\Rightarrow \quad \dfrac{x}{5} = 35 \Rightarrow x = 175$
∴ Required number $= 175 \times \dfrac{4}{5}$
$= 35 \times 4 = 140$

15. (c) Let a man have ₹ x.
According to the question,
$\dfrac{(100-25)}{100} \times \dfrac{(100-20)}{100} \times x = 480$
$\Rightarrow \quad x = \dfrac{480 \times 100 \times 100}{80 \times 75}$ = ₹ 800

16. (a) Let the 3rd number be 100.
III   II   I
100 150 120
∴ Required percentage $= \dfrac{150}{120} \times 100\% = 125\%$

17. (b) Let number of boys = B
and number of girls = G
Then, 10% of $B = \dfrac{1}{4}$ of G
$\Rightarrow \quad \dfrac{B}{10} = \dfrac{G}{4} \Rightarrow \dfrac{B}{G} = \dfrac{10}{4} = \dfrac{5}{2}$
$\Rightarrow \quad B : G = 5 : 2$

18. (c) Let 1st number be x and 2nd number be y.
According to the question,
48% of x = 60% y
$\Rightarrow \quad x \times \dfrac{48}{100} = y \times \dfrac{60}{100}$
$\Rightarrow \quad \dfrac{x}{y} = \dfrac{60}{100} \times \dfrac{100}{48} = \dfrac{5}{4}$
∴ $x : y = 5 : 4$

19. (c) Let the positive number be x.
According to the question,
$x \times (15 + 20)\% = 126 \Rightarrow x \times 35 = 126 \times 100$
∴ $x = \dfrac{12600}{35} = 360$
So, one-third of the number $= \dfrac{360}{3} = 120$

20. (d) B's marks = C's marks + 5% of 400
= 300 + 20 = 320
Now, A's marks = B's marks + 10% of 400
$= 320 + \dfrac{10}{100} \times 400$
= 320 + 40 = 360

21. (d) Let total number of staff be 100.
∴ Female staff = 40
Male staff = (100 − 40) = 60
Votes cast by females $= \dfrac{40}{100} \times 40 = 16$
Votes cast by males $= \dfrac{60}{100} \times 60 = 36$
Votes cast by both (males + females)
= 16 + 36 = 52
∴ Percentage votes obtained = 52%

22. (d) Total marks scored by Mathew in all
subjects = 42 + 51 + 58 + 35 + 48 = 234
∵ Maximum marks is 60 in any subject.
∴ Maximum marks = 60 × 5 = 300
∴ Percentage of Mathew's marks
$= \dfrac{\text{Marks obtained}}{\text{Maximum marks}} \times 100\%$
$= \dfrac{234}{300} \times 100 = \dfrac{234}{3} = 78\%$

23. (a) Number of sweets received by each
student = 15% of 40 $= \dfrac{40 \times 15}{100} = 6$
∴ Number of sweets received by 40 students
= 40 × 6 = 240
Number of sweets received by each teacher
= 20% of 40 $= \dfrac{40 \times 20}{100} = 8$
∴ Number of sweets received by 5 teachers
= 8 × 5 = 40
∴ Total number of sweets = 240 + 40 = 280

24. (b) Increase in percentage of longer pipe
compared to shorter pipe
$= \dfrac{70 - 30}{30} \times 100\%$
$= \dfrac{40}{30} \times 100\% = \dfrac{400}{3}\%$

25. (a) Let the prices of two articles be $3x$ and $4x$, respectively.
Then, $\dfrac{110\% \text{ of } 3x}{4x + 4} = \dfrac{3}{4} \Rightarrow \dfrac{1.1x}{x+1} = 1$

Percentage / 193

$\Rightarrow \quad 1.1x = x + 1$
$\Rightarrow \quad 0.1x = 1 \Rightarrow x = 10$
Thus, original price of the second article
$= 4 \times 10 = ₹ 40$

26. (c) Let the total number of voters be x.
According to the question,
75% of 98% of 75% of $x = 9261$
[∵ 2% votes were rejected]
$\Rightarrow \dfrac{x \times 75 \times 98 \times 75}{100 \times 100 \times 100} = 9261$
$\therefore x = \dfrac{9261 \times 100 \times 100 \times 100}{75 \times 75 \times 98}$
$= 16800$

27. (a) Last year, number of boys in school = 610
∴ Number of boys in school this year
$= 610 - 610 \times \dfrac{20}{100} = 610 - 122 = 488$
∴ Number of girls in school this year
= Number of boys in school this year × 175%
$= 488 \times \dfrac{175}{100} = 488 \times \dfrac{7}{4} = 122 \times 7 = 854$
Hence, number of girls in school this year is 854.

28. (d) According to the question,
$\dfrac{(100 + P) \times 20}{100} = \dfrac{(100 - P) \times 60}{100}$
$\Rightarrow (100 + P) = (100 - P) \times 3$
$\Rightarrow P + 3P = 300 - 100$
$\Rightarrow P = \dfrac{200}{4} = 50\%$
Then, 50% of 70 $= \dfrac{50 \times 70}{100} = 35$

29. (c) Here, percentage of students getting first class in class X
$= \dfrac{24}{30} \times 100\% = 8 \times 10\% = 80\%$
Percentage of students getting first class in class Y $= \dfrac{28}{35} \times 100 = 80\%$
∴ Both classes have equal percentage of students getting first class.

30. (b) Let total money of man have ₹x.
Money he spend $= 7\dfrac{1}{2}\% = \dfrac{15}{2}\%$
∴ Rest money he has
$= x - x \times \dfrac{15}{2} \times \dfrac{1}{100} = ₹\dfrac{185x}{200}$
Now, according to the question,
$\dfrac{185x}{200} - \dfrac{185x}{200} \times \dfrac{75}{100} = 370$
$\Rightarrow \dfrac{185x}{200}\left(1 - \dfrac{75}{100}\right) = 370$

$\Rightarrow \dfrac{185x}{200} \times \dfrac{25}{100} = 370$
$\Rightarrow \dfrac{185x}{200} \times \dfrac{1}{4} = 370$
$\Rightarrow x = \dfrac{370 \times 4 \times 200}{185}$
$\Rightarrow x = ₹ 1600$

31. (a) Let maximum marks be x.
According to the question,
$780 - 728 = 5\%$ of x
$\Rightarrow 52 = 5\%$ of x
$\therefore x = \dfrac{5200}{5} = 1040$

32. (c) Let the original salary be ₹ x.
Then, increased salary $= \left(\dfrac{110}{100}\right)x = ₹\dfrac{11x}{10}$
∵ He received the same salary even after increment.
Amount of salary he did not recieve
$= \dfrac{11x}{10} - x = ₹\dfrac{x}{10}$
Percentage amount of salary he did not receive $= \left(\dfrac{x}{10} / \dfrac{11x}{10}\right) \times 100$
$= \dfrac{x}{11x} \times 100\% = \dfrac{100}{11}\%$

33. (d) Let Priya's marks in Geography = G
Then, Priya's marks in History = G + 9
According to the question,
Marks in History
= Marks in (History + Geography) $\times \dfrac{56}{100}$
$\Rightarrow G + 9 = (G + 9 + G) \times \dfrac{14}{25}$
$\Rightarrow 25G + 225 = 28G + 126$
$\Rightarrow 28G - 25G = 225 - 126$
$\Rightarrow 3G = 99$
$\Rightarrow G = 33$
Hence, Priya's marks in Geography is 33.

34. (b) After removing 25% of blue balls, total blue balls left = 75% of 100
$= \dfrac{75 \times 100}{100} = 75$
After removing 50% of red balls, total red balls left = 50% of 50 $= \dfrac{50 \times 50}{100} = 25$
∴ Required percentage
$= \dfrac{50}{(75 + 25 + 50)} \times 100$
$= \dfrac{50}{150} \times 100 = 33\dfrac{1}{3}\%$

35. (d) Let usual speed and usual time taken by the jogger be x and t, respectively.
Let his new speed be x'.
Then, $xt = x' \cdot \frac{3}{4}t \Rightarrow x = \frac{3}{4}x' \Rightarrow x' = \frac{4}{3}x$
Thus, he has to increase his speed by
$\dfrac{\frac{4}{3}x - x}{x} \times 100\%$, i.e. $33\frac{1}{3}\%$

36. (d) Let sale in 2008 = 100
Sale in 2009 = 20, Sale in 2010 = 100
∴ Required increase $= \dfrac{100 - 20}{20} \times 100$
$= \dfrac{80}{20} \times 100 = 400\%$

37. (b) Given, $x = 15$, $y = 20$ and $z = 25$
According to the formula,
$z\%$ of $A = \left(\dfrac{yz}{x}\right)\%$ of B [by Technique 1]
$\Rightarrow 25\%$ of $A = \dfrac{20 \times 25}{15}\%$ of B
$= \dfrac{100}{3}\%$ of B
$= 33\dfrac{1}{3}\%$ of B

38. (d) Monthly income of a person = ₹ 5000
Increment in income $= \dfrac{30}{100} \times 5000 = 1500$
∴ New income = 5000 + 1500 = ₹ 6500
Fast Track Method
Here, x = ₹ 5000 and y = 30%
According to the formula,
Required new income $= \dfrac{100 + y}{100} \times x$
[by Technique 2]
$= \dfrac{100 + 30}{100} \times 5000$
$= \dfrac{130}{100} \times 5000 = ₹\ 6500$

39. (b) Given that, x = ₹ 15000 and y = 8
According to the formula,
Required new price $= \dfrac{100 - y}{100} \times x$
[by Technique 2]
$= \dfrac{100 - 8}{100} \times 15000$
$= \dfrac{92}{100} \times 15000 = ₹\ 13800$
Alternate Method
New price = 92% of 15000
$= \dfrac{92}{100} \times 15000 = ₹\ 13800$

40. (b) Given, $a = 25\%$
∴ Required percentage
$= \dfrac{a}{100 + a} \times 100\%$
[by Technique 3(i)]
$= \dfrac{25}{100 + 25} \times 100\%$
$= \dfrac{25}{125} \times 100\% = 20\%$

41. (d) Given, $a = 20\%$
∴ Required percentage
$= \dfrac{a}{100 - a} \times 100\%$
[by Technique 3 (ii)]
$= \dfrac{20}{100 - 20} \times 100\%$
$= \dfrac{20}{80} \times 100\% = 25\%$

42. (d) Let Kamal's marks be 100.
∴ Shantanu's marks = 60
Kamal has 40 marks more than Shantanu.
∴ Required percentage
$= \dfrac{40}{60} \times 100\% = 66\dfrac{2}{3}\%$
Fast Track Method
Here, $a = 40\%$
According to the formula,
Required percentage
$= \left(\dfrac{a}{100 - a} \times 100\right)\%$
[by Technique 3 (iii)]
$= \left(\dfrac{40}{100 - 40} \times 100\right)\% = \left(\dfrac{40}{60} \times 100\right)\%$
$= \dfrac{400}{6}\% = 66\dfrac{2}{3}\%$

43. (d) In such case, there is always decrease.
Given, a = (common increase or decrease) = 7%
According to the formula,
Decreased percentage $= \dfrac{a^2}{100}\%$
[by Technique 4]
$= \dfrac{7^2}{100}\% = 0.49\%$

44. (e) Net effect $= \left[-a - b + \dfrac{(-a)(-b)}{100}\right]\%$
[by Technique 5]
$= \left[-12 - 4 + \dfrac{(-12)(-4)}{100}\right]\%$
$= (-16 + 0.48)\% = -15.52\%$
∴ Loss of 15.52% in revenue.

Percentage / 195

45. (c) Given that, $a = 15\%$ and $b = 10\%$
According to the formula,
Required profit $= \left(a + b + \dfrac{ab}{100}\right)\%$
[by Technique 5]
$= \left(15 + 10 + \dfrac{15 \times 10}{100}\right)\%$
$= (25 + 1.5)\% = 26.5\%$

46. (e) According to the question,
Change in length of rectangle $= -50\% = a$
Change in breadth of rectangle $= +100\% = b$
$\therefore$ Net effect $= \left(a + b + \dfrac{ab}{100}\right)\%$
[by Technique 5]
$= \left(-50 + 100 + \dfrac{(-50)(100)}{100}\right)\% = 0\%$
$\therefore$ Both the areas remains same.

47. (a) Net effect on area
$= \left(a + b + \dfrac{ab}{100}\right)\%$ [by Technique 5]
$= \left[-20 + 10 + \dfrac{(-20)(10)}{100}\right]\%$
$= (-10 - 2)\% = -12\%$
Now, after this mistake, new area
$= (100 - 12)\%$ of 200
$= \dfrac{88}{100} \times 200 = 176$ sq cm

48. (a) Here, $a = 20\%$
Required percentage loss
$= \dfrac{a}{100 + a} \times 100\%$ [by Technique 6]
$= \dfrac{20}{100 + 20} \times 100\%$
$= \dfrac{20}{120} \times 100\% = 16\dfrac{2}{3}\%$

49. (a) Let original price of sugar be ₹ 100.
Then, increased price = ₹ 125
$\therefore$ Reduction in consumption
$= \left(\dfrac{125 - 100}{125} \times 100\right)\%$
$= \left(\dfrac{25}{125} \times 100\right)\% = 20\%$

Fast Track Method
Here, $a = 25\%$
According to the formula,
Reduction in consumption
$= \left(\dfrac{a}{100 + a} \times 100\right)\%$ [by Technique 6]
$= \left(\dfrac{25}{125} \times 100\right)\% = 20\%$

50. (a) Let the maximum marks be x.
According to the question,
$\dfrac{40x}{100} = 40 + 40 \Rightarrow \dfrac{40x}{100} = 80$
$\therefore \quad x = 200$

Fast Track Method
Here, $P = 40\%, R = 40$ and $F = 40$
According to the formula,
Maximum marks $= \dfrac{100(R + F)}{P}$
[by Technique 7]
$= \dfrac{100(40 + 40)}{40} = \dfrac{100 \times 80}{40} = 200$

51. (c) Let the maximum marks be M.
$\therefore$ Passing marks $= x\%$ of
$M + a$ and $y\%$ of $M - b$
Hence, $x\%$ of $M + a = y\%$ of $M - b$
$\Rightarrow \dfrac{xM}{100} + a = \dfrac{yM}{100} - b$
$\Rightarrow M\left(\dfrac{y}{100} - \dfrac{x}{100}\right) = a + b$
$\therefore \quad M = \dfrac{100(a + b)}{y - x}$

52. (c) Students failed in Mathematics $= 34\%$
Students failed in English $= 42\%$
Students failed in both subjects $= 20\%$
$\Rightarrow$ Students failed in either subject
$= (34 + 42 - 20)\% = 56\%$
$\therefore$ Students passed in both subjects
$= (100 - 56)\% = 44\%$

53. (a) Here, $x = 35, y = 600$
$\therefore$ Total number of votes casted
$= \dfrac{100 \times y}{100 - 2x}$ [by Technique 11]
$= \dfrac{100 \times 600}{100 - 2 \times 35} = \dfrac{100 \times 600}{30} = 2000$

54. (d) Population after 2 yr
$= P\left(1 + \dfrac{R}{100}\right)^2$ [by Technique 12 (i)]
$= 250000\left(1 + \dfrac{2}{100}\right)^2$
$= 250000 \times \dfrac{51}{50} \times \dfrac{51}{50} = 260100$
$\therefore$ Growth $= 260100 - 250000 = 10100$

55. (c) Let population in year 2011 be x.
Then, population in year 2012
$= x + \dfrac{x \times 8}{100} = \dfrac{27x}{25}$
Now, population in year 2013
$= \dfrac{27x}{25} + \dfrac{27x}{25} \times \dfrac{15}{100} = \dfrac{621x}{500}$

According to the question,
$$\frac{621x}{500} = 93771$$
$$\Rightarrow x = \frac{93771 \times 500}{621}$$
$$\Rightarrow x = 151 \times 500$$
$$\therefore x = 75500$$

56. (b) Let the population at the beginning of the first year be x.
Then, according to the question,
$$x\left(1 + \frac{5}{100}\right) \times \left(1 - \frac{5}{100}\right) = 47880$$
[by Technique 13]
$$\Rightarrow x \times \frac{105}{100} \times \frac{95}{100} = 47880$$

$$\therefore x = 47880 \times \frac{100}{105} \times \frac{100}{95} = 48000$$

57. (b) Given, $P = 5000$, $R_1 = 10\%$ (increase)
$R_2 = 20\%$ (decrease) and $R_3 = 30\%$ (increase)
$\therefore$ Population at the end of third year
$$= P\left(1 + \frac{R_1}{100}\right)\left(1 - \frac{R_2}{100}\right)\left(1 + \frac{R_3}{100}\right)$$
[by Technique 13]
$$= 5000\left(1 + \frac{10}{100}\right)\left(1 - \frac{20}{100}\right)\left(1 + \frac{30}{100}\right)$$
$$= 5000 \times \frac{110}{100} \times \frac{80}{100} \times \frac{130}{100}$$
$$= 5 \times 11 \times 8 \times 13 = 5720$$

Exercise 2 Higher Skill Level Questions

1. (b) Percentage error $= \frac{1.55 - 1.5}{1.5} \times 100$
$$= \frac{0.05}{1.5} \times 100 = \frac{5 \times 100}{150}$$
$$= \frac{10}{3} = 3.33\%$$

2. (d) According to the question,
$$1 - \left(\frac{1}{4} + \frac{1}{5}\right) = 1 - \frac{9}{20} = \frac{11}{20}$$
Answer he found $= 0.45 = \frac{45}{100} = \frac{9}{20}$
$\therefore$ Required error percentage
$$= \frac{\frac{11}{20} - \frac{9}{20}}{\frac{11}{20}} \times 100 = \frac{\frac{2}{20}}{\frac{11}{20}} \times 100$$
$$= \frac{2}{11} \times 100 = \frac{200}{11}\%$$

3. (c) We have, $a\%$ of $a + b\%$ of $B = 2\%$ of ab
$$\Rightarrow \frac{a}{100} \times a + \frac{b}{100} \times b = \frac{2}{100} \times ab$$
$$\Rightarrow a^2 + b^2 = 2ab$$
$$\Rightarrow a^2 - 2ab + b^2 = 0$$
$$\Rightarrow (a - b)^2 = 0$$
$$\Rightarrow a = b$$
$\therefore$ Required percentage $= \frac{b}{a} \times 100$
$$= \frac{a}{a} \times 100 = 100\%$$

4. (c) Let Mohan's initial amount be ₹x.
He gave to Ram $= x \times \frac{25}{100} = \frac{x}{4}$

Now, according to the question,
$$\frac{x}{4} - \frac{x}{4} \times \frac{20}{100} - \frac{x}{4} \times \frac{35}{10} = 2700$$
$$\Rightarrow \frac{x}{4} - \frac{20x}{400} - \frac{35x}{400} = 2700$$
$$\Rightarrow \frac{100x - 20x - 35x}{400} = 2700$$
$$\Rightarrow \frac{100x - 55x}{400} = 2700$$
$$\Rightarrow \frac{45x}{400} = 2700$$
$$\Rightarrow x = \frac{400 \times 2700}{45} = 24000$$
$$\Rightarrow x = ₹24000$$
Hence, Mohan's initial amount = ₹24000

5. (a) Total salary spent on insurance, shopping and household expenses
$$= 15 + 55 = 70\%$$
$\therefore$ Savings $= 100 - 70 = 30\%$
Let the total salary be x.
$\therefore$ 30% of $x = ₹12750$
$$\Rightarrow \frac{30}{100} \times x = ₹12750$$
$$\therefore x = \frac{12750 \times 100}{30} = ₹42500$$

6. (c) Let Mary's monthly salary be ₹x.
Then, paid on EMI $= x \times \frac{15}{100} = ₹\frac{15x}{100}$
$\therefore$ Rest salary $= x - \frac{15x}{100} = ₹\frac{85x}{100}$
Now, paid on internet bill
$$= \frac{85x}{100} \times \frac{10}{100} = ₹\frac{85x}{1000}$$

Percentage / 197

and paid on rent = $\frac{85x}{100} \times \frac{20}{100} = ₹\frac{170x}{1000}$

Now, according to the question,

$x - \frac{15x}{100} - \frac{85x}{1000} - \frac{170x}{1000} = 24990$

$\Rightarrow \frac{1000x - 150x - 85x - 170x}{1000} = 24990$

$\Rightarrow 1000x - 405x = 1000 \times 24990$

$\Rightarrow 595x = 1000 \times 24990$

$\Rightarrow x = \frac{1000 \times 24990}{595}$

$\Rightarrow x = 1000 \times 42 = 42000$

Hence, Mary's monthly salary = ₹42000

7. (c) Let total amount be ₹ x.
Total expenditure = $(23 + 33 + 19 + 16)\%$ of $x = 91\%$ of x
Remaining money
$= (100 - 91)\%$ of $x = 9\%$ of x
According to the question,
9% of $x = 504$

$\therefore x = \frac{504}{9} \times 100 = ₹5600$

Now, total money (food + insurance)%
$= (23 + 33)\%$ of $x = 56\%$ of x
$= 56\%$ of 5600
$= \frac{56}{100} \times 5600 = ₹3136$

8. (d) Let the marks of first students be x.
Then, marks of second students be $(x + 10)$.
According to the question,

$(x + 10) = \frac{75}{100}(2x + 10)$

$\Rightarrow x + 10 = \frac{3}{4}(2x + 10)$

$\Rightarrow x + 10 = \frac{3}{2}(x + 5)$

$\Rightarrow 2x + 20 = 3x + 15$

$\therefore x = 5$ and $x + 10 = 15$

Hence, the marks obtained by them are 15 and 5.

9. (b) Let the number of boys and girls be x and y respectively.
Total sum of ages of boys = $20x$
and total sum of ages of girls = $16y$

Given that, $\frac{20x + 16y}{x + y} = 18$

$\Rightarrow 20x + 16y = 18x + 18y$

$\Rightarrow 2x = 2y \Rightarrow x = y$

Hence, percentage of boys in group is 50%.

10. (a) Let the original income be ₹ x,
i.e. $I = ₹x$

$\therefore$ Saving income $(s_1) = x \times 10\% = ₹\frac{x}{10}$

Initial expenditure = $x - \frac{x}{10}\% = ₹\frac{9x}{10}$

2 yr later, when income has increased by 20%.

Then, new income = $x \times 120\% = ₹\frac{12x}{10}$

and new saving income $(s_2) = ₹\frac{x}{10}$

$\therefore$ New expenditure = New income − New saving income

$= \frac{12x}{10} - \frac{x}{10} = ₹\frac{11x}{10}$

$\therefore$ Required percentage

$= \frac{\text{New expenditure} - \text{Initial expenditure}}{\text{Initial expenditure}} \times 100\%$

$= \frac{\frac{11x}{10} - \frac{9x}{10}}{\frac{9x}{10}} \times 100\% = \frac{200}{9}\% = 22\frac{2}{9}\%$

11. (e) Let the total population be x.

Then, number of females = $\frac{40}{100} \times x = \frac{4x}{10}$

Hence, number of males = $x - \frac{4x}{10} = \frac{6x}{10}$

Number of literate males = $\frac{75}{100} \times \frac{6x}{10} = \frac{9x}{20}$

According to the question, $\frac{6x}{10} = \frac{9x}{20} + 225$

$\Rightarrow \frac{6x}{10} - \frac{9x}{20} = 225$

$\Rightarrow \frac{12x - 9x}{20} = 225 \Rightarrow \frac{3x}{20} = 225$

$\Rightarrow x = \frac{20 \times 225}{3} = 20 \times 75$

$\Rightarrow x = 1500$

Therefore, number of females

$= \frac{4x}{10} = \frac{4 \times 1500}{10} = 600$

Number of literate females

$= \frac{70}{100} \times 600 = 420$

$\therefore$ Number of illiterate females
$= 600 - 420 = 180$

12. (c) Let selling price of goods by Shyam be ₹ 100.

$\therefore$ Selling price of goods by Ram = ₹ 75

Now, according to the question,
125% of selling price of goods by Hari = ₹ 75

$\therefore$ Selling price of goods by Hari

$= \frac{100}{125} \times 75 = ₹60$

198 / Fast Track Objective Arithmetic

So, Hari's goods are cheaper than Shyam's goods by
$$\frac{40}{100} \times 100\% = 40\%$$

13. (b) Let the original price per egg be x.

Then, increased price = ₹$\left(\frac{130}{100}x\right)$

According to the question,
$$\frac{7.80}{x} - \frac{7.80}{\frac{130x}{100}} = 6$$

$\Rightarrow \quad \dfrac{7.80}{x} - \dfrac{780}{130x} = 6$

$\Rightarrow \quad 1014 - 780 = 6 \times 130x$

$\Rightarrow \quad 780x = 234$

$\therefore \quad x = \dfrac{234}{780} = 0.3$

$\therefore$ Present price per dozen
$$= ₹\left(12 \times \frac{130}{100} \times 0.3\right) = ₹\,4.68$$

14. (c) Total number of students = 480
Percentage of total students passed
$$= 75\% \text{ of total student}$$
$$= \frac{75 \times 480}{100} = 360 \text{ students}$$

Now, using the condition from the question.
Let the number of boys be x.
Then, 70% of x + 85% of (480 − x) = 360

$\Rightarrow \quad \dfrac{70 \times x}{100} + \dfrac{85 \times (480 - x)}{100} = 360$

$\Rightarrow \quad 70x - 85x + 40800 = 36000$

$\Rightarrow \quad 40800 - 36000 = 85x - 70x$

$\Rightarrow \quad 4800 = 15x$

$\Rightarrow \quad x = \dfrac{4800}{15} = 320$

Thus, there are 320 boys who appeared for the examination.

15. (a) Let the original price of rice be ₹x per kg.

Increased price of rice = $\dfrac{120x}{100} = ₹\dfrac{5}{4}x$

According to the question,
$$\frac{400}{x} - \frac{400}{\frac{5}{4}x} = 20$$

$\Rightarrow \quad \dfrac{400}{x} - \dfrac{1600}{5x} = 20$

$\Rightarrow \quad 2000 - 1600 = 100x$

$\Rightarrow \quad x = ₹4 \text{ per kg}$

Now, new price of rice = ₹$\dfrac{5}{4}x$ = ₹$\dfrac{5}{4} \times 4$
$$= ₹5 \text{ per kg}$$

16. (b) Quantity of alcohol in 5 L of solution
$$= \frac{40}{100} \times 5 = 2 \text{ L}$$
Quantity of alcohol in 6 L of solution = 2 L
$\therefore$ Strength of alcohol in new solution
$$= \left(\frac{2}{6} \times 100\right)\% = 33\frac{1}{3}\%$$

17. (d) Old price of wheat = ₹ 24 per kg
New price of wheat = ₹ 27 per kg
$\therefore$ Increase in price of wheat
$$= ₹(27 - 24) = ₹\,3$$
Now, reduction in consumption, so as to kept the expenditure fixed
$$= \frac{3}{27} \times 100 = 11.1\%$$

18. (c) There are 100 persons in which 60 men and 40 women.
Number of men who cleared the qualifying test = $70 \times \dfrac{60}{100} = 42$

Number of women who cleared the qualifying test = $75 \times \dfrac{40}{100} = 30$

Number of men who get success in final test = $\dfrac{80}{100} \times 42 = \dfrac{3360}{100} = 33.6$ men

Number of women who get success in final test = $30 \times \dfrac{70}{100} = 21$ women

$\therefore$ More men cleared the examination than women.

19. (b) 6% of household's income > ₹200000
22% of household's income > ₹100000
Therefore, (22 − 6) = 16% of households income is between ₹100000 to ₹200000.
So, 12% + 990 + 16% + 6% = 100%
$\Rightarrow \quad 990 = (100 - 34)\%$
$\Rightarrow \quad 990 = 66\%$

Now, $16\% = \dfrac{990}{66} \times 16 = 240$

Hence, 240 households earn between ₹100000 and ₹200000.

20. (b) Let Sia's initial money be ₹x.

She gave to her brother = $x \times \dfrac{1}{4} = ₹\dfrac{x}{4}$

According to the question,
$$\frac{x}{4} - \left(\frac{x}{8} \times \frac{1}{4} + \frac{x}{4} \times \frac{1}{4}\right) = 125$$

$\Rightarrow \quad \dfrac{x}{4} - \dfrac{x}{32} - \dfrac{x}{16} = 125$

$\Rightarrow \quad \dfrac{8x - x - 2x}{32} = 125$

Percentage / 199

$\Rightarrow \quad \dfrac{5x}{32} = 125 \Rightarrow x = \dfrac{125 \times 32}{5}$

$\Rightarrow \quad x = 25 \times 32 = 800 \Rightarrow x = ₹800$

Hence, Sia's initial money is ₹800.

21. (c) After paying EMI, total amount with Rashmi $= \dfrac{7}{3} \times 15000 = ₹35000$

$\therefore$ Total salary of March
$= \dfrac{35000}{70} \times 100 = ₹50000$

$\therefore$ Total salary of April
$= 50000 + \dfrac{12}{100} \times 50000$
$= 50000 + 6000 = ₹56000$

22. (a) Let the marks awarded to the student in exam be x.
According to the question,
After re-evaluation his score,

$x - x \times \dfrac{40}{100} = 96 \Rightarrow x - \dfrac{2x}{5} = 96$

$\Rightarrow \quad \dfrac{5x - 2x}{5} = 96 \Rightarrow \dfrac{3x}{5} = 96$

$\Rightarrow \quad x = 5 \times \dfrac{96}{3} = 5 \times 32 \Rightarrow x = 160$

So, a student was awarded by 160 marks in examination.

$\therefore$ Marks lose after re-evaluation
$= 160 - 96 = 64$

23. (c) Let the number of students appeared be x.
Students who cleared all sections = 10%
Students who cleared none of sections = 5%
$\therefore$ Rest students = $x - 15\%$ of $x = 0.85x$
Now, according to the question,

$0.85x \times \dfrac{40}{100} = 1020$

$\Rightarrow \quad x = \dfrac{1020 \times 100}{40 \times 0.85} = 3000$

24. (c) Let the population of village be x.
$\therefore \quad$ Males $= \dfrac{5}{9}x$

$\therefore$ Married males = 30% of $\left(\dfrac{5}{9}x\right)$

$= \dfrac{30}{100} \times \dfrac{5}{9}x = \dfrac{x}{6}$

$\therefore$ Married females = Married males $= \dfrac{x}{6}$

Now, total females $= x - \dfrac{5x}{9} = \dfrac{4}{9}x$

$\therefore$ Unmarried females $= \dfrac{4}{9}x - \dfrac{x}{6} = \dfrac{5}{18}x$

$\therefore$ Required percentage
$= \dfrac{\left(\dfrac{5x}{18}\right) \times 100}{x} = \dfrac{500}{18}\% = \dfrac{250}{9}\% = 27\dfrac{7}{9}\%$

25. (a) Let price of ghee before increment be ₹x.
Consumption = 10 kg
Then, expenditure on ghee = ₹$10x$
After increment,
Expenditure on ghee = 110% of $10x = 11x$
Price of ghee = 132% of $x = x \times \dfrac{132}{100}$

$= \dfrac{33x}{25}$ per kg

$\therefore$ New consumption $= \dfrac{11x \times 25}{33x}$ kg $= 8\dfrac{1}{3}$ kg

26. (b) Let Ridhi's monthly salary be ₹x.
Spend on various expenses = 34% of x
$= x \times \dfrac{34}{100} = ₹\dfrac{34x}{100}$

$\therefore$ Remaining salary $= x - \dfrac{34x}{100} = ₹\dfrac{66x}{100}$

Given to her brother $= \dfrac{1}{6} \times \dfrac{66x}{100} = ₹\dfrac{11x}{100}$

Given to her sister $= \dfrac{2}{3} \times \dfrac{66x}{100} = ₹\dfrac{44x}{100}$

and savings $= \dfrac{66x}{100} - \dfrac{11x}{100} - \dfrac{44x}{100} = ₹\dfrac{11x}{100}$

Now, according to the question,
$\dfrac{44x}{100} - \dfrac{11x}{100} = 10560 \Rightarrow \dfrac{33x}{100} = 10560$

$\Rightarrow \quad x = \dfrac{100 \times 10560}{33}$

$\Rightarrow \quad x = 100 \times 320$

$\Rightarrow \quad x = ₹32000$

$\therefore$ Ridhi's savings $= \dfrac{11x}{100} = \dfrac{11 \times 32000}{100}$
$= ₹3520$

27. (e) Let the number of registered voters be x.
$\therefore$ Number of votes casted $= \dfrac{70}{100} \times x = \dfrac{7x}{10}$

Let A received y votes and B received $\left(\dfrac{7x}{10} - y\right)$ votes.

According to the question,
$y = \left(\dfrac{7x}{10} - y\right) + 400$

$\Rightarrow \quad 2y = \dfrac{7x}{10} + 400$

$\Rightarrow \quad y = \dfrac{7x}{20} + 200$ ...(i)

Also, $y - \dfrac{12.5}{100} \times y = \dfrac{7x}{10} - y$

$\Rightarrow \quad y - \dfrac{y}{8} = \dfrac{7x}{10} - y \Rightarrow \dfrac{15y}{8} = \dfrac{7x}{10}$

$\Rightarrow \quad y = \dfrac{56x}{150}$ ...(ii)

From Eqs. (i) and (ii), we get

$\dfrac{56x}{150} = \dfrac{7x}{20} + 200 \Rightarrow \dfrac{56x}{150} - \dfrac{7x}{20} = 200$

$\Rightarrow \dfrac{112x - 105x}{300} = 200 \Rightarrow 7x = 60000$

$\Rightarrow x = \dfrac{60000}{7} = 8571.4 \approx 8572$

28. (d) Let $A = 100, B = 50$, then
$P = 40\%$ of $100 + 65\%$ of 50
$= \dfrac{40}{100} \times 100 + \dfrac{65}{100} \times 50$
$= 40 + 32.5 = 72.5$
$Q = 50\%$ of $100 + 50\%$ of 50
$= \dfrac{50}{100} \times 100 + \dfrac{50}{100} \times 50$
$= 50 + 25 = 75$
Here, $P < Q$
Again, let $A = 101, B = 100$, then
$P = 40\%$ of $101 + 65\%$ of 100
$= \dfrac{40}{100} \times 101 + \dfrac{65}{100} \times 100$
$= 40.4 + 65 = 105.4$
$Q = 50\%$ of $101 + 50\%$ of 100
$= \dfrac{50}{100} \times 101 + \dfrac{50}{100} \times 100$
$= 50.5 + 50 = 100.5$
Here, $P > Q$
Therefore, none of the relation can be concluded with certainty.

29. (b) Let monthly salary of Suri be ₹ x.
So, Suri gave her mother $= x \times \dfrac{25}{100} = ₹\dfrac{x}{4}$
Rent paid $= \dfrac{3x}{4} \times \dfrac{15}{100} = ₹\dfrac{9x}{80}$
Kept for monthly expenses
$= \dfrac{3x}{4} \times \dfrac{25}{100} = ₹\dfrac{3x}{16}$
∴ Deposit in bank $= \left(x - \dfrac{x}{4}\right) - \left(\dfrac{9x}{80} + \dfrac{3x}{16}\right)$
$= \dfrac{3x}{4} - \dfrac{9x}{80} - \dfrac{3x}{16}$
$= \dfrac{60x - 9x - 15x}{80} = \dfrac{36x}{80} = \dfrac{9x}{20}$
Now, according to the question,
$\dfrac{x}{4} + \dfrac{9x}{20} = 42000$
$\Rightarrow \dfrac{5x}{20} + \dfrac{9x}{20} = 42000 \Rightarrow \dfrac{14x}{20} = 42000$
$\Rightarrow x = \dfrac{20 \times 42000}{14} = 20 \times 3000$
$\Rightarrow x = 60000$
Hence, monthly salary of Suri = ₹60000

30. (d) In 2003, total population of the village
$P = 4800$
40% of females $= \dfrac{40}{100} \times 4800 = 1920$
and number of males
$= 4800 - 1920 = 2880$
In 2004, total population of the village
$= 4800 \times \dfrac{110}{100} = 5280$
Total number of females = 1920
Number of males in 2004 = $5280 - 1920$
$= 3360$
Number of males increased in 2004, then
$2003 = 3360 - 2880 = 480$
$\Rightarrow$ Total increase in percentage of males in 2004 as compared to 2003
$= \dfrac{480}{2880} \times 100 = \dfrac{100}{6} = 16\dfrac{4}{6} = 16\dfrac{2}{3}\%$

31. (b) Let the number of votes casted for B be x.
∴ Votes casted for $A = \dfrac{140}{100}x = 1.4x$
and votes casted for $C = \dfrac{120}{100}x = 1.2x$
According to the question,
$1.4x - 1.2x = 4000$
$\Rightarrow 0.2x = 4000 \Rightarrow x = 20000$
Total voted casted $= x + 1.4x + 1.2x$
$= 3.6x = 3.6 \times 20000 = 72000$
∴ Total number of voters in voting list
$= \dfrac{72000}{90} \times 100 = 80000$

32. (c) Let Puneet has total money ₹ 100.
Money given to his wife $= 100 \times \dfrac{20}{100} = ₹20$
Money kept with himself
$= 100 \times \dfrac{22}{100} = ₹22$
Rest amount $= 100 - (20 + 22)$
$= 100 - (42) = ₹58$
Now, money given to his two sons
$= 58 \times \dfrac{60}{100} = ₹34.80$
Money given to his daughters
$= 58 - 34.80 = ₹23.20$
Now, difference between the amount given to his daughter and kept with himself
$= ₹ 2940$
But, $(23.20 - 22)\% = 2940$
$\Rightarrow 1.20\% = 2940$
∴ $100\% = \dfrac{2940}{1.20} \times 100 = ₹ 245000$
∴ Required money given to his sons
$= 245000 \times \dfrac{34.80}{100} = ₹ 85260$

Chapter 12
Profit and Loss

Profit and loss are the terms related to monetary transactions in trade and business. Whenever a purchased article is sold, then either profit is earned or loss is incurred.

Cost Price (CP) The price at which an article is purchased or manufactured, is called the cost price.

Selling Price (SP) The price at which an article is sold, is called the selling price.

Overhead Charges Such charges are the extra expenditures on purchased goods apart from actual cost price. Overhead charges include freight charges, rent, salary of employees, repairing cost on purchased articles etc.

* If the overhead charges are not specified in the question, then they are not considered.

Profit (SP > CP) When an article is sold at a price more than its cost price, then profit or gain is earned.

Loss (CP > SP) When an article is sold at a price lower than its cost price, then loss is incurred.

Basic Formulae Related to Profit and Loss

1. Profit or gain = SP − CP
2. Loss = CP − SP
3. Gain% = $\dfrac{\text{Gain }(G)}{\text{CP}} \times 100\%$
4. Loss% = $\dfrac{\text{Loss }(L)}{\text{CP}} \times 100\%$
5. SP = $\left(\dfrac{100 + \text{Gain\%}}{100}\right) \times \text{CP}$
6. SP = $\left(\dfrac{100 - \text{Loss\%}}{100}\right) \times \text{CP}$
7. CP = $\left(\dfrac{100}{100 + \text{Gain\%}}\right) \times \text{SP}$
8. CP = $\left(\dfrac{100}{100 - \text{Loss\%}}\right) \times \text{SP}$

MIND IT!

1. Profit and loss percentage are always calculated on cost price unless otherwise stated in the question.
2. If an article is sold at a certain gain (say 45%), then SP = 145% of CP
3. If an article is sold at a certain loss (say 25%), then SP = 75% of CP

Ex. 1 A man buys an article for ₹ 300 and sells it for ₹ 900. Find profit/loss.

Sol. Here, SP > CP
∴ Profit is earned.
According to the formula,
Profit = SP − CP = 900 − 300 = ₹ 600

Ex. 2 Raman purchased a car for ₹ 5 lakh and sold it for ₹ 4 lakh. Find profit/loss in this transaction.

Sol. Here, SP < CP
∴ Loss is incurred in this case.
According to the formula,
Loss = CP − SP = (5 − 4) lakh = ₹ 1 lakh

Ex. 3 A person buys a toy for ₹ 50 and sells it for ₹ 75. What will be his gain per cent?

Sol. Given, CP = ₹ 50 and SP = ₹ 75
∴ Gain = SP − CP = 75 − 50 = ₹ 25
According to the formula,
$$\text{Gain\%} = \frac{\text{Gain}}{\text{CP}} \times 100\% = \frac{25}{50} \times 100\% = 50\%$$

Ex. 4 A person buys a cycle for ₹ 450 but because of certain urgency, he sells it for ₹ 350. Find his loss per cent.

Sol. Given, CP = ₹ 450 and SP = ₹ 350
∴ Loss = CP − SP = 450 − 350 = ₹ 100
According to the formula,
$$\text{Loss\%} = \frac{\text{Loss}}{\text{CP}} \times 100\% = \frac{100}{450} \times 100\% = \frac{200}{9}\% = 22\frac{2}{9}\%$$

Ex. 5 Find the SP, when CP is ₹ 80 and gain is 20%.

Sol. Given, CP = ₹ 80 and gain = 20%
∴ $$SP = \left(\frac{100 + \text{Gain\%}}{100}\right) \times CP$$
$$= \left(\frac{100 + 20}{100}\right) \times 80 = \frac{120}{100} \times 80 = 12 \times 8 = ₹ 96$$

Ex. 6 Find the SP, when CP is ₹ 80 and loss is 20%.

Sol. Given, CP = ₹ 80 and loss = 20%
∴ $$SP = \left(\frac{100 - \text{Loss\%}}{100}\right) \times CP$$
$$= \left(\frac{100 - 20}{100}\right) \times 80 = 8 \times 8 = ₹ 64$$

Ex. 7 Find the CP, when SP is ₹ 40 and gain is 15%.

Sol. Given, SP = ₹ 40 and gain = 15%
∴ $$CP = \left(\frac{100}{100 + \text{Gain\%}}\right) \times SP$$
$$= \left(\frac{100}{100 + 15}\right) \times 40 = \frac{100}{115} \times 40 = ₹ 34.78$$

Profit and Loss / 203

Ex. 8 Find the CP, when SP is ₹ 200 and loss is 35%.

Sol. Given, SP = ₹ 200 and loss = 35%

$$\therefore \quad CP = \left(\frac{100}{100 - Loss\%}\right) \times SP = \left(\frac{100}{100 - 35}\right) \times 200 = \frac{100}{65} \times 200 = ₹ 307.6$$

Ex. 9 A toy is bought for ₹ 150 and sold at a gain of 8%. Find its selling price.

Sol. Given, CP = ₹ 150 and gain = 8%

$$\therefore \quad SP = \left(\frac{100 + Gain\%}{100}\right) \times CP = \frac{108}{100} \times 150 = 54 \times 3 = ₹ 162$$

Ex. 10 A table is bought for ₹1500 and sold at a loss of 6%. Find its selling price.

Sol. Given, CP = ₹ 1500 and loss = 6%

$$\therefore \quad SP = \left(\frac{100 - Loss\%}{100}\right) \times CP = \frac{94}{100} \times 1500 = ₹ 1410$$

Ex. 11 By selling a watch for ₹ 1440, a man losses 10%. At what price should he sell it to gain 10%?

Sol. Given, SP = ₹ 1440 and loss = 10%

$$\therefore \quad CP = \left(\frac{100}{100 - Loss\%}\right) \times SP = \frac{100}{90} \times 1440 = ₹ 1600$$

Now, CP = ₹ 1600 and gain = 10%

$$\therefore \text{ Required } SP = \left(\frac{100 + Gain\%}{100}\right) \times CP = 1600 \times \frac{110}{100} = ₹ 1760$$

Ex. 12 Ravish lost 20% by selling a radio set for ₹ 3072. What per cent will he gain by selling it for ₹ 4080?

Sol. Given, SP = ₹ 3072 and loss = 20%

$$\therefore \quad CP = \left(\frac{100}{100 - Loss\%}\right) \times SP = \frac{100}{80} \times 3072 = ₹ 3840$$

Now, CP = ₹ 3840 and SP = ₹ 4080

$$\therefore \quad Gain = SP - CP = 4080 - 3840 = ₹ 240$$

$$\therefore \quad Gain\% = \frac{Gain}{CP} \times 100\% = \frac{240}{3840} \times 100\% = 6.25\%$$

Ex. 13 A vendor sells apples at 10 for ₹ 1, gaining 40%. How many apples did he buy for ₹ 1?

Sol. Given, SP of 10 apples = ₹ 1 and gain = 40%

$$CP \text{ of 10 apples} = \left(\frac{100}{100 + Gain\%}\right) \times SP = \frac{100}{140} \times 1 = ₹ \frac{5}{7}$$

Since, ₹ $\frac{5}{7}$ yield 10 apples.

$$\therefore ₹ 1 \text{ will yield } 10 \times \frac{7}{5} = 14 \text{ apples}$$

Ex. 14 A grocer buys 160 kg of rice at ₹ 27 per kg and mixes it with 240 kg of rice available at ₹ 32 per kg. At what rate per kg should he sell the mixture to gain 20% on the whole?

Sol. ∵ CP of 400 kg of rice = 160 × 27 + 240 × 32 = 4320 + 7680 = ₹ 12000

Now, CP = ₹ 12000 and gain = 20% (given)

204 / Fast Track Objective Arithmetic

$$\therefore \quad SP = \left(\frac{100 + Gain\%}{100}\right) \times CP = \frac{120}{100} \times 12000 = ₹\,14400$$

$$\therefore \quad SP \text{ per kg} = \frac{14400}{400} = ₹\,36$$

Ex. 15 A man purchases a certain number of apples at 3 per rupee and the same number of apples at 4 per rupee. He mixes them together and sells them at 3 per rupee. What is his gain or loss per cent?

Sol. CP of 1 apple for 1st rate = $₹\,\dfrac{1}{3}$

CP of 1 apple for 2nd rate = $₹\,\dfrac{1}{4}$

CP of 2 apples after mixing = $\dfrac{1}{3} + \dfrac{1}{4} = ₹\,\dfrac{7}{12}$

CP of 1 apple after mixing = $\dfrac{7}{12 \times 2} = ₹\,\dfrac{7}{24}$

SP of 1 apple after mixing = $₹\,\dfrac{1}{3}$

$$\therefore \quad \text{Gain} = \dfrac{1}{3} - \dfrac{7}{24} = \dfrac{8-7}{24} = ₹\,\dfrac{1}{24}$$

$$\therefore \quad \text{Gain\%} = \left(\dfrac{1}{24} \times \dfrac{24}{7} \times 100\right)\% = \dfrac{100}{7}\% = 14\dfrac{2}{7}\%$$

Fast Track
Techniques
to solve the QUESTIONS

Technique 1 If a person sells two different articles at the same selling price, one at a gain of $a\%$ and another at a loss of $a\%$, then the seller always incurs a loss which is given by

$$\text{Loss\%} = \left(\dfrac{a}{10}\right)^2 \%$$

♦ In this case, SP is immaterial.

Ex. 16 A man sold two radios for ₹ 2000 each. On one he gains 16% and on the other he losses 16%. Find his gain or loss per cent in the whole transaction.

Sol. Here, $a = 16\%$

According to the formula,

$$\text{Loss\%} = \left(\dfrac{a}{10}\right)^2\% = \left(\dfrac{16}{10}\right)^2\% = \dfrac{256}{100}\%$$

$$= 2.56\%$$

Profit and Loss / 205

Technique 2 — If ath part of some items is sold at $x\%$ loss, then required gain per cent in selling rest of the items in order that there is neither gain nor loss in whole transaction, is $\dfrac{ax}{1-a}\%$.

Ex. 17 A medical store owner purchased medicines worth ₹ 6000 from a company. He sold 1/3 part of the medicine at 30% loss. On which gain he should sell his rest of the medicines, so that he has neither gain nor loss?

Sol. Given, $a = \dfrac{1}{3}$ and $x = 30\%$

According to the formula,

Required gain% $= \dfrac{ax}{1-a}\% = \dfrac{\frac{1}{3} \times 30}{1 - \frac{1}{3}}\% = \dfrac{10 \times 3}{2}\% = 15\%$

Technique 3 — A businessman sells his items at a profit/loss of $a\%$. If he had sold it for ₹ R more, he would have gained/lost $b\%$. Then,

CP of items $= \dfrac{R}{b \pm a} \times 100$

'−' = When both are either profit or loss
'+' = When one is profit and other is loss.

Ex. 18 A person sold a table at a profit of $6\dfrac{1}{2}\%$. If he had sold it for ₹ 1250 more, he would have gained 19%. Find the CP of the table.

Sol. In this problem, it is clearly stated that for ₹ 1250 more the gain will rise to 19%.

$6\dfrac{1}{2}\%$ profit + ₹ 1250 = 19% profit

$\Rightarrow$ ₹ $1250 = \left(19 - \dfrac{13}{2}\right)\%$ of CP

[∵ profit% is calculated on CP, hence we can write it % of CP]

$\Rightarrow$ ₹ $1250 = \left(\dfrac{38-13}{2}\right)\%$ of CP

$\Rightarrow$ ₹ $1250 = \dfrac{25}{2}\%\left(\text{or }\dfrac{1}{8}\right)$ of CP

∴ CP = ₹ $1250 \times 8 = $ ₹ 10000

Fast Track Method

Here, $a = 6\dfrac{1}{2}\% = \dfrac{13}{2}\%$, $b = 19\%$ and $R = $ ₹ 1250

According to the formula,

CP of table $= \dfrac{R}{b-a} \times 100$ [∵ both are profits]

$= \dfrac{1250}{19 - \dfrac{13}{2}} \times 100$

$= \dfrac{1250 \times 2}{25} \times 100$

$= $ ₹ 10000

Technique 4 If the cost price of 'a' articles is equal to the selling price of 'b' articles, then Profit percentage $= \dfrac{a-b}{b} \times 100\%$

Ex. 19 If the cost price of 20 articles is equal to the selling price of 18 articles, then find the profit per cent.

Sol. Let CP of 1 article = ₹ x
Then, CP of 20 articles = ₹ $20x$ and SP of 18 articles = ₹ $20x$
$\therefore$ SP of 1 article $= \dfrac{20}{18} x = \dfrac{10x}{9}$

Profit $= \dfrac{10x}{9} - x = \dfrac{x}{9}$

$\therefore$ Profit% $= \dfrac{x/9}{x} \times 100 = \dfrac{100}{9} = 11\dfrac{1}{9}\%$

Direct Approach
Since, 2 articles are gained on selling 18 articles.
$\therefore$ Profit% $= \left(\dfrac{2}{18} \times 100\right)\% = 11\dfrac{1}{9}\%$

Fast Track Method
Here, $a = 20$ and $b = 18$
According to the formula,
Profit% $= \left(\dfrac{a-b}{b} \times 100\right)\% = \left(\dfrac{20-18}{18} \times 100\right)\% = \dfrac{100}{9}\% = 11\dfrac{1}{9}\%$

Technique 5 If a man purchases m items for ₹ x and sells n items for ₹ y, then profit or loss per cent is given by $\dfrac{my - nx}{nx} \times 100\%$.
[here, positive result means profit and negative result means loss]

Ex. 20 If Karan purchases 10 oranges for ₹ 25 and sells 9 oranges for ₹ 25, then find the gain percentage.

Sol. CP of 10 oranges = ₹ 25
Then, CP of 1 orange = ₹ $\dfrac{25}{10}$
Now, SP of 9 oranges = ₹ 25
$\therefore$ SP of 1 orange = ₹ $\dfrac{25}{9}$

$\therefore$ Gain% $= \dfrac{SP - CP}{CP} \times 100\% = \dfrac{\dfrac{25}{9} - \dfrac{25}{10}}{\dfrac{25}{10}} \times 100\%$

$= \dfrac{25 \times 10 - 25 \times 9}{9 \times 10} \times \dfrac{10}{25} \times 100\%$

$= \dfrac{25}{9 \times 25} \times 100\% = \dfrac{100}{9}\% = 11\dfrac{1}{9}\%$

Direct Approach
Let the total number of oranges be 90 (LCM of 10 and 9).
Then, SP of 90 oranges = 25×10 = ₹ 250
and CP of 90 oranges = 25×9 = 225

Profit and Loss / 207

$$\therefore \text{Profit per cent} = \frac{SP - CP}{CP} \times 100\% = \frac{250 - 225}{225} \times 100\%$$
$$= \frac{25}{225} \times 100\% = \frac{100}{9}\% = 11\frac{1}{9}\%$$

Fast Track Method
Here, $m = 10$, $x = 25$, $n = 9$ and $y = 25$
$$\therefore \text{Profit per cent} = \frac{my - nx}{nx} \times 100\% = \frac{25 \times 10 - 9 \times 25}{9 \times 25} \times 100\%$$
$$= \frac{250 - 225}{225} \times 100\% = \frac{25}{225} \times 100\% = \frac{100}{9}\% = 11\frac{1}{9}\%$$

Ex. 21 Shakshi bought pens at a rate of 10 pens for ₹ 11 and sold them at a rate of 11 pens for ₹ 10. Then, find the profit or loss per cent.

Sol. Given, CP of 10 pens = ₹ 11. Then, CP of 1 pen = ₹ $\frac{11}{10}$

Similarly, SP of 11 pens = ₹ 10

$\therefore$ SP of 1 pen = ₹ $\frac{10}{11}$

Here, SP < CP

$$\therefore \text{Loss per cent} = \frac{CP - SP}{CP} \times 100\% = \frac{\frac{11}{10} - \frac{10}{11}}{\frac{11}{10}} \times 100\%$$
$$= \frac{121 - 100}{10 \times 11} \times \frac{10}{11} \times 100\% = \frac{21}{121} \times 100\% = \frac{2100}{121}\% = 17\frac{43}{121}\%$$

Fast Track Method
Here, $m = 10$, $x = 11$, $n = 11$ and $y = 10$
$$\therefore \text{Profit or loss per cent} = \frac{my - nx}{nx} \times 100\% = \frac{10 \times 10 - 11 \times 11}{11 \times 11} \times 100\%$$
$$= \frac{-21}{121} \times 100\% = -\frac{2100}{121}\%$$
$$= -17\frac{43}{121}\% \qquad [\because \text{negative sign indicates a loss}]$$

Technique 6 If A sold an article to B at a profit (loss) of $r_1\%$ and B sold this article to C at a profit (loss) of $r_2\%$, then cost price of article for C is given by

$$\text{Cost price for } A \times \left(1 \pm \frac{r_1}{100}\right)\left(1 \pm \frac{r_2}{100}\right).$$

✦ Use positive sign for profit and negative sign for loss.

Ex. 22 Nikunj sold a machine to Sonia at a profit of 30%. Sonia sold this machine to Anu at a loss of 20%. If Nikunj paid ₹ 5000 for this machine, then find the cost price of machine for Anu.

Sol. $\because$ Nikunj sold the machine for a profit of 30%.

$\therefore$ CP of machine for Sonia = 130% of CP of machine for Nikunj = $\frac{130 \times 5000}{100}$ = ₹ 6500

Now, Sonia sold the machine to Anu at a loss of 20%.

$\therefore$ CP of machine for Anu = 80% of CP of machine for Sonia = $\frac{80 \times 6500}{100}$ = ₹ 5200

Hence, CP of machine for Anu is ₹ 5200.

Fast Track Method

Here, $r_1 = 30\%$ (profit) and $r_2 = 20\%$ (loss)

CP of a machine for Nikunj = ₹ 5000

$\therefore$ CP of machine for Anu = CP of machine for Nikunj $\left(1 + \dfrac{r_1}{100}\right)\left(1 - \dfrac{r_2}{100}\right)$

$= 5000\left(1 + \dfrac{30}{100}\right)\left(1 - \dfrac{20}{100}\right) = 5000 \times \dfrac{130}{100} \times \dfrac{80}{100} = ₹ 5200$

Technique 7 If a dishonest trader professes to sell his items at CP but uses false weight, then

$$\text{Gain\%} = \dfrac{\text{Error}}{\text{True value} - \text{Error}} \times 100\% = \dfrac{\text{True weight} - \text{False weight}}{\text{False weight}} \times 100\%$$

◆ Here, while calculating gain or profit per cent, we have taken false weight as a base. Because CP is what is paid when an item is purchased or manufactured. In this case, dishonest trader is telling false weight to be the CP and he is gaining only when sells at false weight.

Ex. 23 A dishonest dealer professes to sell his goods at cost price but he uses a weight of 930 g for 1 kg weight. Find his gain per cent.

Sol. 70 g is gained on 930 g.

$\therefore \quad \text{Gain\%} = \left(\dfrac{70}{930} \times 100\right)\% = 7\dfrac{49}{93}\%$

Fast Track Method

According to the formula,

$\text{Gain\%} = \dfrac{\text{Error}}{\text{True value} - \text{Error}} \times 100\% = \dfrac{70}{1000 - 70} \times 100\% = \dfrac{70}{930} \times 100\% = 7\dfrac{49}{93}\%$

Technique 8 If a shopkeeper sells his goods at $a\%$ loss on cost price but uses b g instead of c g, then his percentage profit or loss is $\left[(100 - a)\dfrac{c}{b} - 100\right]\%$.

◆ Positive result indicates profit and negative result indicates loss.

Ex. 24 A dealer sells goods at 6% loss on cost price but he uses 14 g instead of 16 g. What is his percentage profit or loss?

Sol. Let the CP of 16g of goods = ₹ 100

$\therefore$ CP of 14g of goods = $\dfrac{100}{16} \times 14 = ₹ 87.5$

Now, SP of 16g of goods = $\dfrac{94}{100} \times 100 = ₹ 94$

Required profit percentage = $\left(\dfrac{94 - 87.5}{87.5}\right) \times 100 = \dfrac{6.5}{87.5} \times 100 = 7\dfrac{3}{7}\%$

Fast Track Method

Given, $a = 6\%$, $b = 14$ g and $c = 16$ g

According to the formula,

Required answer = $\left[(100 - 6)\dfrac{16}{14} - 100\right]\% = \left(94 \times \dfrac{8}{7} - 100\right)\%$

$= \left(\dfrac{752 - 700}{7}\right)\% = \dfrac{52}{7}\%$

$= 7\dfrac{3}{7}\%$ gain   ['+ve' sign shows that there is a gain]

Profit and Loss / 209

Technique 9 — If a dealer sells his goods at *a*% profit or loss on cost price and uses *b*% less weight, then his percentage profit or loss will be $\dfrac{(b \pm a)}{100 - b} \times 100\%$.

✦ Positive result indicates profit and negative result indicates loss.

Ex. 25 A dealer sells his goods at 20% loss on cost price but uses 40% less weight. What is his percentage profit or loss?

Sol. Let the marked weight = 1 kg = 1000 g
Real weight = 60% of 1000 g = 600 g
Let CP of 1 g = ₹ 1
∴ CP of 600 g = ₹ 600
and CP of 1000 g = ₹ 1000
Then, $\quad$ SP = $\dfrac{80}{100} \times 1000$ = ₹ 800

Now, required profit percentage = $\dfrac{(800 - 600)}{600} \times 100$

$= \dfrac{2}{6} \times 100 = 33\dfrac{1}{3}\%$

Fast Track Method

Given, $a = 20\%$ and $b = 40\%$
According to the formula,

Required answer = $\dfrac{(b-a)}{100-b} \times 100\% = \dfrac{(40-20)}{60} \times 100\% = \dfrac{2}{6} \times 100\%$

$= \dfrac{100}{3}\% = 33\dfrac{1}{3}\%$ profit

Technique 10 — If '*a*' part of an article is sold at *x*% profit/loss, '*b*' part at *y*% profit/loss and *c* part at *z*% profit/loss and finally there is a profit/loss of ₹ *R*, then

Cost price of entire article = ₹ $\dfrac{R \times 100}{ax + by + cz}$, where *x* or *y* or *z* is negative, if it indicates a loss.

Ex. 26 If $\dfrac{2}{3}$ part of an article is sold at 30% profit, $\dfrac{1}{4}$ part at 16% profit and remaining part at 12% profit and finally, there is a profit of ₹ 75, then find the cost price of an article.

Sol. Let CP of article be $12x$.

Now, SP of $\dfrac{2}{3}$ part = $\dfrac{130}{100} \times \dfrac{2}{3}(12x)$

and SP of $\dfrac{1}{4}$ part = $\dfrac{116}{100} \times \dfrac{1}{4}(12x)$

∴ SP of remaining part = $\dfrac{112}{100}\left(\dfrac{1}{12}\right)(12x)$

Now, profit = SP − CP

$\Rightarrow \quad \dfrac{130}{100} \times 8x + \dfrac{116}{100} \times 3x + \dfrac{112}{100}x - 12x = 75$

$\Rightarrow \quad 3x = 75$ or $x = 25$

∴ $\quad$ CP of article = $12x = 12 \times 25$ = ₹ 300

Fast Track Method

Here, $a = \dfrac{2}{3}$, $x = 30\%$, $b = \dfrac{1}{4}$, $y = 16\%$, $z = 12\%$ and $R = ₹\,75$

Then, remaining part $(c) = 1 - \left(\dfrac{2}{3} + \dfrac{1}{4}\right) = \dfrac{1}{12}$

$\therefore$ Cost price of entire article $= \dfrac{R \times 100}{ax + by + cz}$

$= \dfrac{75 \times 100}{\dfrac{2}{3} \times 30 + \dfrac{1}{4} \times 16 + \dfrac{1}{12} \times 12}$

$= \dfrac{75 \times 100}{20 + 4 + 1} = \dfrac{7500}{25} = ₹\,300$

Technique 11 If there are two successive profits or losses at $a\%$ and $b\%$ respectively, then a resultant profit or loss $= \left(a + b + \dfrac{ab}{100}\right)\%$, where a or b is negative, if it indicates a loss.

✦ If result is positive (+), then there is a profit and if result is a negative (−), then there is a loss.

Ex. 27 The price of a commodity is diminished by 15% and its demand increases by 20%. Find the resultant profit/loss.

Sol. Let the sales of the commodity be ₹ 100.

Then, new price $= \dfrac{(100 - 85)}{100} \times 100 = ₹\,85$

Now, demand increases by 20%

$\therefore$ Sales $= \dfrac{120}{100} \times 85 = ₹\,102$

Resultant profit percentage $= \dfrac{(102 - 100)}{100} \times 100 = 2\%$

Fast Track Method

Here $a = -15\%$, $b = 20\%$

$\therefore$ Resultant profit $= \left(a + b + \dfrac{ab}{100}\right)\%$

$= -15 + 20 + \dfrac{(-15) \times 20}{100}$

$= -15 + 20 - 3 = 2\%$

Fast Track Practice

Exercise 1 — Base Level Questions

1. Anita purchased a bicycle at a cost of ₹ 3200. She sold it at a loss of ₹ 240. At what price did she sell the bicycle? **[IOB Clerk 2011]**
   (a) ₹ 2960 (b) ₹ 2690
   (c) ₹ 3440 (d) ₹ 3360
   (e) None of these

2. A person buys a book for ₹ 200 and sells it for ₹ 225. What will be his gain per cent? **[SSC LDC 2011]**
   (a) 13% (b) 14% (c) 18.4% (d) 12.5%

3. A person buys a watch for ₹ 500 and sells it for ₹ 300. Find his loss per cent.
   (a) 30% (b) 40% (c) 35% (d) 45%
   (e) None of these

4. A gold bracelet is sold for ₹ 14500 at a loss of 20%. What is the cost price of the gold bracelet? **[SBI PO 2012]**
   (a) ₹ 18125 (b) ₹ 17400
   (c) ₹ 15225 (d) ₹ 16800
   (e) None of these

5. If a shopkeeper purchases cashew nut at ₹ 250 per kg and sells it at ₹ 10 per 50 g, then he will have **[MAT 2014]**
   (a) 20% profit (b) 25% profit
   (c) 20% loss (d) 25% loss

6. Find the percentage loss when the cost price and selling price of an article are in the ratio of 5 : 3.
   (a) 40% (b) 35% (c) 45% (d) 26%
   (e) None of these

7. If the cost price and selling price of an article are in the ratio of 10 : 11, then the percentage profit is **[SSC CGL 2010]**
   (a) 10% (b) 9% (c) 3% (d) 1%

8. The ratio of cost price and selling price of an article 25 : 26. The per cent of profit will be **[SSC CGL (Mains) 2016]**
   (a) 26% (b) 25% (c) 1% (d) 4%

9. The cost price of an item is two-third of its selling price. What is the gain or loss per cent on that item? **[Bank Clerk 2010]**
   (a) 45% (b) 50% (c) 35% (d) 54%
   (e) None of these

10. A trader buys some goods for ₹ 150. If the overhead expenses be 12% of the cost price, then at what price should it be sold to earn 10%? **[RRB ASM 2007]**
    (a) ₹ 184.80 (b) ₹ 185.80
    (c) ₹ 187.8 (d) ₹ 188.80

11. Kamlesh purchased 120 reams of paper at ₹ 100 per ream and the expenditure on transport was ₹ 480. He had to pay an octroi duty of 50 paise per ream and the coolie charges were ₹ 60. What should he charge per ream to gain 40%?
    (a) ₹ 155 (b) ₹ 147 (c) ₹ 138 (d) ₹ 165
    (e) None of these

12. A calculator is bought for ₹ 350 and sold at a gain of 15%. What will be the selling price of calculator? **[Hotel Mgmt. 2010]**
    (a) ₹ 385 (b) ₹ 375
    (c) ₹ 472 (d) ₹ 402.50

13. A salesman expects a gain of 13% on his cost price. If in a month, his sale was ₹ 791000, then what was his profit? **[SSC (10+2) 2012]**
    (a) ₹ 91000 (b) ₹ 97786
    (c) ₹ 85659 (d) ₹ 88300

14. The owner of a furniture shop charges his customer 18% more than the CP. If a customer paid ₹ 10207 for a dining table, then find its original price. **[LIC ADO 2010]**
    (a) ₹ 9240 (b) ₹ 8650
    (c) ₹ 9840 (d) ₹ 7670
    (e) None of these

15. A man sells calculator at the rate of ₹ 250 each which includes a profit of 14%. What amount of profit will he earn in 19 days, if he sells seven calculators per day? **[IDBI SO 2012]**
    (a) ₹ 4665 (b) ₹ 4565 (c) ₹ 4545 (d) ₹4655
    (e) None of these

16. Meera purchased 23 bracelets at the rate of ₹ 160 per bracelet. At what rate per bracelet should she sell the bracelets, so that profit earned is 15%? **[SBI Clerk 2012]**
    (a) ₹ 184 (b) ₹ 186 (c) ₹ 192 (d) ₹ 198
    (e) None of these

17. A rice trader buys 22 quintals of rice for ₹ 3344. 24% rice is lost in transportation. At what rate should he sell to earn 30% profit? **[SSC (10+2) 2017]**
    (a) ₹ 88.86 per quintal
    (b) ₹ 197.6 per quintal
    (c) ₹ 269.2 per quintal
    (d) ₹ 260 per quintal

18. A man bought 30 defective machines for ₹ 1000. He repaired and sold them at ₹ 300 per machine. He got profit of ₹ 150 per machine. How much did he spend on repairs? **[SSC CGL (Pre) 2016]**
    (a) ₹ 5500   (b) ₹ 4500
    (c) ₹ 3500   (d) ₹ 2500

19. A woman bought eggs at ₹ 30 per dozen. The selling price per hundred so as to gain 12% will be
    (a) ₹ 280   (b) ₹ 250
    (c) ₹ 300   (d) ₹ 360
    (e) None of these

20. The difference between the CP and SP of an article is ₹ 240. If the profit is 20%, then the selling price is **[SSC (10+2) 2011]**
    (a) ₹ 1440   (b) ₹ 1400
    (c) ₹ 1240   (d) ₹ 1200

21. The owner of a cell phone shop charges his customer 28% more than the cost price. If the customer paid ₹ 8960 for the cell phone, then what was the cost price of the cell phone? **[OBC Clerk 2009]**
    (a) ₹ 7800   (b) ₹ 7000
    (c) ₹ 6900   (d) ₹ 6850
    (e) None of these

22. A man loses 10% by selling an article for ₹ 180. At what price should he sell it to gain 10%? **[SSC FCI 2012]**
    (a) ₹ 220 (b) ₹ 217.80 (c) ₹ 200 (d) ₹ 216

23. By selling a cell phone for ₹ 2400, a shopkeeper makes a profit of 25%. Then, his profit percentage, if he had sold it for ₹ 2040, is **[SSC (10+2) 2012]**
    (a) 10%   (b) 6.25%   (c) 6.5%   (d) 15%

24. A man sold an article for ₹ 322, gaining 1/6th of his outlay. Find the cost price of the article.
    (a) ₹ 300 (b) ₹ 376 (c) ₹ 175 (d) ₹ 276
    (e) None of these

25. If a saree is sold for ₹ 2880 the seller will face 10% loss, at what price should he sell to gain 20% profit? **[SSC CGL (Pre) 2017]**
    (a) ₹ 4830   (b) ₹ 3840
    (c) ₹ 3480   (d) ₹ 4380

26. Rajdeep loses 20% by selling a radio for ₹ 768. What per cent will he gain by selling it for ₹ 1020?
    (a) 7.25%   (b) 5.25%
    (c) 6.25%   (d) 8.25%
    (e) None of these

27. Ravi sells an article at a gain of $12\frac{1}{2}$%. If he had sold it at ₹ 22.50 more, he would have gained 25%. The cost price of the article is **[RRB Clerk (Pre) 2017]**
    (a) ₹ 162 (b) ₹ 140 (c) ₹ 196 (d) ₹ 180
    (e) None of these

28. The profit earned after selling a pair of shoes for ₹ 2033 is same as the loss incurred after selling the same pair of shoes for ₹ 1063. What is the cost price of the shoes? **[IOB Clerk 2009]**
    (a) ₹ 1650   (b) ₹ 1548   (c) ₹ 1532
    (d) Cannot be determined
    (e) None of these

29. The profit earned after selling an article for ₹ 625 is same as the loss incurred after selling the article for ₹ 435. The cost price of the article is **[SSC (10+2) 2012]**
    (a) ₹ 520 (b) ₹ 530 (c) ₹ 540 (d) ₹ 550

30. The profit earned by a shopkeeper by selling a bucket at a gain of 8% is ₹ 28 more than when he sells it at a loss of 8%. The cost price of the bucket is **[SSC CGL (Mains) 2016]**
    (a) ₹ 170   (b) ₹ 190
    (c) ₹ 175   (d) ₹ 165

31. A person sold an article for ₹ 3600 and got a profit of 20%. Had he sold the article for ₹ 3150, how much profit would he have got? **[CDS 2013]**
    (a) 4%   (b) 5%   (c) 6%   (d) 10%

32. Neeta got profit of 10% on selling an article in ₹ 220. To get the profit of 30%, she should sell the article in how many rupees? **[SSC CGL 2012]**
    (a) ₹ 220 (b) ₹ 230 (c) ₹ 260 (d) ₹ 280

33. An article is sold for ₹ 300 at a profit of 20%. Had it been sold for ₹ 235, the loss percentage would have been **[SSC CGL 2013]**
    (a) 3%   (b) 5%   (c) 6%   (d) 16%

34. Charu purchased a dinner set at 3/10th of its selling price and sold it at 10% more than its CP. Find the gain per cent. **[Hotel Mgmt. 2008]**
    (a) 15%   (b) 5%
    (c) 9%   (d) 10%

Profit and Loss / 213

35. A person sold an article for ₹ 136 and got 15% loss. Had he sold it for ₹ x, he would have got a profit of 15%. Which one of the following is correct? **[CDS 2012 (I)]**
(a) 190 < x < 200 (b) 180 < x < 190
(c) 170 < x < 180 (d) 160 < x < 170

36. A vendor sells lemons at the rate of 5 for ₹ 14, gaining thereby 40%. For how much did he buy a dozen lemon? **[SSC CGL 2010]**
(a) ₹ 20 (b) ₹ 21 (c) ₹ 24 (d) ₹ 28

37. Meena purchased two fans each at ₹ 1200. She sold one fan at the loss of 5% and other at the gain of 10%. Find total gain or loss per cent. **[SBI Clerk 2009]**
(a) 1.2% loss (b) 1.2% gain
(c) 2.5% gain (d) 2.5% loss
(e) None of these

38. A man gains 10% by selling an article for a certain price. If he sells it at double the price, then the profit made is **[SSC CGL 2013]**
(a) 120% (b) 20% (c) 40% (d) 100%

39. A man purchases a certain number of oranges at 4 a rupee and the same number of oranges at 5 a rupee. He mixes them together and sells them at 4 a rupee. What is his gain or loss per cent?
(a) $11\frac{2}{9}$% gain (b) $11\frac{2}{9}$% loss
(c) $11\frac{1}{9}$% gain (d) $11\frac{1}{9}$% loss

40. Two lots of onions with equal quantity, one costing ₹ 10 per kg and the other costing ₹ 15 per kg, are mixed together and whole lot is sold at ₹ 15 per kg. What is the profit or loss? **[CDS 2013 (I)]**
(a) 10% loss (b) 10% profit
(c) 20% profit (d) 20% loss

41. A producer of tea blends two varieties of tea from tea gardens one costing ₹ 18 per kg and another ₹ 20 per kg in the ratio 5 : 3. If he sells the blended variety at ₹ 21 per kg, then his gain percentage is **[SSC FCI 2012]**
(a) 18% (b) 8% (c) 10% (d) 12%

42. A dealer sold three-fourth of his articles at a gain of 24% and the remaining at the cost price. Percentage of gain in the whole transaction is **[SSC CPO 2011]**
(a) 15 (b) 18 (c) 24 (d) 32

43. Mr. Kapoor purchased two toy cycles for ₹ 750 each. He sold these cycles, gaining 6% on one and losing 4% on the other. The gain or loss per cent in the whole transaction is **[SSC CGL (Mains) 2016]**
(a) 1% loss (b) 1% gain
(c) 1.5% loss (d) 1.5 gain

44. Pankaj purchased an item for ₹ 7500 and sold it at the gain of 24%. From that amount he purchased another item and sold it at the loss of 20%. What is his over all gain/loss? **[Union Bank Clerk 2011]**
(a) Loss of ₹ 140
(b) Gain of ₹ 60
(c) Loss of ₹ 60
(d) Neither gain nor loss
(e) None of the above

45. Raj sold an item for ₹ 6384 and incurred a loss of 30%. At what price should he have sold the item to have gained a profit of 30%? **[SBI Clerk 2011]**
(a) ₹ 14656 (b) ₹ 11856 (c) ₹ 13544
(d) Cannot be determined
(e) None of these

46. Gopal bought a cell phone and sold it to Ram at 10% profit. Then, Ram wanted to sell it back to Gopal at 10% loss. What will be Gopal's position if he agreed? **[UPSC CSAT 2017]**
(a) Neither loss nor gain
(b) Loss 1%
(c) Gain 1%
(d) Gain 0.5%

47. The price of a land passing through three hands, rises on the whole by 65%. If the first and second sellers earned 20% and 25% profit, respectively. Find the profit earned by third seller. **[SSC (10+2) 2007]**
(a) 20% (b) 55% (c) 10% (d) 25%

48. If the difference between the selling prices of an article at profit of 6% and 4% is ₹ 3, then the cost price of the article should be **[SSC (10+2) 2010]**
(a) ₹ 100 (b) ₹ 150 (c) ₹ 175 (d) ₹ 200

49. A person sold his watch for ₹ 75 and got a percentage profit equal to the cost price. The cost price of the watch is **[SSC Multitasking 2014]**
(a) ₹ 40 (b) ₹ 45 (c) ₹ 50 (d) ₹ 55

50. A fruit seller buys 700 oranges at the rate of ₹ 500 for 100 oranges and another variety of 500 oranges at the rate of ₹ 700 for 100 oranges and sells them at ₹ 84 per dozen. The profit per cent is **[SSC Multitasking 2014]**
(a) 20% (b) 40% (c) 30% (d) 10%

51. By selling 32 oranges for ₹ 30 a man loses 25%. How many oranges should be sold for ₹ 24 so as to gain 20% in the transaction?
   (a) 16   (b) 24   (c) 32   (d) 28
   (e) None of these

52. By selling 100 pens, a shopkeeper gains the selling price of 40 pens. Find his gain per cent.
   (a) 4.47% (b) 6.67% (c) 8.8% (d) 5.59%
   (e) None of these

53. A furniture seller sells two tables at ₹ 1500 each. He earned a profit of 20% on one table and suffered a loss of 20% on the another table. Net profit or loss in this deal is   [CGPSC 2013]
   (a) 4% loss   (b) 4% profit
   (c) Neither profit nor loss
   (d) 10% loss   (e) 10% profit

54. A man sold two houses for ₹ 96000 each. In the sale of the first house, he incurred 20% profit and in the sale of the second, he incurred 20% loss. What is the gain or loss percentage in total?   [SSC CGL (Mains) 2012]
   (a) 6% gain   (b) 6% loss
   (c) 4% gain   (d) 4% loss

55. A shopkeeper purchased some books from a publication worth ₹ 750. Because of some reasons, he had to sell two-fifth part of the book at a loss of 15%. On which gain he should sell his rest of the books, so that he gets neither gain nor loss?
   (a) 10% (b) 9% (c) 12% (d) 15%
   (e) 18%

56. A man sold his watch at a loss of 5%. Had he sold it for ₹ 56.25 more, he would have gained 10%. What is the cost price (in ₹) of the watch?   [SSC (10+2) 2014]
   (a) ₹ 370 (b) ₹ 365 (c) ₹ 375 (d) ₹ 390

57. A man sold an article at a loss of 20%. If he sells the article for ₹ 12 more, he would have gained 10%. The cost price of the article is
   (a) ₹ 60 (b) ₹ 40 (c) ₹ 30 (d) ₹ 22
   (e) None of these

58. A person sold a watch at a profit of 10%. If he had sold it for ₹ 2000 more, he would have gained 20%. Find the CP of watch.
   (a) ₹ 15000   (b) ₹ 10000
   (c) ₹ 20000   (d) ₹ 25000

59. If an article is sold at a gain of 6% instead of at a loss of 6%, then the seller gets ₹ 6 more. The cost price of the article is   [SSC CGL 2013]
   (a) ₹ 106 (b) ₹ 50 (c) ₹ 94 (d) ₹ 100

60. The selling price of 20 articles is equal to the cost price of 22 articles. The gain percentage is   [SSC CGL 2013]
   (a) 12% (b) 9% (c) 10% (d) 11%

61. If the cost price of 16 tables is equal to the selling price of 12 tables, then the gain per cent is   [SSC Multitasking 2014; SSC CPO 2013]
   (a) $33\frac{1}{3}$% (b) 20% (c) 30% (d) 15%

62. A merchant buys p apples for ₹ q and sells q apples for ₹ p. If $p < q$, then in the whole outlay, he makes   [SSC CGL 2007]
   (a) $10\left(\frac{p^2+q^2}{q^2}\right)$% gain
   (b) $100\left(\frac{q^2-p^2}{q^2}\right)$% loss
   (c) $100\left(\frac{p^2+q^2}{p^2}\right)$% loss
   (d) $100\left(\frac{q^2-p^2}{p^2}\right)$% gain

63. A fruit-seller buys lemons at 2 for a rupee and sells them at 5 for three rupees. What is his gain per cent?   [CDS 2011 (I)]
   (a) 10%   (b) 15%
   (c) 20%   (d) 25%

64. Some apples are bought at 5 for ₹ 10 and sold at 6 for ₹ 15. What is the gain per cent?
   (a) 35% (b) 45% (c) 20% (d) 25%
   (e) None of these

65. A person buys some pencils at 5 for ₹ 1 and sells them at 3 for ₹ 1. Its gain per cent will be   [SSC CGL 2016]
   (a) $66\frac{2}{3}$%   (b) $76\frac{2}{3}$%
   (c) $56\frac{2}{3}$%   (d) $46\frac{2}{3}$%

66. The retail price of a water geyser is ₹1265. If the manufacturer gains 10%, the wholesale dealer gains 15% and the retailer gains 25%, then the cost of the product is   [RRB Clerk (Pre) 2017]
   (a) ₹ 800 (b) ₹ 900 (c) ₹ 700 (d) ₹ 600
   (e) None of these

67. A sold a watch to B at 40% gain and B sold it to C at a loss of 20%. If C bought the watch for ₹ 432, at what price did A purchase it?
   (a) ₹ 385.71 (b) ₹ 216 (c) ₹ 250 (d) ₹ 550

Profit and Loss / 215

68. A dishonest dealer professes to sell his goods at cost price but he uses a weight of 920 g for 1 kg weight. Find his gain per cent.
(a) $7\frac{16}{23}$% (b) $8\frac{16}{23}$% (c) $5\frac{16}{23}$% (d) $3\frac{16}{23}$%
(e) None of these

69. A dealer sells goods at 4% loss on cost price but he uses 28 g instead of 32 g. What is his per cent profit or loss?
(a) $9\frac{5}{7}$% gain (b) $14\frac{3}{7}$% loss
(c) $16\frac{3}{7}$% gain (d) $16\frac{3}{7}$% loss
(e) None of these

70. A dishonest dealer sells articles at 10% loss on cost price but uses the weight of 16 g instead of 18 g. What is his profit or loss per cent?
(a) $1\frac{1}{4}$% gain (b) $1\frac{1}{4}$% loss
(c) $3\frac{1}{4}$% loss (d) $5\frac{1}{4}$% gain
(e) None of these

71. A dishonest dealer sells his goods at 10% loss on cost price but uses 20% less weight. What is his profit or loss per cent?
(a) 12% loss (b) 22.5% gain
(c) 13.9% loss (d) 12.5% gain
(e) None of these

72. A man sells rice at 10% profit and uses weight 30% less than the actual measure. His gain per cent is
(a) $57\frac{2}{8}$% (b) $57\frac{1}{7}$%
(c) $57\frac{2}{5}$% (d) $57\frac{3}{7}$%
(e) None of these

73. A trader sells wheat at 20% profit and uses weight 20% less than the actual measure. His gain per cent is
(a) 35% (b) 38%
(c) 48% (d) 50%
(e) None of these

74. A dishonest dealer sells his goods at 10% loss on cost price and uses 30% less weight. What is his profit or loss per cent?
(a) $28\frac{4}{7}$% loss (b) $28\frac{3}{7}$% profit
(c) $28\frac{3}{7}$% loss (d) $28\frac{4}{7}$% profit
(e) None of these

75. Raghu sells 1/2 part of his agricultural land at 30% profit, 1/4 part at 20% loss and remaining at 36% profit. Finally, there is a profit of ₹ 38000, then the cost of land was
(a) ₹ 200000 (b) ₹ 300000
(c) ₹ 380000 (d) ₹ 150000

76. A shopkeeper marked 15% excess on an article. Due to decrease in demand, he reduces the price by 10%. He will get
(a) loss of 3.5% (b) profit of 3.5%
(c) loss of 3% (d) profit of 3%

Exercise 2 Higher Skill Level Questions

1. A fruit seller buys 240 apples for ₹ 600. Some of these apples are rotten and are thrown away. He sells the remaining apples at ₹ 3.50 each and makes a profit of ₹ 198. The percentage of apples thrown away are **[SSC CPO 2015]**
(a) 5% (b) 7%
(c) 8% (d) 6%

2. A dealer bought 80 cricket bats for ₹ 50 each. He sells 20 of them at a gain of 5%. What must be the gain percentage of the remaining bats, so as to get 10% gain on the whole? **[SSC (10+2) 2012]**
(a) $3\frac{2}{11}$% (b) $12\frac{1}{2}$%
(c) $11\frac{2}{3}$% (d) ₹ 3350

3. A man sold 3/5th of his articles at a gain of 20% and the remaining at cost price. Find the gain earned in the transaction.
[SNAP 2016]
(a) 8% (b) 10% (c) 12% (d) 14%

4. A merchant has 1000 kg of sugar, part of which he sells at 8% profit and the rest at 18% profit. He gains 14% on the whole. The quantity sold at 18% profit is
[SSC CGL 2010]
(a) 500 kg (b) 600 kg (c) 400 kg (d) 640 kg

5. A dealer buys an article marked at ₹ 25000 with 20% and 5% off. He spends ₹ 2000 on its repair and sells it for ₹ 25000. What is his gain or loss per cent? **[SSC FCI 2012]**
(a) 21% loss (b) 10.50% loss
(c) 19.05% gain (d) 25% gain

6. On selling an article at ₹ 530, the gain is 20% more than the loss incurred on selling it at ₹ 475. In order to gain 20%, the selling price will be
   (a) ₹ 900   (b) ₹ 600   (c) ₹ 700   (d) ₹ 500
   (e) None of these

7. By selling an umbrella for ₹30, a shopkeeper gains 20%. During a clearance sale, the shopkeeper allows a discount of 10% of the marked price. His gain percentage during the sale season is [SSC CGL (Mains) 2012]
   (a) 7%   (b) 7.5%   (c) 8%   (d) 9%

8. A person sold a table at a gain of 15%. Had he bought it for 25% less and sold it for ₹ 60 less, he would have made a profit of 32%. The cost price of table was [SSC (10+2) 2013]
   (a) ₹ 300   (b) ₹ 350   (c) ₹ 375   (d) ₹ 400

9. Seema purchased an item for ₹ 9600 and sold it for a loss of 5%. From that money she purchased another item and sold it for a gain of 5%. What is her overall gain/loss? [BOB PO 2011]
   (a) Loss of ₹ 36   (b) Profit of ₹ 24
   (c) Loss of ₹ 54   (d) Profit of ₹ 36
   (e) None of these

10. A man sold two articles-A (at a profit of 40%) and B (at a loss of 20%). He incurred a total profit of ₹ 18 in the whole deal. If article-A costs ₹ 140 less than article-B, what is the price of article-B? [SBI PO (Pre) 2017]
    (a) ₹ 380   (b) ₹ 280   (c) ₹ 340   (d) ₹ 370
    (e) ₹ 300

11. An article passing through two hands is sold at a profit of 40% at the original cost price. If the 1st dealer makes a profit of 20%, then the profit per cent made by the second is
    (a) $15\frac{2}{3}\%$ (b) $16\frac{2}{3}\%$ (c) $13\frac{2}{3}\%$ (d) $11\frac{2}{3}\%$
    (e) None of these

12. A bookseller sells a book at a gain of 10%. If he had bought it at 4% less and sold it for ₹ 6 more, he would have gained $18\frac{3}{4}\%$. The CP of the book is [SSC CGL 2007]
    (a) ₹ 130   (b) ₹ 140   (c) ₹ 150   (d) ₹ 160

13. A man sells an article at 5% above its cost price. If he had bought it at 5% less than what he had paid for it and sold it at ₹ 2 less, he would have gained 10%. The cost price of the article is [SSC CGL 2015]
    (a) ₹ 100   (b) ₹ 300   (c) ₹ 400   (d) ₹ 200

14. A vendor purchased 40 dozen bananas for ₹ 250. Out of these 30 bananas were rotten and could not be sold. At what rate per dozen should he sell the remaining bananas to make a profit of 20%? [SSC (10+2) 2015]
    (a) ₹ 10   (b) ₹ 6   (c) ₹ 12   (d) ₹ 8

15. The total cost of a washing machine with an electric chimney was ₹ 57750. The electric chimney was sold at a profit of 34% and the washing machine at a loss of 24%. If the sale price was the same in both the items, then the cost price of the cheaper item was [SSC (10+2) 2017]
    (a) ₹ 26850   (b) ₹ 20900
    (c) ₹ 28875   (d) ₹ 25850

16. A mobile phone and a tablet were sold at a profit of 10% and at a loss of 8%, respectively. If the cost price of the mobile is 1.5 times that of the tablet, then what is the overall profit percentage earned by selling both the articles? [IBPS SO 2016]
    (a) 3.2%   (b) 2.8%   (c) 5%   (d) 4%
    (e) 1.6%

17. How many kilogram of salt at 42 paise per kg must a man mix with 25 kg of salt at 24 paise per kg, so that he may, on selling the mixture at 40 paise per kg gain 25% on the outlay? [SBI Clerk (Mains) 2016]
    (a) 15 kg   (b) 18 kg   (c) 20 kg   (d) 24 kg
    (e) 26 kg

18. A merchant fixed the selling price of his articles at ₹ 700 after adding 40% profit to the cost price. As the sale was very low at this price level, he decided to fix the selling price at 10% profit. Find the new selling price. [SSC (10+2) 2012]
    (a) ₹ 450   (b) ₹ 490   (c) ₹ 500   (d) ₹ 550

19. A trader purchases a watch and a wall clock for ₹ 390. He sells them making a profit of 10% on the watch and 15% on the wall clock. He earns a profit of ₹ 51.50. The difference between the original prices of the wall clock and the watch is equal to [SSC (10+2) 2012]
    (a) ₹ 110   (b) ₹ 100   (c) ₹ 80   (d) ₹ 120

20. A shopkeeper sells a transistors at 15% above its cost price. If he had bought it at 5% more than what he paid for it and sold it for ₹ 6 more, he would have gained 10%. The cost price of the transistor is [SSC CGL (Mains) 2012]
    (a) ₹ 800   (b) ₹ 1000
    (c) ₹ 1200   (d) ₹ 1400

Profit and Loss / 217

21. A merchant bought two mobiles which together cost him ₹ 480. He sold one of them at a loss of 15% and other at a gain of 19%. If the selling prices of both the mobiles are equal, then find the cost of the lower priced mobile. **[MAT 2014]**
(a) ₹ 150 (b) ₹ 180
(c) ₹ 200 (d) ₹ 220

22. A merchant earns a profit of 20% by selling a basket containing 80 apples whose cost is ₹ 240 but he gives one-fourth of it to his friend at cost price and sells the remaining apples. In order to earn the same profit, at what price must he sell each apple? **[CDS 2012]**
(a) ₹ 3.00 (b) ₹ 3.60
(c) ₹ 3.80 (d) ₹ 4.80

23. Two mobile phones were purchased at the same price. One was sold at a profit of 20% and second was sold at a price, which was ₹ 1520 less than the price at which the first was sold. If the overall profit earned by selling both the mobile phones was 1%, then what was the cost price of one mobile phone? **[IBPS SO 2016]**
(a) ₹ 6000 (b) ₹ 5200
(c) ₹ 4800 (d) ₹ 4000
(e) ₹ 5000

24. A car was sold at a loss of 18%. There would have been a profit of 15%, if the car would have been sold for ₹ 99000 more. At what price, car should be sold to make a profit of 10%? **[CMAT 2016]**
(a) ₹ 250000 (b) ₹ 330000
(c) ₹ 450000 (d) ₹ 300000

25. The cost price of item B is ₹ 150 more than the cost price of item A. Item A was sold at a profit of 10% and item B was sold at a loss of 20%. If the respective ratio of selling price of items A and B is 11: 12, then what is the cost price of item B? **[SBI PO (Pre) 2016]**
(a) ₹ 450 (b) ₹ 420
(c) ₹ 400 (d) ₹ 350
(e) ₹ 480

26. Cost price of article A is ₹ 200 more than the cost price of article B. Article A was sold at 10% loss and article B was sold at 25% profit. If the overall profit earned after selling both the articles is 4%, then what is the cost price of article B? **[LIC AAO 2016]**
(a) ₹ 450 (b) ₹ 550 (c) ₹ 400 (d) ₹ 500
(e) ₹ 300

27. A shopkeeper sells notebooks at the rate of ₹ 45 each and earns a commission of 4%. He also sells pencil box at the rate of ₹ 80 each and earns a commission of 20%. How much amount of commission will be earn in two weeks, if he sells 10 notebooks and 6 pencil boxes a day? **[LIC AAO 2007]**
(a) ₹ 1956 (b) ₹ 1586
(c) ₹ 1496 (d) ₹ 1596
(e) None of these

28. Cost of a packet of coffee powder and a litre of milk are ₹ 20 and ₹ 30, respectively. 10 cups of coffee is made with one packet coffee powder and for each cup 200 mL of milk is used. If coffee is sold at 25% profit, then the selling price of each cup of coffee is **[SSC CPO 2013]**
(a) ₹12.50 (b) ₹ 6.25 (c) ₹ 8 (d) ₹ 10

29. Assuming that profit of a shopkeeper in a particular commodity is a linear expression of transportation charge (t) and the quantity of commodity (q), he earns a profit of ₹ 10000 by selling 20 units at the transport charge of ₹ 400. He also earns a profit of ₹ 12000 by selling 25 units at the transport charge of ₹ 600. What is the linear expression in t and q? **[CDS 2012]**
(a) $600q - 5t$ (b) $500q - 4t$
(c) $600q - 4t$ (d) $500q - 5t$

30. A man bought 500 m of electronic wire at 50 paise per metre. He sold 50% of it at a profit of 5%. At what per cent should he sell the remaining so as to gain 10% on the whole transaction? **[SSC CGL (Mains) 2016]**
(a) 13% (b) 12.5% (c) 15% (d) 20%

31. Cost price of B is 200 more than cost-price of A. B is sold at 10% profit and A is sold at 40% loss and selling price of A and B are in the ratio 4 : 11. If A is sold at 20% loss, then what will be selling price of A? **[RRB PO (Pre) 2017]**
(a) 320 (b) 400 (c) 240 (d) 160
(e) 360

32. The respective ratio of cost price of articles A and B is 7 : 9. Article A was sold at a profit of 40% and article B was sold at a profit of 10%. If the total profit earned after selling both the articles (A and B) is ₹ 148, what is the difference between cost price of articles A and B? **[IBPS Clerk (Pre) 2016]**
(a) ₹ 120 (b) ₹ 160 (c) ₹ 200 (d) ₹ 80
(e) ₹ 40

Answer with Solutions

Exercise 1 Base Level Questions

1. (a) We know that, Loss = CP − SP
$\Rightarrow$ 240 = 3200 − SP
$\therefore$ SP = 3200 − 240 = ₹ 2960

2. (d) Given that, CP = ₹ 200 and SP = ₹ 225
Profit = SP − CP = 225 − 200 = ₹ 25
$\therefore$ Gain% = $\dfrac{Gain}{CP} \times 100\%$
= $\dfrac{25}{200} \times 100\% = 12.5\%$

3. (b) Given that, CP = ₹ 500 and SP = ₹ 300
Loss = CP − SP = 500 − 300 = ₹ 200
$\therefore$ Loss% = $\dfrac{Loss}{CP} \times 100\%$
= $\dfrac{200}{500} \times 100\% = 40\%$

4. (a) CP of bracelet = $\left(\dfrac{100}{100 - Loss\%}\right) \times SP$
= $\dfrac{100}{80} \times 14500 = ₹ 18125$

5. (c) Here, CP = ₹ 250 per kg
and SP = ₹ 10 per 50 g
= 10 × 20 per kg = ₹ 200 per kg
$\because$ CP > SP
$\therefore$ Loss% = $\dfrac{Loss}{CP} \times 100 = \dfrac{250 - 200}{250} \times 100$
= $\dfrac{50}{250} \times 100 = 20\%$

6. (a) Let CP = $5x$ and SP = $3x$
Loss = CP − SP = $5x - 3x = 2x$
$\therefore$ Loss% = $\dfrac{2x}{5x} \times 100\% = 40\%$

7. (a) Let CP = $10x$ and SP = $11x$
$\therefore$ Gain% = $\dfrac{11x - 10x}{10x} \times 100\%$
= $\dfrac{1}{10} \times 100\% = 10\%$

8. (d) CP = ₹ $25x$ and SP = ₹ $26x$
$\therefore$ Profit percentage = $\dfrac{(26x - 25x)}{25x} \times 100$
= $\dfrac{x}{25x} \times 100 = \dfrac{100}{25} = 4\%$

9. (b) Let SP = x. Then, CP = $\dfrac{2x}{3}$
Gain = $\left(x - \dfrac{2x}{3}\right) = \dfrac{x}{3}$

$\therefore$ Gain% = $\dfrac{x/3}{2x/3} \times 100\%$
= $\dfrac{x}{3} \times \dfrac{3}{2x} \times 100\% = 50\%$

10. (a) Total CP = (CP + Overhead expenses)
= 150 + 12% of 150
= $150 + \dfrac{12}{100} \times 150 = ₹ 168$
Given that, gain = 10%
$\therefore$ SP = $\dfrac{110}{100} \times 168 = ₹ 184.80$

11. (b) Total cost price = ₹ (120 × 100) = ₹ 12000
Total expenditure
= $480 + \dfrac{1}{2} \times 120 + 60 = ₹ 600$
Total cost price = 12000 + 600 = ₹ 12600
Gain = 40%
SP of 120 reams = $12600 \times \dfrac{140}{100} = ₹ 17640$
$\therefore$ SP per ream = $\dfrac{17640}{120} = ₹ 147$

12. (d) Given, CP = ₹ 350 and gain = 15%
$\therefore$ SP = $\left(\dfrac{100 + Gain\%}{100}\right) \times CP$
= $\dfrac{115}{100} \times 350 = \dfrac{115 \times 7}{2} = ₹ 402.50$

13. (a) Given, total sale = ₹ 791000
$\therefore$ Cost price = $\left(\dfrac{100}{100 + Gain\%}\right) \times SP$
= $\dfrac{100}{113} \times 791000 = ₹ 700000$
$\therefore$ Required profit
= 791000 − 700000 = ₹ 91000

14. (b) Original CP = $\left(\dfrac{100}{100 + Gain\%}\right) \times SP$
= $\dfrac{100}{118} \times 10207 = ₹ 8650$

15. (e) Profit on one calculator = SP − CP
= $250\left(1 - \dfrac{100}{114}\right) = ₹ 30.70$
$\therefore$ Total amount of the profit
= $19 \times 7 \times 30.70 = ₹ 4083.10$

16. (a) $\because$ Cost price of one bracelet = ₹ 160
Profit earned = 15%

Profit and Loss / 219

∴ Selling price of one bracelet
= Cost price + Profit earned
$= 160 + 160 \times \dfrac{15}{100}$
$= 160 + \dfrac{16 \times 15}{10} = 160 + \dfrac{240}{10}$
$= 160 + 24 = ₹ 184$
Hence, Meera should sell her bracelet at ₹ 184 per piece.

17. (d) Balance quantity of rice
$= \dfrac{22 \times (100 - 24)}{100} = \dfrac{22 \times 76}{100}$
$= \dfrac{418}{25}$ quintals
∴ Cost price of rice $= \dfrac{3344}{\dfrac{418}{25}}$
$= \dfrac{3344 \times 25}{418} = ₹ 200$ per quintal
Now, to gain 30% profit, sale price of rice
$= \dfrac{200 \times (100 + 30)}{100} = 2 \times 130$
$= ₹ 260$ per quintal

18. (c) CP of 30 machines = ₹ 1000
SP of 30 machines $= 30 \times 300 = ₹ 9000$
Total profit on 30 machines
$= 30 \times 150 = ₹ 4500$
∴ Total CP of 30 machines
$= 9000 - 4500 = ₹ 4500$
Now, amount spend on repair
$= 4500 - 1000 = ₹ 3500$

19. (a) Cost price of 12 eggs = ₹ 30
Then, cost price of 1 egg $= \dfrac{30}{12} = ₹ 2.5$
∴ Cost price of 100 eggs $= 2.5 \times 100 = ₹ 250$
Now, let the SP of 100 eggs be ₹ x.
Then, $\dfrac{SP - CP}{CP} \times 100 = $ Profit%
$\Rightarrow \dfrac{x - 250}{250} \times 100 = 12$
$\Rightarrow \dfrac{(x - 250)}{5} \times 2 = 12$
$\Rightarrow x - 250 = \dfrac{12 \times 5}{2} = 30$
∴ $x = 250 + 30 = ₹ 280$

20. (a) Given, profit% = 20%
Difference between SP and CP = 240
Now, profit% $= \dfrac{SP - CP}{CP} \times 100$
$\Rightarrow 20 = \dfrac{240}{CP} \times 100$
$\Rightarrow CP = 240 \times 5 = ₹ 1200$
∴ SP = 1200 + 240 [∵ SP − CP = 240]
= ₹ 1440

21. (b) $CP = \left(\dfrac{100}{100 + \text{Gain\%}}\right) \times SP$
$= 8960 \times \dfrac{100}{128} = 70 \times 100 = ₹ 7000$

22. (a) CP of the article $= 180 \times \dfrac{100}{90} = ₹ 200$
To gain 10%,
SP of article = 110% of 200 = ₹ 220

23. (b) ∵ SP of cell phone = ₹ 2400
∴ CP of cell phone $= 2400 \times \dfrac{100}{100 + 25}$
$= ₹ 1920$
∴ Required percentage profit
$= \dfrac{2040 - 1920}{1920} \times 100 = 6.25\%$

24. (d) Let CP = x, then gain $= \dfrac{x}{6}$
According to the question,
$x + \dfrac{x}{6} = 322 \Rightarrow \dfrac{7x}{6} = 322$
∴ $x = \dfrac{322 \times 6}{7} = ₹ 276$

25. (b) Here, Loss = 10%, Sale price = ₹ 2880
∴ Cost price of saree
$= 2880 \times \dfrac{100}{(100 - 10)} = 2880 \times \dfrac{100}{90}$
$= 32 \times 100 = ₹ 3200$
Now, sale price of saree to get 20% profit
$= $ Cost price $\times \dfrac{(100 + \text{Profit\%})}{100}$
$= 3200 \times \dfrac{(100 + 20)}{100} = 32 \times 120$
$= ₹ 3840$

26. (c) Given, SP = ₹ 768 and loss = 20%
∴ $CP = \dfrac{100}{80} \times 768 = \dfrac{5}{4} \times 768 = ₹ 960$
Now, CP = ₹ 960 and SP = ₹ 1020
Gain = 1020 − 960 = ₹ 60
∴ Gain% $= \dfrac{60}{960} \times 100\% = 6.25\%$

27. (d) According to the question,
$\left(25 - \dfrac{25}{2}\right)\% \; CP = 22.50$
$\Rightarrow \dfrac{25}{2} \times \dfrac{1}{100} \times CP = 22.50$
$\Rightarrow CP = ₹ 180$

28. (b) Let the cost price of the shoes be ₹ x.
According to the question,
$2033 - x = x - 1063$
$\Rightarrow 2x = 2033 + 1063 = 3096$
∴ $x = \dfrac{3096}{2} = ₹ 1548$

29. (b) Let CP be ₹ x.

According to the question,

$$625 - x = x - 435$$
$$\Rightarrow 2x = 1060$$
$$\Rightarrow x = \frac{1060}{2} = ₹ 530$$

30. (c) Let CP of a bucket = ₹x

According to the question,

$$\frac{108x}{100} - \frac{92x}{100} = 28 \Rightarrow \frac{16x}{100} = 28$$
$$\Rightarrow x = \frac{28 \times 100}{16} \Rightarrow x = ₹ 175$$

31. (b) Let the cost price of the article be ₹ x.

Then, $x + \frac{20x}{100} = 3600 \Rightarrow \frac{120x}{100} = 3600$

$\therefore \quad x = 3000$

Now, profit percentage when the article is sold for ₹ 3150 $= \frac{3150 - 3000}{3000} \times 100$

$= \frac{150}{3000} \times 100 = 5\%$

32. (c) Cost price of an article

$= 220 \times \frac{100}{110} = ₹ 200$

For getting the profit of 30%, selling price of an article = 130% of ₹ 200

$= \frac{130}{100} \times 200 = ₹ 260$

33. (c) Let the cost price of article = ₹ x

Then, 120% of $x = 300$

$\Rightarrow \frac{120 \times x}{100} = 300$

$\therefore \quad x = \frac{300 \times 100}{120} = ₹ 250$

Now, SP = ₹ 235

Then, loss percentage

$= \frac{250 - 235}{250} \times 100$

$= \frac{15}{250} \times 100 = 6\%$

34. (d) Let SP = ₹ 100

$\therefore \quad$ CP $= \frac{3}{10} \times 100 = ₹ 30$

New SP $= 30 \times \frac{110}{100} = ₹ 33$

Gain $= 33 - 30 = ₹ 3$

$\therefore$ Gain% $= \frac{3}{30} \times 100\% = \frac{300}{30}\% = 10\%$

35. (b) Cost price $= \frac{\text{Selling price} \times 100}{(100 - \text{Loss}\%)}$

$= \frac{136 \times 100}{85} = ₹ 160$

[$\because$ loss = 15%]

Selling price $(x) = \frac{160 \times (100 + 15)}{100}$

$= \frac{160 \times 115}{100} = ₹ 184$

$\therefore$ Option (b) is correct, because $180 < x < 190$.

36. (c) SP of 1 lemon = ₹ $\frac{14}{5}$

Let CP of 1 lemon = x

According to the question,

$x \times \frac{140}{100} = \frac{14}{5}$

$\therefore \quad x = ₹ 2$

$\therefore$ CP for 12 lemons $= 12 \times 2 = ₹ 24$

37. (c) Total CP $= 2 \times 1200 = ₹ 2400$

SP at 5% loss $= \frac{95}{100} \times 1200 = ₹ 1140$

SP at 10% gain $= \frac{110}{100} \times 1200 = ₹ 1320$

Total SP $= 1140 + 1320 = ₹ 2460$

$\therefore \quad$ Gain $= 2460 - 2400 = ₹ 60$

$\therefore \quad$ Gain% $= \frac{60}{2400} \times 100\% = 2.5\%$

38. (a) Let CP of the article be ₹ x.

Then, SP = 110% of $x = 1.1x$

If SP be double, i.e. $2.2x$, then

Profit% $= \frac{2.2x - x}{x} \times 100\% = 120\%$

39. (c) CP of 1 orange for 1st rate = ₹ $\frac{1}{4}$

CP of 1 orange for 2nd rate = ₹ $\frac{1}{5}$

CP of 2 oranges after mixing

$= \frac{1}{4} + \frac{1}{5} = ₹ \frac{9}{20}$

CP of 1 orange after mixing

$= \frac{9}{20 \times 2} = ₹ \frac{9}{40}$

SP of 1 orange after mixing = ₹ $\frac{1}{4}$

$\therefore \quad$ Gain $= \frac{1}{4} - \frac{9}{40} = ₹ \frac{1}{40}$

$\therefore$ Gain% $= \left(\frac{1}{40} \times \frac{40}{9} \times 100\right)\%$

$= \frac{100}{9}\% = 11\frac{1}{9}\%$

40. (c) Let each lot of onion, contains x kg onion, then total cost price of these two lots together $= 10x + 15x = 25x$

Selling price of whole lot

$= 15 \times (x + x) = 15 \times 2x = 30x$

Profit and Loss / 221

$\therefore$ Profit% $= \dfrac{30x - 25x}{25x} \times 100\%$

$= \dfrac{5x}{25x} \times 100\% = 20\%$

41. (d) Let quantities of two types of tea costing ₹ 18 per kg and ₹ 20 per kg be $5x$ kg and $3x$ kg, respectively.
Then, CP of tea $= 18 \times 5x + 20 \times 3x$

$= 90x + 60x = 150x$

SP of tea $= 21 \times 8x = 168x$

$\therefore$ Profit% $= \dfrac{168x - 150x}{150x} \times 100\%$

$= \dfrac{18}{150} \times 100\% = 12\%$

42. (b) Let CP = ₹ 100

$\therefore$ SP $= \dfrac{3}{4} \times 100 \times \dfrac{124}{100} + 25$

$= 75 \times \dfrac{124}{100} + 25 = 93 + 25 =$ ₹ 118

Clearly, gain = 18%

43. (b) CP of both the cycles $= 2 \times 750 =$ ₹ 1500
Now, SP of both the cycles

$= \dfrac{750 \times 106}{100} + \dfrac{750 \times 96}{100}$

$= 795 + 720 =$ ₹ 1515

Here, SP > CP

$\therefore$ Profit percentage $= \dfrac{(1515 - 1500)}{1500} \times 100$

$= \dfrac{15 \times 100}{1500} = 1\%$

44. (c) $\because$ CP$_1$ = ₹ 7500

$\therefore$ SP$_1$ = $7500 \times \dfrac{124}{100} =$ ₹ 9300 = CP$_2$

and SP$_2$ = $9300 \times \dfrac{80}{100} =$ ₹ 7440

Here, CP$_1$ > SP$_2$
Hence, loss is incurred in this transaction.
$\therefore$ Required loss = (CP$_1$ − SP$_2$)

$= 7500 - 7440 =$ ₹ 60

45. (b) CP $= \dfrac{100}{70} \times 6384 =$ ₹ 9120

To gain 30%, SP = 130% of 9120

SP $= \dfrac{130}{100} \times 9120 =$ ₹ 11856

46. (c) Let CP of cellphone be ₹ 100.
Then, SP for Ram $= 100 \left(1 + \dfrac{10}{100}\right)\left(1 - \dfrac{10}{100}\right)$

$= 100 \times \dfrac{110}{100} \times \dfrac{90}{100} =$ ₹ 99

Therefore, required gain

$= \dfrac{100 - 99}{100} \times 100 = 1\%$

47. (c) Let price of land be ₹ x.

SP for I = Cost price for II $= \dfrac{120x}{100}$

SP for II = Cost price for III $= \dfrac{125}{100} \times \dfrac{120x}{100}$

Let profit earn by III be $a\%$.

Then, $x + \dfrac{65}{100} \times x = \dfrac{125 \times 120}{100 \times 100} \dfrac{x}{1} \times \left(\dfrac{100 + a}{100}\right)$

$\Rightarrow \dfrac{165x}{100} \times \dfrac{100 \times 100 \times 100}{125 \times 120x} = 100 + a$

$\Rightarrow 110 = 100 + a$

$\therefore a = 10\%$

48. (b) Let CP = x
According to the question,

$\dfrac{106x}{100} - \dfrac{104x}{100} = 3 \Rightarrow \dfrac{2x}{100} = 3$

$\therefore x = \dfrac{300}{2} =$ ₹ 150

Alternate Method
According to the question,

$(6\% - 4\%) = 3 \Rightarrow 2\% = 3$

$\Rightarrow (2 \times 50)\% = 3 \times 50$

$\therefore 100\% =$ ₹ 150

49. (c) Let CP of the watch be ₹ x.
According to the question,

$\dfrac{75 - x}{x} \times 100 = x \Rightarrow (75 - x) \times 100 = x^2$

$\Rightarrow x^2 + 100x - 7500 = 0$

$\Rightarrow x^2 + 150x - 50x - 7500 = 0$

$\Rightarrow x(x + 150) - 50(x + 150) = 0$

$\Rightarrow (x + 150)(x - 50) = 0 \Rightarrow x =$ ₹ 50

50. (a) CP of 700 oranges

$= \dfrac{500}{100} \times 700 =$ ₹ 3500

CP of 500 oranges of another variety

$= \dfrac{700}{100} \times 500 =$ ₹ 3500

Total CP = 3500 + 3500 = ₹ 7000
and total number of oranges purchased

$= 700 + 500 = 1200$

Total SP $= \dfrac{84}{12} \times 1200 =$ ₹ 8400

$\therefore$ Profit% $= \dfrac{8400 - 7000}{7000} \times 100\%$

$= \dfrac{1400 \times 100}{7000} = 20\%$

51. (a) Let the cost price be ₹ x.

SP of 1 orange $=$ ₹ $\dfrac{30}{32} =$ ₹ $\dfrac{15}{16}$

According to the question, $\dfrac{75x}{100} = \dfrac{15}{16}$

$\therefore \quad x = \dfrac{15 \times 100}{75 \times 16} = ₹\dfrac{5}{4}$

SP of 1 orange with 20% profit

$= ₹\left(\dfrac{5}{4} \times \dfrac{120}{100}\right) = ₹\dfrac{3}{2}$

$\therefore$ In $₹\dfrac{3}{2}$, the number of oranges sold = 1

$\therefore$ In ₹ 24, the number of oranges sold

$= \dfrac{2}{3} \times 24 = 16$

52. (e) (SP of 100 pens) − (CP of 100 pens)
= SP of 40 pens
$\Rightarrow$ SP of 60 pens = CP of 100 pens
Let CP of each pen = ₹ 1.
Then, CP of 60 pens = ₹ 60
SP of 60 pens = ₹ 100
Gain = 100 − 60 = ₹ 40

$\therefore$ Gain% $= \dfrac{40}{60} \times 100\% = 66.67\%$

53. (a) Loss% $= \left(\dfrac{a}{10}\right)^2 \% = \left(\dfrac{20}{10}\right)^2 \%$

$= 4\%$ [by Technique 1]
[here, a = 20%]

Thus, the seller gets 4% loss in the deal.

54. (d) Here, a = 20%

$\therefore$ Total loss% $= \left(\dfrac{a}{10}\right)^2 \% = \left(\dfrac{20}{10}\right)^2 \% = 4$

[by Technique 1]

55. (a) Here, $a = \dfrac{2}{5}$ and $x = 15\%$.

According to the formula,

Gain% $= \dfrac{ax}{1-a}\%$ [by Technique 2]

$= \dfrac{\dfrac{2}{5} \times 15}{1 - \dfrac{2}{5}}\% = \dfrac{2 \times 15}{3}\% = 10\%$

56. (c) Let CP of the watch be ₹ x.
According to the question,

$\dfrac{x \times (100 + 10)}{100} - \dfrac{x \times (100 - 5)}{100} = 56.25$

$\Rightarrow \dfrac{110x}{100} - \dfrac{95x}{100} = 56.25$

$\Rightarrow \dfrac{15x}{100} = 56.25$

$\Rightarrow x = \dfrac{56.25 \times 100}{15} = \dfrac{5625}{15} = ₹ 375$

Fast Track Method
Here, a = 5%, b = 10% and R = ₹ 56.25

$\therefore$ CP of the watch $= \dfrac{R}{b+a} \times 100$

[by Technique 3]

$= \dfrac{56.25}{10+5} \times 100$

$= \dfrac{5625}{15} = ₹ 375$

57. (b) Let the CP be ₹ x.
According to the question,

$\dfrac{110x}{100} - \dfrac{80x}{100} = 12$

$\Rightarrow \dfrac{11x}{10} - \dfrac{4x}{5} = 12 \Rightarrow 11x - 8x = 120$

$\Rightarrow x = \dfrac{120}{3} = ₹ 40$

Fast Track Method
Here, a = 20%, b = 10% and R = 12

$\therefore$ CP of article $= \dfrac{R \times 100}{b+a}$ [by Technique 3]

$= \dfrac{12 \times 100}{20+10} = ₹ 40$

58. (c) Here, a = 10%, b = 20% and R = 2000
According to the formula,

CP of watch $= \dfrac{R}{b-a} \times 100$

[by Technique 3]

$= \dfrac{2000}{20-10} \times 100$

$= \dfrac{2000 \times 100}{10} = ₹ 20000$

59. (b) Let CP of the article be ₹ x.
Then, 106% of x − 94% of x = 6
$\Rightarrow$ 12% of x = 6

$\therefore \quad x = \dfrac{6 \times 100}{12} = ₹ 50$

Fast Track Method
Here, a = 6%, b = 6% and R = ₹ 6

$\therefore$ CP of an article $= \dfrac{R \times 100}{b+a}$

[by Technique 3]

$= \dfrac{6}{6+6} \times 100 = \dfrac{6}{12} \times 100 = ₹ 50$

60. (c) Here, a = 22 and b = 20

$\therefore$ Gain% $= \dfrac{a-b}{b} \times 100\%$ [by Technique 4]

$= \dfrac{22-20}{20} \times 100\% = 10\%$

61. (a) Here, a = 16 and b = 12

$\therefore$ Required gain% $= \left(\dfrac{a-b}{b}\right) \times 100\%$

[by Technique 4]

$= \dfrac{16-12}{12} \times 100\%$

$= \dfrac{1}{3} \times 100\% = 33\dfrac{1}{3}\%$

Profit and Loss / 223

62. (b) Given, CP of p apples = q

$\therefore$ CP of 1 apple = $\dfrac{q}{p}$ and SP of q apples = p

$\therefore$ SP of 1 apple = $\dfrac{p}{q}$

Given, $p < q$ [$\because$ CP > SP]

$\therefore$ Loss = $\dfrac{q}{p} - \dfrac{p}{q} = \dfrac{q^2 - p^2}{pq}$

and loss% = $\dfrac{\text{loss}}{\text{CP}} \times 100 = \dfrac{\dfrac{q^2 - p^2}{pq}}{q/p} \times 100$

$= \dfrac{(q^2 - p^2)}{q^2} \times 100$

Therefore, when $p < q$, then the person had a loss which is given by

$100 \times \left(\dfrac{q^2 - p^2}{q^2}\right)$ % loss.

63. (c) Since, CP of 2 lemons is ₹ 1.

$\therefore$ CP of 1 lemon is = $\dfrac{1}{2}$ = ₹ 0.5

SP of 5 lemons is ₹ 3.

$\therefore$ SP of 1 lemon is = $\dfrac{3}{5}$ = ₹ 0.6

$\therefore$ Gain% = $\dfrac{0.6 - 0.5}{0.5} \times 100$

$= \dfrac{0.1 \times 100}{0.5} = 20\%$

Fast Track Method

Here, $m = 2, x = 1, n = 5$ and $y = 3$

$\therefore$ Required gain = $\dfrac{my - nx}{nx} \times 100\%$

[by Technique 5]

$= \dfrac{2 \times 3 - 5 \times 1}{5 \times 1} \times 100$

$= \dfrac{1}{5} \times 100 = 20\%$

64. (d) Let number of apples bought

$= 5 \times 6 = 30$ [LCM of 5 and 6]

Then, CP = $\dfrac{10}{5} \times 30$ = ₹ 60

SP = $\dfrac{15}{6} \times 30$ = ₹ 75

Now, gain = SP − CP = 75 − 60 = ₹ 15

$\therefore$ Gain% = $\left(\dfrac{15}{60} \times 100\right)\%$ = 25%

Fast Track Method

Here, $m = 5, x = 10, n = 6$ and $y = 15$

Profit% = $\dfrac{my - nx}{nx} \times 100\%$ [by Technique 5]

$= \dfrac{(5 \times 15 - 6 \times 10)}{6 \times 10} \times 100\%$

$= \dfrac{75 - 60}{60} \times 100\%$

$= \dfrac{15}{60} \times 100\% = 25\%$

65. (a) Here, $m = 5, x = 1, n = 3$ and $y = 1$

$\therefore$ Required gain = $\dfrac{my - nx}{nx} \times 100\%$

[by Technique 5]

$= \dfrac{5 \times 1 - 3 \times 1}{3 \times 1} \times 100\%$

$= \dfrac{2}{3} \times 100\% = 66\dfrac{2}{3}\%$

66. (a) CP of retailer = CP of geyser

$\times \left(1 + \dfrac{r_1}{100}\right)\left(1 + \dfrac{r_2}{100}\right)\left(1 + \dfrac{r_3}{100}\right)$

[by Technique 6]

Here, CP of retailer = ₹ 1265

$r_1 = 10, r_2 = 15, r_3 = 25$

$\therefore$ CP of geyser = $1265 \times \dfrac{100}{110} \times \dfrac{100}{115} \times \dfrac{100}{125}$

= ₹ 800

67. (a) Let A purchased the watch for ₹ x.

Then, CP for $A \times \left(1 + \dfrac{r_1}{100}\right)\left(1 - \dfrac{r_2}{100}\right)$

= CP for C

[by Technique 6]

$\Rightarrow \qquad x \times \dfrac{140}{100} \times \dfrac{80}{100} = 432$

$\therefore \qquad x = \dfrac{432 \times 100}{14 \times 8}$ = ₹ 385.71

68. (b) According to the formula,

Gain% = $\left(\dfrac{\text{Error}}{\text{True value} - \text{Error}}\right) \times 100\%$

[by Technique 7]

$= \dfrac{80}{1000 - 80} \times 100\%$

$= \dfrac{80}{920} \times 100\% = 8\dfrac{16}{23}\%$

Alternate Method

Since, 80 g is gained on 920 g.

$\therefore$ Gain% = $\dfrac{80}{920} \times 100\% = 8\dfrac{16}{23}\%$

69. (a) Given, $a = 4\%, b = 28$ g and $c = 32$ g

Required answer

$= \left[(100 - a)\dfrac{c}{b} - 100\right]\%$ [by Technique 8]

$= \left[(100 - 4)\dfrac{32}{28} - 100\right]\%$

$= \left[96 \times \dfrac{32}{28} - 100\right]\%$

$= \left[96 \times \dfrac{8}{7} - 100\right]\% = \left[\dfrac{768}{7} - 100\right]\%$

$= \left[\dfrac{768 - 700}{7}\right]\% = \dfrac{68}{7}\% = 9\dfrac{5}{7}\%$ gain

[+ve sign shows that profit is earned here]

70. (a) Given that, $a = 10\%$, $b = 16$ g
and $c = 18$ g
∴ Required answer
$= \left[(100 - a)\dfrac{c}{b} - 100\right]\%$ [by Technique 8]
$= \left[(100 - 10)\dfrac{18}{16} - 100\right]\%$
$= \left[90 \times \dfrac{18}{16} - 100\right]\% = \left[90 \times \dfrac{9}{8} - 100\right]\%$
$= \left[\dfrac{810}{8} - 100\right]\% = \left[\dfrac{810 - 800}{8}\right]\%$
$= \dfrac{10}{8}\% = \dfrac{5}{4}\% = 1\dfrac{1}{4}\%$ gain

[+ve sign shows that profit in earned here]

71. (d) Here, $a = 10\%$ and $b = 20\%$
According to the formula,
Required answer $= \left[\dfrac{b - a}{100 - b}\right] \times 100\%$

[by Technique 9]

$= \dfrac{(20 - 10)}{(100 - 20)} \times 100\% = \dfrac{10}{80} \times 100\%$
$= 12.5\%$ gain

[+ve sign shows that profit is earned here]

72. (b) Let the marked weight = 1 kg = 1000 g
Real weight = 70% of 1000 = 700 g
Let   CP of 1 g = ₹ 1
∴   CP of 700 g = ₹ 700
CP of 1000 g = ₹ 1000
SP = 110% of 1000 = $\dfrac{110}{100} \times 1000$ = ₹ 1100
Now,   gain = 1100 − 700 = ₹ 400
∴   Gain% = $\dfrac{400}{700} \times 100\% = 57\dfrac{1}{7}\%$

Fast Track Method
Here, $a = 10\%$ and $b = 30\%$
∴ Required Profit = $\dfrac{b + a}{100 - b} \times 100\%$

[by Technique 9]

$= \dfrac{30 + 10}{100 - 30} \times 100\%$
$= \dfrac{40}{70} \times 100\% = 57\dfrac{1}{7}\%$

73. (d) Let the marked weight = 1 kg = 1000 g
Real weight = 80% of 1000 = 800 g
Let the CP of 1 g = ₹ 1
∴   CP of 800 g = ₹ 800

CP of 1000 g = ₹ 1000
SP = 120% of 1000 = $\dfrac{120}{100} \times 1000$ = ₹ 1200
Now, gain = 1200 − 800 = ₹ 400
∴ Gain% = $\dfrac{400}{800} \times 100\% = 50\%$

Fast Track Method
Here, $a = 20\%$ and $b = 20\%$
∴ Required profit
$= \left(\dfrac{b + a}{100 - b}\right) \times 100\%$ [by Technique 9]
$= \dfrac{20 + 20}{100 - 20} \times 100 = \dfrac{40}{80} \times 100 = 50\%$

74. (d) Given that, $a = 10\%$ and $b = 30\%$.
According to the formula,
Required per cent $= \left[\dfrac{b - a}{100 - b}\right] \times 100\%$

[by Technique 9]

$= \left[\dfrac{30 - 10}{100 - 30}\right] \times 100\%$
$= \left[\dfrac{20}{70}\right] \times 100\% = 28\dfrac{4}{7}\%$ profit

75. (a) Here, $a = \dfrac{1}{2}$, $b = \dfrac{1}{4}$, $c = 1 - \dfrac{1}{2} - \dfrac{1}{4} = \dfrac{1}{4}$
and $x = 30\%$, $y = -20\%$, $z = 36\%$,
$R = 38000$
∴ Cost of land $= \dfrac{R \times 100}{ax + by + cz}$

[by Technique 10]

$= \dfrac{38000 \times 100}{\dfrac{1}{2} \times 30 + \dfrac{1}{4}(-20) + \dfrac{1}{4} \times 36}$
$= \dfrac{38000 \times 100}{15 - 5 + 9} = \dfrac{38000 \times 100}{19}$ = ₹ 200000

76. (b) Here, $a = +15\%$ ['+' for excess]
and $b = -10\%$ ['−' for reduce]
∴ Profit / Loss $= \left(a + b + \dfrac{ab}{100}\right)\%$

[by Technique 11]

$= \left[15 + (-10) + \dfrac{(15) \times (-10)}{100}\right]\%$
$= \left[15 - 10 - \dfrac{15 \times 10}{100}\right]\%$
$= \left(5 - \dfrac{3}{2}\right)\% = \dfrac{7}{2}\% = 3.5\%$

Here, resultant is positive (+), so he will get a profit of 3.5%.

Profit and Loss / 225

Exercise 2 — Higher Skill Level Questions

1. (a) Total cost price = ₹ 600
Selling price = 600 + 198 = ₹ 798
Selling price of each apple = ₹ 3.5
∴ Number of apples = $\dfrac{798}{3.5}$ = 228
Hence, required percentage
$= \dfrac{240 - 228}{240} \times 100$
$= \dfrac{12}{240} \times 100 = 5\%$

2. (c) Let required percentage profit be $x\%$.
According to the question,
10% of (80 × 50) = 5 % of (20 × 50) + x % of (60 × 50)
$\Rightarrow \dfrac{80 \times 50 \times 10}{100} = \dfrac{20 \times 50 \times 5}{100} + \dfrac{60 \times 50 \times x}{100}$
$\Rightarrow \quad 80 = 10 + 6x$
∴ $\quad x = \dfrac{70}{6} = 11\dfrac{2}{3}\%$

3. (c) Let the total cost of article be ₹ 100.
∴ Earning on $\dfrac{3}{5}$th of articles
$= \dfrac{3}{5} \times 100 + 20\%$ of $\dfrac{3}{5}$ of 100 = ₹ 72
On rest of the article, there is no profit no loss.
∴ Overall profit $= \dfrac{(72 + 40) - 100}{100} \times 100\%$
$= \dfrac{12}{100} \times 100\% = 12\%$

Alternate Method
Let the number of articles be 5 and cost of each article be ₹10.
∴ CP of $\dfrac{3}{5} \times 5 = 3$ articles = ₹30
Since, these 3 articles are sold at profit of 20%.
∴ SP = 30 + $\dfrac{20}{100} \times 30$ = ₹ 36
SP of remaining two articles = ₹ 20
Total selling price = ₹ 56
∴ Profit% = $\dfrac{56 - 50}{50} \times 100$
$= \dfrac{6}{50} \times 100 = 12\%$

4. (b) Let the sugar sold at 8% gain be x.
Then, the sugar sold at 18% gain is $(1000 - x)$.
Let CP of sugar = ₹ y per kg
∴ Total CP = ₹ $1000 y$

∴ $\left(\dfrac{108}{100} \times xy\right) + \dfrac{118}{100}(1000 - x)y$
$= \dfrac{114}{100} \times 1000y$
$\Rightarrow 108xy + 118000y - 118xy = 114000y$
$\Rightarrow \qquad 10x = 4000$
∴ $\qquad x = 400$
∴ Quantity sold at 18% profit
$= 1000 - 400 = 600$ kg

5. (c) Total CP = (95% of 80% of ₹ 25000) + 2000
$= \left(\dfrac{95}{100} \times \dfrac{80}{100} \times 25000\right) + 2000$
$= 19000 + 2000 =$ ₹ 21000
∴ CP = ₹ 21000 and SP = ₹ 25000
∵ Gain = 25000 − 21000 = ₹ 4000
∴ Gain% = $\dfrac{4000}{21000} \times 100\% = \dfrac{400}{21}\%$
$= 19.05\%$ (approx.)

6. (b) Let CP be ₹ x.
Then, according to the question,
$\dfrac{120}{100}(x - 475) = (530 - x)$
$\Rightarrow \quad \dfrac{6}{5}(x - 475) = (530 - x)$
$\Rightarrow \quad (6x - 2850) = (2650 - 5x)$
$\Rightarrow \qquad 11x = 5500$
$\Rightarrow \qquad x =$ ₹ 500
∴ Required SP = $500 \times \dfrac{120}{100}$ = ₹ 600

7. (c) Given, selling price of an umbrella = ₹ 30 and profit = 20%.
∴ Cost price of an umbrella
$= \dfrac{30 \times 100}{120} =$ ₹25
During the clearance sale, selling price of an umbrella = $\dfrac{30 \times 90}{100} =$ ₹ 27
∴ Required profit percentage
$= \dfrac{27 - 25}{25} \times 100 = 8\%$

8. (c) Let the cost price of table be ₹ x.
Then, selling price with 15% gain
$= \dfrac{(100 + \text{Gain}\%) \times \text{CP}}{100}$
$= \dfrac{(100 + 15)\% \times x}{100}$
$=$ ₹ $\dfrac{115x}{100}$

New CP = $\dfrac{(100-25) \times x}{100} = ₹ \dfrac{75x}{100}$

New SP = $₹ \left(\dfrac{115x}{100} - 60\right)$

Now, according to the question,

$$\dfrac{\left(\dfrac{115x}{100} - 60\right) - \dfrac{75x}{100}}{\dfrac{75x}{100}} \times 100 = 32$$

$\Rightarrow \dfrac{\dfrac{115x - 6000 - 75x}{100}}{\dfrac{75x}{100}} \times 100 = 32$

$\Rightarrow \dfrac{40x - 6000}{75x} \times 100 = 32$

$\Rightarrow \dfrac{40x - 6000}{3x} \times 4 = 32$

$\Rightarrow 160x - 24000 = 96x$

$\Rightarrow 160x - 96x = 24000$

$\Rightarrow 64x = 24000$

$\Rightarrow x = \dfrac{24000}{64} = ₹ 375$

Hence, the cost price of table is ₹ 375.

9. (e) First selling price

$= 9600 \times \left(\dfrac{100 - 5}{100}\right)$

$= \dfrac{9600 \times 95}{100} = ₹ 9120$

Second selling price

$= \dfrac{9120 \times (100 + 5)}{100}$

$= \dfrac{9120 \times 105}{100} = 9576$

$\therefore$ Loss = CP − SP

$= 9600 − 9576 = ₹ 24$

10. (d) Let price of article B = ₹ x

Then, price of article-A = ₹ $(x - 140)$

According to the question,

$(x - 140) \times \dfrac{40}{100} + x \times \left(\dfrac{-20}{100}\right) = 18$

$\Rightarrow \dfrac{40x - 5600}{100} + \dfrac{(-20x)}{100} = 18$

$\Rightarrow 40x - 5600 - 20x = 1800$

$\Rightarrow 20x = 1800 + 5600$

$\Rightarrow 20x = 7400$

$\Rightarrow x = 370$

Hence, price of article B is ₹ 370.

11. (b) Let CP = ₹ 100, then SP = ₹ 140

Let the profit made by the 2nd dealer be x %.

Then, $(100 + x)$% of 120% of ₹ 100 = ₹ 140

$\Rightarrow \dfrac{100 + x}{100} \times \dfrac{120}{100} \times 100 = 140$

$\Rightarrow 6(100 + x) = 700$

$\Rightarrow 600 + 6x = 700$

$\Rightarrow 6x = 100$

$\therefore x = \dfrac{100}{6}\% = \dfrac{50}{3}\% = 16\dfrac{2}{3}\%$

12. (c) Let CP be ₹ x.

Then, SP = $\dfrac{110x}{100} = ₹ \dfrac{11x}{10}$

New CP = 96% of $x = \dfrac{96x}{100} = ₹ \dfrac{24x}{25}$

SP = $₹ \left(\dfrac{11x}{10} + 6\right)$

Now, according to the question,

$\therefore \left(\dfrac{11x}{10} + 6\right) = \left(\dfrac{100 + 18\dfrac{3}{4}}{100}\right)$ of $\dfrac{24x}{25}$

$\left[\because \text{SP} = \dfrac{(100 + \text{Gain\%})}{100} \times \text{CP}\right]$

$\Rightarrow \dfrac{11x + 60}{10} = \dfrac{475}{400} \times \dfrac{24x}{25} = \dfrac{57x}{50}$

$\Rightarrow 550x + 3000 = 570x$

$\Rightarrow 20x = 3000$

$\Rightarrow x = \dfrac{3000}{20} = ₹ 150$

13. (c) Let CP be ₹ x.

Then, SP = ₹ $1.05x$

Now, new CP = ₹ $0.95x$

and new SP = ₹ $(1.05x - 2)$

$\therefore (1.05x - 2) = 0.95x + 10 \times \dfrac{0.95x}{100}$

$\Rightarrow 1.05x - 2 - 0.95x = \dfrac{0.95x}{10}$

$\Rightarrow 0.1x = 2 + \dfrac{0.95x}{10}$

$\Rightarrow 0.1x = \dfrac{20 + 0.95x}{10}$

$\Rightarrow x = 20 + 0.95x$

$\Rightarrow x - 0.95x = 20$

$\Rightarrow 0.05x = 20$

$\therefore x = \dfrac{20}{0.05} = ₹ 400$

14. (d) Price of 40 dozen bananas = ₹ 250

i.e. price of 480 bananas = ₹ 250

Now, 30 bananas were rotten.

$\therefore$ Price of 450 bananas = ₹ 250

To make a profit of 20%,

Profit and Loss / 227

SP of 1 banana = $\dfrac{250}{450} \times \dfrac{120}{100} = ₹\dfrac{2}{3}$

SP of 12 bananas, i.e. 1 dozen
$= \dfrac{2}{3} \times 12 = ₹8$ per dozen

15. **(b)** Let the CP of washing machine be ₹ x.
Then, the CP of the chimney will be ₹ $(57750 - x)$.
∴ SP of washing machine = SP of Chimney
$\Rightarrow \dfrac{(100-24)}{100} \times x = \dfrac{(100+34)}{100} \times (57750 - x)$
$\Rightarrow \dfrac{76}{100} \times x = \dfrac{134}{100}(57750 - x)$
$\Rightarrow 76x = 134 \times 57750 - 134x$
$\Rightarrow 210x = 134 \times 57750$
$\Rightarrow x = \dfrac{134 \times 57750}{210}$
$\Rightarrow x = ₹36850$
So, the cheaper item will be chimney.
CP of chimney = ₹ $(57750 - 36850)$
$= ₹20900$

16. **(b)** Let CP of tablet = ₹ x
Then, CP of mobile = ₹ $1.5x$
Now, SP of tablet
$= \dfrac{x \times (100 - 8)}{100} = ₹\dfrac{92x}{100}$
and SP of mobile
$= \dfrac{1.5x \times (100 + 10)}{100}$
$= \dfrac{1.5x \times 110}{100} = ₹\dfrac{165x}{100}$
Total CP of both articles
$= x + 1.5x = ₹2.5x$
and total SP of both articles
$= \dfrac{92x}{100} + \dfrac{165x}{100} = ₹\dfrac{257x}{100}$
Now, profit $= \dfrac{257x}{100} - 2.5x = ₹\dfrac{7x}{100}$
∴ Profit percentage $= \dfrac{\dfrac{7x}{100}}{\dfrac{250x}{100}} \times 100$
$= \dfrac{7}{250} \times 100 = \dfrac{14}{5} = 2.8\%$

17. **(c)** Let the required amount of salt be x kg.
Then, CP $= x \times 42p + 25 \times 24p$
$= (42x + 600)p$
and SP $= (x + 25) \times 40p$
$= (40x + 1000)p$
∴ Gain% $= \dfrac{SP - CP}{CP} \times 100$
$\Rightarrow 25 = \dfrac{(40x + 1000) - (42x + 600)}{42x + 600} \times 100$

$\Rightarrow \dfrac{1}{4} = \dfrac{400 - 2x}{42x + 600}$
$\Rightarrow 42x + 600 = 1600 - 8x$
$\Rightarrow 50x = 1000$
∴ $x = 20$ kg

18. **(d)** Let the CP be ₹ x.
According to the question,
$x \times \dfrac{(100 + 40)}{100} = 700$
$\Rightarrow x = \dfrac{700 \times 100}{140} = 500$
∴ New selling price $= \dfrac{500 \times (100 + 10)}{100}$
$= 5 \times 110 = ₹550$

19. **(a)** Let cost price of a watch be ₹ x.
∴ Cost price of a wall clock = ₹ $(390 - x)$
According to the question,
$\dfrac{x \times 10}{100} + \dfrac{(390 - x) \times 15}{100} = 51.50$
$\Rightarrow 390 \times 15 - 5x = 51.50 \times 100$
$\Rightarrow 5x = 5850 - 5150 = 700$
$\Rightarrow x = \dfrac{700}{5} = 140$
∴ Required difference $= (390 - x) - x$
$= 390 - 2x$ [∵ $x = 140$]
$= 390 - 280 = ₹110$

20. **(c)** Let the cost price of transistor be ₹ x.
According to the question,
New CP of transistor $= x \times \dfrac{105}{100}$
and SP of transistor $= \left(\dfrac{115x}{100} + 6\right)$
∴ Profit percentage $= \dfrac{SP - CP}{CP} \times 100$
$\Rightarrow 10 = \dfrac{\dfrac{115x}{100} + 6 - \dfrac{105x}{100}}{\dfrac{105x}{100}} \times 100$
$\Rightarrow 10 = \dfrac{(10x + 600) 100}{105x}$
$\Rightarrow 105x = 100x + 6000 \Rightarrow 5x = 6000$
∴ $x = ₹1200$

21. **(c)** Let CP of one mobile be ₹ x
and CP of 2nd mobile be ₹ $(480 - x)$.
According to the question,
$x \times \dfrac{85}{100} = \dfrac{(480 - x) \times 119}{100}$
$\Rightarrow 85x = 480 \times 119 - 119x$
$\Rightarrow 85x + 119x = 480 \times 119$
$\Rightarrow 204x = 480 \times 119$

$\Rightarrow \quad x = \dfrac{480 \times 119}{204} = ₹280$

∴ 2nd mobile cost price $= 480 - x$
$= 480 - 280 = ₹200$

So, the cost of lower priced mobile is ₹200.

22. (c) ∵ CP of 80 apples $= ₹240$ [given]

∴ CP of 1 apple $= ₹3$

∴ CP of 20 apples $= ₹60$

To earn a profit of 20%,
SP $= 120\%$ of $240 = ₹288$

But he sells $\dfrac{1}{4}$ of his apples, i.e. 20 apples for ₹60.

∴ SP of remaining 60 apples
$= 288 - 60 = ₹228$

∴ SP of 1 apple $= \dfrac{228}{60} = ₹3.80$

23. (d) Let CP of each mobile phone $= ₹x$.

Then, according to the question,

$\left(x + \dfrac{x \times 20}{100}\right) + \left(x + \dfrac{x \times 20}{100} - 1520\right)$
$= (x + x) + \dfrac{(x + x) \times 1}{100}$

$\Rightarrow x + \dfrac{20x}{100} + x + \dfrac{20x}{100} - 1520 = 2x + \dfrac{2x}{100}$

$\Rightarrow \dfrac{40x}{100} - \dfrac{2x}{100} = 1520$

$\Rightarrow \dfrac{38x}{100} = 1520$

$\Rightarrow x = \dfrac{1520 \times 100}{38} = 40 \times 100$

∴ $x = ₹4000$

Hence, CP of each mobile phone is ₹4000.

24. (b) Let cost price be ₹ x.

Selling price at 18% loss $= ₹0.82x$
and selling price at 15% profit $= ₹1.15x$

Now, according to the question,
$1.15x - 0.82x = 99000 \Rightarrow 0.33x = 99000$

∴ $x = \dfrac{99000}{0.33} = ₹300000$

∴ Selling price when car sold at 10% profit
$= \dfrac{300000 \times 110}{100} = ₹330000$

25. (a) Let CP of item $A = ₹x$

Then, CP of item $B = ₹(150 + x)$

Now, SP of item $A = x \times \dfrac{110}{100} = ₹\dfrac{11x}{10}$

and SP of item $B = (x + 150) \times \dfrac{80}{100}$
$= ₹\dfrac{(8x + 1200)}{10}$

Now, according to the question,

$\dfrac{\dfrac{11x}{10}}{(8x + 1200)} = \dfrac{11}{12}$

$\Rightarrow \dfrac{11x}{(8x + 1200)10} = \dfrac{11}{12}$

$\Rightarrow \dfrac{x}{(8x + 1200)} = \dfrac{1}{12}$

$\Rightarrow 12x = 8x + 1200$
$\Rightarrow 12x - 8x = 1200$
$\Rightarrow 4x = 1200$
$\Rightarrow x = 300$

∴ CP of item $B = ₹(150 + x)$
$= ₹(150 + 300) = ₹450$

26. (c) Let the cost price of article B be ₹ x.

Then, the cost price of article A will be ₹ $(x + 200)$.

∴ Selling price of article A
$= (x + 200) - \dfrac{10}{100}(x + 200)$
$= (x + 200) - \dfrac{10x}{100} - \dfrac{2000}{100}$
$= x - \dfrac{x}{10} + 200 - 20$
$= \dfrac{9x}{10} + 180$

Selling price of article B
$= x + \dfrac{25}{100} \times x = \dfrac{5x}{4}$

According to the question,
$\dfrac{9x}{10} + 180 + \dfrac{5x}{4}$
$= (2x + 200) + \dfrac{4}{100}(2x + 200)$

$\Rightarrow \dfrac{9x}{10} + 180 + \dfrac{5x}{4} = 2x + 200 + \dfrac{8x}{100} + 8$

$\Rightarrow \dfrac{9x}{10} + \dfrac{5x}{4} - \dfrac{8x}{100} - 2x = 208 - 180$

$\Rightarrow \dfrac{90x + 125x - 8x - 200x}{100} = 28$

$\Rightarrow \dfrac{7x}{100} = 28$

$\Rightarrow x = 400$

Hence, the cost price of article B is ₹400.

27. (d) 10 notebooks in a day then in two weeks (i.e. 14 days) $= 14 \times 10 = 140$ notebooks.

Commission earned $= 45 \times \dfrac{4}{100} \times 140$
$= ₹252$

6 pencil boxes in a day then in two weeks (i.e. 14 days) $= 14 \times 6$
$= 84$ pencil boxes

Profit and Loss / 229

Commission earned
$$= 80 \times \frac{20}{100} \times 84 = ₹ 1344$$

Total commission earned
$$= 252 + 1344 = ₹ 1596$$

28. (d) Cost of coffee powder used in one cup
$$= \frac{20}{10} = ₹ 2$$

Cost of milk used in one cup
$$= \frac{30}{1000} \times 200 = ₹ 6$$

∴ Cost of each cup of coffee
$$= 2 + 6 = ₹ 8$$

To gain 25% profit, sale price of each cup of coffee = 125% of 8 = ₹ 10

29. (a) By hit and trial,
Case I $t = ₹ 400$ and $q = 20$ units
Then, from option (a),
Total profit = $600q - 5t$
$= 600 \times 20 - 5 \times 400$
$= 12000 - 2000 = ₹ 10000$
Case II $t = ₹ 600$ and $q = 25$ units
Total profit = $600q - 5t$
$= (600 \times 25) - (5 \times 600)$
$= 600(25 - 5)$
$= 600 \times 20 = ₹ 12000$
Hence, the option (a) is correct.

30. (c) According to the question,
$$250 \times 0.5 \times \frac{105}{100} + 250 \times 0.5 \times \frac{x}{100}$$
$$= 500 \times 0.5 \times \frac{110}{100}$$
$$\Rightarrow \frac{250 \times 0.5}{100}(105 + x) = \frac{500 \times 0.5 \times 110}{100}$$
$$\Rightarrow (105 + x) = \frac{500 \times 110}{250}$$
$$\Rightarrow 105 + x = 220$$
$$\Rightarrow x = 220 - 105$$
$$\Rightarrow x = 115$$
∴ Required percentage = 115 − 100 = 15%

31. (a) Let CP of $A = x$, then CP of $B = 200 + x$
According to the question,
$$\frac{\frac{110}{100}(x + 200)}{\frac{60}{100}x} = \frac{11}{4}$$
$$\Rightarrow \frac{x + 200}{6x} = \frac{1}{4} \Rightarrow x = 400$$

If A is sold at 20% loss, then selling price
$$= \frac{80}{100} \times 400 = 320$$

32. (d) Let the cost price of article A be CP_1 and cost price of article B be CP_2.

Cost price = $\dfrac{\text{Profit}}{\text{Profit\%}} \times 100$

$$\Rightarrow CP_1 = \frac{P_1}{40} \times 100 = \frac{P_1}{4} \times 10$$

and $CP_2 = \dfrac{P_2}{10} \times 100$

$$\Rightarrow CP_2 = P_2 \times 10$$

Hence, $\dfrac{CP_1}{CP_2} = \dfrac{10 P_1}{4} \times \dfrac{1}{10 P_2} = \dfrac{P_1}{4 P_2}$

$$\Rightarrow \frac{P_1}{4 P_2} = \frac{7}{9} \quad \left[\because \frac{CP_1}{CP_2} = \frac{7}{9}\text{(given)}\right]$$

$$\Rightarrow P_1 = \frac{28 P_2}{9}$$

Now, $P_1 + P_2 = 148 \Rightarrow \dfrac{28 P_2}{9} + P_2 = 148$

$$\Rightarrow 37 P_2 = 148 \times 9$$
$$\Rightarrow P_2 = 4 \times 9 = ₹ 36$$

and $P_1 = \dfrac{28 P_2}{9} = \dfrac{28}{9} \times 36 = ₹ 112$

∴ $CP_1 = \dfrac{P_1}{40} \times 100 = \dfrac{112}{40} \times 100 = ₹ 280$

and $CP_2 = \dfrac{P_2}{10} \times 100 = \dfrac{36}{10} \times 100 = ₹ 360$

∴ Difference of cost price of Article A and B
$= CP_2 - CP_1$
$= (360 - 280) = ₹ 80$

Chapter 13

Discount

Discount is defined as the amount of rebate given on a fixed price (called as marked price) of an article. It is given by merchants/shopkeepers to increase their sales by attracting customers.

Discount = Marked price − Selling price

Marked Price (List Price)

The price on the label of an article/product is called the marked price or list price. This is the price at which product is intended to be sold. However, there can be some discount given on this price and actual selling price of the product may be less than the marked price. It is generally denoted by MP.

Basic Formulae Related to Discount

(i) Discount% = $\dfrac{\text{Marked price} - \text{Selling price}}{\text{Marked price}} \times 100$

(ii) Discount% = $\dfrac{\text{Discount}}{\text{Marked price}} \times 100$

(iii) Selling price = Marked price $\times \left(1 - \dfrac{r}{100}\right)$

where, $r\%$ is the rate of discount allowed.

(iv) Selling price = Marked price − Discount

✦ Discount is always calculated with respect to marked price of an article.

Successive Discount

When a series of discounts (one after the other) are allowed on the marked price of an article, then these discounts are called successive discounts.

Let $r_1\%, r_2\%, r_3\%, \ldots$ be the series of discounts on an article with marked price of ₹ P, then the selling price of the article after all the discounts is given as

$$P \times \left(1 - \dfrac{r_1}{100}\right) \times \left(1 - \dfrac{r_2}{100}\right) \times \left(1 - \dfrac{r_3}{100}\right) \times \ldots$$

Discount / 231

Ex. 1 An item is sold for ₹ 680 by allowing a discount of 15% on its marked price. Find the marked price of an item.

Sol. Here, selling price of an item = ₹ 680
and rate of discount, $r = 15\%$

∴ Selling price = Marked price $\left(1 - \dfrac{r}{100}\right) \Rightarrow 680 = MP\left(1 - \dfrac{15}{100}\right)$

$\Rightarrow 680 = MP\left(\dfrac{100-15}{100}\right) \Rightarrow MP = \dfrac{680 \times 100}{85} = ₹ 800$

Ex. 2 A shopkeeper on the eve of Diwali allowed a series of discount on television sets. Find the selling price of a television set, if the marked price of television is ₹ 1000 and successive discounts are 10% and 5%.

Sol. Selling price of a television = Marked price $\times \left(1 - \dfrac{r_1}{100}\right)\left(1 - \dfrac{r_2}{100}\right)$

Here, marked price = ₹ 1000, $R_1 = 10\%$ and $R_2 = 5\%$

∴ Selling price = $1000\left(1 - \dfrac{10}{100}\right)\left(1 - \dfrac{5}{100}\right) = 1000\left(\dfrac{100-10}{100}\right)\left(\dfrac{100-5}{100}\right)$

$= 1000 \times \dfrac{90}{100} \times \dfrac{95}{100} = ₹ 855$

Fast Track Techniques
to solve the QUESTIONS

Technique 1 Single discount equivalent to two successive discounts $r_1\%$ and $r_2\%$
$$= \left(r_1 + r_2 - \dfrac{r_1 \times r_2}{100}\right)\%$$

Ex. 3 What will be a single equivalent discount for successive discounts of 10% and 5% on marked price of an article?

Sol. Let the marked price of the article be ₹ 100.
Then, selling price = Marked price $\left(1 - \dfrac{r_1}{100}\right)\left(1 - \dfrac{r_2}{100}\right) = 100\left(\dfrac{90}{100}\right)\left(\dfrac{95}{100}\right) = ₹ 85.5$

∴ Equivalent discount = $\dfrac{(100 - 85.5)}{100} \times 100 = 14.5\%$

Fast Track Method

Given, $r_1 = 10\%$ and $r_2 = 5\%$

Single equivalent discount = $\left(r_1 + r_2 - \dfrac{r_1 \times r_2}{100}\right)\%$

$= \left(10 + 5 - \dfrac{10 \times 5}{100}\right)\% = \left(15 - \dfrac{50}{100}\right)\%$

$= (15 - 0.5)\% = 14.5\%$

Hence, the single equivalent discount is 14.5%.

Technique 2 Single discount equivalent to three successive discounts $r_1\%$, $r_2\%$ and $r_3\%$

$$= \left[1 - \left(1 - \frac{r_1}{100}\right)\left(1 - \frac{r_2}{100}\right)\left(1 - \frac{r_3}{100}\right)\right] \times 100\%$$

Ex. 4 What is the single equivalent discount for successive discounts of 10%, 20% and 15% on marked price for motor bike?

Sol. Given, $r_1 = 10\%$, $r_2 = 20\%$ and $r_3 = 15\%$

∴ Single equivalent discount for three successive discounts

$$= \left[1 - \left(1 - \frac{r_1}{100}\right)\left(1 - \frac{r_2}{100}\right)\left(1 - \frac{r_3}{100}\right)\right] \times 100\%$$

$$= \left[1 - \left(1 - \frac{10}{100}\right)\left(1 - \frac{20}{100}\right)\left(1 - \frac{15}{100}\right)\right] \times 100\%$$

$$= \left[1 - \left(\frac{90}{100}\right)\left(\frac{80}{100}\right)\left(\frac{85}{100}\right)\right] \times 100\% = \left[1 - \frac{9}{10} \times \frac{4}{5} \times \frac{17}{20}\right] \times 100\%$$

$$= \left[\frac{1000 - 612}{1000}\right] \times 100\% = \frac{388}{1000} \times 100\% = 38.8\%$$

Technique 3 If a shopkeeper wants a profit of $R\%$ after allowing a discount of $r\%$, then

Marked Price (MP) of the item $= CP\left(\dfrac{100 + R}{100 - r}\right)$

or Cost Price (CP) of the item $= MP\left(\dfrac{100 - r}{100 + R}\right)$

Ex. 5 The marked price of a bicycle is ₹ 1100. A shopkeeper allows a discount of 10% and get a profit of 10%. Find the cost price of the bicycle.

Sol. Let CP of the bicycle be ₹ x.

Then, SP of the bicycle $= \dfrac{110x}{100} = ₹ \dfrac{11x}{10}$

Now, MP of the bicycle = ₹ 1100

∴ SP of the bicycle $= MP \times \left(1 - \dfrac{\text{Discount}\%}{100}\right) = \dfrac{1100 \times 90}{100} = ₹ 990$

According to the question, $\dfrac{11x}{10} = 990 \Rightarrow x = \dfrac{990 \times 10}{11} = ₹ 900$

Fast Track Method

Here, MP = ₹ 1100, $r = 10\%$ and $R = 10\%$

∴ $CP = MP\left(\dfrac{100 - r}{100 + R}\right) = 1100 \times \dfrac{(100 - 10)}{(100 + 10)} = \dfrac{1100 \times 90}{110} = ₹ 900$

Technique 4 A merchant fixes the marked price of an article in such a way that after allowing a discount of $r\%$, he earns a profit of $R\%$. Then, marked price of the article is $\left(\dfrac{r + R}{100 - r} \times 100\right)\%$ more than its cost price.

Discount / 233

Ex. 6 A shopkeepper allows a discount of 10% on the marked price of calculator, then by what per cent higher than cost price, should the marked price be, so as to gain 20% on selling it at the discount?

Sol. Let the cost price of calculator be ₹ 100.
Then, selling price = 120% of 100   [∵ profit per cent is 20%]

$$= \frac{120 \times 100}{100} = ₹\ 120$$

Now, the selling price of calculator should be 10% less than marked price, because of the discount of 10%.

∴ Marked price $= \dfrac{100 \times SP}{100 - 10} = \dfrac{100 \times 120}{90} = \dfrac{400}{3}$

∴ Required percentage at which calculator is marked higher than cost price

$$= \frac{MP - CP}{CP} \times 100\% = \frac{\frac{400}{3} - 100}{100} \times 100$$

$$= \frac{(400 - 300)}{3 \times 100} \times 100 = \frac{100}{3} = 33\frac{1}{3}\%$$

Fast Track Method

Here, $r = 10\%$ and $R = 20\%$

∴ Required percentage $= \left(\dfrac{r + R}{100 - r} \times 100\right)\% = \dfrac{10 + 20}{100 - 10} \times 100$

$$= \frac{30 \times 100}{90} = \frac{100}{3} = 33\frac{1}{3}\%$$

Technique 5 If a shopkeeper allows a discount of r_1% on an article and marked price of the article is r% more than the cost price, then

$$\text{Profit /Loss per cent in this transaction} = \frac{r \times (100 - r_1)}{100} - r_1$$

✦ Positive value shows a profit, while negative value shows a loss.

Ex. 7 A shopkeeper marked the price 10% more than its cost price. If he allows a discount of 20%, then find his loss per cent.

Sol. Let the cost price of the item = ₹ 100

∴ Marked price $= \dfrac{(100 + 10)}{100} \times 100 = ₹\ 110$

and selling price $= \dfrac{(100 - 20)}{100} \times 110 = ₹\ 88$

∴ Required loss percentage $= \dfrac{(100 - 88)}{100} \times 100 = 12\%$

Fast Track Method

Here, $r_1 = 20\%$ and $r = 10\%$

∴ Loss or profit per cent $= \dfrac{r \times (100 - r_1)}{100} - r_1$

$$= \frac{10 \times (100 - 20)}{100} - 20$$

$$= \frac{10 \times 80}{100} - 20 = -12\%$$

Hence, the loss per cent is 12%.

Fast Track Practice

1. If the marked price of a fan is ₹ 700 and a discount of 10% is given on it, then what is the selling price of the fan?
   (a) ₹ 500 (b) ₹ 575
   (c) ₹ 610 (d) ₹ 630

2. On a 20% discount sale, an article costs ₹ 596. What was the original price of the article?
   (a) ₹ 720 (b) ₹ 735
   (c) ₹ 745 (d) ₹ 775

3. A shopkeeper has announced 14% rebate on marked price of an article. If the selling price of the article is ₹ 645, then the marked price of the article will be
   [SSC CPO 2013]
   (a) ₹ 800 (b) ₹ 810
   (c) ₹ 750 (d) ₹ 775

4. What is the maximum percentage discount (approximately) that a merchant can offer on his marked price, so that he ends up selling at no profit or loss, if he initially marked his goods up by 40%?
   [SSC CGL 2013]
   (a) 60% (b) 28.5%
   (c) 33.5% (d) No discount

5. Marked price of a cycle is ₹ 2000. It sells by two successive discounts of 20% and 10%, respectively. On cash payment, additional 5% discount is also given. On cash payment, SP of the cycle would be
   (a) ₹ 1368 (b) ₹ 1648
   (c) ₹ 1568 (d) ₹ 1668

6. A shopkeeper offers 15% discount on all plastic toys. He offers a further discount of 4% on the reduced price to those customers who pay cash. What does a customer have to pay in cash for a toy of ₹ 200? [SSC CGL (Mains) 2016]
   (a) ₹ 133.7 (b) ₹ 129.8
   (c) ₹ 163.2 (d) ₹ 153.3

7. The difference between a discount of 30% on ₹ 2000 and two successive discounts of 25% and 5% on the same amount is
   [SSC FCI 2012]
   (a) ₹ 30 (b) ₹ 35
   (c) ₹ 25 (d) ₹ 40

8. A shopkeeper bought a table marked at ₹ 200 at successive discount of 10% and 15%, respectively. He spent ₹ 7 on transport and sold the table for ₹ 208. What will be his profit percentage?
   [IBPS PO 2015]
   (a) 30% (b) 40%
   (c) 55% (d) 45%
   (e) 32%

9. A retailer offers the following discount schemes for buyers on an article.
   I. Two successive discounts of 10%.
   II. A discount of 12% followed by a discount of 8%.
   III. Successive discounts of 15% and 5%.
   IV. A discount of 20%.
   The selling price will be minimum under the scheme [SSC CGL 2012]
   (a) I (b) II (c) III (d) IV

10. If on a sale, there is 40% discount on the marked price of ₹ 1000, but the sale is done at ₹ 510 only, then what additional discount did the customer get?
    [SSC CGL (Pre) 2017]
    (a) 25% (b) 15%
    (c) 10% (d) 30%

11. A dinner set is quoted for ₹ 1500. A customer pays ₹ 1173 for it. If the customer got a series of two discounts and the rate of first discount is 15%, then the rate of second discount was
    [SSC CGL (Mains) 2016]
    (a) 15% (b) 7%
    (c) 9% (d) 8%

12. If on a marked price, the difference of selling prices with a discount of 30% and two successive discounts of 20% and 10% is ₹ 72, then the marked price (in ₹) is
    [SSC CGL 2010]
    (a) 3600 (b) 3000
    (c) 2500 (d) 2400

13. The successive discounts of 10%, 20% and 30% is equivalent to single discount of [SSC CGL 2010]
    (a) 60% (b) 49.6%
    (c) 40.5% (d) 36%

14. A shopkeeper earns a profit of 12% on selling a book at 10% discount on the printed price. The ratio of the cost price and the printed price of the book is
    [SSC CPO 2013]
    (a) 45 : 56 (b) 8 : 11
    (c) 47 : 56 (d) 3 : 4

Discount / 235

15. A manufacturer marked an article at ₹ 50 and sold it allowing 20% discount. If his profit was 25%, then the cost price of the article was [SSC CGL 2012]
(a) ₹ 40 (b) ₹ 35
(c) ₹ 32 (d) ₹ 30

16. The cost price of an article is ₹ 800. After allowing a discount of 10%, a gain of 12.5% was made. Then, the marked price of the article is [SSC CGL 2011]
(a) ₹ 1000 (b) ₹ 1100
(c) ₹ 1200 (d) ₹ 1300

17. The marked price of a radio is ₹ 480. The shopkeeper allows a discount of 10% and gains 8%. If no discount is allowed, his gain per cent would be [SSC CGL 2011]
(a) 18% (b) 18.5%
(c) 20.5% (d) 20%

18. A shopkeeper allows a discount of 10% to his customers and still gains 20%, the marked price of the article which costs ₹ 450, is [SSC CGL 2013]
(a) ₹ 600 (b) ₹ 540
(c) ₹ 660 (d) ₹ 580

19. A photographer allows a discount of 10% on the advertised price of a camera. The price that must be marked on the camera, which cost him ₹ 600, to make a profit of 20%, would be
[SSC CGL (Mains) 2016]
(a) ₹ 650 (b) ₹ 800
(c) ₹ 700 (d) ₹ 850

20. After allowing a discount of 16%, there was still a gain of 5%. Then, the percentage of marked price over the cost price is [SSC CPO 2011]
(a) 15% (b) 18%
(c) 21% (d) 25%

21. A shopkeeper sold an article at 20% discount and earned a profit of 4%. By what per cent the marked price of the article more than the cost price of the article? [SBI Clerk (Pre) 2016]
(a) 20% (b) 15%
(c) 40% (d) 25%
(e) 30%

22. A merchant purchases a wrist watch for ₹ 450 and fixes its list price in such a way that after allowing a discount of 10%, he earns a profit of 20%. Then, the list price of the watch is
[SSC Multitasking 2013; SSC CPO 2010]
(a) ₹ 600 (b) ₹ 650
(c) ₹ 700 (d) ₹ 550

23. A merchant marked the price on his goods 20% more than its cost price and allows a discount of 15%. His profit per cent is [SSC CGL 2011]
(a) 1% (b) 2% (c) 10% (d) 15%

24. The printed price of an article is 40% higher than its cost price. Then, the rate of discount, so that he gains 12% profit, is [SSC CGL 2015]
(a) 20% (b) 15%
(c) 21% (d) 18%

25. A dealer marked the price of an item 40% above the cost price. Once he gave successive discounts of 20% and 25% to a particular customer. As a result, he incurred a loss of ₹ 448. At what price did he sell the item to the mentioned customer? [IBPS PO 2015]
(a) ₹ 2416 (b) ₹ 2352
(c) ₹ 2268 (d) ₹ 2152
(e) ₹ 2578

26. A shopkeeper increases the cost price of an item by 20% and offers a discount of 10% on this marked price. What is his gain percentage? [CDS 2016 (II)]
(a) 15% (b) 12%
(c) 10% (d) 8%

27. If an electricity bill is paid before due date, one gets a reduction of 4% on the amount of the bill. By paying the bill before due date, a person got a reduction of ₹ 13. The amount of his electricity bill was [SSC CGL 2013]
(a) ₹ 125 (b) ₹ 225
(c) ₹ 325 (d) ₹ 425

28. A shopkeeper marks an article at ₹ 60 and sells at a discount of 15%. He also gives a gift worth ₹ 3. If he still makes 20% profit, then the cost price is
[SSC CPO 2011]
(a) ₹ 22 (b) ₹ 32
(c) ₹ 40 (d) ₹ 42

29. A merchant has announced 25% rebate on prices of readymade garments at the time of sale. If a purchaser needs to have a rebate of ₹ 400, then how many shirts, each costing ₹ 320, should he purchase?
[SSC CGL 2010]
(a) 10 (b) 7 (c) 6 (d) 5

30. A dozen pair of socks quoted at ₹ 80 are available at a discount of 10%. How many pair of socks can be bought for ₹ 24? [SSC Multitasking 2013]
(a) 4 (b) 5 (c) 3 (d) 6

31. If a shopkeeper sold a book with 20% profit after giving a discount of 10% on marked price. The ratio of cost price and marked price of the book is **[SSC Constable 2012]**
    (a) 6 : 5
    (b) 5 : 6
    (c) 3 : 4
    (d) 2 : 3

32. By selling an article at 3/4th of the marked price, there is a gain of 25%. The ratio of marked price and the cost price is **[SSC CGL 2012]**
    (a) 5 : 3
    (b) 3 : 5
    (c) 3 : 4
    (d) 4 : 3

33. Rita bought a television set with 20% discount on the labelled price. She made a profit of ₹ 800 by selling it for ₹ 16800. The labelled price of the set was **[SSC CGL 2010]**
    (a) ₹ 10000
    (b) ₹ 20000
    (c) ₹ 20800
    (d) ₹ 24000

34. Articles are marked at a price which gives a profit of 25%. After allowing a certain discount, the profit reduces to $12\frac{1}{2}$%. The discount per cent is **[SSC CGL (Mains) 2015]**
    (a) $12\frac{1}{2}$%
    (b) 12%
    (c) 10%
    (d) 11.1%

35. Glenn labelled the price of an article in such a way so as to earn 25% profit. However, while selling he offered 6% discount on the labelled price. If he sold it for ₹ 10340, then what was the cost price of the article? **[SBI Clerk 2015]**
    (a) ₹ 9200
    (b) ₹ 8000
    (c) ₹ 8800
    (d) ₹ 8600
    (e) ₹ 8400

36. A trader sells an item to a retailer at 20% discount, but charges 10% on the discounted price, for delivery and packaging. The retailer sells it for ₹ 2046 more, thereby earning a profit of 25%. At what price had the trader marked the item? **[RBI Officer Grade 2015]**
    (a) ₹ 9400
    (b) ₹ 9000
    (c) ₹ 8000
    (d) ₹ 12000
    (e) ₹ 9300

37. In order that there may be a profit of 20% after allowing a discount of 10% on the marked price, the cost price of an article has to be increased by **[SSC CGL 2012]**
    (a) 30%
    (b) 33%
    (c) $33\frac{1}{3}$%
    (d) $33\frac{2}{3}$%

38. The marked price of a clock is ₹ 3200. It is to be sold at ₹ 2448 at two successive discounts. If the first discount is 10%, then the second discount is **[SSC CGL 2010]**
    (a) 5%
    (b) 10%
    (c) 15%
    (d) 20%

39. If the price of an item is increased by 30% and then allows two successive discounts of 10% and 10%. In last the price of an item is **[SSC CGL 2011]**
    (a) increased by 10%
    (b) increased by 5.3%
    (c) decreased by 3%
    (d) decreased by 5.3%

40. A dealer fixed the price of an article 40% above the cost of production. While selling it he allows a discount of 20% and makes a profit of ₹ 48. The cost of production (in ₹) of the article is **[SSC CGL (Mains) 2015]**
    (a) 400
    (b) 360
    (c) 320
    (d) 420

41. A man bought an article listed at ₹ 1500 with a discount of 20% offered on the list price. What additional discount must be offered to the man to bring the net price to ₹ 1104? **[SSC CGL 2011]**
    (a) 8%
    (b) 10%
    (c) 12%
    (d) 15%

42. At 9% discount, the selling price of a washing machine is ₹ 14000, what is the selling price, if the discount is 22%? **[SSC (10+2) 2017]**
    (a) ₹ 12000
    (b) ₹ 9360
    (c) ₹ 10202.4
    (d) ₹ 13322.4

43. While selling, a businessman allows 40% discount on the marked price and there is a loss of 30%. If it is sold at the marked price, then profit per cent will be **[SSC CGL 2012]**
    (a) 10%
    (b) 20%
    (c) $16\frac{2}{3}$%
    (d) $16\frac{1}{3}$%

44. The cost price of an article is 64% of the marked price. The gain percentage after allowing a discount of 12% on the marked price is **[SSC CGL 2013]**
    (a) 37.5%
    (b) 48%
    (c) 50.5%
    (d) 52%

45. By selling an article at 80% of its marked price, a trader makes a loss of 10%, what will be a profit percentage, if he sells it at 95% of its marked price? **[SSC CGL 2012]**
    (a) 6.9%
    (b) 5%
    (c) 5.9%
    (d) 12.5%

Discount / 237

46. The Maximum Retail Price (MRP) of a product is 55% above its manufacturing cost. The product is sold through a retailer, who earns 23% profit on his purchase price. What is the profit percentage (expressed in nearest integer) for the manufacturer, who sells his product to the retailer, if the retailer gives 10% discount on MRP? [XAT 2015]
(a) 31%   (b) 22%   (c) 15%   (d) 13%
(e) 11%

47. If a commission of 10% is given on the written price of an article, the gain is 20%. The gain per cent, when the commission is increased to 20%, will be [SSC FCI 2012]
(a) $6\frac{2}{3}$%   (b) 5%   (c) 8%   (d) $5\frac{1}{3}$%

48. A shopkeeper sells notebooks at the rate of ₹ 457 each and earns a commission of 4%. He also sells pencil boxes at the rate of ₹ 80 each and earns a commission of 20%. How much amount of commission will he earn in two weeks, if he sells 10 notebooks and 6 pencil boxes a day? [CBI PO 2010]
(a) ₹ 1956   (b) ₹ 1586
(c) ₹ 1496   (d) ₹ 1596
(e) None of these

49. A shopkeeper allows 23% commission on his advertised price and still makes a profit of 10%. If he gains ₹ 56 on one item, then his advertised price of the item (in ₹) is [SSC CGL 2013]
(a) 820   (b) 780   (c) 790   (d) 800

50. Amit and Roshan, two shopkeepers, buy an article for ₹ 1000 and ₹ 2000, respectively. Roshan marks his article up by $2x$% and offers a discount of x%, while Amit marks his article up by x%. If both make the same profit, then what is the value of x?
(a) 40%   (b) 37.5%
(c) 12.5   (d) 25%
(e) 50%

51. The cost of price of article B is 20% more than that of article A. Articles A and B were marked up by 50% and 25%, respectively. Article A was sold at a discount of 4% and article B was sold at a discount of 0.5%. If the selling price of article A was ₹ 30 less than the selling price of article B, what was the cost price of article A? [IBPS Clerk (Mains) 2017]
(a) ₹ 600   (b) ₹ 400
(c) ₹ 500   (d) ₹ 570
(e) ₹ 400

Answer with Solutions

1. (d) Given, marked price = ₹ 700 and discount (r) = 10%
∴ Selling price = Marked price $\left(1 - \frac{r}{100}\right)$
$= 700\left(1 - \frac{10}{100}\right) = ₹ 630$

2. (c) Let the original price be ₹ x.
Since, after discount of 20% article costs ₹ 596.
∴ $596 = \frac{(100-20)}{100} \times x$
⇒ $x = \frac{596 \times 100}{80} = ₹ 745$

3. (c) Here, $r = 14$% and SP = ₹ 645
∴ Marked price of an item $= \frac{SP \times 100}{100 - r}$
$= \frac{645 \times 100}{100 - 14} = \frac{64500}{86} = ₹ 750$

4. (b) Let cost price = ₹ 100 and marked price = 100 + 40 = ₹ 140

Let required discount be x%.
According to the question,
$140 \times \left(\frac{100 - x}{100}\right) = 100$
⇒ $100 - x = \frac{100 \times 100}{140}$
∴ $x = 100 - \frac{100 \times 100}{140} = \frac{40 \times 100}{140}$
$= 28.5$% (approx.)

5. (a) Here, marked price of cycle = ₹ 2000,
$r_1 = 20$%, $r_2 = 10$% and $r_3 = 5$%
∴ SP of cycle = Marked price
$\times \left(1 - \frac{r_1}{100}\right)\left(1 - \frac{r_2}{100}\right)\left(1 - \frac{r_3}{100}\right)$
$= 2000\left(1 - \frac{20}{100}\right)\left(1 - \frac{10}{100}\right)\left(1 - \frac{5}{100}\right)$
$= 2000 \times \frac{80}{100} \times \frac{90}{100} \times \frac{95}{100}$
$= \frac{2 \times 8 \times 9 \times 95}{10} = \frac{144 \times 19}{2} = 72 \times 19$
$= ₹ 1368$

6. (c) Single discount against two discounts
$$= \left(15 + 4 - \frac{15 \times 4}{100}\right)\%$$
$= (19 - 0.6)\% = 18.4\%$ [by Technique 1]
∴ Payment against ₹ 200 toy
$$= \left(200 - 200 \times \frac{18.4}{100}\right)$$
$= (200 - 36.8) = ₹ 163.20$

7. (c) ∵ Single equivalent discount per cent to 25% and 5%
$$= \left(r_1 + r_2 - \frac{r_1 \times r_2}{100}\right)\% \quad \text{[by Technique 1]}$$
$= 25 + 5 - \frac{25 \times 5}{100} = 30 - 1.25 = 28.75\%$
∴ Required difference $= \frac{(30 - 28.75) \times 2000}{100}$
$= \frac{1.25 \times 2000}{100} = ₹ 25$

8. (a) Single discount corresponding to 10% and 15% $= \left(r_1 + r_2 - \frac{r_1 \times r_2}{100}\right)\%$
[by Technique 1]
$= 10 + 15 - \frac{10 \times 15}{100} = 23.5\%$
Given, MP = ₹ 200
Discount = 23.5%
Money spent on transport = ₹ 7
∴ CP = 76.5% of 200 + 7
$= \frac{76.5}{100} \times 200 + 7 = ₹ 160$
SP = ₹ 208
∴ Percentage of profit $= \frac{208 - 160}{160} \times 100$
$= \frac{48 \times 100}{160} = 30\%$

9. (d) I. Equivalent single discount to 10% and 10% $= \left(10 + 10 - \frac{10 \times 10}{100}\right)\% = 19\%$
[by Technique 1]
II. Equivalent single discount to 12% and 8%
$= 12 + 8 - \frac{12 \times 8}{100}$
$= 20 - 0.96 = 19.04\%$
III. Equivalent single discount to 15% and 5%
$= 15 + 5 - \frac{15 \times 5}{100}$
$= 20 - 0.75$
$= 19.25\%$
IV. Equivalent single discount to 20% = 20%
So, the selling price will be minimum under the scheme IV as in this scheme, the discount is maximum.

10. (b) Here, marked price = ₹ 1000
and discount = 40%
∴ Selling price $= 1000 - 1000 \times \frac{40}{100}$
$= 1000 - 400 = ₹ 600$
Real selling price = ₹ 510
∴ Additional discount $= \frac{600 - 510}{600} \times 100$
$= \frac{90}{600} \times 100 = \frac{90}{6} = 15\%$

11. (d) Here, SP = ₹ 1173,
marked price = ₹ 1500
and $r_1 = 15\%$
∴ $SP = MP\left(1 - \frac{r_1}{100}\right)\left(1 - \frac{r_2}{100}\right)$
$\Rightarrow 1173 = 1500\left(1 - \frac{15}{100}\right)\left(1 - \frac{r_2}{100}\right)$
$\Rightarrow 1173 = 15 \times 85 \times \frac{(100 - r_2)}{100}$
$\Rightarrow (100 - r_2) = \frac{1173 \times 100}{15 \times 85} = 92$
∴ $r_2 = 100 - 92 = 8\%$

12. (a) Let the marked price be ₹ x.
The discount equivalent to successive discounts of 20% and 10%
$= \left(r_1 + r_2 - \frac{r_1 r_2}{100}\right)\%$ [by Technique 1]
where, $r_1 = 20$ and $r_2 = 10$
$= \left(20 + 10 - \frac{20 \times 10}{100}\right)$
$= 30 - 2 = 28\%$
According to the question,
$\frac{(100 - 28)x}{100} - \frac{(100 - 30)x}{100} = 72$
$\Rightarrow \frac{72x - 70x}{100} = 72$
∴ $x = \frac{72 \times 100}{2} = ₹ 3600$

13. (b) Here, $r_1 = 10\%$, $r_2 = 20\%$ and $r_3 = 30\%$
∴ Required discount
$= \left[1 - \left(1 - \frac{r_1}{100}\right)\left(1 - \frac{r_2}{100}\right)\left(1 - \frac{r_3}{100}\right)\right]$
$\times 100\%$
[by Technique 2]
$= \left[1 - \left(1 - \frac{10}{100}\right)\left(1 - \frac{20}{100}\right)\left(1 - \frac{30}{100}\right)\right]$
$\times 100\%$
$= \left(1 - \frac{9}{10} \times \frac{4}{5} \times \frac{7}{10}\right) \times 100\%$
$= (1 - 0.504) \times 100\% = 49.6\%$

Discount / 239

14. (a) Let the CP of book be ₹ x.
Then, SP of book = $\dfrac{(100+12) \times x}{100} = \dfrac{112x}{100}$
Now, the printed price = ₹ y
Then, after discount,
$$SP = \dfrac{(100-10) \times y}{100} = \dfrac{90y}{100}$$
Since, both SP are same.
Then, $\dfrac{112x}{100} = \dfrac{90y}{100} \Rightarrow \dfrac{x}{y} = \dfrac{45}{56} = 45:56$

Fast Track Method
Here, $R = 12\%$ and $r = 10\%$
According to the formula,
$\dfrac{CP}{MP} = \dfrac{100-r}{100+R}$ [by Technique 3]
$= \dfrac{100-10}{100+12} = \dfrac{90}{112} = \dfrac{45}{56}$

15. (c) ∵ Marked price of the article = ₹ 50
∴ SP of the article = $\dfrac{50 \times (100-20)}{100}$
$= \dfrac{50 \times 80}{100} = ₹ 40$
Hence, cost price of the article = $\dfrac{40 \times 100}{(100+25)}$
$= \dfrac{40 \times 100}{125} = ₹ 32$

Fast Track Method
Here, MP = ₹50, $r = 20\%$ and $R = 25\%$
∴ CP = MP $\left(\dfrac{100-r}{100+R}\right)$ [by Technique 3]
$= 50 \left(\dfrac{100-20}{100+25}\right)$
$= \dfrac{50 \times 80}{125} = ₹ 32$

16. (a) Let marked price of the article be ₹ x.
∴ SP of the article after a discount of 10%
$= x \times \left(\dfrac{100-10}{100}\right) = ₹ \dfrac{9x}{10}$
⇒ CP of the article with a profit of 12.5%
$= \dfrac{9x}{10} \times \dfrac{100}{100+12.5}$
$= ₹ \dfrac{9x}{10} \times \dfrac{100}{112.5}$
But CP of the article = ₹ 800
⇒ $\dfrac{9x}{10} \times \dfrac{100}{112.5} = 800$
∴ $x = \dfrac{800 \times 112.5 \times 10}{9 \times 100} = ₹ 1000$

Fast Track Method
Here, CP = ₹ 800, $r = 10\%$ and $R = 12.5\%$

∴ Marked Price (MP) = $\dfrac{CP \times (100+R)}{(100-r)}$
[by Technique 3]
$= \dfrac{800 \times (100+12.5)}{(100-10)}$
$= \dfrac{800 \times 112.5}{90} = ₹ 1000$

17. (d) ∵ Marked price of a radio = ₹ 480
∴ SP of a radio = $480 \times \left(\dfrac{100-10}{100}\right)$
$= \dfrac{480 \times 90}{100} = ₹ 432$
⇒ CP of a radio = $\dfrac{432 \times 100}{100+8}$
$= \dfrac{432 \times 100}{108} = ₹ 400$
Profit when article is sold at MP
$= 480 - 400 = ₹ 80$
Hence, profit per cent = $\dfrac{80}{400} \times 100\%$
$= 20\%$

Fast Track Method
Here, MP = ₹ 480, $r = 10\%$ and $R = 8\%$
∴ CP = $\dfrac{MP \times (100-r)}{100+R}$ [by Technique 3]
$= \dfrac{480 \times (100-10)}{100+8}$
$= \dfrac{480 \times 90}{108} = ₹ 400$
∴ Profit = $480 - 400 = ₹ 80$
Hence, profit per cent
$= \dfrac{80}{400} \times 100\% = 20\%$

18. (a) Here, $r = 10\%$, $R = 20\%$ and CP = ₹450
∴ MP = CP $\times \left(\dfrac{100+R}{100-r}\right)$
[by Technique 3]
$= 450 \times \left(\dfrac{100+20}{100-10}\right) = 450 \times \dfrac{120}{90}$
$= ₹ 600$

19. (b) Here, CP = ₹ 600, profit = 20%,
discount = 10%
Let marked price = ₹ x
∴ SP = $x - \dfrac{x \times 10}{100} = \dfrac{90x}{100} = ₹ \dfrac{9x}{10}$
Now, SP = CP $\dfrac{(100+Profit\%)}{100}$
⇒ $\dfrac{9x}{10} = 600 \times \dfrac{(100+20)}{100}$

$\Rightarrow \quad \dfrac{9x}{10} = 6 \times 120$

$\therefore \quad x = \dfrac{6 \times 120 \times 10}{9} = ₹\, 800$

Fast Track Method

Here, CP = 600, R = 20 and r = 10

$\therefore$ MP = CP $\left(\dfrac{100 + R}{100 - r}\right)$ [by Technique 3]

$= 600 \left(\dfrac{120}{90}\right) = ₹\, 800$

20. (d) Here, r = 16% and R = 5%

$\therefore$ Required percentage

$= \left(\dfrac{r + R}{100 - r}\right) \times 100\,\%$ [by Technique 4]

$= \dfrac{16 + 5}{100 - 16} \times 100\% = \dfrac{21 \times 100}{84}\% = 25\%$

21. (e) Let MP of an article be ₹ x.

Then, SP $= x - \dfrac{x \times 20}{100} = \dfrac{80x}{100}$

Now, CP $= \dfrac{SP \times 100}{(100 + \text{Profit}\%)} = \dfrac{80x}{100} \times \dfrac{100}{(100 + 4)}$

$= \dfrac{80x}{104}$

$\therefore$ Required percentage $= \dfrac{x - \dfrac{80x}{104}}{\dfrac{80x}{104}} \times 100$

$= \dfrac{24x}{80x} \times 100 = 30\%$

Fast Track Method

Here, R = 4% and r = 20%

$\therefore$ Required percentage

$= \left(\dfrac{R + r}{100 - r}\right) \times 100\,\%$ [by Technique 4]

$= \left(\dfrac{4 + 20}{100 - 20}\right) \times 100\,\%$

$= \left(\dfrac{24}{80} \times 100\right)\% = 30\%$

22. (a) CP of watch = ₹ 450

SP of watch $= 450 \left(\dfrac{100 + 20}{100}\right) = ₹\, 540$

$\therefore$ List price of watch $= 540 \times \dfrac{100}{(100 - 10)}$

$= ₹\, 600$

Fast Track Method

We know that, to get a profit of R% after allowing a discount of r%, the marked price should be marked $\left(\dfrac{r + R}{100 - r}\right) \times 100\,\%$ more than its cost price. [by Technique 4]

Here, r = 10% and R = 20%

$\therefore$ Required per cent

$= \left(\dfrac{10 + 20}{100 - 10}\right) \times 100\,\% = \dfrac{100}{3}\%$

and CP of the watch = ₹ 450

Hence, MP of the watch

$= 450 \left(\dfrac{100 + \dfrac{100}{3}}{100}\right)$

$= \dfrac{450 \times 400}{300} = ₹\, 600$

23. (b) Let the cost price of item be ₹ x.

$\because$ Marked price of item $= \dfrac{x \times (100 + 20)}{100}$

$= ₹\, \dfrac{6x}{5}$

Then, selling price of item

$= \dfrac{6x}{5} \times \left(\dfrac{100 - 15}{100}\right)$

$= \dfrac{6x \times 17}{5 \times 20} = ₹\, \dfrac{102}{100} x$

$\Rightarrow$ Profit $= \left(\dfrac{102x}{100} - x\right) = ₹\, \dfrac{2}{100} x$

$\therefore$ Gain per cent $= \dfrac{\dfrac{2x}{100} \times 100\%}{x} = 2\%$

Fast Track Method

Here, r = 20% and r_1 = 15%

$\therefore$ Required profit per cent $= \dfrac{r(100 - r_1)}{100} - r_1$

[by Technique 5]

$= \dfrac{20(100 - 15)}{100} - 15$

$= 17 - 15 = 2\%$

24. (a) Let the cost price of the article be ₹ 100.

Then, printed price = ₹ 140

and selling price $= 100 + 100 \times \dfrac{12}{100} = ₹\,112$

Discount = 140 − 112 = ₹ 28

$\therefore$ Discount per cent $= \dfrac{28}{140} \times 100\% = 20\%$

Fast Track Method

Here, r = 40%, r_1 = ? and profit = 12%

$\therefore$ Profit per cent $= \dfrac{r \times (100 - r_1)}{100} - r_1$

[by Technique 5]

$\Rightarrow \quad 12 = \dfrac{40 \times (100 - r_1)}{100} - r_1$

$\Rightarrow \quad 1200 = 4000 - 40 r_1 - 100 r_1$

$\Rightarrow \quad 140 r_1 = 2800$

$\Rightarrow \quad r_1 = 20\%$

25. (b) Let the cost price of the item be ₹ 100.
Then,

$$100 \xrightarrow{+40\%\uparrow} 140 \xrightarrow{-20\%\downarrow} 112 \xrightarrow{-25\%\downarrow} 84$$
$$\text{CP} \qquad\qquad \text{MP} \qquad\qquad\qquad \text{SP}$$

∴ Loss = 16% and given loss = ₹ 448

Now, CP = $\dfrac{448 \times 100}{16}$ = ₹ 2800

and SP = $\dfrac{2800 \times 84}{100}$ = ₹ 2352

Fast Track Method
Single discount equivalent to 20% and 25%
= $20 + 25 - \dfrac{20 \times 25}{100}$ = 40%

Here, r = 40%, r_1 = 40%

∴ Loss per cent = $\dfrac{r(100 - r_1)}{100} - r_1$

[by Technique 5]

= $\dfrac{40(100 - 40)}{100} - 40$

= $\dfrac{40 \times 60}{100} - 40 = -16\%$

Now, loss = 16% and given, loss = ₹448

∴ CP = $448 \times \dfrac{100}{16}$ = ₹2800

and SP = $2800 \times \dfrac{84}{100}$ = ₹2352

26. (d) Let the cost price of an item = ₹100
Cost price of an item after increasing by 20% = ₹120
Selling price after 10% discount of = ₹120
= ₹(120 − 12) = ₹108
∴ Percentage gain = (108 − 100)% = 8%

Fast Track Method
Here, r = 20% and r_1 = 10%

∴ Gain % = $\dfrac{r(100 - r_1)}{100} - r_1$ [by Technique 5]

= $\dfrac{20(100 - 10)}{100} - 10$

= $\dfrac{20 \times 90}{100} - 10 = 18 - 10 = 8\%$

27. (c) Let the amount of electricity bill be ₹ x.

∴ 4% of x = 13 ⇒ $\dfrac{4 \times x}{100}$ = 13

⇒ x = 13 × 25 = ₹ 325

28. (c) ∵ MP of the article = ₹ 60

and SP of the article = $\dfrac{60 \times (100 - 15)}{100}$

= $\dfrac{60 \times 85}{100}$ = ₹ 51

Thus, actual SP of the article
= (51 − 3) = ₹ 48

Hence, CP of the article = $\dfrac{48 \times 100}{100 + 20}$

= $\dfrac{48 \times 100}{120}$ = ₹ 40

29. (d) ∵ Marked price of a shirt = ₹ 320

and discount on a shirt = $\dfrac{320 \times 25}{100}$ = ₹ 80

∴ Number of shirts that has to be purchased to get a rebate of ₹ 400 = $\dfrac{400}{80}$ = 5

30. (a) ∵ MP of one dozen pairs of socks = ₹ 80

∴ SP of one dozen pairs of socks
= $\dfrac{80 \times (100 - 10)}{100}$

= $\dfrac{80 \times 90}{100}$ = ₹ 72

Hence, required number of pairs of socks
purchased for ₹ 24 = $\dfrac{12 \times 24}{72}$ = 4

31. (c) Let the marked price of book be ₹ x.
Then, selling price after 10% discount
= $\dfrac{90x}{100}$ = ₹ $\dfrac{9}{10}x$

Profit = 20%
∴ Cost price of book
= $\dfrac{9}{10} \times \dfrac{100x}{120}$ = ₹ $\dfrac{3x}{4}$

Hence, required ratio = $\dfrac{3x}{4} : x = 3 : 4$

32. (a) Let MP of an article be ₹ x.

∴ SP of an article = ₹ $\dfrac{3}{4}x$

and CP of an article = $\dfrac{3x}{4} \times \dfrac{100}{100 + 25}$

= $\dfrac{3x}{4} \times \dfrac{100}{125}$ = ₹ $\dfrac{3x}{5}$

∴ Required ratio = $x : \dfrac{3x}{5} = 5 : 3$

33. (b) Let labelled price of TV be ₹x.

∴ CP of the TV = $\dfrac{x \times (100 - 20)}{100}$ = ₹ $\dfrac{4x}{5}$

But 16800 − 800 = $\dfrac{4x}{5}$

∴ x = $\dfrac{16000 \times 5}{4}$ = ₹ 20000

34. (c) Let the cost price be ₹ 100.

∴ Marked price = ₹ 125

For a profit of $12\dfrac{1}{2}$%,

Selling price = $100 + \dfrac{25}{2}$ = ₹112.5

∴ Discount percentage

$$= \frac{125 - 112.5}{125} \times 100\%$$

$$= \frac{12.5}{125} \times 100\% = 10\%$$

35. (c) Let CP of the article be ₹ x.
Then, MP of the article
$$= 125\% \text{ of } x = ₹\, 1.25x$$
∴ SP of the article $= 0.94 \times 1.25x$
According to the question,
$$0.94 \times 1.25x = 10340$$
$$\Rightarrow \quad x = \frac{10340}{0.94 \times 1.25} = ₹\, 8800$$

36. (e) Let the marked price of the item be ₹ 100.
Then, $100 \xrightarrow{-20\%} 80 \xrightarrow{+10\%} 88$

∴ $88 + 88 \times \dfrac{125}{100} - 88 = 2046$

$\Rightarrow \quad 110 - 88 = 2046$
$\Rightarrow \quad 22 = 2046 \Rightarrow 1 = 93$
$\Rightarrow$ ₹ 1 is equivalent to ₹ 93.
∴ ₹ 100 is equivalent to ₹ 9300.
∴ Marked price = ₹ 9300
Alternate Method
Let the marked price be ₹ x.
Then, discount = 20%
$\Rightarrow$ Selling price $= x - \dfrac{20}{100} \times x = ₹\, \dfrac{4x}{5}$
∵ Charges = 10%
∴ Selling price $= \dfrac{4x}{5} + \dfrac{10}{100}\left(\dfrac{4x}{5}\right) = ₹\, \dfrac{44x}{50}$
According to the question,
$$\dfrac{44x}{50} + \dfrac{25}{100} \times \dfrac{44x}{50} = \dfrac{44x}{50} + 2046$$
$\Rightarrow \quad \dfrac{11x}{50} = 2046$
$\Rightarrow \quad x = 186 \times 50 = ₹\, 9300$

37. (c) Let MP be ₹ x.
∴ SP $= x \times \left(\dfrac{100-10}{100}\right) = ₹\, \dfrac{9}{10}x$
∴ CP $= \dfrac{9}{10}x \left(\dfrac{100}{100+20}\right) = ₹\, \dfrac{3}{4}x$
Thus, CP has to be increased
$$= \dfrac{x - \dfrac{3x}{4}}{\dfrac{3x}{4}} \times 100\% = \dfrac{\dfrac{x}{4}}{\dfrac{3x}{4}} \times 100\% = 33\dfrac{1}{3}\%$$

38. (c) Let rate of second discount be $r\%$.
Now, marked price of the clock = ₹ 3200
∴ SP of the clock after first discount
$$= \dfrac{3200 \times (100 - 10)}{100} = \dfrac{3200 \times 90}{100}$$

= ₹ 2880
SP of the clock after second discount of $r\%$
$$= \dfrac{2880 \times (100 - r)}{100}$$
But SP of the clock after second discount
= 2448
∴ $\quad 2448 = \dfrac{2880 \times (100-r)}{100}$
$\Rightarrow \quad 100 - r = \dfrac{2448 \times 100}{2880} = 85$
∴ $\quad r = 100 - 85 = 15\%$
Alternate Method
Let rate of second discount be $r\%$.
Here, $x = ₹\, 3200, y = ₹\, 2448, r_1 = 10\%$
and $r_2 = r\%$
∴ $\quad y = x \times \left(\dfrac{100 - r_1}{100}\right)\left(\dfrac{100 - r_2}{100}\right)$
$\Rightarrow 2448 = \dfrac{3200 \times (100-10) \times (100-r)}{100 \times 100}$
$\Rightarrow \quad \dfrac{2448 \times 100 \times 100}{3200 \times 90} = 100 - r$
∴ $\quad r = 100 - 85 = 15\%$

39. (b) Let the cost price of item be ₹ x.
∴ Marked price of item
$$= \dfrac{x \times (100 + 30)}{100} = ₹\, \dfrac{13x}{10}$$
Selling price of the item after two successive discounts
$$= \dfrac{13x}{10} \times \left(\dfrac{100-10}{100}\right) \times \left(\dfrac{100-10}{100}\right)$$
$$= \dfrac{13x}{10} \times \dfrac{90}{100} \times \dfrac{90}{100} = ₹\, \dfrac{1053}{1000}x$$
∴ Increment in the price of item
$$= \dfrac{\dfrac{1053x}{1000} - x}{x} \times 100\% = \dfrac{53}{10}\% = 5.3\%$$

40. (a) Let the cost price be ₹x.
∴ $\quad x \times \dfrac{140}{100} \times \dfrac{80}{100} - x = 48$
$\Rightarrow \quad 1.12x - x = 48$
$\Rightarrow \quad 0.12x = 48$
$\Rightarrow \quad x = \dfrac{48}{0.12} = ₹\, 400$

41. (a) ∵ Listed price of an article = ₹ 1500
∴ Price after first discount
$$= 1500 \times \left(1 - \dfrac{20}{100}\right) = 1500 \times \dfrac{4}{5} = ₹\, 1200$$
Now, second discount = 1200 − 1104 = ₹ 96
Hence, required percentage
$$= \dfrac{96}{1200} \times 100\% = 8\%$$

Discount / 243

Alternate Method
Let rate of second discount be $r_2\%$.
Here, $x = 1500$, $r_1 = 20\%$ and $y = ₹1104$
According to the question,

$$y = x\left(1 - \frac{r_1}{100}\right)\left(1 - \frac{r_2}{100}\right)$$

$\Rightarrow 1104 = 1500 \times \left(1 - \frac{20}{100}\right)\left(1 - \frac{r_2}{100}\right)$

$\Rightarrow \frac{1104}{1500} \times \frac{5}{4} \times 100 = 100 - r_2$

$\Rightarrow 92 = 100 - r_2$

$\therefore r_2 = 100 - 92 = 8\%$

42. (a) Let marked price of a machine be ₹x.
According to the question,

$x - \frac{x \times 9}{100} = 14000$

$\Rightarrow \frac{91x}{100} = 14000 \Rightarrow x = \frac{14000 \times 100}{91}$

Now, selling of price of machine at 22% discount

$= \frac{14000 \times 100}{91} - \frac{14000 \times 100}{91} \times \frac{22}{100}$

$= \frac{14000}{91}(100 - 22)$

$= \frac{14000 \times 78}{91} = \frac{14000 \times 6}{7}$

$= 2000 \times 6 = ₹12000$

43. (c) Let MP of an article be ₹ x.

$\therefore$ SP of an article $= \frac{x \times (100 - 40)}{100} = ₹\frac{3x}{5}$

CP of an article $= \frac{3x}{5} \times \left(\frac{100}{100 - 30}\right)$

$= \frac{3x}{5} \times \frac{100}{70} = ₹\frac{6x}{7}$

$\therefore$ Profit when sold at MP $= \left(x - \frac{6x}{7}\right) = ₹\frac{x}{7}$

Hence, profit per cent $= \frac{x/7}{6x/7} \times 100\%$

$= \frac{50}{3}\% = 16\frac{2}{3}\%$

44. (a) Let MP of the article be ₹ x.

$\therefore$ CP of the article $= \frac{x \times 64}{100} = ₹\frac{16x}{25}$

and SP of the article $= \frac{x \times (100 - 12)}{100}$

$= ₹\frac{x \times 22}{25}$

$\Rightarrow$ Profit $= \left(\frac{22x}{25} - \frac{16x}{25}\right) = ₹\frac{6x}{25}$

$\therefore$ Profit per cent $= \frac{6x/25}{16x/25} \times 100$

$= \frac{6 \times 100}{16} = 37.5\%$

45. (a) Let marked price = ₹100
and selling price = ₹80
In condition of 10% loss, the cost price of
article $= \frac{80 \times 100}{90} = ₹\frac{800}{9}$

According to the question,
When SP = 95, then
Required profit percentage

$= \frac{95 - \frac{800}{9}}{\frac{800}{9}} \times 100 = \frac{55}{8}$

$= 6.9\%$ (approx.)

46. (d) Let the manufacturing cost of product be ₹ 100.
$\therefore$ Maximum Retail Price (MRP) = ₹ 155
Since, retailer gives 10% discount on MRP, then selling price of product

$= 155 - 155 \times \frac{10}{100} = ₹139.50$

Let the purchase price be ₹ x and it is given that, retailer earns 23% profit on his purchase price.

Therefore, $\frac{123x}{100} = 139.50$

$\therefore x = ₹113.41$

So, profit per cent of manufacturer
$= 113.41 - 100$
$= 13.41\% \approx 13\%$

47. (a) Let MP of the article be ₹x.

$\therefore$ SP of the article $= \frac{x \times (100 - 10)}{100} = ₹\frac{9x}{10}$

$\Rightarrow$ CP of the article $= \frac{9x}{10} \times \frac{100}{100 + 20}$

$= \frac{9x \times 10}{120} = ₹\frac{3x}{4}$

Now, new SP of the article

$= \frac{x \times (100 - 20)}{100} = ₹\frac{4x}{5}$

New profit $= \left(\frac{4x}{5} - \frac{3x}{4}\right)$

$= \frac{16x - 15x}{20} = ₹\frac{x}{20}$

Hence, profit per cent

$= \frac{x/20}{3x/4} \times 100\% = \frac{4 \times 100}{3 \times 20}\%$

$= 6\frac{2}{3}\%$

48. (e) ∵ SP of the notebook = ₹ 457

∴ Commission on one notebook = ₹ $\dfrac{4 \times 457}{100}$

and commission on 10 notebooks
$= \dfrac{10 \times 4 \times 457}{100} = ₹\,182.80$

Now, SP of the pencil box = ₹ 80

∴ Commission on 1 pencil box = ₹ $\dfrac{80 \times 20}{100}$

Commission on 6 pencil boxes
$= \dfrac{80 \times 20 \times 6}{100} = ₹\,96$

Hence, total commission of 1 day
$= (182.80 + 96) = ₹\,278.80$

Thus, total commission of 2 weeks
$= 278.80 \times 14 = ₹\,3903.20$

49. (d) Let advertised price of the item be ₹x.

∴ Commission on advertised price of the article = ₹ $\dfrac{x \times 23}{100}$

Price after commission = $x - \dfrac{23x}{100} = ₹\,\dfrac{77x}{100}$

∴ CP of the item = $\dfrac{77x}{100} \times \left(\dfrac{100}{100+10}\right)$

$= \dfrac{77x}{110} = ₹\,\dfrac{7x}{10}$

Now, profit = $\dfrac{77x}{100} - \dfrac{7x}{10} = ₹\,\dfrac{7x}{100}$

$\Rightarrow 56 = \dfrac{7x}{100}$ [given]

∴ $x = \dfrac{56 \times 100}{7} = ₹\,800$

50. (d) CP of amit = ₹1000

CP of Roshan = ₹2000

According to the question,

SP for Roshan

$= 2000 \times \dfrac{(100+2x)}{100} \times \left(\dfrac{100-x}{100}\right)$

$= \dfrac{(100+2x)(100-x)}{5}$

and SP for Amit = $1000 \times \dfrac{100+x}{100}$

$= 10(100+x)$

Now, profit of amit = profit of Roshan

$\Rightarrow 10(100+x) - 1000$

$\quad = \dfrac{(100+2x)(100-x)}{5} - 2000$

$\Rightarrow 1000 + 10x - 1000$

$\quad = \dfrac{10000 - 100x + 200x - 2x^2 - 10000}{5}$

$\Rightarrow 50x = -2x^2 + 100x$
$\Rightarrow 2x^2 - 50x = 0$
$\Rightarrow x(2x-50) = 0 \Rightarrow x = 25\%$

51. (d) Let CP of an article A = ₹x

Then, CP of an article B

$= ₹\left(x + x \times \dfrac{20}{100}\right)$

$= ₹\,\dfrac{120x}{100}$

Now, MP of an article A

$= x + x \times \dfrac{50}{100} = ₹\,\dfrac{150x}{100}$

and MP of an article B

$= \dfrac{120x}{100} + \dfrac{120x}{100} \times \dfrac{25}{100}$

$= ₹\,\dfrac{150x}{100}$

∴ SP of an article A

$= \dfrac{150x}{100} - \dfrac{150x}{100} \times \dfrac{4}{100}$

$= \dfrac{15x}{10} - \dfrac{3x}{50}$

$= \dfrac{75x - 3x}{50}$

$= ₹\,\dfrac{72x}{50}$

and SP of an article B

$= \dfrac{150x}{100} - \dfrac{150x}{100} \times \dfrac{0.5}{100}$

$= \dfrac{15x}{10} - \dfrac{15x}{10} \times \dfrac{1}{200}$

$= \dfrac{3000x - 15x}{2000}$

$= \dfrac{2985x}{2000}$

Now, according to the question,

$\dfrac{2985x}{2000} - \dfrac{72x}{50} = 30$

$\Rightarrow \dfrac{2985x - 2880x}{2000} = 30$

$\Rightarrow \dfrac{105x}{2000} = 30$

$\Rightarrow x = \dfrac{30 \times 2000}{105}$

$\Rightarrow x = 571.42$

∴ $x \approx ₹\,570$

Chapter 14

Simple Interest

When a person borrows some amount of money from another person or organisation (bank), then the person borrowing money (borrower) pays some extra money during repayment, that extra money during repayment is called **Interest**.

For example If A takes ₹ 50 from B and after using ₹ 50, A returns ₹ 55 to B, then A pays (55 – 50), i.e. ₹ 5 as interest.

Let us know the following terms, which will be used in this chapter

Principal (P) Principal is the money borrowed or deposited for a certain time.

Amount (A) The sum of principal and interest is called amount.

∴ Amount = Principal + Simple Interest (SI)

Rate of Interest (R) It is the rate at which the interest is charged on principal. It is always specified in percentage terms.

Time (T) The period, for which the money is borrowed or deposited, is called time.

Simple Interest (SI)

If the interest is calculated on the original principal for any length of time, then it is called simple interest.

$$\text{Simple Interest (SI)} = \frac{\text{Principal }(P) \times \text{Rate }(R) \times \text{Time }(T)}{100}$$

Basic Formulae Related to Simple Interest

- $P = \dfrac{100 \times A}{100 + RT}$
- $SI = \dfrac{ART}{100 + RT}$
- $A = P\left(1 + \dfrac{RT}{100}\right)$

where, SI = Simple Interest, P = Principal, R = Rate of interest, T = Time and A = Amount.

Ex. 1 Find the simple interest on ₹ 200 for 5 yr at 6% per annum.

Sol. Here, $P = ₹ 200$, $T = 5$ yr, $R = 6\%$

$\therefore \quad SI = \dfrac{P \times R \times T}{100} = \dfrac{200 \times 6 \times 5}{100} = ₹ 60$

Ex. 2 A sum at simple interest of 4% per annum amounts to ₹ 3120 in 5 yr. Find the sum.

Sol. Here, $T = 5$ yr, $R = 4\%$, $A = ₹ 3120$

We know that, $P = \dfrac{100 \times A}{100 + RT} = \dfrac{100 \times 3120}{100 + 4 \times 5} = \dfrac{100 \times 3120}{120} = ₹ 2600$

Alternate Method

Let the sum be ₹ P.

Then, $SI = \dfrac{P \times R \times T}{100} = \dfrac{P \times 4 \times 5}{100} = ₹ \dfrac{P}{5}$

$\therefore$ Amount $(A) = P + SI = P + \dfrac{P}{5} = ₹ \dfrac{6P}{5}$

According to the question,

Amount $= 3120 \Rightarrow \dfrac{6P}{5} = 3120$

$\therefore \quad P = \dfrac{3120 \times 5}{6} = ₹ 2600$

Ex. 3 Amit takes some loan from Akash for 2 yr at the rate of 5% per annum and after 2 yr he gave back ₹ 6600 to Akash and completed the payment of his loan. Find the interest paid by Amit.

Sol. Here, $T = 2$ yr, $R = 5\%$, $A = ₹ 6600$

We know that, $SI = \dfrac{ART}{100 + RT} = \dfrac{6600 \times 5 \times 2}{100 + 5 \times 2} = \dfrac{6600 \times 10}{110} = ₹ 600$

Fast Track Method

Here, $SI = \dfrac{P \times R \times T}{100} = \dfrac{P \times 5 \times 2}{100} = ₹ \dfrac{P}{10}$

We know that, $A = P + SI$

$\Rightarrow \quad P + \dfrac{P}{10} = 6600 \Rightarrow \dfrac{11P}{10} = 6600$

$\Rightarrow \quad 11P = 66000 \Rightarrow P = 6000$

$\therefore$ Required interest = Amount − Principal = 6600 − 6000 = ₹ 600

MIND IT!

1. If rate of interest is half-yearly, then rate $= \left(\dfrac{R}{2}\right)\%$ and time $= 2T$

2. If rate of interest is quarterly, then rate $= \left(\dfrac{R}{4}\right)\%$ and time $= 4T$

3. If rate of interest is monthly, then rate $= \left(\dfrac{R}{12}\right)\%$ and time $= 12T$

4. To calculate interest, the day on which amount is deposited, is not counted but the day on which amount is withdrawn, is counted.

Ex. 4 A takes ₹ 3000 from B for 2 yr at the rate of 10% half-yearly interest. What amount will be paid by A to B after the end of 2 yr?

Sol. Here, $P = ₹ 3000, T = 2$ yr $= 4$ half-years, $R = 10\% = \dfrac{10}{2}\%$

Now, $\text{SI} = \dfrac{P \times R \times T}{100} = \dfrac{3000 \times 4 \times 10}{100 \times 2} = ₹ 600$

∴ Amount paid to $B = P + \text{SI} = 3000 + 600 = ₹ 3600$

Instalments

When a borrower pays the total money in some equal parts (i.e. not in a single amount), then we say that he/she is paying in **Instalments**.

For example A borrowed ₹ 100 from B and he pays back it to B in several parts, i.e. ₹ 20 in 5 times or ₹ 50 in 2 times etc. The important point is that borrower has to pay the interest for using the borrowed sum or purchased article. In general, the value of each instalment is kept constant. If a loan of ₹ A at R% interest per annum is to be repaid in n equal yearly instalments, then

$$A = \left[x + \left(x + \dfrac{x \times R \times 1}{100}\right) + \left(x + \dfrac{x \times R \times 2}{100}\right) + \left(x + \dfrac{x \times R \times 3}{100}\right) + \ldots + \left(x + \dfrac{x \times R(n-1)}{100}\right)\right]$$

where, A = Total amount paid, x = Value of each instalment

Also, $A = P + \dfrac{P \times n \times R}{100}$

where, P = Principal, n = Number of instalments and R = Rate of interest

Ex. 5 A scooty is sold by an automobile agency for ₹ 19200 cash or for ₹ 4800 cash down payment together with five equal monthly instalments. If the rate of interest charged by the company is 12% per annum, then find the value of each instalment.

Sol. ∵ Balance of the price to be paid through instalments,

$P = 19200 - 4800 = ₹ 14400$

Now, according to the formula,

$$A = \left[x + \left(\dfrac{x \times R \times 1}{100}\right) + \left(x + \dfrac{x \times R \times 2}{100}\right) + \ldots + \left(x + \dfrac{x \times R \times 4}{100}\right)\right]$$

where, $A = P + \dfrac{P \times n \times R}{100}$

$\Rightarrow \left(14400 + \dfrac{14400 \times 12 \times 5}{100 \times 12}\right) = \left[x + \left(x + \dfrac{12x}{12 \times 100}\right) + \left(x + \dfrac{12x \times 2}{12 \times 100}\right)\right.$

$\left. + \ldots + \left(x + \dfrac{12x \times 4}{12 \times 100}\right)\right]$

$\Rightarrow 15120 = 5x + \dfrac{x}{10} \Rightarrow x = \dfrac{151200}{51} = ₹ 2964.70$

✦ In the left hand side and right hand side, given amounts are equal. Each amount is equal to the total amount payable after 5 months.

Fast Track Techniques to solve the QUESTIONS

Technique ① If a sum of money becomes n times in T yr at simple interest, then formula for calculating rate of interest will be given as

$$R = \frac{100(n-1)}{T}\%$$

Ex. 6 A sum of money becomes four times in 20 yr at SI. Find the rate of interest.

Sol. Given, $T = 20$ yr

Let sum = ₹ P. Then, the sum after 20 yr = $4P$

∴ SI = $4P - P = 3P$

Now, $3P = \dfrac{PRT}{100} = \dfrac{P \times R \times 20}{100} \Rightarrow 3 = \dfrac{20R}{100} = \dfrac{R}{5}$

∴ $R = 15\%$

Fast Track Method

Here, $T = 20$ yr and $n = 4$

∴ $R = \dfrac{100(n-1)}{T} = \dfrac{100(4-1)}{20} = \dfrac{100 \times 3}{20} = 15\%$

Technique ②

(i) If a sum of money at a certain rate of interest becomes n times in T_1 yr and m times in T_2 yr, then

$$T_2 = \left(\frac{m-1}{n-1}\right) \times T_1$$

(ii) If a sum of money in a certain time becomes n times at R_1 rate of interest and m times at R_2 rate of interest,

$$R_2 = \left(\frac{m-1}{n-1}\right) \times R_1$$

Ex. 7 A sum becomes two times in 5 yr at a certain rate of interest. Find the time in which the same amount will be 8 times at the same rate of interest.

Sol. Let the sum = ₹ P

Then, for 5 yr, SI = $2P - P = P$

∵ SI = $\dfrac{P \times R \times T}{100}$

∴ $P = \dfrac{P \times R \times 5}{100} = \dfrac{PR}{20} \Rightarrow R = 20\%$

Again, for another time (T), SI = $8P - P = ₹ 7P$

Simple Interest / 249

$$\therefore \quad 7P = \frac{P \times 20 \times T}{100} = \frac{20\,TP}{100} = \frac{TP}{5}$$

Now, $T = 7 \times 5 = 35$ yr

Fast Track Method

Here, $n = 2, m = 8, T_1 = 5$ and $T_2 = ?$

$$\therefore \quad T_2 = \left(\frac{m-1}{n-1}\right) \times T_1 = \left(\frac{8-1}{2-1}\right) \times 5 = 35 \text{ yr}$$

Ex. 8 In a certain time, a sum becomes 3 times at the rate of 5% per annum. At what rate of interest, the same sum becomes 6 times in same duration?

Sol. Let sum = ₹ P

Then, for 5% rate of interest, SI = $3P - P = 2P$

$$\therefore \quad \text{SI} = \frac{P \times R \times T}{100} \Rightarrow 2P = \frac{P \times 5 \times T}{100} = \frac{PT}{20} \Rightarrow T = 40 \text{ yr}$$

Again, for another rate (R), SI = $6P - P = 5P \Rightarrow 5P = \frac{P \times R \times 40}{100} = \frac{2PR}{5}$

$$\therefore \quad R = \frac{25}{2} = 12.5\%$$

Fast Track Method

Here, $n = 3, m = 6, R_1 = 5\%$ and $R_2 = ?$

$$\therefore \quad R_2 = \left(\frac{m-1}{n-1}\right) \times R_1 = \left(\frac{6-1}{3-1}\right) \times 5 = \frac{5}{2} \times 5 = \frac{25}{2} = 12.5\%$$

Technique 3 If a certain sum P in a certain time T amounts to ₹ A_1 at the rate of $R_1\%$ and the same sum amounts to ₹ A_2 at the rate of $R_2\%$ in same time, then

$$P = \left(\frac{A_2 R_1 - A_1 R_2}{R_1 - R_2}\right) \text{ and } T = \left(\frac{A_1 - A_2}{A_2 R_1 - A_1 R_2}\right) \times 100.$$

✦ If the rate of interest is uniform in above mentioned condition and time is variable, then

$$P = \left(\frac{A_2 T_1 - A_1 T_2}{T_1 - T_2}\right) \text{ and } R = \left(\frac{A_1 - A_2}{A_2 T_1 - A_1 T_2}\right) \times 100$$

where, T_1 and T_2 are the time for first and second conditions, respectively.

Ex. 9 A certain sum in certain time becomes ₹ 500 at the rate of 8% per annum SI and the same sum amounts to ₹ 200 at the rate of 2% per annum SI in the same duration. Find the sum and time.

Sol. According to the question, $A_1 - A_2 = 500 - 200$

$$\Rightarrow \quad \left(P + \frac{P \times 8 \times T}{100}\right) - \left(P + \frac{P \times 2 \times T}{100}\right) = 500 - 200$$

$$\Rightarrow \quad \frac{6PT}{100} = 300 \Rightarrow PT = \frac{300 \times 100}{6} = 5000$$

Again, for 8% rate, SI = $\frac{P \times R \times T}{100} = \frac{5000 \times 8}{100} = ₹ 400$ [$\because PT = ₹5000$]

$$\therefore \quad \text{Sum } (P) = 500 - 400 = ₹ 100 \quad [\because P = \text{Amount} - \text{SI}]$$

We have, $PT = ₹ 5000$

$$\therefore \quad T = \frac{5000}{P} = \frac{5000}{100} = 50 \text{ yr}$$

250 / Fast Track Objective Arithmetic

Fast Track Method
Here, $R_1 = 8\%$, $R_2 = 2\%$, $A_1 = ₹500$ and $A_2 = ₹200$
Now, according to the formula,
$$P = \frac{A_2R_1 - A_1R_2}{R_1 - R_2} = \frac{200 \times 8 - 500 \times 2}{8 - 2} = \frac{1600 - 1000}{6} = \frac{600}{6} = ₹100$$

and time, $T = \frac{A_1 - A_2}{A_2R_1 - A_1R_2} \times 100 = \frac{500 - 200}{200 \times 8 - 500 \times 2} \times 100 = \frac{300}{600} \times 100 = 50$ yr

Ex. 10 A certain sum at a certain rate of SI amounts to ₹1125 in 4 yr and ₹1200 in 7 yr. Find the sum and rate of interest.

Sol. According to the question,
$$\left(P + \frac{P \times R \times 7}{100}\right) - \left(P + \frac{P \times R \times 4}{100}\right) = 1200 - 1125$$
$$\Rightarrow \frac{7PR}{100} - \frac{4PR}{100} = 75 \Rightarrow \frac{3PR}{100} = 75 \Rightarrow PR = 2500$$

For 4 yr, SI $= \frac{P \times R \times T}{100} = \frac{2500 \times 4}{100} = ₹100$

∴ Sum $= 1125 - 100 = ₹1025$

Again, we have $PR = ₹2500$

∴ $R = \frac{2500}{1025} = 2.43\%$

Fast Track Method
Here, $A_1 = ₹1125$, $A_2 = ₹1200$, $T_1 = 4$ yr and $T_2 = 7$ yr
According to the formula,
$$P = \frac{A_2T_1 - A_1T_2}{T_1 - T_2} = \frac{1200 \times 4 - 1125 \times 7}{4 - 7} = \frac{4800 - 7875}{-3} = \frac{-3075}{-3} = ₹1025$$

We know that, $R = \frac{A_1 - A_2}{A_2T_1 - A_1T_2} \times 100\% = \frac{1125 - 1200}{1200 \times 4 - 1125 \times 7} \times 100$

$= \frac{-7500}{4800 - 7875} = \frac{-7500}{-3075} = 2.43\%$

Technique 4 If SI for a certain sum P_1 for time T_1 and rate of interest R_1 is I_1 and SI for another sum P_2 for time T_2 and rate of interest R_2 is I_2, then

Difference of SI $= I_2 - I_1 = \frac{P_2R_2T_2 - P_1R_1T_1}{100}$

✦ In the above mentioned condition, if all the parameters are constant but time is variable, then $I_2 - I_1 = \frac{PR(T_2 - T_1)}{100}$.

✦ When only rate of interest is variable, then $I_2 - I_1 = \frac{PT(R_2 - R_1)}{100}$.

✦ When only sum is variable, then $I_2 - I_1 = \frac{RT(P_2 - P_1)}{100}$.

✦ When only one parameter remains constant and the remaining are variables, then

(i) $I_2 - I_1 = \frac{P(R_2T_2 - R_1T_1)}{100}$ (ii) $I_2 - I_1 = \frac{R(P_2T_2 - P_1T_1)}{100}$ (iii) $I_2 - I_1 = \frac{T(P_2R_2 - P_1R_1)}{100}$

Simple Interest / 251

Ex. 11 The simple interest on a certain sum of money at 5% per annum for 4 yr and 3 yr differ by ₹ 42. Find the sum (in ₹).

Sol. According to the question,

$$\frac{P \times 5 \times 4}{100} - \frac{P \times 5 \times 3}{100} = 42$$

$$\Rightarrow \quad 20P - 15P = 4200$$

$$\therefore \quad P = \frac{4200}{5} = ₹ 840$$

Fast Track Method

Here, $I_2 - I_1 = 42$, $T_2 = 4$ yr, $T_1 = 3$ yr and $R = 5\%$

According to the formula,

$$I_2 - I_1 = \frac{PR(T_2 - T_1)}{100}$$

$$\Rightarrow \quad 42 = \frac{P \times 5(4 - 3)}{100}$$

$$\therefore \quad P = 42 \times 20 = ₹ 840$$

Ex. 12 Simple interest for a sum of ₹ 1550 for 2 yr is ₹ 20 more than the simple interest for ₹ 1450 for the same duration. Find the rate of interest.

Sol. Given, $P_1 = ₹ 1550$, $T_1 = 2$ yr, $P_2 = ₹ 1450$ and $T_2 = 2$ yr

According to the question,

$$\frac{P_1 \times R \times T_1}{100} - \frac{P_2 \times R \times T_2}{100} = 20$$

$$\Rightarrow \quad \frac{1550 \times 2 \times R}{100} - \frac{1450 \times 2 \times R}{100} = 20 \Rightarrow \frac{200R}{100} = 20 \Rightarrow R = 10\%$$

Fast Track Method

Here, $I_1 - I_2 = ₹ 20$, $P_1 = ₹ 1550$, $P_2 = ₹ 1450$ and $T = 2$ yr

According to the formula,

$$I_2 - I_1 = \frac{RT(P_2 - P_1)}{100}$$

$$\Rightarrow \quad -20 = \frac{R \times 2(1450 - 1550)}{100} \qquad \begin{bmatrix} \because I_1 - I_2 = 20 \\ \therefore I_2 - I_1 = -20 \end{bmatrix}$$

$$\Rightarrow \quad -10 = \frac{-100R}{100} \Rightarrow R = 10\%$$

Ex. 13 For a certain sum, the simple interest in 2 yr at 8% per annum is ₹ 90 more than the simple interest in 1.5 yr at the rate of 10% per annum for the same sum. Find the sum.

Sol. According to the question,

$$\frac{P \times 8 \times 2}{100} - \frac{P \times 10 \times 1.5}{100} = 90 \Rightarrow \frac{P}{100} = 90 \Rightarrow P = ₹ 9000$$

Fast Track Method

Here, $I_1 - I_2 = ₹ 90$, $T_1 = 2$ yr, $R_1 = 8\%$, $R_2 = 10\%$ and $T_2 = 1.5$ yr

According to the formula, $I_2 - I_1 = \dfrac{P(R_2 T_2 - R_1 T_1)}{100}$

$$\Rightarrow \quad -90 = \frac{P}{100}(10 \times 1.5 - 8 \times 2) \Rightarrow P = \frac{-90 \times 100}{15 - 16} = ₹ 9000$$

Technique 5 If $\frac{1}{x}$ part of a certain sum P is lent out at $R_1\%$ SI, $\frac{1}{y}$ part is lent out at $R_2\%$ SI and the remaining $\frac{1}{z}$ part at $R_3\%$ SI and this way the interest received be I, then $P = \dfrac{I \times 100}{\dfrac{R_1}{x} + \dfrac{R_2}{y} + \dfrac{R_3}{z}}$

Ex. 14 Alok lent out a certain sum. He lent 1/3 part of his sum at 7% SI, 1/4 part at 8% SI and remaining part at 10% SI. If ₹ 510 is his total interest, then find the money lent out.

Sol. Let entire sum = ₹P
Now, according to the question,

$$\frac{1}{3}P \times 7\% + \frac{1}{4}P \times 8\% + \left[1 - \left(\frac{1}{3} + \frac{1}{4}\right)\right] \times P \times 10\% = 510$$

$\Rightarrow \quad \dfrac{\frac{1}{3}P \times 7}{100} + \dfrac{\frac{1}{4}P \times 8}{100} + \dfrac{\frac{5}{12}P \times 10}{100} = 510$

$\Rightarrow \quad \dfrac{7P}{3} + 2P + \dfrac{25P}{6} = 510 \times 100$

$\Rightarrow \quad \dfrac{14P + 12P + 25P}{6} = 510 \times 100$

$\therefore \quad P = \dfrac{510 \times 100 \times 6}{51} = ₹\,6000$

Fast Track Method

Here, $R_1 = 7\%$, $R_2 = 8\%$, $R_3 = 10\%$

and $\dfrac{1}{x} = \dfrac{1}{3}$, $\dfrac{1}{y} = \dfrac{1}{4}$, $I = ₹\,510$

$\therefore \quad \dfrac{1}{z} = \left[1 - \left(\dfrac{1}{3} + \dfrac{1}{4}\right)\right] = \dfrac{5}{12}$

According to the formula,

$P = \dfrac{I \times 100}{\dfrac{R_1}{x} + \dfrac{R_2}{y} + \dfrac{R_3}{z}} = \dfrac{510 \times 100}{\dfrac{7}{3} + \dfrac{8}{4} + \dfrac{50}{12}} = \dfrac{51000}{\dfrac{7}{3} + 2 + \dfrac{25}{6}}$

$= \dfrac{51000}{\dfrac{14 + 12 + 25}{6}} = \dfrac{51000}{\dfrac{51}{6}} = \dfrac{51000}{51} \times 6 = ₹\,6000$

Technique 6 A sum of ₹ P is lent out in n parts in such a way that the interest on first part at $R_1\%$ for T_1 yr, the interest on second part at $R_2\%$ for T_2 yr and the interest on third part at $R_3\%$ for T_3 yr and so on, are equal, then the ratio in which the sum was divided in n parts is given by

$$\dfrac{1}{R_1 T_1} : \dfrac{1}{R_2 T_2} : \dfrac{1}{R_3 T_3} : \ldots : \dfrac{1}{R_n T_n}$$

Simple Interest / 253

Ex. 15 A sum of ₹ 7700 is lent out in two parts in such a way that the interest on one part at 20% for 5 yr is equal to that on another part at 9% for 6 yr. Find the two sums.

Sol. Let the first sum be ₹ x. Then, second sum = ₹ $(7700 - x)$
Now, according to the question,
$$\frac{x \times 20 \times 5}{100} = \frac{(7700 - x) \times 9 \times 6}{100}$$
$\Rightarrow \quad 50x = (7700 - x) \times 27 \Rightarrow 50x = 7700 \times 27 - 27x$
$\Rightarrow \quad 77x = 7700 \times 27 \Rightarrow x = ₹ 2700$
$\therefore$ Second part = $(7700 - x) = 7700 - 2700 = ₹ 5000$

Fast Track Method
Here, sum = ₹ 7700, $R_1 = 20\%$, $T_1 = 5$ yr, $R_2 = 9\%$ and $T_2 = 6$ yr
Ratio of two sums = $\dfrac{1}{R_1 T_1} : \dfrac{1}{R_2 T_2} = \dfrac{1}{20 \times 5} : \dfrac{1}{9 \times 6}$
$= \dfrac{1}{10 \times 5} : \dfrac{1}{9 \times 3} = \dfrac{1}{50} : \dfrac{1}{27} = 27 : 50$

$\therefore$ First part = $\dfrac{27}{27 + 50} \times 7700 = \dfrac{27}{77} \times 7700 = ₹ 2700$
and Second part = $\dfrac{50}{27 + 50} \times 7700 = \dfrac{50}{77} \times 7700 = ₹ 5000$

Technique 7 The annual payment that will discharge a debt of ₹ P due in T yr at the rate of interest $R\%$ per annum is given by
$$\frac{100\,P}{100\,T + \dfrac{RT(T-1)}{2}}.$$

Ex. 16 What annual payment will discharge a debt of ₹ 848 in 8 yr at 8% per annum?

Sol. Here, $P = ₹ 848$, $T = 8$ yr and $R = 8\%$
According to the formula,
Annual payment = $\dfrac{100\,P}{100\,T + \dfrac{RT(T-1)}{2}}$
$= \dfrac{100 \times 848}{100 \times 8 + \dfrac{8 \times 8 (8-1)}{2}}$
$= \dfrac{848 \times 100}{800 + 32 \times 7} = \dfrac{84800}{1024} = ₹ 82.8125$

Multi Concept Questions

1. A person invested some amount at the rate of 12% simple interest and the remaining at 10%. He received yearly an interest of ₹ 130. Had he interchanged the amount invested, he would have received an interest of ₹ 134. How much money did he invest at different rates?
   (a) ₹ 500 at 10%, ₹ 800 at 12%
   (b) ₹ 700 at 10%, ₹ 600 at 12%
   (c) ₹ 800 at 10%, ₹ 400 at 12%
   (d) ₹ 700 at 10%, ₹ 500 at 12%

   ➙ (d) Let ₹ x be invested at 12% per annum and ₹ y be invested at 10% per annum.
   According to the question,
   $$12\% \text{ of } x + 10\% \text{ of } y = 130$$
   $$\Rightarrow 12x + 10y = 13000 \qquad \ldots(i)$$
   and after interchanging the amount,
   $$10\% \text{ of } x + 12\% \text{ of } y = 134$$
   $$\Rightarrow 10x + 12y = 13400 \qquad \ldots(ii)$$
   On solving Eqs. (i) and (ii) and evaluating x and y, we get
   $$x = 500 \text{ and } y = 700$$
   Hence, ₹ 500 are invested at 12% per annum and ₹ 700 are invested at ₹ 10% per annum.

2. A private finance company A claims to be lending money at simple interest. But the company includes the interest every 6 months for calculating principal. If company A is charging an interest of 10%, the effective rate of interest after 1 yr becomes
   (a) 10.25%   (b) 12.50%   (c) 11.25%   (d) 10.75%

   ➙ (a) Let the sum be ₹ 100.
   Then, SI for first 6 months $= \dfrac{100 \times 10 \times 1}{100 \times 2} = ₹ 5$
   Now, principal becomes $100 + 5 = 105$.
   $\therefore$ SI for last 6 months $= \dfrac{105 \times 10 \times 1}{100 \times 2} = ₹ 5.25$
   Hence, amount at the end of 1 yr $= 105 + 5.25 = ₹ 110.25$
   $\therefore$ Effective SI $= 110.25 - 100 = ₹ 10.25$
   Effective rate, $R = \dfrac{100 \times SI}{P \times T} = \dfrac{100 \times 10.25}{100 \times 1} = ₹ 10.25\%$

3. The rates of simple interest in two banks x and y are in the ratio of 10 : 8. Rajni wants to deposit his total savings in two banks in such a way that she receive equal half-yearly interest from both. She should deposit the savings in banks x and y in the ratio of
   (a) 4 : 5   (b) 3 : 5   (c) 5 : 4   (d) 2 : 1

   ➙ (a) Here, $R_1 = 10x$, $R_2 = 8x$ and $T_1 = T_2 = \dfrac{1}{2}$ yr
   Let the savings be P and Q and rates of simple interest be $10x$ and $8x$, respectively.
   Then, $\dfrac{P \times R_1 \times T_1}{100} = \dfrac{Q \times R_2 \times T_2}{100}$
   $\Rightarrow P \times 10x \times \dfrac{1}{2} \times \dfrac{1}{100} = Q \times 8x \times \dfrac{1}{2} \times \dfrac{1}{100} \Rightarrow 10P = 8Q$
   $\Rightarrow \dfrac{P}{Q} = \dfrac{8}{10} = \dfrac{4}{5}$
   $\therefore P : Q = 4 : 5$

Fast Track Practice

Exercise 1 Base Level Questions

1. What would be the simple interest obtained on an amount of ₹ 8930 at the rate of 8% per annum after 5 yr?
   [NICL AO 2015]
   (a) ₹ 5413   (b) ₹ 2678
   (c) ₹ 3572   (d) ₹ 4752
   (e) None of these

2. What will be simple interest for 1 yr and 4 months on a sum of ₹ 25800 at the rate of 14% per annum? [Delhi Police SI 2007]
   (a) ₹ 4816   (b) ₹ 2580
   (c) ₹ 4816.75 (d) ₹ 4815

3. A sum at simple interest of $13\frac{1}{2}$% per annum amounts to ₹ 3080 in 4 yr. Find the sum.
   (a) ₹ 1550   (b) ₹ 1680
   (c) ₹ 2000   (d) ₹ 1850
   (e) None of these

4. The sum which amounts to ₹ 364.80 in 8 yr at 3.5% simple interest per annum is   [CDS 2011]
   (a) ₹ 285 (b) ₹ 280 (c) ₹ 275 (d) ₹ 270

5. A sum of ₹ 2668 amounts to ₹ 4669 in 5 yr at the rate of simple interest. Find the rate per cent. [SSC CGL 2008]
   (a) 15.2% (b) 14.9% (c) 16% (d) 15%

6. Kriya deposits an amount of ₹ 65800 to obtain a simple interest at the rate of 14% per annum for 4 yr. What total amount will Kriya get at the end of 4 yr?
   [SBI Clerk 2009]
   (a) ₹ 102648   (b) ₹ 115246
   (c) ₹ 125578   (d) ₹ 110324
   (e) None of these

7. What is the ratio of simple interest earned on certain amount at the rate of 12% per annum for 9 yr and that for 12 yr?   [RRB (Non-tech) 2016]
   (a) 1 : 2   (b) 2 : 3
   (c) 3 : 4   (d) 4 : 5

8. Mr. Deepak invested an amount of ₹ 21250 for 6 yr. At what rate of simple interest, will he obtain the total amount of ₹ 26350 at the end of 6 yr? [SBI Clerk 2012]
   (a) 6% per annum   (b) 5% per annum
   (c) 8% per annum   (d) 12% per annum
   (e) None of these

9. The simple interest on ₹ 4000 in 3 yr at the rate of x% per annum equals to the simple interest on ₹ 5000 at the rate of 12% per annum in 2 yr. The value of x is
   [SSC CGL 2013]
   (a) 6%   (b) 8%   (c) 9%   (d) 10%

10. The simple interest on a sum of money is 1/144 of the principal and the number of years is equal to the rate per cent per annum. What will be the rate per cent per annum?
    (a) $\frac{3}{5}$% (b) $\frac{5}{6}$% (c) $\frac{7}{6}$% (d) $\frac{1}{6}$%
    (e) None of these

11. How long will a sum of money invested at 5% per annum SI take to increase its value by 50%?
    (a) 10 yr (b) 12 yr (c) 15 yr (d) 7 yr
    (e) None of these

12. Suresh borrowed ₹ 800 at 6% and Naresh borrowed ₹ 600 at 10%. After how much time, will they both have equal debts?
    [SSC CGL 2008]
    (a) $15\frac{1}{3}$ yr   (b) $14\frac{1}{2}$ yr
    (c) $18\frac{1}{3}$ yr   (d) $16\frac{2}{3}$ yr

13. Jim invested ₹ 1700 and ₹ 2300 in schemes A and B respectively for 3 yr. If schemes A and B offer simple interest at 6% per annum and at 9% per annum respectively, what is the total interest earned by Jim from both the schemes together after 3 yr?   [IBPS Clerk (Pre) 2017]
    (a) ₹ 927 (b) ₹ 933 (c) ₹ 949 (d) ₹ 935
    (e) ₹ 945

14. Raju lent ₹ 400 to Ajay for 2 yr and ₹ 100 to Manoj for 4 yr and received from both ₹ 60 as collective interest. Find the rate of interest, simple interest being calculated.
    [SSC CGL 2008]
    (a) 5%   (b) 6%   (c) 8%   (d) 9%

15. Rashmi lent ₹ 600 to Geeta for 2 yr and ₹ 150 to Seeta for 4 yr and received altogether ₹ 80 as simple interest from both. Find the rate of interest.
    (a) $3\frac{4}{9}$% (b) $2\frac{4}{9}$% (c) $5\frac{4}{9}$% (d) $4\frac{4}{9}$%
    (e) None of these

256 / Fast Track Objective Arithmetic

16. In 4 yr, ₹ 6000 amounts to ₹ 8000. In what time at the same rate, will ₹ 525 amount to ₹ 700? **[SNAP 2012]**
    (a) 2 yr (b) 3 yr
    (c) 4 yr (d) 5 yr

17. Harsha makes a fixed deposit of ₹ 20000 in Bank of India for a period of 3 yr. If the rate of interest be 13% SI per annum charged half-yearly, what amount will he get after 42 months? **[SSC CPO 2007]**
    (a) ₹ 27800 (b) ₹ 28100
    (c) ₹ 29100 (d) ₹ 30000

18. The effective annual rate of interest corresponding to a nominal rate of 22% per annum payable half-yearly is **[SSC (10+2) 2017]**
    (a) 44% (b) 23.21%
    (c) 46.42% (d) 22%

19. Vikram borrowed ₹ 6450 at 5% simple interest repayable in 4 equal instalments. What will be the annual instalment payable by him?
    (a) ₹ 1710 (b) ₹ 1800
    (c) ₹ 1910 (d) ₹ 1860

20. In what time, does a sum of money become four fold at the simple interest rate of 10% per annum?
    (a) 40 yr (b) 30 yr (c) 15 yr (d) 25 yr
    (e) None of these

21. A certain sum becomes 8 fold in 15 yr at simple interest. What will be the rate of interest? **[SSC (10+2) 2012]**
    (a) $46\frac{5}{3}$% (b) $46\frac{2}{3}$% (c) $46\frac{5}{8}$% (d) $46\frac{11}{12}$%

22. A certain sum becomes 3 fold at 4% annual rate of interest. At what rate, it will become 6 fold? **[CBI Clerk 2008]**
    (a) 10% (b) 12% (c) 8% (d) 9%
    (e) None of these

23. In a certain time, a sum becomes 4 times at the rate of 5% per annum. At what rate of simple interest, the same sum becomes 8 times in the same duration?
    (a) $12\frac{2}{3}$% (b) $11\frac{3}{5}$% (c) $11\frac{2}{3}$% (d) $12\frac{3}{5}$%
    (e) None of these

24. A sum becomes 6 fold at 5% per annum. At what rate, the sum becomes 12 fold? **[LIC AAO 2009]**
    (a) 10% (b) 12% (c) 9% (d) 11%
    (e) None of these

25. A sum becomes two fold in 6 yr at a certain rate of interest. Find the time, in which the same amount will be 10 fold at the same rate of interest.

    (a) 35 yr (b) 49 yr (c) 59 yr (d) 54 yr
    (e) None of these

26. At simple interest, a sum becomes 3 times in 20 yr. Find the time, in which the sum will be double at the same rate of interest. **[RRB 2007]**
    (a) 8 yr (b) 10 yr (c) 12 yr (d) 14 yr

27. A sum of money invested at simple interest triples itself in 8 yr. In how many years it become 8 times itself at the same rate? **[MAT 2014]**
    (a) 24 (b) 28 (c) 30 (d) 21

28. A certain sum becomes ₹ 600 in a certain time at the rate of 6% simple interest. The same sum amounts to ₹ 200 at the rate of 1% simple interest in the same duration. Find the sum and time.
    (a) ₹120 and $66\frac{2}{3}$ yr
    (b) ₹150 and $66\frac{2}{3}$ yr
    (c) ₹130 and $66\frac{2}{3}$ yr
    (d) ₹160 and $66\frac{2}{3}$ yr
    (e) None of the above

29. At a simple interest, a sum amounts to ₹ 1012 in $2\frac{1}{2}$ yr and becomes ₹ 1067.20 in 4 yr. What is the rate of interest? **[SSC (10+2) 2009]**
    (a) 2.5% (b) 3% (c) 4% (d) 5%

30. A sum was lent out for a certain time. The sum amounts to ₹ 400 at 10% annual interest rate. When the sum was lent out at 4% annual interest rate, it amounts to ₹ 200. Find the sum. **[SSC (10+2) 2008]**
    (a) ₹ $\frac{200}{3}$ (b) ₹ 100
    (c) ₹ $\frac{400}{3}$ (d) ₹ 500

31. A certain sum at simple interest amounts to ₹ 1350 in 5 yr and to ₹ 1620 in 8 yr. What is the sum? **[CDS 2011]**
    (a) ₹ 700 (b) ₹ 800
    (c) ₹ 900 (d) ₹ 1000

32. A principal amounts to ₹ 944 in 3 yr and to ₹ 1040 in 5 yr, each sum being invested at the same simple interest. The principal was **[SSC CGL 2013]**
    (a) ₹ 800 (b) ₹ 991
    (c) ₹ 750 (d) ₹ 900

33. The difference between the simple interests received from two different

Simple Interest / 257

banks on ₹ 500 in 2 yr is ₹ 2.5. The difference between their rate of interest is
(a) 0.10% (b) 0.25% (c) 0.50% (d) 1.00%

34. Simple interest for the sum of ₹ 1230 for 2 yr is ₹ 10 more than the simple interest for ₹ 1130 for the same duration. Find the rate of interest. [SSC Multitasking 2012]
(a) 5% (b) 6% (c) 8% (d) 2%

35. For a certain sum, the simple interest in 2 yr at 8% per annum is ₹ 110 more than the simple interest in 1 yr at the rate of 5% per annum for the same sum. Find the sum. [SSC Multitasking 2013]
(a) ₹ 5000 (b) ₹ 1000
(c) ₹ 1050 (d) ₹ 950

36. A sum was invested for 4 yr at a certain rate of simple interest. If it had been invested at 2% more annual rate of interest, then ₹ 56 more would have been obtained. What is the sum? [LIC AAO 2007]
(a) ₹ 680 (b) ₹ 700 (c) ₹ 720 (d) ₹ 820
(e) None of these

37. The difference of simple interest from two banks for ₹ 1000 in 2 yr is ₹ 20. Find the difference in rates of interest. [UP Police 2008]
(a) 2% (b) 1.5% (c) 1% (d) 2.5%

38. 2/3 part of my sum is lent out at 3%, 1/6 part is lent out at 6% and remaining part is lent out at 12%. All the three parts are lent out at simple interest. If the annual income is ₹ 25, what is the sum? [RBI Clerk 2009]
(a) ₹ 500 (b) ₹ 650
(c) ₹ 600 (d) ₹ 450
(e) None of these

39. A sum of ₹ 1521 is lent out in two parts in such a way that the interest on one part at 10% for 5 yr is equal to that of another part at 8% for 10 yr. What will be the two parts of sum? [Hotel Mgmt. 2007]
(a) ₹ 926 and ₹ 595 (b) ₹ 906 and ₹ 615
(c) ₹ 916 and ₹ 605 (d) ₹ 936 and ₹ 585

40. What annual payment will discharge a debt of ₹ 1696 in 4 yr at 4% per annum?
(a) ₹ 525 (b) ₹ 425
(c) ₹ 325 (d) ₹ 400
(e) None of these

41. What annual payment will discharge a debt of ₹ 1092 due in 2 yr at 12% simple interest?
(a) ₹ 725 (b) ₹ 325
(c) ₹ 515 (d) ₹ 900
(e) ₹ 400

Exercise 2 Higher Skill Level Questions

1. The simple interest on a sum of money will be ₹ 200 after 5 yr. In the next 5 yr, principal is tripled. What will be the total interest at the end of the 10th yr?
(a) ₹ 650 (b) ₹ 850 (c) ₹ 800 (d) ₹ 750

2. Mr. Pawan invests an amount of ₹ 24200 at the rate of 4% per annum for 6 yr to obtain a simple interest, later he invests the principal amount as well as the amount obtained as simple interest for another 4 yr at the same rate of interest. What amount of simple interest will be obtained at the end of the last 4 yr? [Dena Bank Clerk 2009]
(a) ₹ 4800 (b) ₹ 4850.32
(c) ₹ 4801.28 (d) ₹ 4700
(e) None of these

3. The simple interest on a certain sum for 8 months at 4% per annum is ₹ 129 less than the simple interest on the same sum for 15 months at 5% per annum. What is the sum? [SBI Clerk (Mains) 2016]
(a) ₹ 2580 (b) ₹ 2400
(c) ₹ 2529 (d) ₹ 3600
(e) ₹ 2900

4. The interest earned when ₹ 'P' is invested for five years in a scheme offering 12% per annum simple interest is more than the interest earned when the same sum (₹ P) is invested for two years in another scheme offering 8% per annum simple interest, by ₹1100. What is the value of P? [IBPS SO 2016]
(a) ₹ 2500 (b) ₹ 2000
(c) ₹ 4000 (d) ₹ 3500
(e) ₹ 3000

5. A sum was put at simple interest at a certain rate for 3 yr. Had it been put at 1% higher rate, it would have fetched ₹ 5100 more. The sum is [RRB Clerk (Pre) 2017]
(a) ₹ 170000 (b) ₹ 150000
(c) ₹ 125000 (d) ₹ 120000
(e) None of these

6. ₹ *XYZ* was deposited at simple interest at a specific rate for 3 yr. Had it been

deposited at 2% higher rate, it would have fetched ₹ 360 more. Find ₹ XYZ [SNAP 2016]
(a) ₹ 5500   (b) ₹ 5000
(c) ₹ 6000   (d) ₹ 4500

7. Neeta borrowed some money at the rate of 6% per annum for the first 3 yr, at the rate of 9% per annum for the next 5 yr and at the rate of 13% per annum for the period beyond 8 yr. If she pays a total interest of ₹ 8160 at the end of 11 yr, how much money did she borrow? [Bank PO 2008]
(a) ₹ 12000 (b) ₹ 10000 (c) ₹ 8000
(d) Data is inadequate
(e) None of these

8. If a sum of money at a certain rate of simple interest per year doubles in 5 yr and at a different rate of simple interest per year becomes three times in 12 yr, then the difference in the two rates of simple interest per year is [CDS 2016 (I)]
(a) 2% (b) 3% (c) $3\frac{1}{3}$% (d) $4\frac{1}{3}$%

9. A person invests ₹ 12000 as fixed deposit at a bank at the rate of 10% per annum simple interest. But due to some pressing needs, he has to withdraw the entire money after 3 yr, for which the bank allowed him a lower rate of interest. If he gets ₹ 3320 less than, what he would have got at the end of 5 yr, the rate of interest allowed by bank is [SSC (10+2) 2012]
(a) $7\frac{8}{9}$% (b) $8\frac{7}{9}$% (c) $7\frac{8}{9}$% (d) $7\frac{4}{9}$%

10. The simple interest obtained when a sum of money is invested for 4 yr at 18% per annum is ₹ 427 more than the simple interest obtained if the same sum of money is invested for 2 yr at 22% per annum. What is the amount obtained when the same sum of money is invested for 4 yr at 18% per annum? [SBI Clerk 2015]
(a) ₹ 2130 (b) ₹ 2623 (c) ₹ 1096 (d) ₹ 1854
(e) ₹ 2475

11. An equal amount of sum is invested in two schemeqs for 4 yr each, both offering simple interest. When invested in scheme A at 8% per annum, the sum amounts to ₹ 5280. In scheme B, invested at 12% per annum, it amounts to ₹ 5920. What is the total sum invested? [RBI Clerk 2015]
(a) ₹ 4000   (b) ₹ 3500
(c) ₹ 4200   (d) ₹ 8000
(e) None of these

12. A sum of ₹ 1550 was lent partly at 5% and partly at 8% per annum simple interest.

The total interest received after 4 yr was ₹ 400. The ratio of the money lent at 5% to that lent at 8% is [SSC CPO 2013]
(a) 16 : 15 (b) 17 : 15
(c) 16 : 13 (d) 16 : 19

13. The interest earned on ₹ 4000 when invested in scheme A for two years at 7% per annum simple interest is half of the interest earned when ₹ X is invested for five years in the same scheme at the same rate of interest. What is the value of X? [IBPS SO 2016]
(a) ₹ 2000   (b) ₹ 3000
(c) ₹ 3600   (d) ₹ 2400
(e) ₹ 3200

14. An equal amount of sum, ₹ P is invested in scheme A and Scheme B. Both the schemes A and B offer simple interest at the rate of 12% and 9% respectively. If at the end of two years total amount received from both the schemes together was ₹ 21780, what is the value of P? [SBI Clerk (Pre) 2016]
(a) ₹ 9000   (b) ₹ 9600
(c) ₹ 12000 (d) ₹ 8400
(e) ₹ 8000

15. A took a certain sum as loan from bank at a rate of 8% simple interest per annum. A lends the same amount to B at 12% simple interest per annum. If at the end of five years, A made profit of ₹ 800 from the deal, how much was the original sum? [SBI PO 2015]
(a) ₹ 6500   (b) ₹ 4000   (c) ₹ 6200
(d) ₹ 6000   (e) ₹ 4500

16. Sum of money was invested for 'T' yr in Scheme A offering simple interest. The amount received after 'T' yr was twice the sum of money invested in the scheme. What will be amount received for Scheme A, when a sum of ₹ 5450 is invested for '2T' yr? [IBPS Clerk (Mains) 2017]
(a) ₹ 16290 (b) ₹ 15500
(c) ₹ 15050 (d) ₹ 16350
(e) ₹ 16500

17. Reena had ₹ 10000 with her. Out of this money, she lent some money to Akshay for 2 yr at 15% simple interest. She lent remaining money to Brijesh for an equal number of years at the rate of 18%. After 2 yr, Reena found that Akshay had given her ₹ 360 more as interest as compared to Brijesh. The amount of money which Reena had lent to Brijesh must be
(a) ₹ 4000   (b) ₹ 2500
(c) ₹ 3500   (d) ₹ 4200
(e) None of these

Answer with Solutions

Exercise 1 Base Level Questions

1. (c) Here, $P = ₹8930$, $R = 8\%$ and $T = 5$ yr

and $SI = \dfrac{P \times R \times T}{100}$

$= \dfrac{8930 \times 8 \times 5}{100} = \dfrac{357200}{100}$

$= ₹3572$

2. (a) Here, $P = ₹25800$, $R = 14\%$

and $T = 1$ yr 4 months

$= \left(1 + \dfrac{4}{12}\right) = \left(1 + \dfrac{1}{3}\right) = \dfrac{4}{3}$ yr

According to the formula,

$SI = \dfrac{P \times R \times T}{100} = \dfrac{25800 \times 14 \times \dfrac{4}{3}}{100}$

$= \dfrac{258 \times 14 \times 4}{3} = ₹4816$

3. (c) Let the sum be $₹P$.

Here, $R = 13\dfrac{1}{2}\%$ and $T = 4$ yr

Then, $SI = \dfrac{P \times R \times T}{100}$

$= \dfrac{P \times \dfrac{27}{2} \times 4}{100} = \dfrac{P \times 54}{100} = \dfrac{27P}{50}$

$\therefore$ Amount $= P + \dfrac{27P}{50} = \dfrac{77P}{50}$

According to the question,

$\dfrac{77P}{50} = 3080$

$\therefore P = \dfrac{3080 \times 50}{77} = ₹2000$

Fast Track Method

Here, $T = 4$ yr, $R = 13\dfrac{1}{2}\% = \dfrac{27}{2}\%$,

$A = ₹3080$, $P = ?$

According to the formula,

$P = \dfrac{100 \times A}{100 + RT}$

$= \dfrac{100 \times 3080}{100 + \dfrac{27}{2} \times 4}$

$= \dfrac{100 \times 3080}{154}$

$= 100 \times 20 = ₹2000$

4. (a) Given, $t = 8$ yr, $r = 3.5\%$, $A = ₹364.80$

Let amount $= ₹P$

$\because A = P\left(1 + \dfrac{RT}{100}\right)$

$\therefore 364.80 = P\left(1 + 3.5 \times \dfrac{8}{100}\right)$

$\Rightarrow 364.80 = P\left(1 + \dfrac{35 \times 8}{1000}\right)$

$\Rightarrow \dfrac{3648}{10} = P\left(\dfrac{128}{100}\right)$

$\Rightarrow P = \dfrac{36480}{128} = ₹285$

5. (d) Here, $P = ₹2668$, $T = 5$ yr, $A = ₹4669$

We know that,

Amount (A) = Principal (P) + Simple Interest (SI)

$4669 = 2668 + SI$

$\Rightarrow SI = 4669 - 2668 = ₹2001$

Again, $SI = \dfrac{P \times R \times T}{100}$

$\therefore 2001 = \dfrac{2668 \times R \times 5}{100}$

$\Rightarrow R = \dfrac{2001 \times 100}{2668 \times 5}$

$= \dfrac{2001 \times 5}{667} = 15\%$

Fast Track Method

Here, $A = ₹4669$, $P = ₹2668$, $T = 5$, $R = ?$

According to the formula,

$P = \dfrac{100 \times A}{100 + RT}$

$\Rightarrow 2668 = \dfrac{100 \times 4669}{100 + 5R}$

$\Rightarrow 266800 + 13340R = 466900$

$\Rightarrow 13340R = 466900 - 266800$

$\therefore R = \dfrac{200100}{13340} = 15\%$

6. (a) Here, $P = ₹65800$, $R = 14\%$, $T = 4$ yr

Now, $SI = \dfrac{65800 \times 14 \times 4}{100} = ₹36848$

$\therefore$ Required amount $= P + SI$

$= 65800 + 36848 = ₹102648$

Fast Track Method
Here, $P = ₹65800$, $R = 14\%$, $T = 4$ yr
According to the formula,
$$A = P\left(1 + \frac{RT}{100}\right)$$
$$= 65800\left(1 + \frac{14 \times 4}{100}\right)$$
$$= \frac{65800 \times 156}{100}$$
$$= 658 \times 456 = ₹102648$$

7. (c) Let the certain amount be ₹ P.
Then, first SI $= \dfrac{P \times 12 \times 9}{100} = \dfrac{₹108P}{100}$
and second SI $= \dfrac{P \times 12 \times 12}{100} = \dfrac{₹144P}{100}$
$\therefore$ Required ratio $= \dfrac{\text{First SI}}{\text{Second SI}}$
$= \dfrac{108P}{144P} = \dfrac{108}{144} = \dfrac{3}{4} = 3 : 4$

8. (e) SI $= 26350 - 21250 = ₹5100$
$\therefore$ Rate $= \dfrac{SI \times 100}{\text{Principal} \times \text{Time}}$
$= \dfrac{5100 \times 100}{21250 \times 6} = 4\%$

9. (d) Since, the two simple interests are equal.
Then, $\dfrac{4000 \times 3 \times x}{100} = \dfrac{5000 \times 12 \times 2}{100}$
$\therefore \quad x = 10\%$

10. (b) Let the principal be ₹ P.
Then, according to the question,
$\dfrac{P \times R \times R}{100} = \dfrac{P}{144}$
$\qquad$ [$\because$ time and rate are equal]
$\Rightarrow \qquad R^2 = \dfrac{100}{144}$
$\Rightarrow \qquad R = \dfrac{10}{12}$
$\therefore \qquad R = \dfrac{5}{6}\%$

11. (a) Let the sum be ₹ P.
Then, 50% of $P = \dfrac{P}{2} = SI$
Now, $\dfrac{P}{2} = \dfrac{P \times 5 \times T}{100}$ [as rate $= 5\%$]
$\Rightarrow \dfrac{P}{2} = \dfrac{5PT}{100} \Rightarrow \dfrac{1}{2} = \dfrac{T}{20}$
$\therefore \qquad T = 10$ yr

12. (d) Given, $R_1 = 6\%$, $R_2 = 10\%$
According to the question,
$800 + \dfrac{800 \times 6 \times T}{100} = 600 + \dfrac{600 \times 10 \times T}{100}$
$\Rightarrow \qquad 800 + 48T = 600 + 60T$
$\qquad 12T = 200 \Rightarrow 3T = 50$
$\therefore \qquad T = \dfrac{50}{3} = 16\dfrac{2}{3}$ yr

13. (a) $\because$ Jim invested ₹1700 at 6% per annum for 3 yr.
$\therefore \quad SI_1 = \dfrac{1700 \times 6 \times 3}{100} = ₹306$
Also, Jim invested ₹2300 at 9% per annum for 3 yr.
$\therefore \quad SI_2 = \dfrac{2300 \times 9 \times 3}{100} = ₹621$
$\therefore$ Total of $SI_1 + SI_2 = 306 + 621 = ₹927$

14. (a) According to the question,
$\dfrac{R \times 400 \times 2}{100} + \dfrac{R \times 100 \times 4}{100} = 60$
$\Rightarrow \qquad 12R = 60$
$\therefore \qquad R = \dfrac{60}{12} = 5\%$

15. (d) Given, $T_1 = 2$ yr and $T_2 = 4$ yr,
$P_1 = 600$, $P_2 = 150$ and $SI_1 + SI_2 = 80$
According to the question,
$\dfrac{600 \times R \times 2}{100} + \dfrac{150 \times R \times 4}{100} = 80$
$\Rightarrow 120R + 60R = 800 \Rightarrow 180R = 800$
$\therefore \quad R = \dfrac{800}{180} = \dfrac{80}{18} = \dfrac{40}{9} = 4\dfrac{4}{9}\%$

16. (c) Amount $= ₹8000$
Time $(T) = 4$ yr, Principal $(P) = ₹6000$
Simple Interest $(SI) = (A - P)$
$= $ Amount $-$ Principal
$= 8000 - 6000 = ₹2000$
Rate $(R) = ?$
According to the formula,
$SI = \dfrac{P \times R \times T}{100} \Rightarrow 2000 = \dfrac{6000 \times R \times 4}{100}$
$\Rightarrow \qquad R = \dfrac{2000 \times 100}{6000 \times 4} = \dfrac{25}{3}\%$
Again, amount $(A) = ₹700$,
Principal $(P) = ₹525$,
Rate $(R) = \dfrac{25}{3}\%$, Time $(T) = ?$
$\therefore \qquad SI = A - P = 700 - 525 = ₹175$
Using formula, $SI = \dfrac{P \times R \times T}{100}$

Simple Interest / 261

$\Rightarrow \quad 175 = \dfrac{525 \times \dfrac{25}{3} \times T}{100}$

$\Rightarrow \quad T = \dfrac{175 \times 100 \times 3}{525 \times 25} = 4$ yr

17. (c) Given, time = 42 months

$= \dfrac{42}{12}$ yr $= 3\dfrac{1}{2}$ yr,

$T = \dfrac{7}{2} \times 2 = 7$ half-yearly

and rate $= \dfrac{13}{2}$% half-yearly

Now, SI $= \dfrac{20000 \times 13 \times 7}{100 \times 2} = ₹ 9100$

$\therefore$ Amount $(A) = 20000 + 9100 = ₹ 29100$

18. (b) Rate of interest $= \dfrac{22}{2} = 11$% half-yearly

Effective annual rate

$= 11 + 11 + \dfrac{11 \times 11}{100} = 22 + \dfrac{121}{100}$

$= 22 + 1.21 = 23.21$%

19. (b) Here, $A = ₹6450$, $R = 5$%, $n = 4$

$\therefore \quad A = P\left(1 + \dfrac{nR}{100}\right)$

$= 6450\left(1 + \dfrac{4 \times 5}{100}\right) = ₹7740$

Let the annual instalment payable by him be ₹ x.

According to the question,

$A = x + \left(x + \dfrac{x \times R \times 1}{100}\right) + \left(x + \dfrac{x \times R \times 2}{100}\right)$

$\qquad + \left(x + \dfrac{x \times R \times 3}{100}\right)$

$\Rightarrow 7740 = 4x + \left(\dfrac{x \times 5 \times 1}{100}\right) + \left(\dfrac{x \times 5 \times 2}{100}\right)$

$\qquad + \left(\dfrac{x \times 5 \times 3}{100}\right)$

$\Rightarrow \quad 7740 = 4x + \dfrac{30x}{100}$

$\Rightarrow \quad 774000 = 400x + 30x$

$\Rightarrow \quad x = \dfrac{774000}{430} = ₹1800$

20. (b) As P becomes $4P$ in time T.

$\therefore \quad$ SI $= (4P - P) = ₹ 3P$

Now, $3P = \dfrac{P \times R \times T}{100}$

$\Rightarrow \quad 3 = \dfrac{10 \times T}{100} \Rightarrow T = 30$ yr

Fast Track Method

Here, $n = 4$, $R = 10$, $T = ?$

According to the formula,

$R = \dfrac{100(n-1)}{T}$ [by Technique 1]

$\Rightarrow \quad 10 = \dfrac{100(4-1)}{T} = \dfrac{100 \times 3}{T}$

$\therefore \quad T = 30$ yr

21. (b) Let sum $= P$

Then, after 15 yr,

Sum $= ₹ 8P$

$\therefore \quad$ SI $= 8P - P = ₹ 7P$

Now, $7P = \dfrac{P \times R \times 15}{100} \Rightarrow 7 = \dfrac{15R}{100} = \dfrac{3R}{20}$

$\therefore \quad R = \dfrac{20 \times 7}{3} = \dfrac{140}{3} = 46\dfrac{2}{3}$%

Fast Track Method

Here, $n = 8$, $T = 15$ yr

$\therefore \quad R = \dfrac{100(n-1)}{T}$ [by Technique 1]

$= \dfrac{100(8-1)}{15} = \dfrac{100 \times 7}{15}$

$= \dfrac{20 \times 7}{3} = \dfrac{140}{3} = 46\dfrac{2}{3}$%

22. (a) Let the sum be ₹ P.

Then, for 4% rate, SI $= (3P - P) = 2P$

$\therefore \quad 2P = \dfrac{P \times 4 \times T}{100} \Rightarrow 1 = \dfrac{2T}{100} = \dfrac{T}{50}$

$\therefore \quad T = 50$ yr

Again, for another rate (R),

SI $= (6P - P) = 5P$

$\Rightarrow \quad 5P = \dfrac{P \times R \times 50}{100} = \dfrac{PR}{2}$

$\therefore \quad R = 10$%

Fast Track Method

Here, $n = 3$, $m = 6$, $R_1 = 4$%

$\therefore \quad R_2 = \dfrac{m-1}{n-1} \times R_1$ [by Technique 2]

$= \dfrac{6-1}{3-1} \times 4 = \dfrac{5}{2} \times 4 = 10$%

23. (c) Let the sum be P.

Then, for $R = 5$%

SI $= (4P - P) = 3P$

$\therefore \quad 3P = \dfrac{P \times 5 \times T}{100} = \dfrac{PT}{20}$

$\Rightarrow \quad T = 60$ yr

Again, for another rate (R),

SI $= (8P - P) = 7P$

$\therefore \quad 7P = \dfrac{P \times R \times 60}{100}$

$\Rightarrow \quad R = \dfrac{7 \times 100}{60} = \dfrac{35}{3}$% $= 11\dfrac{2}{3}$%

Fast Track Method
Here, $n = 4$, $m = 8$, $R_1 = 5\%$, $R_2 = ?$
$\therefore \quad R_2 = \dfrac{m-1}{n-1} \times R_1$ [by Technique 2]
$= \dfrac{8-1}{4-1} \times 5 = \dfrac{7 \times 5}{3} = \dfrac{35}{3} = 11\dfrac{2}{3}\%$

24. (d) SI at 5% $= 6P - P = ₹ 5P$
$\therefore \quad 5P = \dfrac{P \times 5 \times T}{100} \Rightarrow T = 100$ yr
Now, for new rate (R),
$\qquad$ SI $= 12P - P = ₹ 11P$
$\Rightarrow \quad 11P = \dfrac{P \times R \times 100}{100}$
$\therefore \quad R = 11\%$
Fast Track Method
Hence, $R_1 = 5\%$, $n = 6$, $m = 12$
$\therefore \quad R_2 = \dfrac{m-1}{n-1} \times R_1$ [by Technique 2]
$= \dfrac{12-1}{6-1} \times 5 = \dfrac{11}{5} \times 5 = 11\%$

25. (d) Let sum $= ₹ P$
Then, for 6 yr
$\qquad$ SI $= (2P - P) = ₹ P$
$\therefore \quad P = \dfrac{P \times R \times 6}{100} = \dfrac{3PR}{50}$
$\therefore \quad R = \dfrac{50}{3}\%$
Again, for another time (T),
$\qquad$ SI $= 10P - P = ₹ 9P$
$\therefore \quad 9P = \dfrac{P \times \dfrac{50}{3} \times T}{100}$
$\Rightarrow \quad 9P = \dfrac{P \times 50 \times T}{300} \Rightarrow 9 = \dfrac{T}{6}$
$\therefore \quad T = 54$ yr
Fast Track Method
Here, $n = 2$, $m = 10$, $T_1 = 6$ yr, $T_2 = ?$
$\therefore \quad T_2 = \dfrac{m-1}{n-1} \times T_1$ [by Technique 2]
$= \dfrac{10-1}{2-1} \times 6 = 54$ yr

26. (b) Let the sum $= ₹ P$.
Then, for 20 yr,
$\qquad$ SI $= (3P - P) = ₹ 2P$
$\therefore \quad 2P = \dfrac{P \times R \times 20}{100} \Rightarrow 2 = \dfrac{R}{5}$
$\Rightarrow \quad R = 10\%$
Again, for another time (T),
$\qquad$ SI $= (2P - P) = P$

$\therefore \quad P = \dfrac{P \times 10 \times T}{100}$
$\therefore \quad T = 10$ yr
Fast Track Method
Here, $n = 3$, $m = 2$, $T_1 = 20$ yr
$\therefore \quad T_2 = \dfrac{m-1}{n-1} \times T_1$ [by Technique 2]
$= \dfrac{2-1}{3-1} \times 20 = 10$ yr

27. (b) Let the principal be $₹ P$.
Then, amount $= ₹ 3P$
Given, time $= 8$ yr
Simple Interest (SI) $= 3P - P = 2P$
Now, $\quad$ SI $= \dfrac{P \times R \times T}{100}$
$\therefore \quad 2P = \dfrac{P \times R \times 8}{100} \Rightarrow R = 25\%$
Again, new amount $= ₹ 8P$
$\therefore \quad$ SI $= 8P - P = 7P \Rightarrow$ SI $= \dfrac{P \times R \times T}{100}$
Thus, $\quad 7P = \dfrac{P \times 25 \times t}{100} \Rightarrow t = \dfrac{100 \times 7P}{P \times 25}$
$\therefore \quad t = 28$ yr
Fast Track Method
Here, $n = 3$, $m = 8$ and $T_1 = 8$ yr
$\therefore \quad T_2 = \dfrac{m-1}{n-1} \times T_1$ [by Technique 2]
$\Rightarrow \quad T_2 = \dfrac{7}{2} \times 8 = 28$ yr

28. (a) According to the question,
$\left(P + \dfrac{P \times 6 \times T}{100} \right) - \left(P + \dfrac{P \times 1 \times T}{100} \right)$
$\qquad\qquad\qquad\qquad\qquad\qquad = 600 - 200$
$\Rightarrow \quad \dfrac{5PT}{100} = 400 \Rightarrow PT = 8000$
Again, for 6% rate, SI $= \dfrac{PTR}{100} = \dfrac{8000 \times 6}{100}$
$\qquad\qquad\qquad\qquad\quad = ₹ 480$
$\therefore$ Sum $= 600 - 480 = ₹ 120$
As we have, $PT = 8000$
$\therefore \quad T = \dfrac{8000}{120} = \dfrac{200}{3} = 66\dfrac{2}{3}$ yr
Fast Track Method
Here, $R_1 = 6\%$, $R_2 = 1\%$, $A_1 = ₹ 600$
and $A_2 = ₹ 200$
$\therefore \quad P = \dfrac{A_2 R_1 - A_1 R_2}{R_1 - R_2}$ [by Technique 3]
$= \dfrac{200 \times 6 - 600 \times 1}{6 - 1}$
$= \dfrac{1200 - 600}{5} = \dfrac{600}{5} = ₹ 120$

Simple Interest / 263

Time, $T = \dfrac{A_1 - A_2}{A_2 R_1 - A_1 R_2} \times 100$

$= \dfrac{600 - 200}{1200 - 600} \times 100$

$= \dfrac{400 \times 100}{600} = \dfrac{200}{3} = 66\dfrac{2}{3}$ yr

29. (c) Given, $T_1 = 2\dfrac{1}{2}$ yr, $T_2 = 4$ yr

According to the question,

$\left(P + \dfrac{P \times R \times 4}{100}\right) - \left(P + \dfrac{P \times R \times 2.5}{100}\right)$

$= 1067.20 - 1012 = 55.2$

$\Rightarrow \dfrac{1.5 PR}{100} = 55.2$

$\Rightarrow PR = \dfrac{552 \times 100}{15} = 3680$

For 4 yr, SI $= \dfrac{PRT}{100} = \dfrac{3680 \times 4}{100} = ₹147.2$

$\therefore$ Sum $(P) = 1067.2 - 147.2 = ₹920$

We have, $PR = 3680$

$\therefore R = \dfrac{3680}{P} = \dfrac{3680}{920} = 4\%$

Fast Track Method

Here, $T_1 = 2\dfrac{1}{2}$ yr $= 2.5$ yr, $T_2 = 4$ yr,

$A_2 = ₹1067.20$, $A_1 = 1012$

$\therefore R = \dfrac{A_1 - A_2}{A_2 T_1 - A_1 T_2} \times 100$

[by Technique 3]

$= \dfrac{1012 - 1067.20}{(1067.2 \times 2.5) - (1012 \times 4)} \times 100$

$= \dfrac{-55.20}{2668 - 4048} \times 100$

$= \dfrac{-5520}{-1380} = 4\%$

30. (a) According to the question,

$\left(P + \dfrac{P \times 10 \times T}{100}\right) - \left(P + \dfrac{P \times 4 \times T}{100}\right)$

$= 400 - 200$

$\Rightarrow \dfrac{6PT}{100} = 200$

$\Rightarrow PT = \dfrac{200 \times 100}{6} = \dfrac{10000}{3}$

Again, for 10% rate,

SI $= \dfrac{P \times 10 \times T}{100} = \dfrac{\dfrac{10000}{3} \times 10}{100} = \dfrac{1000}{3}$

$\therefore$ Sum $(P) = 400 - \dfrac{1000}{3} = \dfrac{1200 - 1000}{3}$

$= ₹\dfrac{200}{3}$

Fast Track Method

Given, $R_1 = 10\%$, $R_2 = 4\%$, $A_1 = ₹400$, $A_2 = ₹200$

$\therefore P = \dfrac{A_2 R_1 - A_1 R_2}{R_1 - R_2}$ [by Technique 3]

$= \dfrac{200 \times 10 - 400 \times 4}{10 - 4}$

$= \dfrac{2000 - 1600}{6} = \dfrac{400}{6} = ₹\dfrac{200}{3}$

31. (c) Given, $A_1 = ₹1350$, $A_2 = ₹1620$,

$T_1 = 5$ yr and $T_2 = 8$ yr

Let the principal amount be $₹P$.

$\therefore$ In time $8 - 5 = 3$ yr,

Simple interest $= 1620 - 1350 = ₹270$

$\therefore P = \dfrac{A_2 T_1 - A_1 T_2}{T_1 - T_2}$

$= \dfrac{1620 \times 5 - 1350 \times 8}{5 - 8}$

[by Technique 3]

$= \dfrac{8100 - 10800}{-3} = \dfrac{2700}{3} = ₹900$

32. (a) According to the question,

$\left(P + \dfrac{P \times 5 \times R}{100}\right) - \left(P + \dfrac{P \times 3 \times R}{100}\right)$

$= 1040 - 944$

$\Rightarrow 2 PR = 9600$

or $PR = 4800$ ...(i)

Now, for $T = 3$ yr,

$P + \dfrac{P \times R \times 3}{100} = 944$

$\Rightarrow P + \dfrac{4800 \times 3}{100} = 944$ [from Eq. (i)]

$\therefore P = 944 - 144 = ₹800$

33. (b) Let two rates be $R_1\%$ and $R_2\%$ per annum, respectively.

Then, difference

$= \dfrac{P \times R_2 \times T}{100} - \dfrac{P \times R_1 \times T}{100}$

$\Rightarrow \dfrac{500 \times R_2 \times 2}{100} - \dfrac{500 \times R_1 \times 2}{100} = 2.5$

$\Rightarrow 10(R_2 - R_1) = 2.5$

$\therefore R_2 - R_1 = \dfrac{2.5}{10} = 0.25\%$ per annum

Fast Track Method

Here, $I_2 - I_1 = ₹2.5$, $T = 2$ yr and $P = ₹500$

$\therefore\ I_2 - I_1 = \dfrac{PT(R_2 - R_1)}{100}$ [by Technique 4]

$\Rightarrow\ 2.5 = \dfrac{500 \times 2}{100}[R_2 - R_1]$

$\therefore\ R_2 - R_1 = \dfrac{2.5}{10} = 0.25\%$ per annum

34. (a) According to the question,

$\dfrac{1230 \times 2 \times R}{100} - \dfrac{1130 \times 2 \times R}{100} = 10$

$\Rightarrow\ \dfrac{200}{100}R = 10$

$\therefore\ R = 5\%$

Fast Track Method

Here, $I_1 - I_2 = ₹10,\ P_1 = ₹1230,$
$P_2 = ₹1130,\ T = 2$ yr

$\therefore\ I_2 - I_1 = \dfrac{RT(P_2 - P_1)}{100}$ [by Technique 4]

$\Rightarrow\ -10 = \dfrac{R \times 2(1130 - 1230)}{100}$

$\Rightarrow\ -10 = \dfrac{-200R}{100}$

$\therefore\ R = 5\%$

35. (b) Given, $SI_1 - SI_2 = 110,\ T_1 = 2$ yr,
$T_2 = 1$ yr, $R_1 = 8\%,\ R_2 = 5\%$
According to the question,

$\dfrac{P \times 8 \times 2}{100} - \dfrac{P \times 5 \times 1}{100} = 110 \Rightarrow \dfrac{11P}{100} = 110$

$\therefore\ P = ₹1000$

Fast Track Method

Given, $I_1 - I_2 = ₹110,\ T_1 = 2$ yr
$R_1 = 8\%,\ R_2 = 5\%,\ T_2 = 1$ yr

$\therefore\ I_2 - I_1 = \dfrac{P(R_2T_2 - R_1T_1)}{100}$

[by Technique 4]

$\Rightarrow\ -110 = \dfrac{P(5 \times 1 - 8 \times 2)}{100}$

$\Rightarrow\ -110 = \dfrac{P(5 - 16)}{100} = \dfrac{-11P}{100}$

$\therefore\ P = ₹1000$

36. (b) According to the question,

$\dfrac{P \times (R + 2) \times 4}{100} - \dfrac{P \times R \times 4}{100} = 56$

$\Rightarrow\ \dfrac{4PR + 8P - 4PR}{100} = 56 \Rightarrow \dfrac{8P}{100} = 56$

$\therefore\ P = \dfrac{56 \times 100}{8} = ₹700$

Fast Track Method

Given, $I_2 - I_1 = ₹56,\ R_2 - R_1 = 2\%$
and $T = 4$ yr

$\therefore\ I_2 - I_1 = \dfrac{PT(R_2 - R_1)}{100}$

[by Technique 4]

$\Rightarrow\ 56 = \dfrac{P \times 4 \times 2}{100}$

$\therefore\ P = \dfrac{56 \times 100}{8} = ₹700$

37. (c) Let the two rates be R_1 and R_2.
According to the question,

$\dfrac{1000 \times 2 \times R_1}{100} - \dfrac{1000 \times 2 \times R_2}{100} = 20$

$\Rightarrow\ \dfrac{2000(R_1 - R_2)}{100} = 20$

$\therefore\ (R_1 - R_2) = \dfrac{20}{20} = 1\%$

Fast Track Method

Here, $P = ₹1000,\ T = 2$ yr, $I_2 - I_1 = ₹20$

$\therefore\ I_2 - I_1 = \dfrac{PT(R_2 - R_1)}{100}$ [by Technique 4]

$\Rightarrow\ 20 = \dfrac{1000 \times 2(R_2 - R_1)}{100}$

$\Rightarrow\ (R_2 - R_1) = \dfrac{20}{20} = 1\%$

38. (a) Let entire sum be ₹ P.
According to the question,

$\dfrac{2}{3}P \times 3\% + \dfrac{1}{6}P \times 6\%$

$+ \left[1 - \left(\dfrac{2}{3} + \dfrac{1}{6}\right)\right]P \times 12\% = 25$

$\Rightarrow\ \dfrac{2P}{3} \times \dfrac{3}{100} + \dfrac{P}{6} \times \dfrac{6}{100}$

$+ \left[1 - \dfrac{4+1}{6}\right]\dfrac{12P}{100} = 25$

$\Rightarrow\ \dfrac{2P}{100} + \dfrac{P}{100} + \dfrac{2P}{100} = 25$

$\Rightarrow\ 5P = 2500 \Rightarrow P = ₹500$

Fast Track Method

Given, $R_1 = 3\%,\ R_2 = 6\%,\ R_3 = 12\%,$

$\dfrac{1}{x} = \dfrac{2}{3},\ \dfrac{1}{y} = \dfrac{1}{6},\ \dfrac{1}{z} = 1 - \left(\dfrac{2}{3} + \dfrac{1}{6}\right)$

$= \dfrac{6 - (4+1)}{6} = \dfrac{6-5}{6} = \dfrac{1}{6}$ and $I = ₹25$

$\therefore\ P = \dfrac{I \times 100}{\dfrac{R_1}{x} + \dfrac{R_2}{y} + \dfrac{R_3}{z}}$ [by Technique 5]

$= \dfrac{25 \times 100}{3 \times \dfrac{2}{3} + 6 \times \dfrac{1}{6} + 12 \times \dfrac{1}{6}} = \dfrac{2500}{2 + 1 + 2}$

$= ₹500$

Simple Interest / 265

39. (d) Given, $T_1 = 5$ yr, $R_1 = 10\%$,
$T_2 = 10$ yr and $R_2 = 8\%$
Let the first part $= x$
Then, second part $= (1521 - x)$
Now, according to the question,
$$\frac{x \times 5 \times 10}{100} = \frac{(1521 - x) \times 10 \times 8}{100}$$
$\Rightarrow \quad 5x = 12168 - 8x$
$\Rightarrow \quad 13x = 12168 \Rightarrow x = ₹936$
and second part $= 1521 - 936 = ₹585$

Fast Track Method
Given, $T_1 = 5$ yr, $T_2 = 10$ yr, $R_1 = 10\%$ and $R_2 = 8\%$
$\therefore$ Ratio of two parts $= \dfrac{1}{R_1 T_1} : \dfrac{1}{R_2 T_2}$
[by Technique 6]
$= \dfrac{1}{10 \times 5} : \dfrac{1}{8 \times 10} = \dfrac{1}{50} : \dfrac{1}{80} = \dfrac{1}{5} : \dfrac{1}{8} = 8 : 5$

First part $= \dfrac{8}{8+5} \times 1521 = \dfrac{8}{13} \times 1521$
$= 8 \times 117 = ₹936$
$\therefore$ Second part $= 1521 - 936 = ₹585$

40. (d) Given, debt $(P) = ₹1696$,
$R = 4\%$ and $T = 4$ yr
$\therefore$ Annual payment
$= \dfrac{100P}{100 \times T + \dfrac{RT(T-1)}{2}}$ [by Technique 7]
$= \dfrac{1696 \times 100}{4 \times 100 + \dfrac{4 \times 3 \times 4}{2}}$
$= \dfrac{1696 \times 100}{400 + 24} = \dfrac{1696 \times 100}{424} = ₹400$

41. (c) Given, debt $(P) = ₹1092$,
$R = 12\%$ and $T = 2$ yr
$\therefore$ Annual payment $= \dfrac{100P}{100T + \dfrac{RT(T-1)}{2}}$
[by Technique 7]
$= \dfrac{1092 \times 100}{100 \times 2 + \dfrac{24(2-1)}{2}} = \dfrac{1092 \times 100}{212}$
$= ₹515.09 \approx ₹515$

Exercise 2 Higher Skill Level Questions

1. (c) According to the question,
SI for 5 yr $= ₹200$
When principal is tripled, then SI for 5 yr will also be triple and hence SI for next 5 yr will be ₹ (200×3), i.e. ₹600.
$\therefore$ Total SI for 10 yr $= 200 + 600 = ₹800$

2. (c) In case I,
SI $= \dfrac{P \times R \times T}{100} = \dfrac{24200 \times 4 \times 6}{100} = ₹5808$
$\therefore$ Amount = Principal + SI
SI $= 24200 + 5808 = 30008$
In case II, SI $= \dfrac{30008 \times 4 \times 4}{100}$
$= ₹4801.28$

3. (d) According to the question,
$\dfrac{P \times 8 \times 4}{12 \times 100} = \dfrac{P \times 15 \times 5}{12 \times 10} - 129$
$\Rightarrow \quad 32P = 75P - 154800$
$\Rightarrow \quad 43P = 154800$
$\Rightarrow \quad P = ₹3600$

4. (a) According to the question,
$\dfrac{P \times 12 \times 5}{100} - \dfrac{P \times 8 \times 2}{100} = 1100$
$\Rightarrow \quad \dfrac{60P}{100} - \dfrac{16P}{100} = 1100$

$\Rightarrow \quad \dfrac{44P}{100} = 1100$
$\Rightarrow \quad P = \dfrac{1100 \times 100}{44}$
$\Rightarrow \quad P = ₹2500$

5. (a) Let the principle be ₹ P and rate of interest be $R\%$.
Then, according to the question,
$\dfrac{P(R+1)3}{100} - \dfrac{3PR}{100} = 5100$
$\Rightarrow \quad 3P = 5100 \times 100$
$\therefore \quad$ Principal $= ₹170000$

6. (c) Given, principal $= ₹XYZ$
Let the rate of interest be $r\%$.
$\therefore$ New rate of interest $= (r+2)\%$
Then, according to the question,
$\dfrac{XYZ \times (r+2) \times 3}{100} - \dfrac{XYZ \times r \times 3}{100} = 360$
$\Rightarrow \quad \dfrac{3XYZ}{100}(r+2-r) = 360$
$\Rightarrow \quad \dfrac{6XYZ}{100} = 360$
$\Rightarrow \quad XYZ = \dfrac{360 \times 100}{6}$
$\therefore \quad XYZ = ₹6000$

7. (c) Let the sum borrowed be ₹ P.
Then, according to the question,
$$\frac{P \times 6 \times 3}{100} + \frac{P \times 9 \times 5}{100} + \frac{P \times 13 \times 3}{100} = 8160$$
$$\Rightarrow \frac{18P + 45P + 39P}{100} = 8160$$
$$\Rightarrow \frac{102P}{100} = 8160 \Rightarrow P = \frac{8160 \times 100}{102}$$
$$\therefore P = ₹ 8000$$

8. (c) Let the principal be ₹P.
Then, amount of money = ₹ 2P
$$\therefore SI = 2P - P = ₹P$$
Now, $P = \frac{P \times r \times 5}{100} \Rightarrow r = 20\%$
Amount of money after 12 yr = ₹ 3P
$$\therefore SI = 3P - P = 2P$$
Now, $2P = \frac{P \times R \times 12}{100} \Rightarrow R = \frac{50}{3}\%$
∴ Difference between two interest rates
$$= (r - R)\% = \left(20 - \frac{50}{3}\right)\% = \frac{10}{3}\% = 3\frac{1}{3}\%$$

9. (d) Let the rate of interest allowed by bank be r%.
According to the question,
$$\frac{12000 \times 5 \times 10}{100} - \frac{12000 \times 3 \times r}{100} = 3320$$
$$\Rightarrow 6000 - 360r = 3320$$
$$\Rightarrow 360r = 6000 - 3320 = 2680$$
$$\Rightarrow r = \frac{2680}{360} = 7\frac{4}{9}\%$$

10. (b) Let the sum of money invested be ₹ P.
Then, $\frac{P \times 18 \times 4}{100} - \frac{P \times 22 \times 2}{100} = 427$
$$\Rightarrow 72P - 44P = 42700$$
$$\Rightarrow P = \frac{42700}{28} = ₹ 1525$$
Therefore, amount obtained by the sum in 4 yr at 18% per annum
$$= P + \frac{Prt}{100} \quad [\because A = P + SI]$$
$$= 1525 + \frac{1525 \times 18 \times 4}{100}$$
$$= 1525 + 1098 = ₹ 2623$$

11. (d) Let the equal invested sum be ₹x.
Then, for scheme A, $x + \frac{x \times 4 \times 8}{100} = 5280$
$$[\because A = P + SI]$$
$$\Rightarrow x + \frac{8x}{25} = 5280$$
$$\Rightarrow 33x = 5280 \times 25$$
$$\Rightarrow x = \frac{5280 \times 25}{33} = ₹ 4000$$
Now, we need not to calculate the sum invested in scheme B, as sum invested in both the schemes is equals.
For scheme B, x = ₹ 4000
∴ Total sum = 4000 + 4000 = ₹8000

Alternate Method
Difference in interests
$$= 5920 - 5280 = 640$$
Difference in rate = (12 − 8)% = 4%,
Time = 4 yr
In 4 yr, rate of interest = 4% × 4 = 16%
16% = 640
$$100\% = \frac{640 \times 100}{16} = ₹4000$$
∴ Total sum invested
$$= 4000 + 4000 = ₹ 8000$$

12. (a) Let the sum lent at 5% = P
∴ Sum lent at 8% = (1550 − P)
Then, $\frac{P \times 5 \times 4}{100} + \frac{(1550 - P) \times 8 \times 4}{100}$
$$= 400$$
$$\Rightarrow 20P - 32P + 1550 \times 32 = 40000$$
$$\Rightarrow -12P + 49600 = 40000$$
$$\Rightarrow -12P = -9600 \Rightarrow P = ₹ 800$$
Sum lent at 8% = 1550 − 800 = ₹ 750
∴ Required ratio = 800 : 750 = 16 : 15

13. (e) $SI = \frac{4000 \times 7 \times 2}{100} = ₹560$
Since, this interest is half of the interest earned when ₹ x is invested for same rate and scheme. So, interest earned by ₹ x is same scheme for 5 yr will be double.
∴ Interest earned by invested of ₹ x
$$= 2 \times 560 = ₹1120$$
$$\therefore SI = \frac{x \times 7 \times 5}{100}$$
$$\Rightarrow 1120 = \frac{x \times 7}{20}$$
$$\Rightarrow x = \frac{1120 \times 20}{7} = 160 \times 20$$
$$\Rightarrow x = ₹3200$$

14. (a) Here, $R_1 = 12\%, R_2 = 9\%, T = 2$ yr
Now, according to the question,
$$\left(P + \frac{P \times R_1 \times T}{100}\right) \div \left(P + \frac{P \times R_2 \times T}{100}\right)$$
$$= 21780$$

Simple Interest / 267

$\Rightarrow \left(P + \dfrac{P \times 12 \times 2}{100}\right) + \left(P + \dfrac{P \times 9 \times 2}{100}\right)$
$\qquad = 21780$

$\Rightarrow \left(\dfrac{100P + 24P}{100}\right) + \left(\dfrac{100P + 18P}{100}\right)$
$\qquad = 21780$

$\Rightarrow \dfrac{124P}{100} + \dfrac{118P}{100} = 21780$

$\Rightarrow \dfrac{242P}{100} = 21780 \Rightarrow P = \dfrac{100 \times 21780}{242}$

$\therefore \quad P = 100 \times 90 = ₹9000$

15. (b) Let the original sum A took from bank
$\qquad = ₹100$
Interest after 5 yr to be paid to bank by A
$= \dfrac{100 \times 8 \times 5}{100} = ₹40$

Interest received by A from B for 5 yr
$= \dfrac{100 \times 12 \times 5}{100} = ₹60$

Profit made by $A = 60 - 40 = ₹20$
For ₹100, profit made by A is ₹20.
Given that, profit made by A is ₹800.
Hence, the original sum A took from
bank $= \dfrac{800}{20} \times 100 = ₹4000$

16. (d) From, $SI = \dfrac{P \times R \times T}{100}$

$\Rightarrow \quad P = \dfrac{P \times R \times T}{100}$

$\Rightarrow \quad RT = 100 \qquad \ldots(i)$

Now, principal = ₹5450, Time = $2T$ yr

$\therefore \quad SI = \dfrac{P \times R \times T}{100} = \dfrac{5450 \times R \times 2T}{100}$
$= \dfrac{5450 \times RT}{50}$
$= 109 \times 100 \qquad [\because RT = 100]$
$= ₹10900$

$\therefore$ Total amount received from scheme A
$= P + SI$
$= 5450 + 10900 = ₹16350$

17. (a) Let the money lent to Akshay = ₹x
Then, money lent to Brijesh = ₹$(10000 - x)$
$\qquad [\because \text{total amount} = ₹10000]$

SI for Akshay = $\dfrac{x \times 15 \times 2}{100} = \dfrac{3x}{10}$

SI for Brijesh = $\dfrac{(10000 - x) \times 18 \times 2}{100}$
$= \dfrac{9}{25}(10000 - x)$

According to the given condition,

$\dfrac{3x}{10} - \dfrac{9}{25}(10000 - x) = 360$
$[\text{as SI (Akshay)} - \text{SI (Brijesh)} = 360]$

$\Rightarrow \dfrac{3x}{10} - 3600 + \dfrac{9x}{25} = 360$

$\Rightarrow \dfrac{3x}{10} + \dfrac{9x}{25} = 360 + 3600 = 3960$

$\Rightarrow \dfrac{33x}{50} = 3960 \Rightarrow x = \dfrac{3960 \times 50}{33}$

$\Rightarrow \quad x = 6000$

$\therefore$ Amount of money lent to Brijesh
$= 10000 - 6000 = ₹4000$

Chapter 15

Compound Interest

As we know that when we borrow some money from bank or any person, then we have to pay some extra money at the time of repaying. This extra money is known as **interest**. If interest accrued on principal, it is known as simple interest. Sometimes, it happens that we repay the borrowed money bit late. After the completion of a specific period, interest is accrued on the principal as well as the interest due on the principal. This interest is known as **Compound Interest** (CI).

∴ Compound Interest (CI) = Amount (A) − Principal (P)

- The amount at the end of each year becomes the principal for the next year.
- In case of simple interest, the principal remains constant for the whole time but in case of compound interest, principal keeps on changing every year.

Basic Formulae Related to Compound Interest

Let principal = P, rate = R% per annum and time = n yr.

1. If the interest is compounded annually, then

$$\text{Amount} = P\left(1 + \frac{R}{100}\right)^n$$

Also, compound interest = Amount (A) − Principal (P)

or $$\text{CI} = P\left[\left(1 + \frac{R}{100}\right)^n - 1\right]$$

2. If the interest is compounded half-yearly, then $R = \frac{R}{2}$ and $n = 2n$.

∴ $$\text{Amount} = P\left(1 + \frac{R}{2 \times 100}\right)^{2n}$$

Compound Interest / 269

3. If the interest is compounded quarterly, then $R = \dfrac{R}{4}$ and $n = 4n$.

∴ $\quad$ Amount $= P\left(1 + \dfrac{R}{4 \times 100}\right)^{4n}$

4. If the interest is compounded annually but time is in fraction (suppose time $= n\dfrac{a}{b}$ yr), then

$$\text{Amount} = P\left(1 + \dfrac{R}{100}\right)^{n} \times \left[1 + \dfrac{\left(\dfrac{a}{b}\right)R}{100}\right]$$

5. If the rates of interest are $R_1\%$, $R_2\%$ and $R_3\%$ for 1st, 2nd and 3rd yr respectively, then

$$\text{Amount} = P\left(1 + \dfrac{R_1}{100}\right)\left(1 + \dfrac{R_2}{100}\right)\left(1 + \dfrac{R_3}{100}\right)$$

Ex. 1 Find the compound interest on ₹ 8000 at 4% per annum for 2 yr, compounded annually.

Sol. Given that, principal $(P) = ₹ 8000$, rate $(R) = 4\%$ and time $(n) = 2$ yr
Now, according to the formula,

$$\text{Amount} = P\left(1 + \dfrac{R}{100}\right)^{n} = 8000\left(1 + \dfrac{4}{100}\right)^{2} = 8000 \times \dfrac{26}{25} \times \dfrac{26}{25} = ₹ 8652.80$$

∴ Compound Interest (CI) = Amount − Principal = ₹ 8652.80 − ₹ 8000 = ₹ 652.80

Ex. 2 Ruchi invested ₹ 1600 at the rate of compound interest for 2 yr. She got ₹ 1764 after the specified period. Find the rate of interest.

Sol. Given that, $P = ₹ 1600$, $n = 2$ yr and $A = ₹ 1764$
Now, according to the formula,

$$\text{Amount} = P\left(1 + \dfrac{R}{100}\right)^{n} \Rightarrow 1764 = 1600\left(1 + \dfrac{R}{100}\right)^{2}$$

$\Rightarrow \quad \dfrac{1764}{1600} = \left(\dfrac{100 + R}{100}\right)^{2} \Rightarrow \dfrac{441}{400} = \left(\dfrac{100 + R}{100}\right)^{2}$

$\Rightarrow \quad \left(\dfrac{21}{20}\right)^{2} = \left(\dfrac{100 + R}{100}\right)^{2} \Rightarrow \dfrac{100 + R}{100} = \dfrac{21}{20}$

$\Rightarrow \quad 100 + R = \dfrac{21}{20} \times 100$

$\Rightarrow \quad 100 + R = 105$

∴ $\quad R = 105 − 100 = 5\%$

Ex. 3 Find the compound interest on ₹ 5000 in 2 yr at 4% per annum, if the interest being compounded half-yearly.

Sol. Given that, principal $(P) = ₹ 5000$, rate $(R) = 4\%$ per annum and time $(n) = 2$ yr
Now, according to the formula,

Amount $(A) = P\left(1 + \dfrac{R}{2 \times 100}\right)^{2n} = 5000\left(1 + \dfrac{4}{200}\right)^{4} = 5000 \times \dfrac{51}{50} \times \dfrac{51}{50} \times \dfrac{51}{50} \times \dfrac{51}{50}$

$\quad = \dfrac{51 \times 51 \times 51 \times 51}{1250} = ₹ 5412.16$

∴ $\quad$ CI $= A − P = 5412.16 − 5000 = ₹ 412.16$

Ex. 4 Find the compound interest on ₹ 8000 at 20% per annum for 9 months, compounded quarterly.

Sol. Given that, $P = ₹ 8000$, $n = 9$ months $= 3/4$ yr and $R = 20\%$ per annum
According to the formula,

$$\text{Amount} = P\left(1 + \frac{R}{4 \times 100}\right)^{4n} = 8000\left(1 + \frac{20}{400}\right)^{3/4 \times 4} = 8000\left(1 + \frac{5}{100}\right)^3$$

$$= 8000 \times \frac{21}{20} \times \frac{21}{20} \times \frac{21}{20} = ₹ 9261$$

∴ CI $= 9261 - 8000 = ₹ 1261$

Ex. 5 Find the compound interest on ₹ 2000 at 15% per annum for 2 yr 4 months, compounded annually.

Sol. Given that, $P = ₹ 2000$, $n = 2$, $\frac{a}{b} = \frac{4}{12} = \frac{1}{3}$ and $R = 15\%$

Now, according to the formula,

$$\text{Amount} = P\left(1 + \frac{R}{100}\right)^n \times \left[1 + \frac{\left(\frac{a}{b}\right)R}{100}\right] = \left[2000\left(1 + \frac{15}{100}\right)^2 \times \left(1 + \frac{\frac{1}{3} \times 15}{100}\right)\right]$$

$$= 2000 \times \frac{23}{20} \times \frac{23}{20} \times \frac{21}{20} = \frac{11109}{4} = 2777.25$$

∴ CI $= 2777.25 - 2000 = ₹ 777.25$

Ex. 6 What sum of money at compound interest will amount to ₹ 4499.04 in 3 yr, if the rate of interest is 3% for the 1st yr, 4% for the 2nd yr and 5% for the 3rd yr?

Sol. Given that, $A = ₹ 4499.04$, $R_1 = 3\%$, $R_2 = 4\%$, $R_3 = 5\%$ and $P = ?$
Now, according to the formula,

$$\text{Amount} = P\left(1 + \frac{R_1}{100}\right)\left(1 + \frac{R_2}{100}\right)\left(1 + \frac{R_3}{100}\right)$$

$$\Rightarrow 4499.04 = P\left(1 + \frac{3}{100}\right)\left(1 + \frac{4}{100}\right)\left(1 + \frac{5}{100}\right)$$

$$\Rightarrow 4499.04 = P(1.03)(1.04)(1.05)$$

∴ $P = \frac{4499.04}{1.03 \times 1.04 \times 1.05} = \frac{4499.04}{1.12476} = ₹ 4000$

Instalments

When a borrower pays the sum in parts, then we say that he/she is paying in instalments II.

∴ $P = \left[\dfrac{x}{\left(1 + \dfrac{R}{100}\right)} + \dfrac{x}{\left(1 + \dfrac{R}{100}\right)^2} + \dfrac{x}{\left(1 + \dfrac{R}{100}\right)^3} + \ldots + \dfrac{x}{\left(1 + \dfrac{R}{100}\right)^n}\right]$

where, $x =$ Value of each instalment

Total amount paid in n instalments, $A = P\left(1 + \dfrac{R}{100}\right)^n$ and $n =$ Number of instalments

Compound Interest / 271

Ex. 7 Sapna borrowed some money on compound interest and returned it in 3 yr in equal annual instalments. If rate of interest is 15% per annum and annual instalment is ₹ 486680, then find the sum borrowed.

Sol. Given that, rate of interest, $R = 15\%$ per annum
Annual instalment, $x = ₹ 486680$
and total number of instalments, $n = 3$

$$\therefore P = \left[\frac{x}{\left(1+\frac{R}{100}\right)} + \frac{x}{\left(1+\frac{R}{100}\right)^2} + \frac{x}{\left(1+\frac{R}{100}\right)^3}\right]$$

$$= x\left[\frac{100}{(100+R)} + \frac{(100)^2}{(100+R)^2} + \frac{(100)^3}{(100+R)^3}\right]$$

$$= 486680\left[\left(\frac{100}{100+15}\right) + \left(\frac{100}{100+15}\right)^2 + \left(\frac{100}{100+15}\right)^3\right]$$

$$= 486680 \times \left[\frac{20}{23} + \left(\frac{20}{23}\right)^2 + \left(\frac{20}{23}\right)^3\right]$$

$$= 486680 \times \frac{20}{23}\left(1 + \frac{20}{23} + \frac{400}{529}\right) = 486680 \times \frac{20}{23}\left(\frac{529 + 460 + 400}{529}\right)$$

$$= 486680 \times \frac{20}{23} \times \frac{1389}{529} = ₹ 1111200$$

Hence, the principal borrowed is ₹ 1111200.

Fast Track Techniques
to solve the QUESTIONS

Technique 1

(i) Difference between CI and SI for 2 yr, $D = P\left(\frac{R}{100}\right)^2 = \frac{SI \times R}{200}$

or $CI - SI = SI \times \frac{R}{200} \Rightarrow CI = SI\left(1 + \frac{R}{200}\right)$

(ii) Difference between CI and SI for 3 yr, $D = P\left(\frac{R}{100}\right)^2\left(\frac{R}{100} + 3\right)$

✦ SI and CI for one year on the same sum and at same rate are equal.

Ex. 8 The difference between compound interest and simple interest for 2 yr at rate of 5% per annum is ₹ 5, then find the sum.

Sol. Given that, $CI - SI = ₹ 5$, rate $(R) = 5\%$
Then, according to the formula,

$$D = P\left(\frac{R}{100}\right)^2 \Rightarrow 5 = P\left(\frac{5}{100}\right)^2 \Rightarrow 5 = \frac{P \times 5 \times 5}{100 \times 100} \Rightarrow P = ₹ 2000$$

Ex. 9 The difference between CI and SI for 3 yr at the rate of 20% per annum is ₹ 152. What is the principal lent?

Sol. Difference between CI and SI for 3 yr = ₹ 152

$$\therefore \quad P\left(\frac{R}{100}\right)^2 \left(\frac{R}{100} + 3\right) = 152$$

$$\Rightarrow \quad P\left(\frac{20}{100}\right)^2 \left(\frac{20}{100} + 3\right) = 152$$

$$\Rightarrow \quad P\left(\frac{1}{25}\right)\left(\frac{16}{5}\right) = 152 \Rightarrow P = \frac{152 \times 25 \times 5}{16}$$

$$\Rightarrow \quad P = 9.5 \times 25 \times 5 = ₹\ 1187.5$$

Ex. 10 If the simple interest for a certain sum for 2 yr at 5% per annum is ₹ 200, then what will be the compound interest for same sum for same period and at the same rate of interest?

Sol. Given that, SI = ₹ 200 and $R = 5\%$

According to the formula,

$$CI = SI\left(1 + \frac{R}{200}\right) = 200\left(1 + \frac{5}{200}\right) = 200 \times \frac{205}{200} = ₹\ 205$$

Technique 2 If a certain sum at compound interest becomes x times in n_1 yr and y times in n_2 yr, then $x^{\frac{1}{n_1}} = y^{\frac{1}{n_2}}$.

Ex. 11 If a certain sum at compound interest becomes double in 5 yr, then in how many years, it will be 16 times at the same rate of interest?

Sol. If sum = x, then

x becomes $2x$ in 5 yr, i.e. $2(2x) = 4x$
$2x$ becomes $4x$ in 10 yr, i.e. $2(4x) = 8x$
$4x$ becomes $8x$ in 15 yr, i.e. $2(8x) = 16x$

Hence, $8x$ becomes $16x$ in 20 yr.

Fast Track Method

Here, $n_1 = 5$ yr, $x = 2$, $y = 16$ and $n_2 = ?$
According to the formula,

$$x^{\frac{1}{n_1}} = y^{\frac{1}{n_2}} \quad \Rightarrow \quad 2^{\frac{1}{5}} = 16^{\frac{1}{n_2}}$$

$$\Rightarrow \quad 2^{\frac{1}{5}} = (2)^{4 \times \frac{1}{n_2}} \quad \Rightarrow \quad \frac{1}{5} = \frac{4}{n_2}$$

$$\therefore \quad n_2 = 5 \times 4 = 20 \text{ yr}$$

Technique 3 If a certain sum at compound interest becomes A_1 in n yr and A_2 in $(n + 1)$ yr, then

(i) Rate of compound interest $= \dfrac{(A_2 - A_1)}{A_1} \times 100\%$

(ii) Sum $= A_1 \left(\dfrac{A_1}{A_2}\right)^n$

Compound Interest / 273

Ex. 12 A sum of money invested at compound interest amounts to ₹ 800 in 2 yr and ₹ 840 in 3 yr. Find the rate of interest per annum and the sum.

Sol. Amount of 2 yr (A_1) = 800, amount of 3 yr (A_2) = 840, $n_1 = 2$ and $n_2 = n_1 + 1 = 3$

Given, $$800 = P\left(1 + \frac{R}{100}\right)^2 \qquad ...(i)$$

and $$840 = P\left(1 + \frac{R}{100}\right)^3 \qquad ...(ii)$$

On dividing Eq. (ii) by Eq. (i), we get

$$\frac{840}{800} = \frac{\left(1 + \frac{R}{100}\right)^3}{\left(1 + \frac{R}{100}\right)^2}$$

$$\Rightarrow \frac{21}{20} = 1 + \frac{R}{100} \Rightarrow \frac{21}{20} - 1 = \frac{R}{100}$$

$$\Rightarrow \frac{1}{20} = \frac{R}{100}$$

$$\therefore R = 5\%$$

Now, let the sum = x, then amount = Sum $\left(1 + \frac{R}{100}\right)^n$

$$\Rightarrow x\left(1 + \frac{5}{100}\right)^2 = 800$$

$$\Rightarrow \frac{105}{100} \times \frac{105}{100} \times x = 800$$

$$\Rightarrow \frac{21}{20} \times \frac{21}{20} \times x = 800$$

$$\therefore x = \frac{800 \times 20 \times 20}{21 \times 21} = \frac{320000}{441} = ₹ 725.62$$

Fast Track Method

Here, $A_1 = ₹ 800$ and $A_2 = ₹ 840$

According to the formula,

Rate of compound interest = $\dfrac{A_2 - A_1}{A_1} \times 100\%$

$$= \frac{840 - 800}{800} \times 100\% = \frac{40}{8}\% = 5\%$$

and $$\text{sum} = A_1 \left(\frac{A_1}{A_2}\right)^n = 800 \times \left(\frac{800}{840}\right)^2$$

$$= 800 \times \frac{800}{840} \times \frac{800}{840} = \frac{320000}{441} = ₹ 725.62$$

Multi Concept Questions

1. Suneeta borrowed certain sum from Reena for 2 yr at simple interest. Suneeta lent this sum to Venu at the same rate for 2 yr at compound interest. At the end of 2 yr, she received ₹ 110 as compound interest but paid ₹ 100 as simple interest. Find the sum and the rate of interest, respectively.
  (a) ₹ 250, 10% per annum
  (b) ₹ 250, 20% per annum
  (c) ₹ 250, 25% per annum
  (d) ₹ 250, 30% per annum

↪ **(b)** Let the sum borrowed be ₹ x and rate of interest be $R\%$.
Given, Compound Interest (CI) = ₹ 110, Simple Interest (SI) = ₹ 100 and time $(t) = 2$ yr
By using the formula,
$$CI - SI = \frac{SI \times R}{200} \Rightarrow 110 - 100 = \frac{100 \times R}{200} \Rightarrow R = \frac{10 \times 200}{100} = 20\%$$
Again, using the same formula,
$$CI - SI = \frac{PR^2}{100^2} \Rightarrow 110 - 100 = \frac{x \times 20 \times 20}{100 \times 100} \Rightarrow x = \frac{10 \times 100 \times 100}{20 \times 20} \Rightarrow x = 250$$
∴ Sum borrowed = ₹ 250 and rate of interest = 20%

2. Find the least number of complete year in which a sum of money put out at 20% compound interest will be more than double.
  (a) 3 yr   (b) 4 yr   (c) 5 yr   (d) 8 yr

↪ **(b)** Let the sum be P.
According to the question, $P\left(1 + \frac{20}{100}\right)^n > 2P$ or $\left(\frac{6}{5}\right)^n > 2$; so $n > 4$
By Hit and Trial This is true for $n = 4$; as $\left(\frac{6}{5}\right)^4 = \left(\frac{6}{5} \times \frac{6}{5} \times \frac{6}{5} \times \frac{6}{5}\right) > 2$

3. The half life of Uranium-233 is 160000 yr, i.e. Uranium-233 decays at a constant rate in such a way that it reduces to 50% in 160000 yr. In how many years, will it reduce to 25%?
  (a) 200000 yr   (b) 150000 yr   (c) 280000 yr   (d) 320000 yr

↪ **(d)** Let the initial amount of Uranium-233 be 1 unit and the rate of decay be $R\%$ per unit per year.
∴ Amount of Uranium-233 after 160000 yr $= 1 \times \left(1 - \frac{R}{100}\right)^{160000}$
$$\Rightarrow \frac{50}{100} \times 1 = 1 \times \left(1 - \frac{R}{100}\right)^{160000} \quad \left[\because 50\% = \frac{1}{2}\right] \quad ...(i)$$
Assume that Uranium-233 reduces to 25% in n yr, then
$$\frac{25}{100} = 1 \times \left(1 - \frac{R}{100}\right)^n \Rightarrow \left(\frac{1}{2}\right)^2 = \left(1 - \frac{R}{100}\right)^n \quad ...(ii)$$
From Eqs. (i) and (ii), we get
$$\left[\left(1 - \frac{R}{100}\right)^{160000}\right]^2 = \left(1 - \frac{R}{100}\right)^n \Rightarrow \left(1 - \frac{R}{100}\right)^{320000} = \left(1 - \frac{R}{100}\right)^n \Rightarrow n = 320000 \text{ yr}$$
Thus, Uranium-233 will reduce to 25% in 320000 yr.

Fast Track Practice

Exercise 1 — Base Level Questions

1. What will be the compound interest on a sum of ₹ 50000 after 3 yr at the rate of 12% per annum?
   (a) ₹ 80000   (b) ₹ 70246.40
   (c) ₹ 20246.40   (d) ₹ 70000
   (e) None of these

2. In 3 yr, ₹ 3000 amounts to ₹ 3993 at x% compounded interest, compounded annually. The value of x is [SSC CGL (Mains) 2016]
   (a) 10   (b) 18   (c) 5   (d) 8

3. A sum of money amounts to ₹ 6655 at the rate of 10% compounded annually for 3 yr. The sum of money is
   [SSC CGL (Pre) 2016]
   (a) ₹ 5000   (b) ₹ 5500
   (c) ₹ 6000   (d) ₹ 6100

4. The compound interest on ₹ 30000 at 7% per annum for a certain time is ₹ 4347. The time is
   (a) 2 yr   (b) 2.5 yr
   (c) 3 yr   (d) 4 yr

5. The sum for 2 yr gives a compound interest of ₹ 3225 at 15% rate. Then, sum is [SSC CGL (Mains) 2016]
   (a) ₹ 10000   (b) ₹ 20000
   (c) ₹ 15000   (d) ₹ 32250

6. Find the compound interest on ₹ 31250 at 16% per annum compounded quarterly for 9 months.
   (a) ₹ 4000   (b) ₹ 3902
   (c) ₹ 3500   (d) ₹ 4200
   (e) None of these

7. What amount will be received on a sum of ₹ 15000 in $1\frac{1}{4}$ yr at 12% per annum, if interest is compounded quarterly?
   [SSC LDC 2008]
   (a) ₹ 16596.88   (b) ₹ 16789.08
   (c) ₹ 17630.77   (d) ₹ 17389.10

8. What amount will be received on a sum of ₹ 1750 in 2½ yr, if the interest is compounded at the rate of 8% per annum? [SBI PO 2009]
   (a) ₹ 2125   (b) ₹ 2122.85
   (c) ₹ 2100   (d) ₹ 2200
   (e) ₹ 2300

9. Find the compound interest for a sum of ₹ 9375 in 2 yr, if the rate of interest for the first year is 2% and for the second year is 4%. [RRB 2007]
   (a) ₹ 570   (b) ₹ 670   (c) ₹ 770   (d) ₹ 760

10. A sum is invested for 3 yr at compound interest at 5%, 10% and 20% respectively. In three years, if the sum of amounts to ₹ 16632, then find the sum.
    [RRB Clerk (Pre) 2017]
    (a) ₹ 11000   (b) ₹ 12000
    (c) ₹ 13000   (d) ₹ 14000
    (e) None of these

11. The income of Shantanu was ₹ 4000. In the first 2 yr, his income decreased by 10% and 5% respectively but in the third year, the income increased by 15%. What was his income at the end of third year?
    (a) ₹ 3933   (b) ₹ 4000
    (c) ₹ 3500   (d) ₹ 3540
    (e) None of these

12. The principal amount which yields a compound interest of ₹ 208 in the second year at 4%, is [SSC CGL 2012]
    (a) ₹ 5000   (b) ₹ 10000
    (c) ₹ 130000   (d) ₹ 6500

13. Raja invested ₹ 15000 at the rate of 10% per annum for 1 yr. If the interest is compounded half-yearly, then find the amount received by Raja at the end of the year.
    (a) ₹ 16537.50   (b) ₹ 18000
    (c) ₹ 19000.50   (d) ₹ 20000
    (e) None of these

14. At what per cent annual compound interest rate, a certain sum amounts to its 27 times in 3 yr? [RRB 2007]
    (a) 100%   (b) 150%
    (c) 75%   (d) 200%

15. A certain amount of money earns ₹ 540 as simple interest in 3 yr. If it earns a compound interest of ₹ 376.20 at the same rate of interest in 2 yr, then find the amount. [SSC CPO 2015]
    (a) ₹ 1600   (b) ₹ 2000
    (c) ₹ 1800   (d) ₹ 2100

276 / Fast Track Objective Arithmetic

16. A sum of ₹ 400 amounts to ₹ 441 in 2 yr. What will be its amount, if the rate of interest is increased by 5%? [SSC CGL 2008]
 (a) ₹ 484   (b) ₹ 560
 (c) ₹ 512   (d) ₹ 600

17. A sum of ₹ 8400 was taken as a loan. This is to be paid in two equal instalments. If the rate of interest is 10% per annum, compounded annually, then the value of each instalment is [CDS 2017 (I)]
 (a) ₹ 4200   (b) ₹ 4480
 (c) ₹ 4840   (d) None of these

18. A man borrows ₹ 5100 to be paid back with compound interest at the rate of 4% per annum by the end of 2 yr in two equal yearly instalments. How much will each instalment be?
 (a) ₹ 2704   (b) ₹ 2800
 (c) ₹ 3000   (d) ₹ 2500
 (e) None of these

19. A borrowed sum was paid in the two annual instalments of ₹ 121 each. If the rate of compound interest is 10% per annum, what sum was borrowed?
 (a) ₹ 217.80   (b) ₹ 210
 (c) ₹ 220   (d) ₹ 200

20. A sum of ₹ 11000 was taken as loan. This is to be paid in two equal annual instalments. If the rate of interest is 20% compounded annually, then the value of each instalment is [SSC CPO 2013]
 (a) ₹ 7500   (b) ₹ 7000
 (c) ₹ 7100   (d) ₹ 7200

21. The difference between compound interest and simple interest at the same rate of interest R% per annum on an amount of ₹ 15000 for 2 yr is ₹ 96. What is the value of R? [CDS 2015 (I)]
 (a) 8%   (b) 10%   (c) 12%   (d) 14%

22. Equal sums of money are deposited in two different banks by M/s Enterprises, one at compounds interest, compounded annually and the other at simple interest, both at 5% per annum. If after 2 yr, the difference in the amounts comes to ₹ 200, what are the amounts deposited with each bank? [MAT 2015]
 (a) ₹ 64000   (b) ₹ 72000
 (c) ₹ 80000   (d) ₹ 84000

23. The difference between the simple and the compound interest on a certain sum of money at 4% per annum in 2 yr is ₹ 10. What is the sum? [CDS 2017 (I)]
 (a) ₹ 5000   (b) ₹ 6000
 (c) ₹ 6250   (d) ₹ 7500

24. The simple interest on a certain sum for 2 yr is ₹ 120 and compound interest is ₹ 129. Find the rate of interest. [LIC ADO 2009]
 (a) 14%   (b) 15%
 (c) 12%   (d) $12\frac{1}{2}$%
 (e) None of these

25. The simple interest for certain sum in 2 yr at 4% per annum is ₹ 80. What will be the compound interest for the same sum, if conditions of rate and time period are same? [SSC DEO 2009]
 (a) ₹ 91.60   (b) ₹ 81.60
 (c) ₹ 71.60   (d) ₹ 80

26. What is the difference between compound interest and simple interest for 2 yr on the sum of ₹ 1250 at 4% per annum? [RBI 2008]
 (a) ₹ 3   (b) ₹ 4
 (c) ₹ 2   (d) ₹ 8
 (e) None of these

27. What is the difference between the compound interest and simple interest calculated on an amount of ₹ 16200 at the end of 3 yr at 25% per annum? (Rounded off to two digits after decimal) [IBPS Clerk 2011]
 (a) ₹ 3213.44   (b) ₹ 3302.42
 (c) ₹ 3495.28   (d) ₹ 3290.63
 (e) None of these

28. The difference between compound and simple rates of interest on ₹ 10000 for 3 yr at 5% per annum is [SSC CGL (Mains) 2012]
 (a) ₹ 76.25   (b) ₹ 76.75
 (c) ₹ 76.50   (d) ₹ 76

29. An amount at compound interest doubles itself in 4 yr. In how many years will the amount become 8 times itself? [SSC CGL (Pre) 2016]
 (a) 8 yr   (b) 12 yr
 (c) 16 yr   (d) 24 yr

30. A sum, at the compound rate of interest, becomes $2\frac{1}{2}$ times in 6 yr. The same sum becomes what times in 18 yr? [Hotel Mgmt. 2007]
 (a) $\frac{5}{2}$ (b) $\frac{25}{4}$ (c) $\frac{125}{8}$ (d) $\frac{625}{16}$

31. A sum amounts to ₹ 2916 in 2 yr and ₹ 3149.28 in 3 yr at compound interest. The sum is [SSC (10+2) 2012]
 (a) ₹ 1500   (b) ₹ 2500
 (c) ₹ 2000   (d) ₹ 3000

Compound Interest / 277

Exercise 2 — Higher Skill Level Questions

1. What will be the present worth of ₹ 169 due in 2 yr at 4% per annum compound interest? **[SSC LDC 2008]**
   (a) ₹ 156.25 (b) ₹ 160
   (c) ₹ 150.50 (d) ₹ 154.75

2. What is the difference between the compound interests on ₹ 1000 for 1 yr at 10% per annum compounded yearly and half-yearly? **[SSC CGL (Pre) 2017]**
   (a) ₹ 1.5 (b) ₹ 0.5 (c) ₹ 2.5 (d) ₹ 3.5

3. The simple interest on a certain sum of money for 3 yr at 8% per annum is half the compound interest on ₹ 8000 for 2 yr at 10% per annum. Find the sum on which simple interest is calculated.
   (a) ₹ 3500 (b) ₹ 3800
   (c) ₹ 4000 (d) ₹ 3600
   (e) None of these

4. The simple interest, at the rate of 6% per annum received on a principal of ₹ X, was ₹ 482.40 when invested for 3 yr in scheme A. If scheme B offered compound interest (compounded annually) at 10% per annum, what was the interest received by investing ₹ 'X-680' for 2 yr in scheme B? **[SBI Clerk (Pre) 2016]**
   (a) ₹ 420 (b) ₹ 490
   (c) ₹ 530 (d) ₹ 540
   (e) ₹ 650

5. Poonam invests ₹ 4200 in scheme A which offers 12% per annum simple interest. She also invests ₹ 4200 − P in scheme B offering 10% per annum compound interest (compounded annually). The difference between the interests Poonam earns from both the schemes at the end of 2 yr is ₹ 294, what is the value of P? **[SBI PO (Pre) 2017]**
   (a) 1500 (b) 800 (c) 600 (d) 1000
   (e) Other than those gives as options

6. An amount is invested in a bank at compound rate of interest. The total amount, including interest, after first and third year is ₹ 1200 and ₹ 1587, respectively. What is the rate of interest? **[SSC CGL (Mains) 2012]**
   (a) 10% (b) 3.9% (c) 12% (d) 15%

7. There is 60% increase in an amount in 6 yr at simple interest. What will be the compound interest on ₹ 12000 after 3 yr at the same rate of interest? **[CDS 2015 (I)]**
   (a) ₹ 2160 (b) ₹ 3120
   (c) ₹ 3972 (d) ₹ 6240

8. Scheme A offers compound interest (compounded annually) at a certain rate of interest (per cent per annum). When a sum was invested in the scheme, it amounted to ₹ 14112 after 2 yr and ₹ 16934.40 after 3 yr. What was the sum of money invested? **[SBI Clerk 2015]**
   (a) ₹ 9000 (b) ₹ 10200
   (c) ₹ 8800 (d) ₹ 9400
   (e) ₹ 9800

9. A certain sum is invested for 2 yr in scheme A at 10 % per annum compound interest (compounded annually). Same sum is also invested for 3 yr in scheme B at X% per annum simple interest, interest earned from scheme A is half of that earned from scheme B. What is the value of X? **[LIC AAO 2016]**
   (a) 20% (b) 9%
   (c) 12% (d) 18%
   (e) 14%

10. A man invested ₹ P for 2 yr in scheme A A which offered 20% per annum compound interest (compounded annually). He lent the interest earned from scheme A to Shubh, at the rate of 7.5% per annum simple interest. If at the end of 2 yr, Shubh gave ₹ 3036 to Raman and thereby repaid the whole amount (actual loan + interest), what is the value of P? **[SBI PO (Pre) 2016]**
    (a) ₹ 6000 (b) ₹ 5800
    (c) ₹ 6800 (d) ₹ 5400
    (e) ₹ 6400

11. A man borrowed some money and agreed to pay-off by paying ₹ 3150 at the end of the first year and ₹ 4410 at the end of the second year. If the rate of compound interest is 5% per annum, then the sum is **[SSC CGL (Mains) 2016]**
    (a) ₹ 5000 (b) ₹ 6500
    (c) ₹ 7000 (d) ₹ 9200

12. ₹ 260200 is divided between Ram and Shyam so that the amount that Ram receives in 4 yr is the same as that Shyam receives in 6 yr. If the interest is compounded annually at the rate of 4% per annum, then Ram's share is **[SSC CGL (Mains) 2016]**
    (a) ₹ 125000 (b) ₹ 135200
    (c) ₹ 152000 (d) ₹ 108200

13. A sum of ₹ 8448 is to be divided between A and B who are respectively 18 and 19 yr old, in such a way that if their shares be invested at 6.25% per annum at compound interest, they will receive equal amounts on attaining the age of 21 yr. The present share of A is
   (a) ₹ 4225
   (b) ₹ 4352
   (c) ₹ 4096
   (d) ₹ 4000

14. The simple interest on a sum of money for 3 yr is ₹ 240 and the compound interest on the same sum, at the same rate for 2 yr is ₹ 170. The rate of interest is   [SSC CPO 2016]
   (a) $5\dfrac{5}{17}$ %
   (b) 8%
   (c) $29\dfrac{1}{6}$ %
   (d) $12\dfrac{1}{2}$ %

15. In the beginning of the year 2004, a person invests some amount in a bank. In the beginning of 2007, the accumulated interest is ₹ 10000 and in the beginning of the year 2010, the accumulated interest becomes ₹ 25000. The interest rate is compounded annually and the annual interest rate is fixed. The principal amount is   [XAT 2015]
   (a) ₹ 16000
   (b) ₹ 18000
   (c) ₹ 20000
   (d) ₹ 25000
   (e) None of these

Answer with Solutions

Exercise 1 Base Level Questions

1. (c) Given, $P = ₹ 50000$, $R = 12\%$ and $n = 3$ yr
According to the formula,
Amount $= P\left(1 + \dfrac{R}{100}\right)^n$
$= 50000\left(1 + \dfrac{12}{100}\right)^3$
$= 50000 \times \dfrac{28}{25} \times \dfrac{28}{25} \times \dfrac{28}{25}$
$= \dfrac{16 \times 28 \times 28 \times 28}{5}$
$= \dfrac{351232}{5} = ₹ 70246.40$
∴ CI $= 70246.40 - 50000 = ₹ 20246.40$

2. (a) $r = x\%$, $P = ₹ 3000$, $A = ₹ 3993$, $t = 3$ yr
∴ $A = P\left(1 + \dfrac{r}{100}\right)^t$
⇒ $3993 = 3000\left(1 + \dfrac{x}{100}\right)^3$
⇒ $\dfrac{3993}{3000} = \left(1 + \dfrac{x}{100}\right)^3$
⇒ $\left(1 + \dfrac{x}{100}\right)^3 = \dfrac{1331}{1000} = \left(\dfrac{11}{10}\right)^3$
⇒ $1 + \dfrac{x}{100} = \dfrac{11}{10}$ ⇒ $\dfrac{x}{100} = \dfrac{11}{10} - 1$
⇒ $\dfrac{x}{100} = \dfrac{1}{10}$ ⇒ $x = \dfrac{100}{10} = 10$

3. (a) Here, $r = 10\%$, $n = 3$ yr,
$A = ₹ 6655$ and $P = ?$
∴ $A = P\left(1 + \dfrac{r}{100}\right)^n$
⇒ $6655 = P\left(1 + \dfrac{10}{100}\right)^3$
⇒ $6655 = P\left(\dfrac{11}{10}\right)^3$
⇒ $P = 6655 \times \left(\dfrac{10}{11}\right)^3$
⇒ $P = 6655 \times \dfrac{1000}{1331} = ₹ 5000$

4. (a) Given, CI $= ₹ 4347$, $P = ₹ 30000$ and $R = 7\%$
By formula, CI $= P\left[\left(1 + \dfrac{R}{100}\right)^n - 1\right]$
⇒ $4347 = 30000\left[\left(1 + \dfrac{7}{100}\right)^n - 1\right]$
⇒ $4347 = 30000\left[\left(\dfrac{107}{100}\right)^2 - 1\right]$
⇒ $\left(\dfrac{107}{100}\right)^n = \dfrac{4347}{30000} + 1$
⇒ $\left(\dfrac{107}{100}\right)^n = \dfrac{34347}{30000} = \dfrac{11449}{10000}$
⇒ $\left(\dfrac{107}{100}\right)^n = \left(\dfrac{107}{100}\right)^2$ ⇒ $n = 2$

Compound Interest / 279

5. (a) Here, CI = ₹ 3225, $r = 15\%$ and $t = 2$ yr

$\therefore \quad \text{CI} = P\left[\left(1 + \dfrac{r}{100}\right)^t - 1\right]$

$\Rightarrow \quad 3225 = P\left[\left(1 + \dfrac{15}{100}\right)^2 - 1\right]$

$\Rightarrow \quad 3225 = P\left[\left(1 + \dfrac{3}{20}\right)^2 - 1\right]$

$\Rightarrow \quad 3225 = P\left[\left(\dfrac{23}{20}\right)^2 - 1\right]$

$\Rightarrow \quad 3225 = P\left(\dfrac{529}{400} - 1\right)$

$\Rightarrow \quad 3225 = P \times \dfrac{129}{400} \Rightarrow P = \dfrac{3225 \times 400}{129}$

$\Rightarrow \quad P = 25 \times 400 = ₹\,10000$

6. (b) Given, $P = ₹\,31250$,
$\quad n = 9$ months $= 3$ quarters
and $\quad R = 16\%$ per annum
$\quad = 4\%$ per quarter
According to the formula,

Amount $(A) = P\left(1 + \dfrac{R}{100}\right)^n$

$= 31250\left(1 + \dfrac{4}{100}\right)^3$

$= 31250 \times \dfrac{26}{25} \times \dfrac{26}{25} \times \dfrac{26}{25} = ₹\,35152$

$\therefore$ CI $= A - P = 35152 - 31250 = ₹\,3902$

7. (d) Given, $P = ₹\,15000$, $R = 12\%$
and $n = 1\dfrac{1}{4} = \dfrac{5}{4}$ yr
According to the formula,

Amount $= P\left(1 + \dfrac{R}{100 \times 4}\right)^{4n}$

$= 15000\left(1 + \dfrac{12}{100 \times 4}\right)^{4 \times \frac{5}{4}}$

$= 15000\left(\dfrac{412}{400}\right)^5 = 15000\left(\dfrac{103}{100}\right)^5$

$= 15000 \times \dfrac{103}{100} \times \dfrac{103}{100} \times \dfrac{103}{100} \times \dfrac{103}{100} \times \dfrac{103}{100}$

$= \dfrac{15 \times 103 \times 103 \times 103 \times 103 \times 103}{10000000}$

$= ₹\,17389.111 = ₹\,17389.10$ (approx.)

8. (b) Given, $P = ₹\,1750, R = 8\%, n = 2$
and $\dfrac{a}{b} = \dfrac{1}{2}$.
According to the formula,

Amount $= P\left(1 + \dfrac{R}{100}\right)^n \times \left(1 + \dfrac{a/b \times R}{100}\right)$

$= 1750\left(1 + \dfrac{8}{100}\right)^2\left(1 + \dfrac{\frac{1}{2} \times 8}{100}\right)$

$= 1750\left(\dfrac{27}{25}\right)^2 \times \dfrac{26}{25} = 1750 \times \dfrac{27}{25} \times \dfrac{27}{25} \times \dfrac{26}{25}$

$= ₹\,2122.848 = ₹\,2122.85$ (approx).

9. (a) Given, $P = ₹\,9375$, $R_1 = 2\%$ and $R_2 = 4\%$
According to the formula,

$A = P\left(1 + \dfrac{R_1}{100}\right)\left(1 + \dfrac{R_2}{100}\right)$

$= 9375\left(1 + \dfrac{2}{100}\right)\left(1 + \dfrac{4}{100}\right)$

$= 9375 \times \dfrac{51}{50} \times \dfrac{26}{25}$

$= 7.5 \times 51 \times 26 = ₹\,9945$

Now, CI $= A - P = 9945 - 9375 = ₹\,570$

10. (b) Let P be the sum.

$\therefore 16632 = P\left(1 + \dfrac{5}{100}\right)\left(1 + \dfrac{10}{100}\right)\left(1 + \dfrac{20}{100}\right)$

$\Rightarrow 16632 = P \times \dfrac{21}{20} \times \dfrac{11}{10} \times \dfrac{6}{5}$

$\Rightarrow \quad P = ₹\,12000$

11. (a) Given, $P = ₹\,4000$, $R_1 = 10\%$ (decreases),
$R_2 = 5\%$ (decreases) and $R_3 = 15\%$ (increases)
According to the formula,
Income
$= P\left(1 - \dfrac{R_1}{100}\right)\left(1 - \dfrac{R_2}{100}\right)\left(1 + \dfrac{R_3}{100}\right)$

$= 4000\left(1 - \dfrac{10}{100}\right)\left(1 - \dfrac{5}{100}\right)\left(1 + \dfrac{15}{100}\right)$

$= 4000 \times \dfrac{9}{10} \times \dfrac{19}{20} \times \dfrac{23}{20}$

$= 9 \times 19 \times 23 = ₹\,3933$

12. (a) Let the principal be ₹ x.
Amount after first year

$= x\left(1 + \dfrac{4}{100}\right)^1 = \dfrac{x \times 104}{100}$

and interest after second year

$= \dfrac{x \times 104}{100} \times \dfrac{4}{100}$

According to the question,

$\dfrac{x \times 104 \times 4}{100 \times 100} = 208$

$\Rightarrow \quad x = ₹\,5000$

13. (a) Given, $P = ₹15000, R = 10\%$ and $n = 1$ yr
According to the formula,

Amount $= P\left(1 + \dfrac{R}{2 \times 100}\right)^{2n}$

$= 15000\left(1+\dfrac{5}{100}\right)^2 = 15000 \times \left(\dfrac{21}{20}\right)^2$

$= 15000 \times \dfrac{21}{20} \times \dfrac{21}{20} = ₹\, 16537.50$

14. (d) Let sum be P and given that $n = 3$ yr and amount $= 27P$
According to the formula,

$\text{Amount} = P\left(1+\dfrac{R}{100}\right)^n$

$\Rightarrow 27P = P\left(1+\dfrac{R}{100}\right)^3$

$\Rightarrow 27 = \left(1+\dfrac{R}{100}\right)^3$

$\Rightarrow \left(1+\dfrac{R}{100}\right)^3 = (3)^3$

$\Rightarrow 1+\dfrac{R}{100} = 3$

$\Rightarrow \dfrac{R}{100} = 3-1 = 2$

$\therefore R = 200\%$

15. (b) Given, SI = ₹ 540 and $t = 3$ yr
Let amount of money be ₹ P.

$\because \quad \text{SI} = \dfrac{P \times r \times t}{100}$

$\Rightarrow \dfrac{P \times 3 \times r}{100} = 540$

$\Rightarrow r = \dfrac{540 \times 100}{3P} = \dfrac{18000}{P}$

Now, CI = ₹ 376.20, $t = 2$ yr

$\because \quad \text{CI} = P\left[\left(1+\dfrac{r}{100}\right)^t - 1\right]$

$\therefore 376.20 = P\left[\left(1+\dfrac{18000}{100P}\right)^2 - 1\right]$

$= P\left[\left(1+\dfrac{180}{P}\right)^2 - 1\right]$

$= P\left[\left(\dfrac{P+180}{P}\right)^2 - 1\right]$

$= P\left[\dfrac{P^2 + 32400 + 360P - P^2}{P^2}\right]$

$\Rightarrow 376.20 = \dfrac{32400 + 360P}{P}$

$\Rightarrow 376.20 P = 32400 + 360P$

$\Rightarrow 16.20 P = 32400 \Rightarrow P = ₹\, 2000$

16. (a) According to the given condition,

$441 = 400\left(1+\dfrac{R}{100}\right)^2 \Rightarrow \dfrac{441}{400} = \left(1+\dfrac{R}{100}\right)^2$

$\Rightarrow \left(\dfrac{21}{20}\right)^2 = \left(1+\dfrac{R}{100}\right)^2$

$\Rightarrow \dfrac{21}{20} = 1+\dfrac{R}{100}$

$\Rightarrow \dfrac{21}{20} - 1 = \dfrac{R}{100}$

$\Rightarrow \dfrac{R}{100} = \dfrac{1}{20}$

$\Rightarrow R = 5\%$

$\therefore$ New rate $= 5 + 5 = 10\%$

$\therefore \text{Amount} = 400\left(1+\dfrac{10}{100}\right)^2$

$= 400 \times \dfrac{11}{10} \times \dfrac{11}{10} = ₹\, 484$

17. (c) Let each instalment be ₹ x.
Then, according to the question,

$\dfrac{x}{\left(1+\dfrac{10}{100}\right)} + \dfrac{x}{\left(1+\dfrac{10}{100}\right)^2} = 8400$

$\Rightarrow \dfrac{10}{11}x + \dfrac{100}{121}x = 8400$

$\Rightarrow \dfrac{10}{11}\left(\dfrac{21}{11}x\right) = 8400$

$\therefore x = \dfrac{8400 \times 121}{10 \times 21} = ₹\, 4840$

18. (a) Let each instalment be ₹ x.
Then, according to the question,

$\dfrac{x}{\left(1+\dfrac{4}{100}\right)} + \dfrac{x}{\left(1+\dfrac{4}{100}\right)^2} = 5100$

$\Rightarrow \dfrac{25x}{26} + \dfrac{625x}{676} = 5100$

$\Rightarrow 1275x = 5100 \times 676$

$\therefore x = \dfrac{5100 \times 676}{1275} = ₹\, 2704$

19. (b) According to the question,

Borrowed sum $= \dfrac{121}{\left(1+\dfrac{10}{100}\right)} + \dfrac{121}{\left(1+\dfrac{10}{100}\right)^2}$

$= \dfrac{121}{\dfrac{11}{10}} + \dfrac{121}{\dfrac{11}{10} \times \dfrac{11}{10}} = 110 + 100 = ₹\, 210$

20. (d) Let value of each instalment be ₹ x.
Then, $\dfrac{x}{\left(1+\dfrac{20}{100}\right)} + \dfrac{x}{\left(1+\dfrac{20}{100}\right)^2} = 11000$

$\Rightarrow x\left(\dfrac{5}{6} + \dfrac{25}{36}\right) = 11000 \Rightarrow x\left(\dfrac{55}{36}\right) = 11000$

$\therefore \quad x = ₹\, 7200$

21. (a) Given, principal $(P) = ₹\, 15000$,
Time $(n) = 2$ yr and rate $= R\%$

$\therefore \text{SI} = \dfrac{\text{Principal} \times \text{Time} \times \text{Rate}}{100}$

Compound Interest / 281

$\Rightarrow \quad SI = \dfrac{15000 \times 2 \times R}{100}$

$\therefore \quad SI = 300R$ ...(i)

Now, $CI = P\left[\left(1+\dfrac{R}{100}\right)^n - 1\right]$

$= 15000\left[\left(1+\dfrac{R}{100}\right)^2 - 1\right]$ ...(ii)

According to the question,
$CI - SI = ₹96$

$\Rightarrow 15000\left[\left(1+\dfrac{R}{100}\right)^2 - 1\right] - 300R = 96$

[from Eqs. (i) and (ii)]

$\Rightarrow 15000\left(1+\dfrac{R}{100}\right)^2 - 15000 - 300R = 96$

$\Rightarrow 15000\dfrac{(100+R)^2}{10000} - 15000 - 300R = 96$

$\Rightarrow \dfrac{3(100+R)^2}{2} - 300R = 96 + 15000 = 15096$

$\Rightarrow 3(10000 + R^2 + 200R) - 600R = 30192$

$\Rightarrow 600R + 30000 + 3R^2 - 600R = 30192$

$\Rightarrow 3R^2 = 30192 - 30000$

$\Rightarrow 3R^2 = 192 \Rightarrow R^2 = 64 \Rightarrow R = 8$

Hence, the value of R is 8%.

Fast Track Method
Given, principal, $P = ₹15000$
Rate, $R = ?$
Difference, $D = ₹96$
For 2 yr difference, $D = \dfrac{PR^2}{(100)^2}$

[by Technique 1(i)]

$\Rightarrow 96 = \dfrac{15000 \times R^2}{100 \times 100} \Rightarrow R^2 = 64$

$\therefore \quad R = 8\%$

22. (c) Let the amount deposited with each bank be ₹ P.
For compound interest,
Amount = ₹ P, time $(t) = 2$ yr
and rate of interest $(r) = 5\%$

We know that, $CI = P\left[\left(1+\dfrac{r}{100}\right)^t - 1\right]$

$\Rightarrow CI = P\left[\left(1+\dfrac{5}{100}\right)^2 - 1\right]$

$= P\left[\left(\dfrac{21}{20}\right)^2 - 1\right] = ₹\dfrac{41P}{400}$

For simple interest,
Amount = ₹ P, time $(t) = 2$ yr
and rate of interest $(r) = 5\%$

$\because \quad SI = \dfrac{P \times r \times t}{100}$

$\Rightarrow \quad SI = \dfrac{P \times 5 \times 2}{100} = ₹\dfrac{P}{10}$

Now, according to the question,
$CI - SI = ₹200$

$\Rightarrow \dfrac{41P}{400} - \dfrac{P}{10} = 200 \Rightarrow P\left[\dfrac{41-40}{400}\right] = 200$

$\therefore \quad P = 200 \times 400 = ₹80000$

Fast Track Method
Here, $CI - SI = ₹200$, $R = 5\%$

$\because \quad CI - SI = \dfrac{PR^2}{100^2}$

$\Rightarrow 200 = \dfrac{P \times 5^2}{100^2}$ [by Technique 1(i)]

$\Rightarrow P = \dfrac{200 \times 100 \times 100}{5 \times 5} = ₹80000$

23. (c) Here, $R = 4\%$ and $T = 2$ yr
Now, difference between compound interest and simple interest of 2 yr

$= P\left(\dfrac{R}{100}\right)^2$ [by Technique 1(i)]

where P is sum.

$\therefore \quad 10 = P\left(\dfrac{4}{100}\right)^2$

$\Rightarrow \quad 10 = P\left(\dfrac{1}{25}\right)^2$

$\Rightarrow \quad P = 10(25)^2 = ₹6250$

24. (b) Given, $SI = ₹120$, $n = 2$ yr and $CI = ₹129$
By formula,
$SI = \dfrac{P \times R \times t}{100} \Rightarrow 120 = \dfrac{P \times R \times 2}{100}$

$\Rightarrow \quad PR = 6000$

Now, $CI = P\left[\left(1+\dfrac{R}{100}\right)^n - 1\right]$

$\Rightarrow 129 = P\left[\left(1+\dfrac{R}{100}\right)^2 - 1\right]$

$\Rightarrow 1290000 = P[(100+R)^2 - 100^2]$

$\Rightarrow 1290000 = P[100^2 + R^2 + 2R \times 100 - 100^2]$

$\Rightarrow 1290000 = P(R^2 + R \times 200)$

$\Rightarrow 1290000 = \dfrac{6000}{R}(R^2 + 200R)$

$\left[\because P = \dfrac{6000}{R}\right]$

$\Rightarrow \quad 1290 = 6(R + 200)$

$\Rightarrow \quad 1290 = 6R + 1200$

$\Rightarrow \quad 6R = 1290 - 1200$

$\therefore \quad R = \dfrac{90}{6} = 15\%$

282 / Fast Track Objective Arithmetic

Fast Track Method
Here, $n = 2$ yr, SI = ₹ 120 and CI = ₹ 129
By formula,
$$CI = SI\left(1 + \frac{R}{200}\right) \quad \text{[by Technique 1(i)]}$$
$$\Rightarrow 129 = 120\left(1 + \frac{R}{200}\right) \Rightarrow \frac{129}{120} = 1 + \frac{R}{200}$$
$$\Rightarrow \frac{129}{120} - 1 = \frac{R}{200} \Rightarrow \frac{9}{120} = \frac{R}{200}$$
$$\therefore R = \frac{9 \times 200}{120} = 15\%$$

25. (b) Given, $n = 2$ yr, $R = 4\%$ and SI = 80
According to the formula,
$$CI = SI\left(1 + \frac{R}{100}\right) \quad \text{[by Technique 1(i)]}$$
$$= 80\left(1 + \frac{4}{200}\right) = \frac{80 \times 51}{50} = ₹ 81.60$$

26. (c) Given, $P = ₹ 1250$, $n = 2$ yr and $R = 4\%$
According to the formula,
Difference between compound interest and
simple interest = $\frac{PR^2}{100^2}$ [by Technique 1(i)]
$$\therefore \text{Required difference} = \frac{1250 \times 4 \times 4}{100 \times 100} = ₹ 2$$

27. (d) Let required difference be ₹ D.
By formula, $D = P\left(\frac{R}{100}\right)^2\left(3 + \frac{R}{100}\right)$
[by Technique 1(i)]
$$= 16200 \times \left(\frac{25}{100}\right)^2 \times \left(3 + \frac{25}{100}\right)$$
$$= 16200 \times \frac{625}{10000} \times \frac{13}{4}$$
$$= \frac{162 \times 625 \times 13}{4 \times 100} = ₹ 3290.63$$

28. (a) Required difference
$$= P\left(\frac{R}{100}\right)^2\left(\frac{300 + R}{100}\right)$$
[by Technique 1(ii)]
$$= 10000\left(\frac{5}{100}\right)^2\left(\frac{305}{100}\right) = 76.25$$

29. (b) Let the amount invested by P.
Then, according to the question,
$$2P = P\left(1 + \frac{R}{100}\right)^4 \Rightarrow (2)^{1/4} = \left(1 + \frac{R}{100}\right) \quad ...(i)$$
and $8P = P\left(1 + \frac{R}{100}\right)^n \Rightarrow 8 = \left(1 + \frac{R}{100}\right)^n$
$$\Rightarrow 8 = (2^{1/4})^n \quad \text{[from Eq. (i)]}$$
$$\Rightarrow 2^3 = 2^{n/4}$$
On comparing the powers, we get
$$\frac{n}{4} = 3 \Rightarrow n = 12 \text{ yr}$$

Fast Track Method
Here, $x = 2$ times, $y = 8$ times
$n_1 = 4$ yr and $n_2 = ?$
$$\because x^{\frac{1}{n_1}} = y^{\frac{1}{n_2}} \quad \text{[by Technique 2]}$$
$$\Rightarrow 2^{\frac{1}{4}} = 8^{\frac{1}{n_2}}$$
$$\Rightarrow 2^{\frac{1}{4}} = 2^{\frac{3}{n_2}}$$
On comparing the powers, we get
$$\frac{3}{n_2} = \frac{1}{4}$$
$$\Rightarrow n_2 = 12 \text{ yr}$$

30. (c) If sum is x, then x becomes $\frac{5}{2}x$ in 6 yr,
$\frac{5}{2}x$ becomes $\frac{25}{4}x$ in 12 yr and
$\frac{25}{4}x$ becomes $\frac{125}{8}x$ in 18 yr.
Thus, the sum becomes $\frac{125}{8}$ times in 18 yr.

Fast Track Method
Here, $n_1 = 6$ yr, $x = 2\frac{1}{2} = \frac{5}{2}$, $n_2 = 18$ yr
According to the formula, $x^{1/n_1} = y^{1/n_2}$
[by Technique 2]
$$\Rightarrow \left(\frac{5}{2}\right)^{1/6} = y^{1/18}$$
$$\Rightarrow y = \left(\frac{5}{2}\right)^{18/6}$$
$$\Rightarrow y = \left(\frac{5}{2}\right)^3 = \frac{5 \times 5 \times 5}{2 \times 2 \times 2} = \frac{125}{8}$$

31. (b) Let the required amount be ₹ P.
According to the question,
$$2916 = P\left(1 + \frac{R}{100}\right)^2 \quad ...(i)$$
and $3149.28 = P\left(1 + \frac{R}{100}\right)^3 \quad ...(ii)$
On dividing Eq. (ii) by Eq. (i), we get
$$1 + \frac{R}{100} = \frac{3149.28}{2916} \Rightarrow \frac{R}{100} = \frac{3149.28}{2916} - 1$$
$$\Rightarrow R = \frac{233.28}{2916} \times 100 = 8\%$$
From Eq. (i),
$$P = \frac{2916 \times 100 \times 100}{108 \times 108} = ₹ 2500$$

Fast Track Method
$$\text{Sum} = A_1\left(\frac{A_1}{A_2}\right)^n \quad \text{[By Technique 3 (ii)]}$$
$$= 2916\left(\frac{2916}{3149.28}\right)^2 = ₹ 2500$$

Compound Interest / 283

Exercise 2 *Higher Skill Level Questions*

1. (*a*) Given, $R = 4\%$, $n = 2$ yr, $A = ₹169$
and $P = ?$
According to the formula,

Amount $= P\left(1 + \dfrac{R}{100}\right)^n$

$\Rightarrow 169 = P\left(1 + \dfrac{4}{100}\right)^2 \Rightarrow 169 = P\left(\dfrac{26}{25}\right)^2$

$\therefore P = \dfrac{169 \times 25 \times 25}{26 \times 26} = \dfrac{105625}{676}$

$= ₹156.25$

2. (*c*) Here, rate = 10% per annum
and principal = ₹1000
$\therefore$ Difference in interest

$= 1000\left[\left\{\left(1 + \dfrac{10}{2 \times 100}\right)^2 - 1\right\}\right.$

$\left. - \left\{\left(1 + \dfrac{10}{100}\right)^1 - 1\right\}\right]$

$= 1000\left[\left\{\left(\dfrac{21}{20}\right)^2 - 1\right\} - \left\{\dfrac{11}{10} - 1\right\}\right]$

$= 1000\left[\left(\dfrac{441}{400} - 1\right) - \dfrac{1}{10}\right]$

$= 1000\left[\dfrac{41}{400} - \dfrac{1}{10}\right]$

$= 1000\left(\dfrac{41 - 40}{400}\right)$

$= 1000 \times \dfrac{1}{400} = \dfrac{10}{4} = ₹2.5$

3. (*a*) CI $= 8000 \times \left(1 + \dfrac{10}{100}\right)^2 - 8000$

$= 8000 \times \dfrac{11}{10} \times \dfrac{11}{10} - 8000$

$= 9680 - 8000 = ₹1680$

$\therefore$ Sum $= \left(\dfrac{840 \times 100}{3 \times 8}\right) = ₹3500$

$\left[\begin{array}{l}\because \text{SI is half of CI.} \\ \therefore \text{SI} = \dfrac{1680}{2} = 840\end{array}\right]$

4. (*a*) Here, SI = ₹482.40, $R = 6\%$, $T = 3$ yr
and $P = X$

$\therefore \quad \text{SI} = \dfrac{P \times R \times T}{100}$

$\Rightarrow \quad 482.40 = \dfrac{X \times 6 \times 3}{100} \Rightarrow 18X = 48240$

$\Rightarrow \quad X = \dfrac{48240}{18} \Rightarrow X = 2680$

Now, for scheme B,
$P = X - 680 = 2680 - 680$
$\Rightarrow P = ₹2000, T = 2\text{yr}, R = 10\%$

$\therefore \quad \text{CP} = P\left[\left(1 + \dfrac{R}{100}\right)^T - 1\right]$

$= 2000\left[\left(1 + \dfrac{10}{100}\right)^2 - 1\right] = 2000\left[\left(\dfrac{11}{10}\right)^2 - 1\right]$

$= 2000 \times \left(\dfrac{121}{100} - 1\right) = 2000\left(\dfrac{121 - 100}{100}\right)$

$= 2000 \times \dfrac{21}{100} = 20 \times 21 = ₹420$

Hence, the interest received in Scheme B is ₹420.

5. (*b*) Here, SI $= \dfrac{4200 \times 12 \times 2}{100} = 1008$

and CI $= (4200 - P)\left[\left(1 + \dfrac{10}{100}\right)^2 - 1\right]$

$= (4200 - P)\left(\dfrac{121}{100} - 1\right)$

$= (4200 - P) \times \dfrac{21}{100} = 882 - \dfrac{21P}{100}$

Now, SI − CI = 294 [given]

$\Rightarrow \quad 1008 - \left(882 - \dfrac{21P}{100}\right) = 294$

$\Rightarrow \quad 1008 - 882 + \dfrac{21P}{100} = 294$

$\Rightarrow \quad 126 + \dfrac{21P}{100} = 294$

$\Rightarrow \quad \dfrac{21P}{100} = 294 - 126$

$\Rightarrow \quad \dfrac{21P}{100} = 168$

$\Rightarrow \quad P = \dfrac{168 \times 100}{21}$

$\therefore \quad P = 8 \times 100 = ₹800$

6. (*d*) Let amount be ₹x and rate of interest be $r\%$ annually.
According to the question,
Amount after 1st yr = ₹1200

$\Rightarrow \quad x\left(1 + \dfrac{R}{100}\right) = 1200$ ...(i)

Amount after 3rd yr = 1587

$\Rightarrow \quad x\left(1 + \dfrac{R}{100}\right)^3 = 1587$ ...(ii)

On dividing Eq. (ii) by Eq. (i), we get

$\left(1 + \dfrac{R}{100}\right)^2 = \dfrac{1587}{1200} = \dfrac{529}{400}$

$\Rightarrow \quad 1 + \dfrac{R}{100} = \dfrac{23}{20} \Rightarrow \dfrac{R}{100} = \dfrac{3}{20}$

$\Rightarrow \quad R = 15\%$

7. (c) Let the principal be ₹ x.

Then, SI = ₹ $\dfrac{60x}{100}$

$\because \quad \text{SI} = \dfrac{\text{Principal} \times \text{Rate} \times \text{Time}}{100}$

$\Rightarrow \quad \dfrac{60x}{100} = \dfrac{x \times r \times 6}{100}$ [∵ time = 6 yr]

$\therefore \quad r = 10\%$

Again, principal $(P) = ₹\ 12000$,
time $(n) = 3$ yr

Amount $= P\left(1 + \dfrac{r}{100}\right)^n$

$= 12000\left(1 + \dfrac{10}{100}\right)^3 = 12000\left(\dfrac{11}{10}\right)^3$

$= 12000 \times \dfrac{11 \times 11 \times 11}{1000}$

$= 12 \times 121 \times 11 = ₹\ 15972$

$\therefore \quad \text{CI} = A - P$
$= 15972 - 12000 = ₹\ 3972$

8. (e) Let the money invested be ₹ P and the rate of interest be $r\%$.

Then, according to the question,

$P\left(1 + \dfrac{r}{100}\right)^2 = 14112$ ...(i)

and $P\left(1 + \dfrac{r}{100}\right)^3 = 16934.40$ ...(ii)

On dividing Eq. (ii) by Eq. (i), we get

$\left(1 + \dfrac{r}{100}\right) = \dfrac{16934.40}{14112}$ ...(iii)

$\Rightarrow \quad 1 + \dfrac{r}{100} = 1.2$

$\therefore \quad r = 20\%$

On putting the value of r in Eq. (i), we get

$P\left(1 + \dfrac{20}{100}\right)^2 = 14112$

$\Rightarrow \quad P \times \dfrac{6}{5} \times \dfrac{6}{5} = 14112$

$\Rightarrow \quad P = \dfrac{14112 \times 5 \times 5}{6 \times 6} = ₹\ 9800$

9. (e) Let the sum invested be ₹ P.

Then, according to the question,

$P\left(1 + \dfrac{10}{100}\right)^2 - P = \dfrac{1}{2}\left(\dfrac{P \times X \times 3}{100}\right)$

$\Rightarrow \quad P\left(\dfrac{11}{10}\right)^2 - P = \dfrac{3PX}{200}$

$\Rightarrow \quad \dfrac{121P - 100P}{100} = \dfrac{3PX}{200}$

$\Rightarrow \quad \dfrac{21}{100} P = \dfrac{3PX}{200}$

$\therefore \quad X = \dfrac{21 \times 200}{3 \times 100} = 7 \times 2 = 14\%$

10. (a) Here, $r = 20\%$ and $n = 2$ yr

$\therefore \quad \text{CI} = P\left[\left(1 + \dfrac{r}{100}\right)^n - 1\right]$

$= P\left[\left(1 + \dfrac{20}{100}\right)^2 - 1\right]$

$= P\left[\left(1 + \dfrac{1}{5}\right)^2 - 1\right]$

$= P\left[\left(\dfrac{6}{5}\right)^2 - 1\right] = P\left(\dfrac{36}{25} - 1\right)$

$= P \times \dfrac{11}{25} \Rightarrow \text{CI} = \dfrac{11P}{25}$

Now, CI is lent at 7.5% per annum on SI for 2 yr.

$\therefore \quad \text{SI} = \dfrac{11P}{25} \times 7.5 \times \dfrac{2}{100} = \dfrac{165P}{2500}$

Now, according to the question,

$\dfrac{11P}{25} + \dfrac{165P}{2500} = 3036$

$\Rightarrow \quad \dfrac{1100P + 165P}{2500} = 3036$

$\therefore \quad P = \dfrac{3036 \times 2500}{1265} = ₹\ 6000$

11. (c) Here, $r = 5\%$, $A_1 = ₹\ 3150$, $A_2 = ₹\ 4410$

Now, $A_1 = P_1\left(1 + \dfrac{r}{100}\right)^t$

$\Rightarrow \quad 3150 = P_1\left(1 + \dfrac{5}{100}\right)^1$

$\Rightarrow \quad 3150 = P_1 \times \dfrac{21}{20} \Rightarrow P_1 = \dfrac{3150 \times 20}{21}$

$\Rightarrow \quad P_1 = ₹\ 3000$

Now, $A_2 = P_2\left(1 + \dfrac{r}{100}\right)^t$

$\Rightarrow \quad 4410 = P_2\left(1 + \dfrac{5}{10}\right)^2$

$\Rightarrow \quad 4410 = P_2 \times \left(\dfrac{21}{20}\right)^2$

$\Rightarrow \quad P_2 = \dfrac{4410 \times 400}{441} = ₹\ 4400$

$\therefore$ Total sum $= P_1 + P_2$
$= 3000 + 4000 = ₹\ 7000$

12. (b) Let amount received by Ram = ₹ x

Then, amount received by Shyam
$= ₹\ (260200 - x)$

Compound Interest / 285

According to the question, $x\left(1+\dfrac{4}{100}\right)^4$

$= (260200 - x)\left(1+\dfrac{4}{100}\right)^6$

$\Rightarrow x = (260200 - x)\left(1+\dfrac{1}{25}\right)^2$

$\Rightarrow x = (260200 - x) \times \left(\dfrac{26}{25}\right)^2$

$\Rightarrow 625x = 260200 \times 676 - 676x$
$\Rightarrow 625x + 676x = 260200 \times 676$
$\Rightarrow 1301x = 260200 \times 676$
$\Rightarrow x = 200 \times 676 = ₹135200$

13. (c) Let shares of A and B be $₹x$ and $₹(8448 - x)$, respectively.
Amount got by A after 3 yr
= Amount got by B after 2 yr

$\Rightarrow x\left(1+\dfrac{6.25}{100}\right)^3 = (8448 - x) \times \left(1+\dfrac{6.25}{100}\right)^2$

$\Rightarrow 1 + \dfrac{6.25}{100} = \dfrac{8448 - x}{x}$

$\Rightarrow 1 + \dfrac{1}{16} = \dfrac{8448 - x}{x} \Rightarrow \dfrac{17}{16} = \dfrac{8448 - x}{x}$

$\Rightarrow 17x = 135168 - 16x$
$\Rightarrow 33x = 13516$
$\therefore x = ₹4096$

14. (d) Let the principal be $₹P$.
Given, SI $= ₹240$ and time $(t) = 3$ yr

Then, SI $= \dfrac{P \times R \times t}{100}$

$\Rightarrow 240 = \dfrac{P \times R \times 3}{100}$

$\Rightarrow P = \dfrac{8000}{R}$ ...(i)

and CI $= P\left[\left(1+\dfrac{R}{100}\right)^n - 1\right]$

[$\because n = 2$ yr and CI $= ₹170$]

$\Rightarrow 170 = \dfrac{8000}{R}\left[\left(1+\dfrac{R}{100}\right)^2 - 1\right]$

$\left[\text{from Eq. (i)}, P = \dfrac{8000}{R}\right]$

$\Rightarrow \dfrac{8000}{R}\left[\left(\dfrac{100+R}{100}\right)^2 - 1\right] = 170$

$\Rightarrow \dfrac{8000}{R}\left[\dfrac{10000 + R^2 + 200R - 10000}{10000}\right] = 170$

$\Rightarrow \dfrac{8000}{R}\left[\dfrac{R^2 + 200R}{10000}\right] = 170$

$\Rightarrow \dfrac{8R(R + 200)}{10R} = 170$

$\Rightarrow R + 200 = \dfrac{1700}{8}$

$\Rightarrow R = \dfrac{1700}{8} - 200 = \dfrac{1700 - 1600}{8}$

$\Rightarrow R = \dfrac{100}{8} \Rightarrow R = 12\dfrac{1}{2}\%$

15. (c) Let the principal amount be $₹P$ and annual interest rate be $r\%$.
Then, according to the question,

$P\left[\left(1+\dfrac{r}{100}\right)^3 - 1\right] = 10000$ ...(i)

and $P\left[\left(1+\dfrac{r}{100}\right)^6 - 1\right] = 25000$ ...(ii)

Let $\left(1+\dfrac{r}{100}\right) = a$, then the equation become

$P(a^3 - 1) = 10000$ ...(iii)
and $P(a^6 - 1) = 25000$ ...(iv)

On dividing Eq. (iii) by Eq. (iv), we get

$\dfrac{a^3 - 1}{a^6 - 1} = \dfrac{10}{25} \Rightarrow \dfrac{a^3 - 1}{(a^3)^2 - 1} = \dfrac{10}{25}$

$\Rightarrow \dfrac{a^3 - 1}{(a^3 - 1)(a^3 + 1)} = \dfrac{10}{25} \Rightarrow (a^3 + 1) = \dfrac{5}{2}$

$[\because a^2 - b^2 = (a - b)(a + b)]$

$\Rightarrow a^3 = \dfrac{5}{2} - 1 \Rightarrow a^3 = \dfrac{3}{2}$

On putting the value of a^3 in Eq. (iii), we get

$P\left(\dfrac{3}{2} - 1\right) = 10000 \Rightarrow P = 20000$

Hence, the principal amount is ₹20000.

Chapter 16

True Discount and Banker's Discount

True Discount

If a person borrows certain money from another person for a certain period and the borrower wants to clear-off the debt right now, then for paying back the debt, the borrower gets certain discount which is called True Discount (TD).

For example If a person has to pay ₹ 160 after 4 yr and the rate of interest is 15% per annum. It is clear that ₹ 100 at 15% will amount to ₹ 160 in 4 yr. Therefore, the payment of ₹ 100 now will clear-off the debt of ₹ 160 due 4 yr hence at 15% per annum.

Thus, following points are the outcomes

1. Sum due is equal to ₹ 160 due 4 yr hence.
2. Present Worth (PW) is ₹ 100.
3. True Discount (TD) = 160 − 100 = ₹ 60

Present Worth (PW) The money to be paid back before due date to clear-off a debt is called the present worth.

Amount (A) The sum due is called amount, i.e. A = PW + TD

True Discount (TD) It is the difference between the amount and the present worth, i.e. TD = A − PW.

MIND IT!
1. True discount is the interest on present worth.
2. Interest is reckoned on PW and TD is reckoned on amount.
   According to the definition,
   $$TD = A - PW$$

True Discount and Banker's Discount

Banker's Discount

The difference between the amount shown on a bond that is bought by a bank from a customer and the amount that the customer actually receives from the bank is called Banker's Discount (BD).

For example If a person X buys goods having value of ₹ 5000 from another person Y at a credit of 6 months, then X prepares a bill called 'Bill of Exchange' (Hundi). The 'Bill of Exchange' is prepared by Y and X signing this bill (bill of exchange) to allow Y to withdraw the amount from his (X's) bank account on a date that falls exactly after 6 months. This is the date which is known as 'Nominally Due Date'.

If we add some more days (grace days/period), we get 'Legally Due Date'.

Suppose Y wants to have the money before the legally due date. Then, he can have the money from the banker or a broker, who deducts SI on face value (₹ 5000 in this case) for the period from the date on which the bill was discounted (i.e. paid by the banker) and the legally due date. This amount is known as Banker's discount.

Now, following outcomes can be written as

Banker's Discount (BD) = Interest on bill for remaining time (unexpired time)

$$= \frac{\text{Value of bill} \times \text{Rate} \times \text{Remaining time}}{100} = \frac{A \times R \times T}{100}$$

where, A = Face value, T = Remaining time and R = Rate%

Also, Banker's Gain (BG) = BD − TD (unexpired time)

✦ Banker's Discount (BD) is little bit more than True Discount (TD).
✦ When the date of bill is not given, then the grace days are not added.

Ex. 1 Find the banker's discount at a bill of ₹ 25500 due 2 months hence when rate of interest is 3% per annum.

Sol. Given that, A = ₹ 25500, R = 3%, T = 2 months = $\frac{2}{12}$ yr = $\frac{1}{6}$ yr

According to the formula, BD = $\frac{A \times R \times T}{100} = \frac{25500 \times 3 \times \frac{1}{6}}{100}$ = ₹ 127.5

Fast Track Formulae to solve the QUESTIONS

▶ **Formula 1**

If the rate of interest is R% per annum, time is T yr and present worth is PW, then

True Discount (TD) = $\frac{PW \times R \times T}{100}$

Ex. 2 What will be the true discount for the present worth of ₹ 6000 for a period of 9 months at 12% per annum?

Sol. Given that, T = 9 months = $\frac{9}{12}$ yr, R = 12% and PW = ₹ 6000

∴ TD = $\frac{PW \times R \times T}{100} = \frac{6000 \times 12 \times \frac{9}{12}}{100}$ = 60 × 9 = ₹ 540

Formula 2

If the true discount on a certain sum of money due certain year hence and the simple interest on the same sum for the same time and at the same rate is given, then

$$\text{Sum due } (A) = \frac{SI \times TD}{SI - TD}$$

[here, SI = Simple Interest and TD = True Discount]

Ex. 3 The true discount on a certain sum of money due 10 yr hence is ₹ 68 and the simple interest on the same sum for the same time and at the same rate of interest is ₹ 102. Find the sum due.

Sol. Given that, TD = ₹ 68 and SI = ₹ 102

$$\therefore \text{ Sum due } (A) = \frac{SI \times TD}{SI - TD}$$

$$= \frac{102 \times 68}{102 - 68} = \frac{102 \times 68}{34} = ₹ 204$$

Formula 3

If the true discount on a certain sum of money due T yr hence and the simple interest on the same sum for the same time and at the same rate of interest R% per annum are given, then

$$SI - TD = \frac{TD \times R \times T}{100}$$

Ex. 4 The true discount on a certain sum of money due 16 yr hence is ₹ 300 and the simple interest on the same sum for the same time and at the same rate of interest is ₹ 900, find the rate per cent.

Sol. Given that, SI = ₹ 900, TD = ₹ 300, T = 16 yr and R = ?

According to the formula, $SI - TD = \dfrac{TD \times R \times T}{100}$

$$\Rightarrow \quad 900 - 300 = \frac{300 \times R \times 16}{100}$$

$$\Rightarrow \quad 600 = \frac{300 \times 4 \times R}{25}$$

$$\Rightarrow \quad 50 = 4R$$

$$\therefore \quad R = \frac{50}{4} = \frac{25}{2} = 12.5\%$$

Alternate Method

According to the formula,

$$\text{Sum due} = \frac{SI \times TD}{SI - TD} = \frac{900 \times 300}{600} = ₹ 450$$

Since, TD is SI on amount. So, ₹ 900 is SI on ₹ 450 for 16 yr.

$$\therefore \quad 900 = \frac{450 \times R \times 16}{100}$$

$$\Rightarrow \quad R = \frac{900 \times 100}{450 \times 16} = 12.5\% \qquad \left[\because SI = \frac{P \times r \times t}{100}\right]$$

True Discount and Banker's Discount / 289

▶ Formula 4

When the sum is put at compound interest, then PW = $\dfrac{A}{\left(1+\dfrac{R}{100}\right)^T}$.

Ex. 5 What will be the present worth of ₹ 4840 due 2 yr hence, when the interest is compounded at 10% per annum? Also, find the true discount.

Sol. Given that, A = ₹ 4840, T = 2 yr and R = 10%

According to the formula,

$$PW = \dfrac{A}{\left(1+\dfrac{R}{100}\right)^T} = \dfrac{4840}{\left(1+\dfrac{10}{100}\right)^2} = \dfrac{4840}{\left(\dfrac{11}{10}\right)^2} = \dfrac{4840 \times 10 \times 10}{11 \times 11} = ₹4000$$

∴ TD = A − PW = 4840 − 4000 = ₹ 840

▶ Formula 5

If the face value of a bill due T yr hence is A and rate is R%, then

Banker's Gain (BG) = $\dfrac{A(R \times T)^2}{100(100 + R \times T)}$ or BG = $\dfrac{(TD)^2}{PW}$ ⇒ TD = $\sqrt{BG \times PW}$

Ex. 6 The face value of a bill due 3 yr hence is ₹ 13800. If the rate of simple interest is 5% per annum, what will be the banker's gain at the bill?

Sol. Given that, A = ₹ 13800, R = 5% and T = 3 yr

According to the formula,

$$BG = \dfrac{A(R \times T)^2}{100(100 + R \times T)} = \dfrac{13800(5 \times 3)^2}{100(100 + 5 \times 3)} = \dfrac{13800 \times 225}{100 \times 115} = ₹ 270$$

Ex. 7 The present worth of a sum due sometime hence is ₹ 400 and the true discount is ₹ 80. Find the banker's gain.

Sol. Given, SI on ₹ 400 = ₹ 80

∴ SI on ₹ 480 = $\dfrac{80}{400}$ × 480 = ₹ 96

Now, TD = ₹ 80 and BD = ₹ 96

∴ BG = BD − TD = 96 − 80 = ₹ 16

Fast Track Method

Given that, PW = ₹ 400 and TD = ₹ 80

According to the formula,

$$TD = \sqrt{(PW) \times (BG)}$$

⇒ $80 = \sqrt{400 \times BG} = 20\sqrt{BG}$

∴ BG = 4 × 4 = ₹ 16   [squaring on both sides]

Ex. 8 If the true discount at a bill of ₹ 3720 due 8 months hence is ₹ 120, then find the banker's gain.

Sol. ∵ Present Worth (PW) of bill = 3720 − 120 = ₹ 3600

∴ Banker's Gain (BG) = $\dfrac{(TD)^2}{PW} = \dfrac{120 \times 120}{3600} = \dfrac{144}{36} = ₹ 4$

Formula 6

If the true discount at a bill due T yr hence is TD and the annual rate of interest is $R\%$, then

$$\text{Banker's Gain (BG)} = \frac{TD \times R \times T}{100}$$

Ex. 9 If the banker's gain is ₹ 180 at a bill due 6 yr hence at 10% per annum, then find the banker's discount.

Sol. Given that, BG = ₹ 180, R = 10% and T = 6 yr
According to the formula,

$$BG = \frac{TD \times R \times T}{100} \Rightarrow 180 = \frac{TD \times 10 \times 6}{100}$$

∴ $$TD = \frac{100 \times 180}{10 \times 6} = ₹ 300$$

Now, BG = BD − TD
⇒ BD = BG + TD = 180 + 300 = ₹ 480

Formula 7

If the banker's gain on a bill due T yr hence at $R\%$ rate of simple interest is BG, then

$$\text{Present Worth (PW)} = BG \left(\frac{100}{R \times T}\right)^2$$

Ex. 10 The banker's gain of a sum due 4 yr hence at 12% per annum is ₹ 72. Find the present worth of that bill.

Sol. Given that, BG = ₹ 72, R = 12% and T = 4 yr
According to the formula,

$$PW = BG \left(\frac{100}{R \times T}\right)^2 = 72 \left(\frac{100}{12 \times 4}\right)^2 = 72 \times \frac{100}{48} \times \frac{100}{48}$$

$$= 72 \times \frac{25}{12} \times \frac{25}{12} = \frac{625}{2} = ₹ 312.50$$

Formula 8

If BD and TD have been given on a bill, then

$$\text{Amount of bill (A)} = \frac{BD \times TD}{BD - TD} = \frac{BD \times TD}{BG}$$

Ex. 11 If BD and TD on a sum due certain time hence at a certain rate are ₹ 72 and ₹ 60 respectively, find the sum.

Sol. Given that, BD = ₹ 72 and TD = ₹ 60
According to the formula,

$$\text{Required sum } (A) = \frac{BD \times TD}{BD - TD} = \frac{72 \times 60}{72 - 60} = \frac{72 \times 60}{12} = ₹ 360$$

Fast Track Practice

1. What will be the true discount for the present worth of ₹ 600 for a period of 4 yr at 4% per annum rate of interest?
   (a) ₹ 100 (b) ₹ 32
   (c) ₹ 86   (d) ₹ 96
   (e) None of these

2. If the true discount for a sum of ₹ 50000 for a period of 4 yr at a certain rate of interest per annum is ₹ 2000, find the rate of interest.
   (a) 1% (b) 1.5% (c) 2% (d) 5%
   (e) None of these

3. What will be the present worth of ₹ 3720 due 3 yr hence at 8% per annum? Also, find the true discount.
   (a) PW = ₹ 3000 and TD = ₹ 720
   (b) PW = ₹ 15000 and TD = ₹ 720
   (c) PW = ₹ 3000 and TD = ₹ 320
   (d) PW = ₹ 1000 and TD = ₹ 420
   (e) None of the above

4. Find the sum due, if the true discount on a sum due 2 yr hence at 7% is ₹ 672.
   (a) ₹ 5500 (b) ₹ 5425
   (c) ₹ 5472 (d) ₹ 5300
   (e) None of these

5. The true discount on a certain sum of money due 4 yr hence is ₹ 75 and the simple interest on the same sum for the same time and at the same rate of interest is ₹ 225. Find the rate per cent.
   (a) 25% (b) 50% (c) 31% (d) 45%
   (e) None of these

6. If the present worth of a certain sum due 2 yr hence at 10% per annum compound interest is ₹ 2000, then find the amount.
   (a) ₹ 2100 (b) ₹ 2000
   (c) ₹ 2300 (d) ₹ 2420

7. Find the banker's discount at a bill of ₹ 12750 due four months hence when rate of interest is 6% per annum.
   (a) ₹ 250 (b) ₹ 120
   (c) ₹ 255 (d) ₹ 300
   (e) None of these

8. The present worth of a sum due sometime, hence is ₹ 576 and the banker's gain is ₹ 9. Find the true discount.
   (a) ₹ 8 (b) ₹ 70 (c) ₹ 95 (d) ₹ 72
   (e) None of these

9. The face value of a bill due 5 yr hence is ₹ 13800. If the rate of simple interest is 5% per annum, what will be the banker's gain at the bill?
   (a) ₹ 690 (b) ₹ 600 (c) ₹ 590 (d) ₹ 625
   (e) None of these

10. The true discount at a bill of ₹ 7440 due 16 months, hence is ₹ 240. Find the banker's gain.
    (a) ₹ 10 (b) ₹ 6 (c) ₹ 4 (d) ₹ 8
    (e) None of these

11. The present worth of a bill due sometime, hence is ₹ 2200 and true discount on the bill is ₹ 220. Find the banker's discount and banker's gain.
    (a) BG = ₹ 22 and BD = ₹ 242
    (b) BG = ₹ 242 and BD = ₹ 22
    (c) BG = ₹ 11 and BD = ₹ 121
    (d) BG = ₹ 31 and BD = ₹ 343
    (e) None of the above

12. The true discount on a certain sum due 1 yr hence at 30% per annum, is ₹ 240. What is the banker's discount on the same sum for the same time and at the same rate?
    (a) ₹ 400 (b) ₹ 212
    (c) ₹ 312 (d) ₹ 445
    (e) None of these

13. The true discount on a bill 24 months, hence at 24% per annum, is ₹ 144. What will be the banker's discount?
    (a) ₹ 213.12 (b) ₹ 415
    (c) ₹ 515.12 (d) ₹ 616
    (e) None of these

14. If rate of interest and time on a certain bill are numerically equal and true discount is 81 times of banker's gain, find the rate of interest.
    (a) $2\frac{9}{13}\%$ (b) $1\frac{2}{9}\%$
    (c) $1\frac{7}{9}\%$ (d) $1\frac{1}{9}\%$
    (e) None of these

15. The banker's gain of a sum due 8 yr, hence at 24% per annum, is ₹ 144. Find the present worth of that bill.
    (a) ₹ 39.06 (b) ₹ 45
    (c) ₹ 38.06 (d) ₹ 50
    (e) None of these

16. If the banker's discount and banker's gain on a certain bill, are ₹ 196 and ₹ 28, respectively. Find the amount of bill.
   (a) ₹ 1200   (b) ₹ 1376
   (c) ₹ 1176   (d) ₹ 1400
   (e) None of these

17. The banker's discount on a certain sum due 4 yr, hence is $\frac{11}{10}$ of the true discount. Find the rate per cent per annum.
   (a) 2.5%   (b) 5%
   (c) 5.5%   (d) 1.5%
   (e) None of these

18. Kailash wants to sell his television. There are two offers, one at ₹ 10000 cash and the other at a credit of ₹ 11200 to be paid after 8 months, money being at 18% per annum. Which one is better offer?
   (a) ₹ 11200 at credit
   (b) ₹ 10000 in cash
   (c) Both are equally good
   (d) Cannot be determined

19. Jagatram, a trader, owes a merchant Maganlal ₹ 5014 due 1 yr hence. Jagatram wants to settle the account after 3 months. If the rate of interest is 12% per annum, how much cash should Jagatram pay?
   (a) ₹ 9200
   (b) ₹ 5600
   (c) ₹ 4600
   (d) ₹ 6600
   (e) None of the above

20. Aarti has to pay ₹ 440 to Babita after 1 yr. Babita asks Aarti to pay ₹ 220 in cash and defer the payment of ₹ 220 for 2 yr. Aarti agrees to it. If the rate of interest is 10% per annum, find Aarti's gain or loss.
   (a) Aarti gains ₹ 3.33
   (b) Aarti gains ₹ 8
   (c) Aarti losses ₹ 9
   (d) Aarti losses ₹ 3.33
   (e) None of the above

21. X owes Y ₹ 3146 payable $1\frac{1}{2}$ yr hence. Also, Y owes X, ₹ 2889 payable 6 months hence. If they want to settle the account forthwith, keeping 14% the rate of interest, who should pay and how much?
   (a) Y should pay ₹ 100
   (b) Y should pay ₹ 50
   (c) X should pay ₹ 100
   (d) X should pay ₹ 50
   (e) None of the above

22. Vandana bought a watch for ₹ 600 and sold it the same day for ₹ 688.50 at a credit of 9 months and this way she gained 2%. Find the rate of interest per annum.
   (a) $16\frac{2}{3}\%$ (b) $15\frac{2}{3}\%$
   (c) $11\frac{2}{3}\%$ (d) $5\frac{2}{3}\%$
   (e) None of these

Answer with Solutions

1. (d) Given that, $T = 4$ yr, $R = 4\%$,
PW = ₹ 600 and TD = ?
According to the formula,
$$TD = \frac{PW \times R \times T}{100} \quad \text{[by Formula 1]}$$
$$= \frac{600 \times 4 \times 4}{100}$$
$$= 6 \times 4 \times 4 = ₹ 96$$

2. (a) Given that, PW = ₹ 50000,
TD = ₹ 2000, $T = 4$ yr and $R = ?$
According to the formula,
$$TD = \frac{PW \times R \times T}{100} \quad \text{[by Formula 1]}$$

$$\Rightarrow 2000 = \frac{50000 \times R \times 4}{100}$$
$$\therefore R = \frac{2000 \times 100}{50000 \times 4} = 1\%$$

3. (a) Let PW be ₹ x.
$$\therefore 3720 - x = \frac{x \times 8 \times 3}{100} = \frac{6x}{25}$$
$$\Rightarrow 31x = 3720 \times 25$$
$$\therefore x = \frac{3720 \times 25}{31} = ₹ 3000$$
Now, TD = Amount − PW
$$= 3720 - 3000 = ₹ 720$$

True Discount and Banker's Discount / 293

4. (c) Let PW be x.
Then, $672 = \dfrac{x \times 7 \times 2}{100} = \dfrac{14x}{100}$

$\therefore \quad x = \dfrac{672 \times 100}{14} = ₹\, 4800$

$\therefore$ Amount $(A) = 4800 + 672 = ₹\, 5472$

5. (b) Given that, SI = ₹ 225, TD = ₹ 75,
$T = 4$ yr and $R = ?$
According to the formula,

$SI - TD = \dfrac{TD \times R \times T}{100}$ [by Formula 3]

$\Rightarrow \quad 225 - 75 = \dfrac{75 \times R \times 4}{100}$

$\Rightarrow \quad 150 = \dfrac{75 \times R \times 4}{100}$

$\Rightarrow \quad 3R = 150 \Rightarrow R = 50\%$

6. (d) Given that, PW = ₹ 2000, $T = 2$ yr, $R = 10\%$ and $A = ?$
According to the formula,

$PW = \dfrac{A}{\left(1 + \dfrac{R}{100}\right)^T}$ [by Formula 4]

$\Rightarrow \quad 2000 = \dfrac{A}{\left(1 + \dfrac{10}{100}\right)^2}$

$\Rightarrow \quad 2000 = \dfrac{A \times 10 \times 10}{11 \times 11}$

$\therefore \quad A = \dfrac{2000 \times 11 \times 11}{10 \times 10} = 20 \times 121 = ₹\, 2420$

7. (c) Given that, $A = ₹\, 12750$, $R = 6\%$
and $T = 4$ months $= \dfrac{4}{12} = \dfrac{1}{3}$ yr, BD $= ?$
According to the formula,

$BD = \dfrac{A \times R \times T}{100}$

$= \dfrac{12750 \times 6 \times \dfrac{1}{3}}{100}$

$= \dfrac{12750 \times 2}{100} = ₹\, 255$

8. (d) Given that, PW = ₹ 576, BG = ₹ 9 and TD = ?
According to the formula,
$TD = \sqrt{PW \times BG}$ [by Formula 5]
$= \sqrt{576 \times 9} = 24 \times 3 = ₹\, 72$

9. (a) Given that, $A = ₹\, 13800$, $R = 5\%$, $T = 5$ yr and BG $= ?$
According to the formula,

$BG = \dfrac{A(R \times T)^2}{100(100 + R \times T)}$ [by Formula 5]

$= \dfrac{13800 \, (5 \times 5)^2}{100 \, (100 + 5 \times 5)}$

$= \dfrac{13800 \times 625}{100 \times 125} = ₹\, 690$

10. (d) Present worth of bill = ₹ (7440 − 240)
$= ₹\, 7200$
According to the formula,

Banker's Gain (BG) $= \dfrac{(TD)^2}{PW}$ [by Formula 5]

$= \dfrac{240 \times 240}{7200} = \dfrac{576}{72} = ₹\, 8$

11. (a) Given that, PW = ₹ 2200, TD = ₹ 220
According to the formula,

$BG = \dfrac{(TD)^2}{PW}$ [by Formula 5]

$= \dfrac{220 \times 220}{2200} = ₹\, 22$

$\therefore$ BD = (TD + BG) = ₹ (220 + 22) = ₹ 242

12. (c) We know that,

$BG = SI$ on $TD = \left(\dfrac{240 \times 30 \times 1}{100}\right) = ₹\, 72$

$\because \quad BG = BD - TD$
$\therefore \quad BD = BG + TD = 72 + 240 = ₹\, 312$

13. (a) Given, $R = 24\%$, $T = 24$ months $= 2$ yr
and TD = ₹ 144
According to the formula,

$BG = \dfrac{TD \times R \times T}{100}$ [by Formula 6]

$= \dfrac{144 \times 24 \times 2}{100} = ₹\, 69.12$

$\therefore BD = TD + BG = 144 + 69.12 = ₹\, 213.12$

14. (d) Given that, $R = T$ and TD = 81 BG
According to the formula,

$BG = \dfrac{TD \times R \times T}{100}$ [by Formula 6]

$\therefore \quad \dfrac{R^2}{100} = \dfrac{BG}{TD} = \dfrac{1}{81}$

$\Rightarrow \quad R = \sqrt{\dfrac{100}{81}} \Rightarrow R = \dfrac{10}{9}\% = 1\dfrac{1}{9}\%$

15. (a) Given that, BG = ₹ 144, $R = 24\%$,
$T = 8$ yr and PW $= ?$
According to the formula,

$PW = BG \left(\dfrac{100}{R \times T}\right)^2$ [by Formula 7]

$= 144 \times \left(\dfrac{100}{24 \times 8}\right)^2 = 144 \times \dfrac{100}{192} \times \dfrac{100}{192}$

$= ₹\, 39.06$

16. (c) Given that, BD = ₹ 196 and BG = ₹ 28

∴ TD = BD − BG = 196 − 28 = ₹ 168

According to the formula,

$$A = \frac{BD \times TD}{BG} \quad \text{[by Formula 8]}$$

$$= \frac{196 \times 168}{28} = 196 \times 6 = ₹ 1176$$

17. (a) Let TD = x, then BD = $\frac{11x}{10}$

According to the formula,

$$\text{Sum} = \frac{BD \times TD}{BD - TD} \quad \text{[by Formula 8]}$$

$$= \frac{\frac{11x}{10} \times x}{\frac{11x}{10} - x} = \frac{\frac{11x^2}{10}}{\frac{x}{10}} = 11x$$

∵ SI on ₹ $11x$ for 4 yr is ₹ $\frac{11x}{10}$.

∴ Rate = $\left(\frac{100 \times \frac{11x}{10}}{11x \times 4} \right)$ %

= 2.5% per annum

18. (c) PW of ₹ 11200 due 8 months hence

$$= \frac{11200 \times 100}{100 + 18 \times \frac{8}{12}}$$

$$= \frac{11200 \times 100}{112} = ₹ 10000$$

Clearly, both are equally good.

19. (c) Required cash = PW of ₹ 5014 due 9 months hence

$$= \frac{5014 \times 100}{100 + \left(12 \times \frac{9}{12}\right)} = \frac{5014 \times 100}{100 + 9}$$

$$= \frac{5014 \times 100}{109} = ₹ 4600$$

20. (d) Money to be paid by Aarti

= PW of ₹ 440 due 1 yr hence

$$= \frac{440 \times 100}{100 + (10 \times 1)} = \frac{440 \times 100}{110}$$

= 4 × 100 = ₹ 400

Aarti actually pays ₹ 220 + PW of ₹ 220 due 2 yr hence

$$= 220 + \frac{220 \times 100}{100 + (10 \times 2)}$$

$$= 220 + \frac{220 \times 100}{120}$$

= 220 + 183.33 = ₹ 403.33

∴ Loss of Aarti = 403.33 − 400 = ₹ 3.33

21. (a) X owes PW of ₹ 3146 due $\frac{3}{2}$ yr hence

$$= \frac{3146 \times 100}{100 + \left(14 \times \frac{3}{2}\right)} = \frac{3146 \times 100}{100 + 21}$$

$$= \frac{314600}{121} = ₹ 2600$$

Y owes PW of ₹ 2889 due 6 months hence

$$= \frac{2889 \times 100}{100 + \left(14 \times \frac{1}{2}\right)} = \frac{2889 \times 100}{100 + 7}$$

$$= \frac{2889 \times 100}{107} = ₹ 2700$$

∴ Y must pay ₹ (2700 − 2600), i.e. ₹ 100 to X.

22. (a) SP = 102% of ₹ 600 = $\frac{102}{100} \times 600$ = ₹ 612

Now, PW = ₹ 612 and sum = ₹ 688.50

∴ TD = 688.50 − 612 = ₹ 76.50

Thus, SI on ₹ 612 for 9 months is ₹ 76.50.

∴ Rate = $\frac{100 \times 76.50}{612 \times \frac{3}{4}}$

$$= \frac{100 \times 76.50}{153 \times 3} = 16\frac{2}{3}\%$$

Chapter 17

Ratio and Proportion

Ratio

When two or more similar quantities are compared, then to represent this comparison, ratios are used.

or

Ratio of two quantities is the number of times one quantity contains another quantity of same kind.

The ratio between x and y can be represented as $x:y$, where x is called **antecedent** and y is called **consequent**.

i.e. $\dfrac{x}{y}$ or $x:y$

For example The ratio of ₹ 100 and ₹ 500 can be possible but the ratio of ₹ 100 and 500 apples cannot be possible. Hence, the units of quantity for the comparison of ratio should be same.

Types of Ratios

There are various types of ratios, which are as follow

1. **Duplicate Ratio** If two numbers are in ratio, then the ratio of their squares is called duplicate ratio. If x and y are two numbers, then the duplicate ratio of x and y would be $x^2:y^2$.

   For example Duplicate ratio of $3:4 = 3^2:4^2 = 9:16$

2. **Sub-duplicate Ratio** If two numbers are in ratio, then the ratio of their square roots is called sub-duplicate ratio. If x and y are two numbers, then the sub-duplicate ratio of x and y would be $\sqrt{x}:\sqrt{y}$.

   For example Sub-duplicate ratio of $2:3$ is $\sqrt{2}:\sqrt{3}$.

3. **Triplicate Ratio** If two numbers are in ratio, then the ratio of their cubes is called triplicate ratio. If x and y are two numbers, then the triplicate ratio of x and y would be $x^3:y^3$.

   For example Triplicate ratio of $2:3 = 2^3:3^3 = 8:27$

4. **Sub-triplicate Ratio** If two numbers are in ratio, then the ratio of their cube roots is called sub-triplicate ratio. If x and y are two numbers, then the sub-triplicate ratio of x and y would be $\sqrt[3]{x} : \sqrt[3]{y}$.

   For example Sub-triplicate ratio of $1 : 125 = \sqrt[3]{1} : \sqrt[3]{125} = 1 : 5$

5. **Inverse Ratio** If two numbers are in ratio and their antecedent and consequent are interchanged, then the ratio obtained is called inverse ratio. If x and y are two numbers and their ratio is $x : y$, then its inverse ratio will be $y : x$.

   For example Inverse ratio of $4 : 5$ is $5 : 4$.

6. **Compound Ratio** If two or more ratios are given and the antecedent of one is multiplied with antecedent of other and respective consequents are also multiplied, then the ratio obtained is called compound ratio. If $a : b$, $c : d$ and $e : f$ are three ratios, then their compound ratio will be $ace : bdf$.

   For example Compound ratio of $2 : 5$, $6 : 7$ and $9 : 13 = \dfrac{2 \times 6 \times 9}{5 \times 7 \times 13} = \dfrac{108}{455}$

* If the antecedent is greater than the consequent, then the ratio is known as the ratio of greater inequality such as $7 : 5$.

* If the antecedent is less than the consequent, then the ratio is called the ratio of less inequality such as $5 : 7$.

Comparison of Ratios

Rules used to compare the different ratios are as follow

Rule ❶

If the given ratios are $a : b$ and $c : d$, then
(i) $a : b > c : d$, if $ad > bc$ (ii) $a : b < c : d$, if $ad < bc$ (iii) $a : b = c : d$, if $ad = bc$

Ex. 1 Which is greater $\dfrac{5}{8}$ or $\dfrac{9}{14}$?

Sol. Let $\dfrac{a}{b} = \dfrac{5}{8}$ and $\dfrac{c}{d} = \dfrac{9}{14}$; $ad = 5 \times 14 = 70$ and $bc = 8 \times 9 = 72$

$\because$ $ad < bc$

$\therefore$ $\dfrac{a}{b} < \dfrac{c}{d} = \dfrac{5}{8} < \dfrac{9}{14}$

Rule ❷

If two ratios are given for comparison, convert each ratio in such a way that both ratios have same denominator, then compare their numerators, the fraction with greater numerator will be greater.

Ex. 2 Find the greater ratio between $2 : 3$ and $4 : 5$.

Sol. Here, $\dfrac{2 \times 5}{3 \times 5} = \dfrac{10}{15}$ and $\dfrac{4 \times 3}{5 \times 3} = \dfrac{12}{15}$ $\left[\because \text{LCM of } \dfrac{2}{3} \text{ and } \dfrac{4}{5}, \text{ i.e. } 3 \text{ and } 5 \text{ is } 15 \right]$

$\Rightarrow$ $\dfrac{12}{15} > \dfrac{10}{15}$

$\therefore$ $4 : 5 > 2 : 3$

Ratio and Proportion / 297

Rule ③

If two ratios are given for comparison, convert each ratio in such a way that both ratios have same numerator, then compare their denominators. The fraction with lesser denominator will be greater.

Ex. 3 Find the least fraction between $\frac{6}{7}$ and $\frac{7}{9}$.

Sol. Here, $\frac{6 \times 7}{7 \times 7} = \frac{42}{49}$ and $\frac{7 \times 6}{9 \times 6} = \frac{42}{54}$ $\left[\because \text{LCM of } \frac{6}{7} \text{ and } \frac{7}{9}, \text{ i.e. 6 and 7 is 42} \right]$

$\Rightarrow \quad \frac{42}{49} > \frac{42}{54} \Rightarrow \frac{6}{7} > \frac{7}{9}$

Proportion

An equality of two ratios is called the **proportion**. If $\frac{a}{b} = \frac{c}{d}$ or $a:b = c:d$, then we can say that a, b, c and d are in proportion and can be written as $a:b::c:d$, where symbol '::' represents proportion and it is read as 'a is to b' as 'c is to d'.
Here, a and d are called '**extremes**' and b and c are called as '**means**'.

Basic Rules of Proportion

Rule ①

If $a:b::b:c$, then c is called third proportional to a and b, which are in continued proportion and will be calculated as

$$a:b::b:c \Rightarrow a:b = b:c \Rightarrow a \times c = b \times b \Rightarrow b^2 = ac \Rightarrow c = \frac{b^2}{a}$$

Ex. 4 Calculate the 3rd proportional to 16 and 32.

Sol. Let 3rd proportional be x.
Then, $\quad 16:32::32:x$
$\Rightarrow \quad \frac{16}{32} = \frac{32}{x}$
$\therefore \quad x = \frac{32 \times 32}{16} = 64$

Rule ②

If $a:b::c:d$, then d is called the 4th proportional to a, b and c, d will be calculated as

$$a:b::c:d \Rightarrow a:b = c:d \Rightarrow a \times d = c \times b \Rightarrow d = \frac{bc}{a}$$

Ex. 5 Find the 4th proportional to 3, 7 and 9.

Sol. Let 4th proportional be x.
Then, $\quad 3:7::9:x \Rightarrow \frac{3}{7} = \frac{9}{x}$
$\Rightarrow \quad 3x = 9 \times 7 \Rightarrow x = \frac{9 \times 7}{3} = 21$

Rule 3

Mean proportional between a and b is $\sqrt{ab}$.
Let the mean proportional between a and b be x.
Then, $\quad a : x :: x : b \Rightarrow ab = x^2 \Rightarrow x = \sqrt{ab}$

Ex. 6 What will be the mean proportional between 4 and 25?

Sol. Let the mean proportional be x.
Then, $\quad 4 : x :: x : 25$
$\Rightarrow \quad 4 \times 25 = x \times x$
$\Rightarrow \quad x = \sqrt{4 \times 25} = 10$

Invertendo, Alternendo, Componendo and Dividendo

If $\dfrac{a}{b} = \dfrac{c}{d}$, then

(i) Invertendo : $\dfrac{b}{a} = \dfrac{d}{c}$ 	(ii) Alternendo : $\dfrac{a}{c} = \dfrac{b}{d}$

(iii) Componendo : $\dfrac{a+b}{b} = \dfrac{c+d}{d}$ 	(iv) Dividendo : $\dfrac{a-b}{b} = \dfrac{c-d}{d}$

(v) Componendo and dividendo : $\left(\dfrac{a+b}{a-b}\right) = \left(\dfrac{c+d}{c-d}\right)$

(vi) If $\dfrac{a}{b} = \dfrac{c}{d} = \dfrac{i}{j} = \ldots = k$, then

(a) $\dfrac{a+c+i+\ldots}{b+d+j+\ldots} = k$ 	(b) $\dfrac{pa+qc+ri+\ldots}{pb+qd+rj+\ldots} = k$, where $p, q, r, \ldots =$ constant

Ex. 7 Find the value of $\dfrac{a+b}{a-b}$, if $\dfrac{a}{b} = \dfrac{5}{3}$.

Sol. Given, $\dfrac{a}{b} = \dfrac{5}{3}$

By componendo and dividendo,

$\dfrac{a+b}{a-b} = \dfrac{5+3}{5-3} = \dfrac{8}{2} = \dfrac{4}{1}$

Ex. 8 If $\dfrac{3}{a} = \dfrac{18}{b} = \dfrac{24}{c} = \dfrac{9}{5}$, then find the value of $a+b+c$.

Sol. We know that, if $\dfrac{a}{b} = \dfrac{c}{d} = \dfrac{e}{f} = k$. Then, $\dfrac{(a+c+e)}{(b+d+f)} = k$

$\therefore \quad \dfrac{3+18+24}{a+b+c} = \dfrac{9}{5}$ or $(a+b+c) = \dfrac{45 \times 5}{9} = 25$

Ratio and Proportion / 299

Fast Track Techniques to solve the QUESTIONS

Technique 1
(i) If $A:B = a:b$ and $B:C = m:n$, then
$$A:B:C = am:mb:nb \text{ and } A:C = am:bn$$
(ii) If $A:B = a:b, B:C = c:d$ and $C:D = e:f$, then
$$A:B:C:D = ace:bce:bde:bdf$$

Ex. 9 If $a:b = 5:14$ and $b:c = 7:3$, then find $a:b:c$.

Sol. $a:b:c = (5 \times 7) : (7 \times 14) : (14 \times 3)$
$= 35:98:42 = 5:14:6$ [dividing the ratio by 7]

Ex. 10 The ratio of $A:B = 1:3, B:C = 2:5$ and $C:D = 2:3$. Find the value of $A:B:C:D$.

Sol. Given, $A:B = 1:3, B:C = 2:5, C:D = 2:3$
$\therefore A:B:C:D = (1 \times 2 \times 2) : (3 \times 2 \times 2) : (3 \times 5 \times 2) : (3 \times 5 \times 3) = 4:12:30:45$

Technique 2
(i) If x is divided in $a:b$, then 1st part $= \dfrac{ax}{a+b}$; 2nd part $= \dfrac{bx}{a+b}$

(ii) If x is divided in $a:b:c$, then
1st part $= \dfrac{ax}{a+b+c}$; 2nd part $= \dfrac{bx}{a+b+c}$; 3rd part $= \dfrac{cx}{a+b+c}$

Ex. 11 Divide 1111 in the ratio of 8 : 3.

Sol. Let 1st part be $8x$ and 2nd part be $3x$.
According to the question,
$$8x + 3x = 1111 \Rightarrow 11x = 1111$$
$\therefore \quad x = \dfrac{1111}{11} = 101$

Now, 1st part $= 8x = 8 \times 101 = 808$
and   2nd part $= 3x = 3 \times 101 = 303$

Fast Track Method
1st part $= \dfrac{8}{8+3} \times 1111 = \dfrac{8}{11} \times 1111 = 8 \times 101 = 808$
and   2nd part $= \dfrac{3}{8+3} \times 1111 = \dfrac{3}{11} \times 1111 = 3 \times 101 = 303$

Ex. 12 Divide 2324 in the ratio of 35 : 28 : 20.

Sol. 1st part $= \dfrac{35}{35+28+20} \times 2324 = \dfrac{35}{83} \times 2324 = 35 \times 28 = 980$;

2nd part $= \dfrac{28}{83} \times 2324 = 28 \times 28 = 784$; 3rd part $= \dfrac{20}{83} \times 2324 = 20 \times 28 = 560$

Ex. 13 The sum of three numbers is 315. If the ratio between 1st and 2nd is 2 : 3 and the ratio between 2nd and 3rd is 4 : 5, then find the 2nd number.

Sol. 1st number : 2nd number = 2 : 3 = (2 × 4) : (3 × 4) = 8 : 12;
2nd number : 3rd number = 4 : 5 = (4 × 3) : (5 × 3) = 12 : 15
∴ 1st number : 2nd number : 3rd number = 8 : 12 : 15
Now, 2nd number = $\dfrac{12}{8 + 12 + 15} \times 315 = \dfrac{12}{35} \times 315 = 108$

Technique 3 The incomes of two persons are in ratio of $a : b$ and their expenditures are in the ratio of $c : d$. If each of them saves ₹ X, then their incomes are given by $\dfrac{X(d-c)}{ad-bc} \times a$ and $\dfrac{X(d-c)}{ad-bc} \times b$, respectively and their expenditures are given by $\dfrac{X(b-a)}{ad-bc} \times c$ and $\dfrac{X(b-a)}{ad-bc} \times d$, respectively.

Ex. 14 The ratio of incomes of Raman and Gagan is 4 : 3 and ratio of their expenditures is 3 : 2. If each person saves ₹ 2500, then find their incomes and expenditures.

Sol. Let the income of Raman be ₹ $4x$ and that of Gagan be ₹ $3x$.
Expenditure of Raman = ₹ $(4x - 2500)$; Expenditure of Gagan = ₹ $(3x - 2500)$
According to the question,
$$\dfrac{4x - 2500}{3x - 2500} = \dfrac{3}{2}$$
$\Rightarrow \quad 8x - 5000 = 9x - 7500$
$\Rightarrow \quad x = 7500 - 5000 = 2500$
Income of Raman = $4x = 4 \times 2500 = $ ₹ 10000
Income of Gagan = $3x = 3 \times 2500 = $ ₹ 7500
Expenditure of Raman = $4x - 2500 = 10000 - 2500 = $ ₹ 7500
Expenditure of Gagan = $3x - 2500 = 7500 - 2500 = $ ₹ 5000

Fast Track Method
Here, $a = 4, b = 3, c = 3, d = 2$ and $X = $ ₹ 2500
Income of Raman = $\dfrac{X(d-c)}{ad-bc} \times a$
$= \dfrac{2500(2-3)}{8-9} \times 4 = $ ₹ 10000
Income of Gagan = $\dfrac{X(d-c)}{ad-bc} \times b$
$= \dfrac{2500(2-3)}{8-9} \times 3 = $ ₹ 7500
Expenditure of Raman = $\dfrac{X(b-a)}{ad-bc} \times c = \dfrac{2500(3-4)}{8-9} \times 3 = $ ₹ 7500
Expenditure of Gagan = $\dfrac{X(b-a)}{ad-bc} \times d$
$= \dfrac{2500(3-4)}{8-9} \times 2 = $ ₹ 5000

Ratio and Proportion / 301

Technique 4 If two numbers are in ratio $a:b$ and x is added to the numbers, then the ratio becomes $c:d$. Two numbers will be $\dfrac{xa(c-d)}{ad-bc}$ and $\dfrac{xb(c-d)}{ad-bc}$, respectively.

Ex. 15 Two numbers are in the ratio of $2:3$. If 15 is added to both the numbers, then the ratio between two numbers becomes $\dfrac{11}{14}$. Find the greater number.

Sol. Let the numbers be $2x$ and $3x$.

According to the question,
$$\frac{2x+15}{3x+15} = \frac{11}{14}$$
$$\Rightarrow 14(2x+15) = 11(3x+15)$$
$$\Rightarrow 28x + 210 = 33x + 165$$
$$\Rightarrow 33x - 28x = 210 - 165$$
$$\Rightarrow 5x = 45$$
$$\Rightarrow x = \frac{45}{5} = 9$$

∴ Greater number $= 3x = 3 \times 9 = 27$

Fast Track Method

Here, $a=2, b=3, c=11, d=14$ and $x=15$

∴ 1st number $= \dfrac{xa(c-d)}{ad-bc}$
$$= \frac{15 \times 2(11-14)}{2 \times 14 - 3 \times 11} = \frac{30 \times (-3)}{28-33} = \frac{30 \times (-3)}{-5} = 18$$

and 2nd number $= \dfrac{xb(c-d)}{ad-bc}$
$$= \frac{15 \times 3(11-14)}{2 \times 14 - 3 \times 11} = \frac{45 \times (-3)}{28-33} = \frac{45 \times (-3)}{-5} = 27$$

Hence, the greater number is 27.

Technique 5 Two numbers are in ratio $a:b$ and x is subtracted from the numbers, then the ratio becomes $c:d$. The two numbers will be $\dfrac{xa(d-c)}{ad-bc}$ and $\dfrac{xb(d-c)}{ad-bc}$, respectively.

Ex. 16 Two numbers are in the ratio of $3:5$. If 9 is subtracted from each, the ratio becomes $12:23$. Find the greater number.

Sol. Here, $a=3, b=5, c=12, d=23$ and $x=9$

Then, 1st number $= \dfrac{xa(d-c)}{ad-bc} = \dfrac{9 \times 3(23-12)}{3 \times 23 - 5 \times 12} = \dfrac{27 \times 11}{69-60} = \dfrac{297}{9} = 33$

and 2nd number $= \dfrac{xb(d-c)}{ad-bc} = \dfrac{9 \times 5(23-12)}{3 \times 23 - 5 \times 12} = \dfrac{45 \times 11}{69-60} = \dfrac{45 \times 11}{9} = 55$

Hence, the greater number is 55.

Multi Concept Questions

1. A sum of ₹ 430 has been distributed among 45 people consisting of men, women and children. The total amounts given to men, women and children are in the ratio 12 : 15 : 16. But the amounts received by each man, woman and child are in the ratio 6 : 5 : 4. Find what each man, woman and child receives (in ₹).
   (a) 12, 10, 8 (b) 18, 15, 12 (c) 120, 150, 160 (d) 60, 75, 80

   ➡ (a) ∵ Total amount = ₹ 430 and total people = 45
   Ratio of personal shares = 6 : 5 : 4
   Ratio of the amounts = 12 : 15 : 16
   ∴ Ratio of men, women and children = $\frac{12}{6} : \frac{15}{5} : \frac{16}{4} = 2 : 3 : 4$
   Sum of these ratios = 2 + 3 + 4 = 9
   Number of men = $\left(\frac{45 \times 2}{9}\right)$ = 10 and number of women = $\left(\frac{45 \times 3}{9}\right)$ = 15
   Also, number of children = 45 − (10 + 15) = 20
   Now, divide ₹ 430 in the ratio 12 : 15 : 16.
   Total amount of men's share = $\frac{430 \times 12}{43}$ = ₹ 120
   Total amount of women's share = $\frac{430 \times 15}{43}$ = ₹ 150
   Total amount of children's share = 430 − (120 + 150) = ₹ 160
   ∴ Each man's share = $\frac{120}{10}$ = ₹12
   Each woman's share = $\frac{150}{15}$ = ₹10
   Each child's share = $\frac{160}{20}$ = ₹8

2. The ratio between the number of passengers travelling by 1st and 2nd class between the two railway stations is 1 : 50, whereas the ratio of 1st and 2nd class fares between the same stations is 3 : 50. If on a particular day, ₹ 1325 were collected from the passengers travelling between these stations, then what was the amount collected from the 2nd class passengers?
   (a) ₹ 750 (b) ₹ 850 (c) ₹ 1000 (d) ₹ 1250

   ➡ (d) Let the number of passengers in 1st class be x and the number of passengers in 2nd class be $50x$.
   Then, total amount of 1st class = $3x$ and total amount of 2nd class = $50x$
   Ratio of the amounts collected from the 1st class and the 2nd class passengers = 3 : 50
   ∴ Amount collected from the 2nd class passengers = $\frac{b}{a+b} \times x$
   where, x = total amount, a = 3, b = 50
   ∴ Amount = $\frac{50}{53} \times 1325$ = ₹ 1250

Fast Track *Practice*

Exercise 1 *Base Level Questions*

1. What will be the duplicate ratio of 2 : 7?
   (a) 4 : 49
   (b) 49 : 4
   (c) 4 : 14
   (d) 8 : 343
   (e) None of these

2. Find the sub-duplicate ratio of 81 : 64.
   (a) 8 : 9
   (b) 4 : 9
   (c) 9 : 8
   (d) 7 : 8
   (e) None of these

3. Find the triplicate ratio of 7 : 5.
   (a) 125 : 343
   (b) 343 : 125
   (c) 344 : 125
   (d) 343 : 126
   (e) None of these

4. Calculate the sub-triplicate ratio of 512 : 729.
   (a) 9 : 8
   (b) 4 : 9
   (c) 7 : 8
   (d) 8 : 9
   (e) None of these

5. What will be the inverse ratio of 17 : 19?
   (a) 19 : 17
   (b) 18 : 17
   (c) 17 : 18
   (d) 19 : 5
   (e) None of these

6. Find the compound ratio of 2 : 7, 5 : 3 and 4 : 7.
   (a) 147 : 40
   (b) 40 : 147
   (c) 147 : 30
   (d) 30 : 147
   (e) None of these

7. What is the third proportional to 9 and 45? [SSC CGL (Pre) 2017]
   (a) 405
   (b) 225
   (c) 5
   (d) 81

8. Find the 4th proportional to 4, 16 and 7.
   (a) 28 (b) 29 (c) 22 (d) 25
   (e) None of these

9. Find the mean proportional between 9 and 64.
   (a) 25 (b) 24 (c) 27 (d) 35
   (e) None of these

10. What is the mean proportional between $(15+\sqrt{200})$ and $(27-\sqrt{648})$? [CDS 2012 (II)]
    (a) 4
    (b) $14\sqrt{7}$
    (c) $3\sqrt{5}$
    (d) $4\sqrt{5}$

11. If a, b, c, d and e are in continued proportion, then a/e is equal to [CDS 2013 (I)]
    (a) a^3/b^3
    (b) a^4/b^4
    (c) b^3/a^3
    (d) b^4/a^4

12. If $A : B = 2 : 3$, $B : C = 5 : 7$ and $C : D = 3 : 10$, then what is $A : D$ equal to? [CDS 2014 (I)]
    (a) 1 : 7 (b) 2 : 7 (c) 1 : 5 (d) 5 : 1

13. If $A : B = 3 : 4$ and $B : C = 8 : 9$, then find the value of $A : B : C$. [SSC CGL 2010]
    (a) 3 : 4 : 5
    (b) 1 : 2 : 3
    (c) 7 : 12 : 17
    (d) 6 : 8 : 9

14. If $\frac{1}{2}$ of $A = \frac{2}{5}$ of $B = \frac{1}{3}$ of C, then $A : B : C$ is equal to [SSC (10+2) 2013]
    (a) 4 : 5 : 6
    (b) 6 : 4 : 5
    (c) 5 : 4 : 6
    (d) 4 : 6 : 5

15. If $4a = 5b$ and $7b = 9c$, then $a : b : c$ is equal to
    (a) 45 : 36 : 28
    (b) 44 : 33 : 28
    (c) 28 : 36 : 45
    (d) 36 : 28 : 45
    (e) None of these

16. If $A = \frac{1}{4}B$ and $B = \frac{1}{2}C$, then find the value of $A : B : C$. [SSC (10+2) 2010]
    (a) 8 : 4 : 1
    (b) 4 : 2 : 1
    (c) 1 : 4 : 8
    (d) 1 : 2 : 4

17. If $P : Q : R = 2 : 3 : 4$, then find $\frac{P}{Q} : \frac{Q}{R} : \frac{R}{P}$.
    (a) 8 : 9 : 24
    (b) 9 : 8 : 24
    (c) 24 : 8 : 9
    (d) 8 : 24 : 9
    (e) None of these

18. If $\frac{a}{7} = \frac{b}{9} = \frac{c}{11}$, then find $a : b : c$. [Hotel Mgmt. 2007]
    (a) 11 : 9 : 7
    (b) 9 : 7 : 11
    (c) 7 : 9 : 11
    (d) 11 : 7 : 9

19. If $\frac{1}{x} : \frac{1}{y} : \frac{1}{z} = 2 : 3 : 5$, then determine $x : y : z$.
    (a) 6 : 15 : 10
    (b) 3 : 15 : 10
    (c) 15 : 3 : 10
    (d) 15 : 10 : 6
    (e) None of these

20. If $A:B = 8:15$, $B:C = 5:8$ and $C:D = 4:5$, then $A:B:C:D$ is equal to
(a) 8 : 15 : 24 : 30   (b) 24 : 30 : 15 : 8
(c) 30 : 7 : 15 : 4   (d) 8 : 24 : 15 : 30

21. If $2A = 3B = 4C$, then find $A:B:C$.
[SSC (10+2) 2010]
(a) 2 : 3 : 4   (b) 4 : 3 : 2
(c) 6 : 4 : 3   (d) 3 : 4 : 6

22. If 10% of $(A + B) = 50\%$ of $(A - B)$, then find $A:B$.
(a) 1 : 2 (b) 5 : 2 (c) 2 : 3 (d) 3 : 2
(e) None of these

23. If $a:b = b:c$, then ratio $a^4:b^4$ is equal to
[SSC (10+2) 2010]
(a) $ac:b^2$ (b) $a^2:c^2$
(c) $c^2:a^2$ (d) $b^2:ac$

24. If $xy = 36$, then which of the following is correct?
(a) $x:9 = 4:y$ (b) $9:x = 4:y$
(c) $x:17 = y:7$ (d) $x:6 = y:6$
(e) None of these

25. If $x:y = 7:5$, then what is the value of $(5x - 2y):(3x + 2y)$? [CDS 2012 (I)]
(a) 5/4   (b) 6/5   (c) 25/31   (d) 31/42

26. If $\dfrac{x}{2y} = \dfrac{6}{7}$, then find the value of $\dfrac{x-y}{x+y} + \dfrac{14}{19}$.
(a) 5   (b) 1   (c) 4   (d) 3
(e) None of these

27. If $P^2 + 4Q^2 = 4PQ$, then determine $P:Q$.
(a) 1 : 3 (b) 3 : 1 (c) 2 : 1 (d) 1 : 2
(e) None of these

28. The quantity that must be added to each term of $a:b$, so as to make it $c:d$, is
[SSC Multitasking 2013]
(a) $\dfrac{ab - cd}{a - b}$ (b) $\dfrac{ac + bd}{c + a}$
(c) $\dfrac{ad + bc}{c + d}$ (d) $\dfrac{ad - bc}{c - d}$

29. If $(x^3 - y^3):(x^2 + xy + y^2) = 5:1$ and $(x^2 - y^2):(x - y) = 7:1$, then the ratio $2x:3y$ equals [SSC CGL 2015]
(a) 3 : 2 (b) 4 : 1 (c) 2 : 3 (d) 4 : 3

30. If $a + b:b + c:c + a = 6:7:8$ and $a + b + c = 14$, then find c.
(a) 6   (b) 7   (c) 8   (d) 10
(e) 14

31. If $a:b = c:d = e:f = 1:2$, then find $(3a + 5c + 7e):(3b + 5d + 7f)$.
(a) 1 : 2   (b) 2 : 1
(c) 3 : 1   (d) 1 : 3
(e) None of these

32. If $(3x - 2y):(2x + 3y) = 5:6$, then one of the value of $\dfrac{\sqrt[3]{x} + \sqrt[3]{y}}{\sqrt[3]{x} - \sqrt[3]{y}}$ is
[SSC CGL (Mains) 2015]
(a) 5   (b) 25
(c) $\dfrac{1}{25}$ (d) $\dfrac{1}{5}$

33. If $(a + b):(a - b) = 5:3$, then find $(a^2 + b^2):(a^2 - b^2)$. [SSC CPO 2011]
(a) 17 : 15   (b) 25 : 9
(c) 4 : 1   (d) 16 : 1

34. One-half of a certain number is equal to 65% of the 2nd number. Find the ratio of 1st to 2nd number.
[BOB Clerk 2010]
(a) 10 : 13   (b) 8 : 13
(c) 13 : 8   (d) 13 : 10
(e) None of these

35. If 182 is divided in the ratio of 3 : 5 : 4 : 1 in four parts, then minimum part is
[SSC (10+2) 2017]
(a) 28   (b) 15
(c) 14   (d) 7

36. The total number of students in a school is 2140. If the number of girls in the school is 1200, then what is the ratio of the total number of boys to the total number of girls in the school?
[PNB Clerk 2009]
(a) 26 : 25   (b) 47 : 60
(c) 18 : 13   (d) 31 : 79
(e) None of these

37. 35% of a number is two times 75% of another number. What is the ratio between the first and the second number, respectively? [SBI Clerk 2012]
(a) 35 : 6   (b) 31 : 7
(c) 23 : 7   (d) 32 : 9
(e) None of these

38. Divide 27 into two parts, so that 5 times the first and 11 times the second together equal to 195. Then, ratio of first and second part is [SBI Clerk (Pre) 2016]
(a) 17 : 10   (b) 3 : 2
(c) 2 : 7   (d) 5 : 4
(e) 5 : 2

39. In a college union, there are 48 students. The ratio of the number of boys to the number of girls is 5 : 3. The number of girls to be added in the union, so that the number of boys to girls in 6 : 5, is
[SSC CGL (Mains) 2016]
(a) 6 (b) 7 (c) 12 (d) 17

40. A sum of money is divided amongst A, B, C and D in the ratio of 3 : 7 : 9 : 13. If the share of B is ₹ 4872, then what will be the total amount of money of A and C together? [PNB Clerk 2008]
(a) ₹ 8352 (b) ₹ 6998
(c) ₹ 9784 (d) ₹ 7456
(e) None of these

41. In a class, the number of boys and girls is in the ratio of 4 : 5. If 10 more boys join the class, the ratio of numbers of boys and girls becomes 6 : 5. How many girls are there in the class? [UBI Clerk 2009]
(a) 20
(b) 30
(c) 25
(d) Cannot be determined
(e) None of the above

42. The respective ratio of Sita's, Riya's and Kunal's monthly incomes is 84 : 76 : 89. If Riya's annual income is ₹ 456000, then what is the sum of Sita's and Kunal's annual incomes? (In some cases monthly income and in some cases annual income is used.) [IBPS Clerk 2011]
(a) ₹ 195000 (b) ₹ 983500
(c) ₹ 1130000 (d) ₹ 1038000
(e) None of these

43. In a class of 49 students, the ratio of girls to boys is 4 : 3. If 4 girls leave the class, the ratio of girls to boys would be
[CDS 2017 (I)]
(a) 11 : 7 (b) 8 : 7
(c) 6 : 5 (d) 9 : 8

44. The speeds of three cars are in the ratio of 2 : 3 : 4. Find the ratio between the time taken by these cars to cover the same distance. [Canara Bank PO 2008]
(a) 2 : 3 : 4 (b) 4 : 3 : 2
(c) 4 : 3 : 6 (d) 6 : 4 : 3
(e) None of these

45. A person distributes his pens among four friends A, B, C and D in the ratio $\frac{1}{3} : \frac{1}{4} : \frac{1}{5} : \frac{1}{6}$. What is the minimum number of pens that the person should have? [SSC CGL 2013]

(a) 75 (b) 45 (c) 57 (d) 65

46. The respective ratio between two positive numbers (x and y) is 3 : 5. When 2 is added to both the numbers, the ratio between x and y becomes 5 : 8. What is the difference between both the numbers? [IBPS Clerk (Pre) 2016]
(a) 2 (b) 12 (c) 9 (d) 6
(e) 3

47. A certain number is divided into two parts such that 5 times the first part added to 11 times the second part makes 7 times the whole. The ratio of the first part to the second part is [SSC (10+2) 2013]
(a) 2 : 1 (b) 5 : 11
(c) 1 : 2 (d) 2 : 3

48. There are ₹ 225 consisting of ₹ 1, 50 paise and 25 paise coins. The ratio of their numbers in that order is 8 : 5 : 3. The number of ₹1 coins is [SSC CGL (Mains) 2016]
(a) 80 (b) 112 (c) 160 (d) 172

49. If the positions of the digits of a two-digit number are interchanged, the number newly formed is smaller than the original number by 45. Also, the ratio of the new number to the original number is 3 : 8. What is the original number?
(a) 61 (b) 72 (c) 94
(d) Cannot be determined
(e) None of these

50. The marks of 3 students A, B and C are in the ratio 10 : 12 : 15. If the maximum marks of the paper are 100, then the marks of B cannot be in the range of
[SSC CGL (Mains) 2013]
(a) 20-30 (b) 40-50
(c) 70-80 (d) 80-90

51. A certain amount of money is to be divided among P, Q and R in the ratio of 3 : 5 : 7, respectively. If the amount received by R is ₹ 4000 more than the amount received by Q, what will be the total amount received by P and Q together? [SBI Clerk (Mains) 2016]
(a) ₹ 8000 (b) ₹ 12000
(c) ₹ 1000 (d) ₹ 16000
(e) ₹ 20000

52. A sum of ₹ 300 is divided among P, Q and R in such a way that Q gets ₹ 30 more than P and R gets ₹ 60 more than Q. Then, ratio of their shares is
[SSC CGL 2013]
(a) 2 : 3 : 5 (b) 3 : 2 : 5
(c) 2 : 5 : 3 (d) 5 : 3 : 2

53. An amount of money is to be distributed among P, Q and R in the ratio of 2 : 7 : 9. The total of P's and Q's share is equal to R's share. What is the difference between the shares of P and Q?
 [SSC CGL (Mains) 2016]
(a) ₹ 5000 (b) ₹ 7500
(c) ₹ 9000
(d) Information inadequate

54. ₹ 5625 are divided among A, B and C, so that A receives 1/2 as much as B and C together receive and B receives 1/4 as much as A and C together receive. Find the sum of shares of A and B.
(a) ₹ 5000 (b) ₹ 3000
(c) ₹ 15000 (d) ₹ 9000
(e) None of these

55. The ratio of the ages of a father to that of his son is 5 : 2. If the product of their ages (in years) is 1000, then find the father's age after 10 yr. **[SSC CGL 2010]**
(a) 50 yr (b) 60 yr (c) 80 yr (d) 100 yr

56. From each of two given numbers, half the smaller number is subtracted. After such subtraction, the larger number is 4 times as large as the smaller number. What is the ratio of the numbers?
 [SSC CGL 2012]
(a) 5 : 2 (b) 1 : 4
(c) 4 : 1 (d) 4 : 5

57. In a certain school, the ratio of boys to girls is 7 : 5. If there are 2400 students in the school, then how many girls are there in the school? **[CDS 2012 (II)]**
(a) 500 (b) 700 (c) 800 (d) 1000

58. Amit and Sudesh have invested in the ratio of 4 : 7. If both invested a total amount of ₹ 49500, then find the investment of Sudesh. **[SBI Clerk 2010]**
(a) ₹ 31500 (b) ₹ 1800
(c) ₹ 31000 (d) ₹ 18500
(e) None of these

59. If a sum of ₹ 1664 is divided between P and Q in the ratio of 1/3 : 1/5, then find P's share.
(a) ₹ 1085 (b) ₹ 1015
(c) ₹ 1090 (d) ₹ 1040
(e) None of these

60. Divide ₹ 990 into 3 parts in such a way that half of the first part, one-third of the second part and one-fifth of the third part are equal. **[SSC FCI 2013]**
(a) 198, 494, 298
(b) 198, 297, 495
(c) 200, 300, 490
(d) 196, 298, 496

61. A sum of ₹ 7000 is divided among A, B and C in such a way that the shares of A and B are in the ratio 2 : 3 and those of B and C are in the ratio 4 : 5. The share of B is **[SSC (10+2) 2012]**
(a) ₹ 1600 (b) ₹ 2000
(c) ₹ 2400 (d) ₹ 3000

62. Weekly incomes of two persons are in the ratio of 7 : 3 and their weekly expenses are in the ratio of 5 : 2. If each of them saves ₹ 300 per week, then the weekly income of the first person is **[SNAP 2012]**
(a) ₹ 7500 (b) ₹ 4500
(c) ₹ 6300 (d) ₹ 5400

63. The monthly incomes of X and Y are in the ratio of 4 : 3 and their monthly expenses are in the ratio of 3 : 2. However, each saves ₹ 6000 per month. What is their total monthly income? **[UPSC CSAT 2017]**
(a) ₹ 28000 (b) ₹ 42000
(c) ₹ 56000 (d) ₹ 84000

64. Two numbers are in the ratio of 2 : 3. If 9 is added to each number, they will be in the ratio 3 : 4. What is the product of the two numbers? **[CDS 2012]**
(a) 360 (b) 480
(c) 486 (d) 512

Exercise 2 Higher Skill Level Questions

1. In an office, one-third of the workers are women, half of the women are married and one-third of the married women have children. If three-fourth of the men are married and one-third of the married men have children, then what is the ratio of married women to married men? **[CDS 2016 (II)]**

(a) 1 : 2 (b) 2 : 1
(c) 3 : 1 (d) 1 : 3

2. By increasing the price of entry ticket to a fair in the ratio 3 : 7 the number of visitors to the fair has decreased in the ratio 16 : 13. In what ratio has the total collection increased or decreased?
 [SSC (10+2) 2017]

(a) Decreased in the ratio 91 : 48
(b) Increased in the ratio 48 : 91
(c) Increased in the ratio 39 : 112
(d) Decreased in the ratio 112 : 39

3. In a town, 80% of the population are adults of which the men and women are in the ratio of 9 : 7, respectively. If the number of adult women is 4.2 lakh, what is the total population of the town? **[UBI Clerk 2009]**
(a) 12 lakh (b) 9.6 lakh
(c) 9.8 lakh (d) 11.6 lakh
(e) None of these

4. Mr. Shrimant inherits 2505 gold coins and divides them among his three sons; Bharat, Parat and Marat; in a certain ratio. Out of the total coins received by each of them, Bharat sells 30 coins, Parat donates his 30 coins and Marat losses 25 coins. Now, the ratio of gold coins with them is 46 : 41 : 34, respectively. How many coins did Parat receive from his father? **[BOB Clerk 2009]**
(a) 705 (b) 950 (c) 800 (d) 850
(e) None of these

5. Salary of Mr. X is 80% of the salary of Mr. Y and the salary of Mr. Z is 120% of the salary of Mr. X. What is the ratio between the salaries of X, Y and Z, respectively? **[SBI Clerk 2009]**
(a) 4 : 6 : 5 (b) 4 : 5 : 6
(c) 16 : 24 : 25 (d) 16 : 25 : 24
(e) None of these

6. ₹ 710 were divided among A, B and C in such a way that A had ₹ 40 more than B and C had ₹ 30 more than A. How much was C's share?
(a) ₹ 270 (b) ₹ 300 (c) ₹ 135 (d) ₹ 235
(e) None of these

7. A sum of money is to be divided equally among P, Q and R in the respective ratio of 5 : 6 : 7 and another sum of money is to be divided between S and T equally. If S got ₹ 2100 less than P, then how much amount did Q receive? **[SBI Clerk 2008]**
(a) ₹ 2500
(b) ₹ 2000
(c) ₹ 1500
(d) Cannot be determined
(e) None of the above

8. Nandita scores 80% marks in five subjects together, viz. Hindi, Science, Mathematics, English and Sanskrit, where in the maximum marks of each subject were 105. How many marks did Nandita score in Science, if she scored 89 marks in Hindi, 92 marks in Sanskrit, 98 marks in Mathematics and 81 marks in English? **[IBPS Clerk 2011]**
(a) 60 (b) 75 (c) 65 (d) 70
(e) None of these

9. Salaries of Akash, Bablu and Chintu are in the ratio of 2 : 3 : 5. If their salaries were increased by 15%, 10% and 20% respectively, then what will be the new ratio of their salaries? **[SSC FCI 2013]**
(a) 3 : 3 : 10 (b) 23 : 33 : 60
(c) 20 : 22 : 40 (d) None of these

10. A's income is ₹ 140 more than B's income and C's income is ₹ 80 more than D's. If the ratio of A's and C's income is 2 : 3 and the ratio of B's and D's income is 1 : 2, then the incomes of A, B, C and D are respectively. **[SSC CGL (Mains) 2016]**
(a) ₹ 260, ₹ 120, ₹ 320 and ₹ 240
(b) ₹ 300, ₹ 160, ₹ 600 and ₹ 520
(c) ₹ 400, ₹ 260, ₹ 600 and ₹ 520
(d) ₹ 320, ₹ 180, ₹ 480 and ₹ 360

11. Every month, Mr. Duggabati spends 24% of his monthly income in paying rent and 30% on shopping of groceries. Out of the remaining, he invests in fixed deposit and the lottery in the respective ratio of 9 : 7 respectively. If in a year he deposited a total of ₹ 124200 in fixed deposit, how much did he pay as rent in year? **[IBPS Clerk (Mains) 2017]**
(a) ₹ 124000 (b) ₹ 110400
(c) ₹ 117600 (d) ₹ 124560
(e) ₹ 115200

12. The average age of boys in the class is twice the number of girls in the class. The ratio of boys and girls in the class of 50 is 4 : 1. What is the total of the ages (in yr) of the boys in the class?
[NICL AO 2015]
(a) 2000 (b) 2500 (c) 800 (d) 400
(e) 500

13. The respective ratio between the monthly salaries of Rene and Som is 5 : 3. Out of her monthly salary Rene gives 1/6th as rent, 1/5th to her mother, 30% as her education loan and keeps 25% aside for miscellaneous expenditure. Remaining ₹ 5000 she keeps as savings. What is Som's monthly salary? **[IBPS Clerk 2015]**
(a) ₹ 21000 (b) ₹ 24000
(c) ₹ 27000 (d) ₹ 36000
(e) ₹ 18000

14. A batsman played three matches in a tournament. The respective ratio between the scores of 1st and 2nd match was 5 : 4 and that between the scores of 2nd and 3rd match was 2 : 1. The difference between the 1st and 3rd match was 48 runs. What was the batsman's average score in all the three matches?
 [LIC AAO 2016]
 (a) 44 (b) $58\frac{2}{3}$
 (c) 70 (d) $40\frac{2}{3}$
 (e) $50\frac{1}{4}$

15. Out of two sections A and B, 10 students of section B shift to A, as a result strength of A becomes 3 times the strength of B. But, if 10 students shift over from A to B, both A and B become equal in strength. Ratio of the number of students in section A that of section B is
 [SSC CPO 2013]
 (a) 2 : 1 (b) 5 : 3
 (c) 3 : 1 (d) 9 : 4

16. A cat takes 5 leaps for every 4 leaps of a dog but 3 leaps of the dog are equal to 4 leaps of the cat. What is the ratio of the speeds of the cat to that of the dog?
 [Delhi Police SI 2007]
 (a) 11 : 15 (b) 15 : 11
 (c) 16 : 15 (d) 15 : 16

17. Brothers A and B had some savings in the ratio 4 : 5. They decided to buy a gift for their sister, sharing the cost in the ratio 3 : 4. After they bought, A spent two-third of his amount, while B is left with ₹ 145. Then, the value of gift is
 [SSC CGL (Mains) 2013]
 (a) ₹ 70 (b) ₹ 105
 (c) ₹ 140 (d) ₹ 175

18. The electricity bill of a certain establishment is partly fixed and partly varies as the number of units of electricity consumed. When in a certain month 540 units are consumed, the bill is ₹ 1800. In another month, 620 units are consumed and the bill is ₹ 2040. In yet another month 500 units are consumed. The bill for that month would be
 (a) ₹ 1560 (b) ₹ 1680
 (c) ₹ 1840 (d) ₹ 1950

19. In a factory, the ratio of the numbers of employees of three types A, B and C is 9 : 13 : 18 and their wages are in the ratio of 10 : 7 : 4. If number of employees of type C is 54 and wages of every employee of type B is ₹ 1400, then find the total wages of all the employees of type A.
 (a) ₹ 51000
 (b) ₹ 54000
 (c) ₹ 56000
 (d) ₹ 57000
 (e) ₹ 59000

20. Out of 120 applications for a post, 70 are males and 80 have a driver's license. What is the ratio between the minimum to maximum number of males having driver's license? [UPSC CSAT 2013]
 (a) 1 : 2
 (b) 2 : 3
 (c) 3 : 7
 (d) 5 : 7
 (e) None of the above

21. In a certain examination, the number of those who passed was 4 times the number of those who failed. If there had been 35 fewer candidates and 9 more had failed, the ratio of passed and failed candidates would have been 2 : 1, then the total number of candidates was
 [SSC (10+2) 2013]
 (a) 135 (b) 155
 (c) 145 (d) 150

22. In a coloured picture of blue and yellow colour, blue and yellow colour is used in the ratio of 4 : 3 respectively. If in upper half, blue : yellow is 2 : 3, then in the lower half blue : yellow is
 [SSC CGL (Mains) 2016]
 (a) 1 : 1 (b) 2 : 1
 (c) 26 : 9 (d) 9 : 26

23. The cost of a diamond varies directly as the square of its weight. A diamond broke into four pieces with their weights in the ratio of 1 : 2 : 3 : 4. If the loss in total value of the diamond was ₹ 70000, what was the price of the original diamond? [CDS 2017 (I)]
 (a) ₹ 100000 (b) ₹ 140000
 (c) ₹ 150000 (d) ₹ 175000

Answer with Solutions

Exercise 1 Base Level Questions

1. (a) Required duplicate ratio of 2 : 7
   $= 2^2 : 7^2 = 4 : 49$

2. (c) Required sub-duplicate ratio of 81 : 64
   $= \sqrt{81} : \sqrt{64} = 9 : 8$

3. (b) Required triplicate ratio of 7 : 5
   $= 7^3 : 5^3 = 343 : 125$

4. (d) Required sub-triplicate ratio of 512 : 729
   $= \sqrt[3]{512} : \sqrt[3]{729} = 8 : 9$

5. (a) Required inverse ratio of 17 : 19
   $= \dfrac{1}{17} : \dfrac{1}{19} = 19 : 17$

6. (b) Required compound ratio
   $= \dfrac{2 \times 5 \times 4}{7 \times 3 \times 7} = \dfrac{40}{147} = 40 : 147$

7. (b) Let third proportion of 9 and 45 be x.
   Then, $9 : 45 :: 45 : x$
   $\Rightarrow 9 \times x = 45 \times 45$
   $\Rightarrow x = \dfrac{45 \times 45}{9}$
   $\Rightarrow x = 5 \times 45 = 225$

8. (a) Let the 4th proportional be x.
   Then, $4 : 16 :: 7 : x$
   $\Rightarrow \dfrac{4}{16} = \dfrac{7}{x}$
   $\Rightarrow 4x = 7 \times 16$
   $\therefore x = \dfrac{7 \times 16}{4} = 7 \times 4 = 28$

9. (b) Required mean proportional
   $= \sqrt{9 \times 64} = 3 \times 8 = 24$

10. (c) Here, $a = 15 + \sqrt{200}$, $b = 27 - \sqrt{648}$
    $\therefore$ Mean proportional between two numbers
    $= \sqrt{ab} = \sqrt{(15 + \sqrt{200})(27 - \sqrt{648})}$
    $= \sqrt{\begin{array}{c}(15 \times 27) - 15 \times \sqrt{648} + 27 \times \sqrt{200} \\ - (\sqrt{200} \times \sqrt{648})\end{array}}$
    $= \sqrt{\begin{array}{c}405 - (15 \times 18\sqrt{2}) + (27 \times 10\sqrt{2}) \\ - (10\sqrt{2} \times 18\sqrt{2})\end{array}}$
    $= \sqrt{405 - 270\sqrt{2} + 270\sqrt{2} - 180 \times 2}$
    $= \sqrt{405 - 180 \times 2}$
    $= \sqrt{405 - 360} = \sqrt{45} = 3\sqrt{5}$

11. (b) Since, a, b, c, d and e are in continued proportion.
    $\therefore \dfrac{a}{b} = \dfrac{b}{c} = \dfrac{c}{d} = \dfrac{d}{e}$
    $\Rightarrow \dfrac{e}{d} = \dfrac{d}{c} = \dfrac{c}{b} = \dfrac{b}{a}$
    $\Rightarrow c = \dfrac{b^2}{a}$ $\left[\because \dfrac{c}{b} = \dfrac{b}{a}\right]$
    $d = \dfrac{c^2}{b} = \dfrac{b^4}{a^2} \cdot \dfrac{1}{b} = \dfrac{b^3}{a^2}$
    $e = \dfrac{d^2}{c} = \dfrac{b^6}{a^4} \cdot \dfrac{a}{b^2} = \dfrac{b^4}{a^3}$
    $\therefore \dfrac{a}{e} = \dfrac{a}{(b^4/a^3)} = \dfrac{a^4}{b^4}$

12. (a) Here, $A : B = 2 : 3$, $B : C = 5 : 7$ and $C : D = 3 : 10$
    $\therefore \dfrac{A}{D} = \dfrac{A}{B} \times \dfrac{B}{C} \times \dfrac{C}{D} = \dfrac{2}{3} \times \dfrac{5}{7} \times \dfrac{3}{10} = \dfrac{1}{7}$
    $= 1 : 7$

13. (d) Given that,
    $A : B = 3 : 4 = (3 \times 2) : (4 \times 2) = 6 : 8$
    and $B : C = 8 : 9$ [by Technique 1(i)]
    $\therefore A : B : C = 6 : 8 : 9$
    [as consequent of the first ratio is equal to the antecedent of second ratio]

14. (a) Given, $\dfrac{1}{2}$ of $A = \dfrac{2}{5}$ of $B = \dfrac{1}{3}$ of C
    $\Rightarrow \dfrac{1}{2} \times A = \dfrac{2}{5} \times B = \dfrac{1}{3} \times C$
    Let $\dfrac{A}{2} = \dfrac{2B}{5} = \dfrac{C}{3} = k$.
    Then, $A = 2k$, $B = \dfrac{5k}{2}$, $C = 3k$
    $\therefore A : B : C = 2k : \dfrac{5k}{2} : 3k = 4 : 5 : 6$

15. (a) Given that, $4a = 5b$
    $\therefore \dfrac{a}{b} = \dfrac{5}{4}$
    Also, $7b = 9c \Rightarrow \dfrac{b}{c} = \dfrac{9}{7}$
    $\therefore a : b = 5 : 4 = (5 \times 9) : (4 \times 9) = 45 : 36$
    $b : c = 9 : 7 = (9 \times 4) : (7 \times 4) = 36 : 28$
    $\therefore a : b : c = 45 : 36 : 28$

16. (c) Given that, $A = \dfrac{1}{4} B$

∴ $\dfrac{A}{B} = \dfrac{1}{4} \Rightarrow A : B = 1 : 4$

Also, $B = \dfrac{1}{2} C \Rightarrow \dfrac{B}{C} = \dfrac{1}{2}$

⇒ $B : C = 1 : 2$

Here, $a = 1, b = 4, m = 1, n = 2$

∴ $A : B : C = am : mb : nb$
[by Technique 1(i)]
$= (1 \times 1) : (1 \times 4) : (2 \times 4) = 1 : 4 : 8$

17. (a) Given that, $P : Q : R = 2 : 3 : 4$

Let $P = 2K, Q = 3K, R = 4K$

∴ $\dfrac{P}{Q} = \dfrac{2K}{3K} = \dfrac{2}{3}, \dfrac{Q}{R} = \dfrac{3K}{4K} = \dfrac{3}{4}$ and $\dfrac{R}{P} = \dfrac{4K}{2K} = \dfrac{2}{1}$

∴ $\dfrac{P}{Q} : \dfrac{Q}{R} : \dfrac{R}{P} = \dfrac{2}{3} : \dfrac{3}{4} : \dfrac{2}{1}$

[∵ LCM of 3, 4, 1 = 12]

$= \dfrac{2 \times 4}{3 \times 4} : \dfrac{3 \times 3}{4 \times 3} : \dfrac{2 \times 12}{1 \times 12}$

$= 2 \times 4 : 3 \times 3 : 24$

$= 8 : 9 : 24$

18. (c) Let $\dfrac{a}{7} = \dfrac{b}{9} = \dfrac{c}{11} = K$

Then, $a = 7K, b = 9K, c = 11K$

∴ $a : b : c = 7K : 9K : 11K = 7 : 9 : 11$

19. (d) Let $\dfrac{1}{x} = 2K, \dfrac{1}{y} = 3K$ and $\dfrac{1}{z} = 5K$.

Then, $x = \dfrac{1}{2K}, y = \dfrac{1}{3K}$ and $z = \dfrac{1}{5K}$

∴ $x : y : z = \dfrac{1}{2K} : \dfrac{1}{3K} : \dfrac{1}{5K} = \dfrac{1}{2} : \dfrac{1}{3} : \dfrac{1}{5}$

$= 15 : 10 : 6$
[take LCM of denominators]

20. (a) Given, $A : B = 8 : 15, B : C = 5 : 8$ and $C : D = 4 : 5$

Here, $a = 8, b = 15, c = 5, d = 8, e = 4$ and $f = 5$

∴ $A : B : C : D = ace : bce : bde : bdf$
[by Technique 1(ii)]

$= 8 \times 5 \times 4 : 15 \times 5 \times 4 : 15 \times 8 \times 4 : 15 \times 8 \times 5$

$= 160 : 300 : 480 : 600$

$= 8 : 15 : 24 : 30$

21. (c) Given, $2A = 3B = 4C$

Now, $2A = 3B \Rightarrow \dfrac{A}{B} = \dfrac{3}{2}$

⇒ $A : B = 3 : 2 = (3 \times 2) : (2 \times 2) = 6 : 4$

Again, $3B = 4C \Rightarrow \dfrac{B}{C} = \dfrac{4}{3}$

⇒ $B : C = 4 : 3$

∴ $A : B : C = 6 : 4 : 3$

22. (d) Given, 10% of $(A + B) = 50\%$ of $(A - B)$

⇒ $\dfrac{A + B}{A - B} = \dfrac{50}{10} = \dfrac{5}{1}$

⇒ $A + B = 5A - 5B$

⇒ $5A - A = B + 5B$

⇒ $4A = 6B \Rightarrow \dfrac{A}{B} = \dfrac{6}{4} = \dfrac{3}{2}$

∴ $A : B = 3 : 2$

23. (b) Given that, $\dfrac{a}{b} = \dfrac{b}{c}$

⇒ $b^2 = ac$

∴ $a^4 : b^4 = a^4 : a^2c^2 = a^2 : c^2$

24. (a) Given, $xy = 36$

∴ $xy = 4 \times 9 \Rightarrow \dfrac{x}{9} = \dfrac{4}{y} \Rightarrow x : 9 = 4 : y$

25. (c) Given, $\dfrac{x}{y} = \dfrac{7}{5}$

∴ $\dfrac{5x - 2y}{3x + 2y} = \dfrac{(5 \times 7 - 2 \times 5)}{(3 \times 7 + 2 \times 5)} = \dfrac{35 - 10}{21 + 10} = \dfrac{25}{31}$

Alternate Method

We have, $\dfrac{5x - 2y}{3x + 2y}$

On dividing numerator and denominator by y, we get

$= \dfrac{5 \dfrac{x}{y} - 2}{3 \dfrac{x}{y} + 2} = \dfrac{5 \times \dfrac{7}{5} - 2}{3 \times \dfrac{7}{5} + 2} = \dfrac{7 - 2}{\dfrac{21 + 10}{5}} = \dfrac{25}{31}$

26. (b) Given, $\dfrac{x}{2y} = \dfrac{6}{7} \Rightarrow \dfrac{x}{y} = \dfrac{12}{7}$

By componendo and dividendo,

$\dfrac{x + y}{x - y} = \dfrac{12 + 7}{12 - 7} = \dfrac{19}{5}$

∴ $\dfrac{x - y}{x + y} + \dfrac{14}{19} = \dfrac{5}{19} + \dfrac{14}{19} = \dfrac{19}{19} = 1$

27. (c) ∵ $P^2 + 4Q^2 = 4PQ$

⇒ $P^2 + 4Q^2 - 4PQ = 0 \Rightarrow (P - 2Q)^2 = 0$

⇒ $P - 2Q = 0 \Rightarrow P = 2Q$

⇒ $\dfrac{P}{Q} = \dfrac{2}{1}$

∴ $P : Q = 2 : 1$

28. (d) Let the quantity be x.

Then, $\dfrac{a + x}{b + x} = \dfrac{c}{d}$

⇒ $(a + x) d = (b + x) c$

⇒ $ad + dx = bc + cx$

⇒ $ad - bc = cx - dx$

⇒ $ad - bc = x (c - d)$

∴ $x = \dfrac{ad - bc}{c - d}$

Ratio and Proportion / 311

29. (b) Given, $\dfrac{x^3 - y^3}{x^2 + xy + y^2} = \dfrac{5}{1}$

∴ $\dfrac{(x-y)(x^2 + xy + y^2)}{x^2 + xy + y^2} = \dfrac{5}{1}$

⇒ $x - y = 5$ ...(i)

and $\dfrac{x^2 - y^2}{x - y} = 7 \Rightarrow x + y = 7$ ...(ii)

From Eqs. (i) and (ii), we get $x = 6, y = 1$

∴ Required ratio
$= 2x : 3y = 2 \times 6 : 3 \times 1 = 12 : 3 = 4 : 1$

30. (a) Let $a + b = 6K$
$b + c = 7K$ ...(i)
$c + a = 8K$
and $a + b + c = 14$ [given] ...(ii)

From Eq. (i),
$a + b + b + c + c + a = 6K + 7K + 8K$
⇒ $2(a + b + c) = 21K$
⇒ $2 \times 14 = 21K \Rightarrow K = \dfrac{28}{21} = \dfrac{4}{3}$

∴ $a + b = 6K \Rightarrow a + b = 6 \times \dfrac{4}{3} = 8$ ...(iii)

On subtracting Eq. (iii) from Eq. (ii), we get
$(a + b + c) - (a + b) = 14 - 8 = 6$

31. (a) Given that, $\dfrac{a}{b} = \dfrac{c}{d} = \dfrac{e}{f} = \dfrac{1}{2}$

⇒ $a = \dfrac{b}{2}, c = \dfrac{d}{2}, e = \dfrac{f}{2}$

∴ $\left(\dfrac{3a + 5c + 7e}{3b + 5d + 7f}\right) = \dfrac{\left(\dfrac{3b}{2} + \dfrac{5d}{2} + \dfrac{7f}{2}\right)}{(3b + 5d + 7f)}$

$= \dfrac{\dfrac{1}{2}(3b + 5d + 7f)}{(3b + 5d + 7f)} = \dfrac{1}{2}$

∴ Required ratio = $1 : 2$

32. (a) Given, $\dfrac{3x - 2y}{2x + 3y} = \dfrac{5}{6}$

⇒ $18x - 12y = 10x + 15y$

⇒ $8x = 27y \Rightarrow \dfrac{y}{x} = \dfrac{8}{27} \Rightarrow \sqrt[3]{\dfrac{y}{x}} = \dfrac{2}{3}$

Now, $\dfrac{\sqrt[3]{x} + \sqrt[3]{y}}{\sqrt[3]{x} - \sqrt[3]{y}} = \dfrac{\left[\sqrt[3]{x}\left(1 + \sqrt[3]{\dfrac{y}{x}}\right)\right]}{\left[\sqrt[3]{x}\left(1 - \sqrt[3]{\dfrac{y}{x}}\right)\right]}$

$= \dfrac{\left(1 + \dfrac{2}{3}\right)}{\left(1 - \dfrac{2}{3}\right)} = \dfrac{\left(\dfrac{5}{3}\right)}{\left(\dfrac{1}{3}\right)} = 5$

33. (a) ∵ $\dfrac{a+b}{a-b} = \dfrac{5}{3} \Rightarrow 3a + 3b = 5a - 5b$

⇒ $2a = 8b \Rightarrow a = 4b \Rightarrow \dfrac{a}{b} = \dfrac{4}{1}$

Now, $\dfrac{(a^2 + b^2)}{(a^2 - b^2)} = \dfrac{\dfrac{a^2}{b^2} + 1}{\dfrac{a^2}{b^2} - 1} = \dfrac{\left(\dfrac{a}{b}\right)^2 + 1}{\left(\dfrac{a}{b}\right)^2 - 1}$

$= \dfrac{\left(\dfrac{4}{1}\right)^2 + 1}{\left(\dfrac{4}{1}\right)^2 - 1} = \dfrac{16 + 1}{16 - 1} = \dfrac{17}{15}$

∴ $(a^2 + b^2) : (a^2 - b^2) = 17 : 15$

34. (d) Let 1st number = x and 2nd number = y

According to the question,
$\dfrac{1}{2}$ of $x = 65\%$ of $y \Rightarrow \dfrac{x}{2} = \dfrac{65y}{100}$

⇒ $\dfrac{x}{y} = \dfrac{130}{100} = \dfrac{13}{10}$

∴ $x : y = 13 : 10$

35. (c) Required minimum part
$= \dfrac{1}{(3 + 5 + 4 + 1)} \times 182$
$= \dfrac{1}{13} \times 182 = 14$

36. (b) Number of boys = $(2140 - 1200) = 940$

∴ Boys : Girls = $940 : 1200 = 47 : 60$

37. (e) Let first number be x and second number be y.

Then, according to the question,
$x \times 35\% = 2 \times y \times 75\%$

⇒ $x \times \dfrac{35}{100} = 2y \times \dfrac{75}{100}$

⇒ $\dfrac{x}{y} = \dfrac{2 \times 75}{35} = \dfrac{2 \times 15}{7} = \dfrac{30}{7}$

∴ $x : y = 30 : 7$

38. (a) Let the first part = x

Then, second part becomes = $(27 - x)$
Now, according to the question,
$5x + 11(27 - x) = 195$
⇒ $5x + 297 - 11x = 195$
⇒ $-6x = 195 - 297 \Rightarrow -6x = -102$
⇒ $x = \dfrac{102}{6} = 17 \Rightarrow x = 17$

∴ First part = $x = 17$
and second part = $(27 - x) = (27 - 17) = 10$
∴ Ratio of first and second part = $17 : 10$

39. (b) Let the number of boys in union = $5x$
and number of girls in union = $3x$
According to the question,
$5x + 3x = 48$
⇒ $8x = 48 \Rightarrow x = 6$

∴ Number of boys = $5x = 5 \times 6 = 30$
Number of girls = $3x = 3 \times 6 = 18$
Let G girls should be added to make the ratio $6 : 5$.

According to the question,
$$\frac{30}{18+G} = \frac{6x}{5x} = \frac{6}{5}$$
$\Rightarrow \quad (18+G) \times 6 = 5 \times 30$
$\Rightarrow \quad 18 + G = 25$
$\Rightarrow \quad G = 25 - 18 = 7$

40. (a) Let A's share $= 3x$, B's share $= 7x$
C's share $= 9x$ and D's share $= 13x$
According to the question,
$7x = 4872$
$\therefore \quad x = \frac{4872}{7} = 696$
Share of A and $C = 3x + 9x = 12x$
$= 12 \times 696 = ₹\,8352$

41. (c) Let the original number of boys be $4x$ and number of girls be $5x$.
According to the question,
$$\frac{4x+10}{5x} = \frac{6}{5}$$
$\Rightarrow \quad 30x = 20x + 50$
$\Rightarrow \quad 10x = 50 \Rightarrow x = \frac{50}{10} = 5$
$\therefore$ Number of girls $= 5x = 5 \times 5 = 25$

42. (d) Let monthly income of Sita, Riya and Kunal be $84x, 76x$ and $89x$, respectively.
Given, annual income of Riya $= ₹\,456000$
$\therefore$ Monthly income of Riya
$= \frac{456000}{12} = ₹\,38000$
According to the question,
$76x = 38000 \Rightarrow x = 500$
So, sum of monthly incomes of Sita and Kunal $= 84x + 89x = 173x$
$= 173 \times 500 = 86500$
Therefore, annual income $= 86500 \times 12$
$= ₹\,1038000$

43. (b) Here, number of boys $= \frac{3}{4+3} \times 49 = 21$
and number of girls $= \frac{4}{4+3} \times 49 = 28$
If 4 girls leave the class, then remaining girls $= 28 - 4 = 24$
$\therefore$ Required ratio $= 24 : 21 = 8 : 7$

44. (d) Ratio of speeds $= 2 : 3 : 4$ [given]
$\therefore$ Time $\propto \frac{1}{\text{Speed}}$
$\therefore$ Ratio of time taken $= \frac{1}{2} : \frac{1}{3} : \frac{1}{4} = 6 : 4 : 3$

45. (c) Ratio among A, B, C and D
$= \frac{1}{3} : \frac{1}{4} : \frac{1}{5} : \frac{1}{6}$
On rearranging the ratio, we get
$\frac{60}{3} : \frac{60}{4} : \frac{60}{5} : \frac{60}{6} = 20 : 15 : 12 : 10$
So, minimum number of pens can be when the common ratio is 1.

So, minimum number of pen
$= 20 + 15 + 12 + 10 = 57$

46. (b) Let the two positive numbers be x and y.
Then, $\frac{x}{y} = \frac{3}{5}$ [given]
$\Rightarrow \quad x = \frac{3y}{5}$ ...(i)
and $\frac{x+2}{y+2} = \frac{5}{8}$ [given]
$\Rightarrow \quad 8(x+2) = 5(y+2)$
$\Rightarrow \quad 8x + 16 = 5y + 10$
$\Rightarrow \quad 8x - 5y = 10 - 16$
$\Rightarrow \quad 8\left(\frac{3y}{5}\right) - 5y = -6 \quad \left[\text{put } x = \frac{3y}{5}\right]$
$\Rightarrow \quad 24y - 25y = -30$
$\Rightarrow \quad -y = -30 \Rightarrow y = 30$
and $x = \frac{3y}{5} = \frac{3 \times 30}{5} = 18$
Difference between both the numbers
$= 30 - 18 = 12$

47. (a) Let first and second part of the number be x and y, respectively.
Then, $5x + 11y = 7(x+y)$
$\Rightarrow \quad 11y - 7y = 7x - 5x$
$\Rightarrow \quad 4y = 2x \Rightarrow x : y = 2 : 1$

48. (c) Let the number of coins of ₹ 1, 50 paise and 25 paise be $8x, 5x$ and $3x$, respectively.
According to the question,
$(1)(8x) + (0.50)(5x) + (0.25)(3x) = 225$
$\Rightarrow \quad 8x + \frac{5}{2}x + \frac{3}{4}x = 225$
$\Rightarrow \quad 32x + 10x + 3x = 225 \times 4$
$\Rightarrow \quad 45x = 225 \times 4$
$\Rightarrow \quad x = \frac{225 \times 4}{45} = 20$
Therefore, number of coins of ₹$1 = 8x$
$= 8 \times 20 = 160$

49. (b) Let the two numbers be $8x$ and $3x$, respectively.
According to the question,
$8x - 3x = 45 \Rightarrow 5x = 45$
$\therefore \quad x = \frac{45}{5} = 9$
Hence, original number $= 8x = 8 \times 9 = 72$

50. (d) Let the marks of A, B and C are $10x, 12x$ and $15x$, respectively.
Let $x = 6$
$\therefore$ Maximum marks of C can be
$= 15 \times 6 = 90$
So, maximum marks of B can be
$= 12 \times 6 = 72$
As the marks are fixed and they cannot exceed the maximum marks.
So, the marks of B cannot be in the range of (80-90), i.e. B cannot score above 80.

Ratio and Proportion / 313

51. **(d)** Let the amount received by P, Q and R be $3x, 5x$ and $7x$, respectively.
Now, according to the question,
$5x + 4000 = 7x$
$\Rightarrow \quad 2x = 4000 \Rightarrow x = 2000$
$\therefore$ Total amount received by P and Q
$= 3x + 5x = 8x$
$= 8 \times 2000 = ₹16000$

52. **(a)** Let the share of $P = x$
Then, Q's share $= x + 30$
and R's share $= (x + 30) + 60 = x + 90$
$\because$ Sum of money with P, Q and $R = 300$
$\therefore \quad x + x + 30 + x + 90 = 300$
$\Rightarrow \quad 3x + 120 = 300$
$\therefore \quad x = \dfrac{300 - 120}{3} = 60$
$\therefore$ Required ratio $= 60 : (60 + 30) : (60 + 90)$
$= 60 : 90 : 150 = 2 : 3 : 5$

53. **(d)** Let the amount to be distributed be ₹x.
$P : Q : R = 2 : 7 : 9$
Sum of the ratios $= 2 + 7 + 9 = 18$
$\therefore$ P's share $= \dfrac{2}{18} \times x = \dfrac{x}{9}$, Q's share $= \dfrac{7}{18} x$
and R's share $= \dfrac{9x}{18} = \dfrac{x}{2}$
As given, $\dfrac{x}{9} + \dfrac{7x}{18} = \dfrac{x}{2}$
Thus, we get no conclusion. Amount should necessarily be known.

54. **(b)** According to the question,
$A = \dfrac{1}{2}(B + C) \Rightarrow B + C = 2A$
$\Rightarrow A + B + C = 3A$ [adding A on both sides]
$\Rightarrow \quad 3A = 5625$
$\Rightarrow \quad A = \dfrac{5625}{3} = ₹1875$ and $B = \dfrac{1}{4}(A + C)$
$\Rightarrow \quad A + C = 4B \Rightarrow A + B + C = 5B$
[adding B on both sides]
$\Rightarrow 5B = 5625 \Rightarrow B = \dfrac{5625}{5} = ₹1125$
$\therefore \quad A + B = 1875 + 1125 = ₹3000$

55. **(b)** Let the age of father $= 5x$ yr
and the age of son $= 2x$ yr
According to the question,
$5x \times 2x = 1000 \Rightarrow 10x^2 = 1000$
$\Rightarrow \quad x^2 = 100 \Rightarrow x = \sqrt{100} = 10$
$\therefore$ Father's age after 10 yr $= (5x + 10)$
$= 5 \times 10 + 10 = 60$ yr

56. **(a)** Let the smaller number $= x$
and the greater number $= y$
According to the question,
$\left(y - \dfrac{x}{2}\right) = 4\left(x - \dfrac{x}{2}\right)$

$\Rightarrow \quad y - \dfrac{x}{2} = 4 \cdot \dfrac{x}{2}$
$\Rightarrow \quad y = 2x + \dfrac{x}{2} \Rightarrow y = \dfrac{5x}{2}$
$\therefore \quad y : x = 5 : 2$

57. **(d)** Let the number of boys and girls be $7x$ and $5x$, respectively.
Given, total number of students $= 2400$
$\Rightarrow 7x + 5x = 2400$
$\Rightarrow \quad 12x = 2400 \Rightarrow x = 200$
$\therefore$ Required number of girls
$= 5x = 5 \times 200 = 1000$
Fast Track Method
Here, $a = 7, b = 5, x = 2400$
Number of girls $= \dfrac{bx}{a + b}$ [by Technique 2(i)]
$= \dfrac{5 \times 2400}{7 + 5} = \dfrac{5}{12} \times 2400$
$= 1000$

58. **(a)** Given, total amount $= ₹49500$
Let part of Amit's investment $= ₹4x$
and part of Sudesh's investment $= ₹7x$
According to the question,
$4x + 7x = 49500 \Rightarrow 11x = 49500$
$\Rightarrow \quad x = \dfrac{49500}{11} \Rightarrow x = 4500$
Hence, investment of Sudesh $= 7x$
$= 7 \times 4500 = ₹31500$
Fast Track Method
Here, $x = 49500, a = 4$ and $b = 7$
$\therefore$ Investment of Sudesh $= \dfrac{bx}{a + b}$
$= \dfrac{7 \times 49500}{7 + 4}$ [by Technique 2(i)]
$= 7 \times 4500 = ₹31500$

59. **(d)** Given that, P's share : Q's share
$= \dfrac{1}{3} : \dfrac{1}{5} = 5 : 3$
Here, $a = 5, b = 3, x = 1664$
$\therefore P$'s share $= \dfrac{ax}{a + b}$ [by Technique 2(i)]
$= \dfrac{5 \times 1664}{5 + 3} = \dfrac{5}{8} \times 1664 = ₹1040$

60. **(b)** If three parts be x, y and z, then
$\dfrac{x}{2} = \dfrac{y}{3} = \dfrac{z}{5} \Rightarrow x : y : z = 2 : 3 : 5$
$\therefore \quad x = \dfrac{2}{10} \times 990 = 198$ [by Technique 2(ii)]
$y = \dfrac{3}{10} \times 990 = 297$
and $z = \dfrac{5}{10} \times 990 = 495$

61. **(c)** Given, $A : B = 2 : 3$
and $\qquad B : C = 4 : 5$

314 / **Fast Track** Objective Arithmetic

Then, $A : B : C = 8 : 12 : 15$

∴ Share of $B = \dfrac{12}{8 + 12 + 15} \times 7000$

$= \dfrac{12 \times 7000}{35} = ₹ 2400$

62. (c) Let the incomes of two persons be $7x$ and $3x$, respectively.
Expenditure of first person $= 7x - 300$
Expenditure of second person $= 3x - 300$
According to the question,
$\dfrac{7x - 300}{3x - 300} = \dfrac{5}{2}$

$\Rightarrow 14x - 600 = 15x - 1500 \Rightarrow x = 900$

∴ Income of first person $= 7x$
$= 7 \times 900 = ₹ 6300$

Fast Track Method
Ratio of incomes $= 7 : 3$
Ratio of expenses $= 5 : 2$
So, $a = 7, b = 3, c = 5, d = 2$ and $x = ₹ 300$
∴ Income of 1st person
$= \dfrac{xa(d - c)}{(ad - bc)}$ [by Technique 3]
$= \dfrac{300 \times 7(2 - 5)}{14 - 15} = ₹ 6300$

63. (b) Let the monthly incomes of X and Y be $4x$ and $3x$, respectively and their monthly expenses be $3y$ and $2y$, respectively
According to the question,
$4x - 3y = 6000$ and $3x - 2y = 6000$
$\Rightarrow 4x - 3y = 3x - 2y \Rightarrow x = y$
∴ $x = 6000$

Hence, total monthly income
$= ₹ 7x = ₹ 7 \times 6000 = ₹ 42000$

Fast Track Method
Given, ratio of incomes $= 4 : 3$
and ratio of expenses $= 3 : 2$
Here, $a = 4, b = 3, c = 3, d = 2$ and $x = 6000$
∴ Income of 1st person
$= \dfrac{6000(4)(2 - 3)}{(4 \times 2) - (3 \times 3)}$ [by Technique 3]
$= ₹ 24000$
Income of 2nd person
$= \dfrac{6000 \times (3)(2 - 3)}{(4 \times 2) - (3 \times 3)} = ₹ 18000$

Hence, total monthly income
$= ₹ (24000 + 18000) = ₹ 42000$

64. (c) Let the two numbers be $2x$ and $3x$.
According to the question,
$\dfrac{2x + 9}{3x + 9} = \dfrac{3}{4}$

$\Rightarrow 8x + 36 = 9x + 27 \Rightarrow x = 9$
∴ Product of the two numbers $= 2x \times 3x$
$= 6x^2 = 6(9)^2 = 486$

Fast Track Method
Here, $a = 2, b = 3, c = 3, d = 4$ and $x = 9$
∴ First number $= \dfrac{9 \times 2(3 - 4)}{2 \times 4 - 3 \times 3}$
[by Technique 4]
$= 18$
and second number $= \dfrac{9 \times 3(3 - 4)}{2 \times 4 - 3 \times 3} = 27$
∴ Product of two numbers $= 18 \times 27 = 486$

Exercise 2 Higher Skill Level Questions

1. (d) Here, $\dfrac{1}{3}$ of workers are women and $\dfrac{2}{3}$ of workers are men.

∴ $\dfrac{1}{2}$ of the women are married $= \dfrac{1}{2} \times \dfrac{1}{3} = \dfrac{1}{6}$

and $\dfrac{3}{4}$ of the men are married $= \dfrac{3}{4} \times \dfrac{2}{3} = \dfrac{1}{2}$

∴ Ratio of married women to married men are $\dfrac{1}{6} : \dfrac{1}{2} \Rightarrow 1 : 3$

2. (b) Total collection = Price of each entry ticket × Number of visitors
Increase in ratio $= 16 \times 3 : 13 \times 7 = 48 : 91$

3. (a) Let the number of adult men be x.
Then, ratio of the numbers of adult men and adult women $9 : 7 = x : 4.2$
$\Rightarrow x = \dfrac{9}{7} \times 4.2 = 5.4$ lakh
Total adult population $= 4.2 + 5.4 = 9.6$ lakh

If the population of the town be y lakh, then
$\dfrac{80y}{100} = 9.6 \Rightarrow y = \dfrac{9.6 \times 100}{80} = 12$ lakh

4. (d) According to the question,
$46x + 30 + 41x + 30 + 34x + 25 = 2505$
$\Rightarrow 121x = 2505 - 85 = 2420$
$\Rightarrow x = \dfrac{2420}{121} = 20$

∴ Number of coins received by Parat
$= 41x + 30 = 41 \times 20 + 30 = 850$

5. (e) Let Y's salary $= 100$
∴ X's salary $= 80$
and Z's salary $= \dfrac{80 \times 120}{100} = 96$

∴ Required ratio $= 80 : 100 : 96$
$= 20 : 25 : 24$

6. (a) Let B gets $= ₹ x$.
Then, A gets $(x + 40)$ and C gets $(x + 70)$.

Ratio and Proportion / 315

According to the question,
$x + 40 + x + x + 70 = 710$
$\Rightarrow \quad 3x = 710 - 110 = 600$
$\therefore \quad x = \dfrac{600}{3} = 200$
$\therefore$ C's share $= 200 + 70 = ₹\,270$

7. (d) Cannot be determined, since the total amount of money is not given in either of the case.

8. (a) Total of maximum marks of all subjects
$= 105 \times 5 = 525$
$\therefore$ 80% of $525 = \dfrac{525 \times 80}{100} = 420$
Marks obtained in four subjects (Hindi, Sanskrit, Mathematics and English)
$= 89 + 92 + 98 + 81 = 360$
$\therefore$ Marks obtained in Science
$= 420 - 360 = 60$

9. (b) Ratio of salaries of Akash, Babloo and Chintu $= 2 : 3 : 5$
Let the common ratio be x.
Then, salaries of Akash, Babloo and Chintu will be $2x, 3x$ and $5x$, respectively.
Now, 15% increase in Akash's salary
$= 15\%$ of $2x = \dfrac{15 \times 2x}{100} = 0.3x$
$\therefore$ New salary $= 2x + 0.3x = 2.3x$
Also, 10% increase in Babloo's salary
$= 10\%$ of $3x = \dfrac{10 \times 3x}{100} = 0.3x$
$\therefore$ New salary $= 3x + 0.3x = 3.3x$
Again, 20% increase in Chintu's salary
$= 20\%$ of $5x = \dfrac{20 \times 5x}{100} = 1x$
$\therefore$ New salary $= 5x + x = 6x$
$\because$ New ratio $=$ Ratio of new salaries
$= 2.3x : 3.3x : 6x$
On multiplying with 10 and dividing by x, then the ratio will be $23 : 33 : 60$.

10. (c) Let the income of A, B, C and D be ₹ A, ₹ B, ₹ C and ₹ D, respectively.
According to the question,
$\quad A = B + 140 \qquad \qquad \ldots(i)$
and $\quad C = D + 80 \qquad \qquad \ldots(ii)$
Also, $\quad \dfrac{A}{C} = \dfrac{2}{3}$
$\Rightarrow \quad \dfrac{B + 140}{D + 80} = \dfrac{2}{3} \qquad \ldots(iii)$
Also, $\quad \dfrac{B}{D} = \dfrac{1}{2} \Rightarrow D = 2B$
On putting $D = 2B$ in Eq. (iii), we get
$3(B + 140) = 2(2B + 80) \Rightarrow B = ₹\,260$
Then, $D = ₹\,520$, $A = ₹\,400$ and $C = ₹\,600$

11. (e) Let the monthly salary of Mr. Duggabati $= ₹\,x$
Expenses or rent and domestic items
$= x \times (24 + 30)\% = x \times 54\%$
$= x \times \dfrac{54}{100} = ₹\,\dfrac{54x}{100}$
$\therefore$ Rest amount from salary
$= x - \dfrac{54x}{100} = ₹\,\dfrac{46x}{100}$
Now, investments in fixed deposit and lottery are $9a$ and $7a$, respectively.
According to the question,
$\Rightarrow \quad \dfrac{46x}{100} \times \dfrac{9a}{9a + 7a} = 124200$
$\Rightarrow \quad \dfrac{46x}{100} \times \dfrac{9a}{16a} = 124200$
$\Rightarrow \quad x = \dfrac{124200 \times 16 \times 100}{46 \times 9}$
$\Rightarrow \quad x = ₹\,480000$
Now, rent paid by Duggabati in year
$= 480000 \times \dfrac{24}{100} = ₹\,115200$

12. (c) Let total age of boys in the class be x yr.
Now, number of boys
$= \dfrac{4}{(4+1)} \times 50 = \dfrac{4}{5} \times 50 = 40$
$\therefore$ Number of girls $= 50 - 40 = 10$
Now, according to the question,
$\dfrac{x_1 + x_2 + \ldots + x_{40}}{40} = 2 \times 10$
$\Rightarrow x_1 + x_2 + \ldots + x_{40} = 800 \Rightarrow x = 800$

13. (d) Let the monthly salaries of Rene and Som be $5x$ and $3x$, respectively.
Then, according to the question,
$5x - \left[5x \times \dfrac{1}{6} + 5x \times \dfrac{1}{5} + 5x \times \dfrac{3}{10} + \dfrac{5x}{4} \right]$
$= 5000$
$\Rightarrow 5x - \left[\dfrac{5x}{6} + x + \dfrac{3x}{2} + \dfrac{5x}{4} \right] = 5000$
$\Rightarrow 5x - \left(\dfrac{10x + 12x + 18x + 15x}{12} \right) = 5000$
$\Rightarrow \quad 5x - \left(\dfrac{55x}{12} \right) = 5000$
$\Rightarrow 60x - 55x = 60000 \Rightarrow x = 12000$
Hence, Som's monthly salary $= 3x$
$= 3 \times 12000 = ₹\,36000$

14. (b) Let the scores of three matches be a, b and c, respectively.
Then, $\dfrac{a}{b} = \dfrac{5}{4}$ and $\dfrac{b}{c} = \dfrac{2}{1}$
$\Rightarrow \quad 4a = 5b$ and $b = 2c$
Also, $a - c = 48$
$\Rightarrow \quad \dfrac{5b}{4} - \dfrac{b}{2} = 48 \Rightarrow \dfrac{5b - 2b}{4} = 48$
$\Rightarrow \quad 3b = 4 \times 48 \Rightarrow b = 4 \times 16 = 64$
$\therefore \quad a = \dfrac{5b}{4} = 5 \times 16 = 80$
and $\quad c = \dfrac{b}{2} = \dfrac{64}{2} = 32$

∴ Average score
$$= \frac{80 + 64 + 32}{3} = \frac{176}{3} = 58\frac{2}{3}$$

15. (b) ∵ $A + 10 = 3(B - 10)$
⇒ $A - 3B = -30 - 10$
⇒ $A - 3B = -40$ ...(i)
and $(A - 10) = (B + 10)$
⇒ $A - B = 20$ ...(ii)
On subtracting Eq. (ii) from Eq. (i), we get
$A - 3B - A + B = -40 - 20$
⇒ $-2B = -60 \Rightarrow B = 30$
and $A = 20 + B = 20 + 30 = 50$
∴ Ratio of number of students of A and B
$$= \frac{50}{30} = \frac{5}{3} = 5 : 3$$

16. (d) 4 leaps of cat = 3 leaps of dog
⇒ 1 leap of cat = 3/4 leap of dog
Cat takes 5 leaps for every 4 leaps of a dog.
∴ Required ratio
= (5 × Cat's leap) : (4 × Dog's leap)
$= \left(5 \times \frac{3}{4} \text{ Dog's leap}\right) : (4 \times \text{Dog's leap})$
= 15 : 16

17. (c) Let the savings of A and B are $4x$, $5x$ and the share in cost of gift are $3y$, $4y$, respectively.
According to the question,
For A, $4x - 3y = \frac{2}{3} \times 4x \Rightarrow x = \frac{9y}{4}$ ...(i)
For B, $5x - 4y = 145$
⇒ $5 \times \frac{9y}{4} - 4y = 145$ [from Eq. (i)]
⇒ $y = 20$
∴ Cost of gift = $3y + 4y = 7 \times 20 = ₹140$

18. (b) Let the fixed amount be ₹ x and the cost of each unit be ₹ y.
Then, $540y + x = 1800$ ...(i)
and $620y + x = 2040$ ...(ii)
On subtracting Eq. (i) from Eq. (ii), we get
$80y = 240 \Rightarrow y = 3$.
On putting $y = 3$ in Eq. (i), we get
$540 \times 3 + x = 1800$
⇒ $x = (1800 - 1620) = 180$
∴ Fixed charges = ₹ 180,
Charge per unit = ₹ 3
Total charges for consuming 500 units
= ₹ $(180 + 500 \times 3)$ = ₹ 1680

19. (b) Given, ratio of employees = 9 : 13 : 18
Let number of A's employees = $9x$
Number of B's employees = $13x$
and number of C's employees = $18x$
According to the question,
$18x = 54 \Rightarrow x = 3$
∴ Number of employees of type A
$= 9 \times 3 = 27$

Similarly, wages of every employee of type A
$= \frac{10}{7} \times 1400 = ₹ 2000$
∴ Required wages = $27 \times 2000 = ₹ 54000$

20. (c) Since, there are 70 males out of 120 applicants, there must be 50 females. For the minimum number of males to have a driver's license all 50 females must have a driver's license. Thus, the number of males having a driver's license, will be $80 - 50 = 30$. The maximum possible number of males having a driver's license is 70.
∴ Ratio between the minimum and the maximum number of males having driver's license = 30 : 70 = 3 : 7

21. (b) Let the number of failed and passed candidates be x and $4x$, respectively. Therefore, total number of candidates was $5x$.
According to the question,
If total number of students had been $5x - 35$,
then $\frac{(5x - 35) - (x + 9)}{x + 9} = \frac{2}{1}$
⇒ $4x - 44 = 2(x + 9)$
⇒ $4x - 2x = 18 + 44$
⇒ $2x = 62 \Rightarrow x = 31$
Thus, total number of candidates was 31×5, i.e. 155.

22. (c) Let the total coloured part be 70
Part of picture which is coloured blue
$= \frac{4}{7} \times 70 = 40$
∴ Part of picture coloured yellow = 30
For upper half of picture
Part coloured blue = $\frac{2}{5} \times 35 = 14$
Part coloured yellow = $\frac{3}{5} \times 35 = 21$
Now, remaining part for blue colour
$= 40 - 14 = 26$
and remaining part for yellow colour
$= 30 - 21 = 9$
and required ratio = 26 : 9.

23. (a) Let the weights of the pieces of diamond are $x, 2x, 3x, 4x$.
∴ Total weight = $10x$
∴ Total cost = $(10x)^2 = 100x^2$ ...(i)
Cost of each piece = $x^2, 4x^2, 9x^2, 16x^2$
∴ Total cost pieces = $30x^2$
∴ Total loss = $100x^2 - 30x^2 = 70x^2$
But total loss = ₹ 70000
∴ $70x^2 = 70000$
⇒ $x^2 = 1000$
∴ Total cost of original diamond = $100x^2$
$= 100 \times 1000$ [∵ $x^2 = 1000$]
= ₹ 100000

Chapter 18

Mixture or Alligation

Mixture

The new product obtained by mixing two or more ingredients in a certain ratio is called a mixture.

or

Combination of two or more quantities is known as mixture.

Mean Price

The cost price of a unit quantity of the mixture is called the **mean price**. It will always be higher than cost price of cheaper quantity and lower than cost price of dearer quantity.

Rule of Mixture or Alligation

It is the rule that enables us to find the ratio in which two or more ingredients at the given price must be mixed to produce a mixture of a desired price.

According to this rule, $\dfrac{n_1}{n_2} = \dfrac{A_2 - A_w}{A_w - A_1}$

where, n_1/n_2 is the ratio, in which two quantities should be mixed, while A_1, A_2 and A_w are the cheaper price, dearer price and mean price, respectively.

The above rule can be represented pictorially as shown below

```
       Cheaper price        Dearer price
           A₁                    A₂
              ↘  Mean price  ↙
                    Aw
              ↙              ↘
         A₂ − Aw              Aw − A₁
```

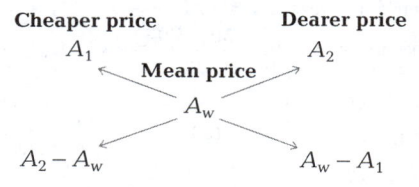

[remember, $A_1 < A_w < A_2$]

Amount of cheaper : Amount of dearer $(n_1 : n_2) = \dfrac{A_2 - A_w}{A_w - A_1}$

MIND IT! The rule is also applicable for solving questions based on average, speed, percentage, price, ratio etc., and not for absolute values. In other words, we can use this method whenever per cent, per hour, per kg etc., are being compared.

Ex. 1 In what proportion, must wheat at ₹ 6.20 per kg be mixed with wheat at ₹ 7.20 per kg, so that the mixture be worth ₹ 6.50 per kg?

Sol. Given, cost price of cheaper quantity = ₹ 6.20 per kg
Cost price of dearer quantity = ₹ 7.20 per kg
and mean price = ₹ 6.50 per kg
According to the rule of alligation,

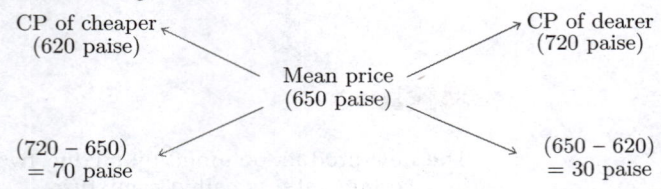

```
    CP of cheaper                    CP of dearer
    (620 paise)                       (720 paise)
                  Mean price
                  (650 paise)
    (720 − 650)                      (650 − 620)
    = 70 paise                       = 30 paise
```

∴ Required ratio = 70 : 30 = 7 : 3

Ex. 2 A mixture of a certain quantity of milk with 16 L of water is worth ₹ 0.75 per litre. If pure milk is worth ₹ 2.25 per litre, then how much milk is there in the mixture?

Sol. Water is available free of cost, so its cost price = ₹ 0
Now, according to the rule of alligation,

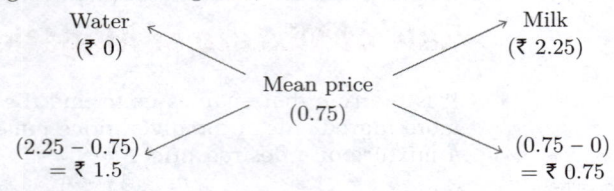

```
      Water                             Milk
      (₹ 0)                            (₹ 2.25)
                  Mean price
                   (0.75)
    (2.25 − 0.75)                    (0.75 − 0)
    = ₹ 1.5                          = ₹ 0.75
```

∴ Water : Milk = 1.5 : 0.75 = 2 : 1
Clearly, quantity of milk = $\frac{1}{2}$ of water = $\frac{1}{2} \times 16 = 8$ L

Ex. 3 If 50 L of milk solution has 40% milk in it, then how much milk should be added to make it 60% in the solution?

Sol. Given, total quantity = 50 L
∴ Quantity of milk = $\frac{50 \times 40}{100} = 20$ L
Let x L milk should be added.
Then, $(20 + x) = \frac{(50 + x) \times 60}{100}$

or $\frac{20 + x}{50 + x} = \frac{6}{10}$

$\Rightarrow$ $200 + 10x = 300 + 6x$
$\Rightarrow$ $4x = 100 \Rightarrow x = 25$ L

Fast Track Techniques to solve the QUESTIONS

Technique 1 If a container initially contains a units of liquid and b units of liquid is taken out and it is filled with b units of another liquid, then after n operations, the final quantity of the original liquid in the container is given as $\left[a\left(1-\dfrac{b}{a}\right)^n\right]$ units.

Ex. 4 A container contains 40 L of milk. From this container, 4 L of milk was taken out and replaced by water. This process was further repeated two times. How much milk is now there in the container?

Sol. Given, original quantity of milk = 40 L
Since, 4 L of milk was taken out.
∴ Quantity of milk in the new mixture = 40 − 4 = 36 L.
Now, when 4 L of this mixture taken out.
Quantity of milk taken out = $4 \times \dfrac{36}{40} = 3.6$ L
∴ Quantity of milk left = 36 − 3.6 = 32.4 L
Similarly, quantity of milk taken out in third step
$$= \dfrac{4 \times 32.4}{40} = 3.24 \text{ L}$$
∴ Quantity of milk left = 32.4 − 3.24 = 29.16 L

Fast Track Method
Here, $a = 40$ L, $b = 4$ L and $n = 3$
According to the formula, after n operations, quantity of milk = $\left[a\left(1-\dfrac{b}{a}\right)^n\right]$

∴ Quantity of milk left = $\left[40\left(1-\dfrac{4}{40}\right)^3\right] = \left(40 \times \dfrac{9}{10} \times \dfrac{9}{10} \times \dfrac{9}{10}\right)$

$$= \dfrac{40 \times 729}{10 \times 10 \times 10} = 29.16 \text{ L}$$

Technique 2 In a container, milk and water are present in the ratio $a : b$.

(i) If x L of water is added to this mixture, the ratio becomes $a : c$. Then, quantity of milk in original mixture = $\dfrac{ax}{c-b}$ L and quantity of water in original mixture = $\dfrac{bx}{c-b}$ L

(ii) If x L of milk is added to this mixture, the ratio becomes $c : b$. Then, quantity of milk in original mixture = $\dfrac{ax}{c-a}$ and quantity of water in original mixture = $\dfrac{bx}{c-a}$

Ex. 5 In a container, milk and water are present in the ratio 7 : 5. If 15 L water is added to this mixture, the ratio of milk and water becomes 7 : 8. Find the quantity of water in the new mixture.

Sol. Let the quantity of milk and water in initial mixture be $7x$ L and $5x$ L.
Then, according to the question,
$$\frac{7x}{5x + 15} = \frac{7}{8} \Rightarrow 7x \times 8 = 7(5x + 15)$$
$$\Rightarrow \qquad 56x = 35x + 105 \Rightarrow 56x - 35x = 105$$
$$\Rightarrow \qquad 21x = 105 \Rightarrow x = \frac{105}{21} = 5$$
∴ Quantity of water in initial mixture = $5 \times 5 = 25$ L
and quantity of water in new mixture = $25 + 15 = 40$ L

Fast Track Method
Here, $a = 7, b = 5, c = 8$ and $x = 15$ L
According to the formula,
Quantity of water in original mixture = $\dfrac{bx}{c - b} = \dfrac{5 \times 15}{8 - 5} = 25$ L
∴ Quantity of water in new mixture = $25 + 15 = 40$ L

Ex. 6 A vessel contain liquid A and liquid B in the ratio 5 : 7. If 10 L of liquid A is added to the vessel, the new ratio becomes 6 : 7. Find the quantity of liquid A in the original mixture.

Sol. Let the quantity of liquid A and B in the vessel initially be $5x$ and $7x$, respectively.
Then, according to the question, $\dfrac{5x + 10}{7x} = \dfrac{6}{7} \Rightarrow 5x + 10 = 6x \Rightarrow x = 10$
∴ Quantity of liquid A in original mixture = $5 \times 10 = 50$ L.

Fast Track Method
Here $a = 5, b = 7, c = 6, x = 10$
Quantity of liquid A in original mixture = $\dfrac{ax}{c - a}$ L $= \dfrac{5 \times 10}{6 - 5} = 50$ L

Technique 3 A container has milk and water in the ratio $a : b$, a second container has milk and water in the ratio $c : d$. If both the mixtures are emptied into a third container, then the ratio of milk to water in third container is given by $\left(\dfrac{a}{a + b} + \dfrac{c}{c + d}\right) : \left(\dfrac{b}{a + b} + \dfrac{d}{c + d}\right)$.

Ex. 7 Two containers have milk and water in the ratio 2 : 1 and 3 : 1, respectively. If both containers are emptied into a bigger container, then find the ratio of milk to water in bigger container?

Sol. Given, ratio of milk and water in 1st container = 2 : 1
∴ Quantity of milk in 1st container = $\dfrac{2}{3}$
and quantity of water in 1st container = $\dfrac{1}{3}$
Similarly, ratio of milk and water in 2nd container = 3 : 1

Mixture or **Alligation** / 321

∴ Quantity of milk in 2nd container = $\dfrac{3}{4}$

and quantity of water in 2nd container = $\dfrac{1}{4}$

Now, after pouring both mixture in one container.

Quantity of milk = $\dfrac{2}{3} + \dfrac{3}{4} = \dfrac{17}{12}$

and quantity of water = $\dfrac{1}{4} + \dfrac{1}{3} = \dfrac{7}{12}$

Hence, required ratio = $\dfrac{17}{12} : \dfrac{7}{12} = 17 : 7$

Fast Track Method

Here, $a = 2, b = 1, c = 3$ and $d = 1$

∴ Ratio of milk to water in bigger container

$$= \left(\dfrac{a}{a+b} + \dfrac{c}{c+d}\right) : \left(\dfrac{b}{a+b} + \dfrac{d}{c+d}\right)$$

$$= \left(\dfrac{2}{2+1} + \dfrac{3}{3+1}\right) : \left(\dfrac{1}{2+1} + \dfrac{1}{3+1}\right)$$

$$= \dfrac{17}{12} : \dfrac{7}{12} = 17 : 7$$

Multi Concept
Questions

1. Jagatram, a milk seller has certain quantity of milk to sell. In what ratio, he should mix water to gain 5% by selling the mixture at the cost price?
   (a) 1 : 10 (b) 1 : 5 (c) 1 : 20 (d) 1 : 15

   (c) Let the cost price of milk be ₹ 1 per litre.

   ∴ SP of 1 L of mixture = ₹ 1; gain = 5%

   ∴ CP of 1 L of mixture = $\dfrac{100}{105} \times 1 = ₹ \dfrac{20}{21}$

   According to the rule of alligation,

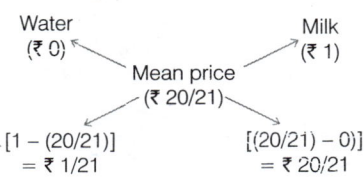

   Water   Milk
   (₹ 0)   (₹ 1)
      Mean price
      (₹ 20/21)

   [1 − (20/21)]   [(20/21) − 0)]
   = ₹ 1/21   = ₹ 20/21

   ∴ Required ratio = $\dfrac{1}{21} : \dfrac{20}{21} = 1 : 20$

2. If the price of three types of rice are ₹ 480, ₹ 576 and ₹ 696 per quintal, then find the ratio in which these types of rice should be mixed, so that the resultant mixture cost ₹ 564 per quintal?
(a) 22 : 11 : 9   (b) 28 : 14 : 7   (c) 14 : 7 : 21   (d) 11 : 77 : 7

(d) Let amount of 3 types of rice be x, y and z, respectively.
Using rule of alligation for x and y,

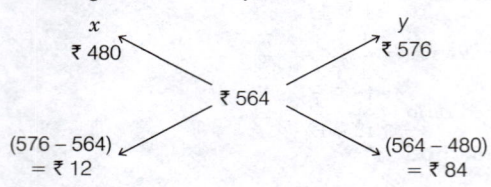

$$
\begin{array}{ccc}
x & & y \\
₹\ 480 & & ₹\ 576 \\
& ₹\ 564 & \\
(576 - 564) & & (564 - 480) \\
= ₹\ 12 & & = ₹\ 84
\end{array}
$$

∴ $x : y = 12 : 84 = 1 : 7$

Similarly, using the rule of alligation for x and z,

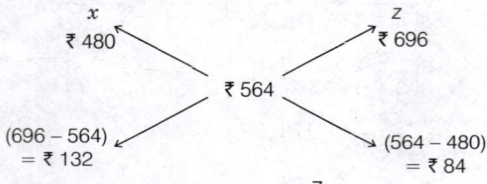

$$
\begin{array}{ccc}
x & & z \\
₹\ 480 & & ₹\ 696 \\
& ₹\ 564 & \\
(696 - 564) & & (564 - 480) \\
= ₹\ 132 & & = ₹\ 84
\end{array}
$$

∴ $x : z = 132 : 84 = 11 : 7 = 1 : \dfrac{7}{11}$

On combining the two ratios, we get

$$x : y : z = 1 : 7 : \dfrac{7}{11} = 11 : 77 : 7$$

Fast Track **Practice**

Exercise ❶ Base Level Questions

1. In what proportion must a grocer mix wheat at ₹ 2.04 per kg and ₹ 2.88 per kg so as to make a mixture of worth ₹ 2.52 per kg? **[PNB Clerk 2008]**
   (a) 2 : 3 (b) 3 : 2 (c) 5 : 3 (d) 3 : 4
   (e) None of these

2. A mixture of certain quantity of milk with 8 L of water is worth 45 paise per litre. If pure milk is worth 54 paise per litre, how much milk is there in the mixture? **[Hotel Mgmt. 2010]**
   (a) 40 L (b) 35 L (c) 25 L (d) 45 L

3. A merchant has 2000 kg of rice, one part of which he sells at 36% profit and the rest at 16% profit. He gains 28% on the whole. Find the quantity sold at 16%.
   [SBI Clerk 2011]
   (a) 400 kg (b) 300 kg (c) 900 kg (d) 800 kg
   (e) None of these

4. A person has ₹ 8400. He lent a part of it at 4% and the remaining at $3\frac{1}{3}$% simple interest. His total annual income was ₹ 294. Find the sum he lent at 4%.
   (a) ₹ 2310   (b) ₹ 2110
   (c) ₹ 2500   (d) ₹ 2100
   (e) None of these

5. A merchant has 50 kg of pulse. He sells one part at a profit of 10% and other at 5% loss. Overall he had a gain of 7%. Find the quantity of pulses, which he sold at 10% profit and 5% loss.
   (a) 40 kg, 10 kg (b) 40 kg, 15 kg
   (c) 40 kg, 12 kg (d) 40 kg, 9 kg

6. A milkman bought 15 L of milk and mixed 3 L of water in it. If the price per kg of the mixture becomes ₹ 22, what is cost price of the milk per litre? **[CDS 2012]**
   (a) ₹ 28.00   (b) ₹ 26.40
   (c) ₹ 24.00   (d) ₹ 22.60

7. In a mixture of 150 L, the ratio of milk to water is 2 : 1. What amount of water should be further added to the mixture so as to make the ratio of the milk to water 1 : 2 respectively? **[NICL AO (Pre) 2017]**
   (a) 145 L (b) 160 L (c) 180 L (d) 150 L
   (e) Other than those given as options

8. 300 g of sugar solution has 40% of sugar in it. How much sugar should be added to make it 50% in the solution?
   [SSC CGL (Mains) 2015]
   (a) 80 g (b) 60 g (c) 120 g (d) 40 g

9. 25 kg of alloy X is mixed with 125 kg of alloy Y. If the amount of lead and tin in the alloy X is in the ratio 1 : 2 and the amount of lead and tin in the alloy Y is in the ratio 2 : 3, then what is the ratio of lead to tin in the mixture? **[CDS 2017 (I)]**
   (a) 1 : 2 (b) 2 : 3 (c) 3 : 5 (d) 7 : 11

10. A vessel of 160 L is filled with milk and water. 70% of milk and 30% of water is taken out of the vessel. It is found that the vessel is vacated by 55%. Find the quantity of milk and water in this mixture.
    (a) Milk = 100 L; Water = 60 L
    (b) Milk = 50 L; Water = 110 L
    (c) Milk = 70 L; Water = 90 L
    (d) Milk = 60 L; Water = 100 L
    (e) None of the above

11. A vessel contains liquids P and Q in the ratio 5 : 3. If 16 L of the mixture are removed and the same quantity of liquid Q is added, the ratio becomes 3 : 5. What quantity does the vessel hold?
    [RRB Clerk (Pre) 2017]
    (a) 35 L (b) 45 L (c) 40 L (d) 50 L
    (e) None of these

12. Jar A contains 'X' L of pure milk only. A 27 L mixture of milk and water in the respective ratio of 4 : 5, is added to jar A. The new mixture thus formed in jar A contains 70% milk, what is the value of X? **[IBPS Clerk (Pre) 2016]**
    (a) 23 L (b) 30 L (c) 27 L (d) 48 L
    (e) 28 L

13. A jar has 40 L milk. From the jar, 8 L of milk was taken out and replaced by an equal quantity of water. If 8 L of the newly formed mixture is taken out of the jar, what is the final quantity of milk left in the jar? **[IBPS SO 2016]**
    (a) 32.5 L   (b) 30 L
    (c) 25.6 L   (d) 24.2 L
    (e) 24 L

14. Tea worth ₹ 126 per kg and ₹ 135 per kg are mixed with a third variety in the ratio 1 : 1 : 2. If the mixture is worth ₹ 153 per kg, the price of the third variety per kg will be [SSC (10+2) 2012]

(a) ₹ 169.50 (b) ₹ 170.0
(c) ₹ 175.50 (d) ₹ 180.0

15. From a container having pure milk, 20% is replaced by water and the process is repeated thrice. At the end of the third operation, purity of the milk is [SSC CGL 2014]

(a) 45% (b) 56% (c) 51.2% (d) 48.8%

16. The respective ratio of milk and water in the mixture is 4 : 3 respectively. If 6 L of water is added to this mixture, the respective ratio of milk and water becomes 8 : 7. What is the quantity of milk in the original mixture? [IBPS Clerk 2015]

(a) 96 L (b) 36 L (c) 84 L (d) 48 L
(e) None of these

17. Two containers of equal capacity are full of mixture of oil and water. In the first, the ratio of oil to water is 4 : 7 and in the second, it is 7 : 11. Now, both the mixtures are mixed in a bigger container. What is the resulting ratio of oil to water? [XAT 2015]

(a) 149 : 247 (b) 247 : 149
(c) 143 : 241 (d) 241 : 143
(e) None of these

Exercise 2 *Higher Skill Level Questions*

1. A butler stole wine from a butt of sherry which contained 80% of spirit and he replaced it by wine containing only 32% spirit. Then, the butt was of 48% strength only. How much of the butt did he steal? [UP Police 2007]

(a) 1/4 (b) 3/5 (c) 2/5 (d) 2/3

2. How many kilograms of tea worth ₹ 25 per kg must be blended with 30 kg of tea worth ₹ 30 per kg, so that by selling the blended variety at ₹ 30 per kg, there should be a gain of 10%?

(a) 36 kg (b) 40 kg (c) 32 kg (d) 42 kg

3. A trader mixes 14 kg rice of variety A which costs ₹ 60 per kg with 18 kg of quantity of type B rice. He sells the mixture at ₹ 65 per kg and earns a profit of $\dfrac{100}{3}$%. Then, what was the cost price of type B rice? [RRB PO (Pre) 2017]

(a) 30 (b) 20 (c) 40 (d) 50
(e) 45

4. Jar A has 'X' L of mixture of apple juice and water in the respective ratio of 5 : 1 and jar B has 'X' L of mixture of mango juice and water in the respective ratio of 2 : 1. 30 L mixture was taken from jar A and jar B each and mixed in jar C. If 12 L of mixture was taken out from jar what was the final quantity of water in jar C? [IBPS Clerk (Mains) 2017]

(a) 5 L (b) 6 L (c) 4 L (d) 9 L
(e) 12 L

5. A milkman claims to sell milk at its cost price only but he is making a profit of 20%, since he has mixed some amount of water in the milk. What is the percentage of milk in the mixture? [CDS 2015 (I)]

(a) 80% (b) $\dfrac{250}{3}$% (c) 75% (d) $\dfrac{200}{3}$%

6. A container is filled with liquid, 6 part of which are water and 10 part milk. How much of the mixture must be drawn off and replaced with water so that the mixture may be half water and half milk?

(a) $\dfrac{1}{3}$ (b) $\dfrac{1}{7}$ (c) $\dfrac{1}{5}$ (d) $\dfrac{1}{8}$

7. A jar contains a mixture of milk and water in the respective ratio of 3 : 1. When 4 L of the mixture is taken out and thereafter 3 L of milk is added to the remaining mixture, the respective ratio of milk and water in the resultant mixture thus formed is 4 : 1. What was the initial quantity of water in the mixture? [IBPS SO 2016]

(a) 1 L (b) 6 L (c) 4 L (d) 2 L
(e) 3 L

8. A vessel contains a mixture of milk and water in the respective ratio of 10 : 3. 26 L of this mixture was taken out and replaced with 10 L of water. If the resultant respective ratio of milk and water in the mixture was 5 : 2, what was the initial quantity of mixture in the vessel? [SBI PO (Pre) 2016]

(a) 143 L (b) 182 L (c) 169 L (d) 156 L
(e) 130 L

Mixture or Alligation / 325

9. A jar contains mixture of milk and water in the respective ratio of 3 : 1. 24 L of the mixture is taken out and 24 L of water was added to it. If the resultant ratio between milk and water in the jar was 2 : 1, what was the initial quantity of mixture in the jar? **[SBI PO (Pre) 2017]**
   (a) 160 L (b) 180 L (c) 200 L (d) 250 L
   (e) 216 L

10. There are three bottles of mixture of syrup and water of ratios 2 : 3, 3 : 4 and 7 : 5. 10 L of first and 21 L of second bottles are taken. How much quantity from third bottle is to be taken so that final mixture from three bottles will be of ratios 1 : 1? **[SSC CGL (Mains) 2016]**
    (a) 25 L (b) 20 L (c) 35 L (d) 30 L

11. A vessel contains 100 L mixture of milk and water in the respective ratio of 22 : 3. 40 L of the mixture is taken out from the vessel and 4.8 L of each pure milk and pure water is added to the mixture. By what per cent is the quantity of water in the final mixture less than the quantity of milk? **[IBPS Clerk 2015]**
    (a) $78\frac{1}{2}$% (b) $79\frac{1}{8}$% (c) $72\frac{5}{6}$% (d) 76%
    (e) $77\frac{1}{2}$%

12. A vessel was containing 80 L of pure milk. 16 L of pure milk was taken out and replaced with equal amount of water. 16 L of newly formed mixture of water and milk was taken out and then 24 L of water was added to the mixture. What is the respective ratio between the quantity of milk and water in the final mixture? **[NICL AO 2016]**
    (a) 34 : 23 (b) 34 : 21
    (c) 28 : 23 (d) 32 : 21
    (e) 32 : 23

Answer with Solutions

Exercise 1 — Base Level Questions

1. (d) According to the rule of alligation,

   Cheaper price (₹ 2.04) — Mean price (₹ 2.52) — Dearer price (₹ 2.88)

   (2.88 − 2.52) = ₹ 0.36   (2.52 − 2.04) = ₹ 0.48

   ∴ Required ratio = 0.36 : 0.48 = 3 : 4

2. (a) As water is available free of cost, so its cost price = ₹ 0
   According to the rule of alligation,

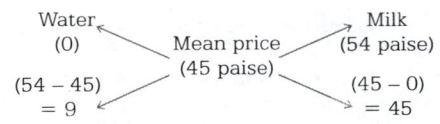

   Water (0) — Mean price (45 paise) — Milk (54 paise)

   (54 − 45) = 9   (45 − 0) = 45

   Water : Milk = 9 : 45 = 1 : 5
   ∴ Quantity of milk = 5 × 8 = 40 L

3. (d) According to the rule of alligation,

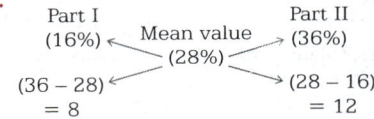

   Part I (16%) — Mean value (28%) — Part II (36%)

   (36 − 28) = 8   (28 − 16) = 12

   Part I (16%) : Part II (36%) = 8 : 12 = 2 : 3
   ∴ Quantity sold at 16% profit
   $= \dfrac{a}{a+b} \times$ Total quantity $= \dfrac{2}{5} \times 2000 = 800$ kg

4. (d) SI on ₹ 8400 for 1 yr = ₹ 294
   ∴ Rate of interest $= \dfrac{100 \times 294}{8400 \times 1} = \dfrac{7}{2} = 3\dfrac{1}{2}$%

   According to the rule of alligation,

   Rate I (4%) — Mean price (7/2%) — Rate II (10/3%)

   [7/2 − (10/3)] = 1/6   (4 − 7/2) = 1/2

   Rate I (4%) : Rate II $\left(\dfrac{10}{3}\%\right) = \dfrac{1}{6} : \dfrac{1}{2} = 1 : 3$

   ∴ Money lent at 4% $= \dfrac{1}{4} \times 8400 = ₹ 2100$

5. (a) According to the rule of alligation,

   Pulse sold at profit (10)% — Mean value 7% — Pulse sold at loss (− 5)%

   7 − (− 5)% = 12%   (10 − 7)% = 3%

   [− ve sign indicates loss]

∴ Ratio of pulses sold at 10% profit and 5% loss = 12 : 3 = 4 : 1
∴ Quantity of pulse sold at 10% profit
$= \frac{4}{4+1} \times 50$
$= \frac{4}{5} \times 50 = 40$ kg
and quantity of pulse sold at 5% loss
$= \frac{1}{4+1} \times 50$
$= \frac{1}{5} \times 50 = 10$ kg

6. (b) Let the cost of milk be ₹ x per litre and the cost of water is ₹ 0 per litre.
Then, according to the rule of alligation,

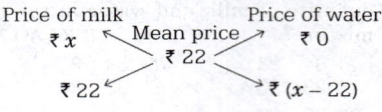

Price of milk   Price of water
₹ x Mean price   ₹ 0
       ₹ 22
₹ 22   ₹ $(x - 22)$

∴ $22 : (x - 22) = 15 : 3$
⇒ $\frac{22}{x - 22} = \frac{15}{3}$
⇒ $\frac{22}{x - 22} = 5$
⇒ $22 = 5x - 110$
⇒ $5x = 132$
∴ $x = ₹\ 26.40$

7. (d) Given, quantity of mixture = 150 L
Quantity of milk in mixture
$= \frac{2}{3} \times 150 = 100$ L
∴ Quantity of water = 150 − 100 L = 50 L
Now, for new mixture 1 part = 100 L
∴ 2 part = 200 L = water in new mixture
∴ Water to be added = (200 − 50) L
                     = 150 L

8. (b) By alligation rule,
Mixture I (300 g)   Mixture II
sugar 40%   Pure sugar
         50%   100%
         (Mean)
(100 − 50)%   (50 − 40)%
= 50%   = 10%
∴ Required ratio = 50 : 10 = 5 : 1 = 300 : 60
Hence, 60 g sugar should be added.
Alternate Method
40% sugar is in 300 g of sugar solution.
∴ Quantity of sugar = $\frac{300 \times 40}{100} = 120$ g
Let x g sugar should be added.
According to the question,
$120 + x = \frac{(300 + x) \times 50}{100}$

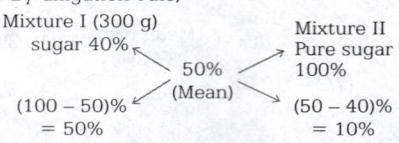

⇒ $\frac{120 + x}{300 + x} = \frac{1}{2}$ ⇒ $300 + x = 240 + 2x$
∴ $x = 60$ g

9. (d) Amount of lead in $X = \frac{1}{1+2} \times 25 = \frac{25}{3}$ g
Amount of tin in $X = \frac{2}{1+2} \times 25 = \frac{50}{3}$ kg
Amount of lead in $Y = \frac{2}{2+3} \times 125$
$= \frac{2 \times 125}{5} = 50$ kg
Amount of tin in $Y = \frac{3}{2+3} \times 125$
$= \frac{3}{5} \times 125 = 75$ kg
When X and Y are mixed, then
Amount of lead $= \frac{25}{3} + 50 = \frac{175}{3}$ kg
Amount of tin $= \frac{50}{3} + 75 = \frac{275}{3}$ kg
∴ Ratio of lead to tin in the mixture
$= \frac{175}{3} : \frac{275}{3}$
$= 175 : 275 = 7 : 11$

10. (a) Here, the percentage value of water and milk that is taken from the vessel should be taken into consideration.
Percentage of milk   Percentage of water
(30)   (70)
             45
(70 − 45) = 25   (45 − 30) = 15
Milk : Water = 25 : 15 = 5 : 3
∴ Quantity of milk in the mixture
$= \frac{5}{8} \times 160 = 100$ L
and quantity of water $= \frac{3}{8} \times 160 = 60$ L

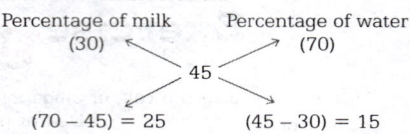

11. (c) Let the quantity of liquids P and Q be $5x$ and $3x$ litres respectively.
Quantity of P removed $= \frac{5}{5+3} \times 16 = 10$ L
Quantity of Q removed $= \frac{3}{5+3} \times 16 = 6$ L
Now, $\frac{5x - 10}{3x - 6 + 16} = \frac{3}{5}$
⇒ $25x - 50 = 9x + 30$
⇒ $16x = 80$
⇒ $x = 5$
∴ Quantity that vessel hold = 8 × 5 = 40 L

12. (a) Quantity of milk in 27 L mixture
$= \frac{4}{4+5} \times 27 = \frac{4}{9} \times 27 = 12$ L

and quantity of water in 27 L mixture
$$= \frac{5}{4+5} \times 27 = \frac{5}{9} \times 27 = 15 \text{ L}$$
Now, in jar A, milk = 70%
and in jar A, water = 30%
Now, according to the question, in jar A,
$$\frac{X+12}{15} = \frac{70}{30}\% = \frac{7}{3}$$
$\Rightarrow \quad 3X + 36 = 105$
$\Rightarrow \quad 3X = 105 - 36$
$\Rightarrow \quad 3X = 36 \Rightarrow X = 23 \text{ L}$

13. (c) In 40 L mixture of milk and water,
Quantity of milk = 32 L
and quantity of water = 8 L
∴ Ratio of milk to water in mixture
= 32 : 8 = 4 : 1
Now, 8 L mixture is taken out from the jar.
∴ Quantity of milk taken out from jar
$$= \frac{4}{5} \times 8 = 6.4 \text{ L}$$
Now, final quantity of milk left in the jar
= 32 − 6.4 = 25.6 L

14. (c) Let the cost of third variety tea will be ₹ x per kg.
According to the question,
$$\frac{126 + 135 + 2x}{4} = 153$$
$\Rightarrow \quad 261 + 2x = 4 \times 153$
$\Rightarrow \quad 2x = 612 - 261 = 351$
$\therefore \quad x = \frac{351}{2} = ₹ 175.50$
Hence, the cost of third variety tea is ₹ 175.50 per kg.

15. (c) Let in begining container having quantity of milk = 100 L
∴ 20% of 100 L = 20 L
Here, a = 100 L, b = 20 L, n = 3
∴ Net amount of milk at last
$$= a\left(1 - \frac{b}{a}\right)^n = 100\left(1 - \frac{20}{100}\right)^3$$
[by Technique 1]
$$= 100 \times \left(\frac{80}{100}\right)^3 = 100 \times \left(\frac{4}{5}\right)^3$$
$$= \frac{100 \times 4 \times 4 \times 4}{5 \times 5 \times 5} = \frac{256}{5} = 51.2\%$$

16. (d) Let quantity of milk and water in original mixture be 4x and 3x, respectively. 6 L of water is added and the ratio of milk and water becomes 8 : 7.
$\therefore \quad \dfrac{4x}{3x+6} = \dfrac{8}{7}$
$\Rightarrow \quad 28x = 24x + 48$
$\Rightarrow \quad 4x = 48 \Rightarrow x = 12$

∴ Quantity of milk in original mixture
= 4x = 4 × 12 = 48 L
Fast Track Method
Here, a : b = 4 : 3 = 8 : 6, x = 6 L
and a : c = 8 : 7
∴ Quantity of milk in the original mixture
$$= \frac{ax}{c-b} \text{ L} \quad \text{[by Technique 2]}$$
$$= \frac{8 \times 6}{7-6} = 48 \text{ L}$$

17. (a) Given, ratio of oil and water in Ist container = 4 : 7
∴ Quantity of oil in Ist container = $\dfrac{4}{11}$
and quantity of water in Ist container = $\dfrac{7}{11}$
Similarly, ratio of oil and water in IInd container = 7 : 11
∴ Quantity of oil in IInd container = $\dfrac{7}{18}$
and quantity of water in IInd container = $\dfrac{11}{18}$
Now, after pouring both mixtures in one container, quantity of oil = $\dfrac{4}{11} + \dfrac{7}{18}$
$$= \frac{72 + 77}{198} = \frac{149}{198}$$
and quantity of water = $\dfrac{7}{11} + \dfrac{11}{18}$
$$= \frac{126 + 121}{198} = \frac{247}{198}$$
∴ Required ratio = $\dfrac{149}{198} : \dfrac{247}{198}$
= 149 : 247
Fast Track Method
Here, a = 4, b = 7, c = 7 and d = 11
∴ Ratio of oil to water in bigger container
$$= \left(\frac{a}{a+b} + \frac{c}{c+d}\right) : \left(\frac{b}{a+b} + \frac{d}{c+d}\right)$$
[by Technique 3]
$$= \left(\frac{4}{4+7} + \frac{7}{7+11}\right) : \left(\frac{7}{4+7} + \frac{11}{7+11}\right)$$
$$= \left(\frac{4}{11} + \frac{7}{18}\right) : \left(\frac{7}{11} + \frac{11}{18}\right)$$
$$= \left(\frac{72+77}{198}\right) : \left(\frac{126+121}{198}\right)$$
$$= \frac{149}{198} : \frac{247}{198} = 149 : 247$$

Exercise 2 — Higher Skill Level Questions

1. (d) According to the rule of alligation,

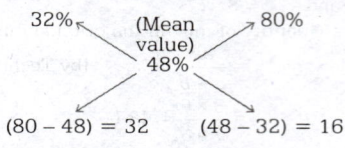

```
    32%              80%
        \  (Mean   /
         \ value) /
          \ 48%  /
         /      \
   (80 – 48) = 32    (48 – 32) = 16
```

∴ Required ratio = 32 : 16 = 2 : 1

Clearly, 1/3 of the butt of sherry was left and the butler stole 2/3 of the butt.

2. (a) SP = ₹ 30 per kg, gain = 10%

∴ CP = $\frac{30 \times 100}{110}$ = ₹ $\frac{300}{11}$ per kg.

By alligation rule,

```
   Type 1           Type 2
    25    (Mean)     30
          300/11
    30/11           25/11
```

Type 1 (₹ 25 per kg) : Type 2 (₹ 30 per kg)
= $\frac{30}{11} : \frac{25}{11}$ = 6 : 5

Now, 5 parts = 30 kg (Quantity of tea worth ₹ 30 per kg)

∴ 1 part = 6 kg
and 6 parts = 36 kg
∴ Quantity of tea worth ₹ 25 per kg = 36 kg

3. (c) Let cost price of mixture be y.

$$\left[\because \text{profit} = \frac{100}{3}\%\right]$$

So, $\frac{4}{3}y = 65$

⇒ $y = 48.75$

From mixture and alligation,

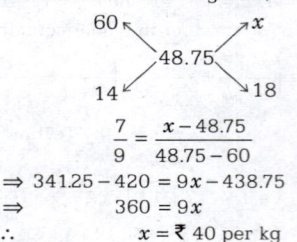

```
    60              x
       \          /
        48.75
       /          \
    14              18
```

$\frac{7}{9} = \frac{x - 48.75}{48.75 - 60}$

⇒ 341.25 − 420 = 9x − 438.75
⇒ 360 = 9x
∴ x = ₹ 40 per kg

4. (e) Given, ratio of jar A = 5 : 1 and ratio of jar B = 2 : 1
From jar A and jar B, 30 L of mixture was taken.
So, amount of apple juice = 25 L
amount of mango juice = 20 L
and amount of water = 5 L + 10 L = 15 L

Now, 12 L of mixture was taken out from jar C.

∴ Amount of water taken out = $\frac{15}{60} \times 12$ = 3 L

Hence, final quantity of water in jar C
= (15 − 3) L = 12 L

5. (b) Let CP of 1 L of milk be ₹ x.
∴ SP of 1 L of milk = $x \times 120\%$ = ₹ 1.2x
Now, as in ₹ 1.2x, the quantity of milk sold
= 1 L
∴ In ₹ x, quantity of milk sold
= $\frac{1}{1.2} \times x = \frac{5}{6}$ L

According to the question,
CP of milk and SP of mixture are same, therefore in mixture, quantity of milk must be $\frac{5}{6}$ L.

Hence, the required percentage
= $\frac{5}{6} \times 100\% = \frac{250}{3}\%$

6. (c) Let the container initially contains 16 L of liquid.

Let a L of water be added to the mixture.
Quantity of water in the new mixture
= $\left(6 - \frac{6a}{16} + a\right)$ L

Quantity of milk in the new mixture
= $\left(10 - \frac{10a}{16}\right)$ L

According to the question,
$6 - \frac{6a}{16} + a = 10 - \frac{10a}{16}$

⇒ 96 − 6a + 16a = 160 − 10a
⇒ 96 + 10a = 160 − 10a
⇒ 20a = 64 ⇒ $a = \frac{64}{20} = \frac{16}{5}$

∴ Part of mixture replaced = $\frac{1}{16} \times \frac{16}{5} = \frac{1}{5}$

7. (c) Let quantity of milk and water be 3x and x, respectively.
∴ Total mixture in jar = 3x + x = 4x
Now, quantity of milk in 4 L mixture
= $4 \times \frac{3x}{4x} = 3$ L

and quantity of water = 4 − 3 = 1 L
According to the question,
$\frac{3x - 3 + 3}{x - 1} = \frac{4}{1}$

⇒ 3x = 4x − 4
⇒ 4x − 3x = 4 ⇒ x = 4

Hence, the initial quantity of water in the mixture is 4 L.

Mixture or Alligation / 329

8. (d) Given, ratio of mixture = 10 : 3

∴ Quantity of milk in 26 L of mixture
$= \dfrac{10}{(10+3)} \times 26 = 20$ L

and quantity of water in 26 L of mixture
$= 26 - 20 = 6$ L

Let quantity of milk initially $= 10k$
and quantity of water initially $= 3k$
Now, 26 L of mixture is taken out and 10 L water is added in the mixture $= 10k - 20$
and quantity of water in mixture
$= 3k - 6 + 10 = (3k + 4)$
Now, according to the question,
$\dfrac{10k - 20}{(3k + 4)} = \dfrac{5}{2}$

$\Rightarrow \quad 20k - 40 = 15k + 20$
$\Rightarrow \quad 20k - 15k = 20 + 40$
$\Rightarrow 5k = 60 \Rightarrow k = 12$

∴ Total quantity of mixture $= 10k + 3k$
$= 13k = 13 \times 12 = 156$ L

9. (e) Let quantity of milk in mixture $= 3x$
and quantity of water in mixture $= x$
Now, 24 L of mixture is taken out from jar.

∴ Quantity of milk taken out
$= 24 \times \dfrac{3x}{(3x + x)} = \dfrac{24 \times 3x}{4x} = 18$ L

and quantity of water taken out
$= \dfrac{24 \times x}{3x + x} = \dfrac{24 \times x}{4x} = 6$ L

According to the question,
$\dfrac{3x - 18}{x - 6 + 24} = \dfrac{2}{1} \Rightarrow \dfrac{3x - 18}{x + 18} = \dfrac{2}{1}$

$\Rightarrow \qquad 3x - 18 = 2x + 36$
$\Rightarrow \qquad 3x - 2x = 36 + 18 \Rightarrow x = 54$

∴ Initial quantity of mixture in jar
$= 3x + x = 4x = 4 \times 54 = 216$ L

10. (d) In first bottle, syrup $= \dfrac{2}{5} \times 10 = 4$ L

Water $= \dfrac{3}{5} \times 10 = 6$ L

In second bottle, syrup $= \dfrac{3}{7} \times 21 = 9$ L

Water $= \dfrac{4}{7} \times 21 = 12$ L

Let mixture taken out from third bottle $= x$ L

∴ In third bottle, syrup $= \dfrac{7}{12} \times x = \dfrac{7x}{12}$ L

Water $= \dfrac{5}{12} \times x = \dfrac{5x}{12}$ L

According to the question,
$\dfrac{4 + 9 + \dfrac{7x}{12}}{6 + 12 + \dfrac{5x}{12}} = \dfrac{1}{1}$

$\Rightarrow \quad 13 + \dfrac{7x}{12} = 18 + \dfrac{5x}{12}$

$\Rightarrow \quad \dfrac{7x}{12} - \dfrac{5x}{12} = 18 - 13$

$\Rightarrow \quad \dfrac{2x}{12} = 5 \Rightarrow x = \dfrac{5 \times 12}{2}$

$\Rightarrow \qquad x = 30$

11. (b) Given, total quantity of mixture = 100 L
Ratio of milk and water = 22 : 3
Quantity of milk after removing 40 L of mixture and adding 4.8 L milk
$= 100 \times \dfrac{22}{25} - 40 \times \dfrac{22}{25} + 4.8$
$= 88 + 4.8 - 35.2$
$= 92.8 - 35.2 = 57.6$ L

Quantity of water
$= 100 \times \dfrac{3}{25} - 40 \times \dfrac{3}{25} + 4.8$
$= 12 + 4.8 - 4.8 = 12$ L

∴ Percentage of quantity of water less than the quantity of milk
$= \dfrac{57.6 - 12}{57.6} \times 100\% = \dfrac{45.6}{57.6} \times 100$
$= 79.166\% \approx 79\dfrac{1}{8}\%$

12. (e) Given, total quantity of pure milk = 80 L
When 16 L of pure milk is taken out, then quantity of pure milk = 64 L
and quantity of water = 16 L

∴ Ratio = 64 : 16 = 4 : 1
Now, quantity of milk in new mixture
$= 80 \times \dfrac{4}{5} - 16 \times \dfrac{4}{5} = \dfrac{320 - 64}{5} = \dfrac{256}{5}$ L

and quantity of water in new mixture
$= 80 \times \dfrac{1}{5} - 16 \times \dfrac{1}{5} + 24$
$= \dfrac{64}{5} + 24 = \dfrac{184}{5}$

∴ Liquid ratio $= \dfrac{256}{5} : \dfrac{184}{5} = 32 : 23$

Chapter 19

Partnership

When two or more persons make an association and invest money for running a certain business and after certain time receive profit in the ratio of their invested money and time period of investment, then such an association is called **partnership** and the persons involved in the partnership are called **partners**.

Types of Partnership

There are two types of partnership, which are as follow

1. Simple Partnership

If all partners invest their different capitals (money) for the same time period or same capital for different time period, then their profit or loss is in the ratio of their investments or time period of investment, then such a partnership is called simple partnership.

2. Compound Partnership

If all partners invest their different capitals (money) for different time period, then their profit not only depends on their investments but also on the time period of their investment, then such a partnership is called compound partnership.

Types of Partners

There are two types of partners, which are as follow

1. Active or Working Partner

A partner who not only invests money, but also take part in the business activities for which he draws a defined salary or gets some share from profit before its division is called an active partner.

2. Sleeping Partner

A partner who only invests money and does not take part in business activities is called sleeping partner.

Case I Ratio of Profit/Loss in Case of Simple Partnership

(i) When the investments made by all the partners $X_1, X_2, X_3, \ldots$ are for the same time period, then profit or loss is distributed amongst them in the ratio of their investments,

i.e. Ratio of profit/loss = Ratio of investments

$\Rightarrow \quad P_1 : P_2 : P_3 : \ldots = X_1 : X_2 : X_3 : \ldots$

(ii) When the amount of capital invested by different partners is same for different time periods, $t_1, t_2, t_3, \ldots$, then profit or loss is distributed amongst them in the ratio of their time periods.

i.e. Ratio of profit/loss = Ratio of time period for which the capital is invested

$\Rightarrow \quad P_1 : P_2 : P_3 : \ldots = t_1 : t_2 : t_3 : \ldots$

Ex. 1 A and B start a business by investing ₹ 4000 and ₹ 12000, respectively. Find the ratio of their profits after 1 yr.

Sol. The gain will be distributed amongst A and B in the ratio of their investments.

i.e. $\dfrac{\text{Investment of } A}{\text{Investment of } B} = \dfrac{\text{Profit of } A}{\text{Profit of } B}$

Hence, ratio of their profits $= \dfrac{4000}{12000} = \dfrac{1}{3} = 1 : 3$

Ex. 2 A and B jointly start a business. The investment of A is equal to three times the investment of B. Find the share of A in the annual profit of ₹ 52000.

Sol. Let the share of $B = ₹\ x$. Then, the share of $A = ₹\ 3x$

∴ $\dfrac{\text{Profit of } A}{\text{Profit of } B} = \dfrac{\text{Investment of } A}{\text{Investment of } B}$

Hence, ratio of their profit $= \dfrac{3x}{x} = \dfrac{3}{1} = 3 : 1$

According to the question, A's share $= \dfrac{3}{3+1} \times 52000 = \dfrac{3}{4} \times 52000 = ₹\ 39000$

Ex. 3 A and B invested ₹ 24000 and ₹ 8000 for a period of 2 yr. After 2 yr, they earned ₹ 48000. What will be the shares of A and B out of this earning?

Sol. Here, A's share : B's share = A's investment : B's investment = 24000 : 8000 = 3 : 1

Now, let A's share $= 3x$ and B's share $= x$

According to the question, $3x + x = 48000 \Rightarrow 4x = 48000$

∴ $x = 12000$

Clearly, share of $B = x = ₹\ 12000$ and share of $A = 3x = 3 \times 12000 = ₹\ 36000$

Alternate Method

Here, A's share : B's share = A's investment : B's investment = 24000 : 8000 = 3 : 1

Now, A's share $= \dfrac{3}{3+1} \times 48000 = 3 \times 12000 = ₹\ 36000$

and B's share $= \dfrac{1}{3+1} \times 48000 = 1 \times 12000 = ₹\ 12000$

Ex. 4 A, B and C start a business with investment of ₹ 50000 each. A remains in partnership for 9 months, B for 6 months and C for 12 months. Then, find the ratio of their profits.

Sol. Since, the ratio of profits of A, B and C will be in the ratio of time period of investment.

So, A's profit : B's profit : C's profit = 9 : 6 : 12 = 3 : 2 : 4

Case II Ratio of Profit/Loss in Case of Compound Partnership

When capital invested by the partners is given as $X_1, X_2, X_3, \ldots$ for different time period $t_1, t_2, t_3, \ldots$ in a business, then

Ratio of their profits $P_1 : P_2 : P_3 : \ldots$ = Amount of capital invested × Time period for which the capital is invested = $X_1 t_1 : X_2 t_2 : X_3 t_3 : \ldots$

To find the amount of rent paid by the person for using the piece of land or any other property for different time periods, where the total amount of rent is given, then the concept of compound partnership is used.

Ex. 5 A starts a business with ₹ 2000 and B joins him after 3 months with ₹ 8000. Find the ratio of their profits at the end of the year.

Sol. Here, A invested for 1 yr (12 months) and B invested 3 months later. It means B invested for (12 − 3) months.

∵ Ratio of profit A and B = Capital invested × Time period of investment
∴ A's share : B's share = 2000 × 12 : 8000 × (12 − 3)
= 2 × 12 : 8 × 9 = 24 : 72 = 1 : 3

Ex. 6 There are twelve friends A, B, C, D, E, F, G, H, I, J, K and L who invested money in some business in the ratio of 1 : 2 : 3 : 4 : 5 : 6 : 7 : 8 : 9 : 10 : 11 : 12 and the duration for which they invested the money is in the ratio of 12 : 11 : 10 : 9 : 8 : 7 : 6 : 5 : 4 : 3 : 2 : 1, respectively. Who will get the maximum profit at the end of the year?   [CDS 2016 (II)]

Sol. We have, A, B, C, D, E, F, G, H, I, J, K and L invested in 1 : 2 : 3 : 4 : 5 : 6 : 7 : 8 : 9 : 10 : 11 : 12.

Ratio of duration in an year are 12 : 11 : 10 : 9 : 8 : 7 : 6 : 5 : 4 : 3 : 2 : 1.

∴ Ratio of total money invested in duration in any year
= 1 × 12 : 2 × 11 : 3 × 10 : 4 × 9 : 5 × 8 : 6 × 7 : 7 × 6 : 8 × 5 : 9 × 4
: 10 × 3 : 11 × 2 : 12 × 1
= 12 : 22 : 30 : 36 : 40 : 42 : 42 : 40 : 36 : 30 : 22 : 12

Thus, F and G get the maximum profit at the end of year.

Ex. 7 A starts a business with ₹ 4000 and B joins the business 4 months later with an investment of ₹ 5000. After 1 yr, they earn a profit of ₹ 22000. Find the share of A and B.

Sol. A's share : B's share = 4000 × 12 : 5000 × (12 − 4) = 4 × 12 : 5 × 8 = 6 : 5
Now, let the share of A = 6x and the share of B = 5x
According to the question, 6x + 5x = 22000
⇒ 11x = 22000 ⇒ x = ₹ 2000
∴ Share of A = 6x = 6 × 2000 = ₹ 12000
and share of B = 5x = 5 × 2000 = ₹ 10000

Alternate Method

Here, A's share : B's share = 4000 × 12 : 5000 × 8 = 6 : 5

Now, A's share = $\dfrac{6}{6+5}$ × 22000 = ₹ 12000

and B's share = $\dfrac{5}{6+5}$ × 22000 = ₹ 10000

Fast Track Techniques to solve the QUESTIONS

Technique 1 If $x_1 : x_2 : x_3 : ...$ is the ratio of investments and $P_1 : P_2 : P_3 : ...$ is the ratio of profits, then the ratio of time periods of investment is given by

$$\frac{P_1}{x_1} : \frac{P_2}{x_2} : \frac{P_3}{x_3} : ....$$

Ex. 8 A, B and C invested capitals in the ratio of 2 : 3 : 5. At the end of the business terms, they received the profit in the ratio of 5 : 3 : 12. Find the ratio of time for which they contributed their capitals.

Sol. Let required ratio of time be $t_1 : t_2 : t_3$.
Then, ratio of product of time and investment = Ratio of profits
$\Rightarrow \qquad 2t_1 : 3t_2 : 5t_3 = 5 : 3 : 12$

Taking first two terms of the ratio, we get

$$\frac{2t_1}{3t_2} = \frac{5}{3} \Rightarrow \frac{t_1}{t_2} = \frac{5}{2} = \frac{25}{10} \Rightarrow t_1 : t_2 = 25 : 10 \qquad ...(i)$$

Taking last two terms of the ratio, we get

$$\frac{3t_2}{5t_3} = \frac{3}{12} \Rightarrow \frac{t_2}{t_3} = \frac{5}{12} = \frac{10}{24}$$

$\Rightarrow \qquad t_2 : t_3 = 10 : 24 \qquad ...(ii)$

From Eqs. (i) and (ii), we get

$$t_1 : t_2 : t_3 = 25 : 10 : 24$$

which is the required ratio.

Fast Track Method
Here, $P_1 : P_2 : P_3 = 5 : 3 : 12$ and $x_1 : x_2 : x_3 = 2 : 3 : 5$

$\therefore$ Required ratio $= \dfrac{P_1}{x_1} : \dfrac{P_2}{x_2} : \dfrac{P_3}{x_3}$

$= \dfrac{5}{2} : \dfrac{3}{3} : \dfrac{12}{5} = \dfrac{5}{2} : 1 : \dfrac{12}{5}$

$= \dfrac{5 \times 10}{2} : 1 \times 10 : \dfrac{12 \times 10}{5}$

$= 25 : 10 : 24$

Technique 2 If $P_1 : P_2 : P_3 : ...$ is the ratio of profits and $t_1 : t_2 : t_3 : ...$ is the ratio of time periods, then ratio of investments is given by

$$\frac{P_1}{t_1} : \frac{P_2}{t_2} : \frac{P_3}{t_3} : ....$$

Ex. 9 A, B and C each does certain investments for time periods in the ratio of 5 : 6 : 8. At the end of the business terms, they received the profits in the ratio of 5 : 3 : 12. Find the ratio of investments of A, B and C.

Sol. Let the required ratio of investments be $I_1 : I_2 : I_3$.
∴ Ratio of product of time and investment = Ratio of profits
$$\Rightarrow \quad 5I_1 : 6I_2 : 8I_3 = 5 : 3 : 12$$
Taking first two terms of ratio, we get
$$\frac{5I_1}{6I_2} = \frac{5}{3} \Rightarrow \frac{I_1}{I_2} = \frac{2}{1}$$
$$\Rightarrow \quad I_1 : I_2 = 2 : 1 \qquad \ldots(i)$$
Taking last two terms of ratio, we get
$$\frac{6I_2}{8I_3} = \frac{3}{12} \Rightarrow \frac{I_2}{I_3} = \frac{1}{3}$$
$$\Rightarrow \quad I_2 : I_3 = 1 : 3 \qquad \ldots(ii)$$
From Eqs. (i) and (ii), we get
$$I_1 : I_2 : I_3 = 2 : 1 : 3$$

Fast Track Method

Here, $t_1 : t_2 : t_3 = 5 : 6 : 8$ and $P_1 : P_2 : P_3 = 5 : 3 : 12$

∴ Required ratio = $\dfrac{P_1}{t_1} : \dfrac{P_2}{t_2} : \dfrac{P_3}{t_3}$

$$= \frac{5}{5} : \frac{3}{6} : \frac{12}{8} = 1 : \frac{1}{2} : \frac{3}{2} = 2 : 1 : 3$$

Multi Concept Questions

1. A started a business with ₹ 52000 and after 4 months, B joined him with ₹ 39000. At the end of the year, out of the total profit B recieved total ₹ 20000 including 25% of the profits as commission for managing the business. What amount did A receive?
(a) ₹ 20500 (b) ₹ 21000 (c) ₹ 20000 (d) ₹ 30000

➥ (c) Ratio of profits of A and B = Amount of capital invested × Time period of investment
= 52000 × 12 : 39000 × 8 = 2 : 1
Let the total profit = ₹ x
B receives 25% as commission for managing business. Then, the remaining 75% of the total profit x is shared between A and B in the ratio 2 : 1.
Hence, B will get $\frac{1}{3}$rd part of 75% in addition to his commission.
Hence, B's total earning = 25% of total profit + $\frac{1}{3}$ × 75% of total profit
$$= 0.25x + \frac{1}{3} \times (0.75x)$$
⇒ $20000 = 0.25x + 0.25x$
⇒ $0.5\,x = 20000$
∴ $x = 40000$
So, total profit = ₹ 40000
Hence, A's total earning = Total profit − B's total earning = 40000 − 20000 = ₹ 20000
∴ A's profit = ₹ 20000

2. A, B and C started a business with their investments in the ratio 1 : 2 : 4. After 6 months, A invested the half amount more as before and B invested twice the amount more as before, while C withdraws $\frac{1}{4}$th of their investments. Find the ratio of their profits at the end of the year.
(a) 5 : 10 : 15 (b) 15 : 20 : 25 (c) 5 : 16 : 20 (d) 5 : 16 : 14

➥ (d) Let us assume their initial investments be $x, 2x$ and $4x$, respectively.
According to the question,
Investment of A for 1yr = x for 6 months + $\left(x + \frac{x}{2}\right)$ for 6 months
$$= 6x + \frac{3x}{2} \times 6 = 6x + 9x = 15x$$
Investment of B for 1yr = $2x$ for 6 months + $(2x + 4x)$ for 6 months
$$= 12x + 36x = 48x$$
Investment of C for 1yr = $4x$ for 6 months + $\left(4x - \frac{4x}{4}\right)$ for 6 months
$$= 24x + (3x \times 6)$$
$$= 24x + 18x = 42x$$
∴ A : B : C = $15x : 48x : 42x$ = 15 : 48 : 42 = 5 : 16 : 14
Hence, the ratio of their profits are 5 : 16 : 14.

Fast Track **Practice**

Exercise ❶ Base Level Questions

1. Ravi and Kavi start a business by investing ₹ 8000 and ₹ 72000, respectively. Find the ratio of their profits at the end of year.
   (a) 2 : 9   (b) 5 : 9
   (c) 7 : 9   (d) 1 : 9
   (e) None of these

2. P and Q entered a partnership for 3 yr. At the start of the business, they invested ₹ 13000 and ₹ 25000, respectively. At the end of 3 yr, their total profit was ₹ 76000. What will be share of Q out of this profit?
   (a) ₹ 50000   (b) ₹ 26000
   (c) ₹ 55000   (d) ₹ 21000
   (e) None of these

3. Rajan and Sajan started a business initially with ₹ 14200 and ₹ 15600, respectively. If total profits at the end of year is ₹ 74500, then what is the Rajan's share in the profit?   [Bank Clerk 2009]
   (a) ₹ 39000   (b) ₹ 39600
   (c) ₹ 35000   (d) ₹ 35500
   (e) None of these

4. A, B and C invested ₹ 45000, ₹ 90000 and ₹ 90000, respectively to start a business. At the end of two years, they earned a profit of ₹ 164000. What will be B's share in the total profit?   [IBPS Clerk 2015]
   (a) ₹ 56000   (b) ₹ 36000
   (c) ₹ 72000   (d) ₹ 65600
   (e) ₹ 59000

5. Srikant and Vividh started a business investing amounts of ₹ 185000 and ₹ 225000, respectively. If Vividh's share in the profit earned by them is ₹ 9000, then what is the total profit earned by them together?   [Bank Clerk 2009]
   (a) ₹ 17400   (b) ₹ 16400
   (c) ₹ 16800   (d) ₹ 17800
   (e) None of these

6. P, Q and R start a business. P invests 3 times as much as Q invests and Q invests $\frac{2}{3}$rd as much as R invests. Find the ratio of capitals of P, Q and R.
   (a) 3 : 2 : 6   (b) 2 : 6 : 3
   (c) 6 : 2 : 3   (d) 5 : 2 : 3

7. P, Q and R start a business jointly. Twice the capital of P is equal to thrice the capital of Q and the capital of Q is four times the capital of R. Find the share of Q in an annual profit of ₹ 148500.
   (a) ₹ 54000   (b) ₹ 64000
   (c) ₹ 56000   (d) ₹ 55000
   (e) None of these

8. A and B invested in a business in the ratio 5 : 2. If 9% of total profit goes to charity and A's share is ₹ 650, then the total profit is   [WBSSC CGL 2014]
   (a) ₹ 1000   (b) ₹ 1200
   (c) ₹ 900   (d) ₹ 1500

9. A and B enter into a partnership by making investments in the ratio 1 : 2, 5% of the total profit goes to charity. If B's share is ₹ 760, then what is the total profit earned?   [SBI Clerk (Mains) 2016]
   (a) ₹ 1200   (b) ₹ 1800
   (c) ₹ 2400   (d) ₹ 1560
   (e) ₹ 2000

10. A starts a business with ₹ 9000 and B joins him after 6 months with an investment of ₹ 45000. What will be the ratio of the profits of A and B at the end of year?   [Hotel Mgmt. 2008]
    (a) 1 : 5   (b) 5 : 2
    (c) 2 : 5   (d) 5 : 1

11. Ajay started a business investing ₹ 25000. After 3 months, Vijay joined him with a capital of ₹ 30000. At the end of the year, they made a profit of ₹ 38000. What will be the Ajay's share in the profit?
    (a) ₹ 10000   (b) ₹ 18000
    (c) ₹ 15000   (d) ₹ 20000

12. A and B invest in the ratio of 3 : 5, respectively. After 6 months, C enters the business with the investment of the capital equal to that of B. What will be the ratio of the profits of A, B and C at the end of year?   [SSC LDC 2008]
    (a) 6 : 10 : 5   (b) 3 : 5 : 5
    (c) 3 : 5 : 2   (d) 6 : 2 : 3

Partnership / 337

13. Ramesh and Priya started a business initially with ₹ 5100 and ₹ 6600, respectively. Investments done by both the persons are for different time periods. If the total profit is ₹ 5460, then what is the profit of Ramesh? **[Bank PO 2010]**
    (a) ₹ 1530   (b) ₹ 1600
    (c) ₹ 1400   (d) Data inadequate
    (e) None of these

14. A started a business with an investment of ₹ 5000. After 2 months, B and C joined with ₹ 2500 and ₹ 3500, respectively. If total annual profit was ₹ 4800, what was B's share in the annual profit?
    [IBPS SO 2016]
    (a) ₹ 1150   (b) ₹ 1000
    (c) ₹ 1050   (d) ₹ 1820
    (e) ₹ 1200

15. M, N and P invest ₹ 50000 for a business. M invests ₹ 4000 more than N and N invests ₹ 5000 more than P. Out of the total profit of ₹ 70000, what is the share received by M?
    (a) ₹ 29400   (b) ₹ 30000
    (c) ₹ 35000   (d) ₹ 40000
    (e) None of these

16. A, B and C together start a business. B invests 1/6 of the total capital while investments of A and C are equal. If the annual profit on this investment is ₹ 33600, then find the difference between the profits of B and C. **[SSC CGL 2007]**
    (a) ₹ 8400   (b) ₹ 7200
    (c) ₹ 6000   (d) ₹ 9600

17. A and B started a joint business. A's investment was thrice the investment of B and the period of his investment was twice the period of investment of B. If B got ₹ 6000 as profit, then what will be the 20% of total profit? **[Hotel Mgmt. 2007]**
    (a) ₹ 5000   (b) ₹ 8400
    (c) ₹ 3500   (d) ₹ 4500

18. A and B entered into a partnership investing ₹ 16000 and ₹ 12000, respectively. After 3 months, A withdrew ₹ 5000 while B invested ₹ 5000 more. After 3 more months, C joins the business with a capital of ₹ 21000. The share of B exceeds that of C, out of a total profit of ₹ 26400 after 1 yr by **[SSC CGL 2015]**
    (a) ₹ 1200   (b) ₹ 2400
    (c) ₹ 4800   (d) ₹ 3600

19. A, B and C enter into a partnership. A invests some amount at the beginning, B invests double the amount of A after 6 months and C invests thrice the amount of A after 8 months. If the annual profit is ₹ 54000, then find the C's share.
    (a) ₹ 3000   (b) ₹ 18000
    (c) ₹ 15000   (d) ₹ 21000
    (e) None of these

20. A, B and C invested capitals in the ratio of 4 : 6 : 9. At the end of the business term, they received the profit in the ratio of 2 : 3 : 5. Find the ratio of their time for which they contributed their capitals.
    [Delhi Police SI 2008]
    (a) 1 : 1 : 9   (b) 2 : 2 : 9
    (c) 10 : 10 : 9 (d) 9 : 9 : 10

21. A, B and C do certain investments for time periods in the ratio of 2 : 1 : 8. At the end of the business term, they received the profits in the ratio of 3 : 4 : 2. Find the ratio of investments of A, B and C.
    (a) 6 : 16 : 1   (b) 2 : 5 : 1
    (c) 6 : 17 : 1   (d) 6 : 19 : 3

Exercise 2 *Higher Skill Level Questions*

1. Sonu invested 10% more than the investment of Mona and Mona invested 10% less than the investment of Raghu. If the total investment of all the three persons is ₹ 5780, then find the investment of Raghu. **[Bank PO 2010]**
    (a) ₹ 2010
    (b) ₹ 2000
    (c) ₹ 2100
    (d) ₹ 2210
    (e) None of the above

2. Aarti, Vinita and Kamla became partners in a business by investing money in the ratio of 5 : 7 : 6. Next year, they increased their investments by 26%, 20% and 15%, respectively. In what ratio should profit earned during 2nd year be distributed?
    (a) 21 : 28 : 23   (b) 23 : 28 : 21
    (c) 28 : 23 : 21   (d) 35 : 41 : 7

3. Avinash, Manoj and Arun started a business in partnership investing in the ratio of 3 : 2 : 5 respectively. At the end of the year, they earned a profit of ₹ 45000 which is 15% of their total investment. How much did Manoj invest?
    (a) ₹ 60000   (b) ₹ 180000
    (c) ₹ 30000   (d) ₹ 90000
    (e) None of these

4. A, B and C started a business by investing ₹ 20000, ₹ 28000 and ₹ 36000, respectively. After 6 months, A and B withdrew an amount of ₹ 8000 each and C invested an additional amount of ₹ 8000. All of them invested for equal period of time. If at the end of the year, C got ₹ 12550 as his share of profit, then what was the total profit earned?
[IBPS Clerk 2015]
(a) ₹ 25100 (b) ₹ 26600
(c) ₹ 24300 (d) ₹ 22960
(e) ₹ 21440

5. A and B started a business by investing ₹ 35000 and ₹ 20000, respectively. After 5 months, B left the business and C joined the business with a sum of ₹ 15000. The profit earned at the end of year is ₹ 84125. What is the share of B in profit? [Bank Clerk 2011]
(a) ₹ 14133
(b) ₹ 15000
(c) ₹ 13460
(d) Cannot be determined
(e) None of the above

6. Amitabh, Brijesh and Kamlesh enter into a partnership with shares in the ratio of 7/2 : 4/3 : 6/5. After 4 months, Amitabh increases his share by 50%. If the total profit at the end of the year is ₹ 21600, then what will be the share of Brijesh in the profit? [SSC (10+2) 2013]
(a) ₹ 8000 (b) ₹ 12000
(c) ₹ 4000 (d) ₹ 7000

7. A and B started a business with initial investments in the respective ratio of 18 : 7. After 4 months from the start of the business, A invest ₹ 2000 more and B invested ₹ 7000 more. At the end of one year, if the profit was distributed among them in the ratio of 2 : 1 respectively, what was the total initial investment with which A and B started the business? [IBPS Officer Grade 2015]
(a) ₹ 50000 (b) ₹ 25000
(c) ₹ 150000 (d) ₹ 75000
(e) ₹ 125000

8. A, B and C started a business together. The respective ratio of investments of A and B was 3 : 5 and the respective ratio of investments of B and C was 10 : 13. If at the end of the year, C received ₹ 5876 as his share of annual profit, what was the total annual profit earned by all of them together? [SBI Clerk 2015]
(a) ₹ 13108 (b) ₹ 12756
(c) ₹ 13224 (d) ₹ 12984
(e) ₹ 12188

9. A and B started a business with an investment of ₹ 2800 and ₹ 5400, respectively. After 4 months, C joined with ₹ 4800. If the difference between C's share and A's share in the annual profit was ₹ 400, what was the total annual profit? [IBPS PO 2016]
(a) ₹ 13110 (b) ₹ 12540
(c) ₹ 17100 (d) ₹ 11400
(e) ₹ 14250

10. A started a business by investing ₹ 25000. At the end of 4th month from the start of the business, B joined with ₹ 15000 and at the end of 6th month from the start of the business, C joined with ₹ 20000. If A's share in profit at the end of year was ₹ 7750, what was the total profit received? [SBI PO (Pre) 2016]
(a) ₹ 13950 (b) ₹ 13810
(c) ₹ 13920 (d) ₹ 12780
(e) ₹ 14040

11. Aayush and Babloo are partners in a business. Aayush contributes 1/4 of the capital for 15 months and Babloo received 2/3 of the profit. How long Babloo's money was used?
(a) 10 months (b) 9 months
(c) 11 months (d) 7 months
(e) None of these

12. A began a business with ₹ 2250 and was joined afterwards by B with ₹ 2700. If the profits at the end of the year were divided in the ratio of 2 : 1, after how much time B joined the business?
(a) 5 months (b) 6 months
(c) 3 months (d) 7 months
(e) None of these

13. Pihu and Rani start a business together by investing ₹ 9000 and ₹ 6300, respectively. After a certain period, Pihu withdrew from the business completely. If at the end of 2 yr, the total profit earned was ₹ 13050 and Pihu's share was ₹ 6750, after how many months did Pihu withdraw from business?
[SBI Clerk (Pre) 2016]
(a) 23 months (b) 14 months
(c) 20 months (d) 18 months
(e) 12 months

14. A started a business with an initial investment of ₹ 1200. 'X' month after the start of business, B joined A with an initial investment of ₹ 1500. If total profit was ₹ 1950 at the end of year and B's share of profit was ₹ 750. Find 'X'.
[RRB PO (Pre) 2017]

(a) 5 months (b) 6 months
(c) 7 months (d) 8 months
(e) 9 months

15. X and Y entered into partnership with ₹ 700 and ₹ 600 respectively. After 3 months, X withdrew 2/7 of his stock but after another 3 months, he puts back 3/5 of what he had withdrawn. The profit at the end of the year is ₹ 726. How much of this should X receive?

(a) ₹ 336 (b) ₹ 366 **[CDS 2016 (II)]**
(c) ₹ 633 (d) ₹ 663

16. A, B and C entered into partnership in a business. A got 3/5 of the profit. B and C distributed the remaining profit equally. If C got ₹ 400 less than A, then the total profit was **[SSC CPO 2013]**

(a) ₹ 1600 (b) ₹ 1200
(c) ₹ 1000 (d) ₹ 800

17. X and Y make a partnership. X invests ₹ 8000 for 8 months and Y remains in the business for 4 months. Out of the total profit, Y claims 2/7 of the profit. How much money was contributed by Y?

(a) ₹ 5000 (b) ₹ 5400
(c) ₹ 7400 (d) ₹ 6400

18. A, B and C start a small business. A contributes one-fifth of the total capital invested in the business. B contributes as much as A and C together. Total profit at the end of the year was ₹ 5200. What was C's profit share? **[SBI PO 2015]**

(a) ₹ 1510 (b) ₹ 2510.
(c) ₹ 1500 (d) ₹ 2560
(e) ₹ 1560

19. A and B invest ₹ 3000 and ₹ 4000, respectively in a business. A receives ₹ 10 per month out of the profit as a remuneration for running the business and the rest of the profit is divided in proportion to the investments. If in a year, A totally receives ₹ 390, what does B receive? **[SSC CPO 2008]**

(a) ₹ 630 (b) ₹ 360
(c) ₹ 480 (d) ₹ 380

20. Anil is an active and Vimal is a sleeping partner in a business. Anil invests ₹ 12000 and Vimal invests ₹ 20000. Anil receives 10% profit for managing, the rest being divided in proportion to their capitals. Out of the total profit of ₹ 9000, the money received by Anil is **[Hotel Mgmt. 2010]**

(a) ₹ 4500 (b) ₹ 4800
(c) ₹ 4600 (d) ₹ 3937.50

21. Tom started a business with ₹ 52000 and after 4 months, Harry joined with ₹ 39000. At the end of the year out of the total profits, Harry received total ₹ 20000 including one-fourth of the profits as commission for managing the business. What profit did Tom receive? **[NICL AO (Pre) 2017]**

(a) ₹ 40000 (b) ₹ 20000
(c) ₹ 30000 (d) ₹ 50000
(e) Other than those given as options

22. A and B started a business by investing ₹ 18000 and ₹ 24000, respectively. At the end of 4th month from the start of the business, C joins with ₹ 15000. At the end of 8th month, B quits at which time C invests ₹ 3000 more. At the end of 10th month, B rejoins with the same investment. If profit at the end of the year is ₹ 12005, what is B's share of profit? **[LIC AAO 2014]**

(a) ₹ 4000 (b) ₹ 4440
(c) ₹ 4360 (d) ₹ 4900
(e) ₹ 3920

23. A and B started a business with ₹ 20000 and ₹ 35000, respectively. They agreed to share the profit in the ratio of their capital. C joins the partnership with the condition that A, B and C will share profit equally and pays ₹ 220000 as premium for this, to be shared between A and B. This is to be divided between A and B in the ratio of **[SSC CGL 2013]**

(a) 10 : 1 (b) 1 : 10
(c) 9 : 10 (d) 10 : 9

24. A sum of ₹ 15525 is divided among Sunil, Anil and Jamil such that if ₹ 22, ₹ 35 and ₹ 48 are diminished from their shares respectively, their remaining sums shall be in the ratio 7 : 10 : 13. What would have been the ratio of their sums in ₹ 16, ₹ 77 and ₹ 37 respectively were added to their original shares? **[SSC CGL (Mains) 2016]**

(a) 9 : 13 : 17 (b) 18 : 26 : 35
(c) 36 : 52 : 67 (d) None of these

Answer with Solutions

Exercise 1 — Base Level Questions

1. (d) Ratio of profits = Ratio of investments
= 8000 : 72000 = 1 : 9

2. (a) P's share : Q's share
= Ratio of their investments
= 13000 : 25000 = 13 : 25
Now, let P's share = $13x$
and Q's share = $25x$
According to the question,
$13x + 25x = 76000$
$\Rightarrow x = \dfrac{76000}{38} = 2000$
∴ Q's share = $25x = 25 \times 2000 = ₹ 50000$

Alternate Method
P's share : Q's share
= Ratio of their investments
= 13000 : 25000 = 13 : 25
∴ Q's share = $\dfrac{25}{13+25} \times 76000$
= $\dfrac{25}{38} \times 76000 = ₹ 50000$

3. (d) Ratio of profits = Ratio of investments
∴ Rajan's share : Sajan's share
= 14200 : 15600 = 142 : 156 = 71 : 78
Let Rajan's share = $71x$
and Sajan's share = $78x$
According to the question,
$71x + 78x = 74500$
$\Rightarrow 149x = 74500$
∴ $x = \dfrac{74500}{149} = 500$
∴ Rajan's share = $71x = 71 \times 500 = ₹ 35500$

Alternate Method
Ratio of profits of Rajan and Sajan
= Ratio of their investments
= 14200 : 15600 = 142 : 156 = 71 : 78
∴ Rajan's share = $\dfrac{71}{71+78} \times 74500$
= $\dfrac{71}{149} \times 74500 = ₹ 35500$

4. (d) This is the case of simple partnership.
Here, $R_1 = ₹ 45000$, $R_2 = ₹ 90000$
and $R_3 = ₹ 90000$
∴ Ratio of profits = Ratio of investments
$\Rightarrow P_1 : P_2 : P_3 = R_1 : R_2 : R_3$
= 45000 : 90000 : 90000
= 45 : 90 : 90 = 1 : 2 : 2

∴ Sum of profit ratios = 1 + 2 + 2 = 5
Given, total profit = ₹ 164000
∴ B's share = $\dfrac{\text{Part of } B}{\text{Sum of profit ratios}}$
= $\dfrac{2}{5} \times 164000 = ₹ 65600$

5. (b) As we know,
Ratio of profits = Ratio of investments
∴ Srikant's share : Vividh's share
= 185000 : 225000 = 37 : 45
Let Srikant's share = $37x$
and Vividh's share = $45x$
According to the question,
$45x = 9000$
∴ $x = \dfrac{9000}{45} = 200$
∴ Total profit = $37x + 45x = 82x$
= $82 \times 200 = ₹ 16400$

6. (c) Let investment of R = x
Then, investment of Q = $\dfrac{2x}{3}$
and investment of P = $2x$
∴ Ratio of capitals of P, Q and R
= $2x : \dfrac{2x}{3} : x = 6x : 2x : 3x = 6 : 2 : 3$

7. (a) Let R's capital = 1
Then, Q's capital = 4
and 2 (P's capital) = 3 (Q's capital)
= $3 \times 4 = 12$
∴ P's capital = $\dfrac{12}{2} = 6$
∴ P's share : Q's share : R's share = 6 : 4 : 1
Thus, Q's share profit
= $\dfrac{4}{6+4+1} \times 148500$
= $\dfrac{4}{11} \times 148500 = 4 \times 13500 = ₹ 54000$

8. (a) Let total profit be ₹ x.
After charity, remaining profit
= $x - \dfrac{x \times 9}{100} = ₹ \dfrac{91x}{100}$
Now, A's share = $\dfrac{5}{(5+2)} \times \dfrac{91x}{100}$
[∵ ratio of profits = ratio of investments]

According to the question,
$$650 = \frac{5}{7} \times \frac{91x}{100} \Rightarrow x = \frac{650 \times 7 \times 100}{5 \times 91}$$
$$= 1000$$

9. So, total profit is ₹ 1000.

10. (a) Let total profit earned = ₹ x
Given, B's share of profit = ₹ 760
$$\Rightarrow \frac{95x}{100} \times \frac{2}{3} = 760 \Rightarrow x = ₹ 1200$$

11. (c) A's share : B's share
$$= 9000 \times 12 : 45000 \times (12 - 6)$$
$$= 12 : 5 \times 6 = 2 : 5$$

(d) Ajay's share : Vijay's share
$$= 25000 \times 12 : 30000 \times (12 - 3)$$
$$= 25 \times 12 : 30 \times 9 = 10 : 9$$
$$\therefore \text{Ajay's share} = \frac{10}{10 + 9} \times 38000$$
$$= \frac{10}{19} \times 38000 = 10 \times 2000 = ₹ 20000$$

12. (a) Let investment of $A = 3x$
∴ Investment of $B = 5x$
    Investment of $C = 5x$
As we know,
Ratio of profits = Capital invested
                 × Time period of investment
∴ A's share : B's share : C's share
$$= 3x \times 12 : 5x \times 12 : 5x \times (12 - 6)$$
$$= 3x \times 12 : 5x \times 12 : 5x \times 6 = 6 : 10 : 5$$

13. (d) Time of investment is not given in the question. So, we cannot find the profit of Ramesh. Hence, the data is inadequate.

14. (b) Ratio of profits of A, B and C
$$= 5000 \times 12 : (12 - 2) \times 2500$$
$$: (12 - 2) \times 3500$$
$$= 60000 : 25000 : 35000$$
$$= 60 : 25 : 35 = 12 : 5 : 7$$
Now, share of B in annual profit
$$= \frac{5}{12 + 5 + 7} \times 4800$$
$$= \frac{5 \times 4800}{24} = ₹ 1000$$

15. (a) Let investment of $P = ₹ x$
Then, investment of $N = ₹ (x + 5000)$
and investment of $M = (x + 5000) + 4000$
$$= ₹ (x + 9000)$$
According to the question,
$$x + (x + 5000) + (x + 9000) = 50000$$
$$\Rightarrow 3x + 14000 = 50000$$
$$\Rightarrow 3x = 50000 - 14000 = 36000$$
$$\therefore x = \frac{36000}{3} = 12000$$
Clearly, investment of $P = ₹ 12000$

Investment of $N = (x + 5000)$
$$= 12000 + 5000 = ₹ 17000$$
Investment of $M = (x + 9000)$
$$= 12000 + 9000 = ₹ 21000$$
M's share : N's share : P's share
$$= 21000 : 17000 : 12000 = 21 : 17 : 12$$
Hence, M's share
$$= \frac{21}{21 + 17 + 12} \times 70000$$
$$= \frac{21}{50} \times 70000 = ₹ 29400$$

16. (a) Given, investment of B
$$= \frac{1}{6} \text{ of total capital}$$
∴ Investments of A and C each
$$= \frac{1}{2}\left(1 - \frac{1}{6}\right) \text{ of total capital}$$
$$= \frac{1}{2} \times \frac{5}{6} \text{ of total capital}$$
$$= \frac{5}{12} \text{ of total capital}$$
Now, A's share : B's share : C's share
$$= \frac{5}{12} : \frac{1}{6} : \frac{5}{12} = 5 : 2 : 5$$
Let   A's share = $5x$,
       B's share = $2x$
and   C's share = $5x$
According to the question,
$$5x + 2x + 5x = 33600 \Rightarrow 12x = 33600$$
$$\Rightarrow x = \frac{33600}{12} = 2800$$
∴ Difference in the profits of B and C
$$= 5x - 2x = 3x = 3 \times 2800$$
$$= ₹ 8400$$

17. (b) Let investment of B be x for y months.
Then, A's investment = $3x$ for $2y$ months
∴ $A : B = (3x \times 2y) : (x \times y)$
$$= 6xy : xy = 6 : 1$$
Let the total profit = m.
Then, $m \times \frac{1}{7} = 6000$
∴ $m = 6000 \times 7 = ₹ 42000$
∴ 20% of 42000 = ₹ 8400

18. (d) Ratio of investments of A, B and C
$$= (16000 \times 3 + 11000 \times 9)$$
$$: (12000 \times 3 + 17000 \times 9) : (21000 \times 6)$$
$$= 147000 : 189000 : 126000$$
$$= 147 : 189 : 126 = 7 : 9 : 6$$
Given, the total profit = ₹ 26400
Sum of profit ratio = 7 + 9 + 6 = 22
So, the share of B exceeds that of C by
$$\frac{3}{22} \times 26400 = ₹ 3600$$

342 / **Fast Track** Objective Arithmetic

19. (b) Let investment of $A = x$,
Investment of $B = 2x$
and investment of $C = 3x$
∴ A's share : B's share : C's share
$= (x \times 12) : (2x \times 6) : (3x \times 4)$
$= 12x : 12x : 12x = 1 : 1 : 1$
∴ C's share $= \dfrac{1}{1+1+1} \times 54000$
$= \dfrac{1}{3} \times 54000 = ₹ 18000$

20. (d) Let required ratio of time be $t_1 : t_2 : t_3$.
Then, ratio of product of time investments
= Ratio of profits
$\Rightarrow 4t_1 : 6t_2 : 9t_3 = 2 : 3 : 5$
Taking first two terms of the ratio,
$\dfrac{4t_1}{6t_2} = \dfrac{2}{3}$
$\Rightarrow \dfrac{t_1}{t_2} = \dfrac{12}{12} = \dfrac{1}{1}$
$\Rightarrow t_1 : t_2 = 9 : 9$
Taking last two terms of the ratio,
$\dfrac{6t_2}{9t_3} = \dfrac{3}{5} \Rightarrow \dfrac{t_2}{t_3} = \dfrac{9}{10}$
$\Rightarrow t_2 : t_3 = 9 : 10$
∴ $t_1 : t_2 : t_3 = 9 : 9 : 10$

Fast Track Method
Here, $P_1 : P_2 : P_3 = 2 : 3 : 5$
and $x_1 : x_2 : x_3 = 4 : 6 : 9$
∴ Required ratio
$= \dfrac{P_1}{x_1} : \dfrac{P_2}{x_2} : \dfrac{P_3}{x_3}$ [by Technique 1]
$= \dfrac{2}{4} : \dfrac{3}{6} : \dfrac{5}{9} = \dfrac{1}{2} : \dfrac{1}{2} : \dfrac{5}{9} = 9 : 9 : 10$

21. (a) Let the required ratio of investments be
$I_1 : I_2 : I_3$.
Ratio of product of time and investments
= Ratio of profits
$= 2I_1 : I_2 : 8I_3 = 3 : 4 : 2$
Taking first two terms of the ratio,
$\dfrac{2I_1}{I_2} = \dfrac{3}{4} \Rightarrow \dfrac{I_1}{I_2} = \dfrac{3}{8} = \dfrac{6}{16}$
$\Rightarrow I_1 : I_2 = 6 : 16$
Taking last two terms of the ratio,
$\dfrac{I_2}{8I_3} = \dfrac{4}{2} \Rightarrow \dfrac{I_2}{I_3} = \dfrac{16}{1}$
$\Rightarrow I_2 : I_3 = 16 : 1$
∴ $I_1 : I_2 : I_3 = 6 : 16 : 1$

Fast Track Method
Here, $t_1 : t_2 : t_3 = 2 : 1 : 8$
and $P_1 : P_2 : P_3 = 3 : 4 : 2$
∴ Required ratio $= \dfrac{P_1}{t_1} : \dfrac{P_2}{t_2} : \dfrac{P_3}{t_3}$
[by Technique 2]
$= \dfrac{3}{2} : \dfrac{4}{1} : \dfrac{2}{8} = \dfrac{3 \times 8}{2} : \dfrac{4 \times 8}{1} : \dfrac{2 \times 8}{8}$
$= 12 : 32 : 2 = 6 : 16 : 1$

Exercise 2 Higher Skill Level Questions

1. (b) Let share of Raghu be 100.
Then, share of Mona = 90
and share of Sonu = 99
Sonu's investment : Mona's investment
: Raghu's investment = 99 : 90 : 100
∴ Investment of Raghu
$= \dfrac{5780}{99 + 90 + 100} \times 100$
$= \dfrac{5780}{289} \times 100 = ₹ 2000$

2. (a) Let investment of Aarti during first year
$= 5x$
Investment of Vinita during first year = $7x$
Investment of Kamla during first year = $6x$
Then, their investments during second year
$= (126\% \text{ of } 5x) : (120\% \text{ of } 7x) : (115\% \text{ of } 6x)$
$= \left(\dfrac{126}{100} \times 5x\right) : \left(\dfrac{120}{100} \times 7x\right) : \left(\dfrac{115}{100} \times 6x\right)$
$= 630 : 840 : 690$
$= 21 : 28 : 23$

3. (a) Total investment
$= \dfrac{100}{15} \times 45000 = ₹ 300000$
∵ Investment of Avinash : Investment of Manoj : Investment of Arun = 3 : 2 : 5
∴ Investment of Manoj
$= \dfrac{2}{10} \times 300000$
$= ₹ 60000$

4. (a) ∵ Ratio of investment by A, B and C
$= (20000 \times 6 + 12000 \times 6)$
$: (28000 \times 6 + 20000 \times 6)$
$: (36000 \times 6 + 44000 \times 6)$
$= (120 + 72) : (168 + 120)$
$: (216 + 264)$
$= 192 : 288 : 480 = 8 : 12 : 20$
$= 2 : 3 : 5$

Partnership / 343

Let the total profit be ₹ x.

∵ C's share = $\dfrac{5}{2+3+5}$ × Total profit

According to the question,

$12550 = \dfrac{5}{10} \times x$

⇒ $x = \dfrac{12550 \times 10}{5}$

= ₹ 25100

5. (c) Ratio of equivalent profits of A, B and C
= (35000 ×12) : (20000 × 5) : (15000 × 7)
= 35 × 12 : 20 × 5 : 15 × 7
= 84 : 20 : 21

Let A's share = $84x$, B's share = $20x$
and C's share = $21x$

According to the question,

$84x + 20x + 21x = 84125$

⇒ $125x = 84125$

⇒ $x = \dfrac{84125}{125} = 673$

∴ B's share = $20x = 20 \times 673$ = ₹ 13460

6. (c) Given ratio = $\dfrac{7}{2} : \dfrac{4}{3} : \dfrac{6}{5}$

= $\dfrac{7}{2} \times 30 : \dfrac{4}{3} \times 30 : \dfrac{6}{5} \times 30 = 105 : 40 : 36$

Let investment of Amitabh = $105x$
Investment of Brijesh = $40x$
Investment of Kamlesh = $36x$
Ratio of their investments
= ($105x \times 4$ + 150% of $105x \times 8$)
                   : ($40x \times 12$) : ($36x \times 12$)

= $\left(420x + \dfrac{150}{100} \times 105x \times 8\right) : (480x) : (432x)$

= $(1680x) : (480x) : (432x) = 35 : 10 : 9$

∴ Brijesh's share = $\left(21600 \times \dfrac{10}{35 + 10 + 9}\right)$

= $\left(21600 \times \dfrac{10}{54}\right)$ = ₹ 4000

7. (a) Let the initial investments of A and B be $18x$ and $7x$.

Then, total investment by A
= $18x \times 4 + (18x + 2000) \times 8$
= $72x + 144x + 16000 = 216x + 16000$

Total investment by B
= $7x \times 4 + (7x + 7000) \times 8$
= $28x + 56x + 56000 = 84x + 56000$

According to the question,

$\dfrac{216x + 16000}{84x + 56000} = \dfrac{2}{1}$

⇒ $216x + 16000 = 168x + 112000$
⇒ $216x - 168x = 112000 - 16000$

⇒ $48x = 96000$ ⇒ $x = \dfrac{96000}{48} = 2000$

∴ Total initial investment
= $(18 + 7)x = (18 + 7) \times 2000$ = ₹ 50000

8. (a) Given, $A : B = 3 : 5 = 6 : 10$
and $B : C = 10 : 13$
∴ $A : B : C = 6 : 10 : 13$

Ratio of share of annual profits of A, B and C
= Ratio of investments of A, B and C
= 6 : 10 : 13

Let shares of annual profits of A, B and C
be $6x$, $10x$ and $13x$, respectively, Then,

$13x = 5876$ ⇒ $x = 452$

∴ Total amount of profit earned by them
= $6x + 10x + 13x = 29x$
= 452×29 = ₹ 13108

9. (d) Ratio of profits of A, B and C
= $2800 \times 12 : 5400 \times 12 : 4800 \times (12 - 4)$
= $28 \times 12 : 54 \times 12 : 48 \times 8 = 14 : 27 : 16$

Let total annual profit be ₹ x and profits of A, B and C be $14x$, $27x$ and $16x$ respectively.

Now, according to the question,

$\dfrac{16x}{(14 + 27 + 16)} - \dfrac{14x}{(14 + 27 + 16)} = 400$

⇒ $\dfrac{16x}{57} - \dfrac{14x}{57} = 400$

⇒ $\dfrac{2x}{57} = 400$

⇒ $x = \dfrac{400 \times 57}{2}$

= $200 \times 57 = 11400$

Hence, the annual profit is ₹ 11400.

10. (a) Let total profit = ₹ x

∴ Ratio of profits of A, B and C
= $25000 \times 12 : 15000 \times 8 : 20000 \times 6$
= $5 : 2 : 2$

A's share in profit = $\dfrac{5}{(5 + 2 + 2)} \times x$

⇒ $7750 = \dfrac{5x}{9}$ ⇒ $x = \dfrac{7750 \times 9}{5}$

⇒ $x = 1550 \times 9$ ⇒ x = ₹ 13950

∴ Total profit = ₹ 13950

11. (a) Let the total capital be ₹ x and Babloo's money is used for y months.

Capital of Aayush = $\dfrac{x}{4}$

∴ Capital of Babloo = $\left(x - \dfrac{x}{4}\right) = \dfrac{3x}{4}$

Ratio of investments of Aayush and Babloo

= $\left(\dfrac{x}{4} \times 15\right) : \left(\dfrac{3x}{4} \times y\right) = \dfrac{15}{4} : \dfrac{3y}{4} = 5 : y$

Ratio of profits of Aayush and Babloo
$= \frac{1}{3} : \frac{2}{3} = 1 : 2 \Rightarrow \frac{5}{y} = \frac{1}{2}$
∴ $y = 10$ months
Hence, Babloo's money was used for 10 months.

12. (d) Let B remained in the business for x months.
Then, $A : B = (2250 \times 12) : (2700 \times x)$
$\Rightarrow 2 : 1 = (27000 : 2700x) = (10 : x)$
∴ $\frac{10}{x} = \frac{2}{1}$ [∵ ratio of profit is $2 : 1$]
$\Rightarrow 2x = 10 \Rightarrow x = \frac{10}{2} = 5$
Clearly, B joined after $(12 - 5) = 7$ months.

13. (d) Let Pihu withdraw from business after x months.
Total profit earned $= ₹ 13050$
and Pihu's share $= ₹ 6750$
∴ Rani's share $= ₹ (13050 - 6750) = ₹ 6300$
Now, by using the formula,
Ratio of investments = Ratio of profits
$\Rightarrow 9000 \times x : 6300 \times 24 = 6750 : 6300$
$\Rightarrow \frac{9000 \times x}{6300 \times 24} = \frac{6750}{6300}$
$\Rightarrow \frac{9000 \times x}{24} = \frac{6750}{1}$
$\Rightarrow x = \frac{24 \times 6750}{9000} = \frac{24 \times 675}{900} = \frac{16200}{900}$
$\Rightarrow x = 18$ months
Hence, Pihu withdraws from business after 18 months.

14. (b) Ratio of profits of A and $B = 1200 : 750$
$= 24 : 15 = 8 : 5$
So, $\frac{1200 \times 12}{1500 \times X} = \frac{8}{5} \Rightarrow X = 6$ months
Hence, B joined 6 months after the start of business.

15. (b) Investment by X : Investment by Y
$= [700 \times 3 + (700 - 200) \times 3$
$\quad + (500 + 120) \times 6] : (600 \times 12)$
$= (2100 + 1500 + 3720) : (7200)$
$= 7320 : 7200 = 61 : 60$
∵ X's profit : Y's profit $= 61 : 60$
∴ Profit of $X = ₹ \frac{61}{121} \times 726 = ₹ 366$

16. (c) Let the total profit be $₹ x$.
Then, A's share in profit $= ₹ \frac{3}{5} x$
Remaining profit $= x - \frac{3}{5}x = \frac{5x - 3x}{5} = \frac{2x}{5}$
∴ B's share in profit $= ₹ \frac{x}{5}$

C's share in profit $= ₹ \frac{x}{5}$
According to the question,
$\left(\frac{3x}{5} - \frac{x}{5}\right) = 400$
$\Rightarrow \frac{2x}{5} = 400 \Rightarrow x = \frac{400 \times 5}{2} = ₹ 1000$

17. (d) Y gets $\frac{2}{7}$ of the profit.
∴ X gets $\left(1 - \frac{2}{7}\right) = \frac{5}{7}$ of the profit
∴ $X : Y = \frac{5}{7} : \frac{2}{7} = 5 : 2$
Let the contribution of Y be a.
Then, ratio of investments by $X : Y$ is
$8000 \times 8 : a \times 4 = 5 : 2$
$\Rightarrow \frac{64000}{4a} = \frac{5}{2} \Rightarrow 20a = 128000$
∴ $a = \frac{128000}{20} = ₹ 6400$

18. (e) Let the total amount invested by A, B and C together be $₹ 100$.
∴ Amount invested by $A = ₹ \frac{1}{5} \times 100$
i.e. $A = ₹ 20$ ...(i)
∴ Amount invested by B and C together
$= 100 - 20$
i.e. $B + C = ₹ 80$ ...(ii)
Given, B contributes as much as A and C together.
i.e. $B = A + C$ ...(iii)
Using Eqs. (i), (ii) and (iii), we have
$B - C = 20$ ...(iv)
On solving Eqs. (iv) and (ii), we have
$B - C = 20$
$B + C = 80$
$\overline{2B = 100}$
$\Rightarrow B = ₹ 50$
∴ $C = ₹ 30$
Now, ratio of investments of A, B and C
$= 20 : 50 : 30 = 2 : 5 : 3$
Ratio of profits = Ratio of investments
$= 2 : 5 : 3$
Given, total profit $= ₹ 5200$
∴ Profit shared by $C = \frac{3}{10} \times 5200 = ₹ 1560$

19. (b) Let the annual profit be $₹ x$. Then, $₹ (x - 120)$ will be distributed between A and B as their shares of profit.
∴ Ratio of profits = Ratio of investments
So, $A : B = 3000 : 4000 = 3 : 4$
∴ A's share $= 120 + (x - 120) \times \frac{3}{7}$

∴ A's share = $120 + (x - 120) \times \dfrac{3}{7}$

According to the question,

$$120 + (x - 120) \times \dfrac{3}{7} = 390$$

$\Rightarrow (x - 120) \times \dfrac{3}{7} = 390 - 120 = 270$

$\Rightarrow x - 120 = 270 \times \dfrac{7}{3} = 630$

∴ B's share = $\dfrac{4}{7} \times (x - 120)$

$= \dfrac{4}{7} \times 630 = ₹ 360$

20. (d) For management, money received by Anil = 10% of 9000 = ₹ 900

Balance = ₹ (9000 − 900) = ₹ 8100

Now, ratio of investments
= 12000 : 20000 = 3 : 5

∴ Anil's share = $8100 \times \dfrac{3}{3 + 5}$

$= 8100 \times \dfrac{3}{8} = ₹ 3037.50$

∴ Amount received by Anil
= 900 + 3037.50 = ₹ 3937.50

21. (b) Ratio of profit of Tom and Harry
= 52000 × 12 : 39000 × 8
= 52 × 12 : 39 × 8 = 2 : 1

Let the total profit earned be ₹ x.

Now, according to the question,

$\dfrac{1}{3}\left(\dfrac{3x}{4}\right) + \dfrac{x}{4} = 20000$

$\Rightarrow \dfrac{x}{4} + \dfrac{x}{4} = 20000$

$\Rightarrow x = ₹ 40000$

∴ Share of Tom = $\dfrac{2}{3} \times \dfrac{3}{4} \times 40000 = ₹ 20000$

22. (d) Ratio of profits of A, B and C
= Investment by them × Time period
= 18000 × 12 : (24000 × 8 + 24000 × 2)
: (15000 × 4 + 18000 × 4)
= 18 × 12 : 24 × 10 : (15 × 4 + 18 × 4)
= 216 : 240 : 132 = 18 : 20 : 11

∴ B's share of profit

$= \dfrac{20}{(18 + 20 + 11)} \times 12005$

$= \dfrac{240100}{49} = ₹ 4900$

23. (a) Ratio of total capitals of A and B
= 20000 × 12 : 35000 × 12
= 240000 : 420000

Now, C gives ₹ 220000 to both to make the capital equal.

∴ A's capital : B's capital
= 240000 : 420000
+ 200000 : 20000
―――――――――
440000 : 440000

If A takes ₹ 200000 and B takes ₹ 20000 from C, then both have the equal capital.

∴ Required ratio of divided amount
= 200000 : 20000
= 20 : 2 = 10 : 1

Hence, A and B should divide the amount in the ratio of 10 : 1.

24. (c) Given, (7k + 10k + 13k)
− (22 + 35 + 48) = 15525

$\Rightarrow$ 30k − 105 = 15525
$\Rightarrow$ 30k = 15525 + 105
$\Rightarrow$ 30k = 15630
$\Rightarrow$ k = 521

∴ Share of Sunil = (7k − 22)
= 7 × 521 − 22
= 3647 − 22
= 3625

Share of Anil = (10k − 35)
= 10 × 521 − 35
= 5210 − 35 = 5175

and share of Jamil = (13k − 48)
= 13 × 521 − 48
= 6773 − 48
= 6725

If ₹ 16, ₹ 77 and ₹ 37 respectively added to their original shares, then

Required ratio
= (3625 + 16) : (5175 + 77) : (6725 + 37)
= 3641 : 5252 : 6762
≈ 36 : 52 : 67

Chapter 20

Unitary Method

Unitary method is a fundamental tool to solve arithmetic problems based on variation in quantities. The method endorses a simple technique to find the amount related to unit quantity.

This method can be applied in questions based on time and work, speed and distance, work and wages etc.

Direct Proportion

Two quantities are said to be in direct proportion to each other, if on increasing (or decreasing) a quantity, the other quantity also increases (or decreases) to the same extent

i.e. (Quantity 1) $\propto$ (Quantity 2)

For example Number of men $\propto$ Volume of work done

(time constant)

i.e. if number of men increases, the volume of work done also increases. Similarly, if volume of work increases, the number of men required to finish the work also increases.

Ex. 1 If the price of 8 bananas is ₹ 40, then find out the price of 12 bananas.

Sol. $\because$ Price of 8 bananas = ₹ 40

$\therefore$ Price of 1 banana = ₹ $\dfrac{40}{8}$

$\therefore$ Price of 12 bananas = $\dfrac{40}{8} \times 12 = 5 \times 12 =$ ₹ 60

Alternate Method

Let the price of 12 bananas be ₹ x.

More bananas, More cost (Direct proportion)

Now, $8 : 12 :: 40 : x \Rightarrow \dfrac{8}{12} = \dfrac{40}{x} \Rightarrow 8 \times x = 12 \times 40$

$\therefore \quad x = \dfrac{12 \times 40}{8} =$ ₹ 60

Unitary Method / 347

Ex. 2 Ramesh walks 160 m everyday, how many kilometres will he walk in 4 weeks?

Sol. Given, total days in 4 weeks = $4 \times 7 = 28$ days
Since, he walks 160 m in 1 day.
$\therefore$ Total walking distance in 4 weeks = $28 \times 160 = 4480$ m = 4.480 km [$\because 1$ km = 1000 m]

Alternate Method
Let the walking distance in 4 weeks or 28 days be x m.
More days, More walking distance (Direct proportion)
Hence, $1 : 28 :: 160 : x \Rightarrow 1 \times x = 28 \times 160$
$\therefore$ Total distance $(x) = 28 \times 160 = 4480$ m = 4.48 km

Ex. 3 If the wages of 12 men for 30 days is ₹ 4200, find out the wages of 18 men for 24 days.

Sol. Let the required wages be ₹ x.
More men, More wages (Direct proportion)
Less days, Less wages (Direct proportion)

$$\left. \begin{array}{l} \text{Men} \quad 12 : 18 \\ \text{Days} \quad 30 : 24 \end{array} \right\} :: 4200 : x$$

$\Rightarrow \quad 12 \times 30 \times x = 18 \times 24 \times 4200 \Rightarrow x = \dfrac{18 \times 24 \times 4200}{12 \times 30} = ₹ 5040$

$\therefore$ Required wages = ₹ 5040

Indirect Proportion

Two quantities are said to be in indirect proportion to each other, if on increasing (or decreasing) a quantity, the other quantity decreases (or increases) to the same extent,

i.e. $\quad$ (Quantity 1) $\propto \dfrac{1}{\text{(Quantity 2)}}$

For example The time taken by a vehicle in covering a certain distance is inversely proportional to the speed of the vehicle,

i.e. $\quad$ Speed $\propto \dfrac{1}{\text{Time}}$

Ex. 4 If Karan travels at a speed of 60 km/h and covers a distance in 9 h, how much time will he take to travel the same distance at a speed of 90 km/h?

Sol. $\because$ Speed = $\dfrac{\text{Distance}}{\text{Time}}$ or Distance = Speed $\times$ Time

Given, speed = 60 km/h, time = 9 h
$\therefore \quad$ Distance = $60 \times 9 = 540$ km
Now, to cover the same distance at speed of 90 km/h.
$\therefore \quad$ Time taken = $\dfrac{\text{Distance}}{\text{Speed}} = \dfrac{540}{90} = 6$ h

Alternate Method
Let the time taken to cover the distance be x.
More speed, Less time (Indirect proportion)
$60 : 90 :: x : 9 \Rightarrow \dfrac{60}{90} = \dfrac{x}{9} \Rightarrow \dfrac{60 \times 9}{90} = x \Rightarrow x = 6$ h

Ex. 5 A man can purchase 50 kg of rice for ₹ 1000. If the price of rice is reduced by 10%, then how many kilograms of rice can now be bought for the same amount of money?

Sol. ∵ Price of rice per kg = ₹ $\dfrac{1000}{50}$

∴ Price of rice = ₹ 20 per kg

Price of rice after 10% reduction = $\dfrac{90}{100} \times 20 =$ ₹ 18 per kg

Now, let the quantity of rice purchase be x kg.

∵ Quantity of rice purchase ∝ $\dfrac{1}{\text{Price of rice}}$ (Indirect proportion)

∴ $50 : x :: 18 : 20 \Rightarrow \dfrac{50}{x} = \dfrac{18}{20}$

∴ $x = \dfrac{50 \times 20}{18} = 55.55$ kg

Hence, the man can purchase 55.55 kg of rice after 10% reduction in price of rice.

🢂 MIND IT!

(i) If M_1 persons can do W_1 work in D_1 days and M_2 persons can do W_2 work in D_2 days, then we have a general formula, $M_1 D_1 W_2 = M_2 D_2 W_1$.

(ii) If we include working hours (say T_1 and T_2) for the two groups, then the relationship is $M_1 D_1 W_2 T_1 = M_2 D_2 W_1 T_2$.

Ex. 6 If 30 men working 18 h per day can reap a field in 32 days, in how many days can 36 men reap the field working 16 h per day?

Sol. Let the required number of days be x.

More men, Less days (Indirect proportion)

Less hours, More days (Indirect proportion)

$\left. \begin{array}{lcc} \text{Men} & 36 : 30 \\ \text{Hours per day} & 16 : 18 \end{array} \right\} :: 32 : x$

$\Rightarrow \quad 36 \times 16 \times x = 30 \times 18 \times 32$

∴ $x = \dfrac{30 \times 18 \times 32}{36 \times 16} = 30$

Hence, the required number of days is 30.

Fast Track Method

Here, $M_1 = 30$, $D_1 = 32$, $T_1 = 18$ h, $M_2 = 36$ and $T_2 = 16$ h

∵ $M_1 D_1 T_1 = M_2 D_2 T_2$

$\Rightarrow \quad 30 \times 32 \times 18 = 36 \times 16 \times D_2$

∴ $D_2 = \dfrac{30 \times 32 \times 18}{36 \times 16} = 30$ days

Fast Track Practice

Exercise 1 Base Level Questions

1. If cost of 24 oranges is ₹ 72, then find out the cost of 120 oranges. [Bank Clerk 2009]
   (a) ₹ 180 (b) ₹ 360 (c) ₹ 172 (d) ₹ 500
   (e) None of these

2. If cost of 15 eggs is ₹ 75, then find out the cost of 4 dozen eggs. [IOB Clerk 2010]
   (a) ₹ 240 (b) ₹ 300 (c) ₹ 150 (d) ₹ 185
   (e) None of these

3. If 16 dozen bananas cost ₹ 360, then how many bananas can be bought in ₹ 60?
   (a) 16 (b) 48 (c) 32 (d) 50
   (e) None of these

4. A worker makes a toy in every 2 h. If he works for 80 h, then how many toys will he make? [Hotel Mgmt. 2008]
   (a) 40 (b) 54 (c) 45 (d) 39

5. If price of m articles is ₹ n, then what is the price of 5 articles?
   (a) ₹ $\frac{5n}{m}$ (b) ₹ $\frac{mn}{5}$ (c) ₹ $\frac{m}{n}$ (d) ₹ $\frac{5m}{n}$
   (e) None of these

6. If 45 m of a uniform rod weighs 171 kg, then what will be the weight of 12 m of the same rod? [MBA 2008]
   (a) 49 kg   (b) 42.5 kg
   (c) 55 kg   (d) 45.6 kg

7. Maganlal, a worker, makes an article in every $\frac{2}{3}$ h. If he works for $7\frac{1}{2}$ h, how many articles will he make?
   (a) $11\frac{1}{4}$ (b) $11\frac{1}{3}$ (c) $11\frac{1}{6}$ (d) $11\frac{2}{5}$
   (e) None of these

8. Shantanu completes 5/8 of a job in 20 days. At this rate, how many more days will he take to finish the job?
   (a) 6 (b) 18 (c) 5 (d) 12
   (e) None of these

9. 20 men can build 56 m long wall in 6 days. What length of a similar wall can be built by 70 men in 3 days?
   (a) 100 m (b) 98 m (c) 48 m (d) 85 m
   (e) None of these

10. A tree is 12 m tall and casts an 8 m long shadow. At the same time, a flag pole casts a 100 m long shadow. How long is the flag pole?
    (a) 150 m (b) 200 m (c) 125 m (d) 115 m
    (e) None of these

11. A motorcycle gives an average of 45 km/L. If the cost of petrol is ₹ 20 per litre. The amount required to complete a journey of 540 km is [SSC CGL (Pre) 2016]
    (a) ₹ 120   (b) ₹ 360
    (c) ₹ 200   (d) ₹ 240

12. If 12 persons working 16 h per day earn ₹ 33600 per week, then how much will 18 persons earn working 12 h per day?
    (a) ₹ 40000   (b) ₹ 35000
    (c) ₹ 28800   (d) ₹ 37800
    (e) None of these

13. 12 monkeys can eat 12 bananas in 12 min. In how many minutes can 4 monkeys eat 4 bananas? [SSC CGL 2015]
    (a) 10   (b) 4
    (c) 8   (d) 12

14. If 10 spiders can catch 10 flies in 10 min, then how many flies can 200 spiders catch in 200 min? [MBA 2010]
    (a) 2000 (b) 5000 (c) 4000 (d) 3000

15. 12 men can do a piece of work in 24 days. How many days are needed to complete the work, if 8 men are engaged in the same work? [IDBI Clerk 2010]
    (a) 28 (b) 36 (c) 48 (d) 52
    (e) None of these

16. 22 men can complete a job in 16 days. In how many days, will 32 men complete that job? [IOB Clerk 2009]
    (a) 14 (b) 12 (c) 16 (d) 9
    (e) None of these

17. If 30 men working 9 h per day can reap a field in 16 days, in how many days, will 36 men reap the field working 8 h per day?
    (a) 15 (b) 25 (c) 18 (d) 10
    (e) None of these

18. A garrison of 1000 men had provisions for 48 days. However, a reinforcement of 600 men arrived. How long will now food last for?
    (a) 35 days   (b) 30 days
    (c) 25 days   (d) 45 days
    (e) None of these

19. 2000 soldiers in a fort had enough food for 20 days. But some soldiers were transferred to another fort and the food lasted for 25 days. How many soldiers were transferred? [SSC (10+2) 2013]
    (a) 400 (b) 450 (c) 525 (d) 500

20. If in a hostel, food is available for 45 days for 50 students. For how many days will this food be sufficient for 75 students? [CGPSC 2013]
(a) 25 (b) 28
(c) 30 (d) 40
(e) None of these

21. In a garrison, there was food for 1000 soldiers for one month. After 10 days, 1000 more soldiers joined the garrison. How long would the soldiers be able to carry on with the remaining food? [UPSC CSAT 2013]
(a) 25 days (b) 20 days
(c) 15 days (d) 10 days

22. A garrison is provided with ration for 72 soldiers to last for 54 days. How long would the same amount of food last for 90 soldiers, if the individual ration is reduced by 10%? [SSC CGL 2013]
(a) 48 days (b) 72 days
(c) 54 days (d) 126 days

Exercise 2 Higher Skill Level Questions

1. If m persons can paint a house in d days, how many days will it take for $(m+2)$ persons to paint the same house? [CDS 2015 (II)]
(a) $md + 2$ (b) $md - 2$
(c) $\dfrac{m+2}{md}$ (d) $\dfrac{md}{m+2}$

2. 3 men can do a piece of work in 18 days. 6 boys can also do the same work in 18 days. In how many days, 4 men and 4 boys together will finish the work? [OBC Clerk 2010]
(a) 10 (b) 6 (c) 12 (d) 9
(e) None of these

3. A man and a boy working together can complete a work in 24 days. If for the last 6 days, the man alone does the work, then it is completed in 26 days. How long will the boy take to complete the work alone?
(a) 72 days (b) 73 days
(c) 49 days (d) 62 days
(e) None of these

4. 25 men can reap a field in 20 days. When should 15 men leave the work, if the whole field is to be reaped in $37\dfrac{1}{2}$ days after they leave the work?
(a) After 5 days (b) After 10 days
(c) After 9 days (d) After 7 days
(e) None of these

5. 10 men and 8 women can together complete a work in 5 days. Work done by a woman is equal to the half of the work done by a man. In how many days will 4 men and 6 women complete that work? [RBI Clerk 2010]
(a) 12 (b) 10 (c) $8\dfrac{2}{3}$ (d) $9\dfrac{3}{4}$
(e) None of these

6. If 12 engines consume 30 metric tonne of coal when each is running 18 h per day, how much coal will be required for 16 engines, each running 24 h per day, it being given that 6 engines of former type consume as much as 8 engines of latter type?
(a) 10 tonne (b) 5 tonne
(c) 25 tonne (d) 40 tonne
(e) None of these

7. If 12 men or 18 women can do a piece of work in 14 days, then how long will 8 men and 16 women take to finish the work? [RRB 2007]
(a) 9 days (b) 10 days
(c) 12 days (d) 14 days

8. 8 men can complete a work in 12 days, 4 women can complete it in 48 days and 10 children can complete the same work in 24 days. In how many days can 10 men, 4 women and 10 children complete the same work?
(a) 10 (b) 5 (c) 7 (d) 6

9. A typist types one page in 4 min. If between 1 pm to 2 pm, 1080 pages have to be typed. How many typist would be needed? [BSSC CGL (Pre) 2015]
(a) 108 (b) 60 (c) 90 (d) 72

10. 40 men complete one-third of a work in 40 days. How many more men should be employed to finish the rest of the work in 50 more days?
(a) 12 (b) 20 (c) 18 (d) 24
(e) None of these

11. 20 men can complete a piece of work in 16 days. After 5 days from the start of the work, some men left. If the remaining work was completed by the remaining men in $18\dfrac{1}{3}$ days, then how many men left after 5 days from the start of the work? [LIC ADO 2015]
(a) 4 (b) 10 (c) 8 (d) 5
(e) 6

Unitary Method / 351

12. 24 men can finish a piece of work in 18 days, while 30 women can finish the same piece of work in 12 days. In how many days 16 men and 24 women together can finish the same piece of work? **[LIC ADO 2015]**

(a) $8\dfrac{5}{14}$ (b) $10\dfrac{11}{14}$

(c) $9\dfrac{9}{14}$ (d) $9\dfrac{1}{7}$

(e) $8\dfrac{8}{7}$

Answer with Solutions

Exercise 1 — Base Level Questions

1. (b) ∵ Cost of 24 oranges = ₹ 72

∴   Cost of 1 orange = ₹ $\dfrac{72}{24}$

∴   Cost of 120 oranges = $\dfrac{72}{24} \times 120$
$= 3 \times 120 =$ ₹ 360

Alternate Method
Let the required cost be ₹ x.
More oranges, More cost
                     (Direct proportion)
$24 : 120 :: 72 : x$

∴ $x = \dfrac{120 \times 72}{24} =$ ₹ 360

2. (a) ∵ Cost of 15 eggs = ₹ 75

∴   Cost of 1 egg = ₹ $\dfrac{75}{15}$

∴ Cost of 4 dozen $(4 \times 12 = 48)$ eggs
$= \dfrac{75}{15} \times 48 = 5 \times 48 =$ ₹ 240

3. (c) Let the required number of bananas be x.
16 dozen bananas
$= 16 \times 12 = 192$ bananas
Less bananas, Less cost
                     (Direct proportion)
$360 : 60 :: 192 : x$

∴ $x = \dfrac{60 \times 192}{360} = 32$ bananas

4. (a) Let number of toys be x.
More hours, More toys (Direct proportion)
$2 : 80 :: 1 : x \;\Rightarrow\; x = \dfrac{80}{2} = 40$ toys

5. (a) ∵ Price of m articles = ₹ n

∴   Price of 1 article = ₹ $\dfrac{n}{m}$

∴   Price of 5 articles = ₹ $\dfrac{5n}{m}$

6. (d) ∵ Weight of 45 m rod = 171 kg

∴ Weight of 1 m rod = $\dfrac{171}{45}$ kg

∴ Weight of 12 m rod
$= \dfrac{171}{45} \times 12 = 45.6$ kg

Alternate Method
Let the required weight be x kg.
Less length, Less weight
                     (Direct proportion)
$45 : 12 :: 171 : x$

∴ $x = \dfrac{12 \times 171}{45} = 45.6$ kg

7. (a) ∵ In $\dfrac{2}{3}$ h, 1 article is made.

∴ In 1h, $\dfrac{3}{2}$ articles are made.

∴ In $7\dfrac{1}{2} = \dfrac{15}{2}$ h, $\dfrac{3}{2} \times \dfrac{15}{2} = \dfrac{45}{4}$ articles are made.

∴   Required articles = $\dfrac{45}{4} = 11\dfrac{1}{4}$

8. (d) Let the required number of days be x.

∴   Remaining work = $1 - \dfrac{5}{8} = \dfrac{3}{8}$

Less work, Less days   (Direct proportion)
$\dfrac{5}{8} : \dfrac{3}{8} :: 20 : x \;\Rightarrow\; x = \dfrac{3}{8} \times 20 \times \dfrac{8}{5}$

∴ $x = 12$ days

Alternate Method

∵ $\dfrac{5}{8}$ work is done in 20 days.

∴ 1 work will be done in $\dfrac{20 \times 8}{5} = 32$ days.

∴ Required number of days to complete the remaining work = $(32 - 20) = 12$ days

9. (b) Let the required length be x m.
More men, More length
                     (Direct proportion)

Less days, Less length (Direct proportion)

Men   20 : 70   } :: 56 : x
Days   6 : 3

$\therefore$ $(20 \times 6 \times x) = (70 \times 3 \times 56)$

$\therefore$ $x = \dfrac{70 \times 3 \times 56}{20 \times 6} = 98$ m

Fast Track Method

Here, $M_1 = 20, D_1 = 6, W_1 = 56,$
$M_2 = 70, D_2 = 3$ and $W_2 = ?$

$\because$ $\dfrac{M_1 D_1}{W_1} = \dfrac{M_2 D_2}{W_2} \Rightarrow \dfrac{20 \times 6}{56} = \dfrac{70 \times 3}{W_2}$

$\therefore$ $W_2 = \dfrac{70 \times 3 \times 56}{20 \times 6} = 98$ m

10. (a) $\because$ 8 m shadow means original height
$= 12$ m

$\therefore$ 1 m shadow means original height
$= \dfrac{12}{8}$ m

$\therefore$ 100 m shadow means original height
$= \dfrac{12}{8} \times 100$ m $= \dfrac{6}{4} \times 100$
$= 6 \times 25 = 150$ m

11. (d) Petrol required to drive 45 km = 1 L

Petrol required to drive 1 km $= \dfrac{1}{45}$ L

Petrol required to drive 540 km
$= \dfrac{1}{45} \times 540$ L $= 12$ L

Cost of 1 L petrol = ₹ 20

$\therefore$ Cost of 12 L petrol $= 12 \times 20 =$ ₹ 240

12. (d) Let the required earnings be ₹ x.
More persons, More earnings
                      (Direct proportion)
Less hours per day, Less earnings
                      (Direct proportion)

Persons   12 : 18   } :: 33600 : x
Hours per day   16 : 12

$\therefore$ $(12 \times 16 \times x) = (18 \times 12 \times 33600)$

$\Rightarrow$ $x = \dfrac{18 \times 12 \times 33600}{12 \times 16}$

$= 18 \times 2100 =$ ₹ 37800

Fast Track Method

Here, $M_1 = 12, H = 16, W_1 = 33600,$
$M_2 = 18, H = 12$ and $W_2 = ?$

By using the formula, $\dfrac{M_1 H_1}{W_1} = \dfrac{M_2 H_2}{W_2}$

$\Rightarrow \dfrac{12 \times 16}{33600} = \dfrac{18 \times 12}{W_2} \Rightarrow W_2 = \dfrac{33600 \times 18}{16}$

$\therefore$ $W_2 = 2100 \times 18 =$ ₹ 37800

13. (d) Let the required time be x min.
More monkeys, More bananas
                      (Direct proportion)
More time, More bananas
                      (Direct proportion)

Monkeys   12 : 4   } :: 12 : 4
Time (in min)   12 : x

$\therefore$ $12 \times 12 \times 4 = 4 \times x \times 12$

$\Rightarrow$ $x = \dfrac{12 \times 12 \times 4}{12 \times 4} = 12$ min

14. (c) Let the required number of flies be x.
More spiders, More flies
                      (Direct proportion)
More time, More flies   (Direct proportion)

Spiders   10 : 200   } :: 10 : x
Time   10 : 200

$\therefore$ $(10 \times 10 \times x) = (200 \times 200 \times 10)$

$\Rightarrow$ $x = \dfrac{200 \times 200 \times 10}{10 \times 10} = 4000$ flies

15. (b) Let the required number of days be x.
Less men, More days (Indirect proportion)

$8 : 12 :: 24 : x \Rightarrow x = \dfrac{12 \times 24}{8} = 36$ days

Fast Track Method

Here, $M_1 = 12, D_1 = 24, M_2 = 8, D_2 = ?$

$\because$ $M_1 D_1 = M_2 D_2 \Rightarrow 12 \times 24 = 8 \times D_2$

$\therefore$ $D_2 = \dfrac{12 \times 24}{8} = 36$ days

16. (e) $\because$ 22 men do the work in 16 days.
$\therefore$ 1 man will do the work in 16×22 days.
$\therefore$ 32 men will do the job in
$\dfrac{16 \times 22}{32}$ i.e. 11 days.

17. (a) Let the required number of days be x.
More men, Less days   (Indirect proportion)
Less hours, More days
                      (Indirect proportion)

Men   36 : 30   } :: 16 : x
Hours / day   8 : 9

$\therefore$ $(36 \times 8 \times x) = (30 \times 9 \times 16)$

$\therefore$ $x = \dfrac{30 \times 9 \times 16}{36 \times 8} = 15$ days

Fast Track Method

Here, $M_1 = 30, D_1 = 16, T_1 = 9, M_2 = 36$
and $T_2 = 8$
By using the formula,
$M_1 D_1 T_1 = M_2 D_2 T_2$

$\Rightarrow 30 \times 16 \times 9 = 36 \times D_2 \times 8$

$\therefore$ $D_2 = \dfrac{30 \times 16 \times 9}{36 \times 8} = 15$ days

Unitary Method / 353

18. (b) ∵ For 1000 men, provision lasts for 48 days.
∴ For 1 man, provision lasts for (48×1000) days.
∴ For $(1000 + 600)$ men, provision will last for $\dfrac{48 \times 1000}{1600}$ days.
∴ Required number of days
$= \dfrac{48 \times 1000}{1600} = 3 \times 10 = 30$

19. (a) Let the number of soldiers transferred be x.
Now, the food would last for 25 days for $(2000 - x)$ soldiers.
Less men, More days (Indirect proportion)
$25 : 20 :: 2000 : 2000 - x$
$\Rightarrow 2000 - x = \dfrac{2000 \times 20}{25}$
$\Rightarrow 2000 - x = 1600$
∴ $x = 2000 - 1600 = 400$

Fast Track Method
Here, $M_1 = 2000$, $D_1 = 20$
Let the soldiers leaving the fort be x.
Then, $M_2 = 2000 - x$, $D_2 = 25$
By using formula, $M_1 D_1 = M_2 D_2$
$2000 \times 20 = (2000 - x) 25$
$\Rightarrow \dfrac{2000 \times 20}{25} = 2000 - x$
$\Rightarrow 1600 = 2000 - x \Rightarrow x = 400$

20. (c) ∵ For 50 students, food is sufficient for 45 days.
∴ For 1 student, food is sufficient for 45×50 days.

∴ For 75 students, food is sufficient for $\dfrac{45 \times 50}{75}$ days, i.e. for 30 days.

21. (d) Let us assume that each soldier eats one unit of food per day. Thus, total units of food at the begining will be
$1000 \times 30 = 30000$.
After 10 days, 1000 soldiers would have eaten $1000 \times 10 = 10000$ units of food. Thus, food left after 10 days equals 20000 units. Now, there are total of 2000 soldiers who eat one unit of food every day. So, the number of days that 20000 units of food will serve 2000 soldiers are $\dfrac{20000}{2000} = 10$ days.

Alternate Method
∵ For 1000 soldiers, remaining food is sufficient for $30 - 10 = 20$ days.
∴ For $1000 + 1000 = 2000$ soldiers, food will be sufficient for
$\dfrac{1000}{2000} \times 20 = 10$ days

22. (a) Let the required number of days be x.
More men, Less days
(Indirect proportion)

Soldiers 90 72
Ration $\dfrac{9}{10}$ 1 $\Bigg\} :: 54 : x$

∴ $x = \dfrac{72 \times 54 \times 10}{90 \times 9} = 48$ days

Exercise 2 Higher Skill Level Questions

1. (d) ∵ m persons paint a house in d days.
∴ 1 person paints a house in $(m \times d)$ days and $m + 2$ persons paint a house in $\left(\dfrac{md}{m+2}\right)$ days.

2. (d) 3 men ≡ 6 boys
$\Rightarrow$ 1 man ≡ 2 boys
∴ 4 men + 4 boys ≡ 4 men + 2 men
= 6 men
∵ 3 men can do a work in 18 days.
∴ 1 man can do a work in 18×3 days.
∴ 6 men can do the work in $\dfrac{18 \times 3}{6}$ days.
∴ Required number of days
$= \dfrac{18 \times 3}{6} = 9$ days

Alternate Method
3 men ≡ 6 boys
$\Rightarrow$ 1 man ≡ 2 boys
∴ 4 men + 4 boys ≡ 4 men + 2 men
= 6 men
More men, Less days (Indirect proportion)
$6 : 3 :: 18 : x$
$\Rightarrow x = \dfrac{3 \times 18}{6}$
∴ $x = 3 \times 3 = 9$ days

3. (a) Let man's 1 day's work $= \dfrac{1}{m}$
and boy's 1 day's work $= \dfrac{1}{n}$
1 day's work of man and boy $= \dfrac{1}{24}$

354 / Fast Track Objective Arithmetic

Man's 6 days work = $\dfrac{6}{m}$

Now, for 20 days, both man and boy do the work and for last 6 days, only man does the work.
According to the question,

$$\dfrac{1}{m} + \dfrac{1}{n} = \dfrac{1}{24} \qquad \ldots(i)$$

and $\quad 20\left(\dfrac{1}{m} + \dfrac{1}{n}\right) + \dfrac{6}{m} = 1$

$\Rightarrow \quad \left(20 \times \dfrac{1}{24}\right) + \dfrac{6}{m} = 1$ [from Eq. (i)]

$\Rightarrow \quad \dfrac{6}{m} = \left(1 - \dfrac{20}{24}\right) = \dfrac{4}{24} = \dfrac{1}{6} \Rightarrow \dfrac{1}{m} = \dfrac{1}{36}$

Now, from Eq. (i), we have

$\dfrac{1}{m} + \dfrac{1}{n} = \dfrac{1}{24} \Rightarrow \dfrac{1}{36} + \dfrac{1}{n} = \dfrac{1}{24}$

$\Rightarrow \quad \dfrac{1}{n} = \left(\dfrac{1}{24} - \dfrac{1}{36}\right) = \dfrac{1}{72}$

Hence, the boy alone can do the work in 72 days.

4. *(a)* Let 25 men work for m days.

Work done in 1 day's by 25 men = $\dfrac{m}{20}$

Remaining work = $\left(1 - \dfrac{m}{20}\right)$

25 men's 1 day's work = $\dfrac{1}{20}$

1 man's 1 day's work = $\dfrac{1}{20} \times \dfrac{1}{25} = \dfrac{1}{500}$

10 men's 1 day's work = $\dfrac{1}{500} \times 10 = \dfrac{1}{50}$

10 men's $\dfrac{75}{2}$ day's work

$= \dfrac{1}{50} \times \dfrac{75}{2} = \dfrac{75}{100} = \dfrac{3}{4}$

∴ $\left(1 - \dfrac{m}{20}\right) = \dfrac{3}{4} \Rightarrow \dfrac{m}{20} = \dfrac{1}{4}$

$\Rightarrow \quad m = \dfrac{1}{4} \times 20 = 5$

Clearly, 15 men leave after 5 days.

5. *(b)* 1 man ≡ 2 women

10 men + 8 women
≡ 20 women + 8 women
= 28 women

4 men + 6 women ≡ 8 women + 6 women
= 14 women

∵ 28 women do the work in 5 days.

∴ 1 woman can do the same work in (28×5) days.

∴ 14 women will do the same work in $\left(\dfrac{28 \times 5}{14}\right)$ days.

∴ Required number of days

$= \dfrac{28 \times 5}{14} = 2 \times 5 = 10$

Alternate Method

1 woman ≡ $\dfrac{1}{2}$ man

10 men + 8 woman ≡ 10 men + 4 men
= 14 men

4 men + 6 women ≡ 4 + 3 = 7 men

More men, Less days (Indirect proportion)

7 : 14 :: 5 : x

∴ $x = \dfrac{14 \times 5}{7} = 10$ days

6. *(d)* Let the required quantity of coal consumed be x.

More engines, More coal consumption
(Direct proportion)

More hours, More coal consumption
(Direct proportion)

Less rate of consumption, Less coal consumption (Direct proportion)

Engines	12 : 16	
Working hours	18 : 24	:: 30 : x
Rate of consumption	$\dfrac{1}{6} : \dfrac{1}{8}$	

∴ $12 \times 18 \times \dfrac{1}{6} \times x = 16 \times 24 \times \dfrac{1}{8} \times 30$

$\Rightarrow 36x = 1440 \Rightarrow x = \dfrac{1440}{36} = 40$

Hence, quantity of coal consumed will be 40 tonne.

7. *(a)* 12 men ≡ 18 women

$\Rightarrow$ 1 man ≡ $\dfrac{18}{12}$ women ≡ $\dfrac{3}{2}$ women

∴ 8 men ≡ $\dfrac{3}{2} \times 8 = 12$ women

∴ 8 men + 16 women
= 12 women + 16 women
= 28 women

∵ 18 women can do the work in 14 days.

∴ 1 woman can do the same work in (14×18) days.

∴ 28 women will do the same work in $\left(\dfrac{14 \times 18}{28}\right)$ days.

∴ Required number of days

$= \dfrac{14 \times 18}{28} = 9$ days

Unitary Method / 355

8. (d) 1 man can finish the work in
$(8 \times 12) = 96$ days
1 woman can finish the work in
$(4 \times 48) = 192$ days
1 child can finish the work in
$(10 \times 24) = 240$ days
1 man's 1 day's work $= \dfrac{1}{96}$
1 woman's 1 day's work $= \dfrac{1}{192}$
1 child's 1 day's work $= \dfrac{1}{240}$
$\therefore$ (10 men + 4 women + 10 children)'s 1 day's work
$= \left(\dfrac{10}{96} + \dfrac{4}{192} + \dfrac{10}{240}\right) = \left(\dfrac{5}{48} + \dfrac{1}{48} + \dfrac{1}{24}\right)$
$= \left(\dfrac{5+1+2}{48}\right) = \dfrac{8}{48} = \dfrac{1}{6}$

Hence, they will finish the work in 6 days.

9. (d) More pages $\propto$ More typist
(Direct proportion)
More working time $\propto \dfrac{1}{\text{Typist}}$
(Indirect proportion)
Let the required typist be x.

Pages	Time (in min)	Typist
1	4 ↑	1
1080 ↓	60	x ↓

Here, direction of arrow depicts the proportion.
$\left.\begin{array}{c} 1080 : 1 \\ 4 : 60 \end{array}\right\} :: x : 1$
$\Rightarrow \quad 1 \times 60 \times x = 1080 \times 4 \times 1$
$\therefore \quad x = \dfrac{1080 \times 4}{60} = 72$

10. (d) Work done $= \dfrac{1}{3}$
Remaining work $= \left(1 - \dfrac{1}{3}\right) = \left(\dfrac{3-1}{3}\right) = \dfrac{2}{3}$
Let the number of additional men be x.
More work, More men (Direct proportion)
More days, Less men (Indirect proportion)
$\left.\begin{array}{lcc} \text{Work} & \dfrac{1}{3} : \dfrac{2}{3} \\ \text{Days} & 50 : 40 \end{array}\right\} :: 40 : (40 + x)$

$\therefore \quad \dfrac{1}{3} \times 50 \times (40 + x) = \dfrac{2}{3} \times 40 \times 40$
$\Rightarrow \quad 5 \times (40 + x) = 2 \times 40 \times 4$
$\Rightarrow \quad 200 + 5x = 320$
$\Rightarrow \quad 5x = 320 - 200 = 120$
$\therefore \quad x = \dfrac{120}{5} = 24$

$\therefore$ Required number of men = 24

11. (c) Let x men left after 5 days.
$\therefore \quad \dfrac{w}{20 \times 11} = \dfrac{w}{(20-x) \times \dfrac{55}{3}}$
$\Rightarrow \quad (20 - x) \times \dfrac{55}{3} = 20 \times 11$
$\Rightarrow \quad \dfrac{1100}{3} - \dfrac{55}{3}x = 220$
$\Rightarrow \quad \dfrac{1100}{3} - 220 = \dfrac{55x}{3}$
$\Rightarrow \quad \dfrac{440}{3} = \dfrac{55x}{3}$
$\Rightarrow \quad x = 8$ men

Hence, 8 men left after 5 days.

12. (c) $\because$ 24 men can finish a piece of work in 18 days.
$\therefore$ 1 man can finish the same work in
18×24 days = 432 days
Again, 30 women can finish same work in 12 days.
$\therefore$ 1 woman can finish the same work in
$30 \times 12 = 360$ days
Now, 432 men days = 360 women days
or 6 men = 5 women
$\therefore$ 1 man $= \dfrac{5}{6}$ women
$\therefore$ 16 men $= \dfrac{5}{6} \times 16$ women $= \dfrac{40}{3}$ women

Again, 1 woman can finish the work in 360 days.
$\therefore \left(\dfrac{40}{3} + 24 = \dfrac{112}{3}\right)$ women can finish the work in
$\dfrac{360 \times 3}{112} = \dfrac{45 \times 3}{14}$
$= \dfrac{135}{14} = 9\dfrac{9}{14}$ days

Chapter 21

Problems Based on Ages

Age is defined as a period of time that a person has lived or a thing has existed. Age is measured in **months, years, decades** and so on.

Problems based on ages generally consists of information of ages of two or more persons and a relation between their ages in present/future/past. Using the information, it is asked to calculate the ages of one or more persons in present/future/past.

Important Rules for Problems Based on Ages

Rule ❶

If the ratio of present ages of A and B is $x : y$ and n yr ago, the ratio of their ages was $p : q$, then $\dfrac{kx - n}{ky - n} = \dfrac{p}{q}$, where k is a constant.

Ex. 1 The ratio of the ages of A and B at present is $3 : 1$. Four years earlier, the ratio was $4 : 1$. Find the present age of A. **[SSC CPO 2015]**

Sol. Let the present ages of A and B be $3x$ yr and x yr, respectively.

Now, 4 yr ago, age of $A = (3x - 4)$ yr

and age of $B = (x - 4)$ yr

According to the question,

$$\dfrac{3x - 4}{x - 4} = \dfrac{4}{1}$$

$\Rightarrow \qquad 3x - 4 = 4x - 16$

$\Rightarrow \qquad 4x - 3x = -4 + 16$

$\Rightarrow \qquad x = 12$

∴ Present age of $A = 3x = 3 \times 12 = 36$ yr

Problems Based on Ages / 357

Rule 2

If the ratio of present ages of A and B is $x : y$ and after n yr, the ratio of their ages will be $p : q$, then $\dfrac{kx + n}{ky + n} = \dfrac{p}{q}$, where k is a constant.

Ex. 2 At present, the ratio of the ages of Maya and Chhaya is 6:5 and fifteen years from now, the ratio will get changed to 9:8. Find the present age of Maya. **[SSC CGL 2011]**

Sol. Let the present ages of Maya and Chhaya be $6x$ yr and $5x$ yr, respectively.

According to the question,

After 15 yr, $\dfrac{6x + 15}{5x + 15} = \dfrac{9}{8}$

$\Rightarrow \quad 48x + 120 = 45x + 135$
$\Rightarrow \quad 3x = 15 \Rightarrow x = 5$

Hence, present age of Maya $= 5 \times 6 = 30$ yr

MIND IT! Mostly questions on ages can be solved with the use of linear equations. So, the method to solve linear equations is important for this chapter which is discussed in chapter equations.

Fast Track Techniques to solve the QUESTIONS

Technique 1 If t yr after, age of one person is n times the age of another person and at present the age of first person is m times the age of another person, then

$$\text{Age of first person} = tm\left(\dfrac{n-1}{m-n}\right) \text{ yr}$$

and $\text{age of second person} = t\left(\dfrac{n-1}{m-n}\right)$ yr

Ex. 3 The present age of Karan is 5 times the age of Shivam. After 10 yr, Karan will be 3 times as old as Shivam. What are the present ages of Karan and Shivam?

Sol. Let present age of Shivam $= x$ yr

Then, present age of Karan $= 5x$ yr

After 10 yr, the ratio of ages will be 3 : 1.

According to the question,

$\dfrac{5x + 10}{x + 10} = \dfrac{3}{1}$

358 / Fast Track Objective Arithmetic

$$\Rightarrow \quad 5x + 10 = 3(x + 10)$$
$$\Rightarrow \quad 5x + 10 = 3x + 30$$
$$\Rightarrow \quad 5x - 3x = 30 - 10$$
$$\Rightarrow \quad 2x = 20$$
$$\Rightarrow \quad x = \frac{20}{2} = 10$$

∴ Karan's present age = $5 \times 10 = 50$ yr and Shivam's present age = 10 yr

Fast Track Method

Here, $t = 10$, $m = 5$ and $n = 3$

Karan's present age = $tm \left(\dfrac{n - 1}{m - n} \right)$

$$= 10 \times 5 \left(\frac{3 - 1}{5 - 3} \right) = 50 \times \frac{2}{2} = 50 \text{ yr}$$

Shivam's present age = $t \left(\dfrac{n - 1}{m - n} \right)$

$$= 10 \left(\frac{3 - 1}{5 - 3} \right)$$
$$= 10 \times \frac{2}{2} = 10 \text{ yr}$$

Technique 2 If t_1 yr before, age of a person was m times the age of another person. After t_2 yr, age of a person will be n times the age of second person, then

Age of first person = $\dfrac{t_2 m (n - 1) + t_1 n(m - 1)}{m - n}$ yr

and age of second person = $\dfrac{t_2 (n - 1) + t_1 (m - 1)}{m - n}$ yr

Ex. 4 Mukesh told his granddaughter Sailee that five years earlier, he was seven times as old as she was. After 15 yr, he will be thrice as old as she will be. Find the sum of their present ages. **[NIFT 2015]**

Sol. Let Mukesh's present age be y yr and Sailee's present age be x yr.
According to the question,

$$y - 5 = 7(x - 5)$$
$$\Rightarrow \quad y - 5 = 7x - 35$$
$$\Rightarrow \quad 7x - y = 30 \quad \ldots(i)$$

Again,
$$(y + 15) = 3(x + 15)$$
$$\Rightarrow \quad y + 15 = 3x + 45$$
$$\Rightarrow \quad y - 3x = 30 \quad \ldots(ii)$$

On solving Eqs. (i) and (ii), we get
$$x = 15 \text{ and } y = 75$$

∴ Sum of their present ages = $75 + 15 = 90$ yr

Fast Track Method

Here, $t_1 = 5, m = 7, t_2 = 15$ and $n = 3$

$\therefore$ Mukesh's age $= \dfrac{t_2 m (n-1) + t_1 n (m-1)}{m - n}$

$= \dfrac{15 \times 7 (3-1) + 5 \times 3 (7-1)}{7 - 3}$

$= \dfrac{15 \times 14 + 15 \times 6}{4} = \dfrac{210 + 90}{4} = \dfrac{300}{4} = 75$ yr

and Sailee's age $= \dfrac{t_2 (n-1) + t_1 (m-1)}{m - n}$

$= \dfrac{15 (3-1) + 5 (7-1)}{7 - 3}$

$= \dfrac{15 \times 2 + 5 \times 6}{4} = \dfrac{30 + 30}{4} = \dfrac{60}{4} = 15$ yr

$\therefore$ Sum of their present ages $= 75 + 15 = 90$ yr

Technique 3 If M is as elder to N as he is younger to P and sum of ages of N and P is t yr, $N < M < P$, then

M's age $= \dfrac{\text{Sum of ages of } N \text{ and } P}{2} = \dfrac{S}{2}$, where S = Sum.

Ex. 5 If Akshay is as much elder than Vinay as he is younger to Kartik and sum of ages of Vinay and Kartik is 48 yr, then find the age of Akshay.

Sol. Let present age of Akshay be x yr and he is younger to Kartik by y yr.

Then, Kartik's age $= (x + y)$ yr and Vinay's age $= (x - y)$ yr

Now, according to the question,

Sum of ages of Kartik and Vinay $= 48$

$\Rightarrow \qquad (x + y) + (x - y) = 48$

$\Rightarrow \qquad\qquad\qquad 2x = 48$

$\therefore \qquad\qquad\qquad x = \dfrac{48}{2} = 24$ yr

Hence, present age of Akshay is 24 yr.

Fast Track Method

Present age of Akshay $= \dfrac{\text{Sum of ages}}{2} = \dfrac{48}{2} = 24$ yr

Fast Track Practice

1. Raju decided to marry 3 yr after he gets a job. He was 17 yr old when he passed class 12th. After passing class 12th, he had completed his graduation course in 3 yr and PG course in 2 yr. He got the job exactly 1 yr after completing his PG course. At what age will he get married?
   [Allahabad Bank Clerk 2010]
   (a) 27 yr (b) 26 yr (c) 28 yr (d) 23 yr
   (e) None of these

2. A father is nine times as old as his son and the mother is eight times as old as the son. The sum of the father's and the mother's age is 51 yr. What is the age of the son?   **[UPSC CSAT 2015]**
   (a) 7 yr (b) 5 yr (c) 4 yr (d) 3 yr

3. A father's age is three times the sum of the ages of his two children, but 20 yr hence his age will be equal to the sum of their ages. Then, the father's age is
   [RRB Clerk (Pre) 2017]
   (a) 30 yr (b) 40 yr (c) 35 yr (d) 45 yr
   (e) None of these

4. The average of the present ages of Sachin and Saurabh is 36 yr. If Sachin is 8 yr older than Saurabh, what is the Saurabh's present age?   **[Allahabad Bank Clerk 2010]**
   (a) 30 yr (b) 34 yr (c) 32 yr (d) 40 yr
   (e) None of these

5. Raman's present age is three times his daughter's and nine-thirteenth of his mother's present age. The sum of the present ages of all three of them is 125 yr. What is the difference between the present ages of Raman's daughter and Raman's mother?   **[Allahabad Bank PO 2009]**
   (a) 45 yr (b) 40 yr (c) 50 yr
   (d) Cannot be determined
   (e) None of these

6. A man born in the year 1896 AD. If in the year AD x^2, his age is $(x-4)$. The value of x is   **[SSC CPO 2015]**
   (a) 44 yr (b) 36 yr
   (c) 42 yr (d) 40 yr

7. The ratio of Smita's age to her mother is 3 : 7 respectively and the difference in their ages is 32 yr. What will be the ratio of their ages 4 yr hence?
   [NICL AO (Pre) 2017]
   (a) 4 : 19 (b) 5 : 14 (c) 3 : 20 (d) 7 : 15
   (e) 3 : 8

8. The ratio of the present age of Manoj to that of Wasim is 3 : 11. Wasim is 12 yr younger than Rehana. Rehana's age after 7 yr will be 85 yr. What is the present age of Manoj's father, who is 25 yr older than Manoj?   **[IOB PO 2011]**
   (a) 43 yr (b) 67 yr (c) 45 yr (d) 69 yr
   (e) None of these

9. The ratio between the present ages of Indira and Lizzy is 3 : 8, respectively. After 8 yr, Indira's age will be 20 yr. What was Lizzy's age 5 yr ago?   **[BOI Clerk 2010]**
   (a) 37 yr (b) 27 yr
   (c) 28 yr (d) 38 yr
   (e) None of these

10. Joe's present age is 2/7th of his father's present age. Joe's brother is 3 yr older than Joe. The respective ratio between present ages of Joe's father and Joe's brother is 14 : 5. What is Joe's present age?   **[SBI PO 2015]**
    (a) 6 yr (b) 15 yr (c) 12 yr (d) 18 yr
    (e) 20 yr

11. Two years hence, the respective ratio between A's age at that time and B's age at that time will be 6 : 5. A's age thirteen years ago was half of B's present age. What is A's present age?
    [IBPS Clerk (Pre) 2016]
    (a) 16 yr (b) 40 yr (c) 28 yr (d) 22 yr
    (e) 34 yr

12. Four years ago, the respective ratio between 1/2 of A's age at that time and four times of B's age at that time was 5 : 12. Eight years hence 1/2 of A's age at that time will be less than B's age at that time by 2 yr. What is B's present age?
    [IBPS Clerk 2015]
    (a) 10 yr (b) 14 yr (c) 12 yr (d) 5 yr
    (e) 8 yr

13. The ratio between the present ages of A and B is 2 : 3 respectively. B's age sixteen years hence will be twice of A's age four years hence. What is the difference between the present ages of A and B?   **[LIC AAO 2016]**
    (a) 6 yr (b) 12 yr (c) 8 yr (d) 4 yr
    (e) 15 yr

14. The present age of Charu is 2.5 times the present age of Harsh. Had Harsh been two years younger and Charu been 13 years older, Charu's age would have

been 3.5 times Harsh's age. What is Harsh's present age? **[SBI Clerk (Pre) 2016]**
(a) 48 yr (b) 25 yr (c) 20 yr (d) 28 yr
(e) 36 yr

15. At present, Ami's age is twice Rio's age and Cami is two years older than Ami. Two years ago, the respective ratio between Rio's age at that time and Cami's age at that time was 4 : 9. What will be Ami's age after four years?
[SBI PO (Pre) 2016]
(a) 40 yr (b) 30 yr (c) 42 yr (d) 36 yr
(e) 48 yr

16. At present, Ron is eight years younger to Emma. Harry is two years younger to Emma. If the respective ratio between the present age of Ron and that of Harry is 3 : 4. What is Harry's present age?
[IBPS SO 2016]
(a) 20 yr (b) 8 yr (c) 12 yr (d) 24 yr
(e) 18 yr

17. A woman is 5 yr younger than her husband and 3 times as old as her daughter. If the daughter attains 21 yr of age after 6 yr, what is the present age of the husband? **[IB PA 2016]**
(a) 50 yr (b) 55 yr (c) 40 yr (d) 45 yr

18. Ram's age after 17 yr will be 6 yr less than twice his present age. The respective ratio between Vivan's present age and Amit's present age is 21 : 19. If Amit's age 31 yr hence will be same as three times Ram's present age, what will be Vivan's age 4 yr hence? **[IBPS Clerk (Mains) 2017]**
(a) 44 yr (b) 45yr (c) 46 yr (d) 48 yr
(e) Other than those given as options

19. The difference between the present ages of Trisha and Shalini is 14 yr. Seven years ago, the ratio of their ages was 5 : 7, respectively. What is Trisha's present age? **[IBPS Clerk 2015]**
(a) 49 yr (b) 56 yr (c) 63 yr (d) 35 yr
(e) 40 yr

20. Five years ago, the respective ratio between the age of Opi and that of Mini was 5 : 3. Nikki is 5 yr younger to Opi. Nikki is 5 yr older to Mini. What is Nikki's present age? **[IBPS SO 2016]**
(a) 35 yr (b) 25 yr (c) 20 yr (d) 10 yr
(e) 30 yr

21. The age of Mr. X last year was the square of a number and it would be the cube of a number next year. What is the least number of years he must wait for his age to become the cube of a number again? **[UPSC CSAT 2017]**
(a) 42 (b) 38 (c) 25 (d) 16

22. A is eighteen years older to B. The respective ratio of B's age six years hence and C's present age is 3 : 2. If at present A's age is twice the age of C, what was B's age four years ago? **[SBI PO (Pre) 2017]**
(a) 24 yr (b) 28 yr (c) 29 yr (d) 20 yr
(e) 26 yr

23. Leela got married 6 yr ago. Today her age is $1\frac{1}{4}$ times her age at the time of her marriage. Her son's age is $\frac{1}{10}$ times her age. What is the present age of her son?
[CDS 2017 (I)]
(a) 1 yr (b) 2 yr (c) 3 yr (d) 4 yr

24. At present, Anil is 1.5 times of Purvi's age. 8 yr hence, the respective ratio between Anil and Purvi's ages will be 25 : 18. What is Purvi's present age?
[IBPS Clerk 2011]
(a) 50 yr (b) 28 yr (c) 42 yr (d) 36 yr
(e) None of these

25. At present, Meena is eight times her daughter's age. 8 yr from now, the ratio of the ages of Meena and her daughter will be 10 : 3, respectively. What is Meena's present age? **[IDBI PO 2009]**
(a) 32 yr
(b) 40 yr
(c) 36 yr
(d) Cannot be determined
(e) None of the above

26. Before 7 yr, the ratio of ages of A and B was 3 : 4. After 9 yr, ratio of their ages will be 7 : 8. The present age of B will be
[SBI PO 2012]
(a) 16 yr (b) 19 yr (c) 28 yr (d) 23 yr
(e) None of these

27. At present, Mani is 4 times the age of Vaibhav. After 3 yr, Mani will be 2 times as old as Vaibhav. What is the present age of Mani?
(a) 10 yr (b) 6 yr (c) 12 yr (d) 8 yr

28. John's grandfather was five times older to him 5 yr ago. He would be two times of his age after 25 yr from now. What is the ratio of John's age to that of his grandfather? **[SNAP 2016]**
(a) 7 : 11 (b) 5 : 11 (c) 3 : 11 (d) 4 : 11

29. Kashish is as elder to Arun as he is younger to Manish. The sum of age of Arun and Manish is 26 yr. Then, the age of Kashish is **[SSC (10+2) 2014]**
(a) 12 yr (b) 14 yr (c) 13 yr (d) 15 yr

Answer with Solutions

1. (b) ∵ Age of Raju when he got the job
$= 17 + 3 + 2 + 1 = 23$ yr
∴ Age of Raju at the time of marriage
$= 23 + 3 = 26$ yr

2. (d) Let the present age of the son be x yr.
Then, present age of father $= 9x$ yr
and present age of mother $= 8x$ yr
Now, according to the question,
$9x + 8x = 51 \Rightarrow 17x = 51$
$\Rightarrow \quad x = 3$ yr

3. (a) Let the father's present age be x yr and present age of his children be a and b yr.
Then, according to the question,
∴ $(a + b) = \dfrac{x}{3}$
and $(a + b) + 20 + 20 = x + 20$
$\Rightarrow \dfrac{x}{3} + 20 = x$
$\Rightarrow \dfrac{2x}{3} = 20$
$\Rightarrow x = 10 \times 3 = 30$
Hence, the father's age is 30 yr.

4. (c) ∵ Total age of Sachin and Saurabh
$= 36 \times 2 = 72$ yr
Let age of Saurabh $= x$ yr
Then, age of Sachin $= (x + 8)$ yr
According to the question,
$x + x + 8 = 72 \Rightarrow 2x = 64$
∴ $x = 32$ yr

5. (c) Let age of Raman's daughter $= x$ yr
Then, age of Raman $= 3x$ yr
and age of Raman's mother
$= \dfrac{13}{9} \times 3x = \dfrac{13}{3} x$ yr
According to the question,
$x + 3x + \dfrac{13}{3} x = 125$
$\Rightarrow 3x + 9x + 13x = 125 \times 3$
∴ $x = \dfrac{125 \times 3}{25} = 15$
Hence, required difference
$= \dfrac{13}{3} x - x = \dfrac{10x}{3}$
$= \dfrac{10 \times 15}{3} = 50$ yr

6. (a) Here, $1896 + (x - 4) = x^2$
$\Rightarrow 1896 = x^2 - (x - 4)$
$\Rightarrow x^2 - x - 1892 = 0$
$\Rightarrow x^2 - 44x + 43x - 1892 = 0$

$\Rightarrow x(x - 44) + 43(x - 44) = 0$
$\Rightarrow (x + 43)(x - 44) = 0$
$\Rightarrow x = 44$ yr $\quad [\because x \ne -43]$

7. (d) Let Smita's present age $= S$
and her mother's present age $= M$
According to the question,
$\dfrac{S}{M} = \dfrac{3}{7}$
$\Rightarrow \quad 3M = 7S$ ...(i)
and $\quad M - S = 32$ ...(ii)
$\Rightarrow \quad 3M - 3S = 96$
$\Rightarrow \quad 7S - 3S = 96 \quad [\because 3M = 7S]$
$\Rightarrow \quad 4S = 96 \Rightarrow S = 24$ yr
∴ $\quad 3M = 7S = 7 \times 24$
$\Rightarrow \quad M = \dfrac{7 \times 24}{3} = 56$ yr
∴ Ratio of their ages after 4 yr
$= \dfrac{S + 4}{M + 4} = \dfrac{24 + 4}{56 + 4}$
$= \dfrac{28}{60} = \dfrac{7}{15} = 7 : 15$

8. (a) Let the present ages of Manoj and Wasim be $3x$ yr and $11x$ yr, respectively.
According to the question,
$11x = 85 - 7 - 12$
$\Rightarrow \quad 11x = 66 \Rightarrow x = 6$
∴ Present age of Manoj $= 3 \times 6 = 18$ yr
Hence, present age of Manoj's father
$= 18 + 25 = 43$ yr

9. (b) Let present ages of Indira and Lizzy be $3x$ yr and $8x$ yr, respectively.
Then, $3x + 8 = 20 \Rightarrow 3x = 12 \Rightarrow x = 4$
Hence, Lizzy's required age $= 8x - 5$
$= 32 - 5 = 27$ yr

10. (c) Let the Joe's father present age be x yr.
Joe's present age $= \dfrac{2}{7} x$ yr
Joe's elder brother present age
$= \left(\dfrac{2}{7} x + 3\right)$ yr
According to the question,
$\dfrac{x}{3 + \dfrac{2}{7} x} = \dfrac{14}{5} \Rightarrow 5x = 14\left(3 + \dfrac{2}{7} x\right)$
$\Rightarrow \quad 5x = 42 + 4x \Rightarrow 5x - 4x = 42$
$\Rightarrow \quad x = 42$ yr
∴ Joe's present age $= \dfrac{2}{7} x = \dfrac{2}{7} \times 42 = 12$ yr

11. (d) Let the present ages of A and B be x yr and y yr, respectively.

Problem Based on Ages / 363

According to the question,
$$\frac{x+2}{y+2} = \frac{6}{5}$$
$$\Rightarrow \quad 5x + 10 = 6y + 12$$
$$\Rightarrow \quad 5x - 6y = 12 - 10$$
$$\Rightarrow \quad 5x - 6y = 2$$
$$\Rightarrow \quad 5x - 2 = 6y$$
$$\Rightarrow \quad y = \frac{5x-2}{6} \quad \ldots(i)$$
$$x - 13 = \frac{1}{2}y$$
$$\Rightarrow \quad x = \frac{1}{2}y + 13$$
$$\Rightarrow \quad 2x = y + 26$$
$$\Rightarrow \quad 2x = \frac{5x-2}{6} + 26 \quad \left[\because y = \frac{5x-2}{6}\right]$$
$$\Rightarrow \quad 12x = 5x - 2 + 156$$
$$\Rightarrow \quad 12x - 5x = 154$$
$$\Rightarrow \quad 7x = 154$$
$$\Rightarrow \quad x = 22$$
Thus, the present age of A is 22 yr.

12. (a) Let the present ages of A and B be x yr and y yr.
4 yr ago, age of A = $(x-4)$ yr and age of B = $(y-4)$ yr
According to the question,
$$\frac{(x-4)}{\frac{2}{4(y-4)}} = \frac{5}{12}$$
$$\Rightarrow \quad \frac{(x-4)}{8(y-4)} = \frac{5}{12} \Rightarrow \frac{x-4}{y-4} = \frac{10}{3}$$
$$\Rightarrow \quad 3x - 12 = 10y - 40$$
$$\Rightarrow \quad 10y - 3x = 28 \quad \ldots(i)$$
After 8 yr, age of A = $(x+8)$ yr and age of B = $(y+8)$ yr
Now, $\quad \frac{x+8}{2} + 2 = y + 8$
$$\Rightarrow \quad \frac{x}{2} + 4 + 2 = y + 8$$
$$\Rightarrow \quad y - \frac{x}{2} = -2$$
$$\Rightarrow \quad 2y - x = -4 \quad \ldots(ii)$$
On multiplying Eq. (ii) by 3 and then subtracting it from Eq. (i), we get
$$10y - 3x = 28$$
$$6y - 3x = -12$$
$$\underline{-\quad +\quad\quad +}$$
$$4y = 40$$
$$\Rightarrow \quad y = 10$$
Hence, the present age of B is 10 yr.

13. (c) Let the present ages of A and B be x yr and y yr.
Then, according to the question,
$$\frac{x}{y} = \frac{2}{3}$$

$$\Rightarrow \quad 3x = 2y$$
and $\quad \frac{x+4}{y+16} = \frac{1}{2}$
$$\Rightarrow \quad 2x + 8 = y + 16$$
$$\Rightarrow \quad 2x = y + 8$$
$$\Rightarrow \quad 2x = \frac{3x}{2} + 8$$
$$\Rightarrow \quad 4x - 3x = 16 \Rightarrow x = 16$$
∴ A's present age = 16 yr
and B's present age
$$y = \frac{3x}{2} = \frac{3 \times 16}{2} = 24 \text{ yr}$$
∴ Difference of present ages of A and B
$= 24 - 16 = 8$ yr

14. (c) Let present age of Harsh = x yr and present age of Charu = y yr
Then, according to the question,
$$2.5x = y \quad \ldots(i)$$
and $\quad 3.5(x-2) = (y+13) \quad \ldots(ii)$
$$\Rightarrow \quad 3.5(x-2) = 2.5x + 13 \quad \text{[from Eq. (i)]}$$
$$\Rightarrow \quad 3.5x - 7 = 2.5x + 13$$
$$\Rightarrow \quad 3.5x - 2.5x = 13 + 7 \Rightarrow x = 20$$
Hence, the present age of Harsh is 20 yr.

15. (a) Let Ami's present age = x yr
Rio's present age = y yr
and Cami's present age = z yr
According to the question,
$$x = 2y \text{ and } z = (x+2)$$
Also, $\quad \frac{y-2}{z-2} = \frac{4}{9} \Rightarrow \frac{\left(\frac{x}{2} - 2\right)}{(x+2) - 2} = \frac{4}{9}$
$$\Rightarrow \quad \frac{x-4}{2x} = \frac{4}{9} \Rightarrow 9x - 36 = 8x$$
$$\Rightarrow \quad 9x - 8x = 36 \Rightarrow x = 36$$
∴ Ami's present age = 36 yr
and Ami's age after 4 yr
$= 36 + 4 = 40$ yr

16. (d) Let present age of Ron = R
Present age of Harry = H
and present age of Emma = E
According to the question,
$$R = E - 8 \quad \ldots(i)$$
$$H = E - 2 \quad \ldots(ii)$$
and $\quad \frac{R}{H} = \frac{3}{4} \quad \ldots(iii)$
Now, putting the values of R and H from Eqs. (i) and (ii) in Eq. (iii), we get
$$\frac{E-8}{E-2} = \frac{3}{4} \Rightarrow 4E - 32 = 3E - 6$$
$$\Rightarrow \quad 4E - 3E = -6 + 32$$
$$\Rightarrow \quad E = 26$$
Now, putting the value of E in Eq. (ii), we get
$$H = 26 - 2 = 24$$
Hence, present age of Harry is 24 yr.

17. (a) Let the present ages of woman, husband and daughter be W yr, H yr and D yr, respectively.
According to the question,
$W = H - 5$ ...(i)
$W = 3D$ ...(ii)
and $D + 6 = 21$
$\Rightarrow D = 21 - 6 = 15$ yr
From Eq. (ii), we get $W = 3 \times 15 = 45$ yr
From Eq. (i), we get $H = W + 5$
$\Rightarrow H = 45 + 5 = 50$ yr
Hence, the husband's present age is 50 yr.

18. (c) Let present age of Ram = x yr
Given, $x + 17 = 2x - 6$
$\Rightarrow x = 23$
$\therefore$ Present age of Ram = 23 yr
Now, let present ages of Vivan and Amit be $21a$ and $19a$.
Then, $19a + 31 = 3x$
$\Rightarrow 19a + 31 = 3 \times 23$
$\Rightarrow 19a = 69 - 31$
$\Rightarrow 19a = 38$
$\Rightarrow a = 2$
$\therefore$ Present age of Vivan
$= 21a = 21 \times 2$
$= 42$ yr
Now, age of Vivan after 4 yr
$= 42 + 4 = 46$ yr

19. (b) Let the present age of Shalini be S yr and present age of Trisha be T yr.
7 yr ago, age of Shalini = $(S - 7)$ yr and age of Trisha = $(T - 7)$ yr
According to the question,
$T - S = 14$ ...(i)
and $\dfrac{S-7}{T-7} = \dfrac{5}{7} \Rightarrow \dfrac{S-7}{14+S-7} = \dfrac{5}{7}$
[from Eq. (i)]
$\Rightarrow \dfrac{S-7}{S+7} = \dfrac{5}{7} \Rightarrow 7S - 49 = 5S + 35$
$\Rightarrow 2S = 84 \Rightarrow S = 42$ yr
From Eq. (i), $T = 42 + 14 = 56$ yr
Hence, the present age of Trisha is 56 yr.

20. (b) Let present age of Nikki = x yr
Present age of Opi = y yr
and present age of Mini = z yr
According to the question,
$\dfrac{y-5}{z-5} = \dfrac{5}{3}$ ...(i)
Then, $x = y - 5$ ...(ii)
and $x = z + 5$
$\Rightarrow z = x - 5$ ...(iii)
Now, putting the values of $(y - 5)$ and z from Eqs. (ii) and (iii) to Eq. (i), we get
$\dfrac{x}{x-5-5} = \dfrac{5}{3} \Rightarrow \dfrac{x}{x-10} = \dfrac{5}{3}$

$\Rightarrow 5x - 50 = 3x \Rightarrow 5x - 3x = 50$
$\Rightarrow 2x = 50 \Rightarrow x = 25$
Hence, the present age of Nikki is 25 yr.

21. (b) Let the present age of Mr. X be x yr.
Then, according to the question,
$(x - 1) = y^2$ and $(x + 1) = z^3$
$\Rightarrow y^2 + 1 = z^3 - 1$
$\Rightarrow z^3 - y^2 = 2$
By hit and trial method, $z = 3$ and $y = 5$.
$\therefore$ Present age of Mr. $X = y^2 + 1 = 26$
Now, next cube after 27 is 64.
Hence, Mr. X have to wait $64 - 26 = 38$ yr for his age to become cube of a number again.

22. (e) Given, $B + 18 = A \Rightarrow B = A - 18$
and $\dfrac{B+6}{C} = \dfrac{3}{2}$ and $A = 2C$
Now, $\dfrac{B+6}{C} = \dfrac{3}{2}$
$\Rightarrow \dfrac{A-18+6}{C} = \dfrac{3}{2}$ $[\because B = A - 18]$
$\Rightarrow \dfrac{A-12}{C} = \dfrac{3}{2}$
$\Rightarrow \dfrac{2C-12}{C} = \dfrac{3}{2}$ $[\because A = 2C]$
$\Rightarrow 4C - 24 = 3C \Rightarrow 4C - 3C = 24$
$\Rightarrow C = 24$ yr
$\therefore A = 2 \times 24 = 48$ yr
and $B = 48 - 18 = 30$ yr
Hence, B's age four years ago
$= (30 - 4)$ yr $= 26$ yr

23. (c) Let the age of Leela at her marriage was x yr.
$\therefore$ Present age of Leela = $(x + 6)$ yr
According to the question,
$x + 6 = \dfrac{5}{4}x$
$\Rightarrow \dfrac{1}{4}x = 6 \Rightarrow x = 24$
$\because$ Present age of Leela = $(24 + 6) = 30$ yr
$\therefore$ Leela's son age = $\dfrac{1}{10}$ Age of Leela
$= \dfrac{1}{10} \times 30 = 3$ yr

24. (b) Let age of Purvi = x yr
Then, age of Anil = $1.5x$ yr
According to the question,
$\dfrac{1.5x + 8}{x + 8} = \dfrac{25}{18}$
$\Rightarrow 27x + 144 = 25x + 200$
$\therefore x = \dfrac{56}{2} = 28$
Hence, the present age of Purvi is 28 yr.

25. (a) Let age of daughter $= x$ yr
Then, age of Meena $= 8x$ yr
According to the question,
$$\frac{8x+8}{x+8} = \frac{10}{3}$$
$\Rightarrow \quad 24x + 24 = 10x + 80$
$\Rightarrow \quad 14x = 56 \Rightarrow x = 4$
Hence, Meena's present age $= 4 \times 8 = 32$ yr

26. (d) Let the ages of A and B before 7 yr were $3x$ yr and $4x$ yr, respectively.
$\therefore$ Present age of $A = 3x + 7$
and present age of $B = 4x + 7$
Now, according to the question,
$$\frac{3x+7+9}{4x+7+9} = \frac{7}{8}$$
$\Rightarrow \quad 24x + 128 = 28x + 112 \Rightarrow 4x = 16$
$\therefore \quad x = 4$
Hence, present age of $B = 4 \times 4 + 7$
$= 16 + 7 = 23$ yr

27. (b) Let Mani's and Vaibhav's present ages be M yr and V yr, respectively.
According to the question,
$\quad M = 4V$ ...(i)
and $\quad M + 3 = 2(V + 3)$ ...(ii)
$\Rightarrow \quad 4V + 3 = 2V + 6$ [from Eq. (i)]
$\Rightarrow \quad 2V = 3$
$\Rightarrow \quad V = 1.5$ yr
$\therefore$ Mani's present age $= 4V = 4 \times 1.5 = 6$ yr

Fast Track Method
Here, $t = 3$, $m = 4$, $n = 2$
$\therefore$ Mani's age $= tm\left(\dfrac{n-1}{m-n}\right)$
[by Technique 1]
$= 3 \times 4 \left(\dfrac{2-1}{4-2}\right) = \dfrac{12 \times 1}{2} = 6$ yr

28. (c) Let the age of John be x yr and age of grandfather be y yr.
Now, 5 yr ago, John's age $= (x - 5)$ yr
and 5 yr ago, grandfather's age $= (y - 5)$ yr
According to the question,
$\quad y - 5 = 5(x - 5)$
$\Rightarrow \quad y - 5 = 5x - 25$
$\Rightarrow \quad y - 5x = -20$ ...(i)

After 25 yr, John's age $= (x + 25)$ yr
and grandfather's age $= (y + 25)$ yr
Then, $2(x + 25) = y + 25$
$\Rightarrow \quad 2x + 50 = y + 25$
$\Rightarrow \quad 2x - y = -25$ ...(ii)
From Eqs. (i) and (ii), we get
$\quad x = 15$, $y = 55$
$\therefore$ Required ratio of ages $= 15:55 = 3:11$

Fast Track Method
Here, $t_1 = 5$, $t_2 = 25$, $n = 2$ and $m = 5$
$\therefore$ Grandfather's age
$= \dfrac{t_2 m(n-1) + t_1 n(m-1)}{(m-n)}$
[by Technique 2]
$= \dfrac{25 \times 5(2-1) + 5 \times 2(5-1)}{5-2}$
$= \dfrac{125 + 40}{3} = \dfrac{165}{3} = 55$ yr
and John's age $= \dfrac{t_2(n-1) + t_1(m-1)}{(m-n)}$
$= \dfrac{25(2-1) + 5(5-1)}{5-2}$
$= \dfrac{25 + 20}{3} = \dfrac{45}{3} = 15$ yr
$\therefore$ Required ratio $= \dfrac{\text{John's age}}{\text{Grandfather's age}}$
$= \dfrac{15}{55} = \dfrac{3}{11} = 3:11$

29. (c) Let present age of Kashish be x yr and he is younger to Manish by y yr.
Then, Manish's age $= (x + y)$ yr
and Arun's age $= (x - y)$ yr
According to the question,
Sum of ages of Manish and Arun $= 26$
$\Rightarrow \quad x + y + x - y = 26$
$\Rightarrow \quad 2x = 26 \Rightarrow x = 13$
Hence, the present age of Kashish is 13 yr.

Fast Track Method
Age of Kashish
$= \dfrac{\text{Sum of ages of Arun and Manish}}{2}$
$= \dfrac{26}{2} = 13$ yr [by Technique 3]

Chapter 22

Work and Time

In this chapter, we will study techniques to solve problems based on work and its completion time as well as number of persons required to finish the given work in stipulated time. Suppose you are a contractor and you got a contract to construct a flyover in a certain time. For this, you need to calculate the number of men required to finish the work according to their work efficiency.

Some Important Relations

1. **Work and Person** Directly proportional (more work, more men and conversely more men, more work).
2. **Time and Person** Inversely proportional (more men, less time and conversely more time, less men).
3. **Work and Time** Directly proportional (more work, more time and conversely more time, more work).
   + While solving these types of problems, the work done is always supposed to be equal to 1.

Basic Rules Related to Work and Time

Rule ❶

If a person can do a piece of work in n days (hours), then that person's 1 day's (hour's) work $= \dfrac{1}{n}$.

Ex. 1 Vandana completes a work in 35 days. What work will she do in 1 day?

Sol. We know that, if a person can do a piece of work in n days, then person's 1 day's work $= \dfrac{1}{n}$

Here, $n = 35$

∴ Required work done $= \dfrac{1}{35}$

Work and Time / 367

Rule 2

If a person's 1 day's (hour's) work = $\frac{1}{n}$, then the person will complete the work in n days (hours).

Ex. 2 Kavi completes $\frac{1}{13}$ part of a certain work in 1 day. In how many days, will he complete the whole work?

Sol. We know that, if a person's 1 day's work = $\frac{1}{n}$, then the person will complete the whole work in n days.

Here, $\frac{1}{n} = \frac{1}{13}$

∴ Required number of days = 13

Rule 3

If a person is n times efficient than the second person, then work done by

First person : Second person = n : 1

and time taken to complete a work by

First person : Second person = 1 : n

Ex. 3 P can do a work 3 times faster than Q and therefore takes 40 days less than Q. Find the time in which P and Q can complete the work individually.

Sol. We know that, if a person is n times efficient than the second person, then time taken to complete a work by

First person : Second person = 1 : n

∴ Time taken to complete the work by $P : Q = 1 : 3$

According to the question,

Time taken by Q − Time taken by $P = 40$

⇒ $3K - K = 40$ ⇒ $2K = 40$

⇒ $K = 20$

∴ Number of days required by $P = 20$ and number of days required by $Q = 60$

Rule 4

If ratio of number of men required to complete a work is $m : n$, then the ratio of time taken by them will be $n : m$.

Ex. 4 If 12 men can finish a work in 20 days, then find the number of days required to complete the same work by 15 men.

Sol. We know that if ratio of numbers of men required to complete a work is $m : n$, then ratio of time taken by them will be $n : m$.

According to the question,

Ratio of numbers of men = 12 : 15 = 4 : 5

∴ Ratio of time taken = 5 : 4

Let us suppose 15 men can finish a work in x days.

Then, $20 : x = 5 : 4$

⇒ $x = 16$

∴ Required number of days = 16

Fast Track Techniques to solve the QUESTIONS

Technique 1

(i) If A can do a piece of work in x days and B can do the same work in y days, then $(A + B)$'s 1 day's work $= \dfrac{1}{x} + \dfrac{1}{y} = \dfrac{x+y}{xy}$

Time taken by $(A + B)$ to complete the work $= \dfrac{xy}{x+y}$ days

(ii) If A can do a piece of work in x days, B can do the same work in y days and C can do the same work in z days, then

$(A + B + C)$'s 1 day work $= \dfrac{1}{x} + \dfrac{1}{y} + \dfrac{1}{z} = \dfrac{yz + xz + xy}{xyz}$

Time taken by $(A + B + C)$ to complete the work $= \dfrac{xyz}{xy + yz + zx}$ days

Ex. 5 A can do a piece of work in 10 days and B can do the same work in 12 days. How long will they take to finish the work, if both work together?

Sol. A's 1 day's work $= \dfrac{1}{10}$ and B's 1 day's work $= \dfrac{1}{12}$

$(A + B)$'s 1 day's work $= \dfrac{1}{10} + \dfrac{1}{12} = \dfrac{6+5}{60} = \dfrac{11}{60}$ day

$\therefore (A + B)$ complete the whole work in $\dfrac{60}{11}$ days $= 5\dfrac{5}{11}$ days.

Fast Track Method

Here, $x = 10$ and $y = 12$

$\therefore$ Number of days taken by A and B $= \dfrac{xy}{x+y} = \dfrac{10 \times 12}{10 + 12} = 5\dfrac{5}{11}$ days

Ex. 6 If A can do a piece of work in 4 days, B can do the same work in 8 days and C can do the same work in 12 days, then working together, how many days will they take to complete the work?

Sol. A's 1 day's work $= \dfrac{1}{4}$; B's 1 day's work $= \dfrac{1}{8}$; C's 1 day's work $= \dfrac{1}{12}$

According to the question,

$(A + B + C)$'s 1 day work $= \dfrac{1}{4} + \dfrac{1}{8} + \dfrac{1}{12} = \dfrac{6+3+2}{24} = \dfrac{11}{24}$

$\therefore (A + B + C)$ complete the whole work in $\dfrac{24}{11}$ days $= 2\dfrac{2}{11}$ days.

Fast Track Method

Here, $x = 4, y = 8$ and $z = 12$

$\therefore$ Time taken by $(A + B + C)$ to complete the work $= \dfrac{xyz}{xy + yz + zx} = \dfrac{4 \times 8 \times 12}{32 + 96 + 48}$

$= \dfrac{4 \times 8 \times 12}{176} = \dfrac{24}{11} = 2\dfrac{2}{11}$ days

Work and Time / 369

Technique 2 If A and B can complete a work in x days and A alone can complete that work in y days, then the number of days required to complete the work by $B = \dfrac{xy}{y-x}$ days.

Ex. 7 A and B together can do a piece of work in 12 days and A alone can do it in 18 days. In how many days can B alone do it?

Sol. $(A+B)$'s 1 day's work $= \dfrac{1}{12}$ and A's 1 day's work $= \dfrac{1}{18}$

$\therefore$ B's 1 day's work $= (A+B)$'s 1 day's work $- A$'s 1 day's work

$= \dfrac{1}{12} - \dfrac{1}{18} = \dfrac{3-2}{36} = \dfrac{1}{36}$

$\therefore$ Time taken by B to complete the work alone $= 36$ days

Fast Track Method

Here, $x = 12$ and $y = 18$

$\therefore$ Time taken by $B = \dfrac{xy}{y-x} = \dfrac{12 \times 18}{18-12} = 36$ days

Technique 3 If M_1 persons can do W_1 work in D_1 days working T_1 h in a day and M_2 persons can do W_2 work in D_2 days working T_2 h in a day, then the relationship between them is

$$M_1 D_1 T_1 W_2 = M_2 D_2 T_2 W_1.$$

Ex. 8 10 persons can make 20 toys in 12 days working 12 h per day. Then, in how many days can 24 persons make 32 toys working 16 h per day?

Sol. Given that, $M_1 = 10, M_2 = 24, D_1 = 12, D_2 = ?, T_1 = 12, T_2 = 16, W_1 = 20$ and $W_2 = 32$

According to the formula, $M_1 D_1 T_1 W_2 = M_2 D_2 T_2 W_1$

$\Rightarrow 10 \times 12 \times 12 \times 32 = 24 \times D_2 \times 16 \times 20$ $\Rightarrow D_2 = \dfrac{10 \times 12 \times 12 \times 32}{24 \times 16 \times 20} = \dfrac{12}{2} = 6$ days

Technique 4 If A and B can do a piece of work in x days, B and C can do the same work in y days and A and C can do it in z days, then working together A, B and C can do that work in $\dfrac{2xyz}{xy+yz+zx}$ days.

Ex. 9 A and B can do a piece of work in 3 days. B and C can do the same work in 9 days, while C and A can do it in 12 days. Find the time in which A, B and C can finish the work, working together.

Sol. Here, $(A+B)$'s 1 day's work $= \dfrac{1}{3}$, $(B+C)$'s 1 day's work $= \dfrac{1}{9}$

and $(C+A)$'s 1 day's work $= \dfrac{1}{12}$

Now, $2(A+B+C)$'s 1 day's work $= \dfrac{1}{3} + \dfrac{1}{9} + \dfrac{1}{12} = \dfrac{12+4+3}{36} = \dfrac{19}{36}$

$\therefore$ $(A+B+C)$'s 1 day's work $= \dfrac{19}{36 \times 2} = \dfrac{19}{72}$

Hence, $(A+B+C)$ complete the work in $\dfrac{72}{19}$ days or in $3\dfrac{15}{19}$ days.

Fast Track Method
Here, $x = 3, y = 9$ and $z = 12$
According to the formula,
Required time taken by A, B and $C = \dfrac{2xyz}{xy + yz + zx} = \dfrac{2 \times 3 \times 9 \times 12}{3 \times 9 + 9 \times 12 + 12 \times 3}$

$= \dfrac{2 \times 3 \times 9 \times 12}{27 + 108 + 36} = \dfrac{2 \times 3 \times 9 \times 12}{171} = \dfrac{72}{19} = 3\dfrac{15}{19}$ days

Technique 5 If a_1 men or b_1 women can finish a work in D days, then time taken by a_2 men and b_2 women to complete the work in $\dfrac{D(a_1 b_1)}{(a_2 b_1 + a_1 b_2)}$ days.

Ex. 10 If 6 men or 8 women can reap a field in 86 days, how long will 14 men and 10 women take to reap it?

Sol. ∵ 6 men = 8 women ⇒ 1 man = $\dfrac{8}{6} = \dfrac{4}{3}$ women

∴ 14 men = $\dfrac{4}{3} \times 14$ women = $\dfrac{56}{3}$ women

14 men + 10 women = $\left(\dfrac{56}{3} + 10\right)$ women = $\dfrac{86}{3}$ women

Now, $M_1 = 8, M_2 = \dfrac{86}{3}, W_1 = W_2 = 1, D_1 = 86$ and $D_2 = ?$

According to the formula, $M_1 D_1 W_2 = M_2 D_2 W_1$

⇒ $8 \times 86 \times 1 = \dfrac{86}{3} \times D_2 \times 1 \Rightarrow D_2 = \dfrac{8 \times 86 \times 3}{86} = 24$ days

Fast Track Method
Here, $a_1 = 6, b_1 = 8, a_2 = 14, b_2 = 10$ and $D = 86$

∴ Number of days = $\dfrac{D(a_1 b_1)}{(a_2 b_1 + a_1 b_2)} = \dfrac{86 \times 6 \times 8}{14 \times 8 + 6 \times 10} = \dfrac{86 \times 6 \times 8}{172} = 24$ days

Technique 6 If A can do a work in x days and B can do y% faster than A, then B will complete the work in $\dfrac{100 x}{100 + y}$ days.

Ex. 11 Kamal can do a work in 15 days and Vimal is 50% more expert than Kamal to complete the same work, then find total time taken to complete the work by Vimal.

Sol. Let Vimal takes $2x$ days to complete the work. Then, Kamal will take $3x$ days to complete the same work.
According to the question,
$3x = 15 \Rightarrow x = 5$
∴ Vimal will take = $2x = 2 \times 5 = 10$ days

Fast Track Method
Here, $x = 15$ days and $y = 50$%

Now, time taken by Vimal = $\dfrac{100 x}{100 + y} = \dfrac{100 \times 15}{100 + 50}$

$= \dfrac{1500}{150} = 10$ days

Work and Time / 371

Technique 7   If a men can do a piece of work in x days and b boys can do the same work in y days, then time taken to complete the same work by c men and d boys will be $\dfrac{1}{\dfrac{c}{ax} + \dfrac{d}{by}}$ days.

Ex. 12   If 5 men can do a work in 2 days and 3 boys can do the same work in 5 days, then find the time taken to complete same work by 10 men and 3 boys.

Sol. $\because$ 1 day work of 5 men $= \dfrac{1}{2}$

$\therefore$ 1 day work of 1 man $= \dfrac{1}{2 \times 5} = \dfrac{1}{10}$

and 1 day work of 3 boys $= \dfrac{1}{5}$

$\therefore$ 1 day work of 1 boy $= \dfrac{1}{3 \times 5} = \dfrac{1}{15}$

Hence, 1 day work of 10 men and 3 boys $= 10 \times \dfrac{1}{10} + 3 \times \dfrac{1}{15} = 1 + \dfrac{1}{5} = \dfrac{6}{5}$

$\therefore$ Time taken by 10 men and 3 boys to complete the work $= \dfrac{5}{6}$ days

Fast Track Method

Given, $a = 5, b = 3, x = 2, y = 5, c = 10$ and $d = 3$

Time taken by 10 men and 3 boys $= \dfrac{1}{\dfrac{c}{ax} + \dfrac{d}{by}} = \dfrac{1}{\dfrac{10}{5 \times 2} + \dfrac{3}{3 \times 5}} = \dfrac{1}{1 + \dfrac{1}{5}} = \dfrac{5}{6}$ days

Technique 8   If a_1 men and b_1 boys can complete a work in x days, while a_2 men and b_2 boys can complete the same work in y days, then

$$\dfrac{\text{1 day work of 1 man}}{\text{1 day work of 1 boy}} = \dfrac{(yb_2 - xb_1)}{(xa_1 - ya_2)}$$

Ex. 13   If 12 men and 16 boys can finish a work in 5 days, while 13 men and 24 boys can finish the same work in 4 days. Compare the one day's work of 1 man and 1 boy.

Sol. $\because$ 1 day work of $12 M + 16 B = \dfrac{1}{5}$ ...(i)

and 1 day work of $13 M + 24 B = \dfrac{1}{4}$ ...(ii)

On solving Eqs. (i) and (ii), we get

1 day work of 1 man $= \dfrac{1}{100}$ and 1 day work of 1 boy $= \dfrac{1}{200}$

$\therefore \quad \dfrac{\text{1 day work of 1 man}}{\text{1 day work of 1 boy}} = \dfrac{200}{100} = \dfrac{2}{1}$

Fast Track Method

Here, $a_1 = 12$, $b_1 = 16$, $x = 5$, $a_2 = 13$, $b_2 = 24$ and $y = 4$

$$\therefore \frac{1 \text{ day work of 1 man}}{1 \text{ day work of 1 boy}} = \frac{(yb_2 - xb_1)}{(xa_1 - ya_2)}$$

$$= \frac{4 \times 24 - 5 \times 16}{5 \times 12 - 4 \times 13} = \frac{96 - 80}{60 - 52} = \frac{16}{8} = \frac{2}{1}$$

Technique 9 If X takes a days more to complete a work than the time taken by $(X + Y)$ to do same work and Y takes b days more than the time taken by $(X + Y)$ to do the same work, then $(X + Y)$ do the work in $\sqrt{ab}$ days.

Ex. 14 When X alone does a piece of work, he takes 16 days more than the time taken by $(X + Y)$ to complete the work, while Y alone takes 9 days more than the time taken by $(X + Y)$ to finish the work. What time X and Y together will take to finish this work?

Sol. Let time taken by X and Y to complete the work $= z$ days
According to the question,

$$\frac{1}{16 + z} + \frac{1}{9 + z} = \frac{1}{z}$$

$\Rightarrow \quad z(9 + z + 16 + z) = (16 + z)(9 + z)$
$\Rightarrow \quad 9z + z^2 + 16z + z^2 = 144 + 16z + 9z + z^2$
$\Rightarrow \quad z^2 - 144 = 0 \Rightarrow z = 12$

Hence, the time taken by X and Y to complete the work is 12 days.

Fast Track Method
Here, $a = 16$ and $b = 9$
According to the formula, required time $= \sqrt{ab} = \sqrt{16 \times 9} = 4 \times 3 = 12$ days

Technique 10 A and B, each alone can do a piece of work in a and b days, respectively. Both begin together and if

(i) A leaves the work x days before its completion, then total time taken for completion of work will be given as $T = \dfrac{(a + x)b}{(a + b)}$ days.

(ii) B leaves the work x days before its completion, then total time taken for completion of work will be given as $T = \dfrac{(b + x)a}{(a + b)}$ days.

Ex. 15 A can do a piece of work in 10 days while B can do it in 15 days. They begin together but 5 days before the completion of the work, B leaves off. Find the total number of days for the work to be completed.

Sol. $\because$ A's 5 days work $= \dfrac{5}{10} = \dfrac{1}{2}$, then remaining work $= 1 - \dfrac{1}{2} = \dfrac{1}{2}$

$(A + B)$'s 1 day work $= \dfrac{1}{10} + \dfrac{1}{15} = \dfrac{3 + 2}{30} = \dfrac{5}{30} = \dfrac{1}{6}$

$\therefore$ $(A + B)$ finish $\dfrac{1}{6}$ work in 1 day.

So, $(A + B)$ will finish $\dfrac{1}{2}$ work in $\dfrac{1}{2} \times 6$ days or in 3 days.

∴ Required time = 5 + 3 = 8 days

Fast Track Method

Here, $a = 10$ days, $b = 15$ days, $x = 5$ and $T = ?$

According to the formula,

Required time $= \dfrac{(b + x)a}{(a + b)} = \dfrac{(15 + 5)10}{10 + 15} = \dfrac{20 \times 10}{25} = 4 \times 2 = 8$ days

Technique 11 A and B do a piece of work in a and b days, respectively. Both begin together but after some days, A leaves off and the remaining work is completed by B in x days. Then, the time after which A left, is given by

$$T = \dfrac{(b - x)a}{a + b}$$

Ex. 16 A and B can do a piece of work in 40 days and 50 days, respectively. Both begin together but after a certain time, A leaves off. In this case, B finishes the remaining work in 20 days. After how many days did A leave?

Sol. ∵ B's 20 days work $= \dfrac{20}{50} = \dfrac{2}{5}$, then remaining work $= 1 - \dfrac{2}{5} = \dfrac{3}{5}$

$(A + B)$'s 1 day work $= \dfrac{1}{40} + \dfrac{1}{50} = \dfrac{5 + 4}{200} = \dfrac{9}{200}$

∴ $(A + B)$ do $\dfrac{9}{200}$ work in 1 day.

So, $(A + B)$ will do $\dfrac{3}{5}$ work in $\dfrac{3}{5} \times \dfrac{200}{9} = \dfrac{40}{3} = 13\dfrac{1}{3}$ days.

Fast Track Method

Here, $a = 40$ days, $b = 50$ days, $x = 20$ and $T = ?$

∴ Required time $= \dfrac{(b - x)a}{a + b} = \dfrac{(50 - 20) \times 40}{(40 + 50)}$

$= \dfrac{30 \times 40}{90} = \dfrac{40}{3} = 13\dfrac{1}{3}$ days

Fast Track Practice

Exercise 1 Base Level Questions

1. A and B working together can finish a piece of work in 12 days while B alone can finish it in 30 days. In how many days can A alone finish the work? [CDS 2017 (I)]
   (a) 18 days (b) 20 days
   (c) 24 days (d) 25 days

2. Praful has done 1/2 of a job in 30 days, Sarabjit completes the rest of the job in 10 days. In how many days can they together do the job? [SSC (10+2) 2017]
   (a) 30 days (b) 45 days
   (c) 15 days (d) 60 days

3. A can do a piece of work in x days and B can do the same work in $3x$ days. To finish the work together, they take 12 days. What is the value of x? [CDS 2012]
   (a) 8 (b) 10 (c) 12 (d) 16

4. A completes 1/3 part of a work in 5 days, while B completes 2/5 part of the same work in 10 days. In how many days will they complete doing the work together? [SSC (10+2) 2017]
   (a) $7\frac{3}{4}$ days (b) $9\frac{3}{8}$ days
   (c) $8\frac{4}{5}$ days (d) 10 days

5. A can do a piece of work in 4 days and B can complete the same work in 12 days. What is the number of days required to do the same work together? [CDS 2013]
   (a) 2 (b) 3 (c) 4 (d) 5

6. A can do a piece of work in 8 days, B can do it in 10 days and C can do it in 20 days. In how many days can A, B and C together complete the work?
   (a) $3\frac{7}{11}$ (b) $3\frac{5}{11}$ (c) $3\frac{2}{11}$ (d) $3\frac{9}{11}$
   (e) None of these

7. A and B together can do a piece of work in 12 days, while B alone can finish it in 30 days. A alone can finish the work in [SSC (10+2) 2012]
   (a) 15 days (b) 18 days
   (c) 20 days (d) 25 days

8. X can complete a job in 12 days. If X and Y work together, they can complete the job in $6\frac{2}{3}$ days. Y alone can complete the job in
   (a) 10 days (b) 12 days
   (c) 15 days (d) 18 days

9. A, B and C can do a job working alone in 12, 16 and 24 days respectively. In how many days they can do the job, if they worked together? [SSC CGL (Pre) 2017]
   (a) $\frac{16}{3}$ (b) $\frac{15}{4}$ (c) $\frac{17}{3}$ (d) $\frac{19}{4}$

10. A, B and C can complete a work in 2 h. If A does the job alone in 6 h and B in 5 h, then how long will it take for C to finish the job alone? [SSC Multitasking 2014]
    (a) $5\frac{1}{2}$ h (b) $7\frac{1}{2}$ h (c) 9 h (d) $4\frac{1}{2}$ h

11. A and B together can complete a work in 3 days. They started together but after 2 days, B left the work. If the work is completed after 2 more days, B alone could do the work in how many days? [SSC CGL 2007]
    (a) 5 (b) 6 (c) 7 (d) 10

12. A can do a piece of work in 10 days, while B can do it in 6 days. B worked at it for 4 days. How long will A take to finish the remaining work?
    (a) $3\frac{1}{3}$ days (b) $3\frac{2}{3}$ days
    (c) $3\frac{2}{5}$ days (d) $3\frac{5}{7}$ days
    (e) None of these

13. A and B can complete a job in 24 days working together. A alone can complete it in 32 days. Both of them worked together for 8 days and then A left. The number of days B will take to complete the remaining job is [SSC CGL 2012]
    (a) 16 (b) 32 (c) 64 (d) 128

14. A, B and C can do a work separately in 16, 32 and 48 days, respectively. They started the work together but B leaving off 8 days and C, 6 days before the completion of the work. In what time is the work finished? [SSC CGL 2015]
    (a) 9 days (b) 10 days
    (c) 14 days (d) 12 days

Work and Time / 375

15. A can do a work in 20 days and B can do in 10 days. A starts the work and works alone for 5 days. Then, B joins A and they finish the work. In how many days the work gets finished? **[NIFT 2015]**
    (a) 10 (b) 12 (c) 9 (d) 8

16. A, B and C can do a piece of work in 24, 30 and 40 days, respectively. They began the work together but C left 4 days before completion of the work. In how many days was the work done? **[SSC CGL 2015]**
    (a) 11 (b) 12 (c) 13 (d) 14

17. A can complete a work in 20 days and B in 30 days. A worked alone for 4 days and then B completed the remaining work along with C in 18 days. In how many days can C working alone complete the work? **[SSC CGL 2012]**
    (a) 12 (b) 68 (c) 72 (d) 90

18. A, B and C can do a piece of work individually in 8, 12 and 15 days, respectively. A and B start working but A quits after working for 2 days. After this, C joins B till the completion of work. In how many days will the work be completed? **[CDS 2014]**
    (a) $5\frac{8}{9}$ (b) $4\frac{6}{7}$ (c) $6\frac{7}{13}$ (d) $3\frac{3}{4}$

19. A and B can complete a work in 8 days, working together. B alone can do it in 12 days. After working for 4 days, B left the work. How many days will A take to complete the remaining work?
    (a) 16 (b) 18 (c) 20 (d) 22

20. A and B working separately can do a piece of work in 9 and 15 days, respectively. If they work for a day alternately, with B beginning, then the work will be completed in **[SSC (10+2) 2015]**
    (a) 10 days (b) 11 days
    (c) 12 days (d) 13 days

21. Piyush has done 1/3rd of a job in 30 days, Sanjeev completes the rest of the job in 60 days. In how many days can they together do the job? **[SSC (10+2) 2017]**
    (a) 15 days (b) 45 days
    (c) 30 days (d) 10 days

22. A mason can build a tank in 12 h. After working for 6 h, he took the help of a boy and finished the work in another 5 h. The time that the boy will take alone to complete the work is **[CDS 2013]**
    (a) 30 h (b) 45 h (c) 60 h (d) 64 h

23. A and B can do a job together in 12 days. A is 2 times as efficient as B. In how many days can B alone complete the work? **[SSC (10+2) 2012]**
    (a) 36 (b) 12 (c) 18 (d) 9

24. A takes twice as much time as B and C takes thrice as much time as B to finish a work. Working together, they can finish the work in 12 days. Find the number of days needed for A to do the work alone. **[SSC CGL 2011]**
    (a) 20 (b) 22 (c) 33 (d) 44

25. A is thrice as good a workman as B and therefore is able to finish a job in 30 days less than B. How many days will they take to finish the job working together? **[RRB 2007]**
    (a) $10\frac{1}{4}$ (b) $11\frac{3}{4}$ (c) $11\frac{1}{4}$ (d) $11\frac{1}{3}$

26. A takes twice the time taken by B and thrice the time taken by C to do a particular piece of work. Working together, they can complete the work in 2 days. Find the number of days taken by A, B and C respectively to complete the work alone. **[LIC ADO 2007]**
    (a) 12, 6 and 4 (b) 18, 9 and 6
    (c) 24, 12 and 8 (d) 6, 3 and 2
    (e) None of these

27. A, B and C, each working alone, can finish a piece of work in 27, 33 and 45 days, respectively. If A starts by working alone for 12 days, then B takes over from A and works for 11 days. At this stage, C takes over from B and completes the remaining work. In how many days the whole work was completed? **[SBI Clerk (Pre) 2016]**
    (a) 33 days (b) 31 days
    (c) 39 days (d) 35 days
    (e) 37 days

28. A is thrice as efficient as B. A started working and after 4 days, he was replaced by B. B then worked for 15 days and left. If A and B together finished 75% of the total work, in how many days B alone can finish the whole work? **[SBI PO (Pre) 2016]**
    (a) 27 days (b) 45 days
    (c) 24 days (d) 36 days
    (e) 42 days

29. A can complete a task in 15 days. B is 50% more efficient than A. Both A and B started working together on the task and after few days, B left task and A finished the remaining 1/3 of the given work. For how many days A and B worked together? **[RRB Clerk (Pre) 2017]**
    (a) 3 (b) 5 (c) 4 (d) 6
    (e) 2

30. 5 men start working to complete a work in 15 days. After 5 days, 10 women are accompanied by them to complete the work in next 5 days. If the work is to be done by women only, when could the work be over, if 10 women have started it? **[Bank Clerk 2007]**
(a) 10 days (b) 18 days
(c) 15 days (d) 12 days
(e) None of these

31. 12 men can finish a piece of work in 20 days. 8 men started working and after 10 days were replaced by 18 women. These 18 women finished the remaining work in 16 days. In how many days 18 women can finish the whole work? **[LIC AAO 2016]**
(a) 32 (b) 18 (c) 28 (d) 24
(e) 21

32. P can do 1/4th of work in 10 days, Q can do 40% of work in 40 days and R can do 1/3rd of work in 13 days. Who will complete the work first?
[SSC CGL (Mains) 2017]
(a) P (b) Q
(c) R (d) Both P and R

33. 6 boys can complete a piece of work in 16 h. In how many hours will 8 boys complete the same work? **[Bank Clerk 2011]**
(a) 10 (b) 8 (c) 12 (d) 14
(e) None of these

34. In a hostel, there are 120 students and food stock is for 45 days. If 30 new students join the hostel, in how many days will the complete stock be exhausted? **[SSC FCI 2012]**
(a) 38 (b) 40 (c) 32 (d) 36

35. A certain number of men can do a piece of work in 60 days. If there were 6 men more, the work can be finished 20 days earlier. The number of men working is
[SSC CGL (Pre) 2016]
(a) 6 (b) 12 (c) 18 (d) 24

36. 20 women can complete a piece of work in 7 days. If 8 more women are put on the job. In how many days will they complete the work? **[DMRC CRA 2012]**
(a) 4.5 (b) 5 (c) 5.5 (d) 4.5

37. 12 men can do a piece of work in 24 days. How many days are needed to complete the work, if 8 men do this work?
[IDBI PO 2010]
(a) 28 (b) 36 (c) 48 (d) 52
(e) None of these

38. 40 men can build a wall 200 m long in 12 days, working 8 h per day. What will be the number of days that 30 men will take to build a similar wall 300 m long, working 6 h per day? **[SSC FCI 2012]**
(a) 32 (b) 18 (c) 36 (d) 9

39. If m men working m h per day, can do m units of work in m days, then n men working n h per day would be able to complete how many units of work in n days?
(a) $\frac{n^3}{m^2}$ (b) $\frac{m^3}{n^2}$ (c) $\frac{m^4}{n^2}$ (d) $\frac{n^4}{m^3}$

40. 15 men complete a work in 16 days. If 24 men are employed, then the time required to complete that work will be **[CDS 2014]**
(a) 7 days (b) 8 days
(c) 10 days (d) 12 days

41. 20 workers working for 5 h per day complete a work in 10 days. If 25 workers are employed to work 10 h per day, what is the time required to complete the work? **[CDS 2013]**
(a) 4 days (b) 5 days
(c) 6 days (d) 8 days

42. A certain number of men can do a piece of work in 80 days. If there were 10 men less, it could be finished in 20 days more. How many men are there in the starting?
(a) 45 (b) 50 (c) 40 (d) 60

43. A stock of food is enough for 240 men for 48 days. How long will the same stock last for 160 men? **[CDS 2012]**
(a) 54 days (b) 60 days
(c) 64 days (d) 72 days

44. 45 people take 18 days to dig a pond. If the pond would have to be dug in 15 days, then the number of people to be employed will be **[CDS 2012]**
(a) 50 (b) 54 (c) 60 (d) 72

45. In a school, Mid-Day Meal food is sufficient for 250 students for 33 days, if each student is given 125 g meals. 80 more students joined the school. If same amount of meal is given to each student, then the food will last for
[CLAT 2013]
(a) 20 days (b) 40 days
(c) 30 days (d) 25 days

46. 90 men are engaged to do a piece of work in 40 days but it is found that in 25 days, 2/3 work is completed. How many men should be allowed to go off, so that the work may be finished in time?
[SSC (10+2) 2008]
(a) 10 (b) 15 (c) 20 (d) 25

Work and Time / 377

47. It is given that 16 men working 18 h per day can build a wall 36 m long, 4 m broad and 24 m high in 20 days. How many men will be required to build a wall 64 m long, 6 m broad and 18 m high working 12 h per day in 16 days?
(a) 60   (b) 20   (c) 30   (d) 35

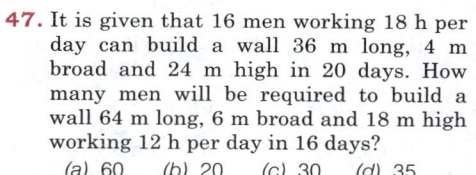

48. If 4 men working 4 h per day for 4 days complete 4 units of work, then how many units of work will be completed by 2 men working for 2 h per day in 2 days? **[CDS 2015 (II)]**
(a) 2   (b) 1   (c) $\frac{1}{2}$ (d) $\frac{1}{8}$

49. 10 men and 8 women together can complete a work in 5 days. Work done by one woman in a day is equal to half the work done by a man in 1 day. How many days will it take for 4 men and 6 women to complete that work? **[Bank Clerk 2009]**
(a) 12   (b) 10   (c) $8\frac{2}{3}$ (d) $4\frac{3}{4}$
(e) None of these

50. If 15 men take 21 days of 8 h each to do a piece of work, then what is the number of days of 6 h each that 21 women would take, if 3 women would do as much work as 2 men? **[CDS 2017 (I)]**
(a) 18   (b) 20   (c) 25   (d) 30

51. A project manager hired 16 men to complete a project in 38 days. However, after 30 days, he realised that only 5/9th of the work is complete. How many more men does he need to hire to complete the project on time? **[SBI PO 2015]**
(a) 48   (b) 24   (c) 32   (d) 16
(e) 36

52. If one man or two women or three boys can finish a work in 88 days, then how many days will one man, one woman and one boy together take to finish the same work? **[Bank Clerk 2009]**
(a) 46   (b) 54   (c) 48   (d) 44
(e) 60

53. 4 men or 6 women or 10 children can paint a house in 5 days. The painting is given to a couple and their 5 sons. They finish the job in **[SSC Multitasking 2014]**
(a) $\frac{11}{60}$ days   (b) $5\frac{5}{11}$ days
(c) $5\frac{6}{11}$ days   (d) $11\frac{1}{5}$ days

54. A contractor undertook to finish a certain work in 124 days and employed 120 men. After 64 days, he found that he had already done 2/3 of the work. How many men can be discharged now, so that the work may finish in time? **[SSC CGL 2013]**
(a) 40   (b) 50   (c) 48   (d) 56

55. A contract is to be completed in 92 days and 234 men were set to work, each working 16 h per day. After 66 days, 4/7 of the work is completed. How many additional men may be employed, so that the work may be completed in time, each man now working 18 h per day?
(a) 162   (b) 234   (c) 262   (d) 81
(e) None of these

56. A contract is to be completed in 50 days and 105 men were set to work, each working 8 h per day. After 25 days, 2/5th of the work is finished. How many additional men be employed, so that the work may be completed on time, each man now working 9 h per day? **[SNAP 2012]**
(a) 34   (b) 36   (c) 35   (d) 37

57. A and B can do a piece of work in 72 days. B and C can do it in 120 days. A and C can do it in 90 days. In what time can A alone do it? **[SSC CGL 2011]**
(a) 80 days   (b) 100 days
(c) 120 days   (d) 150 days

58. A and B can do a piece of work in 10 h. B and C can do it in 15 h, while A and C take 12 h to complete the work. B independently can complete the work in **[CDS 2012]**
(a) 12 h   (b) 16 h   (c) 20 h   (d) 24 h

59. If 3 men or 4 women can build a wall in 43 days, in how many days can 7 men and 5 women build this wall? **[SSC (10+2) 2012]**
(a) 16   (b) 25   (c) 21   (d) 12

60. 3 men can do a piece of work in 18 days. 6 children can also do that work in 18 days. 4 men and 4 children together will finish the work in how many days? **[Bank Clerk 2010]**
(a) 10   (b) 6   (c) 12   (d) 9
(e) None of these

61. 4 goats or 6 sheeps can graze a field in 50 days. 2 goats and 9 sheeps can graze the field in **[CDS 2013]**
(a) 100 days   (b) 75 days
(c) 50 days   (d) 25 days

62. A does 20% less work than B. If A can complete a piece of work in $7\frac{1}{2}$ h, then B can do it in **[SSC CGL 2013]**
(a) 4 h   (b) 6 h
(c) 8 h   (d) 10 h

63. If 5 men can do a piece of work in 10 days and 12 women can do the same work in 15 days, the number of days required to complete the work by 5 men and 6 women is [CDS 2017 (I)]
    (a) $7\frac{1}{2}$ days (b) 8 days
    (c) $9\frac{1}{2}$ days (d) 12 days

64. 15 men can finish a work in 20 days, however it takes 24 women to finish it in 20 days. If 10 men and 8 women undertake to complete the work, then they will take [SSC (10+2) 2015]
    (a) 15 days (b) 30 days
    (c) 10 days (d) 20 days

65. 10 men can complete a piece of work in 6 days and 6 women can complete the same piece of work in 12 days. In how many days will 15 men and 10 women together complete the work? [IBPS SO 2016]
    (a) $3\frac{1}{2}$ days (b) $2\frac{4}{7}$ days
    (c) $3\frac{1}{4}$ days (d) $2\frac{1}{2}$ days
    (e) 3 days

66. If the work done by 8 men and 4 boys in 1 day is 7 times the work done by 1 man and 1 boy, then compare the work done by 1 man and 1 boy in 1 day. [CDS 2013]
    (a) 1 (b) 2 (c) 3 (d) 1/2

67. When A alone does a piece of work, he takes 25 days more than the time taken by (A + B) to complete that particular work, while B alone takes 49 days more than the time taken by (A + B) to finish the same work. A and B together will take what time to finish this work?
    (a) 35 days (b) 25 days
    (c) 15 days (d) 45 days
    (e) None of these

68. A can do a piece of work in 10 days and B in 20 days. They begin together but A leaves 2 days before the completion of the work. The whole work will be done in [SSC (10+2) 2012]
    (a) 8 days (b) $7\frac{2}{3}$ days
    (c) 7 days (d) 6 days

69. A and B can do a piece of work in 30 and 36 days, respectively. They began the work together but A leaves after some days and B finished the remaining work in 25 days. After how many days did A leave? [SSC CGL 2015]
    (a) 6 (b) 11 (c) 10 (d) 5

70. A and B can do a piece of work in 60 days and 75 days, respectively. Both begin together but after a certain time, A leaves off. In such case, B finishes the remaining work in 30 days. After how many days did A leave?
    (a) 25 (b) 21 (c) 20 (d) 24

71. Ajay can do a piece of work in 25 days and Sanjay can finish it in 20 days. They work together for 5 days and then Ajay goes away. In how many days will Sanjay finish the remaining work? [DMRC CRA 2012]
    (a) 11 (b) 12 (c) 14 (d) None of these

72. A and B each working alone can do a work in 15 days and 25 days, respectively. They started the work together but B left after some time and A finished the remaining work in 7 days. After how many days from the start did B leave? [SSC CGL 2012]
    (a) 3 (b) 5 (c) 7 (d) 9

73. Dinu and Tinu can do a piece of work in 45 and 40 days, respectively. They began the work together, but Dinu leaves after some days and Tinu finished the remaining work in 23 days. After how many days did Dinu leave? [XAT 2015]
    (a) 7 days (b) 8 days
    (c) 9 days (d) 11 days
    (e) None of these

Exercise 2 Higher Skill Level Questions

1. The time taken by 4 men to complete a job is double the time taken by 5 children to complete the same job. Each man is twice as fast as a woman. How long will 12 men, 10 children and 8 women take to complete a job, given that a child would finish the job in 20 days?
    [SSC CGL (Mains) 2016]
    (a) 4 days (b) 3 days
    (c) 2 days (d) 1 day

2. 3 women and 18 children together take 2 days to complete a piece of work. How many days will 9 children alone take to complete the piece of work, if 6 women alone can complete the piece of work in 3 days? [IBPS Clerk 2011]
    (a) 9 (b) 7
    (c) 5 (d) 6
    (e) None of these

3. A can build up structure in 8 days and B can break it in 3 days. A has worked for 4 days and then B joined to work with A for another 2 days. In how many days will A alone build up the remaining part of the structure? [SBI Clerk (Mains) 2016]
   (a) 10   (b) 9   (c) 12   (d) 8
   (e) 7

4. 16 men and 10 women together can complete a project in 10 days. If 12 women can complete the project in 25 days, in how many days 10 men complete the same project? [IBPS SO 2016]
   (a) 28 days   (b) 24 days
   (c) 18 days   (d) 26 days
   (e) 10 days

5. The efficiency of P is twice that of Q, whereas the efficiency of P and Q together is three times that of R. If P, Q and R work together on a job, in what ratio should they share their earnings?   [CDS 2015 (I)]
   (a) 2 : 1 : 1   (b) 4 : 2 : 1
   (c) 4 : 3 : 2   (d) 4 : 2 : 3

6. P and Q together can do a job in 6 days. Q and R can finish the same job in 60/7 days. P started the work and worked for 3 days. Q and R continued for 6 days. Then, the difference of days in which R and P can complete the job, is [SSC CGL 2015]
   (a) 15   (b) 10   (c) 8   (d) 12

7. 6 men can do a piece of work in 12 days while 8 women can do the same work in 18 days. The same work can be done by 18 children in 10 days. 4 men, 12 women and 20 children work together for 2 days. If only men have to complete remaining work in 1 day, then find the required number of men.   [Bank PO 2010]
   (a) 36   (b) 24   (c) 18
   (d) Cannot be determined
   (e) None of these

8. A can do 50% more work than B in the same time. B alone can do a piece of work in 30 h. B starts working and had already worked for 12 h, when A joins him. How many hours should B and A work together to complete the remaining work?   [CDS 2016 (II)]
   (a) 6 h   (b) 12 h
   (c) 4.8 h   (d) 7.2 h

9. Consider the following diagrams x men, working at constant speed, do a certain job in y days.

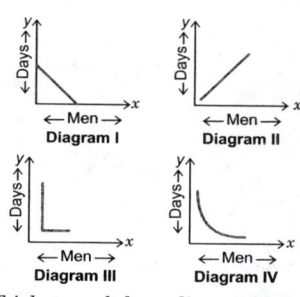

   Diagram I
   Diagram II
   Diagram III
   Diagram IV

   Which one of these diagrams shows the relation between x and y? [UPSC CSAT 2013]
   (a) Diagram I   (b) Diagram II
   (c) Diagram III   (d) Diagram IV

10. 18 men can complete a piece of work in 24 days and 12 women can complete the same piece of work in 32 days. 18 men start working and after few days, 4 men leave the job and 8 women join. If the remaining work is completed in $15\frac{15}{23}$ days, after how many days did the four men leave?   [NICL AO 2015]
    (a) 8   (b) 5   (c) 6   (d) 4
    (e) 2

11. The work done by a woman in 8 h is equal to the work done by a man in 6 h and by a boy in 12 h. If working 6 h per day, 9 men can complete a work in 6 days, then in how many days 12 women, 12 men and 12 boys together finish the same work working 8 h per day? [FCI Assist. Grade 2015]
    (a) $1\frac{1}{3}$ (b) $3\frac{2}{3}$
    (c) $1\frac{1}{2}$ (d) None of these

12. 18 women complete a project in 24 days and 24 men complete the same project in 15 days. 16 women worked for 3 days and then they left. 20 men work for next 2 days and then they are joined by 16 women. In how many days will they complete the remaining work?   [IBPS Clerk (Mains) 2017]
    (a) $6\frac{1}{5}$ (b) $7\frac{3}{5}$ (c) $5\frac{2}{5}$ (d) $9\frac{1}{5}$
    (e) Other than those given as options

13. Time taken by A alone to finish a piece of work is 60% more than that taken by A and B together to finish the same piece of work. C is twice as efficient as B. If B and C together can complete the same piece of work in $13\frac{1}{3}$ days, in how many days can A alone finish the same piece of work?   [SBI PO (Pre) 2017]
    (a) 36   (b) 24   (c) 16   (d) 28
    (e) Other than those given as options

Answer with Solutions

Exercise 1 Base Level Questions

1. (b) ∵ A and B do the work in 12 days.

∴ One day work of A and B = $\dfrac{1}{12}$

B alone do the work in 30 days.

∴ One day work of B = $\dfrac{1}{30}$

∴ One day work of A

$= \dfrac{1}{12} - \dfrac{1}{30} = \dfrac{5-2}{60} = \dfrac{3}{60} = \dfrac{1}{20}$

Hence, A can do the work along in 20 days.

2. (c) Praful complete full job in

$30 \times 2 = 60$ days

Sarabjit completes full job in

$10 \times 2 = 20$ days

Their combined one day work

$= \dfrac{1}{60} + \dfrac{1}{20} = \dfrac{4}{60} = \dfrac{1}{15}$

So, they both take 15 days to complete full job.

3. (d) Here, A's 1 day's work = $\dfrac{1}{x}$

and B's 1 day's work = $\dfrac{1}{3x}$

∴ (A + B)'s 1 day's work = $\dfrac{1}{x} + \dfrac{1}{3x} = \dfrac{4}{3x}$

Given, 1 day's work of both (A + B) = $\dfrac{1}{12}$

$\Rightarrow \dfrac{4}{3x} = \dfrac{1}{12} \Rightarrow 3x = 48$

$\Rightarrow x = 16$

4. (b) ∵ 5 days work of A = $\dfrac{1}{3}$

∴ 1 day work of A = $\dfrac{1}{3 \times 5} = \dfrac{1}{15}$

and 10 days work of B = $\dfrac{2}{5}$

∴ 1 day work of B = $\dfrac{2}{5 \times 10} = \dfrac{1}{25}$

∴ 1 day work of A and B = $\dfrac{1}{15} + \dfrac{1}{25}$

$= \dfrac{5+3}{75} = \dfrac{8}{75}$

∴ Time taken by both together

$= \dfrac{75}{8}$ days $= 9\dfrac{3}{8}$ days

5. (b) Here, A's 1 day's work = $\dfrac{1}{4}$

and B's 1 day's work = $\dfrac{1}{12}$

∴ (A + B)'s 1 day's work = $\dfrac{1}{4} + \dfrac{1}{12}$

$= \dfrac{3+1}{12} = \dfrac{4}{12} = \dfrac{1}{3}$

∴ A and B together do the work in 3 days.

Fast Track Method

Here, $x = 4$ and $y = 12$

∴ Required number of days

$= \dfrac{xy}{x+y} = \dfrac{4 \times 12}{4 + 12}$ [by Technique 1(i)]

$= \dfrac{4 \times 12}{16} = 3$

6. (a) A's 1 day's work = $\dfrac{1}{8}$

B's 1 day's work = $\dfrac{1}{10}$

C's 1 day's work = $\dfrac{1}{20}$

(A + B + C)'s 1 day's work

$= \dfrac{1}{8} + \dfrac{1}{10} + \dfrac{1}{20}$

$= \dfrac{5+4+2}{40} = \dfrac{11}{40}$

∴ (A + B + C) can finish the work in

$\dfrac{40}{11}$ days or $3\dfrac{7}{11}$ days.

Fast Track Method

Here, $x = 8, y = 10, z = 20$

According to the formula,

Time taken by A, B and C to complete

the work = $\dfrac{xyz}{xy + yz + zx}$

[by Technique 1(ii)]

$= \dfrac{8 \times 10 \times 20}{8 \times 10 + 10 \times 20 + 20 \times 8}$

$= \dfrac{1600}{80 + 200 + 160}$

$= \dfrac{1600}{440} = 3\dfrac{7}{11}$ days

Work and Time / 381

7. (c) $(A + B)$'s 1 day's work $= \dfrac{1}{12}$

and B's 1 day's work $= \dfrac{1}{30}$

∴ A's 1 day's work $= \dfrac{1}{12} - \dfrac{1}{30} = \dfrac{5-2}{60} = \dfrac{1}{20}$

Hence, A can finish this work in 20 days.

Fast Track Method

Here, $x = 12$ and $y = 30$

∴ Required number of days

$= \dfrac{xy}{y-x} = \dfrac{12 \times 30}{30 - 12}$ [by Technique 2]

$= \dfrac{12 \times 30}{18} = 20$ days

∴ A alone can finish the work in 20 days.

8. (c) X's 1 day's work $= \dfrac{1}{12}$

$(X + Y)$'s 1 day's work $= \dfrac{3}{20}$

∴ Y's 1 day's work $= \dfrac{3}{20} - \dfrac{1}{12} = \dfrac{4}{60} = \dfrac{1}{15}$

∴ Number of days taken by Y to complete the work = 15 days

Fast Track Method

Here, $x = \dfrac{20}{3}$, $y = 12$

∴ Number of days taken by $y = \dfrac{xy}{y-x}$

[by Technique 2]

$= \dfrac{\dfrac{20}{3} \times 12}{12 - \dfrac{20}{3}} = 15$ days

9. (a) A's one day work $= \dfrac{1}{12}$

B's one day work $= \dfrac{1}{16}$

C's one day work $= \dfrac{1}{24}$

∴ One day work of A, B and C

$= \dfrac{1}{12} + \dfrac{1}{16} + \dfrac{1}{24} = \dfrac{4+3+2}{48} = \dfrac{9}{48} = \dfrac{3}{16}$

∴ Total time taken by A, B and C together

$= \dfrac{16}{3}$ days

10. (b) Let C alone can finish the job in x h. According to the question,

Work done by A, B and C together in 1 h $= \dfrac{1}{2}$

⇒ $\dfrac{1}{6} + \dfrac{1}{5} + \dfrac{1}{x} = \dfrac{1}{2}$

⇒ $\dfrac{1}{x} = \dfrac{1}{2} - \dfrac{1}{6} - \dfrac{1}{5} = \dfrac{15 - 5 - 6}{30}$

⇒ $\dfrac{1}{x} = \dfrac{4}{30} = \dfrac{2}{15}$

∴ $x = \dfrac{15}{2} = 7\dfrac{1}{2}$ h

11. (b) $(A + B)$'s 2 days work $= 2 \times \dfrac{1}{3} = \dfrac{2}{3}$

Remaining work $= 1 - \dfrac{2}{3} = \dfrac{1}{3}$

A will complete $\dfrac{1}{3}$ work in 2 days.

A will complete 1 work in 6 days.

A's 1 day's work $= \dfrac{1}{6}$

B's 1 day's work $= \dfrac{1}{3} - \dfrac{1}{6} = \dfrac{1}{6}$

Hence, B will take 6 days to complete the work alone.

12. (a) B's 4 days work $= \dfrac{1}{6} \times 4 = \dfrac{2}{3}$

∴ Remaining work $= 1 - \dfrac{2}{3} = \dfrac{1}{3}$

A's 1 day's work $= \dfrac{1}{10}$

∴ $\dfrac{1}{3}$ work is finished by A in

$\left(10 \times \dfrac{1}{3}\right) = 3\dfrac{1}{3}$ days

13. (c) Let B will take x days to complete the remaining job.

According to the question,

$\dfrac{1}{A} + \dfrac{1}{B} = \dfrac{1}{24}$ and $\dfrac{1}{A} = \dfrac{1}{32}$

∴ $\dfrac{1}{B} = \dfrac{1}{24} - \dfrac{1}{32} = \dfrac{1}{96}$ ⇒ $B = 96$ days

According to the question,

$8\left(\dfrac{1}{A} + \dfrac{1}{B}\right) + x \times \dfrac{1}{B} = 1$

⇒ $8 \times \dfrac{1}{24} + \dfrac{x}{96} = 1$

⇒ $\dfrac{1}{3} + \dfrac{x}{96} = 1 \Rightarrow \dfrac{x}{96} = 1 - \dfrac{1}{3}$

∴ $x = \dfrac{2 \times 96}{3} = 64$

Hence, B complete the remaining job in 64 days.

14. (d) Let the total work = LCM of (16, 32, 48)
= 96 units

∴ A's work = 6 units/day
B's work = 3 units/ day

and C's work = 2 units/day

Let the work be finished in x days.

$$\therefore \quad 6x + 3(x-8) + 2(x-6) = 96$$
$$\Rightarrow \quad 6x + 3x - 24 + 2x - 12 = 96$$
$$\Rightarrow \quad 11x = 132$$
$$\therefore \quad x = 12 \text{ days}$$

15. (a) Here, A can do a work in 20 days and B can do a work in 10 days.

Now, A's 1 days, work = $\dfrac{1}{20}$

$\therefore$ A's 5 day work = $\dfrac{1}{20} \times 5 = \dfrac{1}{4}$

$\therefore$ Rest work = $\left(1 - \dfrac{1}{4}\right) = \dfrac{3}{4}$

Now, $(A + B)$'s 1 day work
$$= \dfrac{1}{20} + \dfrac{1}{10} = \dfrac{1+2}{20} = \dfrac{3}{20}$$

$\because (A + B)$ completes $\dfrac{3}{20}$ th of work in 1 day.

$\Rightarrow (A + B)$ completes 1 work in $\dfrac{1 \times 20}{3}$ days.

$\therefore (A + B)$ completes $\dfrac{3}{4}$ th of work in $\dfrac{20}{3} \times \dfrac{3}{4}$, i.e. 5 days.

Hence, total number of days = 5 + 5 = 10

16. (a) Work done by A in one day = $\dfrac{1}{24}$

Work done by B in one day = $\dfrac{1}{30}$

Work done by C in one day = $\dfrac{1}{40}$

Work done by $(A + B + C)$ in one day
$$= \dfrac{1}{24} + \dfrac{1}{30} + \dfrac{1}{40} = \dfrac{5+4+3}{120} = \dfrac{12}{120}$$

Work done by $(A + B)$ in one day
$$= \dfrac{1}{24} + \dfrac{1}{30} = \dfrac{5+4}{120} = \dfrac{9}{120}$$

Work done by $(A + B)$ in last 4 days
$$= \dfrac{9}{120} \times 4 = \dfrac{36}{120}$$

Remaining work = $1 - \dfrac{36}{120} = \dfrac{84}{120}$

$\therefore$ Time taken by $(A + B + C)$ to complete the $\dfrac{84}{120}$ work = $\dfrac{84}{120} \times \dfrac{120}{12} = 7$ days

$\therefore$ Total time taken to complete the work
$$= 7 + 4 = 11 \text{ days}$$

17. (d) Work done by A in 4 days = $\dfrac{4}{20} = \dfrac{1}{5}$

$\therefore$ Remaining work = $1 - \dfrac{1}{5} = \dfrac{4}{5}$

Let C working alone can complete the work in x days.

According to the question,
$$\dfrac{18}{30} + \dfrac{18}{x} = \dfrac{4}{5}$$

$\Rightarrow 18(x + 30) \times 5 = 4 \times 30x$
$\Rightarrow 90x + 2700 = 120x \Rightarrow 30x = 2700$
$\therefore \quad x = 90$ days

18. (a) Work done by A and B in 1 day
$$= \dfrac{1}{8} + \dfrac{1}{12} = \dfrac{5}{24}$$

2 days work of A and B = $\dfrac{10}{24}$

After 2 days, A left the work.

$\therefore$ Remaining work = $1 - \dfrac{10}{24} = \dfrac{14}{24}$

One day's work of B and C together
$$= \dfrac{1}{12} + \dfrac{1}{15} = \dfrac{9}{60}$$

So, the number of days required by B and C to finish work = $\dfrac{14/24}{9/60} = \dfrac{14}{24} \times \dfrac{60}{9} = \dfrac{35}{9}$

$\therefore$ Total days to complete the work
$$= 2 + \dfrac{35}{9} = \dfrac{53}{9} = 5\dfrac{8}{9} \text{ days}$$

19. (a) $(A + B)$'s 1 day's work = $\dfrac{1}{8}$

B's 1 day's work = $\dfrac{1}{12}$

$\therefore$ A's 1 day's work = $\dfrac{1}{8} - \dfrac{1}{12} = \dfrac{3-2}{24} = \dfrac{1}{24}$

$\therefore$ A can complete the work in 24 days.

Now, B's 4 days work = $\dfrac{4}{12} = \dfrac{1}{3}$

$\therefore$ Remaining work = $1 - \dfrac{1}{3} = \dfrac{2}{3}$

As, time taken by A to complete the whole work is 24 days.

$\therefore$ Time taken by A to do $\dfrac{2}{3}$ of the work
$$= \dfrac{2}{3} \times 24 = 16 \text{ days}$$

20. (c) Work done by A and B in 2 days
$$= \dfrac{1}{9} + \dfrac{1}{15} = \dfrac{5+3}{45} = \dfrac{8}{45}$$

Since, they both work for a day alternately.

$\therefore$ Work done by both in 10 days
$$= \dfrac{10}{2} \times \dfrac{8}{45} = \dfrac{40}{45}$$

Now, remaining work = $1 - \dfrac{40}{45} = \dfrac{1}{9}$

Next day, B will do the work.

$\therefore$ Work done by B on 11th day
$$= \dfrac{1}{15}$$

Remaining work = $\dfrac{1}{9} - \dfrac{1}{15} = \dfrac{2}{45}$

∴ A will complete the remaining work on 12th day.
∴ Total time taken by both A and B to complete the work
= (11 + 1) = 12 days

21. (b) Time taken by Piyush to complete 1/3 work = 30 days
∴ Time taken by Piyush to complete 1 work
= $30 \times \dfrac{1}{1/3} = 30 \times 3 = 90$ days
∴ One day work of Piyush = $\dfrac{1}{90}$
Now, rest work = $1 - \dfrac{1}{3} = \dfrac{2}{3}$
Time taken by Sanjeev to complete 2/3 work = 60 days
∴ Time taken by Sanjeev to complete 1 work
= $60 \times \dfrac{1}{2/3} = \dfrac{60 \times 3}{2} = 90$ days
∴ One day work of Sanjeev = $\dfrac{1}{90}$
∴ One day work of both
= $\dfrac{1}{90} + \dfrac{1}{90} = \dfrac{2}{90} = \dfrac{1}{45}$
∴ Time taken to complete the work by both
= $\dfrac{1}{1/45} = 45$ days

22. (c) Mason's 1 h work = $\dfrac{1}{12}$
Mason's 6 h work = $\dfrac{6}{12} = \dfrac{1}{2}$
Remaining work = $1 - \dfrac{1}{2} = \dfrac{1}{2}$
Remaining work can be finished in 5 h.
Total work can be finished in $2 \times 5 = 10$ h.
Now, $\dfrac{1}{12} + \dfrac{1}{B} = \dfrac{1}{10} \Rightarrow \dfrac{1}{B} = \dfrac{1}{10} - \dfrac{1}{12} = \dfrac{1}{60}$
Hence, boy can complete the work in 60 h.

23. (a) Let time taken by A to complete a work
= x days
∴ Time taken by B to complete the same work = 2x days
According to the question,
$\dfrac{1}{x} + \dfrac{1}{2x} = \dfrac{1}{12} \Rightarrow \dfrac{3}{2x} = \dfrac{1}{12}$
∴ $x = \dfrac{3 \times 12}{2} = 18$
Time taken by B to complete the same work
= $2x = 2 \times 18 = 36$ days

24. (d) Let time taken by B = x
Then, time taken by A = 2x
and time taken by C = 3x

According to the question,
$\dfrac{1}{x} + \dfrac{1}{2x} + \dfrac{1}{3x} = \dfrac{1}{12} \Rightarrow 1 + \dfrac{1}{2} + \dfrac{1}{3} = \dfrac{x}{12}$
$\Rightarrow \dfrac{6 + 3 + 2}{6} = \dfrac{x}{12} \Rightarrow 11 = \dfrac{x}{2} \Rightarrow x = 22$
∴ Required number of days = 2x
= $2 \times 22 = 44$ days

25. (c) Let A takes x and B takes 3x days to finish the job.
According to the question,
$3x - x = 30$
$\Rightarrow x = \dfrac{30}{2} = 15$ days
∴ B's time to complete the work
= $3x = 3 \times 15 = 45$ days
∴ (A + B)'s 1 day's work
= $\dfrac{1}{15} + \dfrac{1}{45} = \dfrac{3+1}{45} = \dfrac{4}{45}$
∴ (A + B) will finish the work in $\dfrac{45}{4}$ days.
Hence, required time = $\dfrac{45}{4} = 11\dfrac{1}{4}$ days.

26. (a) Let A, B and C take $x, \dfrac{x}{2}$ and $\dfrac{x}{3}$ days respectively to complete the work.
(A + B + C)'s 1 day's work = $\dfrac{1}{x} + \dfrac{2}{x} + \dfrac{3}{x} = \dfrac{6}{x}$
According to the question, $\dfrac{6}{x} = \dfrac{1}{2}$
$\left[\text{as, } (A + B + C)\text{'s 1 day's work} = \dfrac{1}{2} \right]$
$\Rightarrow x = 12$
∴ Time taken by A to complete the work
= 12 days
Time taken by B to complete the work
= $\dfrac{12}{2} = 6$ days
Time taken by C to complete the work
= $\dfrac{12}{3} = 4$ days

27. (a) A's 12 days work = $\dfrac{12}{27} = \dfrac{4}{9}$
∴ Remaining work = $1 - \dfrac{4}{9} = \dfrac{5}{9}$
B's 11 days work = $\dfrac{11}{33} = \dfrac{1}{3}$
∴ Remaining work = $\dfrac{5}{9} - \dfrac{1}{3} = \dfrac{5-3}{9} = \dfrac{2}{9}$
Now, 1 work completed by C in 45 days.
∴ $\dfrac{2}{9}$ work completed by C in $45 \times \dfrac{2}{9}$,
i.e. 10 days.

∴ Whole work will be completed in (12 + 11 + 10), i.e. 33 days.

28. (d) Let A and B complete the work in x and $3x$ days, respectively.
Then, according to the question,
$$\frac{4}{x} + \frac{15}{3x} = \frac{75}{100} \Rightarrow \frac{4}{x} + \frac{5}{x} = \frac{3}{4} \Rightarrow \frac{9}{x} = \frac{3}{4}$$
$$\Rightarrow x = \frac{9 \times 4}{3} = 12 \Rightarrow x = 12$$
∴ Time taken by B to complete the work
$= 3x = 3 \times 12$ days $= 36$ days

29. (c) ∵ A complete the work in 15 days.
∴ B will complete the work in 10 days.
They together will complete whole work
$$= \frac{15 \times 10}{25} = 6 \text{ days}$$
∴ A and B together worked
$$= 6 \times \frac{2}{3} = 4 \text{ days}$$

30. (c) According to the question,
5 men's 1 day's work $= \dfrac{1}{15}$ ...(i)
∴ 5 men's 5 days work $= \dfrac{1}{15} \times 5 = \dfrac{1}{3}$
∴ Remaining work $= 1 - \dfrac{1}{3} = \dfrac{2}{3}$
5 men + 10 women can do $\dfrac{2}{3}$ of the work in 5 days.
∴ 5 men + 10 women can do the whole of the work in $\dfrac{15}{2}$ days.
∴ (5 men + 10 women's) 1 day's work
$$= \frac{2}{15}$$ ...(ii)
10 women's 1 day's work
$$= \frac{2}{15} - 5 \text{ men's 1 day's work}$$
$$= \frac{2}{15} - \frac{1}{15} = \frac{1}{15}$$
So, 10 women can finish the work in 15 days.

31. (d) 1 man's 1 day's work $= \dfrac{1}{12 \times 20} = \dfrac{1}{240}$
Work done by 8 men in 10 days $= \dfrac{10 \times 8}{240} = \dfrac{1}{3}$
∴ Remaining work $= 1 - \dfrac{1}{3} = \dfrac{2}{3}$
Two-third work is completed by 18 women in 16 days.
∴ 1 work is completed by 18 women in
$\dfrac{16 \times 3}{2} = 8 \times 3 = 24$ days

32. (c) Given, time taken to complete the 1/4 work by $P = 10$ days
∴ Time taken to complete 1 work by P
$$= \frac{10}{1/4} = 40 \text{ days}$$
Time taken to complete $\dfrac{40}{100}$ work by Q
$= 40$ days
∴ Time taken to complete 1 work by Q
$$= \frac{40}{40/100} = 100 \text{ days}$$
Time taken to complete the 1/3 work by R
$= 13$ days
∴ Time taken to complete 1 work by R
$$= \frac{13}{1/3} = 39 \text{ days}$$
It is clear from above that R will complete the work first.

33. (c) Given, $M_1 = 6$, $M_2 = 8$, $T_1 = 16$ h,
and $T_2 = ?$, $W_1 = W_2 = 1$
According to the formula,
$M_1 T_1 W_2 = M_2 T_2 W_1$ [by Technique 3]
∴ $6 \times 16 \times 1 = 8 \times T_2 \times 1$
∴ $T_2 = \dfrac{16 \times 6}{8} = 2 \times 6 = 12$ h

34. (d) Let food stock will be exhausted in x days.
Given, $M_1 = 120$, $D_1 = 45$
$M_2 = 120 + 30 = 150$ and $D_2 = x$
According to the formula,
$M_1 D_1 = M_2 D_2$ [by Technique 3]
$\Rightarrow 120 \times 45 = 150 \times x$
∴ $x = \dfrac{120 \times 45}{150} = 36$

35. (b) Let number of men's working $= x$
Then, by formula, $M_1 D_1 = M_2 D_2$
[by Technique 3]
$\Rightarrow x \times 60 = (x + 6) \times (60 - 20)$
$\Rightarrow x \times 60 = (x + 6) \times 40$
$\Rightarrow x \times 3 = (x + 6) \times 2 \Rightarrow 3x = 2x + 12$
$\Rightarrow 3x - 2x = 12 \Rightarrow x = 12$

36. (b) Given, $M_1 = 20$, $D_1 = 7$,
$M_2 = 20 + 8 = 28$ and $D_2 = ?$
According to the formula,
$M_1 D_1 = M_2 D_2$ [by Technique 3]
$\Rightarrow 20 \times 7 = 28 \times D_2$
∴ $D_2 = \dfrac{20 \times 7}{28} = 5$ days

37. (b) Given, $M_1 = 12$, $M_2 = 8$, $D_1 = 24$, $W_1 = 1$
$W_2 = 1$, $D_2 = ?$

Work and Time / 385

According to the formula,
$M_1 D_1 W_2 = M_2 D_2 W_1$ [by Technique 3]
$\Rightarrow 12 \times 24 \times 1 = 8 \times D_2 \times 1$
$\therefore D_2 = \dfrac{12 \times 24}{8} = 12 \times 3 = 36$ days

38. (a) Given, $M_1 = 40, W_1 = 200, T_1 = 8$,
$D_1 = 12, M_2 = 30, W_2 = 300, T_2 = 6$ and $D_2 = ?$
According to the formula,
$M_1 T_1 D_1 W_2 = M_2 T_2 D_2 W_1$ [by Technique 3]
$\Rightarrow 40 \times 8 \times 12 \times 300 = 30 \times 6 \times D_2 \times 200$
$\therefore D_2 = \dfrac{40 \times 8 \times 12 \times 300}{30 \times 6 \times 200} = 32$ days

39. (a) Let required number of units of work = x
$W_1 = m$ and $W_2 = x$
According to the formula,
$M_1 T_1 D_1 W_2 = M_2 T_2 D_2 W_1$ [by Technique 3]
$\Rightarrow m \times m \times m \times x = n \times n \times n \times m$
$\therefore x = \dfrac{m \times n^3}{m^3} = \dfrac{n^3}{m^2}$

40. (c) Here, $M_1 = 15, D_1 = 16, W_1 = W_2 = 1$
$M_2 = 24, D_2 = ?$
According to the formula,
$M_1 D_1 W_2 = M_2 D_2 W_1$ [by Technique 3]
$\Rightarrow 15 \times 16 \times 1 = 24 \times D_2 \times 1$
$\therefore D_1 = \dfrac{15 \times 16}{24} = 10$ days
Therefore, 10 days are required to complete the work.

41. (a) Given, $M_1 = 20, M_2 = 25, T_1 = 5$,
$T_2 = 10, D_1 = 10$ and $D_2 = ?$
According to the formula,
$M_1 T_1 D_1 = M_2 T_2 D_2$ [by Technique 3]
$\Rightarrow 20 \times 5 \times 10 = 25 \times 10 \times D_2$
$\therefore D_2 = \dfrac{20 \times 5 \times 10}{25 \times 10} = 4$ days

42. (b) Let original number of men = x
Time taken by x men = 80 days
Now, $(x - 10)$ men can finish the work in $(80 + 20) = 100$ days
Here, $M_1 = x, M_2 = (x - 10), D_1 = 80$ and $D_2 = 100$
According to the formula,
$M_1 D_1 = M_2 D_2$ [by Technique 3]
$\Rightarrow x \times 80 = 100 \times (x - 10)$
$\Rightarrow 8x = 10x - 100$
$\Rightarrow 10x - 8x = 100 \Rightarrow x = 50$

43. (d) Given, $M_1 = 240, D_1 = 48, M_2 = 160$ and $D_2 = ?$

According to the formula,
$M_1 D_1 = M_2 D_2$ [by Technique 3]
$\Rightarrow 240 \times 48 = 160 \times D_2$
$\therefore D_2 = \dfrac{240 \times 48}{160} = 72$ days

44. (b) Given, $M_1 = 45, D_1 = 18, M_2 = ?, D_2 = 15$
According to the formula,
$M_1 D_1 = M_2 D_2$ [by Technique 3]
$\therefore M_2 = \dfrac{M_1 D_1}{D_2}$
$\Rightarrow M_2 = \dfrac{45 \times 18}{15} = 3 \times 18$
$\Rightarrow M_2 = 54$

45. (d) $M_1 = 250, D_1 = 33$ day,
Per day meal; $W_1 = W_2 = 125$ g
$M_2 = (250 + 80) = 330$ and $D_2 = ?$
According to the formula,
$M_1 D_1 W_2 = M_2 D_2 W_1$ [by Technique 3]
$\Rightarrow 250 \times 33 \times 125 = 330 \times D_2 \times 125$
$\Rightarrow D_2 = \dfrac{250 \times 33}{330}$
$\therefore D_2 = 25$ days

46. (b) Let x men be allowed to go off.
Then, $M_1 = 90, D_1 = 25, D_2 = 15, W_1 = \dfrac{2}{3}$,
$W_2 = 1 - \dfrac{2}{3} = \dfrac{1}{3}, M_2 = (90 - x)$
According to the formula,
$M_1 D_1 W_2 = M_2 D_2 W_1$ [by Technique 3]
$\Rightarrow (90 \times 25)\left(\dfrac{1}{3}\right) = (90 - x) \times 15 \times \dfrac{2}{3}$
$\Rightarrow 90 \times 25 \times \dfrac{1}{3} = 10 (90 - x)$
$\Rightarrow 75 = 90 - x$
$\therefore x = 90 - 75 = 15$

47. (a) Given, $M_1 = 16, T_1 = 18$,
$W_1 = 36 \times 4 \times 24, D_1 = 20$ and $M_2 = ?$,
$T_2 = 12, W_2 = 64 \times 6 \times 18, D_2 = 16$
According to the formula,
$\dfrac{M_1 D_1 T_1}{W_1} = \dfrac{M_2 D_2 T_2}{W_2}$ [by Technique 3]
$\Rightarrow \dfrac{16 \times 20 \times 18}{36 \times 4 \times 24} = \dfrac{M_2 \times 12 \times 16}{64 \times 6 \times 18}$
$\Rightarrow \dfrac{15}{36 \times 24} = \dfrac{M_2 \times 2}{64 \times 6 \times 18}$
$\therefore M_2 = 60$ men

48. (c) 4 men working 4 h per day for 4 days to complete 4 units of work.
i.e. 4 units of work completed in
$(4 \times 4 \times 4) = 64$ h

386 / Fast Track Objective Arithmetic

$\therefore$ 1 unit of work completed $= \dfrac{64}{4} = 16$ h

Now, 2 men working for 2 h per day in 2 days.
i.e. $(2 \times 2 \times 2) = 8$ h

In 8 h $= \dfrac{8}{16}$ units of work completed

$= \dfrac{1}{2}$ unit of work completed

Fast Track Method
Here, $M_1 = 4, T_1 = 4, D_1 = 4, W_1 = 4$
and $M_2 = 2, T_2 = 2, D_2 = 2, W_2 = ?$
According to the formula,
$M_1 D_1 T_1 W_2 = M_2 D_2 T_2 W_1$ [by Technique 3]
$\Rightarrow 4 \times 4 \times 4 \times W_2 = 2 \times 2 \times 2 \times 4$
$\Rightarrow W_2 = \dfrac{2 \times 2 \times 2 \times 4}{4 \times 4 \times 4} = \dfrac{1}{2}$

49. (b) According to the question,
1 man = 2 women
$\therefore$ 10 men + 8 women = (20 + 8) women
= 28 women
4 men + 6 women = (8 + 6) women
= 14 women
Given, $M_1 = 28, M_2 = 14, D_1 = 5, D_2 = ?$
and $W_1 = W_2 = 1$
According to the formula,
$M_1 D_1 W_2 = M_2 D_2 W_1$ [by Technique 3]
$\Rightarrow 28 \times 5 \times 1 = 14 \times D_2 \times 1$
$\therefore D_2 = \dfrac{28 \times 5}{14} = 10$ days

50. (d) We have, $M_1 = 15, D_1 = 21, H_1 = 8,$
$M_2 = 14, D_2 = ?, H_2 = 6$
[$\because 3W = 2M \Rightarrow 21W = 14M$]
$\therefore \dfrac{M_1 \times D_1 \times H_1}{W_1} = \dfrac{M_2 \times D_2 \times H_2}{W_2}$
[by Technique 3]
$\Rightarrow \dfrac{15 \times 21 \times 8}{W} = \dfrac{14 \times 6 \times D_2}{W}$
$\Rightarrow D_2 = \dfrac{15 \times 21 \times 8}{14 \times 6} = 30$

51. (c) Here, $M_1 = 16, D_1 = 30, W_1 = \dfrac{5}{9},$
$M_2 = x, D_2 = 8$ and $W_2 = \dfrac{4}{9}$
According to the formula,
$\dfrac{M_1 D_1}{W_1} = \dfrac{M_2 D_2}{W_2}$ [by Technique 3]
$\Rightarrow \dfrac{16 \times 30}{\dfrac{5}{9}} = \dfrac{x \times 8}{\dfrac{4}{9}} \Rightarrow x = \dfrac{16 \times 30 \times \dfrac{4}{9}}{8 \times \dfrac{5}{9}}$

$\Rightarrow x = 48$
$\therefore$ More men needed to hire to complete the project on time
$= 48 - 16 = 32$ men

52. (c) 1 man = 2 women = 3 boys
$\therefore$ 1 man + 1 woman + 1 boy
$= \left(3 + \dfrac{3}{2} + 1\right)$ boys $= \dfrac{11}{2}$ boys

Here, $M_1 = 3, M_2 = \dfrac{11}{2}, D_1 = 88$ and $D_2 = ?$
According to the formula,
$M_1 D_1 = M_2 D_2$ [by Technique 3]
$\Rightarrow 3 \times 88 = \dfrac{11}{2} \times D_2$
$\Rightarrow D_2 = \dfrac{3 \times 2 \times 88}{11} = 3 \times 2 \times 8 = 48$
$\therefore$ Required number of days is 48.

53. (b) 4 men = 6 women = 10 children
$\Rightarrow$ 1 man $= \dfrac{5}{2}$ children
and 1 woman $= \dfrac{5}{3}$ children
Now, 1 couple + 5 children
= 1 man + 1 woman + 5 children
$= \left(\dfrac{5}{2} + \dfrac{5}{3} + 5\right) = \dfrac{55}{6}$ children
According to the formula,
$M_1 D_1 = M_2 D_2$ [by Technique 3]
$\Rightarrow 10 \times 5 = \dfrac{55}{6} \times D_2$
$\therefore D_2 = \dfrac{60}{11} = 5\dfrac{5}{11}$ days

54. (d) Given, $M_1 = 120, M_2 = (120 - x),$
$D_1 = 64, D_2 = 60, W_1 = \dfrac{2}{3}$ and $W_2 = \dfrac{1}{3}$
According to the formula,
$\dfrac{M_1 D_1}{W_1} = \dfrac{M_2 D_2}{W_2}$ [by Technique 3]
$\Rightarrow \dfrac{120 \times 64}{2/3} = \dfrac{(120 - x) \times 60}{1/3}$
$\Rightarrow \dfrac{120 \times 64}{2 \times 60} = (120 - x)$
$\Rightarrow (120 - x) = 64$
$\therefore x = 120 - 64 = 56$
Hence, 56 men can be discharged to finish the work in time.

55. (a) Remaining work $= \left(1 - \dfrac{4}{7}\right) = \dfrac{3}{7}$
Remaining period = (92 - 66) = 26 days

Let the number of additional men $= x$
Given, $M_1 = 234$, $D_1 = 66$, $T_1 = 16$,
$W_1 = \dfrac{4}{7}$, $M_2 = (234 + x)$, $D_2 = 26$,
$T_2 = 18$, $W_2 = \dfrac{3}{7}$

According to the formula,
$M_1 W_2 T_1 D_1 = M_2 W_1 T_2 D_2$ [by Technique 3]

$\Rightarrow 234 \times \dfrac{3}{7} \times 16 \times 66 = (234 \times x) \times \dfrac{4}{7} \times 18 \times 26$

$\Rightarrow \quad 234 + x = \dfrac{3 \times 66 \times 16 \times 234}{4 \times 26 \times 18}$

$\Rightarrow \quad 234 + x = 36 \times 11 = 396$

$\therefore \quad x = 396 - 234 = 162$

So, additional men to be employed $= 162$

56. (c) Given, $M_1 = 105$, $D_1 = 25$, $T_1 = 8$, $W_1 = \dfrac{2}{5}$

Now, let the additional men be x.
Then, $\quad M_2 = 105 + x$, $T_2 = 9$, $D_2 = 25$
and $\quad W_2 = 1 - \dfrac{2}{5} = \dfrac{3}{5}$

According to the formula,
$\dfrac{M_1 D_1 T_1}{W_1} = \dfrac{M_2 D_2 T_2}{W_2}$ [by Technique 3]

$\Rightarrow \dfrac{105 \times 25 \times 8}{2/5} = \dfrac{(105 + x) \times 25 \times 9}{3/5}$

$\Rightarrow \dfrac{105 \times 8}{2} = \dfrac{(105 + x) \times 9}{3}$

$\Rightarrow 105 \times 4 = (105 + x) \times 3$

$\Rightarrow 105 \times 4 = 105 \times 3 + 3x \Rightarrow 3x = 105$

$\therefore \quad x = 35$ men

57. (c) $(A + B)$'s 1 day's work $= \dfrac{1}{72}$

$(B + C)$'s 1 day's work $= \dfrac{1}{120}$

$(A + C)$'s 1 day's work $= \dfrac{1}{90}$

$2(A + B + C)$'s 1 day's work $= \dfrac{1}{72} + \dfrac{1}{120} + \dfrac{1}{90}$

$\therefore (A + B + C)$'s 1 day's work
$= \dfrac{5 + 3 + 4}{360 \times 2} = \dfrac{12}{360 \times 2} = \dfrac{1}{60}$

$\therefore$ A's 1 day's work $= (A + B + C)$'s 1 day's work $- (B + C)$'s 1 day's work
$= \dfrac{1}{60} - \dfrac{1}{120} = \dfrac{2 - 1}{120} = \dfrac{1}{120}$

$\therefore$ A alone can finish the work in 120 days.

Fast Track Method
Here, $x = 72$, $y = 120$ and $z = 90$
$\therefore$ Time taken by A, B and C

$= \dfrac{2xyz}{xy + yz + zx}$ [by Technique 4]

$= \dfrac{2 \times 72 \times 120 \times 90}{72 \times 120 + 120 \times 90 + 72 \times 90}$

$= \dfrac{2 \times 72 \times 120 \times 90}{25920} = 60$ days

Now, A's 1 day's work
$= (A + B + C)$'s 1 day's work
$\qquad - (B + C)$'s 1 day's work
$= \dfrac{1}{60} - \dfrac{1}{120} = \dfrac{1}{120}$

Hence, A can finish the work in 120 days.

58. (d) A's and B's 1 h work $= \dfrac{1}{10}$

B's and C's 1 h work $= \dfrac{1}{15}$

and A's and C's 1 h work $= \dfrac{1}{12}$

$\therefore$ A's, B's and C's 1 h work
$= \dfrac{1}{2}\left(\dfrac{1}{10} + \dfrac{1}{15} + \dfrac{1}{12}\right) = \dfrac{1}{2} \times \dfrac{1}{4} = \dfrac{1}{8}$

Hence, B's work in 1 h $= \dfrac{1}{8} - \dfrac{1}{12} = \dfrac{1}{24}$

$\therefore$ B independently can complete the work in 24 h.

Fast Track Method
Here, $x = 10$, $y = 15$ and $z = 12$
$\therefore$ Time taken by A, B and C

$= \dfrac{2xyz}{xy + yz + zx}$ [by Technique 4]

$= \dfrac{2 \times 10 \times 15 \times 12}{10 \times 15 + 15 \times 12 + 12 \times 10}$

$= \dfrac{2 \times 10 \times 15 \times 12}{450} = 8$ days

Now, B's 1 hour's work
$= \dfrac{1}{8} - \dfrac{1}{12} = \dfrac{1}{24}$.

Hence, B can complete the work in 24 h.

59. (d) Here, $a_1 = 3$, $b_1 = 4$, $D = 43$, $a_2 = 7$
and $b_2 = 5$
$\therefore$ Required number of days

$= \dfrac{D a_1 b_1}{a_2 b_1 + a_1 b_2}$ [by Technique 5]

$= \dfrac{43 \times 3 \times 4}{7 \times 4 + 3 \times 5} = \dfrac{43 \times 3 \times 4}{28 + 15}$

$= \dfrac{43 \times 3 \times 4}{43} = 12$

60. (d) 3 men $=$ 6 children
$\Rightarrow$ 1 man $=$ 2 children
$\therefore$ 4 men $+$ 4 children

388 / Fast Track Objective Arithmetic

$= 4 \text{ men} + \dfrac{4}{2} \text{ men} = 6 \text{ men}$

Given, $M_1 = 3$, $M_2 = 6$, $D_1 = 18$,
$W_1 = W_2 = 1$ and $D_2 = ?$
According to the formula,
$M_1 D_1 W_2 = M_2 D_2 W_1$
[by Technique 3]
$\Rightarrow 3 \times 18 \times 1 = 6 \times D_2 \times 1$
$\therefore D_2 = \dfrac{3 \times 18}{6} = 9 \text{ days}$

Fast Track Method
Here, $a_1 = 3$, $b_1 = 6$, $D = 18$, $a_2 = 4$
and $b_2 = 4$
$\therefore$ Required number of days
$= \dfrac{D a_1 b_1}{a_2 b_1 + a_1 b_2}$ [by Technique 5]
$= \dfrac{18 \times 3 \times 6}{4 \times 6 + 3 \times 4} = \dfrac{18 \times 3 \times 6}{36} = 9$

61. (d) Part of field grazed by 4 goats in 1 day
$= \dfrac{1}{50}$
Part of field grazed by 1 goat in 1 day
$= \dfrac{1}{50 \times 4} = \dfrac{1}{200}$
Also, $4g = 6s$
[here, g = goats and s = sheeps]
$\Rightarrow 1s = \dfrac{4}{6} g = \dfrac{2}{3} g$
Now, $2g + 9s = 2g + 9 \times \dfrac{2}{3} g$
$= 2g + 6g = 8g$
$\therefore$ 8 goats can graze the field in
$\dfrac{1}{8/200} = 25 \text{ days}$

Fast Track Method
Here, $a_1 = 4$, $b_1 = 6$, $a_2 = 2$, $b_2 = 9$
and $D = 50$
$\therefore$ Required number of days $= \dfrac{D a_1 b_1}{a_2 b_1 + a_1 b_2}$
[by Technique 5]
$= \dfrac{50 \times 4 \times 6}{2 \times 6 + 4 \times 9} = \dfrac{50 \times 4 \times 6}{12 + 36} = \dfrac{50 \times 4 \times 6}{48} = 25$

62. (b) Let time taken by $B = x$
Efficiency $\propto \dfrac{1}{\text{Time taken}}$
So, if B is 100% efficient, then A is 80% efficient.
So, $\dfrac{80}{100} = \dfrac{x}{15/2} \Rightarrow x = 6 \text{ h}$

Fast Track Method
Let B completes 100% work.
Then, A completes 80% work.
$\Rightarrow$ B is 25% faster than A.
Here, $x = \dfrac{15}{2}$ and $y = 25$
$\therefore B$ will complete the work
$= \dfrac{100 x}{100 + y}$ [by Technique 6]
$= \dfrac{100 \times \dfrac{15}{2}}{125} = 6 \text{ h}$

63. (a) We have, $a = 5$, $b = 12$, $x = 10$, $y = 15$, $c = 5$ and $d = 6$
$\therefore$ Number of days required to complete the work by 5 men and 6 women
$= \dfrac{1}{\dfrac{c}{ax} + \dfrac{d}{by}} = \dfrac{1}{\dfrac{5}{5 \times 10} + \dfrac{6}{12 \times 15}}$
[by Technique 7]
$= \dfrac{1}{\dfrac{1}{10} + \dfrac{1}{30}} = \dfrac{1}{\dfrac{3+1}{30}}$
$= \dfrac{30}{4} = 7\dfrac{1}{2} \text{ days}$

64. (d) 15 men $\times 20 = 24$ women $\times 20$
$\Rightarrow$ 15 men = 24 women
$\Rightarrow$ 5 men = 8 women
Now, (10 men + 8 women) $\times D$
$= 24$ women $\times 20$
$\Rightarrow$ (16 women + 8 women) $\times D$
$= 24$ women $\times 20$
$\Rightarrow$ 24 women $\times D = 24$ women $\times 20$
$\therefore D = 20 \text{ days}$

Fast Track Method
Here, $x = 20$, $a = 15$, $c = 10$, $b = 24$, $y = 20$ and $d = 8$
$\therefore$ Required number of days
$= \dfrac{1}{\dfrac{c}{ax} + \dfrac{d}{by}}$ [by Technique 7]
$= \dfrac{1}{\dfrac{10}{15 \times 20} + \dfrac{8}{24 \times 20}} = \dfrac{1}{\dfrac{1}{30} + \dfrac{1}{3 \times 20}}$
$= \dfrac{1}{\dfrac{2+1}{60}} = \dfrac{60}{3} = 20 \text{ days}$

65. (b) Here, $a = 10$, $x = 6$, $b = 6$, $y = 12$, $c = 15$ and $d = 10$
Time taken to complete the work by 15 men and 10 women

Work and Time / 389

$$= \frac{1}{\frac{c}{ax} + \frac{d}{by}} \quad \text{[by Technique 7]}$$

$$= \frac{1}{\frac{15}{10 \times 6} + \frac{10}{6 \times 12}} = \frac{1}{\frac{3}{2 \times 6} + \frac{5}{3 \times 12}}$$

$$= \frac{1}{\frac{3}{12} + \frac{5}{36}} = \frac{1}{\frac{9+5}{36}} = \frac{1}{\frac{14}{36}}$$

$$= \frac{36}{14} = \frac{18}{7} = 2\frac{4}{7} \text{ days}$$

66. (c) Here, $a_1 = 8$, $b_1 = 4$, $a_2 = 1$, $b_2 = 1$,
$y = 7$ and $x = 1$
$$\therefore \frac{\text{Work done by 1 man}}{\text{Work done by 1 boy}} = \frac{(yb_2 - xb_1)}{(xa_1 - ya_2)}$$
[by Technique 8]
$$= \frac{7 \times 1 - 4}{8 - 7 \times 1} = \frac{3}{1}$$

67. (a) According to the formula,
Required time $= \sqrt{ab}$ [by Technique 9]
where, $a = 25$ and $b = 49$
$\therefore$ Required time $= \sqrt{25 \times 49}$
$= 5 \times 7 = 35$ days

68. (a) Let the required days be x.
A works for $(x - 2)$ days, while B works for x days.
According to the question,
$$\frac{x-2}{10} + \frac{x}{20} = 1$$
$\Rightarrow \quad 2x - 4 + x = 20 \Rightarrow 3x = 24$
$\therefore \quad x = 8$ days

Fast Track Method
Here, $a = 10$, $b = 20$ and $x = 2$
$\therefore$ Required time $= \frac{(a + x) b}{(a + b)}$
[by Technique 10 (i)]
$$= \frac{(10 + 2) \times 20}{30} = 8 \text{ days}$$

69. (d) $\because$ A's one day work $= \frac{1}{30}$
and B's one day work $= \frac{1}{36}$
$\therefore (A + B)$'s one day work $= \frac{1}{30} + \frac{1}{36} = \frac{11}{180}$
25 days work of $B = \frac{25}{36}$
$\therefore$ Remaining work $= 1 - \frac{25}{36} = \frac{11}{36}$
A and B complete $\frac{11}{36}$ work in

$$\frac{11}{36} \div \frac{11}{180} = 5 \text{ days}$$

So, A leaves after 5 days.

Fast Track Method
Here, $a = 30$, $b = 36$, $x = 25$, $T = ?$
$$\therefore \quad T = \frac{(b - x) a}{a + b} \quad \text{[by Technique 11]}$$
$$= \frac{(36 - 25) \times 30}{(30 + 36)}$$
$$= \frac{11 \times 30}{66} = 5 \text{ days}$$

Hence, A leaves after 5 days.

70. (c) B's 30 days work $= \frac{30}{75} = \frac{2}{5}$

Remaining work $= 1 - \frac{2}{5} = \frac{3}{5}$

$(A + B)$'s 1 day's work $= \frac{1}{60} + \frac{1}{75}$
$$= \frac{5 + 4}{300} = \frac{9}{300} = \frac{3}{100}$$

$(A + B)$ do $\frac{3}{100}$ work in 1 day.

$\therefore (A + B)$ will do $\frac{3}{5}$ work in
$$\frac{3}{5} \times \frac{100}{3} = 20 \text{ days}$$

$\therefore A$ left the work after 20 days.

Fast Track Method
Here, $a = 60$ days, $b = 75$ days, $x = 30$, $T = ?$
$\therefore$ Required time $= \frac{(b - x) a}{a + b}$
[by Technique 11]
$$= \frac{(75 - 30) \times 60}{60 + 75} = \frac{45 \times 60}{135} = 20 \text{ days}$$

71. (a) Ajay's 1 day's work $= \frac{1}{25}$

Sanjay's 1 day's work $= \frac{1}{20}$

Ajay's and Sanjay's together 1 day's work
$$= \frac{1}{25} + \frac{1}{20} = \frac{4 + 5}{100} = \frac{9}{100}$$

Their 5 days work together
$= 5 \times 1$ day's work
$$= 5 \times \frac{9}{100} = \frac{45}{100}$$

Remaining work $= 1 - 5$ days work
$$= 1 - \frac{45}{100} = \frac{55}{100}$$

Now, this remaining work is done by Sanjay.
Let Sanjay takes x days to complete it.

Then, $\dfrac{1}{20} \times x = \dfrac{55}{100}$

$\Rightarrow \qquad x = \dfrac{55 \times 20}{100}$

$\therefore \qquad x = 11$ days

So, remaining work is done in 11 days by Sanjay.

Fast Track Method

Here, $a = 25$ days, $b = 20$ days,
$x = ?$ and $T = 5$

According to the formula,

$T = \dfrac{(b-x)a}{a+b}$ [by Technique 11]

$\Rightarrow \quad 5 = \dfrac{(20-x) \times 25}{45}$

$\Rightarrow \quad 45 = 100 - 5x$

$\Rightarrow \quad 5x = 55$

$\therefore \qquad x = 11$ days

72. (b) Let B left the work after x days from the start.

According to the question,

$\dfrac{x}{25} + \dfrac{x+7}{15} = 1 \Rightarrow \dfrac{3x + 5x + 35}{75} = 1$

$\Rightarrow \qquad 8x = 75 - 35 \Rightarrow x = \dfrac{40}{8}$

$\therefore \qquad x = 5$

Fast Track Method

Here, $a = 25$, $b = 15$ and $x = 7$

$\therefore$ Required number of days

$= \dfrac{(b-x)a}{a+b}$ [by Technique 11]

$= \dfrac{(15-7)25}{25+15} = \dfrac{8 \times 25}{40} = 5$

73. (c) Let total work to be done
= LCM of time taken by Dinu and Tinu

Dinu (45) — 8
              360 (Total work)
Tinu (40) — 9

Amount of work done in 1 day by Dinu

$= \dfrac{360}{45} = 8$ units

Amount of work done in 1 day by Tinu

$= \dfrac{360}{40} = 9$ units

Work finished by Tinu in 23 days
$= 23 \times 1$ day work of time
$= 23 \times 9 = 207$ units

$\Rightarrow$ Units of work done by them together
$= 360 - 207 = 153$ units

$\therefore$ Number of days they worked together

$= \dfrac{153}{17} = 9$ days

Alternate Method

Dinu's one day's work $= \dfrac{1}{45}$

Tinu's one day's work $= \dfrac{1}{40}$

Dinu's + Tinu's one day's work

$= \dfrac{1}{45} + \dfrac{1}{40}$

$= \dfrac{40+45}{1800} = \dfrac{85}{1800} = \dfrac{17}{360}$

Tinu's 23 days work $= \dfrac{23}{40}$

Remaining work $= 1 - \dfrac{23}{40} = \dfrac{17}{40}$

$\dfrac{17}{40}$ th work was done by Dinu and Tinu together.

$\therefore \dfrac{17}{40}$ work was completed in

$\dfrac{17}{40} \times \dfrac{360}{17} = 9$ days

Fast Track Method

Here, $a = 45$, $b = 40$ and $x = 23$

$\therefore$ Required number of days

$= \dfrac{(b-x)a}{a+b}$ [by Technique 11]

$= \dfrac{(40-23) \times 45}{40+45} = \dfrac{17 \times 45}{85} = 9$

Exercise 2 Higher Skill Level Questions

1. (d) 4 men takes double the time taken by 5 children to do a job.

So, 5 children = 8 men ...(i)

Also, it is given that 1 man = 2 women ...(ii)

Now, 12 men, 10 children and 8 women
= 16 men + 10 children
   [from Eqs. (i) and (ii)]
= 20 children

Now, 1 child can do that job in 20 days.

$\therefore$ Time taken by 20 children $= \dfrac{20}{20}$ days

$= 1$ day

2. (d) Since, 3 women + 18 children complete work in 2 days.

Work and Time / 391

Therefore, (3 × 2) women + (18 × 2) children complete work in 1 day, i.e. 6 women + 36 children complete work in 1 day.

Work of 36 children for 1 day $= 1 - \frac{1}{3} = \frac{2}{3}$

$\left[\because \text{ work of 6 women for 1 day} = \frac{1}{3}\right]$

$\therefore$ 36 children do $\frac{2}{3}$ part of the work in 1 day.

$\Rightarrow$ 36 children can do the work in $\frac{3}{2}$ days.

$\Rightarrow$ 9 children can do the work in

$\left(\frac{3}{2} \times 4\right) = 6$ days

3. (d) Time taken by A to build a structure
= 8 days
Time taken by B to break a structue = 3 days
Work done in 6 days

$= \frac{6}{8} - \frac{2}{3} = \frac{18 - 16}{24} = \frac{1}{12}$

Remaining work $= 1 - \frac{1}{12} = \frac{11}{12}$

$\therefore$ Required time $= 8 \times \frac{11}{12} \approx 8$ days

4. (b) One day work of

$(16M + 10W) = \frac{1}{10}$ ...(i)

and one day work of $12\,W = \frac{1}{25}$ ...(ii)

Now, multiplying Eq. (i) by 12 and Eq. (ii) by 10 and then subtracting Eq. (ii) from Eq. (i), we get

$192M = \frac{12}{10} - \frac{10}{25}$

$\Rightarrow \quad 192M = \frac{60 - 20}{50} = \frac{40}{50} = \frac{4}{5}$

$\Rightarrow \quad M = \frac{4}{5 \times 192} = \frac{1}{5 \times 48} = \frac{1}{240}$

So, 1 man's one day's work $= \frac{1}{240}$

So, 1 man will complete the work in 240 days.
Hence, 10 men will complete the work in $\frac{240}{10}$, i.e. 24 days.

5. (a) Let time taken by P to complete the work be x days and time taken by Q to complete the work be $2x$ days.

$\therefore \quad$ Q's one day's work $= \frac{1}{2x}$

and P's one day's work $= \frac{1}{x}$

Now, (P + Q)'s one day's work

$= \frac{1}{x} + \frac{1}{2x} = \frac{3}{2x}$

$\therefore$ (P + Q) will complete the whole work in $\frac{2x}{3}$ days.

Now, according to the question,
R will complete this work in $2x$ days.

$\therefore$ R's one day's work $= \frac{1}{2x}$

Required ratio $= \frac{1}{x} : \frac{1}{2x} : \frac{1}{2x}$

$= 1 : \frac{1}{2} : \frac{1}{2} = 2 : 1 : 1$

6. (b) Work done by Q and R in 6 days

$= \frac{6 \times 7}{60} = \frac{7}{10}$

$\therefore \quad$ Remaining $= 1 - \frac{7}{10} = \frac{3}{10}$

P does 3/10 work in 3 days.

$\Rightarrow$ 1 work is done by P in $\frac{3 \times 10}{3} = 10$ days

Since, P alone can do the work in 10 days.

Q's one day's work $= \frac{1}{6} - \frac{1}{10} = \frac{10 - 6}{60} = \frac{1}{15}$

Q alone can do the work in 15 days.

$\therefore$ R's one day's work $= \frac{7}{60} - \frac{1}{15} = \frac{7 - 4}{60} = \frac{1}{20}$

$\therefore$ R alone can do the work in 20 days.
So, required difference = 20 − 10 = 10 days

7. (a) 6 × 12 men = 8 × 18 women
= 18 × 10 children
$\Rightarrow \quad$ 12 men = 24 women = 30 children
$\Rightarrow \quad$ 2 men = 4 women = 5 children
Now, 4 men + 12 women + 20 children
= 4 men + 6 men + 8 men = 18 men
Time to do remaining work for 1 man
= (6 × 12 − 18 × 2) = 36 days
$\therefore$ Required number of men to finish the work in 1 day = 36

8. (d) We have, A can do 50% more work than B in same hour.
B alone can do a piece of work in 30 h.
$\therefore$ A alone can do a piece of work in 20 h.

In 1 h, B can do $\frac{1}{30}$ piece of work and A can do $\frac{1}{20}$ piece of work.

$\therefore$ In 1 h, A and B can do $\left(\frac{1}{30} + \frac{1}{20}\right)$

$= \frac{1}{12}$ piece of work.

∴ A and B both can do piece of work in 12 h.
But B has already worked for 12 h.

∴ B has finished $\frac{12}{30}$ part of work.

Now, remaining work = $1 - \frac{12}{30} = \frac{18}{30}$

For completing $\frac{18}{30}$ work by A and B combining, hours required

$= \frac{18}{30} \times 12 = 7.2$ h

9. (d) As the number of men increase the number of days taken to do the work must decrease which means that diagrams II and III are ruled out. Moreover, the men work at constant speed which means that no case will the work be done in zero days no matter, how many men are put to work. Thus, diagram I is ruled out because, here the graph touches the zero line on both the axes. Thus, the right answer is diagram IV.

10. (d) Here, 18 men's 24 days work
    = 12 women's 32 days work
    ∴ $18M \times 24 = 12W \times 32$
    ⇒ $18M \times 2 = W \times 32$
    ⇒ 9 men = 8 women

Let 18 men works for x days or 4 men leave the work after x days.
Then, according to the question,

$\frac{18 \times x}{\frac{x}{24}} = (14 + 9) \text{ men} \times \frac{360}{23} \times \frac{1}{\left(1 - \frac{x}{24}\right)}$

⇒ $18 \times 24 = \frac{360}{\left(1 - \frac{x}{24}\right)}$ ⇒ $x = 4$ days

Hence, 4 men leaves the job after 4 days.

11. (c) Here, efficiency of a woman, a man and a boy = $\frac{1}{8} : \frac{1}{6} : \frac{1}{12} = 3 : 4 : 2$

Total units of work = $9 \times 6 \times 6 \times 4 = 1296$

Work completed by 12 women, 12 men and 12 boys working 8 h per day

$= \frac{1296}{12 \times 8 \times 3 + 12 \times 8 \times 4 + 12 \times 8 \times 2}$

$= \frac{1296}{96(3 + 4 + 2)} = \frac{1296}{96 \times 9}$

$= \frac{12}{8} = 1\frac{1}{2}$ days

12. (e) 18 women complete the work = 24 days
    ∴ 1 day work of 1 woman = $\frac{1}{18 \times 24}$
    and 24 men complete the work = 15 days

∴ 1 day work of 1 man = $\frac{1}{24 \times 15}$

Now, 3 days work of 16 women

$= 16 \times 3 \times \frac{1}{18 \times 24} = \frac{1}{9}$

∴ Rest work = $1 - \frac{1}{9} = \frac{8}{9}$

Now, 2 days work of 20 men

$= 2 \times 20 \times \frac{1}{24 \times 15} = \frac{1}{9}$

∴ Rest work = $\frac{8}{9} - \frac{1}{9} = \frac{7}{9}$

1 day work of 20 M + 16 W

$= \frac{20}{24 \times 15} + \frac{16}{18 \times 24}$

$= \frac{1}{18} + \frac{1}{27} = \frac{3 + 2}{54} = \frac{5}{54}$

∴ Number of days required to complete $\frac{7}{9}$th work = $\frac{7}{9} \times \frac{54}{5} = \frac{42}{5} = 8\frac{2}{5}$ days

13. (b) Let $(A + B)$ take x days to complete the work.

∴ One day work of $(A + B) = \frac{1}{x}$

and one day work of $(B + C) = \frac{1}{13\frac{1}{3}}$

$= \frac{1}{40/3} = \frac{3}{40}$

and A takes 60% more time than $(A + B)$

i.e. time taken by A = $x \times \frac{160}{100} = \frac{8x}{5}$

∴ One day work of A = $\frac{5}{8x}$

Given, C = 2B

∴ $B + C = \frac{3}{40}$ ⇒ $B + 2B = \frac{3}{40}$

⇒ $3B = \frac{3}{40}$ ⇒ $B = \frac{1}{40}$

Now, $A + B = \frac{1}{x}$ ⇒ $\frac{5}{8x} + \frac{1}{40} = \frac{1}{x}$

⇒ $\frac{1}{x} - \frac{5}{8x} = \frac{1}{40}$ ⇒ $\frac{8-5}{8x} = \frac{1}{40}$

⇒ $\frac{3}{8x} = \frac{1}{40}$ ⇒ $x = \frac{3 \times 40}{8}$

⇒ $x = 15$

∴ One day work of A = $\frac{5}{8x} = \frac{5}{8 \times 15} = \frac{1}{24}$

∴ Time taken by A alone to complete the work = 24 days

Chapter 23

Work and Wages

Activity involving physical efforts, done in order to achieve a result is known as **Work**.

Money received by a person for a certain work is called the **Wages** of the person for that particular work.

In other words, we can find the entire wages of any person by the following formula

> Entire wages = Total number of days
> × Wages of 1 day of any person

For example If Arjun's monthly wages is ₹ 4200 and he worked for all 30 days, then his daily wages will be calculated as

Total wages = Number of days × Daily wages
$\Rightarrow$ 4200 = 30 × Daily wages
$\therefore$ Daily wages = $\dfrac{4200}{30}$ = ₹ 140

Some Important Points

(i) Wages is directly proportional to the work done. It means, more money will be received for more work and less money will be received for less work.

(ii) Wages is indirectly proportional to the time taken by the individual.

(iii) Wages is directly proportional to 1 day work of each individual.
For example If Karan can do a piece of work in 10 days and Arun can do the same piece of work in 15 days. Then, ratio of Karan and Arun's wages will be 15 : 10, i.e. 3 : 2.

(iv) If X, Y and Z can do a piece of work in d_1, d_2 and d_3 days respectively, then ratio of their shares is
$d_2 d_3 : d_3 d_1 : d_1 d_2$.

Fast Track Formulae to solve the QUESTIONS

▶ Formula 1

If A and B can do a piece of work in x and y days respectively, the ratio of their wages will be $y : x$. Then, the wages earned by A and B will be

$$A\text{'s wages} = \frac{\text{Total wages}}{x+y} \times y; \quad B\text{'s wages} = \frac{\text{Total wages}}{x+y} \times x$$

Ex. 1 Akanksha can do a piece of work in 6 days, while Vasudha can do the same work in 5 days. If the total amount to be given for this work is ₹ 660, then what will be the share of Vasudha, if both work together?

Sol. Time taken by Akanksha = 6 days

∴ 1 day's work of Akanksha = $\frac{1}{6}$

Time taken by Vasudha = 5 days

∴ 1 day's work of Vasudha = $\frac{1}{5}$

Given, total amount earned = ₹ 660

∵ Ratio of their incomes = $\frac{1}{6} : \frac{1}{5} = 5 : 6$

∴ Vasudha's share = $\frac{660}{5+6} \times 6 = ₹ 360$

▶ Formula 2

If A, B and C can do a piece of work in x, y and z days respectively, the ratio of their wages will be $yz : xz : xy$. Then, wages earned by A, B and C respectively will be

$$A\text{'s wages} = \frac{\text{Total wages}}{(yz+xz+xy)} \times yz; \quad B\text{'s wages} = \frac{\text{Total wages}}{(yz+xz+xy)} \times xz;$$

$$C\text{'s wages} = \frac{\text{Total wages}}{(yz+xz+xy)} \times xy$$

Ex. 2 A, B and C take ₹ 535 for doing a piece of work together. If working alone, each takes 5 days, 6 days and 7 days respectively, then find the share of each.

Sol. Given, total wages = ₹ 535

A can do a work in 5 days; B can do a work in 6 days; C can do a work in 7 days.

∴ A's share = $\frac{535}{(7 \times 6)+(7 \times 5)+(6 \times 5)} \times (7 \times 6)$

$= \frac{535}{42+35+30} \times 42 = \frac{535}{107} \times 42 = 42 \times 5 = ₹ 210$

B's share = $\frac{535}{107} \times 35 = 5 \times 35 = ₹ 175$

and C's share = $\frac{535}{107} \times 30 = 5 \times 30 = ₹ 150$

Work and Wages / 395

▶Formula 3

Total wages earned by certain persons in doing certain work
= (1 person's 1 day wages) × (Number of persons) × (Number of days)

∴ Required number of persons

$$= \frac{\text{Total wages}}{\text{Number of days} \times \text{1 person's 1 day wages}}$$

Ex. 3 Wages of 45 women for 48 days amount to ₹ 31050. How many men must work for 16 days to receive ₹ 11500, if the daily wages of a man being double those of a woman?

Sol. 1 day's wages of a woman = $\dfrac{31050}{45 \times 48} = ₹ \dfrac{115}{8}$

1 day's wages of a man = $\dfrac{115}{8} \times 2 = ₹ \dfrac{115}{4}$

∴ Required number of men = $\dfrac{11500}{16 \times \dfrac{115}{4}}$ = 25 men

▶Formula 4

A can do a piece of work in x days. With the help of B, A can do the same work in y days. If they get ₹ a for that work, then

$$\text{Share of } A = ₹ \left(\frac{ay}{x}\right) \text{ and share of } B = ₹ \left[\frac{a(x-y)}{x}\right]$$

Ex. 4 Suresh can do a work in 20 days. Suresh and Surendra together do the same work in 15 days. If they got ₹ 400 for that work, then find the share of Suresh and Surendra.

Sol. Suresh's 1 day's wages = $\dfrac{\text{Total wages}}{\text{Total number of days}} = \dfrac{400}{20} = ₹ 20$

Suresh's 15 days wages = $20 \times 15 = ₹ 300$

∴ Surendra's share in wages = ₹ (400 − 300) = ₹ 100

Fast Track Method

Here, $x = 20$, $y = 15$ and $a = 400$

Now, according to the formula,

Share of Suresh = $\dfrac{a \times y}{x} = \dfrac{400 \times 15}{20} = ₹ 300$

Share of Surendra = $\dfrac{a(x-y)}{x} = \dfrac{400 \times (20-15)}{20} = \dfrac{400 \times 5}{20} = ₹ 100$

▶Formula 5

X, Y and Z undertake to do a work for ₹ R. If X and Y together do only m/n of the work and rest is done by Z alone, then the share of Z is given by

$$R\left(1 - \frac{m}{n}\right).$$

Ex. 5 A, B and C undertake to do a work for ₹ 480. A and B together do 1/4 of the work and rest is done by the C alone. How much should C get?

Sol. Here, $R = ₹ 480, m = 1$ and $n = 4$
Now, according to the formula,
Share of $C = R\left(1 - \dfrac{m}{n}\right) = 480\left(1 - \dfrac{1}{4}\right) = 480 \times \dfrac{3}{4} = ₹ 360$

Ex. 6 X and Y contracted a piece of work for ₹ 1600. X alone can do it in 6 days, while Y alone can do that work in 8 days. They completed the work in 3 days taking help of A. Find the share of A.

Sol. $\because$ Time taken by $X = 6$ days. So, X's 1 day work $= \dfrac{1}{6}$
and time taken by $Y = 8$ days. So, Y's 1 day work $= \dfrac{1}{8}$
If A can do a work in n days, then A's 1 day work $= \dfrac{1}{n}$
According to the question, $(X + Y + A)$ complete the work in 3 days.
$\therefore$ X's 3 days work + Y's 3 days work + A's 3 days work = 1
$\Rightarrow \quad \dfrac{3}{6} + \dfrac{3}{8} + \dfrac{3}{n} = 1 \Rightarrow \dfrac{3}{n} = \left(1 - \dfrac{3}{6} - \dfrac{3}{8}\right) = 1 - 3\left(\dfrac{4+3}{24}\right)$
$\Rightarrow \quad \dfrac{3}{n} = 1 - \dfrac{7}{8} = \dfrac{1}{8} \Rightarrow n = 3 \times 8 \Rightarrow n = 24$
$\therefore$ X's share : Y's share : A's share $= \dfrac{1}{6} : \dfrac{1}{8} : \dfrac{1}{24} = \dfrac{12}{72} : \dfrac{9}{72} : \dfrac{3}{72} = 12 : 9 : 3 = 4 : 3 : 1$
Now, A's share $= \dfrac{1}{4+3+1} \times 1600 = \dfrac{1}{8} \times 1600 = ₹ 200$

Ex. 7 A can do a piece of work in 10 days, while B alone can do it in 15 days. They work together for 5 days and rest of the work is done by M in 2 days. If they get ₹ 9000 for the whole work, then how should they divide the money?

Sol. Given, time taken by $A = 10$ days, then A's 1 day work $= \dfrac{1}{10}$
and time taken by $B = 15$ days, B's 1 day work $= \dfrac{1}{15}$
Now, $(A + B)$'s 5 days work $= 5\left(\dfrac{1}{10} + \dfrac{1}{15}\right) = 5\left(\dfrac{3+2}{30}\right) = 5 \times \dfrac{5}{30} = \left(5 \times \dfrac{1}{6}\right) = \dfrac{5}{6}$
Remaining work $= 1 - \dfrac{5}{6} = \dfrac{6-5}{6} = \dfrac{1}{6}$
$\therefore$ M's 2 days work $= \dfrac{1}{6}$
Now, (A's 5 days work) : (B's 5 days work) : (M's 2 days work)
$= \dfrac{5}{10} : \dfrac{5}{15} : \dfrac{1}{6} = \dfrac{1}{2} : \dfrac{1}{3} : \dfrac{1}{6} = \dfrac{3}{6} : \dfrac{2}{6} : \dfrac{1}{6} = 3 : 2 : 1$
Let A's share $= 3x$, B's share $= 2x$ and M's share $= x$
According to the question, $3x + 2x + x = 9000$
$\Rightarrow \quad 6x = 9000 \Rightarrow x = \dfrac{9000}{6} = 1500$
$\therefore$ A's share $= 3x = 3 \times 1500 = ₹ 4500$, B's share $= 2x = 2 \times 1500 = ₹ 3000$
and M's share $= x = ₹ 1500$

Fast Track Practice

1. A alone can do a piece of work in 8 days while B alone can do it in 10 days. If they together complete this work and get ₹ 900 as their remuneration, then find the shares of both the persons.
   (a) A = ₹ 800 and B = ₹ 100
   (b) A = ₹ 500 and B = ₹ 400
   (c) A = ₹ 600 and B = ₹ 300
   (d) A = ₹ 300 and B = ₹ 600
   (e) None of the above

2. Shantanu can do a piece of work in 12 days and Manu can do the same work in 10 days. If they work together, then in what ratio Shantanu and Manu will receive their wages?
   (a) 5 : 6   (b) 3 : 2   (c) 1 : 6   (d) 5 : 7

3. A can do a piece of work in 9 days and B can do the same work in 15 days. If they work together, then in what ratio A and B will receive their wages? [Bank Clerk 2008]
   (a) 3 : 5   (b) 5 : 3   (c) 2 : 5   (d) 5 : 2
   (e) None of these

4. A and B can complete a piece of work in 15 days and 10 days, respectively. They contracted to complete the work for ₹30000. The share of A in the contracted money will be [SSC CGL (Pre) 2016]
   (a) ₹ 18000   (b) ₹ 16500
   (c) ₹ 12500   (d) ₹ 12000

5. A person can do a piece of work in 26 days and another person can do the same work in 39 days. If they work together, then by what per cent the wages of 1st person is more than that of 2nd person? [UP Police 2007]
   (a) 25%   (b) 35%   (c) 15%   (d) 50%

6. A alone can finish a work in 2 days, while B alone can finish it in 3 days. If they work together to finish it, then out of total wages of ₹ 6000, what will be the 20% of A's share? [Hotel Mgmt. 2009]
   (a) ₹ 720   (b) ₹ 350   (c) ₹ 820   (d) ₹ 420

7. A, B and C completed a work costing ₹ 1800. A worked for 6 days, B worked for 4 days and C worked for 9 days. If their daily wages are in the ratio of 5 : 6 : 4, then how much amount will be received by A? [SSC CPO 2007]
   (a) ₹ 600   (b) ₹ 500   (c) ₹ 900   (d) ₹ 450

8. A, B and C can do a work in 6, 8 and 12 days, respectively. If they do the work together and earn ₹ 2700, what is the share of C in that amount? [SNAP 2016]
   (a) ₹ 600   (b) ₹ 900   (c) ₹ 1000   (d) ₹ 700

9. A can finish a work in 15 days, B in 20 days and C in 25 days. All these three worked together and earned ₹ 4700. The share of C is [CDS 2012]
   (a) ₹ 1200   (b) ₹ 1500
   (c) ₹ 1800   (d) ₹ 2000

10. A and B undertaken to do a piece of work for ₹ 1200. A alone can do it in 8 days, while B can do it in 6 days. With the help of C, they complete it in 3 days. Find C's share. [SSC CGL (Mains) 2012]
    (a) ₹ 450   (b) ₹ 300   (c) ₹ 150   (d) ₹ 100

11. A, B and C get ₹ 5400 for doing a work in 36 days. A and C get ₹ 1880 for doing the same work in 20 days, while B and C get ₹ 3040 for doing the same work in 40 days. Find the amount received by C per day.
    (a) ₹ 50   (b) ₹ 25   (c) ₹ 30   (d) ₹ 20

12. Vikas can do a work in 16 days. Vikas and Vikram can together do the work in 12 days. If they got ₹ 500 for that work, then find the share of Vikas.
    (a) ₹ 375   (b) ₹ 400   (c) ₹ 270   (d) ₹ 350

13. P, Q and R enter into a contract for a piece of work for ₹ 1100. P and Q together are supposed to do 7/22 of the work. How much does R get?
    (a) ₹ 750   (b) ₹ 350   (c) ₹ 4751   (d) ₹ 900
    (e) None of these

14. X, Y and Z undertake to do a work for ₹ 6000. X and Y together do 3/4 of the work and rest is done by Z alone. How much should be share of Z?
    (a) ₹ 1350   (b) ₹ 1200
    (c) ₹ 1500   (d) ₹ 1450

15. P works thrice as fast as Q, whereas P and Q together can work four times as fast as R. If P, Q and R together work on a job, in what ratio should they share the earnings? [UPSC CSAT 2017]
    (a) 3 : 1 : 1   (b) 3 : 2 : 4
    (c) 4 : 3 : 4   (d) 3 : 1 : 4

16. A man and a boy received ₹ 1400 as wages for 10 days for the work they did together. The man's efficiency in the work was six times that of the boy. What is the daily wages of the boy?
(a) ₹ 10 (b) ₹ 15
(c) ₹ 20 (d) ₹ 30

17. A sum of money is sufficient to pay A's wages for 21 days and B's wages for 28 days. The same money is sufficient to pay the wages of both for [SSC CGL 2013]
(a) $24\frac{1}{2}$ days (b) 12 days
(c) $12\frac{1}{4}$ days (d) 14 days

18. The labourers A, B, C were given a contract of ₹ 750 for doing a certain piece of work. All the three together can finish the work in 8 days. A and C together can do it in 12 days, while A and B together can do it in $13\frac{1}{3}$ days. The money will be divided in the ratio   [SSC CGL (Mains) 2016]
(a) 4 : 5 : 6 (b) 4 : 7 : 5
(c) 5 : 7 : 4 (d) 5 : 6 : 8

19. A, B and C can do a piece of work in 20, 24 and 30 days, respectively. They undertook to do the piece of work for ₹ 5400. They begin the work together but B left 2 days before the completion of work and C left 5 days before the completion of work. The share of A from the assured money is [SSC CPO 2013]
(a) ₹ 2700 (b) ₹ 540
(c) ₹ 1800 (d) ₹ 600

20. A can do a piece of work in 16 days and B in 24 days. They take the help of C and three together finish the work in 6 days. If the total remuneration for the work is ₹ 400, then the amount (in ₹) each will receive, in proportion to do the work is [SSC CGL 2014]
(a) A : 150, B : 100, C : 150
(b) A : 100, B : 150, C : 150
(c) A : 150, B : 150, C : 100
(d) A : 100, B : 150, C : 100

21. Total wages of 3 men, 2 women and 4 boys is ₹ 26. If the wages of 3 men is equal to that of 4 women and the wages of 2 women is equal to that of 3 boys, then find out the total wages of 4 men, 3 women and 2 boys.
(a) ₹ 29 (b) ₹ 35
(c) ₹ 65 (d) ₹ 20

22. A person was appointed for a 50 days job on a condition that he will be paid ₹ 12 for every working day but he will be fined ₹ 6 for everyday he remains absent. After the completion of the work, he got ₹ 420. For how many days, he did not work?
(a) 15 days (b) 5 days
(c) 10 days (d) 20 days

23. 4 men and 6 women get ₹ 1600 by doing a piece of work in 5 days. 3 men and 7 women get ₹ 1740 by doing the same work in 6 days. In how many days, 7 men and 6 women can complete the same work getting ₹ 3760? [SSC CGL 2009]
(a) 6 days (b) 8 days
(c) 10 days (d) 12 days

24. Men, women and children are employed to do a work in the proportion of 3 : 2 : 1 and their wages as 5 : 3 : 2. When 90 men are employed, then total daily wages of all amounts to ₹ 10350. Find the daily wages of a man. [SSC CGL (Mains) 2012]
(a) ₹ 45 (b) ₹ 57
(c) ₹ 115 (d) ₹ 75

25. 2 men and 1 woman can do a piece of work in 14 days, while 4 women and 2 men can do the same work in 8 days. If a man gets ₹ 90 per day, then what should be the wages per day of a woman? [CDS 2013]
(a) ₹ 48 (b) ₹ 60
(c) ₹ 72 (d) ₹ 135

26. A alone can do a piece of work in 6 days and B alone in 8 days. A and B undertook to do it for ₹ 3200. With the help of C, they completed the work in 3 days. How much is to be paid to C? [IB PA 2016]
(a) ₹ 375 (b) ₹ 400
(c) ₹ 600 (d) ₹ 800

Answer with Solutions

1. (b) ∵ A's 1 day work = $\dfrac{1}{8}$

and B's 1 day work = $\dfrac{1}{10}$

∴ A's share : B's share
$= \dfrac{1}{8} : \dfrac{1}{10} = \dfrac{5}{40} : \dfrac{4}{40} = 5:4$

Let A's share = $5x$ and B's share = $4x$
According to the question,
$5x + 4x = 900 \Rightarrow 9x = 900 \Rightarrow x = 100$
∴ A's share = $5x = 5 \times 100 =$ ₹ 500
and B's share = $4x = 4 \times 100 =$ ₹ 400

Fast Track Method
Ratio of wages of A and B = 10 : 8
A's wages = $\dfrac{A}{A+B} \times$ Total wages
[by Formula 1]
$= \dfrac{10}{10+8} \times 900$
= ₹ 500
Similarly,
B's wages = $\dfrac{8}{10+8} \times 900 =$ ₹ 400

2. (a) ∵ Shantanu's 1 day work = $\dfrac{1}{12}$

and Manu's 1 day work = $\dfrac{1}{10}$

∴ Shantanu's share : Manu's share
$= \dfrac{1}{12} : \dfrac{1}{10} = \dfrac{5}{60} : \dfrac{6}{60} = 5:6$

3. (b) Here, A's 1 day work = $\dfrac{1}{9}$

and B's 1 day work = $\dfrac{1}{15}$

∴ A's share : B's share = $\dfrac{1}{9} : \dfrac{1}{15} = \dfrac{5}{45} : \dfrac{3}{45}$
$= 5:3$

4. (d) Here, A's 1 day's work = $\dfrac{1}{15}$

and B's 1 day's work = $\dfrac{1}{10}$

∴ Ratio = $\dfrac{1}{15} : \dfrac{1}{10} = 2:3$

Sum of the ratios = $2 + 3 = 5$
∴ A's share = ₹ $\dfrac{2}{5} \times 30000 =$ ₹12000

5. (d) Let 1st person be x and 2nd person be y.
Then, x's 1 day's work = $\dfrac{1}{26}$

y's 1 day's work = $\dfrac{1}{39}$

x's share : y's share = $\dfrac{1}{26} : \dfrac{1}{39}$
$= \dfrac{3}{78} : \dfrac{2}{78} = 3:2$

Difference of ratio = $3 - 2 = 1$
∴ Required percentage = $\dfrac{1}{2} \times 100\% = 50\%$

∴ 1st person's wages is 50% more than the 2nd person's wages.

6. (a) A's 1 day's work = $\dfrac{1}{2}$

B's 1 day's work = $\dfrac{1}{3}$

A's share : B's share = $\dfrac{1}{2} : \dfrac{1}{3} = \dfrac{3}{6} : \dfrac{2}{6} = 3:2$

A's share = $\dfrac{3}{5} \times 6000 = 3 \times 1200 =$ ₹ 3600

∴ 20% of A's share = $3600 \times \dfrac{20}{100} =$ ₹ 720

Fast Track Method
Here, $x = 2$ and $y = 3$
∴ A's share = $\dfrac{\text{Total wages}}{(x+y)} \times y$
[by Formula 1]
$= \dfrac{6000}{5} \times 3 =$ ₹3600

∴ 20% of A's share = $3600 \times \dfrac{20}{100} =$ ₹720

7. (a) Ratio of the wages of A, B and C
= 5 : 6 : 4
A's share : B's share : C's share
= $(6 \times 5) : (4 \times 6) : (9 \times 4)$
= 30 : 24 : 36 = 5 : 4 : 6
∴ A's share = $\dfrac{5}{15} \times 1800 =$ ₹ 600

8. (a) Work done by A in a day = $\dfrac{1}{6}$

Work done by B in a day = $\dfrac{1}{8}$

Work done by C in a day = $\dfrac{1}{12}$

∴ Ratio of amount shared = $\dfrac{1}{6} : \dfrac{1}{8} : \dfrac{1}{12}$
= 4 : 3 : 2

∴ Share of C = $\dfrac{2}{9} \times 2700$ = ₹ 600

Fast Track Method
Here, $x = 6, y = 8$ and $z = 12$
∴ C's share = $\dfrac{\text{Total wages}}{xy + yz + xz} \times xy$

[by Formula 2]
= $\dfrac{2700}{48 + 96 + 72} \times 48 = \dfrac{2700}{216} \times 48$ = ₹ 600

9. (a) A's 1 day work = $\dfrac{1}{15}$
B's 1 day work = $\dfrac{1}{20}$
C's 1 day work = $\dfrac{1}{25}$
∵ A, B and C worked together.
∴ Ratio of amount shared
= $\dfrac{1}{15} : \dfrac{1}{20} : \dfrac{1}{25}$
= $\dfrac{20}{300} : \dfrac{15}{300} : \dfrac{12}{300}$ = 20 : 15 : 12
∴ Share of C = $\dfrac{12}{47} \times 4700$ = ₹ 1200

Fast Track Method
Here, $x = 15, y = 20$ and $z = 25$
∴ C's share = $\dfrac{\text{Total wages}}{xy + yz + xz} \times xy$

[by Formula 2]
= $\dfrac{4700}{(15 \times 20 + 20 \times 25 + 15 \times 25)} \times 300$
= $\dfrac{4700}{(300 + 500 + 375)} \times 300$
= $\dfrac{4700}{1175} \times 300$ = ₹1200

10. (c) According to the question,
$\dfrac{1}{A} + \dfrac{1}{B} + \dfrac{1}{C} = \dfrac{1}{3} \Rightarrow \dfrac{1}{8} + \dfrac{1}{6} + \dfrac{1}{C} = \dfrac{1}{3}$
$\Rightarrow \dfrac{1}{C} = \dfrac{1}{3} - \left(\dfrac{1}{8} + \dfrac{1}{6}\right) = \dfrac{1}{3} - \left(\dfrac{3+4}{24}\right)$
= $\dfrac{1}{3} - \dfrac{7}{24} = \dfrac{8-7}{24} = \dfrac{1}{24}$
∴ Ratio in shares of A, B and C
= $\dfrac{1}{8} : \dfrac{1}{6} : \dfrac{1}{24} = \dfrac{3}{24} : \dfrac{4}{24} : \dfrac{1}{24}$
= 3 : 4 : 1
∴ C's share = $\dfrac{1}{3 + 4 + 1} \times 1200$ = ₹ 150

11. (d) Amount received by (A + B + C) per
day = $\dfrac{5400}{36}$ = ₹ 150
∴ A + B + C = ₹ 150 ...(i)
Similarly, amount received by (A + C) per
day = $\dfrac{1880}{20}$ = ₹ 94
∴ A + C = ₹ 94 ...(ii)
Amount received by (B + C) per day
= $\dfrac{3040}{40}$ = ₹ 76
∴ B + C = ₹ 76 ...(iii)
From Eqs. (i) and (iii), we get
A + 76 = ₹ 150 ⇒ A = 150 − 76 = ₹ 74
On putting the value of A in Eq. (ii), we get
74 + C = ₹ 94
∴ C = 94 − 74 = ₹ 20
Hence, amount received by C per day
= ₹ 20

12. (a) Given, $x = 16, y = 12, a$ = ₹ 500
∴ Share of Vikas = ₹ $\left(\dfrac{ay}{x}\right)$ [by Formula 4]
= ₹ $\left(\dfrac{500 \times 12}{16}\right)$ = ₹ 375

13. (a) Work done by (P + Q) = $\dfrac{7}{22}$
Work done by R = $\left(1 - \dfrac{7}{22}\right) = \dfrac{22-7}{22} = \dfrac{15}{22}$
∴ (P + Q)'s share : R's share
= $\dfrac{7}{22} : \dfrac{15}{22}$ = 7 : 15
R's share = $\dfrac{15}{22} \times 1100 = 15 \times 50$ = ₹ 750

Fast Track Method
∵ R's share = Total wages $\left(1 - \dfrac{m}{n}\right)$
[by Formula 5]
Here, $\dfrac{m}{n} = \dfrac{7}{22}$
∴ R's share = $1100 \times \left(1 - \dfrac{7}{22}\right)$
= $1100 \times \dfrac{15}{22}$ = ₹ 750

14. (c) Given, $\dfrac{m}{n} = \dfrac{3}{4}, R$ = ₹ 6000
∴ Share of C = $R\left(1 - \dfrac{m}{n}\right)$ [by Formula 5]
= $6000\left(1 - \dfrac{3}{4}\right) = 6000\left(\dfrac{1}{4}\right)$ = ₹ 1500

Work and Wages / 401

15. (a) Given, $P = 3Q$ and $P + Q = 4R$
$\Rightarrow \quad 3Q + Q = 4R \Rightarrow Q = R$
$\therefore$ Ratio of share $= P : Q : R$
$\qquad = 3Q : Q : Q = 3 : 1 : 1$

16. (c) $\because$ Ratio of efficiency of man to boy
$\qquad = 6 : 1$
We know that,
$\qquad$ Efficiency $\propto$ Wages
$\therefore$ Boy's share $= \dfrac{1}{1+6} \times 1400$
$\qquad = \dfrac{1}{7} \times 1400 = ₹ 200$
Now, they worked for 10 days.
$\therefore$ Daily wages of a boy $= \dfrac{200}{10} = ₹ 20$

17. (b) $\because$ A's 1 day work $= \dfrac{1}{21}$
and B's 1 day work $= \dfrac{1}{28}$
$\therefore$ Same money is sufficient to pay the wages of both for
$= \dfrac{1}{\dfrac{1}{21} + \dfrac{1}{28}} = \dfrac{21 \times 28}{21 + 28} = \dfrac{21 \times 28}{49}$
$= 3 \times 4 = 12$ days

18. (a) $(A + B + C) = \dfrac{1}{8}$ ...(i)
$A + C = \dfrac{1}{12}$ ...(ii)
and $A + B = \dfrac{1}{13\dfrac{1}{3}} = \dfrac{1}{\dfrac{40}{3}} = \dfrac{3}{40}$ ...(iii)
Now, solving Eqs. (i), (ii) and (iii), we get
$A = \dfrac{1}{30}, B = \dfrac{1}{24}$ and $C = \dfrac{1}{20}$
$\therefore$ Required ratio $= \dfrac{1}{30} : \dfrac{1}{24} : \dfrac{1}{20}$
$= \dfrac{4}{120} : \dfrac{5}{120} : \dfrac{6}{120} = 4 : 5 : 6$

19. (a) Let the number of days to complete the work be x.
According to the question,
$\dfrac{x}{20} + \dfrac{x-2}{24} + \dfrac{x-5}{30} = 1$
$\Rightarrow \dfrac{6x + 5(x-2) + 4(x-5)}{120} = 1$
$\Rightarrow 6x + 5x + 4x = 120 + 10 + 20$
$\Rightarrow 15x = 150 \Rightarrow x = 10$
$\therefore$ Work done by $A = \dfrac{10}{20} = \dfrac{1}{2}$

Share of A from the assured money
$\qquad = \dfrac{1}{2} \times 5400 = ₹ 2700$

20. (a) Work done by A in 1 day $= \dfrac{1}{16}$
Work done by B in 1 day $= \dfrac{1}{24}$
Work done by A, B and C in 1 day $= \dfrac{1}{6}$
Then, work done by C in 1 day
$\qquad = (A + B + C)$'s 1 day's work
$\qquad - A$'s 1 day's work $- B$'s 1 day's work
$\Rightarrow \dfrac{1}{C} = \dfrac{1}{6} - \dfrac{1}{16} - \dfrac{1}{24}$
$\Rightarrow \dfrac{1}{C} = \dfrac{8 - 3 - 2}{48} = \dfrac{3}{48} = \dfrac{1}{16}$
Ratio of wages of A, B and C
$= \dfrac{1}{A} : \dfrac{1}{B} : \dfrac{1}{C}$
$= \dfrac{1}{16} : \dfrac{1}{24} : \dfrac{1}{16} = \dfrac{3}{48} : \dfrac{2}{48} : \dfrac{3}{48}$
$= 3 : 2 : 3$ [taking LCM]
Share of $A = \dfrac{3}{3+2+3} \times 400$
$\qquad = \dfrac{3}{8} \times 400 = ₹ 150$
Share of $B = \dfrac{2}{3+2+3} \times 400 = ₹ 100$
Share of $C = \dfrac{3}{3+2+3} \times 400 = ₹ 150$

21. (a) Let the wages of 1 man, 1 woman and 1 boy be ₹ x, ₹ y and ₹ z, respectively.
According to the question,
$\qquad 3x + 2y + 4z = 26$ ...(i)
$\qquad 3x = 4y$ ...(ii)
and
$\qquad 2y = 3z$ ...(iii)
From Eqs. (i) and (ii), we get
$\qquad 4y + 2y + 4z = 26$
$\Rightarrow \qquad 6y + 4z = 26$ ...(iv)
From Eqs. (iii) and (iv), we get
$\qquad 9z + 4z = 26 \Rightarrow 13z = 26 \Rightarrow z = 2$
From Eqs. (ii) and (iii), we get
$\qquad y = 3$ and $x = 4$
$\therefore$ Wages of 4 men, 3 women and 2 boys
$\qquad = 4x + 3y + 2z = 4 \times 4 + 3 \times 3 + 2 \times 2$
$\qquad = 16 + 9 + 4 = ₹ 29$

22. (c) Let the person did not work for x days.
It means that he worked for $(50 - x)$ days.
$\therefore$ Fine for being absent $= ₹ 6x$
Wages for working days $= 12(50 - x)$

According to the question,
Received wages = $12(50 - x) - 6x = 420$
$\Rightarrow \quad 600 - 12x - 6x = 420$
$\Rightarrow \quad 18x = 600 - 420 = 180$
$\therefore \quad x = \dfrac{180}{18} = 10$ days

23. (b) Let man's 1 day wages = ₹x
and woman's 1 day wages = ₹y
According to the question,
$5(4x + 6y) = 1600$
$\Rightarrow \quad 4x + 6y = 320 \qquad ...(i)$
and $6(3x + 7y) = 1740$
$\Rightarrow \quad 3x + 7y = 290 \qquad ...(ii)$
On solving Eqs. (i) and (ii), we get
$x = 50$ and $y = 20$
Now, let 7 men and 6 women complete the work in d days.
Then, $\quad d(7x + 6y) = 3760$
$\Rightarrow \quad 7x + 6y = \dfrac{3760}{d}$
$\Rightarrow \quad 7 \times 50 + 6 \times 20 = \dfrac{3760}{d}$
$\Rightarrow \quad 470 = \dfrac{3760}{d} \Rightarrow d = \dfrac{3760}{470} = 8$ days

24. (d) Let the number of men, women and children be $3y$, $2y$ and y, respectively.
Given, $3y = 90 \Rightarrow y = 30$
Number of women = 60
and number of children = 30
Let the men's, women's and children's daily wages be ₹$5x$, ₹$3x$ and ₹$2x$, respectively.
According to the question,
Total daily wages = ₹ 10350
$\Rightarrow 90 \times 5x + 60 \times 3x + 30 \times 2x = 10350$
$\Rightarrow \quad x(450 + 180 + 60) = 10350$
$\therefore \quad x = \dfrac{10350}{690} = 15$
$\therefore$ Daily wages of a man = $15 \times 5 =$ ₹ 75

25. (b) Let man be represent by m and woman be represented by w.
$\because \quad 2m + 1w = \dfrac{1}{14}$
$\Rightarrow \quad 14(2m + 1w) = 1 \qquad ...(i)$
and $\quad 4w + 2m = \dfrac{1}{8}$
$\Rightarrow \quad 8(4w + 2m) = 1 \qquad ...(ii)$
On equating Eqs. (i) and (ii), we get
$14(2m + 1w) = 8(4w + 2m)$
$\Rightarrow \quad 28m + 14w = 32w + 16m$
$\Rightarrow \quad 28m - 16m = 32w - 14w$
$\Rightarrow \quad 12m = 18w$
$\Rightarrow \quad \dfrac{m}{w} = \dfrac{18}{12} = \dfrac{3}{2}$
Now, efficiency of 1 man and 1 woman is 3 : 2.
So, their wages must be in the same ratio,
i.e. $\dfrac{90}{x} = \dfrac{3}{2}$ [here, x = wages of a woman]
$\therefore \quad x = \dfrac{90 \times 2}{3} =$ ₹ 60

26. (b) Let C can alone complete the work in x days.
Then, $\quad \dfrac{1}{x} + \dfrac{1}{8} + \dfrac{1}{6} = \dfrac{1}{3}$
$\Rightarrow \quad \dfrac{1}{x} = \dfrac{1}{3} - \dfrac{1}{8} - \dfrac{1}{6}$
$= \dfrac{8 - 3 - 4}{24} = \dfrac{1}{24}$
$\therefore$ C can alone do the work in 24 days.
Work done by C in 3 days = $\dfrac{3}{24} = \dfrac{1}{8}$
$\therefore$ Amount paid to $C = \dfrac{1}{8} \times 3200 =$ ₹400

Chapter 24

Pipes and Cisterns

Problems on Pipes and Cisterns are based on the basic concept of time and work. Pipes are connected to a tank or cistern and are used to fill or empty the tank or cistern. In pipe and cistern, the work is done in form of filling or emptying a cistern/tank.

Inlet pipe It fills a tank/cistern/reservoir.
Outlet pipe It empties a tank/cistern/reservoir.

Important Facts Related to Pipes and Cisterns

(i) If a pipe can fill/empty a tank in 'm' h, then the part of tank filled/emptied in 1 h $= \dfrac{1}{m}$.

   For example If a pipe can fill the tank in 7 h, then the volume of tank filled in 1 h $= \dfrac{1}{7}$.

(ii) If a pipe can fill/empty '$1/m$' part of a tank in 1 h, then it can fill/empty the whole tank in 'm' h.

   For example If a pipe can fill 1/5 part of a tank in 1 h, then it can fill the whole tank in 5 h.

(iii) Time taken to fill a tank is taken positive (+ve) and time taken to empty a tank is taken negative (–ve).

(iv) If a pipe fills a tank in m h and another pipe fills in n h. Then, part filled by both pipes in 1 h $= \dfrac{1}{m} + \dfrac{1}{n}$.

Ex. 1 An outlet pipe can empty a cistern in 5 h. In what time will the pipe empty $\dfrac{2}{5}$ part of the cistern?

Sol. ∵ Time taken to empty full cistern = 5 h

∴ Time taken to empty $\dfrac{2}{5}$ part of the cistern $= \dfrac{2}{5} \times 5 = 2$ h

Ex. 2 If a pipe can fill a tank in 2 h and another pipe can fill the same tank in 6 h, then what part of a tank will be filled by both the pipes in 1 h, if they are opened simultaneously?

Sol. In 1 h, part filled by 1st pipe $= \dfrac{1}{m} = \dfrac{1}{2}$

In 1 h, part filled by 2nd pipe $= \dfrac{1}{n} = \dfrac{1}{6}$

$\therefore$ In 1 h, part filled by both the pipes together $= \left(\dfrac{1}{m} + \dfrac{1}{n}\right) = \left(\dfrac{1}{2} + \dfrac{1}{6}\right)$

$= \dfrac{3+1}{6} = \dfrac{4}{6} = \dfrac{2}{3}$ part

Ex. 3 If a pipe can fill a tank in 5 h and another pipe can empty the tank in 10 h, then what part of a tank will be filled by both pipes in 1 h, if both pipes are open simultaneously?

Sol. In 1 h, part filled by 1st pipe $= \dfrac{1}{m} = \dfrac{1}{5}$

In 1 h, part emptied by 2nd pipe $= \dfrac{1}{n} = \dfrac{1}{10}$

$\therefore$ In 1 h, part filled by both pipes when open simultaneously

$= \dfrac{1}{m} - \dfrac{1}{n}$ [– ve sign is used, as 2nd pipe empties the tank]

$= \dfrac{1}{5} - \dfrac{1}{10} = \dfrac{2-1}{10} = \dfrac{1}{10}$ part

Fast Track Techniques to solve the QUESTIONS

Technique 1 A pipe can fill/empty a tank in 'm' h and another pipe can fill/empty the same tank in 'n' h.

(i) If both pipes either fill or empty the tank, then the time taken to fill or empty the tank when both pipes are opened, is $t = \dfrac{mn}{m+n}$.

(ii) If first pipe fills the tank and second pipe empties the tank, then the time taken to fill the tank when both pipes are opened, is $t = \dfrac{mn}{m-n}$, where $m > n$.

(iii) If first pipe fills the tank and second pipe empties the tank, then the time taken to empty the tank when both pipes are opened, is $t = \dfrac{mn}{n-m}$, where $n > m$.

Ex. 4 Two pipes A and B can fill a tank in 18 h and 12 h, respectively. If both the pipes are opened simultaneously, then how much time will be taken to fill the tank?

Pipes and Cisterns / 405

Sol. ∵ Part filled by A alone in 1 h = $\dfrac{1}{18}$ and part filled by B alone in 1 h = $\dfrac{1}{12}$

∴ Part filled by $(A + B)$ in 1 h = $\dfrac{1}{18} + \dfrac{1}{12} = \dfrac{2 + 3}{36} = \dfrac{5}{36}$

Hence, both the pipes together will fill the tank in $\dfrac{36}{5}$ h, i.e. $7\dfrac{1}{5}$ h.

Fast Track Method

Time taken by both pipes to fill the tank = $\dfrac{mn}{m + n}$, where m and n are time taken to fill the tank by individual pipes.

Here, $m = 18$ and $n = 12$

∴ Time taken to fill the tank = $\dfrac{mn}{m + n} = \dfrac{18 \times 12}{18 + 12}$

$= \dfrac{18 \times 12}{30} = \dfrac{3 \times 12}{5} = \dfrac{36}{5} = 7\dfrac{1}{5}$ h

Ex. 5 A pipe can fill a tank in 5 h, while another pipe can empty it in 6 h. If both the pipes are opened simultaneously, then how much time will be taken to fill the tank?

Sol. ∵ Part filled by 1st pipe in 1 h = $\dfrac{1}{5}$ and part filled by 2nd pipe in 1 h = $\dfrac{1}{6}$

∴ Part filled in 1 h by both pipes = $\dfrac{1}{5} - \dfrac{1}{6} = \dfrac{6 - 5}{30} = \dfrac{1}{30}$

Hence, the tank will be filled completely in 30 h.

Fast Track Method

Here, $m = 5$ h and $n = 6$ h

∴ Time taken to fill the tank = $\dfrac{mn}{n - m} = \dfrac{5 \times 6}{6 - 5} = \dfrac{30}{1} = 30$ h

Ex. 6 A pipe can fill a tank in 10 h. Due to a leak in the bottom, it fills the tank in 20 h. If the tank is full, then how much time will the leak take to empty it?

Sol. Let the leak empties full tank in x h, then part emptied in 1 h by leak = $\dfrac{1}{x}$

Also, part filled by inlet pipe in 1 h = $\dfrac{1}{10}$

According to the question, $\dfrac{1}{10} + \dfrac{1}{x} = \dfrac{1}{20}$

⇒ $\dfrac{1}{x} = \dfrac{1}{20} - \dfrac{1}{10} = \dfrac{1 - 2}{20} = -\dfrac{1}{20}$ [−ve sign means leak empties the tank]

Hence, leak will empty the full tank in 20 h.

Fast Track Method

Here, $m = 10$ and $n = 20$

According to the formula,

Required time taken to empty the tank = $\dfrac{mn}{n - m} = \dfrac{10 \times 20}{20 - 10} = \dfrac{200}{10} = 20$ h

Technique 2 If three pipes can fill a tank separately in m, n and p h respectively, then part of tank filled in 1 h by all the three pipes is given by $\left(\dfrac{1}{m} + \dfrac{1}{n} + \dfrac{1}{p}\right)$ and total time taken to fill the tank is given by $\dfrac{mnp}{np + mp + mn}$ h.

MIND IT ! If any one of the three pipes is used to empty the tank, then time taken by that particular pipe will be negative (–ve). Suppose, third pipe is used to empty the tank. Then, the above formulae takes the form as

$$\left(\dfrac{1}{m} + \dfrac{1}{n} - \dfrac{1}{p}\right) \text{ and } \dfrac{mnp}{np + mp - mn} \text{ h.}$$

Ex. 7 Three pipes m, n and p can fill a tank separately in 4, 5 and 10 h, respectively. Find the time taken by all the three pipes to fill the tank when the pipes are opened together.

Sol. Here, part filled by pipe m in 1 h = $\dfrac{1}{4}$; part filled by pipe n in 1 h = $\dfrac{1}{5}$

and part filled by pipe p in 1 h = $\dfrac{1}{10}$

$\therefore$ Part filled by $(m + n + p)$ pipes in 1 h = $\dfrac{1}{4} + \dfrac{1}{5} + \dfrac{1}{10} = \dfrac{5 + 4 + 2}{20} = \dfrac{11}{20}$

$\therefore$ Required time to fill the tank = $\dfrac{20}{11}$ h = $1\dfrac{9}{11}$ h

Fast Track Method

Here, $m = 4, n = 5$ and $p = 10$

$\therefore$ Required time to fill the tank = $\dfrac{mnp}{np + mp + mn} = \dfrac{4 \times 5 \times 10}{5 \times 10 + 4 \times 10 + 4 \times 5}$

$= \dfrac{4 \times 5 \times 10}{50 + 40 + 20} = \dfrac{4 \times 5 \times 10}{110} = \dfrac{20}{11} = 1\dfrac{9}{11}$ h

Ex. 8 Pipe A can fill a tank in 20 h while pipe B alone can fill it in 10 h and pipe C can empty the full tank in 30 h. If all the pipes are opened together, then how much time will be needed to make the tank full?

Sol. Part filled by pipe A alone in 1 h = $\dfrac{1}{20}$; part filled by pipe B alone in 1 h = $\dfrac{1}{10}$

and part emptied by pipe C alone in 1 h = $\dfrac{1}{30}$

Net part filled by $(A + B + C)$ in 1 h = $\left(\dfrac{1}{20} + \dfrac{1}{10} - \dfrac{1}{30}\right) = \left(\dfrac{3 + 6 - 2}{60}\right) = \dfrac{7}{60}$ h

$\therefore$ Required time to fill the tank = $\dfrac{60}{7}$ h = $8\dfrac{4}{7}$ h

Fast Track Method

Here, $m = 20, n = 10$ and $p = 30$

$\therefore$ Required time to fill the tank = $\dfrac{mnp}{np + mp - mn} = \dfrac{20 \times 10 \times 30}{10 \times 30 + 20 \times 30 - 20 \times 10}$

$= \dfrac{6000}{300 + 600 - 200} = \dfrac{6000}{700} = \dfrac{60}{7} = 8\dfrac{4}{7}$ h

Pipes and Cisterns / 407

Technique 3 Two pipes A and B together can fill a tank in time t. If time taken by A alone is more than t by a and time taken by B alone is more than t by b, then $t = \sqrt{ab}$.

Ex. 9 Two pipes A and B are opened together to fill a tank. Both the pipes fill the tank in time t. If A separately takes 4 min more time than t to fill the tank and B takes 64 min more time than t to fill the tank, then find the value of t.

Sol. Time taken by pipe $A = (4 + t)$ min

Time taken by pipe $B = (64 + t)$ min

According to the question,

$$\frac{1}{(4+t)} + \frac{1}{(64+t)} = \frac{1}{t}$$

$$\Rightarrow \frac{64 + t + 4 + t}{(4+t)(64+t)} = \frac{1}{t}$$

$$\Rightarrow t(68 + 2t) = 256 + 4 + 64t + t^2$$
$$\Rightarrow 68t + 2t^2 = t^2 + 68t + 256$$
$$\Rightarrow t^2 = 256$$
$$\Rightarrow t = 16 \text{ min}$$

Fast Track Method

We know that, time taken by both pipes to fill the tank $(t) = \sqrt{ab}$

$\therefore \qquad t = \sqrt{4 \times 64} = 2 \times 8 = 16$ min [here, $a = 4$ and $b = 64$]

Technique 4 A full tank gets emptied in 'a' h due to presence of a leak in it. If a tap which fills it at a rate of 'b' L/h, is opened, then it gets emptied in 'c' h.

Therefore, volume of tank $= \dfrac{abc}{c-a}$.

Ex. 10 A full tank gets emptied in 6 min due to presence of an orifice in it. On opening a tap which can fill the tank at the rate of 8 L/min, the tank gets emptied in 10 min. Find the capacity of tank.

Sol. Let the time taken to fill the tank be x min at the rate of 8 L/min.

According to the question,

$$\frac{-1}{6} + \frac{1}{x} = \frac{-1}{10}$$

$$\Rightarrow \frac{1}{x} = \frac{-1}{10} + \frac{1}{6}$$

$$\Rightarrow \frac{1}{x} = \frac{-6 + 10}{60}$$

$$\Rightarrow \frac{1}{x} = \frac{4}{60} \Rightarrow x = 15 \text{ min}$$

$\therefore$ Capacity of tank $= 8 \times 15 = 120$ L

Fast Track Method

Here, $a = 6$, $b = 8$ and $c = 10$

$\therefore$ Capacity of tank $= \dfrac{abc}{c-a} = \dfrac{6 \times 8 \times 10}{10 - 6} = 120$ L

Technique 5 If two taps A and B, which can fill a tank, such that efficiency of A is n times of B and takes t min less/more than B to fill the tank, then

(i) time taken to fill the tank by both pipes together $= \dfrac{nt}{n^2 - 1}$ min

(ii) time taken to fill the tank by faster tap $= \dfrac{t}{n-1}$ min

(iii) time taken to fill the tank by slower tap $= \dfrac{nt}{n-1}$ min

Ex. 11 Tap A can fill a tank 3 times faster than tap B and takes 28 min less than tap B to fill the tank. If both the taps are opened simultaneously, then find the time taken to fill the tank.

Sol. Let the time taken by tap A to fill the tank be x min.
Then, time taken by tap B to fill the tank be $3x$ min.
Then, according to the question, $3x - x = 28 \Rightarrow 2x = 28 \Rightarrow x = \dfrac{28}{2} = 14$ min

$\therefore$ Time taken by tap $A = 14$ min; part filled by tap A in 1 min $= \dfrac{1}{14}$

Time taken by tap $B = 3 \times 14 = 42$ min
Part filled by tap B in 1 min $= \dfrac{1}{42}$

$\therefore$ Part filled by both taps in 1 min $= \dfrac{1}{14} + \dfrac{1}{42} = \dfrac{3+1}{42} = \dfrac{4}{42} = \dfrac{2}{21}$

So, time taken by both taps to fill the tank working together $= \dfrac{21}{2}$ min

Fast Track Method
Here, $n = 3$ and $t = 28$
According to the formula,

Time taken to fill the tank by both taps together $= \dfrac{nt}{n^2 - 1} = \dfrac{28 \times 3}{(3)^2 - 1}$

$= \dfrac{28 \times 3}{8} = \dfrac{21}{2}$ min

Ex. 12 Pipe A is 3 times slower than the pipe B and takes 16 min more time than the pipe B. How much time will pipe B take to fill the tank?

Sol. Let the time taken by pipe A to fill the tank be $3x$ min. Then, time taken by pipe B to fill the tank be x min.
Then, according to the question,
$$3x - x = 16$$
$\Rightarrow \qquad 2x = 16 \Rightarrow x = 8$
$\therefore$ Time taken by pipe B to fill the tank $= 8$ min

Fast Track Method
Here, $n = 3$ and $t = 16$
According to the formula,

Time taken by pipe B to fill the tank $= \dfrac{t}{n-1} = \dfrac{16}{3-1} = \dfrac{16}{2} = 8$ min

Ex. 13 Pipe A is 2 times slower than pipe B and takes 9 min more time than the pipe B. How much time will pipe A take to fill the tank?

Sol. Let the time taken by pipe A to fill the tank be $2x$ min. Then, time taken by pipe B to fill the tank is x min.
Then, according to the question, $2x - x = 9$
$$\Rightarrow \qquad x = 9 \text{ min}$$
∴ Time taken by pipe A to fill the tank $= 2 \times 9 = 18$ min.

Fast Track Method
Here, $n = 2$ and $t = 9$ min
According to the formula,
Time taken to fill the tank by pipe $A = \dfrac{nt}{n-1} = \dfrac{2 \times 9}{2-1} = \dfrac{18}{1} = 18$ min

Technique 6 Two pipes A and B can fill a tank in x min and y min, respectively. If both the pipes are opened simultaneously, then the time after which pipe B should be closed, so that the tank is full in t min, is $\left[y\left(1 - \dfrac{t}{x}\right)\right]$ min.

Ex. 14 Two pipes A and B can fill a tank in 12 min and 16 min, respectively. If both the pipes are opened simultaneously, then after how much time should B be closed so that the tank is full in 9 min?

Sol. In 1 min, part filled by pipe $A = \dfrac{1}{12}$
and part filled by pipe $B = \dfrac{1}{16}$
Let the pipe B closed after x min.
According to the question,
$$x\left(\dfrac{1}{12} + \dfrac{1}{16}\right) + (9-x)\left(\dfrac{1}{12}\right) = 1$$
$$\Rightarrow \qquad x\left(\dfrac{4+3}{48}\right) + \dfrac{9}{12} - \dfrac{x}{12} = 1$$
$$\Rightarrow \qquad 7x + 36 - 4x = 48$$
$$\Rightarrow \qquad 3x = 12 \Rightarrow x = 4 \text{ min}$$
Hence, pipe B should be closed after 4 min.

Fast Track Method
Here, $x = 12$, $y = 16$ and $t = 9$
∴ Required time after which B should be closed
$$= y\left(1 - \dfrac{t}{x}\right) = 16\left(1 - \dfrac{9}{12}\right)$$
$$= 16 \times \dfrac{3}{12} = 4 \text{ min}$$

Fast Track Practice

Exercise 1 Base Level Questions

1. A pipe can fill a cistern in 6 h. Due to a leak in its bottom, it is filled in 7 h. When the cistern is full, in how much time will it be emptied by the leak?
   [RRB PO (Pre) 2017]
   (a) 42 h (b) 40 h (c) 43 h (d) 45 h
   (e) None of these

2. There are two tanks A and B to fill up a water tank. The tank can be filled in 40 min, if both taps are on. The same tank can be filled in 60 min, if tap A alone is on. How much time will tap B alone take to fill up the same tank?
   [CDS 2012]
   (a) 64 min (b) 80 min
   (c) 96 min (d) 120 min

3. Through an inlet, a tank takes 8 h to get filled up. Due to a leak in the bottom, it takes 2 h more to get it filled completely. If the tank is full, then how much time will the leak take to empty it? [SSC CGL 2010]
   (a) 16 h (b) 20 h (c) 32 h (d) 40 h

4. A tank can be filled by pipe X in 2 h and pipe Y in 6 h. At 10 am, pipe X was opened. At what time will the tank be filled, if pipe Y is opened at 11 am?
   [CDS 2016 (II)]
   (a) 12 : 45 pm (b) 5 : 00 pm
   (c) 11 : 45 am (d) 11 : 50 am

5. Two pipes A and B can independently fill a tank completely in 20 min and 30 min, respectively. If both the pipes are opened simultaneously, how much time will they take to fill the tank completely?
   [UPSC CSAT 2015]
   (a) 10 min (b) 12 min
   (c) 15 min (d) 25 min

6. A cistern can be filled up in 4 h by an inlet A. An outlet B can empty the cistern in 8 h. If both A and B are opened simultaneously, then after how much time will the cistern get filled?
   [Bank Clerk 2009]
   (a) 5 h (b) 7 h (c) 8 h (d) 6 h
   (e) None of these

7. A tap can fill an empty tank in 12 h and a leakage can empty the tank in 20 h. If tap and leakage both work together, then how long will it take to fill the tank?
   [Bank Clerk 2010]
   (a) 25 h (b) 40 h
   (c) 30 h (d) 35 h
   (e) None of these

8. A pipe can fill a tank in 20 h. Due to a leak in the bottom, it is filled in 40 h. If the tank is full, then how much time will the leak take to empty it? [CDS 2013]
   (a) 40 h (b) 30 h
   (c) 50 h (d) 30 h

9. A tank (of capacity 160 L) has one inlet, A and one outlet, B. Inlet A, alone can fill the empty tank in 6 h and outlet B alone can empty the full tank in 24 h. In how many hours will inlet A fill 16 L water in the tank when outlet B is also open?
   [IBPS Clerk (Pre) 2016]
   (a) 1 (b) $\frac{4}{5}$
   (c) $\frac{11}{5}$ (d) $1\frac{1}{2}$
   (e) $\frac{1}{5}$

10. A, B and C are three pipes connected to a tank. A and B together fill the tank in 6 h, B and C together fill the tank in 10 h and A and C together fill the tank in 12 h. In how much time A, B and C fill up the tank together?
    (a) 9 h (b) $5\frac{3}{7}$ h (c) $5\frac{2}{7}$ h (d) $5\frac{5}{7}$ h
    (e) None of these

11. Two pipes A and B can fill a tank in 1 h and 75 min, respectively. There is also an outlet C. If all the three pipes are opened together, then the tank is full in 50 min. How much time will be taken by C to empty the full tank?
    (a) 100 min (b) 150 min
    (c) 200 min (d) 125 min
    (e) None of these

12. A tank is filled in 5 h by three pipes A, B and C. The pipe C is twice as fast as B and B is twice as fast as A. How much time will pipe A alone take to fill the tank? [SBI Clerk (Mains) 2016]
    (a) 20 h (b) 25 h
    (c) 35 h (d) 40 h
    (e) 30 h

13. Pipe A can fill a tank in 30 min, while pipe B can fill the same tank in 10 min and pipe C can empty the full tank in 40 min. If all the pipes are opened together, then how much time will be needed to make the tank full?
    [Hotel Mgmt. 2010]
    (a) $9\frac{3}{13}$ h (b) $9\frac{4}{13}$ h
    (c) $9\frac{7}{13}$ h (d) $9\frac{9}{13}$ h

14. Pipes A and B can fill a tank in 5 and 6 h, respectively. Pipe C can fill it in 30 h. If all the three pipes are opened together, then in how much time the tank will be filled up? [Bank PO 2007]
    (a) $3\frac{3}{14}$ h (b) $2\frac{1}{2}$ h
    (c) $3\frac{9}{14}$ h (d) $2\frac{1}{14}$ h
    (e) None of these

15. Three taps are fitted in a cistern. The empty cistern is filled by the first and the second taps in 3 h and 4 h, respectively. The full cistern is emptied by the third tap in 5 h. If all three taps are opened simultaneously, then the empty cistern will be filled up in
    [SSC CGL 2013]
    (a) $1\frac{14}{23}$ h (b) $2\frac{14}{23}$ h
    (c) 2 h 40 min (d) 1 h 56 min

16. Three taps A, B and C together can fill an empty cistern in 10 min. The tap A alone can fill it in 30 min and the tap B alone can fill it in 40 min. How long will the tap C alone take to fill it? [SSC CPO 2010]
    (a) 16 min (b) 24 min
    (c) 32 min (d) 40 min

17. Two pipes A and B are opened together to fill a tank. Both pipes fill the tank in a certain time. If A separately takes 16 min more than the time taken by (A + B) and B takes 9 min more than the time taken by (A + B). Then, find the time taken by A and B to fill the tank when both the pipes are opened together.
    (a) 10 min (b) 12 min
    (c) 15 min (d) 8 min
    (e) None of these

18. A tank has a leak which would empty it in 8 h. A tap is turned on which admits 3 L a min into the tank and it is now emptied in 12 h. How many litres does the tank hold?

    (a) 4320 (b) 4000
    (c) 2250 (d) 4120

19. A full tank get emptied in 30 min due to leakage. If a tap which fills it at a rate of 10 L/min is opened, then it get emptied in 50 min. Find the capacity of tank.
    (a) 720 L (b) 750 L
    (c) 710 L (d) 760 L

20. Pipe A is 4 times faster than pipe B and takes 30 min less than the pipe B to fill the tank. When will the cistern be full, if both pipes are opened together?
    (a) 6 min (b) 8 min
    (c) 12 min (d) 9 min

21. If tap A can fill a tank 2 times faster than tap B and takes 20 min less than tap B to fill the tank. If both the taps are opened simultaneously, then find the time taken to fill the tank by tap A.
    (a) 20 min (b) 25 min
    (c) 15 min (d) None of these

22. Pipe A is 4 times faster than pipe B and takes 24 min less than the pipe B to fill the tank. Find how much time will pipe B take to fill the tank?
    (a) 28 min (b) 32 min
    (c) 16 min (d) 40 min

23. Two pipes P and Q can fill a cistern in 12 min and 15 min, respectively. If both pipes are opened together and at the end of 3 min, the first is closed, then how much longer will the cistern take to fill?
    [SSC CGL 2013]
    (a) $8\frac{1}{4}$ min (b) $8\frac{3}{4}$ min
    (c) 5 min (d) $8\frac{1}{2}$ min

24. Pipe A alone can fill a tank in 8 h. Pipe B alone can fill the same tank in 10 h. If both the pipes are opened but after 2 h, pipe A is closed, then pipe B will fill the tank in [BSSC CGL 2015]
    (a) 6 h (b) $3\frac{1}{2}$ h
    (c) 4 h (d) $5\frac{1}{2}$ h

25. Two pipes A and B can fill a tank in 24 min and 32 min, respectively. If both the pipes are opened together, then after how much time pipe B should be closed so that the tank is full in 9 min?
    (a) 40 min (b) 30 min
    (c) 10 min (d) 20 min
    (e) None of these

Exercise 2 Higher Skill Level Questions

1. A tap having diameter d can empty a tank in 40 min. How long another tap having diameter $2d$ take to empty the same tank?
   (a) 5 min (b) 20 min
   (c) 10 min (d) 40 min
   (e) 80 min

2. Pipe A can fill an empty tank in 6 h and pipe B in 8 h. If both the pipes are opened and after 2 h, pipe A is closed, then how much time pipe B will take to fill the remaining tank? [SSC CGL 2015]
   (a) $7\frac{1}{2}$ h (b) $3\frac{1}{3}$ h
   (c) $2\frac{2}{5}$ h (d) $2\frac{1}{3}$ h

3. A large tanker can be filled by two pipes A and B in 60 min and 40 min, respectively. How many minutes will it take to fill the tanker from empty state, if B is used for half the time and A and B fill it together for the other half? [FCI Assist. Grade 2015]
   (a) 30 min (b) 15 min
   (c) 20 min (d) None of these

4. If two pipes function together, then the tank will be filled in 12 h. One pipe fills the tank in 10 h faster than the other. How many hours does the faster pipe take to fill up the tank?
   (a) 20 (b) 60 (c) 15 (d) 25
   (e) None of these

5. Two pipes can fill a cistern in 14 h and 16 h, respectively. The pipes are opened simultaneously and it is found that due to leakage in the bottom, it took 92 min more to fill the cistern. When the cistern is full, in what time will the leak empty it?
   (a) $43\frac{19}{23}$ h (b) $43\frac{17}{23}$ h
   (c) $43\frac{13}{23}$ h (d) $43\frac{19}{23}$ h
   (e) None of these

6. A pipe can fill a cistern in 12 min and another pipe can fill it in 15 min, but a third pipe can empty it in 6 min. The first two pipes are kept open for 5 min in the beginning and then the third pipe is also opened. Time taken to empty the cistern is [SSC CGL (Mains) 2013]
   (a) 38 min (b) 22 min
   (c) 42 min (d) 45 min

7. There are three pipes connected with a tank. The first pipe can fill 1/2 part of the tank in 1 h, second pipe can fill 1/3 part of the tank in 1 h. Third pipe is connected to empty the tank. If after opening all the three pipes, 7/12 part of the tank can be filled in 1 h, then how long will third pipe take to empty the full tank? [SSC CGL 2007]
   (a) 3 h (b) 4 h
   (c) 5 h (d) 6 h

8. Two pipes can fill a tank in 20 min and 24 min, respectively and a waste pipe can empty 6 gallon/min. All the three pipes working together can fill the tank in 15 min. Find the capacity of the tank.
   (a) 210 gallon
   (b) 50 gallon
   (c) 150 gallon
   (d) 240 gallon
   (e) None of the above

9. There are two inlets A and B connected to a tank. A and B can fill the tank in 16 h and 10 h, respectively. If both the pipes are opened alternately for 1 h, starting from A, then how much time will the tank take to be filled?
   (a) $13\frac{1}{4}$ h (b) $11\frac{6}{8}$ h
   (c) $12\frac{2}{5}$ h (d) $12\frac{1}{4}$ h
   (e) None of these

10. Two pipes X and Y can fill a cistern in 6 min and 7 min, respectively. Starting with pipe X, both the pipes are opened alternately, each for 1 min. In what time will they fill the cistern?
    (a) $6\frac{2}{7}$ min (b) $6\frac{3}{7}$ min
    (c) $6\frac{5}{7}$ min (d) $6\frac{1}{7}$ min
    (e) None of these

11. Pipe A can fill a tank in 4 h and pipe B can fill it in 6 h. If they are opened on alternate hours and if pipe A is opened first, then in how many hours, the tank will be full? [SSC CGL 2015]
    (a) $4\frac{1}{2}$ (b) $4\frac{2}{3}$
    (c) $3\frac{1}{2}$ (d) $3\frac{1}{4}$

12. A cistern has three pipes A, B and C. Pipes A and B can fill it in 3 h and 4 h, respectively, while pipe C can empty the completely filled cistern in 1 h. If the pipes are opened in order at 3 : 00 pm, 4 : 00 pm and 5 : 00 pm respectively, then at what time will the cistern be empty? **[SSC (10+2) 2007]**
    (a) 6 : 15 pm (b) 7 : 12 pm
    (c) 8 : 12 pm (d) 8 : 35 pm

13. A pipe P can fill a tank in 12 min and another pipe R can fill it in 15 min. But, the third pipe M can empty it in 6 min. The first two pipes P and R are kept open for double the 2.5 min in the beginning and then the third pipe is also opened. In what time is the tank emptied?
    (a) 30 min (b) 25 min
    (c) 45 min (d) 35 min
    (e) None of these

14. Three taps A, B and C can fill a tank in 12 h, 15 h and 20 h, respectively. If A is open all the time and B and C are open for one hour each alternately, in what time will the tank be full? **[NICL AO 2015]**
    (a) 9 h (b) 7 h (c) 8 h (d) 10 h
    (e) 11 h

15. Capacity of tap B is 80% more than that of A. If both the taps are opened simultaneously, then they take 45 h to fill the tank. How long will B take to fill the tank alone?
    (a) 72 h (b) 48 h (c) 66 h (d) 70 h
    (e) None of these

16. There are 7 pipes attached with a tank out of which some are inlets and some are outlets. Every inlet can fill the tank in 10 h and every outlet can empty the tank in 15 h. When all the pipes are opened simultaneously, the tank is filled up in $2\frac{8}{11}$ h. Find the number of inlets and outlets.
    (a) 5, 2 (b) 6, 1 (c) 4, 3 (d) 3, 4
    (e) None of these

17. Three pipes A, B and C can fill a tank in 30 min, 20 min and 10 min, respectively. When the tank is empty, all the three pipes are opened. If A, B and C discharge chemical solutions P, Q and R respectively, then the part of solution R in the liquid in the tank after 3 min is **[SSC (10+2) 2013]**
    (a) $\frac{8}{11}$ (b) $\frac{5}{11}$
    (c) $\frac{6}{11}$ (d) $\frac{7}{11}$

18. There are three taps of diameters 1 cm, $\frac{4}{3}$ cm and 2 cm, respectively. The ratio of the water flowing through them is equal to the ratio of the square of their diameters. The biggest tap can fill the tank alone in 61 min. If all the taps are opened simultaneously, then how long will the tank take to be filled?
    (a) 44 min (b) 45 min
    (c) $44\frac{1}{4}$ min (d) 46 min
    (e) None of these

19. Taps A, B and C are attached with a tank and velocity of water coming through them are 42 L/h, 56 L/h and 48 L/h, respectively. A and B are inlets and C is outlet. If all the taps are opened simultaneously, then tank is filled in 16 h. What is the capacity of the tank?
    (a) 2346 L (b) 1600 L
    (c) 800 L (d) 960 L
    (e) None of these

20. The weight of a container completely filled with water is 2.25 kg. The container weighs 0.77 kg when its 0.2 part is filled with water. The weight of the container when 0.4 part of it is filled with water, is **[SSC CGL 2015]**
    (a) 0.74 kg (b) 1.14 kg
    (c) 1.88 kg (d) 0.40 kg

Answer with Solutions

Exercise 1 — Base Level Questions

1. (a) In 1 h, $\frac{1}{6}$ of the cistern can be filled.

   In 1 h, only $\frac{1}{7}$ of the cistern can be filled due to leak in its bottom.

   ∴ In 1 h, $\frac{1}{6} - \frac{1}{7} = \frac{1}{42}$ of the cistern is empty.

   Hence, the whole cistern will be emptied in 42 h.

2. (d) Part filled by tap A in 1 min = $\frac{1}{60}$

   Let tap B fills the tank in x min.

   Then, part filled by tap B in 1 min = $\frac{1}{x}$

   According to the question,

   $\frac{1}{60} + \frac{1}{x} = \frac{1}{40} \Rightarrow \frac{1}{x} = \frac{1}{40} - \frac{1}{60}$

   $\Rightarrow \frac{1}{x} = \frac{3-2}{120} = \frac{1}{120}$

   ∴ Tap B can fill the tank in 120 min.

3. (d) Let the leak takes x h to empty the tank.

   Now, part filled by inlet in 1 h = $\frac{1}{8}$

   Part filled in 1 h when both tap and leak work together = $\frac{1}{8+2} = \frac{1}{10}$

   According to the question,

   $\frac{1}{x} = \frac{1}{8} - \frac{1}{10} = \frac{5-4}{40} = \frac{1}{40} \Rightarrow x = 40$ h

4. (c) ∵ Part of tank filled by X in 1 h = $\frac{1}{2}$

   and part of tank filled by Y in 1 h = $\frac{1}{6}$

   ∴ Part of tank filled by $(X+Y)$ in 1 h

   $= \frac{1}{2} + \frac{1}{6} = \frac{2}{3}$

   During 10:00 am to 11:00 am, part of tank filled by pipe $X = \frac{1}{2}$

   ∴ Remaining part to be filled = $1 - \frac{1}{2} = \frac{1}{2}$

   Time taken by $(X+Y)$ to filled $\frac{1}{2}$ part of tank

   $= \frac{3}{2} \times \frac{1}{2} = \frac{3}{4}$ h = 45 min

   Hence, the tank will be filled at 11:45 am.

5. (b) Tank filled in 1 min by pipe $A = \frac{1}{20}$

   Tank filled in 1 min by pipe $B = \frac{1}{30}$

   Now, tank filled by both pipes in 1 min

   $= \frac{1}{20} + \frac{1}{30} = \frac{3+2}{60} = \frac{5}{60} = \frac{1}{12}$

   Hence, the time taken to fill the tank by both pipes is 12 min.

   Fast Track Method

   Here, $m = 20$ and $n = 30$

   ∴ Required time = $\frac{mn}{m+n}$

   [by Technique 1 (i)]

   $= \frac{20 \times 30}{20+30} = \frac{600}{50} = 12$ min

6. (c) Part filled by A in 1 h = $\frac{1}{4}$

   Part emptied by B in 1 h = $\frac{1}{8}$

   Part filled by $(A+B)$ in 1 h

   $= \frac{1}{4} + \left(-\frac{1}{8}\right) = \frac{1}{4} - \frac{1}{8} = \frac{2-1}{8} = \frac{1}{8}$

   ∴ Required time to fill the cistern is 8 h.

   $\left[\frac{-1}{8}\right.$ has been taken, because it empties the tank.$\left.\right]$

   Fast Track Method

   Here, $m = 8$ and $n = 4$

   ∴ Required time = $\frac{mn}{m-n} = \frac{8 \times 4}{8-4} = 8$ h

   [by Technique 1(ii)]

7. (c) Part filled by tap in 1 h = $\frac{1}{12}$

   Part emptied by leak in 1 h = $-\frac{1}{20}$

   Net part filled in 1 h when both (tap and leakage) work

   $= \frac{1}{12} - \frac{1}{20} = \frac{5-3}{60}$

   $= \frac{2}{60} = \frac{1}{30}$

   ∴ Required time to fill the tank is 30 h.

Pipes and Cisterns / 415

Fast Track Method
Here, $m = 20$ and $n = 12$
$\therefore \quad t = \dfrac{mn}{m-n}$ [by Technique 1(ii)]
$= \dfrac{12 \times 20}{8} = 30$ h

8. (a) Let the leak empties the full tank in x h, then
Part emptied in 1 h by leak $= \dfrac{1}{x}$
$\therefore$ Part filled by inlet in 1 h $= \dfrac{1}{20}$
According to the question, $\dfrac{1}{20} + \dfrac{1}{x} = \dfrac{1}{40}$
$\therefore \quad \dfrac{1}{x} = \dfrac{1}{40} - \dfrac{1}{20} = \dfrac{1-2}{40} = -\dfrac{1}{40}$
[–ve sign indicates emptying]
Clearly, leak will empty the full tank in 40 h.

Fast Track Method
Here, $m = 20$ and $n = 40$
$\therefore$ Required time $= \dfrac{mn}{n-m}$
[by Technique 1 (iii)]
$= \dfrac{20 \times 40}{40 - 20} = 40$ h

9. (b) Given, capacity of the tank = 160 L
Let the inlet (pipe) A fill the full tank in 'm' hours = 6 h
and the outlet (pipe) B empty the full tank in n hours = 24 h.
Then, time taken to fill the full tank
$= \dfrac{mn}{n-m}$ [by Technique 1(iii)]
$= \dfrac{6 \times 24}{24 - 6} = \dfrac{6 \times 24}{18} = 8$ h
160 L (full tank) of water fill in the tank in 8 h, then 160 L → 8 h
$\therefore$ 16 L → $\dfrac{8}{10}$ h $= \dfrac{4}{5}$ h

10. (d) Part filled by $(A + B)$ in 1 h $= \dfrac{1}{6}$
Part filled by $(B + C)$ in 1 h $= \dfrac{1}{10}$
Part filled by $(A + C)$ in 1 h $= \dfrac{1}{12}$
$\therefore$ Part filled by $2(A + B + C)$ in 1 h
$= \dfrac{1}{6} + \dfrac{1}{10} + \dfrac{1}{12}$
$= \dfrac{10 + 6 + 5}{60} = \dfrac{21}{60} = \dfrac{7}{20}$

$\therefore$ Part filled by $(A + B + C)$ in 1 h
$= \dfrac{7}{2 \times 20} = \dfrac{7}{40}$
$\therefore$ Required time $= \dfrac{40}{7} = 5\dfrac{5}{7}$ h

11. (a) Work done by C in 1 min
$= \left(\dfrac{1}{60} + \dfrac{1}{75} - \dfrac{1}{50}\right)$
$= \dfrac{5 + 4 - 6}{300} = \dfrac{3}{300} = \dfrac{1}{100}$
Hence, C can empty the full tank in 100 min.

12. (c) Let pipe A takes x h to fill the tank.
Then, pipe B takes $\dfrac{x}{2}$ h to fill the tank
and pipe C takes $\dfrac{x}{4}$ h to fill the tank.
According to the question,
$\dfrac{1}{x} + \dfrac{2}{x} + \dfrac{4}{x} = \dfrac{1}{5} \Rightarrow \dfrac{7}{x} = \dfrac{1}{5}$
$\Rightarrow \quad x = 35$ h

13. (a) Part filled by A in 1 min $= \dfrac{1}{30}$
Part filled by B in 1 min $= \dfrac{1}{10}$
Part emptied by C in 1 min $= -\dfrac{1}{40}$
Net part filled in 1 h by $(A + B + C)$
$= \left(\dfrac{1}{30} + \dfrac{1}{10} - \dfrac{1}{40}\right) = \dfrac{4 + 12 - 3}{120} = \dfrac{13}{120}$
$\therefore$ Required time to fill the tank
$= \dfrac{120}{13} = 9\dfrac{3}{13}$ h

14. (b) Part filled by A in 1 h $= \dfrac{1}{5}$
Part filled by B in 1 h $= \dfrac{1}{6}$
Part filled by C in 1 h $= \dfrac{1}{30}$
Net part filled by $(A + B + C)$ in 1 h
$= \left(\dfrac{1}{5} + \dfrac{1}{6} + \dfrac{1}{30}\right) = \dfrac{6 + 5 + 1}{30} = \dfrac{12}{30} = \dfrac{2}{5}$
$\therefore$ Required time to fill the tank $= \dfrac{5}{2} = 2\dfrac{1}{2}$ h

Fast Track Method
Here, $m = 5, n = 6$ and $P = 30$
$\therefore$ Required time $= \dfrac{mnP}{mn + nP + mP}$
[by Technique 2]
$= \dfrac{5 \times 6 \times 30}{5 \times 6 + 6 \times 30 + 5 \times 30}$
$= \dfrac{5 \times 6 \times 30}{30 + 180 + 150} = \dfrac{900}{360} = \dfrac{5}{2} = 2\dfrac{1}{2}$ h

15. (b) Part of tank filled by first tap in 1 h = $\dfrac{1}{3}$

Part of tank filled by second tap in 1 h = $\dfrac{1}{4}$

Part of tank emptied by third tap in 1 h = $\dfrac{1}{5}$

Part of the tank filled by all pipes when opened simultaneously in 1 h

$= \dfrac{1}{3} + \dfrac{1}{4} - \dfrac{1}{5}$

$= \dfrac{20 + 15 - 12}{60} = \dfrac{23}{60}$

∴ Time taken by all the taps to fill the tank when it is empty = $\dfrac{60}{23}$ h = $2\dfrac{14}{23}$ h

Fast Track Method

Here, $m = 3$, $n = 4$ and $p = 5$

∴ Required time = $\dfrac{mnp}{np + mp - mn}$

[by Technique 2]

$= \dfrac{3 \times 4 \times 5}{4 \times 5 + 3 \times 5 - 3 \times 4}$

$= \dfrac{60}{20 + 15 - 12}$

$= \dfrac{60}{23} = 2\dfrac{14}{23}$ h

16. (b) Part filled by $(A + B + C)$ in 1 min = $\dfrac{1}{10}$

Part filled by A in 1 min = $\dfrac{1}{30}$

Part filled by B in 1 min = $\dfrac{1}{40}$

Part filled by $(A + B)$ in 1 min

$= \dfrac{1}{30} + \dfrac{1}{40} = \dfrac{4 + 3}{120} = \dfrac{7}{120}$

∴ Part filled by C in 1 min

$= \dfrac{1}{10} - \dfrac{7}{120} = \dfrac{12 - 7}{120} = \dfrac{5}{120} = \dfrac{1}{24}$

Hence, tap C will fill the cistern in 24 min.

17. (b) Here, $a = 16$ and $b = 9$

∴ Required time = $\sqrt{ab}$ [by Technique 3]

$= \sqrt{16 \times 9} = 4 \times 3 = 12$ min

18. (a) Work done by the inlet in 1 h

$= \left(\dfrac{1}{8} - \dfrac{1}{12}\right) = \dfrac{1}{24}$

Work done by the inlet in 1 min

$= \dfrac{1}{24} \times \dfrac{1}{60} = \dfrac{1}{1440}$

∵ Capacity of $\dfrac{1}{1440}$ part = 3 L

∴ Capacity of the whole

$= 3 \times 1440 = 4320$ L

Fast Track Method

Here, $a = 8$, $b = 3$ L/min = $3 \times 60 = 180$ and $c = 12$

∴ Capacity of tank = $\dfrac{abc}{c - a}$ [by Technique 4]

$= \dfrac{180 \times 8 \times 12}{4}$

$= 4320$ L

19. (b) Here, $a = 30$, $b = 10$ and $c = 50$

∴ Capacity of tank = $\dfrac{abc}{c - a}$ [by Technique 4]

$= \dfrac{30 \times 10 \times 50}{50 - 30}$

$= \dfrac{30 \times 10 \times 50}{20} = 750$ L

20. (b) Let the time taken by pipe A to fill the tank be x min. Then, the time taken by pipe B to fill the tank is $4x$ min.

Then, according to the question,

$4x - x = 30$

$\Rightarrow 3x = 30 \Rightarrow x = 10$

∴ Time taken by pipe $A = 10$ min

Part filled by pipe A in 1 min = $\dfrac{1}{10}$

Time taken by pipe $B = 4 \times 10 = 40$ min

Part filled by pipe B in 1 min = $\dfrac{1}{40}$

Now, part filled by both the pipes A and B in 1 min = $\dfrac{1}{10} + \dfrac{1}{40} = \dfrac{4 + 1}{40} = \dfrac{5}{40} = \dfrac{1}{8}$

∴ Time taken by both the pipes A and B to fill the tank = 8 min

Fast Track Method

Here, $n = 4$ and $t = 30$

Then, time taken to fill the tank by both the pipes together

$= \dfrac{nt}{n^2 - 1}$ [by Technique 5(i)]

$= \dfrac{4 \times 30}{4^2 - 1} = \dfrac{4 \times 30}{16 - 1} = \dfrac{4 \times 30}{15} = 8$ min

21. (a) Let the time taken by tap A to fill the tank be x min. Then, time taken by tap B to fill the tank is $2x$ min.

Then, according to the question,

$2x - x = 20 \Rightarrow x = 20$

∴ Time taken by tap $A = 20$ min

Fast Track Method

Here, $n = 2$ and $t = 20$

∴ Time taken by faster tap $A = \dfrac{t}{n - 1}$

[by Technique 5(ii)]

$= \dfrac{20}{2 - 1} = 20$ min

Pipes and Cisterns / 417

22. (b) Let the time taken by pipe A to fill the tank be x min. Then, the time taken by pipe B to fill the tank is $4x$ min.
Then, according to the question,
$$4x - x = 24 \Rightarrow 3x = 24$$
$$\Rightarrow \quad x = \frac{24}{3} = 8 \text{ min}$$
∴ Time taken by pipe B to fill the tank
$$= 4 \times 8 = 32 \text{ min}$$

Fast Track Method
Here, $n = 4$, $t = 24$
Then, time taken by pipe B to fill the tank
$$= \frac{nt}{n-1} = \frac{4 \times 24}{4-1} \quad \text{[by Technique 5(iii)]}$$
$$= \frac{4 \times 24}{3} = 4 \times 8 = 32 \text{ min}$$

23. (a) Part filled by pipe P in 1 min $= \frac{1}{12}$

Part filled by pipe Q in 1 min $= \frac{1}{15}$

Part filled by both pipes in 1 min
$$= \frac{1}{12} + \frac{1}{15} = \frac{5+4}{60} = \frac{9}{60}$$

Now, part filled by both pipes in 3 min
$$= \frac{3 \times 9}{60} = \frac{27}{60} = \frac{9}{20}$$

∴ Remaining part $= 1 - \frac{9}{20} = \frac{11}{20}$

Let the remaining part be filled by pipe Q in x min.

Then, $x \times \frac{1}{15} = \frac{11}{20}$

$$\Rightarrow \quad x = \frac{15 \times 11}{20} = \frac{33}{4} = 8\frac{1}{4} \text{ min}$$

Fast Track Method
Here, $x = 15$, $y = 12$ and $t = ?$
So, the time in which P is closed
$$= y\left(1 - \frac{t}{x}\right) \quad \text{[by Technique 6]}$$
$$\Rightarrow \quad 3 = 12\left(1 - \frac{t}{15}\right)$$
$$\Rightarrow \quad 3 = 12 - \frac{4t}{5} \Rightarrow 9 = \frac{4t}{5}$$
$$\Rightarrow \quad t = \frac{45}{4}$$
∴ Required time $= \frac{45}{4} - 3 = \frac{33}{4} = 8\frac{1}{4} \text{ min}$

24. (d) Part of tank filled by pipe A in
$$1 \text{ h} = \frac{1}{8}$$
Part of tank filled by pipe B in 1 h $= \frac{1}{10}$

∴ Part of tank filled by both pipes in 1 h
$$= \frac{1}{8} + \frac{1}{10} = \frac{10+8}{80} = \frac{18}{80} = \frac{9}{40}$$
Part of tank filled by both pipes in 2h
$$= \frac{2 \times 9}{40} = \frac{9}{20}$$
Remaining part of tank to be filled
$$= 1 - \frac{9}{20} = \frac{11}{20}$$
∴ Time taken by pipe B to fill this part
$$= \frac{11/20}{1/10} = \frac{11}{2} = 5\frac{1}{2} \text{ h}$$

Fast Track Method
Here, $x = 10$, $y = 8$ and $t = ?$
Time after which A is closed $= y\left(1 - \frac{t}{x}\right)$
[by Technique 6]
$$\Rightarrow \quad 2 = 8\left(1 - \frac{t}{10}\right)$$
$$\Rightarrow \quad \frac{1}{4} = 1 - \frac{t}{10} \Rightarrow \frac{3}{4} = \frac{t}{10} \Rightarrow t = \frac{15}{2} \text{ h}$$
∴ Required time $= \frac{15}{2} - 2 = \frac{11}{2} = 5\frac{1}{2} \text{ h}$

25. (d) Part filled by A in 1 min $= \frac{1}{24}$

and part filled by B in 1 min $= \frac{1}{32}$

Let B be closed after x min. Then,
[Part filled by $(A+B)$ in x min]
+ [Part filled by A in $(9-x)$ min] = 1

∴ $x\left(\frac{1}{24} + \frac{1}{32}\right) + (9-x) \times \frac{1}{24} = 1$

$$\Rightarrow \quad x\left(\frac{4+3}{96}\right) + \frac{(9-x)}{24} = 1$$
$$\Rightarrow \quad \frac{7x}{96} + \frac{(9-x)}{24} = 1$$
$$\Rightarrow \quad \frac{7x + 4(9-x)}{96} = 1$$
$$\Rightarrow \quad 7x + 4(9-x) = 96$$
$$\Rightarrow \quad 7x + 36 - 4x = 96$$
$$\Rightarrow \quad 7x - 4x = 96 - 36$$
$$\Rightarrow \quad 3x = 60 \Rightarrow x = \frac{60}{3} = 20$$

Hence, B must be closed after 20 min.

Fast Track Method
Here, $x = 24$ min, $y = 32$ min, $t = 9$ min
∴ Required time $= y\left(1 - \frac{t}{x}\right)$
[by Technique 6]
$$= 32\left(1 - \frac{9}{24}\right)$$
$$= 32 \times \frac{15}{24} = 20 \text{ min}$$

418 / Fast Track Objective Arithmetic

Exercise 2 — Higher Skill Level Questions

1. (c) Area of tap ∝ Work done by pipe
When diameter is doubled, then area will be four times. So, it will work four times faster.
Hence, required time taken to empty the tank $= 40 \times \dfrac{1}{4} = 10$ min

2. (b) A's one hour's work $= \dfrac{1}{6}$
and B's one hour's work $= \dfrac{1}{8}$
$(A + B)$'s two hour's work $= 2\left(\dfrac{1}{8} + \dfrac{1}{6}\right) = \dfrac{28}{48}$
Remaining work $= 1 - \dfrac{28}{48} = \dfrac{20}{48}$
∴ Time taken by pipe B to complete remaining work
$= \dfrac{20/48}{1/8} = \dfrac{20}{48} \times 8 = \dfrac{20}{6} = 3\dfrac{1}{3}$ h

3. (a) Let the time taken to fill the tank be t min.
According to the question,
$\dfrac{t}{2}\left(\dfrac{1}{60} + \dfrac{1}{40}\right) + \dfrac{t}{2}\left(\dfrac{1}{40}\right) = 1$
$\Rightarrow \quad \dfrac{t}{2}\left(\dfrac{1}{24}\right) + \dfrac{t}{2}\left(\dfrac{1}{40}\right) = 1$
$\Rightarrow \quad \dfrac{t}{2}\left(\dfrac{5+3}{120}\right) = 1$
$\Rightarrow \quad t = \dfrac{120}{8} \times 2 = 30$ min

4. (a) Let one pipe takes m h to fill the tank. Then, the other pipe takes $(m - 10)$ h.
According to the question,
$\dfrac{1}{m} + \dfrac{1}{(m-10)} = \dfrac{1}{12}$
$\Rightarrow \quad \dfrac{m - 10 + m}{m(m - 10)} = \dfrac{1}{12}$
$\Rightarrow \quad 12(m - 10 + m) = m(m - 10)$
$\Rightarrow \quad m^2 - 34m + 120 = 0$
$\Rightarrow \quad m^2 - 30m - 4m + 120 = 0$
$\Rightarrow \quad (m - 30)(m - 4) = 0$
∴ $m = 30$ or 4   but $m \neq 4$
∴ Faster pipe will take $(30 - 10)$ h, i.e. 20 h to fill the tank.

5. (a) Part filled by 1st pipe in 1 h $= \dfrac{1}{14}$
Part filled by 2nd pipe in 1 h $= \dfrac{1}{16}$

Part filled by the two pipes in 1 h
$= \left(\dfrac{1}{14} + \dfrac{1}{16}\right) = \dfrac{8+7}{112} = \dfrac{15}{112}$
∴ Time taken by these two pipes to fill the cistern $= \dfrac{112}{15}$ h = 7 h 28 min
Due to leakage, the time taken
$= 7$ h 28 min + 92 min = 9 h
∴ Work done by (two pipes + leak) in 1 h
$= 1/9$
Work done by the leak in 1 h
$= \dfrac{1}{9} - \dfrac{15}{112} = \dfrac{112 - 135}{1008} = -\dfrac{23}{1008}$
∴ Time taken by leak to empty the full cistern $= \dfrac{1008}{23} = 43\dfrac{19}{23}$ h

6. (d) Let the number of minutes taken to empty the cistern be x min.
According to the question,
$\dfrac{x}{6} - \dfrac{x+5}{12} - \dfrac{x+5}{15} = 0$
$\Rightarrow \quad \dfrac{x}{6} - \dfrac{x}{12} - \dfrac{5}{12} - \dfrac{x}{15} - \dfrac{5}{15} = 0$
$\Rightarrow \quad \dfrac{x}{6} - \dfrac{x}{12} - \dfrac{x}{15} = \dfrac{5}{12} + \dfrac{5}{15}$
$\Rightarrow \quad \dfrac{10x - 5x - 4x}{60} = \dfrac{25 + 20}{60}$
$\Rightarrow \quad \dfrac{x}{60} = \dfrac{45}{60} \Rightarrow x = 45$ min

7. (b) Since, 1st pipe takes 1 h to fill $\dfrac{1}{2}$ part of the tank. So, time taken to fill the whole tank $(m) = 2$ h and 2nd pipe takes 1 h to fill $\dfrac{1}{3}$ part of the tank. So, time taken to fill the whole tank $(n) = 3$ h
Let 3rd pipe takes x h to empty the whole tank.
According to the question,
$\dfrac{1}{m} + \dfrac{1}{n} - \dfrac{1}{x} = \dfrac{7}{12}$
$\Rightarrow \quad \dfrac{1}{2} + \dfrac{1}{3} - \dfrac{1}{x} = \dfrac{7}{12}$
$\Rightarrow \quad \dfrac{1}{x} = \dfrac{6 + 4 - 7}{12} = \dfrac{3}{12} = \dfrac{1}{4} \Rightarrow x = 4$ h

8. (d) Part filled by 1st pipe in 1 min $= \dfrac{1}{20}$
Part filled by 2nd pipe in 1 min $= \dfrac{1}{24}$
Part filled by all the pipes in 1 min $= \dfrac{1}{15}$

Pipes and Cisterns / 419

Work done by the waste pipe in 1 min

$= \dfrac{1}{15} - \left(\dfrac{1}{20} + \dfrac{1}{24}\right)$

$= \dfrac{1}{15} - \left(\dfrac{6+5}{120}\right)$

$= \dfrac{1}{15} - \dfrac{11}{120}$

$= \dfrac{8-11}{120} = \left(-\dfrac{3}{120}\right) = \left(-\dfrac{1}{40}\right)$

[–ve sign indicates emptying]

Now, capacity of $\dfrac{1}{40}$ part = 6 gallon

∴ Capacity of whole tank = 40 × 6
= 240 gallon

9. (c) Part filled by A in 1 h = $\dfrac{1}{16}$

and part filled by B in 1 h = $\dfrac{1}{10}$

Part filled by (A + B) in 2 h

$= \dfrac{1}{16} + \dfrac{1}{10} = \dfrac{13}{80}$

∴ Part filled by (A + B) in 12 h

$= \dfrac{6 \times 13}{80} = \dfrac{78}{80}$

∴ Remaining part = $1 - \dfrac{78}{80} = \dfrac{2}{80} = \dfrac{1}{40}$

Now, it is the turn of A.

Time taken by A to fill $\dfrac{1}{40}$ part of the tank

$= \dfrac{1}{40} \times 16 = \dfrac{2}{5}$ h

∴ Total time taken = $\left(12 + \dfrac{2}{5}\right)$ h = $12\dfrac{2}{5}$ h

10. (b) Part filled by X in 1st min and Y in

2nd min = $\left(\dfrac{1}{6} + \dfrac{1}{7}\right) = \dfrac{13}{42}$

Part filled by (X + Y) working alternately in

6 min = $\dfrac{1}{2} \times \dfrac{13}{42} \times 6 = \dfrac{13}{14}$

∴ Remaining part = $\left(1 - \dfrac{13}{14}\right) = \dfrac{1}{14}$

Now, it is the turn of X.

∵ One-sixth part is filled in 1 min.

∴ One-fourteenth part is filled in

$\left(6 \times \dfrac{1}{14}\right)$ min = $\dfrac{3}{7}$ min

Now, required time = $\left(6 + \dfrac{3}{7}\right) = 6\dfrac{3}{7}$ min

11. (b) Part of tank filled by pipe A in 1 h = $\dfrac{1}{4}$

Part of tank filled by pipe B in 1 h = $\dfrac{1}{6}$

Part of tank filled by (A + B) in 2 h

$= \dfrac{1}{4} + \dfrac{1}{6} = \dfrac{3+2}{12} = \dfrac{5}{12}$

∴ Part of tank filled by (A + B) in 4 h

$= \dfrac{5}{12} \times 2 = \dfrac{5}{6}$

Remaining part = $1 - \dfrac{5}{6} = \dfrac{1}{6}$

Now, it is the turn of A.

Time taken by A to fill $\dfrac{1}{6}$ part of the tank

$= \dfrac{1}{6} \times 4 = \dfrac{2}{3}$ h

∴ Total time taken = $\left(4 + \dfrac{2}{3}\right)$ h = $4\dfrac{2}{3}$ h

12. (b) Let the cistern gets emptied in m h after 3 : 00 pm.

Work done by A in m h, by B in (m − 1) h and by C in (m − 2) h = 0

$\Rightarrow \dfrac{m}{3} + \dfrac{m-1}{4} - (m-2) = 0$

$\Rightarrow 4m + 3(m-1) - 12(m-2) = 0$

$\Rightarrow 5m = 21$

$\Rightarrow m = \dfrac{21}{5} = 4.2$ h

∴ m = 4 h 12 min

So, the required time = 7 : 12 pm

13. (c) According to the question,
Double the 2.5 min = 5 min
Now, part filled in 5 min

$= 5\left(\dfrac{1}{12} + \dfrac{1}{15}\right) = 5\left(\dfrac{5+4}{60}\right)$

$= 5 \times \dfrac{9}{60} = \dfrac{3}{4}$

Part emptied in 1 min when P, R and M, all are opened

$= \dfrac{1}{6} - \left(\dfrac{1}{12} + \dfrac{1}{15}\right) = \dfrac{1}{6} - \left(\dfrac{5+4}{60}\right)$

$= \dfrac{1}{6} - \dfrac{3}{20} = \dfrac{1}{60}$

∵ 1/60th part is emptied in 1 min.

∴ Three-fourth part will be emptied in

$60 \times \dfrac{3}{4} = 15 \times 3 = 45$ min

14. (b) Part of tank filled in 1 h by tap A = $\dfrac{1}{12}$

Part of tank filled in 1 h by tap B = $\dfrac{1}{15}$

Part of tank filled in 1 h by tap C = $\dfrac{1}{20}$

Part of tank filled in 1 h by taps A and B

$= \dfrac{1}{12} + \dfrac{1}{15} = \dfrac{9}{60} = \dfrac{3}{20}$

Part of tank filled in 1 h by A and C
$$= \frac{1}{12} + \frac{1}{20} = \frac{5+3}{60} = \frac{8}{60} = \frac{2}{15}$$
Tank filled in first 2 h $= \frac{3}{20} + \frac{2}{15} = \frac{17}{60}$
Tank filled in 6 h $= \frac{17}{60} \times 3 = \frac{51}{60}$
$\therefore$ Remaining part $= 1 - \frac{51}{60} = \frac{3}{20}$
Now in the 7th hour, tank is filled by taps A and B. So, time taken by A and B to fill $\frac{3}{20}$ part of tank $= \frac{3/20}{3/20} = 1$ h
$\therefore$ Total time to fill the tank $= 6 + 1 = 7$ h

15. (d) Let time taken by B to fill the tank,
$$m = x \text{ h}$$
$\therefore$ Time taken by A to fill the tank,
$$n = x + \frac{x \times 80}{100} = \frac{9x}{5} \text{ h}$$
According to the formula,
Time taken by both the taps to fill the tank
$$t = \frac{mn}{m+n} \quad \text{[by Technique 1(i)]}$$
$$\Rightarrow \quad 45 = \frac{x \times \frac{9x}{5}}{x + \frac{9x}{5}} \Rightarrow 45 \times \frac{14x}{5} = \frac{9x^2}{5}$$
$$\therefore \quad x = \frac{45 \times 14}{9} = 70 \text{ h}$$

16. (a) Let the number of outlets be x.
$\therefore$ Number of inlets $= (7 - x)$
Time taken to fill the tank when all the pipes are opened $= \frac{30}{11}$ h
Part of tank filled in 1 h when all the pipes are opened $= \frac{11}{30}$ h
According to the question,
$$\frac{7-x}{10} - \frac{x}{15} = \frac{11}{30}$$
$$\Rightarrow \quad \frac{3(7-x) - 2x}{30} = \frac{11}{30}$$
$$\Rightarrow \quad 21 - 3x - 2x = 11$$
$$\Rightarrow \quad 5x = 10 \Rightarrow x = 2$$
Hence, number of outlets $= 2$
and number of inlets $= 7 - 2 = 5$

17. (c) Total quantity of solutions P, Q and R from A, B and C respectively, after 3 min
$$= \frac{3}{30} + \frac{3}{20} + \frac{3}{10} = 3\left(\frac{2+3+6}{60}\right)$$
$$= \frac{3 \times 11}{60} = \frac{11}{20}$$

Quantity of solution R in liquid in 3 min
$$= \frac{3}{10}$$
$\therefore$ Part of solution $R = \frac{3/10}{11/20} = \frac{3 \times 20}{10 \times 11} = \frac{6}{11}$

18. (e) Time taken to fill the tank by the tap having 2 cm diameter $= 61$ min
$\therefore$ Time taken to fill the tank by the tap having 1 cm diameter
$$= 61 \times \left(\frac{2}{1}\right)^2 = 244 \text{ min}$$
Similarly, time taken to fill the tank by the tap having $\frac{4}{3}$ cm diameter
$$= 61 \times \left(\frac{2}{4/3}\right)^2 = 61 \times \frac{9}{4}$$
$$= \frac{549}{4} \text{ min}$$
$\therefore$ Part of the tank filled by all the three pipes in 1 min $= \frac{1}{61} + \frac{1}{244} + \frac{1}{549/4}$
$$= \frac{36 + 9 + 16}{2196} = \frac{61}{2196} = \frac{1}{36}$$
Hence, required time taken $= 36$ min

19. (c) Quantity of water admitted by tap 1 in 1 h $= 42$ L
Quantity of water admitted by tap 2 in 1 h $= 56$ L
Quantity of water removed by tap 3 in 1 h $= 48$ L
So, quantity of water filled in the tank in 1 h $= (42 + 56 - 48)$ L $= 50$ L
$\therefore$ Quantity of water filled in 16 h
$$= 16 \times 50 = 800 \text{ L}$$
Hence, the capacity of tank is 800 L.

20. (b) Let the weight of container be x kg.
$\therefore$ Weight of water $= (2.25 - x)$ kg
Now, weight of 0.2 part of filled container with water $= 0.77$ kg
and 0.2 part or $\frac{1}{5}$ part water's weight
$$= (0.77 - x)$$
$\therefore$ Weight of water $= 5(0.77 - x)$
$\Rightarrow \quad 2.25 - x = 5(0.77 - x)$
$\Rightarrow \quad 4x = 3.85 - 2.25 = 1.6$
$\Rightarrow \quad x = 0.4$ kg
So, weight of water filled in container
$$= 2.25 - 0.4 = 1.85 \text{ kg}$$
$\therefore$ Weight of 0.4 part of filled container
$$= (1.85 \times 0.4) + 0.4 = 1.14 \text{ kg}$$

Chapter 25

Speed, Time and Distance

Speed
The rate at which a body or an object travels to cover a certain distance is called **speed** of that body.

Time
The duration in hours, minutes or seconds spent to cover a certain distance is called the **time**.

Distance
The length of the path travelled by any object or a person between two places is known as **distance**.

- Units of speed, time and distance should be in the same metric system.

Relation between Speed, Time and Distance

Speed is the distance covered by an object in unit time. It is calculated by dividing the distance travelled by the time taken.

1. Speed = $\dfrac{\text{Distance}}{\text{Time}}$ 2. Time = $\dfrac{\text{Distance}}{\text{Speed}}$

3. Distance = Speed × Time

Ex. 1 If a car covers 125 km in 5 h, then find the speed of the car.

Sol. We know that,

$$\text{Speed} = \dfrac{\text{Distance}}{\text{Time}}$$

∴ Required speed = $\dfrac{125}{5}$ = 25 km/h

Ex. 2 A train covers a distance of 200 km with a speed of 10 km/h. What time is taken by the train to cover this distance?

Sol. Given, speed = 10 km/h and distance = 200 km

∴ Time = $\dfrac{\text{Distance}}{\text{Speed}} = \dfrac{200}{10} = 20$ h

Ex. 3 A bike crosses a bridge with a speed of 108 km/h. What will be the length of the bridge, if the bike takes 8 h to cross the bridge?

Sol. Given, speed = 108 km/h, time = 8 h

∴ Length of the bridge = Speed × Time = Distance travelled by bike in 8 h
= 108 × 8 = 864 km

Basic Formulae Related to Speed, Time and Distance

▶ Formula 1

Conversion of units a km/h = $\dfrac{a \times 1000 \text{ m}}{3600 \text{ s}} = \dfrac{5a}{18}$ m/s

a m/s = $\dfrac{a \times 1/1000 \text{ km}}{1/3600 \text{ h}} = \dfrac{18a}{5}$ km/h

Ex. 4 Convert 72 km/h into m/s.

Sol. We know that, a km/h = $\left(a \times \dfrac{5}{18}\right)$ m/s

∴ 72 km/h = $\left(72 \times \dfrac{5}{18}\right)$ m/s = 4 × 5 = 20 m/s

Ex. 5 Convert 25 m/s to km/h.

Sol. We know that, a m/s = $\left(a \times \dfrac{18}{5}\right)$ km/h

∴ 25 m/s = $\left(25 \times \dfrac{18}{5}\right)$ = 5 × 18 = 90 km/h

▶ Formula 2

If speed is kept constant, then the distance covered by an object is proportional to time.

i.e. Distance ∝ Time (speed constant) or $\dfrac{D_1}{T_1} = \dfrac{D_2}{T_2}$

Ex. 6 A person covers $20\dfrac{2}{5}$ km in 3 h. What distance will he cover in 5 h?

Sol. Here, speed is kept constant. Therefore, according to the formula,

$\dfrac{D_1}{T_1} = \dfrac{D_2}{T_2}$

Given that, $D_1 = 20\dfrac{2}{5} = \dfrac{102}{5}$ km, $T_1 = 3$ h, $T_2 = 5$ h and $D_2 = ?$

∴ $\dfrac{102/5}{3} = \dfrac{D_2}{5} \Rightarrow D_2 = \dfrac{102 \times 5}{5 \times 3} = 34$ km

Hence, the distance covered by the object in 5 h is 34 km.

Formula 3

If time is kept constant, then the distance covered by an object is proportional to speed, i.e. distance ∝ speed (time constant) or $\dfrac{D_1}{S_1} = \dfrac{D_2}{S_2}$.

Ex. 7 A person covers a distance of 12 km, while walking at a speed of 4 km/h. How much distance he would cover in same time, if he walks at a speed of 6 km/h?

Sol. Given, $D_1 = 12$ km, $S_1 = 4$ km/h, $D_2 = ?$ and $S_2 = 6$ km/h

Since, the time is kept constant.

Therefore, according to the formula, $\dfrac{D_1}{S_1} = \dfrac{D_2}{S_2} \Rightarrow \dfrac{12}{4} = \dfrac{D_2}{6} \Rightarrow D_2 = 18$ km

Hence, the person will cover 18 km.

Formula 4

If distance is kept constant, then the speed of a body is inversely proportional to time, i.e. speed $\propto \dfrac{1}{\text{time}}$ (distance constant)

or $\quad S_1 T_1 = S_2 T_2 = S_3 T_3 = ...$

+ If the ratio of speeds of two objects is $x : y$, then to cover same distance, the ratio of time taken will be $y : x$.

Ex. 8 A person covers a certain distance with a speed of 18 km/h in 8 min. If he wants to cover the same distance in 6 min, what should be his speed?

Sol. We know that, Speed $= \dfrac{\text{Distance}}{\text{Time}} \Rightarrow 18 = \dfrac{\text{Distance} \times 60}{8}$ $\left[\because 8 \text{ min} = \dfrac{8}{60} \text{ h} \right]$

∴ Distance $= \dfrac{18 \times 8}{60} = \dfrac{12}{5}$ km

∴ Speed to cover $\dfrac{12}{5}$ km in 6 min $= \dfrac{\text{Distance}}{\text{Time}} = \dfrac{12/5}{1/10}$ $\left[\because 6 \text{ min} = \dfrac{6}{60} \text{ h} = \dfrac{1}{10} \text{ h} \right]$

$= \dfrac{12}{5} \times 10 = 24$ km/h

Alternate Method

We know that, if distance is same, then speed is inversely proportional to time.

Given, $S_1 = 18$ km/h, $S_2 = ?$, $T_1 = \dfrac{8}{60}$ h and $T_2 = \dfrac{6}{60}$ h

According to the formula, $S_1 T_1 = S_2 T_2$

where, S_1 and S_2 are speeds and T_1 and T_2 are times.

∴ $\quad 18 \times \dfrac{8}{60} = S_2 \times \dfrac{6}{60} \Rightarrow S_2 = \dfrac{18 \times 8}{6} = 24$ km/h

Formula 5

When two bodies A and B are moving with speeds a km/h and b km/h respectively, then the relative speed of two bodies is
(i) $(a + b)$ km/h (if they are moving in opposite directions)
(ii) $(a - b)$ km/h (if they are moving in same direction)

Ex. 9 Two persons are moving in the directions opposite to each other. The speeds of the both persons are 5 km/h and 3 km/h, respectively. Find the relative speed of the two persons in respect of each other.

Sol. We know that, if two persons are moving in opposite direction, then sum of their speeds is the required relative speed.

∴ Required relative speed = 5 + 3 = 8 km/h

Ex. 10 Two buses are running in the same direction. The speeds of two buses are 5 km/h and 15 km/h, respectively. What will be the relative speed of second bus with respect to first?

Sol. We know that, if two buses are running in same direction, then difference in speeds is the required relative speed.

∴ Required relative speed = 15 − 5 = 10 km/h

▶ Formula 6

When a body travels with different speeds for different durations, then average speed of that body for the complete journey is defined as the total distance covered by the body divided by the total time taken to cover the distance,

i.e. Average speed = $\dfrac{\text{Total distance covered by a body}}{\text{Total time taken by the body}}$

Ex. 11 A person covers a distance of 20 km by bus in 35 min. After deboarding the bus, he took rest for 20 min and covers another 10 km by a taxi in 20 min. Find his average speed for the whole journey.

Sol. Total distance covered = (20 + 10) km = 30 km

Total time taken = (35 + 20 + 20) min = 75 min = $\dfrac{75}{60}$ h = $\dfrac{5}{4}$ h

According to the formula,

Average speed = $\dfrac{\text{Total distance covered}}{\text{Total time taken}} = \dfrac{30}{5/4} = 24$ km/h

So, the average speed of person for the whole journey is 24 km/h.

Ex. 12 If a person covers 40 km at a speed of 10 km/h by a cycle, 25 km at 5 km/h on foot and another 100 km at 50 km/h by bus. Then, find his average speed for the whole journey.

Sol. Here, total distance covered by the person = (40 + 25 + 100) km = 165 km

Time taken to cover 40 km = $\dfrac{40}{10}$ = 4 h

Time taken to cover 25 km = $\dfrac{25}{5}$ = 5 h

Time taken to cover 100 km = $\dfrac{100}{50}$ = 2 h

∴ Total time taken for whole journey = (4 + 5 + 2) h = 11 h

According to the formula,

Average speed = $\dfrac{\text{Total distance covered}}{\text{Total time taken}} = \dfrac{165}{11} = 15$ km/h

Speed, Time and Distance / 425

> **MIND IT!** If a body covers a distance D_1 at S_1 km/h, D_2 at S_2 km/h, D_3 at S_3 km/h and so on upto D_n at S_n, then
>
> $$\text{Average speed} = \frac{D_1 + D_2 + D_3 + D_4 + \ldots + D_n}{\frac{D_1}{S_1} + \frac{D_2}{S_2} + \frac{D_3}{S_3} + \frac{D_4}{S_4} + \ldots + \frac{D_n}{S_n}}$$
>
> $$\text{Average speed} \neq \frac{S_1 + S_2 + S_3 + S_4 + \ldots + S_n}{n}$$

Ex. 13 A person covers 20 km distance with a speed of 5 km/h, then he covers the next 15 km with a speed of 3 km/h and the last 10 km is covered by him with a speed of 2 km/h. Find out his average speed for the whole journey.

Sol. Here, $P = 20$ km, $Q = 15$ km, $R = 10$ km and $x = 5$ km/h, $y = 3$ km/h, $z = 2$ km/h

$\therefore$ Required average speed $= \dfrac{P + Q + R}{\dfrac{P}{x} + \dfrac{Q}{y} + \dfrac{R}{z}} = \dfrac{20 + 15 + 10}{\dfrac{20}{5} + \dfrac{15}{3} + \dfrac{10}{2}}$

$= \dfrac{45}{4 + 5 + 5} = \dfrac{45}{14} = 3\dfrac{3}{14}$ km/h

Fast Track Techniques to solve the QUESTIONS

Technique 1 When a certain distance is covered at speed A and the same distance is covered at speed B, then the average speed during the whole journey is given by $\dfrac{2AB}{A + B}$.

Ex. 14 Shantanu covers a certain distance by car driving at 35 km/h and he returns back to the starting point riding on a scooter with a speed of 25 km/h. Find the average speed for the whole journey.

Sol. Let us assume that the distance covered by Shantanu in one direction = D km

$\therefore$ Time taken in first case $= \dfrac{D}{35}$ h

and time taken in second case $= \dfrac{D}{25}$ h

$\therefore$ Average speed $= \dfrac{\text{Total distance covered}}{\text{Total time taken}} = \dfrac{D + D}{\dfrac{D}{35} + \dfrac{D}{25}} = \dfrac{2D}{D\left(\dfrac{1}{35} + \dfrac{1}{25}\right)}$

$= \dfrac{2 \times 35 \times 25}{60} = 29.16$ km/h

Fast Track Method

Here, $A = 35$ km/h and $B = 25$ km/h
According to the formula,

Average speed $= \dfrac{2AB}{A + B} = \dfrac{2 \times 35 \times 25}{60} = \dfrac{175}{6} = 29.16$ km/h

426 / **Fast Track** Objective Arithmetic

Technique 2 If a person covers three equal distances at the speed of A km/h, B km/h and C km/h respectively, then the average speed during the whole journey will be $\dfrac{3ABC}{AB + BC + CA}$.

Ex. 15 If a person covers three equal distances at the speed of 30 km/h, 15 km/h and 10 km/h respectively, then find out his average speed during the whole journey.

Sol. Let the distance be D km.

$\therefore$ Required average speed $= \dfrac{3D}{D \times \dfrac{1}{30} + D \times \dfrac{1}{15} + D \times \dfrac{1}{10}}$

$= \dfrac{3D}{D\left(\dfrac{1}{30} + \dfrac{1}{15} + \dfrac{1}{10}\right)} = \dfrac{3 \times 30}{(1 + 2 + 3)}$

$= \dfrac{3 \times 30}{6} = 15$ km/h

Fast Track Method

Here, $A = 30$ km/h, $B = 15$ km/h and $C = 10$ km/h

$\therefore$ Required average speed $= \dfrac{3\,ABC}{AB + BC + CA} = \dfrac{3 \times 30 \times 15 \times 10}{30 \times 15 + 15 \times 10 + 10 \times 30}$

$= \dfrac{3 \times 30 \times 15 \times 10}{450 + 150 + 300} = \dfrac{3 \times 30 \times 15 \times 10}{900} = 15$ km/h

Technique 3 If a person covers P part of his total distance with speed of x, Q part of total distance with speed of y and R part of total distance with speed of z and so on, then the average speed of a person for the whole journey,

Average speed $= \dfrac{1}{\dfrac{P}{x} + \dfrac{Q}{y} + \dfrac{R}{z} + \ldots}$

Ex. 16 Mr. Sharma travels by car and covers 25% of his journey with a speed of 10 km/h, 45% of his journey with a speed of 5 km/h and remaining 30% of his journey with a speed of 15 km/h. What will be the average speed of Mr. Sharma for the whole journey?

Sol. Let the total distance travelled be D km.

Now, 25% distance $= \dfrac{25}{100} \times D = \dfrac{D}{4}$ km, 45% distance $= \dfrac{45}{100} \times D = \dfrac{9}{20} D$ km

and 30% distance $= \dfrac{30}{100} \times D = \dfrac{3D}{10}$ km

$\therefore$ Required average speed $= \dfrac{D}{\dfrac{D}{4} \times \dfrac{1}{10} + \dfrac{9}{20} D \times \dfrac{1}{5} + \dfrac{3}{D} D \times \dfrac{1}{15}}$

$= \dfrac{D}{D\left(\dfrac{1}{40} + \dfrac{9}{100} + \dfrac{1}{50}\right)}$

$= \dfrac{1}{\dfrac{5 + 18 + 4}{200}} = \dfrac{200}{27} = 7.40$ km/h

Speed, Time and Distance / 427

Fast Track Method

Here, $P = 25\% = \dfrac{1}{4}$, $x = 10$ km/h, $Q = 45\% = \dfrac{45}{100} = \dfrac{9}{20}$, $y = 5$ km/h,

$R = 30\% = \dfrac{30}{100} = \dfrac{3}{10}$ and $z = 15$ km/h

∴ Required average speed $= \dfrac{1}{\dfrac{P}{x} + \dfrac{Q}{y} + \dfrac{R}{z}} = \dfrac{1}{\dfrac{1}{4 \times 10} + \dfrac{9}{20 \times 5} + \dfrac{3}{10 \times 15}}$

$= \dfrac{1}{\dfrac{1}{40} + \dfrac{9}{100} + \dfrac{1}{50}} = \dfrac{1}{\dfrac{5 + 18 + 4}{200}}$

$= \dfrac{200}{27} = 7.40$ km/h

Technique 4 When a person covers a certain distance between two places with speed 'a', but he reaches his destination late by time t_1 but when he covers the same distance with speed 'b', he reaches his destination t_2 time earlier. In this case, the distance between two places is given by

$$D = \dfrac{ab(t_1 + t_2)}{b - a}$$

Ex. 17 Aashutosh covers a certain distance between his home and college by cycle. Having an average speed of 30 km/h, he is late by 20 min. However, with a speed of 40 km/h, he reaches his college 10 min earlier. Find the distance between his house and college.

Sol. Let the required distance be x.

Difference between the time taken $= 20 + 10 = 30$ min $= \dfrac{30}{60}$ h

According to the question,

$\dfrac{x}{30} - \dfrac{x}{40} = \dfrac{30}{60} \Rightarrow \dfrac{x}{30} - \dfrac{x}{40} = \dfrac{1}{2} \Rightarrow 4x - 3x = 60$

∴ $x = 60$ km

Fast Track Method

Here, $a = 30$, $b = 40$, $t_1 = \dfrac{20}{60}$ and $t_2 = \dfrac{10}{60}$

According to the formula,

Required distance, $D = \dfrac{ab(t_1 + t_2)}{b - a}$

$= \dfrac{30 \times 40}{40 - 30} \times \left(\dfrac{20 + 10}{60}\right) = \dfrac{30 \times 40}{10} \times \dfrac{30}{60} = 60$ km

✦ t_1 time late and t_2 time earlier make a difference of $(t_1 + t_2)$.

Technique 5 When a person reaches a certain distance with speed 'a', he gets late by t_1 time and when he increases his speed by 'b' to cover the same distance, then he still gets late by t_2 time. In this case, the distance is calculated by $D = (t_1 - t_2)(a + b)\dfrac{a}{b}$.

Ex. 18 A boy walking at a speed of 20 km/h reaches his school 30 min late. Next time, he increases his speed by 4 km/h but still he is late by 10 min. Find the distance of the school from his home.

Sol. Let the required distance be x.

Here, the difference in time = $30 - 10 = 20$ min = $\frac{20}{60} = \frac{1}{3}$ h

Speed during next journey = $(20 + 4) = 24$ km/h

According to the question,

$$\frac{x}{20} - \frac{x}{24} = \frac{1}{3} \Rightarrow \frac{6x - 5x}{120} = \frac{1}{3} \Rightarrow x = 40 \text{ km}$$

Fast Track Method

Here, $a = 20$ km/h, $b = 4$ km/h, $t_1 = 30$ min and $t_2 = 10$ min

According to the formula, required distance = $(t_1 - t_2)(a + b)\frac{a}{b}$

$$= \frac{(30 - 10)}{60}(20 + 4)\frac{20}{4} = \frac{20}{60} \times 24 \times \frac{20}{4} = 40 \text{ km}$$

Technique 6 When two persons A and B travel distance D between two points P to Q, with speeds 'a' and 'b', respectively and B reaches Q first, returns immediately and meets A at R, then

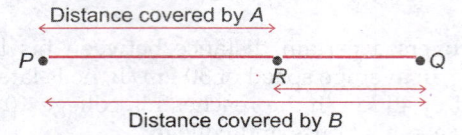

Distance travelled by A (PR) = $2 \times D\left(\frac{a}{a+b}\right)$

Distance travelled by B (PQ + QR) = $2 \times D\left(\frac{b}{a+b}\right)$

Ex. 19 Sonu and Monu travel from point P to Q, a distance of 42 km at 6 km/h and 8 km/h, respectively. Monu reaches Q first and returns immediately and meets Sonu at R. Find the distance from points P to R.

Sol. Here, speed of Sonu = 6 km/h and speed of Monu = 8 km/h

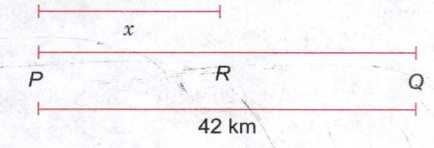

Let $PR = x$ km

According to the question,

Time taken by Sonu to cover PR = Time taken by Monu to cover $(PQ + QR)$

$\Rightarrow \quad \frac{x}{6} = \frac{42 + (42 - x)}{8} \Rightarrow 4x = 3(2 \times 42 - x)$

$\Rightarrow \quad 4x = 252 - 3x \Rightarrow 7x = 252 \Rightarrow x = 36$ km

Speed, Time and Distance / 429

Fast Track Method
Given that, $D = 42$ km, $a = 6$ km/h and $b = 8$ km/h
According to the formula,
Distance travelled by Sonu $= PR = 2D \times \dfrac{a}{a+b} = 2 \times 42 \times \dfrac{6}{6+8}$
$= 2 \times 42 \times \dfrac{6}{14} = 36$ km

Technique 7 A policeman sees a thief at a distance of d. He starts chasing the thief who is running at a speed of 'a' and policeman is chasing with a speed of 'b' $(b > a)$. In this case, the distance covered by the thief when he is caught by the policeman, is given by $d\left(\dfrac{a}{b-a}\right)$.

Ex. 20 A policeman sees a chain snatcher at a distance of 50 m. He starts chasing the chain snatcher who is running with a speed of 2 m/s, while the policeman chasing him with a speed of 4 m/s. Find the distance covered by the chain snatcher when he is caught by the policeman.

Sol. Let us assume that the time taken by the policeman to catch the thief be t hours.
Hence, after time t, the thief will be at a distance of $(t \times 2)$ m from where he started.
At the same time, he will be at a distance, if $(t \times 2 + 50)$ m from where the police started.
Now, distance run by the police in time $t = (t \times 4)$ m
According to the question,
$$4t = 2t + 50 \Rightarrow 4t - 2t = 50 \Rightarrow 2t = 50 \Rightarrow t = 25 \text{ s}$$
∴ Distance covered by the thief $= 2 \times t = 2 \times 25 = 50$ m

Fast Track Method
Here, $d = 50$ m, $a = 2$ m/s and $b = 4$ m/s
According to the formula, required distance $= d\left(\dfrac{a}{b-a}\right) = 50 \times \dfrac{2}{4-2} = 50$ m

Technique 8 Two persons A and B start running at the same time in opposite directions from two points and after passing each other, they complete their journeys in 'x' h and 'y' h, respectively.
Then, A's speed : B's speed $= \sqrt{y} : \sqrt{x}$

Ex. 21 A man sets out to cycle from points P to Q and at the same time, another man starts to cycle from points Q to P. After passing each other, they complete their journeys in 9 h and 4 h, respectively. Find the ratio of speeds of 1st man to that of 2nd man.

Sol. Let the total distance between P and Q be d km and time taken by P and Q to meet at R be t h.
Then, the speed of P and Q are S_p and S_q, respectively.

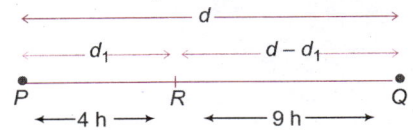

430 / Fast Track Objective Arithmetic

According to the question,
$$t = \frac{d_1}{S_p} = \frac{d - d_1}{S_q}$$

$\Rightarrow \qquad \dfrac{S_p}{S_q} = \dfrac{d_1}{d - d_1}$ ...(i)

and $d_1 = S_q \times 4$ and $d - d_1 = S_p \times 9$.
On putting the above values in Eq. (i), we get
$$\frac{S_p}{S_q} = \frac{4 S_q}{9 S_p} \Rightarrow \frac{S_p^2}{S_q^2} = \frac{4}{9}$$

$\Rightarrow \qquad \dfrac{S_p}{S_q} = \dfrac{2}{3}$ or $S_p : S_q = 2 : 3$

Fast Track Method

Given that, $x = 9$ h and $y = 4$ h
According to the formula,
1st man's speed : 2nd man's speed $= \sqrt{y} : \sqrt{x} = \sqrt{4} : \sqrt{9} = 2 : 3$

Technique 9 If a man changes his speed to $\left(\dfrac{x}{y}\right)$ of his usual speed and gets late by t min or reaches early by t min, then the usual time taken by him
$= \dfrac{tx}{(y - x)}$, if $(y > x)$ and $\dfrac{tx}{(x - y)}$, if $(x > y)$.

Ex. 22 If a man increases his speed to 7/5 times of his original speed and reaches his office 20 min before to fixed time, then find the usual time taken by him.

Sol. Let the original speed be s, distance be d and time taken be t.
Then, $\qquad d = st$ ...(i)
and $\qquad d = \dfrac{7}{5} s \, (t - 20)$

$\Rightarrow \qquad st = \dfrac{7}{5} s \, (t - 20)$

$\Rightarrow \qquad 5t = 7t - 140$

$\Rightarrow \qquad 2t = 140$

$\Rightarrow \qquad t = 70$ min

Fast Track Method

Given that, $\dfrac{x}{y} = \dfrac{7}{5}$, $x = 7$, $y = 5$ and $t = 20$ min

Now, required time $= \dfrac{t \times x}{(x - y)} = \dfrac{20 \times 7}{(7 - 5)} = \dfrac{20 \times 7}{2} = 70$ min $\qquad [\because x > y]$

Multi Concept
Questions

1. Distance between A and B is 72 km. Two men started walking from A and B at the same time towards each other. The person who started from A travelled uniformly with average speed 4 km/h while the other man travelled with varying speed as follows In first hour, his speed was 2 km/h, in the second hour, it was 2.5 km/h, in the third hour, it was 3 km/h and so on. When will they meet each other?
   (a) 7 h   (b) 10 h   (c) 35 km from A   (d) Midway between A and B

   ↪ (d) Both of them are moving in opposite direction, hence relative speed will be sum of their speeds. It means that in first hour, they will travel 6 km; in second hour, they will travel 6.5 km; in third hour, they will travel 7 km and so on.
   Now, relative speeds of both of them in consecutive hours forms an arithmetic progression, i.e. 6, 6.5, 7, 7.5, ..., so on
   where, $a = 6$ and $d = 6.5 - 6 = 0.5$, n = number of required hours to travel 72 km
   Now, according to the formula of AP,
   Total distance covered $= \frac{n}{2}[2a + (n-1)d] \Rightarrow 72 = \frac{n}{2}[2 \times 6 + (n-1)(0.5)]$
   $\Rightarrow \quad 144 = n\left[12 + \frac{(n-1)}{2}\right] \Rightarrow 144 = \frac{n[24 + n - 1]}{2}$
   $\Rightarrow \quad 288 = 24n + n^2 - n$
   $\Rightarrow \quad n^2 + 23n - 288 = 0$
   $\Rightarrow \quad (n-9)(n+32) = 0$
   $\Rightarrow \quad n = 9, n \neq -32$
   Hence, a distance of 72 km will be covered by them in 9 h.
   Now, the distance travelled by A in 9 h $= 9 \times 4 = 36$ km.
   Hence, both of them would meet midway between A and B.

2. Two cars start together in the same direction from the same place. The first goes with a uniform speed of 10 km/h. The second goes at a speed of 8 km/h in the first hour and increase the speed by $\frac{1}{2}$ km each succeeding hour. After how many hours will the second car overtake the first, if both go non-stop?
   (a) 9   (b) 5   (c) 7   (d) 8

   ↪ (a) Let the second car overtakes the first car after n h.
   ∴ Distance covered by first car = Distance covered by second car
   $\Rightarrow \quad 10n = 8 + \left(8 + \frac{1}{2}\right) + \left(8 + \frac{2}{2}\right) + ... + \left(8 + \frac{n-1}{2}\right)$
   $\Rightarrow \quad 10n = 8n + \frac{1}{2}[1 + 2 + ... + (n-1)]$
   $\Rightarrow \quad 10n = 8n + \frac{1}{2} \cdot \frac{n(n-1)}{2}$
   $\Rightarrow \quad 2n = \frac{1}{4}(n^2 - n)$
   $\Rightarrow \quad 8n = n^2 - n$
   $\Rightarrow \quad 8 = n - 1 \Rightarrow n = 9$
   So, the second car overtakes the first car after 9 h.

Fast Track Practice

Exercise 1 Base Level Questions

1. If speed of $3\frac{1}{3}$ m/s is converted to km/h, then it would be [SSC CGL 2012]
   (a) 8 km/h   (b) 9 km/h
   (c) 10 km/h   (d) 12 km/h

2. A missile travels at 1350 km/h. How many metres does it travel in one second? [SSC (10+2) 2017]
   (a) 369 m   (b) 375 m
   (c) 356 m   (d) 337 m

3. A car covers 300 km in 15 h. Find the speed of the car.
   (a) 20 km/h   (b) 25 km/h
   (c) 15 km/h   (d) 24 km/h
   (e) None of these

4. To cover a distance of 315 km in 2.8 h, what should be the average speed of the car? [SSC (10+2) 2017]
   (a) 112.5 m/s   (b) 56.25 m/s
   (c) 62.5 m/s   (d) 31.25 m/s

5. A bus covers a distance of 400 km with a speed of 20 km/h. What time is taken by the bus to cover this distance?
   (a) 25 h (b) 5 h (c) 21 h (d) 20 h
   (e) None of these

6. The speed of a bus is 72 km/h. The distance covered by the bus in 5 s is [SSC (10+2) 2012]
   (a) 50 m   (b) 74.5 m
   (c) 100 m   (d) 60 m

7. A person riding a bike crosses a bridge with a speed of 54 km/h. What is the length of the bridge, if he takes 4 min to cross the bridge? [Hotel Mgmt. 2010]
   (a) 3600 m   (b) 2800 m
   (c) 3500 m   (d) 4500 m

8. A man covered a distance of 12 km in 90 min by cycle. How much distance will he cover in 3 h, if he rides the cycle at a uniform speed? [Bank Clerk 2010]
   (a) 36 km   (b) 24 km
   (c) 30 km   (d) 27 km
   (e) None of these

9. Two men start together to walk a certain distance, one at 4 km/h and another at 3 km/h. The former arrives half an hour before the later. Find the distance. [SSC (10+2) 2012]
   (a) 6 km (b) 9 km (c) 8 km (d) 7 km

10. A bullock cart has to cover a distance of 80 km in 10 h. If it covers half of the journey in $\frac{3}{5}$th time, what should be its speed to cover the remaining distance in the left time?
    (a) 5 km/h   (b) 10 km/h
    (c) 15 km/h   (d) 18 km/h
    (e) 20 km/h

11. A car runs at the speed of 40 km/h when not serviced and runs at 65 km/h when serviced. After servicing, the car covers a certain distance in 5 h. How much approximate time will the car take to cover the same distance when not serviced? [Bank Clerk 2009]
    (a) 10 h (b) 7 h (c) 12 h (d) 8 h
    (e) 6 h

12. A certain distance is covered at a certain speed. If half of the distance is covered in double time, the ratio of the two speeds is [Bank Clerk 2011]
    (a) 4 : 1 (b) 1 : 4 (c) 1 : 2 (d) 2 : 1
    (e) 1 : 1

13. John started from A to B and Vinod from B to A. If the distance between A and B is 125 km and they meet at 75 km from A, what is the ratio of John's speed to that of Vinod's speed? [SSC CGL 2008]
    (a) 2 : 3   (b) 3 : 2
    (c) 4 : 3   (d) 5 : 4

14. The speeds of three cars are in the ratio of 2 : 3 : 5. Find the ratio of the time taken by the above cars to travel the same distance.
    (a) 15 : 10 : 6   (b) 6 : 10 : 15
    (c) 10 : 15 : 6   (d) 10 : 6 : 15
    (e) None of these

15. Moving 6/7 of its usual speed, a train is 10 min late. Find its usual time to cover the journey. [Hotel Mgmt. 2008]
    (a) 25 min   (b) 15 min
    (c) 35 min   (d) 60 min

16. A certain distance is covered at a certain speed. If half of this distance is covered in 4 times of the time, then find the ratio of the two speeds.
    (a) 1 : 8 (b) 1 : 4 (c) 4 : 1 (d) 8 : 1
    (e) None of these

Speed, Time and Distance / 433

17. The speeds of three cars are in the ratio 2 : 3 : 4. What is the ratio between the time taken by these cars to travel the same distance? **[CDS 2016 (II)]**
    (a) 4 : 3 : 2
    (b) 2 : 3 : 4
    (c) 4 : 3 : 6
    (d) 6 : 4 : 3

18. A is twice as fast as B and B is thrice as fast as C. The journey covered by C in 56 min will be covered by A in **[Bank PO 2010]**
    (a) $5\frac{1}{3}$ min
    (b) $2\frac{1}{3}$ min
    (c) $7\frac{1}{3}$ min
    (d) $9\frac{1}{3}$ min
    (e) None of these

19. The ratio of the speeds of A and B is 3 : 4. A takes 20 min more than the time taken by B to reach a particular place. Find the time taken by A and B, respectively to reach that place. **[LIC ADO 2010]**
    (a) 40 min and 30 min
    (b) 80 min and 60 min
    (c) 90 min and 45 min
    (d) 90 min and 50 min
    (e) None of the above

20. Nilu covers a distance by walking for 6 h, while returning, his speed decreases by 2 km/h and he takes 9 h to cover the same distance. What was her speed while returning?
    (a) 2 km/h
    (b) 5 km/h
    (c) 4 km/h
    (d) 7 km/h
    (e) None of these

21. A car covers a distance of 200 km in 2 h 40 min whereas a jeep covers the same distance in 2 h. What is the ratio of their speeds? **[Bank Clerk 2007]**
    (a) 3 : 4
    (b) 4 : 3
    (c) 4 : 5
    (d) 5 : 4
    (e) None of these

22. Two cars A and B start simultaneously from a certain place at the speed of 30 km/h and 45 km/h, respectively. The car B reaches the destination 2 h earlier than A. What is the distance between the starting point and destination? **[CDS 2013]**
    (a) 90 km
    (b) 180 km
    (c) 270 km
    (d) 360 km

23. A boy goes to his school from his house at a speed of 3 km/h and returns at a speed of 2 km/h. If he takes 5 h in going and coming, the distance between his house and school is **[SSC CGL 2016]**
    (a) 6 km
    (b) 5 km
    (c) 5.5 km
    (d) 6.5 km

24. Aashutosh can cover a certain distance in 84 min by covering 2/3rd of distance at 4 km/h and the rest at 5 km/h. Find the total distance.
    (a) 6 km (b) 8 km (c) 9 km (d) 15 km
    (e) None of these

25. Raj while going by bus from home to airport (without any halt) takes 20 min less than time taken when the bus halts for some time. The average speed is 8 km/h more than the average speed of the bus when it halts. If the distance from home to airport is 60 km, what is the speed of the bus when it is travelled? **[IBPS Clerk (Mains) 2017]**
    (a) 24 km/h
    (b) 46 km/h
    (c) 20 km/h
    (d) 49 km/h
    (e) 42 km/h

26. A bus travels at the rate of 54 km/h without stoppages and it travels at 45 km/h with stoppages. How many minutes does the bus stop on an average per hour? **[SSC CGL 2010]**
    (a) 8 (b) 10 (c) 12 (d) 4

27. A person can walk a certain distance and drive back in 6 h. He can also walk both ways in 10 h. How much time will he take to drive both ways? **[CDS 2013]**
    (a) 2 h
    (b) $2\frac{1}{2}$ h
    (c) $5\frac{1}{2}$ h
    (d) 4 h

28. The ratio of speeds of a train and a car is 16 : 15, respectively and a bus covered a distance of 480 km in 8 h. The speed of the bus is 3/4th of the speed of train. What distance will be covered by car in 6 h? **[Bank PO 2010]**
    (a) 450 km
    (b) 480 km
    (c) 360 km
    (d) Cannot be determined
    (e) None of the above

29. The ratio between the speeds of two buses is 5 : 3. If the 1st bus runs 400 km in 8 h, then find the speed of the 2nd bus. **[SSC (10+2) 2010]**
    (a) 30 km/h
    (b) 15 km/h
    (c) 27 km/h
    (d) 37 km/h

30. A man travels some distance at a speed of 12 km/h and returns at a speed of 9 km/h. If the total time taken by him is 2 h 20 min, the distance is **[SSC CGL (Mains) 2016]**
    (a) 35 km
    (b) 21 km
    (c) 9 km
    (d) 12 km

31. A car reached Raipur from Sonagarh in 35 min with an average speed of 69 km/h. If the average speed is increased by 36 km/h, how long will it take to cover the same distance?
   [Bank Clerk 2009]
   (a) 24 min   (b) 27 min
   (c) 23 min   (d) 29 min
   (e) None of these

32. A truck covers a distance of 368 km at a certain speed in 8 h. How much time would a car take at an average speed which is 18 km/h more than that of the speed of the truck to cover a distance which is 16 km more than that travelled by the truck?   [SBI Clerk 2012]
   (a) 7 h   (b) 5 h   (c) 6 h   (d) 8 h
   (e) None of these

33. Sumit drove at the speed of 45 km/h from home to a resort. Returning over the same route, he got stuck in traffic and took an hour longer. Also, he could drive only at the speed of 40 km/h. How many kilometres did he drive each way?
   [DMRC CRA 2012]
   (a) 250   (b) 360
   (c) 375   (d) None of these

34. Ram and Shyam are moving in the directions opposite to each other. The speeds of both persons are 10 km/h and 6 km/h, respectively. Find the speed of Ram with respect of Shyam.
   (a) 6 km/h   (b) 16 km/h
   (c) 4 km/h   (d) 8 km/h
   (e) None of these

35. A and B are 15 km apart and when travelling towards each other, meet after half an hour whereas they meet two and a half hours later, if they travel in the same direction. The faster of the two travels at the speed of
   [SSC CGL (Mains) 2016]
   (a) 15 km/h   (b) 18 km/h
   (c) 10 km/h   (d) 8 km/h

36. Two donkeys are standing 400 m apart. First donkey can run at a speed of 3 m/s and the second can run at 2 m/s. If two donkeys run towards each other, after how much time will they bump into each other?   [SSC CGL (Mains) 2016]
   (a) 60 s   (b) 80 s
   (c) 400 s   (d) 40 s

37. Raj and Prem walk in opposite directions at the rate of 3 and 2 km/h, respectively. How far will they be from each other after 2 h?   [SSC CGL 2015]
   (a) 8 km   (b) 10 km
   (c) 2 km   (d) 6 km

38. The distance between two places A and B is 110 km. 1st car departs from place A to B, at a speed of 40 km/h at 11 am, 2nd car departs from place B to A at a speed of 50 km/h at 1 pm. At what time will both the cars meet each other?   [IBPS SO 2016]
   (a) 1 : 50 pm   (b) 1 : 20 pm
   (c) 2 : 00 pm   (d) 2 : 30 pm
   (e) 2 : 15 pm

39. A car travels a distance of 75 km at the speed of 25 km/h. It covers the next 25 km of its journey at the speed of 5 km/h and the last 50 km of its journey at the speed of 25 km/h. What is the average speed of the car?   [Bank Clerk 2008]
   (a) 40 km/h   (b) 25 km/h
   (c) 15 km/h   (d) 12.5 km/h
   (e) None of these

40. A person covers 9 km with a speed of 3 km/h, 25 km with a speed of 5 km/h and 30 km with a speed of 10 km/h. Find out the average speed of person.
   [LIC ADO 2007]
   (a) $5\frac{9}{11}$ km/h   (b) $11\frac{5}{9}$ km/h
   (c) $9\frac{5}{11}$ km/h   (d) $5\frac{5}{11}$ km/h
   (e) None of these

41. The average speed of a car is 75 km/h. The driver first decreases its average speed by 40% and then increases it by 50%. What is the new average speed now?   [Hotel Mgmt. 2010]
   (a) 67.5 km/h   (b) 60 km/h
   (c) 90 km/h   (d) 60.5 km/h

42. Two cities A and B are 360 km apart. A car goes from A to B with a speed of 40 km/h and returns to A with a speed of 60 km/h. What is the average speed of the car?   [UPSC CSAT 2015]
   (a) 45 km/h   (b) 48 km/h
   (c) 50 km/h   (d) 55 km/h

43. A car covers a distance from town A to town B at the speed of 58 km/h and covers the distance from town B to town A at the speed of 52 km/h. What is the approximate average speed of the car?
   [Bank Clerk 2009]
   (a) 55 km/h   (b) 52 km/h
   (c) 48 km/h   (d) 60 km/h
   (e) None of these

44. A man covers half of his journey at 6 km/h and the remaining half at 3 km/h. Find his average speed. [SSC (10+2) 2007]
    (a) 3 km/h (b) 4 km/h
    (c) 4.5 km/h (d) 9 km/h

45. A person goes from one point to another point with a speed of 5 km/h and comes back to starting point with a speed of 3 km/h. Find the average speed for the whole journey. [SSC CGL 2010]
    (a) 4.5 km/h (b) 4 km/h
    (c) 4.25 km/h (d) 3.75 km/h

46. If you travel 39 km at a speed of 26 km/h, another 39 km at a speed of 39 km/h and again 39 km at a speed of 52 km/h, what is your average speed for the entire journey? [SSC CGL 2008]
    (a) 39 km/h (b) 37 km/h
    (c) 33.33 km/h (d) 36 km/h

47. A man divides his total route of journey into three equal parts and decides to travel the three parts with the speed of 15 km/h, 10 km/h and 5 km/h, respectively. Find his average speed during journey.
    (a) 2.28 km/h (b) 9 km/h
    (c) 14 km/h (d) 8.18 km/h
    (e) None of these

48. A car travels the first one-third of a certain distance with a speed of 10 km/h, the next one-third distance with a speed of 20 km/h and the last one-third distance with a speed of 60 km/h. The average speed of the car for the whole journey is [CDS 2015 (I)]
    (a) 18 km/h (b) 24 km/h
    (c) 30 km/h (d) 36 km/h

49. Mr. Bundda travels in his car and covers 1/4 part of his journey with 8 km/h, 3/5 part with 6 km/h and remaining 3/20 part with a speed of 10 km/h. Find out his average speed during the whole journey.
    (a) 6.83 km/h (b) 9 km/h
    (c) 4 km/h (d) 8.5 km/h
    (e) None of these

50. If a man walks at the rate of 5 km/h, he misses a train by 7 min. However, if he walks at the rate of 6 km/h, he reaches the station 5 min before the arrival of the train. The distance covered by him to reach the station is [SSC CGL (Mains) 2015]
    (a) 6 km (b) 4 km
    (c) 7 km (d) 6.25 km

51. A student walks from his house at $2\frac{1}{2}$ km/h and reaches his school late by 6 min. Next day, he increases his speed by 1 km/h and reaches 6 min before school time. How far is the school from his house? [SSC CGL 2007]
    (a) $\frac{5}{4}$ km (b) $\frac{7}{4}$ km
    (c) $\frac{9}{4}$ km (d) $\frac{11}{4}$ km

52. Rubi goes to a multiplex at the speed of 3 km/h to see a movie and reaches 5 min late. If she travels at the speed of 4 km/h, she reaches 5 min early. Then, the distance of the multiplex from her starting point is [SSC CGL (Mains) 2016]
    (a) 2 km (b) 5 km
    (c) 2 m (d) 5 m

53. A person travels a certain distance at 3 km/h and reaches 15 min late. If he travels at 4 km/h, he reaches 15 min earlier. The distance he has to travel is
    (a) 4.5 km (b) 6 km [CDS 2013]
    (c) 7.2 km (d) 12 km

54. A man riding a bicycle from his house at 10 km/h and reaches his office late by 6 min. He increases his speed by 2 km/h and reaches 6 min before. How far is the office from his house? [SSC CGL 2012]
    (a) 6 km (b) 7 km
    (c) 12 km (d) 16 km

55. A boy walking at a speed of 15 km/h reaches his school 20 min late. Next time, he increases his speed by 5 km/h but still he late by 5 min. Find the distance of the school from his home.
    (a) 5 km (b) 10 km
    (c) 15 km (d) 20 km

56. Two men A and B travel from point P to Q, a distance of 84 km at 12 km/h and 16 km/h, respectively. B reaches Q and returns immediately and meets A at R. Find the distance from P to R.
    (a) 72 km (b) 76 km
    (c) 78 km (d) 68 km
    (e) None of these

57. A thief is spotted by a policeman from a distance of 100 m. When the policeman starts the chase, the thief also starts running. If the speed of the thief is 8 km/h and that of the policeman is 10 km/h, then how far will the thief have to run before he is overtaken? [CDS 2017 (I)]
    (a) 200 m (b) 300 m
    (c) 400 m (d) 500 m

58. Dalbir Singh, a policeman, is 114 m behind a thief. Dalbir Singh runs 21 m and the thief 15 m in a minute. In what time will Dalbir Singh catch the thief?
(a) 19 min (b) 16 min
(c) 21 min (d) 23 min

59. A person sets out to cycle from A to B and at the same time, another person starts from B to A. After passing each other, they complete their journeys in 16 h and 25 h, respectively. Find the ratio of speeds of the 1st man to that of the 2nd man.
(a) 5 : 4 (b) 5 : 3 (c) 4 : 5 (d) 3 : 5
(e) None of these

60. A bus driver decreases his bus speed to $\frac{7}{9}$ times of his original speed and reaches his office 40 min late to fixed time. Find the usual time taken by him.
(a) 140 min (b) 120 min
(c) 100 min (d) 180 min

Exercise 2 *Higher Skill Level Questions*

1. A man started 20 min late and travelling at a speed of $1\frac{1}{2}$ times of his usual speed, reaches his office in time. The time taken by the man to reach his office at his usual speed is **[SSC CGL (Mains) 2012]**
(a) 40 min (b) 1 h 20 min
(c) 1 h (d) 30 min

2. In a flight of 600 km, an aircraft was slowed down due to bad weather. Its average speed for the trip was reduced by 200 km/h and the time of flight increased by 30 min. The duration of the flight is **[CDS 2015 (I)]**
(a) 1 h (b) 2 h (c) 3 h (d) 4 h

3. With a uniform speed, a car covers a distance in 8 h. Had the speed been increased by 4 km/h, the same distance could have been covered in 7 h 30 min. What is the distance covered?
[CDS 2015 (I)]
(a) 420 km (b) 480 km
(c) 520 km (d) 640 km

4. Car M takes 5 h to travel from point A to B. It would have taken 6 h, if the same car had travelled the same distance at a speed which was 15 km/h less than its original speed. What is the distance between points A and B?
[SBI PO (Pre) 2017]
(a) 350 km (b) 300 km
(c) 450 km (d) 420 km
(e) 400 km

5. A cyclist moves non-stop from A to B, a distance of 14 km, at a certain average speed. If his average speed reduces by 1 km/h, he takes 20 min more to cover the same distance. The original average speed of the cyclist is **[CDS 2016 (I)]**
(a) 5 km/h (b) 6 km/h
(c) 7 km/h (d) None of these

6. Two persons A and B start simultaneously from two places c km apart and walk in the same direction. If A travels at the rate of p km/h and B travels at the rate of q km/h, then A has travelled before he overtakes B a distance of **[CDS 2015 (I)]**
(a) $\frac{qc}{p+q}$ km (b) $\frac{pc}{p-q}$ km
(c) $\frac{qc}{p-q}$ km (d) $\frac{pc}{p+q}$ km

7. A man decides to travel 80 km in 8 h partly by foot and partly on a bicycle. If his speed on foot is 8 km/h and on bicycle is 16 km/h, what distance would he travel on foot? **[IB ACIO 2012]**
(a) 20 km (b) 30 km
(c) 48 km (d) 60 km

8. A man covers a certain distance on scooter. Had he moved 3 km/h faster, he would have taken 40 min less. If he had moved 2 km/h slower, he would have taken 40 min more. The distance (in km) is **[SSC Multitasking 2013]**
(a) 42.5 (b) 36 (c) 37.5 (d) 40

9. Shantanu started cycling along the boundaries of a square field ABCD from corner point A. After 1/2 h, he reached the corner point C, diagonally opposite to point A. If his speed was 16 km/h, find the area of the field.
(a) 8 sq km (b) 9 sq km
(c) 32 sq km (d) 19 sq km
(e) None of these

10. Two men A and B run 4 km race on a course 0.25 km round. If their speeds are in the ratio 5 : 4, how often does the winner pass the another? **[CDS 2016 (II)]**
(a) Once (b) Twice
(c) Thrice (d) Four times

Speed, Time and Distance / 437

11. A car driver covers a distance between two cities at a speed of 60 km/h and on the return, his speed is 40 km/h. He goes again from the 1st to the 2nd city at twice the original speed and returns at half the original return speed. Find his average speed for the entire journey.
[LIC AAO 2007]
(a) 55 km/h (b) 50 km/h
(c) 48 km/h (d) 40 km/h
(e) None of these

12. A car starts running with the initial speed of 40 km/h with its speed increasing every hour by 5 km/h. How many hours will it take to cover a distance of 385 km? [SSC Multitasking 2014]
(a) $8\frac{1}{2}$ h (b) $9\frac{1}{2}$ h (c) 9 h (d) 7 h

13. A thief is noticed by a policeman from a distance of 200 m. The thief starts running and the policeman chases him. The thief and the policeman run at the speed of 10 km/h and 11 km/h, respectively. What is the distance between them after 6 min? [SSC CGL 2013]
(a) 100 m (b) 120 m
(c) 150 m (d) 160 m

14. In a journey of 48 km performed by Rickshaw, Bus and Auto, the distance covered by the three ways in that order is in the ratio of 8 : 1 : 3 and charges per kilometre in that order are in the ratio of 8 : 1 : 4. If the Rickshaw charges being 24 paise per km, what is the total cost of the journey? [MAT 2015]
(a) ₹ 9.24 (b) ₹ 10.00
(c) ₹ 12.00 (d) None of these

15. A started at 8 : 30 am from a place P go to a place Q. 30 min later, B left P for Q and caught up with A at 11 am. B reached Q at 12 noon. A will reach Q at [BSSC CGL 2014]
(a) 12 : 30 pm (b) 12 :15 pm
(c) 1 : 00 pm (d) 12 : 20 pm

Answer with Solutions

Exercise 1 Base Level Questions

1. (d) $\because 1$ m/s $= \frac{18}{5}$ km/h
$\therefore 3\frac{1}{3}$ m/s $= \frac{10}{3}$ m/s $= \frac{10}{3} \times \frac{18}{5}$ km/h
$= 12$ km/h

2. (b) We know that to convert km/h in m/s, multiply it by $\frac{5}{18}$.
$\therefore 1350$ km/h $= 1350 \times \frac{5}{18}$ m/s $= 375$ m/s
Hence, 375 m/s means that a missile travels a distance of 375 m in one second.

3. (a) We know that,
Speed $= \frac{\text{Distance}}{\text{Time}}$
$\therefore$ Required speed $= \frac{300}{15} = 20$ km/h

4. (d) Total distance = 315 km = 315000 m
Total time = $2.8 \times 60 \times 60$ s
$\therefore$ Average speed $= \frac{\text{Total distance}}{\text{Total time}}$
$= \frac{315000}{2.8 \times 60 \times 60}$
$= 31.25$ m/s

5. (d) We know that,
Required time $= \frac{\text{Distance}}{\text{Speed}} = \frac{400}{20} = 20$ h

6. (c) Given, speed of bus = 72 km/h
$= 72 \times \frac{5}{18} = 20$ m/s
$\therefore$ Distance travelled in 5 s
$= 5 \times 20 = 100$ m

7. (a) Length of the bridge = Distance travelled by the person in 4 min
= Speed × Time
Speed = 54 km/h $= 54 \times \frac{5}{18} = 3 \times 5 = 15$ m/s
Time = 4 min = $4 \times 60 = 240$ s
$\therefore$ Required length = $15 \times 240 = 3600$ m

8. (b) We know that,
Speed $= \frac{\text{Distance}}{\text{Time}} = \frac{12}{\frac{90}{60}} = \frac{12 \times 60}{90} = 8$ km/h
$\therefore$ Distance covered in 3 h = Speed × Time
$= 8 \times 3 = 24$ km

9. (a) Let total distance = x km
According to the question,
$\frac{x}{3} - \frac{x}{4} = \frac{1}{2} \Rightarrow \frac{x}{12} = \frac{1}{2} \Rightarrow x = 6$ km

10. (b) Total distance covered in 10 h = 80 km
But it covers 40 km in $\frac{3}{5}$ th of time.
i.e. $\frac{3}{5} \times 10 = 6$ h
$\therefore$ Required time = $10 - 6 = 4$ h
and remaining distance = 40 km
Thus, required speed = $\frac{40}{4} = 10$ km/h

11. (d) When serviced, the distance covered by the car = $65 \times 5 = 325$ km.
When not serviced, the time taken by the car to cover 325 km
$= \frac{325}{40} = 8.125 = 8$ h (approx.)

12. (a) Let x km distance be covered in y h.
So, speed in first case = $\frac{x}{y}$ km/h
and speed in second case = $\frac{x/2}{2y} = \frac{x}{4y}$ km/h
$\therefore$ Required ratio = $\frac{x}{y} : \frac{x}{4y} = 1 : \frac{1}{4} = 4 : 1$

13. (b) John's speed : Vinod's speed
$= 75 : (125 - 75) = 75 : 50 = 3 : 2$

14. (a) We know that speed and time are inversely proportional.
$\therefore$ Ratio of time taken
$= \frac{1}{2} : \frac{1}{3} : \frac{1}{5} = \frac{30}{2} : \frac{30}{3} : \frac{30}{5}$
[$\because$ LCM of 2, 3, 5 = 30]
$= 15 : 10 : 6$

15. (d) New speed = $\frac{6}{7}$ of usual speed
$\left[\because \text{speed} \propto \frac{1}{\text{time}}\right]$
Now, time taken = $\frac{7}{6}$ of usual time
$\Rightarrow \left(\frac{7}{6} \text{ of the usual time}\right)$
$-$ (usual time) = 10 min
$\Rightarrow \frac{1}{6}$ of the usual time = 10 min
$\therefore$ Usual time = 60 min

16. (d) Let x km be covered in y h.
Then, first speed = $\frac{x}{y}$ km/h
Again, $\frac{x}{2}$ km is covered in $4y$ h.

$\therefore$ New speed = $\left(\frac{x}{2} \times \frac{1}{4y}\right) = \left(\frac{x}{8y}\right)$ km/h
Ratio of speeds = $\frac{x}{y} : \frac{x}{8y} = 1 : \frac{1}{8} = 8 : 1$

17. (d) Let the speed of three cars be $2x, 3x$ and $4x$ respectively.
Time taken by these cars to travel a distance of D are $\frac{D}{2x}, \frac{D}{3x}$ and $\frac{D}{4x}$, respectively.
$\therefore$ Ratio between time taken by these cars
$= \frac{D}{2x} : \frac{D}{3x} : \frac{D}{4x} = \frac{1}{2} : \frac{1}{3} : \frac{1}{4}$
$= 6 : 4 : 3$

18. (d) Let time taken by $A = y$
Let speed of $C = x$
Then, speed of $B = 3x$
$\therefore$ Speed of $A = 6x$
Now, ratio of speeds of A and C
= Ratio of time taken by C and A
$\Rightarrow 6x : x = 56 : y \Rightarrow \frac{6x}{x} = \frac{56}{y}$
$\therefore y = \frac{56}{6} = 9\frac{2}{6} = 9\frac{1}{3}$ min

19. (b) Ratio of speeds of A and $B = 3 : 4$
Let time taken by A and B be $t + 20$ and t, respectively.
Then, according to the question,
$\frac{t}{t+20} = \frac{3}{4} \Rightarrow 4t = 3t + 60 \Rightarrow t = 60$ min
Hence, time taken by $A = 60 + 20 = 80$ min
and time taken by $B = 60$ min

20. (c) Let the speed in return journey = x
According to the question,
$6(x+2) = 9x \Rightarrow 6x + 12 = 9x$
$\Rightarrow 9x - 6x = 12 \Rightarrow 3x = 12$
$\therefore x = \frac{12}{3} = 4$ km/h

21. (a) Speed of car = $\frac{\text{Distance}}{\text{Time}} = \frac{200}{2\frac{40}{60}}$
$= \frac{600}{8} = 75$ km/h
Speed of jeep = $\frac{200}{2} = 100$ km/h
$\therefore$ Required ratio = $\frac{75}{100} = \frac{3}{4} = 3 : 4$

Alternate Method
Speed of car : Speed of jeep
= Time taken by jeep : Time taken by car
= 120 min : 160 min = 3 : 4

22. (b) Let the time taken by car B to reach destination be x h.

Speed, Time and Distance / 439

Then, time taken by car A to reach destination is $(x + 2)$ h.
Now, $S_1T_1 = S_2T_2$
$\Rightarrow \quad 30 \times (x + 2) = 45 \times x$
$\Rightarrow \quad 30x + 60 = 45x \Rightarrow 15x = 60$
$\therefore \quad x = 4$ h
Now, distance between starting point and destination = $S_2T_2 = 45 \times 4 = 180$ km

23. (a) Let the distance between the school and house be d km and time taken to reach school from house be t h.
According to the question,
$3 \times t = 2 \times (5 - t) \Rightarrow 3t = 10 - 2t$
$\Rightarrow \quad 5t = 10 \Rightarrow t = 2$ h
$\therefore \quad d = 3 \times t = 3 \times 2 = 6$ km

24. (a) Let the total distance = x
Then, according to the question,
$$\frac{\frac{2}{3}x}{4} + \frac{\left(1 - \frac{2}{3}\right)x}{5} = \frac{84}{60}$$
$\Rightarrow \quad \frac{\frac{2}{3}x}{4} + \frac{\frac{1}{3}x}{5} = \frac{84}{60} \Rightarrow \frac{x}{6} + \frac{x}{15} = \frac{84}{60}$
$\Rightarrow \quad \frac{5x + 2x}{30} = \frac{84}{60} \Rightarrow 7x = \frac{84}{60} \times 30$
$\Rightarrow \quad 7x = 42 \Rightarrow x = \frac{42}{7} = 6$ km

25. (e) Let speed of bus (without halt) = x km/h
Then, $\quad T_1 = \frac{d}{x}$ [without halt]
$\Rightarrow \quad T_1 = \frac{60}{x}$ ...(i)
and $\quad T_2 = \frac{d}{x - 8}$ [with halt]
$\Rightarrow \quad T_2 = \frac{60}{x - 8}$ ...(ii)
Now, according to the question,
$T_2 - T_1 = 20$ min
$\Rightarrow \quad \frac{60}{x - 8} - \frac{60}{x} = \frac{20}{60} \Rightarrow 60\left(\frac{1}{x - 8} - \frac{1}{x}\right) = \frac{1}{3}$
$\Rightarrow \quad \frac{1}{x - 8} - \frac{1}{x} = \frac{1}{180} \Rightarrow \frac{x - x + 8}{x(x - 8)} = \frac{1}{180}$
$\Rightarrow \quad x^2 - 8x = 1440$
$\Rightarrow \quad x^2 - 8x - 1440 = 0$ ...(iii)
Now, solving Eq. (iii), we have
$x \approx 42$
Hence, speed of bus = 42 km/h

26. (b) Due to stoppages, bus covers 9 km less per hour.
Time taken to cover 9 km
$= \left(\frac{9}{54} \times 60\right) = 10$ min

Hence, the train stops on an average 10 min per hour.

27. (a) Since, he takes 10 h to walk both ways, therefore the number of hours to walk one way for him will be 5. If he walks one way and drives back the same way, it takes him 6 h which means that the number of hours taken to drive one way is $6 - 5 = 1$ h.
Thus, number of hours to drive both ways will be 2.
Alternate Method
Given that, $\quad W + D = 6$ ...(i)
[$\because W$ = time taken while walking and D = time taken while driving]
Also, $2W = 10 \Rightarrow W = 5$
From Eq. (i), $5 + D = 6 \Rightarrow D = 1$
$2D = 2 \times 1 = 2$
$\therefore$ He will take 2 h to drive both ways.

28. (a) Let speed of train = $16x$
Speed of car = $15x$
Speed of bus = $\frac{480}{8} = 60$ km/h
According to the question,
$60 = (16 x) \times \frac{3}{4} \Rightarrow x = 5$
$\therefore$ Speed of car = $15x = 15 \times 5 = 75$ km/h
$\therefore$ Required distance = Speed $\times$ Time
$= 75 \times 6 = 450$ km

29. (a) Let speed of 1st bus = $5x$
and speed of 2nd bus = $3x$
According to the question,
$5x = \frac{400}{8} = 50 \Rightarrow x = \frac{50}{5} = 10$
$\therefore$ Speed of 2nd bus = $3x = 3 \times 10 = 30$ km/h

30. (d) Let total distance covered by man = d
and time = 2 h 20 min
$= 120 + 20 = 140$ min
$= \frac{140}{60}$ h $= \frac{7}{3}$ h
According to the question,
$\frac{d}{12} + \frac{d}{9} = \frac{7}{3}$
$\Rightarrow \quad \frac{3d + 4d}{36} = \frac{7}{3} \Rightarrow \frac{7d}{36} = \frac{7}{3}$
$\Rightarrow \quad d = \frac{7 \times 36}{3 \times 7} = 12$ km
Hence, distance covered by man = 12 km

31. (c) Distance between Sonagarh and Raipur
= Average speed $\times$ Time
$= \frac{69 \times 35}{60}$ km $= \frac{161}{4}$ km
New speed = $(69 + 36)$ km/h = 105 km/h
$\therefore$ Required time = $\frac{\text{Distance}}{\text{Speed}} = \frac{161}{4 \times 105}$ h
$= \frac{161 \times 60}{4 \times 105}$ min = 23 min

32. (c) Speed of truck = $\dfrac{\text{Distance}}{\text{Time}} = \dfrac{368}{8}$
= 46 km/h
Now, speed of car
= (Speed of truck + 18) km/h
= (46 + 18) = 64 km/h
and distance travelled by car
= 368 + 16 = 384 km
∴ Time taken by car = $\dfrac{\text{Distance}}{\text{Speed}} = \dfrac{384}{64} = 6$ h

33. (b) Let distance he drove be x.
From given condition,
$\dfrac{x}{40} - \dfrac{x}{45} = 1 \Rightarrow \dfrac{45x - 40x}{40 \times 45} = 1$
$\Rightarrow \quad 5x = 40 \times 45 \Rightarrow x = 360$ km

34. (b) We know that for relative speed, the two speeds will be added, if the motions of the two objects are in opposite directions.
∴ Required speed = 10 + 6 = 16 km/h

35. (b) Let speed of $A = a$ km/h
and speed of $B = b$ km/h
When both are in same direction, then
Relative speed = $\dfrac{\text{Distance}}{\text{Time}}$
$\Rightarrow \quad a - b = \dfrac{15}{\frac{5}{2}} \Rightarrow a - b = 6 \quad …(i)$
When both are in opposite direction, then
Relative speed = $\dfrac{\text{Distance}}{\text{Time}}$
$\Rightarrow \quad a + b = \dfrac{15}{\frac{1}{2}} \Rightarrow a + b = 30 \quad …(ii)$
Solving Eqs. (i) and (ii), we get
$a = 18$ km/h and $b = 12$ km/h
Hence, speed of faster man = 18 km/h

36. (b) Distance = 400 m
and relative speed of both the donkeys
= 3 + 2 = 5 m/s
∴ Time to meet both the donkeys
= $\dfrac{400}{5} = 80$ s

37. (b) ∵ Speed of Raj = 3 km/h
and speed of Prem = 2 km/h
∵ Raj and Prem are walking in opposite directions.
∴ Relative speed = 3 + 2 = 5 km/h
Distance covered = Speed × Time
= 5 × 2 = 10 km

38. (b)

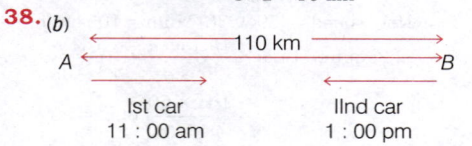

A ←——————110 km——————→ B
Ist car   IInd car
11 : 00 am   1 : 00 pm

Total time from 11 : 00 am to 1 : 00 pm = 2 h
Total distance covered by Ist car upto
1 : 00 pm = 40 × 2 = 80 km
∴ Distance between both the cars
= 110 − 80 = 30 km
Now, meeting time of both cars
= $\dfrac{\text{Distance between both the cars}}{\text{Average speed of both cars}}$
= $\dfrac{30}{40 + 50} = \dfrac{30}{90} = \dfrac{3}{9} = \dfrac{1}{3}$ h
= $\dfrac{1}{3} \times 60$ min = 20 min
Hence, both car will meet at
1 : 00 + 0 : 20 = 1 : 20 pm

39. (c) Time taken to cover first 75 km
= $\dfrac{75}{25} = 3$ h
Time taken to cover next 25 km = $\dfrac{25}{5} = 5$ h
Time taken to cover last 50 km of its journey = $\dfrac{50}{25} = 2$ h
Total distance = 75 + 25 + 50 = 150 km
Total time taken = 3 + 5 + 2 = 10 h
∴ Required average speed
= $\dfrac{\text{Total distance}}{\text{Total time taken}} = \dfrac{150}{10} = 15$ km/h

40. (a) Here, $D_1 = 9$ km, $D_2 = 25$ km, $D_3 = 30$ km,
$S_1 = 3$ km/h, $S_2 = 5$ km/h and $S_3 = 10$ km/h
∴ Required average speed
= $\dfrac{D_1 + D_2 + D_3}{\dfrac{D_1}{S_1} + \dfrac{D_2}{S_2} + \dfrac{D_3}{S_3}} = \dfrac{9 + 25 + 30}{\dfrac{9}{3} + \dfrac{25}{5} + \dfrac{30}{10}}$
= $\dfrac{64}{3 + 5 + 3} = \dfrac{64}{11} = 5\dfrac{9}{11}$ km/h

41. (a) Required average speed
= $75 \times \dfrac{60}{100} \times \dfrac{150}{100} = 75 \times \dfrac{3}{5} \times \dfrac{3}{2}$
= 67.5 km/h

42. (b) Given, distance between the cities A and B = 360 km
Time taken to cover 360 km with the speed of 40 km/h = $\dfrac{360}{40}$ h = 9 h
and time taken to cover 360 km with the speed of 60 km/h = $\dfrac{360}{60}$ h = 6 h
∴ Average speed of the cars
= $\dfrac{\text{Total distance covered}}{\text{Total time taken}}$
= $\dfrac{360 + 360}{9 + 6}$
= $\dfrac{720}{9 + 6} = \dfrac{720}{15} = 48$ km/h

Speed, Time and Distance / 441

Fast Track Method
Here, $x = 40$ km/h and $y = 60$ km/h
∴ Average speed $= \dfrac{2xy}{x+y}$ [by Technique 1]
$= \dfrac{2 \times 40 \times 60}{40 + 60} = \dfrac{80 \times 60}{100} = 48$ km/h

43. (a) We know that, if two equal distances are covered at two different speeds A and B, then
Average speed $= \dfrac{2AB}{A+B} = \dfrac{2 \times 58 \times 52}{58 + 52}$
[by Technique 1]
$= 54.8$ km/h $= 55$ km/h (approx.)

44. (b) Given, $A = 6$ km/h and $B = 3$ km/h
According to the formula,
Average speed $= \dfrac{2AB}{A+B}$ [by Technique 1]
∴ Required average speed
$= \dfrac{2 \times 6 \times 3}{6 + 3} = \dfrac{36}{9} = 4$ km/h

45. (d) Average speed $= \dfrac{2AB}{A+B}$
$= \dfrac{2 \times 5 \times 3}{5 + 3}$ [by Technique 1]
$= \dfrac{30}{8} = 3.75$ km/h

46. (d) Required average speed
$= \dfrac{\text{Total distance covered}}{\text{Total time taken}}$
$= \dfrac{3 \times 39}{\dfrac{39}{26} + \dfrac{39}{39} + \dfrac{39}{52}} = \dfrac{3 \times 39}{\dfrac{3}{2} + 1 + \dfrac{3}{4}}$
$= \dfrac{3 \times 39}{\dfrac{6 + 4 + 3}{4}} = \dfrac{3 \times 39 \times 4}{13}$
$= 3 \times 3 \times 4 = 36$ km/h

Fast Track Method
Here, $A = 26$ km/h, $B = 39$ km/h, $C = 52$ km/h
Then, speed $= \dfrac{3 \times 26 \times 39 \times 52}{26 \times 39 + 39 \times 52 + 26 \times 52}$
[by Technique 2]
$= \dfrac{3 \times 26 \times 39 \times 52}{1014 + 2028 + 1352}$
$= \dfrac{3 \times 26 \times 39 \times 52}{4394} = 36$ km/h

47. (d) We know that, if three equal distances are covered at different speeds of A, B and C km/h respectively, then average speed during whole journey
$= \dfrac{3ABC}{AB + BC + CA}$ km/h [by Technique 2]
$= \dfrac{3 \times 15 \times 10 \times 5}{15 \times 10 + 10 \times 5 + 15 \times 5}$
$= \dfrac{3 \times 15 \times 10 \times 5}{150 + 50 + 75} = \dfrac{3 \times 15 \times 10 \times 5}{275}$
$= 8.18$ km/h

48. (a) Let total distance of AB be x km.

For distance AC,
$t_1 = \dfrac{x/3}{10} = \dfrac{x}{30}$ h $\left[\because \text{time} = \dfrac{\text{distance}}{\text{speed}}\right]$

For distance CD, $t_2 = \dfrac{x/3}{20} = \dfrac{x}{60}$ h

For distance DB, $t_3 = \dfrac{x/3}{60} = \dfrac{x}{180}$ h

Total time taken
$= t_1 + t_2 + t_3 = \dfrac{x}{30} + \dfrac{x}{60} + \dfrac{x}{180}$
$= \dfrac{6x + 3x + x}{180} = \dfrac{10x}{180} = \dfrac{x}{18}$

∴ Average speed $= \dfrac{\text{Total distance}}{\text{Total time taken}}$
$= \dfrac{x}{\dfrac{x}{18}} = 18$ km/h

Fast Track Method
We know that, if three equal distances are covered at different speeds of A, B and C km/h, then average speed during whole journey $= \dfrac{3ABC}{AB + BC + CA}$ km/h
[by Technique 2]
Here, $A = 10$, $B = 20$, $C = 60$
∴ Required average speed
$= \dfrac{3 \times 10 \times 20 \times 60}{10 \times 20 + 20 \times 60 + 10 \times 60}$
$= \dfrac{3 \times 12000}{200 + 1200 + 600} = \dfrac{3 \times 12000}{2000} = 18$ km/h

49. (a) Here, $P = \dfrac{1}{4}$, $Q = \dfrac{3}{5}$, $R = \dfrac{3}{20}$
and $x = 8$ km/h, $y = 6$ km/h, $z = 10$ km/h
∴ Required average speed $= \dfrac{1}{\dfrac{P}{x} + \dfrac{Q}{y} + \dfrac{R}{z}}$
[by Technique 3]
$= \dfrac{1}{\dfrac{1}{4 \times 8} + \dfrac{3}{5 \times 6} + \dfrac{3}{20 \times 10}}$
$= \dfrac{1}{\dfrac{1}{32} + \dfrac{1}{10} + \dfrac{3}{200}} = \dfrac{1}{\dfrac{25 + 80 + 12}{800}}$
$= \dfrac{800}{117} = 6.83$ km/h

50. (a) Let the distance covered by man to reach the station be D km.
Now, according to the question,
$$\frac{D}{5} - \frac{D}{6} = \frac{12}{60}$$
$$\Rightarrow \frac{6D - 5D}{30} = \frac{12}{60} \Rightarrow D = 6 \text{ km}$$

Fast Track Method
Here, $a = 5, b = 6, t_1 = \frac{7}{60}, t_2 = \frac{5}{60}$
$$\therefore D = \frac{ab(t_1 + t_2)}{b - a} \quad \text{[by Technique 4]}$$
$$= \frac{5 \times 6 \left(\frac{7}{60} + \frac{5}{60}\right)}{(6 - 5)}$$
$$= 30 \times \frac{12}{60} = 6 \text{ km}$$

51. (b) Let the required distance = x
According to the question,
$$\frac{x}{5/2} - \frac{x}{7/2} = \frac{12}{60}$$
[$\because$ difference between two times = 6 + 6 = 12 min]
$$\Rightarrow \frac{2x}{5} - \frac{2x}{7} = \frac{1}{5} \Rightarrow 14x - 10x = 7$$
$$\Rightarrow 4x = 7 \Rightarrow x = \frac{7}{4} \text{ km}$$

Fast Track Method
Here, $a = 2\frac{1}{2} = \frac{5}{2}, b = \frac{7}{2}, t_1 = 6$ and $t_2 = 6$
According to the formula,
Required distance $= \frac{ab(t_1 + t_2)}{b - a}$
[by Technique 4]
$$= \frac{\frac{5}{2} \times \frac{7}{2} \left(\frac{6 + 6}{60}\right)}{\frac{7}{2} - \frac{5}{2}} = \frac{35}{4} \times \frac{12}{60} = \frac{7}{4} \text{ km}$$

52. (a) Here, $a = 3$ km/h and $b = 4$ km/h
$t_1 = 5$ min $= \frac{5}{60} = \frac{1}{12}$ h
and $t_2 = 5$ min $= \frac{5}{60} = \frac{1}{12}$ h
$\therefore$ Distance between starting point of Rubi and multiplex $= \frac{ab}{(b-a)}(t_1 + t_2)$
[by Technique 4]
$$= \frac{3 \times 4}{(4 - 3)} \times \left(\frac{1}{12} + \frac{1}{12}\right)$$
$$= \frac{3 \times 4}{1} \times \frac{2}{12} = 2 \text{ km}$$

53. (b) Let the certain distance be d and time t.
Now, by given condition,
$$\frac{d}{3} = (t + 15) \text{ min} = \frac{(t+15)}{60} \text{ h}$$
$$\Rightarrow 20d = t + 15 \Rightarrow t = 20d - 15 \quad \ldots(i)$$
and $\frac{d}{4} = (t - 15) \text{ min} = \frac{(t-15)}{60}$ h
$$\Rightarrow 15d = t - 15 \Rightarrow t = 15d + 15 \quad \ldots(ii)$$
From Eqs. (i) and (ii), we get
$$20d - 15 = 15d + 15 \Rightarrow 5d = 30$$
$$\Rightarrow d = 6 \text{ km}$$

Fast Track Method
Here, $a = 3, b = 4, t_1 = 15, t_2 = 15$
$$\therefore d = \frac{ab(t_1 + t_2)}{(b-a)} \quad \text{[by Technique 4]}$$
$$= \frac{3 \times 4 \left(\frac{15}{60} + \frac{15}{60}\right)}{4 - 3} = 3 \times 4 \times \frac{30}{60} = 6 \text{ km}$$

54. (c) Let the distance be d and time be t.
As per first condition, $\frac{d}{10} = t + \frac{6}{60}$
$$\Rightarrow d = 10t + 1 \quad \ldots(i)$$
As per second condition,
$$\frac{d}{12} = t - \frac{6}{60} \Rightarrow d = 12t - \frac{6}{5} \quad \ldots(ii)$$
From Eqs. (i) and (ii), we have
$$10t + 1 = 12t - \frac{6}{5} \Rightarrow 2t = 1 + \frac{6}{5} = \frac{11}{5} \Rightarrow t = \frac{11}{10}$$
$\therefore$ From Eq. (i), $d = 10 \times \frac{11}{10} + 1$
$$\Rightarrow d = 11 + 1 = 12$$
$$\Rightarrow d = 12 \text{ km}$$

Fast Track Method
Here, $a = 10$ km/h, $b = (10 + 2) = 12$ km/h,
$t_1 = 6$ min $= \frac{1}{10}$ h and $t_2 = \frac{1}{10}$ h
$\therefore$ Required distance $= \frac{ab(t_1 + t_2)}{b - a}$
[by Technique 4]
$$= \frac{10 \times 12 \times \left(\frac{1}{10} + \frac{1}{10}\right)}{12 - 10}$$
$$= \frac{10 \times 12 \times 1}{5 \times 2} = 12 \text{ km}$$

55. (c) Let the distance = x
Here, difference in time
= 20 − 5 = 15 min
$$= \frac{15}{60} = \frac{1}{4} \text{ h}$$
Speed during next journey
= 15 + 5 = 20 km/h
According to the question,
$$\frac{x}{15} - \frac{x}{20} = \frac{1}{4} \Rightarrow \frac{4x - 3x}{60} = \frac{1}{4}$$

Speed, Time and Distance / 443

$\Rightarrow \qquad x = \dfrac{60}{4} = 15$

$\therefore \qquad x = 15$ km

Fast Track Method
Here, $a = 15$ km/h, $b = 5$ km/h,
$t_1 = 20$ min and $t_2 = 5$ min
$\therefore$ Required distance $= (t_1 - t_2)(a+b)\dfrac{a}{b}$
[by Technique 5]
$= \dfrac{(20-5)}{60}(15+5)\dfrac{15}{5} = 15$ km

56. (a) Given, $D = 84$ km, $a = 12$ km/h and $b = 16$ km/h
According to the formula,
Distance travelled by A
$= PR = 2D \times \dfrac{a}{a+b}$ [by Technique 6]
$= 2 \times 84 \times \dfrac{12}{12+16} = 2 \times 84 \times \dfrac{12}{28}$
$= 2 \times 6 \times 6 = 72$ km

57. (c) Let us assume that the time taken by the policeman to catch the thief be t hours. Hence, after time t, the thief will be at a distance of $(t \times 8)$ km from where he started. At the same time, he will be at a distance of $(t \times 8 + 0.1)$ km from where the police started.
Now, distance run by police in time $t = (10 \times t)$ km.
According to the question,
$10t = 8t + 0.1$
$\Rightarrow \qquad 2t = 0.1 \Rightarrow t = 0.05$ h
$\therefore$ Distance run by thief in 0.05 h
$= 8 \times 0.05$ km $= 0.4$ km $= 400$ m

Fast Track Method
Here, $d = 100$ m,
$a = 8$ km/h $= 8 \times \dfrac{5}{18} = \dfrac{20}{9}$ m/s
and $b = 10$ km/h $= 10 \times \dfrac{5}{18} = \dfrac{25}{9}$ m/s
According to the formula,
Required distance $= d \times \left(\dfrac{a}{b-a}\right)$
[by Technique 7]
$= 100 \times \dfrac{20/9}{\dfrac{25}{9} - \dfrac{20}{9}} = 100 \times \dfrac{20}{5} = 400$ m

58. (a) In this case, $(21-15)$ m or 6 m is covered by Dalbir Singh in 1 min.
$\therefore$ 114 m will be covered in
$\left(\dfrac{1}{6} \times 114\right)$ min $= 19$ min

59. (a) Given, $x = 16$ and $y = 25$
According to the formula,
1st man's speed : 2nd man's speed
$= \sqrt{y} : \sqrt{x}$ [by Technique 8]
$= \sqrt{25} : \sqrt{16} = 5 : 4$

60. (a) Given that, $\dfrac{x}{y} = \dfrac{7}{9}, x = 7, y = 9$
and $t = 40$ min
Now, required time $= \dfrac{t \times x}{y - x}$ [$\because y > x$]
[by Technique 9]
$= \dfrac{40 \times 7}{9-7} = \dfrac{40 \times 7}{2} = 140$ min

Exercise 2 Higher Skill Level Questions

1. (c) Let the usual speed of a man be v and the time be t h.
By using the formula,
$v_1 t_1 = v_2 t_2 \Rightarrow vt = 1\dfrac{1}{2}v\left(t - \dfrac{20}{60}\right)$
$\Rightarrow \qquad t = \dfrac{3}{2}\left(t - \dfrac{1}{3}\right) \Rightarrow \dfrac{3}{2}t - t = \dfrac{1}{2}$
$\therefore \qquad t = 1$ h
Hence, the time taken by the man to reach his office at his usual speed is 1 h.

2. (a) Let the original speed of an aircraft be x km/h and its reduced speed
$= (x - 200)$ km/h
Condition I Time taken by aircraft to cover 600 km by the original speed $= \dfrac{600}{x}$ h
Condition II Time taken by aircraft to cover 600 km by the reduced speed $= \dfrac{600}{(x-200)}$ h

According to the question,
$\dfrac{600}{x-200} - \dfrac{600}{x} = \dfrac{1}{2}$ $\left[\because 30 \text{ min} = \dfrac{1}{2}\text{h}\right]$
On dividing both sides by 600, we get
$\dfrac{1}{x-200} - \dfrac{1}{x} = \dfrac{1}{600 \times 2}$
$\Rightarrow \dfrac{x - x + 200}{x(x-200)} = \dfrac{1}{1200}$
$\Rightarrow \dfrac{200}{x(x-200)} = \dfrac{1}{1200}$
$\Rightarrow \qquad x^2 - 200x - 240000 = 0$
$\Rightarrow \qquad (x - 600)(x + 400) = 0$
$\Rightarrow \qquad x = 600$ km/h
[$\because x \neq -400$ km/h]
$\therefore$ Required time $= \dfrac{600}{600} = 1$ h

3. (b) Let the distance between A and B be x km.

$A \xleftarrow{\hspace{1cm} x \text{ km} \hspace{1cm}} B$

Case I Distance = x km, speed = V km/h (let)
and time = 8 h

$\because$ Speed = $\dfrac{\text{Distance}}{\text{Time}}$

$\therefore \quad V = \dfrac{x}{8}$ ...(i)

Case II Speed = $(V + 4)$ km/h

and time = $7\dfrac{1}{2}$ h = $\dfrac{15}{2}$ h

$\therefore \quad V + 4 = \dfrac{x}{15/2} \Rightarrow V + 4 = \dfrac{2x}{15}$

$\Rightarrow \quad \dfrac{x}{8} + 4 = \dfrac{2x}{15}$ [from Eq. (i)]

$\Rightarrow \quad \dfrac{2x}{15} - \dfrac{x}{8} = 4 \Rightarrow \dfrac{x}{120} = 4$

$\therefore \quad x = 480$ km

4. (c) Let distance between points A and B be d km and speed of car be x km/h.

From condition I, $x = \dfrac{d}{5}$ ...(i)

and from condition II, $(x - 15) = \dfrac{d}{6}$ ...(ii)

Now, putting the value of x in Eq. (ii) from Eq. (i), we have

$\dfrac{d}{5} - 15 = \dfrac{d}{6} \Rightarrow \dfrac{d}{5} - \dfrac{d}{6} = 15$

$\Rightarrow \dfrac{6d - 5d}{30} = 15 \Rightarrow \dfrac{d}{30} = 15 \Rightarrow d = 450$ km

Hence, distance between points A and $B = 450$ km

5. (c) Let the original average speed of the cyclist be x km/h.

Then, time taken to cover the distance by original average speed = $\dfrac{14}{x}$ h

$A \xleftarrow{\hspace{1cm} 14 \text{ km} \hspace{1cm}} B$

When original average speed is decreased by 1 km/h. Then, time taken to cover the distance by reduced average speed = $\dfrac{14}{(x-1)}$ h

Now, according to the question,

$\dfrac{14}{(x-1)} - \dfrac{14}{x} = \dfrac{20}{60}$

$\Rightarrow \dfrac{14}{x} + \dfrac{1}{3} = \dfrac{14}{(x-1)} \Rightarrow \dfrac{42 + x}{3x} = \dfrac{14}{(x-1)}$

$\Rightarrow (42 + x)(x - 1) = 42x$

$\Rightarrow 42x - 42 + x^2 - x = 42x$

$\Rightarrow x^2 - x - 42 = 0 \Rightarrow (x - 7)(x + 6) = 0$

$\therefore \quad x = 7$ km/h $[\because x \neq -6]$

Hence, the original average speed of the cyclist is 7 km/h.

6. (b) Let A and B will meet after t h at point E.

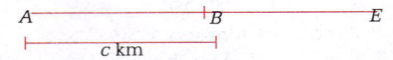

$A \xleftarrow{\hspace{0.5cm} c \text{ km} \hspace{0.5cm}} B \xrightarrow{\hspace{2cm}} E$

Distance travelled by $A = pt$ h
[$\because$ distance = speed $\times$ time]
and distance travelled by $B = qt$ h
According to the question,

$pt = qt + c$

$\Rightarrow \quad pt - qt = c \Rightarrow t(p - q) = c$

$\Rightarrow \quad t = \dfrac{c}{p - q}$ ...(i)

$\therefore$ Distance travelled by $A = pt = \dfrac{pc}{p - q}$ km

[from Eq. (i)]

7. (c) Here, $S_1 = 8$ km/h and $S_2 = 16$ km/h

$\therefore \quad d_1 = S_1 \times t_1 = 8t_1$ ...(i)

and $d_2 = S_2 \times t_2 = 16t_2$ ...(ii)

We know that, $t_1 + t_2 = 8$ ...(iii)

and $d_1 + d_2 = 80$ [given] ...(iv)

Put the values of d_1 and d_2 from Eqs. (i) and (ii) in Eq. (iv), we get

$8t_1 + 16t_2 = 80$

$\Rightarrow \quad 8t_1 + 8t_2 + 8t_2 = 80$

$\Rightarrow \quad 8(t_1 + t_2) + 8t_2 = 80$

$\Rightarrow \quad 8 \times 8 + 8t_2 = 80$ [from Eq. (iii)]

$\Rightarrow \quad 8t_2 = 80 - 64 = 16$

$\Rightarrow \quad t_2 = \dfrac{16}{8} = 2$ h, $t_1 = 8 - 2 = 6$ h

$\therefore$ Distance travelled by foot
$= d_1 = 8 \times 6 = 48$ km

8. (d) Let distance and original speed of the man be d km and s km/h.
Then,

$\dfrac{d}{s} - \dfrac{d}{s + 3} = \dfrac{40}{60} = \dfrac{2}{3} \Rightarrow \dfrac{d(s + 3 - s)}{s(s + 3)} = \dfrac{2}{3}$

$\Rightarrow \quad 9d = 2s(s + 3)$ ...(i)

and $\dfrac{d}{s - 2} - \dfrac{d}{s} = \dfrac{40}{60} = \dfrac{2}{3}$

$\Rightarrow \dfrac{d(s - s + 2)}{s(s - 2)} = \dfrac{2}{3}$

$\Rightarrow \quad 3d = s(s - 2)$ ...(ii)

From Eqs. (i) and (ii), we get

$3s(s - 2) = 2s(s + 3)$

$\Rightarrow 3s^2 - 6s = 2s^2 + 6s \Rightarrow s^2 = 12s \Rightarrow s = 12$

From Eq. (ii), we get $3d = 12(12 - 2)$

$\therefore \quad d = 40$ km

9. (e) Distance = $(AB + BC) = 2x$

$\Rightarrow \quad 2x = 16 \times \dfrac{1}{2} = 8$ km

$\therefore \quad x = \dfrac{8}{2} = 4$ km

$\therefore$ Area of the field = 4×4
$= 16$ sq km

Speed, Time and Distance / 445

10. (c) When A completes 5 rounds and B completes 4 rounds. Then, A will overtake B when he completes 5 rounds.
∴ Distance travelled by A when crossing B
$= 5 \times 0.25 = 1.25$ km
Since, there are three 1.25 km in 4 km, hence A will pass B in 3 times.

11. (d) Required average speed
$$= \frac{4}{\frac{1}{60} + \frac{1}{40} + \frac{1}{120} + \frac{1}{20}}$$
$$= \frac{4}{\frac{2 + 3 + 1 + 6}{120}} = \frac{4 \times 120}{12} = 40 \text{ km/h}$$

12. (d) Required number of hours is the number of terms of the series $40 + 45 + 50 + \ldots$ as speed increases every hour.
Given, sum of the series is 385.
∴ $a = 40, d = 5, S = 385$ and $n = ?$
Using $S = \frac{n}{2}[2a + (n-1)d]$,
$385 = \frac{n}{2}[80 + 5n - 5]$
⇒ $770 = 5n^2 + 75n$
⇒ $n^2 + 15n - 154 = 0$
⇒ $n^2 + 22n - 7n - 154 = 0$
⇒ $n(n + 22) - 7(n + 22) = 0$
⇒ $(n + 22)(n - 7) = 0$
∴ $n = 7$ [∵ $n \ne 22$]

13. (a) Given, speed of thief = 10 km/h
$= \frac{10 \times 1000}{60}$ m/min $= \frac{500}{3}$ m/min
and speed of policeman = 11 km/h
$= \frac{11 \times 1000}{60}$ m/min $= \frac{550}{3}$ m/min
Now, distance travelled by thief in 6 min
$= \frac{500}{3} \times 6 = 1000$ m
[∵ distance = speed × time]
and distance travelled by policeman in 6 min $= \frac{550}{3} \times 6 = 1100$ m
∴ Difference = (1100 − 1000) = 100 m

Alternate Method
Relative speed of policeman with respect to thief = (11 − 10) = 1 km/h.
Now, relative distance between the policeman and thief in 6 min
$=$ Speed × Time
$= 1 \times \frac{6}{60} = \frac{1}{10}$ km $= 100$ m

14. (a) Given, 48 km perform by Rickshaw, Bus and Auto, the distance covered by the three ways in ratio = 8 : 1 : 3

∴ Rickshaw : Bus : Auto = 8 : 1 : 3
Distance travelled by Rickshaw
$= \frac{8}{12} \times 48 = 32$ km
Distance travelled by bus $= \frac{1}{12} \times 48 = 4$ km
and distance travelled by Auto $= \frac{3}{12} \times 48$
$= 12$ km
∴ Charges per kilometre in that order are in the ratio = 8 : 1 : 4
Rickshaw charge being 24 paise per kilometre.
Let total charges being per kilometre be ₹ x.
∴ $\frac{8}{13} \times x = 24 \Rightarrow x = 39$ paise
∴ Bus charges being per kilometre
$= \frac{1}{13} \times 39 = 3$ paise
and auto charges being per kilometre
$= \frac{4}{13} \times 39 = 12$ paise
Hence, the total cost of journey
$= 32 \times 24 + 4 \times 3 + 12 \times 12$
$= 768 + 12 + 144$
$= 924$ paise = ₹ 9.24

15. (b) Let total distance = d
A's starting time = 8 : 30 am
B's starting time = 9 : 00 am
They both meet at 11 : 00 am.
∴ Time taken by A to cover this distance
$= 11 : 00 - 8 : 30 = 2.5$ h and time taken by B to cover this distance $= 11 : 00 - 9 : 00 = 2$ h
∴ Speed of $A = \frac{d}{2.5}$ and speed of $B = \frac{d}{2}$
B reached at 12 : 00 noon.
So, total time taken by B to cover entire distance = 12 : 00 − 9 : 00 = 3 h
∴ Total distance of travelling
$=$ Speed of B × Time taken
$= \frac{d}{2} \times 3 = \frac{3d}{2}$
Now, total time taken by A to cover entire distance $= \frac{\text{Total distance}}{\text{Speed of } A}$
$= \frac{3d/2}{d/2.5} = \frac{3d \times 2.5}{2 \times d} = \frac{7.5}{2}$ h
$= 3$ h 45 min
Hence, A will reach Q at 8 : 30 am + 3 h 45 min, i.e. at 12 : 15 pm.

Chapter 26

Problems Based on Trains

Problems based on trains are same as the problems related to 'Speed, Time and Distance' and some concepts of 'Speed, Time and Distance' are also applicable to these problems. The only difference is that the length of the moving object (train) is taken into consideration in these types of problems.

Basic Rules Related to Problems Based on Trains

Rule ❶

Speed of train $(S) = \dfrac{\text{Distance covered }(d)}{\text{Time taken }(t)}$ or $S = \dfrac{d}{t}$

Here, unit of speed is m/s or km/h.

(i) a km/h $= \left(a \times \dfrac{5}{18}\right)$ m/s (ii) a m/s $= \left(a \times \dfrac{18}{5}\right)$ km/h

Ex. 1 Convert 360 km/h into m/s.

Sol. ∵ 1 km/h $= \dfrac{5}{18}$ m/s

∴ 360 km/h $= 360 \times \dfrac{5}{18}$ m/s $= 100$ m/s

Ex. 2 Convert 150 m/s into km/h.

Sol. ∵ 1 m/s $= \dfrac{18}{5}$ km/h

∴ 150 m/s $= 150 \times \dfrac{18}{5}$ km/h

$= (30 \times 18)$ km/h $= 540$ km/h

Rule 2

The distance covered by a train in passing a pole or a standing man or a signal post or any other object (of negligible length) is equal to the length of the train.

Ex. 3 A train covers 85 m in passing a signal post. What is the length of the train?

Sol. We know that the distance covered by a train in passing a pole or a standing man or a signal post or any other object (of negligible length) is equal to the length of the train. So, in this case, train covers 85 m to pass a signal post.

∴ Length of the train = 85 m

Rule 3

If a train passes a stationary object (bridge, platform etc.) having some length, then the distance covered by train is equal to the sum of the length of train and that particular stationary object which it is passing.

Ex. 4 A 29 m long train passes a platform which is 100 m long. Find the distance covered by the train in passing the platform.

Sol. We know that when a train passes a stationary object having some length, then the distance covered by train is equal to the sum of the length of train and that particular stationary object. In this case, stationary object is 100 m long platform.

∴ Required distance = Length of train + Length of platform = (29 + 100) m = 129 m

Rule 4

If two trains are moving in opposite directions, then their relative speed is equal to the sum of the speeds of both the trains.

Ex. 5 Two trains are moving in opposite directions with speeds of 4 m/s and 8 m/s, respectively. Find their relative speed.

Sol. Given, speed of first train = 4 m/s and speed of second train = 8 m/s

We know that when two trains are moving in opposite directions, then their relative speed = Sum of speeds of both the trains

∴ Required relative speed of trains = (4 + 8) m/s = 12 m/s

Rule 5

If two trains are moving in the same direction, then their relative speed is equal to the difference of speeds of both the trains.

Ex. 6 Two trains are moving in the same direction with speeds of 19 km/h and 25 km/h, respectively. What will be the relative speed of the train running at 25 km/h in respect of the train running at 19 km/h?

Sol. Given, speed of first train = 25 km/h and speed of second train = 19 km/h

We know that when two trains are running in the same direction, then

Their relative speed = Difference of speeds of both the trains

∴ Required relative speed of trains = (25 − 19) km/h = 6 km/h

Rule 6

If two trains of lengths x and y are moving in opposite directions with speeds of u and v respectively, then time taken by the trains to cross each other is equal to $\dfrac{(x+y)}{(u+v)}$.

Ex. 7 Two trains of lengths 80 m and 90 m are moving in opposite directions at 10 m/s and 7 m/s, respectively. Find the time taken by the trains to cross each other.

Sol. Here, $x = 80$ m, $y = 90$ m, $u = 10$ m/s and $v = 7$ m/s
∴ Both trains are moving in opposite directions.
According to the formula, required time $= \dfrac{x+y}{u+v} = \dfrac{80+90}{10+7} = \dfrac{170}{17} = 10$ s

Rule 7

If two trains of lengths x and y are moving in the same direction with speeds of u and v respectively, then time taken by the faster train to cross the slower train is equal to $\left(\dfrac{x+y}{u-v}\right)$.

[here, $u > v$]

Ex. 8 Two trains of lengths 75 m and 95 m are moving in the same direction at 9 m/s and 8 m/s, respectively. Find the time taken by the faster train to cross the slower train.

Sol. Here, $x = 75$ m, $y = 95$ m, $u = 9$ m/s and $v = 8$ m/s $(u > v)$
∴ Both trains are moving in the same direction.
According to the formula, required time $= \dfrac{x+y}{u-v} = \dfrac{75+95}{9-8} = 170$ s

Ex. 9 A train passes a standing man in 6 s and 210 m long platform in 16 s. Find the length and the speed of the train.

Sol. Let length of the train be L m.
If the speed of the train remains same while crossing a standing man and platform.
Then, according to the question, $\dfrac{L}{6} = \dfrac{L+210}{16}$ $\left[\because \text{speed} = \dfrac{\text{length}}{\text{time}}\right]$

$16L = 6L + 1260 \Rightarrow 10L = 1260 \Rightarrow L = \dfrac{1260}{10} = 126$ m

∴ Speed of the train $= \dfrac{\text{Length of the train}}{\text{Time taken}} = \dfrac{126}{6} = 21$ m/s

Ex. 10 A 250 m long train is running at 100 km/h. In what time, will it pass a man running at 10 km/h in the same direction in which the train is going?

Sol. Given, length of train = 250 m
Speed of the train = 100 km/h and speed of man = 10 km/h
∵ Train and man are moving in same direction.
∴ Speed of the train relative to man = Speed of the train − Speed of man
$= 100 - 10 = 90$ km/h $= 90 \times \dfrac{5}{18}$ m/s $= 5 \times 5 = 25$ m/s

Problems Based on Trains / 449

Distance covered in passing the man = 250 m

∴ Time taken = $\dfrac{250}{25}$ = 10 s

Ex. 11 A 220 m long train is running at 120 km/h. In what time, will it pass a man running in the direction opposite to that of the train at 12 km/h?

Sol. Given, length of the train = 220 m
Speed of the train = 120 km/h and speed of man = 12 km/h
∵ Train and man are moving in opposite directions.
∴ Speed of the train relative to man = (120 + 12) km/h
$$= 132 \text{ km/h} = \left(132 \times \dfrac{5}{18}\right) \text{m/s} = \dfrac{110}{3} \text{ m/s}$$
Distance covered in passing the man = 220 m

∴ Time taken = $\dfrac{220}{110} \times 3 = 2 \times 3 = 6$ s

Fast Track Techniques
to solve the QUESTIONS

Technique 1 If a train of length L m passes a platform of x m in t_1 s, then time taken by the same train to pass a platform of length y m is given as
$$t_2 = \left(\dfrac{L+y}{L+x}\right) t_1$$

Ex. 12 A train of length 250 m, passes a platform of 350 m length in 50 s. What time will this train take to pass the platform of 230 m length?

Sol. Given, length of train = 250 m ; length of platform = 350 m
and time taken to cover the distance = 50 s
So, to cross the platform, the train has to cover a distance
= Length of train + Length of platform = (250 + 350) m = 600 m

∴ Speed of train = $\dfrac{\text{Distance covered}}{\text{Time taken}} = \dfrac{600}{50} = 12$ m/s

Now, the time taken to cross the platform of length 230 m
$$= \dfrac{\text{Total distance covered}}{\text{Speed of train}}$$
$$= \dfrac{\text{Length of train + Length of platform}}{\text{Speed of train}}$$
$$= \dfrac{250 + 230}{12} = \dfrac{480}{12} = 40 \text{ s}$$

Fast Track Method
Here, $L = 250$ m, $x = 350$ m, $t_1 = 50$ s, $y = 230$ m and $t_2 = ?$

∴ $t_2 = \left(\dfrac{L+y}{L+x}\right) t_1 = \left(\dfrac{250+230}{250+350}\right) \times 50 = \dfrac{480}{600} \times 50 = 40$ s

Technique 2 From stations P and Q, two trains start moving towards each other with the speeds of a and b, respectively. When they meet each other, it is found that one train covers distance d more than that of another train. In such cases, distance between stations P and Q is given as $\left(\dfrac{a+b}{a-b}\right) \times d$.

Ex. 13 From stations A and B, two trains start moving towards each other with the speeds of 150 km/h and 130 km/h, respectively. When the two trains meet each other, it is found that one train covers 20 km more than that of another train. Find the distance between stations A and B.

Sol. Let the trains meet after time t at a distance x from station B.
Then, another train coming from station A covers a distance of $(x + 20)$.
For station A, distance covered by first train $(x + 20) = 150t$
$\Rightarrow \qquad x = 150t - 20$ ...(i)
For station B, distance covered by second train, $x = 130t$ ...(ii)
From Eqs. (i) and (ii), we get
$130t = 150t - 20 \Rightarrow 150t - 130t = 20 \Rightarrow 20t = 20 \Rightarrow t = 1\,h$
$\therefore$ Distance between stations A and $B = 150t + 130t = 280t = 280 \times 1 = 280\,km$

Fast Track Method
Here, $a = 150$ km/h, $b = 130$ km/h and $d = 20$ km
According to the formula,
Distance between stations A and $B = \left(\dfrac{a+b}{a-b}\right) \times d = \left(\dfrac{150+130}{150-130}\right) \times 20 = \dfrac{280}{20} \times 20$
$= 280\,km$

Technique 3 If two trains leave P for Q at time t_1 and t_2 and travel with speeds a and b respectively, then the distance d from P, where the two trains meet, is given as

$d = \text{Difference in time} \times \dfrac{\text{Product of speeds}}{\text{Difference in speeds}} = (t_2 - t_1) \times \dfrac{a \times b}{b - a}$

where, $t_2 > t_1$ and $b > a$.

Ex. 14 Two trains leave Patna for Delhi at 10 : 00 am and 10 : 30 am, respectively and travel at 120 km/h and 150 km/h, respectively. How many kilometres from Patna will the two trains meet?

Sol. Speed of 1st train $= 120$ km/h and speed of 2nd train $= 150$ km/h
Relative speed of 2nd train $= (150 - 120)$ km/h $= 30$ km/h
As 1st train leaves $\dfrac{1}{2}$ h before, hence in $\dfrac{1}{2}$ h, it will cover 60 km as it moves 120 km in 1 h.

Time taken by 2nd train to gain 60 km $= \dfrac{60}{30} = 2\,h$

$\therefore$ Actual distance covered by 2nd train in 2 h $= 2 \times 150 = 300$ km

Fast Track Method
Here, $a = 120$ km/h, $b = 150$ km/h, $t_1 = 10 : 00$ am and $t_2 = 10 : 30$ am
$[\because t_2 > 1 \text{ and } b > a]$

Problems Based on Trains / 451

According to the formula,

Required distance $= (t_2 - t_1) \times \dfrac{a \times b}{b - a} = (10:30 - 10:00) \times \dfrac{(120 \times 150)}{(150 - 120)}$

$= \dfrac{1}{2} \times \dfrac{120 \times 150}{30} = \dfrac{120 \times 150}{60} = 2 \times 150 = 300$ km

Technique 4 If a train overtakes two persons who are walking with speeds of *a* and *b* respectively, in the same direction and passes the two persons completely in t_1 and t_2 time, respectively, then

Length of the train $= \dfrac{\text{Difference in speeds} \times t_1 \times t_2}{(t_2 - t_1)}$, where $t_2 > t_1$.

♦ In case, $t_1 > t_2$, $(t_1 - t_2)$ is taken in place of $(t_2 - t_1)$ in the denominator.

Ex. 15 A train overtakes two persons who are walking at the rate of 4 km/h and 8 km/h in the same direction and passes them completely in 18 and 20 s, respectively. Find the length of the train.

Sol. Let speed of the train be x m/s.

Given, speed of first person $= 4$ km/h $= \dfrac{4 \times 5}{18} = \dfrac{10}{9}$ m/s

and speed of second person $= 8$ km/h $= \dfrac{8 \times 5}{18} = \dfrac{20}{9}$ m/s

∵ Train and two persons are moving in same direction.

∴ Relative speeds are $\left(x - \dfrac{10}{9}\right)$ m/s and $\left(x - \dfrac{20}{9}\right)$ m/s.

For 1st man,

Length of the train $= \left(x - \dfrac{10}{9}\right) \times 18$ m ...(i)

For 2nd man,

Length of the train $= \left(x - \dfrac{20}{9}\right) \times 20$ m ...(ii)

From Eqs. (i) and (ii), we get

$\left(x - \dfrac{10}{9}\right) \times 18 = \left(x - \dfrac{20}{9}\right) \times 20$

$\Rightarrow \left(\dfrac{9x - 10}{9}\right) \times 18 = \left(\dfrac{9x - 20}{9}\right) \times 20$

$\Rightarrow 9(9x - 10) = 10(9x - 20)$

$\Rightarrow 81x - 90 = 90x - 200$

$\Rightarrow 9x = 110 \Rightarrow x = \dfrac{110}{9}$

From Eq. (i), we get

Length of the train $= \left(\dfrac{110}{9} - \dfrac{10}{9}\right) \times 18 = \dfrac{100}{9} \times 18 = 200$ m

Fast Track Method

Here, $t_1 = 18$ s, $t_2 = 20$ s, $a = 4$ km/h $= 4 \times \dfrac{5}{18}$ m/s $= \dfrac{10}{9}$ m/s

and $b = 8$ km/h $= 8 \times \dfrac{5}{18}$ m/s $= \dfrac{20}{9}$ m/s

According to the formula,

$$\text{Length of the train} = \frac{\text{Difference in speeds} \times t_1 \times t_2}{t_2 - t_1}$$

$$= \frac{\left(\frac{20}{9} - \frac{10}{9}\right) \times 18 \times 20}{20 - 18}$$

$$= \frac{\frac{10}{9} \times 18 \times 20}{2} = 10 \times 20 = 200 \text{ m}$$

Technique 5 If two trains A and B start from stations/points P and Q towards Q and P, respectively and after passing each other, they take t_1 and t_2 time to reach Q and P, respectively and speed of train A is given as a, then

$$\text{Speed of train } B = a\sqrt{\frac{t_1}{t_2}}$$

Ex. 16 Two trains x and y start from Mumbai and Delhi towards Delhi and Mumbai, respectively. After passing each other, they take 12 h 30 min and 8 h to reach Delhi and Mumbai, respectively. If the train from Mumbai is moving at 60 km/h, then find the speed of the other train.

Sol. Here, $a = 60$ km/h, $t_1 = 12$ h 30 min $= 12 + \frac{30}{60} = \frac{25}{2}$ h and $t_2 = 8$ h

According to the formula,

$$\text{Speed of } y = a\sqrt{\frac{t_1}{t_2}} = 60 \times \sqrt{\frac{25}{2 \times 8}} = 60 \times \frac{5}{4} = 75 \text{ km/h}$$

Technique 6 The distance between two stations P and Q is d km. A train with a km/h starts from station P towards Q and after a difference of t h another train with b km/h starts from Q towards station P, then both the trains will meet at a certain point after time T. Then,

$$T = \left(\frac{d \pm tb}{a + b}\right)$$

✦ If second train starts after the first train, then t is taken as positive.
✦ If second train starts before the first train, then t is taken as negative.

Ex. 17 The distance between two stations P and Q is 110 km. A train with speed of 20 km/h leaves station P at 7 : 00 am towards station Q. Another train with speed of 25 km/h leaves station Q at 8 : 00 am towards station P. Then, at what time both trains meet?

Sol. First train leaves at 7 : 00 am and second at 8 : 00 am.
So, first train i.e. from P to Q has travelled 1h more.
Distance covered by first train in 1 h $= 20 \times 1 = 20$ km   [∵ distance = speed × time]
So, distance left between the station $= 110 - 20 = 90$ km
Now, trains are travelling in opposite directions.

So, relative speed = (20 + 25) km/h = 45 km/h

Time taken to cover 90 km = $\dfrac{\text{Distance travelled}}{\text{Relative speed}} = \dfrac{90}{45} = 2$ h

∴ The time, at which they will meet, is 2 h after second train left,
i.e. 8 : 00 am + 2 h = 10 : 00 am.

Fast Track Method

Here, $d = 110$ km, $t = 8:00 - 7:00 = 1$ h, $a = 20$ km/h and $b = 25$ km/h

∴ Time taken by trains to meet, $T = \left(\dfrac{d + tb}{a + b}\right) = \dfrac{110 + (1)(25)}{20 + 25} = \dfrac{135}{45} = 3$ h

∴ They will meet at = 7 : 00 am + 3 h = 10 : 00 am

Technique 7 The distance between two stations P and Q is d km. A train starts from P towards Q and another train starts from Q towards P at the same time and they meet at a certain point after t h. If train starting from P travels with a speed of x km/h slower or faster than another train, then

(i) speed of faster train = $\left(\dfrac{d + tx}{2t}\right)$ km/h

(ii) speed of slower train = $\left(\dfrac{d - tx}{2t}\right)$ km/h

Ex. 18 The distance between two stations A and B is 138 km. A train starts from A towards B and another from B to A at the same time and they meet after 6 h. The train travelling from A to B is slower by 7 km/h compared to other train from B to A, then find the speed of the both trains.

Sol. Let the speed of slower train be x km/h.

Then, speed of faster train = $(x + 7)$ km/h
As the trains are moving in opposite directions.
So, the relative speed = $x + (x + 7) = (2x + 7)$ km/h

Time taken = $\dfrac{\text{Distance travelled}}{\text{Relative speed}}$

$\Rightarrow \quad 6 = \dfrac{138}{2x + 7} \Rightarrow 2x + 7 = \dfrac{138}{6} \Rightarrow 2x + 7 = 23$

$\Rightarrow \quad 2x = 23 - 7 \Rightarrow x = \dfrac{16}{2} = 8$ km/h

Hence, speed of slower train is 8 km/h.
and speed of faster train = $(8 + 7)$ km/h = 15 km/h

Fast Track Method

Here, $d = 138$ km, $t = 6$ h and $x = 7$ km/h

∴ Speed of slower train = $\dfrac{d - tx}{2t} = \dfrac{138 - (6)(7)}{2(6)} = \dfrac{138 - 42}{12} = \dfrac{96}{12} = 8$ km/h

and speed of faster train = $\dfrac{d + tx}{2t} = \dfrac{138 + (6)(7)}{2(6)}$

$= \dfrac{138 + 42}{12} = \dfrac{180}{12} = 15$ km/h

Technique 8 A train covers distance d between two stations P and Q in t_1 h. If the speed of train is reduced by a km/h, then the same distance will be covered in t_2 h.

(i) Distance between P and Q, $d = a \left(\dfrac{t_1 t_2}{t_2 - t_1} \right)$ km

(ii) Speed of the train $= \left(\dfrac{a t_2}{t_2 - t_1} \right)$ km/h

Ex. 19 A train covers distance between two stations A and B in 2 h. If the speed of train is reduced by 6 km/h, then it travels the same distance in 3 h. Calculate the distance between two stations and speed of the train.

Sol. Let the initial speed of train be x km/h and distance between stations A and $B = d$ km

Case I With initial speed,

$$\text{Time taken} = \dfrac{\text{Distance}}{\text{Speed}} \Rightarrow 2 = \dfrac{d}{x} \Rightarrow 2x = d \qquad \ldots(i)$$

Case II With decreased speed,

$$\text{Time taken} = \dfrac{\text{Distance}}{\text{Speed}} \Rightarrow 3 = \dfrac{d}{(x-6)} \Rightarrow 3(x-6) = d \qquad \ldots(ii)$$

From Eqs. (i) and (ii), we get

$$2x = 3(x-6) \Rightarrow 2x = 3x - 18 \Rightarrow 3x - 2x = 18 \Rightarrow x = 18 \text{ km/h}$$

On putting the value of x in Eq. (i), we get

$$2 = \dfrac{d}{18} \Rightarrow d = 2 \times 18 = 36 \text{ km}$$

∴ Initial speed of train = 18 km/h and distance between two stations = 36 km.

Fast Track Method

Here, $t_1 = 2$ h, $t_2 = 3$ h, $a = 6$ km/h and $d = ?$

(i) Distance between A and B,

$$d = a \left(\dfrac{t_1 t_2}{t_2 - t_1} \right) \text{ km}$$

$$\Rightarrow \quad d = 6 \left(\dfrac{2 \times 3}{3 - 2} \right)$$

$$\Rightarrow \quad d = 36 \text{ km}$$

(ii) Speed of the train $= \dfrac{a t_2}{t_2 - t_1} = \dfrac{6 \times 3}{3 - 2} = 18$ km/h

Technique 9 Without stoppage, a train travels at an average speed of a and with stoppage, it covers the same distance at an average speed of b, then

$$\text{stoppage time per hour} = \dfrac{\text{Difference in average speeds}}{\text{Speed without stoppage}} = \dfrac{a - b}{a}$$

where, $a > b$.

Ex. 20 Without stoppage, the speed of a train is 54 km/h and with stoppage, it is 45 km/h. For how many minutes, does the train stop per hour?

Sol. Decrease in speed due to stoppage = 54 − 45 = 9 km/h

Problems Based on Trains / 455

Because of stoppage, train covers 9 km less per hour.

∴ Time taken to cover 9 km = $\dfrac{9}{54} = \dfrac{1}{6}$ h = $\dfrac{1}{6} \times 60 = 10$ min

Fast Track Method

Here, $a = 54$ km/h and $b = 45$ km/h

According to the formula,

Required rest time = $\dfrac{a-b}{a} = \dfrac{54-45}{54}$

$= \dfrac{9}{54} = \dfrac{1}{6}$ h $= \dfrac{1}{6} \times 60 = 10$ min

Technique 10 If two trains of equal lengths and different speeds take t_1 and t_2 time to cross a pole, then time taken by them to cross each other is

$$T = \dfrac{2t_1 t_2}{t_2 \pm t_1}.$$

✦ We use '+ve' sign, if trains are moving in opposite directions and '–ve' sign, if they are moving in same direction.

Ex. 21 Two trains of equal lengths take 5 s and 6 s, respectively to cross a pole. If these trains are moving in the same direction, then how long will they take to cross each other?

Sol. Let the length of train be x m.

Then, speed of first train = $\dfrac{\text{Distance covered}}{\text{Time taken}} = \dfrac{x}{5}$ m/s

Similarly, speed of second train = $\dfrac{x}{6}$ m/s

Now, time taken to cross each other, when moving in same direction

$= \dfrac{\text{Sum of lengths of two trains}}{\text{Relative speed of two trains}} = \dfrac{x+x}{\dfrac{x}{5} - \dfrac{x}{6}}$

$= \dfrac{2x}{\dfrac{6x-5x}{30}} = \dfrac{2x \times 30}{x} = 60$ s

Fast Track Method

Here, $t_1 = 5$ s and $t_2 = 6$ s

According to the formula, required time = $\dfrac{2t_1 t_2}{t_2 - t_1} = \dfrac{2 \times 5 \times 6}{6-5} = 60$ s

Multi Concept Questions

1. After travelling 80 km, a train meets with an accident and then proceeds at 3/4 of its former speed and arrives at its destination 35 min late. Had the accident occurred 24 km further, it would have reached the destination only 25 min late. Find the speed of the train.
   (a) 50 km/h   (b) 30 km/h   (c) 48 km/h   (d) 55 km/h

   (c) When we analyse the question minutely, we see that the speeds of the train upto 80 km are the same in both the cases. Further, the speeds after (80 + 24) km/h = 104 km are same in both the cases. Therefore, the difference in time (35 min – 25 min) = 10 min is only because of the difference in speeds for the 24 km journey.
   Now, let speed of the train be x km/h.
   Then, according to the question,
   Difference between time when 24 km are travelled with $\frac{3}{4}$th of speed and with usual speed
   = 10 min

   $\Rightarrow \quad \dfrac{24}{\frac{3x}{4}} - \dfrac{24}{x} = \dfrac{10}{60}$

   $\Rightarrow \quad \dfrac{24 \times 4}{3x} - \dfrac{24}{x} = \dfrac{10}{60} \Rightarrow \dfrac{32}{x} - \dfrac{24}{x} = \dfrac{1}{6}$

   $\Rightarrow \quad \dfrac{8}{x} = \dfrac{1}{6} \Rightarrow x = 48$ km/h

2. A goods train and a passenger train are running on the parallel tracks in the same direction. The driver of the goods train observes that the passenger train coming from behind, overtakes and crosses his train completely in 1 min whereas a passenger on the passenger train marks that he crosses the goods train in 2/3 min. If the speeds of the trains is in the ratio of 1 : 2, then find the ratio of their lengths.
   (a) 4 : 1   (b) 3 : 1   (c) 1 : 4   (d) 2 : 1

   (d) Let the speeds of the two trains be x and $2x$ and length be L_1 and L_2, respectively.
   Case I When driver of goods train observes that passenger train crosses his train, then
   $$\dfrac{L_1 + L_2}{2x - x} = 1 \text{ min} = 60 \text{ s} \quad \ldots(i)$$

   Case II When a passenger on passenger train observes that he crosses the goods train, then
   $$\dfrac{L_1}{2x - x} = \dfrac{2}{3} \text{ min} = \dfrac{2}{3} \times 60 = 40 \text{ s} \quad \ldots(ii)$$

   On dividing Eq. (i) by Eq. (ii), we get
   $\dfrac{L_1 + L_2}{L_1} = \dfrac{60}{40} \Rightarrow \dfrac{L_2}{L_1} + 1 = \dfrac{3}{2}$

   $\Rightarrow \quad \dfrac{L_2}{L_1} = \dfrac{3}{2} - 1 = \dfrac{1}{2} \Rightarrow L_1 : L_2 = 2 : 1$

Fast Track Practice

Exercise 1 Base Level Questions

1. A train covers 90 m in passing a standing man. Find the length of the train.
   (a) 20 m   (b) 87 m
   (c) 71 m   (d) 90 m
   (e) None of these

2. A 220 m long train passes a signal post in 12 s. Find the speed of the train.
   [Bank Clerk 2010]
   (a) 72 km/h   (b) 60 km/h
   (c) 66 km/h   (d) 69 km/h
   (e) None of these

3. A train takes 9 s to cross a pole. If the speed of the train is 48 km/h, then length of the train is [CDS 2014]
   (a) 150 m   (b) 120 m
   (c) 90 m   (d) 80 m

4. A train running at the speed of 72 km/h goes past a pole in 15 s. What is the length of the train? [CDS 2013]
   (a) 150 m   (b) 200 m
   (c) 300 m   (d) 350 m

5. Train A crosses a pole in 25 s and train B crosses the pole in 1 min 15 s. Length of train A is half the length of train B. What is the ratio between the speeds of A and B, respectively? [SBI Clerk (Mains) 2016]
   (a) 3 : 2   (b) 3 : 4
   (c) 5 : 3   (d) 2 : 5
   (e) 4 : 3

6. 150 m long train running with the speed of 90 km/h cross a bridge in 26 s. What is the length of the bridge? [SSC CGL 2012]
   (a) 500 m   (b) 600 m
   (c) 659 m   (d) 550 m

7. A train travelling at a speed of 30 m/s crosses a 600 m long platform in 30 s. Find the length of the train. [SSC CGL 2007]
   (a) 120 m   (b) 150 m
   (c) 200 m   (d) 300 m

8. A train crossed a platform in 43 s. The length of the train is 170 m. What is the speed of the train? [IBPS Clerk 2011]
   (a) 233 km/h
   (b) 243 km/h
   (c) 265 km/h
   (d) Cannot be determined
   (e) None of the above

9. A train crosses a platform in 30 s travelling with a speed of 60 km/h. If the length of the train be 200 m, then the length of the platform is [SSC CPO 2013]
   (a) 420 m   (b) 500 m
   (c) 300 m   (d) 250 m

10. A train crosses a bridge of length 150 m in 15 s and a man standing on it in 9 s. The train is travelling at a uniform speed. Length of the train is
    [SSC CGL (Mains) 2012]
    (a) 225 m   (b) 200 m
    (c) 135 m   (d) 90 m

11. A train moving with uniform speed crosses a pole in 2 s and a 250 m long bridge in 7 s. Find the length of the train. [SSC CGL 2010]
    (a) 150 m   (b) 120 m
    (c) 100 m   (d) 80 m

12. Train A, whose length is 328 m, can cross a 354 m long platform in 11 s. Train B can cross the same platform in 12 s. If the speed of train B is 7/8th of the speed of train A, what is the length of train B?
    [SBI Clerk 2015]
    (a) 321 m   (b) 303 m
    (c) 297 m   (d) 273 m
    (e) 309 m

13. A train travelling with uniform speed crosses two bridges of lengths 300 m and 240 m in 21 s and 18 s, respectively. Find the speed of the train. [SSC CPO 2011]
    (a) 72 km/h   (b) 68 km/h
    (c) 65 km/h   (d) 60 km/h

14. A 280 m long train crosses a platform which is three times of its length, in 6 min 40 s. What is the speed of the train? [IBPS Clerk 2011]
    (a) 3.2 m/s
    (b) 1.4 m/s
    (c) 2.8 m/s
    (d) Cannot be determined
    (e) None of the above

15. A 110 m long train is running at a speed of 60 km/h. How many seconds does it take to cross an another train of length 170 m, which is standing on parallel track? [SSC CGL 2011]
    (a) 15.6   (b) 16.8   (c) 17.2   (d) 18

16. The ratio between the speeds of two train is 8 : 9. Second train covers 360 km in 4 h. Distance covered (in km) by first train in 3 h is [SSC (10+2) 2012]
(a) 240   (b) 480   (c) 120   (d) 60

17. The average speed of a bus is three-fourth the average speed of a train. The train covers 240 km in 12 h. How much distance will the bus cover in 7 h? [IBPS Clerk 2011]
(a) 110 km   (b) 115 km
(c) 105 km   (d) 100 km
(e) None of these

18. P and Q are 27 km away. Two trains will having speeds of 24 km/h and 18 km/h respectively start simultaneously from P and Q and travel in the same direction. They meet at a point R beyond Q. Distance QR is [SSC CGL 2012]
(a) 126 km   (b) 81 km
(c) 48 km   (d) 36 km

19. The relative speed of a train in respect of a car is 90 km/h when train and car are moving opposite to each other. Find the actual speed of train, if car is moving with a speed of 15 km/h. [Bank PO 2011]
(a) 80 km/h   (b) 105 km/h
(c) 75 km/h   (d) 100 km/h
(e) None of these

20. A 225 m long train is running at a speed of 30 km/h. How much time does it take to cross a man running at 3 km/h in the same direction? [CDS 2017 (I)]
(a) 40 s   (b) 30 s
(c) 25 s   (d) 15 s

21. Two trains are moving in the same direction at 1.5 km/min and 60 km/h, respectively. A man in the faster train observes that it takes 27 s to cross the slower train. The length of the slower train is [CDS 2015 (II)]
(a) 225 m   (b) 230 m
(c) 240 m   (d) 250 m

22. A 440 m long train is running at 240 km/h. In what time will it pass a man running in the direction opposite to that of the train at 24 km/h? [Hotel Mgmt. 2009]
(a) 9 s   (b) 6 s   (c) 12 s   (d) 4 s

23. A 400 m long train takes 36 s to cross a man walking at 20 km/h in the direction opposite to that of the train. What is the speed of the train? [IBPS Clerk 2011]
(a) 20 km/h   (b) 30 km/h
(c) 15 km/h   (d) 11 km/h
(e) None of these

24. Two trains of lengths 70 m and 90 m are moving in opposite directions at 10 m/s and 6 m/s, respectively. Find the time taken by trains to cross each other.
(a) 10 s   (b) 8 s   (c) 12 s   (d) 16 s

25. Two trains of length 512 m and 528 m are running towards each other on parallel lines at 84 km/h and 60 km/h, respectively. In what time, will they be clear of each other from the moment they meet?
(a) 26 s   (b) 25 s   (c) 15 s   (d) 27 s
(e) None of these

26. Two trains of lengths 105 m and 90 m, respectively run at the speeds of 45 km/h and 72 km/h, respectively in opposite directions on parallel tracks. Find the time which they take to cross each other. [SSC CGL 2007]
(a) 5 s   (b) 6 s   (c) 7 s   (d) 8 s

27. A train A is 180 m long, while another train B is 240 m long. Train A has a speed of 30 km/h and train B's speed is 40 km/h. If the trains move in opposite directions, then find when will train A pass train B completely? [SSC (10+2) 2012]
(a) 21 s   (b) 21.6 s   (c) 26.1 s   (d) 26 s

28. A train A of length 180 m, running by 72 km/h crosses the another train B which is running in the opposite directions at speed of 108 km/h and length is 120 m, in how much time? [SSC CGL 2012]
(a) 23 s   (b) 12 s   (c) 6 s   (d) 30 s

29. Two trains of same length are running in parallel tracks in opposite directions with speeds 65 km/h and 85 km/h, respectively. They cross each other in 6 s. The length of each train is [SSC (10+2) 2013]
(a) 100 m   (b) 115 m   (c) 125 m   (d) 150 m

30. Two trains running in opposite directions cross a man standing on the platform in 54 s and 34 s, respectively and they cross each other in 46 s. Find the ratio of their speeds.
(a) 3 : 2   (b) 2 : 3   (c) 5 : 3   (d) 3 : 5
(e) None of these

31. Two trains are running 40 km/h and 20 km/h respectively, in the same direction. The fast train completely passes a man sitting in the slow train in 5 s. The length of the fast train is [SSC CGL 2013]
(a) $23\frac{2}{9}$ m   (b) 27 m
(c) $27\frac{7}{9}$ m   (d) 23 m

Problems Based on Trains / 459

32. Two trains of lengths 50 m and 65 m are moving in the same direction at 18 m/s and 17 m/s, respectively. Find the time taken by the faster train to cross the slower train.
(a) 100 s (b) 114 s
(c) 95 s   (d) 115 s
(e) None of these

33. Two trains of equal length are running on parallel lines in the same direction at 46 km/h and 36 km/h, respectively. The faster train passes the slower train in 36 s. The length of each train is
[SSC (10+2) 2012]
(a) 82 m (b) 50 m
(c) 80 m (d) 72 m

34. From stations M and N, two trains start moving towards each other at speed 125 km/h and 75 km/h, respectively. When the two trains meet each other, it is found that one train covers 50 km more than another. Find the distance between M and N.
(a) 190 km (b) 200 km
(c) 145 km (d) 225 km

35. If two trains leave Jalandhar for Delhi at 8 : 00 am and 10 : 30 am, respectively and travel at 150 km/h and 180 km/h, respectively. How many kilometres from Jalandhar will the two trains meet?
(a) 2150 km (b) 2350 km
(c) 2250 km (d) 2500 km

36. A train passes two persons who are walking in the direction opposite to the direction of train at the rate of 10 m/s and 20 m/s, respectively in 12 s and 10 s, respectively. Find the length of the train.
[SSC CGL 2013]
(a) 500 m (b) 900 m
(c) 400 m (d) 600 m

37. Two trains A and B start from Delhi and Patna towards Patna and Delhi, respectively. After passing each other, they take 16 h and 9 h to reach Patna and Delhi, respectively. If the train from Delhi is moving at 90 km/h, then find the speed (in km/h) of the other train.
(a) 120 (b) 190
(c) 125 (d) 145

38. Two trains A and B start from Howrah and Patna towards Patna and Howrah respectively at the same time. After passing each other, they take 4 h 48 min and 3 h 20 min to reach Patna and Howrah, respectively. If the train from Howrah is moving at 45 km/h, then the speed of the other train is
[SSC Multitasking 2014]
(a) 60 km/h (b) 45 km/h
(c) 35 km/h (d) 54 km/h

39. The distance between two stations P and Q is 145 km. A train with speed of 25 km/h leaves station at 8 : 00 am towards station Q. Another train with speed of 35 km/h leaves station Q at 9 : 00 am towards station P. Then, at what time both trains meet?
[SSC Multitasking 2013]
(a) 10 : 00 am (b) 11 : 00 am
(c) 12 : 00 am (d) 11 : 30 am

40. Two stations P and Q are at a distance of 160 km. Two trains start moving from P and Q to Q and P, respectively and meet each other after 4 h. If speed of the train starting from P is more than that of other train by 6 km/h, then find the speed of both the trains, respectively.
(a) 19 km/h, 13 km/h
(b) 13 km/h, 9 km/h
(c) 17 km/h, 23 km/h
(d) 16 km/h, 10 km/h
(e) None of the above

41. A train covers distance between two stations Delhi and Meerut in 2 h. If the speed of the train is reduced by 6 km/h, then it travels the same distance in 5 h. Calculate the distance between two stations and speed of the train.
(a) 30 km, 10 km/h
(b) 20 km, 10 km/h
(c) 25 km, 10 km/h
(d) 30 km, 10 km/h

42. Excluding stoppages, the speed of a train is 108 km/h and including stoppages, it is 90 km/h. For how many minutes does the train stop per hour? [SSC CPO 2012]
(a) 5 (b) 9
(c) 10 (d) 6

43. Two trains of same length take 6 s and 9 s, respectively to cross a pole. If both the trains are running in the same direction, then how long will they take to cross each other?
(a) 30 s (b) 36 s
(c) 40 s (d) 42 s
(e) None of these

Exercise 2 Higher Skill Level Questions

1. Two places P and Q are 162 km apart. A train leaves P for Q and simultaneously another train leaves Q for P. They meet at the end of 6 h. If the former train travels 8 km/h faster than the other, then speed of train from Q is
   [SSC CGL (Mains) 2015]
   (a) $8\frac{1}{2}$ km/h
   (b) $9\frac{1}{2}$ km/h
   (c) $12\frac{5}{6}$ km/h
   (d) $10\frac{5}{6}$ km/h

2. A freight train left Delhi for Mumbai at an average speed of 40 km/h. Two hours later, an express train left Delhi for Mumbai, following the freight train on a parallel track at an average speed of 60 km/h. How far from Delhi would the express train meet the freight train?
   [UPSC CSAT 2017]
   (a) 480 km
   (b) 260 km
   (c) 240 km
   (d) 120 km

3. Two stations, A and B are 827 km apart from each other. One train starts from station A at 5 am and travel towards station B at 62 km/h. Another train starts from station B at 7 am and travel towards station A at 59 km/h. At what time will they meet? [SBI PO (Pre) 2015]
   (a) 1 : 00 pm
   (b) 11 : 45 am
   (c) 12 : 48 : 35 pm
   (d) 11 : 30 : 30 am
   (e) 1 : 37 : 45 am

4. Two stations A and B, 100 km far away to each other. Two trains starts at the same time from station A and station B. The train starts from station A is running with the speed of 50 km/h to station B. The train starts from station B is running with the speed of 75 km/h to station A. At what distance both the trains meet with each other from station A? [SSC CGL 2012]
   (a) 40 km
   (b) 20 km
   (c) 30 km
   (d) None of these

5. Train A, travelling at S m/s, can cross a platform double its length in 21 s. The same train, travelling at $S + 5$ m/s, can cross the same platform in 18 s. What is the value of S? [SBI Clerk (Pre) 2016]
   (a) 27.5 m/s
   (b) 32.5 m/s
   (c) 30 m/s
   (d) 35 m/s
   (d) 25 m/s

6. A train travelling at 48 km/h completely crosses an another train having half length of first train and travelling in opposite directions at 42 km/h in 12 s. It also passes a railway platform in 45 s. What is the length of the platform?
   [CDS 2016 (II)]
   (a) 600 m
   (b) 400 m
   (c) 300 m
   (d) 200 m

7. A passenger train departs from Delhi at 6 pm for Mumbai. At 9 pm, an express train, whose average speed exceeds that of the passenger train by 15 km/h, leaves Mumbai for Delhi. Two trains meet each other mid-route. At what time do they meet, given that the distance between the cities is 1080 km?
   [CDS 2017 (I)]
   (a) 4 pm
   (b) 2 am
   (c) 12 midnight
   (d) 6 am

8. The average speed of a train in the onward journey is 25% more than that in the return journey. The train halts for 2 h on reaching the destination. The total time taken to complete to and fro journey is 32 h, covering a distance of 1600 km. Find the speed of the train in the onward journey.
   (a) 56.25 km/h
   (b) 60 km/h
   (c) 66.50 km/h
   (d) 67 km/h
   (e) None of the above

9. A train overtakes two persons walking along a railway track. The first one walks at 4.5 km/h and the other one walks at 5.4 km/h. The train needs 8.4 s and 8.5 s respectively, to overtake them. What is the speed of the train, if both the persons are walking in the same direction as the train?
   (a) 66 km/h
   (b) 72 km/h
   (c) 78 km/h
   (d) 81 km/h
   (e) None of these

Problems Based on Trains / 461

10. The speed of a car is 6/5 speed of train. These cover the distance of 100 km in same time while car stops for 10 min on the way for refuelling. What is speed of train? **[NIFT 2015]**
(a) 80 km/h (b) 90 km/h
(c) 100 km/h (d) 110 km/h

11. An express train travels 299 km between two cities. During the first 111 km of the trip, the train travelled through mountainous terrain. The train travelled 10 km/h slower through mountainous terrain than through level terrain. If the total time to travel between two cities was 7 h, then what is the speed of the train on level terrain? **[MAT 2015]**
(a) 55 km/h (b) 56 km/h
(c) 47 km/h (d) 88 km/h

12. To travel 720 km, an express train takes 6 h more than duronto. If however, the speed of the express train is doubled, it takes 2 h less than duronto. The speed of duronto is **[SSC CGL (Pre) 2017]**
(a) 60 km/h (b) 72 km/h
(c) 66 km/h (d) 78 km/h

Answer with Solutions

Exercise 1 — Base Level Questions

1. (d) We know that the distance covered by a train in passing a pole or a standing man or a signal post or any other object (of negligible length) is equal to the length of the train. In this case, train covers 90 m to cross a standing man.
∴ Length of the train = 90 m

2. (c) We know that,
$$\text{Speed} = \frac{\text{Distance}}{\text{Time}}$$
Speed of the train $= \frac{220}{12} \times \frac{18}{5}$
$= 66$ km/h

3. (b) Let the length of the train be x m.
Now, speed = 48 km/h $= 48 \times \frac{5}{18}$ m/s
Train takes 9 s to cross a pole.
∴ Length of train = Speed × Time
$= 48 \times \frac{5}{18} \times 9 = 120$ m

4. (c) ∴ Length of train = Speed of train × Time taken to cross the stationary object
$= \frac{72 \times 5 \times 15}{18} = 300$ m

5. (a) Let the lengths of the trains A and B be a and $2a$, respectively.
When a train crosses a pole, it covers the distance equal to its length.
∴ Required ratio of speeds $= \frac{a}{25} : \frac{2a}{75} = 3 : 2$

6. (a) Let length of bridge be x m.
We know that,
$$\text{Speed} = \frac{\text{Distance}}{\text{Time}}$$
According to the question,
$90 \times \frac{5}{18} = \frac{150 + x}{26}$
⇒ $25 \times 26 = 150 + x$
⇒ $650 = 150 + x$
∴ $x = 500$ m

7. (d) Let length of the train be x m.
Then, distance covered $= (x + 600)$ m
According to the question,
$\frac{x + 600}{30} = 30$ ⇒ $x + 600 = 900$
∴ $x = 900 - 600 = 300$ m

8. (d) Length of platform is not given. So, it cannot be determined.

9. (c) Speed of train = 60 km/h
$= 60 \times \frac{5}{18}$ m/s $= \frac{50}{3}$ m/s
Let length of the platform be x m.
According to the question,
$200 + x = \frac{50}{3} \times 30$
⇒ $200 + x = 500$ ⇒ $x = 300$ m

10. (a) Let the length of the train be x m.
According to the question,
$\frac{x}{9} = $ Speed ...(i)
and $\frac{x + 150}{15} = $ Speed ...(ii)

From Eqs. (i) and (ii), we get
$$\frac{x}{9} = \frac{x+150}{15} \Rightarrow \frac{x}{3} = \frac{x+150}{5}$$
$$\Rightarrow 5x = 3x + 450 \Rightarrow x = 225 \text{ m}$$

11. (c) Let length of the train be L m.
According to the question,
$$\frac{L}{2} = \frac{L+250}{7}$$
$$\Rightarrow 7L = 2L + 500$$
$$\Rightarrow 7L - 2L = 500 \Rightarrow 5L = 500$$
$$\therefore L = \frac{500}{5} = 100 \text{ m}$$

12. (c) Given, length of train = 328 m
and length of platform = 354 m
∴ To cross the platform the train has to cover a distance = Length of train + Length of platform = (328 + 354) m = 682 m
Now speed of train $A = \frac{682}{11} = 62$ m/s
∴ Speed of train $B = \frac{7}{8} \times 62$
$$= \frac{7 \times 31}{4} = \frac{217}{4} \text{ m/s}$$
Let the length of train B be l m, then
$$(l + 354) = \frac{217}{4} \times 12$$
[∵ distance = speed × time]
$$\Rightarrow l + 354 = 651 \Rightarrow l = 297 \text{ m}$$

13. (a) Let length of the train = L
According to the question,
$$\frac{L + 300}{21} = \frac{L + 240}{18}$$
$$\Rightarrow \frac{L + 300}{7} = \frac{L + 240}{6}$$
$$\Rightarrow 6L + 1800 = 7L + 1680$$
$$\therefore L = 120 \text{ m}$$
Taking the length of the 2nd bridge into consideration,
Speed of train = $\frac{L + 240}{18} = \frac{120 + 240}{18}$ m/s
$$= \frac{360}{18} \times \frac{18}{5} \text{ km/h}$$
$$= 72 \text{ km/h}$$

14. (c) Length of the train = 280 m
Length of the platform = 280 × 3 = 840 m
Taken time = 6 min 40 s
= (360 + 40) s = 400 s
∴ Required speed = $\frac{\text{Distance}}{\text{Time}}$
$$= \frac{840 + 280}{400} = \frac{1120}{400} = 2.8 \text{ m/s}$$

15. (b) Speed = 60 km/h = $60 \times \frac{5}{18}$ m/s
$$= \frac{50}{3} \text{ m/s}$$
∴ Required time = $\frac{110 + 170}{50/3}$
$$= 280 \times \frac{3}{50} = 16.8 \text{ s}$$

16. (a) Let speed of 1st train be $8x$ and speed of 2nd train be $9x$.
According to the question,
Speed of second train = $\frac{360}{4} = 90$ km/h
∴ $9x = 90 \Rightarrow x = 10$
So, speed of 1st train = $8 \times 10 = 80$ km/h
∴ Required distance = $80 \times 3 = 240$ km

17. (c) Average speed of train
$$= \frac{240}{12} = 20 \text{ km/h}$$
According to the question,
Average speed of bus = $\frac{3}{4} \times 20 = 15$ km/h
∴ Required distance = $15 \times 7 = 105$ km

18. (b) Let distance $QR = x$ km

←——27 km——→
P•————————•————————•R
    24 km/h Q   18 km/h

Then, $\frac{27 + x}{24} = \frac{x}{18}$
[∵ both the trains meet at point R]
$$\Rightarrow \frac{27 + x}{4} = \frac{x}{3} \Rightarrow 81 + 3x = 4x$$
$$\Rightarrow x = 81 \text{ km}$$

19. (c) Relative speed of train
= Speed of train + Speed of car
$$\Rightarrow 90 = \text{Speed of train} + 15$$
∴ Speed of train = 90 − 15 = 75 km/h

20. (b) We have,
Speed of train = 30 km/h
and speed of man = 3 km/h
∴ Relative speed of train = Speed of train − Speed of man
= (30 − 3) km/h = 27 km/h
$$= 27 \times \frac{5}{18} \text{ m/s} = \frac{15}{2} \text{ m/s}$$
To cross the man, train have to cover the distance equal to its length.
∴ Time to cross = $\frac{\text{Distance}}{\text{Speed}}$
$$= \frac{225}{\left(\frac{15}{2}\right)} \text{ s} = 30 \text{ s}$$

Problems Based on Trains / 463

21. (a) Speed of faster train is 1.5 km/min,
i.e. 90 km/h or 25 m/s
and speed of slower train = 60 km/h = $\frac{50}{3}$ m/s
∵ Both trains are moving in same direction.
∴ Relative speed of train = $\left(25 - \frac{50}{3}\right)$ m/s
$= \frac{25}{3}$ m/s
Time taken by the faster train to cross the slower train = 27 s
∴ Distance = Relative speed × Time taken
$= \frac{25}{3} \times 27 = 225$ m
Hence, the length of slower train is 225 m.

22. (b) Speed of the train relative to man
$= (240 + 24)$ km/h $= 264$ km/h
$= 264 \times \frac{5}{18}$ m/s $= \frac{220}{3}$ m/s
Distance covered in passing the man = 440 m
∴ Time taken $= \frac{440}{220} \times 3 = 6$ s

23. (a) Relative speed of train
$= \frac{400}{36}$ m/s $= \frac{400}{36} \times \frac{18}{5} = 40$ km/h
Relative speed of train
= Speed of train + Speed of man
⇒ 40 = Speed of train + 20
∴ Speed of train = 40 − 20 = 20 km/h

24. (a) Here, $x = 70$ m, $y = 90$ m, $u = 10$ m/s
and $v = 6$ m/s
According to the formula,
Required time $= \frac{x+y}{u+v}$
$= \frac{70+90}{10+6} = \frac{160}{16} = 10$ s

25. (a) Relative speed $= (84 + 60)$ km/h
$= 144$ km/h $= 144 \times \frac{5}{18}$ m/s
$= 8 \times 5 = 40$ m/s
Distance covered in passing each other
$= 512 + 528 = 1040$ m
∴ Required time $= \frac{1040}{40} = 26$ s

Fast Track Method
Here, $x = 512$ m, $y = 528$ m,
$u = 84$ km/h $= 70/3$ m/s
and $v = 60$ km/h $= 50/3$ m/s
According to the formula,
∴ Required time $= \frac{x+y}{u+v} = \frac{512+528}{\frac{70}{3}+\frac{50}{3}}$
$= \frac{1040}{120} \times 3 = 26$ s

26. (b) Total length of the train
$= 105 + 90 = 195$ m
Relative speed = 72 + 45 km/h
$= 117$ km/h $= \left(117 \times \frac{5}{18}\right) = \frac{585}{18}$ m/s
∴ Required time $= \left(195 \times \frac{18}{585}\right)$ s $= 6$ s

Fast Track Method
Here, $x = 105$ m, $y = 90$ m,
$u = 45$ km/h $= 12.5$ m/s
$v = 72$ km/h $= 20$ m/s
According to the formula,
Required time $= \frac{x+y}{u+v} = \frac{105+90}{12.5+20}$
$= \frac{195}{32.5} = 6$ s

27. (b) Total distance $= x + y$
$= 180 + 240 = 420$ m
Relative speed $= u + v = (30 + 40) \times \frac{5}{18}$
$= \frac{70 \times 5}{18}$ m/s
∴ Required time $= \left(\frac{x+y}{u+v}\right) = \frac{420 \times 18}{70 \times 5}$
$= 21.6$ s

28. (c) Required time
$= \frac{\text{Distance covered by train}}{\text{Relative speed of train}}$
$= \frac{180+120}{(72+108) \times \frac{5}{18}}$
$= \frac{300 \times 18}{180 \times 5} = 6$ s

29. (c) Let length of each train be x m.
Then, $(65 + 85) \times \frac{5}{18} = \frac{x+x}{6}$
⇒ $\frac{150 \times 5 \times 6}{18} = 2x$
⇒ $2x = 250$ ⇒ $x = 125$ m

30. (a) Let the speeds of two trains be x and y, respectively.
∴ Length of 1st train $= 54x$
and length of the 2nd train $= 34y$
According to the question,
$\frac{54x + 34y}{x+y} = 46$
⇒ $54x + 34y = 46x + 46y$
⇒ $27x + 17y = 23x + 23y$
⇒ $4x = 6y$ ⇒ $2x = 3y$
⇒ $\frac{x}{y} = \frac{3}{2}$ ⇒ $x:y = 3:2$

31. (c) Length of the fast train
= Relative speed × Time
= $(40 - 20) \times \dfrac{5}{18} \times 5 = 27\dfrac{7}{9}$ m

32. (d) Here, $x = 50$ m, $y = 65$ m, $u = 18$ m/s and $v = 17$ m/s
According to the formula,
Required time = $\dfrac{x+y}{u-v}$
$= \dfrac{50+65}{18-17} = \dfrac{115}{1} = 115$ s

33. (b) Let the length of each train be x m.
Relative speed = $46 - 36 = 10$ km/h
$= \dfrac{10 \times 5}{18}$ m/s $= \dfrac{25}{9}$ m/s
Now, $\dfrac{\text{Sum of length of train}}{\text{Relative speed of train}} =$ Time taken
$\Rightarrow \dfrac{2x}{\dfrac{25}{9}} = 36 \Rightarrow 2x = \dfrac{36 \times 25}{9} = 100$
$\therefore x = 50$ m

34. (b) Let the trains meet after time t at a distance x from station N, then another train coming from station M covers a distance of $(x + 50)$.
For station M, $(x + 50) = 125t$
$\Rightarrow \qquad x = 125t - 50 \qquad$ …(i)
For station N, $x = 75t \qquad$ …(ii)
From Eqs. (i) and (ii), we get
$75t = 125t - 50 \Rightarrow t = 1$ h
$\therefore$ Distance between stations M and N
$= 125 t + 75 t$
$= 200 \times 1 = 200$ km
Fast Track Method
Here, $a = 125$ km/h, $b = 75$ km/h and $d = 50$ km
According to the formula,
Distance between the stations M and N
$= \left(\dfrac{a+b}{a-b}\right) \times d \qquad$ [by Technique 2]
$= \left(\dfrac{125+75}{125-75}\right) \times 50 = \dfrac{200}{50} \times 50$
$= 200$ km

35. (c) Here, $a = 150$ km/h, $b = 180$ km/h, $t_1 = 8 : 00$ am and $t_2 = 10 : 30$ am
$\therefore$ Required distance = $(t_2 - t_1) \dfrac{(a \times b)}{(b-a)}$
[by Technique 3]
$= (10 : 30 - 8 : 00) \left(\dfrac{150 \times 180}{180 - 150}\right)$
$= 2 : 30 \times \left(\dfrac{150 \times 180}{30}\right)$

$= \dfrac{5}{2} \times \left(\dfrac{150 \times 180}{30}\right)$
$= \dfrac{150 \times 180}{6 \times 2} = 2250$ km

36. (d) Let the speed of the train be x.
According to the question,
$(x + 10) \times 12 = (x + 20) \times 10$
$\Rightarrow \quad 6x + 60 = 5x + 100$
$\Rightarrow \quad x = 100 - 60 = 40$ m/s
$\therefore$ Length of the train $= (x + 10) \times 12$
$= (40 + 10) \times 12 = 600$ m
Fast Track Method
Here, $t_1 = 12$ s, $t_2 = 10$ s, $a = 10$ m/s and $b = 20$ m/s
According to the formula,
Length of the train
$= \dfrac{\text{Difference in speeds} \times t_1 \times t_2}{t_1 - t_2}$
[by Technique 4]
$= \dfrac{(20-10) \times 10 \times 12}{12 - 10}$
$= \dfrac{10 \times 10 \times 12}{2} = \dfrac{1200}{2} = 600$ m

37. (a) Given, $t_1 = 16$ h, $t_2 = 9$ h and $a = 90$ km/h
According to the formula,
Speed of $B = a\sqrt{\dfrac{t_1}{t_2}} \qquad$ [by Technique 5]
$= 90 \times \sqrt{\dfrac{16}{9}} = 90 \times \dfrac{4}{3}$
$= 30 \times 4 = 120$ km/h

38. (d) Given, $a = 45$ km/h, $t_1 = 4$ h 48 min and $t_2 = 3$ h 20 min
$\therefore$ Speed of train $B = a\sqrt{\dfrac{t_1}{t_2}}$ [by Technique 5]
$= 45 \sqrt{\dfrac{4 \text{ h and } 48 \text{ min}}{3 \text{ h and } 20 \text{ min}}}$
$= 45 \sqrt{\dfrac{\dfrac{24}{5} \text{ h}}{\dfrac{10}{3} \text{ h}}} = 45 \sqrt{\dfrac{24 \times 3}{5 \times 10}}$
$= 45 \times \sqrt{1.44} = 45 \times 1.2 = 54$ km/h

39. (b) First train leaves at 8 : 00 am and second at 9 : 00 am.
So, first train, i.e. from P to Q has covered 25 km distance in 1 h. So, distance left between the station $= 145 - 25 = 120$ km
Now, trains are travelling in opposite directions.
So, relative speed $= 25 + 35 = 60$ km/h

Problems Based on Trains / 465

Time taken to cover 120 km = $\frac{120}{60}$ = 2 h

∴ The time, at which both the trains will meet, is 2 h after second train left,
i.e. 9 : 00 am + 2 h = 11 : 00 am

Fast Track Method
Here, d = 145 km,
t = 9 : 00 am − 8 : 00 am = 1 h,
a = 25 km/h and b = 35 km/h
∴ Time taken by trains to meet

$$T = \left(\frac{d + tb}{a + b}\right) \quad \text{[by Technique 6]}$$

$$= \frac{145 + 1 \times 35}{25 + 35} = \frac{180}{60} = 3h$$

So, they will meet at 8 : 00 am + 3 h
= 11 : 00 am

40. (c) Let the speed of both trains be x km/h and $(x + 6)$ km/h, respectively.
Then, according to the question,
$$160 = x \times 4 + (x + 6) \times 4$$
⇒ $160 = 4x + 4x + 24$
⇒ $40 = x + x + 6$
⇒ $2x + 6 = 40$
⇒ $2x = 34$
∴ $x = 17$

Hence, speed of both the trains are 17 km/h and (17 + 6) km/h, i.e. 23 km/h.

Fast Track Method
Here, d = 160 km, t = 4 h and x = 6 km/h.
∴ Speed of faster train = $\left(\frac{d + tx}{2t}\right)$ km/h

[by Technique 7]

$= \frac{160 + 24}{8} = 23$ km/h

Speed of slower train = $\left(\frac{d - tx}{2t}\right)$

$= \frac{160 - 24}{8} = 17$ km/h

41. (b) Here, $t_1 = 2$ h, $t_2 = 5$ h, $a = 6$ km/h and $d = ?$
Distance between Delhi and Meerut is
$$d = a\left(\frac{t_1 t_2}{t_2 - t_1}\right) \quad \text{[by Technique 8]}$$
$$= 6\left(\frac{2 \times 5}{5 - 2}\right) = \frac{6 \times 10}{3} = 20 \text{ km}$$

Speed of train $= \frac{at_2}{t_2 - t_1} = \frac{6 \times 5}{5 - 2} = \frac{30}{3}$
$= 10$ km/h

42. (c) Because of stoppages, train covers 18 km less per hour.
∴ Time taken to cover 18 km = $\frac{18}{108}$

$= \frac{1}{6}$ h $= \frac{1}{6} \times 60 = 10$ min

Fast Track Method
Here, a = 108 km/h and b = 90 km/h
According to the formula,

Required rest time = $\frac{a - b}{a}$

$= \frac{108 - 90}{108}$ [by Technique 9]

$= \frac{18}{108} = \frac{1}{6}$ h

$= \frac{1}{6} \times 60$ min = 10 min

43. (b) Given that, $t_1 = 6$ s and $t_2 = 9$ s
Then, time taken by the trains to cross each other $= \frac{2t_1 t_2}{t_2 - t_1}$ [by Technique 10]

$= \frac{2 \times 6 \times 9}{9 - 6} = 36$ s

Exercise 2 *Higher Skill Level Questions*

1. (b) Let the speed of slower train be S km/h.
∴ Speed of faster train = (S + 8) km/h
According to the question,
$\frac{162}{S + S + 8} = 6 \Rightarrow 162 = 12S + 48$
⇒ $12S = 114$
⇒ $S = \frac{114}{12} = 9\frac{1}{2}$ km/h

2. (c) Let the distance covered by freight and express train be 'x' km in time 't' hour and $(t - 2)$ hour, respectively.

According to the question,
$40 \times t = 60 \times (t - 2)$
⇒ $40t = 60t - 120$
⇒ $20t = 120$
⇒ $t = 6$ h
∴ Required distance = $60(6 - 2)$
$= 60 \times 4 = 240$ km

3. (c) Let the two trains meet at x h after 5 am
∴ Time taken by second train is $(x - 2)$ h.
Speed of first train = 62 km/h
Speed of second train = 59 km/h
and total distance covered = 827 km

According to the question,
$62x + 59(x-2) = 827$
$62x + 59x - 118 = 827$
$\Rightarrow 121x = 827 + 118 \Rightarrow 121x = 945$
$\Rightarrow x = \dfrac{945}{121} = 7$ h 48 min 35 s

Thus, the trains will meet at 7 h 48 min 35 s after 5 am, i.e. at 12 : 48 : 35 pm.

4. (a) Let both the trains meet each other at x km from station A.

```
        |←————— 100 km —————→|
        |←— x —→|←—(100 – x)—→|
        A ←50 km/h→  ←75 km/h→ B
```

Then, $\dfrac{x}{50} = \dfrac{100 - x}{75}$

$\Rightarrow 3x = 200 - 2x \Rightarrow 5x = 200$
$\Rightarrow x = 40$ km

5. (c) Let length of train = l m
Then, length of platform = $2l$ m

$\therefore$ Speed $= \dfrac{\text{Distance}}{\text{Time}}$

$\Rightarrow S = \dfrac{l + 2l}{21} \Rightarrow S = \dfrac{3l}{21}$

$\Rightarrow S = \dfrac{l}{7} \Rightarrow l = 7S$ ...(i)

Again, when speed is $(S + 5)$ m/s and time = 18 s

Then, Speed $= \dfrac{\text{Distance}}{\text{Time}}$

$\Rightarrow (S + 5) = \dfrac{l + 2l}{18}$

$\Rightarrow (S + 5) = \dfrac{3l}{18} \Rightarrow (S + 5) = \dfrac{l}{6}$

$\Rightarrow (S + 5) = \dfrac{7S}{6}$ [from Eq. (i)]

$\Rightarrow \dfrac{7S}{6} - S = 5 \Rightarrow \dfrac{S}{6} = 5$

$\Rightarrow S = 5 \times 6$
$\Rightarrow S = 30$ m/s

6. (b) Let the length of the first train be x m.
Then, the length of second train is $\left(\dfrac{x}{2}\right)$ m.

$\therefore$ Relative speed = $(48 + 42)$ km/h
$= \left(90 \times \dfrac{5}{18}\right)$ m/s = 25 m/s

According to the question,

$\dfrac{\left(x + \dfrac{x}{2}\right)}{25} = 12 \Rightarrow \dfrac{3x}{2} = 300 \Rightarrow x = 200$ m

$\therefore$ Length of first train = 200 m
Let the length of platform be y m.
Speed of the first train

$= \left(48 \times \dfrac{5}{18}\right)$ m/s $= \dfrac{40}{3}$ m/s

$\therefore$ Time $= \dfrac{\text{Distance}}{\text{Speed}}$

$\Rightarrow (200 + y) \times \dfrac{3}{40} = 45$

$\Rightarrow 600 + 3y = 1800$
$\Rightarrow y = 400$ m

7. (d) Let the speed of passenger train = x km/h
$\therefore$ Speed of express train = $(x + 15)$ km/h

```
  6 pm                          9 pm
   •————540 km————•————540 km————•
   D                              M
```

Then, according to the question,

$\dfrac{540}{x} - \dfrac{540}{x + 15} = 3$

$\Rightarrow 540\left(\dfrac{1}{x} - \dfrac{1}{x + 15}\right) = 3$

$\Rightarrow 540\left[\dfrac{x + 15 - x}{x(x + 15)}\right] = 3$

$\Rightarrow 540 \times 15 = 3x(x + 15)$
$\Rightarrow 2700 = x^2 + 15x$
$\Rightarrow x^2 + 15x - 2700 = 0$
$\Rightarrow (x + 60)(x - 45) = 0$
$\Rightarrow x = 45$ [$\because x \neq -60$]

$\therefore$ Time taken by passenger train to reach the meeting point $= \dfrac{540}{45} = 12$ h

$\therefore$ Both train will meet at (6 pm + 12 h) = 6 am

8. (b) Let speed in the return journey = x km/h
$\therefore$ Speed in onward journey
$= \dfrac{125}{100}x = \left(\dfrac{5}{4}x\right)$ km/h

Average speed $= \dfrac{2 \times \dfrac{5}{4} x \times x}{\dfrac{5}{4}x + x}$

$= \dfrac{10x}{9}$ km/h

$\therefore 1600 \times \dfrac{9}{10x} = 30$ [$\because$ train halts for 2 h]

$\Rightarrow x = \dfrac{1600 \times 9}{30 \times 10} = 48$ km/h

$\therefore$ Speed in onward journey $= \dfrac{5}{4}x = \dfrac{5}{4} \times 48$
$= 60$ km/h

9. (d) Speed of Ist person
$= 4.5$ km/h $= \left(4.5 \times \dfrac{5}{18}\right)$ m/s
$= \dfrac{5}{4}$ m/s $= 1.25$ m/s

Speed of IInd person = 5.4 km/h

$= \left(5.4 \times \dfrac{5}{18}\right)$ m/s $= \dfrac{3}{2}$ m/s = 1.5 m/s

Let the speed of the train be x m/s.
Then, $(x - 1.25) \times 8.4 = (x - 1.5) \times 8.5$
$\Rightarrow \quad 8.4x - 10.5 = 8.5x - 12.75$
$\Rightarrow \quad 0.1x = 2.25$
$\Rightarrow \quad x = 22.5$ m/s
$\therefore$ Speed of the train $= \left(22.5 \times \dfrac{18}{5}\right)$
$\qquad = 81$ km/h

10. (c) Let speed of train be u km/h.

$\therefore$ Speed of the car $= \dfrac{6}{5} u$ km/h

According to the question,

$\dfrac{100}{u} - \dfrac{100}{\frac{6}{5}u} = \dfrac{10}{60}$ $\left[\because \text{time} = \dfrac{\text{distance}}{\text{speed}}\right]$

$\Rightarrow \quad \dfrac{100}{u} - \dfrac{100 \times 5}{6u} = \dfrac{1}{6}$

$\Rightarrow \quad \dfrac{600 - 500}{6u} = \dfrac{1}{6}$

$\Rightarrow \quad \dfrac{100}{6u} = \dfrac{1}{6}$

$\therefore \quad u = 100$ km/h

11. (c) Let the speed of train be v km/h on mountainous terrain.

$\because$ Time taken $= \dfrac{111}{v}$

Distance travelled on level terrain
$\qquad = 299 - 111 = 188$ km
Speed on level terrain $= (v + 10)$ km/h

$\because \quad$ Time taken $= \dfrac{188}{v + 10}$

Total time taken $= \dfrac{111}{v} + \dfrac{188}{v + 10} = 7$

$\Rightarrow \quad (v + 10)\,111 + 188\,v = 7\,(v^2 + 10\,v)$
$\Rightarrow \quad 111v + 1110 + 188v = 7v^2 + 70v$
$\Rightarrow \quad 7v^2 - 229\,v - 1110 = 0$

On comparing with $av^2 + bv + c = 0$, we get
$\qquad a = 7, b = -229$ and $c = -1110$

Now, $v = \dfrac{-b \pm \sqrt{b^2 - 4ac}}{2a}$

$= \dfrac{229 \pm \sqrt{(229)^2 - 4 \times 7 \times (-1110)}}{14}$

$= \dfrac{229 \pm 289}{14}$

On taking positive sign, we get
$\qquad v = \dfrac{229 + 289}{14} = \dfrac{518}{14} = 37$

Again, on taking negative sign, we get
$\qquad v = \dfrac{229 - 289}{14} = \dfrac{-60}{14}$

$\Rightarrow \quad v = -4.2857$

$\therefore \quad v = 37$ km/h [neglecting (– ve) value]

So, speed of the train on level terrain,
$\qquad (v + 10) = 37 + 10$
$\qquad \qquad = 47$ km/h

12. (b) Given, distance = 720 km
Let time taken by duronto = x h
Then, time taken by express train
$\qquad = (x + 6)$ h
Now, speed of duronto train
$\qquad = \dfrac{720}{x}$ km/h ...(i)

and speed of express train
$\qquad = \left(\dfrac{720}{x + 6}\right)$ km/h ...(ii)

Now, speed of express train is doubled, it takes 2 h less from duronto train.
i.e. 2 × speed of express train
$\qquad = \left(\dfrac{720}{x - 2}\right)$ km/h ...(iii)

Now, from Eqs. (ii) and (iii), we have

$\qquad 2 \times \dfrac{720}{x + 6} = \dfrac{720}{x - 2}$

$\Rightarrow \quad \dfrac{2}{x + 6} = \dfrac{1}{x - 2}$

$\Rightarrow \quad 2x - 4 = x + 6$
$\Rightarrow \quad 2x - x = 6 + 4$
$\Rightarrow \quad x = 10$ h

$\therefore$ Speed of duronto $= \dfrac{720}{x} = \dfrac{720}{10} = 72$ km/h

Chapter 27

Boats and Streams

Boats and streams is an application of concepts of speed, time and distance. Speed of river flowing either aides a swimmer (boat), while travelling with the direction of river or it opposes when travelling against the direction of river.

Here, we will explain the following concepts

Still Water If the speed of water of a river is zero, then water is considered to be still water.

Stream If the water of a river is moving at a certain speed, then it is called as stream.

Speed of Boat Speed of boat means speed of boat (swimmer) in still water. In other words, if the speed of a boat (swimmer) is given, then that particular speed is the speed in still water.

Downstream Motion If the motion of a boat (swimmer) is along the direction of stream, then such motion is called downstream motion.

Upstream Motion If the motion of a boat (swimmer) is against the direction of stream, then such motion is called upstream motion.

Basic Formulae Related to Boats and Streams

If the speed of a boat in still water is x km/h and speed of the stream is y km/h, then

1. Speed downstream = $(x + y)$ km/h
2. Speed upstream = $(x - y)$ km/h
3. Speed of a boat in still water,
$$x = \frac{1}{2} \text{ (Speed downstream + Speed upstream)}$$
4. Speed of stream,
$$y = \frac{1}{2} \text{ (Speed downstream - Speed upstream)}$$

Boats and Streams / 469

Ex. 1 A man can row with a speed of 6 km/h in still water. What will be his speed with the stream, if the speed of stream is 2 km/h?

Sol. Given, speed of man in still water, $x = 6$ km/h and speed of stream, $y = 2$ km/h
∴ Speed downstream $= x + y = 6 + 2 = 8$ km/h

Ex. 2 If the speed of a boat in still water is 8 km/h and the rate of stream is 4 km/h, then find upstream speed of the boat.

Sol. Given, speed of a boat, $x = 8$ km/h and speed of stream, $y = 4$ km/h
∴ Speed upstream $= x - y = 8 - 4 = 4$ km/h

Ex. 3 Shantanu can row upstream at 10 km/h and downstream at 18 km/h. Find Shantanu's rate in still water and the rate of the current.

Sol. Given, speed upstream $= 10$ km/h and speed downstream $= 18$ km/h
According to the formula,

Shantanu's rate in still water $= \dfrac{1}{2}$ (Speed downstream + Speed upstream)

$= \dfrac{1}{2}(18 + 10) = \dfrac{28}{2} = 14$ km/h

∴ Speed of current $= \dfrac{1}{2}$ (Speed downstream − Speed upstream)

$= \dfrac{1}{2}(18 - 10) = \dfrac{8}{2} = 4$ km/h

Ex. 4 What time will be taken by a boat to cover a distance of 64 km along the stream, if speed of boat in still water is 12 km/h and speed of stream is 4 km/h?

Sol. Given that, distance $= 64$ km, speed of boat in still water, $(x) = 12$ km/h
and speed of stream, $(y) = 4$ km/h
∵ Downstream speed of boat $= x + y = 12 + 4 = 16$ km/h

∴ Required time $= \dfrac{\text{Distance}}{\text{Speed (downstream)}} = \dfrac{64}{16} = 4$ h

Ex. 5 A boat takes 8 h to row 48 km downstream and 12 h to row the same distance upstream. Find the boat's rate in still water and rate of current.

Sol. ∵ Speed downstream $= \dfrac{\text{Distance}}{\text{Time}} = \dfrac{48}{8} = 6$ km/h

and speed upstream $= \dfrac{48}{12} = 4$ km/h

Now, rate of boat in still water $= \dfrac{(\text{Speed downstream} + \text{Speed upstream})}{2}$

$= \dfrac{6 + 4}{2} = 5$ km/h

and rate of current $= \dfrac{(\text{Speed downstream} - \text{Speed upstream})}{2}$

$= \dfrac{6 - 4}{2} = 1$ km/h

Fast Track Techniques to solve the QUESTIONS

Technique 1

If speed of stream is a and a boat (swimmer) takes n times as long to row up as to row down the river, then

$$\text{Speed of boat (swimmer) in still water} = \frac{a(n+1)}{(n-1)}$$

♦ This formula is applicable for equal distances.

Ex. 6 Rajnish can row 12 km/h in still water. It takes him twice as long to row up as to row down the river. Find the rate of stream.

Sol. Let rate of stream = a km/h and distance travelled = y km
Given, rate of Rajnish in still water = 12 km/h
Then, rate of downstream = $(12 + a)$ km/h
and rate of upstream = $(12 - a)$ km/h
According to the question,
  Time taken to travel downstream = 2 × Time taken to travel upstream

$$\Rightarrow \frac{12 + a}{y} = \frac{2(12 - a)}{y}$$

$$\Rightarrow 12 + a = 2(12 - a)$$
$$\Rightarrow 12 + a = 24 - 2a$$
$$\Rightarrow a + 2a = 24 - 12$$
$$\Rightarrow 3a = 12$$
$$\Rightarrow a = \frac{12}{3} = 4 \text{ km/h}$$

Fast Track Method

Here, speed of Rajnish in still water = 12 km/h, $n = 2$
and speed of stream $(a) = ?$
According to the formula,

$$\text{Speed in still water} = \frac{a(n+1)}{(n-1)}$$

$$\Rightarrow 12 = \frac{a(2+1)}{(2-1)} \Rightarrow 3a = 12$$

$$\therefore a = \frac{12}{3} = 4 \text{ km/h}$$

Technique 2

A person can row at a speed of x in still water. If stream is flowing at a speed of y, it takes time T to row to a place and back, then

$$\text{Distance between two places} = \frac{T(x^2 - y^2)}{2x}$$

Ex. 7 A man can row 12 km/h in still water. When the river is running at 2.4 km/h, it takes him 1 h to row to a place and to come back. How far is the place?

Sol. Here, man's rate downstream = $12 + 2.4 = 14.4$ km/h
and man's rate upstream = $12 - 2.4 = 9.6$ km/h
Let the required distance be x km.
According to the question,
Total time taken to travel x km upstream and downstream = 1 h

$\Rightarrow \quad \dfrac{x}{14.4} + \dfrac{x}{9.6} = 1 \Rightarrow 9.6x + 14.4x = 14.4 \times 9.6$

$\Rightarrow \quad 24x = 138.24$

$\therefore \quad x = \dfrac{138.24}{24} = 5.76$ km

Fast Track Method
Here, speed of man in still water $(x) = 12$ km/h,
Speed of river $(y) = 2.4$ km/h and $T = 1$ h
According to the formula,

Required distance = $\dfrac{T(x^2 - y^2)}{2x} = \dfrac{1[(12)^2 - (2.4)^2]}{2 \times 12} = \dfrac{138.24}{24} = 5.76$ km

Technique 3 A man rows a certain distance downstream in x h and returns the same distance in y h. When the stream flows at the rate of a km/h, then

Speed of the man in still water = $\dfrac{a(x+y)}{(y-x)}$.

Ex. 8 Kamal can row a certain distance downstream in 12 h and can return the same distance in 18 h. If the stream flows at the rate of 6 km/h, then find the speed of Kamal in still water.

Sol. Let the speed of Kamal in still water = x km/h
Then, Kamal's speed downstream = $(x + 6)$ km/h
and Kamal's speed upstream = $(x - 6)$ km/h
According to the question,
Distance travelled downstream = Distance travelled upstream

$\Rightarrow \quad 12(x+6) = 18(x-6)$
$\Rightarrow \quad 2x + 12 = 3x - 18$
$\Rightarrow \quad 3x - 2x = 18 + 12 \Rightarrow x = 30$ km/h

Fast Track Method
Here, $x = 12$ h, $y = 18$ h and rate of stream, $(a) = 6$ km/h
According to the formula,

Speed of Kamal in still water = $\dfrac{a(x+y)}{(y-x)} = \dfrac{6(12+18)}{(18-12)} = \dfrac{6 \times 30}{6} = 30$ km/h

✦ If in case of technique 3, man's speed in still water is b km/h and we are asked to find the speed of stream, then technique 3 takes the form as

Speed of stream = $\dfrac{b(y-x)}{(x+y)}$.

Ex. 9 If in the above example, the speed of Kamal in still water is 12 km/h, then find the speed of the stream.

Sol. Let the speed of stream = x km/h
Then, Kamal's speed downstream = $(12 + x)$ km/h
and Kamal's speed upstream = $(12 - x)$ km/h
According to the question,
Distance travelled downstream = Distance travelled upstream
$\Rightarrow \quad 12(12 + x) = 18(12 - x)$
$\Rightarrow \quad 2(12 + x) = 3(12 - x)$
$\Rightarrow \quad 24 + 2x = 36 - 3x$
$\Rightarrow \quad 3x + 2x = 36 - 24$
$\Rightarrow \quad 5x = 12$
$\therefore \quad x = \dfrac{12}{5} = 2.4$ km/h

Fast Track Method
Here, $b = 12$, $y = 18$ and $x = 12$
According to the formula,
$$\text{Speed of stream} = \dfrac{b(y - x)}{(x + y)} = \dfrac{12(18 - 12)}{18 + 12} = \dfrac{12 \times 6}{30} = \dfrac{12}{5} = 2.4 \text{ km/h}$$

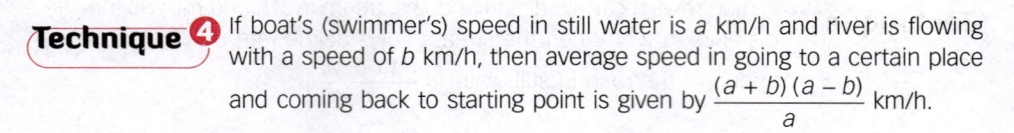

Technique 4 If boat's (swimmer's) speed in still water is a km/h and river is flowing with a speed of b km/h, then average speed in going to a certain place and coming back to starting point is given by $\dfrac{(a + b)(a - b)}{a}$ km/h.

Ex. 10 Ramesh rows in still water with a speed of 4.5 km/h to go to a certain place and to come back. Find his average speed for the whole journey, if the river is flowing with a speed of 1.5 km/h.

Sol. Ramesh's speed upstream = $4.5 - 1.5 = 3$ km/h
Ramesh's speed downstream = $4.5 + 1.5 = 6$ km/h
Let the distance in one direction be x.
Then, time taken in upstream = $\dfrac{x}{3}$ and time taken in downstream = $\dfrac{x}{6}$
$\therefore$ Average speed = $\dfrac{\text{Total distance}}{\text{Total time}} = \dfrac{2x}{\dfrac{x}{3} + \dfrac{x}{6}} = \dfrac{2x \times 18}{6x + 3x} = 4$ km/h

Fast Track Method
Here, $a = 4.5$ km/h and $b = 1.5$ km/h
$\therefore$ Average speed = $\dfrac{(a + b)(a - b)}{a}$
$= \dfrac{(4.5 + 1.5)(4.5 - 1.5)}{4.5}$
$= \dfrac{6 \times 3}{4.5} = \dfrac{18}{4.5} = 4$ km/h

Boats and Streams / 473

Technique 5: When boat's speed (swimmer's speed) in still water is a km/h and river is flowing with a speed of b km/h and time taken to cover a certain distance upstream is T more than the time taken to cover the same distance downstream, then distance $(D) = \dfrac{(a^2 - b^2)T}{2b}$.

Ex. 11 A boat's speed in still water is 10 km/h, while river is flowing with a speed of 2 km/h and time taken to cover a certain distance upstream is 4 h more than time taken to cover the same distance downstream. Find the distance.

Sol. Let the distance be x km.
Then, boat's rate downstream = 10 + 2 = 12 km/h
and boat's rate upstream = 10 − 2 = 8 km/h
According to the question,
Difference between the time = Time taken by boat to travel upstream
− Time taken by boat to travel downstream

$$\Rightarrow \frac{x}{8} - \frac{x}{12} = 4 \Rightarrow \frac{3x - 2x}{24} = 4$$

$$\therefore x = 96 \text{ km}$$

Fast Track Method
Here, $a = 10$ km/h, $b = 2$ km/h and $T = 4$ h
According to the formula,

$$\text{Required distance} = \frac{(a^2 - b^2)}{2b} \times T = \frac{(10^2 - 2^2)}{2 \times 2} \times 4$$

$$= \frac{100 - 4}{4} \times 4 = 100 - 4 = 96 \text{ km}$$

Technique 6: If a man covers l km distance in t_1 h along the direction of river and he covers same distance in t_2 h against the direction of river, then

(i) Speed of man $= \dfrac{1}{2}\left(\dfrac{l}{t_1} + \dfrac{l}{t_2}\right) = \dfrac{l}{2}\left(\dfrac{1}{t_1} + \dfrac{1}{t_2}\right)$

(ii) Speed of stream $= \dfrac{1}{2}\left(\dfrac{l}{t_1} - \dfrac{l}{t_2}\right) = \dfrac{l}{2}\left(\dfrac{1}{t_1} - \dfrac{1}{t_2}\right)$

Ex. 12 A boat covers 20 km in an hour with downstream and covers the same distance in 2 h with upstream. Then, find the speed of boat in still water and speed of stream.

Sol. Here, $l = 20$ km, $t_1 = 1$ h and $t_2 = 2$ h

$$\therefore \text{Speed of boat in still water} = \frac{l}{2}\left(\frac{1}{t_1} + \frac{1}{t_2}\right) = \frac{20}{2}\left(\frac{1}{1} + \frac{1}{2}\right) = 10 \times \frac{3}{2} = 15 \text{ km/h}$$

and $\quad$ Speed of stream $= \dfrac{l}{2}\left(\dfrac{1}{t_1} - \dfrac{1}{t_2}\right) = \dfrac{20}{2}\left(\dfrac{1}{1} - \dfrac{1}{2}\right) = 10 \times \dfrac{1}{2} = 5$ km/h

Fast Track Practice

Exercise 1 Base Level Questions

1. If the speed of a swimmer in still water is 9 km/h. Find the downstream speed of the swimmer, when the river is flowing with the speed of 6 km/h. **[IOB Clerk 2010]**
   (a) 15 km/h (b) 18 km/h
   (c) 3 km/h   (d) 12 km/h
   (e) None of these

2. A boat goes 48 km downstream in 20 h. It takes 4 h more to cover the same distance against the stream. What is the speed of the boat in still water? **[SSC CPO 2013]**
   (a) 2.2 km/h (b) 2 km/h
   (c) 4 km/h   (d) 4.2 km/h

3. A person can row with the stream at 8 km/h and against the stream at 6 km/h. The speed of the current is **[RRB Clerk (Pre) 2017]**
   (a) 1 km/h (b) 2 km/h
   (c) 4 km/h (d) 5 km/h
   (e) None of these

4. A boatman rows 1 km in 5 min along the stream and 6 km in 1 h against the stream. The speed of the stream is **[SSC CGL 2010]**
   (a) 3 km/h   (b) 6 km/h
   (c) 10 km/h (d) 12 km/h

5. The speed of a boat in still water is 500% more than the speed of the current. What is the respective ratio between the speed of the boat downstream and speed of the boat upstream? **[SBI Clerk (Pre) 2016]**
   (a) 9 : 2 (b) 7 : 3 (c) 7 : 5 (d) 9 : 4
   (e) 4 : 3

6. A person can row downstream 20 km in 2 h and upstream 4 km in 2 h. What is the speed of the current? **[CDS 2016 (II)]**
   (a) 2 km/h   (b) 2.5 km/h
   (c) 3 km/h   (d) 4 km/h

7. A man rows 12 km in 5 h against the stream, the speed of current being 4 km/h. What time will be taken by him to row 15 km with the stream? **[SSC CGL 2015]**
   (a) 1 h 26 $\frac{7}{13}$ min (b) 1 h 25 $\frac{7}{13}$ min
   (c) 1 h 24 $\frac{7}{13}$ min (d) 1 h 27 $\frac{7}{13}$ min

8. A boat can travel upstream 13 km and downstream 28 km taking 5 h each time. The velocity of the current is **[IBPS Clerk (Mains) 2017]**
   (a) 0.5 km/h (b) 1 km/h
   (c) 1.5 km/h (d) 2 km/h
   (e) 2.5 km/h

9. A boat can travel with a speed of 16 km/h in still water. If the rate of stream is 5 km/h, then what is the time taken by the boat to cover distance of 84 km downstream? **[SBI Clerk (Mains) 2016]**
   (a) 4 h (b) 5 h (c) 6 h (d) 7 h
   (e) 8 h

10. A motorboat can travel at 10 km/h in still water. It travelled 91 km downstream in a river and then returned to the same place, taking altogether 20 h. The rate of flow of river is **[SSC CGL 2011]**
    (a) 3 km/h (b) 4 km/h
    (c) 2 km/h (d) 5 km/h

11. A man can row against the current three-fourth of a kilometre in 15 min and returns same distance in 10 min, then ratio of his speed to that of current is **[SSC CGL 2010]**
    (a) 3 : 5 (b) 5 : 3 (c) 1 : 5 (d) 5 : 1

12. The speed of the current is 5 km/h. A motorboat goes 10 km upstream and back again to the starting point in 50 min. The speed (in km/h) of the motorboat in still water is **[SSC CPO 2011]**
    (a) 20 (b) 26 (c) 25 (d) 28

13. Speed of motorboat in still water is 45 km/h. If the motorboat travels 80 km along the stream in 1 h 20 min, then the time taken by it to cover the same distance against the stream will be **[SSC CPO 2008]**
    (a) 4 h 20 min (b) 3 h 40 min
    (c) 2 h 40 min (d) 2 h 55 min

14. A boat has to travel upstream 20 km distance from point X of a river to point Y. The total time taken by boat in travelling from point X to Y and Y to X is 41 min 40 s. What is the speed of the boat?

Boats and Streams / 475

(a) 66 km/h (b) 72 km/h
(c) 48 km/h
(d) Cannot be determined

15. A man can row at 10 km/h in still water. If he takes total 5 h to go to a place 24 km away and return, then the speed of the water current is [SSC CGL (Mains) 2012]
(a) 2 km/h (b) 3 km/h
(c) $\frac{1}{2}$ km/h (d) 1 km/h

16. The ratio of speed of a motorboat to that of the current of water is 36 : 5. The motorboat goes along with the current in 5 h 10 min. Find the time to come back of motorboat. [SSC (10+2) 2007]
(a) 5 h 50 min (b) 6 h
(c) 6 h 50 min (d) 12 h 10 min

17. A boat can travel 12.6 km downstream in 35 min. If the speed of the water current is one-fourth the speed of boat in the still water, then what distance the boat can travel upstream in 28 min? [LIC AAO 2016]
(a) 7 km (b) 7.5 km (c) 8.5 km (d) 8 km
(e) 6 km

18. Ashutosh can row 24 km/h in still water. It takes him twice as long as to row up as to row down the river. Find the rate of stream.
(a) 4 km/h (b) 18 km/h
(c) 8 km/h (d) 15 km/h
(e) None of these

19. Pawan can row 24 km/h in still water. When the river is running at 4.8 km/h, it takes him 1 h to row to a place and to come back. How far is the place?
(a) 11.52 km (b) 14 km
(c) 12.52 km (d) 15 km
(e) None of these

20. A man rows in still water with a speed of 6.5 km/h to go to a certain place and come back. Find his average speed for the whole journey, if the river is flowing with a speed of 2.5 km/h.
(a) 6.65 km/h (b) 6.75 km/h
(c) 5.53 km/h (d) 5 km/h

21. A man can row 6 km/h in still water. If the speed of the current is 2 km/h, it takes 3 h more in upstream than in the downstream for the same distance. The distance is [SSC CGL 2012]
(a) 30 km (b) 24 km
(c) 20 km (d) 32 km

22. A boat's speed in still water is 5 km/h, while river is flowing with a speed of 2 km/h and time taken to cover a certain distance upstream is 2 h more than time taken to cover the same distance downstream. Find the distance. [PNB Clerk 2007]
(a) 10.5 km (b) 11 km
(c) 10.9 km (d) 15 km
(e) None of these

23. A boat covers a distance of 30 km downstream in 2 h while it takes 6 h to cover the same distance upstream. What is the speed (in km/h) of the boat? [SNAP 2012]
(a) 5 (b) 7.5 (c) 13 (d) 10

24. A sailor sails a distance of 48 km along the flow of a river in 8 h. If it takes 12 h to return the same distance, then the speed of the flow of the river is [CDS 2013 (I)]
(a) 0.5 km/h (b) 1 km/h
(c) 1.5 km/h (d) 2 km/h

25. A steamer goes downstream from one port to another in 4 h. It covers the same distance upstream in 5 h. If the speed of the stream is 2 km/h, then find the distance between the two ports. [SSC CGL 2007]
(a) 50 km (b) 60 km
(c) 70 km (d) 80 km

Exercise ❷ Higher Skill Level Questions

1. The time taken by a boat to travel 117 km, downstream is 9 h and the same distance upstream is 13 h. The speed of the stream is $\frac{1}{4}$ of the speed of the boat. Find the distance travelled by the boat going upstream in 2 h? [NICL AO (Pre) 2017]
(a) 32 km (b) 14 km
(c) 18 km (d) 20 km
(e) other than these given as options

2. A boat can travel 14.4 km downstream in 32 min. If the speed of the current is 3 km/h, how much time the boat will take to travel 84 km upstream? [IBPS Clerk (Pre) 2016]
(a) 12 h
(b) 8 h
(c) 4 h
(d) 3 h 30 min
(e) 7 h

3. The speed of a boat in still water is 16 km/h and the speed of the current is 2 km/h. The distance travelled by the boat from point A to point B downstream is 12 km more than the distance covered by the same boat from point B to point C upstream in the same time. How much time will the boat take to travel from C to B downstream? **[IBPS SO 2016]**
   (a) 3 h (b) 2 h 30 min
   (c) 3 h 20 min (d) 2 h 20 min
   (e) 2 h

4. The speed of a boat in still water is 14 km/h and the speed of the current is 2 km/h. The time taken by the boat to travel from point A to point B downstream is 1 h less than the time taken to the same boat to travel from point B to point C upstream. If the distance between points A and B is 4 km less than that between points B and C, then what is the distance between points B and C? **[IBPS SO 2016]**
   (a) 30 km (b) 40 km
   (c) 45 km (d) 36 km
   (e) 42 km

5. The speed of the boat in still water is 5 times the speed of the current. It takes 1.1 h to row to point B from point A downstream. The distance between point A and point B is 13.2 km. How much distance will it cover in 312 min upstream? **[IBPS SO 2015]**
   (a) 43.2 km (b) 48 km
   (c) 41.6 km (d) 44.8 km
   (e) 40 km

6. A boat goes from point I to point II and comes back. The speed of water in river is 5 km/h and it takes total time 72 min. If ratio of time from I to II and II to I is 3 : 5, what is the speed of boat in still water? **[NIFT 2015]**
   (a) 25 km/h (b) 20 km/h
   (c) 30 km/h (d) 15 km/h

7. A motorboat travelling at the some speed, can cover 25 km upstream and 39 km downstream in 8 h. At the same speed, it can travel 35 km upstream and 52 km downstream in 11 h. The speed of the stream is **[SSC CGL 2011]**
   (a) 2 km/h (b) 3 km/h
   (c) 4 km/h (d) 5 km/h

8. The speed of boat A is 2 km/h less than the speed of the boat B. The time taken by boat A to travel a distance of 20 km downstream is 30 min more than time taken by B to travel the same distance downstream. If the speed of the current is one-third to the speed of the boat A. What is the speed of boat B? **[LIC AAO 2015]**
   (a) 4 km/h (b) 6 km/h
   (c) 12 km/h (d) 10 km/h
   (e) 8 km/h

9. Two ports A and B are 300 km apart. Two ships leave A for B such that the second leaves 8 h after the first. The ships arrive at B simultaneously. Find the time taken by the slower ship on the trip, if the speed of one of them is 10 km/h higher than that of the other. **[MAT 2015]**
   (a) 25 h (b) 15 h (c) 10 h (d) 20 h

10. A ferry is moving downstream from city A to city B with speed 45 km/h. A passenger jumped into river in middle of cities and starts swimming towards city A. Ferry reached city B and comes to city A with speed 30 km/h and both reached at same time. What is the speed of passenger in river? **[NIFT 2015]**
    (a) 21.75 km/h (b) $18\frac{3}{4}$ km/h
    (c) 11.25 km/h (d) 25.5 km/h

11. The time taken by the boat to cover a distance of 'D-56' km upstream is half of that taken by it to cover a distance of 'D' km downstream. The respective ratio between the speed of the boat downstream and that upstream is 5 : 3. If the time taken to cover 'D-32' km upstream is 4 h, what is the speed of water current? **[SBI PO (Pre) 2017]**
    (a) 5 km/h (b) 3 km/h
    (c) 4 km/h (d) 16 km/h
    (e) 8 km/h

Answer with Solutions

Exercise 1 Base Level Questions

1. (a) Given, swimmer's speed in still water,
$x = 9$ km/h
Rate of stream, $y = 6$ km/h
∴ Speed downstream $= x + y$
$= 9 + 6 = 15$ km/h

2. (a) Speed downstream $= \dfrac{48}{20} = 2.4$ km/h
and speed upstream $= \dfrac{48}{24} = 2$ km/h
∴ Speed of boat in still water
$= \dfrac{1}{2}(2.4 + 2) = \dfrac{4.4}{2} = 2.2$ km/h

3. (a) Given, speed upstream $= 6$ km/h
and speed downstream $= 8$ km/h
∴ Speed of the current $= \dfrac{8-6}{2} = 1$ km/h
Hence, speed of the current is 1 km/h.

4. (a) Downstream speed $= \dfrac{1}{5/60} = 12$ km/h
and upstream speed $= \dfrac{6}{1} = 6$ km/h
∴ Speed of the stream $= \dfrac{12-6}{2} = \dfrac{6}{2}$
$= 3$ km/h

5. (c) Let speed of current $= x$ km/h
Then, speed of boat $= x + x \times \dfrac{500}{100}$
$= x + 5x = 6x$ km/h
∴ Required ratio $= \dfrac{\text{Downstream speed}}{\text{Upstream speed}}$
$= \dfrac{(6x+x)}{(6x-x)} = \dfrac{7x}{5x} = \dfrac{7}{5} = 7:5$

6. (d) Let speed of person $= x$ km/h
and speed of current $= y$ km/h
Distance $= 20$ km in downstream and 4 km in upstream.
According to the question,
$\dfrac{20}{x+y} = 2 \Rightarrow x + y = 10$ ...(i)
and $\dfrac{4}{x-y} = 2 \Rightarrow x - y = 2$ ...(ii)
From Eqs. (i) and (ii), we get
$x = 6$ and $y = 4$
Hence, the speed of current is 4 km/h.

7. (a) Let the speed of boat be x km/h.
Then, speed of upstream $= (x-4)$ km/h

According to the question,
$\dfrac{12}{x-4} = 5 \Rightarrow 12 = 5x - 20 \Rightarrow x = \dfrac{32}{5}$ km/h
∴ Time taken to row 15 km with the stream
$= \dfrac{15}{\dfrac{32}{5}+4} = \dfrac{15 \times 5}{52} = \dfrac{75}{52}$ h
$= 1\dfrac{23}{52} \times 60 = 1$ h $26\dfrac{7}{13}$ min

8. (c) Let the speed of boat in still water
$= a$ km/h
and speed of current $= b$ km/h
Then, $a + b = \dfrac{28}{5}$ ...(i)
and $a - b = \dfrac{13}{5}$ ...(ii)
On solving Eqs. (i) and (ii), we get
$a = 4.1$ and $b = 1.5$
∴ Speed of current $= 1.5$ km/h

9. (a) Given, speed of boat $= 16$ km/h
Speed of stream $= 5$ km/h
Distance covered by downstream $= 84$ km
∴ Speed of downstream
$= 16 + 5 = 21$ km/h
∴ Required time $= \dfrac{84}{21} = 4$ h

10. (a) Given, speed of boat $= 10$ km/h
Let speed of flow of river $= x$ km/h
∴ Upstream speed of boat $= (10-x)$ km/h
and downstream speed of boat
$= (10+x)$ km/h
According to the question,
$\dfrac{91}{10-x} + \dfrac{91}{10+x} = 20$
$\Rightarrow \dfrac{91(10+x+10-x)}{(10-x)(10+x)} = 20$
$\Rightarrow \dfrac{91(20)}{100-x^2} = 20 \Rightarrow 91 = 100 - x^2$
$\Rightarrow x^2 = 9 \Rightarrow x = 3$

11. (d) Let the speed of man and current be x km/h and y km/h, respectively.
The speed upstream $= (x-y)$ km/h
and speed downstream $= (x+y)$ km/h
According to the question,
$\dfrac{3 \times 60}{4 \times 15} = x - y \Rightarrow x - y = 3$ ...(i)
and $\dfrac{3}{4} \times \dfrac{60}{10} = x + y \Rightarrow x + y = \dfrac{9}{2}$ ...(ii)

On adding Eqs. (i) and (ii), we get
$$2x = 3 + \frac{9}{2} \Rightarrow 2x = \frac{6+9}{2} \Rightarrow x = \frac{15}{4}$$
On putting the value of x in Eq. (ii), we get
$$\frac{15}{4} + y = \frac{9}{2} \Rightarrow y = \frac{9}{2} - \frac{15}{4} = \frac{18-15}{4}$$
$$\Rightarrow \quad y = \frac{3}{4}$$
∴ Speed of man, $x = \frac{15}{4}$ km/h
and speed of current, $y = \frac{3}{4}$ km/h
Hence, required ratio $= \frac{15}{4} : \frac{3}{4} = 5 : 1$

12. (c) Let the speed of boat be x km/h.
Given, speed of current = 5 km/h
∴ Upstream speed of boat = $(x - 5)$ km/h
and downstream speed of boat = $(x + 5)$ km/h
According to the question,
$$\frac{10}{x-5} + \frac{10}{x+5} = \frac{50}{60}$$
$$\Rightarrow \quad 10\left(\frac{x+5+x-5}{x^2-25}\right) = \frac{5}{6}$$
$$\Rightarrow \quad 12 \times 2x = x^2 - 25$$
$$\Rightarrow \quad x^2 - 24x - 25 = 0$$
$$\Rightarrow \quad x^2 - 25x + x - 25 = 0$$
$$\Rightarrow \quad (x-25)(x+1) = 0$$
∴ $x = 25$ [∵ $x \ne -1$]
So, the speed of motorboat in still water is 25 km/h.

13. (c) Let speed of stream be x km/h.
Given, speed of motorboat in still water
= 45 km/h
∴ Speed of boat along stream = $(45 + x)$ km/h
According to the question,
$$45 + x = \frac{80}{1\frac{1}{3}} \Rightarrow 45 + x = \frac{80 \times 3}{4}$$
∴ $x = 60 - 45 = 15$ km/h
∴ Speed of boat against stream
= 45 − 15 = 30 km/h
Hence, required time = $\frac{\text{Distance}}{\text{Speed}} = \frac{80}{30}$ h
$= \frac{8}{3} \times 60 = 160$ min = 2 h 40 min

14. (d) Let x be the speed of the boat and y be the speed of the current.
Then, speed upstream = $(x - y)$ km/h
and speed downstream = $(x + y)$ km/h
According to the question,
$$\frac{20}{x-y} + \frac{20}{x+y} = \frac{41}{60} \Rightarrow \frac{40}{3600} = \frac{25}{36} \text{ h}$$
In this equation, there are two variables but only one equation. So, the value of x cannot be determined.

15. (a) Let speed of water current be x km/h.
∴ Speed downstream = $(10 + x)$ km/h
and speed upstream = $(10 − x)$ km/h
According to the question,
$$\frac{24}{10+x} + \frac{24}{10-x} = 5$$
$$\Rightarrow 24(10 - x + 10 + x) = 5(10^2 - x^2)$$
$$\Rightarrow 100 - x^2 = \frac{24 \times 20}{5}$$
$$\Rightarrow 100 - x^2 = 96 \Rightarrow x^2 = 4$$
$$\Rightarrow x = \sqrt{4} = 2 \text{ km/h}$$

16. (c) Let speed of a motorboat be $36x$ km/h.
and speed of the current = $5x$ km/h
∴ Speed downstream = $(36 + 5)x = 41x$ km/h
and speed upstream = $(36 − 5)x = 31x$ km/h
Let the distance be a km.
According to the question,
When boat goes along with the current, then
Distance = Time × Speed
$$\Rightarrow a = \left(5 + \frac{10}{60}\right) \times 41x$$
$$\Rightarrow a = \frac{31}{6} \times 41x \quad \ldots(i)$$
Again, when boat come back
a = Time × Speed
From Eq. (i), $\frac{31}{6} \times 41x$ = Time × 31x
$$\Rightarrow \text{Time} = \frac{41}{6} \Rightarrow \text{Time} = 6 \text{ h } 50 \text{ min}.$$

17. (e) Let the speed of boat in still water be u km/min, then, the speed of water current will be $\frac{u}{4}$ km/min.
Speed in downstream
$$= u + \frac{u}{4} = \left(\frac{5u}{4}\right) \text{ km/min}$$
and speed in upstream
$$= \left(u - \frac{u}{4}\right) = \frac{3u}{4} \text{ km/min}$$
According to the question,
$$\frac{12.6}{35} = \frac{5u}{4} \Rightarrow \frac{12.6 \times 4}{35 \times 5} = u$$
$$\Rightarrow \quad u = 0.288 \text{ km/min}$$
∴ Distance covered by boat upstream in 28 min
$$= \frac{3u}{4} \times 28 = \frac{3 \times 0.288}{4} \times 28$$
$$= 6.048 \text{ km} \approx 6 \text{ km}$$

18. (c) Let the rate of stream be a km/h.
According to the question,
$24 + a = 2(24 − a) \Rightarrow 24 + a = 48 − 2a$
$\Rightarrow 3a = 48 − 24 = 24 \Rightarrow a = \frac{24}{3} = 8$ km/h

Boats and Streams / 479

Fast Track Method
Here, speed of boat in still water
   = 24 km/h and $n = 2$
Rate of stream = a = ?
According to the formula,

Speed of boat in still water = $\dfrac{a(n+1)}{(n-1)}$

[by Technique 1]

$\Rightarrow \quad 24 = \dfrac{a(2+1)}{(2-1)} \Rightarrow 3a = 24$

$\Rightarrow \quad a = \dfrac{24}{3} = 8$ km/h

19. (a) Pawan's speed downstream
   = 24 + 4.8 = 28.8 km/h
Pawan's speed upstream
   = 24 − 4.8 = 19.2 km/h
Let the required distance be x.
According to the question,

$\dfrac{x}{28.8} + \dfrac{x}{19.2} = 1 \Rightarrow \dfrac{19.2x + 28.8x}{552.96} = 1$

$\Rightarrow \quad 19.2x + 28.8x = 552.96$
$\Rightarrow \quad 48x = 552.96$
$\therefore \quad x = \dfrac{552.96}{48} = 11.52$ km

Fast Track Method
$\because x$ = Speed of Pawan in still water = 24 km/h,
  y = Speed of river = 4.8 km/h and $T = 1$ h
According to the formula,

Required distance = $\dfrac{T(x^2 - y^2)}{2x}$

[by Technique 2]

$= \dfrac{1[(24)^2 - (4.8)^2]}{2 \times 24} = \dfrac{576 - 23.04}{2 \times 24}$

$= \dfrac{552.96}{48} = 11.52$ km

20. (c) Man's speed upstream = 6.5 − 2.5
   = 4 km/h
Man's speed downstream = 6.5 + 2.5
   = 9 km/h
Let the distance in one direction be x km.
Then, time taken in upstream = $\dfrac{x}{4}$

and time taken in downstream = $\dfrac{x}{9}$

$\therefore$ Average speed = $\dfrac{\text{Total distance}}{\text{Total time}} = \dfrac{2x}{\dfrac{x}{4} + \dfrac{x}{9}}$

$= \dfrac{2x \times 36}{9x + 4x} = 5.53$ km/h

Fast Track Method
Here, $a = 6.5$ km/h and $b = 2.5$ km/h

$\therefore$ Average speed = $\dfrac{(a+b)(a-b)}{a}$

[by Technique 4]

$= \dfrac{(6.5 + 2.5)(6.5 - 2.5)}{6.5} = \dfrac{9 \times 4}{6.5} = \dfrac{36}{6.5}$

$= 5.53$ km/h

21. (b) Let the total distance be x km.
According to the question,

$\dfrac{x}{6+2} + 3 = \dfrac{x}{6-2} \Rightarrow \dfrac{x}{8} + 3 = \dfrac{x}{4}$

$\Rightarrow \quad \dfrac{x}{4} - \dfrac{x}{8} = 3 \Rightarrow \dfrac{2x - x}{8} = 3$

$\Rightarrow \quad x = 8 \times 3 = 24$ km

Fast Track Method
Here, $a = 6$, $b = 2$ and $T = 3$

$\therefore$ Distance = $\dfrac{(a^2 - b^2)}{2b} \times T$ [by Technique 5]

$= \dfrac{(6^2 - 2^2)}{2 \times 2} \times 3 = \dfrac{(36 - 4)}{4} \times 3 = 24$ km

22. (a) Let the distance be x km.
Then, speed downstream = (5 + 2) = 7 km/h
and speed upstream = (5 − 2) = 3 km/h
According to the question,

$\dfrac{x}{3} - \dfrac{x}{7} = 2 \Rightarrow 7x - 3x = 21 \times 2$

$\therefore \quad x = \dfrac{21 \times 2}{4} = 10.5$ km

Fast Track Method
Here, $a = 5$ km/h, $b = 2$ km/h and $T = 2$ h

$\therefore$ Required distance = $\left(\dfrac{a^2 - b^2}{2b}\right) \times T$

[by Technique 5]

$= \left(\dfrac{5^2 - 2^2}{2 \times 2}\right) \times 2 = \dfrac{25 - 4}{2} = \dfrac{21}{2} = 10.5$ km

23. (d) Speed of boat downstream
   = $\dfrac{\text{Distance covered}}{\text{Time taken}} = \dfrac{30}{2} = 15$ km/h

Now, speed of boat upstream
   = $\dfrac{\text{Distance covered}}{\text{Time taken}} = \dfrac{30}{6} = 5$ km/h

Now, speed of boat
   = $\dfrac{\text{Speed downstream} + \text{Speed upstream}}{2}$

$= \dfrac{15 + 5}{2} = \dfrac{20}{2} = 10$ km/h

Fast Track Method
Here, $l = 30$ km, $t_1 = 2$ h and $t_2 = 6$ h

Then, speed of boat = $\dfrac{1}{2}\left(\dfrac{l}{t_1} + \dfrac{l}{t_2}\right)$

[by Technique 6 (i)]

$= \dfrac{1}{2}\left(\dfrac{30}{2} + \dfrac{30}{6}\right)$

$= \dfrac{1}{2}(15 + 5) = \dfrac{1}{2} \times 20 = 10$ km/h

24. (b) Let the rate of sailing of sailer be x km/h and speed of the flow of water be y km/h.

Downstream speed $(x + y) = \dfrac{48}{8}$

$\Rightarrow \qquad x + y = 6$ ...(i)

and upstream speed $(x - y) = \dfrac{48}{12}$

$\Rightarrow \qquad x - y = 4$ ...(ii)

On solving Eqs. (i) and (ii), we get
$x = 5$ km/h and $y = 1$ km/h

Hence, speed of the flow of river is 1 km/h.

Fast Track Method

Speed of stream $= \dfrac{1}{2}\left(\dfrac{l}{t_1} - \dfrac{l}{t_2}\right)$

[by Technique 6 (ii)]

$= \dfrac{1}{2}\left(\dfrac{48}{8} - \dfrac{48}{12}\right) = \dfrac{1}{2}(6 - 4) = \dfrac{2}{2} = 1$ km/h

25. (d) Let the distance between the two ports be x km.

Then, speed downstream $= \dfrac{x}{4}$ km/h

and speed upstream $= \dfrac{x}{5}$ km/h

$\therefore$ Speed of the stream

$= \dfrac{1}{2}$ (Speed downstream − Speed upstream)

$\Rightarrow 2 = \dfrac{1}{2}\left(\dfrac{x}{4} - \dfrac{x}{5}\right) \Rightarrow 2 = \dfrac{1}{2}\left(\dfrac{5x - 4x}{20}\right)$

$\Rightarrow \dfrac{x}{40} = 2 \Rightarrow x = 80$ km

Fast Track Method

Here, $t_1 = 4$ h and $t_2 = 5$ h

$\therefore$ Speed of stream $= \dfrac{1}{2}\left(\dfrac{1}{t_1} - \dfrac{1}{t_2}\right)$

[by Technique 6 (ii)]

$\Rightarrow \qquad 2 = \dfrac{l}{2}\left(\dfrac{1}{4} - \dfrac{1}{5}\right) \Rightarrow 2 = \dfrac{l}{2}\left(\dfrac{1}{20}\right)$

$\therefore \qquad l = 80$ km

Exercise 2 Higher Skill Level Questions

1. (c) $\because$ Distance travelled = 117 km
and time = 13 h

$\therefore$ Upstream speed $= \dfrac{\text{Distance}}{\text{Time}}$

$= \dfrac{117}{13} = 9$ km/h

Now, distance travelled by boat going upstream in 2 h = speed upstream × 2
$= 9 \times 2 = 18$ km

2. (c) In downstream, a boat can travel a distance of 14.4 km in 32 min or $\dfrac{32}{60}$ h.

$\Rightarrow \qquad d = \text{speed} \times \text{time}$
$\Rightarrow \qquad d = (x + y) \times t$

[here $y \to$ speed of the current and $x \to$ speed of still water]

$\Rightarrow \qquad 14.4 = (x + 3) \times \dfrac{32}{60}$

$\Rightarrow \qquad x + 3 = 27 \Rightarrow x = 27 - 3$
$\Rightarrow \qquad x = 24$ km/h

Now, in upstream a boat can travel a distance of 84 km in time 't' h, i.e.
$d = (x - y) \times t$

$\Rightarrow \qquad 84 = (24 - 3)t \Rightarrow t = \dfrac{84}{21} = 4$ h

3. (d) Given, speed of boat in still water
$= 16$ km/h

and speed of current = 2 km/h

C•────x──→•B $(x+12)$ •A

$\therefore$ Downstream speed of boat
$= 16 + 2 = 18$ km/h

and upstream speed of boat
$= 16 - 2 = 14$ km/h

Now, distance between points B and C
$= x$ km (say)

and distance between points A and B
$= (x + 12)$ km

Now, according to the question,

$\dfrac{x}{14} = \dfrac{x + 12}{18}$

$\Rightarrow 18x = 14x + 168 \Rightarrow 18x - 14x = 168$
$\Rightarrow \qquad 4x = 168 \Rightarrow x = 42$ km

$\therefore$ Time taken by boat to travel from point C to B in downstream motion $= \dfrac{x}{18} = \dfrac{42}{18}$

$= \dfrac{7}{3} = 2\dfrac{1}{3}$ h

$= 2$ h + 20 min = 2 h 20 min

4. (d) Downstream speed of boat
$= 14 + 2 = 16$ km/h

Upstream speed of boat $= 14 - 2 = 12$ km/h

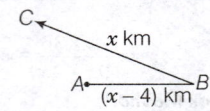

C
 \\ x km
  \\
A•────────•B
  $(x - 4)$ km

Let the distance between B and C be x km.
Then, time taken by boat from point A to B

$= \left(\dfrac{x - 4}{16}\right)$ h

and time taken by boat from point B to C
$$= \left(\frac{x}{12}\right) \text{h}$$
According to the question, $\frac{x-4}{16} + 1 = \frac{x}{12}$
$\Rightarrow \quad 6(x - 4 + 16) = 8x$
$\Rightarrow \quad 6x + 72 = 8x$
$\Rightarrow \quad 2x = 72 \Rightarrow x = 36$
Hence, the distance between points B and C is 36 km.

5. (c) Let the speed of current be x km/h.
Then, speed of boat = $5x$ km/h
Total speed in downstream = $x + 5x = 6x$
Total speed in upstream = $5x - x = 4x$
According to the question,
$1.1 \times 6x = 13.2 \Rightarrow x = 2$ km/h
[∵ time × speed = distance]
∴ Required distance covered in upstream
$= \text{Time} \times \text{Speed} = \frac{312}{60} \times 4x$
$= \frac{312}{60} \times 8 = 5.2 \times 8 = 41.6$ km

6. (b) Let the speed of boat be B km/h and total distance between I and II be x km.
Given, speed of water in river = 5 km/h
Then, speed downstream = $(B + 5)$ km/h
and speed upstream = $(B - 5)$ km/h
Time taken by boat from I to II = $\frac{x}{B+5}$
and time taken by boat from II to I = $\frac{x}{B-5}$
According to the question,
$\frac{\text{Time taken from I to II}}{\text{Time taken from II to I}} = \frac{x}{B+5} \times \frac{B-5}{x} = \frac{3}{5}$
$\Rightarrow \quad \frac{B-5}{B+5} = \frac{3}{5}$
$\Rightarrow \quad 5B - 25 = 3B + 15$
$\Rightarrow \quad 2B = 40$
$\Rightarrow \quad B = 20$ km/h

7. (c) Let the speed of a boat and stream be x km/h and y km/h.
∴ Speed of boat along stream = $(x + y)$ km/h
and speed of boat against stream
$= (x - y)$ km/h
According to the question,
$\frac{25}{x-y} + \frac{39}{x+y} = 8$ ...(i)
and $\frac{35}{x-y} + \frac{52}{x+y} = 11$ ...(ii)
On multiplying Eq. (i) by 4 and Eq. (ii) by 3, then subtracting Eq. (ii) from Eq. (i), we get
$\frac{100}{x-y} - \frac{105}{x-y} = -1 \Rightarrow \frac{5}{x-y} = 1$
$\Rightarrow \quad x - y = 5$ ...(iii)

On substituting the value of $(x - y) = 5$ in Eq. (i), we get
$\frac{25}{5} + \frac{39}{x+y} = 8 \Rightarrow \frac{39}{x+y} = 8 - 5$
$\Rightarrow \quad x + y = \frac{39}{3} \Rightarrow x + y = 13$ ...(iv)
On solving Eqs. (iii) and (iv), we get
$x = 9$ and $y = 4$
Hence, the speed of stream is 4 km/h.

8. (e) Let the speed of boat B be x km/h.
∴ Speed of boat $A = (x - 2)$ km/h
and speed of current $= \left(\frac{x-2}{3}\right)$ km/h
Speed of boat A downstream
= Speed of boat A + Speed of current
$= \left[(x-2) + \frac{(x-2)}{3}\right]$ km/h
and speed of boat B downstream
= Speed of boat B + Speed of current
$= \left(x + \frac{x-2}{3}\right)$ km/h
Now, according to the question,
$\frac{20}{(x-2) + \frac{(x-2)}{3}} = \frac{20}{x + \frac{x-2}{3}} + \frac{30}{60}$
$\Rightarrow \quad \frac{20 \times 3}{3x - 6 + x - 2} = \frac{20 \times 3}{3x + x - 2} + \frac{1}{2}$
$\Rightarrow \quad \frac{60}{4x-8} - \frac{60}{4x-2} = \frac{1}{2}$
$\Rightarrow \quad \frac{60}{4(x-2)} - \frac{60}{2(2x-1)} = \frac{1}{2}$
$\Rightarrow \quad \frac{15}{x-2} - \frac{30}{2x-1} = \frac{1}{2}$
$\Rightarrow \quad \frac{30x - 15 - 30x + 60}{(x-2)(2x-1)} = \frac{1}{2}$
$\Rightarrow \quad \frac{45}{(x-2)(2x-1)} = \frac{1}{2}$
$\Rightarrow \quad (x-2)(2x-1) = 90$
$\Rightarrow \quad 2x^2 - x - 4x + 2 = 90$
$\Rightarrow \quad 2x^2 - 5x + 2 - 90 = 0$
$\Rightarrow \quad 2x^2 - 5x - 88 = 0$
$\Rightarrow \quad 2x^2 - 16x + 11x - 88 = 0$
$\Rightarrow \quad 2x(x-8) + 11(x-8) = 0$
$\Rightarrow \quad (x-8)(2x+11) = 0$
$\Rightarrow \quad 2x + 11 = 0$ and $x - 8 = 0$
$\Rightarrow \quad x = -\frac{11}{2}$ and $x = 8$
∴ $x = 8$ km/h [leaving the negative value]
Hence, the speed of boat B is 8 km/h.

9. (d) Let the speed of the slower ship be v km/h.
∴ Speed of other ship = $(v + 10)$ km/h

Now, it is given that despite of leaving 8 h after one ship departs, another ship reaches on the same time as first ship reaches.

$$\therefore \quad \frac{300}{v} - \frac{300}{v+10} = 8 \Rightarrow 300\left(\frac{1}{v} - \frac{1}{v+10}\right) = 8$$

$$\Rightarrow \quad \frac{v+10-v}{v^2+10v} = \frac{8}{300}$$

$$\Rightarrow \quad 8v^2 + 80v - 3000 = 0$$
$$\Rightarrow \quad v^2 + 10v - 375 = 0$$
$$\Rightarrow \quad v^2 + 25v - 15v - 375 = 0$$
$$\Rightarrow \quad v(v+25) - 15(v+25) = 0$$
$$\Rightarrow \quad v = 15, -25$$
$$\therefore \quad v = 15 \text{ km/h [omitting negative value]}$$

Hence, time spent by slower ship on the trip = $\frac{300}{15}$ = 20 h

10. **(b)** Let the speed of ferry be B km/h and the speed of water be D km/h.

∴ Ferry's speed in downstream
$= B + D = 45$ km/h ...(i)

and Ferry's speed in upstream
$= B - D = 30$ km/h ...(ii)

Now, from Eqs. (i) and (ii), we get
$2B = 75 \Rightarrow B = \frac{75}{2}$ km/h

On putting the value of B in Eq. (i), we get
$D = \frac{15}{2}$ km/h

Let the speed of passenger in river be v km/h and distance between city A and city B be x km.

$A \xleftarrow{\quad x/2 \quad} O \xrightarrow{\quad x/2 \quad} B$
$\xleftarrow{\quad\quad t_2 \quad\quad}$

According to the question, $t_1 + t_2 = t$ where, t_1 is the time of ferry to reached city B and t_2 is the time of ferry reached city B to A.

$$\therefore \quad \frac{x/2}{B+D} + \frac{x}{B-D} = \frac{x/2}{v-D}$$

$$\Rightarrow \quad \frac{x}{2(B+D)} + \frac{x}{B-D} = \frac{x}{2(v-D)}$$

$$\Rightarrow \quad \frac{1}{2\times 45} + \frac{1}{30} = \frac{1}{2(v-D)}$$

$$\Rightarrow \quad \frac{1}{90} + \frac{1}{30} = \frac{1}{2(v-D)}$$

$$\Rightarrow \quad \frac{1+3}{90} = \frac{1}{2(v-D)}$$

$$\Rightarrow \quad \frac{4}{90} = \frac{1}{2(v-D)} \Rightarrow v - D = \frac{45}{4}$$

$$\therefore \quad v = \frac{45}{4} + \frac{15}{2} = \frac{45+30}{4} = \frac{75}{4} = 18\frac{3}{4} \text{ km/h}$$

11. **(c)** Let speed of boat in still water = x km/h and speed of current = y km/h

∴ Downstream speed of boat = $(x+y)$ km/h and upstream speed of boat = $(x-y)$ km/h

Now, according to the question

$$T = \frac{D}{x+y} \quad ...(i)$$

and $\quad \frac{T}{2} = \frac{D-56}{x-y} \Rightarrow T = \frac{2(D-56)}{(x-y)} \quad ...(ii)$

From Eqs. (i) and (ii), we have
$$\frac{D}{x+y} = \frac{2(D-56)}{x-y}$$

$$\Rightarrow \quad \frac{(x-y)}{(x+y)} = \frac{2D-112}{D}$$

$$\Rightarrow \quad \frac{3}{5} = \frac{2D-112}{D}$$

$$\left[\because \frac{x-y}{x+y} = \frac{3}{5} \text{ (given)}\right]$$

$\Rightarrow \quad 10D - 560 = 3D \Rightarrow 10D - 3D = 560$
$\Rightarrow \quad 7D = 560 \Rightarrow D = 80$ km

Again as per the question, $x - y = \frac{D-32}{4}$

$\Rightarrow \quad x - y = \frac{80-32}{4} \Rightarrow x - y = \frac{48}{4} = 12$

$\Rightarrow \quad x - y = 12 \Rightarrow x = (12+y)$

Now, $\quad \frac{x+y}{x-y} = \frac{5}{3}$

$\Rightarrow \quad \frac{12+y+y}{12+y-y} = \frac{5}{3}$ [put $x = 12+y$]

$\Rightarrow \quad \frac{12+2y}{12} = \frac{5}{3}$

$\Rightarrow \quad 36 + 6y = 60 \Rightarrow 6y = 60 - 36$
$\Rightarrow \quad 6y = 24 \Rightarrow y = 4$ km/h

Hence, speed of water current is 4 km/h.

Chapter 28

Races and Games of Skill

A race or a game of skill includes the contestants in a contest and their skill in the concerned contest/game.

Important Terms Related to Races and Games of Skill

Race A race is a contest of speed in running, driving, riding, sailing or rowing.

Race Course The ground/path on which the contest of race is organised in a systematic way, is called a race course.

Starting Point The exact point/place from where a race begins, is called starting point.

Start If two persons A and B are contesting a race and before the start of the race, A is at the starting point and B is ahead of A by x m (say), then it is said that A gives B a start of x m.

For example If A and B are the contestants for a 100 m race and A has to cover 100 m, while B has to cover $(100 - 20) = 80$ m, then A gives B a start of 20 m.

Winning Point (Goal)/**Finishing Point** The exact point/place where a race ends, is known as finishing point or winning point or goal.

Winner A person who reaches the finishing point first, is called the **winner**.

Dead Heat Race A race is said to be a dead heat race, if all the contestants reach the finishing point exactly at the same time.

Game of 100 A game of 100 means that the contestant who scores 100 points first, is declared as the winner.

For example If in a 100 points game, A scores 100 points, while B scores 85 points.

Then, it is said that A can give $(100 - 85) = 15$ points to B.

Some Important Facts about Race

Let the length of the race be L m.

For Two Contestants A and B

1. If A beats B by x m, then

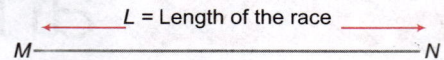

   $\xleftarrow{\quad\quad\quad L = \text{Length of the race} \quad\quad\quad}$
   M————————————————————N

   Distance covered by A (winner) = L m
   Distance covered by B (loser) = $(L - x)$ m

2. If B starts x m ahead of A (or A gives B a start of x m), then

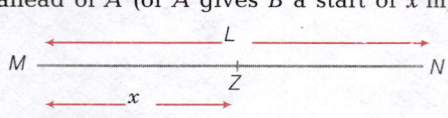

   M————————————Z————N
   $\xleftarrow{\quad x \quad}$

   A starts from M and B starts from Z.
   ∴ Distance covered by $B = (L - x)$ m

3. If A beats B by T s, then

   M————————————————————N

   A and B both start from point M.
   ∴ Time taken by A (winner) = Time taken by B (loser) $- T$
   It means that A completes the race in T s less time than that of B.

4. If B starts the race T s before A starts (or if A gives B a start of T s), then We say that A starts T s after B starts.

5. If both of the contestants get at the finishing point at the same time, then Difference in time of defeat = 0; Difference in distance of defeat = 0

Ex. 1 In a race of 100 m, A gives B a start of 10 m. What distance will be covered by B?

Sol. Since, A gives B a start of 10 m means B is ahead of A by 10 m.
∴ Required distance = $(100 - 10)$ m = 90 m

Ex. 2 In a race, x gives y a start of 30 m making length of race for y a distance of 170 m. Find the total length of race.

Sol. Since, x gives y a start of 30 m, means y is ahead of x by 30 m.
∴ Required length = $(170 + 30) = 200$ m

Ex. 3 In a 100 m race, Ajay runs at the speed of 4 km/h. Ajay gives Brijesh a start of 4 m and still beats him by 15 s. Find the speed of Brijesh.

Sol. Time taken by Ajay to cover 4000 m = 1h = 3600 s
∴ Time taken by Ajay to cover 100 m = $\left(\dfrac{60 \times 60}{4000} \times 100\right)$ s = 90 s

Now, Brijesh covers $(100 - 4)$ m = 96 m in $(90 + 15)$ s = 105 s.
∴ Brijesh's speed = $\dfrac{96}{105}$ m/s = $\dfrac{96}{105} \times \dfrac{18}{5}$ km/h = 3.29 km/h

Races and Games of Skill / 485

Ex. 4 In 1 km race, A beats B by 36 m or 18 s. Find A's time over the course.

Sol. Clearly, B covers 36 m in 18 s.

$\therefore$ B's time over the course $= \dfrac{18}{36} \times 1000 = 500$ s

Now, A's time over the course $= (500 - 18)$ s $= 482$ s

Ex. 5 P covers 1 km in 4 min 40 s, while Q covers the same distance in 5 min. By what distance does P defeat Q?

Sol. Clearly, P beats Q by 20 s.

$\therefore$ Distance covered by Q in 20 s $= \dfrac{1000}{300} \times 20 = 66\dfrac{2}{3}$ m

Hence, P defeat Q by $66\dfrac{2}{3}$ m.

Ex. 6 A can run 1 km in 5 min and B can run the same distance in 6 min. How many metres start can A give to B in 1 km race, so that the race may end in a dead heat?

Sol. Time taken by A to run 1 km $= 300$ s; Time taken by B to run 1 km $= 360$ s.

A can give B a start of $(360 - 300)$ s $= 60$ s

$\because$ In 360 s, B runs 1000 m.

$\therefore$ In 60 s, B runs $\dfrac{1000}{360} \times 60$ m $= \dfrac{1000}{6}$ m $= \dfrac{500}{3}$ m $= 166\dfrac{2}{3}$ m

Hence, A can give a start of $166\dfrac{2}{3}$ m.

For Three Contestants A, B and C

Let A, B and C participate in a race of L m.

Let A comes 1st in the race by beating B by x m and C by y m, respectively.

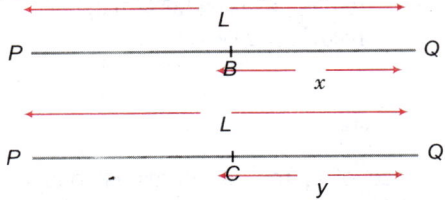

```
           L
    P ─────┼───────── Q
           B      x
```

```
           L
    P ─────┼───────── Q
           C      y
```

Here, the values of x and y will decide 2nd and 3rd positions.

If $x < y$, then B will beat C, i.e. B will get the 2nd position.

If $x > y$, then C will beat B, i.e. C will get the 2nd position.

Ex. 7 A, B and C are three contestants in 1 km race. If A can give B a start of 40 m and A can give C a start of 64 m, then how many metres start can B give C?

Sol. While A covers 1000 m, B covers $(1000 - 40)$ m $= 960$ m

and C will cover $(1000 - 64)$ m $= 936$ m.

So, when B covers 1000 m, C will cover $\left(\dfrac{936}{960} \times 1000\right) = 975$ m

$\therefore$ B can give C a start of $(1000 - 975)$ m, i.e. 25 m.

Fast Track Techniques
to solve the QUESTIONS

Technique 1   If in a race of L m, 1st contestant beats 2nd contestant and 3rd contestant by distances of a_{12} and a_{13}, respectively and the 2nd contestant beats the 3rd contestant by a distance of a_{23}, then we get the relation $(L - a_{12}) a_{23} = L(a_{13} - a_{12})$

Ex. 8 P, Q and R are three contestants in a 2 km race. If P can give Q a start of 100 m and P can give R a start of 138 m, then how many metres start can Q give to R?

Sol. When P covers 2000 m, then Q covers $(2000 - 100)$ m $= 1900$ m
and R covers $(2000 - 138)$ m $= 1862$ m
When Q covers 1900 m, then R covers 1862 m.
When Q covers 2000 m, then R covers $\left(\dfrac{1862}{1900} \times 2000\right)$ m $= 1960$ m

Hence, Q can give R a start of $(2000 - 1960)$ m, i.e. 40 m

Fast Track Method
Here, $a_{12} = 100$ m, $a_{13} = 138$ m, $a_{23} = ?$ and $L = 2000$ m
According to the formula,
$$(L - a_{12}) a_{23} = L(a_{13} - a_{12})$$
$\Rightarrow \quad (2000 - 100) a_{23} = 2000 (138 - 100)$
$\Rightarrow \quad 1900 a_{23} = 2000 \times 38$
$\therefore \quad a_{23} = \dfrac{2000 \times 38}{1900} = \dfrac{760}{19} = 40$ m

Hence, Q can give R a start of 40 m.

Technique 2   If in a race of L_1 m, 1st contestant beats the 2nd contestant by a distance of a_{12}; in a race of L_2 m 2nd contestant beats the 3rd contestant by a distance of a_{23} and in a race of L_3 m 1st contestant beats the 3rd contestant by a distance of a_{13}, then for a race of L m

$$A_{12} = \dfrac{a_{12}}{L_1} \times L; \quad A_{23} = \dfrac{a_{23}}{L_2} \times L; \quad A_{13} = \dfrac{a_{13}}{L_3} \times L$$

Now, we get the following relation
$$(L - A_{12}) A_{23} = L(A_{13} - A_{12})$$

Ex. 9 In a race of 1200 m, A can beat B by 120 m and in a race of 500 m, B can beat C by 100 m. By how many metres will A beat C in a race of 800 m?

Sol. If A runs 1200 m, then B runs 1080 m.
If A runs 800 m, then B runs $\left(\dfrac{1080}{1200} \times 800\right)$ m $= 720$ m

Races and Games of Skill / 487

When B runs 500 m, then C runs 400 m.
When B runs 720 m, then C runs $\left(\dfrac{400}{500} \times 720\right)$ m = 576 m

Hence, A beats C by $(800 - 576)$ m, i.e. 224 m.

Fast Track Method

According to the question, the length of each race is different.

$$a_{12} = 120 \text{ m (for 1200 m)}$$

$\therefore \quad A_{12}$ (for 800 m) $= \dfrac{120}{1200} \times 800 = 80$ m $\quad [\because L_1 = 1200 \text{ m and } L = 800 \text{ m}]$

Similarly, $\quad a_{23} = 100$ m (for 500 m)

A_{23} (for 800 m) $= \dfrac{100}{500} \times 800 = 160$ m and $A_{13} = ?$ $\quad [\because L_2 = 500 \text{ m}]$

According to the formula,

$$(L - A_{12}) A_{23} = L (A_{13} - A_{12})$$

$\Rightarrow \qquad (800 - 80) \, 160 = 800 \, (A_{13} - 80)$
$\Rightarrow \qquad 720 \times 160 = 800 \, A_{13} - 800 \times 80$
$\Rightarrow \qquad 720 \times 16 = 80 \, A_{13} - 6400$
$\Rightarrow \qquad 80 \, A_{13} = 17920$
$\therefore \qquad A_{13} = \dfrac{17920}{80} = 224$ m

Hence, A will beat C by a distance of 224 m.

Multi Concept Questions

1. In 1 km race, the ratio of the speeds of two contestants L and M is 3 : 4. L has a start of 280 m. Then, L wins by how many metres?

    (a) 40   (b) 60   (c) 45   (d) 65

    (a) To reach the winning post, L will have to cover a distance of (1000 − 280) m = 720 m
    When L covers 3 m, then M covers 4 m.
    When L covers 720 m, then M covers $\left(\dfrac{4}{3} \times 720\right)$ m = 960 m
    Thus, when L reaches the winning post, M covers 960 m and therefore, remains 40 m behind.
    ∴ L wins by 40 m.

2. In a game of 80 points, A can give 10 points to B and 20 points to C. Then, how many points can B give C in a game of 70 points?

    (a) 20   (b) 30   (c) 40   (d) 10

    (d) Here, A : B = 80 : 70 and A : C = 80 : 60
    ∴ $\dfrac{B}{C} = \left(\dfrac{B}{A} \times \dfrac{A}{C}\right) = \left(\dfrac{70}{80} \times \dfrac{80}{60}\right) = \dfrac{70}{60} = 70 : 60$
    Hence, in a game of 70 points, B can give C 10 points.

3. In a 500 m race, A gives B a start of 6 s and beat him by 20 m. In another race of 500 m, A beats B by $8\dfrac{1}{7}$ s. Find their speeds.

    (a) A's speed = 11 m/s, B's speed = 10 m/s   (b) A's speed = 11 m/s, B's speed = 9.33 m/s
    (c) A's speed = 12 m/s, B's speed = 9.33 m/s   (d) A's speed = 9.33 m/s, B's speed = 11 m/s

    (b) In a 500 m race, B takes $8\dfrac{1}{7}$ s more time than A.
    In another 500 m race, B takes 6 s more time and run 20 m less distance than A.
    ∴ B can run in $\left(8\dfrac{1}{7} - 6\right)$ s a distance of 20 m; B can run in 1 s a distance of $\dfrac{20}{2\dfrac{1}{7}}$.
    ∴ Speed of B is $\dfrac{28}{3}$ m/s ≈ 9.33 m/s
    Let the speed of A be v_A m/s.
    Then, according to the question,
    $\dfrac{500}{\text{Speed of B}} - \dfrac{500}{\text{Speed of A}} = 8\dfrac{1}{7} \Rightarrow \dfrac{500}{28/3} - \dfrac{500}{v_A} = \dfrac{57}{7}$
    $\Rightarrow \dfrac{1500}{28} - \dfrac{57}{7} = \dfrac{500}{v_A}$
    ∴ $v_A \approx 11$ m/s
    Hence, speed of A is 11 m/s and speed of B is 9.33 m/s.

Fast Track Practice

1. In a race of 150 m, A gives B a start of 20 m. What distance will be covered by B?
   (a) 100 m (b) 130 m
   (c) 170 m (d) 160 m

2. In a race, P gives Q a start of 25 m making length of race for Q a distance of 175 m. Find the total length of race.
   (a) 250 m (b) 200 m
   (c) 225 m (d) 235 m
   (e) None of these

3. In a 800 m race, A gives some start to B and this makes the length of race for B 725 m. What start does B get from A?
   (a) 22 m (b) 35 m (c) 45 m (d) 75 m

4. In a game of 100 points, A scores 100 points while B scores only 65 points. In this game, how much points can A give to B?
   (a) 45 (b) 35 (c) 25 (d) 55
   (e) None of these

5. In a 100 m race, A runs at a speed of $\frac{5}{6}$ m/s. If A gives a start of 4 m to B and still beats him by 12 s, what is the speed of B? **[CDS 2017 (I)]**
   (a) $\frac{5}{4}$ m/s (b) $\frac{7}{5}$ m/s (c) $\frac{4}{3}$ m/s (d) $\frac{6}{5}$ m/s

6. In a 400 m race, A runs at a speed of 16 m/s. If A gives B a start of 16 m and still beats him by 40 s, then what will be the speed of B?
   (a) 6 m/s (b) 8 m/s
   (c) 15 m/s (d) 5.9 m/s

7. In a 200 m race, A runs at a speed of 2 m/s. If A gives B a start of 10 m and still beats him by 5 s, then what will be the speed of B?
   (a) $1\frac{17}{21}$ m/s (b) $3\frac{17}{21}$ m/s
   (c) 21 m/s (d) 2 m/s

8. In a 1000 m race, X beats Y by 140 m or 14 s. What will be the X's time over the course?
   (a) 86 s (b) 90 s (c) 95 s (d) 76 s

9. In 1 km race, P beats Q by 72 m or 12 s. Find the P's time over the course.
   (a) $155\frac{2}{3}$ s (b) 151 s
   (c) $154\frac{2}{3}$ s (d) 160 s

10. X covers 1 km in 8 min 40 s while Y covers the same distance in 10 min. By what distance does X defeat Y?
    (a) $13\frac{1}{3}$ m (b) $133\frac{2}{3}$ m
    (c) $133\frac{2}{5}$ m (d) $133\frac{1}{3}$ m

11. A can run 40 m while B runs 50 m. In 1 km race, B beats A by which of the following distances?
    (a) 175 m (b) 225 m (c) 335 m (d) 200 m
    (e) None of these

12. Raman covers 1 km in 8 min while Suman covers the same distance in 10 min. By what distance does Raman beat Suman?
    (a) 150 m (b) 65 m
    (c) 190 m (d) 200 m
    (e) None of these

13. Yogesh can run 1 km in 6 min 20 s and Vijay can cover the same distance in 6 min 40 s. By what distance can Yogesh beat Vijay?
    (a) 50 m (b) 90 m (c) 45 m (d) 30 m

14. In a game of 200 points, A can give 40 points to B and 56 points to C. How many points can B give to C?
    (a) 20 (b) 15 (c) 10 (d) 5
    (e) None of these

15. In a game of billiards, A can give B, 20 points in the game of 120 points and he can give C, 30 points in the game of 120 points. How many points can B give C in a game of 90?
    (a) 9 (b) 18 (c) 6 (d) 3

16. Arun and Bhaskar start from place P at 6 : 00 am and 7 : 30 am, respectively and run in the same direction. Arun and Bhaskar run at 8 km/h and 12 km/h, respectively. Bhaskar overtakes Arun at **[SSC CGL (Mains) 2012]**
    (a) 10 : 30 am (b) 9 : 00 am
    (c) 11 : 30 am (d) 1 : 00 am

17. A runs $1\frac{2}{3}$ times as fast as B. If A gives B a start of 80 m, how far must the winning post from the starting point be so that A and B might reach it at the same time? **[CDS 2015 (I)]**
    (a) 200 m (b) 300 m
    (c) 270 m (d) 160 m

490 / Fast Track Objective Arithmetic

18. A runs $1\frac{3}{4}$ times as fast as B. If A gives B a start of 30 m, then how far must the winning post be, so that A and B reach it at the same time?
 (a) 52 m (b) 75 m (c) 69 m (d) 70 m

19. In a 200 m race, A can beat B by 50 m and B can beat C by 8 m. In the same race, A can beat C by what distance?
 (a) 60 m (b) 72 m
 (c) 56 m (d) 66 m

20. A, B and C are three contestants in a 500 m race. If A can give B a start of 20 m and A can give C a start of 32 m, then how many metres start can B give to C?
 (a) 12 (b) 14
 (c) 12.5 (d) 13.5

21. In a race of 600 m, A can beat B by 30 m and in a race of 500 m, B can beat C by 25 m. By how many metres will A beat C in a race of 400 m?
 (a) 39 (b) 49 (c) 55 (d) 25

22. A 10 km race is organised at 800 m circular race course. P and Q are the contestants of this race. If the ratio of the speeds of P and Q is 5 : 4, then how many times will the winner overtake the loser?
 (a) 4 (b) 1 (c) 2 (d) 3

23. In a race A, B and C take part. A beats B by 30 m, B beats C by 20 m and A beats C by 48 m.
 I. The length of the race is 300 m.
 II. The speeds of A, B and C are in the ratio 50 : 45 : 42.
 Which of the following is/are correct?
 [CDS 2015 (II)]
 (a) Only I (b) Only II
 (c) Both I and II (d) Neither I nor II

24. A, B and C walk 1 km in 5 min, 8 min and 10 min, respectively. C starts walking from a point at a certain time, B starts from the same point 1 min later and A starts from the same point 2 min later than C. Then, A meet C and B at times [SSC CGL 2013]
 (a) 2 min, 3 min (b) $\frac{4}{3}$ min, 3 min
 (c) 2 min, $\frac{5}{3}$ min (d) 1 min, 2 min

25. In a race of 1000 m, A beats B by 100 m or 10 s. If they start a race of 1000 m simultaneously from the same point and if B gets injured after running 50 m less than half the race length and due to which his speed gets halved, then by how much time will A beat B? [CDS 2016 (I)]
 (a) 65 s (b) 60 s
 (c) 50 s (d) 45 s

Answer with Solutions

1. (b) Required distance = (150 − 20) m
 = 130 m

2. (b) Required length = (175 + 25) m
 = 200 m

3. (d) Start given by A to B
 = (800 − 725) m = 75 m

4. (b) Here, score of A = 100 points
 and score of B = 65 points
 ∴ A can give (100 − 65) = 35 points to B.

5. (c) Let the speed of B be x m/s.
 Then, according to the question,
 $\frac{96}{x} - \frac{100}{(5/3)} = 12 \Rightarrow \frac{96}{x} - 60 = 12$
 $\Rightarrow \frac{96}{x} = 72 \Rightarrow x = \frac{4}{3}$
 Hence, the speed of B is $\frac{4}{3}$ m/s.

6. (d) Time taken by A to cover 400 m
 $= \frac{\text{Distance}}{\text{Speed}} = \frac{400}{16}$ s = 25 s
 ∴ B covers (400 − 16)
 = 384 m in (25 + 40) = 65 s
 ∴ B's speed = $\frac{384}{65}$ = 5.9 m/s

7. (a) Time taken by A to cover 200 m
 $= \frac{200}{2} = 100$ s
 ∴ B covers (200 − 10) m = 190 m in
 (100 + 5) s = 105 s
 ∴ B's speed = $\frac{190}{105} = \frac{38}{21} = 1\frac{17}{21}$ m/s

8. (a) Clearly, Y covers 140 m in 14 s.
 ∴ Y's time over the course
 $= \frac{14}{140} \times 1000 = 100$ s
 ∴ X's time over the course
 = 100 − 14 = 86 s

Races and Games of Skill / 491

Alternate Method

$\because \dfrac{\text{Time taken by } X}{1000-140} = \dfrac{14}{140}$

$\therefore$ Time taken by $X = \dfrac{1}{10} \times 860 = 86$ s

9. (c) Clearly, Q covers 72 m in 12 s.
$\therefore Q$'s time over the course
$= \dfrac{12}{72} \times 1000 = \dfrac{500}{3}$

$\therefore P$'s time over the course
$= \dfrac{500}{3} - 12 = \dfrac{500-36}{3} = \dfrac{464}{3} = 154\dfrac{2}{3}$ s

10. (d) Clearly, X beats Y by 80 s.
Distance covered by Y in 600 s = 1000 m
[$\because$ 10 min = 600 s]
$\therefore$ Distance covered by Y in 80 s
$= \dfrac{1000}{600} \times 80 = \dfrac{400}{3}$ m $= 133\dfrac{1}{3}$ m

11. (d) In a 50 m race, B beats A by 10 m.
In 1 km race, B beats A by
$\left(\dfrac{10}{50} \times 1000\right)$ m = 200 m

12. (d) Raman covers 1 km in 8 min
and Suman cover 1 km in 10 min.
If they starts together, then distance covered by Suman in 8 min
$= \dfrac{1000}{10} \times 8 = 800$ m

$\therefore$ Raman will beat Suman by
$(1000-800)$ m = 200 m

13. (a) Clearly, Yogesh beats Vijay by 20 s.
Distance covered by Vijay in 400 s
= 1000 m   [$\because$ 6 min 40 s = 400 s]

Distance covered by Vijay in 20 s
$= \left(\dfrac{1000}{400} \times 20\right)$ m $= 50$ m

$\therefore$ Yogesh beats Vijay by 50 m.

14. (a) $A:B = 200:160$, $A:C = 200:144$

$\therefore \dfrac{B}{C} = \dfrac{B}{A} \times \dfrac{A}{C} = \dfrac{160}{200} \times \dfrac{200}{144}$
$= \dfrac{160}{144} = \dfrac{10}{9} = \dfrac{200}{180}$

Hence, B can give C 20 points.

15. (a) If A scores 120 points, then B scores 100 points and C scores 90 points.
When B scores 100 points, then C scores 90 points.
When B scores 90 points, then C scores
$\left(\dfrac{90}{100} \times 90\right)$ points = 81 points

$\therefore B$ can give C, 9 points in a game of 90.

16. (a) Distance between Arun and Bhaskar at 7 : 30 am $= 8 \times 1\dfrac{1}{2} = 12$ km

Time taken by Bhaskar in covering a distance of 12 km $= \dfrac{12}{(12-8)} = 3$ h

[$\because$ relative speed = $12 - 8 = 4$]

$\therefore$ Required time = 7 : 30 + 3 : 00
= 10 : 30 am

17. (a) Let the speed of B be x m/s.
$\therefore$ Speed of $A = 1\dfrac{2}{3}x = \dfrac{5x}{3}$ m/s

Ratio of speed of A and $B = \dfrac{5x}{3} : x = 5:3$

$\because$ 2 m are gained in a race of 5 m.
$\therefore$ 1 m is gained in a race of $\dfrac{5}{2}$ m.

So, 80 m are gained in a race of
$= \dfrac{5}{2} \times 80 = 200$ m

18. (d) Since, A is $1\dfrac{3}{4}$ faster than B.
Then, ratio of the rates of A and $B = 7:4$
3 m are gained in a race of 7 m.
30 m are gained in a race of $\dfrac{7}{3} \times 30 = 70$ m

So, winning post is 70 m away from the starting point.

19. (c) According to the question,
$A:B = 200:150$
and $B:C = 200:192$

$\therefore A:C = \left(\dfrac{A}{B} \times \dfrac{B}{C}\right) = \left(\dfrac{200}{150} \times \dfrac{200}{192}\right) = \dfrac{200}{144}$

So, A beats C by $(200-144)$ m = 56 m.

Fast Track Method

Here, $L = 200$ m, $a_{12} = 50$ m, $a_{23} = 8$ m
and $a_{13} = ?$
According to the formula,
$(L-a_{12})a_{23} = L(a_{13}-a_{12})$ [by Technique 1]
$\Rightarrow (200-50) \times 8 = 200(a_{13}-50)$
$\Rightarrow \dfrac{150 \times 8}{200} = a_{13} - 50$
$\Rightarrow a_{13} = 6 + 50 = 56$ m

20. (c) Given that, $a_{12} = 20$ m, $a_{13} = 32$ m, $L = 500$ m and $a_{23} = ?$
According to the formula,
$(L-a_{12})a_{23} = L(a_{13}-a_{12})$
[by Technique 1]
$\Rightarrow (500-20)a_{23} = 500(32-20)$
$\Rightarrow 480\, a_{23} = 500 \times 12$
$\Rightarrow a_{23} = \dfrac{500 \times 12}{480} \Rightarrow a_{23} = \dfrac{50}{4} = 12.5$ m

Hence, B can give C a start of 12.5 m.

21. (a) If A runs 600 m, then B runs 570 m.
If A runs 400 m, B then runs
$$\frac{570}{600} \times 400 = 380 \text{ m}$$
When B runs 500 m, then C runs 475 m.
When B runs 380 m, then C runs
$$\frac{475}{500} \times 380 = 361 \text{ m}$$
∴ A beats C by $(400 - 361)$ m $= 39$ m

Fast Track Method
According to the question,
The length of the each race is different.
$a_{12} = 30$ m (for 600 m)
$L_1 = 600$ m, $L = 400$ m
∴ A_{12} (for 400 m) $= \dfrac{a_{12}}{L_1} \times L = \dfrac{30}{600} \times 400$
$= 20$ m
Similarly, $A_{23} = 25$ m (for 500 m)
$L_2 = 500$ m, $L = 400$ m
A_{23} (for 400 m) $= \dfrac{a_{23}}{L_2} \times L = \dfrac{25}{500} \times 400$
$= 20$ m and $A_{13} = ?$
According to the formula,
$(L - A_{12}) A_{23} = L (A_{13} - A_{12})$
[by Technique 2]
$\Rightarrow (400 - 20) 20 = 400 (A_{13} - 20)$
$\Rightarrow 7600 = 400 A_{13} - 8000$
$\Rightarrow 15600 = 400 A_{13} \Rightarrow A_{13} = 39$ m
So, A beats C by 39 m.

22. (c) Speed of P : Speed of $Q = 5 : 4$
Time taken by P to cover 5 rounds
$=$ Time taken by Q to cover 4 rounds
Distance covered by P in 5 rounds
$= 5 \times \dfrac{800}{1000} = 4$ km
Distance covered by Q in 4 rounds
$= 4 \times \dfrac{800}{1000} = \dfrac{16}{5}$ km
In 5 rounds, P will overtake Q everytime.
It means that after covering 4 km, P will overtake Q one time.
∴ After covering 10 km P will overtake Q
$= \dfrac{1}{4} \times 10 = 2\dfrac{1}{2}$ times ≈ 2 times

23. (c) I. Let the length of race be x m.
Then, distance covered by $A = x$ m.
Distance covered by B when A reach the destination $= x - 30$
Distance covered by C when A reach the destination $= x - 48$
and distance covered by C when B reach the destination $= x - 20$

$\Rightarrow \dfrac{x - 30}{x} = \dfrac{x - 48}{x - 20}$
$\Rightarrow x^2 - 50x + 600 = x^2 - 48x$
$\Rightarrow \qquad x = 300$ m
II. The speeds of A, B and C are in the ratio $= 300 : 270 : 252 = 50 : 45 : 42$

24. (c) A walks 1 km in 5 min.
Then, distance covered by A in 1 min
$= \dfrac{1000}{5} = 200$ m
B walks 1 km in 8 min.
Then, distance covered by B in 1 min
$= \dfrac{1000}{8} = 125$ m
C walks 1 km in 10 min.
Then, distance covered by C in 1 min
$= \dfrac{1000}{10} = 100$ m
Let A meets B and C in x and y min, respectively.
Then, according to the question,
Distance covered by C in $(x + 2)$ min
$=$ Distance covered by A in x min
$\Rightarrow \qquad 100 (x + 2) = 200 x$
$\Rightarrow \qquad 100 x + 200 = 200 x$
$\Rightarrow \qquad 200 = 100 x \Rightarrow x = \dfrac{200}{100} = 2$ min
Now, for A and B,
Distance covered by B in $(y + 1)$ min
$=$ Distance covered by A in y min
$\Rightarrow \qquad 125 (y + 1) = 200 \times y$
$\Rightarrow \qquad 125 y + 125 = 200 y$
$\Rightarrow \qquad 125 = 200 y - 125 y$
$\Rightarrow \qquad 125 = 75 y \Rightarrow y = \dfrac{125}{75} = \dfrac{5}{3}$ min

25. (a) Since, either A beats B by 100 m or 10 s. It means that B runs 100 m in 10 s.
∴ Speed of $B = \dfrac{100}{10} = 10$ m/s

```
                                    B              A
          •──────────────────────────•──────────────•
          ←───────── 900 m ─────────→ ←── 100 m ──→
          ←──────────────── 1000 m ──────────────→
```

∵ B gets injured at a distance of 450 m and his speed gets halved.
So, time taken by B to cover
$1000 \text{ m} = \dfrac{450}{10} + \dfrac{550}{5} = 155$ s
∵ Ratio of speed of A and B is equal to ratio of distance covered by A and B.
$A : B = 1000 : 900 = 10 : 9$
Now, speed of $A = \dfrac{10}{9} \times 10 = \dfrac{100}{9}$ m/s
Time taken by A to cover 1000 m
$= \dfrac{1000}{100} \times 9 = 90$ s
Hence, A beat B by length of time
$= (155 - 90)$ s $= 65$ s

Chapter 29

Clock and Calendar

Clock

A clock is an instrument which displays time divided into hours, minutes and seconds.

A clock mainly consists of four components, i.e. dial, hour hand, minute hand and second hand.

A clock is a circular **dial**. The periphery of the dial is numbered 1 through 12 indicating the hours in a 12 h cycle and a short **hour hand** makes two revolutions a day.

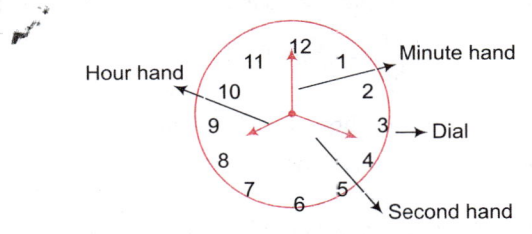

A longer **minute hand** makes one revolution every hour. The face may also include a **second hand** which makes one revolution per minute.

- In 1 h, minute hand covers 60 min spaces, whereas the hour hand covers 5 min spaces.
  Therefore, minute hand gains (60 − 5) = 55 min in 1 h.

Important Points Related to Clock

1. In 1 h, both hour and minute hands coincide once (i.e. 0° apart)
   For example Between 3 and 4 O' clock, hands are together as shown in the adjoining figure.

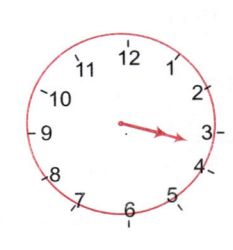

2. In 12 h, both hands coincide 11 times (between 11 and 1 O' clock they coincide once) and in a day both hands coincide 22 times.

   For example Between 11 and 1O' clock, hands are together as shown in the adjoining figure.

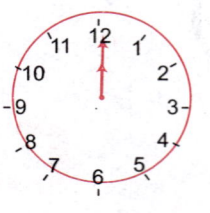

3. If two hands are at 90°, then they are 15 min spaces apart. This happens twice in 1 h. In a period of 12 h, the hands are at right angle 22 times (2 common positions) and in a day both hands are at right angle 44 times.

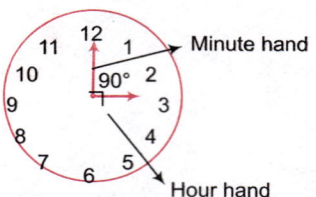

4. If two hands are in opposite direction. (i.e. 180° apart), then they are 30 min spaces apart. This happens once in 1 h. In a period of 12 h both hands are in opposite direction 11 times and in a day both hands are in opposite direction 22 times.

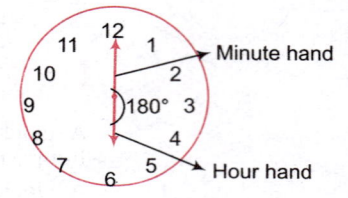

5. Angle covered by minute hand in 1 min

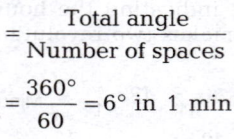

   $$= \frac{\text{Total angle}}{\text{Number of spaces}}$$

   $$= \frac{360°}{60} = 6° \text{ in 1 min}$$

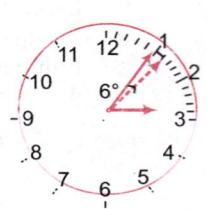

6. Angle covered by hour hand in 1 min.
   As hour hand covers 360° in 12 h. Hence, hour hand covers $\left(\frac{360°}{12}\right) = 30°$ in 1 h.

   Hence, hour hand covers $\left(\frac{30°}{60}\right) = \frac{1°}{2}$ in 1 min.

7. From point 5 and 6, we can say that the minute hand goes ahead by $5\frac{1°}{2}$ in comparison to hour hand.

Concept of Slow or Fast Clocks

If a watch/clock indicates 9 : 15, when the correct time is 9, then it is said to be 15 min too fast. On the other hand, if the watch/clock indicates 6 : 45, when the correct time is 7, then it is said to be 15 min too slow.

Ex. 1 What will be angle between the two hands of a clock at 9 : 50?

Sol. ∵ Angle traced by the hour hand in 12 h = 360°

∴ Angle traced by the hour hand in 1 h = $\frac{360°}{12}$

and angle traced by the hour hand in 9 h and 50 min,

Clock and Calendar / 495

i.e. $\dfrac{59}{6} h = \dfrac{360°}{12} \times \dfrac{59}{6} = 295°$

Similarly, angle traced by the minute hand in 60 min = 360°

∴ Angle traced by minute hand in 50 min = $\dfrac{360°}{60} \times 50 = 300°$

So, the required angle = 300° − 295° = 5°

Ex. 2 A clock gains 10 s in every 3 h. If the clock was set right at 4 : 00 am on Monday morning, then find the correct time when it will indicate on Tuesday evening at 7 : 00 pm.

Sol. Difference of time between 4 : 00 am on Monday to 7 : 00 pm Tuesday

$$= 24 + 12 + 3 = 39 \text{ h}$$

Now, time gained by clock in 3 h = 10 s

Time gained by clock in 1 h = $\dfrac{10}{3}$ s

∴ Time gained by clock in 39 h = $\dfrac{10 \times 39}{3} = 130$ s

So, the correct time is 7 : 02 : 10 pm.

Fast Track Techniques to solve the QUESTIONS

Technique 1 Between n O'clock and (n + 1) O'clock, the two hands of a clock will coincide at $\left(\dfrac{60n}{11}\right)$ min past n.

Ex. 3 At what time between 4 O'clock and 5 O'clock, will the hands of a clock be together?

Sol. At 4 O'clock, the hour hand is at 4 and the minute hand is at 12. It means that they are 20 min spaces apart.

To be together, the minute hand must gain 20 min over the hour hand.

As we know, 55 min is gained by minute hand in 60 min.

∴ 20 min will be gained in $\left(\dfrac{60}{55} \times 20\right)$ min = $\dfrac{60 \times 4}{11} = \dfrac{240}{11}$ min = $21\dfrac{9}{11}$ min

Hence, the hands will coincide at $21\dfrac{9}{11}$ min past 4.

Fast Track Method

Here, $n = 4$ and $(n + 1) = 5$

According to the formula,

The two hands will coincide at $\dfrac{60n}{11}$ min past n.

or $\dfrac{60 \times 4}{11}$ min past 4 or $21\dfrac{9}{11}$ min past 4.

Ex. 4 At what time between 1 O'clock and 2 O'clock, will the hands of a clock be together?

Sol. Given, $n = 1$ and $(n + 1) = 2$

∴ The two hands will coincide at $\left(\dfrac{60n}{11}\right)$ min past 1.

or $\dfrac{60 \times 1}{11}$ min past 1 or $5\dfrac{5}{11}$ min past 1.

Technique 2 Between n O'clock and $(n + 1)$ O'clock, the two hands of a clock will mutually make right angle at $(5n \pm 15) \times \dfrac{12}{11}$ min past n.

Ex. 5 At what time between 7 O'clock and 8 O'clock in the morning, will the both hands of a clock be at right angle?

Sol. At 7 O'clock the minute hand will be 35 min spaces behind the hour hand. Now, when the two hands are at right angle, they are 15 min spaces apart. So, they are at right angles in the following cases.

Case I When minute hand is 15 min spaces behind the hour hand.

In this case, minute hand will have to gain $(35 - 15) = 20$ min spaces.

55 min spaces are gained by it in 60 min.

20 min spaces will be gained by it in $\left(\dfrac{60}{55} \times 20\right)$ min $= \dfrac{240}{11}$ min $= 21\dfrac{9}{11}$ min

Hence, they are at right angle at $21\dfrac{9}{11}$ min past 7.

Case II When the minute hand is 15 min spaces ahead of the hour hand.

To be in this position, the minute hand will have to gain $(35 + 15) = 50$ min spaces.

55 min spaces are gained in 60 min.

50 min spaces are gained in $\left(\dfrac{60}{55} \times 50\right)$ min $= \left(\dfrac{60}{11} \times 10\right)$ min $= 54\dfrac{6}{11}$ min

Hence, they are at right angle at $54\dfrac{6}{11}$ min past 7.

Fast Track Method

Here, $n = 7$ and $(n + 1) = 8$

According to the formula,

The hands will make right angle at

$= (5n \pm 15) \times \dfrac{12}{11}$ min past 7 $= (5 \times 7 \pm 15) \times \dfrac{12}{11}$ min past 7

$= (35 + 15) \times \dfrac{12}{11}$ min past 7 and $(35 - 15) \times \dfrac{12}{11}$ min past 7

$= \dfrac{50 \times 12}{11}$ min past 7 and $\dfrac{20 \times 12}{11}$ min past 7

$= \dfrac{600}{11}$ min past 7 and $\dfrac{240}{11}$ min past 7

Clearly, the two hands will make right angle at $54\dfrac{6}{11}$ min past 7 and $21\dfrac{9}{11}$ min past 7.

Clock and Calendar / 497

Technique 3 Between n O'clock and $(n+1)$ O'clock, the hands of a clock will be in the same straight line (without being together) at

(i) $(5n - 30) \times \dfrac{12}{11}$ min past n, when $n > 6$.

(ii) $(5n + 30) \times \dfrac{12}{11}$ min past n, when $n < 6$.

Ex. 6 At what time between 7 O'clock and 8 O'clock, will the hands of a clock be in the same straight line but not together?

Sol. At 7 O'clock, the hour hand is at 7 and the minute hand is at 12. It means that the two hands are 25 min spaces apart.

To be in the same straight line (but not together), they will be 30 min spaces apart.

∴ The minute hand will have to gain $(30 - 25) = 5$ min spaces over the hour hand.
As we know that 55 min spaces are gained in 60 min.

∴ 5 min will be gained in $\left(\dfrac{60}{55} \times 5\right)$ min $= 5\dfrac{5}{11}$ min

Hence, the hands will be in the same straight line but not together at $5\dfrac{5}{11}$ min past 7.

Fast Track Method

Here, $n = 7$ and $(n + 1) = 8$, also $n > 6$

According to the formula,

Hands will be in the same straight line at $(5n - 30) \times \dfrac{12}{11}$ min past n

$= (5 \times 7 - 30) \times \dfrac{12}{11}$ min past $7 = (35 - 30) \times \dfrac{12}{11}$ min past 7

$= \dfrac{12 \times 5}{11}$ min past $7 = 5\dfrac{5}{11}$ min past 7

Ex. 7 At what time between 3 O'clock and 4 O'clock, will the hands of a clock be in opposite directions?

Sol. At 3 O'clock, the hour hand is at 3 and the minute hand is at 12. It means that the two hands are 15 min spaces apart. But to be in opposite directions, the hands must be 30 min spaces apart. Therefore, the minute hand will have to gain $(30 + 15) = 45$ min spaces over the hour hand.

∵ 55 min spaces are gained in 60 min.

∴ 45 min spaces are gained in $\left(\dfrac{60}{55} \times 45\right)$ min $= \dfrac{60 \times 9}{11}$ min $= \dfrac{540}{11} = 49\dfrac{1}{11}$ min

Hence, the required time $= 49\dfrac{1}{11}$ min past 3.

Fast Track Method

Here, $n = 3$ and $(n + 1) = 4$, also $(n < 6)$

According to the formula,

Hands will be at the same straight line at $(5n + 30) \times \dfrac{12}{11}$ min past n

$= (5 \times 3 + 30) \times \dfrac{12}{11}$ min past $3 = (15 + 30) \times \dfrac{12}{11}$ min past 3

$= \dfrac{45 \times 12}{11}$ min past $3 = 49\dfrac{1}{11}$ min past 3

Clock and Calendar / 499

$\therefore$ Gain in 24 h (one day) $= \left(\dfrac{27}{11} \times \dfrac{60 \times 24}{63}\right)$ min $= \dfrac{4320}{77}$ min $= 56\dfrac{8}{77}$ min

As the result is positive, therefore the clock gains $56\dfrac{8}{77}$ min.

Fast Track Method

Here, $x = 63$ min

According to the formula,

Required result $= \left(\dfrac{720}{11} - x\right)\left(\dfrac{60 \times 24}{x}\right)$ min $= \left(\dfrac{720}{11} - 63\right)\left(\dfrac{60 \times 24}{63}\right)$ min

$= \dfrac{27}{11} \times \dfrac{60 \times 8}{21} = 56\dfrac{8}{77}$ min

As result is positive, therefore the clock gains $56\dfrac{8}{77}$ min.

Calendar

A calendar is a chart or series of pages showing the days, weeks and months of a particular year. A calendar consist of 365 or 366 days divided into 12 months.

Ordinary Year

A year having 365 days is called an ordinary year

(52 complete weeks + 1 extra day = 365 days)

Leap Year

A leap year has 366 days (the extra day is 29th of February) (52 complete weeks + 2 extra days = 366 days.)

A leap year is divisible by 4 except for a century. For a century to be a leap year it must be divisible by 400. For example

* Years like 1988, 2008 are leap years (divisible by 4).
* Centuries like 2000, 2400 are leap years (divisible by 400).
* Years like 1999, 2003 are not leap years (not divisible by 4).
* Centuries like 1700, 1800 are not leap years (not divisible by 400).
* In a century, there are 76 ordinary years and 24 leap years.

Odd Days

Extra days, apart from the complete weeks in a given period are called odd days. An ordinary year has 1 odd day while a leap year has 2 odd days.

To find the numbers of odd days

* Number of days in an ordinary year
  $= 365 = (52 \times 7) + 1 = 52$ weeks + 1 odd day
  Hence, an ordinary year has 1 odd day.
* Number of days in a leap year $= 366 = (52 \times 7) + 2$
  $= 52$ weeks + 2 days
  Hence, a leap year has 2 odd days.
* Number of days in a century (100 yr)
  $= 76$ ordinary years + 24 leap years
  $= 76 \times 1 + 24 \times 2 = 124 = 17 \times 7 + 5 = 17$ week + 5 odd days
  Hence, 100 yr has 5 odd days.

Months	Odd days
January	3
February	0/1 (ordinary/leap)
March	3
April	2
May	3
June	2
July	3
August	3
September	2
October	3
November	2
December	3

- Number of odd days in 200 yr = 5 × 2 = 10 days = 1 week + 3 days = 3 odd days
- Number of odd days in 300 yr = 5 × 3 = 15 days = 2 weeks + 1 day = 1 odd day
- Number of odd days in 400 yr = (5 × 4 + 1) days = 21 days = 3 weeks = 0 odd days
- A 400th is a leap year, therefore 1 more day has been taken.

Similarly, each one of 800 yr, 1200 yr, 1600 yr, 2000 yr, 2400 yr etc., has no odd days. Remember the adjacent table for the number of odd days in different months of an year.

- In an ordinary year, February has no odd days, but in a leap year, February has one odd day.
- The 1st day of a century must be Tuesday, Thursday or Saturday.
- The last day of a century cannot be Tuesday, Thursday or Saturday.

Day Gain/Loss

Ordinary Year (± 1 day)

- When we proceed forward by 1 yr, then 1 day is gained.

  For example 9th August 2013 is Friday, then 9th August 2014 has to be Friday + 1 = Saturday.

- When we move backward by 1 yr, then 1 day is lost.

  For example 24th December 2013 is Tuesday, then 24th December 2012 has to be Tuesday − 1 = Monday.

Leap Year (± 2 days)

- When we proceed forward by 1 leap year, then 2 days are gained.

  For example If it is Wednesday on 25th December 2011, then it would be Friday on 25th December 2012 [Wednesday + 2] because 2012 is a leap year.

- When we move backward by 1 leap year, then 2 days are lost.

  For example If its is Wednesday on 18th December 2012, then it would be Monday on 18th December 2011. [Wednesday − 2] because 2012 is a leap year.

Exception

The day must have crossed 29th February for adding 2 days otherwise 1 day.

For example If 26th January 2011 is Wednesday, 26th January 2012 would be Wednesday + 1 = Thursday (even if 2012 is leap year, we have added + 1 day because 29th February is not crossed).

If 23rd March 2011 is Wednesday, then 23rd March 2012 would be Wednesday + 2 = Friday (+ 2 days because 29th February of leap year is crossed).

To Find a Particular Day on the Basis of Given Day and Date

Following steps are taken into consideration to solve such questions

Step I Firstly, you have to find the number of odd days between the given date and the date for which the day is to be determined.

Step II The day (for a particular date) to be determined, will be that day of the week which is equal to the total number of odd days and this number is counted forward from the given day, in case the given day comes before the day to be determined. But, if the given day comes after the day to be determined, then the same counting is done backward from the given day.

Clock and Calendar / 501

Ex. 10 If 5th January, 1991 was Saturday, what day of the week was it on 4th March, 1992?

Sol. Number of days between 5th January, 1991 and 4th March, 1992
$= (365 - 5)$ days of year 1991 + 31 days of January 1992
+ 29 days of February 1992 + 4 days of March 1992
[as 1992 is completely divisible by 4, hence it is a leap year and that's why February has 29 days]
$= 360 + 31 + 29 + 4 = 424 = 60$ weeks + 4 days
∵ Number of odd days = 4
∴ 4th March, 1992 will be 4th day beyond Saturday.
Hence, the required day will be Wednesday.

Ex. 11 What day of the week was it on 5th November, 1987, if it was Monday on 4th April, 1988?

Sol. Number of days between 5th November, 1987 and 4th April, 1988
$= (30 - 5)$ days of November 1987 + 31 days of December 1987 + 31 days of January 1988 + 29 days of February 1988 + 31 days of March 1988 + 4 days of April 1988
$= 25 + 31 + 31 + 29 + 31 + 4 = 151$ days = 21 weeks + 4 days
∵ Number of odd days = 4   [∵ 1988 is a leap year, so February has 29 days]
∴ 5th November 1987 will be 4 days before Monday.
Hence, the required day is Thursday.

To Find a Particular Day without Given Date and Day

Following steps are taken into consideration to solve such questions

Step I Firstly, you have to find the number of odd days upto the date for which the day is to be determined.

Step II Your required day will be according to the following conditions
  (a) If the number of odd days = 0, then required day is Sunday.
  (b) If the number of odd days = 1, then required day is Monday.
  (c) If the number of odd days = 2, then required day is Tuesday.
  (d) If the number of odd days = 3, then required day is Wednesday.
  (e) If the number of odd days = 4, then required day is Thursday.
  (f) If the number of odd days = 5, then required day is Friday.
  (g) If the number of odd days = 6, then required day is Saturday.

Ex. 12 Find the day of the week on 26th January 1950.

Sol. Number of odd days upto 26th January, 1950.
$=$ Odd days for 1600 yr + Odd days for 300 yr + Odd days for 49 yr
+ Odd days of 26 days of January 1950
$= 0 + 1 + (12 \times 2 + 37) + 5 = 0 + 1 + 61 + 5 = 67$ days
$= 9$ weeks + 4 days = 4 odd days
∴ It was Thursday on 26th January 1950.

◆ 49 yr has 12 leap year and 37 ordinary year.

Ex. 13 Mahatma Gandhi was born on 2 October 1869. What was the day of the week?

Sol. Odd days till the year 1868 = 1600 + 200 + 68
= 0 + 3 + (17 leap + 51 ordinary years)
= 0 + 3 + (17 × 2 + 51 × 1)
= 3 + 85 = 88
= 12 weeks + 4 days

Now, total number of odd days 1869 till October 2, 1869 are

January = 3   February = 0   March = 3
April = 2   May = 3   June = 2
July = 3   August = 3   September = 2
October = 2

Here, 23 days = 3 weeks + 2 days.

∴ Total number of odd days = 4 + 2 = 6 days

Hence, 2 October 1869 was Saturday.

Ex. 14 How many days are there in x weeks x days?

Sol. ∵ Number of days in x weeks = $7 \times x = 7x$

∴ Total number of days in x weeks x days = $7x + x = 8x$ days

Ex. 15 After 2007, which year's calendar will be the same as 2007?

Sol. Count the number of odd days from year 2007 and onwards to get the sum equal to odd days.

Years	2007	2008	2009	2010	2011	2012	2013	2014	2015	2016	2017
Odd days	1	2	1	1	1	2	1	1	1	2	1

Sum = 14 odd days = 0 odd days.

Hence, the calendar for the year 2018 will be same as for the year 2007.

Fast Track Practice

Clock

1. What will be the angle between the hands of a clock when the time is at 4 : 40 pm? **[CMAT 2015]**
   (a) 120° (b) 100°
   (c) 110° (d) 130°

2. At 5 : 30, the hour hand and the minute hand of a clock form an angle of **[CDS 2015 (I)]**
   (a) 30° (b) 15° (c) 70° (d) 45°

3. At what time between 3 O'clock and 4 O'clock, will the hands of a clock be together?
   (a) $16\frac{3}{11}$ min past 3
   (b) $14\frac{3}{11}$ min past 3
   (c) $13\frac{2}{11}$ min past 3
   (d) $16\frac{4}{11}$ min past 3
   (e) None of the above

4. At what point of time after 3 O'clock, hour hand and the minute hand are at right angles for the first time? **[CDS 2014]**
   (a) 9 O' clock (b) 4 h 37 $\frac{1}{6}$ min
   (c) 3 h 30 $\frac{8}{11}$ min (d) 3 h 32 $\frac{8}{11}$ min

5. At what time between 9 O'clock and 10 O'clock, will the hands of a clock be in the same straight line but not together?
   (a) $16\frac{2}{11}$ min past 9
   (b) $16\frac{2}{11}$ min past 10
   (c) $16\frac{4}{11}$ min past 9
   (d) $16\frac{4}{11}$ min past 10
   (e) None of the above

6. Between 6 pm and 7 pm, the minute hand of a clock will be ahead of the hour hand by 3 min at **[UPSC CSAT 2015]**
   (a) 6 : 15 pm (b) 6 : 18 pm
   (c) 6 : 36 pm (d) 6 : 48 pm

7. The minute hand of a clock overtakes the hour hand at intervals of 70 min of the correct time. How much in a day does the clock gain or loss?
   (a) $\frac{7200}{77}$ min gain
   (b) $\frac{7200}{77}$ min loss
   (c) $\frac{7300}{77}$ min loss
   (d) $\frac{7300}{78}$ min gain

8. A clock strikes once at 1 O clock, twice at 2 O'clock and thrice at 3 O'clock and so on. If it takes 8 s to strike at 5 O'clock, then the time taken by it to strike at 10 O'clock is **[CDS 2016 (I); UPSC CSAT 2017]**
   (a) 14 s (b) 16 s
   (c) 18 s (d) None of these

9. A watch which gains uniformly, is 2 min slow at noon on Monday and is 4 min 48 s fast at 2 pm on the following Monday. When was it correct? **[CLAT 2015]**
   (a) 2 pm on Tuesday
   (b) 2 pm on Wednesday
   (c) 3 pm on Thursday
   (d) 1 pm on Friday

10. A watch loses 2 min in every 24 h while another watch gains 2 min in every 24 h. At a particular instant, the two watches showed an identical time. Which of the following statements is correct if 24 h clock is followed? **[UPSC CSAT 2017]**
    (a) The two watches show the identical time again on completion of 30 days
    (b) The two watches show the identical time again on completion of 90 days
    (c) The two watches show the identical time again on completion of 120 days
    (d) None of the above statements is correct.

11. A person goes to a market between 4 pm and 5 pm. When he comes back, he finds that the hours hand and minute hand have interchanged their positions. For how much time (approximately) was he out of his house?
    (a) 55 : 25 min (b) 55 : 30 min
    (c) 55 : 34 min (d) 55 : 38 min

Calendar

12. By which of the following, a leap year must be divisible?
    (a) 9 (b) 6
    (c) 5 (d) 4
    (e) None of these

13. Which of the following is a leap year?
    (a) 2007 (b) 2016
    (c) 2001 (d) 1997
    (e) None of these

14. The last day of a century cannot be
    (a) Thursday (b) Wednesday
    (c) Friday   (d) Monday
    (e) None of these

15. Today is Monday. What will be the day after 64 days?
    (a) Saturday (b) Friday
    (c) Thursday (d) Tuesday
    (e) None of these

16. If it is Saturday on January 1, 2000, then January 1, 2001 would have been [CGPSC 2014]
    (a) Monday   (b) Tuesday
    (c) Friday   (d) Saturday
    (e) None of these

17. January 3, 2007 was Wednesday. What day of the week fell on January 3, 2008?
    (a) Tuesday   (b) Friday
    (c) Thursday (d) Saturday
    (e) None of these

18. What was the day of the week on 2nd Jan, 2010, if it was Sunday on 1st Jan, 2006?
    (a) Saturday (b) Thursday
    (c) Sunday   (d) Friday
    (e) None of these

19. If 5th March, 1999 was Friday, what day of the week was it on 9th March 2000?
    (a) Wednesday (b) Saturday
    (c) Friday   (d) Thursday
    (e) None of these

20. If 1 September, 2014 was a Monday, then what day of the week will 31 December, 2014? [CISF Head Constable 2016]
    (a) Wednesday (b) Thursday
    (c) Friday   (d) Tuesday

21. On 6th March, 2005, Monday falls. What was the day of the week on 7th March, 2004?
    (a) Tuesday   (b) Monday
    (c) Friday   (d) Sunday
    (e) None of these

22. 4th April, 1988 was Monday. What day of the week was it, on 6th November 1987?
    (a) Tuesday   (b) Friday
    (c) Sunday   (d) Saturday
    (e) None of these

23. What was the day of the week on 17th July, 1776?
    (a) Wednesday (b) Thursday
    (c) Monday   (d) Saturday
    (e) None of these

24. What was the day of the week on 17th August, 2010?
    (a) Sunday   (b) Wednesday
    (c) Tuesday   (d) Friday
    (e) None of these

25. Calendar for the year 2008 will be the same for which of the following years?
    (a) 2017 (b) 2019 (c) 2020 (d) 2016
    (e) None of these

Answer with Solutions

1. **(b)** Angle traced by hour hand in 12 h = 360°
   So, angle traced by the hour hand in 4 h 40 min,
   i.e. $\frac{14}{3}$ h = $\left(\frac{360}{12} \times \frac{14}{3}\right)° = 140°$
   Angle traced by the minute hand in 60 min = 360°
   Angle traced by minute hand in 40 min = $\frac{360°}{60} \times 40 = 240°$
   ∴ Required angle = 240° − 140° = 100°

2. **(b)** Angle traced by hour hand in 5 h 30 min
   = $\frac{11}{2}$ h = $\left(\frac{360}{12} \times \frac{11}{2}\right) = 165°$

Clock and Calendar / 505

Angle traced by minute hand in 30 min
$= \left(\dfrac{360}{60} \times 30\right) = 180°$
So, required angle $= 180° - 65° = 15°$

3. **(d)** At 3 O' clock, the hour hand is at 3 and the minute hand is at 12. It means they are 15 min spaces apart.
To be together, the minute hand must gain 15 min over the hour hand.
As we know that 55 min is gained by minute hand in 60 min.
∴ 15 min will be gained in $\left(\dfrac{60}{55} \times 15\right)$ min
$= \dfrac{60 \times 3}{11} = \dfrac{180}{11}$ min $= 16\dfrac{4}{11}$ min
Hence, the hands will coincide at $16\dfrac{4}{11}$ min past 3.

Fast Track Method
Here, $n = 3$ and $(n + 1) = 4$
∴ Two hands will coincide at
$\dfrac{60n}{11}$ min past n [by Technique 1]
$= \dfrac{60 \times 3}{11}$ min past 3
$= \dfrac{180}{11}$ min past 3
$= 16\dfrac{4}{11}$ min past 3

4. **(d)** ∵ Clock will make right angle at
$(5n + 15) \times \dfrac{12}{11}$ min past n.
Given that, $n = 3$
∴ $(5 \times 3 + 15) \times \dfrac{12}{11}$ min past 3
                        [by Technique 2]
$= 30 \times \dfrac{12}{11}$ min past 3
$= 32\dfrac{8}{11}$ min past 3
i.e.   3 h and $32\dfrac{8}{11}$ min

5. **(c)** At 9 O'clock, the hour hand is at 9 and the minute hand is at 12. It means that the two hands are 15 min spaces apart. To be in the same straight line (but not together), they will be 30 min space apart.
∴ The minute hand will have to gain $(30 - 15) = 15$ min spaces over the hour hand.

As we know,
55 min spaces are gained in 60 min.
∴ 15 min will be gained in
$\left(\dfrac{60}{55} \times 15\right)$ min $= \dfrac{180}{11} = 16\dfrac{4}{11}$ min
Hence, the hands will be in the same straight line but not together at $16\dfrac{4}{11}$ min past 9.

Fast Track Method
Here, $n = 9$ and $n + 1 = 10$ $(n > 6)$
The hands will be in the same straight line at $(5n - 30) \times \dfrac{12}{11}$ min past n.
                            [by Technique 3 (i)]
$= (5 \times 9 - 30)\dfrac{12}{11}$ min past 9
$= \dfrac{15 \times 12}{11}$ min past 9
$= \dfrac{180}{11}$ min past 9
$= 16\dfrac{4}{11}$ min past 9

6. **(c)** At 6 O'clock, the minute hand is 30 min spaces behind the hour hand. Given the minute hand is 3 min spaces ahead of the hour hand.
In such case, the minute hand has to gain $(30 + 3)$ min spaces $= 33$ min spaces
∴ 55 min are gained in 60 min
33 min are gained in $\left(\dfrac{60}{55} \times 33\right)$ min
$= \left(\dfrac{60 \times 3}{5}\right)$ min
$= (12 \times 3)$ min $= 36$ min
Hence, the hand will 3 min ahead at 36 min past 6, i.e. 6 : 36 pm.

Fast Track Method
Given that, $n = 6$,
$(n + 1) = 7$ and $x = 3$
According to the formula,
$(5n + x)\dfrac{12}{11}$ min past n
                            [by Technique 4]
$= (5 \times 6 + 3)\dfrac{12}{11}$ min past 6
$= (30 + 3)\dfrac{12}{11}$ min past 6
$= 36$ min past 6
∴ Required time $= 6 : 36$ pm

7. (b) Given that, $x = 70$ min
Required result
$= \left(\dfrac{720}{11} - x\right)\left(\dfrac{60 \times 24}{x}\right)$ min

[by Technique 5]

$= \left(\dfrac{720}{11} - 70\right)\left(\dfrac{60 \times 24}{70}\right)$ min

$= \left(\dfrac{720 - 770}{11}\right)\left(\dfrac{60 \times 24}{70}\right)$ min

$= \left(\dfrac{50}{11} \times \dfrac{6 \times 24}{7}\right)$ min

$= -\dfrac{7200}{77}$ min

[– ve sing indicates that there is a loss]

8. (b) ∵ A clock takes time to strike at 5 O'clock = 8 s
Then, time taken to strike at 1 O'clock
$= \dfrac{8}{5}$ s

∴ Time taken to strike at 10 O'clock
$= \dfrac{8}{5} \times 10 = 16$ s

9. (b) Time from 12 pm on Monday to 2 pm on the following Monday
$= 7$ days 2 h $= 170$ h
So, the watch gains $\left(2 + 4\dfrac{4}{5}\right)$ min or $\dfrac{34}{5}$ min in 170 h.
Now, $\dfrac{34}{5}$ min are gained in 170 h.
∴ 2 min are gained in
$\left(170 \times \dfrac{5}{34} \times 2\right) = 50$ h
So, watch is correct 2 days 2 h after 12 pm on Monday, i.e. it will be correct at 2 pm on Wednesday.

10. (d) Time difference = 4 min in 1 day.
The two clock shows identical time when the time difference is 24 h, i.e. 1440 min.
So, the clocks will show identical time again after $\dfrac{1440}{4} = 360$ days.

11. (d) Since, both of the hands are interchanging their positions. So, sum of the angles formed by both of the hands will be 360°.
Now, suppose that the person was out of the house for t min, then
Angle formed by min hand + Angle formed by hour hand = 360°
$\Rightarrow$ $6t + 0.5t = 360°$

$\Rightarrow$ $6.5t = 360°$
$\Rightarrow$ $t = \dfrac{360°}{6.5} \approx 55{:}38$ min

12. (d) A leap year must be divisible by 4, as every 4th year is a leap year.

13. (b) As 2016 is completely divisible by 4.

14. (a) 100 yr have 5 odd days.
∴ Last day of 1st century is Friday.
200 yr have
$(5 \times 2) = 1$ week $+ 3$ odd days
$= 3$ odd days
∴ Last day of 2nd century is Wednesday,
300 yr have $(5 \times 3) = 2$ week $+ 1$ odd day
$= 1$ odd day
∴ Last day of 3rd century is Monday.
400 yr have 0 odd day.
∴ Last day of 4th century is Sunday.
This cycle is repeated.
∴ Last day of a century cannot be Tuesday or Thursday or Saturday.

15. (d) Each day of the week is repeated after 7 days.
So, after 63 days, it will be Monday.
Hence, after 64 days, it will be Tuesday.

16. (a) ∵ Saturday is on January 1, 2000.
∴ Day on January 1, 2001
= Saturday + 2 = Monday
[∵ 2000 is a leap year]
Hence, the required day is Monday.

17. (c) The year 2007 is an ordinary year, so it has 1 odd day.
3rd day of the year 2007 was Wednesday.
∴ 3rd day of the year 2008 will be one day beyond the Wednesday.
Hence, it will be Thursday.

18. (a) On 31st December, 2005, it was Saturday. Number of odd days from the year 2006 to the year 2009
= 1 odd day of 2006 + 1 odd day of 2007
+ 2 odd days of 2008 (leap year)
+ 1 odd day of 2009
= 1 + 1 + 2 + 1 = 5
∴ On 31st December 2009, it was Thursday. Thus, on 2nd January 2010, it was Saturday.

19. (d) 5th March, 1999 is Friday.
Then, 5th March 2000 = Friday + 2
= Sunday
[∵ 2000 is leap year and it crosses 29th Feb 2000, so 2 is taken as odd day]

Clock and Calendar / 507

∴ 5th March 2000 = Sunday.
Then, 9th March 2000 = Thursday.

20. (a) Total number of days from 1st September, 2014 to 31st December, 2016
$= 29 + 31 + 30 + 31 = 121$
Number of odd days
$= \dfrac{121}{7} = 17$ weeks $+ 2$ days
$= 2$ odd days
∴ Required day = Monday $+ 2$
= Wednesday

21. (b) ∵ 6th March 2005 = Monday
Then, 6th March 2004
= Monday $- 1$ day = Sunday
[∵ 2004 is a leap year but it does not cross 29th February of 2004, so only 1 is taken as odd day]
6th March 2004 = Sunday
7th March 2004 = Monday

22. (b) Number of odd days between 6th November, 1987 and 4th April, 1988.
$= (30 - 6)$ days of November 1987 $+ 31$ days of December 1987 $+ 31$ days of January 1988 $+ 29$ days of February 1988 $+ 31$ days of March 1988 $+ 4$ days of April 1988
$= \{(30 - 6) + 31\} + (31 + 29 + 31 + 4)$
$= 24 + 31 + 29 + 31 + 4 = 150$
$= 21$ weeks $+ 3$ days
$= 3$ odd days
∴ 6th November, 1987 will be three days backward from Monday.
∴ Required day is Friday.

23. (a) Period upto 17th July, 1776
$= (1775$ yr $+$ Period form 1st January 1776 to 17th July, 1776)
Counting the number of odd days
In 1600 yr $= 0$
In 100 yr $= 5$
75 yr = 18 leap years $+ 57$ ordinary years
$= (18 \times 2 + 57 \times 1)$ odd days
$= 93$ odd days $= (13$ weeks $+ 2$ days)
$= 2$ odd days
∴ 1775 yr have $(0 + 5 + 2)$ odd days
$= 7$ odd days $= 1$ week $+ 0$ odd day
$= 0$ odd day
Number of days between 1.1.1776 to 17.7.1776
= January + February + March + April + May + June + July

$= 31 + 29 + 31 + 30 + 31 + 30 + 17$
$= 199$ days $= 28$ weeks $+ 3$ odd days
∴ Total number of odd days $= (3 + 0) = 3$
Hence, the required day is Wednesday.

24. (c) Period upto 17th August, 2010
$= (2009$ yr $+$ Period from 1.1.2010 to 17.8.2010)
Counting the number of odd days.
Odd days in 1600 yr $= 0$
Odd days in 400 yr $= 0$
9 yr = (2 leap years + 7 ordinary years)
$= (2 \times 2 + 7 \times 1) = 1$ week $+ 4$ days
$= 4$ odd days
Number of days between 1.1.2010 to 17.8.2010
= January + February + March + April + May + June + July + August
$= (31 + 28 + 31 + 30 + 31 + 30 + 31 + 17)$ days
$= 229$ days $= 32$ weeks $+ 5$ odd days
Total number of odd days
$= (0 + 0 + 4 + 5)$ days
$= 9$ days $= 1$ week $+ 2$ odd days
Hence, the required day is Tuesday.

25. (b) Count the number of odd days from the year 2008 onwards to get the sum equal to 0 odd day.
Let us see

Years	Odd days
2008	2
2009	1
2010	1
2011	1
2012	2
2013	1
2014	1
2015	1
2016	2
2017	1
2018	1

Sum $= 2 + 1 + 1 + 1 + 2 + 1 + 1 + 1 + 2 + 1 + 1$
$= 14$ days $= 2$ weeks $= 0$ odd days
∴ Calendar for the year 2019 will be the same as for the year 2008.

Chapter 30

Linear Equations

A linear equation is an equation for a straight line. So, the equation which has degree 1, i.e. which has linear power of the variables, is called a **linear equation**.

It is written as $ax + by + c = 0$, where a, b and c are real numbers and a and b both are not zero.

For example $y = 2x + 1$ is a linear equation. The different values of x and y are

x	1	2	0	−1	−2	−3
y	3	5	1	−1	−3	−5

All these values of (x, y), i.e. $(1, 3), (2, 5), (0, 1)$ etc., are the solutions of the given linear equation.

- If same number is added, subtracted or multiplied to both the sides of the equation, then the equality remains the same.
- If both sides are divided by a non-zero number, then the equality remains the same.

Linear Equation in One Variable

A linear equation in which the number of variables is one, is known as linear equation in one variable.

For example $3x + 5 = 10$, $y + 3 = 5$ etc.

Linear Equation in Two Variables

A linear equation in which the number of variables is two, is known as linear equation in two variables.

For example $2x + 5y = 10$, $x + 4y = 8$ etc.

Linear Equations

MIND IT!
1. Linear equation in one variable represents a point in number line.
2. Linear equation in two variable represents a line in XY-plane (cartesian plane).
3. Linear equation in three variables represents a plane in xyz-coordinate system.

Methods of Solving Linear Equations in Two Variables

There are following methods which are useful to solve the linear equations in two variables.

1. Substitution Method

In this method, first represent one variable in the form of another variable, then substitute this value in another equation and solve it. Thus, a value of one variable is obtained and this value is used to find the value of another variable.

Ex. 1 Solve the following equations with substitution method.
$$2x - y = 3, \ 4x - y = 5$$

Sol. Given, $2x - y = 3$ ...(i)
and $4x - y = 5$ ...(ii)
From Eq. (i), $y = 2x - 3$
On putting the value of y in Eq. (ii), we get
$$4x - (2x - 3) = 5 \Rightarrow 2x + 3 = 5$$
$$\therefore \quad x = 1$$
On putting the value of x in Eq. (i), we get
$$2 \times 1 - y = 3 \Rightarrow y = -1$$
Hence, $x = 1$ and $y = -1$.

2. Elimination Method

In this method, the coefficients of one of the variables of each equation are made equal by multiplying the equations by a proper multiple. Now, either add or subtract the equations to get an equation in one variable solve this equation to get the value of the variable. Thus, with the help of this value, we can find the value of another variable.

Ex. 2 Solve the following equations with elimination method.
$$11x - 5y + 61 = 0, \ 3x - 20y - 2 = 0$$

Sol. Given, $11x - 5y + 61 = 0$ ...(i)
and $3x - 20y - 2 = 0$ ...(ii)
Now, multiplying Eq. (i) by 3 and Eq. (ii) by 11 and then subtracting, we get
$$33x - 15y + 183 = 0$$
$$33x - 220y - 22 = 0$$
$$\underline{- \quad + \quad \quad +}$$
$$205y + 205 = 0 \Rightarrow y = -1$$
On putting the value of y in Eq. (ii), we get
$$3x - 20(-1) - 2 = 0 \Rightarrow 3x + 20 - 2 = 0$$
$$\Rightarrow \quad 3x = -18 \Rightarrow x = -6$$
Hence, $x = -6$ and $y = -1$.

3. Cross-Multiplication Method

Let $a_1x + b_1y + c_1 = 0$ and $a_2x + b_2y + c_2 = 0$ are two equations.

∴ By cross-multiplication method, $\dfrac{x}{b_1c_2 - b_2c_1} = \dfrac{y}{c_1a_2 - c_2a_1} = \dfrac{1}{a_1b_2 - a_2b_1}$

$\Rightarrow \quad x = \dfrac{b_1c_2 - b_2c_1}{a_1b_2 - a_2b_1}$ and $y = \dfrac{c_1a_2 - c_2a_1}{a_1b_2 - a_2b_1}$

Ex. 3 Solve the following equations with cross-multiplication method.
$2x - 3y + 1 = 0,\ 3x + 4y - 5 = 0$

Sol. Given, $\quad 2x - 3y + 1 = 0$ ...(i)

and $\quad 3x + 4y - 5 = 0$ ...(ii)

On comparing Eqs. (i) and (ii) with $a_1x + b_1y + c_1 = 0$ and $a_2x + b_2y + c_2 = 0$, we get
$a_1 = 2,\ b_1 = -3,\ c_1 = 1$ and $a_2 = 3,\ b_2 = 4,\ c_2 = -5$

By cross-multiplication method, $x = \dfrac{b_1c_2 - b_2c_1}{a_1b_2 - a_2b_1} = \dfrac{(-3)\times(-5) - 4\times 1}{2\times 4 - 3\times(-3)} = \dfrac{15 - 4}{8 + 9} = \dfrac{11}{17}$

and $\quad y = \dfrac{c_1a_2 - c_2a_1}{a_1b_2 - a_2b_1} = \dfrac{1\times 3 - (-5)\times 2}{2\times 4 - 3\times(-3)}$

$= \dfrac{3 + 10}{8 + 9} = \dfrac{13}{17}$

Consistency for the System of Linear Equations

A set of linear equations is said to be **consistent**, if there exists atleast one solution for these equations. A set of linear equations is said to be **inconsistent**, if there is no solution for the equations.

Consistent System

Let us consider a system of two linear equations as shown below,
$$a_1x + b_1y + c_1 = 0 \quad \text{and} \quad a_2x + b_2y + c_2 = 0$$
The above system will be consistent, if $\dfrac{a_1}{a_2} \neq \dfrac{b_1}{b_2}$ or $\dfrac{a_1}{a_2} = \dfrac{b_1}{b_2} = \dfrac{c_1}{c_2}$.

- If $\dfrac{a_1}{a_2} \neq \dfrac{b_1}{b_2}$, then system has unique solution and represents a pair of intersecting lines.
- If $\dfrac{a_1}{a_2} = \dfrac{b_1}{b_2} = \dfrac{c_1}{c_2}$, then system has infinite solutions and represents overlapping lines.

Inconsistent System

The above system will be inconsistent, if $\dfrac{a_1}{a_2} = \dfrac{b_1}{b_2} \neq \dfrac{c_1}{c_2}$, then the system does not have any solution and represents a pair of parallel lines.

Ex. 4 Check whether the given system is consistent or not. If yes, then find the solution.
$$x - 2y = 0,\ 3x + 4y - 20 = 0$$

Sol. Given, $\quad x - 2y = 0$ ...(i)

and $\quad 3x + 4y - 20 = 0$ ...(ii)

Here, $\dfrac{a_1}{a_2} = \dfrac{1}{3}, \dfrac{b_1}{b_2} = -\dfrac{2}{4} = -\dfrac{1}{2}; \dfrac{c_1}{c_2} = -\dfrac{0}{20} = 0$

$\therefore \quad \dfrac{a_1}{a_2} \ne \dfrac{b_1}{b_2}$

Thus, system is consistent and has a unique solution. To find the solution of the given equations, multiply Eq. (i) by 2 and add it to Eq. (ii), we get

$2x - 4y + 3x + 4y - 20 = 0 \Rightarrow 5x = 20 \Rightarrow x = 4$

On putting the value of x in Eq. (i), we get

$4 - 2y = 0 \Rightarrow y = 4/2 = 2$

Hence, given system has the solution (4, 2).

Ex. 5 For what value of K, the system of equations $Kx - 4y - 8 = 0$ and $8x - 6y - 12 = 0$ has a unique solution?

Sol. For a unique solution, we have

$\dfrac{a_1}{a_2} \ne \dfrac{b_1}{b_2} \Rightarrow \dfrac{K}{8} \ne \dfrac{-4}{-6} \Rightarrow \dfrac{K}{8} \ne \dfrac{2}{3}$

$\therefore \quad K \ne \dfrac{16}{3}$

Ex. 6 For what value of K, the system of equations $2x + 4y + 16 = 0$ and $3x + Ky + 24 = 0$ has an infinite number of solutions?

Sol. For infinite number of solutions, we have

$\dfrac{a_1}{a_2} = \dfrac{b_1}{b_2} = \dfrac{c_1}{c_2} \Rightarrow \dfrac{2}{3} = \dfrac{4}{K} = \dfrac{16}{24}$

$\therefore \quad K = 6$

Ex. 7 For what value of K, the system of equations $Kx - 20y - 6 = 0$ and $6x - 10y - 14 = 0$ has no solution?

Sol. For no solution, we have

$\dfrac{a_1}{a_2} = \dfrac{b_1}{b_2} \ne \dfrac{c_1}{c_2} \Rightarrow \dfrac{K}{6} = \dfrac{-20}{-10} \ne \dfrac{-6}{-14}$

$\therefore \quad K = \dfrac{6 \times 20}{10} = 12$

Ex. 8 If $4x + 9y = 43$ and $6x + 4y = 36$ are the linear equations in x and y, then find the values of x and y.

Sol. Given, $\quad 4x + 9y = 43 \quad$ ...(i)

and $\quad 6x + 4y = 36 \quad$ ...(ii)

On multiplying Eq. (i) by 3 and Eq. (ii) by 2 and then subtracting, we get

$12x + 27y = 129$
$12x + 8y = 72$
$\underline{}$
$\quad\quad 19y = 57 \Rightarrow y = \dfrac{57}{19} = 3$

On putting the value of y in Eq. (i), we get

$4x + 9 \times 3 = 43 \Rightarrow 4x + 27 = 43$

$\Rightarrow \quad 4x = 43 - 27 = 16$

$\therefore \quad x = \dfrac{16}{4} = 4$

Hence, $x = 4$ and $y = 3$.

Ex. 9 The taxi charges in a city consist of a fixed charge together with the charge for the distance covered. For a distance of 20 km, the charge paid is ₹ 205 and for a distance of 25 km, the charge paid is ₹ 255. Find the fixed charges and the charge per km. How much does a person have to pay for covering a distance of 50 km?

Sol. Let the fixed charges be ₹ x and charge per km be ₹ y.
Then, $\quad x + 20y = 205$ ...(i)
and $\quad x + 25y = 255$ ...(ii)
On subtracting Eq. (ii) from Eq. (i), we get

$$x + 20y = 205$$
$$x + 25y = 255$$
$$\underline{- \quad - \quad \quad -}$$
$$-5y = -50 \Rightarrow y = 10$$

On putting the value of y in Eq. (i), we get
$$x + 20 \times 10 = 205 \Rightarrow x + 200 = 205$$
$$\therefore \quad x = 5$$

Hence, amount paid for a distance of 50 km = $x + 50y = 5 + 50 \times 10 = ₹ 505$

Multi Concept Questions

1. Solve the following equations for x and y.
$$\frac{2}{3x+2y} + \frac{3}{3x-2y} = \frac{17}{5}, \frac{5}{3x+2y} + \frac{1}{3x-2y} = 2$$

   (a) 2, 3   (b) 5, 7   (c) 3, 4   (d) 1, 1

➤ (d) Let $\dfrac{1}{3x+2y} = P$ and $\dfrac{1}{3x-2y} = Q$

$\therefore \quad 2P + 3Q = \dfrac{17}{5}$ ...(i)

Similarly, $\quad 5P + Q = 2$ ...(ii)
On multiplying Eq. (i) by 5 and Eq. (ii) by 2 and then subtracting, we get
$\quad 10P + 15Q = 17$ ...(iii)
$\quad 10P + 2Q = 4$ ...(iv)
$\quad \underline{- \quad - \quad \quad -}$
$\Rightarrow \quad 13Q = 13 \Rightarrow Q = 1$
On substituting $Q = 1$ in Eq. (ii), we get
$\quad 5P + 1 = 2 \Rightarrow 5P = 1 \Rightarrow P = \dfrac{1}{5}$

Now, $\quad P = \dfrac{1}{5} = \dfrac{1}{3x+2y}, Q = 1 = \dfrac{1}{3x-2y}$

$\Rightarrow \quad 3x + 2y = 5$ ...(v)
and $\quad 3x - 2y = 1$ ...(vi)
On adding Eqs. (v) and (vi), we get $x = 1$ and $y = 1$

Linear Equations / 513

2. If a, b and c are the positive numbers, then the system of equations
$\frac{x^2}{a^2} + \frac{y^2}{b^2} - \frac{z^2}{c^2} = 1$, $\frac{x^2}{a^2} - \frac{y^2}{b^2} + \frac{z^2}{c^2} = 1$, $\frac{-x^2}{a^2} + \frac{y^2}{b^2} + \frac{z^2}{c^2} = 1$ in x, y and z has

(a) no solution
(b) unique solution
(c) infinitely many solutions
(d) finitely many solutions

↪ **(d)** We have, $\frac{x^2}{a^2} + \frac{y^2}{b^2} - \frac{z^2}{c^2} = 1$ ...(i)

$\frac{x^2}{a^2} - \frac{y^2}{b^2} + \frac{z^2}{c^2} = 1$ ...(ii)

and $\frac{-x^2}{a^2} + \frac{y^2}{b^2} + \frac{z^2}{c^2} = 1$ ...(iii)

On adding Eqs. (i) and (ii), we get
$\frac{x^2}{a^2} = 1 \Rightarrow x = \pm a$

Similarly, adding Eqs. (ii) and (iii), we get
$\frac{z^2}{c^2} = 1 \Rightarrow z = \pm c$

Again, adding Eqs. (i) and (iii), we get $y = \pm b$

Hence, x, y and z have finitely many solutions.

Fast Track Practice

Exercise 1 Base Level Questions

1. The cost of 21 pencils and 9 clippers is ₹ 819. What is the total cost of 7 pencils and 3 clippers together? [DMRC CRA 2012]
(a) ₹ 204 (b) ₹ 409
(c) ₹ 273 (d) ₹ 208

2. Deepak has some hens and some goats. If the total number of animal heads is 90 and the total number of animal feet is 248, what is the total number of goats Deepak has? [PNB Mgmt. Trainee 2010]
(a) 32 (b) 36 (c) 34
(d) Cannot be determined
(e) None of these

3. If $6x - 10y = 10$ and $\frac{x}{x+y} = \frac{5}{7}$, then $(x - y)$ is equal to
(a) 6 (b) 8 (c) 12 (d) 3
(e) None of these

4. Solve $6x + 3y = 7xy$ and $3x + 9y = 11xy$.
(a) $x = 1, y = \frac{3}{2}$ (b) $x = -1, y = \frac{2}{3}$
(c) $x = \frac{3}{2}, y = \frac{1}{2}$ (d) $x = 1, y = -\frac{3}{2}$
(e) None of these

5. In a rare coin collection, there is one gold coin for every three non-gold coins. 10 more gold coins are added to the collection and the ratio of gold coins to non-gold coins would be 1 : 2. Based on the information; the total number of coins in the collection now becomes. [UPSC CSAT 2013]
(a) 90 (b) 80 (c) 60 (d) 50

6. In an examination, a student scores 4 marks for every correct answer and losses 1 mark for every wrong answer. A student attempted all the 200 questions and scored 200 marks. Find the number of questions, he answered correctly. [SSC CGL 2010]
(a) 82 (b) 80
(c) 68 (d) 60

7. If $3^{x+y} = 81$ and $81^{x-y} = 3$, then what is the value of x? [CDS 2012]
(a) 17/16 (b) 17/8
(c) 17/4 (d) 15/4

514 / **Fast Track** Objective Arithmetic

8. Ten chairs and six tables together cost ₹ 6200, three chairs and two tables together cost ₹ 1900. The cost of 4 chairs and 5 tables is [CDS 2013]
   (a) ₹ 3000 (b) ₹ 3300
   (c) ₹ 3500 (d) ₹ 3800

9. If $x + y - 7 = 0$ and $3x + y - 13 = 0$, then what is $4x^2 + y^2 + 4xy$ equal to? [CDS 2013]
   (a) 75 (b) 85
   (c) 91 (d) 100

10. If $999x + 888y = 1332$ and $888x + 999y = 555$. Then, the value of $x + y$ is [SSC CPO 2015]
    (a) 555 (b) 888 (c) 999 (d) 1

11. The pair of linear equations $kx + 3y + 1 = 0$ and $2x + y + 3 = 0$ intersect each other, if [CDS 2017 (I)]
    (a) $k = 6$ (b) $k \neq 6$
    (c) $k = 0$ (d) $k \neq 0$

12. The value of k for which $kx + 3y - k + 3 = 0$ and $12x + ky = k$, have infinite solutions, is [CLAT 2013]
    (a) 0 (b) – 6 (c) 6 (d) 1

13. The system of equations $2x + 4y = 6$ and $4x + 8y = 6$ has
    (a) exactly two solutions
    (b) no solution
    (c) infinitely many solutions
    (d) a unique solution
    (e) None of the above

14. The system of equations $2x + 4y = 6$ and $4x + 8y = 8$ is [CDS 2017 (I)]
    (a) Consistent with a unique solution
    (b) Consistent with infinitely many solutions
    (c) Inconsistent
    (d) None of the above

15. The graphs of $ax + by = c$, $dx + ey = f$ will be
    I. parallel, if the system has no solution.
    II. coincident, if the system has finite numbers of solutions.
    III. intersecting, if the system has only one solution.
    Which of the above statements are correct? [CDS 2012]
    (a) Both I and II
    (b) Both II and III
    (c) Both I and III
    (d) All I, II and III

16. The system of equations $3x + y - 4 = 0$ and $6x + 2y - 8 = 0$ has [CDS 2013]
    (a) a unique solution $x = 1$, $y = 1$
    (b) a unique solution $x = 0$, $y = 4$
    (c) no solution
    (d) infinite solutions

17. The value of K, for which the system of equations $3x - Ky - 20 = 0$ and $6x - 10y + 40 = 0$ has no solution, is [CDS 2016 (I)]
    (a) 10 (b) 6 (c) 5 (d) 3

Exercise ❷ Higher Skill Level Questions

1. The numerator of a fraction is $6x + 1$ and the denominator is $7 - 4x$. x can have any value between – 2 and 2, both included. The values of x for which the numerator is greater than the denominator, are
   (a) $\frac{3}{5} < x \leq 2$ (b) $\frac{3}{5} \leq x \leq 2$
   (c) $0 < x \leq 2$ (d) $-2 \leq x < 2$
   (e) None of these

2. The number of pairs (x, y), where x, y are integers satisfying the equation $21x + 48y = 5$, is [CDS 2015 (II)]
   (a) zero (b) one
   (c) two (d) infinite

3. If $\dfrac{\sqrt{3+x} + \sqrt{3-x}}{\sqrt{3+x} - \sqrt{3-x}} = 2$, then x is equal to [SSC CGL 2010]

   (a) $\dfrac{5}{12}$ (b) $\dfrac{12}{5}$
   (c) $\dfrac{5}{7}$ (d) $\dfrac{7}{5}$

4. If $2^a + 3^b = 17$ and $2^{a+2} - 3^{b+1} = 5$, then
   (a) $a = 2, b = 3$
   (b) $a = -2, b = 3$
   (c) $a = 2, b = -3$
   (d) $a = 3, b = 2$
   (e) None of the above

5. If $\dfrac{p}{x} + \dfrac{q}{y} = m$ and $\dfrac{q}{x} + \dfrac{p}{y} = n$, then what is $\dfrac{x}{y}$ equal to? [CDS 2016 (I)]

   (a) $\dfrac{np + mq}{mp + nq}$ (b) $\dfrac{np + mq}{mp - nq}$
   (c) $\dfrac{np - mq}{mp - nq}$ (d) $\dfrac{np - mq}{mp + nq}$

Linear Equations / 515

6. Consider the following statements
  I. The equation $1990x - 173y = 11$ has no solution in integers for x and y.
  II. The equation $3x - 12y = 7$ has no solution in integers for x and y.
  Which of the above statement(s) is/are correct? **[CDS 2015 (I)]**
  (a) Only I   (b) Only II
  (c) Both I and II   (d) Neither I nor II

7. The ratio of incomes of two persons is 8 : 5 and the ratio of their expenditure is 2 : 1. If each of them manages to save ₹ 1000 per month, find the difference of their monthly income.
  (a) ₹ 2500   (b) ₹ 1500
  (c) ₹ 1000   (d) ₹ 700
  (e) None of these

8. If $\dfrac{3}{x+y} + \dfrac{2}{x-y} = 2$ and $\dfrac{9}{x+y} - \dfrac{4}{x-y} = 1$, then what is the value of $\dfrac{x}{y}$? **[CDS 2012]**
  (a) 3/2   (b) 5   (c) 2/3   (d) 1/5

9. If $\dfrac{a}{b} - \dfrac{b}{a} = \dfrac{x}{y}$ and $\dfrac{a}{b} + \dfrac{b}{a} = x - y$, then what is the value of x? **[CDS 2012]**
  (a) $\dfrac{a+b}{a}$ (b) $\dfrac{a+b}{b}$
  (c) $\dfrac{a-b}{a}$ (d) None of these

10. If $\dfrac{x}{2} + \dfrac{y}{3} = 4$ and $\dfrac{2}{x} + \dfrac{3}{y} = 1$, then what is $x + y$ equal to? **[CDS 2013]**
  (a) 11   (b) 10
  (c) 9   (d) 8

11. The cost of 2.5 kg rice is ₹ 125. The cost of 9 kg rice is equal to that of 4 kg pulses. The cost of 14 kg pulses is equal to that of 1.5 kg tea. The cost of 2 kg tea is equal to that of 5 kg nuts. What is the cost of 11 kg nuts? **[CDS 2016 (II)]**
  (a) ₹ 2310
  (b) ₹ 3190
  (c) ₹ 4070
  (d) ₹ 4620

Answer with Solutions

Exercise 1 Base Level Questions

1. (c) Let cost of 1 pencil and 1 clipper be p and c, respectively.
Now, according to the question,
$21p + 9c = ₹ 819$
$\Rightarrow 3(7p + 3c) = ₹ 819$
$\Rightarrow 7p + 3c = ₹ 273$
∴ Cost of 7 pencils and 3 clippers = ₹ 273

2. (c) Let hens = H, goats = G
According to the question,
$H + G = 90$ ...(i)
and $2H + 4G = 248$ ...(ii)
On multiplying Eq. (i) by 2 and then subtracting from Eq. (ii), we get
$2H + 2G = 180$
$2H + 4G = 248$
$\underline{---}$
$-2G = -68 \Rightarrow G = 34$
So, the total number of goats is 34.

3. (d) Given, $6x - 10y = 10$ ...(i)
and $\dfrac{x}{x+y} = \dfrac{5}{7} \Rightarrow 7x = 5x + 5y$
$\Rightarrow 2x - 5y = 0$ ...(ii)
On multiplying Eq. (ii) by 2 and then subtracting from Eq. (i), we get
$6x - 10y = 10$
$4x - 10y = 0$
$\underline{-+-}$
$2x = 10 \Rightarrow x = 5$
On putting the value of x in Eq. (i), we get
$30 - 10y = 10 \Rightarrow 10y = 20$
$y = 2$
∴ $x - y = 5 - 2 = 3$

4. (a) Given equations are
$6x + 3y = 7xy \Rightarrow \dfrac{6}{y} + \dfrac{3}{x} = 7$ ...(i)
and $3x + 9y = 11xy$
$\Rightarrow \dfrac{3}{y} + \dfrac{9}{x} = 11$ ...(ii)
On multiplying Eq. (ii) by 2 and then subtracting from Eq. (i), we get
$\dfrac{6}{y} + \dfrac{3}{x} = 7$
$\dfrac{6}{y} + \dfrac{18}{x} = 22$
$\underline{---}$
$-\dfrac{15}{x} = -15$
∴ $x = 1$

On putting the value of x in Eq. (i), we get
$$\frac{6}{y} + 3 = 7 \Rightarrow \frac{6}{y} = 4 \Rightarrow y = \frac{6}{4} = \frac{3}{2}$$
Hence, $x = 1$ and $y = \frac{3}{2}$

5. (a) Let the number of gold coins initially be x and the number of non-gold coins be y.
According to the question,
$$3x = y \text{ or } x : y = 1 : 3 \quad …(i)$$
When 10 more gold coins are added total number of gold coins become $x + 10$ and the number of non-gold coins remains the same as y.
Now, we have
$$(10 + x) : y = 1 : 2 \text{ or } 2(10 + x) = y \quad …(ii)$$
On solving Eqs. (i) and (ii),
$$x = 20 \text{ and } y = 60$$
∴ Total number of coins in the collection at the end is
$$x + 10 + y = 20 + 10 + 60 = 90$$

6. (b) Let the number of correct answers be x and number of wrong answers be y.
Then, $4x - y = 200$ …(i)
and $x + y = 200$ …(ii)
On adding Eqs. (i) and (ii), we get
$$\begin{array}{r} 4x - y = 200 \\ x + y = 200 \\ \hline 5x = 400 \end{array}$$
∴ $x = 80$

7. (b) Given, $3^{x+y} = 81 \Rightarrow 3^{x+y} = 3^4$
$\Rightarrow \quad x + y = 4$ …(i)
and $81^{x-y} = 3$
$\Rightarrow (3^4)^{x-y} = 3^1 \Rightarrow (3)^{x-y} = 3^{1/4}$
$\Rightarrow \quad x - y = \frac{1}{4}$ …(ii)
On solving Eqs. (i) and (ii), we get
$$2x = \frac{17}{4} \Rightarrow x = \frac{17}{8}$$

8. (a) Let the cost of one chair be ₹ x and cost of one table be ₹ y.
By given condition,
$10x + 6y = 6200$ …(i)
and $3x + 2y = 1900$
$\Rightarrow 9x + 6y = 5700$ …(ii)
On subtracting Eq. (ii) from Eq. (i), we get
$$x = ₹ 500$$
On putting the value of x in Eq. (i), we get
$5000 + 6y = 6200$
$\Rightarrow \quad 6y = 1200$
∴ $y = ₹ 200$
Cost of 4 chairs and 5 tables
$= 4x + 5y$
$= 4 \times 500 + 5 \times 200$
$= 2000 + 1000 = ₹ 3000$

9. (d) We have, $x + y - 7 = 0$
$\Rightarrow \quad x + y = 7$ …(i)
and $3x + y - 13 = 0$
$\Rightarrow \quad 3x + y = 13$ …(ii)
On subtracting Eq. (i) from Eq. (ii), we get
$$\begin{array}{r} 3x + y = 13 \\ x + y = 7 \\ \hline 2x = 6 \end{array} \Rightarrow x = 3$$
On putting the value of x in Eq. (i), we get
$3 + y = 7 \therefore y = 4$
Now, $4x^2 + y^2 + 4xy$
$= 4 \times (3)^2 + (4)^2 + 4 \times 3 \times 4$
$= 4 \times 9 + 16 + 48$
$= 36 + 16 + 48 = 100$

10. (d) Given, $999x + 888y = 1332$ …(i)
and $888x + 999y = 555$ …(ii)
On adding Eqs. (i) and (ii), we get
$999(x + y) + 888(x + y) = 1887$
$\Rightarrow (x + y)(999 + 888) = 1887$
$\Rightarrow (x + y) = \frac{1887}{1887} = 1$

11. (b) The system of linear equations
$a_1x + b_1y + c_1 = 0$ and $a_2x + b_2y + c_2 = 0$
Has unique solution, it $\frac{a_1}{a_2} \neq \frac{b_1}{b_2}$.
We have, $kx + 3y + 1 = 0$ and $2x + y + 3 = 0$
For unique solution
$$\frac{k}{2} \neq \frac{3}{1} \Rightarrow k \neq 6$$

12. (c) For infinite solution,
$$\frac{a_1}{a_2} = \frac{b_1}{b_2} = \frac{c_1}{c_2} \Rightarrow \frac{k}{12} = \frac{3}{k} = \frac{-k+3}{-k}$$
$\Rightarrow \quad \frac{k}{12} = \frac{3}{k} \Rightarrow k^2 = 36$
∴ $k = \sqrt{36} = 6$

13. (b) Given equations are
$2x + 4y = 6$ and $4x + 8y = 6$
Here, $a_1 = 2, b_1 = 4, c_1 = 6, a_2 = 4$
$b_2 = 8, c_2 = 6$
Then, $\frac{a_1}{a_2} = \frac{2}{4} = \frac{1}{2}; \frac{b_1}{b_2} = \frac{4}{8} = \frac{1}{2}; \frac{c_1}{c_2} = \frac{6}{6} = 1$
∴ $\frac{a_1}{a_2} = \frac{b_1}{b_2} \neq \frac{c_1}{c_2}$
So, given system has no solution.

14. (c) Given system of equations are
$2x + 4y = 6$ and $4x + 8y = 8$.
∴ $a_1 = 2, b_1 = 4, c_1 = (-6)$
and $a_2 = 4, b_2 = 8, c_2 = -8$
Now, $\frac{a_1}{a_2} = \frac{2}{4} = \frac{1}{2}, \frac{b_1}{b_2} = \frac{4}{8} = \frac{1}{2}$
and $\frac{c_1}{c_2} = \frac{-6}{-8} = \frac{3}{4}$

Linear Equations / 517

$\because \quad \dfrac{a_1}{a_2} = \dfrac{b_1}{b_2} \neq \dfrac{c_1}{c_2}$

Hence, the system of equation is inconsistent.

15. (c) As per theory, the graphs of $ax + by = c$, $dx + ey = f$ will be parallel, if the system has no solution; will be coincident, if the system has infinite solution and will be intersecting, if the system has only one solution. Hence, I and III are correct.

16. (d) Given equations of system are
$$3x + y - 4 = 0 \quad \ldots\text{(i)}$$
and
$$6x + 2y - 8 = 0 \quad \ldots\text{(ii)}$$
Here, $a_1 = 3, b_1 = 1, c_1 = -4$,

$a_2 = 6, b_2 = 2$ and $c_2 = -8$,

$\because \quad \dfrac{a_1}{a_2} = \dfrac{b_1}{b_2} = \dfrac{c_1}{c_2} = \dfrac{1}{2}$

So, the system of equations has infinite solutions, because it represents a overlapping line.

17. (c) Given equation are $3x - Ky - 20 = 0$ and $6x - 10y + 40 = 0$.
Here, $a_1 = 3, b_1 = -K, c_1 = -20, a_2 = 6, b_2 = -10$ and $c_2 = 40$

For no solution, $\dfrac{a_1}{a_2} = \dfrac{b_1}{b_2} \neq \dfrac{c_1}{c_2}$

$\Rightarrow \quad \dfrac{3}{6} = \dfrac{-K}{-10} \Rightarrow K = 5$

Exercise 2 Higher Skill Level Questions

1. (a) $6x + 1 > 7 - 4x \Rightarrow x > \dfrac{3}{5}$

$\therefore \quad \dfrac{3}{5} < x \leq 2$

2. (a) Given, $21x + 48y = 5 \Rightarrow 3(7x + 16y) = 5$
If x, y are integers, then LHS of the above equation is multiple of 3, but the RHS of above equation is not multiple of 3.
Hence, no integral values of x and y exist.

3. (b) Given, $\dfrac{\sqrt{3+x} + \sqrt{3-x}}{\sqrt{3+x} - \sqrt{3-x}} = 2$

Let $\sqrt{3+x} = a$ and $\sqrt{3-x} = b$

Then, $\dfrac{a+b}{a-b} = \dfrac{2}{1}$

$\Rightarrow \quad a + b = 2a - 2b \Rightarrow a = 3b$

$\therefore \quad \sqrt{3+x} = 3\sqrt{3-x}$

On squaring both sides, we get
$(\sqrt{3+x})^2 = (3\sqrt{3-x})^2$

$\Rightarrow \quad 3 + x = 9(3-x)$
$\Rightarrow \quad 3 + x = 27 - 9x \Rightarrow 10x = 24$

$\therefore \quad x = \dfrac{12}{5}$

4. (d) Given, $2^a + 3^b = 17$
and $2^{a+2} - 3^{b+1} = 5$
$\Rightarrow \quad 2^a \times 2^2 - 3^b \times 3^1 = 5$
$\Rightarrow \quad 4 \cdot 2^a - 3 \cdot 3^b = 5$
Let $2^a = x$ and $3^b = y$
Then, $x + y = 17 \quad \ldots\text{(i)}$
$4x - 3y = 5 \quad \ldots\text{(ii)}$
On multiplying Eq. (i) by 3 and then adding to Eq. (ii), we get
$3x + 3y = 51$
$4x - 3y = 5$
$\overline{\quad 7x \quad = 56} \Rightarrow x = 8$

On putting the value of x in Eq. (i), we get
$8 + y = 17 \Rightarrow y = 9$
Now, $2^a = x \Rightarrow 2^a = 8 = (2)^3$
$\therefore \quad a = 3$ and $3^b = y = 9 \Rightarrow 3^b = 3^2$
$\Rightarrow \quad b = 2$
Hence, $a = 3$ and $b = 2$.

5. (c) Given, $\dfrac{p}{x} + \dfrac{q}{y} = m$ and $\dfrac{q}{x} + \dfrac{p}{y} = n$

Let $\dfrac{1}{x} = u$ and $\dfrac{1}{y} = v$

Then, $pu + qv = m \quad \ldots\text{(i)}$
and $qu + pv = n \quad \ldots\text{(ii)}$
On multiplying Eq. (i) by q and Eq. (ii) by p and then subtracting, we get
$pqu + q^2v = mq$
$pqu + p^2v = np$
$\overline{\quad q^2v - p^2v = mq - np}$

$\therefore \quad v = \dfrac{mq - np}{q^2 - p^2}$

On putting the value of v in Eq. (i), we get
$u = \dfrac{mp - nq}{p^2 - q^2}$

$\therefore \quad \dfrac{x}{y} = \dfrac{1/u}{1/v} = \dfrac{v}{u} = \dfrac{-(mq - np)}{mp - nq} = \dfrac{np - mq}{mp - nq}$

6. (c) I. Given, $1990x - 173y = 11$
Let x be an integer.
$\therefore \quad 173y = 1990x - 11$

$\Rightarrow \quad y = \dfrac{1990x - 11}{173}$

Here, we substitute the different integer values of x, but we do not get an integer value of y.

518 / Fast Track Objective Arithmetic

II. Given, $3x - 12y = 7$
Let x be an integer.
$\therefore 12y = 3x - 7 \Rightarrow y = \dfrac{3x-7}{12}$
Here, we substitute the different integer values of x, but we do not get an integer value of y.
So, both I and II are correct.

7. (b) Let the incomes of two persons be $8x$ and $5x$ and their expenditure be $2y$ and y, respectively.
$\because$ Saving = Income − Expenditure
$\therefore \qquad 1000 = 8x - 2y$ ...(i)
and $\qquad 1000 = 5x - y$ ...(ii)
On multiplying Eq. (ii) by 2 and subtracting from Eq. (i), we get
$8x - 2y = 1000$
$10x - 2y = 2000$
$\underline{\;-\;\;\;\;\;+\;\;\;\;\;\;\;-\;\;}$
$-2x = -1000 \Rightarrow x = 500$
$\therefore$ Monthly incomes are
$8x = 8 \times 500 = ₹\,4000$
and $5x = 5 \times 500 = ₹\,2500$
$\therefore$ Difference $= 4000 - 2500 = ₹\,1500$

8. (b) Given, $\dfrac{3}{x+y} + \dfrac{2}{x-y} = 2$ ...(i)
and $\dfrac{9}{x+y} - \dfrac{4}{x-y} = 1$ ...(ii)
Let $x + y = a$ and $x - y = b$
On multiplying Eq. (i) by 3 and then subtracting from Eq. (ii), we get
$\dfrac{9}{a} - \dfrac{4}{b} = 1$
$\dfrac{9}{a} + \dfrac{6}{b} = 6$
$\underline{\;-\;\;\;\;\;-\;\;\;\;\;-\;\;}$
$\dfrac{-10}{b} = -5 \Rightarrow b = 2$
Now, putting the value of b in Eq. (i), we get
$\dfrac{3}{a} + \dfrac{2}{2} = 2 \Rightarrow \dfrac{3}{a} = 2 - 1$
$\Rightarrow \qquad a = 3 \Rightarrow x + y = 3$ ...(iii)
and $\qquad x - y = 2$ ...(iv)
On subtracting Eq. (iv) from Eq. (iii), we get
$2y = 1 \Rightarrow y = 1/2$
Now, putting $y = \dfrac{1}{2}$ in Eq. (iii), we get
$x + \dfrac{1}{2} = 3 \Rightarrow x = 3 - \dfrac{1}{2} = \dfrac{6-1}{2} = \dfrac{5}{2}$
$\therefore \dfrac{x}{y} = \dfrac{5/2}{1/2} = \dfrac{5}{2} \times \dfrac{2}{1} = 5$

9. (d) Given equations are
$\dfrac{a}{b} - \dfrac{b}{a} = \dfrac{x}{y} \Rightarrow y = \dfrac{x}{\left(\dfrac{a}{b} - \dfrac{b}{a}\right)}$ ...(i)

and $\dfrac{a}{b} + \dfrac{b}{a} = x - y$ ...(ii)
From Eqs. (i) and (ii), we get
$\dfrac{a}{b} + \dfrac{b}{a} = x - \dfrac{x}{\left(\dfrac{a}{b} - \dfrac{b}{a}\right)}$
$\Rightarrow \left(\dfrac{a}{b} + \dfrac{b}{a}\right)\left(\dfrac{a}{b} - \dfrac{b}{a}\right) = x\left(\dfrac{a}{b} - \dfrac{b}{a} - 1\right)$
$\Rightarrow \left(\dfrac{a^2}{b^2} - \dfrac{b^2}{a^2}\right) = x\left(\dfrac{a^2 - b^2 - ab}{ab}\right)$
$\Rightarrow x = \dfrac{ab}{(a^2 - b^2 - ab)} \times \left(\dfrac{a^4 - b^4}{a^2 b^2}\right)$
$= \dfrac{(a^4 - b^4)}{(a^2 - b^2 - ab)} \times \dfrac{1}{ab}$
$= \dfrac{(a-b)(a+b)(a^2+b^2)}{ab(a^2 - b^2 - ab)}$

10. (b) Given, $\dfrac{x}{2} + \dfrac{y}{3} = 4 \Rightarrow \dfrac{3x + 2y}{6} = 4$
$\Rightarrow 3x + 2y = 24$ ...(i)
and $\dfrac{2}{x} + \dfrac{3}{y} = 1 \Rightarrow \dfrac{2y + 3x}{xy} = 1$
$\Rightarrow 2y + 3x = xy$ ...(ii)
From Eqs. (i) and (ii), we get $xy = 24$
There are 6 possibilities for x and y, respectively.

$2 \times 12 = 24$	$6 \times 4 = 24$
$3 \times 8 = 24$	$8 \times 3 = 24$
$4 \times 6 = 24$	$12 \times 2 = 24$

2 and 12 cannot be the values of x and y as their sum is 14 and it is not given in options. Now, we check both 3 and 8 as well as 4 and 6 as values of x and y or values of y and x. Only 4 as a value of x and 6 as a value of y satisfies the given condition $\dfrac{x}{2} + \dfrac{y}{3} = 4$.
$\therefore x = 4$ and $y = 6$
Hence, $x + y = 4 + 6 = 10$

11. (d) We have, cost of 2.5 kg rice $= ₹\,125$
$\therefore$ Cost of 1 kg rice $= \dfrac{125}{2.5} = ₹\,50$ ...(i)
Let r, p, t and n be the cost of rice, pulse, tea and nuts, respectively.
Given, $\qquad 9r = 4p$ ...(ii)
$\qquad 14p = 1.5t$ ...(iii)
and $\qquad 2t = 5n$ ...(iv)
From Eqs. (ii), (iii) and (iv), we get
$2\left(\dfrac{14}{1.5} \times \dfrac{9}{4}\right)r = 5n \Rightarrow n = \dfrac{42}{5}r$
Now, cost of 1 kg nuts = Cost of $\dfrac{42}{5}$ kg of rice
$= \dfrac{42}{5} \times 50 = ₹\,420$
$\therefore$ Cost of 11 kg nuts $= ₹\,420 \times 11 = ₹\,4620$

Chapter 31

Quadratic Equations

A quadratic equation is an equation in which the highest power of the variable appearing is two. It is written as
$$ax^2 + bx + c = 0$$
where, a and b are coefficients of x^2 and x respectively and c is a constant.

The factor that identifies this expression as quadratic is the exponent 2. The coefficient of x^2, i.e. a cannot be zero ($a \neq 0$).

For example $2x^2 + x - 300 = 0$ is a quadratic equation.

To check whether an equation is quadratic or not, following examples will help to understand it in a better way

S.No.	Equation	Is it quadratic?	Explanation
1.	$3x^3 - 4x + 5 = 0$	No	Here, the first term is raised to the 3rd power. It must be raised to the 2nd power in order to be quadratic.
2.	$5x^2 - 4x + 2 = 0$	Yes	This equation is in the correct form, i.e. $ax^2 + bx + c = 0$
3.	$7x^2 = 49$	Yes	This equation can be rewritten as $$7x^2 - 49 = 0$$ In this equation, b is 0. b or c can be 0, however a cannot be 0.
4.	$2x^2 = 8x - 3$	Yes	This equation can be rewritten as $$2x^2 - 8x + 3 = 0$$ which is in the correct form, i.e. $ax^2 + bx + c = 0$

MIND IT!
1. A quadratic equation has two and only two roots.
2. A quadratic equation cannot have more than two different roots.
3. If α is the root of the quadratic equation $ax^2 + bx + c = 0$, then $(x - \alpha)$ is a factor of $ax^2 + bx + c = 0$.

Methods of Solving the Quadratic Equations

There are following two methods for solving quadratic equations

1. **By Factorisation Method** If $ax^2 + bx + c$ can be factorised as $(x - \alpha)(x - \beta)$, then $ax^2 + bx + c = 0$ is equivalent to $(x - \alpha)(x - \beta) = 0$.
   Thus, $(x - \alpha)(x - \beta) = 0$
   $\Rightarrow (x - \alpha) = 0$ or $(x - \beta) = 0$
   $\Rightarrow x = \alpha$ or $x = \beta$
   Here, α and β are called roots of equation $ax^2 + bx + c = 0$.

2. **By Sridharacharya's Method** Using quadratic formula, write the quadratic equation in the standard form $ax^2 + bx + c = 0$, then roots are
$$\alpha = \frac{-b + \sqrt{b^2 - 4ac}}{2a}, \beta = \frac{-b - \sqrt{b^2 - 4ac}}{2a}$$
Here, $D = b^2 - 4ac$ is called the **discriminant** of quadratic equation.

Important Points Related to Quadratic Equations

1. A real number α is said to be a root of the quadratic equation $ax^2 + bx + c = 0$, if $a\alpha^2 + b\alpha + c = 0$. The zeroes of the quadratic polynomial $ax^2 + bx + c$ and the roots of the quadratic equation $ax^2 + bx + c = 0$ are the same.

2. A quadratic equation $ax^2 + bx + c = 0$ has
   (i) two distinct real roots, if $D > 0$.
   (ii) two equal real roots, if $D = 0$.
   (iii) no real roots, if $D < 0$.
   (iv) reciprocal roots, if $a = c$.
   (v) one root $= 0$, if $c = 0$.
   (vi) negative and reciprocal roots, if $c = -a$.
   (vii) both roots equal to 0, if $b = 0, c = 0$.

3. **Formation of Quadratic Equation** Let α and β be two roots of a quadratic equation. Then, we can form a quadratic equation as
$$x^2 - (\text{Sum of roots})x + (\text{Product of roots}) = 0$$
$\Rightarrow x^2 - (\alpha + \beta)x + (\alpha\beta) = 0 \Rightarrow (x - \alpha)(x - \beta) = 0$
Here, for standard quadratic equation $ax^2 + bx + c = 0$,
Sum of roots $= \alpha + \beta = -\dfrac{b}{a}$; Product of roots $= \alpha\beta = \dfrac{c}{a}$

4. If $ax^2 + bx + c = 0$, where a, b and c are rational, has one root $p + \sqrt{q}$, then the other root will be $p - \sqrt{q}$.
Hence, irrational roots occur in conjugate pair, if the coefficients are rational.

Quadratic Equations / 521

Ex. 1 Solve $5x^2 + 11x + 6 = 0$.

Sol. By factorisation method,
$$5x^2 + 11x + 6 = 0 \Rightarrow 5x^2 + 5x + 6x + 6 = 0$$
$$\Rightarrow 5x(x+1) + 6(x+1) = 0 \Rightarrow (x+1)(5x+6) = 0$$
$$\Rightarrow (x+1) = 0 \text{ or } (5x+6) = 0$$
If $x + 1 = 0$, then $x = -1$ and if $5x + 6 = 0$, then $x = -\dfrac{6}{5}$.

Alternate Method

Given quadratic equation is $5x^2 + 11x + 6 = 0$.
On comparing the given equation by $ax^2 + bx + c = 0$, we get
$$a = 5, b = 11 \text{ and } c = 6$$
By Sridharacharya's method,
$$x = \frac{-b \pm \sqrt{b^2 - 4ac}}{2a} = \frac{-11 \pm \sqrt{121 - 120}}{10} = \frac{-11 \pm 1}{10}$$
Taking positive sign, $\quad x = \dfrac{-11 + 1}{10} = \dfrac{-10}{10} = -1$

Taking negative sign, $\quad x = \dfrac{-11 - 1}{10} = \dfrac{-12}{10} = -\dfrac{6}{5}$

Ex. 2 Find two consecutive odd positive integers, sum of whose squares is 290.

Sol. Let the two consecutive odd positive integers be x and $(x + 2)$.
According to the question,
$$x^2 + (x+2)^2 = 290 \Rightarrow x^2 + x^2 + 4x + 4 = 290 \quad [\because (a+b)^2 = a^2 + b^2 + 2ab]$$
$$\Rightarrow 2x^2 + 4x - 286 = 0 \Rightarrow x^2 + 2x - 143 = 0 \qquad \ldots(i)$$
On comparing Eq. (i) by $ax^2 + bx + c = 0$, we get $a = 1, b = 2$ and $c = -143$
By Sridharacharya's method,
$$x = \frac{-b \pm \sqrt{b^2 - 4ac}}{2a}$$
$$\Rightarrow x = \frac{-2 \pm \sqrt{(2)^2 - [4 \times 1 \times (-143)]}}{2 \times 1} = \frac{-2 \pm \sqrt{4 + 572}}{2}$$
$$\Rightarrow x = \frac{-2 \pm \sqrt{576}}{2} \Rightarrow x = \frac{-2 \pm 24}{2}$$
$$\Rightarrow x = \frac{-2 + 24}{2} = \frac{22}{2} = 11 \text{ and } \frac{-2 - 24}{2} = -\frac{26}{2} = -13$$

Since, x is given to be an odd positive integer.
Therefore, $x \neq -13$
So, two consecutive odd integers are 11 and $11 + 2$, i.e. 11 and 13.

Ex. 3 Two natural numbers are in the ratio of 3 : 5 and their product is 2160. Find the smaller of the numbers.

Sol. Let the numbers be $3x$ and $5x$.
$\because$ Product of the numbers $= 2160 \Rightarrow 3x \times 5x = 2160 \Rightarrow 15x^2 = 2160$
$$\Rightarrow x^2 = 144 \Rightarrow x = \sqrt{144} = \pm 12$$
Since, it is a natural number, so $x = 12$.
$\therefore$ Required smaller number $= 3x = 3 \times 12 = 36$

Ex. 4 The product of two numbers is 24 times the difference of these two numbers. If the sum of these numbers is 14, then find the larger number.

Sol. Let the two numbers be x and y.
According to the question,
$$x \times y = 24(x - y) \Rightarrow xy = 24(x - y) \qquad \ldots(i)$$
and $\quad x + y = 14 \Rightarrow y = 14 - x \qquad \ldots(ii)$
On putting the value of y in Eq. (i), we get
$$x(14 - x) = 24(x - 14 + x) \Rightarrow 14x - x^2 = 24x - 336 + 24x$$
$$\Rightarrow -x^2 + 14x - 24x - 24x + 336 = 0 \Rightarrow -x^2 - 34x + 336 = 0$$
$$\Rightarrow x^2 + 34x - 336 = 0 \Rightarrow x^2 + 42x - 8x - 336 = 0$$
$$\Rightarrow x(x + 42) - 8(x + 42) = 0 \Rightarrow (x + 42)(x - 8) = 0$$
$$\therefore x = 8 \text{ or } -42$$
$$\Rightarrow x = 8 \qquad \text{[ignoring negative value]}$$
On putting the value of x in Eq. (ii), we get
$$y = 6$$
Hence, the larger number is 8.

Ex. 5 Which of the following equations has/have real roots?
(i) $3x^2 + 4x + 5 = 0$ 　　　(ii) $x^2 + x + 4 = 0$
(iii) $(x - 1)(2x - 5) = 0$ 　　(iv) $2x^2 - 3x + 4 = 0$

Sol. Roots of a quadratic equation $ax^2 + bx + c = 0$ are real, if $b^2 - 4ac \geq 0$.
(i) $3x^2 + 4x + 5 = 0$, here $b^2 - 4ac = (4)^2 - 4(3)(5) = -44 < 0$
Hence, the roots of this equation are not real.
(ii) $x^2 + x + 4 = 0$, here $b^2 - 4ac = (1)^2 - 4(1)(4) = 1 - 16 = -15 < 0$
Hence, the roots of this equation are not real.
(iii) $(x - 1)(2x - 5) = 0 \Rightarrow x = 1$ and $x = \dfrac{5}{2}$ So, 1 and $\dfrac{5}{2} > 0$
Hence, the roots of this equation are real.
(iv) $2x^2 - 3x + 4 = 0$, here $b^2 - 4ac = (-3)^2 - 4(2)(4) = 9 - 32 = -23 < 0$
Hence, the roots of this equation are not real.

Ex. 6 If one root of a quadratic equation is $2 + \sqrt{5}$, then find the quadratic equation.

Sol. Since, one root $= 2 + \sqrt{5}$, then another root $= 2 - \sqrt{5}$
$\therefore$ Sum of the roots $= (2 + \sqrt{5}) + (2 - \sqrt{5}) = 4$
Product of the roots $= (2 + \sqrt{5}) \times (2 - \sqrt{5}) = 4 - 5 = -1$
Then, the required quadratic equation is
$$x^2 - (\text{Sum of the roots})\, x + (\text{Product of the roots}) = 0$$
$$\Rightarrow x^2 - 4x - 1 = 0$$

Quadratic Equations / 523

Fast Track Formulae to solve the QUESTIONS

▶ Formula 1

If the equation $ax^2 + bx + c = 0$ has the roots α and β, then the equation having the roots $\dfrac{1}{\alpha}$ and $\dfrac{1}{\beta}$, is $cx^2 + bx + a = 0$.

Ex. 7 If roots of the equation $2x^2 - 6x + 3 = 0$ are α and β, then find the equation having the roots $\dfrac{1}{\alpha}$ and $\dfrac{1}{\beta}$.

Sol. Since, α and β are the roots of the equation $2x^2 - 6x + 3 = 0$.
Then, according to the formula, the equation having the roots $\dfrac{1}{\alpha}$ and $\dfrac{1}{\beta}$ is
$3x^2 - 6x + 2 = 0$.

▶ Formula 2

If the equation $ax^2 + bx + c = 0$ has the roots α and β, then the equation having the roots $\alpha \pm A$ and $\beta \pm A$, is $a(x \mp A)^2 + b(x \mp A) + c = 0$.

Ex. 8 If roots of the equation $x^2 - 5x + 6 = 0$ are α and β, then find the equation having the roots $(\alpha - 1)$ and $(\beta - 1)$.

Sol. Given equation is $x^2 - 5x + 6 = 0$.
Then, the required equation is
$(x + 1)^2 - 5(x + 1) + 6 = 0 \Rightarrow x^2 + 2x + 1 - 5x - 5 + 6 = 0$
$\Rightarrow \qquad x^2 - 3x + 2 = 0$

▶ Formula 3

If α and β are the roots of the equation $ax^2 + bx + c = 0$, then the equation having the roots $A\alpha$ and $A\beta$ is $ax^2 + Abx + A^2c = 0$.

✦ In this formula, the equation having the roots $\dfrac{\alpha}{A}$ and $\dfrac{\beta}{A}$, is $aA^2x^2 + bAx + c = 0$.

Ex. 9 If α and β are the roots of the equation $x^2 - 6x + 5 = 0$, then find the equation having the roots 2α and 2β.

Sol. Given equation is $x^2 - 6x + 5 = 0$.
Then, the required equation is
$x^2 - 2 \times 6x + 2^2 \times 5 = 0$
$\Rightarrow \qquad x^2 - 12x + 20 = 0$

Multi Concept Questions

1. If α and β are the roots of the equation $ax^2 + bx + c = 0$, then find the values of the following expressions in terms of a, b and c.
   (i) $\alpha^2 + \beta^2$
   (ii) $\alpha^4 - \beta^4$

➥ If $ax^2 + bx + c = 0$ and α, β are the roots of the equation, then
$$\alpha + \beta = -\frac{b}{a} \text{ and } \alpha\beta = \frac{c}{a}$$

(i) By using formula $a^2 + b^2 = (a+b)^2 - 2ab$,
$$\alpha^2 + \beta^2 = (\alpha + \beta)^2 - 2\alpha\beta = \left(-\frac{b}{a}\right)^2 - 2\left(\frac{c}{a}\right) = \frac{b^2}{a^2} - \frac{2c}{a} \Rightarrow \frac{b^2 - 2ac}{a^2}$$

(ii) $\alpha^4 - \beta^4 = (\alpha^2 + \beta^2)(\alpha + \beta)(\alpha - \beta) = [(\alpha + \beta)^2 - 2\alpha\beta](\alpha + \beta)\sqrt{(\alpha + \beta)^2 - 4\alpha\beta}$

$[\because a^4 - b^4 = (a^2 + b^2)(a^2 - b^2) = (a^2 + b^2)(a - b)(a + b)]$

$= \left(\frac{b^2 - 2ac}{a^2}\right)\left(\frac{-b}{a}\right)\sqrt{\frac{b^2}{a^2} - 4\frac{c}{a}}$

$= -\frac{b}{a^4}(b^2 - 2ac)\sqrt{b^2 - 4ac}$

2. What are the factors of $x^2 + 4y^2 + 4y - 4xy - 2x - 8$?
   (a) $(x - 2y - 4)(x - 2y + 2)$
   (b) $(x - y + 2)(x - 4y + 4)$
   (c) $(x - y + 2)(x - 4y - 4)$
   (d) $(x + 2y - 4)(x - 2y + 2)$

➥ (a) $x^2 + 4y^2 + 4y - 4xy - 2x - 8$
$= x^2 + 4y^2 - 4xy - 2x + 4y - 8$
$= (x - 2y)^2 - 2(x - 2y) - 8$
$= A^2 - 2A - 8$ [put $x - 2y = A$]
$= A^2 - 4A + 2A - 8$
$= (A - 4)(A + 2)$
$= (x - 2y - 4)(x - 2y + 2)$

3. If the sum of the roots of the equation $5x^2 + (p + q + r)x + pqr$ is zero, then what is the value of $(p^3 + q^3 + r^3)$?
   (a) 4 pqr
   (b) 3 pqr
   (c) 7 pqr
   (d) 8 pqr

➥ (b) Here, $a = 5, b = p + q + r$ and $c = pqr$
Sum of the roots $= \frac{-b}{a} = \frac{-(p+q+r)}{5} = 0$
∴ $p + q + r = 0$
According to the formula, $a^3 + b^3 + c^3 = 3abc$, if $a + b + c = 0$
∴ $p^3 + q^3 + r^3 = 3pqr$ [∵ $p + q + r = 0$]

Fast Track Practice

Exercise 1 Base Level Questions

1. Which of the following is a quadratic equation?
   (a) $x^3 - x^2 - x + 5 = 0$
   (b) $x^4 - 10 = 0$
   (c) $7x^2 = 49$
   (d) $x^4 - x^3 = 9000$
   (e) None of the above

2. Find the roots of the quadratic equation $6x^2 - 11x - 35 = 0$. [SSC (10+2) 2017]
   (a) $\frac{5}{3}, \frac{-7}{2}$
   (b) $\frac{-5}{3}, \frac{7}{2}$
   (c) $\frac{-3}{5}, \frac{2}{7}$
   (d) $\frac{3}{5}, \frac{-2}{7}$

3. What are the roots of the quadratic equation $21x^2 - 37x - 28 = 0$? [SSC (10+2) 2017]
   (a) $\frac{-7}{3}, \frac{4}{7}$
   (b) $\frac{3}{7}, \frac{-7}{4}$
   (c) $\frac{7}{3}, \frac{-4}{7}$
   (d) $\frac{-3}{7}, \frac{7}{4}$

4. If one of the roots of quadratic equation $7x^2 - 50x + k = 0$ is 7, then what is the value of k? [CDS 2012]
   (a) 7
   (b) 1
   (c) $\frac{50}{7}$
   (d) $\frac{7}{50}$

5. Which of the following equations has real roots?
   (a) $2x^2 - 3x + 4 = 0$
   (b) $(x - 1)(2x - 5) = 0$
   (c) $3x^2 + 4x + 5 = 0$
   (d) Cannot be determined
   (e) None of the above

6. If the equation $2x^2 + 3x + p = 0$ has equal roots, then the value of p is
   (a) $\frac{9}{8}$
   (b) $\frac{6}{5}$
   (c) $\frac{4}{3}$
   (d) $\frac{5}{4}$
   (e) $\frac{5}{7}$

7. For what values of k, the equation $x^2 + 2(k - 4)x + 2k = 0$ has equal roots?
   (a) 6 and 4
   (b) 8 and 2
   (c) 10 and 4
   (d) 12 and 2
   (e) None of these

8. Which one of the following is correct? [CDS 2015 (II)]
   (a) $(x + 2)$ is a factor of $x^4 - 6x^3 + 12x^2 - 24x + 32$
   (b) $(x + 2)$ is a factor of $x^4 + 6x^3 - 12x^2 + 24x - 32$
   (c) $(x - 2)$ is a factor of $x^4 - 6x^3 + 12x^2 - 24x + 32$
   (d) $(x - 2)$ is a factor of $x^4 + 6x^3 - 12x^2 + 24x - 32$

9. $(x + 4)$ is a factor of which one of the following expressions? [CDS 2017 (I)]
   (a) $x^2 - 7x + 44$
   (b) $x^2 + 7x - 44$
   (c) $x^2 - 7x - 44$
   (d) $x^2 + 7x + 44$

10. The difference in the roots of the equation $2x^2 - 11x + 5 = 0$ is [CDS 2013]
    (a) 4.5
    (b) 4
    (c) 3.5
    (d) 3

11. The quadratic equation whose roots are 3 and -1, is [CDS 2012]
    (a) $x^2 - 4x + 3 = 0$
    (b) $x^2 - 2x - 3 = 0$
    (c) $x^2 + 2x - 3 = 0$
    (d) $x^2 + 4x + 3 = 0$

12. If one root of
    $(a^2 - 5a + 3)x^2 + (3a - 1)x + 2 = 0$
    is twice the other, then what is the value of a? [CDS 2017 (I)]
    (a) $\frac{2}{3}$
    (b) $\frac{-2}{3}$
    (c) $\frac{1}{3}$
    (d) $\frac{-1}{3}$

Directions (Q. Nos. 13-14) *In these equations, two equations numbered I and II have been given. You have to solve both the equations and choose the correct option.* [NICL AO (Pre) 2017]
   (a) If $x > y$
   (b) If $x \geq y$
   (c) If $x < y$
   (d) If $x \leq y$
   (e) If $x = y$ or relationship between x and y cannot be established

13. I. $x^2 + 10x + 25 = 0$
    II. $y^2 + 29y + 190 = 0$

14. I. $x^2 - 6x - 40 = 0$
    II. $y^2 + 18y - 40 = 0$

Directions (Q. Nos. 15-16) *In each of the following questions, there are two equations. You have to solve both equations and mark the correct answer.*
[SBI Clerk (Mains) 2016]

(a) If $x > y$ (b) If $x \geq y$
(c) If $x < y$ (d) If $x \leq y$
(e) If $x = y$ or relationship cannot be established

15. I. $5x^2 - 44x + 63 = 0$
    II. $15y^2 - 37y + 18 = 0$

16. I. $x^2 = 1296$
    II. $y = 36$

Directions (Q. Nos. 17-18) *In these questions, two equations numbered I and II are given. You have to solve both the equations and mark the correct answer.*
[SBI PO (Pre) 2017]

(a) If $x > y$ (b) If $x \geq y$
(c) If $x < y$ (d) If $x \leq y$
(e) If $x = y$ or relationship cannot be established

17. I. $3x^2 - 4x + 1 = 0$
    II. $15y^2 - 8y + 1 = 0$

18. I. $x^2 + 14x + 45 = 0$
    II. $y^2 + 19y + 88 = 0$

Directions (Q. Nos. 19-20) *In these questions, two equations numbered I and II are given. You have to solve both the equations and choose the appropriate option.*
[LIC AAO 2016]

(a) If $x > y$ (b) If $x \geq y$
(c) If $x < y$ (d) If $x \leq y$
(e) If $x = y$ or the relationship cannot be estabilished

19. I. $3x^2 + 7x + 2 = 0$
    II. $y^2 + 5y + 6 = 0$

20. I. $3x^2 - 10x + 8 = 0$
    II. $2y^2 - 11y + 15 = 0$

Directions (Q. Nos. 21-22) *In these questions, two equations numbered I and II are given. You have to solve both the equations and mark the appropriate option.*
[SBI PO (Pre) 2016]

(a) If $x > y$ (b) If $x < y$
(c) If $x \geq y$ (d) If $x \leq y$
(e) If $x = y$ or relationship between x and y cannot be established

21. I. $6x^2 - 25x + 14 = 0$
    II. $9y^2 - 9y + 2 = 0$

22. I. $8x^2 + 25x + 3 = 0$
    II. $2y^2 + 17y + 30 = 0$

23. If α and β are the roots of the equation $4x^2 - 19x + 12 = 0$, then find the equation having the roots $\dfrac{1}{\alpha}$ and $\dfrac{1}{\beta}$.

(a) $4x^2 + 19x + 12 = 0$
(b) $12x^2 - 19x + 4 = 0$
(c) $12x^2 + 19x + 4 = 0$
(d) $4x^2 + 19x - 12 = 0$
(e) None of the above

24. If α and β are the roots of the equation $x^2 - 11x + 24 = 0$, then find the equation having the roots $\alpha + 2$ and $\beta + 2$.

(a) $x^2 + 15x + 24 = 0$
(b) $x^2 - 15x + 24 = 0$
(c) $x^2 + 15x - 50 = 0$
(d) $x^2 + 15x - 60 = 0$
(e) $x^2 - 15x + 50 = 0$

25. If α and β are the roots of the equation $x^2 + 13x - 30 = 0$, then find the equation having the roots 2α and 2β.

(a) $x^2 + 26x - 120 = 0$
(b) $x^2 + 12x - 60 = 0$
(c) $x^2 + 24x - 12 = 0$
(d) $x^2 + 26x + 120 = 0$
(e) $x^2 - 12x + 60 = 0$

Exercise 2 Higher Skill Level Questions

1. If α and β are the roots of the equation $x^2 + px + q = 0$, then what is $\alpha^2 + \beta^2$ equal to? [CDS 2017 (I)]

(a) $p^2 - 2q$ (b) $q^2 - 2p$
(c) $p^2 + 2q$ (d) $q^2 - q$

2. If $2x^2 - 7xy + 3y^2 = 0$, then the value of $x : y$ is

(a) 3 : 2
(b) 2 : 3
(c) 3 : 1 and 1 : 2
(d) 5 : 6

Quadratic Equations / 527

3. If the sum of the roots of $ax^2 + bx + c = 0$ is equal to the sum of the squares of their reciprocals, then which one of the following relations is correct? **[CDS 2016 (I)]**
(a) $ab^2 + bc^2 = 2a^2c$
(b) $ac^2 + bc^2 = 2b^2a$
(c) $ab^2 + bc^2 = a^2c$
(d) $a^2 + b^2 + c^2 = 1$

4. If $x = \dfrac{\sqrt{a+b} - \sqrt{a-b}}{\sqrt{a+b} + \sqrt{a-b}}$, then what is $bx^2 - 2ax + b$ equal to $(b \neq 0)$? **[CDS 2017 (I)]**
(a) 0
(b) 1
(c) ab
(d) 2ab

5. The number of roots of the equation $3^{2x^2 - 7x + 7} = 9$ is
(a) 1
(b) 2
(c) 3
(d) 4
(e) 5

6. If one of the roots of the equation $x^2 - bx + c = 0$ is the square of the other, then which of the following option is correct? **[CDS 2013]**
(a) $b^3 = 3bc + c^2 + c$
(b) $c^3 = 3bc + b^2 + b$
(c) $3bc = c^3 + b^2 + b$
(d) $3bc = c^3 + b^3 + b^2$

7. In solving a problem, one student makes a mistake in the coefficient of the first degree term and obtains -9 and -1 for the roots. Another student makes a mistake in the constant term of the equation and obtains 8 and 2 for the roots. The correct equation was **[CDS 2013]**
(a) $x^2 + 10x + 9 = 0$
(b) $x^2 - 10x + 16 = 0$
(c) $x^2 - 10x + 9 = 0$
(d) None of the above

8. Aman and Alok attempted to solve a quadratic equation. Aman made a mistake in writing down the constant term and ended up in roots (4, 3). Alok made a mistake in writing down the coefficient of x to get roots (3, 2). The correct roots of the equation are **[CDS 2017 (I)]**
(a) $-4, -3$
(b) 6, 1
(c) 4, 3
(d) $-6, -1$

9. If one root of the equation $\dfrac{x^2}{a} + \dfrac{x}{b} + \dfrac{1}{c} = 0$ is reciprocal of the other, then which one of the following is correct? **[CDS 2012]**
(a) $a = b$
(b) $b = c$
(c) $ac = 1$
(d) $a = c$

10. For which value of k does the pair of equations $x^2 - y^2 = 0$ and $(x - k)^2 + y^2 = 1$ yield a unique positive solution of x? **[CDS 2015 (I)]**
(a) 2
(b) 0
(c) $\sqrt{2}$
(d) $-\sqrt{2}$

11. If $2p + 3q = 12$ and $4p^2 + 4pq - 3q^2 = 126$, then what is the value of $p + 2q$? **[CDS 2016 (II)]**
(a) 5
(b) 21/4
(c) 25/4
(d) 99/16

12. If $\sqrt{3x^2 - 7x - 30} - \sqrt{2x^2 - 7x - 5} = x - 5$ has α and β as its roots, then the value of $\alpha\beta$ is **[CDS 2016 (I)]**
(a) -15
(b) -5
(c) 0
(d) 5

13. If $x = \sqrt{\dfrac{\sqrt{5}+1}{\sqrt{5}-1}}$, then $x^2 - x - 1$ is equal to **[SSC CGL (Mains) 2012]**
(a) 0
(b) 1
(c) 2
(d) 5

14. If α and β are the roots of the quadratic equation $2x^2 + 6x + k = 0$, where $k < 0$, then what is the maximum value of $\dfrac{\alpha}{\beta} + \dfrac{\beta}{\alpha}$? **[CDS 2017 (I)]**
(a) 2
(b) -2
(c) 9
(d) -9

15. Let p and q be non-zero integers. Consider the polynomial $A(x) = x^2 + px + q$. It is given that $(x - m)$ and $(x - km)$ are simple factors of $A(x)$, where m is a non-zero integer and k is a positive integer, $k \geq 2$. Which one of the following is correct? **[CDS 2016 (I)]**
(a) $(k+1)^2 p^2 = kq$
(b) $(k+1)^2 q = kp^2$
(c) $k^2 q = (k+1)p^2$
(d) $k^2 p^2 = (k+1)^2 q$

16. If α and β are the two zeroes of the polynomial $25x^2 - 15x + 2$, then what is a quadratic polynomial whose zeroes are $(2\alpha)^{-1}$ and $(2\beta)^{-1}$? **[CDS 2016 (II)]**
(a) $x^2 + 30x + 2$
(b) $8x^2 - 30x + 25$
(c) $8x^2 - 30x$
(d) $x^2 + 30x$

17. The number of solutions of the equation $\sqrt{x^2 - x + 1} + \dfrac{1}{\sqrt{x^2 - x + 1}} = 2 - x^2$ is

[SSC CGL 2012]

(a) 0   (b) 1
(c) 2   (d) 4

18. In the quadratic equation $x^2 + ax + b = 0$, a and b can take any value from the set {1, 2, 3, 4}. How many pairs of values of a and b are possible in order that the quadratic equation has real roots?

[CDS 2016 (II)]

(a) 6   (b) 7   (c) 8   (d) 16

Answer with Solutions

Exercise 1 Base Level Questions

1. (c) Clearly, $7x^2 = 49$ or $7x^2 - 49 = 0$, which is of the form $ax^2 + bx + c = 0$, where $b = 0$.
Thus, $7x^2 - 49 = 0$ is a quadratic equation.

2. (b) Given, $6x^2 - 11x - 35 = 0$
$\Rightarrow 6x^2 - 21x + 10x - 35 = 0$
$\Rightarrow 3x(2x - 7) + 5(2x - 7) = 0$
$\Rightarrow (3x + 5)(2x - 7) = 0$
$\therefore x = \dfrac{-5}{3}, x = \dfrac{7}{2}$

Hence, the roots are $-5/3$ and $7/2$.

3. (c) Given, $21x^2 - 37x - 28 = 0$
$\Rightarrow 21x^2 - 49x + 12x - 28 = 0$
$\Rightarrow 7x(3x - 7) + 4(3x - 7) = 0$
$\Rightarrow (3x - 7)(7x + 4) = 0$
$\Rightarrow x = \dfrac{7}{3}, \dfrac{-4}{7}$

4. (a) Given equation is $7x^2 - 50x + k = 0$.
Here, $a = 7, b = -50, c = k$
Since, $\alpha + \beta = \dfrac{-b}{a}$
$\therefore \alpha + \beta = \dfrac{50}{7} \Rightarrow \beta = \dfrac{50}{7} - 7$
$\Rightarrow \beta = \dfrac{1}{7}$ [$\because \alpha = 7$, given]
and $\alpha\beta = \dfrac{c}{a} \Rightarrow 7 \times \dfrac{1}{7} = \dfrac{k}{7} \Rightarrow k = 7$

Alternate Method
Given equation is $7x^2 - 50x + k = 0$.
If one root is 7, then it will satisfy the equation, i.e. putting $x = 7$ in equation
$7 \times (7)^2 - 50 \times 7 + k = 0$
$\Rightarrow 7 \times 49 - 350 + k = 0$
$\Rightarrow 343 - 350 + k = 0$
$\therefore k = 7$

5. (b) $(x - 1)(2x - 5) = 0 \Rightarrow x = 1, \dfrac{5}{2}$
So, its roots are real.

6. (a) Given quadratic equation is
$2x^2 + 3x + p = 0$.
Since, the equation has equal roots.
$\therefore D = b^2 - 4ac = 0$
$\Rightarrow 9 - 4(2)(p) = 0$
$\Rightarrow 8p = 9 \Rightarrow p = \dfrac{9}{8}$

7. (b) Given equation is $x^2 + 2(k - 4)x + 2k = 0$.
On comparing with $ax^2 + bx + c = 0$
Here, $a = 1, b = 2(k - 4), c = 2k$
Since, the roots are equal, we have $D = 0$.
$\therefore b^2 - 4ac = 0$
$\Rightarrow 4(k - 4)^2 - 8k = 0$
$\Rightarrow 4(k^2 + 16 - 8k) - 8k = 0$
$\Rightarrow 4k^2 + 64 - 32k - 8k = 0$
$\Rightarrow 4k^2 - 40k + 64 = 0$
$\Rightarrow k^2 - 10k + 16 = 0$
$\Rightarrow k^2 - 8k - 2k + 16 = 0$
$\Rightarrow k(k - 8) - 2(k - 8) = 0$
$\Rightarrow (k - 8)(k - 2) = 0$
Hence, the value of k is 8 or 2.

8. (c) (a) $p(x) = x^4 - 6x^3 + 12x^2 - 24x + 32$
Now, $p(-2) = (-2)^4 - 6(-2)^3 + 12(-2)^2 - 24(-2) + 32$
$= 16 - 6(-8) + 12(4) + 48 + 32$
$= 16 + 48 + 48 + 48 + 32 \neq 0$
Hence, it is not a factor.

(b) $p(x) = x^4 + 6x^3 - 12x^2 + 24x - 32$
Now, $p(-2) = (-2)^4 + 6(-2)^3 - 12(-2)^2 + 24(-2) - 32$
$= 16 - 48 - 48 - 48 - 32 \neq 0$
Hence, it is not a factor.

Quadratic Equations / 529

(c) $p(x) = x^4 - 6x^3 + 12x^2 - 24x + 32$
Now, $p(2) = 2^4 - 6(2)^3 + 12(2)^2$
$\qquad\qquad - 24(2) + 32$
$= 16 - 48 + 48 - 48 + 32$
$= 48 - 48 = 0$
Hence, it is a factor.

(d) $p(x) = x^4 + 6x^3 - 12x^2 + 24x - 32$
Now, $p(2) = 2^4 + 6(2)^3 - 12(2)^2$
$\qquad\qquad + 24(2) - 32$
$= 16 + 48 - 48 + 48 - 32$
$= 64 - 32 \neq 0$
Hence, it is not a factor.

9. (c) We have, $(x + 4)$ as a factor.
∴ Put $x = -4$ in all the options.
(a) $(-4)^2 - 7(-4) + 44$
$= 16 + 28 + 44 = 88 \neq 0$
(b) $(-4)^2 + 7(-4) - 44$
$= 16 - 28 - 44 = -56 \neq 0$
(c) $(-4)^2 - 7(-4) - 44 = 16 + 28 - 44 = 0$
(d) $(-4)^2 + 7(-4) + 44 = 16 - 28 + 44 = 32 \neq 0$
Hence, $(x + 4)$ is a factor of $x^2 - 7x - 44$.

10. (a) Let α and β be the roots of the quadratic equation $2x^2 - 11x + 5 = 0$.
∴ $\quad \alpha + \beta = -\dfrac{(-11)}{2} = \dfrac{11}{2}$ ...(i)
and $\quad \alpha \cdot \beta = \dfrac{5}{2}$ ...(ii)
Now, $(\alpha - \beta)^2 = (\alpha + \beta)^2 - 4\alpha\beta$
$= \left(\dfrac{11}{2}\right)^2 - 4\left(\dfrac{5}{2}\right)$
$= \dfrac{121}{4} - \dfrac{20}{2}$
$= \dfrac{121 - 40}{4} = \dfrac{81}{4} = \left(\dfrac{9}{2}\right)^2$
∴ Difference of roots $= (\alpha - \beta) = \dfrac{9}{2} = 4.5$

11. (b) Given roots of the quadratic equation are 3 and -1.
Let $\alpha = 3$ and $\beta = -1$.
Then, sum of roots $= \alpha + \beta = 3 - 1 = 2$
and product of roots $= \alpha \cdot \beta = (3)(-1) = -3$
∴ Required quadratic equation is
$x^2 - (\alpha + \beta)x + \alpha\beta = 0$
$\Rightarrow x^2 - 2x + (-3) = 0 \Rightarrow x^2 - 2x - 3 = 0$

12. (a) Let α and 2α be the roots of the given equation.
∴ $\quad \alpha + 2\alpha = \dfrac{-(3a - 1)}{a^2 - 5a + 3}$
$\Rightarrow \quad 3\alpha = \dfrac{-(3a - 1)}{a^2 - 5a + 3}$

$\Rightarrow \quad \alpha = \dfrac{-(3a - 1)}{3(a^2 - 5a + 3)}$ ...(i)
Also, $(\alpha)(2\alpha) = \dfrac{2}{a^2 - 5a + 3}$
$\Rightarrow \quad \alpha^2 = \dfrac{1}{a^2 - 5a + 3}$ ...(ii)
From Eqs. (i) and (ii), we get
$\left[\dfrac{-(3a - 1)}{3(a^2 - 5a + 3)}\right]^2 = \dfrac{1}{a^2 - 5a + 3}$
$\Rightarrow \dfrac{(3a - 1)^2}{9(a^2 - 5a + 3)^2} = \dfrac{1}{a^2 - 5a + 3}$
$\Rightarrow \quad (3a - 1)^2 = 9(a^2 - 5a + 3)$
$\Rightarrow \quad 9a^2 - 6a + 1 = 9a^2 - 45a + 27$
$\Rightarrow \quad 39a = 26$
$\Rightarrow \quad a = \dfrac{26}{39} = \dfrac{2}{3}$

13. (a) I. Here, $x^2 + 10x + 25 = 0$
$\Rightarrow x^2 + 5x + 5x + 25 = 0$
$\Rightarrow x(x + 5) + 5(x + 5) = 0$
$\Rightarrow (x + 5)(x + 5) = 0$
$\Rightarrow \qquad x = -5, -5$
II. Here, $y^2 + 29y + 190 = 0$
$\Rightarrow y^2 + 19y + 10y + 190 = 0$
$\Rightarrow y(y + 19) + 10(y + 19) = 0$
$\Rightarrow (y + 19)(y + 10) = 0$
$\Rightarrow \qquad y = -19, -10$
∴ $\qquad x > y$

14. (e) I. Here, $x^2 - 6x - 40 = 0$
$\Rightarrow x^2 - 10x + 4x - 40 = 0$
$\Rightarrow x(x - 10) + 4(x - 10) = 0$
$\Rightarrow (x + 4)(x - 10) = 0$
$\Rightarrow \qquad x = -4, 10$
II. Here, $y^2 + 18y - 40 = 0$
$\Rightarrow y^2 + 20y - 2y - 40 = 0$
$\Rightarrow y(y + 20) - 2(y + 20) = 0$
$\Rightarrow (y + 20)(y - 2) = 0$
$\Rightarrow \qquad y = 2, -20$
Hence, relationship between x and y cannot be established.

15. (b) I. ∵ $\qquad 5x^2 - 44x + 63 = 0$
$\Rightarrow 5x^2 - 35x - 9x + 63 = 0$
$\Rightarrow 5x(x - 7) - 9(x - 7) = 0$
$\Rightarrow \qquad (5x - 9)(x - 7) = 0$
$\Rightarrow \qquad x = \dfrac{9}{5}, 7$
II. ∵ $\qquad 15y^2 - 37y + 18 = 0$
$\Rightarrow 15y^2 - 10y - 27y + 18 = 0$
$\Rightarrow 5y(3y - 2) - 9(3y - 2) = 0$
$\Rightarrow \qquad (5y - 9)(3y - 2) = 0$

$\Rightarrow$ $y = \dfrac{9}{5}, \dfrac{2}{3}$

So, $x \geq y$

16. (d) I. $x^2 = 1296 \Rightarrow x = \pm 36$

II. $y = 36$

So, $y \geq x$.

17. (b) I. Here, $3x^2 - 4x + 1 = 0$
$\Rightarrow 3x^2 - 3x - x + 1 = 0$
$\Rightarrow 3x(x-1) - 1(x-1) = 0$
$\Rightarrow (x-1)(3x-1) = 0$
$\Rightarrow x = 1, \dfrac{1}{3}$

II. Here, $15y^2 - 8y + 1 = 0$
$\Rightarrow 15y^2 - 5y - 3y + 1 = 0$
$\Rightarrow 5y(3y-1) - 1(3y-1) = 0$
$\Rightarrow (3y-1)(5y-1) = 0$
$\Rightarrow y = \dfrac{1}{3}, \dfrac{1}{5}$

So, $x \geq y$.

18. (e) I. Here, $x^2 + 14x + 45 = 0$
$\Rightarrow x^2 + 9x + 5x + 45 = 0$
$\Rightarrow x(x+9) + 5(x+9) = 0$
$\Rightarrow (x+5)(x+9) = 0$
$\Rightarrow x = -5, -9$

II. Here, $y^2 + 19y + 88 = 0$
$\Rightarrow y^2 + 11y + 8y + 88 = 0$
$\Rightarrow y(y+11) + 8(y+11) = 0$
$\Rightarrow (y+8)(y+11) = 0$
$\Rightarrow y = -8, -11$

Hence, relationship between x and y cannot be established.

19. (b) I. $\because$ $3x^2 + 7x + 2 = 0$
$\Rightarrow 3x^2 + 6x + x + 2 = 0$
$\Rightarrow 3x(x+2) + 1(x+2) = 0$
$\Rightarrow (x+2)(3x+1) = 0$
$\Rightarrow x = -2, -\dfrac{1}{3}$

II. $\because$ $y^2 + 5y + 6 = 0$
$\Rightarrow y^2 + 3y + 2y + 6 = 0$
$\Rightarrow y(y+3) + 2(y+3) = 0$
$\Rightarrow (y+2)(y+3) = 0$
$\Rightarrow y = -2, -3$

So, $x \geq y$.

20. (c) I. $\because$ $3x^2 - 10x + 8 = 0$
$\Rightarrow 3x^2 - 6x - 4x + 8 = 0$
$\Rightarrow 3x(x-2) - 4(x-2) = 0$
$\Rightarrow (3x-4)(x-2) = 0$
$\Rightarrow x = 2, \dfrac{4}{3}$

II. $\because$ $2y^2 - 11y + 15 = 0$
$\Rightarrow 2y^2 - 6y - 5y + 15 = 0$
$\Rightarrow 2y(y-3) - 5(y-3) = 0$

$\Rightarrow (2y-5)(y-3) = 0$
$\Rightarrow y = \dfrac{5}{2}, 3$

So, $x < y$.

21. (c) I. $\because$ $6x^2 - 25x + 14 = 0$
$\Rightarrow 6x^2 - 21x - 4x + 14 = 0$
$\Rightarrow 3x(2x-7) - 2(2x-7) = 0$
$\Rightarrow (2x-7)(3x-2) = 0$
$\Rightarrow x = \dfrac{7}{2}, \dfrac{2}{3}$

II. $\because$ $9y^2 - 9y + 2 = 0$
$\Rightarrow 9y^2 - 6y - 3y + 2 = 0$
$\Rightarrow 3y(3y-2) - 1(3y-2) = 0$
$\Rightarrow (3y-2)(3y-1) = 0$
$\Rightarrow y = \dfrac{1}{3}, \dfrac{2}{3}$

So, $x \geq y$.

22. (e) I. $\because$ $8x^2 + 25x + 3 = 0$
$\Rightarrow 8x^2 + 24x + x + 3 = 0$
$\Rightarrow 8x(x+3) + 1(x+3) = 0$
$\Rightarrow (x+3)(8x+1) = 0$
$\Rightarrow x = -3, -\dfrac{1}{8}$

II. $\because$ $2y^2 + 17y + 30 = 0$
$\Rightarrow 2y^2 + 12y + 5y + 30 = 0$
$\Rightarrow 2y(y+6) + 5(y+6) = 0$
$\Rightarrow (y+6)(2y+5) = 0$
$\Rightarrow y = -6, -\dfrac{5}{2}$

Hence, relationship between x and y cannot be determined.

23. (b) Given equation is $4x^2 - 19x + 12 = 0$, and the roots are α and β.

The required equation having the roots $\dfrac{1}{\alpha}$ and $\dfrac{1}{\beta}$ is $12x^2 - 19x + 4 = 0$ [by Formula 1]

24. (e) Given equation is
$$x^2 - 11x + 24 = 0$$
Then, required equation is
$$(x-2)^2 - 11(x-2) + 24 = 0$$
[by Formula 2]
$\Rightarrow x^2 - 4x + 4 - 11x + 22 + 24 = 0$
$\therefore x^2 - 15x + 50 = 0$

25. (a) Given equation is $x^2 + 13x - 30 = 0$.
Then, the required equation is
$$x^2 + 2(13x) - 2^2 \times 30 = 0$$
[by Formula 3]
$\therefore x^2 + 26x - 120 = 0$

Quadratic Equations / 531

Exercise 2 — Higher Skill Level Questions

1. (a) We know that α and β are the roots of the equation $x^2 + px + q = 0$.
$\therefore \quad \alpha + \beta = -p$ and $\alpha\beta = q$
Now, $\alpha^2 + \beta^2 = (\alpha + \beta)^2 - 2\alpha\beta$
$\qquad = (-p)^2 - 2(q) = p^2 - 2q$

2. (c) $\because \quad 2x^2 - 7xy + 3y^2 = 0$
$\Rightarrow \quad 2x^2 - 6xy - xy + 3y^2 = 0$
$\Rightarrow \quad 2x(x - 3y) - y(x - 3y) = 0$
$\Rightarrow \quad (2x - y)(x - 3y) = 0$
Either $2x - y = 0$
$\Rightarrow \quad 2x = y \Rightarrow \dfrac{x}{y} = \dfrac{1}{2}$
or $\quad x - 3y = 0 \Rightarrow x = 3y \Rightarrow \dfrac{x}{y} = \dfrac{3}{1}$

3. (a) Let α and β be the roots of equation $ax^2 + bx + c = 0$.
Then, $\quad \alpha + \beta = \dfrac{-b}{a}$
and $\quad \alpha\beta = \dfrac{c}{a}$
Now, according to the question,
$\alpha + \beta = \left(\dfrac{1}{\alpha}\right)^2 + \left(\dfrac{1}{\beta}\right)^2$
$\Rightarrow \quad \alpha + \beta = \dfrac{\beta^2 + \alpha^2}{(\alpha\beta)^2}$
$\Rightarrow \quad (\alpha + \beta) = \dfrac{(\alpha + \beta)^2 - 2\alpha\beta}{(\alpha\beta)^2}$
$\Rightarrow \quad \dfrac{-b}{a} = \dfrac{b^2/a^2 - 2c/a}{c^2/a^2}$
$\Rightarrow \quad ab^2 - 2a^2c = -bc^2$
$\Rightarrow \quad ab^2 + bc^2 = 2a^2c$

4. (a) We have,
$x = \dfrac{\sqrt{a+b} - \sqrt{a-b}}{\sqrt{a+b} + \sqrt{a-b}}$
$\Rightarrow \quad x = \dfrac{\sqrt{a+b} - \sqrt{a-b}}{\sqrt{a+b} + \sqrt{a-b}} \times \dfrac{\sqrt{a+b} - \sqrt{a-b}}{\sqrt{a+b} - \sqrt{a-b}}$
$\Rightarrow \quad x = \dfrac{(\sqrt{a+b} - \sqrt{a-b})^2}{(a+b) - (a-b)}$
$\Rightarrow \quad x = \dfrac{a + b + a - b - 2\sqrt{a^2 - b^2}}{2b}$
$\Rightarrow \quad x = \dfrac{2a - 2\sqrt{a^2 - b^2}}{2b}$
$\Rightarrow \quad x = \dfrac{a - \sqrt{a^2 - b^2}}{b}$

$\Rightarrow \quad bx = a - \sqrt{a^2 - b^2}$
$\Rightarrow a - bx = \sqrt{a^2 - b^2}$
On squaring both the sides, we get
$\qquad (a - bx)^2 = a^2 - b^2$
$\Rightarrow \quad a^2 + b^2x^2 - 2abx = a^2 - b^2$
$\Rightarrow \quad b^2x^2 - 2abx + b^2 = 0$
$\therefore \quad bx^2 - 2ax + b = 0$

5. (b) We have, $3^{2x^2 - 7x + 7} = 9 = 3^2$
On comparing the exponents on both sides, we get $2x^2 - 7x + 7 = 2$
$\Rightarrow \quad 2x^2 - 7x + 5 = 0$
which is a quadratic equation.
$\therefore$ It has two roots.

6. (a) Given that, one root of the equation $x^2 - bx + c = 0$ is square of other root of this equation, i.e. roots (α, α^2).
$\therefore$ Sum of roots $= \alpha + \alpha^2 = -\dfrac{(-b)}{1}$
$\Rightarrow \quad \alpha(\alpha + 1) = b$ ...(i)
and product of roots $= \alpha \cdot \alpha^2 = \dfrac{c}{1}$
$\Rightarrow \quad \alpha^3 = c \Rightarrow \alpha = c^{1/3}$ ...(ii)
From Eqs. (i) and (ii), we get
$c^{1/3}(c^{1/3} + 1) = b$ ...(iii)
On cubing both sides, we get
$c(c^{1/3} + 1)^3 = b^3$
$\Rightarrow \quad c\{c + 1 + 3c^{1/3}(c^{1/3} + 1)\} = b^3$
$\Rightarrow \quad c\{c + 1 + 3b\} = b^3$ [from Eq. (iii)]
$\Rightarrow \quad b^3 = 3bc + c^2 + c$

7. (c) When mistake is done in first degree term, the roots of the equation are -9 and -1.
$\therefore$ Equation is
$\quad (x + 1)(x + 9) = x^2 + 10x + 9$ ...(i)
When mistake is done in constant term, then the roots of equation are 8 and 2.
$\therefore$ Equation is
$\quad (x - 2)(x - 8) = x^2 - 10x + 16$ ...(ii)
$\therefore$ Required equation from Eqs. (i) and (ii)
$\qquad = x^2 - 10x + 9$

Also, we see in both the cases, 1st degree term is same with opposite sign, i.e. in such question, we should take data from given conditions and find the correct equation.

Alternate Method
Let α and β be the roots of the equation
$ax^2 + bx + c = 0$.
When mistake is done in first degree term,
then $\alpha\beta = (-9)(-1)$
$\Rightarrow \quad \alpha\beta = 9$ [not effected by mistake]
When mistake is done in constant term, then
$\alpha + \beta = 8 + 2 = 10$
[not effected by mistake]
$\therefore$ Required equation is
$x^2 - (\alpha + \beta)x + \alpha\beta = 0$
i.e. $x^2 - 10x + 9 = 0$.

8. (b) Let the quadratic equation be
$ax^2 + bx + c = 0$
If α and β are roots, then
$\alpha + \beta = \dfrac{-b}{a}$ and $\alpha\beta = \dfrac{c}{a}$
Since, Aman made a mistake in writing down the constant term.
$\therefore \quad \alpha + \beta = 4 + 3 = 7$
Also, Alok made a mistake in writing down the coefficient of x.
$\therefore \quad \alpha\beta = 3 \times 2 = 6$
So, the equation will be
$x^2 - (\alpha + b)x + \alpha\beta = 0$
$\Rightarrow \quad x^2 - 7x + 6 = 0$
$\Rightarrow \quad (x - 6)(x - 1) = 0$
$\Rightarrow \quad x = 6, 1$

9. (d) Given quadratic equation is
$\dfrac{x^2}{a} + \dfrac{x}{b} + \dfrac{1}{c} = 0$ ...(i)
Now, by condition, the roots of Eq. (i) are α and $\dfrac{1}{\alpha}$.
Now, product of roots $= \dfrac{1/c}{1/a}$
$\Rightarrow \quad \alpha \cdot \dfrac{1}{\alpha} = \dfrac{a}{c} \Rightarrow c = a$
which is the required relation.

10. (c) Given, $x^2 - y^2 = 0 \Rightarrow y^2 = x^2$ ...(i)
and $(x - k)^2 + y^2 = 1$
$\Rightarrow \quad x^2 + k^2 - 2kx + x^2 = 1$ [from Eq. (i)]
$\Rightarrow \quad 2x^2 - 2kx + (k^2 - 1) = 0$
$\because$ Equation has unique solution, then
$D = 0 \Rightarrow b^2 - 4ac = 0$
$\Rightarrow \quad (-2k)^2 - 4 \times 2 \times (k^2 - 1) = 0$
$\Rightarrow \quad 4k^2 - 8k^2 + 8 = 0 \Rightarrow -4k^2 + 8 = 0$
$\Rightarrow \quad -4k^2 = -8 \Rightarrow k = \pm\sqrt{2}$
For positive roots,
$-\dfrac{b}{a} > 0$ and $\dfrac{c}{a} > 0$

$\Rightarrow \quad -\left(\dfrac{-2k}{2}\right) > 0$ and $\left(\dfrac{k^2 - 1}{2}\right) > 0$
$\Rightarrow \quad k > 0$ and $k^2 > 1$
$\therefore \quad k = \sqrt{2}$

11. (d) Given, $2p + 3q = 12$ ...(i)
and $4p^2 + 4pq - 3q^2 = 126$
$\Rightarrow 4p^2 + 6pq - 2pq - 3q^2 = 126$
$\Rightarrow \quad (2p + 3q)(2p - q) = 126$
$\Rightarrow \quad 12(2p - q) = 126$
[from Eq. (i)]
$\Rightarrow \quad 2p - q = \dfrac{21}{2}$ ...(ii)
On solving Eqs. (i) and (ii), we get
$p = \dfrac{87}{16}$ and $q = \dfrac{3}{8}$
$\therefore \quad p + 2q = \dfrac{87}{16} + 2 \times \dfrac{3}{8} = \dfrac{99}{16}$

12. (a) Given,
$\sqrt{3x^2 - 7x - 30} = (x - 5) + \sqrt{2x^2 - 7x - 5}$
On squaring both sides, we get
$3x^2 - 7x - 30 = (x - 5)^2 + (2x^2 - 7x - 5)$
$\qquad + 2(x - 5)\sqrt{2x^2 - 7x - 5}$
$\Rightarrow 3x^2 - 7x - 30 = x^2 + 25 - 10x + 2x^2$
$\qquad -7x - 5 + (2x - 10)\sqrt{2x^2 - 7x - 5}$
$\Rightarrow 3x^2 - 7x - 30 - x^2 + 10x + 7x - 2x^2 - 20$
$\qquad = (2x - 10)\sqrt{2x^2 - 7x - 5}$
$\Rightarrow (10x - 50) = (2x - 10)\sqrt{2x^2 - 7x - 5}$
$\Rightarrow 10(x - 5) = 2(x - 5)\sqrt{2x^2 - 7x - 5}$
$\Rightarrow \sqrt{2x^2 - 7x - 5} = 5$ ...(i)
Again, squaring both sides, we get
$2x^2 - 7x - 5 = 25$
$\Rightarrow 2x^2 - 7x - 30 = 0$ ...(ii)
If α and β are roots of Eq. (ii), then
Products of roots, $\alpha\beta = \dfrac{\text{Constant term}}{\text{Coefficient of } x^2}$
$= \dfrac{-30}{2} = -15$

13. (a) Here, $x = \sqrt{\dfrac{\sqrt{5} + 1}{\sqrt{5} - 1}}$
On rationalising the terms given in square root, we get $x = \sqrt{\dfrac{\sqrt{5} + 1}{\sqrt{5} - 1} \times \dfrac{\sqrt{5} + 1}{\sqrt{5} + 1}} = \dfrac{\sqrt{5} + 1}{2}$
Now, substituting the value of x in
$x^2 - x - 1$.

Quadratic Equations / 533

$\therefore x^2 - x - 1 = \left(\dfrac{\sqrt{5}+1}{2}\right)^2 - \left(\dfrac{\sqrt{5}+1}{2}\right) - 1$

$= \dfrac{5+1+2\sqrt{5}}{4} - \dfrac{\sqrt{5}+1}{2} - 1$

$= \dfrac{6+2\sqrt{5}-2\sqrt{5}-2-4}{4} = 0$

14. (d) We have, $2x^2 + 6x + k = 0$

$\therefore \quad \alpha+\beta = -\dfrac{6}{2} = -3$ and $\alpha\beta = \dfrac{k}{2}$

Now, $\dfrac{\alpha}{\beta} + \dfrac{\beta}{\alpha} = \dfrac{\alpha^2+\beta^2}{\alpha\beta}$

Since, numerator is always positive and denominator will be always negative ($k < 0$).

$\therefore \dfrac{\alpha^2+\beta^2}{\alpha\beta}$ will be a negative value.

So, options (a) and (c) are wrong.

Now, $\dfrac{\alpha^2+\beta^2}{\alpha\beta} = \dfrac{(\alpha+\beta)^2 - 2\alpha\beta}{\alpha\beta}$

$= \dfrac{9-k}{k/2} = \dfrac{2(9-k)}{k}$

Put $\dfrac{2(9-k)}{k} = -2$

$\Rightarrow \quad 9 - k = -k$

$\Rightarrow \quad 9 = 0$, which is not possible.

Put $\dfrac{2(9-k)}{k} = -9$

$\Rightarrow \quad k = -\dfrac{18}{7}$, which is possible.

Hence, the maximum value is -9.

15. (b) We have, $A(x) = x^2 + px + q$

$\because (x - m)$ and $(x - km)$ are the factors of $A(x)$, then m and km are the roots of $A(x)$.

$\therefore \quad m + km = -p \Rightarrow m(k+1) = -p$

$\Rightarrow \quad m = \dfrac{-p}{(k+1)}$ ...(i)

and $m \cdot km = q \Rightarrow m^2 k = q$

$\Rightarrow \quad \dfrac{p^2}{(k+1)^2} \cdot k = q$ [from Eq. (i)]

$\Rightarrow \quad (k+1)^2 q = kp^2$

16. (b) Let $\quad 25x^2 - 15x + 2 = 0$

$\Rightarrow \quad 25x^2 - 10x - 5x + 2 = 0$

[by factorisation]

$\Rightarrow \quad 5x(5x - 2) - 1(5x - 2) = 0$

$\Rightarrow \quad (5x - 1)(5x - 2) = 0$

$\Rightarrow \quad 5x = 1$ or $5x = 2$

$\Rightarrow \quad x = \dfrac{1}{5}$ or $x = \dfrac{2}{5}$

It is given that α and β are the zeroes of $25x^2 - 15x + 2 = 0$.

$\therefore \quad \alpha = \dfrac{1}{5}$ and $\beta = \dfrac{2}{5}$ ...(i)

Now, the quadratic equation whose roots are $(2\alpha)^{-1}$ and $(2\beta)^{-1}$, is given by

$x^2 - [(2\alpha)^{-1} + (2\beta)^{-1}]x + (2\alpha)^{-1} \cdot (2\beta)^{-1} = 0$

$\Rightarrow \quad x^2 - \left[\dfrac{1}{2\alpha} + \dfrac{1}{2\beta}\right]x + \left(\dfrac{1}{2\alpha}\right)\left(\dfrac{1}{2\beta}\right) = 0$

$\Rightarrow x^2 - \left[\dfrac{1}{2\times\dfrac{1}{5}} + \dfrac{1}{2\times\dfrac{2}{5}}\right]x + \left(\dfrac{1}{2\times\dfrac{1}{5}}\right)\left(\dfrac{1}{2\times\dfrac{2}{5}}\right) = 0$

[from Eq. (i)]

$\Rightarrow \quad x^2 - \left[\dfrac{5}{2} + \dfrac{5}{4}\right]x + \left(\dfrac{5}{2}\right)\left(\dfrac{5}{4}\right) = 0$

$\Rightarrow \quad x^2 - \dfrac{15}{4}x + \dfrac{25}{8} = 0$

$\Rightarrow \quad \dfrac{8x^2 - 30x + 25}{8} = 0$

$\therefore \quad 8x^2 - 30x + 25 = 0$

Hence, the required quadratic polynomial is $8x^2 - 30x + 25$.

17. (b) We know that, AM $\geq$ GM

$\therefore \quad \sqrt{a} + \dfrac{1}{\sqrt{a}} \geq 2$

Here, $\sqrt{x^2 - x + 1} + \dfrac{1}{\sqrt{x^2 - x + 1}} \geq 2$

$\Rightarrow \quad 2 - x^2 \geq 2 \Rightarrow x^2 \leq 0 \Rightarrow x = 0$

[$\because x^2 < 0$ is not possible]

Hence, the given equation has only one solution.

18. (b) For real roots,

$B^2 - 4AC \geq 0$

So, by equation, $a^2 - 4b \geq 0$

$\Rightarrow \quad a^2 \geq 4b$

When $b = 1$, then $a^2 \geq 4$

$\Rightarrow \quad a^2 - 4 \geq 0$

$\therefore \quad b = 1$ and $a = 2, 3, 4$

When $b = 2$, then $a^2 - 8 \geq 0$

$\therefore \quad b = 2$ and $a = 3, 4$

When $b = 3$, then $a^2 - 12 \geq 0$

$\therefore \quad b = 3$ and $a = 4$

When $b = 4$, then $a^2 - 16 \geq 0$

$\therefore \quad a = 4$ and $b = 4$

Hence, 7 pairs of values of a and b are possible.

Chapter 32

Permutations and Combinations

In our day-to-day life, we are interested in knowing the number of ways, in which a particular work can be done. For this, we will have to know all the possible ways to do that work and it can be done with the help of **permutation** and **combination**.

Factorial

Factorial of a number can be defined as the product of all natural numbers upto that number,

i.e. $n! = n \times (n-1) \times (n-2) \times (n-3) \times (n-4) \times \ldots \times 1 = n \times (n-1)!$

e.g. $4! = 4 \times 3 \times 2 \times 1 = 4 \times 3!$

✦ Factorial of negative number and integers is not defined.
$$0! = 1, \quad 1! = 1$$

Permutation

Each of the different arrangements which can be made by taking some or all of a given number of things or objects at a time, is called a permutation. Permutation implies arrangement, where order of the things is important.

For example The permutations of three items a, b and c taken two at a time are ab, ba, ac, ca, cb and bc. Since, the order in which the items are taken, is important, ab and ba are counted as two different permutations.

Let r and n be positive integers such that $1 \leq r \leq n$. Then, the number of permutations of n different things, taken r at a time, is denoted by ^nP_r or $P(n, r)$.

Formula for permutation, $^nP_r = \dfrac{n!}{(n-r)!}$.

✦ $^nP_n = n!$, $^nP_0 = 1$

Ex. 1 If $^nP_4 = 360$, then find n.

Sol. Given, $^nP_4 = 360 \Rightarrow \dfrac{n!}{(n-4)!} = 360$

$\Rightarrow \dfrac{(n)(n-1)(n-2)(n-3)(n-4)!}{(n-4)!} = 360$ $\left[\because {}^nP_r = \dfrac{n!}{(n-r)!}\right]$

$\Rightarrow n(n-1)(n-2)(n-3) = 360 = 6 \times 5 \times 4 \times 3$

$\therefore n = 6$

Cases of Permutation

There are several cases of permutation

1. Formation of Numbers with Given Digits

In these type of questions, it is asked to form numbers with some different digits. These digits can be used with repetition or without repetition.

Ex. 2 How many numbers of four digits can be formed with the digits 1, 2, 3, 4 and 5? (Repetition of digits is not allowed)

Sol. There are five numbers and number of places to be filled up = 4

So, the required number of four digits is

$$^5P_4 = \dfrac{5!}{(5-4)!} = \dfrac{5 \times 4 \times 3 \times 2 \times 1}{1} = 120$$

Ex. 3 How many numbers between 400 and 1000 can be made with the digits 2, 3, 4, 5, 6 and 0? (Repetition of digits is not allowed)

Sol. Here, it is mentioned that repetition of digits, is not allowed.

Now, for any number of three digits to be in between 400 and 1000, its hundred place must be occupied by 4 or 5 or 6 (because if it will start with 0, 2 and 3, then it will not lie between 400 to 1000).

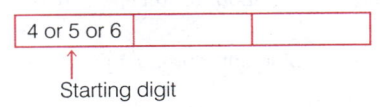

| 4 or 5 or 6 | | |

↑
Starting digit

So, remaining two places can be filled up by five digits (since, six digits 2, 3, 4, 5, 6, 0 are given and if the first place is occupied by 4, then remaining two can be filled up by five digits 2, 3, 5, 6 and 0).

So, the number of ways to fill first place = 3

and the number of ways to fill remaining two places = $^5P_2$

$\therefore$ Required numbers = $3 \times {}^5P_2$

$= 3 \times \dfrac{5!}{(5-2)!} = \dfrac{3 \times 5 \times 4 \times 3 \times 2 \times 1}{3 \times 2 \times 1} = 60$

2. Formation of Words with Given Letters

These questions are very much similar to previous case questions but here in place of numbers, word are formed from a set of English alphabets given in the form of a word.

* Number of permutations of n objects out of which p are alike and are of first type, q are alike and are of second type and r are alike and are of third type, is $\dfrac{n!}{p!\,q!\,r!}$.

Ex. 4 In how many ways, can the letters of the word 'DIRECTOR' be arranged, so that the three vowels are never together?

Sol. Here, total number of letters = 8 and total number of vowels = 3
Since, R occurs two times.

∴ Total number of arrangements when there is no repetition = $\frac{8!}{2!}$ = 20160, but when three vowels are together, regarding them as one letter, we have only 5 + 1 = 6 letters.

These 6 letters can be arranged in $\frac{6!}{2!}$ ways, since R occurs twice.

Also, three vowels can be arranged among themselves in 3! ways.

Hence, number of arrangements when the three vowels are together = $3! \times \frac{6!}{2!}$ = 2160

Number of arrangements, so that the three vowels are never together
= 20160 − 2160 = 18000

3. Arrangement of Persons in a Row or at a Round Table

These type of questions are based on arrangement of person (boy or girls etc.,) in a straight line facing some direction or around some circular object like table etc.

+ Number of permutations of n objects taken all at a time is $n!$, when repetition is not allowed.

Ex. 5 In how many different ways 5 girls can be seated in a row?

Sol. Number of ways in which 5 girls can be seated in a row = $5! = 5 \times 4 \times 3 \times 2 \times 1 = 120$

4. Arrangement of Books on a Shelf

In such questions, arrangement of books is done into a shelf in a row or one over the other.

+ Questions based on sending invitation to different persons are similar to questions based on arrangement of books.
+ Number of permutations of n different objects taken r at a time, when repetition is allowed = n^r.

Ex. 6 In how many ways, 3 books can be given away to 7 boys, when each boy is eligible for any of the books?

Sol. ∵ First book can be given to any of the 7 boys, second book can be given to any of the 7 boys, third book can be given to any of the 7 boys.
∴ Required number of ways = 7^3 = 343

Ex. 7 A gentleman has 6 friends to invite. In how many ways, can he send invitation cards to them, if he has three servants to carry the cards?

Sol. There are 6 friends and each friend can be invited by any one of the three servants.
So, number of ways a friend can be invited = 3
∴ Number of ways six friends can be invited = 3^6 = 729

Combination

Combination of things means selection of things. Here, order of things has no importance.

For example The combination of two letters from the group of three letters A, B and C would be AB, BC, AC.

In combination, we make groups. So, AB or BA as a group is same.

Let r and n be the positive integers such that $1 \le r \le n$. Then, the number of combination of n different things, when r things are taken into consideration, is denoted by ^nC_r or $C(n, r)$.

Formula for combination, $^nC_r = \dfrac{n!}{r!(n-r)!}$.

Some Important Points

- $^nC_n = {^nC_0} = 1$
- $^nC_r = {^nC_{n-r}}$ or $^nC_r = (^nP_r)/r!$
- $^nC_{r-1} + {^nC_r} = {^{n+1}C_r}$
- $^nC_1 + {^nC_2} + \ldots + {^nC_n} = 2^n - 1$

Ex. 8 Find the value of the following.

(i) $^{15}C_{11}$ (ii) $^{10}C_4$

Sol. (i) $^{15}C_{11} = \dfrac{15!}{11!(15-11)!}$ $\qquad \left[\because {^nC_r} = \dfrac{n!}{(n-r)!\, r!}\right]$

$= \dfrac{15 \times 14 \times 13 \times 12 \times 11!}{11! \times 4!} = \dfrac{15 \times 14 \times 13 \times 12}{4 \times 3 \times 2} = 1365$

(ii) $^{10}C_4 = \dfrac{10!}{4!(10-4)!} = \dfrac{10 \times 9 \times 8 \times 7 \times 6!}{4! \times 6!} = \dfrac{10 \times 9 \times 8 \times 7}{4 \times 3 \times 2 \times 1} = 210$

Cases of Combination

There are several cases of combination

1. Formation of Committee from a Given Set of Persons

These questions are based on formation of a committee consisting of some members (male and/or female) from a group of persons following a certain condition.

Ex. 9 In how many ways can 5 members form a committee out of 10 be selected, so that

(i) two particular members must be included?
(ii) two particular members must not be included?

Sol. (i) When two particular members are included, then we have to select $5 - 2 = 3$ members out of $10 - 2 = 8$.

$\therefore$ Required number of ways $= C(8, 3) = {^8C_3}$

$= \dfrac{8!}{3!(8-3)!} = \dfrac{8 \times 7 \times 6 \times 5!}{3! \times 5!} = \dfrac{8 \times 7 \times 6}{6} = 56$

(ii) When two particular members are not included, then we have to select 5 members out of $10 - 2 = 8$.

$\therefore$ Required number of ways $= C(8, 5) = {^8C_5} = \dfrac{8!}{5!(8-5)!} = \dfrac{8 \times 7 \times 6 \times 5!}{5! \times 3!}$

$= \dfrac{8 \times 7 \times 6}{6} = 56$

2. Selection of Questions from Question Paper

In such questions, a question paper is given with one or more parts and the different ways in which some specified number of questions can be attempted is asked.

Ex. 10 A question paper has two parts, part A and part B, each containing 10 questions. If the student has to choose 8 from part A and 5 from part B, then in how many ways can he choose the questions?

Sol. Required number of ways
$$= C(10, 8) \cdot C(10, 5) = {}^{10}C_8 \times {}^{10}C_5 = \frac{10!}{8!(10-8)!} \times \frac{10!}{5!(10-5)!}$$
$$= \frac{10!}{8!2!} \times \frac{10!}{5!5!} = \frac{10 \times 9}{2} \times \frac{10 \times 9 \times 8 \times 7 \times 6}{5 \times 4 \times 3 \times 2} = 5 \times 9 \times 3 \times 2 \times 7 \times 6 = 11340$$

Fundamental Principles of Counting

There are two fundamental principles of counting, which are as follows:

Multiplication Principle

If an operation can be performed in m different ways, following which a second operation can be performed in n different ways, then the two operations in succession can be performed in $m \times n$ ways.

This can be extended to any finite number of mutually inclusive operations.

In general, if there are n jobs to perform and each can be performed in m_i ways $(i = 1, 2, 3, \ldots, n)$, then number of ways of doing all things simultaneously is $m_1 \times m_2 \times m_3 \times \ldots \times m_n$.

Here, the jobs performed are mutually inclusive.

Ex. 11 A hall has 12 gates. In how many ways, can a man enter the hall through one gate and come out through a different gate?

Sol. Since, there are 12 ways of entering into the hall, the man can come out through a different gate in 11 ways.

Hence, by the fundamental principle of multiplication, total number of ways is $12 \times 11 = 132$.

Addition Principle

If an operation can be performed in m different ways and another operation, which is independent of the first operation, can be performed in n different ways, then either of the two operations can be performed in $(m + n)$ ways. This can be extended to any finite number of mutually exclusive operations.

In general, if there are n independent jobs, each of which can be performed in m_i ways, then the total number of ways of performing all things simultaneously is $m_1 + m_2 + m_3 + \ldots + m_n$.

Here, the jobs performed are mutually exclusive.

Ex. 12 There are 25 students in a class with 15 boys and 10 girls. The class teacher selects either a boy or a girl for monitor post of the class. In how many ways, the class teacher can make this selection?

Sol. As there are 15 boys and 10 girls and monitor selected can be anyone from the given students.

Hence, required number of ways = $15 + 10 = 25$

Permutations and Combinations / 539

Fast Track Formulae to solve the QUESTIONS

▶ Formula 1

If $^nC_x = {}^nC_y$, then either $x = y$ or $x + y = n$

Ex. 13 If $^{15}C_{3r} = {}^{15}C_{r+3}$, then find r.

Sol. $\because \quad {}^{15}C_{3r} = {}^{15}C_{r+3}$

$\Rightarrow$ Either $3r = r + 3$ or $3r + r + 3 = 15$

$\Rightarrow$ Either $r = \dfrac{3}{2}$ or $r = 3$

Since, r cannot be a fraction, so $r = 3$

▶ Formula 2

Number of circular permutations of n different objects $= (n-1)!$

Ex. 14 Find the number of ways, in which 10 boys can form a ring?

Sol. Let us assume the boys be $B_1, B_2, B_3, B_4, B_5, B_6, B_7, B_8, B_9$ and B_{10}.
If we assume B_1 fixed, then other 9 boys can be arranged in 9! ways.
$\therefore$ Total number of ways $= 9! = 362880$

Fast Track Method

Given, total number of boys $= 10$
According to the formula,
Total number of ways in which ring can be formed $= (10-1)! = 9! = 362880$

▶ Formula 3

In a circular permutation, if clockwise and anti-clockwise arrangements are considered to be same, then the number of circular permutations of n objects $= \dfrac{(n-1)!}{2}$

Ex. 15 Find the total number of ways, in which 10 beads can be strung into a necklace.

Sol. Given, total number of beads $= 10$
According to the formula,
Required number of ways $= \dfrac{(10-1)!}{2} = \dfrac{9!}{2} = \dfrac{362880}{2} = 181440$

▶ Formula 4

Number of ways to declare the result, where 'n' match are played $= 2^n$

Ex. 16 In a cricket tournament 5 matches were played, then in how many ways result can be declared?

Sol. Total ways to declare the result $= 2^n = 2^5 = 32$

Formula 5

Let there are n persons in a hall. If every person shakes his hand with every other person only once, then total number of handshakes

$$= {}^nC_2 = \frac{n(n-1)}{2}$$

✦ If in place of handshakes each person gives a gift to another person, then formula changes to
$$= n(n-1)$$

✦ Number of diagonals in a polygon of n sides $= {}^nC_2 - n$

Ex. 17 In a party, every person shakes his hand with every other person only once. If total number of handshakes is 210, then find the number of persons.

Sol. Let the number of persons be n. Then, according to the question, ${}^nC_2 = 210$

$$\Rightarrow \frac{n(n-1)}{2} = 210$$

$$\Rightarrow n(n-1) = 420 = 21 \times 20 \Rightarrow n = 21$$

Formula 6

If there are n non-collinear points in a plane, then
(i) Number of straight lines formed $= {}^nC_2$
(ii) Number of triangles formed $= {}^nC_3$
(iii) Number of quadrilaterals formed $= {}^nC_4$

Ex. 18 In a plane, there are 16 non-collinear points. Find the number of straight lines formed.

Sol. Here, $n = 16$
∴ Required number of straight lines formed $= {}^nC_2$

$$= {}^{16}C_2 = \frac{16!}{2!(16-2)!} = \frac{16 \times 15 \times 14!}{2 \times 14!}$$

$$= 8 \times 15 = 120$$

Formula 7

If there are n points in a plane out of which m are collinear, then
(i) Number of straight lines formed $= {}^nC_2 - {}^mC_2 + 1$
(ii) Number of triangles formed $= {}^nC_3 - {}^mC_3$

Ex. 19 In a plane, there are 11 points, out of which 5 are collinear. Find the number of triangles made by these points.

Sol. Here, $n = 11, m = 5$
Then, required number of triangles $= {}^nC_3 - {}^mC_3 = {}^{11}C_3 - {}^5C_3$

$$= \frac{11!}{3!(11-3)!} - \frac{5!}{3!(5-3)!}$$

$$= \frac{11 \times 10 \times 9}{3 \times 2 \times 1} - \frac{5 \times 4 \times 3}{3 \times 2 \times 1}$$

$$= 165 - 10 = 155$$

Fast Track Practice

Exercise 1 Base Level Questions

1. If $(1 \times 2 \times 3 \times 4 \times ... \times n) = n!$, then $(14! - 13! - 12!)$ is equal to [SSC (10+2) 2012]
   (a) $14 \times 12 \times (12!)$ (b) $14 \times 12 \times (13!)$
   (c) $14 \times 13 \times (13!)$ (d) $13 \times 12 \times (12!)$

2. Find the value of $^5P_2$.
   (a) 15   (b) 18
   (c) 20   (d) 122

3. If $^nP_3 = 9240$, then find the value of n.
   (a) 20   (b) 21
   (c) 22   (d) 23

4. There is a 7-digit telephone number with all different digits. If the digit at extreme right and extreme left are 5 and 6 respectively, then how many such telephone numbers are possible? [IB ACIO 2012]
   (a) 120   (b) 100000
   (c) 6720   (d) 30240

5. In how many different ways, can the letters of the word 'INHALE' be arranged? [SBI Clerk 2012]
   (a) 720   (b) 360   (c) 120   (d) 650
   (e) None of these

6. In how many ways, the letters of the word 'ARMOUR' can be arranged? [Bank PO 2010]
   (a) 720   (b) 300   (c) 640   (d) 350
   (e) None of these

7. In how many ways, the letters of the word 'BANKING' can be arranged? [Bank PO 2010]
   (a) 5040   (b) 2540   (c) 5080   (d) 2520
   (e) None of these

8. In how many ways, the letters of the word 'STRESS' can be arranged? [Bank PO 2010]
   (a) 360   (b) 240   (c) 720   (d) 120
   (e) None of these

9. In how many different ways, the letters of the word 'FINANCE' can be arranged? [NABARD 2010]
   (a) 5040   (b) 2040   (c) 2510   (d) 4080
   (e) None of these

10. In how many different ways, can the letters of the word 'VENTURE' be arranged? [IBPS Clerk 2011]
    (a) 840   (b) 5040
    (c) 1260   (d) 2520
    (e) None of these

11. In how many ways, can the letters of the word 'ASSASSINATION' be arranged, so that all the S are together?
    (a) 10!   (b) 14!/(4!)
    (c) 151200   (d) 3628800

12. How many different signals, can be made by 5 flags from 8 flags of different colours?
    (a) 6270   (b) 1680   (c) 20160   (d) 6720

13. In how many ways, 12 balls can be divided between 2 boys, one receiving 5 and the other 7 balls?
    (a) 1784   (b) 1584   (c) 1854   (d) 1560
    (e) None of these

14. If 2 boys and 2 girls are to be arranged in a row so that the girls are not next to each other, how many possible arrangements are there? [UPSC CSAT 2017]
    (a) 3   (b) 6   (c) 12   (d) 24

15. In how many ways, a cricket team of 11 players can be made from 15 players, if a particular player is always chosen?
    (a) 1835   (b) 1001   (c) 1635   (d) 1365

16. In how many ways, a cricket team of 11 players can be made from 15 players, if a particular player is never chosen?
    (a) 364   (b) 480   (c) 1365   (d) 640

17. A committee of 5 members is going to be formed from 3 trainees, 4 professors and 6 research associates. How many ways can they be selected, if [Bank PO 2010]
    (i) in committee, there are 2 trainees and 3 research associates?
       (a) 15   (b) 45   (c) 60   (d) 9
       (e) None of these
    (ii) there are 4 professors and 1 research associate or 3 trainees and 2 professors?
       (a) 12   (b) 13   (c) 24   (d) 52
       (e) None of these

18. From among 36 teachers in a school, one principal and one vice-principal are to be appointed. In how many ways can this be done? [RRB Clerk (Pre) 2017]
   (a) 1260 (b) 1250 (c) 1240 (d) 1800
   (e) None of these

19. In how many different ways, 5 boys and 5 girls can sit on a circular table, so that the boys and girls are alternate?
   (a) 2880 (b) 2800 (c) 2680 (d) 2280

20. An examination paper contains 8 questions of which 4 have 3 possible answers each, 3 have 2 possible answers each and the remaining one question has 5 possible answers. The total number of possible answers to all the questions is [MAT 2015]
   (a) 1728 (b) 1278 (c) 1306 (d) 3240

21. If $^{50}C_r = {}^{50}C_{r+2}$, then find r.
   (a) 24 (b) 23 (c) 22 (d) 21

22. 20 persons were invited to a party. In how many ways, they and the host can be seated at a circular table?
   (a) 18!   (b) 19!
   (c) 20!   (d) 25!

23. In how many ways, can 24 persons be seated around a circular table, if there are 13 seats?
   (a) $\dfrac{24!}{13 \times 11!}$ (b) $\dfrac{22!}{14 \times 12!}$
   (c) $\dfrac{23!}{13 \times 11!}$ (d) $\dfrac{24!}{12 \times 12!}$

24. Find the number of ways, in which 12 different beads can be arranged to form a necklace.
   (a) $\dfrac{11!}{2}$ (b) $\dfrac{10!}{2}$ (c) $\dfrac{12!}{2}$
   (d) Cannot be determined

25. In a cricket tournament 9 matches were played, then in how many ways result can be declared?
   (a) 1024 (b) 512
   (c) 256   (d) 128

26. In a meeting between two countries, each country has 12 delegates. All the delegates of one country shake hands with all delegates of the other country. Find the number of handshakes possible? [SSC CGL 2008]
   (a) 72   (b) 144
   (c) 288 (d) 234

27. In a "Kavi Sammelan" every poet shakes his hand with other poet only once. If total number of handshakes is 300, then find the number of poets.
   (a) 25 (b) 24
   (c) 23 (d) 22

28. There are five lines in a plane, no two of which are parallel. The maximum number of points in which they can intersect is [CDS 2016 (I)]
   (a) 4   (b) 6
   (c) 10 (d) None of these

29. There is a polygon of 12 sides. How many triangles can be drawn using the vertices of polygon?
   (a) 200 (b) 220 (c) 240 (d) 260
   (e) 280

30. There are 10 points in a plane, out of which 5 are collinear. Find the number of straight lines formed by joining them.
   (a) 36 (b) 45
   (c) 30 (d) 35

31. There are 14 points in a plane, out of which 4 are collinear. Find the number of triangles made by these points.
   (a) 364 (b) 360 (c) 368 (d) 365

Exercise 2 — Higher Skill Level Questions

1. If $^{56}P_{r+6} : {}^{54}P_{r+3} = 30800$, then find $^{r}P_2$.
   (a) 1840 (b) 2640
   (c) 1640 (d) 820

Directions (Q. Nos. 2-5) Find the number of permutations that can be made from the letters of the word 'OMEGA'.

2. O and A occupying end places.
   (a) 12 (b) 14 (c) 20 (d) 18

3. E being always in the middle.
   (a) 18 ways
   (b) 24 ways
   (c) 48 ways
   (d) 20 ways

4. Vowels occupying odd places.
   (a) 12 ways   (b) 16 ways
   (c) 6 ways   (d) 20 ways

5. Vowels being never together.
   (a) 36 ways   (b) 84 ways
   (c) 120 ways (d) 10 ways

Permutations and Combinations / 543

6. A question paper consists of two sections having 3 and 5 questions respectively. The following note is given on the paper. 'It is not necessary to attempt all the questions'. One question from each section is compulsory. In how many ways, a candidate can select the question?
   (a) 38 (b) 217 (c) 256 (d) 320

7. How many necklaces of 12 beads can be made from 18 beads of various colours?
   (a) $\dfrac{118 \times 13!}{2}$ (b) $\dfrac{110 \times 14!}{2}$
   (c) $\dfrac{119 \times 13!}{2}$ (d) $\dfrac{110 \times 12!}{2}$

8. There are 5 tasks and 5 persons. Task 1 cannot be assigned to either person 1 or person 2. Task 2 must be assigned to either person 3 or person 4. Every person is to be assigned one task. In how many ways can the assignment be done? **[UPSC CSAT 2015]**
   (a) 6 (b) 12
   (c) 24 (d) 144

9. The number of ways in which a committee of 3 ladies and 4 gentlemen can be appointed from a group consisting of 8 ladies and 7 gentlemen, given Mrs. X refuses to serve in a committee if Mr. Y is its member, is
   (a) 1960 (b) 3240
   (c) 1540 (d) 2065

10. A selection is to be made for one post of Principal and two posts of Vice-Principal. Amongst the six candidates called for the interview, only two are eligible for the post of Principal while they all are eligible for the post of Vice-Principal. The number of possible combinations of selection is **[UPPSC 2015]**
    (a) 4 (b) 12
    (c) 18 (d) None of these

11. The figure below shows the network connecting cities A, B, C, D, E and F. The arrows indicate permissible direction of travel. What is the number of distinct paths from A to F?

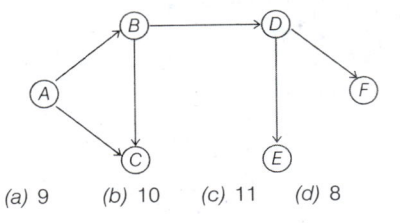

    (a) 9 (b) 10 (c) 11 (d) 8

12. In how many ways, can 15 people be seated around two round tables with seating capacities of 7 and 8 people? **[IB ACIO 2013]**
    (a) 15!/(8!) (b) 7!/88!
    (c) $^{15}C_8 \times 6! \times 7!$ (d) $^{15}C_8 \times 8!$
    (e) None of these

13. There are 10 stations on a railway line. The number of different journey tickets that are required by the authorities, is **[SNAP 2012]**
    (a) 92 (b) 90
    (c) 91 (d) 93

14. A new flag is to be designed with six vertical stripes using some or all of the colours yellow, green, blue and red. Then, the number of ways this can be made such that no two adjacent stripes have the same colour is
    (a) 12×81
    (b) 16×192
    (c) 20×125
    (d) 24×216

15. In the given figure, the lines represent one way roads allowing travel only Northwards or only Westwards. Along how many distinct routes can a car reach point B from point A?

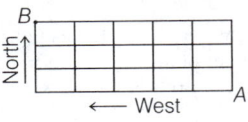

    (a) 15 (b) 56
    (c) 120 (d) 336

16. An intelligence agency forms a code of two distinct digits selected from 0, 1, 2, ..., 9 such that the first digit of code is non-zero. The code, handwritten on a slip, can however potentially create confusion when read upside down, for example the code 91 may appear as 16. How many codes are there for which no such confusion can arise?
    (a) 80 (b) 78
    (c) 71 (d) 69

17. If the last 6 digits of $[(M)!-(N)!]$ are 999000, which of the following option is not possible for $M \times (M-N)$? (Both M and N are positive integers and $M > N \cdot (M)!$ is factorial M). **[XAT 2015]**
    (a) 150 (b) 180
    (c) 200 (d) 225
    (e) 234

Answer with Solutions

Exercise 1 Base Level Questions

1. (a) $14! - 13! - 12!$
$= 14 \times 13 \times 12! - 13 \times 12! - 12!$
$= 12!(14 \times 13 - 13 - 1)$
$= 12!(182 - 14)$
$= 168 \times 12!$
$= 14 \times 12 \times 12!$

2. (c) We know that, $^nP_r = \dfrac{n!}{(n-r)!}$
$\therefore \ ^5P_2 = \dfrac{5!}{(5-2)!} = \dfrac{5!}{3!} = 5 \times 4 = 20$

3. (c) Given, $^nP_3 = 9240 \Rightarrow \dfrac{n!}{(n-3)!} = 9240$
$\Rightarrow n(n-1)(n-2) = 9240$
$\Rightarrow n(n-1)(n-2) = 22 \times 21 \times 20$
$\therefore n = 22$

4. (c) There is a 7-digit telephone number but extreme right and extreme left positions are fixed. i.e. $6 \times \times \times \times \times 5$
$\therefore$ Required number of ways
$= 8 \times 7 \times 6 \times 5 \times 4 = 6720$
[as there are 10 digits out of which 2 are fixed]

5. (a) The word 'INHALE' has 6 distinct letters.
$\therefore$ Number of arrangements $= n! = 6!$
$= 6 \times 5 \times 4 \times 3 \times 2 \times 1 = 720$

6. (e) Number of arrangements $= \dfrac{n!}{p!\,q!\,r!}$
Total letters = 6, but R has come twice.
So, required number of arrangements
$= \dfrac{6!}{2!} = \dfrac{6 \times 5 \times 4 \times 3 \times 2!}{2!} = 360$

7. (d) Total letters = 7, but N has come twice.
So, required number of arrangements
$= \dfrac{7!}{2!} = \dfrac{7 \times 6 \times 5 \times 4 \times 3 \times 2!}{2!} = 2520$

8. (d) Required number of arrangements
$= \dfrac{6!}{3!}$ [∵ S has come thrice]
$= \dfrac{6 \times 5 \times 4 \times 3!}{3!} = 120$

9. (e) Total number of letters = 7, but N has come twice.
So, required number of arrangements
$= \dfrac{7!}{2!} = \dfrac{7 \times 6 \times 5 \times 4 \times 3 \times 2!}{2!} = 2520$

10. (d) Required number of arrangements
$= \dfrac{7!}{2!} = \dfrac{7 \times 6 \times 5 \times 4 \times 3 \times 2!}{2!} = 2520$

11. (c) When all 4S are taken together and consider as 1. Then, 10 letters in total can be arranged in 10! ways.
But, here are 3 'A', 2 'I' and 2 'N'.
$\therefore$ Required number of ways $= \dfrac{10!}{3! \times 2! \times 2!}$
$= 151200$

12. (d) Number of ways taking 5 flags out of 8 flags $= ^8P_5 = \dfrac{8!}{(8-5)!} = \dfrac{8!}{3!}$
$= \dfrac{8 \times 7 \times 6 \times 5 \times 4 \times 3!}{3!} = 6720$

13. (b) Here, order is important. Then, the number of ways in which 12 different balls can be divided between two boys who receive 5 and 7 balls respectively
$= \dfrac{12!}{5!\,7!} \times 2! = 1584$

14. (c) Total possible arrangements $= 4! = 24$
Total possible arrangements when two girls sit together $= 3! \times 2! = 12$
$\therefore$ Required arrangements $= 24 - 12 = 12$

15. (b) Since, particular player is always chosen. It means that $11 - 1 = 10$ players are to be selected out of the remaining $15 - 1 = 14$ players.
$\therefore$ Required number of ways
$= ^{14}C_{10} = \dfrac{14!}{10! \times 4!}$
$= \dfrac{14 \times 13 \times 12 \times 11}{4 \times 3 \times 2 \times 1}$
$= 7 \times 13 \times 11 = 91 \times 11 = 1001$

16. (a) Since, particular player is never chosen. It means that 11 players are to be selected out of $15 - 1 = 14$ players.
$\therefore$ Required number of ways
$= ^{14}C_{11} = \dfrac{14!}{11!\,3!} = \dfrac{14 \times 13 \times 12}{3 \times 2 \times 1} = 364$

17. (i) (c) Required number $= ^3C_2 \times ^6C_3$
$= \dfrac{3!}{2!\,(3-2)!} \times \dfrac{6!}{3!\,(6-3)!}$
$= \dfrac{3 \times 2 \times 1}{2 \times 1 \times 1} \times \dfrac{6 \times 5 \times 4 \times 3 \times 2 \times 1}{3 \times 2 \times 1 \times 3 \times 2 \times 1} = 60$

(ii) (a) Required number
$= {}^4C_4 \times {}^6C_1 + {}^3C_3 \times {}^4C_2$
$= \dfrac{4!}{4!(4-4)!} \times \dfrac{6!}{1!(6-1)!}$
$\quad + \dfrac{3!}{3!(3-3)!} \times \dfrac{4!}{2!(4-2)!}$
$= 1 \times 6 + 1 \times 6 = 12$

18. (a) Since, one principal can be appointed in 36 ways.
Then, one vice-principal appointed in remaining 35 ways.
∴ Total number of ways = 36 × 35 = 1260

19. (a) After fixing up one boy on the table, the remaining can be arranged in 4! ways, but boys and girls have to be alternate. There will be 5 places, one place each between two boys. These 5 places can be filled by 5 girls in 5! ways.

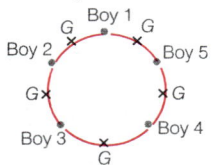

Hence, by the principle of multiplication, the required number of ways = 4! × 5! = 2880

20. (d) Number of ways (m) to answer 4 questions having three possible answer
$= 3^4$
Number of ways (n) to answer 3 question having two possible answer = 2^3
∴ Number of ways (p) to answer 1 question having 5 possible answer = 5
Hence, total number of possible answers will be counted by fundamental principle of counting = $m \times n \times p = 3^4 \times 2^3 \times 5$
$= 81 \times 8 \times 5 = 3240$

21. (a) By using Formula 1, ${}^nC_x = {}^nC_y$
$\Rightarrow x = y$ or $x + y = n$
Now, ${}^{50}C_r = {}^{50}C_{r+2}$
$\Rightarrow r + r + 2 = 50$ or $r = r + 2$
$\Rightarrow 2r = 48$ [∵ $r = r + 2$ is not possible]
∴ $r = 24$

22. (c) Total persons on the circular table
= 20 guests + 1 host = 21
∴ Required number of ways
$= (n-1)!$ [by Formula 2]
$= (21-1)!$ [here, $n = 21$]
$= 20!$

23. (a) First, we select 13 persons out of 24 persons in ${}^{24}C_{13}$ ways. Now, these 13 persons can be seated in 12! ways around a table. So, required number of ways

$= {}^{24}C_{13} \times 12! = \dfrac{24!}{13!(24-13)!} \times 12!$
[by Formula 2]
$= \dfrac{24!}{13!\,11!} \times 12! = \dfrac{24!}{13 \times 11!}$

24. (a) Number of beads is 12. It is not mentioned that whether the beads are arranged in clockwise or anti-clockwise direction.
∴ Required number of arrangements
$= \dfrac{1}{2}(n-1)!$ [by Formula 3]
$= \dfrac{(12-1)!}{2} = \dfrac{11!}{2}$

25. (b) Total number of ways to declare the result = $2^n = 2^9 = 512$ [by Formula 4]

26. (b) Total number of handshakes
$= 12 \times 12 = 144$

27. (a) Let number of persons be n.
Then, according to the question,
${}^nC_2 = 300$ [by Formula 5]
$\Rightarrow \dfrac{n!}{2!(n-2)!} = 300 \Rightarrow \dfrac{n(n-1)(n-2)!}{2(n-2)!} = 300$
$\Rightarrow n(n-1) = 600 \Rightarrow n(n-1) = 25 \times 24$
$\Rightarrow n = 25$
Hence, the number of poets is 25.

28. (c) We know that intersection point is formed by the intersection of two lines.
∴ Number of intersection points = Number of ways of selecting 2 lines out of the given 5 non-parallel lines
∴ Required number of points
$= {}^5C_2 = \dfrac{5 \times 4}{2} = 10$

29. (b) Required number of triangles
$= {}^nC_3 = {}^{12}C_3$ [by Formula 6]
$= \dfrac{12 \times 11 \times 10}{6} = 220$

30. (a) Required number of straight lines
$= {}^nC_2 - {}^mC_2 + 1$ [by Formula 7]
$= {}^{10}C_2 - {}^5C_2 + 1$ [here, $n = 10$, $m = 5$]
$= \dfrac{10 \times 9}{2} - \dfrac{5 \times 4}{2} + 1 = 45 - 10 + 1 = 36$

31. (b) Required number of triangles
$= {}^nC_3 - {}^mC_3$ [by Formula 7]
$= {}^{14}C_3 - {}^4C_3$ [here, $n = 14$, $m = 4$]
$= \dfrac{14 \times 13 \times 12 \times 11!}{3! \times 11!} - \dfrac{4!}{3! \times 1!}$
$= \dfrac{14 \times 13 \times 12}{6} - \dfrac{4}{1}$
$= 14 \times 26 - 4 = 364 - 4 = 360$

Exercise 2 Higher Skill Level Questions

1. (c) $\because \dfrac{^{56}P_{r+6}}{^{54}P_{r+3}} = \dfrac{30800}{1}$

$\Rightarrow \dfrac{56!}{(50-r)!} \times \dfrac{(51-r)!}{54!} = \dfrac{30800}{1}$

$\Rightarrow \dfrac{56 \times 55 \times 54!(51-r)(50-r)!}{(50-r)!\,54!} = \dfrac{30800}{1}$

$\Rightarrow 51 - r = \dfrac{30800}{56 \times 55} = 10$

$\Rightarrow r = 51 - 10 = 41$

$\therefore\ ^rP_2 = {}^{41}P_2 = \dfrac{41!}{39!} = 41 \times 40 = 1640$

2. (a) When O and A occupy end places. Then, the three letters (M, E, G) can be arranged among themselves in 3! = 6 ways and two letters (O, A) can be arranged among themselves in 2! = 2 ways.
$\therefore$ Total number of ways = $6 \times 2 = 12$

3. (b) When E is fixed in the middle, then there are four places left to be filled by four remaining letters O, M, G and A and this can be done in 4! ways.
$\therefore$ Total number of ways = 4! = 24

4. (a) Three vowels (O, E, A) can be arranged at the odd places in 3! ways (1st position, 3rd position, 5th position) and two consonants (M, G) can be arranged at the even places in 2! ways (2nd place and 4th place).
$\therefore$ Total number of ways = $3! \times 2! = 12$

5. (b) Total number of ways = 5! = 120

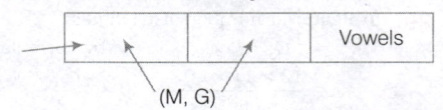

[Diagram: four boxes with "Vowels" label, arrows pointing to (M, G)]

Combining the vowels at one place (OEA) with remaining 2 letters MG, letters can be arranged in 3! ways. Also, three vowels can be arranged in 3! ways. So, when vowels are together, then number of words
$= 3! \times 3! = 36$
$\therefore$ Required number of ways, when vowels are never together = $120 - 36 = 84$

6. (b) Here, we have two sections A and B (say). Section A has 3 questions and B has 5 questions and one question from each section is compulsory according to the given condition.
$\therefore$ Number of ways selecting one or more than one question from section A
$= 2^3 - 1 = 7$

Similarly, number of ways of selecting one or more than one question from section B $= 2^5 - 1 = 31$
According to the rule of multiplication, the required number of ways in which a candidate can select the questions
$= 7 \times 31 = 217$

7. (c) First, we can select 12 beads out of 18 beads in $^{18}C_{12}$ ways. Now, these 12 beads can make a necklace in $\dfrac{11!}{2}$ ways as clockwise and anti-clockwise arrangements are same.
So, required number of ways = $^{18}C_{12} \cdot \dfrac{11!}{2!}$

$= \dfrac{18 \times 17 \times 16 \times 15 \times 14 \times 13 \times 11!}{6 \times 5 \times 4 \times 3 \times 2 \times 1 \times 2!}$

$= \dfrac{17 \times 7 \times 13!}{2!} = \dfrac{119 \times 13!}{2}$

8. (c) Task 2 must be assigned to P_4 or P_3
$= {}^2C_1 = 2$

P_1	P_2	P_3	P_4	P_5
T_1	T_2	T_3	T_4	T_5

$\therefore$ For task 1, only 2 persons are available i.e. P_3 or P_4 and $P_5 = {}^2C_1 = 2$
For task 3, only 3 persons are available
$= {}^3C_1 = 3$
For task 4, only 2 persons are available
$= {}^2C_1 = 2$
For task 5, only 1 person is left = 1
$\therefore$ Total number of ways
$= 2 \times 2 \times 3 \times 2 \times 1 = 24$

9. (d) If Mrs. X is selected among the ladies in the committee, then Mr. Y is not selected or if Mrs. X is not selected then Mr. Y can be there in the committee.
So, required number of ways
$= {}^8C_3 \times {}^6C_4 + {}^7C_3 \times {}^7C_4$
$= \dfrac{8 \times 7 \times 6}{3 \times 2} \times \dfrac{6 \times 5}{2 \times 1} + \dfrac{7 \times 6 \times 5}{3 \times 2} \times \dfrac{7 \times 6 \times 5}{3 \times 2}$
$= 840 + 1225 = 2065$

10. (d) Out of the six candidates called for interview, two are eligible for Principal.
$\therefore$ Selecting one candidate out of these two can be done in $^2C_1$ ways.
Given that all six are eligible for Vice-Principal post. So, out of the remaining 5 candidates, two can be selected in $^5C_2$ ways.

Probability / 549

Possible Outcomes
All possibilities related to an event are known as possible outcomes.

Tossing a Coin When a coin is tossed, there are two possible outcomes. So, we can say that the probability of getting H is 1/2 or the probability of getting T is 1/2.

Throwing a Die When a single die is thrown, there are six possible outcomes 1, 2, 3, 4, 5 and 6.

The probability of getting anyone of these numbers is $\frac{1}{6}$.

Ex. 1 There are 5 marbles in a bag. 3 of them are red and 2 of them are blue. What is the probability that a blue marble will be picked?

Sol. ∵ Number of favourable outcomes = 2   [∵ there are 2 blue marbles]
and total number of outcomes = 5   [∵ there are 5 marbles in total]
So, required probability = $\frac{2}{5}$ = 0.4

Event

Event is the single result of an experiment. *For example* Getting a head is an event related to tossing of a coin.

Types of Events
Various types of events are as follows

1. Certain and Impossible Events
A **certain event** is certain to occur. *For example* S (sample space) is a certain event. Probability of certain event is 1, i.e. $P(S) = 1$.

An **impossible event** has no chance of occurring, i.e. ϕ is the impossible event. Probability of impossible event is 0, i.e. $P(\phi) = 0$.

Ex. 2 A teacher chooses a student at random from a class of 30 boys. What is the probability that the student chosen is a boy?

Sol. Since, all the students are boys, so chosen may be any one, i.e.
Favourable cases = Total cases = 30
∴ Required probability = $\frac{30}{30}$ = 1

Ex. 3 A bag contains 20 black marbles. If a marble is picked at random from the bag, then find the probability that marble picked is of red colour.

Sol. Since, the bag contains 20 black marbles and there is no red marble in the bag.
So, favourable cases = 0 and total outcomes = 20
∴ Required probability = $\frac{0}{20}$ = 0

2. Equally Likely Events
Events related to an experiment are said to be equally likely events, if probability of occurrence of each event is same.

For example When a die is rolled, then possible outcomes of getting an odd number
= possible outcomes of getting an even number = 3
So, getting an even number or odd number are equally likely events.

3. Complement of an Event

The complement of an event A is the set of all outcomes in the sample space that are not included in the outcomes of event A. The complement of event A is represented by $\overline{A}$ (read as A bar). The probability of complement of an event can be found by subtracting the given probability from 1.

$$\therefore \quad P(\overline{A}) = 1 - P(A)$$

Ex. 4 A single card is chosen at random from a standard deck of 52 playing cards. What is the probability of choosing a card that is not a king?

Sol. Since, a standard deck contains 4 king.

So, probability of getting a king = $\dfrac{4}{52}$

Now, probability of not getting a king = 1 − Probability of getting a king

$$= 1 - \dfrac{4}{52} = 1 - \dfrac{1}{13} = \dfrac{13 - 1}{13} = \dfrac{12}{13}$$

4. Mutually Exclusive and Exhaustive Events

(i) Two events E_1 and E_2 related to an experiment E, having sample space S are known as **mutually exclusive**, if the probability of occurrence of both events simultaneously is zero, i.e. $P(E_1 \cap E_2) = 0$.

For example When a coin is tossed either head or tail will appear. Head and tail cannot occur simultaneously. Therefore, occurrence of a head or a tail are two mutually exclusive events.

(ii) Two events E_1 and E_2 related to an experiment E, having sample space S are known as **mutually exhaustive**, if the probability of occurrence of event E_1 or E_2 is 1, i.e. $P(E_1 \cup E_2) = 1$.

For example Let A be probability of getting an even number when a die is rolled and B be the probability of getting an odd number. The probability of occurrence of event A or event B is 1, i.e. any of the events can occur, so they are mutually exhaustive.

✦ Events $E_1, E_2, E_3, \ldots, E_n$ related to S are known as
  (i) mutually exclusive, if $P(E_1 \cap E_2 \cap E_3 \cap \ldots \cap E_n) = 0$ or $E_1 \cap E_2 \cap E_3 \cap \ldots \cap E_n = 0$.
  (ii) mutually exhaustive, if $P(E_1 \cup E_2 \cup E_3 \cup \ldots \cup E_n) = 1$ or $E_1 \cup E_2 \cup E_3 \cup \ldots \cup E_n = S$.

5. Dependent Events

Two events are called dependent, if the outcomes or occurrence of the first affects the outcomes or occurrence of the second, so that the probability is changed.

Ex. 5 A card is chosen at random from a standard deck of 52 playing cards. Without replacing it, a second card is chosen. What is the probability that the first card chosen is a queen and the second card chosen is a jack?

Sol. Here, $P(\text{Queen on first pick}) = \dfrac{4}{52}$

and $P(\text{Jack on 2nd pick given that Queen on 1st pick}) = \dfrac{4}{51}$

[∵ one card is already picked]

$$\therefore \quad P(\text{Queen and Jack}) = \dfrac{4}{52} \times \dfrac{4}{51} = \dfrac{4}{663}$$

6. Independent Events

Two events A and B are called independent, if occurring or non-occurring of A does not affect the occurring or non-occurring of B.
If A and B are independent events, then
$$P(A \text{ and } B) = P(A \cap B) = P(A) \cdot P(B)$$
For example Getting head after tossing a coin and getting a 5 on rolling single 6-sided die are independent events.
In general we can say that, if events $E_1, E_2, \ldots, E_n$ related to an experiment are independent, then
$$P(E_1 \cap E_2 \cap E_3 \cap \ldots \cap E_n) = P(E_1) P(E_2) P(E_3) P(E_4) \ldots P(E_n)$$
or
$$P(E_1 \cup E_2 \cup E_3 \cup \ldots \cup E_n) = 1 - P(\overline{E_1}) P(\overline{E_2}) P(\overline{E_3}) \ldots P(\overline{E_n}).$$

Ex. 6 A coin is tossed and a single 6-sided die is rolled. Find the probability of getting the head side of the coin and getting a 3 on the die.

Sol. Probability of getting a head when a coin is tossed = $\dfrac{1}{2}$

Probability of getting a 3 when a die is rolled = $\dfrac{1}{6}$

Now, the required probability that both occurs at the same time = $\dfrac{1}{2} \times \dfrac{1}{6} = \dfrac{1}{12}$.

Rules/Theorems Related to Probability

The various theorems related to probability are discussed below

Addition Rule of Probability

When two events A and B are mutually exclusive, the probability that A or B will occur, is the sum of the probability of each event.
$$P(A \text{ or } B) = P(A) + P(B), \text{ i.e. } P(A \cup B) = P(A) + P(B)$$
But when two events A and B are non-mutually exclusive, the probability that A or B will occur, is
$$P(A \text{ or } B) = P(A) + P(B) - P(A \text{ and } B)$$
i.e.
$$P(A \cup B) = P(A) + P(B) - P(A \cap B)$$

Multiplication Theorem of Probability

When two events A and B are mutually exclusive, the probability that A and B will occur simultaneously is given as $P(A \cap B) = P(A) \cdot P(B/A)$
or
$$P(A \cap B) = P(A) \cdot P(B) \qquad [\because A \text{ and } B \text{ are independent events}]$$

Ex. 7 From a well-shuffled pack of 52 cards, a card is drawn at random. Find the probability that it is either a heart or a queen.

Sol. Let A be the probability of getting a heart card and B be the probability of getting a queen card.

Then, $P(A) = \dfrac{13}{52}$, $P(B) = \dfrac{4}{52}$, $P(A \cap B) = \dfrac{1}{52}$ $\quad [\because \text{one heart card is a queen}]$

∴ Required probability = $P(A \cup B) = P(A) + P(B) - P(A \cap B)$

$$= \dfrac{13}{52} + \dfrac{4}{52} - \dfrac{1}{52} = \dfrac{13 + 4 - 1}{52} = \dfrac{16}{52} = \dfrac{4}{13}$$

Ex. 8 Eight persons A, B, C, D, E, F, G and H appeared for an interview. Find the probability that both A and D are selected in the interview.

Sol. Probability that A is selected, $P(A) = \dfrac{1}{8}$

Probability that D is selected, $P(D) = \dfrac{1}{8}$

∴ Required probability that both are selected $= \dfrac{1}{8} \times \dfrac{1}{8} = \dfrac{1}{64}$

Law of Total Probability

If $E_1, E_2, E_3, \ldots, E_n$ are n mutually exclusive events related to an experiment, then probability of an event A which occurs with E_1 or E_2 or $E_3 \ldots E_n$ is given by

$$P(A) = P(E_1) P(A/E_1) + P(E_2) P(A/E_2) + \ldots + P(E_n) P(A/E_n)$$

Conditional Probability

The conditional probability of an event B in relationship to an event A is the probability that event B occurs given that event A has already been occurred. The notation for conditional probability is $P(B/A)$. It is pronounced as the probability of happening of an event B given that A has already been happened.

$$P(A/B) = \dfrac{P(A \cap B)}{P(B)} \text{ and } P(B/A) = \dfrac{P(A \cap B)}{P(A)}$$

Ex. 9 A Mathematics teacher conducted two tests in her class. 25% of the students passed both tests and 42% of the students passed the first test. What per cent of the students passed the second test given that they have already passed the first test?

Sol. This problem describes a conditional probability, since it asks us to find the probability that the second test was passed given that the first test was passed. This can be solved by multiplication rule.

i.e. $P(B/A) = \dfrac{P(A \text{ and } B)}{P(A)}$

According to the formula,

$$P\left(\dfrac{\text{Second}}{\text{First}}\right) = \dfrac{P(\text{First and Second})}{P(\text{First})} = \dfrac{0.25}{0.42} \approx 0.60 \approx 60\%$$

✦ The probability of r success in n trials of an event is given as ${}^nC_r p^r q^{n-r}$, where p is the probability of success of that event in single trial and q is the probability of failure of that event.

Types of Questions

Various types of questions asked on probability are as follow

Based on Coins

This types of questions are based on tossing of coin (s) and obtaining a particular face (Head / Tail) or obtaining same face on two or more coins.

Probability / 553

Ex. 10 What is the probability of each outcome, when a coin is tossed?

Sol. Here, $S = \{H, T\}$, i.e. $n(S) = 2$

∴ $P(\text{Head}) = \dfrac{1}{2}$; $P(\text{Tail}) = \dfrac{1}{2}$

Ex. 11 A coin is tossed twice, then find the probability that a head is obtained atleast once.

Sol. When a coin is tossed twice, then possible outcome, $n(S) = \{HH, HT, TH, TT\} = 4$

Way in which a head is obtained atleast once, $n(P) = \{HH, HT, TH\} = 3$

∴ Required probability $= \dfrac{n(P)}{n(S)} = \dfrac{3}{4}$

Based on Dice

This type of questions are based on rolling of one or more dice and getting a particular number on the face or a particular sum on faces of the dice etc.

Ex. 12 A single 6-sided die is rolled. What is the probability of getting an even number and getting an odd number?

Sol. Since, the possible outcomes are 1, 2, 3, 4, 5 and 6.

Even numbers are 2, 4 and 6, i.e. 3 outcomes.

So, the number of favourable (even) outcomes = 3

Similarly, odd numbers are 1, 3 and 5, i.e. 3 outcomes.

So, the number of favourable (odd) outcomes = 3

and total number of outcomes = 6

So, the probability of getting an even number $= \dfrac{3}{6} = \dfrac{1}{2}$

Similarly, probability of getting an odd number $= \dfrac{3}{6} = \dfrac{1}{2}$

Ex. 13 A single 6-sided die is rolled. What is the probability of getting either 2 or 5?

Sol. ∵ Total outcomes when a die is rolled = 6 [given]

Probability of getting a number in a single throw of die $= \dfrac{1}{6}$

So, $P(2) = \dfrac{1}{6}$ and $P(5) = \dfrac{1}{6}$

∴ Required probability of getting either 2 or 5

$= P(2 \text{ or } 5) = P(2) + P(5) = \dfrac{1}{6} + \dfrac{1}{6}$

$= \dfrac{1+1}{6} = \dfrac{2}{6} = \dfrac{1}{3}$

554 / *Fast Track* Objective Arithmetic

Based on Playing Cards
There are total of 52 cards in a deck of playing cards, which are explained below

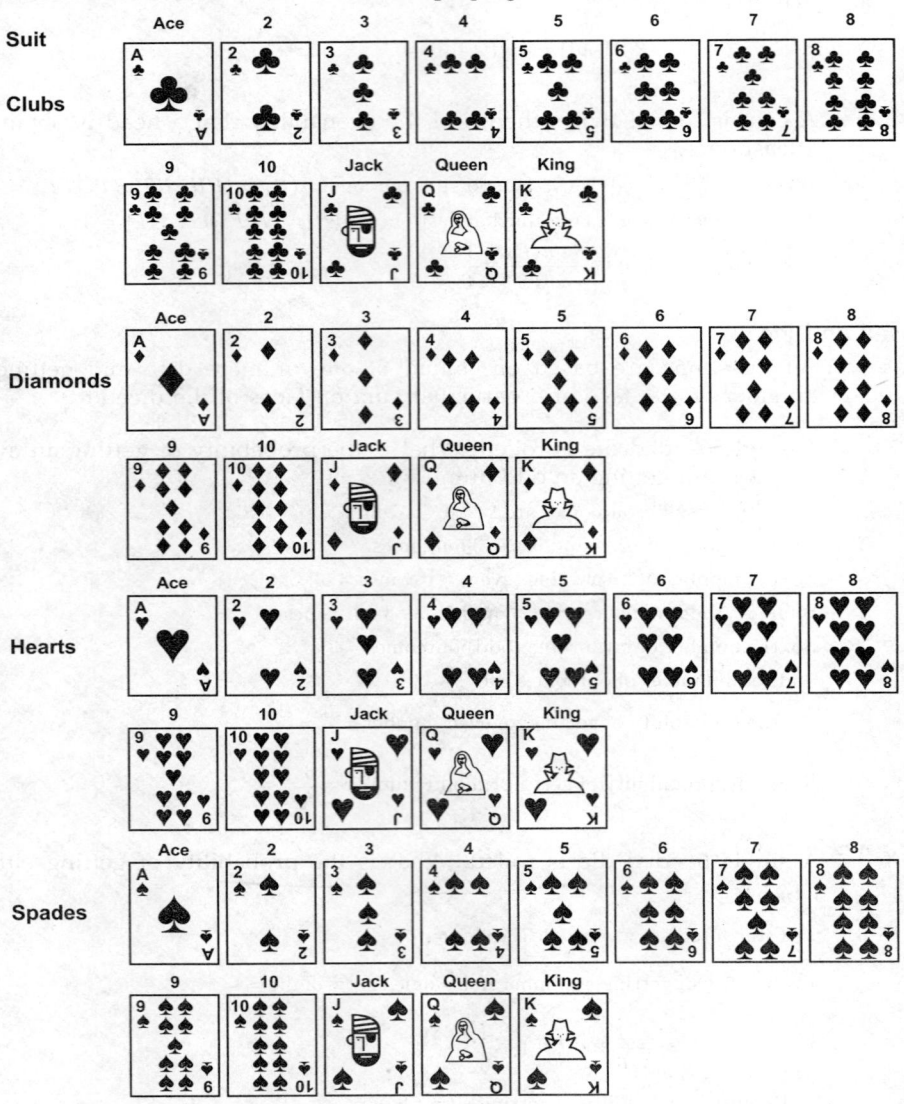

- There are 13 cards of each suit clubs, diamonds, hearts and spades.
- There are 4 aces, 4 jacks, 4 queens and 4 kings.
- There are 26 red and 26 black cards.
- There are 12 face cards; king, queen and jacks in each suit.

Ex. 14 A total of five cards are chosen at random from a standard deck of 52 playing cards. What is the probability of choosing 5 aces?

Sol. There are only 4 aces, so we cannot choose 5 aces. Thus, it is impossible event.

$\therefore$ Probability (5 aces) = $\dfrac{0}{52}$ = 0

Ex. 15 A single card is chosen at random from a standard deck of 52 playing cards. What is the probability of choosing a king or a club?

Sol. There are 4 king in a standard deck, 13 club cards and 1 king of club card.

So, probability of getting a king = $\dfrac{4}{52}$

Probability of getting a club = $\dfrac{13}{52}$

Probability of getting a king of club = $\dfrac{1}{52}$

$\therefore$ Required probability of getting a king or a club

$= \dfrac{4}{52} + \dfrac{13}{52} - \dfrac{1}{52} = \dfrac{4 + 13 - 1}{52} = \dfrac{16}{52} = \dfrac{4}{13}$

Based on Marbles or Balls

These types of questions are based on choosing a ball or a marble of particular colour from one or more bags containing different coloured balls or marbles.

Ex. 16 A glass jar contains 1 red, 3 green, 2 blue and 4 yellow marbles. If a single marble is chosen at random from the jar, what is the probability that it is yellow or green?

Sol. Total marbles = 1 + 3 + 2 + 4 = 10, i.e. $n(S) = 10$

Now, probability of getting a yellow marble = $\dfrac{4}{10}$

Probability of getting a green marble = $\dfrac{3}{10}$

Since, the events are mutually exclusive.

$\therefore \quad P$ (yellow or green) = P (yellow) + P (green) = $\dfrac{4}{10} + \dfrac{3}{10} = \dfrac{7}{10}$

Ex. 17 A person has 2 bags. He has 3 black and 4 white balls in one bag and 4 black and 3 white balls in another bag. Find the probability of getting a black ball.

Sol. 1st bag contains 3 black and 4 white balls and 2nd bag contains 4 black and 3 white balls.

Case I If 1st bag was chosen among the two bags and ball drawn is black, then

Required probability = $\dfrac{1}{2} \times \dfrac{3}{7} = \dfrac{3}{14}$

[$\because$ 1/2 represents the probability of choosing a bag out of two]

Case II If 2nd bag was chosen among the two bags and ball drawn is black, then

Required probability = $\dfrac{1}{2} \times \dfrac{4}{7} = \dfrac{4}{14}$

∴ Required probability of choosing the black ball among the two bags

$$= P(\text{1st bag}) \text{ or } P(\text{2nd bag})$$
$$= \frac{3}{14} + \frac{4}{14} = \frac{3+4}{14} = \frac{7}{14} = \frac{1}{2}$$

Miscellaneous (Choosing a Student, Hitting a Target, etc.)

Ex. 18 In a Mathematics class of 30 students, 17 are boys and 13 are girls. In a unit test, 4 boys and 5 girls made an A grade. If a student is chosen at random from the class, what is the probability of choosing a girl or an 'A grade student'?

Sol. Here, total number of boys = 17 and total number of girls = 13

Also, girls getting A grade = 5 and boys getting A grade = 4

Probability of choosing a girl = $\frac{13}{30}$

Probability of choosing a A grade student = $\frac{9}{30}$ [∵ 4 + 5 = 9]

Now, A grade student chosen can be a girl.

So, probability of choosing it = $\frac{5}{30}$

∴ Required probability of choosing a girl or an A grade student

$$= \frac{13}{30} + \frac{9}{30} - \frac{5}{30}$$
$$= \frac{22-5}{30} = \frac{17}{30}$$

Ex. 19 A person can hit a target 4 out of 7 shots. If he fires 10 shots, what is the probability that he hit the target twice?

Sol. Here, $n = 10$ and $r = 2$

Now, success $(p) = \frac{4}{7}$ and failure $(q) = 1 - \frac{4}{7} = \frac{3}{7}$

∴ P (hit the target twice) = $^{10}C_2 \left(\frac{4}{7}\right)^2 \left(\frac{3}{7}\right)^8$

$$= \frac{10!}{2!(10-2)!} \left(\frac{4}{7}\right)^2 \left(\frac{3}{7}\right)^8 = 45 \cdot \frac{(4)^2 (3)^8}{7^{10}}$$

Ex. 20 The probability that it is Friday and a student is absent, is 0.03. Since, there are 5 school days in a week, the probability that it is Friday is 0.2. What is the probability that a student is absent, given that today is Friday?

Sol. Probability of knowing that it is Friday and a student is absent = 0.03

Also, probability of that day will be a Friday = 0.2

Now, if it is given that today is Friday then it is a conditional probability.

∴ $P(\text{Absent}/\text{Friday}) = \dfrac{P \text{ (Friday and absent)}}{P \text{ (Friday)}} = \dfrac{0.03}{0.2} = 0.15$

Fast Track Practice

Exercise 1 Base Level Questions

1. The probability that a leap year selected at random contains 53 Sunday, is
   (a) $\frac{7}{366}$ (b) $\frac{26}{183}$ (c) $\frac{1}{7}$ (d) $\frac{2}{7}$
   (e) None of these

2. Let E be the set of all integers with 1 at their unit places. The probability that a number chosen from {2, 3, 4, ... , 50} is an element of E, is
   (a) $\frac{5}{49}$ (b) $\frac{4}{49}$ (c) $\frac{3}{49}$ (d) $\frac{2}{49}$
   (e) None of these

3. A bag contains 63 cards (numbered 1, 2, 3, ..., 63). Two cards are picked at random from the bag (one after another and without replacement), what is the probability that the sum of number of both cards drawn is even? [SBI PO (Pre) 2017]
   (a) $\frac{11}{21}$ (b) $\frac{34}{63}$ (c) $\frac{7}{11}$ (d) $\frac{11}{63}$
   (e) Other than those given as options

4. What is the probability of choosing three distinct numbers randomly from (1, 2, 3,..., 100) such that all are divisible by both 2 and 3? [CMAT 2015]
   (a) $\frac{4}{1125}$ (b) $\frac{5}{1126}$ (c) $\frac{3}{1125}$ (d) $\frac{4}{1155}$

5. A bag contains 20 tickets numbered from 1 to 20. Two tickets are drawn at random. What is the probability that both numbers are prime? [SBI Clerk (Mains) 2016]
   (a) $\frac{8}{20}$ (b) $\frac{14}{95}$ (c) $\frac{7}{20}$ (d) $\frac{21}{190}$
   (e) $\frac{21}{95}$

6. The probability that a man will be alive for 10 more years is $\frac{1}{4}$ and the probability that his wife will alive for 10 more years is $\frac{1}{3}$. The probability that none of them will be alive for 10 more years, is
   (a) $\frac{5}{12}$ (b) $\frac{1}{2}$ (c) $\frac{7}{12}$ (d) $\frac{11}{12}$
   (e) None of these

7. In a lottery, 10000 tickets are sold and ten prizes are awarded. What is the probability of not getting a prize, if you buy one ticket?
   (a) 9/10000 (b) 9/10
   (c) 999/1000 (d) 9999/10000

8. Two persons A and B appear in an interview for two vacancies. If the probabilities of their selections are 1/4 and 1/6 respectively, then the probability that none of them is selected, is
   (a) $\frac{5}{8}$ (b) $\frac{5}{12}$
   (c) $\frac{1}{12}$ (d) $\frac{1}{24}$
   (e) None of these

9. The probabilities of solving a problem by three students A, B and C are $\frac{1}{2}, \frac{1}{3}$ and $\frac{1}{4}$, respectively. The probability that the problem will be solved, is
   (a) $\frac{1}{4}$ (b) $\frac{1}{2}$ (c) $\frac{3}{4}$ (d) $\frac{1}{3}$
   (e) None of these

10. Two events A and B have probabilities 0.25 and 0.50, respectively. The probability that both A and B occur simultaneously is 0.12. Then, the probability that neither A nor B occurs is [MAT 2015]
    (a) 0.13 (b) 0.38 (c) 0.63 (d) 0.37

11. If three unbiased coins are tossed simultaneously, then the probability of exactly two heads is
    (a) 1/8 (b) 2/8 (c) 3/8 (d) 4/8
    (e) None of these

12. Five coins are tossed at a time. Then, the probability of obtaining atleast one tail is
    (a) $\frac{31}{32}$ (b) $\frac{1}{32}$ (c) $\frac{1}{5}$ (d) $\frac{5}{32}$
    (e) None of these

13. The probability of getting a composite number when a six-faced unbiased die is tossed, is
    (a) $\frac{1}{4}$ (b) $\frac{1}{3}$ (c) $\frac{1}{2}$ (d) 1

14. Two dice are thrown. If the total on the faces of the two dice is 6, then find the probability that there are two odd numbers on the faces. [MAT 2015]
    (a) $\frac{2}{5}$ (b) $\frac{1}{5}$ (c) $\frac{5}{9}$ (d) $\frac{3}{5}$

15. A card is drawn from a well-shuffled pack of cards. The probability of getting a queen of club or a king of heart is
    (a) $\frac{1}{52}$ (b) $\frac{1}{26}$ (c) $\frac{1}{13}$ (d) $\frac{1}{39}$
    (e) None of these

16. What is the probability that a card drawn at random from a pack of 52 cards is either a king or a spade? [Bank Clerk 2010]
    (a) $\frac{17}{52}$ (b) $\frac{4}{13}$ (c) $\frac{3}{13}$ (d) $\frac{13}{52}$
    (e) None of these

17. Two card are drawn at random from a well-shuffled pack of 52 cards. What is the probability of getting two hearts or two diamonds? [RRB Clerk (Pre) 2017]
    (a) $\frac{3}{26}$ (b) $\frac{2}{17}$ (c) $\frac{1}{26}$ (d) $\frac{4}{13}$
    (e) None of these

18. A bag contains 4 red balls, 6 green balls and 5 blue balls. If three balls are picked at random, what is the probability that two of them are green and one of them is blue? [SBI PO 2015]
    (a) $\frac{20}{91}$ (b) $\frac{10}{91}$ (c) $\frac{15}{91}$ (d) $\frac{5}{91}$
    (e) $\frac{25}{91}$

19. A bag contains 6 red, 5 green and 4 yellow coloured balls. 2 balls are drawn at random after one another without replacement, then what is the probability that atleast one ball is green. [RRB PO (Pre) 2017]
    (a) $\frac{2}{3}$ (b) $\frac{4}{5}$ (c) $\frac{3}{8}$ (d) $\frac{4}{7}$
    (e) $\frac{2}{7}$

20. There are 5 red balls, 4 yellow balls and 3 green balls in a basket. If 3 balls are drawn at random, what is the probability that atleast 2 of them are green in colour? [NICL AO 2015]
    (a) $\frac{1}{11}$ (b) $\frac{13}{55}$ (c) $\frac{3}{11}$ (d) $\frac{11}{55}$
    (e) $\frac{7}{55}$

Directions (Q. Nos. 21-22) *Study the given information carefully and answer the questions that follow.* [Bank PO 2010]
A basket contains 4 red, 5 blue and 3 green marbles.

21. If three marbles are picked at random, what is the probability that atleast one is blue?
    (a) $\frac{7}{12}$ (b) $\frac{37}{44}$ (c) $\frac{5}{12}$ (d) $\frac{7}{44}$
    (e) None of these

22. If three marbles are picked at random, what is the probability that either all are green or all are red?
    (a) $\frac{7}{44}$ (b) $\frac{7}{12}$ (c) $\frac{5}{12}$ (d) $\frac{1}{44}$
    (e) None of these

Directions (Q. Nos. 23-25) *Read the following information carefully and answer the given questions.* [NICL AO 2015]
There are 5 red balls, 6 green balls and 7 blue balls in a box.

23. If three balls are drawn randomly, what is the probability that one of them is red and the remaining two are blue?
    (a) $\frac{5}{34}$ (b) $\frac{25}{136}$ (c) $\frac{35}{272}$ (d) $\frac{45}{272}$
    (e) $\frac{15}{136}$

24. If one ball is drawn randomly, what is the probability that it is either red or blue?
    (a) $\frac{5}{6}$ (b) $\frac{5}{9}$ (c) $\frac{3}{4}$ (d) $\frac{2}{3}$
    (e) $\frac{1}{3}$

25. If two balls are drawn randomly, what is the probability that atleast one of them is green?
    (a) $\frac{31}{51}$ (b) $\frac{29}{51}$ (c) $\frac{35}{51}$ (d) $\frac{24}{51}$
    (e) $\frac{26}{51}$

Directions (Q. Nos. 26-29) *Study the given information carefully and answer the questions that follow.* [Bank PO 2010]
An urn contains 6 red, 4 blue, 2 green and 3 yellow marbles.

26. If three marbles are picked at random, what is the probability that two are blue and one is yellow?

Probability / 559

(a) $\dfrac{3}{91}$ (b) $\dfrac{1}{5}$ (c) $\dfrac{18}{455}$ (d) $\dfrac{7}{15}$
(e) None of these

27. If four marbles are picked at random, what is the probability that atleast one is blue?
(a) $\dfrac{4}{15}$ (b) $\dfrac{69}{91}$ (c) $\dfrac{11}{15}$ (d) $\dfrac{22}{91}$
(e) None of these

28. If two marbles are picked at random, what is the probability that either both are green or both are yellow?
(a) $\dfrac{5}{91}$ (b) $\dfrac{1}{35}$ (c) $\dfrac{1}{3}$ (d) $\dfrac{4}{105}$
(e) None of these

29. If four marbles are picked at random, what is the probability that one is green, two are blue and one is red?
(a) $\dfrac{24}{455}$ (b) $\dfrac{13}{35}$ (c) $\dfrac{11}{15}$ (d) $\dfrac{7}{91}$
(e) None of these

30. A box contains 24 marbles, some are green and others are blue. If a marble is drawn at random from the box, the probability that it is green, is 2/3. The number of blue marbles in the box is
(a) 13 (b) 12 (c) 16 (d) 8

31. There are two bags containing white and black balls. In the first bag, there are 8 white and 6 black balls and in the second bag, there are 4 white and 7 black balls. One ball is drawn at random from any of these two bags. Find the probability of this ball being black.
(a) $\dfrac{5}{9}$ (b) $\dfrac{7}{19}$ (c) $\dfrac{41}{77}$ (d) $\dfrac{9}{17}$
(e) None of these

32. There are 6 red balls and 4 yellow balls in a bag. Two balls are simultaneously drawn at random. What is the probability that both the balls are of same colour?   **[LIC AAO 2016]**
(a) 7/15 (b) 5/12 (c) 7/11 (d) 7/8
(e) 1/8

33. In a bag, there are 6 red balls and 9 green balls. Two balls are drawn at random. What is the probability that atleast one of the balls drawn is red? **[SBI PO (Pre) 2016]**
(a) $\dfrac{29}{35}$ (b) $\dfrac{7}{15}$ (c) $\dfrac{23}{35}$ (d) $\dfrac{17}{35}$
(e) $\dfrac{19}{35}$

34. A committee of 3 members is to be selected out of 3 men and 2 women. What is the probability that the committee has atleast one woman?   **[Bank PO 2008]**

(a) $\dfrac{1}{10}$ (b) $\dfrac{9}{20}$ (c) $\dfrac{9}{10}$ (d) $\dfrac{1}{20}$
(e) None of these

35. Out of 13 applicants for a job, there are 5 women and 8 men. It is desired to select 2 persons for the job. The probability that atleast one of the selected persons will be a woman, is
(a) $\dfrac{25}{39}$ (b) $\dfrac{14}{35}$ (c) $\dfrac{5}{13}$ (d) $\dfrac{10}{13}$
(e) None of these

36. From 4 children, 2 women and 4 men, 4 persons are selected. The probability that there are exactly 2 children among the selected persons, is
(a) $\dfrac{11}{21}$ (b) $\dfrac{9}{21}$ (c) $\dfrac{10}{21}$ (d) $\dfrac{5}{21}$
(e) None of these

Directions (Q. Nos. 37-40) *Study the following information carefully to answer the questions that follow.*   **[Bank PO 2009]**

A box contains 2 blue caps, 4 red caps, 5 green caps and 1 yellow cap.

37. If two caps are picked at random, what is the probability that both are blue?
(a) $\dfrac{1}{6}$ (b) $\dfrac{1}{10}$ (c) $\dfrac{1}{12}$ (d) $\dfrac{1}{45}$
(e) None of these

38. If four caps are picked at random, what is the probability that none is green?
(a) $\dfrac{7}{99}$ (b) $\dfrac{5}{99}$ (c) $\dfrac{7}{12}$ (d) $\dfrac{5}{12}$
(e) None of these

39. If three caps are picked at random, what is the probability that two are red and one is green?
(a) $\dfrac{9}{22}$ (b) $\dfrac{6}{19}$ (c) $\dfrac{1}{6}$ (d) $\dfrac{3}{22}$
(e) None of these

40. If one cap is picked at random, what is the probability that it is either blue or yellow?
(a) $\dfrac{2}{9}$ (b) $\dfrac{1}{4}$ (c) $\dfrac{3}{8}$ (d) $\dfrac{6}{11}$
(e) None of these

41. An elevator starts with 5 passengers and stops at 8 different floors of the house. Find out the probability of all the 5 passengers alighting at different floors.
(a) $\dfrac{101}{512}$ (b) $\dfrac{105}{512}$ (c) $\dfrac{107}{512}$ (d) $\dfrac{109}{512}$
(e) None of these

Exercise 2 Higher Skill Level Questions

1. The probability that a man can hit a target is 3/4. He tries 5 times. The probability that he will hit the target atleast three times, is
   (a) $\dfrac{291}{364}$ (b) $\dfrac{371}{464}$ (c) $\dfrac{471}{502}$ (d) $\dfrac{459}{512}$
   (e) None of these

2. In a ward-robe, Nitish has 3 trousers. One of them is black, second is blue and third is brown. In this ward-robe, he has 4 shirts also. One of them is black and the other 3 are white. He opens his ward-robe in the dark and picks out one shirt-trouser pair without examining the colour. What is the likelihood that neither the shirts nor the trousers are black?
   (a) $\dfrac{1}{12}$ (b) $\dfrac{1}{6}$ (c) $\dfrac{1}{4}$ (d) $\dfrac{1}{2}$
   (e) None of these

3. Ramesh plans to order a birthday gift for his friend from an online retailer. However, the birthday coincides with the festival season during which there is a huge demand for buying online goods and hence deliveries are often delayed. He estimates that the probabilities of receiving the gift, on time, from the retailers A, B, C and D would be 0.6, 0.8, 0.9 and 0.5, respectively. Playing safe, he orders from all four retailers simultaneously. What would be the probability that his friend would receive the gift on time? **[XAT 2015]**
   (a) 0.004 (b) 0.006
   (c) 0.216 (d) 0.994
   (e) 0.996

4. Four boys and three girls stand in a queue for an interview. The probability that they stand in alternate position, is
   (a) $\dfrac{1}{34}$ (b) $\dfrac{1}{35}$ (c) $\dfrac{1}{17}$ (d) $\dfrac{1}{68}$
   (e) None of these

5. If a 4-digit number is formed at random using the digits 1, 3, 5, 7 and 9 without repetition, then the probability that it is divisible by 5, is
   (a) $\dfrac{4}{5}$ (b) $\dfrac{3}{5}$
   (c) $\dfrac{1}{5}$ (d) $\dfrac{2}{3}$
   (e) None of these

6. A die is rolled three times and sum of three numbers appearing on the uppermost face is 15. The chance that the first roll was four, is
   (a) 2/5 (b) 1/5
   (c) 1/6 (d) None of these

7. Murari has 9 pairs of dark blue socks and 9 pairs of black socks. He keeps them all in the same bag. If he picks out three socks at random, what is the probability that he will get a matching pair?
   (a) $\dfrac{{}^9C_3 \times {}^9C_1}{{}^{18}C_3}$ (b) $\dfrac{2 \times {}^9C_2 \times {}^9C_1}{{}^{18}C_3}$
   (c) 1 (d) $\dfrac{4}{7}$
   (e) None of these

8. A speaks the truth 3 out of 4 times and B speaks 5 out of 6 times. What is the probability that they will contradict each other in stating the same fact?
   (a) $\dfrac{2}{3}$ (b) $\dfrac{1}{3}$
   (c) $\dfrac{5}{6}$ (d) $\dfrac{1}{2}$

9. A player can take a maximum of 4 chances to hit a bottle with a flying disc. The probability of hitting the bottle at the first, second, third and fourth shots are 0.1, 0.2, 0.35 and 0.45, respectively. What is the probability that the player hits the bottle with the flying disc? **[SNAP 2014]**
   (a) 0.6573 (b) 0.2574
   (c) 0.7426 (d) None of these

Answer with Solutions
Exercise 1 Base Level Questions

1. **(d)** In a leap year, there are 366 days. It means 52 full weeks + 2 odd days. These two days can be (Mon, Tues), (Tues, Wed), (Wed, Thurs), (Thurs, Fri), (Fri, Sat), (Sat, Sun) or (Sun, Mon).
   ∴ Required probability = $\dfrac{2}{7}$

2. **(b)** Here, $n(S) = 49$
   Favourable numbers are 11, 21, 31, 41.
   ∴ Required probability = $\dfrac{4}{49}$

3. **(e)** Total number of cards in bag
   $= 63\{1, 2, 3, \ldots, 63\}$
   Total number of even number of cards
   $= 31\{2, 4, \ldots, 62\}$
   ∴ Required probability = $\dfrac{^{31}C_2}{^{63}C_2}$
   $= \dfrac{\dfrac{31!}{2!\,29!}}{\dfrac{63!}{2!\,61!}} = \dfrac{31 \times 30}{1 \times 2} \cdot \dfrac{1 \times 2}{63 \times 62}$
   $= \dfrac{31 \times 30}{63 \times 62} = \dfrac{30}{63 \times 2} = \dfrac{15}{63} = \dfrac{5}{21}$

4. **(d)** Number of numbers divisible by both 2 and 3 between 1 to 100 = 16
   (i.e. 6, 12, 18, 24, 30, 36, 42, 48, 54, 60, 66, 72, 78, 84, 90, 96)
   Hence, required probability = $\dfrac{^{16}C_3}{^{100}C_3}$
   $= \dfrac{16!}{3!\,13!} \times \dfrac{3! \times 97!}{100!}$
   $= \dfrac{16 \cdot 15 \cdot 14}{3!} \times \dfrac{3!}{100 \cdot 99 \cdot 98} = \dfrac{4}{1155}$

5. **(b)** Prime numbers from 1 to 20
   $= 2, 3, 5, 7, 11, 13, 17, 19$
   ∴ Required probability
   $= \dfrac{^8C_2}{^{20}C_2} = \dfrac{8 \times 7}{2 \times 1} \times \dfrac{2 \times 1}{20 \times 19} = \dfrac{14}{95}$

6. **(b)** Required probability = $P(\overline{A}) \times P(\overline{B})$
   $= \left(1 - \dfrac{1}{4}\right) \times \left(1 - \dfrac{1}{3}\right) = \dfrac{3}{4} \times \dfrac{2}{3} = \dfrac{1}{2}$

7. **(c)** Total lottery tickets = 10000
   Total prizes in the lottery = 10
   ∴ Probability of getting a prize
   $= \dfrac{10}{10000} = \dfrac{1}{1000}$

   Now, probability of not getting a prize
   $= 1 -$ Probability of getting a prize
   $= 1 - \dfrac{1}{1000} = \dfrac{999}{1000}$

8. **(a)** Required probability = $P(\overline{A}) \times P(\overline{B})$
   $= \left(1 - \dfrac{1}{4}\right)\left(1 - \dfrac{1}{6}\right) = \dfrac{3}{4} \times \dfrac{5}{6} = \dfrac{5}{8}$

9. **(c)** First, we find the probability of not solving the problem.
   $P(\overline{A}) \times P(\overline{B}) \times P(\overline{C})$
   $= \left(1 - \dfrac{1}{2}\right) \times \left(1 - \dfrac{1}{3}\right) \times \left(1 - \dfrac{1}{4}\right)$
   $= \dfrac{1}{2} \times \dfrac{2}{3} \times \dfrac{3}{4} = \dfrac{1}{4}$
   ∴ Required probability = $1 - \dfrac{1}{4} = \dfrac{3}{4}$

10. **(d)** Given, $P(A) = 0.25, P(B) = 0.50$
    and $P(A \cap B) = 0.12$
    Now, $P(A \cup B) = P(A) + P(B) - P(A \cap B)$
    $= 0.25 + 0.50 - 0.12$
    $= 0.63$
    ∴ Probability that neither A nor B occurs,
    $P(\overline{A} \cap \overline{B}) = 1 - P(A \cup B)$
    $= 1 - 0.63 = 0.37$

11. **(c)** $n(S) = \{(HHH), (HTH), (THH), (HHT), (THT), (TTH), (HTT), (TTT)\} = 2^3 = 8$
    Let E = Event of getting exactly two heads
    $= \{(H, H, T), (H, T, H), (T, H, H)\}$
    $\Rightarrow n(E) = 3$
    ∴ Required probability = $\dfrac{3}{8}$

12. **(a)** Total events = $n(S) = 2^5 = 32$
    $n(\overline{E})$ of getting heads on all the three coins
    $= 1$ and $P(\overline{E}) = \dfrac{1}{32}$
    ∴ $P(E) = 1 - P(\overline{E}) = 1 - \dfrac{1}{32} = \dfrac{31}{32}$

13. **(b)** Here, $n(S) = 6, n(E) = (4, 6) = 2$
    ∴ $P(E) = \dfrac{2}{6} = \dfrac{1}{3}$

14. **(d)** Total possible pairs of numbers those sums to 6 = 5
    i.e. (1, 5), (3, 3), (2, 4), (4, 2), (5, 1)
    Number of pairs having two odd numbers on the faces = 3, i.e. (1, 5), (5, 1), (3, 3)
    Hence, required probability = $\dfrac{3}{5}$

15. (b) Total number of ways = 52

There is one queen of club and one king of heart.

∴ Favourable number of ways = 1 + 1 = 2

∴ Required probability = $\dfrac{2}{52} = \dfrac{1}{26}$

16. (b) Probability of getting a king, $P(A) = \dfrac{4}{52}$

Probability of getting a spade, $P(B) = \dfrac{13}{52}$

Probability of getting a king of spade,
$P(A \cap B) = \dfrac{1}{52}$

∴ Probability of either a king or a spade,
$P(A \cup B) = \dfrac{4}{52} + \dfrac{13}{52} - \dfrac{1}{52} = \dfrac{16}{52} = \dfrac{4}{13}$

17. (b) ∴ Required probability
$= \dfrac{^{13}C_2 + {}^{13}C_2}{^{52}C_2} = \dfrac{78 + 78}{1326} = \dfrac{156}{1326} = \dfrac{2}{17}$

18. (c) Total number of balls in a bag
= 4 + 6 + 5 = 15

Total number of ways of drawing 3 balls
$= {}^{15}C_3$

Probability that two of them are green and one of them is blue

$= \dfrac{^{6}C_2 \times {}^{5}C_1}{^{15}C_3} = \dfrac{\dfrac{6!}{2! \times 4!} \times \dfrac{5!}{1! \times 4!}}{\dfrac{15!}{3! \times 12!}} = \dfrac{15}{91}$

19. (d) Probability that no ball is green
$= \dfrac{^{10}C_1 \times {}^{9}C_1}{15 \times 14} = \dfrac{90}{15 \times 14} = \dfrac{3}{7}$

∴ Required probability $= 1 - \dfrac{3}{7} = \dfrac{4}{7}$

20. (e) Here, total balls in the basket
= 5 red + 4 yellow + 3 green = 12 balls

Now, 3 balls chosen at random in which atleast two balls are green in colour.

Selecting 3 balls $= {}^{9}C_1 \times {}^{3}C_2 + {}^{3}C_3$

Sample space $= {}^{12}C_3$

∴ P (getting atleast 2 green balls)
$= \dfrac{^{9}C_1 \times {}^{3}C_2 + {}^{3}C_3}{^{12}C_3} = \dfrac{9 \times 3 + 1}{220} = \dfrac{28}{220} = \dfrac{7}{55}$

21. (b) Total number of possible outcomes
$= {}^{12}C_3 = 220$

Number of events which do not contain blue marbles (3 marbles out of 7 marbles)
$= {}^{7}C_3 = 35$

∴ Required probability $= 1 - \dfrac{35}{220} = \dfrac{37}{44}$

22. (d) Total number of ways of selection of 3 marbles out of 12 = $n(S) = {}^{12}C_3 = 220$

Total number of favourable events
$= n(E) = {}^{3}C_3 + {}^{4}C_3 = 1 + 4 = 5$

∴ Required probability $= \dfrac{5}{220} = \dfrac{1}{44}$

23. (c) Total number of balls in a box
= 5 red + 6 green + 7 blue = 18

Required probability
$= \dfrac{^{5}C_1 \times {}^{7}C_2}{^{18}C_3} = \dfrac{5 \times 21}{816} = \dfrac{35}{272}$

24. (d) Required probability
$= \dfrac{^{5}C_1 + {}^{7}C_1}{^{18}C_1} = \dfrac{5+7}{18} = \dfrac{12}{18} = \dfrac{2}{3}$

25. (b) Required probability $= \dfrac{^{6}C_1 \times {}^{12}C_1 + {}^{6}C_2}{^{18}C_2}$

$= \dfrac{6 \times 12 + 15}{\dfrac{18 \times 17}{2}} = \dfrac{72 + 15}{153} = \dfrac{87}{153} = \dfrac{29}{51}$

26. (c) Ways of selection of two blue marbles
$= {}^{4}C_2$

Ways of selection of one yellow marble = ${}^{3}C_1$

Ways of selection of three marbles = ${}^{15}C_3$

∴ Required probability $= \dfrac{^{4}C_2 \times {}^{3}C_1}{^{15}C_3}$

$= \dfrac{\dfrac{4!}{2!(4-2)!} \times \dfrac{3!}{1!(3-1)!}}{\dfrac{15!}{3!(15-3)!}} = \dfrac{18}{455}$

27. (b) Probability that none is blue from four marbles $= \dfrac{^{15-4}C_4}{^{15}C_4} = \dfrac{^{11}C_4}{^{15}C_4} = \dfrac{22}{91}$

So, probability that atleast one is blue from four marbles $= 1 - \dfrac{22}{91} = \dfrac{69}{91}$

28. (d) Ways of selection of two green marbles
$= {}^{2}C_2 = 1$

Ways of selection of two yellow marbles
$= {}^{3}C_2 = 3$

So, probability (both are green) $= \dfrac{1}{^{15}C_2}$ ...(i)

Probability (both are yellow) $= \dfrac{3}{^{15}C_2}$ ...(ii)

Then, required probability
$= \dfrac{1}{^{15}C_2} + \dfrac{3}{^{15}C_2} = \dfrac{4}{^{15}C_2} = \dfrac{4}{105}$

29. (a) Ways of selection of 4 marbles,
$n(S) = {}^{15}C_4$

Probability / 563

Ways of selection of one green marble,
$n(E_1) = {}^2C_1$
Ways of selection of two blue marbles,
$n(E_2) = {}^4C_2$
Ways of selection of one red marble,
$n(E_3) = {}^6C_1$
∴ Required probability
$$= \frac{{}^2C_1 \times {}^4C_2 \times {}^6C_1}{{}^{15}C_4} = \frac{24}{455}$$

30. (d) Let the number of green marbles = x
Then, probability of getting a green marble,
$$\frac{{}^xC_1}{{}^{24}C_1} = \frac{2}{3} \Rightarrow \frac{x}{24} = \frac{2}{3} \therefore x = 16$$
So, number of blue marbles = 24 − 16 = 8

31. (c) Probability of selecting first bag out of two bags and getting one black ball from first bag = $\frac{1}{2} \times \frac{{}^6C_1}{{}^{14}C_1} = \frac{3}{14}$

Probability of selecting second bag and getting one black ball from it
$$= \frac{1}{2} \times \frac{{}^7C_1}{{}^{11}C_1} = \frac{7}{22}$$

Hence, required probability = $\frac{3}{14} + \frac{7}{22} = \frac{41}{77}$

32. (a) Number of ways of drawing two balls from a collection of 10 balls
$$= {}^{10}C_2 = \frac{10 \times 9}{1 \times 2} = 45$$
∴ Required probability
$$= \frac{{}^6C_2 + {}^4C_2}{45} = \frac{\frac{6 \times 5}{1 \times 2} + \frac{4 \times 3}{1 \times 2}}{45}$$
$$= \frac{15 + 6}{45} = \frac{21}{45} = \frac{7}{15}$$

33. (c) Required probability
$$= \frac{{}^6C_2 + {}^9C_1 \times {}^6C_1}{{}^{15}C_2} = \frac{\frac{6 \times 5}{1 \times 2} + \frac{9 \times 6}{1}}{\frac{15 \times 14}{1 \times 2}}$$
$$= \frac{15 + 54}{105} = \frac{69}{105} = \frac{23}{35}$$

34. (c) Required probability
$$= \frac{{}^2C_1 \times {}^3C_2 + {}^2C_2 \times {}^3C_1}{{}^5C_3} = \frac{9}{10}$$

35. (a) Total number of ways = ${}^{13}C_2$
Favourable number of ways of selecting men only = ${}^8C_2$
∴ Probability of selecting no woman
$$= \frac{{}^8C_2}{{}^{13}C_2} = \frac{14}{39}$$

∴ Probability of selecting atleast one woman
$$= 1 - \frac{14}{39} = \frac{25}{39}$$

36. (b) Total number of cases = ${}^{10}C_4$
Favourable number of cases = ${}^4C_2 \cdot {}^6C_2$
[since, we have to select 2 children out of 4 and remaining 2 persons are to be selected from remaining 6 persons (2W + 4M)]
∴ Required probability
$$= \frac{{}^4C_2 \cdot {}^6C_2}{{}^{10}C_4} = \frac{\frac{4 \times 3}{2 \times 1} \times \frac{6 \times 5}{2 \times 1}}{\frac{10 \times 9 \times 8 \times 7}{4 \times 3 \times 2 \times 1}} = \frac{90}{210} = \frac{9}{21}$$

37. (e) Total number of caps
= 2 + 4 + 5 + 1 = 12
Total number of outcomes,
$n(S) = {}^{12}C_2 = 66$
Favourable number of outcomes,
$n(E) = {}^2C_2 = 1$
∴ Required probability = $\frac{1}{66}$

38. (a) Total number of caps = 12
Total number of results, $n(S) = {}^{12}C_4 = 495$
Out of 5 caps, number of ways not to pick a green cap, $n(E_1) = {}^5C_0 = 1$
and out of 7 caps, number of ways to pick 4 caps,
$n(E_2) = {}^7C_4 = 35$
∴ Required probability = $\frac{1 \times 35}{495} = \frac{7}{99}$

39. (d) Total number of caps = 12
∴ $n(S) = {}^{12}C_3 = 220$
Out of 4 red caps, number of ways to pick 2 caps, $n(E_1) = {}^4C_2 = 6$
and out of 5 green caps, number of ways to pick one green cap,
$n(E_2) = {}^5C_1 = 5$
∴ $P(E) = \frac{n(E_1) \times n(E_2)}{n(S)} = \frac{6 \times 5}{220} = \frac{3}{22}$

40. (b) Total number of caps = 12
∴ $n(S) = {}^{12}C_1 = 12$
Out of (2 blue + 1 yellow) caps, number of ways to pick one cap, $n(E) = {}^3C_1 = 3$
∴ Required probability,
$$P(E) = \frac{n(E)}{n(S)} = \frac{3}{12} = \frac{1}{4}$$

41. (b) Required probability
$$= \frac{\text{Favourable number of cases}}{\text{Total number of cases}} = \frac{{}^8P_5}{8^5} = \frac{105}{512}$$

Exercise 2 Higher Skill Level Questions

1. **(d)** Given, $n = 5$ and $r = 3$
   Then, success, $p = \dfrac{3}{4}$
   and failure, $q = 1 - \dfrac{3}{4} = \dfrac{1}{4}$
   $\therefore$ Required probability that man hit the target thrice
   $= {}^5C_3\left(\dfrac{3}{4}\right)^3\left(\dfrac{1}{4}\right)^2 + {}^5C_4\left(\dfrac{3}{4}\right)^4\left(\dfrac{1}{4}\right) + {}^5C_5\left(\dfrac{3}{4}\right)^5$
   $= \dfrac{270}{1024} + \dfrac{405}{1024} + \dfrac{243}{1024}$
   $= \dfrac{918}{1024} = \dfrac{459}{512}$

2. **(d)** Probability that trousers are not black
   $= \dfrac{2}{3}$
   Probability that shirts are not black $= \dfrac{3}{4}$
   $\therefore$ Required probability $= \dfrac{2}{3} \times \dfrac{3}{4} = \dfrac{1}{2}$

3. **(e)** The probabilities of not receiving the gift on time from retailers A, B, C and D are 0.4, 0.2, 0.1 and 0.5, respectively.
   $\therefore$ Probability of receiving gift on time from atleast one retailer
   $= \{1 - (0.4 \times 0.2 \times 0.1 \times 0.5)\} = 0.996$

4. **(b)** Total number of possible arrangements for 4 boys and 3 girls in a queue $= 7!$
   When they occupy alternate position, then the arrangement would be like BGBGBGB. Thus, total number of possible arrangements $= 4! \times 3!$
   $\therefore$ Required probability $= \dfrac{4! \times 3!}{7!}$
   $= \dfrac{4 \times 3 \times 2 \times 3 \times 2}{7 \times 6 \times 5 \times 4 \times 3 \times 2} = \dfrac{1}{35}$

5. **(c)** Number of 4-digit numbers which are formed with 1, 3, 5, 7, 9,
   $n(S) = {}^5P_4 = 5 \times 4 \times 3 \times 2 = 120$
   Number of 4-digit numbers which are formed with 1, 3, 5, 7, 9 and are divisible by 5,
   $n(E) = {}^4P_3 = 4 \times 3 \times 2 = 24$
   $\therefore P(E) = \dfrac{n(E)}{n(S)} = \dfrac{24}{120} = \dfrac{1}{5}$

6. **(d)** Total number of favourable outcomes, $n(S) = 6^3 = 216$
   Combinations of outcomes for getting sum of 15 on uppermost face $= \{(4, 5, 6), (5, 4, 6), (6, 5, 4), (5, 6, 4), (4, 6, 5), (6, 4, 5), (5, 5, 5), (6, 6, 3), (6, 3, 6), (3, 6, 6)\}$
   Now, outcomes on which first roll was a four, $n(E) = (4, 5, 6), (4, 6, 5)$
   $\therefore P(E) = \dfrac{n(E)}{n(S)} = \dfrac{2}{216} = \dfrac{1}{108}$

7. **(c)** Since, there are only two types of socks in the bag. So, if Murari picks up 3 socks, then certainly two of them are of same type. Thus, this is a certain event. Hence, required probability $= 1$

8. **(b)** State of contradiction will appear in two ways as follows
   (i) A speaks truth and B does not speak truth $= \dfrac{3}{4} \times \left(1 - \dfrac{5}{6}\right)$
   $= \dfrac{3}{4} \times \dfrac{1}{6} = \dfrac{3}{24}$
   (ii) A does not speak truth and B speaks truth $= \left(1 - \dfrac{3}{4}\right) \times \left(\dfrac{5}{6}\right)$
   $= \dfrac{1}{4} \times \dfrac{5}{6} = \dfrac{5}{24}$
   Both cases in combination will give resultant probability that both are contradicting each other.
   $\therefore$ Required probability
   $= \dfrac{3}{24} + \dfrac{5}{24} = \dfrac{8}{24} = \dfrac{1}{3}$

9. **(c)** Probability of not hitting the bottle at first shot $= 0.9$
   Probability of not hitting at the bottle at second shot $= 0.8$
   Probability of not hitting at the bottle at third shot $= 0.65$
   Probability of not hitting at the bottle at fourth shot $= 0.55$
   Then, probability of player hits the bottle with the flying disc
   $= (1 - 0.9 \times 0.8 \times 0.65 \times 0.55)$
   $= (1 - 0.2574) = 0.7426$

Chapter 34

Area and Perimeter

Area and perimeter are measuring parameters related to various two-dimensional figures like triangle, rectangle, square etc.

Area

Total space enclosed by the boundary of a plane figure is called the area of that particular figure.

In another words, the area of a figure is a measure associated with the part of plane enclosed in the figure. Area is measured in square unit, like square metre, square centimetre etc.

For example If length (l) of a rectangle is 5 cm and breadth (b) is 2 cm, then

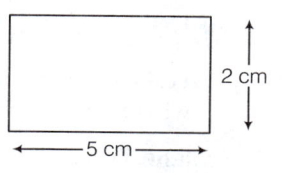

Area of rectangle = $l \times b = 5 \times 2 = 10$ sq cm

Perimeter

Perimeter is the length of border around any enclosed plane. Therefore, sum of the sides of a plane figure is the perimeter of that particular figure.

Unit of perimeter is same as the unit of sides of a given figure, like metre, centimetre etc.

For example If the sides of a triangle are 2 cm, 8 cm and 4 cm, respectively, then

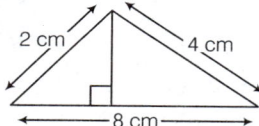

Perimeter of the triangle = 2 cm + 8 cm + 4 cm = 14 cm

Triangle

A figure enclosed by three sides is known as a triangle. A triangle has three angles with their sum equal to 180°. Adjoining figure represents a triangle with sides AB, BC, CA. ∠A, ∠B, ∠C are the three angles of the triangle.

Types of Triangles

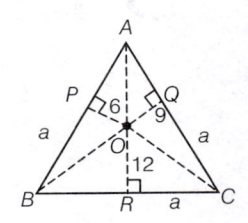

Various types of triangles are discussed below

1. Equilateral Triangle

A triangle whose all three sides are equal and each angle is equal to 60°, is called an equilateral triangle.

(i) Area $= \dfrac{\sqrt{3}}{4} a^2 \approx 0.433 a^2$ (ii) Height $= \dfrac{\sqrt{3}}{2} a \approx 0.866 a$

(iii) Perimeter $= 3a$, where a = Side of the triangle

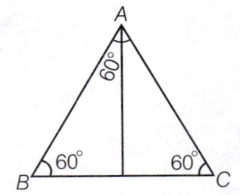

2. Isosceles Triangle

A triangle whose two sides and two angles are equal and altitude drawn on non-equal side bisect it, is called an isosceles triangle.

(i) Area $= \dfrac{b}{4} \sqrt{4a^2 - b^2}$

(ii) Height $= \sqrt{a^2 - \left(\dfrac{b}{2}\right)^2} = \dfrac{1}{2}\sqrt{4a^2 - b^2}$

(iii) Perimeter $= a + a + b = 2a + b$

where, a = Each of two equal sides; b = Third side

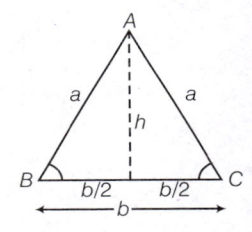

3. Scalene Triangle

A triangle whose all three sides are unequal, is called a scalene triangle.

(i) Area $= \sqrt{s(s-a)(s-b)(s-c)}$ [Heron's formula]

where, $s = \dfrac{a+b+c}{2}$ and a, b and c are the sides of the triangle and s is the semi-perimeter.

(ii) Perimeter $= a + b + c$

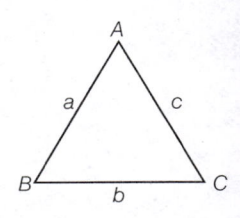

4. Right Angled Triangle

A triangle having one of its angles equal to 90°, is called a right angled triangle.

(i) Area $= \dfrac{1}{2} \times$ Base $\times$ Height $= \dfrac{1}{2} \times b \times p$

(ii) Perimeter $= p + b + h$

(iii) $h^2 = p^2 + b^2 \Rightarrow h = \sqrt{p^2 + b^2}$

where, p = Perpendicular, b = Base and h = Hypotenuse

✦ This is known as Pythagoras theorem.

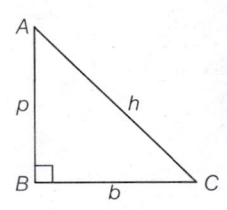

5. Isosceles Right Angled Triangle

It is a triangle with one angle equal to 90° and two sides containing the right angle are equal.

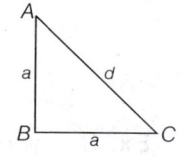

(i) Area = $\dfrac{1}{2} \times a^2$

(ii) Perimeter = $a + a + d = 2a + d$

where, a = Each of the equal sides; d = Hypotenuse.

Properties of Triangles

(i) Sum of any two sides of a triangle is greater than the third side.
(ii) Side opposite to the greatest angle will be the greatest and side opposite to the smallest angle will be the smallest.
(iii) Among all the triangles that can be formed with a given perimeter, the equilateral triangle will have the maximum area. 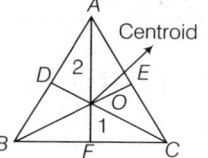
(iv) The lines joining the mid-points of sides of a triangle to the opposite vertex are called **medians**. In the adjoining figure, AF, BE and CD are medians.
(v) The median of a triangle divides it into two triangles of equal areas.
(vi) Radius of an incircle of an equilateral triangle of side a is $\dfrac{a}{2\sqrt{3}}$ and area is $\dfrac{\pi a^2}{12}$.
(vii) Radius of circumcircle of an equilateral triangle of side a will be $\dfrac{a}{\sqrt{3}}$ and area is $\dfrac{\pi a^2}{3}$.

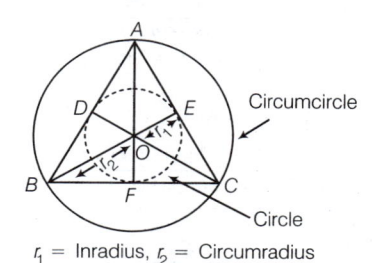

r_1 = Inradius, r_2 = Circumradius

(viii) The area of the triangle formed by joining the mid-points of the sides of a given triangle is 1/4th of the area of the given triangle.

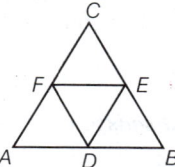

Ex. 1 Find the perimeter of a triangle with sides equal to 3 cm, 8 cm and 5 cm.
Sol. Required perimeter = Sum of the sides = $(3 + 8 + 5)$ cm = 16 cm

Ex. 2 The perimeter of an equilateral triangle is 45 cm. Find its area.
Sol. Given that, perimeter of an equilateral triangle is 45 cm.
Let each side of triangle be a cm.

Then, sum of sides = 45 cm $\Rightarrow 3a = 45$

$\therefore \quad a = \dfrac{45}{3} = 15$ cm

$\therefore \quad$ Area $= \dfrac{\sqrt{3}}{4} a^2 = \dfrac{\sqrt{3}}{4} \times 15^2 = \dfrac{225\sqrt{3}}{4}$ sq cm

Ex. 3 The perimeter of an isosceles triangle is 32 cm while its equal sides together measure 18 cm. Find the third side and each of the equal sides.

Sol. Let the third side be x.
According to the question,
Perimeter = Sum of all three sides $= x + 2a$
$\Rightarrow \quad x + 2a = 32 \Rightarrow x + 18 = 32$ $\qquad [\because 2a = 18,\text{ given}]$
$\therefore \quad x = 32 - 18 = 14$ cm
So, the third side = 14 cm and each equal side $= \dfrac{18}{2} = 9$ cm

Ex. 4 The area of a right angled triangle is 42 sq cm. If its perpendicular is equal to 10 cm, find its base.

Sol. Given that, area = 42 sq cm, perpendicular (p) = 10 cm, base (b) = ?
According to the question,
$\qquad$ Area $= \dfrac{1}{2} \times$ Base $\times$ Height (Perpendicular) $= \dfrac{1}{2} \times b \times p$
$\Rightarrow \quad 42 = \dfrac{1}{2} \times b \times 10 \Rightarrow b = \dfrac{2 \times 42}{10} = \dfrac{84}{10} = 8.4$ cm

Ex. 5 Find the area of a triangle whose sides are 26 cm, 28 cm and 30 cm.

Sol. Given that, $a = 26$ cm, $b = 28$ cm, $c = 30$ cm
From Heron's formula, area $= \sqrt{s(s-a)(s-b)(s-c)}$
Now, $\qquad s = \dfrac{a+b+c}{2} = \dfrac{26+28+30}{2} = \dfrac{84}{2} = 42$
and $\qquad (s - a) = 42 - 26 = 16$ cm,
$\qquad (s - b) = 42 - 28 = 14$ cm and $(s - c) = 42 - 30 = 12$ cm
$\therefore$ Required area $= \sqrt{s(s-a)(s-b)(s-c)} = \sqrt{42 \times 16 \times 14 \times 12}$
$\qquad = \sqrt{14 \times 3 \times 16 \times 14 \times 4 \times 3} = 14 \times 4 \times 2 \times 3 = 336$ sq cm

Quadrilateral

A figure enclosed by four sides is called a quadrilateral. A quadrilateral has four angles and sum of these angles is equal to 360°.

Types of Quadrilaterals

Various types of quadrilaterals are discussed below

1. Parallelogram
A quadrilateral, in which opposite sides are parallel, is called a parallelogram.
(i) Area = Base × Height $= b \times h$
(ii) Perimeter $= 2(a + b)$

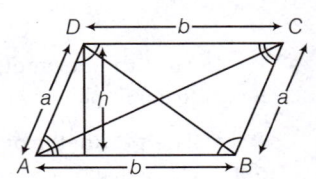

Properties of Parallelogram
(i) Diagonals of a parallelogram bisect each other.
(ii) Each diagonal of a parallelogram divides it into two triangles of equal area.
(iii) A parallelogram and a rectangle have equal areas, if they are on the same base and between the same parallel lines.
(iv) The opposite angles of parallelogram are equal, i.e. $\angle A = \angle C$ and $\angle B = \angle D$. (from the above figure)

2. Square
It is a parallelogram with all 4 sides equal and each angle is equal to 90°.

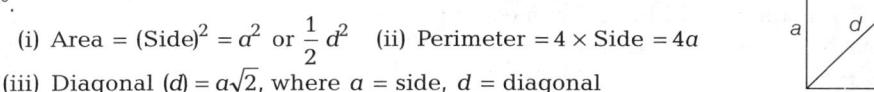

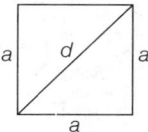

(i) Area = $(Side)^2 = a^2$ or $\frac{1}{2}d^2$ (ii) Perimeter = $4 \times$ Side = $4a$

(iii) Diagonal $(d) = a\sqrt{2}$, where a = side, d = diagonal

Properties of Square
(i) Diagonals of a square are equal and bisect each other at right angle (90°).
(ii) Diagonal is the diameter of the circumscribing circle that circumscribes the square and circumradius = $\frac{a}{\sqrt{2}}$.

3. Rectangle
It is a parallelogram with opposite sides equal and each angle is equal to 90°. The diagonals of a rectangle are of equal lengths and they bisect each other.

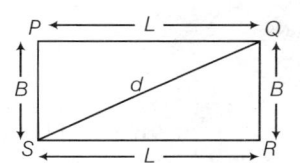

(i) Area = Length × Breadth = $L \times B$
(ii) Perimeter = $2(L + B)$
(iii) Diagonal $(d) = \sqrt{L^2 + B^2}$

✦ All rectangles are parallelograms but converse is not true.

4. Trapezium
It is a quadrilateral with one pair of opposite sides parallel.

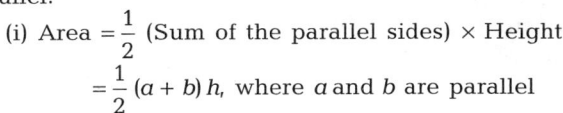

(i) Area = $\frac{1}{2}$ (Sum of the parallel sides) × Height

= $\frac{1}{2}(a + b)h$, where a and b are parallel sides and h is the height or perpendicular distance between a and b.

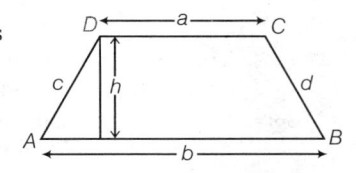

(ii) Perimeter = $AB + BC + CD + AD$

(iii) Area of trapezium, when the lengths of parallel and non-parallel sides are given = $\frac{a+b}{k}\sqrt{s(s-k)(s-c)(s-d)}$, where $k = b - a$ and $s = \frac{k+c+d}{2}$

(iv) Perpendicular distance 'h' between the two parallel sides
= $\frac{2}{k}\sqrt{s(s-k)(s-c)(s-d)}$

5. Rhombus

It is a parallelogram with all 4 sides equal. The opposite angles in a rhombus are equal but they are not right angle. A rhombus has unequal diagonal and they bisect each other at right angle (90°).

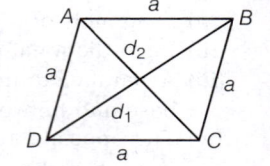

(i) Area $= \frac{1}{2} \times d_1 \times d_2$ (ii) Perimeter $= 4a$

(iii) Side $(a) = \frac{1}{2}\sqrt{d_1^2 + d_2^2}$ (iv) $4a^2 = d_1^2 + d_2^2$

where, a = side, d_1 and d_2 are diagonals.

Ex. 6 A square field has its area equal to 289 sq m. Find its side and perimeter.

Sol. Given, area = 289 sq m $\Rightarrow a^2 = 289$ [$\because a$ = side of square field]
$\therefore \quad a = \sqrt{289} = 17$ cm
$\therefore$ Perimeter $= 4a = 4 \times 17 = 68$ m

Ex. 7 The area of a rectangular field is 400 sq m. If the breadth of the field is 16 m, find the length of the field.

Sol. Given, $B = 16$ m, $L = ?$, area = 400 sq m
According to the question, $L \times B = 400$ [$\because L$ = length, B = breadth]
$\Rightarrow \quad L \times 16 = 400$
$\therefore \quad L = \frac{400}{16} = 25$ m

Ex. 8 The length and breadth of a rectangle are 6 cm and 4 cm, respectively. What will be its diagonal?

Sol. Given that, $L = 6$ cm, $B = 4$ cm, $d = ?$
According to the formula,
$d = \sqrt{L^2 + B^2} = \sqrt{6^2 + 4^2} = \sqrt{36 + 16} = \sqrt{52}$
$= \sqrt{13 \times 4} = 2\sqrt{13}$ cm

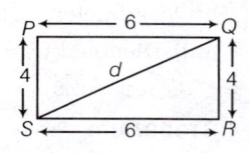

Ex. 9 The base of a parallelogram is twice its height. If the area of the parallelogram is 144 sq cm, find its height.

Sol. Let height of the parallelogram = a
$\therefore$ Base of parallelogram = $2a$
We know that,
Area of parallelogram = Base × Height
$\Rightarrow \quad 144 = 2a \times a \Rightarrow a^2 = \frac{144}{2} = 72$
$\therefore \quad a = \sqrt{72} = 6\sqrt{2}$ cm

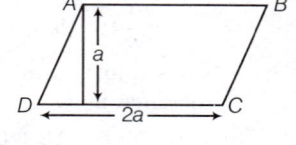

Ex. 10 The difference between two parallel sides of a trapezium is 8 cm. The perpendicular distance between them is 38 cm. If the area of the trapezium is 950 cm², find the length of the parallel sides.

Sol. Let the two parallel sides be a and b.
According to the question, $a - b = 8$ ...(i)
We know that,
Area $= \frac{1}{2} \times$ (Sum of parallel sides) × Height

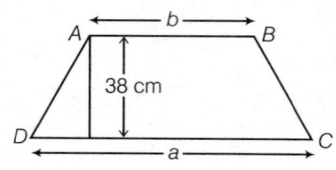

$\Rightarrow \quad \dfrac{1}{2}(a+b) \times 38 = 950$ [∵ height = 38 cm]

$\Rightarrow \quad a + b = \dfrac{1900}{38} = 50$ ...(ii)

From Eqs. (i) and (ii), we get $2a = 58$

∴ $\quad a = \dfrac{58}{2} = 29$ cm

Now, from Eq. (i), $a - b = 8$

$\Rightarrow \quad b = a - 8 = 29 - 8 = 21$ cm

Hence, $a = 29$ cm, $b = 21$ cm.

Regular Polygon

In regular polygons, all sides and all interior angles are equal. A polygon is called pentagon, hexagon, heptagon, octagon, nanogon and decagon accordingly, if it contains 5, 6, 7, 8, 9 and 10 sides, respectively.

If each side of a regular polygon of n sides = a, then

(i) Area of regular pentagon $= 5a^2 \dfrac{\sqrt{3}}{4}$

(ii) Area of regular hexagon $= 6a^2 \dfrac{\sqrt{3}}{4}$

(iii) Area of regular octagon $= 2(\sqrt{2}+1)a^2$

(iv) Area of n sided polygon $= \dfrac{1}{2} na^2 \sin\left(\dfrac{2\pi}{n}\right)$

(v) Perimeter of n sided polygon $= n \times$ Side

(vi) Each exterior angle $= \dfrac{360°}{n}$

(vii) Each interior angle $= 180°$ − Exterior angle

(viii) Number of diagonals $= \left\{\dfrac{n(n-1)}{2} - n\right\}$

Regular Hexagon

Ex. 11 Find the area of a regular hexagon whose side measures 18 cm.

Sol. Given that, $a = 18$ cm

∴ Area of regular hexagon $= 6 \dfrac{\sqrt{3}}{4} a^2 = \dfrac{6 \times \sqrt{3} \times 18 \times 18}{4}$

$= 6 \times 81\sqrt{3} = 486\sqrt{3}$ sq cm

Ex. 12 An equilateral triangle of side 6 cm has its corners cut-off to form a regular hexagon. What is the area of this regular hexagon?

Sol. Side of the regular hexagon $= \dfrac{1}{3} \times 6 = 2$ cm

∴ Area of the hexagon $= 6 \times \dfrac{\sqrt{3}}{4} (\text{Side})^2$

$= 6 \times \dfrac{\sqrt{3}}{4} \times (2)^2$

$= 6\sqrt{3}$ sq cm

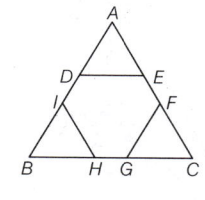

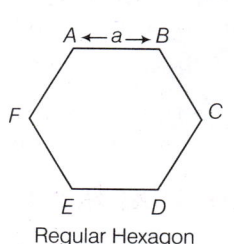

Circle

It is a plane figure enclosed by a line on which every point is equidistant from a fixed point (centre) inside the curve.

(i) Area = πr^2

(ii) Circumference (perimeter) = $2\pi r$

(iii) Diameter = $2r$

(iv) Length of the arc $(l) = \dfrac{\pi r \theta}{180°}$

(v) Area of sector $AOB = \dfrac{\pi r^2 \theta}{360°}$

where, θ = angle enclosed between two radii

✦ Sector is a part of area of circle between two radii.

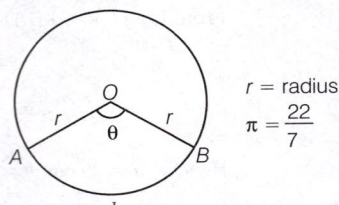

r = radius

$\pi = \dfrac{22}{7}$

Semi-circle

A circle when separated into two parts along its diameter, then each half part is known as semi-circle.

(i) Area of semi-circle = $\dfrac{1}{2}\pi r^2$

(ii) Perimeter = $\pi r + 2r$

where, r = radius of semi-circle

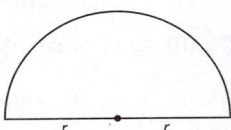

Circular Ring

(i) Area = $\pi (R^2 - r^2)$

(ii) Difference in circumference of both the rings
$= (2\pi R - 2\pi r) = 2\pi (R - r)$

where, R = radius of bigger ring
and r = radius of smaller ring

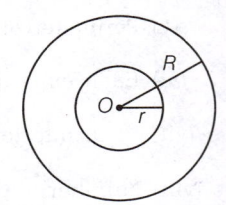

Ex. 13 The inner circumference of a 7 m wide circular race track is 220 m. Find the radius of the outer circle.

Sol. Let R = radius of outer circle, r = radius of inner circle
According to the question,
$2\pi r = 220 \Rightarrow r = \dfrac{220}{2\pi} = \dfrac{220 \times 7}{44} = 35$ m

Clearly, $R - r = 7 \Rightarrow R - 35 = 7$

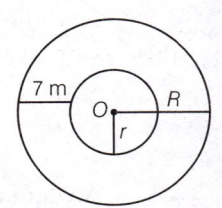

∴ $R = 35 + 7 = 42$ m

Ex. 14 A wheel makes 2000 revolutions in covering a distance of 88 km. Find the radius of wheel.

Sol. Distance covered in 1 revolution = $\dfrac{\text{Total distance}}{\text{Total revolutions}} = \dfrac{88 \times 1000}{2000} = 44$ m

Now, circumference of wheel = $2\pi r = 44$

$\Rightarrow 2 \times \dfrac{22}{7} \times r = 44 \Rightarrow r = 44 \times \dfrac{7}{44} = 7$ m

Area and Perimeter / 573

Fast Track Techniques to solve the QUESTIONS

Technique 1 If the length and breadth of a rectangle are increased by $a\%$ and $b\%$, respectively, then area will be increased by $\left(a + b + \dfrac{ab}{100}\right)\%$.

Ex. 15 If the length and breadth of a rectangle are increased by 10% and 8%, respectively, then by what per cent will the area of that rectangle be increased?

Sol. Let the length of the rectangle be l and its breadth be b.

Then, new length $= \dfrac{110}{100} \times l$ and new breadth $= \dfrac{108}{100} \times b$

$\therefore$ Required percentage $= \left(\dfrac{\dfrac{110}{100} \times l \times \dfrac{108}{100} \times b - lb}{lb}\right) \times 100$

$= \left(\dfrac{110 \times 108 - 10000}{10000}\right) \times 100 = \dfrac{1880}{100} = 18.8\%$

Fast Track Method

Given that, $a = 10$, $b = 8$

According to the formula,

Percentage increase in area $= \left(10 + 8 + \dfrac{10 \times 8}{100}\right)\%$

$= \left(18 + \dfrac{80}{100}\right)\% = \left(18 + \dfrac{4}{5}\right)\% = 18\dfrac{4}{5}\% = 18.8\%$

✦ If any one or both the sides of rectangle is/are decreased, then put negative sign for that value in Technique 1.

Ex. 16 If the length of a rectangle is increased by 5% and the breadth of the rectangle is decreased by 6%, then find the percentage change in area.

Sol. Given that, $a = 5$, $b = -6$

According to the formula,

Percentage change in area $= \left(5 - 6 - \dfrac{5 \times 6}{100}\right)\% = -1 - \dfrac{30}{100} = -1 - 0.30$

$= -1.3\%$ (decrease)

✦ Negative value shows that there is a decrease in area.

Technique 2 If all the sides of any two-dimensional figure are changed by $a\%$, then its area will be changed by $\left(2a + \dfrac{a^2}{100}\right)\%$. In case of circle, radius (or diameter) is increased in place of sides.

Ex. 17 If sides of a square are increased by 5%, by what per cent, its area will be increased?

Sol. Let the side of square be s. Then, area of square $= s^2$

$\therefore$ New side of square $= \dfrac{105}{100} \times s$ and new area $= \left(\dfrac{105}{100} \times s\right)^2$

Required percentage $= \left\{\dfrac{\left(\dfrac{105}{100}s\right)^2 - s^2}{s^2}\right\} \times 100 = \left(\dfrac{105 \times 105 - 10000}{10000}\right) \times 100 = 10.25\%$

Fast Track Method
Given that, $a = 5$
According to the formula,

Percentage increase in area $= \left(2 \times 5 + \dfrac{5^2}{100}\right)\% = \left(10 + \dfrac{25}{100}\right)\% = \left(10 + \dfrac{1}{4}\right)\% = 10.25\%$

Technique 3
If the length of a rectangle is increased by $a\%$, then its breadth will have to be decreased by $\left(\dfrac{100 \times a}{100 + a}\right)$ in order to maintain the same area of rectangle.

Ex. 18 The length of rectangle is increased by 25%. By what per cent should its breadth be decreased so as to maintain the same area?

Sol. Let the length and breadth of the rectangle be l and b, respectively.

Then, new length $= \dfrac{125}{100} l$ and new breadth $= \dfrac{(100 - x) b}{100}$

According to the question, area of the rectangle remains same.

$\therefore lb = \dfrac{125}{100} l \times \dfrac{(100 - x)}{100} b \Rightarrow 125(100 - x) = 10000 \Rightarrow 100 - x = 80$

$\therefore x = 20\%$

Hence, breadth of the rectangle must be decreased by 20%.

Fast Track Method
Here, $a = 25$
According to the formula,

Breadth of rectangle must be decreased by $\left(\dfrac{100 \times 25}{100 + 25}\right)\% = 20\%$

Technique 4
If all the measuring sides of any two-dimensional figure are changed (increased or decreased) by $a\%$, then its perimeter also changes by $a\%$. In case of circle such change takes place because of the change in radius (or diameter).

Ex. 19 If diameter of a circle is increased by 12.5%, find the percentage increase in its circumference.

Sol. Given that, $a = 12.5$
According to the formula,
Required percentage increase $= 12.5\%$

Area and Perimeter / 575

Technique 5 If area of a square is a sq units, then the area of the circle formed by the same perimeter is given by $\dfrac{4a}{\pi}$ sq units.

Ex. 20 If area of a square is 44 sq cm, find the area of the circle formed by the same perimeter.

Sol. Let the side of the square be s and radius of the circle be r.
Then, area of square = (side)^2 = 44 sq cm or side = $\sqrt{44} = 2\sqrt{11}$ cm
Now, perimeter of square = circumference of circle
$$\Rightarrow \quad 4 \times 2\sqrt{11} = 2\pi r \quad \Rightarrow \quad r = \dfrac{4\sqrt{11}}{\pi}$$
Now, area of circle = $\pi r^2 = \pi \left(\dfrac{4\sqrt{11}}{\pi}\right)^2 = \dfrac{16 \times 11}{22} \times 7 = 56$ sq cm

Fast Track Method
Given that, $a = 44$ sq cm
According to the formula,
Required area = $\dfrac{4a}{\pi} = \dfrac{4 \times 44}{22/7} = \dfrac{4 \times 44 \times 7}{22} = 4 \times 2 \times 7 = 56$ sq cm

Technique 6 Area of a square inscribed in a circle of radius r is equal to $2r^2$.

Ex. 21 Find the area of a square inscribed in a circle of radius 5 cm.

Sol. From the figure, Diameter of circle = Diagonal of square
$\therefore$ Diagonal of square = 10 cm $[\because d = 2r]$
So, area of square = $\dfrac{1}{2} \times$ (Diagonal)$^2 = \dfrac{1}{2} \times (10)^2 = \dfrac{1}{2} \times 100 = 50$ sq cm

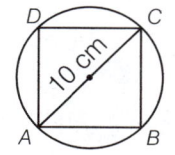

Fast Track Method
Here, $r = 5$ cm
Then, area of square = $2r^2 = 2 \times (5)^2 = 2 \times 25 = 50$ sq cm

Technique 7 The area of the largest triangle inscribed in a semi-circle of radius r is equal to r^2.

Ex. 22 The largest triangle is inscribed in a semi-circle of radius 7 cm. Find the area inside the semi-circle which is not occupied by triangle.

Sol. Given that, radius = 7 cm, diameter = 14 cm
According to the formula,
Area of the largest triangle = $7^2 = 49$ sq cm
Area of semi-circle = $\dfrac{\pi r^2}{2} = \dfrac{\dfrac{22}{7} \times 7^2}{2} = 77$ sq cm 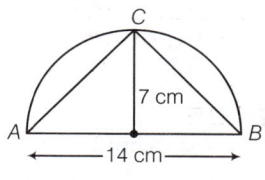
$\therefore$ Required area = Area of semi-circle − Area of the largest triangle
$= 77 - 49 = 28$ sq cm

Technique 8 If the length and breadth of a rectangle are l and b, then area of circle of maximum radius inscribed in that rectangle is $\dfrac{\pi b^2}{4}$.

Ex. 23 Find the area of circle with maximum radius that can be inscribed in the rectangle of length 12 cm and breadth 8 cm.

Sol. From the figure,
Diameter of largest circle that can be inscribed in a rectangle is equal to its breadth.

∴ Diameter of circle = 8 cm ⇒ Radius = $\frac{8}{2}$ = 4 cm

∴ Area of circle = $\pi r^2 = \pi \times (4)^2 = 16\pi$ sq cm

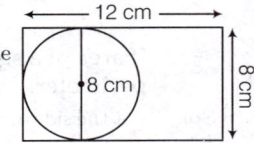

Fast Track Method
Here, $l = 12$ and $b = 8$

∴ Area of circle with maximum radius = $\frac{\pi b^2}{4} = \frac{\pi(8)^2}{4} = \frac{64\pi}{4} = 16\pi$ sq cm

Technique 9 If a pathway of width x is made inside or outside a rectangular plot of length l and breadth b, then area of pathway is
(i) $2x(l + b + 2x)$, if path is made outside the plot.
(ii) $2x(l + b - 2x)$, if path is made inside the plot.

Ex. 24 There is a garden of 140 m × 120 m and a gravel path is to be made of an equal width all around it, so as to take up just one-fourth of the garden. What must be the width of the path?

Sol. Since, path covers $\frac{1}{4}$th area of the garden, that means path is inside the garden.

Given, $l = 140$ m, $b = 120$ m, $x = ?$
According to the question, $2x(l + b - 2x) = \frac{1}{4} \times l \times b$

⇒ $2x(140 + 120 - 2x) = \frac{1}{4} \times 140 \times 120$

⇒ $x(260 - 2x) = 2100 \Rightarrow x^2 - 130x + 1050 = 0 \Rightarrow x = \frac{-b \pm \sqrt{b^2 - 4ac}}{2a}$

Here, $a = 1, b = -130, c = 1050$

⇒ $x = \frac{130 \pm \sqrt{(130)^2 - 4 \times 1050}}{2} = \frac{130 \pm \sqrt{12700}}{2} = \frac{130 \pm 112.7}{2} \Rightarrow x = 8.65$ or 121.3

Leaving 121.3, since width of the path cannot be greater than breadth.
∴ Width of the park = 8.65 m

Technique 10 If two paths, each of width x are made parallel to length (l) and breadth (b) of the rectangular plot in the middle of the plot, then area of the paths is $x(l + b - x)$.

Ex. 25 A rectangular grass plot 80 m × 60 m has two roads, each 10 m wide, running in the middle of it, one parallel to length and the other parallel to breadth. Find the area of the roads.

Sol. Given, area of the grass plot = 80 × 60 = 4800 sq m
and width of the road = 10 m
New length without road = (80 − 10) = 70 m
New breadth without roads = (60 − 10) = 50 m
Area of park without roads = 70 × 50 = 3500 sq m
Now, area of the roads = Area of plot − Area of plot without roads
= 4800 − 3500 = 1300 sq m

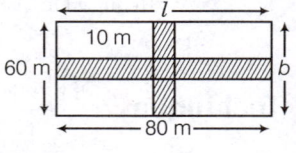

Multi Concept Questions

1. In the adjoining figure, the side of square ABCD is 7 cm. What is the area of the shaded portion formed by the arcs BD of the circles with the centre at C and A? 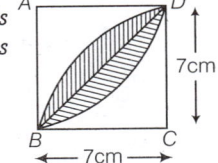
(a) 7 cm^2 (b) 28 cm^2
(c) 14 cm^2 (d) 21cm^2

↪ **(b)** Here, we have two sectors BCD and ABD. Both are symmetrical.
∴ Taking sector BCD.
Area of shaded portion = Area of sector BCD − Area of △BCD
$$= \frac{\pi r^2 \theta}{360°} - \frac{1}{2} \times b \times h$$
$$= \frac{\pi(7)^2 \times 90°}{360°} - \frac{1}{2} \times 7 \times 7$$
$$= \frac{77}{2} - \frac{49}{2} = \frac{28}{2} \text{ sq cm}$$ 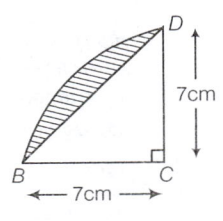
∴ Area of complete shaded portion = $2 \times \frac{28}{2} = 28$ sq cm

2. If the sum of the lengths of the diagonals of a rhombus is 10 m and if its area is 9 m^2, then what is the sum of squares of the diagonals?
(a) 36 m^2 (b) 64 m^2 (c) 80 m^2 (d) 100 m^2

↪ **(b)** Let the two diagonals of rhombus be x and y, respectively.
∴ $x + y = 10$ m   [given] ...(i)
We know that, area of rhombus = $\frac{1}{2} \times$ product of diagonals
⇒ $9 = \frac{1}{2} \times x \times y \Rightarrow xy = 18$ m^2 ...(ii)
Now, $x^2 + y^2 = (x + y)^2 - 2xy$
On putting values from Eqs. (i) and (ii), we get
$x^2 + y^2 = (10)^2 - 2(18) = 100 - 36 = 64$ m^2

3. The quadrants shown in the adjoining figure are each of diameter 12 cm each. What is the area of the shaded portion? 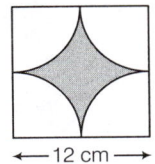
(a) $12(12 - \pi)$ cm^2 (b) $144(4 - \pi)$ cm^2
(c) 36π cm^2 (d) $36(4 - \pi)$ cm^2

↪ **(d)** Area of shaded region
= Area of square of side 12 cm − Area of 4 quadrants of radius 6 cm
$= (12)^2 - 4 \times \frac{1}{4} \times \pi \times (6)^2$
$= 144 - 36\pi = 36(4 - \pi)$ cm^2

578 / **Fast Track** Objective Arithmetic

4. *ABCD is a square, E and F are the mid-points of BC and CD. What is the ratio of area of △AEF to that of the square ABCD.*
 (a) 3 : 8 (b) 5 : 8 (c) 3 : 2 (d) 7 : 4

↪ (*a*) Let the side of the square be *a*.
∴ Area of the square = $a \times a = a^2$ ...(i)
Now, area of △AEF = (Area of square) − (Area of △ABE + Area of △CEF + Area of △ADF)

$$= a^2 - \left(\frac{1}{2} \times a \times \frac{a}{2} + \frac{1}{2} \times \frac{a}{2} \times \frac{a}{2} + \frac{1}{2} \times a \times \frac{a}{2}\right)$$

$$= a^2 - \left(\frac{a^2}{4} + \frac{a^2}{8} + \frac{a^2}{4}\right) = a^2 - \frac{5a^2}{8} = \frac{3a^2}{8} \quad ...(ii)$$

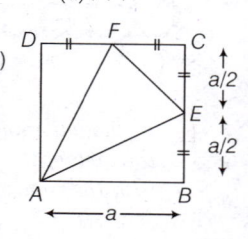

From Eqs. (i) and (ii), we get

Area of △AEF : Area of square ABCD = $\frac{3a^2}{8} : a^2 = 3 : 8$

5. *The three perpendicular distances of three sides of an equilateral triangle from a point which lies inside that triangle are 6 cm, 9 cm and 12 cm, respectively. The perimeter of the triangle is* **[SSC CGL (Mains) 2012]**
 (a) $42\sqrt{2}$ cm (b) $45\sqrt{3}$ cm
 (c) $52\sqrt{2}$ cm (d) $54\sqrt{3}$ cm

↪ (*d*) Let O be the point inside the equilateral △ ABC of side *a* cm.
Then, area of △ABC = Area of △BOC + Area of △AOC + Area of △AOB

∴ $\frac{\sqrt{3}}{4}a^2 = \frac{1}{2}BC \times OR + \frac{1}{2}AC \times OQ + \frac{1}{2}AB \times OP$

⇒ $\frac{\sqrt{3}}{4}a^2 = \frac{1}{2} \times a \times 12 + \frac{1}{2} \times a \times 9 + \frac{1}{2} \times a \times 6$

⇒ $\frac{\sqrt{3}}{4}a^2 = a\left(\frac{12}{2} + \frac{9}{2} + \frac{6}{2}\right)$

⇒ $\frac{\sqrt{3}}{4}a = \left(\frac{12 + 9 + 6}{2}\right)$

⇒ $a = \frac{4}{\sqrt{3}} \times \frac{27}{2} = \frac{54}{\sqrt{3}}$ cm

∴ Perimeter = $3 \times a = 3 \times \frac{54}{\sqrt{3}} = 54\sqrt{3}$ cm

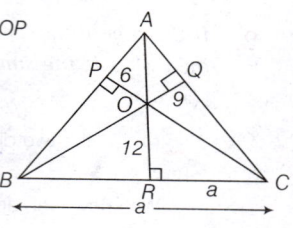

Fast Track Practice

Triangles

1. What is the area of a triangle with sides of length 12 cm, 13 cm and 5 cm? **[CDS 2016 (II)]**
   (a) 30 cm^2 (b) 35 cm^2
   (c) 40 cm^2 (d) 42 cm^2

2. The area of a right angled triangle is 40 sq cm. If its base is equal to 28 cm, find its height. **[Bank Clerk 2010]**
   (a) $3\frac{6}{7}$ cm (b) $4\frac{6}{7}$ cm
   (c) $2\frac{6}{7}$ cm (d) $5\frac{6}{7}$ cm
   (e) None of these

3. The area of an equilateral triangle is $\frac{\sqrt{243}}{4}$ sq cm. Find the length of its side. **[SSC CGL 2011]**
   (a) 3 cm (b) $3\sqrt{3}$ cm
   (c) 9 cm (d) $\sqrt{3}$ cm

4. The perimeter of an isosceles triangle is 26 cm while equal sides together measure 20 cm. The third side and each of the equal sides are respectively
   (a) 6 cm and 10 cm
   (b) 8 cm and 9 cm
   (c) 10 cm and 8 cm
   (d) 14 cm and 6 cm

5. The sides of a triangle are in the ratio of $\frac{1}{2} : \frac{1}{3} : \frac{1}{4}$. If the perimeter is 52 cm, then the length of the smallest side is **[RRB Clerk (Pre) 2017]**
   (a) 9 cm (b) 10 cm (c) 11 cm (d) 12 cm
   (e) None of these

6. The base of a triangular wall is 7 times its height. If the cost of painting the wall at ₹ 350 per 100 sq m is ₹ 1225, then what is the base length?
   (a) 50 m (b) 70 m
   (c) 75 m (d) 100 m

7. The three sides of a triangle are 15, 25 and x units. Which one of the following is correct? **[CDS 2014]**
   (a) $10 < x < 40$ (b) $10 \le x \le 40$
   (c) $10 \le x < 40$ (d) $10 < x \le 40$

8. The sides of a right angled triangle are equal to three consecutive numbers expressed in centimetres. What can be the area of such a triangle? **[CDS 2014]**
   (a) 6 cm^2 (b) 8 cm^2
   (c) 10 cm^2 (d) 12 cm^2

9. If the area of an equilateral triangle is x and its perimeter is y, then which one of the following is correct? **[CDS 2013]**
   (a) $y^4 = 432x^2$ (b) $y^4 = 216x^2$
   (c) $y^2 = 432x^2$ (d) None of these

10. The area of an isosceles $\triangle ABC$ with $AB = AC$ and altitude $AD = 3$ cm is 12 sq cm. What is its perimeter?
    (a) 18 cm (b) 16 cm **[CDS 2013]**
    (c) 14 cm (d) 12 cm

11. The ratio of length of each equal side and the third side of an isosceles triangle is 3 : 4. If the area of the triangle is $18\sqrt{5}$ sq units, the third side is **[SSC (10 + 2) 2012]**
    (a) $8\sqrt{2}$ units (b) 12 units
    (c) 16 units (d) $5\sqrt{10}$ units

12. Three sides of a triangular field are of lengths 15 m, 20 m and 25 m long, respectively. Find the cost of sowing seeds in the field at the rate of ₹ 5 per sq m. **[SSC CGL 2013]**
    (a) ₹ 750 (b) ₹ 150 (c) ₹ 300 (d) ₹ 600

13. If the side of an equilateral triangle is increased by 60%, then by what per cent will the area increase? **[CMAT 2015]**
    (a) 145% (b) 125% (c) 98% (d) 156%

14. Let P, Q, R be the mid-points of sides AB, BC, CA respectively of a $\triangle ABC$. If the area of $\triangle ABC$ is 5 sq units, then the area of $\triangle PQR$ is **[CDS 2017 (I)]**
    (a) $\frac{5}{3}$ sq units (b) $\frac{5}{2\sqrt{2}}$ sq units
    (c) $\frac{5}{4}$ sq units (d) 1 sq unit

15. D and E are points on the sides AB and AC respectively of $\triangle ABC$ such that DE is parallel to BC and $AD : DB = 4 : 5$, CD and BE intersect each other at F. Then, the ratio of the areas of $\triangle DEF$ and $\triangle CBF$ is **[SSC CGL (Mains) 2016]**
    (a) 16 : 25 (b) 16 : 81
    (c) 81 : 16 (d) 4 : 9

16. The inradius of triangle is 4 cm and its area is 34 sq cm. The perimeter of the triangle is [SSC CGL (Pre) 2016]
    (a) 8.5 cm   (b) 17 cm
    (c) 34 cm   (d) 20 cm

17. Lengths of the perpendiculars from a point in the interior of an equilateral triangle on its sides are 3 cm, 4 cm and 5 cm. The area of the triangle is [SSC CGL (Pre) 2016]
    (a) $48\sqrt{3}$ cm^2 (b) $54\sqrt{3}$ cm^2
    (c) $72\sqrt{3}$ cm^2 (d) $80\sqrt{3}$ cm^2

18. A ΔDEF is formed by joining the mid-points of the sides of ΔABC. Similarly, ΔPQR is formed by joining the mid-points of the sides of ΔDEF. If the sides of ΔPQR are of lengths 1, 2 and 3 units, what is the perimeter of ΔABC? [CDS 2013]
    (a) 18 units
    (b) 24 units
    (c) 48 units
    (d) Cannot be determined

19. ABC is a triangle in which D is the mid-point of BC and E is the mid-point of AD. Which of the following statements is/are correct?
    I. The area of ΔABC is equal to four times the area of ΔBED.
    II. The area of ΔADC is twice the area of ΔBED.
    Select the correct answer using the codes given below. [CDS 2016 (I)]
    (a) Only I
    (b) Only II
    (c) Both I and II
    (d) Neither I nor II

20. In an equilateral triangle, another equilateral triangle is drawn inside joining the mid-points of the sides of given equilateral triangle and the process is continued upto 7 times. What is the ratio of area of fourth triangle to that of seventh triangle? [CDS 2016 (II)]
    (a) 256 : 1   (b) 128 : 1
    (c) 64 : 1   (d) 16 : 1

Quadrilaterals and Polygons

21. The ratio between the length and the breadth of a rectangle is 2 : 1. If breadth is 5 cm less than the length, what will be the perimeter of the rectangle?
    (a) 30 cm   (b) 25 cm
    (c) 35 cm   (d) 40 cm
    (e) None of these

22. The perimeter of a rectangle having area equal to 144 cm^2 and sides in the ratio 4 : 9 is [CDS 2013]
    (a) 52 cm   (b) 56 cm
    (c) 60 cm   (d) 64 cm

23. The area of a rectangular field is 15 times the sum of its length and breadth. If the length of that field is 40 m, what is the breadth of that field? [DMRC 2012]
    (a) 24 m (b) 25 m (c) 28 m (d) 32 m

24. The length and perimeter of a rectangle are in the ratio of 5 : 18. What will be the ratio of its length and breadth? [SSC CGL 2012]
    (a) 4 : 3 (b) 3 : 5 (c) 5 : 4 (d) 4 : 7

25. The respective ratio between length and breadth of a rectangle is 8 : 5 and its perimeter is 208 cm. If the side of the square is 40% less than the breadth of the rectangle, what is the perimeter of the square? [IBPS Clerk (Pre) 2016]
    (a) 54 cm   (b) 68 cm
    (c) 88 cm   (d) 96 cm
    (e) 92 cm

26. One side of a rectangular field is 9 m and one of its diagonal is 20 m. Find the area of the field.
    (a) $9\sqrt{319}$ sq m   (b) $7\sqrt{314}$ sq m
    (c) $2\sqrt{319}$ sq m   (d) $5\sqrt{319}$ sq m
    (e) None of these

27. The area of a rectangle lies between 40 cm^2 and 45 cm^2. If one of the sides is 5 cm, then its diagonal lies between [CDS 2014]
    (a) 8 cm and 10 cm
    (b) 9.5 cm and 10 cm
    (c) 10 cm and 12 cm
    (d) 11 cm and 13 cm

28. If the length of a rectangle decreases by 5 m and breadth increases by 3 m, then its area reduces by 9 sq m. If length and breadth of this rectangle increased by 3 m and 2 m respectively, then its area increased by 67 sq m. What is the length of rectangle? [RRB 2012]
    (a) 9 m   (b) 15.6 m
    (c) 17 m   (d) 18.5 m

Area and Perimeter / 581

29. The area of a rectangle whose length is 5 more than twice its width is 75 sq units. What is the perimeter of the rectangle? **[CDS 2012]**
(a) 40 units (b) 30 units
(c) 24 units (d) 20 units

30. If the sides of a rectangle are increased by 5%, find the percentage increase in its diagonals.
(a) 6% (b) 4% (c) 5% (d) 9%
(e) None of these

31. If each of the dimensions of a rectangle is increased by 200%, the area is increased by **[CDS 2017 (I)]**
(a) 300% (b) 400% (c) 600% (d) 800%

32. The length of a rectangle is twice its breadth. If the length is decreased by half of the 10 cm and the breadth is increased by half of the 10 cm, the area of the rectangle is increased by 5 sq cm more than 70 sq cm. Find the length of the rectangle.
(a) 30 cm (b) 40 cm (c) 21 cm (d) 45 cm

33. The area of a rectangle, whose one side is a, is $2a^2$. What is the area of a square having one of the diagonals of the rectangle as side?
(a) $2a^2$ (b) $3a^2$ (c) $4a^2$ (d) $5a^2$

34. The sum of the length and breadth of a rectangle is 6 cm. A square is constructed such that one of its sides is equal to a diagonal of the rectangle. If the ratio of areas of the square and rectangle is 5 : 2, the area of the square (in cm^2) is **[SSC CGL (Mains) 2016]**
(a) 20 cm^2 (b) 10 cm^2
(c) $\sqrt{5}$ cm^2 (d) $10\sqrt{2}$ cm^2

35. A rectangle has 20 cm as its length and 200 sq cm as its area. If the area is increased to $1\frac{1}{5}$ times the original area by increasing its length only, then the perimeter of the rectangle so formed (in cm) is **[SSC CPO 2013]**
(a) 72 (b) 60 (c) 64 (d) 68

36. A rectangle has 30 cm as its length and 720 sq cm as its area. Its area is increased to $1\frac{1}{4}$ times its original area by increasing only its length. Its new perimeter is
(a) 123 cm (b) 125 cm
(c) 119 cm (d) 121 cm
(e) None of these

37. Find the cost of carpeting a room 8 m long and 6 m broad with a carpet 75 cm wide at ₹ 20 per m.
(a) ₹ 1300 (b) ₹ 1500
(c) ₹ 1750 (d) ₹ 1280
(e) None of these

38. A rectangular grassy plot 160 m × 45 m has a gravel path 3 m wide all the four sides inside it. Find the cost of gravelling the path at ₹ 5 per sq m.
(a) ₹ 5970 (b) ₹ 4970
(c) ₹ 6490 (d) ₹ 4970
(e) None of these

39. A ground 100×80 m^2 has two cross roads in its middle. The road parallel to the length is 5 m wide and the other road is 4 m wide, both roads are perpendicular to each other. The cost of laying the bricks at the rate of ₹ 10 per m^2, on the roads, will be **[CLAT 2013]**
(a) ₹ 7000 (b) ₹ 8000
(c) ₹ 9000 (d) ₹ 1000

40. The breadth of a rectangle is 25 m. The total cost of putting a grass bed on this field was ₹ 12375, at the rate of ₹ 15 per sq m. What is the length of the rectangular field? **[Bank Clerk 2011]**
(a) 27 m (b) 30 m (c) 33 m (d) 32 m
(e) None of these

41. A took 15 s to cross a rectangular field diagonally walking at the rate of 52 m/min and B took the same time to cross the same field along its sides walking at the rate of 68 m/min. Find the area of the field.
(a) 45 sq m (b) 35 sq m
(c) 51 sq m (d) 30 sq m
(e) None of these

42. The length and the breadth of a rectangular plot are in the ratio of 5 : 3. The owner spends ₹ 3000 for surrounding it from all the sides at the rate of ₹ 7.5 per metre. What is the difference between the length and breadth of the plot? **[SSC CGL 2008]**
(a) 50 m (b) 100 m
(c) 75 m (d) 60 m

43. The length and breadth of a rectangular plot are in the ratio of 9 : 7. If the cost of fencing the plot at the rate of ₹ 27.75 per metre is ₹ 3552, what is the area of the plot? **[SBI Clerk (Mains) 2016]**
(a) 1236 sq m (b) 1008 sq m
(c) 1152 sq m (d) 1288 sq m
(e) 1056 sq m

44. A rectangular garden of length 12 m is surrounded by a 2 m wide path. If the area of the garden is 84 m^2 and the cost of gravelling is ₹ 8 per m^2, what is the total cost of gravelling the path?
 [SBI Clerk (Pre) 2016]
 (a) ₹ 780 (b) ₹ 742
 (c) ₹ 724 (d) ₹ 775
 (e) ₹ 736

45. What is the area of a square having perimeter 68 cm? [Bank Clerk 2009]
 (a) 361 sq cm (b) 284 sq cm
 (c) 269 sq cm (d) 289 sq cm
 (e) None of these

46. The diagonals of two squares are in the ratio of 3 : 2. Find the ratio of their areas.
 (a) 9 : 4 (b) 9 : 2
 (c) 9 : 5 (d) 9 : 7

47. The diagonal of a square is $4\sqrt{2}$ cm. The diagonal of another square whose area is double that of the first square is
 [SNAP 2008]
 (a) 8 cm (b) $8\sqrt{2}$ cm
 (c) $4\sqrt{2}$ cm (d) 6 cm
 (e) None of these

48. If the sides of a square is increased by 25%, then the area of the square will be increased by [SSC CGL 2013]
 (a) 125% (b) 50%
 (c) 56.25% (d) 53.75%

49. ABCD is a square. If the sides AB and CD are increased by 30%, sides BC and AD are increased by 20%, then the area of the resulting rectangle exceeds the area of the square by [CDS 2015 (II)]
 (a) 50% (b) 52% (c) 54% (d) 56%

50. The perimeter of two squares is 12 cm and 24 cm. The area of the bigger square is how many times that of the smaller? [CTET 2012]
 (a) 2 times (b) 3 times
 (c) 4 times (d) 5 times

51. The perimeters of two squares are 68 cm and 60 cm. Find the perimeter of the third square whose area is equal to the difference of the areas of these two squares. [SSC (10+2) 2011]
 (a) 64 cm (b) 60 cm
 (c) 32 cm (d) 8 cm

52. A cost of cultivating a square field at a rate of ₹ 135 per hectare is ₹ 1215. The cost of putting a fence around it at the rate of 75 paise per metre would be

 (a) ₹ 360 (b) ₹ 810
 (c) ₹ 900 (d) ₹ 1800
 (e) None of these

53. In the given figure, side of each square is 1 cm. The area (in sq cm) of the shaded part is [CTET 2012]

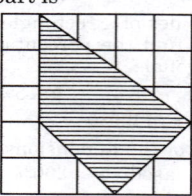

 (a) 8 (b) 9 (c) 10 (d) 11

54. Floor of a square room of side 10 m is to be completely covered with square tiles, each having length 50 cm. The smallest number of tiles needed is [CTET 2012]
 (a) 200 (b) 300
 (c) 400 (d) 500

55. The side of a square is equal to the length of a rectangle, also the side of the square is twice the breadth of the rectangle. If the sum of the areas of the square and rectangle is 48 cm^2, what is the length of the rectangle? [IBPS SO 2016]
 (a) 5 cm (b) 3 cm
 (c) 6 cm (d) 9 cm
 (e) 4 cm

56. The given figure has been obtained by folding a rectangle. The total area of the figure (as visible) is 144 m^2. Had the rectangle not been folded, then current overlapping part would have been a square. What would have been the total area of the original unfolded rectangle? [XAT 2015]

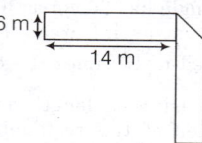

 6 m
 14 m

 (a) 128 m^2 (b) 154 m^2
 (c) 162 m^2 (d) 172 m^2
 (e) None of these

57. The base of a parallelogram is thrice of its height. If the area of the parallelogram is 2187 sq cm, find its height.
 (a) 27 cm (b) 35 cm
 (c) 29 cm (d) 26 cm
 (e) None of these

58. The sides of a parallelogram are 12 cm and 8 cm long and one of the diagonals is 10 cm long. If d is the length of other diagonal, then which one of the following is correct? [CDS 2012]
 (a) $d < 8$ cm
 (b) 8 cm $< d <$ 10 cm
 (c) 10 cm $< d <$ 12 cm
 (d) $d > 12$ cm

59. Find the distance between the two parallel sides of a trapezium, if the area of the trapezium is 500 sq m and the two parallel sides are equal to 30 m and 20 m, respectively. [Bank PO 2008]
 (a) 20 m (b) 15 m (c) 18 m (d) 25 m
 (e) None of these

60. The area of a trapezium is 384 cm^2. If its parallel sides are in the ratio 3 : 5 and the perpendicular distance between them is 12 cm, the smaller of the parallel sides is [SSC CGL 2012]
 (a) 20 cm   (b) 24 cm
 (c) 30 cm   (d) 36 cm

61. The ratio of the length of the parallel sides of a trapezium is 3 : 2. The shortest distance between them is 15 cm. If the area of the trapezium is 450 sq cm, find the sum of the lengths of the parallel sides. [SSC CGL 2011]
 (a) 15 cm   (b) 36 cm
 (c) 42 cm   (d) 60 cm

62. The diagonals of a trapezium $ABCD$ intersect each other at the point O. If $AB = 2CD$, then ratio of the areas of $\triangle AOB$ and $\triangle COD$ is [SSC CGL (Mains) 2016]
 (a) 4 : 1 (b) 1 : 16 (c) 1 : 4 (d) 16 : 1

63. The cross-section of a canal is trapezium in shape. If the canal is 20 m wide at the top and 12 m wide at the bottom and the area of the cross-section is 640 sq m, find the length of the cross-section.
 (a) 80 m (b) 40 m (c) 60 m (d) 70 m
 (e) None of these

64. The difference between two parallel sides of a trapezium is 8 cm. The perpendicular distance between them is 19 cm while the area of trapezium is 760 sq cm. What will be the lengths of the parallel sides?
 (a) 44 cm and 36 cm
 (b) 44 cm and 38 cm
 (c) 36 cm and 25 cm
 (d) 39 cm and 26 cm
 (e) None of the above

65. In a trapezium, the two non-parallel sides are equal in length, each being of 5 units. The parallel sides are at a distance of 3 units apart. If the smaller side of the parallel sides is of length 2 units, then the sum of the diagonals of the trapezium is [CDS 2014]
 (a) $10\sqrt{5}$ units   (b) $6\sqrt{5}$ units
 (c) $5\sqrt{5}$ units   (d) $3\sqrt{5}$ units

66. In a trapezium $ABCD$, AB and DC are parallel sides and $\angle ADC = 90°$. If $AB = 15$ cm, $CD = 40$ cm and diagonal $AC = 41$ cm, then the area of the trapezium $ABCD$ is [SSC CGL (Mains) 2016]
 (a) 245 cm^2 (b) 240 cm^2
 (c) 247.5 cm^2 (d) 250 cm^2

67. The diagonals of a rhombus are 1 m and 1.5 m in lengths. The area of the rhombus is [CLAT 2013]
 (a) 0.75 m^2 (b) 1.5 m^2
 (c) 1.5 m^2 (d) 0.375 m^2

68. If the diagonals of a rhombus are 4.8 cm and 1.4 cm, then what is the perimeter of the rhombus? [CDS 2013]
 (a) 5 cm   (b) 10 cm
 (c) 12 cm   (d) 20 cm

69. The area of a rhombus having one side 10 cm and one diagonal 12 cm, is [SSC CGL (Mains) 2016]
 (a) 48 cm^2 (b) 96 cm^2
 (c) 144 cm^2 (d) 192 cm^2

70. If area of a regular pentagon is $125\sqrt{3}$ sq cm, how long is its each side? [MBA 2008]
 (a) 10 cm   (b) 15 cm
 (c) 16 cm   (d) 25 cm

71. Find each interior angle of a regular pentagon.
 (a) 108°   (b) 105°
 (c) 103°   (d) 101°
 (e) None of these

72. The area of a regular hexagon of side a is equal to [CDS 2017 (I)]
 (a) $\dfrac{\sqrt{2}}{3}a^2$ sq units   (b) $\dfrac{3\sqrt{3}}{2}a^2$ sq units
 (c) $\dfrac{1}{3}a^2$ sq units   (d) $\dfrac{\sqrt{3}}{2}a^2$ sq units

73. Calculate each exterior angle of a regular octagon.
 (a) 35°   (b) 39°
 (c) 40°   (d) 45°

Circles

74. A railing of 288 m is required for fencing a semi-circular park. Find the area of the park. (take π = 22/7) **[SSC Multitasking 2014]**
    (a) 4928 m^2 (b) 9865 m^2
    (c) 8956 m^2 (d) 9856 m^2

75. The area of a sector of a circle of radius 36 cm is 72π cm^2. The length of the corresponding arc of the sector is **[CDS 2014]**
    (a) π cm (b) 2π cm (c) 3π cm (d) 4π cm

76. The inner circumference of a circular race track 7 m wide is 440 m. Find the radius of the outer circle.
    (a) 57 m (b) 68 m (c) 77 m (d) 69 m
    (e) None of these

77. The area of a sector of a circle is 77 sq cm and the angle of the sector is 45°. Find the radius of the circle. **[SSC (10+2) 2012]**
    (a) 7 cm (b) 14 cm (c) 21 cm (d) 28 cm

78. The wheel of an engine turns 350 times round its axle to cover a distance of 1.76 km. The diameter of the wheel is **[SSC FCI 2012]**
    (a) 150 cm (b) 155 cm
    (c) 165 cm (d) 160 cm

79. The areas of two circular fields are in the ratio 16 : 49. If the radius of the bigger field is 14 m, then what is the radius of the smaller field? **[CDS 2017 (I)]**
    (a) 4 m (b) 8 m (c) 9 m (d) 10 m

80. The ratio of the areas of the circumcircle and the incircle of a square is **[SSC CGL (Mains) 2012]**
    (a) 2 : 1 (b) 1 : 2
    (c) $\sqrt{2}$: 1 (d) 1 : $\sqrt{2}$

81. The radii of two concentric circles are 68 cm and 22 cm. The area of the closed figure bounded by the boundaries of the circles is **[SSC CGL (Mains) 2016]**
    (a) 4140π sq cm (b) 4110π sq cm
    (c) 4080π sq cm (d) 4050π sq cm

82. The area of a circle is increased by 22 sq cm when its radius is increased by 1 cm. Find the original radius of the circle. **[SSC CGL 2007]**
    (a) 6 cm (b) 3.2 cm
    (c) 3 cm (d) 3.5 cm

83. From a circular piece of cardboard of radius 3 cm, two sectors of 40° each have been cut-off. The area of the remaining portion is **[CDS 2015 (I)]**
    (a) 11 sq cm (b) 22 sq cm
    (c) 33 sq cm (d) 44 sq cm

84. The radius of a circle is increased, so that its circumference increases by 15%. The area of the circle will increase by **[CDS 2017 (I)]**
    (a) 31.25% (b) 32.25%
    (c) 33.25% (d) 34.25%

85. The radius of a circle is so increased that its circumference increased by 5%. The area of the circle, then increases by **[SNAP 2010]**
    (a) 12.5% (b) 10.25%
    (c) 10.5% (d) 11.25%
    (e) None of these

86. The cost of levelling a circular field at 50 paise per square metre is ₹ 7700. The cost of putting up a fence all round it at ₹ 1.20 per metre is **[SSC CGL (Mains) 2016]**
    (a) ₹ 132 (b) ₹ 264 (c) ₹ 528 (d) ₹ 1056

87. Three circles each of radius 3.5 cm touch one another. The area subtended between them is **[CDS 2017 (I)]**
    (a) $6(\sqrt{3}\pi - 2)$ sq units
    (b) $6(2\pi - \sqrt{3})$ sq units
    (c) $\frac{49}{8}(2\sqrt{3} - \pi)$ sq units
    (d) $\frac{49}{8}(\sqrt{3} - \pi)$ sq units

88. The diameter of the front wheel of an engine is $2x$ cm and that of rear wheel is $2y$ cm. To cover the same distance, what is the number of times the rear wheel revolves when the front wheel revolves n times? **[CDS 2016 (II)]**
    (a) $\frac{n}{xy}$ (b) $\frac{ny}{x}$ (c) $\frac{nx}{y}$ (d) $\frac{xy}{n}$

89. The wheels of a car are of diameter 80 cm each. The car is travelling at a speed of 66 km/h. What is the number of complete revolutions each wheel makes in 10 min? **[CDS 2016 (II)]**
    (a) 4275 (b) 4350 (c) 4375 (d) 4450

90. What is the number of rounds that a wheel of diameter $\frac{5}{11}$ m will make in traversing 7 km? **[CDS 2016 (II)]**
    (a) 3300 (b) 3500 (c) 4400 (d) 4900

Area and Perimeter / 585

91. A man riding a bicycle, completes one lap of a circular field along its circumference at the speed of 14.4 km/h in 1 min 28 s. What is the area of the field?
    (a) 7958 sq m
    (b) 9856 sq m
    (c) 8842 sq m
    (d) Cannot be determined
    (e) None of the above

92. A person observed that he takes 30 s less time to cross a circular ground along its diameter than to cover it once along the boundary. If his speed was 30 m/min, then the radius of the circular ground is (take $\pi = 22/7$) [SSC CGL 2013]
    (a) 10.5 m
    (b) 3.5 m
    (c) 5.5 m
    (d) 7.5 m

93. The circumference of circle A is 75 m more than its diameter. If the radius of circle B is 3.5 m more than the radius of circle A, what is the circumference of circle B? [SBI Clerk (Pre) 2016]
    (a) 110 m
    (b) 140 m
    (c) 163 m
    (d) 96 m
    (e) 132 m

94. Consider the following statements
    I. Area of segment of a circle is less than area of its corresponding sector.
    II. Distance travelled by a circular wheel of diameter $2d$ cm in one revolution is greater than $6d$ cm.
    Which of the above statement(s) is/are correct? [CDS 2012]
    (a) Only I
    (b) Only II
    (c) Both I and II
    (d) Neither I nor II

95. What is the area of the larger segment circle formed by a chord of length 5 cm subtending an angle of 90° at the centre? [CDS 2014]
    (a) $\frac{25}{4}\left(\frac{\pi}{2} + 1\right)$ cm²
    (b) $\frac{25}{4}\left(\frac{\pi}{2} - 1\right)$ cm²
    (c) $\frac{25}{4}\left(\frac{3\pi}{2} + 1\right)$ cm²
    (d) None of these

96. The external fencing of a circular path around a circular plot of land is 33 m more than its interior fencing. The width of the path around the plot is [SSC (10+2) 2014]
    (a) 5.52 m
    (b) 5.25 m
    (c) 2.55 m
    (d) 2.25 m

97. A circle of 3 m radius is divided into three areas by semi-circles of radii 1 m and 2 m as shown in the adjoining figure. The ratio of the three areas A, B and C will be [CDS 2016 (I)] 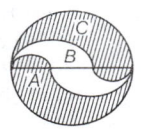
    (a) 2 : 3 : 2
    (b) 1 : 1 : 1
    (c) 4 : 3 : 4
    (d) 1 : 2 : 1

Miscellaneous

98. In a quadrilateral $ABCD$, it is given that $BD = 16$ cm. If $AL \perp BD$ and $CM \perp BD$ such that $AL = 9$ cm and $CM = 7$ cm, then area of quadrilateral $ABCD$ is equal to

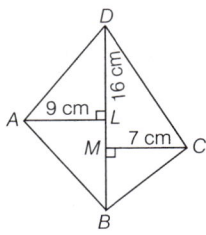

    (a) 256 cm²
    (b) 128 cm²
    (c) 64 cm²
    (d) 96 cm²

99. How many circular plates of diameter d be taken out of a square plate of side $2d$ with minimum loss of material? [CDS 2014]
    (a) 8
    (b) 6
    (c) 5
    (d) 2

100. What is the area of a circle whose area is equal to that of a triangle with sides 7 cm, 24 cm and 25 cm? [CDS 2013]
    (a) 80 cm²
    (b) 84 cm²
    (c) 88 cm²
    (d) 90 cm²

101. The circumference of a circle is 25 cm. Find the side of the square inscribed in the circle.
    (a) $\frac{25}{\pi\sqrt{2}}$ cm
    (b) $\frac{21}{\pi\sqrt{3}}$ cm
    (c) $\frac{23}{\pi\sqrt{2}}$ cm
    (d) $\frac{29}{\pi\sqrt{3}}$ cm
    (e) None of these

102. If area of a square is 64 sq cm, then find the area of the circle formed by the same perimeter. [MBA 2009]
    (a) $\frac{215}{\pi}$ sq cm
    (b) $\frac{216}{\pi}$ sq cm
    (c) $\frac{256}{\pi}$ sq cm
    (d) $\frac{318}{\pi}$ sq cm

103. The side of a square is 5 cm which is 13 cm less than the diameter of a circle. What is the approximate area of the circle? **[Bank Clerk 2011]**
(a) 245 sq cm (b) 235 sq cm
(c) 265 sq cm (d) 255 sq cm
(e) 275 sq cm

104. AB and CD are two diameters of a circle of radius r and they are mutually perpendicular. What is the ratio of the area of the circle to the area of the $\triangle ACD$? **[CDS 2014]**
(a) $\frac{\pi}{2}$ (b) π (c) $\frac{\pi}{4}$ (d) 2π

105. Find the area of the largest triangle that can be inscribed in a semi-circle of radius 9 cm.
(a) 81 sq cm (b) 51 sq cm
(c) 91 sq cm (d) 75 sq cm
(e) None of these

106. Find the area of the largest circle that can be drawn inside a rectangle with sides 18 cm and 14 cm. **[SSC CGL 2007]**
(a) 49 sq cm (b) 154 sq cm
(c) 378 sq cm (d) 1078 sq cm

107. Find the area of circle with maximum radius that can be inscribed in the rectangle of length 16 cm and breadth 4 cm.
(a) 5π cm^2 (b) 6π cm^2
(c) 3π cm^2 (d) 4π cm^2

108. The area of a square is twice the area of a circle. The area of the circle is 392 sq cm. Find the length of the side of the square. **[Bank Clerk 2011]**
(a) 28 cm (b) 26 cm
(c) 24 cm (d) 22 cm
(e) None of these

109. A circle of radius 10 cm has an equilateral triangle inscribed in it. The length of the perpendicular drawn from the centre to any side of the triangle is **[CDS 2014]**
(a) $2.5\sqrt{3}$ cm (b) $5\sqrt{3}$ cm
(c) $10\sqrt{3}$ cm (d) None of these

110. The area of a rectangle is 4 times the area of a square. The area of the square is 729 sq cm and the length of the rectangle is 81 cm. What is the difference between the side of the square and the breadth of the rectangle?
(a) 18 cm (b) 27 cm
(c) 24 cm (d) 9 cm
(e) None of these

111. The circumference of a circle is equal to the perimeter of a rectangle. The length and the breadth of the rectangle are 45 cm and 43 cm, respectively. What is the half the radius of the circle? **[Bank Clerk 2011]**
(a) 56 cm (b) 14 cm
(c) 28 cm (d) 7 cm
(e) None of these

112. What is area of largest triangle inscribed in a semi-circle of radius r units? **[CDS 2016 (II)]**
(a) r^2 sq units
(b) $2r^2$ sq units
(c) $3r^2$ sq units
(d) $4r^2$ sq units

113. If the area of a circle is equal to the area of square with side $2\sqrt{\pi}$ units, what is the diameter of the circle? **[CDS 2012]**
(a) 1 unit (b) 2 units
(c) 4 units (d) 8 units

114. What is the area between a square of side 10 cm and two inverted semi-circular cross-sections each of radius 5 cm inscribed in the square? **[CDS 2013]**
(a) 17.5 cm^2 (b) 18.5 cm^2
(c) 20.5 cm^2 (d) 21.5 cm^2

115. Find the area of shaded portion, if each side of the square is 14 cm.

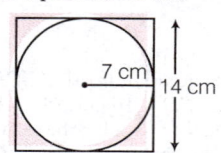

(a) 50 sq cm (b) 45 sq cm
(c) 62 sq cm (d) 42 sq cm

116. What is the total area of three equilateral triangles inscribed in a semi-circle of radius 2 cm? **[CDS 2014]**
(a) 12 cm^2 (b) $\frac{3\sqrt{3}}{4}$ cm^2
(c) $\frac{9\sqrt{3}}{4}$ cm^2 (d) $3\sqrt{3}$ cm^2

117. A copper wire when bent in the form of a square encloses an area of 121 cm^2, if the same wire is bent in the form of a circle, it encloses an area equal to **[CDS 2017 (I)]**
(a) 121 cm^2 (b) 144 cm^2
(c) 154 cm^2 (d) 168 cm^2

118. In the given figure, AC is parallel to ED and AB = DE = 5 cm and BC = 7 cm. What is the area ABDE : area BDE : area BCD equal to? [CDS 2017 (I)]

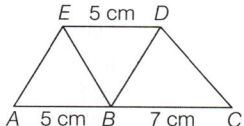

```
     E   5 cm   D
      /\      /\
     /  \    /  \
    /    \  /    \
   A  5cm  B  7cm  C
```

(a) 10 : 5 : 7 (b) 8 : 4 : 7
(c) 2 : 1 : 2 (d) 8 : 4 : 5

119. The length of a side of an equilateral triangle is 8 cm. The area of the region lying between the circumcircle and the incircle of the triangle is
[SSC CGL (Mains) 2016]
(a) $50\frac{1}{7}$ cm^2 (b) $50\frac{2}{7}$ cm^2
(c) $75\frac{1}{7}$ cm^2 (d) $75\frac{2}{7}$ cm^2

120. A square park has each side 50 m. At each corner of the park, there is a flower bed in the form of a quadrant of radius 7 m, as shown in the figure. Find the area of remaining part of the park.

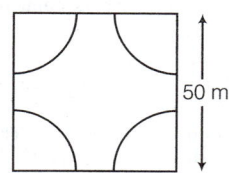

(a) 2346 sq m (b) 2340 sq m
(c) 2250 sq m (d) 2155 sq m

121. Which one of the following is a Pythagorean triplet in which one side differs from the hypotenuse by two units? [CDS 2014]
(a) $(2n + 1, 4n, 2n^2 + 2n)$
(b) $(2n, 4n, n^2 + 1)$
(c) $(2n^2, 2n, 2n + 1)$
(d) $(2n, n^2 - 1, n^2 + 1)$
where, n is a positive real number.

122. The area of a rectangle is equal to the area of a circle with circumference equal to 39.6 m. What is the length of the rectangle, if its breadth is 4.5 m?
[IBPS Clerk 2011]
(a) 33.52 m (b) 21.63 m
(c) 31.77 m (d) 27.72 m
(e) None of these

123. The area of a rectangle is 1.8 times the area of a square. The length of the rectangle is 5 times the breadth. The side of the square is 20 cm. What is the perimeter of the rectangle?
[Bank Clerk 2011]
(a) 145 cm (b) 144 cm
(c) 133 cm (d) 135 cm
(e) None of these

124. A rectangle of maximum area is drawn inside a circle of diameter 5 cm. What is the maximum area of such a rectangle? [CDS 2014]
(a) 25 cm^2 (b) 12.5 cm^2
(c) 12 cm^2 (d) None of these

125. The largest triangle is inscribed in a semi-circle of radius 4 cm. Find the area inside the semi-circle which is not occupied by the triangle.
(a) $8(\pi - 2)$ sq cm (b) $7(\pi - 1)$ sq cm
(c) $8(\pi - 1)^2$ sq cm (d) $6(\pi - 2)$ sq cm
(e) None of these

126. The perimeter of a square is twice the perimeter of a rectangle. If the perimeter of the square is 72 cm and the length of the rectangle is 12 cm, what is the difference between the breadth of the rectangle and the side of the square? [IBPS Clerk 2011]
(a) 9 cm (b) 12 cm (c) 18 cm (d) 3 cm
(e) None of these

127. The area of circle is equal to the area of a rectangle having perimeter of 50 cm and the length is more than its breadth by 3 cm. What is the diameter of the circle? [Bank Clerk 2007]
(a) 7 cm (b) 21 cm
(c) 28 cm (d) 14 cm
(d) None of these

128. A circle and a square have the same perimeter. Which one of the following is correct? [CDS 2016 (II)]
(a) Their areas are equal
(b) The area of the circle is larger
(c) The area of the square is $\pi/2$ times area of circle
(d) The area of the square is π times area of circle

129. If the perimeter of a circle is equal to that of a square, then what is the ratio of area of circle to that of square?
[CDS 2016 (II)]
(a) 22 : 7 (b) 14 : 11
(c) 7 : 22 (d) 11 : 14

588 / Fast Track Objective Arithmetic

130. The radius of a circular field is equal to the side of a square field. If the difference between the perimeter of the circular field and that of the square field is 32 m, what is the perimeter of the square field? **[LIC AAO 2016]**
    (a) 84 m (b) 95 m (c) 56 m (d) 28 m
    (e) 112 m

131. The area of circle inscribed in an equilateral triangle is 154 sq cm. What is the perimeter of the triangle?
    (a) 21 cm   (b) $42\sqrt{3}$ cm
    (c) $21\sqrt{3}$ (d) 42 cm

132. One diagonal of a rhombus is 60% of the other diagonal. Then, area of the rhombus is how many times the square of the length of the larger diagonal?
    (a) $\dfrac{1}{5}$ (b) $\dfrac{2}{5}$ (c) $\dfrac{6}{7}$ (d) $\dfrac{3}{10}$
    (e) None of these

133. The diameters of two circles are the side of a square and the diagonal of the square. The ratio of the areas of the smaller circle and the larger circle is **[SSC CGL 2013]**
    (a) 1 : 4   (b) $\sqrt{2} : \sqrt{3}$
    (c) 1 : $\sqrt{2}$ (d) 1 : 2

134. A regular hexagon is inscribed in a circle of radius 5 cm. If x is the area inside the circle but outside the regular hexagon, then which one of the following is correct? **[CDS 2013]**
    (a) 12 cm² < x < 15 cm²
    (b) 15 cm² < x < 17 cm²
    (c) 17 cm² < x < 19 cm²
    (d) 19 cm² < x < 21 cm²

135. A circular road is constructed outside a square field. The perimeter of the square field is 200 ft. If the width of the road is $7\sqrt{2}$ ft and cost of construction is ₹ 100 per sq ft. Find the lowest possible cost to construct 50% of the total road. **[XAT 2015]**
    (a) ₹ 70400   (b) ₹ 125400
    (c) ₹ 140800   (d) ₹ 235400
    (e) None of these

136. Two circles, each of radius r, with centres P and Q are such that each circle passes through the centre of the other circle. Then, the area common to the circles is less than one-third of the sum of the areas of the two circles by **[CDS 2015 (I)]**
    (a) $\dfrac{\sqrt{3}r^2}{4}$ (b) $\dfrac{\sqrt{3}r^2}{3}$
    (c) $\dfrac{\sqrt{3}r^2}{2}$ (d) $\sqrt{3}r^2$

137. Three equal circles each of diameter d are drawn on a plane in such a way that each circle touches the other two circles. A big circle is drawn in such a manner that it touches each of the small circles internally. The area of the big circle is **[CDS 2015 (I)]**
    (a) πd^2 (b) $\pi d^2 (2 - \sqrt{3})^2$
    (c) $\dfrac{\pi d^2 (\sqrt{3} + 1)^2}{2}$ (d) $\dfrac{\pi d^2 (\sqrt{3} + 2)^2}{12}$

Answer with Solutions

1. **(a)** Given, the length of the sides are
   $a = 12$ cm, $b = 13$ cm and $c = 5$ cm
   Then, $s = \dfrac{a+b+c}{2}$
   $= \dfrac{12+13+5}{2} = 15$ cm
   ∴ Area of triangle $= \sqrt{s(s-a)(s-b)(s-c)}$
   $= \sqrt{15(15-12)(15-13)(15-5)}$
   $= \sqrt{15 \times 3 \times 2 \times 10}$
   $= \sqrt{900} = 30$ cm²

2. **(c)** Given that, area = 40 sq cm,
   base = 28 cm and height
   = perpendicular = ?
   We know that,
   Area $= \dfrac{1}{2} \times$ Base $\times$ Perpendicular
   $\Rightarrow 40 = \dfrac{1}{2} \times 28 \times$ Perpendicular
   ∴ Perpendicular $= \dfrac{40}{14} = \dfrac{20}{7}$
   $= 2\dfrac{6}{7}$ cm

Area and Perimeter / 589

3. (a) According to the question,
$$\frac{\sqrt{3}}{4}a^2 = \frac{\sqrt{243}}{4} \Rightarrow a^2 = \frac{\sqrt{81 \times 3}}{\sqrt{3}} = \frac{9\sqrt{3}}{\sqrt{3}}$$
$$\therefore \quad a = \sqrt{9} = 3 \text{ cm}$$

4. (a) Let the third side be x.
According to the question,
$$x + 20 = 26 \Rightarrow x = 26 - 20 = 6 \text{ cm}$$
$$\therefore \text{ Each equal side} = \frac{20}{2} = 10 \text{ cm}$$

5. (d) Given, sides of a triangle are in ratio $\frac{1}{2} : \frac{1}{3} : \frac{1}{4}$, i.e. $6 : 4 : 3$.
Let the sides be $6k, 4k$ and $3k$, respectively.
$$\therefore \quad 13k = 52 \Rightarrow k = 4$$
Hence, the sides of the triangle are 24 cm, 16 cm and 12 cm, respectively.
$\therefore$ Smallest side = 12 cm

6. (b) Let the height of the triangle be x.
Then, $BC = 7x$
Area of $\triangle ABC$
$$= \frac{1}{2} \times \text{Base} \times \text{Height}$$
$$= \frac{1}{2} \times 7x \times x = \frac{7}{2}x^2$$
Area of painting in ₹ 350 = 100 sq m
$\therefore$ Area of painting in
₹ 1225 = $\frac{100}{350} \times 1225 = 350$ sq m
It means $\frac{7}{2}x^2 = 350 \Rightarrow x = 10$ m
$\therefore$ Base = $7x = 70$ m

7. (a) In a triangle, sum of two sides is always greater than third side,
i.e. $\quad x < 25 + 15 = 40$ ...(i)
Difference of two sides is always less than third side, i.e. $25 - 15 = 10 < x$ ...(ii)
From Eqs. (i) and (ii), we get
$$10 < x < 40$$

8. (a) Since, the triangle is right angled.
$\therefore$ All the three consecutive sides must satisfy Pythagoras theorem.
$\therefore$ 3, 4 and 5 are the sides of triangle which satisfy Pythagoras theorem.
$\because \quad (5^2 = 4^2 + 3^2)$
$\therefore$ Area of triangle = $\frac{1}{2} \times 4 \times 3 = 6$ cm^2

9. (a) Area of equilateral triangle
$$= \frac{\sqrt{3}}{4}a^2 = x \qquad ...(i)$$
and perimeter = $3a = y \Rightarrow a = \frac{y}{3}$ ...(ii)

Now, putting the value of a from Eq. (ii) in Eq. (i), we get
$$\frac{\sqrt{3}\left(\frac{y}{3}\right)^2}{4} = x \Rightarrow x = \frac{\sqrt{3} \times y^2}{9 \times 4}$$
$$\Rightarrow x = \frac{y^2}{3\sqrt{3} \times 4} \Rightarrow x = \frac{y^2}{12\sqrt{3}}$$
$$\Rightarrow \quad 12\sqrt{3}\,x = y^2$$
On squaring both sides, we get $y^4 = 432x^2$

10. (a) Let $AB = CA = a$ cm and base = b cm
Now, area of $\triangle ABC = \frac{1}{2} \times b \times h$

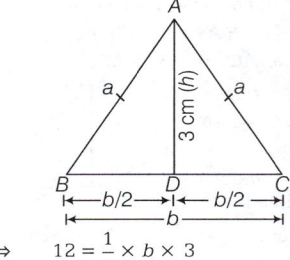

$$\Rightarrow \quad 12 = \frac{1}{2} \times b \times 3$$
$$\therefore \quad b = \frac{12 \times 2}{3} = 8 \text{ cm}$$
Here, $BD = CD = \frac{b}{2} = \frac{8}{2} = 4$ cm
In right angled $\triangle ABD$, by Pythagoras theorem,
$$AB = \sqrt{BD^2 + AD^2}$$
$$\Rightarrow \quad a = \sqrt{4^2 + 3^2}$$
$$= \sqrt{16 + 9} = \sqrt{25} = 5 \text{ cm}$$
Now, perimeter of the isosceles triangle
$$= 2a + b = 2 \times 5 + 8$$
$$= 10 + 8 = 18 \text{ cm}$$

11. (b) Let the sides of isosceles triangle be $3x, 3x$ and $4x$, respectively.
Then, half-perimeter $(s) = \frac{a+b+c}{2}$
$$= \frac{3x + 3x + 4x}{2} = 5x$$
Given, area of isosceles triangle
$$= 18\sqrt{5} \text{ sq units}$$
$$\Rightarrow \quad \sqrt{s(s-a)(s-b)(s-c)} = 18\sqrt{5}$$
$$\Rightarrow \quad \sqrt{5x(5x-3x)(5x-3x)(5x-4x)} = 18\sqrt{5}$$
$$\Rightarrow \quad \sqrt{5x \times 2x \times 2x \times x} = 18\sqrt{5}$$
$$\Rightarrow \quad 2\sqrt{5}x^2 = 18\sqrt{5}$$
$$\Rightarrow \quad x^2 = 9 \Rightarrow x = 3$$
$\therefore$ Third side of isosceles triangle
$$= 4 \times 3 = 12 \text{ units}$$

Alternate Method
Let the sides of isosceles triangle be $3x$, $3x$ and $4x$.
Then, area of isosceles triangle
$$= \frac{b}{4}\sqrt{4a^2 - b^2}$$
$\Rightarrow \quad 18\sqrt{5} = \frac{4x}{4}\sqrt{4(3x)^2 - (4x)^2}$
[here, $a = 3x$, $b = 4x$]
$\Rightarrow \quad 18\sqrt{5} = x\sqrt{36x^2 - 16x^2}$
$\Rightarrow \quad 18\sqrt{5} = 2\sqrt{5}x^2$
$\Rightarrow \quad x = \sqrt{9} = 3$
∴ Third side of isosceles triangle
$= 4 \times 3 = 12$ units

12. (a) Since, $AC^2 = (25)^2 = 625$
and $AB^2 + BC^2$
$= (15)^2 + (20)^2$
$= 225 + 400 = 625$
So, the triangular field is right angled at B.
∴ Area of the field $= \frac{1}{2} \times AB \times BC$
$= \frac{1}{2} \times 15 \times 20 = 150$ m^2
Since, the cost of sowing seed is ₹ 5 per sq m.
∴ Cost of sowing seed for 150 m^2
$= 150 \times 5 = ₹ 750$

13. (d) Let the each side of equilateral triangle be 100.
Area of equilateral triangle
$= \frac{\sqrt{3}}{4} \times (\text{Side})^2 = \frac{\sqrt{3}}{4} \times (100)^2$
$= \frac{\sqrt{3}}{4} \times 10000$
$= 2500\sqrt{3}$
New length of triangle when each side is increased by 60%
$= 100 + 60\%$ of 100
$= 100 + \frac{60}{100} \times 100 = 160$
∴ New area of the equilateral triangle
$= \frac{\sqrt{3}}{4} \cdot (160)^2 = 6400\sqrt{3}$
∴ Required percentage increase in the area
$= \frac{6400\sqrt{3} - 2500\sqrt{3}}{2500\sqrt{3}} \times 100\%$
$= \frac{3900\sqrt{3}}{2500\sqrt{3}} \times 100\% = 156\%$

Fast Track Method
Here, $a = 60\%$
∴ Percentage increase in area
$= \left(2a + \frac{a^2}{100}\right)\%$
$= \left(2 \times 60 + \frac{(60)^2}{100}\right)\%$ [by Technique 2]
$= (120 + 36)\% = 156\%$

14. (c) We know that,
$\frac{\text{ar }(\Delta PQR)}{\text{ar }(\Delta ABC)} = \frac{1}{4}$
$\Rightarrow \quad \text{ar }(\Delta PQR) = \frac{1}{4} \text{ ar }(\Delta ABC)$

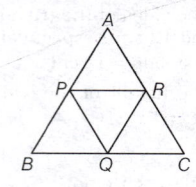

$\Rightarrow \quad \text{ar }(\Delta PQR) = \frac{1}{4} \times 5$
$= \frac{5}{4}$ sq units

15. (b) Here, it is given that
$\frac{AD}{DB} = \frac{4}{5}$
or
$\frac{DB}{AD} = \frac{5}{4}$

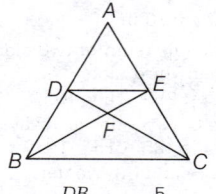

$\Rightarrow \quad \frac{DB}{AD} + 1 = \frac{5}{4} + 1$
$\Rightarrow \quad \frac{AB}{AD} = \frac{9}{4}$ ...(i)
Now, $\Delta ADE \sim \Delta ABC$
∴ $\frac{AB}{AD} = \frac{BC}{DE} = \frac{9}{4}$ [from Eq. (i)]
or let $BC = 9x$, $DE = 4x$
Also, $\Delta DEF \sim \Delta CBF$
∵ $DE \parallel BC$ and DC and BE are transversals.
∴ $\frac{\text{ar }(\Delta DEF)}{\text{ar }(\Delta CBF)} = \frac{(4x)^2}{(9x)^2} = \frac{16}{81}$

Area and Perimeter / 591

16. (b) Area of △ABC = Area of △AOB
+ Area of △BOC + Area of △AOC

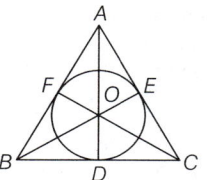

$\Rightarrow 34 = \frac{1}{2} \times AB \times 4 + \frac{1}{2} \times BC \times 4 + \frac{1}{2} \times AC \times 4$

$\Rightarrow 34 = 2 \times$ Perimeter

$\Rightarrow$ Perimeter $= \frac{34}{2} = 17$ cm

17. (a) Let a be the side of equilateral triangle.
Given, $OE = 5$ cm, $OF = 3$ cm
and $OD = 4$ cm

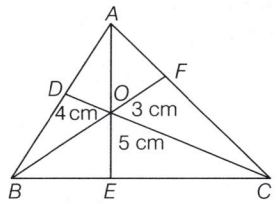

∴ Area of the triangle = Area of △AOC
+ Area of △BOC + Area of △AOB

$\Rightarrow \frac{\sqrt{3}}{4}a^2 = \frac{1}{2} \times AC \times OF + \frac{1}{2} \times BC \times OE$
$\qquad + \frac{1}{2} \times AB \times OD$

$\Rightarrow \frac{\sqrt{3}}{4}a^2 = \frac{1}{2} \times a \times OF + \frac{1}{2} \times a \times OE$
$\qquad + \frac{1}{2} \times a \times OD$

$\Rightarrow \frac{\sqrt{3}}{2}a = OF + OE + OD$

$\Rightarrow a = \frac{2}{\sqrt{3}}(OF + OE + OD)$

$\Rightarrow a = \frac{2}{\sqrt{3}}(3 + 5 + 4) = \frac{24}{\sqrt{3}}$

∴ Area of triangle $= \frac{\sqrt{3}}{4}a^2$

$= \frac{\sqrt{3}}{4} \times \left(\frac{24}{\sqrt{3}}\right)^2 = 48\sqrt{3}$ cm²

18. (b) Perimeter of △PQR = 1 + 2 + 3 = 6 units

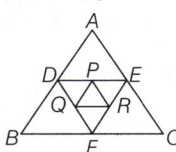

Now, in △DEF,

$\frac{DQ}{DF} = \frac{1}{2} = \frac{PQ}{FE}$

[∵ a line joining the mid-points of two sides of a triangle is parallel to the third side and is half of it]

So, $2PQ = FE$
Similarly, $DF = 2PR$
and $DE = 2QR$

∴ Perimeter of △DEF = 2 × 6 = 12 units
Similarly,
Perimeter of △ABC = 2 × Perimeter of △DEF
$= 2 \times 12 = 24$ units

19. (c) In △ABC, AD is the median and bisects the area of △ABC.

∴ Area of △ABD $= \frac{1}{2} \times$ Area of △ABC

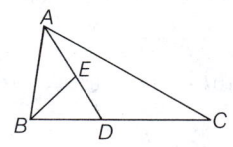

Since, E is the mid-point of AD. Then, BE is the median of △ABD. So, BE bisects the area of △ABD.

I. Area of △BED $= \frac{1}{2} \times$ Area of △ABD

$= \frac{1}{2}\left[\frac{1}{2} \times \text{Area of } \triangle ABC\right]$

$= \frac{1}{4} \times$ Area of △ABC

Hence, Statement I is correct.

II. Area of △ADC = Area of △ABD
$= 2 \times$ Area of △BED

So, Statement II is also correct.

20. (c) Let the side of equilateral triangle be x.
Now, another equilateral triangle is drawn inside joining the mid-points of the sides of given equilateral triangle and process is continued upto 7 times. Then, the side of the fourth triangle is $\frac{x}{2^3}$ and seventh triangle is $\frac{x}{2^6}$.

∴ Required ratio = Area of fourth triangle
 : Area of seventh triangle

$= \frac{\sqrt{3}}{4}\left(\frac{x}{2^3}\right)^2 : \frac{\sqrt{3}}{4}\left(\frac{x}{2^6}\right)^2 = 2^6 : 1 = 64 : 1$

21. (a) Let length $= 2x$ and breadth $= x$
According to the question,
$2x - x = 5 \Rightarrow x = 5$

∴ Required perimeter $= 2(2x + x) = 2 \times 3x$
$= 2 \times 3 \times 5 = 30$ cm

22. (a) Let $l = 9x$ and $b = 4x$
Area of rectangle = $l \times b$
$\Rightarrow \quad 144 = 9x \times 4x$
$\Rightarrow \quad x^2 = \dfrac{144}{36} \Rightarrow x^2 = 4$
$\Rightarrow \quad x = 2$
$\therefore \quad l = 18$ cm and $b = 8$ cm
Hence, perimeter of rectangle = $2(l + b)$
$= 2(18 + 8) = 2 \times 26 = 52$ cm

23. (a) Length of rectangle = 40 m
Let breadth of rectangle = x
Then, according of the question,
Area = 15 (Length + Breadth)
$\Rightarrow \quad (40 + x)15 = 40 \times x$
$\Rightarrow \quad 600 + 15x = 40x \Rightarrow 25x = 600$
$\therefore \quad x = 24$ m

24. (c) According to the question,
$\dfrac{l}{2(l+b)} = \dfrac{5}{18}$
$\Rightarrow \quad 10l + 10b = 18l \Rightarrow 8l = 10b$
$\Rightarrow \quad \dfrac{l}{b} = \dfrac{10}{8} = \dfrac{5}{4}$
$\therefore \quad l : b = 5 : 4$
Hence, ratio of length and breadth of a rectangle is 5 : 4.

25. (d) $\dfrac{\text{Length}}{\text{Breadth}} = \dfrac{8}{5} \Rightarrow \dfrac{l}{b} = \dfrac{8}{5}$
Let the length of the rectangle (l) = $8x$
and breadth of the rectangle (b) = $5x$
According to the question,
$2(l + b) = 208$ [given]
$\Rightarrow \quad 2(8x + 5x) = 208$
$\Rightarrow \quad 26x = 208$
$\Rightarrow \quad x = 8$ cm
Breadth of the rectangle (b)
$= 5x = 5 \times 8 = 40$ cm
Side of the square (s) = 40% less than breadth = $\dfrac{60}{100} \times 40 = 24$ cm
$\therefore$ Perimeter of square = 4×24 cm
$= 96$ cm

26. (a) In $\triangle ABC$, by Pythagoras theorem,
$(AB)^2 = (AC)^2 - (BC)^2$
$\Rightarrow AB = \sqrt{(20)^2 - (9)^2} = \sqrt{319}$

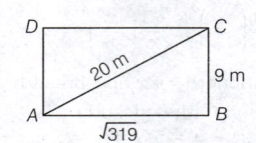

D ———— C
    20 m
        9 m
A ———— B
    √319

$\therefore$ Required area = Length × Breadth
$= 9\sqrt{319}$ sq m

27. (b) It is given that area of rectangle lies between 40 cm^2 and 45 cm^2.
Now, one side = 5 cm
Since, area cannot be less than 40 cm^2.
Hence, other side cannot be less than
$\dfrac{40}{5} = 8$ cm
Since, area cannot be greater than 45 cm^2.
Hence, other side cannot be greater than
$\dfrac{45}{5} = 9$ cm
$\therefore$ Minimum value of diagonal
$= \sqrt{8^2 + 5^2} = \sqrt{89} = 9.43$ cm
and maximum value of diagonal
$= \sqrt{9^2 + 5^2} = \sqrt{106} = 10.3$ cm

28. (c) Let length and breadth of a rectangle be x and y, respectively.
Then, as per first condition,
$(x - 5)(y + 3) = xy - 9$
$\Rightarrow \quad xy - 5y + 3x - 15 = xy - 9$
$\Rightarrow \quad 3x - 5y = 6$ ...(i)
As per second condition,
$(x + 3)(y + 2) = xy + 67$
$\Rightarrow \quad xy + 3y + 2x + 6 = xy + 67$
$\Rightarrow \quad 2x + 3y = 61$ ...(ii)
On multiplying Eq. (i) by 3 and Eq. (ii) by 5, then adding, we get
$9x - 15y = 18$
$10x + 15y = 305$
$\overline{\qquad 19x = 323} \Rightarrow x = \dfrac{323}{19} = 17$
Hence, the length of rectangle is 17 m.

29. (a) Let the width of the rectangle = x units
$\therefore$ Length = $(2x + 5)$ units
According to the question,
Area = $x(2x + 5) \Rightarrow 75 = 2x^2 + 5x$
$\Rightarrow \quad 2x^2 + 5x - 75 = 0$
$\Rightarrow \quad 2x^2 + 15x - 10x - 75 = 0$
$\Rightarrow \quad x(2x + 15) - 5(2x + 15) = 0$
$\Rightarrow \quad (2x + 15)(x - 5) = 0$
$\Rightarrow \quad x = 5$ and $\dfrac{-15}{2}$
Since, width cannot be negative.
$\therefore$ Width = 5 units
and length = $2x + 5 = 2 \times 5 + 5 = 15$ units
$\therefore$ Perimeter of the rectangle
$= 2(15 + 5) = 40$ units

30. (c) Let the length and breadth of rectangle be x and y, respectively.
Then, new length = $x + \dfrac{5}{100} \times x = \dfrac{21x}{20}$
and new breadth = $\dfrac{21y}{20}$

∴ New diagonal = $\sqrt{\left(\dfrac{21x}{20}\right)^2 + \left(\dfrac{21y}{20}\right)^2}$

$= \sqrt{\dfrac{441}{400}(x^2 + y^2)}$

$= \dfrac{21}{20}\sqrt{x^2 + y^2}$

∴ Required percentage increase in diagonals

$= \dfrac{\dfrac{21}{20}\sqrt{x^2 + y^2} - \sqrt{x^2 + y^2}}{\sqrt{x^2 + y^2}} \times 100$

$= \left(\dfrac{21}{20} - 1\right) \times 100$

$= \dfrac{1}{20} \times 100 = 5\%$

31. (d) Let the length and width of the rectangle be x and y, respectively.
Then, area of rectangle = xy
Now, length of rectangle after increasing by 200% = $x + \dfrac{200}{100}x = 3x$
Width of rectangle after increasing by 200%
$= y + \dfrac{200}{100}y = 3y$
∴ New area = $(3x)(3y) = 9xy$
∴ Percentage increase in area
$= \dfrac{9xy - xy}{xy} \times 100$
$= \dfrac{8xy}{xy} \times 100 = 800\%$

32. (b) Given that, $l = 2b$
[here, l = length and b = breadth]
Decrease in length = Half of 10 cm
$= \dfrac{1}{2} \times 10 = 5$ cm
Increase in breadth = Half of 10 cm
$= \dfrac{1}{2} \times 10 = 5$ cm
Increase in the area = $(70 + 5) = 75$ sq cm
According to the question,
$(l - 5)(b + 5) = lb + 75$
⇒ $(2b - 5)(b + 5) = 2b^2 + 75$ [as $l = 2b$]
⇒ $5b - 25 = 75$
⇒ $5b = 100$
∴ $b = \dfrac{100}{5} = 20$ cm
∴ $l = 2b = 2 \times 20 = 40$ cm

33. (d) ∵ Area of rectangle = $2a^2 = l \times b$
⇒ $l \times b = 2a^2 = l \times a$ [∵ $b = a$ units]
⇒ $l = 2a$

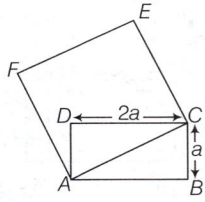

Now in $\triangle ACD$,
$AC^2 = AD^2 + CD^2 = a^2 + 4a^2 = 5a^2$
∴ Side of square (AC) = $a\sqrt{5}$
Hence, area of square = $(a\sqrt{5})^2 = 5a^2$

34. (a) Let the side of square be 's', length and breadth of rectangle be 'l' and 'b' respectively.
∴ diagonal of rectangle = $\sqrt{l^2 + b^2} = s$
⇒ $s^2 = l^2 + b^2$
Also, $\dfrac{s^2}{lb} = \dfrac{5}{2} \Rightarrow \dfrac{l^2 + b^2}{lb} = \dfrac{5}{2}$
⇒ $\dfrac{(l + b)^2 - 2lb}{lb} = \dfrac{5}{2}$ [∵ $l + b = 6$, given]
⇒ $72 = 9lb \Rightarrow lb = 8$ cm^2
Now, $\dfrac{s^2}{lb} = \dfrac{5}{2} \Rightarrow \dfrac{s^2}{8} = \dfrac{5}{2} \Rightarrow s^2 = 20$ cm^2
Hence, the area of square is 20 cm^2.

35. (d) $l_1 = 20$ cm, $A_1 = 200$ sq cm
∴ $b_1 = \dfrac{200}{20} = 10$ cm
Now, $A_2 = 200 \times \dfrac{6}{5} = 240$ sq cm; $b_2 = 10$ cm
∴ $l_2 = \dfrac{240}{10} = 24$ cm
∴ Perimeter of new rectangle = $2(l_2 + b_2)$
$= 2(24 + 10) = 2 \times 34 = 68$ cm

36. (a) Original breadth of rectangle
$= \dfrac{720}{30} = 24$ cm
Now, area of rectangle = $\dfrac{5}{4} \times 720 = 900$ cm^2
∴ New length of rectangle = $\dfrac{900}{24} = 37.5$ cm
∴ New perimeter of rectangle = $2(l + b)$
$= 2(37.5 + 24)$
$= 2 \times 61.5 = 123$ cm

37. (d) Area of the carpet = Area of the room
$= 8 \times 6 = 48$ sq m
Width of the carpet = $\dfrac{75}{100} = \dfrac{3}{4}$ m
Length of the carpet = $\left(48 \times \dfrac{4}{3}\right)$
$= 16 \times 4 = 64$ m
∴ Cost of carpeting = $64 \times 20 = ₹ 1280$

38. (a) Given that, $l = 160$ m, $b = 45$ m and $x = 3$ m

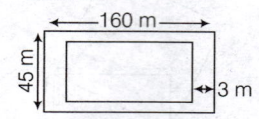

∴ Area of path $= 2x(l + b - 2x)$ [by Technique 9(ii)]
$= 6(160 + 45 - 6)$
$= 6 \times 199 = 1194$ sq m
∴ Required cost $= (1194 \times 5) = ₹ 5970$

39. (b) Area to be paved with bricks
$= 5 \times 100 + 4 \times 80 - 4 \times 5 = 800$ m²

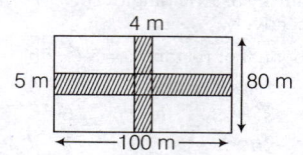

∴ Cost of lying bricks $= 800 \times 10 = ₹ 8000$

40. (c) Area of the rectangular field
$= \dfrac{\text{Total cost}}{\text{Cost per sq m}} = \dfrac{12375}{15} = 825$ sq m

According to the question,
$(L \times B) = 825$
[∵ $L =$ length and $B =$ breadth]
$\Rightarrow L \times 25 = 825$
∴ $L = \dfrac{825}{25} = 33$ m

41. (e) Distance travelled by A in 15 s
$=$ Speed $\times$ Time $= \dfrac{52 \times 15}{60} = 13$ m

∴ Diagonal of the field $= 13$ m
∴ $\sqrt{l^2 + b^2} = 13 \Rightarrow l^2 + b^2 = 169$ ...(i)

Distance travelled by B in 15 s
$= \dfrac{68 \times 15}{60} = 17$ m $\Rightarrow l + b = 17$ m ...(ii)

On solving Eqs. (i) and (ii), we get
$(l + b)^2 = l^2 + b^2 + 2lb$
$\Rightarrow 289 = 169 + 2lb$
$\Rightarrow 120 = 2lb$
$\Rightarrow lb = 60$
$\Rightarrow$ Area $= 60$ sq m

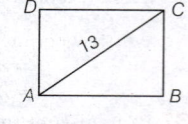

42. (a) Perimeter of the field $= \dfrac{\text{Total cost}}{\text{Cost per metre}}$
$= \dfrac{3000}{7.5} = 400$ m
$\Rightarrow 2(l + b) = 400$
$\Rightarrow 2(5x + 3x) = 400$
$\Rightarrow 8x = 200$

∴ $x = \dfrac{200}{8} = 25$
∴ Required difference $= (5x - 3x)$
$= 2x = 2 \times 25 = 50$ m

43. (b) Let length and breadth of the plot be $9x$ and $7x$, respectively.
∴ Perimeter of the plot $= 2(9x + 7x) = 32x$
Now, cost of fencing $= ₹ 27.75$ per m
∵ Total cost $= ₹ 3552$
Perimeter $= \dfrac{3552}{27.75}$
$\Rightarrow 32x = 128 \Rightarrow x = 4$ m
Length $= 9 \times 4 = 36$ cm
Breadth $= 7 \times 4 = 28$ m
∴ Area of the plot $=$ Length $\times$ Breadth
$= 36 \times 28 = 1008$ m²

44. (e) Given, area of rectangular garden
$= 84$ sq m
Length of garden $(l) = 12$ m
∴ Breadth of garden $= \dfrac{84}{12} = 7$ m

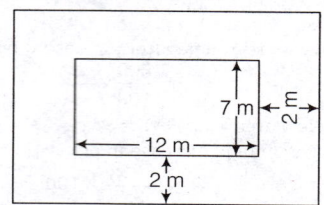

Now, length of park with surrounded path
$= 12 + 2 \times 2 = 12 + 4 = 16$ m
and breadth of park with surrounded path
$= 7 + 2 \times 2 = 7 + 4 = 11$ m
∴ Area of park with surrounded path
$= 16 \times 11 = 176$ m²
∴ Area of surrounded path
$= 176 - 84 = 92$ m²
∴ Cost of gravelling on path
$= 92 \times 8 = ₹ 736$

Fast Track Method
Here, $x = 2$ m, $l = 12$ m, $b = \dfrac{84}{12} = 7$ m
∴ Area of pathway $= 2x(l + b + 2x)$
[by Technique 9(ii)]
$= 2 \times 2(12 + 7 + 2 \times 2)$
$= 4 \times 23 = 92$ m²
∴ Cost of gravelling on path $= 92 \times 8 = ₹ 736$

45. (d) According to the question,
$4a = 68$ [∵ $a =$ side]
∴ $a = \dfrac{68}{4} = 17$ cm
∴ Required area $= a^2 = (17)^2 = 289$ sq cm

Area and Perimeter / 595

46. (a) Let the diagonals of the squares be $3x$ and $2x$.

∴ Ratio of their areas = $\dfrac{\frac{1}{2}(3x)^2}{\frac{1}{2}(2x)^2} = \dfrac{9}{4} = 9:4$

$\left[\because \text{area} = \dfrac{1}{2} \text{ diagonal}\right]$

47. (a) Diagonal of square = $\sqrt{2}a$ [∵ a = side]

$\Rightarrow \quad 4\sqrt{2} = \sqrt{2}a \Rightarrow a = 4$ cm

Now, area of square = $a^2 = (4)^2 = 16$

Side of a square whose area is 2×16.

$a_1^2 = 32 \Rightarrow a_1 = \sqrt{32} \Rightarrow a_1 = 4\sqrt{2}$

Now, diagonal of new square
$= \sqrt{2}a = \sqrt{2} \times 4\sqrt{2} = 8$ cm

48. (c) Here, $a = 25$

∴ Required increment

$= \left(2a + \dfrac{a^2}{100}\right)\%$ [by Technique 2]

$= \left(2 \times 25 + \dfrac{(25)^2}{100}\right) = \left(50 + \dfrac{625}{100}\right)\%$

$= 56.25\%$

49. (d) Let the side of square be x.

∴ Area of square = x^2

Now, increased length = $x\left(1 + \dfrac{30}{100}\right) = \dfrac{13x}{10}$

and increased breadth

$= x\left(1 + \dfrac{20}{100}\right) = \dfrac{12x}{10}$

Now, new area of rectangle

$= \dfrac{13x}{10} \times \dfrac{12x}{10} = \dfrac{156x^2}{100}$

Increase in area $= \dfrac{156x^2}{100} - x^2 = \dfrac{56x^2}{100}$

∴ Percentage increase in area

$= \dfrac{56x^2}{100x^2} \times 100 = 56\%$

Fast Track Method

Here, $a = 30$, $b = 20$

∴ Percentage increase in area

$= \left(a + b + \dfrac{ab}{100}\right)\%$ [by Technique 1]

$= \left(30 + 20 + \dfrac{30 \times 20}{100}\right)\% = 56\%$

50. (c) We know that,
Perimeter of square = $4 \times$ Side

∴ $\quad 4 \times a = 12$ [for smaller square]

$\Rightarrow \quad a = 3$

∴ Area of smaller square
$= 3 \times 3 = 9$ cm^2 ...(i)

Now, $4 \times b = 24$ [for bigger square]

$\Rightarrow \quad b = 6$

∴ Area of bigger square
$= 6 \times 6 = 36$ cm^2 $= 4 \times 9$ cm^2
$= 4 \times$ Area of smaller square
   [from Eq. (i)]

Hence, area of bigger square is 4 times that of smaller square.

51. (c) Here, $a_1 = \dfrac{68}{4} = 17$ cm

and $\quad a_2 = \dfrac{60}{4} = 15$ cm

where, a_1 and a_2 are the sides of the squares.
According to the question,
Area of the third square
$= [(17)^2 - (15)^2]$
$= (17 + 15)(17 - 15)$
$= 32 \times 2 = 64$ sq cm

Let a_3 be the side of the third square.
According to the question,

$(a_3)^2 = 64$ sq cm

$\Rightarrow \quad a_3 = \sqrt{64} = 8$ cm

∴ Perimeter of the third square
$= 4 \times a_3 = 4 \times 8 = 32$ cm

52. (c) Area of the field $= \dfrac{1215}{135} = 9$ hec

$= 90000$ m^2 [∵ 1 hec = 10000 m^2]

∴ Side of the field = $\sqrt{90000} = 300$ m
Perimeter of the field = $4 \times 300 = 1200$ m
Now, cost of putting a fence around field

$= 1200 \times \dfrac{75}{100} = ₹ 900$

53. (c) Here, shaded part $ABCD$ can be visualised as $\triangle ABD + \triangle BCD$.

$\left\{\begin{array}{l}\text{where } BD \text{ is common base and} \\ AB, CE \text{ are respective heights}\end{array}\right\}$

So, area can be calculated by finding the area of two triangles and adding them.

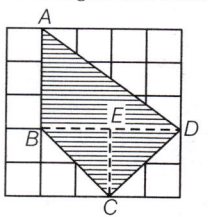

∴ Area of shaded region
$=$ Area of $\triangle ABD +$ Area of $\triangle BCD$

$= \dfrac{1}{2} AB \times BD + \dfrac{1}{2} CE \times BD$

$= \dfrac{1}{2} \times 3 \times 4 + \dfrac{1}{2} \times 2 \times 4 = 6 + 4$

$= 10$ sq cm

596 / *Fast Track* Objective Arithmetic

54. (c) Area of square room = $(10)^2$ = 100 sq m
= $100 \times (100)^2$ sq cm
= $100 \times 100 \times 100$ sq cm
Now, area of tile = $(50)^2 = 50 \times 50$ sq cm
∴ Number of tiles needed
$= \dfrac{\text{Area of square room}}{\text{Area of tile}}$
$= \dfrac{100 \times 100 \times 100}{50 \times 50} = 400$
Hence, 400 tiles will be needed.

55. (c) Let length of rectangle = l cm
and breadth of rectangle = b cm
Given, side of square = length of rectangle
= l cm
and side of square = 2 × breadth of rectangle
$\Rightarrow l = 2 \times b = 2b \Rightarrow b = \dfrac{l}{2}$
Now, according to the question,
$l^2 + (l \times b) = 48 \Rightarrow l^2 + \left(l \times \dfrac{l}{2}\right) = 48$
$\Rightarrow l^2 + \dfrac{l^2}{2} = 48 \Rightarrow \dfrac{3l^2}{2} = 48$
$\Rightarrow l^2 = 32 \Rightarrow l^2 = 32 \approx 36$
$\Rightarrow l^2 = 36$
$\Rightarrow l = 6$ cm
Hence, the length of rectangle is 6 cm.

56. (c) The given figure is as shown below

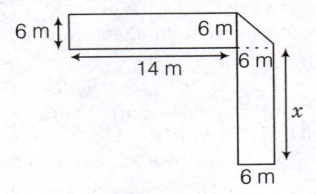

Given, total visible area = 144 m²
$\Rightarrow 6 \times 14 + \dfrac{1}{2}(6 \times 6) + 6x = 144$
$\Rightarrow 84 + 18 + 6x = 144$
$\Rightarrow 6x = 144 - 102 = 42$
∴ $x = 7$ m
Thus, original unfolded area of rectangle
$= 6 \times 14 + 6 \times 6 + 7 \times 6 = 162$ m²

57. (a) We know that,
Area of parallelogram = Base × Height
Let height = a and base = $3a$

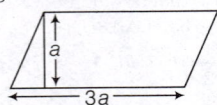

According to the question,
$3a \times a = 2187$

$\Rightarrow 3a^2 = 2187 \Rightarrow a^2 = \dfrac{2187}{3} = 729$
∴ $a = 27$ cm

58. (d) In parallelogram,
$d_1^2 + d_2^2 = 2(l^2 + b^2)$

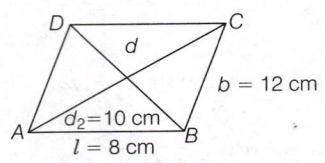

∴ $d^2 + (10)^2 = 2(64 + 144)$
$\Rightarrow d^2 = 2 \times 208 - 100$
$\Rightarrow d^2 = 416 - 100 = 316 \Rightarrow d = \sqrt{316}$
$\Rightarrow d = 17.76$ cm $\Rightarrow d > 12$

59. (a) According to the question,

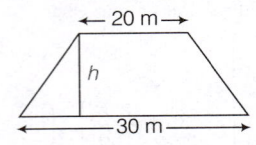

Area $= \dfrac{1}{2}(30 + 20) \times h$
$\Rightarrow 50h = 500 \times 2 \Rightarrow h = 20$ m

60. (b) Let the sides of trapezium be $5x$ and $3x$, respectively.
According to the question,
$\dfrac{1}{2} \times (5x + 3x) \times 12 = 384$
$\Rightarrow 8x = \dfrac{384 \times 2}{12} \Rightarrow x = \dfrac{64}{8} = 8$ cm
Length of smaller of the parallel sides
$= 8 \times 3 = 24$ cm

61. (d) Let the lengths of parallel sides be $3x$ and $2x$.
We know that,
Area of trapezium $= \dfrac{1}{2}$ (Sum of the parallel sides) × Distance between them
$\Rightarrow \dfrac{1}{2}(3x + 2x)15 = 450$
$\Rightarrow 75x = 900$
$\Rightarrow x = \dfrac{900}{75} = 12$ cm
∴ Sum of the parallel sides $= (3x + 2x)$
$= 5x = 5 \times 12 = 60$ cm

62. (a) Given that, $AB = 2CD$
$\Rightarrow \dfrac{AB}{CD} = \dfrac{2}{1}$
∴ $AB = 2x, CD = x$
Also, $\triangle COD \sim \triangle AOB$

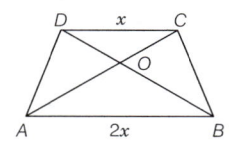

Area and Perimeter / 597

Since, $CD \parallel AB$ and BD and AC are transversals.

∴ $\dfrac{\text{ar}(\triangle AOB)}{\text{ar}(\triangle COD)} = \dfrac{4x^2}{x^2} = 4:1$

63. (b) Let length of cross-section be x.
According to the question,
$\dfrac{1}{2}(20 + 12) \times x = 640$

⇒ $\dfrac{1}{2}(32)x = 640$

⇒ $16x = 640 \Rightarrow x = 40$ m

64. (a) Let the two parallel sides be a and b.

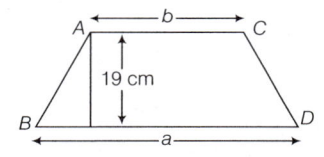

According to the question,
$a - b = 8$ ...(i)
and $\dfrac{1}{2}(a + b) \times 19 = 760$

⇒ $a + b = 40 \times 2$
⇒ $a + b = 80$ ...(ii)

From Eqs. (i) and (ii), we get
$2a = 88 \Rightarrow a = \dfrac{88}{2} = 44$

From Eq. (i), we get $a - b = 8$
∴ $b = a - 8 = 44 - 8 = 36$ cm
∴ $a = 44$ cm and $b = 36$ cm

65. (b) In $\triangle BCF$, by Pythagoras theorem,
$(5)^2 = (3)^2 + (BF)^2 \Rightarrow BF = 4$ units

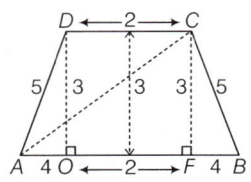

So, $AO = BF = 4$ units
∴ $AB = 2 + 4 + 4 = 10$ units
Now, in $\triangle ACF$, $AC^2 = CF^2 + FA^2$
⇒ $AC^2 = 3^2 + 6^2 \Rightarrow AC = \sqrt{45}$ units
Similarly, $BD = \sqrt{45}$ units
∴ Sum of diagonals = $AC + BD$
$= \sqrt{45} + \sqrt{45} = 2\sqrt{45} = 6\sqrt{5}$ units

66. (c) In right angled $\triangle ADC$,
$AC = 41$ cm, $DC = 40$ cm

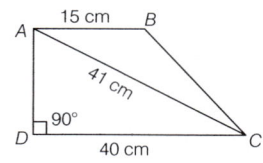

∵ $(AC)^2 = (AD)^2 + (DC)^2$
∴ $(AD)^2 = (AC)^2 - (DC)^2$
⇒ $(AD)^2 = (41)^2 - (40)^2$
⇒ $(AD)^2 = 1681 - 1600 \Rightarrow (AD)^2 = 81$
⇒ $AD = \sqrt{81} \Rightarrow AD = 9$ cm
[which is height of trapezium $ABCD$]
∴ Area of trapezium $ABCD$
$= \dfrac{1}{2} \times$ (Sum of parallel lines) $\times$ (Height)
$= \dfrac{1}{2} \times (AB + CD) \times AD$
$= \dfrac{1}{2} \times (15 + 40) \times 9$
$= \dfrac{1}{2} \times 55 \times 9 = 247.5$ cm^2

67. (a) Area of rhombus $= \dfrac{1}{2} \times d_1 \times d_2$
$= \dfrac{1}{2} \times 1 \times 1.5 = 0.75$ m^2

68. (b) Perimeter of rhombus
$= 2\sqrt{d_1^2 + d_2^2} = 2\sqrt{(4.8)^2 + (1.4)^2}$
$= 2\sqrt{23.04 + 1.96}$
$= 2\sqrt{25} = 2 \times 5 = 10$ cm

69. (b) Side of a rhombus $(a) = 10$ cm
and diagonal $(d_1) = 12$ cm
We know that, $4a^2 = d_1^2 + d_2^2$
⇒ $4 \times (10)^2 = (12)^2 + d_2^2$
⇒ $4 \times 100 = 144 + d_2^2$
⇒ $d_2^2 = 400 - 144 \Rightarrow d_2^2 = 256$
⇒ $d_2 = 16$ cm
Now, area of rhombus
$= \dfrac{1}{2} \times d_1 \times d_2 = \dfrac{1}{2} \times 12 \times 16$
$= 6 \times 16 = 96$ cm^2

70. (a) We know that,
Area of regular pentagon $= 5a^2 \dfrac{\sqrt{3}}{4}$
According to the question,
$\dfrac{5a^2\sqrt{3}}{4} = 125\sqrt{3} \Rightarrow a^2 = \dfrac{125\sqrt{3} \times 4}{5\sqrt{3}} = 100$
∴ $a = 10$ cm

71. (a) Each exterior angle of regular pentagon
$$= \frac{360°}{5} = 72°$$
∴ Each interior angle
$$= 180° - \text{Exterior angle}$$
$$= 180° - 72° = 108°$$

72. (b) In $\triangle OBC$,
$$\angle BOC = 60°$$
[∵ ABCDEF is a regular hexagon]
Also, $OC = OB$
∴ $\angle OCB = \angle OBC = 60°$
[∵ $\angle OBC + \angle OCB + \angle BOC = 180°$]

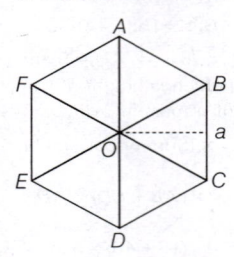

∴ $\triangle OBC$ is an equilateral triangle.
Now, let the side of hexagon be a.
Then, area of equilateral $\triangle OBC = \frac{\sqrt{3}}{4}a^2$
and area of hexagon ABCDEF
$$= 6 \times \text{ar}(\triangle OBC)$$
$$= 6 \times \frac{\sqrt{3}}{4}a^2 = \frac{3\sqrt{3}}{2}a^2$$

73. (d) Each exterior angle of regular octagon
$$= \frac{360°}{8} = 45°$$

74. (a) Let the radius of the park be r, then
$$\pi r + 2r = 288$$
$$\Rightarrow (\pi + 2)r = 288$$
$$\Rightarrow \left(\frac{22}{7} + 2\right)r = 288 \Rightarrow r = \frac{288 \times 7}{36} = 56$$
∴ Area of the park $= \frac{1}{2}\pi r^2$
$$= \frac{1}{2} \times \frac{22}{7} \times 56 \times 56 = 4928 \text{ m}^2$$

75. (d) Given, area of sector $= 72\pi$ cm²
$$\Rightarrow \frac{\pi r^2 \theta}{360°} = 72\pi$$
∴ $\theta = \frac{72 \times 360}{36 \times 36} = 20°$ [∵ $r = 36$ cm]

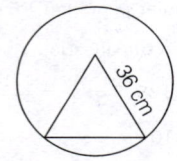

Now, length of arc
$$= \frac{\pi r \theta}{180°} = \frac{\pi \times 36 \times 20°}{180°} = 4\pi \text{ cm}$$

76. (c) Here, R = Radius of outer circle
and r = Radius of inner circle

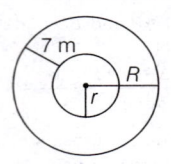

According to the question, $2\pi r = 440$
$$\Rightarrow r = \frac{440}{2\pi} = \frac{440}{2 \times \frac{22}{7}}$$
$$= \frac{440 \times 7}{44} = 70 \text{ m}$$
Clearly, $R - r = 7 \Rightarrow R - 70 = 7$
∴ $R = 70 + 7 = 77$ m

77. (b) Let the radius of circle $= r$ cm
According to the question,
Area of sector $= 77$ cm²
$$\Rightarrow \frac{\theta}{360°} \times \pi r^2 = 77 \Rightarrow \frac{45°}{360°} \times \pi r^2 = 77$$
$$\Rightarrow r^2 = \frac{77 \times 7 \times 8}{22} \Rightarrow r = 14 \text{ cm}$$

78. (d) Distance covered in 1 round
$$= \frac{\text{Total distance}}{\text{Total round}}$$
$$= \frac{1.76 \times 1000}{350} = \frac{176}{35} \text{ m}$$
∴ $2\pi r = \frac{176 \times 100}{35}$ cm
$$\Rightarrow 2r = \text{Diameter} = \frac{17600 \times 7}{22 \times 35}$$
$$= 160 \text{ cm}$$

79. (b) Let r_1 and r_2 be the radii of the circles.
∴ $\frac{\pi r_1^2}{\pi r_2^2} = \frac{16}{49} \Rightarrow \frac{r_1^2}{r_2^2} = \frac{16}{49}$
$$\Rightarrow \frac{r_1}{r_2} = \frac{4}{7} \Rightarrow r_1 = \frac{4}{7}r_2$$
$$\Rightarrow r_1 = \frac{4}{7} \times 14 = 8 \text{ m} \quad [\because r_2 = 14 \text{ m}]$$

80. (a) Ratio of the areas of the circumcircle and incircle of a square
$$= \frac{(\text{Diagonal})^2 \pi}{(\text{Side})^2 \pi} = \frac{(\text{Side} \times \sqrt{2})^2}{(\text{Side})^2} = \frac{2}{1} \text{ or } 2:1$$
$$\left[\because \text{circumradius} = \frac{\text{diagonal}}{2}\right.$$
$$\left. \text{and inradius} = \frac{\text{side}}{2}\right]$$

Area and Perimeter / 599

81. (a) Here, $r_2 = 68$ cm, $r_1 = 22$ cm

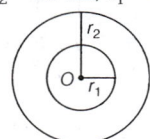

Required area = $\pi(r_2^2 - r_1^2)$ cm^2
= $\pi(68^2 - 22^2)$
= $\pi(68 + 22)(68 - 22)$
= $90 \times 46 \times \pi = 4140\pi$ cm^2

82. (c) Let original radius be r.
Then, according to the question,
$\pi(r+1)^2 - \pi r^2 = 22$
$\Rightarrow \pi \times [(r+1)^2 - r^2] = 22$
$\Rightarrow \dfrac{22}{7} \times (r+1+r)(r+1-r) = 22$
$\Rightarrow 2r + 1 = 7 \Rightarrow 2r = 6$
$\therefore r = \dfrac{6}{2} = 3$ cm

83. (b) Given, radius of circle, $r = 3$ cm
$\therefore$ Area of remaining portion = Area of circle
    $- 2$ (Area of sector $OABCO$)
= $\pi r^2 - \dfrac{2\pi r^2 \theta}{360°} = \pi r^2 \left(1 - \dfrac{2\theta}{360°}\right)$

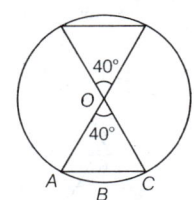

= $\pi r^2 \left(1 - \dfrac{2 \times 40°}{360°}\right)$ [$\because \theta = 40°$]

= $\dfrac{22}{7} \times 9 \left(1 - \dfrac{2}{9}\right) = \dfrac{22}{7} \times 9 \times \dfrac{7}{9}$

= 22 sq cm

84. (b) Let original radius of the circle be r and new radius be r'.
Since, circumference is increased by 15%.
$\therefore \dfrac{2\pi r' - 2\pi r}{2\pi r} \times 100 = 15 \Rightarrow r' = \dfrac{115}{100}r$

Now, percentage increase in area
= $\dfrac{\text{New area} - \text{Old area}}{\text{Old area}} \times 100$

= $\dfrac{\pi r'^2 - \pi r^2}{\pi r^2} \times 100 = \dfrac{r'^2 - r^2}{r^2} \times 100$

= $\dfrac{\left(\dfrac{115}{100}r\right)^2 - r^2}{r^2} \times 100$

= $\dfrac{115^2 - 100^2}{100^2} \times 100$

= $\dfrac{13225 - 10000}{100} = \dfrac{3225}{100} = 32.25\%$

85. (b) Increase in circumference of circle = 5%
$\therefore$ Increase in radius is also 5%.
Now, increase in area of circle
= $\left(2a + \dfrac{a^2}{100}\right)\%$ [by Technique 2]
[where, a = increase in radius]
= $\left(2 \times 5 + \dfrac{5 \times 5}{100}\right)\% = 10.25\%$

86. (c) Area of circular field = $\dfrac{\text{Expenses}}{\text{Rate}}$

$\Rightarrow \pi r^2 = \dfrac{7700}{0.50} \Rightarrow \dfrac{22}{7} \times r^2 = \dfrac{7700}{50} \times 100$

$\Rightarrow \dfrac{22}{7} \times r^2 = 15400 \Rightarrow r^2 = \dfrac{15400 \times 7}{22}$

$\Rightarrow r^2 = 700 \times 7 \Rightarrow r^2 = 4900$
$\Rightarrow r = 70$ m
Now, perimeter of circular field = $2\pi r$
= $2 \times \dfrac{22}{7} \times 70 = 440$ cm
$\therefore$ Cost of putting up a fence all round
= $440 \times 1.20 = ₹ 528$

87. (c) $\because AF = 3.5$ cm and $BF = 3.5$ cm
$\therefore AB = AF + BF = 3.5 + 3.5 = 7$ cm
$\therefore AB = BC = AC = 7$ cm
$\therefore \triangle ABC$ is an equilateral triangle.

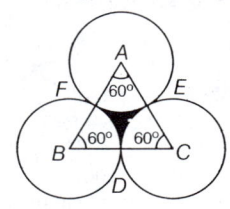

Now, required area = Area of $\triangle ABC$
    $- 3 \times$ Area of sector AFE
= $\dfrac{\sqrt{3}}{4}(7)^2 - 3 \times \dfrac{60}{360} \times \pi(3.5)^2$

= $\dfrac{49\sqrt{3}}{4} - \dfrac{49}{8}\pi = \dfrac{49}{8}(2\sqrt{3} - \pi)$ cm^2

88. (c) Diameter of front wheel = $2x$ cm
Diameter of rear wheel = $2y$ cm
Total distance travelled by front wheel in one revolution = $\pi(2x) = 2\pi x$
Total distance travelled by rear wheel in one revolution = $\pi(2y) = 2\pi y$
Total distance travelled by front wheel in n revolution = $2n x \pi$ ...(i)
Let the same distance is travelled by rear wheel in N revolutions.

∴ Total distance = $2Ny\pi$ ...(ii)
Since, Eqs. (i) and (ii) are equal,
∴ $2nx\pi = 2Ny\pi \Rightarrow N = \dfrac{nx}{y}$

∴ Rear wheel revolves $\dfrac{nx}{y}$ times.

89. (c) Given, radius of wheel = $\dfrac{80}{2}$ cm = 40 cm
Speed = 66 km/h = $\dfrac{6600000}{60}$ cm/min
and time = 10 min
∴ Distance travelled in 10 min
= Speed × Time
= $\dfrac{6600000}{60} \times 10 = 1100000$ cm
Now, number of complete revolutions
= $\dfrac{\text{Distance}}{\text{Circumference of a wheel}}$
= $\dfrac{1100000}{2\pi r} = \dfrac{1100000 \times 7}{2 \times 22 \times 40}$
= $625 \times 7 = 4375$

90. (d) Given, diameter of wheel = $\dfrac{5}{11}$ m
Radius of wheel = $\dfrac{5}{22}$ m
Distance to be covered = 7 km = 7000 m
Now, distance traversed in one round = $2\pi r$
Now, let the wheel traversed 7 km in n rounds, then
7 km = $n \times 2\pi r$
$\Rightarrow 7000 m = n \times 2 \times \dfrac{22}{7} \times \dfrac{5}{22}$
$\Rightarrow 700 \times 7 = n$
∴ $n = 4900$

91. (b) The man takes 3600 s for 14.4 km
The man will take 88 s for
$\dfrac{14.4 \times 88}{3600} = \dfrac{352}{1000}$ km = 352 m
Now, circumference of circular field
= 352 m $\Rightarrow 2\pi r = 352$ m
$\Rightarrow 2 \times \dfrac{22}{7} \times r = 352 \Rightarrow r = 56$ m
Therefore, area of the field = πr^2
= $\dfrac{22}{7} \times 56 \times 56 = 8 \times 22 \times 56$
= 9856 sq m

92. (b) Let the radius of circular field = r m
Speed of person = $\dfrac{30}{60} = \dfrac{1}{2}$ m/s
According to the question,
$\dfrac{2\pi r}{1/2} - \dfrac{2r}{1/2} = 30 \Rightarrow 4\pi r - 4r = 30$
$\left[\because \text{time} = \dfrac{\text{distance}}{\text{speed}}\right]$

$\Rightarrow \left(4 \times \dfrac{22}{7} - 4\right)r = 30$
$\Rightarrow (12.5 - 4)r = 30 \Rightarrow (8.5)r = 30$
$\Rightarrow r = \dfrac{30}{8.5} = 3.5$ m

93. (e) Let radius of circle $A = r_A$
and radius of circle $B = r_B$
Now, according to the question,
$r_B = 3.5 + r_A$ ...(i)
and $2r_A + 75 = 2\pi r_A$
$\Rightarrow 2\pi r_A - 2r_A = 75$
$\Rightarrow 2r_A(\pi - 1) = 75$
$\Rightarrow 2r_A \times \left(\dfrac{22}{7} - 1\right) = 75$
$\Rightarrow 2r_A \times \dfrac{15}{7} = 75 \Rightarrow r_A = \dfrac{7 \times 75}{2 \times 15} = \dfrac{35}{2}$
$\Rightarrow r_A = 17.5$ m
∴ $r_B = 3.5 + 17.5$ [from Eq. (i)]
$\Rightarrow r_B = 21$ m
∴ Circumference of circle $B = 2\pi r_B$
= $2 \times \dfrac{22}{7} \times 21$
= $2 \times 22 \times 3 = 132$ m

94. (c) I. We know that, area of segment PRQP
= Area of sector OPRQO − Area of $\triangle OPQ$
= $\dfrac{\pi r^2 \theta}{360} - \dfrac{1}{2} \times \text{Base} \times \text{Height}$

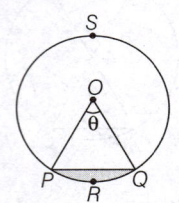

So, the area of a segment of a circle is always less than area of its corresponding sector.

II. Distance travelled by a circular wheel of diameter $2d$ cm in one revolution
= $2 \times 3.14 \times d = 6.28 d$
which is greater than $6d$ cm.

95. (c) In $\triangle AOB$, $AO = OB = r$ [radius of circle]
By Pythagoras theorem,
$AB^2 = OA^2 + OB^2$
$\Rightarrow (5)^2 = r^2 + r^2$
∴ $r^2 = \dfrac{25}{2}$ cm
Now, area of sector AOB
= $\dfrac{\theta}{360°} \times \pi r^2$
= $\dfrac{90°}{360°} \times \pi \times \dfrac{25}{2} = \dfrac{25\pi}{8}$ cm^2

Area and Perimeter / 601

Now, area of minor segment
= Area of sector − Area of triangle
$$= \frac{25\pi}{8} - \frac{r^2}{2} = \frac{25\pi}{8} - \frac{25}{4} = \left(\frac{25\pi - 50}{8}\right)$$
∴ Area of major segment = Area of circle
− Area of minor segment
$$= \pi r^2 - \left(\frac{25\pi - 50}{8}\right) = \frac{25\pi}{2} - \frac{(25\pi - 50)}{8}$$
$$= \frac{100\pi - 25\pi + 50}{8} = \frac{75\pi + 50}{8}$$
$$= \frac{25}{8}(3\pi + 2) = \frac{25}{4}\left(\frac{3\pi}{2} + 1\right) \text{ cm}^2$$

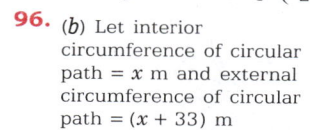

96. (b) Let interior circumference of circular path = x m and external circumference of circular path = $(x + 33)$ m 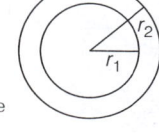
Again, let interior radius be r_1 m and external radius be r_2 m.
Now, $x = 2\pi r_1$ ...(i)
and $(x + 33) = 2\pi r_2$ ...(ii)
On putting the value of x from Eq. (i) in Eq. (ii), we get $2\pi r_1 + 33 = 2\pi r_2$
⇒ $2\pi r_2 - 2\pi r_1 = 33$
⇒ $2\pi (r_2 - r_1) = 33$
⇒ $r_2 - r_1 = \frac{33}{2\pi} = \frac{33}{2 \times (22/7)} = \frac{33 \times 7}{2 \times 22}$
$$= \frac{21}{4} = 5.25 \text{ m}$$
∴ Width of the path around the plot is
= 5.25 m

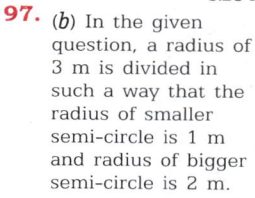

97. (b) In the given question, a radius of 3 m is divided in such a way that the radius of smaller semi-circle is 1 m and radius of bigger semi-circle is 2 m. 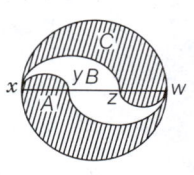
Area of shaded portion A
= Area of semi-circle of radius 3 m
− Area of semi-circle of radius 2 m
+ Area of semi-circle of radius 1 m
$$= \frac{1}{2}\pi (3)^2 - \frac{1}{2}\pi (2)^2 + \frac{1}{2}\pi (1)^2$$
$$= \frac{1}{2}\pi (9 - 4 + 1) = 3\pi \text{ sq m}$$
Area of portion B
= 2 [Area of semi-circle of radius 2 m
− Area of semi-circle of radius 1 m]
$$= 2\left[\frac{1}{2}\pi (2)^2 - \frac{1}{2}\pi (1)^2\right] = 3\pi \text{ sq m}$$
Similarly, area of shaded portion C
= Area of portion A
Hence, the ratio of areas A, B and C is 1 : 1 : 1.

98. (b) ar (quadrilateral ABCD)
= ar (△ABD) + ar (△BCD)
$$= \left(\frac{1}{2} \times BD \times AL\right) + \left(\frac{1}{2} \times BD \times CM\right)$$ 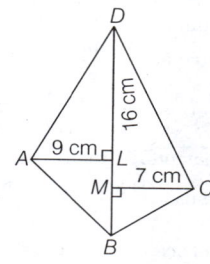
$$= \left(\frac{1}{2} \times 16 \times 9\right) + \left(\frac{1}{2} \times 16 \times 7\right)$$
$$= (72 + 56) = 128 \text{ cm}^2$$

99. (c) Area of square plate = (Side)^2
$$= (2d)^2 = 4d^2$$
Area of circular plate = $\pi \left(\frac{d}{2}\right)^2 = \frac{\pi d^2}{4}$
∴ Number of circular plates
$$= \frac{4d^2}{\pi d^2 / 4} = \frac{4 \times 4}{\pi} \approx 5$$
Since, nearest integer value is 5.

100. (b) Given that, $a = 7$, $b = 24$ and $c = 25$
Semi-perimeter of triangle
$$= \frac{a + b + c}{2} = \frac{7 + 24 + 25}{2} = \frac{56}{2} = 28 \text{ cm}$$
According to the question,
Area of circle = Area of triangle
$$= \sqrt{s(s - a)(s - b)(s - c)}$$
$$= \sqrt{28(28 - 7)(28 - 24)(28 - 25)}$$
$$= \sqrt{28 \times 21 \times 4 \times 3} = \sqrt{7056} = 84 \text{ cm}^2$$

101. (a) Circumference of the circle, $2\pi r = 25$
⇒ $r = \frac{25}{2\pi}$
According to the formula,
Side of inscribed square = $r\sqrt{2}$
$$= \frac{25}{2\pi} \times \sqrt{2} = \frac{25}{\pi\sqrt{2}} \text{ cm}$$

102. (c) Area of square = 64 sq cm
⇒   (Side)^2 = 64
∴   Side = $\sqrt{64}$ = 8 cm
According to the question,
$$2\pi r = 4 \times 8 \Rightarrow r = \frac{4 \times 8}{2\pi} = \frac{16}{\pi}$$

∴ Area of the circle
$$= \pi \times \frac{16}{\pi} \times \frac{16}{\pi} = \frac{256}{\pi} \text{ sq cm.}$$

Fast Track Method
Given that, $a = 64$ sq cm
According to the formula,
Required area $= \frac{4a}{\pi}$ [by Technique 5]
$= \frac{4 \times 64}{\pi} = \frac{256}{\pi}$ sq cm

103. (d) Diameter of the circle = 13 + 5 = 18 cm
∴ Radius $= \frac{\text{Diameter}}{2} = \frac{18}{2} = 9$ cm
Area of the circle $= \pi r^2 = \frac{22}{7} \times 9^2$
$= \frac{22 \times 81}{7} = \frac{1782}{7} = 254.57$
= 255 sq cm (approx.)

104. (b) Required ratio $= \frac{\text{Area of circle}}{\text{Area of } \triangle ACD}$

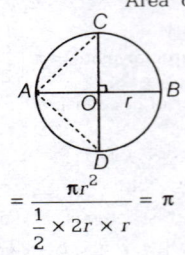

$= \frac{\pi r^2}{\frac{1}{2} \times 2r \times r} = \pi$

105. (a) Given that, $r = 9$ cm

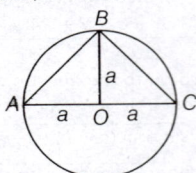

∴ Required area $= \frac{1}{2} \times \text{Base} \times \text{Height}$
$= \frac{1}{2} \times 18 \times 9 = 81$ sq cm

Fast Track Method
Here, $a = 9$ cm
Required area $= a^2$ [by Technique 7]
$= 9^2 = 81$ sq cm

106. (b) Radius of circle $(r) = \frac{1}{2} \times 14 = 7$ cm

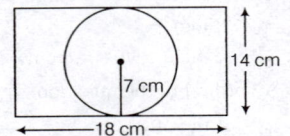

∴ Area of the circle $= \pi r^2 = \frac{22}{7} \times 7 \times 7$
= 154 sq cm

Fast Track Method
Here, $b = 14$ cm
∴ Area of circle $= \frac{\pi b^2}{4}$ [by Technique 8]
$= \frac{\pi \times 14 \times 14}{4} = \frac{22 \times 14 \times 14}{7 \times 4}$
$= 11 \times 14 = 154$ sq cm

107. (d) Here, $b = 4$

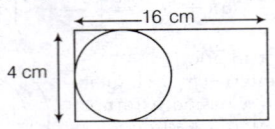

∴ Area of circle $= \frac{\pi b^2}{4}$ [by Technique 8]
$= \frac{\pi (4)^2}{4} = 4\pi \text{ cm}^2$

108. (a) According to the question,
Area of square $= 2 \times 392 \Rightarrow a^2 = 784$
∴ $a = \sqrt{784} = 28$ cm

109. (d) Circumradius $= \frac{2}{3} \times$ Height
∴ Height $= \frac{10 \times 3}{2} = 15$ cm
So, length of perpendicular drawn from centre = 15 − 10 = 5 cm

110. (d) According to the question,
Area of rectangle = 4 × 729
$\Rightarrow$ 81 × B = 4 × 729
∴ B = 4 × 9 = 36 cm
Area of square = 729 $\Rightarrow$ $a^2 = 729$
[∵ a = side of square]
∴ $a = \sqrt{729} = 27$
∴ Required difference = B − a
= 36 − 27 = 9 cm

111. (b) According to the question,
$2\pi r = 2(45 + 43) = 2 \times 88 \Rightarrow r = \frac{2 \times 88}{2\pi}$
∴ $\frac{1}{2}r = \frac{2 \times 88 \times 7}{2 \times 44} = 14$ cm

112. (a) Largest triangle inscribed in a semi-circle is $\triangle ABC$.

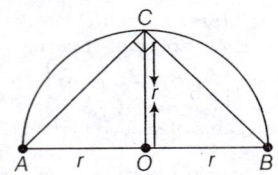

∴ Area of $\triangle ABC = \frac{1}{2} AB \times OC$

$= \frac{1}{2} \times 2r \times r$ [∵ $AB = 2r$, $OC = r$]

$= r^2$ sq units

113. (c) ∵ Area of the circle

= Area of the square = (Side)^2

∴ $\pi r^2 = (2\sqrt{\pi})^2 \Rightarrow \pi r^2 = 4\pi$

$\Rightarrow r^2 = \frac{4\pi}{\pi} = 4 \Rightarrow r = \sqrt{4} = 2$ units

∴ Diameter of circle $(d) = 2 \times r = 2 \times 2$
= 4 units

114. (d) Area between square and semi-circles
= Area of square
  − 2 (Area of semi-circle)

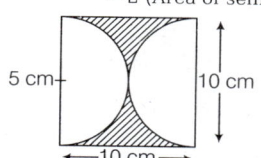

$= (10)^2 - \frac{2}{2} \times \frac{22}{7} (5)^2$

$= 100 - 78.5 = 21.5$ cm^2

115. (d) Area of shaded part
= Area of square − Area of incircle

$= \left[(14)^2 - \left(\frac{22}{7} \times 7 \times 7 \right) \right]$

$= (196 - 154) = 42$ sq cm

116. (d) Since, $\triangle AOB, \triangle BOC$ and $\triangle COD$ are equilateral triangles.

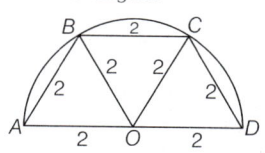

∴ Sides = 2 cm

Now, total area $= 3 \times \frac{\sqrt{3}}{4}$ (side)^2

$= 3 \times \frac{\sqrt{3}}{4} \times 4 = 3\sqrt{3}$ cm^2

117. (c) Let the length of the square be a cm.

∴ $a^2 = 121 \Rightarrow a = 11$ cm

∴ Length of the wire = Perimeter of square of side a

$= 4a = 4 \times 11 = 44$ cm

Let the radius of circle be r cm.

∴ Circumference of circle = Length of wire

$\Rightarrow \quad 2\pi r = 44$

$\Rightarrow \quad r = \frac{44}{2 \times \frac{22}{7}} = 7$ cm

∴ Area enclosed by circle $= \pi r^2$

$= \frac{22}{7} \times (7)^2 = 154$ cm^2

118. (a) Since, $AC \parallel ED$

∴ $EF = BH = DG = x$

Now, in quadrilateral $ABDE$,
$AB \parallel ED$ [∵ $AC \parallel ED \Rightarrow AB \parallel ED$]
and $AB = ED = 5$ cm

∴ Quadrilateral $ABDE$ is a parallelogram.

Now,

ar (parallelogram $ABDE$) $= AB \times EF = 5x$

ar $(\triangle BDE) = \frac{1}{2} \times ED \times BH = \frac{5}{2} x$

and ar $(\triangle BCD) = \frac{1}{2} \times BC \times DG = \frac{7}{2} x$

∴ ar $(ABDE)$: ar (BDE) : ar $(BCD) = 5x : \frac{5}{2}x : \frac{7}{2}x$

= 10 : 5 : 7

119. (b) Given, side of an equilateral triangle
= 8 cm

Now, radius of circumcircle

= Side of an equilateral triangle $\times \frac{1}{\sqrt{3}}$

$= \frac{8}{\sqrt{3}}$ cm

and radius of incircle $=$ Side $\times \frac{1}{2\sqrt{3}}$

$= \frac{4}{\sqrt{3}}$ cm

∴ Required area

$= \left[\left(\frac{8}{\sqrt{3}} \right)^2 - \left(\frac{4}{\sqrt{3}} \right)^2 \right] \pi$ cm^2

$= \left(\frac{64}{3} - \frac{16}{3} \right) \pi$ cm$^2 = \left(\frac{48}{3} \right) \pi$ cm^2

$= 16 \times \frac{22}{7}$ cm$^2 = \frac{352}{7}$ cm$^2 = 50 \frac{2}{7}$ cm^2

120. (a) As four quadrants make a circle.

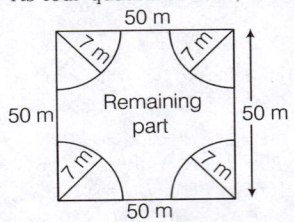

∴ Area of park without flower bed
= Area of square − Area of circle
$$= \left[(50)^2 - \left(\frac{22}{7} \times 7 \times 7\right)\right]$$
$$= [2500 - 154] = 2346 \text{ sq m}$$
∴ Area of remaining part is 2346 sq m.

121. (d) By hit and trial method,
Put $n = 2$ in option (d), we get
$= [(2 \times 2), (2)^2 - 1, (2)^2 + 1] = (4, 3, 5)$
which satisfies Pythagoras theorem and one side differs from hypotenuse by 2 units.

122. (d) Let the radius of the circle be r.
Perimeter of given circle = 39.6 m
$\Rightarrow 2\pi r = 39.6 \Rightarrow r = \frac{7 \times 39.6}{22 \times 2} = 6.3 \text{ m}$
Area of circle $= \frac{22}{7} \times 6.3 \times 6.3$
$= \frac{873.18}{7} = 124.74 \text{ m}^2$
Now, area of a rectangle = Area of a circle
$\Rightarrow l \times b = 124.74 \Rightarrow l \times 4.5 = 124.74$
∴ $l = \frac{124.74}{4.5} = 27.72 \text{ m}$

123. (b) Area of square = (Side)^2 = 20^2
= 400 sq cm
∴ Area of rectangle = 1.8 × 400 = 720 sq cm
Let length and breadth of rectangle be $5x$ and x, respectively.
Then, according to the question,
$5x \times x = 720$
$\Rightarrow 5x^2 = 720 \Rightarrow x^2 = \frac{720}{5} = 144$
∴ $x = \sqrt{144} = 12 \text{ cm}$
Perimeter of rectangle $= 2(5x + x) = 12x$
$= 12 \times 12 = 144 \text{ cm}$

124. (c) Let ABCD be the rectangle inscribed in the circle of diameter 5 cm.

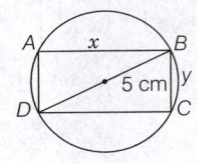

∴ Diameter = Diagonal of rectangle
Now, let x and y be the length and breadth of rectangle, respectively.
In $\triangle ABD$,
$AB^2 + AD^2 = (5)^2 \Rightarrow x^2 + y^2 = 25$
Since, they form a Pythagorean triplet.
So, $x = 4$ and $y = 3$
∴ Area of rectangle $= 3 \times 4 = 12 \text{ cm}^2$

125. (a) According to the formula,

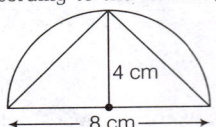

Area of the triangle
$= \frac{1}{2} \times 8 \times 4 = 16 \text{ sq cm}$
Area of semi-circle $= \frac{\pi r^2}{2} = \frac{16\pi}{2} = 8\pi$
∴ Required answer $= 8\pi - 16$
$= 8(\pi - 2)$ sq cm

126. (b) Perimeter of square = 72 cm
∴ Perimeter of rectangle $= \frac{72}{2} = 36 \text{ cm}$
$\Rightarrow 2 \times (l + b) = 36$
[here, l and b are dimensions of rectangle]
$\Rightarrow 2 \times (12 + b) = 36 \Rightarrow 12 + b = 18$
∴ $b = 18 - 12 = 6 \text{ cm}$
and side of square $= \frac{72}{4} = 18 \text{ cm}$
∴ Difference between breadth of the rectangle and side of the square
$= 18 - 6 = 12 \text{ cm}$

127. (d) Let breadth of the rectangle be b.
Then, length $= (b + 3)$
According to the question,
$2(b + b + 3) = 50$
$\Rightarrow 2b + 3 = 25$
$\Rightarrow 2b = 22$
$\Rightarrow b = \frac{22}{2} = 11 \text{ cm}$
∴ Breadth = 11 cm
$\Rightarrow$ Length = (11 + 3) = 14 cm
∴ Area of the circle = Area of the rectangle
$\Rightarrow \pi r^2 = 14 \times 11$
$\Rightarrow r^2 = \frac{14 \times 11 \times 7}{22} = 49$
∴ $r = \sqrt{49} = 7 \text{ cm}$
∴ Diameter $= 2r = 2 \times 7 = 14 \text{ cm}$.

128. (b) Let r and a be the radius of a circle and side of a square respectively.
According to the question, $2\pi r = 4a$

Area and Perimeter / 605

$\Rightarrow \quad a = \dfrac{\pi r}{2}$

Now, area of circle $= \pi r^2 = 3.14 r^2$

and area of square $= a^2 = \dfrac{\pi^2 r^2}{4} = 2.46 r^2$

Clearly, the area of circle is larger than the area of square.

129. (b) Let r be the radius of a given circle and a be the length of each side of a given square.
According to the question,
Perimeter of circle = Perimeter of square
$\Rightarrow \quad 2\pi r = 4a$
$\Rightarrow \quad a = \dfrac{\pi r}{2}$ ...(i)

Now, $\dfrac{\text{Area of circle}}{\text{Area of square}} = \dfrac{\pi r^2}{a^2} = \dfrac{\pi r^2}{\left(\dfrac{\pi r}{2}\right)^2}$

[from Eq. (i)]

$= \dfrac{\pi r^2}{\pi^2 r^2} \times 4 = \dfrac{4}{\pi} = \dfrac{4 \times 7}{22} = \dfrac{14}{11}$

Hence, the required ratio is 14 : 11.

130. (c) Let r be the radius of circular field.
Then, side of square field = r m
According to the question,
$2\pi r - 4r = 32$
$\Rightarrow \quad 2r(\pi - 2) = 32$
$\Rightarrow \quad r\left(\dfrac{22 - 14}{7}\right) = 16$
$\therefore \quad r = \dfrac{16 \times 7}{8} = 14$ m

Hence, perimeter of square field = $4r$
$= 4 \times 14 = 56$ m

131. (b) We know that, the radius of a circle inscribed in an equilateral triangle $= \dfrac{a}{2\sqrt{3}}$

where, a is the length of the side of equilateral triangle.
Given that, area of a circle inscribed in an equilateral triangle = 154 cm²

$\therefore \quad \pi \left(\dfrac{a}{2\sqrt{3}}\right)^2 = 154$

$\Rightarrow \quad \left(\dfrac{a}{2\sqrt{3}}\right)^2 = \dfrac{154 \times 7}{22} = (7)^2$

$\Rightarrow \quad a = 14\sqrt{3}$ cm

$\therefore$ Perimeter of equilateral triangle
$= 3a = 3(14\sqrt{3}) = 42\sqrt{3}$ cm

132. (d) Let one diagonal be x.
Then, other diagonal $= \left(\dfrac{60x}{100}\right) = \dfrac{3x}{5}$

Area of rhombus $= \dfrac{1}{2} \times x \times \dfrac{3x}{5} = \dfrac{3}{10} x^2$

$= \dfrac{3}{10} \times$ Square of larger diagonal

Hence, area of rhombus is $\dfrac{3}{10}$ times.

133. (d) Diagonal of a square $= \sqrt{2} \times$ Side

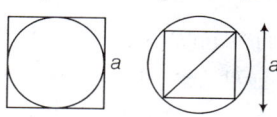

$\therefore$ Ratio of area of smaller circle to larger circle

$= \dfrac{\pi r_1^2}{\pi r_2^2} = \dfrac{\pi \times \left(\dfrac{a}{2}\right)^2}{\pi \times \left(\dfrac{\sqrt{2}a}{2}\right)^2}$

[here, a = side of square]

$= \dfrac{1}{2} = 1 : 2$

134. (a) $\because OB = OA$ = radius
Also, $\angle AOB = 60°$ $\left[\because \dfrac{360°}{6} = 60°\right]$

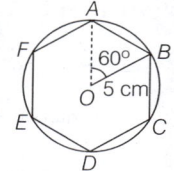

and $\angle OAB = \angle OBA = 60°$
[$\because$ angle opposite to equal sides are equal]
$\therefore \Delta AOB$ is an equilateral triangle.
Then, $AB = 5$ cm
So, required area,
x = Area of circle − Area of hexagon
$= \pi r^2 - \dfrac{3\sqrt{3}(a)^2}{2}$
$= \dfrac{22}{7} \times (5)^2 - \dfrac{3\sqrt{3}}{2} \times (5)^2$ [$\because r = a = 5$]
$= 78.57 - 64.95$
$= 13.62$ cm²

135. (b) Let length of a side of square be a ft.

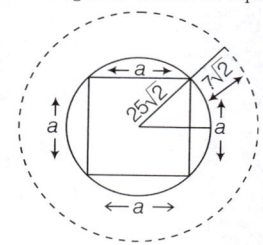

∵ Perimeter of square = 4a
⇒ 4a = 200 ⇒ a = 50 ft
∴ Diagonal of square = $\sqrt{50^2 + 50^2}$
$= \sqrt{5000} = 50\sqrt{2}$ ft
∴ Radius of inner circle $= \dfrac{50\sqrt{2}}{2}$
$= 25\sqrt{2}$ ft
∵ Width of the road = $7\sqrt{2}$ ft [given]
∴ Area of the road = $\pi[(32\sqrt{2})^2 - (25\sqrt{2})^2]$
$= \pi(2048 - 1250)$
$= \dfrac{22}{7} \times 798 = 2508$ sq ft
∴ 50% area of the road
$= \dfrac{50}{100} \times 2508 = 1254$ sq ft
∴ Lowest possible cost to construct 50% of the total road
$= 1254 \times 100 = ₹\ 125400$

136. (c) Given, radius of each circle is r.

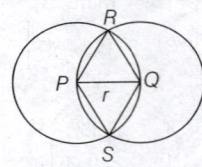

Clearly, PQR is an equilateral triangle.
∴ Area of sector PQRP $= \dfrac{\pi r^2 \theta}{360°}$
$= \dfrac{\pi r^2 \times 60°}{360°} = \dfrac{\pi r^2}{6}$

Now, area of segment PQR
= Area of sector PQRP
       − Area of equilateral triangle
$= \dfrac{\pi r^2}{6} - \dfrac{\sqrt{3}}{4} r^2$

Area of half common circle
$= \dfrac{\pi r^2}{6} + \dfrac{\pi r^2}{6} - \dfrac{\sqrt{3}}{4} r^2$
$= \dfrac{\pi r^2}{3} - \dfrac{\sqrt{3}}{4} r^2$

Area of common circle
$= 2 \times \left(\dfrac{\pi r^2}{3} - \dfrac{\sqrt{3}}{4} r^2\right)$

$= 2r^2 \left(\dfrac{\pi}{3} - \dfrac{\sqrt{3}}{4}\right)$

According to the question,
Required area $= \dfrac{2}{3} \pi r^2 - 2r^2 \left(\dfrac{\pi}{3} - \dfrac{\sqrt{3}}{4}\right)$
$= \dfrac{2\pi r^2}{3} - \dfrac{2\pi r^2}{3} + \dfrac{2\sqrt{3} r^2}{4} = \dfrac{\sqrt{3} r^2}{2}$

137. (d) Given, diameter of each circle = d
∴ Radius of each circle = d/2
⇒ CE = AF = BF = CD = d/2
Here, △ABC is an equilateral triangle.

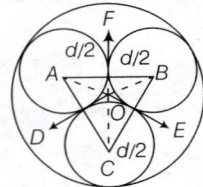

∴ Altitude of equilateral triangle $= \dfrac{\sqrt{3}}{2}$ (Side)
$= \dfrac{\sqrt{3}}{2} (AB) = \dfrac{\sqrt{3}}{2} d$ [∵ AB = AF + FB = d]

∵ Ratio of centroid of equilateral triangle
$= 2 : 1$

∴ Length of OC $= \dfrac{2}{3} \times \dfrac{\sqrt{3}}{2} d = \dfrac{\sqrt{3}}{3} d$

where, O is the centre of big circle.
Now, radius of big circle,
R = OC + CE
$= \dfrac{\sqrt{3}}{3} d + \dfrac{d}{2} = d \left[\dfrac{\sqrt{3}}{3} + \dfrac{1}{2}\right]$
$= \dfrac{d \times \sqrt{3}\,(2 + \sqrt{3})}{6}$

∴ Area of big circle
$= \pi R^2 = \pi \left[d \times \dfrac{\sqrt{3}\,(2+\sqrt{3})}{6}\right]^2$
$= \pi \times \left[\dfrac{d^2 \times 3\,(2+\sqrt{3})^2}{36}\right]$
$= \dfrac{\pi d^2 \,(2+\sqrt{3})^2}{12}$

Chapter 35

Volume and Surface Area

Volume and surface area are related to solids or hollow bodies. These bodies occupy space and have usually three dimensions length, breadth and height.

Volume

The volume of any solid or hollow figure is the amount of space enclosed within its bounding faces. It is always measured in cubic unit like cubic meter, cubic centimetre etc.

For example If length, breadth and height of a box are 9 cm, 4 cm and 2 cm respectively, then
Volume of the box = Length × Breadth × Height
$= 9 \times 4 \times 2 = 72$ cm^3

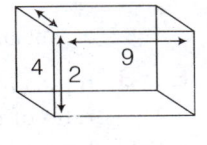

Surface Area

The area of the plane surfaces that bind the solid is called its surface area. Surface area is measured in square units like square centimetre, square metre etc.

For example A cube has 6 surfaces and each surface is in a square like shape. Therefore, its surface area will be $6a^2$ sq units, where a^2 is the area of each surface of the cube.

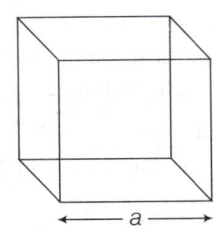

Cube

A solid body having 6 equal faces with equal length, breadth and height is called a **cube**. Infact, each face of a cube is a square.

(i) Volume of the cube = a^3
(ii) Lateral surface area of the cube = $4a^2$
(iii) Total surface area of the cube = $6a^2$
(iv) Diagonal of the cube = $a\sqrt{3}$

where, a = Side (edge) of the cube

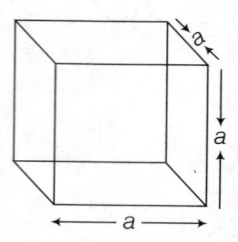

Ex. 1 The diagonal of a cube is $12\sqrt{3}$ cm. Find its volume and surface area.

Sol. Let the edge of cube be a cm.
We know that,
Diagonal of cube = $a\sqrt{3}$
So, $a\sqrt{3} = 12\sqrt{3} \Rightarrow a = 12$
∴ Volume of cube = $a^3 = 12^3 = 1728$ cm^3
Surface area of cube = $6a^2 = 6 \times (12)^2 = 6 \times 144 = 864$ sq cm

Ex. 2 The surface area of a cube is 486 sq cm. Find its volume.

Sol. Let the edge of cube be a cm.
We know that, surface area of cube = $6a^2$
∴ $6a^2 = 486 \Rightarrow a^2 = \dfrac{486}{6} = 81$
$\Rightarrow a = \sqrt{81} = 9$ cm
∴ Volume of cube = $a^3 = 9^3 = 9 \times 9 \times 9 = 729$ cm^3

Ex. 3 If the volumes of two cubical blocks are in the ratio of 8 : 1, what will be the ratio of their edges?

Sol. Let a_1 and a_2 be the sides of two cubes, respectively.
∴ Ratio of volumes = (Ratio of sides)^3
$\Rightarrow \dfrac{8}{1} = \left(\dfrac{a_1}{a_2}\right)^3 \Rightarrow \dfrac{a_1}{a_2} = \sqrt[3]{\dfrac{8}{1}} = \dfrac{2}{1} = 2 : 1$

Hence, the ratio of their edges is 2 : 1.

Cuboid

A rectangular solid body having 6 rectangular faces is called a **cuboid**.

(i) Volume of the cuboid = lbh
(ii) Lateral surface area of a cuboid = $2(l+b)h$
(iii) Total surface area of the cuboid = $2(lb + bh + lh)$
(iv) Diagonal of cuboid = $\sqrt{l^2 + b^2 + h^2}$

where, l = Length, b = Breadth and h = Height

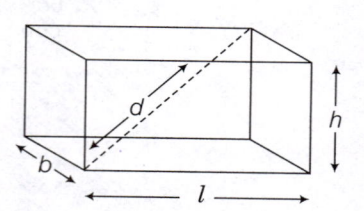

Volume and Surface Area / 609

Some other cube or cuboidal shaped objects are as follow

Room

A rectangular room has 4 walls (surfaces) and opposite walls have equal areas. 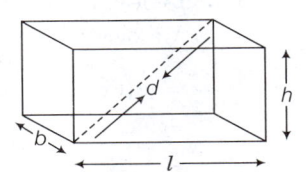
  (i) Total area of walls $= 2(l + b) \times h$
  (ii) Total volume of the room $= lbh$
  (iii) Area of floor or roof $= lb$
      where, l = Length, b = Breadth and h = Height

Box

A box has its shape like cube or cuboid. The amount that a box can hold or contain, is called the capacity of the box. Infact, capacity is the internal volume. 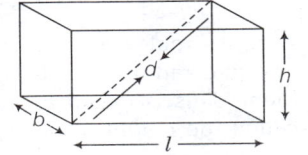
  (i) Surface area of an open box
      $= 2(\text{Length} + \text{Breadth}) \times \text{Height} + \text{Length} \times \text{Breadth}$
      $= 2(l + b)h + lb$
  (ii) Capacity of box $= (l - 2t)(b - 2t)(h - 2t)$
      where, t = Thickness of the box
  (iii) Volume of the material of the box = External volume – Internal volume (capacity)
      $= lbh - (l - 2t)(b - 2t)(h - 2t)$
      where, l = Length, b = Breadth and h = Height

✦ For calculation of any of the parameter, length, breadth and height should be in same units.

Ex. 4 Find the volume and surface area of a cuboid 18 m long, 14 m broad and 7 m high.

Sol. Given, length = 18 m, breadth = 14 m and height = 7 m
∴ Volume of the cuboid = Length × Breadth × Height = $18 \times 14 \times 7 = 1764 \, m^3$
and surface area of cuboid = $2(lb + bh + lh) = 2(18 \times 14 + 14 \times 7 + 18 \times 7)$
$= 2(252 + 98 + 126) = (2 \times 476) = 952$ sq m

Ex. 5 What is the length of the largest pole that can be placed in a room 11 m long, 8 m broad and 9 m high?

Sol. Here, $l = 11 \, m, b = 8 \, m$ and $h = 9 \, m$
∴ Length of the largest pole = Length of the diagonal of room
$= \sqrt{l^2 + b^2 + h^2} = \sqrt{11^2 + 8^2 + 9^2}$
$= \sqrt{121 + 64 + 81} = \sqrt{266} = \sqrt{4 \times 66.5} = 2\sqrt{66.5}$ m

Ex. 6 A wooden box measures 10 cm × 6 cm × 5 cm. Thickness of wood is 2 cm. Find the volume of the wood required to make the box.

Sol. Given, $l = 10 \, cm, b = 6 \, cm, h = 5 \, cm$ and $t = 2 \, cm$
∴ External volume of wooden box = $l \times b \times h = 10 \times 6 \times 5 = 300 \, cm^3$
and internal volume of wooden box = $(l - 2t)(b - 2t)(h - 2t)$
$= (10 - 4) \times (6 - 4) \times (5 - 4) = 6 \times 2 \times 1 = 12 \, cm^3$
∴ Volume of the wood = External volume of wooden box
                                        – Internal volume of wooden box
$= 300 - 12 = 288 \, cm^3$

Ex. 7 The length, breadth and height of a cuboid are in the ratio of 6 : 5 : 4 and its whole surface area is 66600 cm². What is its volume?

Sol. Let length, breadth and height of a cuboid be $6x$, $5x$ and $4x$, respectively.
We know that,
Total surface area of cuboid = $2(lb + bh + hl)$ cm² $\Rightarrow 2(lb + bh + lh) = 66600$
$\Rightarrow 2(6x \times 5x + 5x \times 4x + 6x \times 4x) = 66600 \Rightarrow 148x^2 = 66600$
$\Rightarrow x^2 = \dfrac{66600}{148} = 450 \Rightarrow x = \sqrt{450} = \sqrt{2 \times 225} = 15\sqrt{2}$ cm

∴ Volume of the cuboid = $l \times b \times h = 6x \times 5x \times 4x$
$= 120x^3 = 120(15\sqrt{2})^3 = 810000\sqrt{2}$ cm³

Right Circular Cylinder

A right circular cylinder is a solid or a hollow body with circular ends of equal radius and the line joining their centres perpendicular to them. This is called axis of the cylinder. The length of the axis is called the height of the cylinder.

A cylinder has three surfaces

(i) Curved surface (ii) Bottom (iii) Top

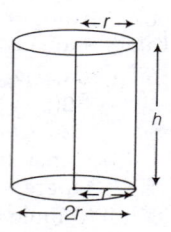

Solid Cylinder

(i) Volume of cylinder = Area of base × Height = $\pi r^2 h$
(ii) Curved surface area = Perimeter of base × Height = $2\pi rh$
(iii) Total surface area = Curved surface area + Area of both the circles (top and bottom surfaces) = $2\pi rh + 2\pi r^2 = 2\pi r(h + r)$

where, r = Radius of base and h = Height

Hollow Cylinder

If the cylinder is hollow, then
(i) Volume of hollow cylinder = Outer volume – Inner volume
$= \pi h(R^2 - r^2)$
(ii) Curved surface area = Curved surface area of outer surface
+ Curved surface area of inner surface
$= 2\pi Rh + 2\pi rh = 2\pi h(R + r)$
(iii) Total surface area of hollow cylinder
= Curved surface area + Area of both top and bottom surface
$= 2\pi h(R + r) + 2\pi(R^2 - r^2)$

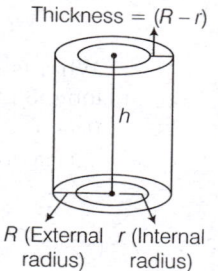

where, R = External radius of base, r = Internal radius of base and h = Height

Ex. 8 Find the volume, curved surface area and the total surface area of a cylinder with diameter of base 14 cm and height 80 cm.

Sol. Given, diameter of base = 14 cm
∴ Radius of base = $\dfrac{14}{2}$ = 7 cm and height of cylinder, h = 80 cm

Now, volume of cylinder = $\pi r^2 h = \dfrac{22}{7} \times 7 \times 7 \times 80 = (22 \times 1 \times 7 \times 80) = 12320$ cm³

Volume and Surface Area / 611

Curved surface area of cylinder = $2\pi rh = 2 \times \dfrac{22}{7} \times \dfrac{14}{2} \times 80 = (44 \times 80) = 3520 \text{ cm}^2$

∴ Total surface area of cylinder
= Curved surface area + Area of both top and bottom surface
= $2\pi rh + 2\pi r^2 = 2\pi r(h+r) = 2 \times \dfrac{22}{7} \times 7 \times (80 + 7) = 3828 \text{ cm}^2$

Ex. 9 How many iron rods each of length 14 m and diameter 4 cm can be made out of 0.88 m³ of iron?

Sol. Given, diameter of rod = 4 cm

∴ Radius of rod = $\dfrac{4}{2} = 2 \text{ cm} = \dfrac{2}{100}$ m and length of rod, $h = 14$ m

∴ Volume of one iron rod = $\pi r^2 h = \dfrac{22}{7} \times \dfrac{2}{100} \times \dfrac{2}{100} \times 14$

$= 22 \times \dfrac{1}{50} \times \dfrac{1}{50} \times 2 = \dfrac{44}{2500} = \dfrac{11}{625} \text{ m}^3$

∵ Volume of iron = 0.88 m³

∴ Number of rods = $\dfrac{\text{Volume of iron}}{\text{Volume of one iron rod}} = \dfrac{0.88}{11/625} = 0.88 \times \dfrac{625}{11} = 50$

Ex. 10 A hollow cylinder made of wood has thickness 1 cm while its external radius is 3 cm. If the height of the cylinder is 8 cm, find the volume, curved surface area and total surface area of the cylinder.

Sol. Given, external radius of hollow cylinder, $R = 3$ cm
and height, $h = 8$ cm

∵ Inner radius of hollow cylinder, r = External radius − Thickness
= 3 − 1 = 2 cm

∴ Required volume = $\pi h(R^2 - r^2) = \dfrac{22}{7} \times 8 \,(3^2 - 2^2)$

$= \dfrac{22}{7} \times 8 \times 5 = \dfrac{880}{7} \text{ cm}^3$

Curved surface area = $2\pi h(R + r) = 2 \times \dfrac{22}{7} \times 8 \times 5 = \dfrac{1760}{7} \text{ cm}^2$

∴ Total surface area = $2\pi h\,(R + r) + 2\pi(R^2 - r^2)$

$= 2 \times \dfrac{22}{7} \times 8 \,(3 + 2) + 2 \times \dfrac{22}{7}\,(3^2 - 2^2)$

$= 2 \times \dfrac{22}{7} \times 8 \times 5 + 2 \times \dfrac{22}{7}\,(5) = \dfrac{1760}{7} + \dfrac{220}{7} = \dfrac{1980}{7} \text{ cm}^2$

Right Circular Cone

Right Circular Cone is a solid or hollow body with a round base and pointed top. It is formed by the rotation of a right angled triangle around its height

(i) Volume = $\dfrac{1}{3} \times$ Base area × Height = $\dfrac{1}{3} \pi r^2 h$

(ii) Slant height $(l) = \sqrt{r^2 + h^2}$

(iii) Curved surface area = $\pi r l = \pi r \sqrt{r^2 + h^2}$

(iv) Total surface area = Curved surface area + Area of base
$= \pi r l + \pi r^2 = \pi r(l + r)$

where, r = Radius of base, h = Height and l = Slant height

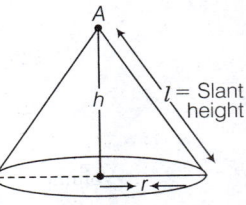

Frustum of Right Circular Cone

If a cone is cut by a plane parallel to the base, so as to divide the cone into two parts, upper part and lower part, then the lower part is called frustum of cone.

(i) Volume of frustum = $\dfrac{\pi h}{3}(r^2 + R^2 + rR)$

(ii) Slant height $(l) = \sqrt{h^2 + (R-r)^2}$

(iii) Curved surface area = $\pi(R+r)l$

(iv) Total surface area = $\pi\{(r+R)l + r^2 + R^2\}$

where, r = Radius of top, R = Radius of base, h = Height and l = Slant height

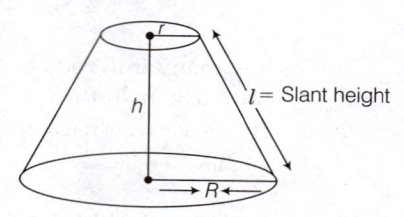

Ex. 11 The diameter of a right circular cone is 14 m and its slant height is 10 m. Find its curved surface area, total surface area and volume.

Sol. Given, diameter of cone = 14 m

$\therefore$ Radius of cone, $r = \dfrac{14}{2} = 7$ m and slant height, $l = 10$ m

$\therefore$ Curved surface area = $\pi r l = \dfrac{22}{7} \times 7 \times 10 = 22 \times 10 = 220\, m^2$

Total surface area = $\pi r(r+l) = \dfrac{22}{7} \times 7(7+10) = 22 \times 17 = 374\, m^2$

$\therefore$ Volume = $\dfrac{1}{3}\pi r^2 h = \dfrac{1}{3}\pi r^2 \times \sqrt{l^2 - r^2}$ $[\because l^2 = h^2 + r^2]$

$= \dfrac{1}{3} \times \dfrac{22}{7} \times (7)^2 \times \sqrt{10^2 - 7^2} = \dfrac{154}{3}\sqrt{51}\, m^3$

Ex. 12 Find the cost of colouring the total surface of the right circular cone in example 12, if the rate of colouring is 14 paise per sq m.

Sol. $\because$ Calculated total surface area = 374 m²   [from Ex. 11]

$\therefore$ Required cost = $374 \times \dfrac{14}{100}$ = ₹ 52.36

Ex. 13 A frustum of a right circular cone has a diameter of base and top 20 cm and 12 cm, respectively and a height of 10 cm. Find the area of its whole surface and volume.

Sol. Given, $R = 10$ cm, $r = 6$ cm and $h = 10$ cm

Slant height $(l) = \sqrt{h^2 + (R-r)^2} = \sqrt{10^2 + (10-6)^2} = \sqrt{100 + 16} = 10.77$ cm

$\therefore$ Whole surface area = $\pi(R^2 + r^2 + Rl + rl)$

$= \dfrac{22}{7}(100 + 36 + 10 \times 10.77 + 6 \times 10.77)$

$= \dfrac{22}{7}(136 + 107.7 + 64.62)$

$= \dfrac{22}{7} \times 308.32 = \dfrac{6783.04}{7} = 969\, cm^2$ (approx.)

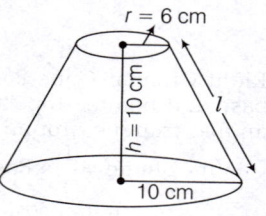

$\therefore$ Volume = $\dfrac{\pi h}{3}(R^2 + r^2 + Rr) = \dfrac{22}{7} \times \dfrac{10}{3}(100 + 36 + 60)$

$= \dfrac{22}{7} \times \dfrac{10}{3} \times 196 = \dfrac{43120}{21} = 2053.33\, cm^3$

Volume and Surface Area / 613

Sphere

A **sphere** is a three-dimensional solid figure, which is made up of all points in the space, which lie at a constant distance from a fixed point. That constant distance and fixed point are respectively called the radius and centre of the sphere. Infact, a sphere is like a solid ball.

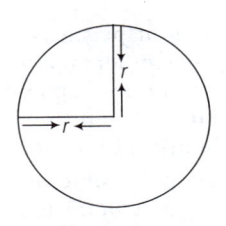

(i) Volume of the sphere $= \dfrac{4}{3}\pi r^3$

(ii) Total surface area $= 4\pi r^2$, where r = Radius

Hollow Sphere or Spherical Shell

Its both external and internal surfaces are spherical and both the surfaces have a common central point.

(i) Volume of hollow sphere $= \dfrac{4}{3}\pi(R^3 - r^3)$

(ii) Internal surface area $= 4\pi r^2$

(iii) External surface area $= 4\pi R^2$

where, R = External radius and r = Internal radius

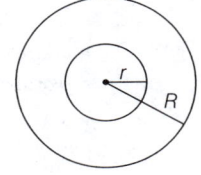

Hemisphere

It is the half part of a sphere.

(i) Volume of the hemisphere $= \dfrac{2}{3}\pi r^3$

(ii) Total surface area $= 3\pi r^2$

(iii) Curved surface area $= 2\pi r^2$, where r = Radius

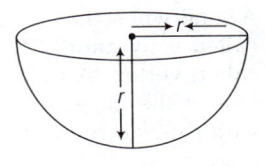

Ex. 14 Find the volume and the surface area of a sphere of diameter 14 cm.

Sol. Given, diameter of sphere = 14 cm

∴ Radius of sphere, $r = \dfrac{14}{2} = 7$ cm

∵ Volume of sphere $= \dfrac{4}{3}\pi r^3 = \dfrac{4}{3} \times \dfrac{22}{7} \times 7 \times 7 \times 7 = \dfrac{4 \times 22 \times 49}{3} = 1437.33$ cm^3

∴ Surface area of sphere $= 4\pi r^2 = 4 \times \dfrac{22}{7} \times 7 \times 7 = 616$ sq cm

Ex. 15 Find the volume, curved surface area and the total surface area of a hemisphere of radius 7 cm.

Sol. Given, radius of hemisphere, $r = 7$ cm

∴ Volume of hemisphere $= \dfrac{2}{3}\pi r^3 = \dfrac{2}{3} \times \dfrac{22}{7} \times 7 \times 7 \times 7 = \dfrac{44 \times 49}{3} = 718.66$ cm^3

Curved surface area of hemisphere $= 2\pi r^2 = 2 \times \dfrac{22}{7} \times 7 \times 7 = 44 \times 7 = 308$ sq cm

and total surface area of hemisphere $= 3\pi r^2 = 3 \times \dfrac{22}{7} \times 7 \times 7 = 462$ sq cm

Prism

A right **prism** is a solid whose top and bottom faces are parallel to each other and are identical polygons that are parallel. The faces joining the top and bottom faces are rectangles and are called **lateral faces**. The distance between the base and the top is called **height** or **length** of the right prism.

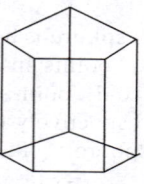

(i) Volume of prism = Area of base × Height of the prism
(ii) Lateral (curved) surface area = Perimeter of base × Height of prism
(iii) Total surface area = Lateral surface area + 2 × Area of base

Ex. 16 The base of a right prism is a square having side of 10 cm. If its height is 8 cm, then find the total surface area and volume of the prism.

Sol. Given, side = 10 cm and height = 3 cm
Now, lateral surface area = Perimeter of the base × Height
= [10 + 10 + 10 + 10] × 8 = 40 × 8 = 320 cm^2 [∵ base of prism is a square]
and area of base = Area of square = 10 × 10 = 100 cm^2
∴ Total surface area = Lateral surface area + 2 × Area of base
= 320 + 2 × 100 = 320 + 200 = 520 cm^2
and Volume of the prism = Area of base × Height of prism = 100 × 8 = 800 cm^3

Pyramid

A solid whose base is a polygon and whose faces are triangles, is called a **pyramid**. The triangular faces meet at a common point called vertex. 'A pyramid whose base is regular polygon and the foot of the perpendicular from the vertex to the base, coincides with the centre of the base, is called a right pyramid.'

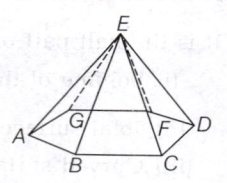

(i) Volume of a pyramid = $\frac{1}{3}$ × Area of base × Height (vertical)
(ii) Lateral surface area = $\frac{1}{2}$ × Perimeter of the base × Slant height
(iii) Total surface area = Lateral surface area + Area of the base

Ex. 17 The base of a pyramid is a square whose side is 10 cm. It's slant and vertical heights are 13 cm and 12 cm, respectively. Then, find the total surface area and volume of the pyramid.

Sol. Given, side = 10 cm, height = 12 cm and slant height = 13 cm
Now, lateral surface area = $\frac{1}{2}$ × Perimeter of the base × Slant height
= $\frac{1}{2}$ × (10 × 4) × 13 = 260 cm^2
[∵ perimeter of base = perimeter of square = 4a]
and area of the base = (side)^2 = (10)^2 = 100 cm^2
∴ Total surface area = Lateral surface area + Area of the base = 260 + 100 = 360 cm^2
and Volume of the pyramid = $\frac{1}{3}$ × Area of base × Height (Vertical)
= $\frac{1}{3}$ × 10 × 10 × 12 = 400 cm^3 [∵ area of base = area of square = (side)^2]

Volume and Surface Area

Fast Track Techniques to solve the QUESTIONS

Technique 1 If length, breadth and height of a cuboid are changed by $x\%$, $y\%$ and $z\%$ respectively, then its volume is increased by

$$\left[x + y + z + \frac{xy + yz + xz}{100} + \frac{xyz}{(100)^2} \right]\%.$$

+ Increment in the value is taken as positive and decrement in value is taken as negative. Positive result shows total increment and negative result shows total decrement.

Ex. 18 If all the dimensions of a cuboid are increased by 100%, by what per cent does the volume of cuboid increase?

Sol. Let the length, breadth and height of the cuboid be l cm, b cm and h cm, respectively.
Then, volume of cuboid = $l \times b \times h = lbh$ cm^3
Now, the dimensions are increased by 100%.
∴ New length = $2l$ cm, new breadth = $2b$ cm and new height = $2h$ cm
Now, new volume of cuboid = $2l \times 2b \times 2h = 8lbh$ cm^3
∴ Required percentage increase = $\left(\dfrac{8lbh - lbh}{lbh} \right) \times 100\% = 700\%$

Fast Track Method
Here, $x = y = z = 100$
According to the formula,

Percentage increase in volume = $\left[x + y + z + \dfrac{xy + yz + zx}{100} + \dfrac{xyz}{(100)^2} \right]\%$

$= \left[100 + 100 + 100 + \dfrac{100 \times 100 + 100 \times 100 + 100 \times 100}{100} + \dfrac{100 \times 100 \times 100}{(100)^2} \right]\%$

$= \left[300 + \dfrac{10000 + 10000 + 10000}{100} + \dfrac{1000000}{10000} \right]\%$

$= \left[300 + \dfrac{30000}{100} + 100 \right]\% = (300 + 300 + 100)\% = 700\%$

Technique 2 If side of a cube or radius (or diameter) of sphere is increased by $x\%$, then its volume increases by $\left[\left(1 + \dfrac{x}{100}\right)^3 - 1 \right] \times 100\%$.

Ex. 19 If side of a cube is increased by 10%, by how much per cent does its volume increase?

Sol. Let the side of the cube be 10 cm.
∴ Volume of cube = $(10)^3 = 1000$ cm^3
Now, each side is increased by 10%.

∴ New side of cube = $\frac{110}{100} \times 10 = 11$ cm

and new volume of cube = $(11)^3 = 1331\,\text{cm}^3$

∴ Required percentage increase = $\left(\frac{1331-1000}{1000}\right) \times 100\% = 33.1\%$

Fast Track Method

Here, $x = 10$

According to the formula,

Percentage increase in volume = $\left[\left(1 + \frac{x}{100}\right)^3 - 1\right] \times 100\%$

$= \left[\left(1 + \frac{10}{100}\right)^3 - 1\right] \times 100\%$

$= \left[\left(1 + \frac{1}{10}\right)^3 - 1\right] \times 100\% = \left[\left(\frac{11}{10}\right)^3 - 1\right] \times 100\%$

$= \left[\frac{1331}{1000} - 1\right] \times 100\%$

$= (1.331 - 1) \times 100\%$

$= (0.331 \times 100)\% = 33.1\%$

Technique ③ If height of a cylinder is changed by $x\%$ and radius remains unchanged, then the volume changes by $x\%$.

Ex. 20 If height of a cylinder is increased by 4%, while radius remains unchanged, by how much per cent volume increases?

Sol. Here, $x = 4$

According to the formula, percentage increase in volume = 4%

Technique ④ If radius of a cylinder is changed by $x\%$ and height remains unchanged, then volume changes by $\left(2x + \frac{x^2}{100}\right)\%$.

♦ If edge of a cube or radius of sphere is changed by $x\%$, then surface area changes by $\left(2x + \frac{x^2}{100}\right)\%$.

Ex. 21 If radius of a cylinder is increased by 10%, while height remains unchanged, by what per cent does the volume of cylinder increase?

Sol. Let the radius and height of cylinder be r and h, respectively.

∴ Volume of cylinder = $\pi r^2 h$

Now, radius is increased by 10%.

∴ New radius = $\frac{110}{100} \times r = 1.1\,r$

and new volume of cylinder = $\pi (1.1\,r)^2 h = 1.21\,\pi r^2 h$

∴ Required percentage change = $\left(\frac{1.21\,\pi r^2 h - \pi r^2 h}{\pi r^2 h}\right) \times 100\% = 21\%$

Volume and Surface Area / 617

Fast Track Method

Here, $x = 10$. According to the formula,

Percentage increase in volume $= \left[2x + \dfrac{x^2}{100}\right]\%$

$= \left(2 \times 10 + \dfrac{10 \times 10}{100}\right)\% = 21\%$

Ex. 22 When the radius of an sphere is increased by 4%, what per cent increase takes place in surface area of sphere?

Sol. Here, $x = 4$

According to the formula,

Percentage increase in area $= \left[2x + \dfrac{x^2}{100}\right]\% = \left[2 \times 4 + \dfrac{(4)^2}{100}\right]\% = \left[8 + \dfrac{16}{100}\right]\% = 8.16\%$

Technique 5 If radius of a cylinder or cone is changed by $x\%$ and height is changed by $y\%$, then volume changes by $\left[2x + y + \dfrac{x^2 + 2xy}{100} + \dfrac{x^2 y}{100^2}\right]\%$.

Ex. 23 If radius of a cylinder is decreased by 4%, while its height is increased by 2%, then what will be effect on volume?

Sol. Let the radius and height of cylinder be r and h, respectively.

$\therefore$ Volume of cylinder $= \pi r^2 h$

Now, radius is decreased by 4% and height is increased by 2%

$\therefore$ New radius $= \dfrac{96}{100} \times r = 0.96 r$

and new height $= \dfrac{102}{100} \times h = 1.02 h$

New volume of cylinder $= \pi (0.96 r)^2 (1.02 h) = 0.940032 \pi r^2 h$

$\therefore$ Required percentage in volume $= \dfrac{(0.940032 - 1)\pi r^2 h}{\pi r^2 h} \times 100 = -5.99\%$ (decrease)

[here, negative sign shows decrease in volume]

Fast Track Method

Here, $x = -4$ and $y = 2$

According to the formula,

Net effect on volume $= \left[2x + y + \dfrac{x^2 + 2xy}{100} + \dfrac{x^2 y}{100^2}\right]\%$

$= \left[2 \times (-4) + 2 + \dfrac{(-4)^2 + 2 \times (-4)2}{100} + \dfrac{(-4)^2 \times 2}{100^2}\right]\%$

[negative sign shows that there is a decrease in radius]

$= \left[-8 + 2 + \dfrac{16 - 16}{100} + \dfrac{32}{100^2}\right]\%$

$= [-6 + 0 + 0.0032]\%$

$= -5.99\%$ (decrease)

618 / Fast Track Objective Arithmetic

Technique 6 If in a cylinder or cone, height and radius both change by $x\%$, then volume changes by $\left[\left(1 \pm \dfrac{x}{100}\right)^3 - 1\right] \times 100\%$.

✦ Take '+ve' sign for increase and '–ve' sign for decrease.

Ex. 24 If in a cylinder, both height and radius increase by 100%, by what per cent does its volume increase?

Sol. Let the height and radius of the cylinder be h and r, respectively.
Then, volume of cylinder $= \pi r^2 h$
Now, both height and radius are increased by 100%.
∴ New height $= 2h$ and new radius $= 2r$
∴ New volume of cylinder $= \pi(2r)^2 (2h) = 8\pi r^2 h$
∴ Required percentage increase $= \left(\dfrac{8\pi r^2 h - \pi r^2 h}{\pi r^2 h}\right) \times 100\% = 700\%$

Fast Track Method
Here, $x = 100$
According to the formula,

Percentage increase in volume $= \left[\left(1 + \dfrac{x}{100}\right)^3 - 1\right] \times 100\%$

$= \left[\left(1 + \dfrac{100}{100}\right)^3 - 1\right] \times 100\%$

$= [(2)^3 - 1] \times 100\%$

$= (8 - 1) \times 100\% = 700\%$

Ex. 25 By what per cent does the volume of cylinder decrease, if there is 100% decrease in its radius and height?

Sol. Here, $x = -100$
According to the formula,

Percentage decrease in volume $= \left[\left\{1 + \dfrac{x}{100}\right\}^3 - 1\right] \times 100\%$

$= \left[\left\{1 + \dfrac{(-100)}{100}\right\}^3 - 1\right] \times 100\%$ [negative signs indicates decrease]

$= [(1-1)^3 - 1] \times 100\% = -100\%$

If two measuring dimensions which are included in the surface area of a cube, cuboid, sphere, cylinder or cone are increased or decreased by $x\%$ and $y\%$, then the surface area of the figure will increase or decrease by $\left(x + y + \dfrac{xy}{100}\right)\%$

[take positive sign for increase and negative sign for decrease]

Volume and Surface Area / 619

Ex. 26 When each side of a cube is increased by 20%, then find the increase in total surface area of a cube.

Sol. Here, $x = y = 20$
According to the formula,

Total surface area $= \left(x + y + \dfrac{xy}{100}\right)\%$

$= \left(20 + 20 + \dfrac{20 \times 20}{100}\right)\%$

$= \left(40 + \dfrac{400}{100}\right)\%$

$= 44\%$

Technique 7 Three cubes of metal whose sides are x, y and z respectively, are melted to form a new (bigger) cube. If there is no loss of weight in this process. Then, side of new cube will be $\sqrt[3]{x^3 + y^3 + z^3}$.

Ex. 27 The sides of three cubes of metal are 30 cm, 40 cm and 50 cm, respectively. Find the side of new cube formed by melting these cubes together.

Sol. Since, three cubes are melted to form a new cube.
∴ Sum of volume of three cubes = Volume of new cube
Let the side of new cube be a cm.
Then, $(30)^3 + (40)^3 + (50)^3 = a^3$

$\Rightarrow \qquad a^3 = (27000 + 64000 + 125000)\,\text{cm}^3$

$\Rightarrow \qquad a = \sqrt[3]{216000} = 60\,\text{cm}$

Hence, the side of new cube is 60 cm.

Fast Track Method

Here, $x = 30$, $y = 40$ and $z = 50$

∴ Side of new cube $= \sqrt[3]{x^3 + y^3 + z^3}$

$= \sqrt[3]{(30)^3 + (40)^3 + (50)^3}$

$= \sqrt[3]{27000 + 64000 + 125000}$

$= \sqrt[3]{216000} = 60\,\text{cm}$

Multi Concept Questions

1. A tank is 7 m long and 4 m wide. At what speed should water run through a pipe 5 cm broad and 4 cm deep, so that in 6 h and 18 min, water level in the tank rises by 4.5 m?
   (a) 12 km/h   (b) 10 km/h   (c) 14 km/h   (d) 18 km/h

   ↪ **(b)** Volume of water flown into the tank
   = Length of tank × Breadth of tank × Height of water rise
   = $7 \times 4 \times 4.5 = 126 \, m^3$

   Time 6 h 18 m = $\left(6 + \dfrac{18}{60}\right)h = \left(6 + \dfrac{3}{10}\right)h = \dfrac{63}{10}h$

   Let the length of water flown through pipe be x m.
   Then, volume of water flown through pipe = Volume of water flown in the tank

   $x \times \dfrac{5}{100} \times \dfrac{4}{100} = 126 \Rightarrow x = \dfrac{126 \times 100 \times 100}{5 \times 4}$

   $\Rightarrow \quad x = 126 \times 5 \times 100 \, m \Rightarrow x = \dfrac{126 \times 5 \times 100}{1000} \, km$

   ∴ Speed of water flowing = $\dfrac{\text{Length of water flown}}{\text{Time taken}}$

   $= \dfrac{126 \times 5 \times 100}{1000 \times \dfrac{63}{10}} = \dfrac{126 \times 5}{63} = 10 \, km/h$

2. A water supply tank as shown below in the figure with given dimensions is used in a house. 5 members use full tank for five days. If two members use the tank full, how long the tank works for proper supply?

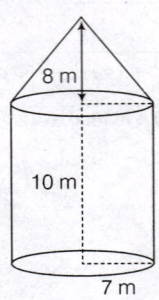

   (a) 10.5 days   (b) 12.5 days
   (c) 15.5 days   (d) 8 days

   ↪ **(b)** Water available in the tank = Volume of cone + Volume of cylinder
   $\Rightarrow \quad W = \dfrac{1}{3}\pi r^2 h_1 + \pi r^2 h_2$

   Here, $r = 7$, h_1 = Height of cone = 8 and h_2 = Height of cylinder = 10

Volume and Surface Area / 621

$\therefore$ Volume of water $= \dfrac{1}{3} \times \dfrac{22}{7} \times (7)^2 \times 8 + \dfrac{22}{7} \times (7)^2 \times 10$

$= \dfrac{1}{3} \times 22 \times 56 + 22 \times 7 \times 10$

$= 410.66 + 1540 = 1950.66 \text{ m}^3$

Since, 5 members use 1950.66 m³ water in five days.

$\therefore$ 5 members use $\dfrac{1950.66}{5}$ m³ water in one day.

$\therefore$ 1 member use water in 1 day $= \dfrac{1950.66}{5 \times 5} = 78.0264 \text{ m}^3$

$\therefore$ Number of days for which two members can use water $= \dfrac{1950.66}{78.0264 \times 2} = 12.5$

Alternate Method

For 5 members, tank lasts for 5 days
For 1 member, tank will last for (5 × 5) days
For 2 members, the tank will last for $\left(\dfrac{5 \times 5}{2}\right)$ dyas = 12.5 days

3. A milk tank of following dimensions used to carry the milk to the milk booth. If 50 tanks used daily, then find the consumption of milk at milk booth.

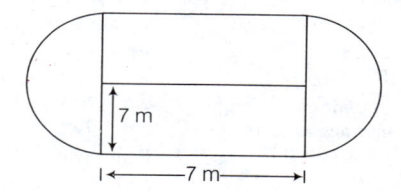

[Figure: capsule-shaped tank with height 7 m and length 7 m]

(a) 185766.66 m³   (b) 105775.66 m³   (c) 155766.66 m³   (d) 125766.66 m³

↪ **(d)** Milk tank capacity = Volume of two hemisphere + Volume of cylinder

$= 2 \times \dfrac{2}{3} \pi r^3 + \pi r^2 h = \dfrac{4}{3} \times \dfrac{22}{7} \times 7 \times 7 \times 7 + \dfrac{22}{7} \times (7)^2 \times 7$

$= \dfrac{4}{3} \times 22 \times 7 \times 7 + 22 \times 7^2 = 22 \times 7^2 \times \dfrac{7}{3} = 2515.33 \text{ m}^3$

$\therefore$ Daily consumption of milk at booth
= Milk in one tank × Number of tanks
= 2515.33 × 50 = 125766.66 m³

4. A cylindrical jar, whose base has a radius of 15 cm, is filled with water upto a height of 20 cm. A solid iron spherical ball of radius 10 cm is dropped in the jar to submerge completely in water. Find the increase in the level of water.   [SSC CPO 2013]

(a) $5\dfrac{17}{27}$ cm   (b) $5\dfrac{5}{7}$ cm   (c) $5\dfrac{8}{9}$ cm   (d) $5\dfrac{25}{27}$ cm

↪ **(d)** Let the level of water will be increased by h cm.

Then, $\pi \times (15)^2 \times h = \dfrac{4}{3} \pi (10)^3$

$\therefore \quad h = \dfrac{4}{3} \times \dfrac{10 \times 10 \times 10}{15 \times 15} = 5\dfrac{25}{27}$ cm

Fast Track Practice

Cube, Cuboid, Room or Box

1. The diagonal of a cube is $2\sqrt{3}$ cm. Find the surface area of the cube.
   (a) 15 sq cm   (b) 18 sq cm
   (c) 25 sq cm   (d) 24 sq cm

2. If the total surface area of a cube is 6 sq units, then what is the volume of the cube? [CDS 2013]
   (a) 1 cu unit   (b) 2 cu units
   (c) 4 cu units (d) 6 cu units

3. If the volume of a cube is 729 cm^3, what is the length of its diagonal? [CDS 2012]
   (a) $9\sqrt{2}$ cm   (b) $9\sqrt{3}$ cm
   (c) 18 cm   (d) $18\sqrt{3}$ cm

4. The surface area of a cube is 726 cm^2. Find the volume of the cube.
   [SSC (10+2) 2008]
   (a) 1331 cm^3 (b) 1232 cm^3
   (c) 1626 cm^3 (d) 1836 cm^3

5. The ratio of volumes of two cubes is 8 : 125. The ratio of their surface area is
   [SSC (10+2) 2014]
   (a) 4 : 25   (b) 2 : 75
   (c) 2 : 15   (d) 4 : 15

6. The volume of a cube is numerically equal to sum of length of its edges. What is the total surface area in square units?
   [CDS 2012]
   (a) 12   (b) 36   (c) 72   (d) 144

7. Three cubes of sides 1 cm, 6 cm and 8 cm are melted to form a new cube. Find half of the surface area of the new cube.
   (a) 243 sq cm   (b) 463 sq cm
   (c) 486 sq cm   (d) 293 sq cm

8. If each side of a cube is decreased by 19%, then decrease in surface area is
   [Hotel Mgmt. 2010]
   (a) 40%   (b) 38.4%
   (c) 35%   (d) 34.39%

9. If the side of a cube is increased by 12%, by how much per cent does its volume increase?
   (a) 40.4928%   (b) 50.5240%
   (c) 60.3292%   (d) 30.4928%
   (e) None of these

10. Find the surface area of a cuboid 10 m long, 5 m broad and 3 m high.
    [Bank Clerk 2008]
    (a) 105 sq m   (b) 104 sq m
    (c) 170 sq m   (d) 190 sq m
    (e) None of these

11. If the length and breadth of a cuboid are increase by 5% and 4%, respectively. Then, the percentage increase in its volume is
    (a) 8%   (b) 7%
    (c) 9.2% (d) 10%

12. The maximum length of a pencil that can be kept in a rectangular box of dimensions 8 cm × 6 cm × 2 cm, is [SNAP 2012]
    (a) $2\sqrt{13}$ cm   (b) $2\sqrt{14}$ cm
    (c) $2\sqrt{26}$ cm   (d) $10\sqrt{2}$ cm

13. The capacity of a cuboid tank of water is 50000 L. Find the breadth of the tank, if its length and depth are 2.5 m and 10 m, respectively.
    (a) 2 m   (b) 4 m
    (c) 9 m   (d) 6 m

14. The area of four walls of a room is 120 m^2. The length of the room is twice its breadth. If the height of the room is 4 m, what is area of the floor? [CDS 2016 (II)]
    (a) 40 m^2 (b) 50 m^2
    (c) 60 m^2 (d) 80 m^2

15. The whole surface area of a rectangular block is 8788 sq cm. If length, breadth and height are in the ratio of 4 : 3 : 2, then find the length. [SSC CGL 2008]
    (a) 26 cm   (b) 52 cm
    (c) 104 cm (d) 13 cm

16. A metal box measures 20 cm × 12 cm × 5 cm. Thickness of the metal is 1 cm. Find the volume of the metal required to make the box.
    (a) 550 cm^3 (b) 656 cm^3
    (c) 660 cm^3 (d) 475 cm^3
    (e) None of these

17. A cube has each edge 2 cm and a cuboid is 1 cm long, 2 cm wide and 3 cm high. The paint in a certain container is sufficient to paint an area equal to 54 cm^2. Which one of the following is correct? [CDS 2014]
    (a) Both cube and cuboid can be painted
    (b) Only cube can be painted

(c) Only cuboid can be painted
(d) Neither cube nor cuboid can be painted

18. If the areas of three adjacent faces of a cuboidial box are 120 sq cm, 72 sq cm and 60 sq cm respectively, then find the volume of the box.
    (a) 820 cm^3 (b) 720 cm^3
    (c) 750 cm^3 (d) 750 cm^3
    (e) None of these

19. From the four corners of a rectangular sheet of dimensions 25×20 cm, square of side 2 cm is cut off from four corners and a box is made. The volume of the box is **[SSC CGL (Mains) 2016]**
    (a) 828 cm^3 (b) 672 cm^3
    (c) 500 cm^3 (d) 1000 cm^3

20. Internal length, breadth and height of a rectangular box are 10 cm, 8 cm and 6 cm, respectively. How many boxes are needed in which cubes whose volume is 6240 cu cm can be packed? **[CTET 2012]**
    (a) 12 (b) 13 (c) 15 (d) 17

21. What are the dimensions (length, breadth and height, respectively) of a cuboid with volume 720 cu cm, surface area 484 sq cm and the area of the base 72 sq cm? **[CDS 2012]**
    (a) 9, 8 and 10 cm (b) 12, 6 and 10 cm
    (c) 18, 4 and 10 cm (d) 30, 2 and 12 cm

22. The diagonals of three faces of a cuboid are 13, $\sqrt{281}$ and 20 linear units. Then, the total surface area of the cuboid is **[CDS 2015 (I)]**
    (a) 650 sq units (b) 658 sq units
    (c) 664 sq units (d) 672 sq units

23. A cubic metre of copper weighing 9000 kg is rolled into a square bar 9 m long. An exact cube is cut off from the bar. How much does the cube weight? **[CDS 2016 (II)]**
    (a) 1000 kg (b) $\frac{1000}{3}$ kg
    (c) 300 kg (d) $\frac{500}{3}$ kg

24. A rectangular block of wood having dimensions 3 m × 2 m × 1.75 m has to be painted on all its faces. The layer of paint must be 0.1 mm thick. Paint comes in cubical boxes having their edges equal to 10 cm. The minimum number of boxes of paint to be purchased is **[CDS 2015 (I)]**
    (a) 5 (b) 4 (c) 3 (d) 2

Cylinder

25. A pillar 14 cm in diameter is 5 m high. How much material was used to construct it? **[Bank Clerk 2010]**
    (a) (77×10^2) cm^3
    (b) (77×10^4) cm^3
    (c) (77×10^5) cm^3
    (d) (77×10^3) cm^3
    (e) None of the above

26. If the lateral surface area of a cylinder is 94.2 sq cm and its height is 5 cm, then find the radius of its base. ($\pi = 3.14$) **[Hotel Mgmt. 2009]**
    (a) 5 cm (b) 8 cm
    (c) 3 cm (d) 4 cm

27. A rod of 2 cm diameter and 30 cm length is converted into a wire of 3 m length of uniform thickness. The diameter of the wire is **[CLAT 2013]**
    (a) 2/10 cm (b) $2/\sqrt{10}$ cm
    (c) $1/\sqrt{10}$ cm (d) 1/10 cm

28. What is the height of a solid cylinder of radius 5 cm and total surface area is 660 sq cm? **[CDS 2012]**
    (a) 10 cm (b) 12 cm
    (c) 15 cm (d) 16 cm

29. The diameter of a roller is 84 cm and its length 120 cm. It takes 500 complete revolutions to move once over to level a playground. Find the area of the playground (in sq m).
    (a) 1632 (b) 1817
    (c) 1532 (d) 1584
    (e) None of these

30. The respective ratio of radii of two right circular cylinders (A and B) is 2 : 3. The respective ratio of volumes of cylinders A and B is 9 : 7, then what are the ratio of heights of cylinders A and B? **[SBI PO (Pre) 2016]**
    (a) 8 : 5 (b) 81 : 28
    (c) 7 : 6 (d) 5 : 4
    (e) 6 : 5

31. The radius of the base and the height of a solid right circular cylinder are in the ratio 2 : 3 and its volume is 1617 cm^3. What is the total surface area of the cylinder? **[CDS 2016 (II)]**
    (a) 462 cm^2 (b) 616 cm^2
    (c) 770 cm^2 (d) 786 cm^2

32. The ratio of the curved surface area to the total surface area of a right circular

cylinder is 1 : 2. If the total surface area is 616 cm², what is the volume of the cylinder? [CDS 2016 (II); SSC CGL 2013]
(a) 539 cm³ (b) 616 cm³
(c) 1078 cm³ (d) 1232 cm³

33. A drainage tile is a cylindrical shell 21 cm long. The inside and outside diameters are 4.5 cm and 5.1 cm, respectively. What is the volume of the clay required for the tile?
(a) 6.96π cm³ (b) 6.76π cm³
(c) 5.76π cm³ (d) None of these

34. The radius of a wire is decreased to one-third. If volume remains the same, length will increase by [SSC CGL (Mains) 2016]
(a) 1.5 times (b) 3 times
(c) 6 times (d) 9 times

35. If the radius of a cylinder is decreased by 8%, while its height is increased by 4%, what will be the effect on volume?
(a) 11.9744% (decrease)
(b) 11.9744% (increase)
(c) 12.4678% (decrease)
(d) 12.4678% (increase)
(e) None of the above

36. The height and the total surface area of a right circular cylinder are 4 cm and 8π sq cm, respectively. The radius of the base of cylinder is [SSC CGL (Mains) 2016]
(a) $(2\sqrt{2} - 2)$ cm (b) $(2 - \sqrt{2})$ cm
(c) 2 cm (d) $\sqrt{2}$ cm

37. The respective ratio between numerical values of curved surface area and volume of right circular cylinder is 1 : 7. If the respective ratio between the diameter and height of the cylinder is 7 : 5, what is the total surface area of the cylinder? [SBI PO (Pre) 2017]
(a) 2992 m² (b) 3172 m²
(c) 2882 m² (d) 3576 m²
(e) 3992 m²

38. 30 metallic cylinders of same size are melted and cast in the form of cones having the same radius and height as those of the cylinders.
Consider the following statements.
Statement I A maximum of 90 cones will be obtained.
Statement II The curved surface of the cylinder can be flattened in the shape of a rectangle but the curved surface of the cone when flattened has the shape of triangle.

Which one of the following is correct in respect of the above? [CDS 2015 (II)]
(a) Both Statement I and Statement II are correct and Statement II is the correct explanation of Statement I
(b) Both Statement I and Statement II are correct and Statement II is not the correct explanation of Statement I
(c) Statement I is correct but Statement II is not correct
(d) Statement I is not correct but Statement II is correct

39. The radius of an open cylindrical milk container is half its height and surface area of the inner part is 616 sq cm. The amount of milk that the container can hold, approximately, is
[use $\sqrt{5} = 2.23$ and $\pi = 22/7$]
[SSC CGL (Mains) 2016]
(a) 1.42 L (b) 1.53 L
(c) 1.71 L (d) 1.82 L

40. Water flows through a cylindrical pipe of internal diameter 7 cm at the rate of 5 m/s. The time, in minutes, the pipe would take to fill an empty rectangular tank 4 m × 3 m × 2.31 m is [CDS 2015 (I)]
(a) 28 (b) 24 (c) 20 (d) 12

41. Three rectangles R_1, R_2 and R_3 have the same area. Their lengths x_1, x_2 and x_3, respectively are such that $x_1 < x_2 < x_3$. If V_1, V_2 and V_3 are the volumes of the cylinders formed from the rectangles R_1, R_2 and R_3, respectively by joining the parallel sides along the breadth, then which one of the following is correct?
[CDS 2015 (I)]
(a) $V_3 < V_2 < V_1$ (b) $V_1 < V_3 < V_2$
(c) $V_1 < V_2 < V_3$ (d) $V_3 < V_1 < V_2$

42. A flask in the shape of a right circular cone of height 24 cm is filled with water. The water is poured in right circular cylindrical flask whose radius is one-third of the radius of the base of the circular cone. Then, the height of the water in the cylindrical flask is
[SSC (10+2) 2014]
(a) 32 cm (b) 24 cm
(c) 48 cm (d) 72 cm

43. Rain water from a roof of 22 m × 20 m drains into a cylindrical vessel having diameter of base 2 m and height 3.5 m. If the vessel is just full, what is the rainfall? [CDS 2016 (II)]
(a) 3.5 cm (b) 3 cm
(c) 2.5 cm (d) 2 cm

Volume and Surface Area / 625

Cone and Frustum of Cone

44. The curved surface area of a right circular cone of radius 14 cm is 440 sq cm. What is the slant height of the cone?
(a) 10 cm (b) 11cm [CDS 2012]
(c) 12 cm (d) 13 cm

45. The diameter of base of a right circular cone is 7 cm and slant height is 10 cm, then what is its lateral surface area?
[CDS 2012]
(a) 110 sq cm (b) 100 sq cm
(c) 70 sq cm (d) 49 sq cm

46. What is the whole surface area of a cone of base radius 7 cm and height 24 cm?
[CDS 2014]
(a) 654 sq cm (b) 704 sq cm
(c) 724 sq cm (d) 964 sq cm

47. The volume of a right circular cone is 100π cm^3 and its height is 12 cm. Find its slant height. [Bank PO 2008]
(a) 13 cm (b) 16 cm (c) 9 cm (d) 26 cm
(e) None of these

48. The diameter of a right circular cone is 14 m, while its slant height is 9 m. Find the volume of the cone.
(a) $\dfrac{49\pi\sqrt{32}}{3}$ m^3 (b) $\dfrac{50\pi\sqrt{32}}{3}$ m^3
(c) $\dfrac{3}{49\pi\sqrt{32}}$ m^3 (d) $\dfrac{\pi\sqrt{32}}{9}$ m^3
(e) None of these

49. The ratio of the radius and height of a cone is 5 : 12. Its volume is $314\dfrac{2}{7}$ cm^3. Its slant height is
(a) 18 cm (b) 13 cm (c) 16 cm (d) 15 cm
(e) None of these

50. If the ratio of volumes of two cones is 2 : 3 and the ratio of the radii of their bases is 1 : 2, then the ratio of their heights will be
(a) 3 : 8 (b) 8 : 3 (c) 9 : 2 (d) 8 : 1
(e) None of these

51. If the volumes of two right circular cones are in the ratio 1 : 3 and their diameters are in the ratio 3 : 5, then the ratio of their heights is [SSC CPO 2013]
(a) 25 : 27 (b) 1 : 5 (c) 3 : 5 (d) 5 : 27

52. A conical cap has the base diameter 24 cm and height 16 cm. What is the cost of painting the surface of the cap at the rate of 70 paise per sq cm? [CDS 2013]
(a) ₹ 520 (b) ₹ 524 (c) ₹ 528 (d) ₹ 532

53. Into a conical tent of radius 8.4 m and vertical height 3.5 m, how many full bags of wheat can be emptied, if space required for the wheat in each bag is 1.96 m^3? [CDS 2016 (II)]
(a) 264 (b) 201
(c) 132 (d) 105

54. Shantanu's cap is in the form of a right circular cone of base radius 7 cm and height 24 cm. Find the area of the sheet required to make 5 such caps.
(a) 5000 sq cm (b) 2750 sq cm
(c) 3000 sq cm (d) 2700 sq cm
(e) None of these

55. The radius of the base of a right circular cone is doubled. To keep the volume fixed, the height of the cone will be
[SSC (10+2) 2012]
(a) half of the previous height
(b) one-third of the previous height
(c) one-fourth of the previous height
(d) $\dfrac{1}{\sqrt{2}}$ times of the previous height

56. If the height of the right circular cone is increased by 200% and the radius of the base is reduced by 50%, then the volume of the cone [CDS 2015 (I)]
(a) increases by 25%
(b) increases by 50%
(c) remains unchanged
(d) decreases by 25%

57. If the radius of a right circular cone is increased by p% without increasing its height, then what is the percentage increase in the volume of the cone?
[CDS 2017 (I)]
(a) p^2 (b) $2p^2$
(c) $\dfrac{p^2}{100}$ (d) $p\left(2 + \dfrac{p}{100}\right)$

58. The frustum of a right circular cone has the diameters of base 10 cm, of top 6 cm and a height of 5 cm. Find its slant height.
(a) $\sqrt{29}$ cm (b) $3\sqrt{3}$ cm
(c) $\sqrt{13}$ cm (d) $4\sqrt{3}$ cm
(e) None of these

59. The frustum of a right circular cone has the radii of base 4 cm, of the top 2 cm and a height of 6 cm. Find the volume of the frustum.
(a) 115 cm^3 (b) 156 cm^3
(c) 185 cm^3 (d) 176 cm^3
(e) None of these

Directions (Q. Nos. 60-61) *The areas of the ends of a frustum of a pyramid are P and Q, where P < Q and H is its thickness.*
[CDS 2012]

60. What is the difference in radii of the ends of the frustum?
    (a) $\dfrac{\sqrt{Q}-\sqrt{P}}{\sqrt{\pi}}$
    (b) $\dfrac{\sqrt{Q}-\sqrt{P}}{\pi}$
    (c) $\sqrt{Q}-\sqrt{P}$
    (d) None of these

61. What is the volume of the frustum?
    (a) $3H(P+Q+\sqrt{PQ})$
    (b) $H(P+Q+\sqrt{PQ})$
    (c) $H(P+Q+\sqrt{PQ})/3$
    (d) $H(P+Q-\sqrt{PQ})/3$

62. A cone of radius r cm and height h cm is divided into two parts by drawing a plane through the middle point of its height and parallel to the base. What is the ratio of the volume of the original cone to the volume of the smaller cone? [CDS 2014]
    (a) 4:1 (b) 8:1 (c) 2:1 (d) 6:1

63. A drinking glass of height 24 cm is in the shape of frustum of a cone and diameters of its bottom and top circular ends are 4 cm and 18 cm, respectively. If we take capacity of the glass as πx cm^3, then what is the value of x? [CDS 2016 (II)]
    (a) 824   (b) 1236
    (c) 1628   (d) 2472

64. Let V be the volume of an inverted cone with vertex at origin and the axis of the cone is along positive Y-axis. The cone is filled with water upto half of its height. The volume of water is [CDS 2015 (II)]
    (a) $\dfrac{V}{8}$ (b) $\dfrac{V}{6}$
    (c) $\dfrac{V}{3}$ (d) $\dfrac{V}{2}$

65. The height of a cone is 60 cm. A small cone is cut off at the top by a plane parallel to the base and its volume is $\dfrac{1}{64}$ the volume of original cone. What is the height from the base at which the section is made? [CDS 2016 (II)]
    (a) 15 cm
    (b) 20 cm
    (c) 30 cm
    (d) 45 cm

66. A solid metal cylinder of 10 cm height and 14 cm diameter is melted and recast into two cones in the proportion of 3 : 4 (volume), keeping the height 10 cm. What would be the percentage change in the flat surface area before and after? [XAT 2014]
    (a) 9%   (b) 16%
    (c) 25%   (d) 50%
    (e) None of these

Sphere

67. What will be the surface area of the sphere having 4 cm radius? [RRB 2010]
    (a) 64π sq cm   (b) 69π sq cm
    (c) 32π sq cm   (d) 35π sq cm

68. What is the volume of a sphere of radius 3 cm? [CDS 2016 (II)]
    (a) 36π cm^3 (b) 18π cm^3
    (c) 9π cm^3 (d) 6π cm^3

69. If the surface area of a sphere is 616 sq cm, what is its volume? [CDS 2012]
    (a) 4312/3 cm^3 (b) 4102/3 cm^3
    (c) 1257 cm^3 (d) 1023 cm^3

70. What is the diameter of the largest circle lying on the surface of a sphere of surface area 616 sq cm? [CDS 2014]
    (a) 14 cm   (b) 10.5 cm
    (c) 7 cm   (d) 3.5 cm

71. If the surface area of a sphere is reduced to one-ninth of the area, its radius reduces to [CDS 2017 (I)]
    (a) one-fourth   (b) one-third
    (c) one-fifth   (d) one-ninth

72. If the ratio of the diameters of two spheres is 3 : 5, then what is the ratio of their surface areas? [CDS 2012]
    (a) 9 : 25   (b) 9 : 10
    (c) 3 : 5   (d) 27 : 125

73. The cost of painting a spherical vessel of diameter 14 cm is ₹ 8008. What is the cost of painting per square centimetre? [CDS 2016 (II)]
    (a) ₹ 8 (b) ₹ 9 (c) ₹ 13 (d) ₹ 14

74. If 64 identical small spheres are made out of a big sphere of diameter 8 cm, what is surface area of each small sphere?
    (a) π cm^2 (b) 2π cm^2
    (c) 4π cm^2 (d) 8π cm^2

75. A metallic sphere of radius 12 cm is melted into three smaller spheres. If the radii of two smaller spheres are 6 cm and 8 cm, the radius of the third is
    (a) 14 cm   (b) 16 cm
    (c) 10 cm   (d) 12 cm

Volume and Surface Area / 627

76. Weight of a solid metallic sphere of radius 4 cm is 4 kg. The weight of a hollow sphere made with same metal, whose outer diameter is 16 cm and inner diameter is 12 cm, is
(a) 20.5 kg (b) 15.5 kg
(c) 16.5 kg (d) 18.5 kg

77. A hemisphere has 28 cm diameter. Find its curved surface area. [SSC CGL 2009]
(a) 1232 sq cm (b) 1236 sq cm
(c) 1238 sq cm (d) 1233 sq cm

78. What will be the difference between total surface area and curved surface area of a hemisphere having 2 cm diameter?
(a) 2π sq cm (b) 3π sq cm
(c) π sq cm (d) 4π sq cm
(e) None of these

79. A hemispherical bowl has 3.5 cm radius. It is to be painted inside as well as outside. Find the cost of painting it at the rate of ₹ 5 per 10 sq cm. [Bank Clerk 2010]
(a) ₹ 50 (b) ₹ 81 (c) ₹ 56 (d) ₹ 77
(e) None of these

80. A sphere and a hemisphere have the same surface area. The ratio of their volumes is [SSC (10+2) 2012]
(a) $\frac{\sqrt{3}}{4} : 1$ (b) $\frac{3\sqrt{3}}{4} : 1$
(c) $\frac{\sqrt{3}}{8} : 1$ (d) $\frac{3\sqrt{3}}{8} : 1$

81. If the radius of a sphere is increased by 10%, then the volume will be increased by [CDS 2015 (I)]
(a) 33.1% (b) 30%
(c) 50% (d) 10%

82. If the radius of a sphere is increased by 3%, then what per cent increase takes place in surface area of the sphere? [Bank Clerk 2010]
(a) 6.09% (b) 7%
(c) 5.06% (d) 9%
(e) None of these

83. If radius of a sphere is decreased by 24%, by what per cent does its surface area decrease? [Bank Clerk 2011]
(a) 44% (b) 49%
(c) 42.24% (d) 46.2%
(e) None of these

84. A solid metal sphere is melted and smaller spheres of equal radii are formed. 10% of the volume of the sphere is lost in the process. The smaller spheres have a radius, that is 1/9th the large sphere. If 10 L of paint was needed to paint the larger sphere, how many litres is need to paint all the smaller spheres? [MAT 2015]
(a) 90 L (b) 81 L
(c) 180 L (d) 324 L

Prism and Pyramid

85. A prism has the base a right angled triangle whose sides adjacent to the right angle are 10 cm and 12 cm long. The height of the prism is 20 cm. The density of the material of the prism is 6 g/cu cm. The weight of the prism is [SSC (10+2) 2012]
(a) 3.4 kg (b) 4.8 kg
(c) 6.4 kg (d) 7.2 kg

86. Find the total surface area of a pyramid having a slant height of 8 cm and a base which is a square of side 4 cm (in cm^2)?
(a) 80 (b) 64
(c) 72 (d) 84

87. The base of a right prism is a right angled isosceles triangle whose hypotenuse is a cm. If the height of the prism is h cm, then its volume is [SSC (10+2) 2012]

(a) $\frac{a^2 h}{4}$ cm^3 (b) $\frac{a^2 h}{6}$ cm^3
(c) $\frac{a^2 h}{8}$ cm^3 (d) $\frac{a^2 h}{12}$ cm^3

88. The perimeter of the triangular base of a right prism is 60 cm and the sides of the base are in the ratio 5 : 12 : 13. Then, its volume will be (height of the prism being 50 cm)
(a) 6000 cm^3 (b) 6600 cm^3
(c) 5400 cm^3 (d) 9600 cm^3

89. A prism and a pyramid have the same base and the same height. Find the ratio of the volumes of the prism and the pyramid.
(a) 1 : 1 (b) 1 : 3 (c) 3 : 1
(d) Cannot be determined

Miscellaneous

90. Ice-cream, completely filled in a cylinder of diameter 35 cm and height 32 cm, is to be served by completely filling identical disposable cones of diameter 4 cm and height 7 cm. The maximum number of persons that can be served in this way is [CDS 2017 (I)]
    (a) 950 (b) 1000
    (c) 1050 (d) 1100

Directions (Q. Nos. 91-92) *A tent of a circus is made of canvas and is in the form of right circular cylinder and right circular cone above it. The height and diameter of the cylindrical part of the tent are 5 m and 126 m, respectively. The total height of the tent is 21 m.* [CDS 2016 (II)]

91. What is the slant height of the cone?
    (a) 60 m (b) 65 m
    (c) 68 m (d) 70 m

92. How many square metres of canvas are used?
    (a) 14450 (b) 14480
    (c) 14580 (d) 14850

93. A solid brass sphere of radius 2.1 dm is converted into a right circular cylindrical rod of same radius and length 7 cm. The ratio of total surface areas of the rod to the sphere is [SSC CGL (Mains) 2017]
    (a) 3 : 1 (b) 2 : 3
    (c) 7 : 3 (d) 3 : 7

94. A ball of radius 1 cm is put into a cylindrical pipe so that it fits inside the pipe. If the length of the pipe is 14 m, what is the curved surface area of the pipe? [CDS 2017 (I)]
    (a) 2200 sq cm (b) 4400 sq cm
    (c) 8800 sq cm (d) 17600 sq cm

95. A solid sphere of radius 3 cm is melted to form a hollow right circular cylindrical tube of length 4 cm and external radius 5 cm. The thickness of the tube is [SSC CGL (Mains) 2016]
    (a) 1 cm (b) 9 cm
    (c) 0.6 cm (d) 1.5 cm

96. Let A be a pyramid on a square base and B be a cube. If a, b and c denote the number of edges, number of faces and number of corners, respectively. Then, the result $a = b + c$ is true for [CDS 2013]
    (a) only A (b) only B
    (c) Both A and B (d) Neither A nor B

97. A conical flask is full of water. The flask has base radius r and height h. This water is poured into a cylindrical flask of base radius mr. The height of water in the cylindrical flask is
    (a) $\frac{h}{2}m^2$ (b) $\frac{2h}{m}$
    (c) $\frac{h}{3m^2}$ (d) $\frac{m}{2h}$

98. Seven equal cubes each of side 5 cm are joined end-to-end. Find the surface area of the resulting cuboid. [Bank PO 2007]
    (a) 750 sq cm
    (b) 1500 sq cm
    (c) 2250 sq cm
    (d) 700 sq cm
    (e) None of the above

99. In a shower, 10 cm of rain falls. What will be the volume of water that falls on 1 hec area of ground? [Delhi Police SI 2008]
    (a) 500 cm^3 (b) 650 cm^3
    (c) 1000 cm^3 (d) 750 cm^3

100. What is the volume of the largest sphere that can be curved out of a cube of edge 3 cm? [CDS 2012]
    (a) 9π cm^3 (b) 6π cm^3
    (c) 4.5π cm^3 (d) 3π cm^3

101. A right circular metal cone (solid) is 8 cm high and the radius is 2 cm. It is melted and recast into a sphere. What is the radius of the sphere? [CDS 2012]
    (a) 2 cm (b) 3 cm
    (c) 4 cm (d) 5 cm

102. Let the largest possible right circular cone and largest possible sphere be fitted into two cubes of same length. If C and S denote the volume of cone and volume of sphere respectively, then which one of the following is correct?
    (a) C = 2S (b) S = 2C [CDS 2012]
    (c) C = S (d) C = 3S

103. A cylindrical box of radius 5 cm contains 10 solid spherical balls, each of radius 5 cm. If the top most ball touches the upper cover of the box, then volume of the empty space in the box is [SSC CPO 2007]
    (a) $\frac{2500}{3}\pi$ cm^3 (b) 5000π cm^3
    (c) 2500π cm^3 (d) $\frac{5000}{3}\pi$ cm^3

Volume and Surface Area / 629

104. A hospital room is to accommodate 56 patients. It should be done in such a way that every patient gets 2.2 m^2 of floor and 8.8 m^3 of space. If the length of the room is 14 m, then breadth and height of the room are respectively
    (a) 8.8 m, 4 m   (b) 8.4 m, 4.2 m
    (c) 8 m, 4 m   (d) 7.8 m, 4.2 m

105. What part of a ditch 48 m long, 16.5 m broad and 4 m deep can be filled by the earth got by digging a cylindrical tunnel of diameter 4 m and length 56 m? [SSC CGL 2007]
    (a) $\dfrac{1}{9}$ (b) $\dfrac{2}{9}$
    (c) $\dfrac{7}{9}$ (d) $\dfrac{8}{9}$

106. The floor of a rectangular hall has a perimeter 250 m. If the cost of painting the four walls at the rate of ₹ 10 per sq m is ₹ 15000, then find the height of the hall. [Hotel Mgmt. 2010]
    (a) 8 m   (b) 6 m
    (c) 9 m   (d) 7 m

107. Consider the following statements
    I. If the height of a cylinder is doubled, the area of the curved surface is doubled.
    II. If the radius of a hemispherical solid is doubled, its total surface area becomes fourfold.
    Which of the above statements is/are correct? [CDS 2015 (II)]
    (a) Only I   (b) Only II
    (c) Both I and II (d) Neither I nor II

108. A solid consists of circular cylinder with exact fitting right circular cone placed on the top. The height of the cone is h. If total volume of the solid is three times the volume of the cone, then the height of the circular cylinder is [SSC CGL 2012]
    (a) $2h$ (b) $\dfrac{2h}{3}$ (c) $4h$ (d) $\dfrac{3h}{2}$

109. A building is in the form of a cylinder surmounted by a hemispherical dome on the diameter of the cylinder. The height of the building is three times the radius of the base of the cylinder. The building contains $67\dfrac{1}{21}$ m^3 of air. What is the height of the building? [CDS 2016 (II)]
    (a) 6 m   (b) 4 m   (c) 3 m   (d) 2 m

110. Water flows at the rate of 10 m/min from a cylindrical pipe 5 mm in diameter. How long will it take to fill up a conical vessel whose diameter at the base is 40 cm and depth is 24 cm?
    (a) 51 min 12 s   (b) 52 min 1 s
    (c) 48 min 15 s   (d) 55 min

111. A hemispherical basin of 150 cm diameter holds water 120 times as much as a cylindrical tube. If the height of the tube is 15 cm, then the diameter of the tube is [SSC CGL 2008]
    (a) 27 cm   (b) 24 cm
    (c) 25 cm   (d) 26 cm

112. A well of inner diameter 14 m is dug to a depth of 15 m. Earth taken out of it has been evenly spread all around it to a width of 7 m to form an embankment. Find the height of embankment so formed. [DMRC 2012]
    (a) 7 m   (b) 5 m
    (c) 14 m   (d) None of these

113. The radius of a sphere is 6 cm. It is melted and drawn into a wire of radius 0.2 cm. The length of the wire is [SSC CGL (Mains) 2016]
    (a) 81 m   (b) 80 m   (c) 75 m   (d) 72 m

Answer with Solutions

1. (d) Given, diagonal $= 2\sqrt{3}$
    $\Rightarrow a\sqrt{3} = 2\sqrt{3}$ [a = edge of the cube]
    $\therefore \quad a = 2$
    $\therefore$ Required surface area $= 6a^2$
    $= 6 \times 2^2 = 24$ sq cm

2. (a) Total surface area of a cube $= 6a^2$
    $\Rightarrow \quad 6 = 6a^2 \Rightarrow a^2 = 1$
    $\therefore \quad a = 1$

    Now, volume of the cube
    $= a^3 = 1^3 = 1$ cu unit

3. (b) Volume of cube $= (\text{Side})^3$
    $\therefore \quad 729 = a^3$
    $\Rightarrow \quad a = 9$ cm
    $\therefore$ Diagonal of cube $= \text{Side} \times \sqrt{3}$
    $= 9 \times \sqrt{3} = 9\sqrt{3}$ cm

4. (a) According to the question,
$6a^2 = 726$ [a = edge of the cube]
$\Rightarrow \quad a^2 = \dfrac{726}{6} = 121$
$\therefore \quad a = \sqrt{121} = 11$ cm
$\therefore$ Required volume $= a^3 = 11^3 = 1331$ cm^3

5. (a) We know that,
Volume of cube $= a^3$
where, a is the side of cube.
$\therefore \quad \dfrac{a_1^3}{a_2^3} = \dfrac{8}{125}$
$\Rightarrow \quad \left(\dfrac{a_1}{a_2}\right)^3 = \left(\dfrac{2}{5}\right)^3$
Comparing on both the sides, we get
$\dfrac{a_1}{a_2} = \dfrac{2}{5}$
Now, ratio of their surface areas
$= \dfrac{6a_1^2}{6a_2^2} = \left(\dfrac{a_1}{a_2}\right)^2$
$= \left(\dfrac{2}{5}\right)^2 = 4:25$

6. (c) Let the edge of a square be x, then
Sum of its edges $= 12x$
Now, by given condition, $x^3 = 12x$
$\Rightarrow \quad x(x^2 - 12) = 0$
$\Rightarrow \quad x^2 = 12 \qquad [\because x \neq 0]\ \ldots(i)$
$\therefore$ Its total surface area $= 6x^2$
$= 6(12) = 72$ sq units

7. (a) Volume (new cube) $= (1^3 + 6^3 + 8^3)$
$= 729$ cm^3
Let a be the side of a new cube.
Then, $a^3 = 729$ [by Technique 7]
$\Rightarrow \quad a = \sqrt[3]{729} \Rightarrow a = 9$ cm
$\therefore$ Surface area of the new cube $= 6a^2$
$= 6 \times 9^2 = 486$ sq cm
$\therefore \dfrac{\text{Surface area}}{2} = \dfrac{486}{2} = 243$ sq cm

8. (d) Here, $x = y = -19\%$
According to the formula,
Percentage decrease in surface area
$= \left[x + y + \dfrac{xy}{100}\right]\%$
$= \left[-19 - 19 + \dfrac{(-19) \times (-19)}{100}\right]\%$
$= \left[-38 + \dfrac{361}{100}\right]\%$
$= [-38 + 3.61]\% = -34.39\%$

9. (a) Here, $x = 12\%$
According to the formula,
Percentage increase in volume
$= \left[\left(1 + \dfrac{x}{100}\right)^3 - 1\right] \times 100\%$
[by Technique 2]
$= \left[\left(1 + \dfrac{12}{100}\right)^3 - 1\right] \times 100\%$
$= [(1.12)^3 - 1] \times 100\%$
$= 0.404928 \times 100\%$
$= 40.4928\%$

10. (d) Given, $l = 10$ m, $b = 5$ m, $h = 3$ m
$\therefore \quad lb = 10 \times 5 = 50,$
$bh = 5 \times 3 = 15$
and $\quad lh = 10 \times 3 = 30$
$\therefore$ Surface area of a cuboid
$= 2(lb + bh + lh)$
$= 2(50 + 15 + 30)$
$= 2 \times 95 = 190$ sq m

11. (c) New length $= l + \dfrac{5}{100} \times l = \dfrac{21l}{20}$
New breadth $= b + \dfrac{4}{100} \times b = \dfrac{26b}{25}$
$\therefore$ New volume $= \dfrac{21l}{20} \times \dfrac{26b}{25} \times h = \dfrac{273}{250} lbh$
Required percentage increase in volume
$= \dfrac{\dfrac{273}{250} lbh - lbh}{lbh} \times 100$
$= \dfrac{23}{250} \times 100$
$= \dfrac{230}{25} = 9.2\%$

Fast Track Method
Here, $x = 5, y = 4$ and $z = 0$
$\therefore$ Required percentage volume
$= \left[x + y + z + \dfrac{xy + yz + zx}{100} + \dfrac{xyz}{(100)^2}\right]\%$
[by Technique 1]
$= \left[5 + 4 + 0 + \dfrac{5 \times 4 + 0}{100} + \dfrac{0}{(100)^2}\right]\%$
$= 9.2\%$

12. (c) Length of largest pencil that can be kept in a box
$=$ Diagonal of box
$= \sqrt{l^2 + b^2 + h^2}$
$= \sqrt{64 + 36 + 4}$
[here, $l = 8$ cm, $b = 6$ cm, $h = 2$ cm]
$= \sqrt{104} = 2\sqrt{26}$ cm

Volume and Surface Area / 631

13. (a) Capacity of tank = 50000 L = 50 m^3
$$\left[\because 1 \text{ L} = \frac{1}{1000} \text{m}^3\right]$$
∴ Breadth = $\frac{50}{2.5 \times 10}$ = 2 m

14. (b) Let l, b and h be the length, breadth and height of the room.
Now, area of four walls = $2(l + b)h$
$\quad\quad 120 = 2(l + b) \times 4$
$\Rightarrow \quad 120 = 8(2b + b) \quad [\because l = 2b]$
$\Rightarrow \quad b = 5$ m
∴ $\quad l = 2 \times 5 = 10$ m
∴ Area of the floor
$\quad = l \times b = (10 \times 5) = 50$ m^2

15. (b) Let length, breadth and height be $4x, 3x$ and $2x$, respectively.
Whole surface area = $2(lb + bh + lh)$
$\Rightarrow (lb + bh + lh) = \frac{8788}{2} = 4394$
$\Rightarrow (4 \times 3 + 3 \times 2 + 2 \times 4) x^2 = 4394$
$\Rightarrow \quad 26x^2 = 4394$
$\Rightarrow \quad x^2 = 169$
$\Rightarrow \quad x = 13$
∴ Length = $4x = 4 \times 13 = 52$ cm

16. (c) External volume
$\quad = 20 \times 12 \times 5$
$\quad = 1200$ cm^3
Internal volume
$\quad = (20 - 2) \times (12 - 2) \times (5 - 2)$
$\quad = 18 \times 10 \times 3$
$\quad = 540$ cm^3
∴ Volume of the metal = External volume − Internal volume
$\quad = 1200 - 540 = 660$ cm^3

17. (a) Surface area of cube which can be painted = $6\,(\text{Side})^2 = 6(2)^2 = 24$ cm^2
Now, surface area of cuboid which can be painted = $2(lb + bh + lh)$
$\quad = 2(2 + 6 + 3) = 22$ cm^2
Total surface area of both cube and cuboid
$\quad = 22 + 24 = 46$ cm^2 < 54 cm^2
Therefore, both cube and cuboid can be painted.

18. (b) Given, $lb = 120$ sq cm, $bh = 72$ sq cm and $lh = 60$ sq cm.
∴ $\quad lb \times bh \times lh = 120 \times 72 \times 60$
$\Rightarrow \quad (lbh)^2 = 120 \times 72 \times 60$
$\Rightarrow \quad lbh = \sqrt{120 \times 72 \times 60}$
$\quad = \sqrt{12 \times 10 \times 12 \times 6 \times 10 \times 6}$
$\quad = 12 \times 10 \times 6 = 720$ cm^3

19. (b) Square of side 2 cm is cut from rectangular sheet as shown in the figure.

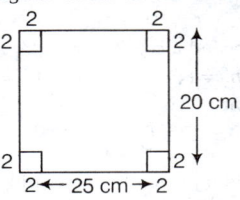

Then, length of the box = $25 - 2 \times 2$
$\quad = 25 - 4 = 21$ cm
and breadth of the box
$\quad = 20 - 2 \times 2 = 20 - 4 = 16$ cm
and height of the box will be 2 cm.
∴ Volume of the box
$\quad$ = Length × Breadth × Height
$\quad = 21 \times 16 \times 2 = 672$ cm^3

20. (b) Volume of rectangular box
$\quad = 10 \times 8 \times 6 = 480$ cm^3
Volume of cubes = 6240 cm^3
∴ Required boxes
$\quad = \dfrac{\text{Volume of cubes}}{\text{Volume of rectangular box}}$
$\quad = \dfrac{6240}{480} = 13$
Hence, 13 boxes are needed.

21. (a) Volume of the cuboid = 720 cm^3
Height of the cuboid
$\quad = \dfrac{\text{Volume of the cuboid}}{\text{Base area of the cuboid}}$
$\quad = \dfrac{720}{72} = 10$ cm
Surface area of the cuboid = 484
$\Rightarrow \quad 484 = 2\,(lb + bh + hl)$
$\Rightarrow \quad 242 = 72 + 10b + 10l$
$\Rightarrow \quad 242 - 72 = 10(l + b)$
$\Rightarrow \quad \dfrac{170}{10} = l + b \Rightarrow l + b = 17$
and $\quad lb = 72$
By Hit and Trial, we get
$\quad l = 9$ and $b = 8$
∴ It is obvious that length, breadth and height of the cuboid is 9 cm, 8 cm and 10 cm.

22. (c) Let the dimensionals of cuboid be l, b and h, respectively.

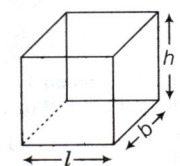

For first face,
$$l^2 + b^2 = 13^2 \quad [\because l^2 + b^2 = d^2]$$
$$\Rightarrow \quad l^2 + b^2 = 169 \quad \ldots(i)$$
For second face,
$$b^2 + h^2 = (\sqrt{281})^2 \Rightarrow b^2 + h^2 = 281 \quad \ldots(ii)$$
For third face,
$$h^2 + l^2 = 20^2$$
$$\Rightarrow \quad h^2 + l^2 = 400 \quad \ldots(iii)$$
On adding Eqs. (i), (ii) and (iii), we get
$$2(l^2 + b^2 + h^2) = 850$$
$$\Rightarrow \quad l^2 + b^2 + h^2 = 425 \quad \ldots(iv)$$
On putting $l^2 + b^2 = 169$ in the above equation, we get
$$169 + h^2 = 425 \Rightarrow h^2 = 425 - 169$$
$$\Rightarrow \quad h^2 = 256 \Rightarrow h = 16 \text{ units}$$
From Eq. (ii), $b^2 + h^2 = 281$
$$\Rightarrow \quad b^2 + 16^2 = 281 \quad [\because h = 16]$$
$$\Rightarrow \quad b^2 + 256 = 281 \Rightarrow b^2 = 25$$
$$\Rightarrow \quad b = 5 \text{ units}$$
From Eq. (iii),
$$h^2 + l^2 = 400 \Rightarrow 16^2 + l^2 = 400$$
$$\Rightarrow \quad l^2 = 400 - 256$$
$$\Rightarrow \quad l^2 = 144 \Rightarrow l = 12 \text{ units}$$
$\therefore$ Total surface area of cuboid
$$= 2(lb + bh + hl)$$
$$= 2(12 \times 5 + 5 \times 16 + 16 \times 12)$$
$$= 2(60 + 80 + 192)$$
$$= 2 \times 332 = 664 \text{ sq units}$$
Hence, the total surface area of the cuboid is 664 sq units.

23. (b) Density of copper $= \dfrac{\text{Mass}}{\text{Volume}}$
$$= \dfrac{9000}{1} \text{ kg/m}^3$$
$$= 9000 \text{ kg/m}^3$$
Let x be the side of cross-section of square bar.
$\therefore$ Volume of square bar = Volume of copper
$$\Rightarrow \quad 9 \times x^2 = 1 \Rightarrow x = \dfrac{1}{3} \text{ m}$$
Volume of cube of side x
$$= x^3 = \left(\dfrac{1}{3}\right)^3 = \dfrac{1}{27} \text{ m}^3$$
$\therefore$ Mass of cube = Volume $\times$ Density
$$= \dfrac{1}{27} \times 9000 = \dfrac{1000}{3} \text{ kg}$$

24. (c) Given, dimensions of a rectangular block of wood are 3 m, 2 m and 1.75 m, i.e. 300 cm, 200 cm and 175 cm. and thickness of layer of paint = 0.1 mm
$$= \dfrac{1}{100} \text{ cm} \quad \left[\because 1 \text{ mm} = \dfrac{1}{10} \text{ cm}\right]$$

$\therefore$ Volume of rectangular block of wood with paint
$$= 2\left[200 \times 300 \times \dfrac{1}{100} + 200 \times 175 \times \dfrac{1}{100}\right.$$
$$\left. + 175 \times 300 \times \dfrac{1}{100}\right]$$
$$= 2[600 + 350 + 525]$$
$$= 2 \times 1475 = 2950 \text{ cm}^3$$
Given, edge of a cubical box = 10 cm
$\therefore$ Volume of cubical box = (Edge)^3
$$= 10^3 = 1000 \text{ cm}^3$$
Now, minimum number of boxes $= \dfrac{2950}{1000}$
$$= 2.95 = 3 \text{ (approx.)}$$
Hence, the minimum number of boxes is 3.

25. (d) Volume of the cylinder $= \pi r^2 h$
$$= \dfrac{22}{7} \times 7 \times 7 \times 500 = 77000$$
$$= (77 \times 10^3) \text{ cm}^3$$

26. (c) Given, lateral surface area = 94.2 sq cm
$$\Rightarrow \quad 2\pi rh = 94.2$$
$$\therefore \quad r = \dfrac{94.2}{2\pi h} = \dfrac{94.2}{2 \times 3.14 \times 5} = 3 \text{ cm}$$
$$[\because h = 5 \text{ cm, given}]$$

27. (b) Given, $r_1 = 1$ cm, $h_1 = 30$ cm
and $h_2 = 300$ cm
$\therefore$ Volume of rod = Volume of wire
$$\Rightarrow \quad \pi r_1^2 h_1 = \pi r_2^2 h_2$$
$$\Rightarrow \quad \pi \times (1)^2 \times 30 = \pi \times r_2^2 \times 300$$
$$\Rightarrow \quad r_2^2 = \dfrac{30}{300} \Rightarrow r_2 = \dfrac{1}{\sqrt{10}} \text{ cm}$$
$\therefore$ Diameter $= 2r_2 = 2 \times \dfrac{1}{\sqrt{10}} = \dfrac{2}{\sqrt{10}}$ cm

28. (d) Let the height and radius of solid cylinder be h and r cm, respectively.
Given that, radius $(r) = 5$ cm
and total surface area = 660 cm^2
$$\Rightarrow \quad 2\pi rh + 2\pi r^2 = 660$$
$$\Rightarrow \quad 2\pi r(h + r) = 660$$
$$\Rightarrow \quad (h + 5) = \dfrac{330}{5\pi} = \dfrac{330}{5} \times \dfrac{7}{22}$$
$$\Rightarrow \quad h = \dfrac{66 \times 7}{22} - 5 = 21 - 5$$
$\therefore$ Required height = 16 cm

29. (d) In one revolution,
Area covered = Curved surface area of roller of playground
$$\Rightarrow \quad 2\pi rh = 2 \times \dfrac{22}{7} \times 42 \times 120$$
$$= 31680 \text{ sq cm}$$

Volume and Surface Area / 633

In 500 revolutions,
Area covered = 31680×500
$= (1584 \times 10^4)$ sq cm
$= \dfrac{1584 \times 10^4}{10^4} = 1584$ sq m

30. (b) Here, $r_A = 2r$, $r_B = 3r$
and $V_A = 9V$, $V_B = 7V$
$\therefore \dfrac{V_A}{V_B} = \dfrac{\pi r_A^2 h_A}{\pi r_B^2 h_B} \Rightarrow \dfrac{9V}{7V} = \dfrac{(2r)^2 \times h_A}{(3r)^2 \times h_B}$
$\Rightarrow \dfrac{9}{7} = \dfrac{4r^2 \times h_A}{9r^2 \times h_B}$
$\Rightarrow \dfrac{h_A}{h_B} = \dfrac{9 \times 9}{4 \times 7} = \dfrac{81}{28}$
$\Rightarrow h_A : h_B = 81 : 28$

31. (c) Let the radius of the base be $2x$ and the height of the cylinder be $3x$.
Given, volume = 1617 cm^3
$\Rightarrow \pi r^2 h = 1617$
$\Rightarrow \dfrac{22}{7}(2x)^2(3x) = 1617$
$\Rightarrow x^3 = \dfrac{1617 \times 7}{22 \times 12} \Rightarrow x = \dfrac{7}{2}$
$\therefore r = 2 \times \dfrac{7}{2} = 7$ cm
and $h = 3 \times \dfrac{7}{2} = \dfrac{21}{2}$
Now, total surface area = $2\pi rh + 2\pi r^2$
$= 2\pi r(h + r)$
$= 2 \times \dfrac{22}{7} \times 7 \left(\dfrac{21}{2} + 7\right)$
$= 2 \times 22 \times \dfrac{35}{2} = 770$ cm^2

32. (c) Let r and h be the radius and height of cylinder.
Then, $\dfrac{\text{Curved surface area}}{\text{Total surface area}} = \dfrac{1}{2}$
$\Rightarrow \dfrac{2\pi rh}{2\pi r(r + h)} = \dfrac{1}{2}$
$\Rightarrow 2h = r + h$
$\Rightarrow h = r$
Now, total surface area = 616 cm^2
$2\pi r(r + r) = 616$ [$\because r = h$]
$\Rightarrow 2 \times \dfrac{22}{7} \times 2 \times r^2 = 616$
$\Rightarrow r^2 = 49 \Rightarrow r = 7$ cm
$\therefore$ Volume of cylinder = $\pi r^2 h$
$= \dfrac{22}{7} \times (7)^3 = 1078$ cm^3

33. (d) Volume of clay required
= External volume – Internal volume
$= \pi \left[\left(\dfrac{5.1}{2}\right)^2 - \left(\dfrac{4.5}{2}\right)^2\right] \times 21$
$= \pi[(2.55)^2 - (2.25)^2] \times 21$
$= \pi[(2.55 + 2.25)(2.55 - 2.25)] \times 21$
$= \pi (0.3 \times 4.8) \times 21 = 30.24 \pi$ cm^3

34. (d) Here, $r_1 = r$, $r_2 = \dfrac{r}{3}$
According to the question,
Volume of new wire = Volume of old wire
$\Rightarrow \pi(r_2)^2 l_2 = \pi(r_1)^2 l_1$
$\Rightarrow \pi \times \left(\dfrac{r}{3}\right)^2 \times l_2 = \pi \times (r)^2 \times l_1$
$\Rightarrow \dfrac{r^2}{9} \times l_2 = r^2 \times l_1$
$\Rightarrow l_2 = 9 l_1$
So, it is clear from above length that will increase by 9 times.

35. (a) Here, $x = -8\%$, $y = 4\%$
According to the formula,
Net effect on volume
$= \left[2x + y + \dfrac{x^2 + 2xy}{100} + \dfrac{x^2 y}{100^2}\right]\%$
[by Technique 5]
$= \left[2 \times (-8) + 4 + \dfrac{(-8)^2 + 2 \times (-8) \times 4}{100}\right.$
$\left. + \dfrac{(-8)^2 \times 4}{100^2}\right]\%$
$= \left[-16 + 4 + \dfrac{64 - 64}{100} + \dfrac{256}{10^4}\right]\%$
$= [-12 + 0 + 0.0256]\%$
$= -11.9744\%$ (decrease)

36. (a) Given, height of cylinder = 4 cm
$\therefore$ Total surface area of cylinder = 8π cm^2
Let radius of the cylinder be r cm.
Then, $2\pi r(h + r) = 8\pi$
$\Rightarrow r^2 + 4r - 4 = 0$
$\Rightarrow r = \dfrac{-4 \pm \sqrt{16 - 4(-4)(1)}}{2(1)} = \dfrac{-4 \pm \sqrt{32}}{2}$
$= \dfrac{-4 \pm 4\sqrt{2}}{2} = -2 \pm 2\sqrt{2}$
$\Rightarrow r = 2\sqrt{2} - 2$
or $r = -2 - 2\sqrt{2}$ [not possible]
$\therefore$ Radius of cylinder is $(2\sqrt{2} - 2)$ cm.

37. (a) Given,
$\dfrac{\text{Curved surface area of cylinder}}{\text{Volume of cylinder}} = \dfrac{1}{7}$

$\Rightarrow \dfrac{2\pi rh}{\pi r^2 h} = \dfrac{1}{7} \Rightarrow \dfrac{2}{r} = \dfrac{1}{7}$

$\Rightarrow r = 14$ m

and $\dfrac{\text{Diameter of cylinder}}{\text{Height of cylinder}} = \dfrac{7}{5}$

$\Rightarrow \dfrac{2r}{h} = \dfrac{7}{5} \Rightarrow \dfrac{2 \times 14}{h} = \dfrac{7}{5}$

$\Rightarrow h = 20$ m

Now, total surface area of cylinder

$= 2\pi r(r+h) = 2 \times \dfrac{22}{7} \times 14(14+20)$

$= 2 \times 22 \times 2 \times 34 = 2992$ m^2

38. (b) Let radius be r and height be h of the cylinder.

Then, according to the question,

$30 \cdot \pi r^2 h = \dfrac{1}{3} \pi r^2 h \times n$

where, n = number of cones

$\therefore \quad n = 30 \times 3 = 90$ cones

Statement I is true. Statement II is also correct. But Statement II is not correct explanation of Statement I.

39. (b) Given, $r = \dfrac{h}{2}$

$\because$ Surface area = 616 sq cm [given]

$\Rightarrow 2\pi rh + \pi r^2 = 616$

$\Rightarrow \dfrac{2\pi h^2}{2} + \dfrac{\pi h^2}{4} = 616$

$\Rightarrow \pi h^2 + \dfrac{\pi h^2}{4} = 616$

$\Rightarrow \pi h^2 \left(1 + \dfrac{1}{4}\right) = 616$

$\Rightarrow h^2 = 616 \times \dfrac{4}{5} \times \dfrac{7}{22} = \dfrac{784}{5}$

$\Rightarrow h = \dfrac{28}{\sqrt{5}} = 12.55$ cm

$\therefore r = \dfrac{h}{2} = \dfrac{12.55}{2} = 6.27$ cm

Hence, volume $= \pi r^2 h$

$= \dfrac{22}{7} \times 6.27 \times 6.27 \times 12.55 \approx 1.53$ L

40. (b) Given, diameter of cylindrical pipe = 7 cm

$\therefore$ Radius of pipe,

$r = \dfrac{7}{2}$ cm $= \dfrac{7}{2 \times 100}$ m $= \dfrac{7}{200}$ m

$\therefore$ Volume of water flows through a cylindrical pipe

$= \pi r^2 h = \dfrac{22}{7} \times \left(\dfrac{7}{200}\right)^2 \times 5 \times 60$ m^3

Now, required time

$= \dfrac{\text{Volume of rectangular tank}}{\text{Volume of cylindrical pipe}}$

$= \dfrac{4 \times 3 \times 2.31 \times 7 \times 200 \times 200}{22 \times 49 \times 5 \times 60} = 24$ min

41. (a) Let the area be A.

Then, breadth of R_1, R_2 and R_3

$= \dfrac{A}{x_1}, \dfrac{A}{x_2}, \dfrac{A}{x_3}$

Cylinder formed by joining parallel side of breadth,

Here, length becomes height

and $2\pi r = $ breadth $= \dfrac{A}{x}$

$\therefore \quad r = \dfrac{A}{2\pi x}$

and $V = \pi r^2 h = \pi \dfrac{A^2}{4\pi^2 x^2} \times x = \dfrac{A^2}{4\pi x}$

Now,

V_1, V_2 and $V_3 = \dfrac{A^2}{4\pi x_1}, \dfrac{A^2}{4\pi x_2}$ and $\dfrac{A^2}{4\pi x_3}$

$\therefore \quad x_1 < x_2 < x_3 \Rightarrow \dfrac{1}{x_1} > \dfrac{1}{x_2} > \dfrac{1}{x_3}$

$\Rightarrow V_1 > V_2 > V_3$

or $V_3 < V_2 < V_1$

42. (d) Let radius of cone be r cm.

and radius of cylinder be R cm.

According to the question,

Radius of cylinder

$(R) = \dfrac{1}{3} \times$ Radius of cone

$\Rightarrow \quad R = \dfrac{r}{3}$ ...(i)

Now, volume of cone = Volume of cylinder

$\Rightarrow \dfrac{1}{3} \pi r^2 h = \pi R^2 H$

$\Rightarrow \dfrac{1}{3} \times r^2 \times 24 = \left(\dfrac{r}{3}\right)^2 \times H$ [$\because h = 24$ cm]

$\Rightarrow r^2 \times 8 = \dfrac{r^2}{9} \times H$

$\therefore \quad H = 9 \times 8 = 72$ cm

43. (c) Given, length and breadth of roof

$= 22$ m $\times 20$ m

radius of cylindrical vessel $= \dfrac{2}{2} = 1$ m

and height of cylindrical vessel = 3.5 m

$\therefore$ Rainfall $= \dfrac{\text{Capacity of cylindrical vessel}}{22 \text{ m} \times 20 \text{ m}}$

$= \dfrac{\pi r^2 h}{22 \text{ m} \times 20 \text{ m}}$

$= \dfrac{22 \times 1 \times 1}{7} \times \dfrac{3.5 \text{ m}^3}{22 \times 20 \text{ m}^2} = \dfrac{1}{40}$ m

$= 2.5$ cm

44. (a) $\because$ Curved surface area of right circular cone $= \pi rl$

$\therefore \quad 440 = \dfrac{22}{7} \times 14 \times l \Rightarrow l = \dfrac{440 \times 7}{22 \times 14}$

$= 10$ cm

Volume and Surface Area / 635

45. (a) Given that,
Diameter of a right circular cone = 7 cm
∴ Radius of a right circular cone = $\frac{7}{2}$ cm
and slant height of a right circular cone
$(l) = 10$ cm
∴ Lateral surface area of a cone = $\pi r l$
$= \frac{22}{7} \times \frac{7}{2} \times 10 = 11 \times 10 = 110$ cm²

46. (b) Slant height, $l = \sqrt{h^2 + r^2}$
$= \sqrt{(24)^2 + (7)^2} = \sqrt{576 + 49}$
$= \sqrt{625} = 25$
Total surface area
$= \pi r(l + r)$
$= \frac{22}{7} \times 7 \times (25 + 7)$
$= \frac{22}{7} \times 7 \times 32$
$= 704$ sq cm

47. (a) ∵ Volume = $\frac{1}{3} \pi r^2 h$
According to the question,
$\frac{1}{3} \pi r^2 h = 100\pi$
$\Rightarrow \frac{1}{3} \pi r^2 \times 12 = 100\pi \Rightarrow r^2 = 25$
∴ $r = \sqrt{25} = 5$ cm
∴ Slant height $(l) = \sqrt{h^2 + r^2} = \sqrt{12^2 + 5^2}$
$= \sqrt{169} = 13$ cm

48. (a) Given, $l = 9$ m and diameter = 14 m
∴ $r = \frac{14}{2} = 7$ m
Now, volume = $\frac{1}{3} \pi r^2 h$
$= \frac{1}{3} \pi \times 49 \times \sqrt{l^2 - r^2}$
$= \frac{1}{3} \pi \times 49 \times \sqrt{81 - 49}$
$= \frac{1}{3} \times 49\pi \times \sqrt{32} = \frac{49\pi\sqrt{32}}{3}$ m³

49. (b) Let radius = $5x$, height = $12x$
According to the question,
$\frac{1}{3} \times \frac{22}{7} \times (5x)^2 \times 12x = \frac{2200}{7}$
$\Rightarrow x^3 = 1 \Rightarrow x = 1$
∴ $r = 5, h = 12$
∴ Slant height $(l) = \sqrt{r^2 + h^2} = \sqrt{25 + 144}$
$= \sqrt{169} = 13$ cm

50. (b) $\frac{\frac{1}{3}\pi r_1^2 h_1}{\frac{1}{3}\pi r_2^2 h_2} = \frac{2}{3} \Rightarrow \left(\frac{r_1}{r_2}\right)^2 \times \frac{h_1}{h_2} = \frac{2}{3}$
$\Rightarrow \left(\frac{1}{2}\right)^2 \times \frac{h_1}{h_2} = \frac{2}{3}$ $\left[\because \frac{r_1}{r_2} = \frac{1}{2}\right]$
$\Rightarrow \frac{h_1}{h_2} = \frac{8}{3}$
∴ $h_1 : h_2 = 8 : 3$

51. (a) Let diameter, radius and height of first cone are d_1, r_1 and h_1, respectively and that of second cone are d_2, r_2 and h_2, respectively.
$\frac{r_1}{r_2} = \frac{d_1}{d_2} = \frac{3}{5}, \frac{h_1}{h_2} = ?$
Given, $\frac{\frac{1}{3}\pi r_1^2 h_1}{\frac{1}{3}\pi r_2^2 h_2} = \frac{1}{3}$
$\Rightarrow \left(\frac{r_1}{r_2}\right)^2 \times \frac{h_1}{h_2} = \frac{1}{3} \Rightarrow \left(\frac{3}{5}\right)^2 \times \frac{h_1}{h_2} = \frac{1}{3}$
$\Rightarrow \frac{h_1}{h_2} = \frac{1}{3} \times \frac{25}{9} = \frac{25}{27}$

52. (c) $l = \sqrt{h^2 + r^2} = \sqrt{16^2 + 12^2}$
$= \sqrt{256 + 144} = \sqrt{400} = 20$ cm

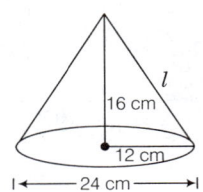

Curved surface area
$= \pi r l = \frac{22}{7} \times 12 \times 20$ cm²
∴ Cost of painting $= \frac{22}{7} \times 12 \times 20 \times 0.70$
$= ₹ 528$

53. (c) Volume of conical tent = $\frac{1}{3}\pi r^2 h$
$= \frac{1}{3} \times \frac{22}{7} \times 8.4 \times 8.4 \times 3.5$
$= 258.72$ m³
and volume of 1 wheat bag = 1.96 m³
∴ Required number of bags
$= \frac{\text{Volume of tent}}{\text{Volume of wheat bag}}$
$= \frac{258.72}{1.96} = 132$

636 / Fast Track Objective Arithmetic

54. (b) Slant height $(l) = \sqrt{r^2 + h^2}$
$= \sqrt{7^2 + 24^2}$
$= \sqrt{49 + 576}$
$= \sqrt{625} = 25$ cm
Curved surface area $= \pi rl = \dfrac{22}{7} \times 7 \times 25$
$= 550$ sq cm
∴ Area of 5 caps $= 550 \times 5 = 2750$ sq cm

55. (c) In first situation,
Radius $= r_1$, height $= h_1$ and volume $= V_1$
In second situation,
Radius $= 2r_1$, height $= h_2$ and volume $= V_2$
If the volume is fixed, then
$V_1 = V_2$
$\Rightarrow \dfrac{1}{3}\pi r_1^2 h_1 = \dfrac{1}{3}\pi (2r_1)^2 h_2 \Rightarrow h_1 = 4h_2$
∴ $h_2 = \dfrac{h_1}{4}$
Therefore, height of the cone will be one-fourth of the previous height.

56. (d) Here, $x = -50\%$, $y = 200\%$
According to the formula,
Net effect
$= \left[2x + y + \dfrac{x^2 + 2xy}{100} + \dfrac{x^2 y}{100^2} \right]\%$
[by Technique 5]
$= \left[-100 + 200 + \dfrac{2500 - 20000}{100} + \dfrac{500000}{10000} \right]\%$
[∵ $x = -50, y = 200$]
$= [100 - 175 + 50] \% = -25\%$

57. (d) Let r and h be the original radius and height of the cone, respectively.
Again, let r' and h' be the new radius and height of the cone, respectively.
Now, according to the question
$r' = r + p\%$ of $r = r + \dfrac{p}{100} r = \left(\dfrac{100+p}{100}\right) r$
and $h' = h$
Now, original volume $= \dfrac{1}{3} \pi r^2 h$
and new volume $= \dfrac{1}{3} \times \pi (r')^2 h'$
∴ Percentage increase in the volume
$= \dfrac{\text{New volume} - \text{Original volume}}{\text{Original volume}} \times 100$
$= \dfrac{\dfrac{1}{3}\pi r'^2 h' - \dfrac{1}{3}\pi r^2 h}{\dfrac{1}{3}\pi r^2 h} \times 100$

$= \dfrac{r'^2 h' - r^2 h}{r^2 h} \times 100$
$= \dfrac{\left(\dfrac{100+p}{100}\right)^2 r^2 h - r^2 h}{r^2 h} \times 100$
$= \dfrac{(100+p)^2 - 100^2}{100}$
$= \dfrac{(200+p)p}{100} = p\left(2 + \dfrac{p}{100}\right)$

58. (a) Slant height $= \sqrt{h^2 + (R-r)^2}$
$= \sqrt{5^2 + (5-3)^2}$
$= \sqrt{25 + 4} = \sqrt{29}$ cm

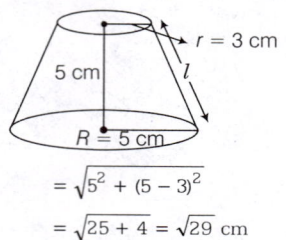

59. (d) Volume $= \dfrac{\pi h}{3}(R^2 + r^2 + Rr)$
$= \dfrac{1}{3} \times \dfrac{22}{7} \times 6 \times (4^2 + 2^2 + 4 \times 2)$
$= \dfrac{22}{7} \times 2 \times 28 = 22 \times 8$
$= 176$ cm^3

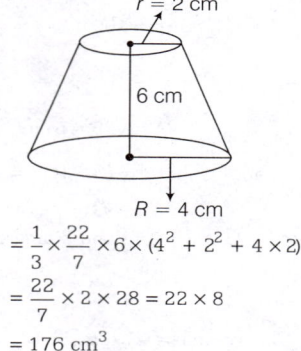

60. (a) Given that, area of first end $= P = \pi r^2$
and area of second end $= Q = \pi R^2$

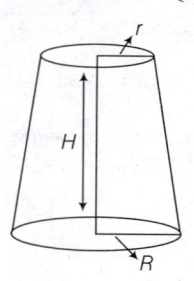

Given, $P < Q$
$\Rightarrow r = \sqrt{\dfrac{P}{\pi}}$ and $R = \sqrt{\dfrac{Q}{\pi}}$

Volume and Surface Area / 637

∴ Difference in radii of the ends of the frustum = $R - r$

$$= \sqrt{\frac{Q}{\pi}} - \sqrt{\frac{P}{\pi}} = \frac{\sqrt{Q} - \sqrt{P}}{\sqrt{\pi}}$$

61. (c) We know that,
Volume of frustum

$$= \frac{\pi H}{3}(R^2 + r^2 + Rr)$$

$$= \frac{\pi}{3}H\left\{\left(\sqrt{\frac{Q}{\pi}}\right)^2 + \left(\sqrt{\frac{P}{\pi}}\right)^2 + \sqrt{\frac{Q}{\pi}} \cdot \sqrt{\frac{P}{\pi}}\right\}$$

$$= \frac{\pi H}{3}\left\{\frac{Q}{\pi} + \frac{P}{\pi} + \frac{\sqrt{PQ}}{\pi}\right\}$$

$$= \frac{H}{3}(P + Q + \sqrt{PQ})$$

62. (b) Let the cone be divided into two parts by a line l.

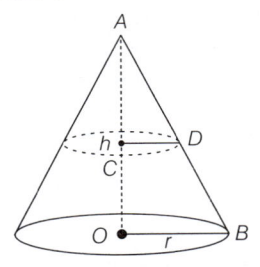

In $\triangle AOB$ and $\triangle ACD$, $\triangle AOB \sim \triangle ACD$
By basic proportionality theorem,

$$\frac{AC}{AO} = \frac{CD}{OB} \Rightarrow \frac{\frac{1}{2}AO}{AO} = \frac{CD}{OB}$$

$$\Rightarrow CD = \frac{OB}{2} = \frac{r}{2}$$

∴ Required ratio = $\dfrac{\text{Volume of original cone}}{\text{Volume of smaller cone}}$

$$= \frac{\frac{1}{3}\pi r^2 h}{\frac{1}{3}\pi \left(\frac{r}{2}\right)^2 \left(\frac{h}{2}\right)} = \frac{8}{1} = 8 : 1$$

63. (a) Here, $h = 24$ cm, $r = \dfrac{4}{2}$ cm = 2 cm

and $R = \dfrac{18}{2}$ cm = 9 cm

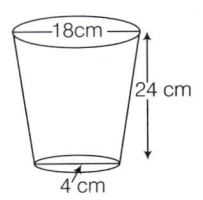

Now, capacity of glass = $\dfrac{\pi h}{3}(r^2 + R^2 + rR)$

$$= \pi \left\{\frac{24}{3}(2^2 + 9^2 + 2 \times 9)\right\}$$

$$= \pi \{8(4 + 81 + 18)\} = 824 \pi \text{ cm}^3$$

According to the question,
Capacity = πx cm^3
On comparing, we get $x = 824$

64. (a) $\dfrac{r}{r_1} = \dfrac{h}{\frac{h}{2}} = \dfrac{2}{1}$

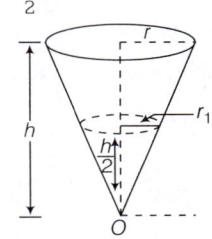

We have, $V = \dfrac{1}{3}\pi r^2 h$

∴ $r^2 = \dfrac{3V}{\pi h}$

Also, $r_1 = \dfrac{r}{2} \Rightarrow r_1^2 = \dfrac{r^2}{4}$

∵ Volume of water = $\dfrac{1}{3}\pi r_1^2 \dfrac{h}{2}$

$$= \frac{1}{6}\pi h \frac{r^2}{4} = \frac{1}{6} \times \pi h \times \frac{1}{4} \times \left(\frac{3V}{\pi h}\right)$$

$$= \frac{1}{24}\pi h \frac{3V}{\pi h} = \frac{3V}{24} = \frac{V}{8}$$

65. (d) In $\triangle AOE$ and $\triangle ADC$,
$\triangle AOE \sim \triangle ADC$

∴ $\dfrac{h_1}{h_1 + h_2} = \dfrac{r_1}{r_2}$ ...(i)

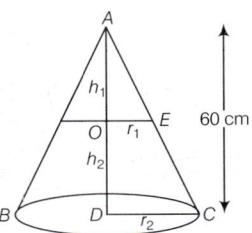

According to the question,

$$\frac{1}{3}\pi r_1^2 h_1 = \frac{1}{64} \times \frac{1}{3}\pi r_2^2 (h_1 + h_2)$$

$$\Rightarrow r_1^2 h_1 = \frac{1}{64} r_2^2 (h_1 + h_2)$$

$$\Rightarrow \left(\dfrac{h_1}{h_1+h_2}\right) = \dfrac{r_2^2}{r_1^2} \times \dfrac{1}{64}$$

[from Eq. (i) and $h_1 + h_2 = 60$ cm (given)]

$$\Rightarrow \dfrac{h_1}{60} = \dfrac{(h_1+h_2)^2}{h_1^2} \times \dfrac{1}{64}$$

$$\Rightarrow h_1^3 = \dfrac{(60)^3}{64}$$

$$\Rightarrow h_1 = 15$$

Hence, required height,
$$h_2 = 60 - h_1 = 45 \text{ cm}.$$

66. (d) Here, radius $(r) = \dfrac{14}{7} = 7$ cm

and height $(h) = 10$ cm

∴ Volume of cylinder
$$= \pi r^2 h = \dfrac{22}{7} \times 7^2 \times 10$$
$$= 1540 \text{ cm}^3$$

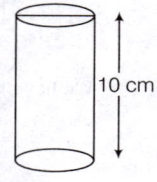

This cylinder is melted and recast into two cones in the proportion of 3 : 4 (volume), keeping the height 10 cm. Let volume of first cone be V_1 cm³ and volume of second cone be V_2 cm³.

Then, $\dfrac{V_1}{V_2} = \dfrac{3}{4}$

$$\Rightarrow \dfrac{\dfrac{1}{3}\pi r_1^2 h}{\dfrac{1}{3}\pi r_2^2 h} = \dfrac{3}{4}$$

$$\Rightarrow \dfrac{r_1^2}{r_2^2} = \dfrac{3}{4}$$

$$\Rightarrow r_1^2 = \dfrac{3}{4} r_2^2 \qquad \ldots(i)$$

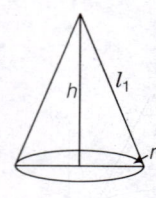

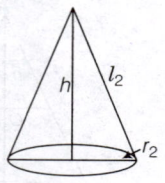

where, h = Height,
r_1 = Radius of 1st cone,
r_2 = Radius of 2nd cone,
l_1 = Slant height of 1st cone

and l_2 = Slant height of 2nd cone

Now, $V_1 + V_2 = 1540$ cm³

$$\Rightarrow \dfrac{1}{3}\pi r_1^2 h + \dfrac{1}{3}\pi r_2^2 h = 1540$$

$$\Rightarrow \dfrac{1}{3}\pi h (r_1^2 + r_2^2) = 1540$$

$$\Rightarrow \dfrac{1}{3} \times \dfrac{22}{7} \times 10(r_1^2 + r_2^2) = 1540$$

$$\Rightarrow r_1^2 + r_2^2 = \dfrac{154 \times 21}{22}$$

$$r_1^2 + r_2^2 = 147 \qquad \ldots(ii)$$

$$\Rightarrow \dfrac{3}{4} r_2^2 + r_2^2 = 147 \qquad \text{[from Eq. (i)]}$$

$$\Rightarrow \dfrac{7 r_2^2}{4} = 147 \Rightarrow r_2^2 = \dfrac{147 \times 4}{7} \Rightarrow r_2^2 = 84$$

$$\Rightarrow r_1^2 = \dfrac{3}{4} \times 84 = 63$$

Now, flat surface area of the cylinder
$$= 2 \times \pi r^2 = 2\pi \times (7)^2 = 98\pi \text{ cm}^2$$

and flat surface area or the cones
$$= \pi r_1^2 + \pi r_2^2$$
$$= 63\pi + 84\pi$$
$$= 147\pi \text{ cm}^2$$

∴ Required percentage change
$$= \dfrac{147\pi - 98\pi}{98\pi} \times 100\%$$
$$= \dfrac{49}{98} \times 100\% = 50\%$$

67. (a) Required surface area = $4\pi r^2$
$$= 4 \times \pi \times 4^2 = 64\pi \text{ sq cm}$$

68. (a) Given, radius of sphere, $r = 3$ cm

∴ Volume of a sphere $= \dfrac{4}{3}\pi r^3 = \dfrac{4}{3}\pi(3)^3$
$$= 4 \times \pi \times 9 = 36\pi \text{ cm}^3$$

69. (a) Curved surface area of the sphere
$$= 4\pi r^2 \quad \text{or} \quad 616 = 4\pi r^2$$

$$\Rightarrow \pi r^2 = \dfrac{616}{4} = 154$$

$$\Rightarrow r^2 = \dfrac{154 \times 7}{22} = 49$$

∴ $r = \sqrt{49} = 7$ cm

∴ Volume of the sphere $= \dfrac{4}{3}\pi r^3$

$$= \dfrac{4}{3} \times \dfrac{22}{7} \times 7 \times 7 \times 7 = \dfrac{4312}{3} \text{ cm}^3$$

70. (a) Surface area of sphere = 616 cm²

∴ $4\pi r^2 = 616 \Rightarrow r^2 = \dfrac{616 \times 7}{4 \times 22}$

$$\Rightarrow r^2 = 7 \times 7 \Rightarrow r = 7 \text{ cm}$$

∴ Diameter of largest circle lying on sphere $= 2 \times r = 2 \times 7 = 14$ cm

Volume and Surface Area / 639

71. (b) Let original and new radius of the sphere are r and r', respectively.
Now, according to the question.
$$4\pi r'^2 = \frac{1}{9} 4\pi r^2$$
$$\Rightarrow \quad r'^2 = \frac{1}{9} r^2$$
$$\Rightarrow \quad r' = \frac{1}{3} r$$

72. (a) Let the diameter's of two sphere be d_1 and d_2, respectively.
$$\therefore \quad d_1 : d_2 = 3 : 5$$
$\therefore$ Ratio of their surface areas $= \dfrac{4\pi r_1^2}{4\pi r_2^2}$
$$= \frac{(2r_1)^2}{(2r_2)^2} = \frac{d_1^2}{d_2^2} = \left(\frac{d_1}{d_2}\right)^2 = \left(\frac{3}{5}\right)^2 = \frac{9}{25}$$
$$= 9 : 25$$

73. (c) Given, diameter of vessel is 14 cm and cost of painting is ₹ 8008.
$\therefore$ Cost of painting per square centimetre
$$= \frac{\text{Total cost}}{\text{Surface area of vessel}} = \frac{8008}{4\pi r^2}$$
$$= \frac{8008 \times 7}{4 \times 22 \times 7 \times 7} = ₹ 13$$

74. (c) Volume of small spheres
$$= \frac{\text{Volume of bigger sphere}}{\text{Number of small spheres}} = \frac{\frac{4}{3}\pi(4)^3}{64}$$
$$= \frac{4}{3} \times \frac{\pi \times 4 \times 4 \times 4}{64} = \frac{4}{3} \pi \text{ cm}^3$$
Let radius of small sphere be r'.
$$\therefore \quad \frac{4}{3}\pi r'^3 = \frac{4}{3}\pi \Rightarrow r' = 1 \text{ cm}$$
Now, surface area of small sphere
$$= 4\pi r'^2 = 4\pi \text{ cm}^2$$

75. (c) Let radius of the third sphere be r.
Then, $\dfrac{4}{3} \pi \times (12)^3 = \dfrac{4}{3} \pi \times (6)^3 + \dfrac{4}{3} \pi$
$$\times (8)^3 + \frac{4}{3}\pi r^3$$
$\Rightarrow \quad (12)^3 = (6)^3 + (8)^3 + r^3$
$\Rightarrow \quad r^3 = 1728 - 216 - 512$
$\Rightarrow \quad r^3 = 1000$
$\therefore \quad r = 10$ cm

76. (d) Volume of solid sphere of radius 4 cm
$$= \frac{4}{3}\pi(4)^3 \text{ cm}^3$$
Volume of hollow sphere
$$= \frac{4}{3}\pi[(8)^3 - (6)^3] \text{ cm}^3$$

$\because$ Weight of $\dfrac{4}{3}\pi(4)^3$ cm^3 = 4 kg
$\therefore$ Weight of $\dfrac{4}{3}\pi[(8)^3 - (6)^3]$ cm^3
$$= \frac{4}{\frac{4}{3}\pi(4)^3} \cdot \frac{4}{3}\pi[(8)^3 - (6)^3]$$
$$= \frac{4(512 - 216)}{4^3} = 18.5 \text{ kg}$$

77. (a) Curved surface area $= 2\pi r^2$
$$= 2\pi \times 14 \times 14$$
$$= 2 \times \frac{22}{7} \times 14 \times 14$$
$$= 2 \times 22 \times 2 \times 14 = 1232 \text{ sq cm}$$

78. (c) Given, diameter = 2 cm
$\therefore \quad r = 1$ cm
Now, total surface area of hemisphere $= 3\pi r^2$
and curved surface area $= 2\pi r^2$
$\therefore$ Required difference $= 3\pi r^2 - 2\pi r^2 = \pi r^2$
$$= \pi \times 1^2 = \pi \text{ sq cm}$$

79. (d) Curved surface area of the hemisphere
$$= 2\pi r^2 = 2 \times \frac{22}{7} \times \frac{7}{2} \times \frac{7}{2} = 77 \text{ sq cm}$$
As bowl is to be painted inside and outside.
$\therefore$ Total surface to be painted
$$= 77 \times 2 = 154 \text{ sq cm}$$
$\therefore$ Cost of painting 154 sq cm
$$= \frac{5}{10} \times 154 = \frac{1}{2} \times 154 = ₹ 77$$

80. (d) According to the question,
Surface area of sphere
$\quad$ = Surface area of hemisphere
$$4\pi r_1^2 = 3\pi r_2^2 \Rightarrow \frac{r_1}{r_2} = \frac{\sqrt{3}}{2}$$
$$\Rightarrow \quad r_1 = \frac{\sqrt{3}}{2} r_2$$
$\therefore$ Ratio of volumes $= \dfrac{\frac{4}{3}\pi r_1^3}{\frac{4}{3}\pi r_2^3} = \left(\dfrac{\frac{\sqrt{3}}{2} r_2}{r_2}\right)^3$
$$= \frac{\frac{3\sqrt{3}}{8}}{1} = \frac{3\sqrt{3}}{8} : 1$$

81. (a) Let the initial radius of sphere be r.
$\therefore$ Volume of sphere $= \dfrac{4}{3}\pi r^3$
According to the question,
If the radius of a sphere is increased by 10%.
Then, new radius $r' = r + r \times 10\%$
$$= r + \frac{r}{10} = \frac{11r}{10}$$

∴ New volume of sphere = $\dfrac{4}{3}\pi r'^3$

$= \dfrac{4\pi}{3}\left(\dfrac{11r}{10}\right)^3 = \dfrac{4}{3}\pi \times \dfrac{1331}{1000}r^3$

Increased volume $= \dfrac{4}{3}\pi \times \dfrac{1331}{1000}r^3 - \dfrac{4}{3}\pi r^3$

$= \dfrac{4}{3}\pi r^3\left[\dfrac{1331}{1000} - 1\right]$

$= \dfrac{4}{3}\pi r^3 \times \dfrac{331}{1000}$

∴ Increased percentage

$= \dfrac{\dfrac{4}{3}\pi r^3 \times \dfrac{331}{1000}}{\dfrac{4}{3}\pi r^3} \times 100\%$

$= \dfrac{331}{1000} \times 100\% = 33.1\%$

Fast Track Method
Here, $x = 10\%$
∴ Percentage increase in volume

$= \left[\left(1 + \dfrac{10}{100}\right)^3 - 1\right] \times 100\%$

[by Technique 2]

$= \left[\left(\dfrac{11}{10}\right)^3 - 1\right] \times 100\%$

$= \left[\dfrac{1331}{1000} - 1\right] \times 100\%$

$= \dfrac{331}{1000} \times 100\% = 33.1\%$

82. (a) According to the formula,
Percentage increase in surface area

$= \left[2x + \dfrac{x^2}{100}\right]\% = \left[2 \times 3 + \dfrac{(3)^2}{100}\right]\%$

[by Technique 4]

$= [6 + 0.09]\% = 6.09\%$

83. (c) According to the formula,
Percentage decrease in surface area

$= \left[2 \times (-24) + \dfrac{(-24) \times (-24)}{100}\right]\%$

[by Technique 4]

$= [-48 + 5.76]\% = -42.24\%$

84. (b) Let volume of larger sphere be V and radius be R after making smaller spheres of equal radii, the volume lost is 10%.
Total volumes of smaller spheres

$= V - \dfrac{1}{10}V = \dfrac{9V}{10} = \dfrac{9}{10}\left(\dfrac{4}{3}\pi r^3\right)$

[where, r = Radius of one smaller sphere]

∵ Radius of the smaller sphere, $r = \dfrac{1}{9}R$

Number of smaller spheres

$= \dfrac{\dfrac{9}{10}\left(\dfrac{4}{3}\pi R^3\right)}{\dfrac{4}{3}\pi\left(\dfrac{1}{9}R\right)^3} = \dfrac{\dfrac{9}{10}\left(\dfrac{4}{3}\pi R^3\right)}{\dfrac{1}{(9)^3}\left(\dfrac{4}{3}\pi R^3\right)}$

$= \dfrac{9}{10} \times (9)^3 = \dfrac{9^4}{10}$

∵ Surface area of larger sphere ($4\pi R^2$) require 10 L of paint.

∴ Surface area of $\dfrac{9^4}{10}$ smaller spheres will require amount of paint

$= \dfrac{9^4}{10} \times 4\pi\left(\dfrac{R}{9}\right)^2$

$= \dfrac{9^4}{10 \times 9^2} \times 4\pi R^2$

$= \dfrac{9^4}{10 \times 9^2} \times 10 = 9^2 = 81$ L

85. (d) Volume of prism
= Area of base × Height
$= \dfrac{1}{2} \times 10 \times 12 \times 20 = 1200$ cm^3

∴ Weight of prism $= 1200 \times 6 = 7200$ g
$= 7.2$ kg

86. (a) Total surface area of the pyramid

$= \left[\dfrac{1}{2}\text{(Perimeter of the base) (Slant height)}\right]$
+ Area of the base

$= \dfrac{1}{2}(4)(4)(8) + 4^2 = 80$ cm^2

87. (a) Let $AB = BC = x$
In $\triangle ABC$, $AB^2 + BC^2 = AC^2$
$\Rightarrow x^2 + x^2 = a^2$
$\Rightarrow 2x^2 = a^2 \Rightarrow x = \dfrac{a}{\sqrt{2}}$

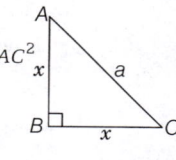

∴ Volume of prism = Area of base × Height

$= \dfrac{1}{2} \cdot \dfrac{a}{\sqrt{2}} \cdot \dfrac{a}{\sqrt{2}} \times h$

$= \dfrac{a^2 h}{4}$ cm^3

88. (a) Let the sides of the base are $5x, 12x$ and $13x$, respectively.
Given, perimeter of base = 60 cm
$\Rightarrow 5x + 12x + 13x = 60$
∴ $x = \dfrac{60}{30} = 2$

So, the sides of base are 10 cm, 24 cm and 26 cm.

∴ Volume of prism $= \dfrac{1}{2} \times 10 \times 24 \times 50$

$= 6000$ cm^3

Volume and Surface Area / 641

89. (c) Volume of the prism
= (Area of the base) × (Height)
Volume of the pyramid
$= \frac{1}{3}$ (Area of the base) × (Height)
∴ Required ratio $= \dfrac{A \times H}{\frac{1}{3} \times A \times H} = 3 : 1$

Therefore, ratio of the volumes of the prism and the pyramid = 3 : 1

90. (c) Number of persons
$= \dfrac{\text{Volume of the cylinder}}{\text{Volume of each cone}}$

$= \dfrac{\pi \left(\frac{35}{2}\right)^2 (32)}{\frac{1}{3}\pi(2)^2(7)}$

$= \dfrac{35 \times 35 \times 32 \times 3}{2 \times 2 \times 2 \times 2 \times 7} = 1050$

91. (b) Given, height of tent = 21 m
and height of cylindrical part (h_1) = 5 m

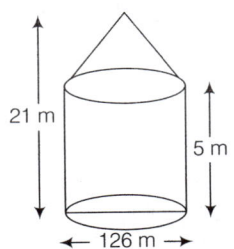

∴ Height of cone (h_2) = (21 – 5) m = 16 m
and radius of cone $= \dfrac{126}{2} = 63$ m

Now, slant height $(l) = \sqrt{r^2 + h_2^2}$
$= \sqrt{(63)^2 + (16)^2}$
$= \sqrt{3969 + 256}$
$= \sqrt{4225} = 65$ m

92. (d) Total canvas used = Curved surface area of cylinder + Curved surface area of cone
$= 2\pi r h_1 + \pi r l = \pi r(2h_1 + l)$
$= \dfrac{22}{7} \times 63(2 \times 5 + 65)$
$= 14850$ m²

93. (b) Required ratio
$= \dfrac{\text{Surface area of rod}}{\text{Surface area of sphere}}$
$= \dfrac{2\pi r(h + r)}{4\pi r^2} = \dfrac{h + r}{2r}$

$= \dfrac{7 + 21}{42} = \dfrac{28}{42}$
$= 2 : 3$ [∵ 1 dm = 10 cm]

94. (c) Diameter of the cylinder = Diameter of spherical ball
= 2 cm
∴ Radium of cylinder = 1 cm
It is given that length of cylinder = 14 m
= 1400 cm
∴ Surface area of pipe $= 2 \times \dfrac{22}{7} \times 1 \times 1400$
= 8800 sq cm

95. (a) Radius of solid sphere, R = 3 cm
Height of cylinder, h = 4 cm
External radius of cylinder, r_2 = 5 cm
Let internal radius of cylinder be r_1 cm.
Now, volume of sphere = Volume of cylinder
∴ $\dfrac{4}{3}\pi R^3 = \pi(r_2^2 - r_1^2) h$
⇒ $\dfrac{4}{3} \times \pi \times (3)^3 = \pi(5^2 - r_1^2) \times 4$
⇒ $9 = 25 - r_1^2 \Rightarrow r_1^2 = 25 - 9$
⇒ $r_1 = 4$ cm
Now, thickness $= (r_2 - r_1)$ cm = (5 – 4) cm
= 1 cm

96. (d) For cube figure,
Edges, a = 12
Faces, b = 6
Corners, c = 8
For pyramid figure,
Edges, a = 8
Faces, b = 5
Corners, c = 5
So, a = b + c is neither true for cube nor for the pyramid.

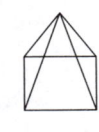

97. (c) Volume of water
= Volume of conical flask $= \dfrac{1}{3}\pi r^2 h$
Now, the water is poured into cylindrical flask.
∴ Volume of cylinder = Volume of water
⇒ $\pi(mr)^2 \times \text{Height} = \dfrac{1}{3}\pi r^2 h$
∴ Height $= \dfrac{h}{3m^2}$

98. (a) Given, l = Length of the cuboid
= 5 × 7 = 35 cm
b = Breadth of the cuboid = 5 cm
and h = Height of the cuboid = 5 cm
∴ Surface area = 2(lb + bh + lh)
= 2[35 × 5 + 5 × 5 + 35 × 5]
= 2[175 + 25 + 175]
= 2 × 375
= 750 sq cm

99. (c) ∵ 1 hec = 10000 m²
∴ Volume of water = Base area × Height
$= 10000 \times \dfrac{10}{100} = 1000 \, m^3$

100. (c) From figure, it is clear that
Diameter of a sphere = Side of the cube
= 3 cm
∴ Radius $= \dfrac{3}{2}$ cm
∴ Volume of the largest sphere $= \dfrac{4}{3} \pi (\text{Radius})^3$
$= \dfrac{4}{3} \pi \left(\dfrac{3}{2}\right)^3$
$= \dfrac{4}{3} \pi \cdot \dfrac{27}{8} = \dfrac{9}{2} \pi = 4.5 \pi \, cm^3$

101. (a) Given that, the height and radius of a right circular metal cone (solid) are 8 cm and 2 cm, respectively.
i.e. $h = 8$ cm and $r = 2$ cm
Let the radius of the sphere be R.
Then, by given condition,
$\dfrac{1}{3} \pi r^2 h = \dfrac{4}{3} \pi R^3 \Rightarrow 4 \times 8 = 4 R^3$
$\Rightarrow R^3 = (2)^3 \Rightarrow R = 2$
∴ Radius of the sphere = 2 cm

102. (b) Let the side of both cube be a, then the height and radius of a cone
$r = \dfrac{a}{2}$ and $h = a$

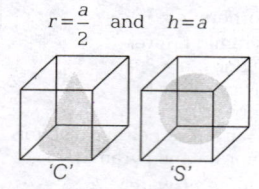

'C' 'S'

and radius of sphere $R = \dfrac{a}{2}$
∴ Volume of cone (C) $= \dfrac{1}{3} \pi r^2 h$
$= \dfrac{1}{3} \pi \left(\dfrac{a}{2}\right)^2 (a) = \dfrac{\pi a^3}{12}$ ...(i)
and volume of sphere (S)
$= \dfrac{4}{3} \pi R^3 = \dfrac{4}{3} \pi \left(\dfrac{a}{2}\right)^3 = \dfrac{\pi a^3}{6}$ ...(ii)
From Eqs. (i) and (ii), we get S = 2C

103. (a) Height of the cylinder
= 10 × Diameter of each ball
= 10 × 10 = 100 cm
∴ Required empty space
= Volume of cylinder
 − 10 × Volume of each ball
$= \pi (5)^2 (100) - \dfrac{4}{3} \pi (5)^3 \times 10$

$= \pi (5)^2 \times 10 \left[10 - \dfrac{20}{3}\right]$
$= \pi (5)^2 \times 10 \left[\dfrac{30 - 20}{3}\right] = \dfrac{2500}{3} \pi \, cm^3$

104. (a) Let the breadth and height of room be b m and h m, respectively.
Then, according to the question,
$l \times b = n \times$ Area occupied by one patient
$\Rightarrow 14 \times b = 56 \times 2.2$
∴ $b = \dfrac{56 \times 2.2}{14} = 8.8$ m
and volume = 8.8 m³
∴ $h \times$ Area of occupied by one patient = 8.8
∴ $h = \dfrac{8.8}{2.2}$
$\Rightarrow h = 4$ m

105. (b) Volume of the earth dugout as a tunnel
$= \pi r^2 h = \dfrac{22}{7} \times 2 \times 2 \times 56 = 704 \, m^3$
Volume of the ditch $= 48 \times \dfrac{33}{2} \times 4$
$= 24 \times 33 \times 4 = 3168 \, m^3$
∴ Part required $= \dfrac{704}{3168} = \dfrac{2}{9}$

106. (b) ∵ Area of 4 walls = Lateral surface area
Perimeter $= 2(l + b) = 250$
$\Rightarrow l + b = \dfrac{250}{2} = 125$ m
Area to be painted
$= \dfrac{\text{Cost}}{\text{Rate}} = \dfrac{15000}{10} = 1500$ sq m
Area of 4 walls $= 2(l + b)h = 250 \, h$
Now, $250 \, h = 1500$
∴ $h = \dfrac{1500}{250} = 6$ m

+ In painting related problems notice whether roof is painted or not.

107. (c) I. Area of curved surface of a cylinder
$= 2 \pi r h$
∵ $2 \pi r (2h) = 4 \pi r h$
Hence, the curved surface area is also doubled, which is correct.
II. Total surface area of hemisphere
$= 3 \pi r^2$
∵ $3 \pi (2r)^2 = 3 \pi \times 4 r^2 = 4 \times 3 \pi r^2$
= 4 times the total surface area which is also correct.

108. (b) Let the height of circular cylinder be H and radius of cylinder or cone be r.
According to the question,
$\dfrac{\text{Total volume of the solid}}{\text{Volume of circular cone}} = 3$

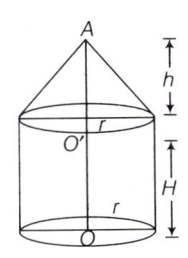

Volume and Surface Area / 643

$$\Rightarrow \quad \frac{\pi r^2 H + \frac{1}{3}\pi r^2 h}{\frac{1}{3}\pi r^2 h} = 3$$

$$\Rightarrow \quad \pi r^2 H + \frac{1}{3}\pi r^2 h = \pi r^2 h$$

$$\Rightarrow \quad \pi r^2 H = \frac{2}{3}\pi r^2 h \quad \Rightarrow \quad H = \frac{2}{3}h$$

109. (a) Let r be the radius of the cylinder. 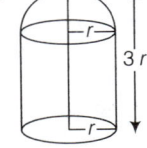
Given, air in building
$= 67\frac{1}{21} \text{ m}^3$
and height of building (h)
$= 3r$.
Now, height of cylindrical part h_1
$= 3r - r = 2r$
Then, air in building = capacity of building

$$67\frac{1}{21} = \pi r^2 h_1 + \frac{2}{3}\pi r^3$$

$$\Rightarrow \quad \frac{1408}{21} = \pi r^2 \left(h_1 + \frac{2}{3}r\right)$$

$$\Rightarrow \quad \frac{1408}{21} = \pi r^2 \left(2r + \frac{2}{3}r\right)$$

$$\Rightarrow \quad \frac{1408}{21} = \pi r^2 \left(\frac{8r}{3}\right) \Rightarrow r^3 = 8 \Rightarrow r = 2$$

∴ Height of building $= 3r = 3 \times 2 = 6$ m

110. (a) Given, radius of pipe
$= \frac{5}{2 \times 10} = \frac{5}{20}$ cm [∵ 1 cm = 10 mm]
Height of pipe = 1000 cm
Radius of vessel = 20 cm and height = 24 cm
Volume of water flow in one minute from cylindrical pipe
$= \pi \left(\frac{5}{20}\right)^2 \times 1000 = \frac{125}{2}\pi \text{ cm}^3$
and volume of conical vessel
$= \frac{1}{3}\pi(20)^2 \times 24 = 3200\pi \text{ cm}^3$
∴ Required time $= \frac{3200\pi \times 2}{125\pi}$
$= 51\frac{1}{5}$ or 51 min 12 s

111. (c) Given, diameter = 150 cm
∴ $R = \frac{150}{2}$ cm
According to the question,
Volume of hemispherical basin
= 120 × volume of cylindrical tube

$$\Rightarrow \quad \frac{2}{3}\pi \left(\frac{150}{2}\right)^3 = 120 \; \pi r^2 \times 15$$

$$\Rightarrow \quad \frac{2}{3} \times \frac{150 \times 150 \times 150}{8} = 120 \times 15 \times r^2$$

$$\Rightarrow \quad r^2 = \frac{150 \times 150 \times 150}{12 \times 120 \times 15}$$

$$\Rightarrow \quad r^2 = \frac{625}{4} \Rightarrow r = \sqrt{\frac{625}{4}} = \frac{25}{2}$$

∴ Diameter $= 2r = 2 \times \frac{25}{2} = 25$ cm

112. (b) Let the height of embankment be h.
Then, volume of earth taken out $= \pi r_1^2 h$
$= \frac{22}{7} \times (7)^2 \times 15$

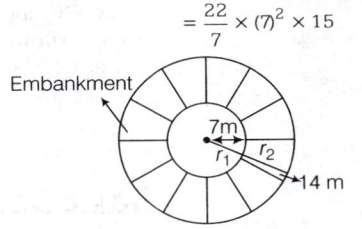

∴ Volume of earth taken out
= Volume of embankment
[an embankment is a heap of stone or mud or earth to stop water]

$\Rightarrow \pi r_1^2 h = \pi (r_2^2 - r_1^2) h \quad [\because r_1 = 7, r_2 = 14]$

$\Rightarrow \frac{22}{7} \times 7^2 \times 15 = \frac{22}{7}(14^2 - 7^2)h$

∴ $h = \frac{7^2 \times 15}{196 - 49} = 5$ m

113. (d) ∵ Radius of sphere = 6 cm
∴ Volume of sphere $= \frac{4}{3}\pi r^3 = \frac{4}{3} \times \pi \times (6)^3$
$= 288\pi \text{ cm}^3$
It is also given that after melting this sphere a wire is made.
Let the length of wire = l cm
and radius of wire = 0.2 cm
∴ Volume of wire of length, $l = \pi r^2 l$
$= \pi \times (0.2)^2 \times l$
According to the question,
Volume of wire = Volume of sphere

$\Rightarrow l = \frac{288}{(0.2)^2} \text{ cm} \Rightarrow l = \frac{288 \times 100}{2 \times 2}$ cm

$\Rightarrow l = 288 \times 25$ cm

$\Rightarrow l = \frac{288 \times 25}{100} = \frac{288}{4} = 72$ m

Chapter 36

Geometry

Geometry is a branch of Mathematics which deals with the questions and concepts of shape, size, relative position of figures, their angles etc.

It can be broadly divided into two parts
Plane Geometry Plane geometry is about flat shapes like line, circle, triangle etc. These are two-dimensional figure which can easily be drawn on paper.
Solid Geometry Solid geometry is about three-dimensional objects like cube, prism, sphere etc. These are three-dimensional figure.

Basic Definitions Related to Geometry

There are various definitions related to geometry, which are given below

Point

A figure of which length, breadth and height cannot be measured is called a point and it is infinitesimal.

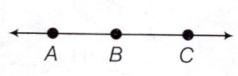

Line

A line is defined by its length but has no breadth. A line contains infinite points and hence can be extended infinitely in both directions.

If three or more than three points lie on the same line, then they are called collinear points, otherwise they are non-collinear points.

> **Plane** It is a flat surface having length and breadth both but no thickness. It is a two-dimensional figure.

Line Segment

A part of a line with two end points is called a line segment. A line segment has a definite length.

Geometry

Ray
A part of line with one end point, is called a ray. A ray has no definite length.

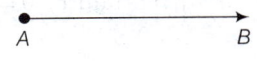

Perpendicular Lines
Two lines which lie in the same plane and intersect each other at right angles are called as perpendicular lines.
Here, OA is perpendicular to the line BC and is denoted by OA ⊥ BC.

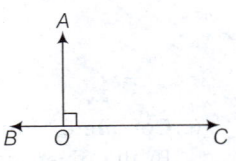

Parallel Lines
Two lines in the same plane are said to be parallel, if they don't have any intersection point, when produced on either side. The distance between the two parallel lines remains constant for the whole length. Here, l and m are called parallel lines and is denoted by $l \parallel m$.

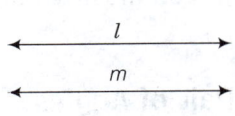

Transversal Lines
A straight line that cuts two or more parallel lines at distinct points is called a transversal line. In the adjacent figure, l and m are parallel lines and p is a transversal line.

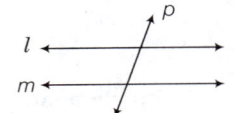

Angle
An angle is formed by two rays with a common initial point. Let O be the initial point, then O is called the vertex of the angle. Generally, angles are measured in degrees. e.g. $\angle AOB = \theta°$.

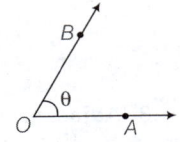

Types of Angles
According to the definition, the angles are of following types

1. **Acute angle** The angle whose value lies between 0° and 90°, is called an acute angle. In the given figure, $0 < \theta < 90°$.

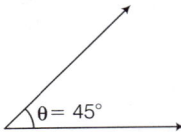

    $\theta = 45°$

2. **Right angle** The angle whose value is 90°, is called a right angle. In the given figure, $\theta = 90°$.

    $\theta = 90°$

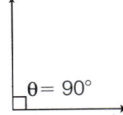

3. **Obtuse angle** The angle whose value lies between 90° and 180°, is called an obtuse angle. In the given figure, $90° < \theta < 180°$.

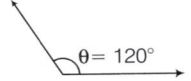

    $\theta = 120°$

4. **Straight angle** The angle whose value is 180°, is called a straight angle. In the given figure, $\theta = 180°$.

$$\theta = 180°$$

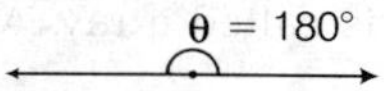

5. **Reflex angle** The angle whose value lies between 180° and 360°, is called a reflex angle. In the given figure, $180° < \theta < 360°$.

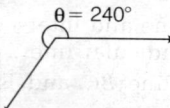

$$\theta = 240°$$

6. **Complete angle** The angle whose value is 360°, is called a complete angle. In the given figure, $\theta = 360°$.

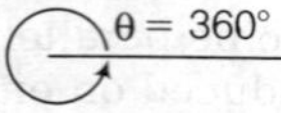

$$\theta = 360°$$

Pair of Angles

1. **Complementary angles** If the sum of two angles is 90°, then they are called complementary angles. Let θ_1 and θ_2 be two angles, then $\theta_1 + \theta_2 = 90°$.

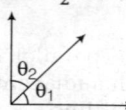

2. **Supplementary angles** If the sum of two angles is 180°, then they are called supplementary angles. Let θ_1 and θ_2 be two angles, then $\theta_1 + \theta_2 = 180°$.

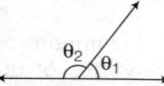

3. **Adjacent angles** Two angles are called adjacent angles, if
   (i) they have a common vertex.
   (ii) they have a common arm.
   (iii) their non-common arms are on different sides of the common arm.

4. **Linear pair of angles** If the non-common arms of two adjacent angles form a line, then these angles are called linear pair of angles. In the adjacent figure, $\angle AOC$ and $\angle BOC$ form a linear pair of angles.
   $\therefore \angle AOB = \angle AOC + \angle BOC$

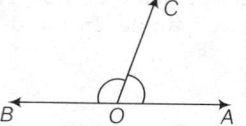

Angle Bisector

A ray which divides an angle into two equal parts is called an angle bisector. An angle bisector can be internal or external. In Fig. (i), OB is the internal angle bisector and $\angle AOB = \angle COB$. In Fig. (ii), OB′ is the external angle bisector and $\angle A'OB' = \angle COB'$.

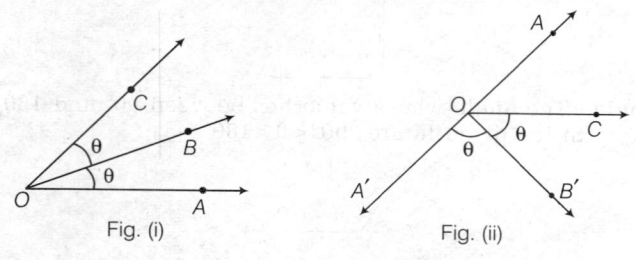

Fig. (i)

Fig. (ii)

Angles Made by a Transversal

Let l and m be parallel lines and n be the transversal which cuts these parallel lines.

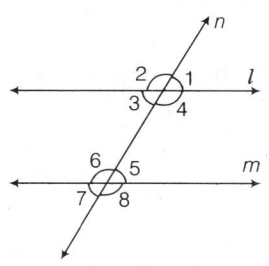

The different angles formed are as follows

1. **Corresponding angles** Pairs of corresponding angles are $\angle 1$ and $\angle 5$, $\angle 2$ and $\angle 6$, $\angle 4$ and $\angle 8$, $\angle 3$ and $\angle 7$. All pairs of corresponding angles are equal.
   i.e. $\quad\quad\quad\quad \angle 1 = \angle 5, \angle 2 = \angle 6, \angle 4 = \angle 8$ and $\angle 3 = \angle 7$

2. **Vertically opposite angles** Pairs of vertically opposite angles are $\angle 1$ and $\angle 3$, $\angle 4$ and $\angle 2$, $\angle 8$ and $\angle 6$, $\angle 5$ and $\angle 7$. All pairs of vertically opposite angles are equal.
   i.e. $\quad\quad\quad\quad \angle 1 = \angle 3, \angle 4 = \angle 2, \angle 8 = \angle 6$ and $\angle 5 = \angle 7$

3. **Alternate interior angles** Pairs of alternate interior angles are $\angle 3$ and $\angle 5$, $\angle 4$ and $\angle 6$. All pairs of alternate interior angles are equal.
   i.e. $\quad\quad\quad\quad \angle 3 = \angle 5$ and $\angle 4 = \angle 6$

4. **Alternate exterior angles** Pairs of alternate exterior angles are $\angle 2$ and $\angle 8$, $\angle 1$ and $\angle 7$. All pairs of alternate exterior angles are equal.
   i.e. $\quad\quad\quad\quad \angle 1 = \angle 7$ and $\angle 2 = \angle 8$

   ✦ The sum of interior angles on the same side of transversal is equal to $180°$.
   i.e. $\quad\quad\quad\quad \angle 3 + \angle 6 = 180°$ and $\angle 4 + \angle 5 = 180°$
   ✦ The sum of exterior angles on the same side of transversal is equal to $180°$.
   i.e. $\quad\quad\quad\quad \angle 2 + \angle 7 = 180°$ and $\angle 1 + \angle 8 = 180°$

Ex. 1 In the given figure, AB and CD are parallel lines. If $\angle EGB = 50°$, find $\angle CHG$.

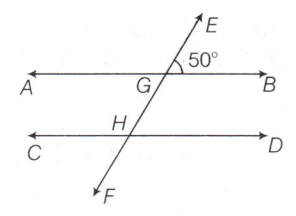

Sol. We have, $\quad\quad\quad \angle AGH = \angle EGB \quad\quad$ [vertically opposite angles]
$\Rightarrow \quad\quad\quad\quad\quad \angle AGH = 50°$
Now, $\quad\quad\quad \angle AGH + \angle CHG = 180° \quad\quad$ [interior angles on the same side of the transversal are supplementary]
$\therefore \quad\quad\quad\quad 50° + \angle CHG = 180°$
$\Rightarrow \quad\quad\quad\quad \angle CHG = 180° - 50° = 130°$

Ex. 2 An angle θ° is one-fourth of its supplementary angle. What is the measure of the angle θ°?

Sol. If the sum of two angles is 180°, then the angles are said to be supplementary angles.
So, the supplementary angle of θ° is (180° − θ°).
Given that, $\theta° = \frac{1}{4}(180° - \theta°) \Rightarrow 4\theta° = 180° - \theta°$
$\Rightarrow \quad 5\theta° = 180° \Rightarrow \theta° = \frac{180°}{5} = 36°$

Ex. 3 In the given figure, find the value of x.

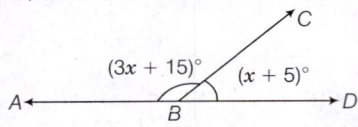

Sol. Given, $\angle ABC = (3x + 15)°$ and $\angle DBC = (x + 5)°$
∵ ABD is a straight line.
∴ $\quad \angle ABD = 180°$
$\Rightarrow \quad \angle ABC + \angle DBC = 180°$
$\Rightarrow \quad (3x + 15)° + (x + 5)° = 180° \Rightarrow 4x = 160° \Rightarrow x = 40°$

Ex. 4 In the figure given below, LOM is a straight line. What is the value of x?

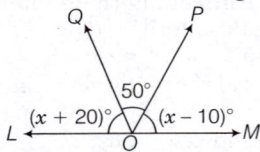

Sol. Given, $\angle LOQ = (x + 20)°$, $\angle POQ = 50°$ and $\angle POM = (x - 10)°$
∵ LOM is a straight line.
∴ $\quad \angle LOM = 180° \Rightarrow \angle LOQ + \angle QOP + \angle POM = 180°$
$\Rightarrow \quad (x + 20°) + 50° + (x - 10°) = 180°$
$\Rightarrow \quad 2x + 60° = 180°$
$\Rightarrow \quad 2x = 120° \Rightarrow x = 60°$

Ex. 5 AB ∥ CD and the line EF cuts these lines at the points M and N, respectively. Bisectors of ∠BMN and ∠MND meet at the point Q. Find the value of ∠MQN.

Sol. Here, $\angle BMN + \angle DNM = 180°$
 [∵ angles on the same side of transversal are supplementary]
On dividing both sides by 2, we get
$\frac{1}{2}\angle BMN + \frac{1}{2}\angle DNM = \frac{180°}{2} = 90°$
$\Rightarrow \quad \angle QMN + \angle MNQ = 90°$ ...(i)
Now, in △MNQ,
$\quad \angle MQN + \angle QMN + \angle MNQ = 180°$ [∵ sum of all angles of a triangle is 180°]
$\Rightarrow \angle MQN = 180° - (\angle QMN + \angle MNQ)$
$= 180° - 90° = 90°$ [from Eq. (i)]

Triangle

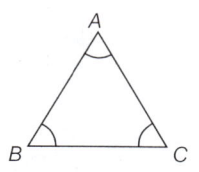

A triangle is a three-sided closed plane figure which is formed by joining three non-collinear points. In the adjoining figure, A, B, C are three non-collinear points which forms $\triangle ABC$.

Now, in $\triangle ABC$, there are
  (i) **Three vertices** A, B and C.
  (ii) **Three sides** AB, BC and AC.
  (iii) **Three angles** $\angle A, \angle B$ and $\angle C$ and sum of these three angles is $180°$.
    i.e. $\angle A + \angle B + \angle C = 180°$

Types of Triangles

1. Based on Sides

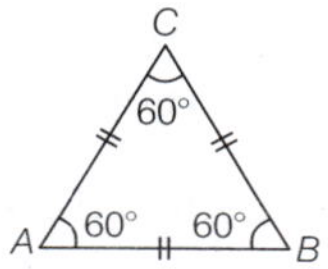

  (i) **Equilateral triangle** A triangle having all sides equal is called an equilateral triangle. In this triangle, all angles are equal to $60°$.
    i.e. $AB = BC = CA$ and $\angle A = \angle B = \angle C = 60°$

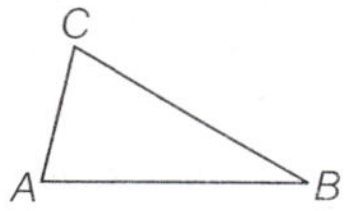

  (ii) **Scalene triangle** A triangle having all sides of different length is called a scalene triangle.
    i.e. $AB \ne BC \ne CA$.

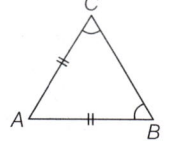

  (iii) **Isosceles triangle** A triangle having two sides equal is called an isosceles triangle. In this triangle, angles opposite to congruent sides are also equal.
    i.e. $AB = AC$
    $\Rightarrow$ $\angle C = \angle B$

2. Based on Angles

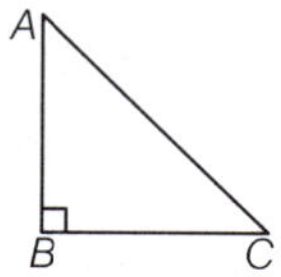

  (i) **Right angled triangle** A triangle one of whose angles measures $90°$ is called a right angled triangle.
    In $\triangle ABC$, $\angle B = 90°$
    So, $\triangle ABC$ is a right angled triangle.

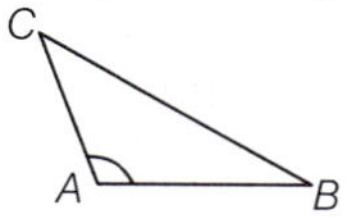

  (ii) **Obtuse angled triangle** A triangle one of whose angles lies between $90°$ and $180°$ is called an obtuse angled triangle.
    In $\triangle ABC$, $\angle A > 90°$
    So, $\triangle ABC$ is an obtuse angled triangle.

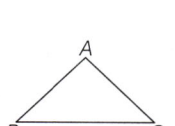

  (iii) **Acute angled triangle** A triangle whose each angle is less than $90°$ is called an acute angled triangle.
    In $\triangle ABC$, $(\angle A, \angle B, \angle C) < 90°$
    So, $\triangle ABC$ is an acute angled triangle.

Properties of Triangles

  (i) The sum of two sides is always more than third side.
  (ii) The difference of two sides is always less than third side.
  (iii) Greater angle has greater side opposite to it and smaller angle has smaller side opposite to it.

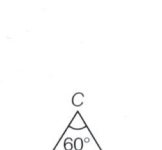

(iv) In the adjoining figure, let p, q and r be the sides of a $\triangle ABC$ and r is the largest side, then
   (a) If $r^2 < p^2 + q^2$, then triangle is acute angled triangle.
   (b) If $r^2 = p^2 + q^2$, then triangle is right angled triangle.
   (c) If $r^2 > p^2 + q^2$, then triangle is obtuse angled triangle.

(v) The exterior angle is equal to the sum of two interior angles not adjacent to it.
$$\angle ACD = \angle BCE = \angle A + \angle B$$

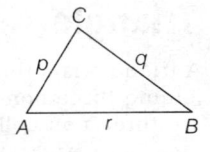

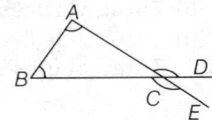

Congruency of Triangles

Two triangles are said to be congruent, if both are exactly of same size, i.e. all angles and sides of one triangle are equal to corresponding angles and sides of other. In congruent triangles, corresponding parts are equal.

Similarity of Triangles

Two triangles are said to be similar to each other, if
   (i) their corresponding sides are proportional.
   (ii) their corresponding angles are equal.

The criteria for similarity of triangles are as given below

1. **AA similarity** If two angles of a triangle are equal to the corresponding two angles of the other triangle, then two triangles are similar.
2. **SAS similarity** If two sides of one triangle are proportional to the corresponding sides of other triangle and the angle included between the sides are equal, then two triangles are similar.
3. **SSS similarity** If three sides of one triangle are proportional to the corresponding three sides of other triangle, then two triangles are similar.

Properties of Similar Triangles

1. Ratio of the areas of two similar triangles is equal to the ratio of the squares of any two corresponding sides.
2. Ratio of the areas of two similar triangles is equal to the ratio of the squares of corresponding altitudes and medians.
3. In two similar $\triangle ABC$ and $\triangle PQR$,

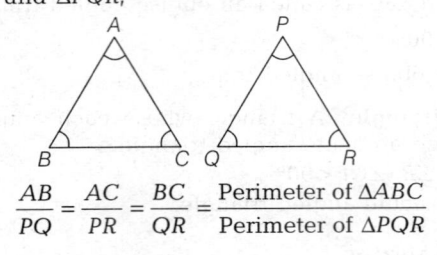

$$\frac{AB}{PQ} = \frac{AC}{PR} = \frac{BC}{QR} = \frac{\text{Perimeter of } \triangle ABC}{\text{Perimeter of } \triangle PQR}$$

Important Points Related to Triangles

1. The internal bisector of an angle of a triangle divides the opposite side internally in the ratio of sides containing the angle.
2. The line joining the mid-point of any two sides of a triangle is parallel to the third side and equal to half of it. **[Mid-point theorem]**

3. Any line parallel to one side of a triangle divides the other two sides proportionally. [**Basic proportionality theorem**]
If DE is drawn parallel to BC, then
$$\frac{AD}{DB} = \frac{AE}{EC} \text{ or } \frac{AD}{AB} = \frac{AE}{AC}$$
or
$$\frac{AD}{DE} = \frac{AB}{BC} \text{ or } \frac{AE}{DE} = \frac{AC}{BC}$$

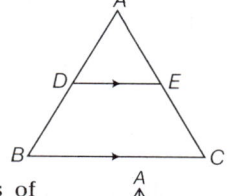

4. **Apollonius theorem** In a triangle, the sum of the squares of any two sides of a triangle is equal to twice the sum of square of the median to the third side and square of half the third side.
In $\triangle ABC$, if AD is median, then $AB^2 + AC^2 = 2(AD^2 + BD^2)$.

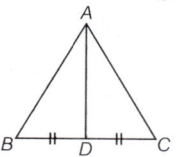

Important Terms Related to Triangles

There are following terms related to triangles

Orthocentre

A perpendicular drawn to a side of a triangle from the vertex opposite to that side, is called an **altitude**. All the three altitudes of a triangle meet a point called the **orthocentre** of the triangle.
In $\triangle ABC$, AD, BE and CF are altitudes and meet at orthocentre O.

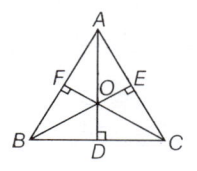

Incentre

All the three angle bisectors of a triangle meet at a point called the **incentre** of the triangle. The incentre is the centre of a circle which can be perfectly **inscribed** in the triangle.
$\therefore$ Inradius $= ID = IE = IF$. Also, $\angle BIC = 90 + \dfrac{\angle A}{2}$.
In $\triangle ABC$, AD, BE and CF are angle bisectors and meet at incentre I.

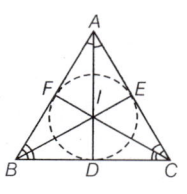

Circumcentre

A line passing through the mid-point of the side of a triangle and perpendicular to it, is called **perpendicular bisector**. All the three perpendicular bisectors of a triangle meet at a point called the **circumcentre** of the triangle. The circumcentre is the centre of a circle which can be perfectly **circumscribed** about the triangle.
Circumradius $= PC = QC = RC$ and $\angle QCR = 2\angle P$
In $\triangle PQR$, the angle bisectors meet at circumcentre C. Also, D, E and F are the mid-points of QR, RP and PQ, respectively.

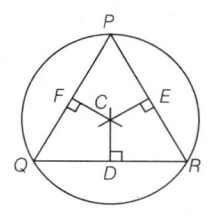

Centroid

A line joining the mid-point of a side of a triangle with the vertex opposite to that side is called **median**. All the three medians of a triangle meet at a point called the **centroid** of the triangle. Centroid divides the median in the ratio 2 : 1.
In $\triangle ABC$, AD, BE and CF are the medians and meet at the centroid G. Also, $\dfrac{AG}{GD} = \dfrac{BG}{GE} = \dfrac{CG}{GF} = \dfrac{2}{1}$ and D, E and F are the mid-points of BC, CA and AB, respectively.

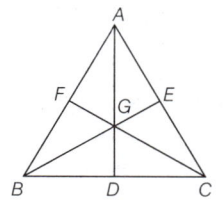

Ex. 6 In the given figure, $\angle BAC : \angle ABC = 2 : 3$. Find the measure of $\angle ABC$.

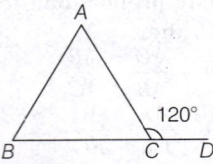

Sol. Let $\angle A = 2x$ and $\angle B = 3x$
Then, $\qquad 2x + 3x = 120°$
$\qquad\qquad$ [∵ exterior angle is equal to the sum of the interior opposite angles]
$\Rightarrow \qquad 5x = 120° \Rightarrow x = 24°$
$\therefore \qquad \angle ABC = 3x = 3 \times 24° = 72°$

Ex. 7 In $\triangle PQR$, if $PQ = 6$ cm, $PR = 8$ cm, $QS = 3$ cm and PS is the bisector of $\angle QPR$, what is the length of QR?

Sol. Since, PS is the angle bisector of $\angle QPR$.
$\therefore \qquad \dfrac{QS}{SR} = \dfrac{PQ}{PR}$
$\Rightarrow \qquad \dfrac{3}{SR} = \dfrac{6}{8} \Rightarrow SR = \left(\dfrac{3 \times 8}{6}\right)$ cm $= 4$ cm
$\therefore \qquad QR = QS + SR = 3 + 4$
$\qquad QR = 7$ cm

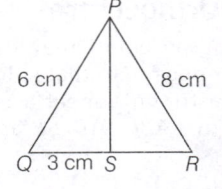

Ex. 8 The angles of triangle are in the ratio $3 : 5 : 7$. The triangle is of which type?

Sol. Let the angles measure $3x$, $5x$ and $7x$.
Then, $\qquad 3x + 5x + 7x = 180°$ $\qquad$ [by angle sum property of triangle]
$\Rightarrow \qquad 15x = 180° \Rightarrow x = 12$
$\therefore$ These angles are $36°$, $60°$ and $84°$.
Hence, the triangle is acute angled triangle.

Ex. 9 In a $\triangle ABC$, $\angle BCA = 90°$ and CD is perendicular to AB. If $AD = 4$ cm and $BD = 9$ cm, then find the value of DC.

Sol. In $\triangle ABC$ and $\triangle ACD$, $\angle A = \angle A$ $\qquad$ [common]
$\qquad \angle ACB = \angle ADC = 90°$
$\therefore \triangle ABC \sim \triangle ACD \Rightarrow \underset{\text{I}}{\dfrac{AC}{AD}} = \underset{\text{II}}{\dfrac{AB}{AC}} = \underset{\text{III}}{\dfrac{BC}{CD}}$

From I and II, $\qquad \dfrac{AC}{13} = \dfrac{4}{AC}$
$\therefore \qquad AC^2 = 4 \times 13 = 52$ cm $\qquad$ ...(i)
Similarly, $\triangle ABC \sim \triangle CBD \Rightarrow \underset{\text{IV}}{\dfrac{AB}{BC}} = \underset{\text{V}}{\dfrac{BC}{BD}} = \underset{\text{VI}}{\dfrac{AC}{CD}}$

From IV and VI, $\qquad \dfrac{AB}{BC} = \dfrac{AC}{CD}$
On squaring both sides, we get
$\qquad \dfrac{(AB)^2}{(BC)^2} = \dfrac{(AC)^2}{(CD)^2}$

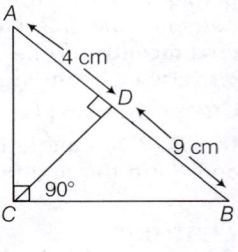

$$\Rightarrow \frac{(AC)^2 + (BC)^2}{(BC)^2(AC)^2} = \frac{1}{CD^2} \Rightarrow \frac{1}{BC^2} + \frac{1}{AC^2} = \frac{1}{CD^2} \quad [\because (AB)^2 = (AC)^2 + (CB)^2]$$

From IV and V,

$$\frac{BC}{BD} = \frac{AB}{BC} \Rightarrow \frac{BC}{13} = \frac{9}{BC}$$

$$\Rightarrow BC^2 = 9 \times 13 = 117 \qquad \ldots(ii)$$

Now, $\frac{1}{CD^2} = \frac{1}{AC^2} + \frac{1}{BC^2} = \frac{1}{52} + \frac{1}{117} = \frac{9+4}{13 \times 4 \times 9}$ [from Eqs. (i) and (ii)]

$$\Rightarrow \frac{1}{CD^2} = \frac{1}{36} \Rightarrow CD = 6 \text{ cm}$$

Quadrilateral

It is a plane figure bounded by four straight lines. It has four sides and four internal angles. The sum of the internal angles of a quadrilateral is equal to 360°.

Types of Quadrilaterals

There are following types of quadrilaterals as given below

1. Parallelogram

A quadrilateral in which the opposite sides are equal and parallel, is called a parallelogram.

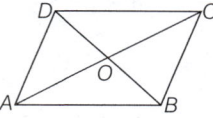

In a parallelogram,
  (i) The diagonals bisect each other.
  (ii) The sum of any two adjacent interior angles is equal to 180°.
  $$\angle A + \angle B = \angle B + \angle C = \angle C + \angle D = \angle D + \angle A = 180°$$
  (iii) The opposite angles are equal in magnitudes $\angle A = \angle C$ and $\angle B = \angle D$.
  (iv) Line joining the mid-points of the adjacent sides of a quadrilateral form a parallelogram.
  (v) Line joining the mid-points of the adjacent sides of a parallelogram is a parallelogram.
  (vi) The parallelogram inscribed in a circle is a rectangle and circumscribed about a circle is a rhombus.
 (vii) $AC^2 + BD^2 = 2(AB^2 + BC^2)$

2. Rhombus

A parallelogram in which all the sides are equal, is called a rhombus.
  (i) The opposite sides are parallel and all the sides are of equal lengths.
  i.e. $AB = BC = CD = DA$
  (ii) The sum of any two adjacent interior angles is equal to 180°.
  i.e. $\angle A + \angle B = \angle B + \angle C = \angle C + \angle D = \angle D + \angle A = 180°$
  (iii) The opposite angles are equal in magnitudes, i.e. $\angle A = \angle C$ and $\angle B = \angle D$.

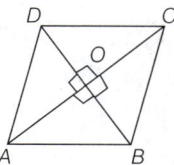

(iv) The diagonals bisect each other at right angles and form four right angled triangles.
(v) Area of the four right triangles are equal,
i.e. ar ($\triangle AOB$) = ar ($\triangle BOC$) = ar ($\triangle COD$) = ar ($\triangle DOA$) and each equals 1/4th the area of the rhombus.
(vi) Figure formed by joining the mid-points of the adjacent sides of a rhombus is a rectangle.

3. Rectangle

A parallelogram in which the adjacent sides are perpendicular to each other and opposite side are equal, is called a rectangle.

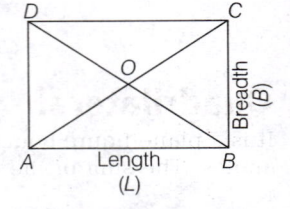

(i) The diagonals of a rectangle are of equal magnitudes and bisect each other, i.e. $AC = BD$ and $OA = OB = OC = OD$.
(ii) The figure formed by joining the mid-points of adjacent sides of a rectangle is a rhombus.
(iii) The quadrilateral formed by the points of intersection of the angle bisectors of a parallelogram is a rectangle.

4. Square

A parallelogram in which all the sides are equal and perpendicular to each other, is called a square.

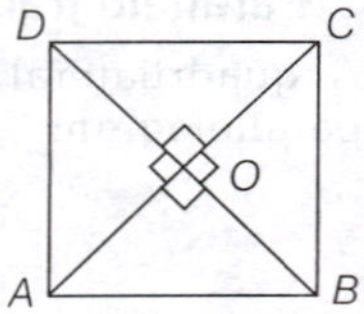

(i) The diagonals bisect each other at right angles and form four isosceles right angled triangles.
(ii) The diagonals of a square are of equal magnitudes, i.e. $AC = BD$.
(iii) The figure formed by joining the mid-points of adjacent sides of a square is again a square.

5. Trapezium

It is a quadrilateral whose only one pair of opposite sides are parallel.
ABCD is a trapezium as $AB \parallel DC$.

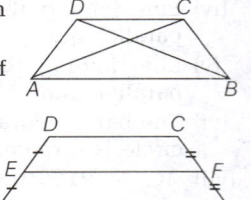

(i) If the non-parallel sides, i.e. (AD and BC) are equal, then diagonals will also be equal to each other.
(ii) Diagonals intersect each other in the ratio of lengths of parallel sides.
(iii) Line joining the mid-points of oblique (non-parallel) sides is half the sum of parallel sides and is called the median, i.e. median, $EF = \frac{1}{2}(AB + DC)$.

Cyclic Quadrilateral

A quadrilateral whose vertices are on the circumference of a circle, is called a cyclic quadrilateral. The opposite angles of a cyclic quadrilateral are supplementary.

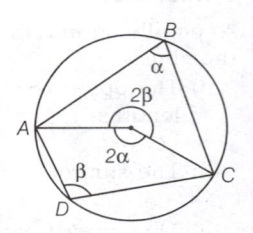

i.e. $\alpha + \beta = 180°$

If the side of a cyclic quadrilateral is produced, then the exterior angle is equal to the interior opposite angle,

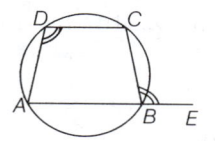

i.e. $\angle ADC = \angle CBE$

Polygons

A polygon is a closed plane figure bounded by straight lines.

Types of Polygons

Polygons are as following four types
1. **Convex polygon** A polygon in which none of its interior angles is more than 180°, is called convex polygon.
2. **Concave polygon** A polygon in which atleast one angle is more than 180°, is called concave, polygon.
3. **Irregular polygon** A polygon in which all the sides or angles are not of the same measure, is called an irregular polygon.
4. **Regular polygon** A regular polygon has all its sides and angles equal.

Number of sides	Name of the polygon	Sum of all the angles	Number of diagonals
3	Triangle	180°	0
4	Quadrilateral	360°	2
5	Pentagon	540°	5
6	Hexagon	720°	9
7	Heptagon	900°	14
8	Octagon	1080°	20
9	Nonagon	1260°	27
10	Decagon	1440°	35

(i) Each exterior angle of a regular polygon $= \dfrac{360°}{\text{Number of sides}}$

(ii) Each interior angle = 180° − Exterior angle

(iii) Sum of all interior angles $= (2n-4) \times 90° = (n-2) \times 180°$

(iv) Sum of all exterior angles = 360°

(v) Number of diagonals of polygon of n sides $= \dfrac{n(n-3)}{2}$

Ex. 10 The angles of a quadrilateral are in the ratio 3 : 4 : 5 : 6. What is the measure of smallest of these angles?

Sol. Let the angles of the quadrilateral be $3x, 4x, 5x$ and $6x$, respectively.
Then, $3x + 4x + 5x + 6x = 360° \Rightarrow 18x = 360° \Rightarrow x = 20°$
∴ Smallest angle $= 3x = 3 \times 20° = 60°$

Ex. 11 In the given figure, ABCD is a parallelogram. E and F are the centroids of $\triangle ABD$ and $\triangle BCD$, respectively. What is EF equal to?

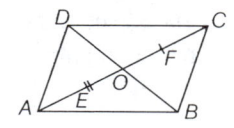

Sol. Since, E is the centroid of $\triangle ABD$ and AO is its median.
Then, $\qquad AE : EO = 2 : 1$
$\therefore \qquad EO = \dfrac{1}{3} OA$
Similarly, $\qquad FO = \dfrac{1}{3} OC$
$\therefore \quad EF = EO + OF = \dfrac{1}{3} OA + \dfrac{1}{3} OC = \dfrac{1}{3} AC = AE \Rightarrow EF = AE$

Ex. 12 $ABCD$ is a rhombus in which $\angle C = 60°$. Then, find the value of $AC : BD$.

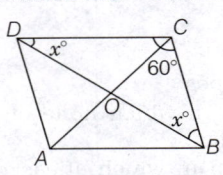

Sol. Since, $ABCD$ is a rhombus. So, its all sides are equal.
Now, $BC = DC \Rightarrow \angle BDC = \angle DBC = x°$
In $\triangle BCD$, $\qquad \angle BCD = 60°$ [given]
$\therefore \qquad x° + x° + 60° = 180° \Rightarrow 2x = 120° \Rightarrow x = 60°$
Now, $\qquad \angle BCD = \angle DBC = \angle BDC = 60°$
So, $\triangle BCD$ is an equilateral triangle.
$\therefore \qquad BD = BC = a$ (let)
In right angled $\triangle AOB$,
$AB^2 = OA^2 + OB^2 \Rightarrow OA^2 = AB^2 - OB^2 = a^2 - \left(\dfrac{a}{2}\right)^2$
$\Rightarrow \qquad OA^2 = a^2 - \dfrac{a^2}{4} = \dfrac{3a^2}{4} \Rightarrow OA = \dfrac{\sqrt{3}a}{2}$
$\Rightarrow \qquad AC = 2 \times OA = \left(2 \times \dfrac{\sqrt{3}a}{2}\right) = \sqrt{3}a$
$\therefore \qquad AC : BD = \sqrt{3}a : a = \sqrt{3} : 1$

Ex. 13 In the given figure, A, B, C and D are the concyclic points. Find the value of x.

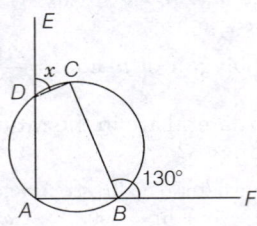

Sol. $\because \angle ABC + \angle CBF = 180°$ [linear pair]
$\therefore \qquad \angle ABC = 180° - 130° = 50°$
In $ABCD$, $\angle ABC + \angle ADC = 180° \Rightarrow \angle ADC = 180° - 50° = 130°$
Now, $\qquad \angle ADC + \angle CDE = 180° \Rightarrow 130° + x = 180°$ [linear pair]
$\Rightarrow \qquad x = 180° - 130° = 50°$

Ex. 14 In the given figure, $AD \parallel BC$. Find the value of x.

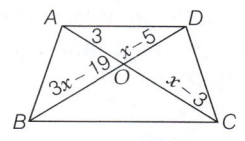

Sol. Here, $AD \parallel BC$ [property of trapezium]

∴ $\dfrac{OA}{OC} = \dfrac{OD}{OB} \Rightarrow \dfrac{3}{x-3} = \dfrac{x-5}{3x-19}$

$\Rightarrow \quad 9x - 57 = x^2 - 8x + 15 \Rightarrow x^2 - 17x + 72 = 0$

$\Rightarrow \quad (x-8)(x-9) = 0 \Rightarrow x = 8, 9$

Circle

A circle is a set of points which are equidistant from a given point. The given point is known as the centre of that circle.

In the given figure, O is the centre of circle and r is the radius of circle. It is represented by $C(O, r)$.

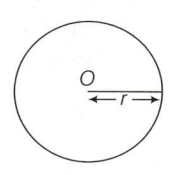

Important Terms Related to Circles

There are following terms related to circles

Chord

A line segment whose end points lie on the circle is called a chord AB in the given figure.

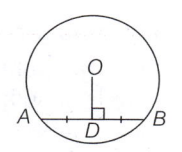

  (i) The line joining the centre of a circle to the mid-point of a chord is perpendicular to the chord. In the given figure, $AD = DB$ and $OD \perp AB$.
 (ii) Equal chords of a circle are equidistant from the centre and *vice-versa*.
(iii) Equal chords subtends equal angles at the centre and *vice-versa*.

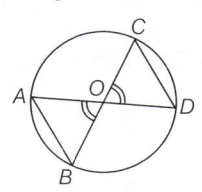

 (iv) In a circle or in congruent circles, equal chords are made by equal arcs.
  (v) If two chords AB and CD of a circle intersect inside a circle (outside the circle when produced at point E), then $AE \times BE = CE \times DE$.

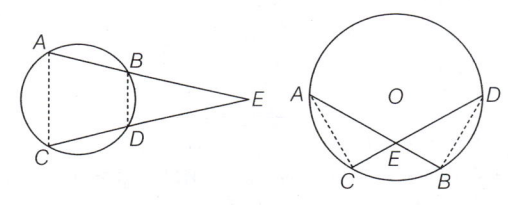

Secant

A line segment which intersects the circle at two distinct points, is called as secant of the circle. In the given figure, PQ is a secant, which intersect circle at two points A and B.

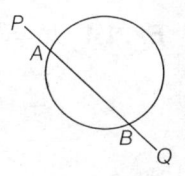

Sector of Circle

The region enclosed by an arc of a circle and its two bounding radii is called a sector of the circle.
In the adjonining figure, OABO is the sector of the circle C (O, r).

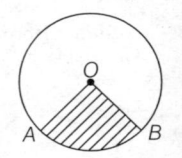

Segment of a Circle

A chord divides the circle into two regions. These two regions are called the segments of a circle.
In the given figure, PQRP is the major segment and PSQP is the minor segment.

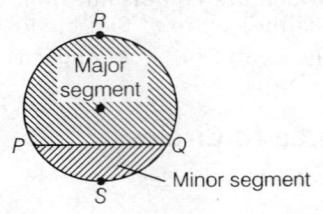

+ Angles made in the same segment by a chord are equal.

Central Angle

The angle subtended at the centre by any two points on the circumference of the circle is called central angle.
  (i) Angle in a semi-circle is a right angle.
  (ii) The angle subtended by an arc at the centre of the circle is twice the angle subtended by the same arc at any point on the remaining part of the circle.
  In the adjoining figure, $\angle AOB$ is the central angle subtended by arc AB and $\angle AO'B$ is angle subtended by the arc AB on the circle.
  So, $\angle AOB = 2\angle AO'B$.

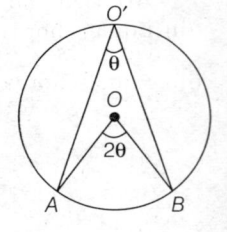

Tangent

A line segment which has only one common point with circumference of a circle, i.e. it touches the circle only at one point is called the tangent to the circle.

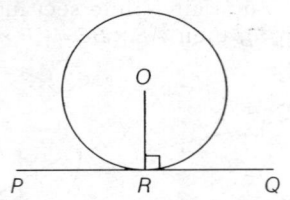

In the adjoining figure, PQ is tangent which touches the circle at point R.
  (i) Radius is always perpendicular to tangent at the point of contact, i.e. $OR \perp PQ$.

(ii) The length of two tangents drawn from the external point to the circle are equal, i.e. $PT = PQ$.

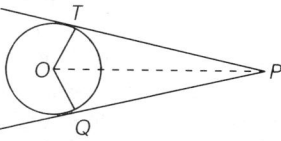

(iii) The angle which a chord makes with a tangent at its point of contact is equal to any angle in the alternate segment. $\angle PTA = \angle ABT$, where AT is the chord and PT is the tangent to the circle.

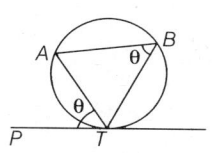

(iv) If PT is a tangent (with P being an external point and T being the point of contact) and PAB is a secant to circle (with A and B as the points, where the secant cuts the circle), then $PT^2 = PA \times PB$.

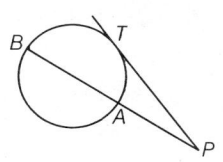

Pair of Circles

(i) (a) When two circles touch externally, then the distance between their centres is equal to the sum of their radii, i.e. $AB = AC + BC$.

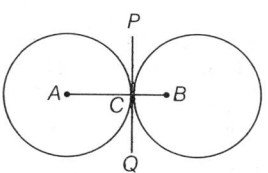

(b) When two circles touch internally then, the distance between their centres is equal to the difference between their radii, i.e. $AB = AC - BC$.

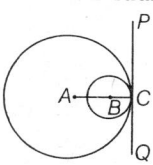

(ii) In a given pair of circles, there are two types of tangents. The direct tangents and the cross (or transverse) tangents. In the given figure, AB and CD are the direct tangents. EH and GF are the transverse tangents.

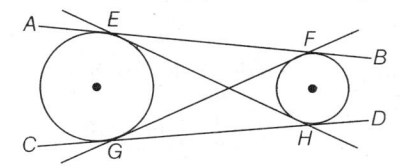

✦ When two circles of radii r_1 and r_2 have their centres at a distance d apart, then

(i) length of the direct common tangent $= \sqrt{d^2 - (r_1 - r_2)^2}$.

(ii) length of transverse tangent $= \sqrt{d^2 - (r_1 + r_2)^2}$.

Ex. 15 A chord AB is drawn in a circle with centre O and radius 5 cm. If the shortest distance between centre and chord is 4 cm, then find the length of chord AB.

Sol. In the adjoining figure, $AO = 5$ cm (radius) 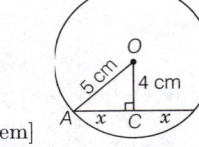
$OC = 4$ cm (shortest distance between centre and chord)
Let length of chord AB be $2x$, then $AC = x$.
 [∵ perpendicular to a chord bisects the chord]
In right angled $\triangle AOC$, $AO^2 = AC^2 + OC^2$
 [by Pythagoras theorem]
$\Rightarrow (5)^2 = x^2 + (4)^2 \Rightarrow 25 = x^2 + 16 \Rightarrow x = \sqrt{25 - 16} = \sqrt{9} = 3$ cm
∴ Length of chord $AB = 2x = 2 \times 3 = 6$ cm

Ex. 16 Find the value of x in the given figure.

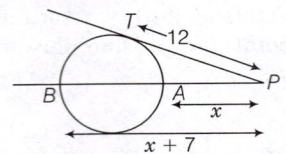

Sol. Since, PT is a tangent and PAB is a secant to the circle.
∴ $PT^2 = PA \cdot PB \Rightarrow 144 = x(x+7) \Rightarrow x^2 + 7x - 144 = 0$
$\Rightarrow (x+16)(x-9) = 0 \Rightarrow x = 9$ [∵ $x = -16$ (not possible)]

Ex. 17 Two circles touch each other internally. Their radii are 2 cm and 3 cm. What is the length of the biggest chord of the greater circle which is outside the inner circle?

Sol. Let O and O' be the centres of greater and smaller circles, respectively. 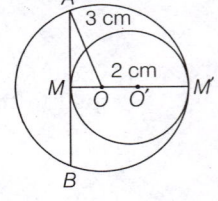
∴ $OM = O'M - OO' = 2 - 1 = 1$
 [∵ $OO' = OM' - O'M' = 3 - 2 = 1$ cm]
In right angled $\triangle AOM$, $AM^2 = OA^2 - OM^2$
$\Rightarrow AM^2 = 9 - 1 = 8 \Rightarrow AM = 2\sqrt{2}$
∴ Length of biggest chord, $AB = 2 \cdot AM = 2 \cdot 2\sqrt{2} = 4\sqrt{2}$ cm

Ex. 18 In the given figure, it is given that O is the centre of the circle and $\angle AOC = 140°$, find $\angle ABC$.

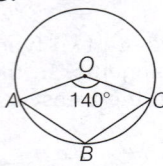

Sol. Join A and C to any point D on the circle. We know that, the angle made by an arc at any point on the circle is half of the angle subtended at the centre.
∴ $\angle ADC = \dfrac{1}{2} \angle AOC = \dfrac{1}{2} \times 140° = 70°$
Since, $ABCD$ is a cyclic quadrilateral. 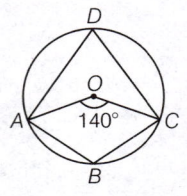
∴ $\angle ADC + \angle ABC = 180°$
 [∵ sum of opposite angles in cyclic quadrilateral is 180°]
$\Rightarrow 70° + \angle ABC = 180°$
∴ $\angle ABC = 110°$

Fast Track Practice

Lines and Angles

1. An angle which is less than 360° and more than 180°, is called
   (a) reflex angle
   (b) a/an straight angle
   (c) acute angle
   (d) obtuse angle
   (e) None of these

2. If D is the number of degrees and R is the number of radians in an angle θ, then which one of the following is correct?
   [CDS 2017 (I)]
   (a) $\pi D = 180R$
   (b) $\pi D = 90R$
   (c) $\pi R = 180D$
   (d) $\pi R = 90D$

3. An angle is twice its complementary angle. What is the measure of the angle?
   [SSC (10+2) 2017]
   (a) 30°
   (b) 90°
   (c) 60°
   (d) 120°

4. In the given figure, straight lines AB and CD intersect at O. If $\angle \beta = 3 \angle p$, then $\angle p$ is equal to

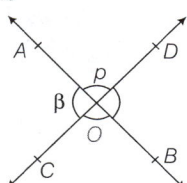

   (a) 40°
   (b) 45°
   (c) 50°
   (d) 55°
   (e) None of these

5. In the given figure, l, m and n are three parallel lines and t_1 and t_2 are two transversal lines which cut l, m and n at A, B, C and P, Q, R, respectively. Which of the following options is correct?

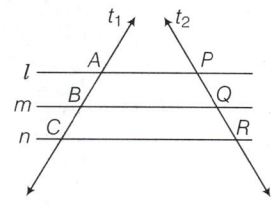

   (a) $\dfrac{BC}{PQ} = \dfrac{AB}{QR}$
   (b) $\dfrac{AP}{BQ} = \dfrac{BQ}{CR}$
   (c) $\dfrac{AB}{BC} = \dfrac{PQ}{QR}$
   (d) $\dfrac{BQ}{AP} = \dfrac{PQ}{AB}$
   (e) None of these

6. In the given figure, $\angle COE = 90°$. Find the value of x.

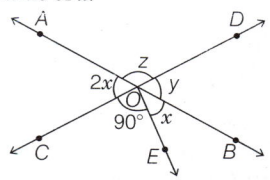

   (a) 120°
   (b) 60°
   (c) 45°
   (d) 30°
   (e) None of these

7. Two transversals S and T cut a set of distinct parallel lines. S cuts the parallel lines in points A, B, C, D and T cuts the parallel lines in points E, F, G and H, respectively. If $AB = 4$, $CD = 3$ and $EF = 12$, then what is the length of GH?
   (a) 4
   (b) 6
   (c) 8
   (d) 9

8. In the given figure, $AB \parallel CD$. If $\angle CAB = 80°$ and $\angle EFC = 25°$, then $\angle CEF$ is equal to

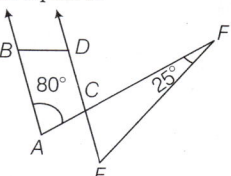

   (a) 65°
   (b) 55°
   (c) 45°
   (d) 75°

9. In the given figure, $AB \parallel CD$ and $EF \parallel GH$. Find the relation between a and b.

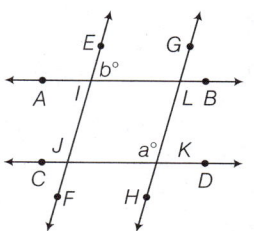

   (a) $2a + b = 180°$
   (b) $a + b = 180°$
   (c) $a - b = 180°$
   (d) $a + 2b = 180°$
   (e) None of these

662 / *Fast Track* Objective Arithmetic

10. In the given figure, if $l \parallel m$, then find the value of x (in degrees).

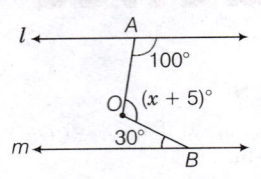

(a) 105° (b) 100°
(c) 110° (d) 115°
(e) None of these

11. In the given figure, $AB \parallel CD$ and they cut PQ and QR at E, F and G, H, respectively. If $\angle PQR = x$, then find the value of x (in degrees).

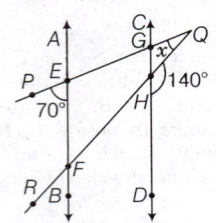

(a) 20° (b) 30°
(c) 24° (d) 32°
(e) None of these

12. In the figure given below, PQ is parallel to RS and PR is parallel to QS. If $\angle LPR = 35°$ and $\angle UST = 70°$, then what is $\angle MPQ$ equal to? [CDS 2017 (I)]

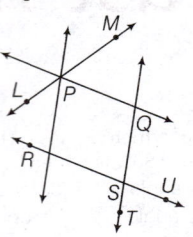

(a) 55° (b) 70° (c) 75° (d) 80°

13. In the given figure, $AB \parallel CD$. If $\angle ABO = 130°$ and $\angle OCD = 110°$, then $\angle BOC$ is equal to

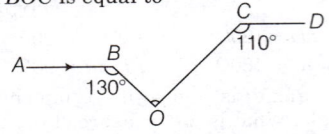

(a) 50° (b) 60° (c) 70° (d) 90°

14. In the figure given below, EC is parallel to AB, $\angle ECD = 70°$ and $\angle BDO = 20°$. What is the value of $\angle OBD$?

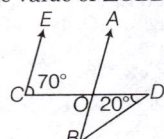

(a) 20° (b) 30° (c) 40° (d) 50°

Triangles

15. In a $\triangle ABC$, $\angle A : \angle B : \angle C = 2 : 4 : 3$. The shortest side and the longest side of the triangle are respectively [SSC CPO 2013]
(a) AC and AB (b) AC and BC
(c) BC and AC (d) AB and AC

16. In a $\triangle ABC$, $\angle A = 90°$, $\angle C = 55°$ and $\overline{AD} \perp \overline{BC}$. What is the value of $\angle BAD$? [SSC CGL 2013]
(a) 60° (b) 45° (c) 55° (d) 35°

17. In the figure given below, ABC is a triangle with $AB = BC$ and D is an interior point of $\triangle ABC$ such that $\angle DAC = \angle DCA$.

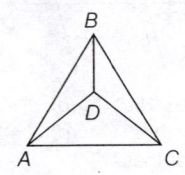

Consider the following statements
I. $\triangle ADC$ is an isosceles triangle.
II. D is the centroid of $\triangle ABC$.
III. $\triangle ABD$ is congruent to $\triangle CBD$.
Which of the above statements are correct? [CDS 2017(I)]
(a) I and II only (b) II and III only
(c) I and III only (d) I, II and III

18. $\triangle ABC$ is right angled at B. BD is an altitude $AD = 4$ cm and $DC = 9$ cm. What is the value of BD? [SSC CGL (Pre) 2017]
(a) 5 cm (b) 4.5 cm
(c) 5.5 cm (d) 6 cm

19. ABC is a right angled triangle such that $AB = a - b$, $BC = a$ and $CA = a + b$. D is a point on BC such that $BD = AB$. The ratio of $BD : DC$ for any values of a and b is given by [CDS 2013]
(a) 3 : 2 (b) 4 : 3
(c) 5 : 4 (d) 3 : 1

Geometry / 663

20. In △ABC, the angle bisector of ∠A cuts BC at E. Find the length of AC, if lengths of AB, BE and EC are 9 cm, 3.6 cm and 2.4 cm? **[SSC (10+2) 2017]**
    (a) 5.4 cm (b) 8 cm
    (c) 4.8 cm (d) 6 cm

21. The bisectors BI and CI of ∠B and ∠C of a △ABC meet in I. What is ∠BIC equal to? **[CDS 2013]**
    (a) $90° - \frac{A}{4}$ (b) $90° + \frac{A}{4}$
    (c) $90° - \frac{A}{2}$ (d) $90° + \frac{A}{2}$

22. In the figure given below, ∠PQR = 90° and QL is a median, PQ = 5 cm and QR = 12 cm. Then, QL is equal to **[CDS 2013]**

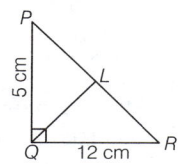

    (a) 5 cm (b) 5.5 cm
    (c) 6 cm (d) 6.5 cm

23. ABC and XYZ are two similar triangles with ∠C = ∠Z, whose areas are respectively 32 cm² and 60.5 cm². If XY = 7.7 cm, then what is AB equal to?
    (a) 5.6 cm (b) 5.8 cm **[CDS 2013]**
    (c) 6.0 cm (d) 6.2 cm

24. The side AC of a △ABC is extended to D such that BC = CD. If ∠ACB is 70°, then what is ∠ADB equal to? **[CDS 2013]**
    (a) 35° (b) 45° (c) 70° (d) 110°

25. E is the mid-point of the median AD of a △ABC. If BE is extended, it meets the side AC at F, then CF is equal to
    (a) AC/3 (b) 2AC/3 **[CDS 2013]**
    (c) AC/2 (d) None of these

26. In the figure given below, ∠A = 80° and ∠ABC = 60°. BD and CD bisect angles B and C, respectively. What are the values of x and y, respectively? **[CDS 2017 (I)]**

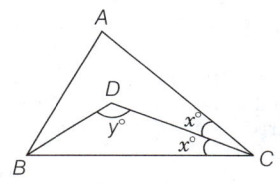

    (a) 10° and 130° (b) 10° and 125°
    (c) 20° and 130° (d) 20° and 125°

27. If O is the orthocentre of a △ABC and ∠BOC = 100°, then measure of ∠BAC is **[SSC CGL (Mains) 2016]**
    (a) 100° (b) 180°
    (c) 80° (d) 200°

28. If AD is the internal angle bisector of △ABC with AB = 3 cm and AC = 1 cm, then what is BD : BC equal to? **[CDS 2014]**
    (a) 1 : 3 (b) 1 : 4
    (c) 2 : 3 (d) 3 : 4

29. In the figure given below, PQR is a non-isosceles right-angled triangle, right angled at Q. If LM and QT are parallel and QT = PT, then what is ∠RLM equal to? **[CDS 2017 (I)]**

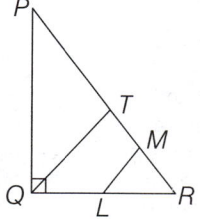

    (a) ∠PQT (b) ∠LRM
    (c) ∠RML (d) ∠QPT

30. ABC is a triangle and D is a point on the side BC. If BC = 12 cm, BD = 9 cm and ∠ADC = ∠BAC, then the length of AC is equal to **[CDS 2017 (I)]**
    (a) 5 cm (b) 6 cm
    (c) 8 cm (d) 9 cm

31. In △ABC, D and E are points on sides AB and AC, such that DE ∥ BC. If AD = x, DB = x − 2, AE = x + 2 and EC = x − 1, then the value of x is **[SSC CPO 2013]**
    (a) 4 (b) 2
    (c) 1 (d) 8

32. In a △ABC, AB = AC and D is a point on AB, such that AD = DC = BC. Then, ∠BAC is equal to **[SSC FCI 2012]**
    (a) 40° (b) 45°
    (c) 30° (d) 36°

33. The mid-points of AB and AC of a △ABC are respectively X and Y. If BC + XY = 12 units, then the value of BC − XY is **[SSC FCI 2012]**
    (a) 6 (b) 8
    (c) 4 (d) 12

34. In the figure given below,
∠ABC = ∠AED = 90°.

Consider the following statements

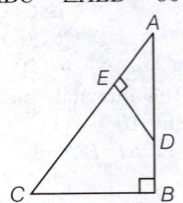

I. ABC and AED are similar triangles.

II. The four points B, C, E and D may lie on a circle.

Which of the above statement(s) is/are correct? [CDS 2012]

(a) Only I
(b) Only II
(c) Both I and II
(d) Neither I nor II

35. In a △ABC, XY is drawn parallel to BC, cutting sides at X and Y, where AB = 4.8 cm, BC = 7.2 cm and BX = 2 cm. What is the length of XY? [CDS 2012]

(a) 4 cm
(b) 4.1 cm
(c) 4.2 cm
(d) 4.3 cm

36. The angles $x°$, $a°$, $c°$ and $(\pi - b)°$ are indicated in the figure given below.

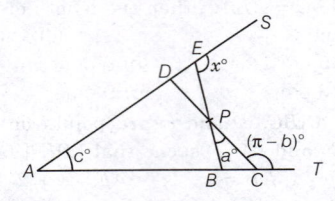

Which one of the following is correct?

(a) $x° = a° + c° - b°$ [CDS 2012]
(b) $x° = b° - a° - c°$
(c) $x° = a° + b° + c°$
(d) $x° = a° - b° + c°$

37. In the figure given below, YZ is parallel to MN, XY is parallel to LM and XZ is parallel to LN. Then, MY is [CDS 2012]

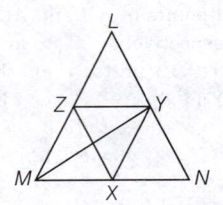

(a) the median of △LMN
(b) the angular bisector of ∠LMN
(c) perpendicular to LN
(d) perpendicular bisector of LN

38. In a △PQR, point X is on PQ and point Y is on PR such that XP = 1.5 units, XQ = 6 units, PY = 2 units and YR = 8 units.

Consider the following statements

I. QR = 5XY

II. QR is parallel to XY.

III. △PYX is similar to △PRQ.

Select the correct answer using the codes given below. [CDS 2016 (I)]

(a) I and II
(b) II and III
(c) I and III
(d) I, II and III

39. ABC is an equilateral triangle and X, Y and Z are the points on BC, CA and AB respectively, such that BX = CY = AZ. Consider the following statements

I. XYZ is an equilateral triangle.

II. △XYZ is similar to △ABC.

Which of the above statement(s) is/are correct? [CDS 2016 (I)]

(a) Only I
(b) Only II
(c) Both I and II
(d) Neither I nor II

40. Let △ABC and △DEF be such that ∠ABC = ∠DEF, ∠ACB = ∠DFE and ∠BAC = ∠EDF. Let L be the mid-point of BC and M be the mid-point of EF.

Consider the following statements

I. △ABL and △DEM are similar.

II. △ALC is congruent to △DMF even, if AC ≠ DF.

Which one of the following is correct in respect of the above statements?
[CDS 2016 (I)]

(a) Both Statements I and II are true and Statement II is the correct explanation of Statement I
(b) Both Statements I and II are true but Statement II is not the correct explanation of Statement I
(c) Statement I is true but Statement II is false
(d) Statement I is false but Statement II is true

41. In △ABC, DE ∥ BC, where D is a point on AB and E is a point on AC. DE divides the area of △ABC into two equal parts. Then, DB : AB is equal to
[SSC CGL (Mains) 2015]

(a) $(\sqrt{2} - 1) : \sqrt{2}$
(b) $\sqrt{2} : (\sqrt{2} - 1)$
(c) $\sqrt{2} : (\sqrt{2} + 1)$
(d) $(\sqrt{2} + 1) : \sqrt{2}$

Quadrilaterals and Polygons

42. Three angles of a quadrilateral are 80°, 95° and 112°. Its fourth angle is
    (a) 78° (b) 73° (c) 85° (d) 100°

43. If one angle of a parallelogram is 24° less than twice the smallest angle, then the largest angle of the parallelogram is
    (a) 68° (b) 102° (c) 112° (d) 136°

44. The external angle of a regular polygon is 72°. Find the sum of all the internal angles of it. [SSC CGL 2012]
    (a) 360° (b) 480° (c) 352° (d) 540°

45. The ratio of the numbers of sides of two regular polygons is 1:2. If each interior angle of the first polygon is 120°, then the measure of each interior angle of the second polygon is [SSC CGL 2012]
    (a) 140°   (b) 135°
    (c) 150°   (d) 160°

46. In the figure given below, $PQRS$ is a parallelogram. PA bisects $\angle P$ and SA bisects $\angle S$. What is $\angle PAS$ equal to? [CDS 2017 (I)]

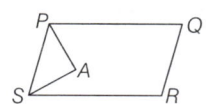

    (a) 60° (b) 75° (c) 90° (d) 100°

47. In quadrilateral $ABCD$ shown in the figure $\angle DAB = \angle DCX = 120°$. If $\angle ABC = 105°$, then what is the value of $\angle ADC$? [CDS 2012] 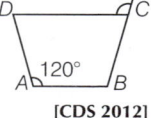
    (a) 45° (b) 60° (c) 75° (d) 95°

48. $ABCD$ is a rectangle. The diagonals AC and BD intersect at O. If $AB = 32$ cm and $AD = 24$ cm, then what is OD equal to? [CDS 2017 (I)]
    (a) 22 cm   (b) 20 cm
    (c) 18 cm   (d) 16 cm

49. In the figure given below, $PQRS$ is a parallelogram. If AP, AQ, CR and CS are the bisectors of $\angle P, \angle Q, \angle R$ and $\angle S$ respectively, then $ABCD$ is a [CDS 2013]

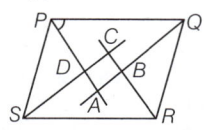

    (a) square   (b) rhombus
    (c) rectangle   (d) None of these

50. In a trapezium $ABCD$, AB is parallel to CD and the diagonals intersect each other at O. What is the ratio of OA to OC equal to? [CDS 2017 (I)]
    (a) Ratio of OB to OD
    (b) Ratio of BC to CD
    (c) Ratio of AD to AB
    (d) Ratio of AC to BD

51. In the figure given below, $ABCD$ is a trapezium. If EF is parallel to AD and BC. Then, $\angle y$ is equal to [CDS 2012]

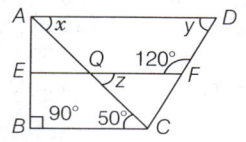

    (a) 30° (b) 45°
    (c) 60° (d) 65°

52. If $ABCD$ is a trapezium in which $AD \parallel BC$ and $AB = DC = 10$ m, then the distance of AD from BC [SSC CGL (Pre) 2017]

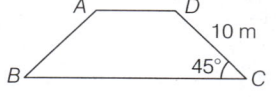

    (a) $10\sqrt{2}$ m   (b) $4\sqrt{2}$ m
    (c) $5\sqrt{2}$ m   (d) $6\sqrt{2}$ m

53. If $ABCD$ is a rectangle and P, Q, R, S are the mid-points of $\overline{AB}, \overline{BC}, \overline{CD}$ and $\overline{DA}$ respectively, then the area of the quadrilateral $PQRS$ is equal to
    (a) 1/3 ar $(ABCD)$ [SSC CGL 2013]
    (b) 3/4 ar $(ABCD)$
    (c) 1/2 ar $(ABCD)$
    (d) ar $(ABCD)$

54. An equilateral $\triangle TQR$ is drawn inside a square $PQRS$. The value of $\angle PTS$ is [SSC (10+2) 2012]
    (a) 75°   (b) 90°
    (c) 120° (d) 150°

55. Let $ABCD$ be a rectangle and P, Q, R, S be the mid-points of sides AB, BC, CD, DA respectively. Then, the quadrilateral $PQRS$ is a [CDS 2017 (I)]
    (a) square
    (b) rectangle but need not be a square
    (c) rhombus but need not be a square
    (d) parallelogram but need not be a rhombus

56. ABCD is a rhombus. If AB is produced to F and BA is produced to E such that AB = AE = BF, then [SSC CGL 2013]
    (a) $ED^2 + CF^2 = EF^2$ (b) ED ∥ CF
    (c) ED > CF (d) ED ⊥ CF

57. ABCD is a quadrilateral such that BC = BA and CD > AD. Which one of the following is correct?
    (a) ∠BAD = ∠BCD (b) ∠BAD < ∠BCD
    (c) ∠BAD > ∠BCD (d) 2∠BAD = ∠BCD

58. ABCD is a square, X is the mid-point of AB and Y is the mid-point of BC. Consider the following statements
    I. Triangles ADX and BAY are congruent.
    II. ∠DXA = ∠AYB.
    III. DX is inclined at an angle 60° with AY.
    IV. DX is not perpendicular to AY.
    Which of the above statements are correct? [CDS 2017 (I)]
    (a) II, III and IV (b) I, II and IV
    (c) I, III and IV (d) I and II

59. PQRA is a rectangle, AP = 22 cm, PQ = 8 cm. △ABC is a triangle, whose vertices lie on the sides of PQRA such that BQ = 2 cm and QC = 16 cm. Then, the length of the line joining the mid-points of the sides AB and BC is
    [SSC CGL (Mains) 2016]
    (a) $4\sqrt{2}$ cm (b) 5 cm
    (c) 6 cm (d) 10 cm

60. An equilateral △BOC is drawn inside a square ABCD. If ∠AOD = 2θ, what is tan θ equal to? [CDS 2015 (II)]
    (a) $2 - \sqrt{3}$ (b) $1 + \sqrt{2}$
    (c) $4 - \sqrt{3}$ (d) $2 + \sqrt{3}$

61. Consider the following statements
    I. If n ≥ 3 and m ≥ 3 are distinct positive integers, then the sum of the exterior angles of a regular polygon of m sides is different from the sum of the exterior angles of a regular polygon of n sides.
    II. If m and n are integers such that m > n ≥ 3. Then, the sum of the interior angles of a regular polygon of m sides is greater than the sum of the interior angles of a regular polygon of n sides and their sum is $(m + n) \pi / 2$.
    Which of the above statement(s) is/are correct? [CDS 2016 (I)]
    (a) Only I (b) Only II
    (c) Both I and II (d) Neither I nor II

Circles

62. R and r are the radii of two circles (R > 1). If the distance between the centres of the two circles is d, then length of common tangent of two circles is
    (a) $\sqrt{r^2 - d^2}$ (b) $\sqrt{d^2 - (R - r)^2}$
    (c) $\sqrt{(R - r)^2 - d^2}$ (d) $\sqrt{R^2 - d^2}$

63. From the circumcentre I of △ABC, perpendicular ID is drawn on BC. If ∠BAC = 60°, then the value of ∠BID is
    [SSC CGL 2012]
    (a) 75° (b) 60°
    (c) 45° (d) 80°

64. In a △ABC, O is its circumcentre and ∠BAC = 50°. The measure of ∠OBC is
    [SSC CPO 2013]
    (a) 60° (b) 30° (c) 40° (d) 50°

65. ABC is an equilateral triangle inscribed in a circle. D is any point on the arc BC. What is ∠ADB equal to? [CDS 2013]
    (a) 90° (b) 60°
    (c) 45° (d) None of these

66. P is a point outside a circle and is 13 cm away from its centre. A secant drawn from the point P intersects the circle at points A and B in such a way that PA = 9 cm and AB = 7 cm. The radius of the circle is [SSC CGL 2012]
    (a) 5.5 cm (b) 5 cm (c) 4 cm (d) 4.5 cm

67. AB and CD are two chords of a circle meeting externally at P.
    Consider the following statements
    I. PA × PD = PC × PB
    II. △PAC and △PDB are similar.
    Which of the above statement(s) is/are correct? [CDS 2014]
    (a) Only I (b) Only II
    (c) Both I and II (d) Neither I nor II

68. The diagonals AC and BD of a cyclic quadrilateral ABCD intersect each other at the point P. Then, it is always true that [SSC CGL 2013]
    (a) AP·CP = BP·DP (b) AP·BP = CP·DP
    (c) AP·CD = AB·CP (d) BP·AB = CD·CP

Geometry / 667

69. In the given figure, PAB is a secant and PT is a tangent to the circle from P. If $PT = 5$ cm, $PA = 4$ cm and $AB = x$ cm, then x is equal to [SSC CGL (Pre) 2016]

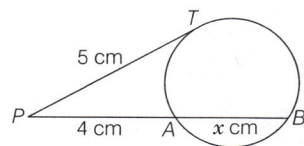

(a) 2.5 cm (b) 2.6 cm
(c) 2.25 cm (d) 2.75 cm

70. $ABCD$ is a cyclic quadrilateral of which AB is the diameter. Diagonals AC and BD intersect at E. If $\angle DBC = 35°$, then $\angle AED$ measures [SSC CGL (Mains) 2016]
(a) 35° (b) 45° (c) 55° (d) 90°

71. If a circle of radius 12 cm is divided into two equal parts by one concentric circle, then radius of inner circle is [SSC CGL (Pre) 2017]
(a) 6 cm (b) 4 cm
(c) $6\sqrt{2}$ cm (d) $4\sqrt{2}$ cm

72. In the given figure, O is the centre of a circle and $\angle OAB = 50°$. Then, $\angle BOD$ is equal to

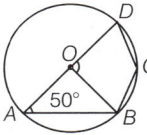

(a) 130° (b) 50° (c) 100° (d) 80°

73. If a quadrilateral has an inscribed circle, then the sum of a pair of opposite sides equals [CDS 2016 (II)]
(a) half the sum of the diagonals
(b) sum of the other pair of opposite sides
(c) sum of two adjacent sides
(d) None of the above

74. If the sides of a quadrilateral $ABCD$ touch a circle and $AB = 6$ cm, $CD = 5$ cm $BC = 7$ cm, then the length (in cm) of AD is [SSC CGL (Mains) 2012]
(a) 4 (b) 6 (c) 8 (d) 9

75. $ABCD$ is a concyclic quadrilateral of a circle $ABCD$ with radius r and centre at O. If AB is the diameter and CD is parallel and half of AB and if the circle completes one rotation about the centre O, then the locus of the mid-point of CD, which is a circle of radius, is [CDS 2016 (II)]
(a) $\dfrac{3r}{2}$ (b) $\dfrac{2r}{3}$ (c) $\dfrac{\sqrt{3}r}{3}$ (d) $\dfrac{\sqrt{3}r}{2}$

76. A, B, C and D are four points on a circle. AC and BD intersect at a point E such that $\angle BEC = 130°$ and $\angle ECD = 20°$. Then, $\angle BAC$ is equal to [SSC CGL 2013]
(a) 90° (b) 100° (c) 110° (d) 120°

77. If two tangents inclined at an angle 60° are drawn to a circle of radius 3 cm, then what is the length of each tangent? [CDS 2016 (II)]
(a) $3\sqrt{3}$ cm (b) $\sqrt{3}$ cm
(c) 6 cm (d) $2\sqrt{2}$ cm

78. Consider the following statements
I. The angular measure (in radian) of a circular arc of fixed length subtending at its centre decreases, if the radius of the arc increases.
II. 1800° is equal to 5π radian.

Which of the statement(s) is/are correct?
(a) Only I (b) Only II
(c) Both I and II (d) Neither I nor II

79. Consider the following statements in respect of two chords XY and ZT of a circle intersecting at P.
I. $PX \cdot PY = PZ \cdot PT$
II. PXZ and PTY are similar triangles.

Which of the statement(s) given above is/are correct? [CDS 2013]
(a) Only I (b) Only II
(c) Both I and II (d) Neither I nor II

80. The diameter of a circle with centre at C is 50 cm. CP is a radial segment of the circle. AB is a chord perpendicular to CP and passes through P. CP produced intersects the circle at D. If $DP = 18$ cm, then what is the length of AB? [CDS 2013]
(a) 24 cm (b) 32 cm
(c) 40 cm (d) 48 cm

81. ABC is triangle right angled at B. If $AB = 6$ cm and $BC = 8$ cm, then what is the length of the circumradius of the $\triangle ABC$? [CDS 2014]
(a) 10 cm (b) 7 cm (c) 6 cm (d) 5 cm

82. O is the centre of a circle. AC and BD are two chords of the circle intersecting each other at P. If $\angle AOB = 15°$ and $\angle APB = 30°$, then $\angle COD$ is equal to [SSC CGL 2012]
(a) 25° (b) 20° (c) 45° (d) 50°

83. Two circles of same radius 5 cm, intersect each other at A and B. If $AB = 8$ cm, then the distance between the centres is [SSC CGL 2013]
(a) 10 cm (b) 4 cm (c) 6 cm (d) 8 cm

668 / *Fast Track* Objective Arithmetic

84. Two parallel chords of a circle whose diameter is 13 cm, are respectively 5 cm and 12 cm, in length. If both chords are on the same side of the diameter, then the distance between these chords is
   [CDS 2017 (I)]
   (a) 5.5 cm (b) 5 cm
   (c) 3.5 cm (d) 3 cm

85. Two concentric circles having common centre O and chord AB of the outer circle intersect the inner circle at points C and D. If distance of chord from the centre is 3 cm, outer radius is 13 cm and inner radius is 7 cm, then length (in cm) of AC is [SSC CGL 2012]
   (a) $8\sqrt{10}$ (b) $6\sqrt{10}$ (c) $4\sqrt{10}$ (d) $2\sqrt{10}$

86. In a circle of radius 3 units, a diameter AB intersects a chord of length 2 units perpendicular at P. If $AP > BP$, then what is the ratio of AP to BP? [CDS 2016 (II)]
   (a) $(3 + \sqrt{10}) : (3 - \sqrt{10})$
   (b) $(3 + \sqrt{8}) : (3 - \sqrt{8})$
   (c) $(3 + \sqrt{3}) : (3 - \sqrt{3})$
   (d) $3 : \sqrt{3}$

87. In the given figure, PQR is a tangent to the circle at Q, whose centre is O and AB is a chord parallel to PR such that $\angle BQR = 70°$. Then, $\angle AQB$ is equal to

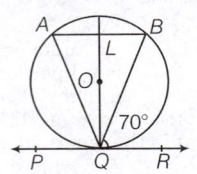

   (a) 20° (b) 35° (c) 40° (d) 45°

88. In the given figure, O is the centre of a circle, BOA is its diameter and the tangent at the point P meets BA extended at T. If $\angle PBO = 30°$, then $\angle PTA$ is equal to

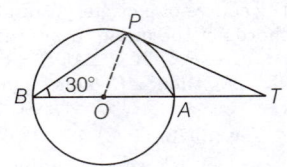

   (a) 60° (b) 30° (c) 15° (d) 45°

89. In the given figure, O is the centre of a circle, PQL and PRM are the tangents at the points Q and R respectively and S is a point on the circle such that $\angle SQL = 50°$ and $\angle SRM = 60°$. Then, $\angle QSR$ is equal to

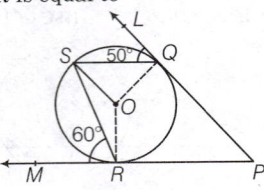

   (a) 40° (b) 50° (c) 60° (d) 70°

90. ABC is an isosceles triangle with $AB = AC$. A circle through B touching AC at the middle point intersects AB at P. Then, $AP : AB$ is equal to [SSC CGL 2013]
   (a) 3 : 5 (b) 1 : 4 (c) 4 : 1 (d) 2 : 3

91. If a chord AB of a circle C_1 of radius $(\sqrt{3} + 1)$ cm touches a circle C_2 of radius $(\sqrt{3} - 1)$ cm, then the length of AB is [SSC CGL 2013]
   (a) $8\sqrt{3}$ cm (b) $4\sqrt[4]{3}$ cm
   (c) $4\sqrt{3}$ cm (d) $2\sqrt[4]{3}$ cm

92. P and Q are two points on a circle with centre at O. R is a point on the minor arc of the circle between the points P and Q. The tangents to the circle from the point S are drawn which touch the circle at P and Q. If $\angle PSQ = 20°$, then $\angle PRQ$ is equal to [SSC CGL 2013]
   (a) 200° (b) 160°
   (c) 100° (d) 80°

93. The length of the diagonal of a square is 8 cm. A circle has been drawn circumscribing the square. The area (in sq cm) of the portion between the circle and the square is [SSC FCI 2012]
   (a) $16\frac{2}{7}$ (b) $18\frac{2}{7}$ (c) $10\frac{2}{7}$ (d) $12\frac{2}{7}$

94. In the given figure, AB and CD are two parallel chords of a circle with centre O and radius 5 cm. Also, $AB = 8$ cm and $CD = 6$ cm. If $OP \perp AB$ and $OQ \perp CD$, then determine the length of PQ. [DMRC 2012]

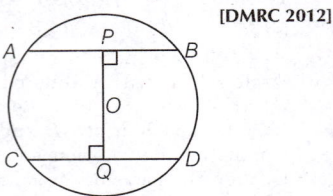

   (a) 7 cm (b) 10 cm
   (c) 8 cm (d) None of these

95. Suppose AB is a diameter of a circle, whose centre is at O and C be any point on the circle. If $CD \perp AB$, $CD = 12$ cm, $AD = 16$ cm, then BD is equal to
 [SSC FCI 2012]
(a) 10 cm (b) 12 cm
(c) 8 cm   (d) 9 cm

96. Three circles of radii 4 cm, 6 cm and 8 cm touch each other pairwise externally. The area of the triangle formed by the line segments joining the centres of the three circles is [SSC (10+2) 2012]
(a) $6\sqrt{6}$ sq cm   (b) $24\sqrt{6}$ sq cm
(c) $144\sqrt{13}$ sq cm   (d) $12\sqrt{105}$ sq cm

97. A, B and C are three points on a circle. The tangent at C meets BA extended at T. Given, $\angle ATC = 36°$ and $\angle ACT = 48°$, the angle subtended by AB at the centre of the circle is [SSC FCI 2012]
(a) 84°   (b) 48°
(c) 96°   (d) 72°

98. In the figure given below, $AO = CD$, where O is the centre of the circle. What is the value of $\angle APB$? [CDS 2012]

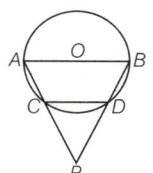

(a) 60°   (b) 50°
(c) 45°   (d) 30°

99. If a square of side x and an equilateral triangle of side y are inscribed in a circle, then what is the ratio of x to y?
 [CDS 2016 (II)]
(a) $\sqrt{\dfrac{2}{3}}$ (b) $\sqrt{\dfrac{3}{2}}$ (c) $\dfrac{3}{\sqrt{2}}$ (d) $\dfrac{\sqrt{2}}{3}$

100. PQ is chord of length 6 cm of a circle of radius 5 cm. Tangents to the circle at P and Q meet at T. Length of TP is
 [SSC CGL 2013]
(a) 4.75 cm   (b) 2.75 cm
(c) 3.75 cm   (d) 4.25 cm

101. AB and CD are two parallel chords of a circle such that $AB = 10$ cm and $CD = 24$ cm. If the chords are on the opposite sides of the centre and distance between them is 17 cm, then the radius of the circle is [SSC CGL 2013]
(a) 12 cm   (b) 13 cm
(c) 10 cm   (d) 11 cm

102. Each of the two circles of same radii a passes through the centre of the other. If the circles cut each other at the points A, B and O, O' are their centres, then area of the quadrilateral $AOBO'$ is [SSC CGL 2013]
(a) $\dfrac{1}{4}a^2$ (b) $\dfrac{1}{2}a^2$
(c) $\dfrac{\sqrt{3}}{2}a^2$ (d) a^2

103. AB is a line segment of length $2a$, with M as mid-point. Semi-circles are drawn on one side with AM, MB and AB as diameters as shown in the figure. A circle with centre O and radius r is drawn such that this circle touches all the three semi-circles. The value of r is
 [CDS 2015 (II)]

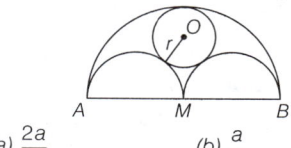

(a) $\dfrac{2a}{3}$ (b) $\dfrac{a}{2}$
(c) $\dfrac{a}{3}$ (d) $\dfrac{a}{4}$

104. Consider a circle with centre at C. Let OP, OQ denote respectively the tangents to the circle drawn from a point O outside the circle. Let R be a point on OP and S be a point on OQ such that $OR \times SQ = OS \times RP$. Consider the following statements

I. If X is the circle with centre at O and radius OR and Y is the circle with centre at O and radius OS, then $X = Y$.

II. $\angle POC + \angle QCO = 90°$

Which of the following statement(s) is/are correct? [CDS 2016 (I)]
(a) Only I   (b) Only II
(c) Both I and II   (d) Neither I nor II

105. In the figure given below, D is the diameter of each circle. What is the diameter of the shaded circle?
 [CDS 2017 (I)]

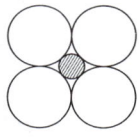

(a) $D(\sqrt{2} - 1)$ (b) $D(\sqrt{2} + 1)$
(c) $D(\sqrt{2} + 2)$ (d) $D(2 - \sqrt{2})$

Answer with Solutions

1. (a) An angle which is less than 360° and more than 180°, is called a reflex angle.

2. (c) We know that,
π Radian = 180°
∴ $\pi R = 180 D$

3. (c) Let an angle be x, then
Complementary angle = $\dfrac{x}{2}$
According to the question,
$\dfrac{2x + x}{2} = 90°$
⇒ $3x = 180° \Rightarrow x = 60°$
∴ Measure of the angle = 60°

4. (b) ∵ $\angle p + \angle \beta = 180°$ [linear pair]
⇒ $\angle p + 3\angle p = 180°$
⇒ $4\angle p = 180°$
∴ $\angle p = 45°$

5. (c) Given that, $l \parallel m \parallel n$ and t_1, t_2 are the transversal lines, therefore $\dfrac{AB}{BC} = \dfrac{PQ}{QR}$.
[if three or more parallel lines are intersected by two transversals, then intercepts made by the transversals are in same proportion]

6. (d) Since, AOB is a straight line.
We have, $\angle COE + \angle EOB + \angle AOC = 180°$
⇒ $90° + x + 2x = 180°$
⇒ $3x = 90°$
∴ $x = \dfrac{90°}{3} = 30°$

7. (d) Let $GH = x$

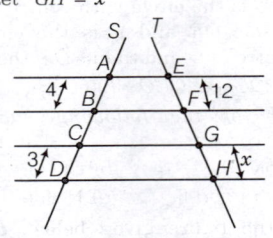

By proportionality law, intercepts cut by transversals are in same proportion.
∴ $\dfrac{AB}{CD} = \dfrac{EF}{GH} \Rightarrow \dfrac{4}{3} = \dfrac{12}{x} \Rightarrow x = 3 \times 3 = 9$

8. (b) Let $\angle CEF = x°$
Now, $AB \parallel CD$ and AF is a transversal.
∴ $\angle DCF = \angle CAB = 80°$ [corresponding angles]
In $\triangle CEF$, side EC has been produced to D.
⇒ $x + 25° = 80°$ [exterior angle theorem]
⇒ $x = 55°$

9. (b) Since, $AB \parallel CD, \angle EIL = \angle IJK$
[corresponding angles]
⇒ $\angle IJK = b$ [∵ $\angle EIL = b$]
Since, $EF \parallel GH$, $\angle IJK + \angle JKL = 180°$
[sum of the interior angles on the same side of the transversal]
⇒ $b + a = 180°$
Hence, the required relation is
$a + b = 180°$.

10. (a) Draw a line n passing through O and parallel to l and m.

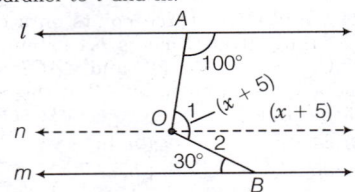

Since, $l \parallel n$
⇒ $\angle 1 + 100° = 180°$
[∵ sum of the interior angles on the same side of the transversal]
⇒ $\angle 1 = 80°$
Since, $n \parallel m$
∴ $\angle 2 = 30°$ [alternate angles]
Now, $\angle AOB = \angle 1 + \angle 2 = (80 + 30)°$
$= 110°$
But $\angle AOB = (x + 5)° = 110°$
⇒ $x = (110 - 5)° = 105°$

11. (b) Since, $AB \parallel CD$ and PQ is transversal.
$\angle PEF = \angle EGH$ [corresponding angles]
⇒ $\angle EGH = 70°$ [∵ $\angle PEF = 70°$]
Now, $\angle EGH + \angle HGQ = 180°$ [linear pair]
⇒ $\angle HGQ = (180° - 70°) = 110°$
Also, $\angle DHQ + \angle GHQ = 180°$ [linear pair]
⇒ $\angle GHQ = (180 - 140)° = 40°$
In $\triangle GQH$,
$\angle GQH + \angle GHQ + \angle HGQ = 180°$
[∵ sum of the angles in a triangle is 180°]
⇒ $x + 40° + 110° = 180°$
⇒ $x + 150° = 180°$
⇒ $x = 180° - 150° = 30°$

12. (c) We have, $\angle UST = 70°$
∴ $\angle QSR = \angle UST = 70°$
[vertically opposite angles]
Again, $\angle SRP + \angle QSR = 180°$
[cointerior angles]
⇒ $\angle SRP = 180° - 70° = 110°$
Again, $\angle RPQ = 180° - \angle SRP$
$= 180° - 110° = 70°$

Now, $\angle LPR + \angle RPQ + \angle MPQ = 180°$
[∵ sum of all angles on a line is 180°]
∴ $35° + 70° + \angle MPQ = 180°$
⇒ $\angle MPQ = 75°$

13. (b) Through O, draw $EOF \parallel AB \parallel CD$.

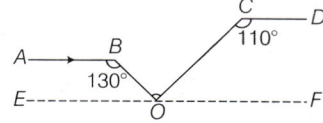

Now, $AB \parallel EO$ and BO is the transversal.
⇒ $\angle ABO + \angle EOB = 180°$
[∵ sum of the interior angles on the same side of the transversal]
⇒ $130° + \angle EOB = 180°$
⇒ $\angle EOB = 50°$
Again, $CD \parallel OF$ and CO is the transversal.
∴ $\angle OCD + \angle COF = 180°$
⇒ $110° + \angle COF = 180°$
⇒ $\angle COF = 70°$
Let $\angle BOC = x°$
Then, $\angle EOB + \angle BOC + \angle COF = 180°$
[straight line]
⇒ $50° + x° + 70° = 180°$
⇒ $x = (180° - 120°) = 60°$
∴ $\angle BOC = 60°$

14. (d) Given that, $EC \parallel AB$
∴ $\angle ECO + \angle AOC = 180°$

⇒ $\angle AOC = 180° - 70° = 110°$
∴ $\angle BOD = \angle AOC = 110°$
[vertically opposite angles]
Now, in $\triangle OBD$,
$\angle BOD + \angle ODB + \angle DBO = 180°$
∴ $110° + 20° + x° = 180°$
⇒ $x° = 50°$

15. (c) Let $\angle A = 2x$, $\angle B = 4x$ and $\angle C = 3x$

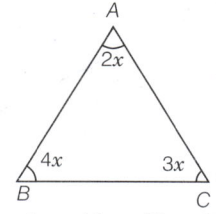

We know that, $\angle A + \angle B + \angle C = 180°$
∴ $2x + 4x + 3x = 180°$

⇒ $9x = 180°$
⇒ $x = 20°$
Now, $\angle A = 40°$, $\angle B = 80°$ and $\angle C = 60°$
Hence, the shortest side of triangle = side opposite to the smallest angle = BC and the longest side of triangle = side opposite to the greatest angle = AC.

16. (c) In $\triangle BAC$,
$\angle B = 180° - (90° + 55°) = 35°$

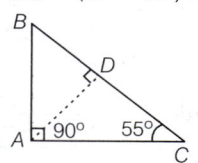

Now, in $\triangle ADB$, $\angle ADB = 90°$
∴ $\angle ADB + \angle DBA + \angle BAD = 180°$
⇒ $\angle BAD = 180° - 90° - 35° = 55°$

17. (c) In $\triangle ADC$,

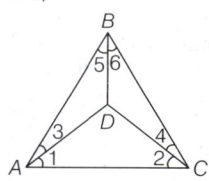

$\angle 1 = \angle 2$ [given]
∴ $AD = CD$
[if two angles of a triangle are equal, then the corresponding sides are also equal]
∴ $\triangle ADC$ is an isosceles triangle'
Now, in $\triangle ABD$ and $\triangle CBD$,
$AB = BC$ [given]
$AD = CD$ [proved above]
$BD = BD$ [common]
∴ $\triangle ABD \cong \triangle CBD$
[by SSS congruence rule]
$\angle 3 = \angle 4$ and $\angle 5 = \angle 6$ [by CPCT]
Hence, D is the intersecting point of angle bisector. So, D is incentre of $\triangle ABC$.

18. (d) In $\triangle ABC$ and $\triangle ADB$,
$\dfrac{AB}{AC} = \dfrac{4}{AB}$
⇒ $AB^2 = 4 \times AC = 4 \times 13 = 52$

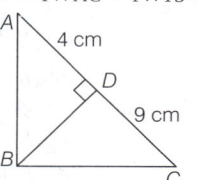

In $\triangle ABC$ and $\triangle BDC$,
$\dfrac{BC}{AC} = \dfrac{9}{BC}$
⇒ $BC^2 = 9 \times AC = 9 \times 13 = 117$

Now, $\dfrac{1}{BD^2} = \dfrac{1}{AB^2} + \dfrac{1}{BC^2} = \dfrac{1}{52} + \dfrac{1}{117}$

$\Rightarrow \dfrac{1}{BD^2} = \dfrac{1}{36} \Rightarrow BD = 6$ cm

19. (d) In right angled $\triangle ABC$,
$(a + b)^2 = (a - b)^2 + a^2$

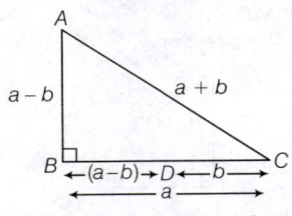

$\Rightarrow a^2 + b^2 + 2ab = a^2 + b^2 - 2ab + a^2$
$\Rightarrow 4ab = a^2 \Rightarrow 4b = a$

Now, $\dfrac{BD}{DC} = \dfrac{a-b}{b} = \dfrac{4b-b}{b} = \dfrac{3b}{b} = \dfrac{3}{1} = 3:1$

20. (d) In $\triangle ABC$, if A is the angle bisector, then according to the angle bisector theorem,

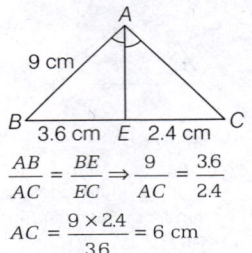

$\dfrac{AB}{AC} = \dfrac{BE}{EC} \Rightarrow \dfrac{9}{AC} = \dfrac{3.6}{2.4}$

$\Rightarrow AC = \dfrac{9 \times 2.4}{3.6} = 6$ cm

21. (d) Given that, BI and CI are angle bisectors of $\angle B$ and $\angle C$, respectively.

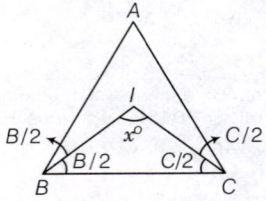

Now, in $\triangle BIC$,

$x° + \dfrac{B}{2} + \dfrac{C}{2} = 180°$

[let $\angle BIC = x°$]

$\Rightarrow x° = 180° - \dfrac{1}{2}(B + C)$

$\Rightarrow x° = 180° - \dfrac{1}{2}(180° - A)$

[$\because$ in $\triangle ABC$, $\angle A + \angle B + \angle C = 180°$]

$\Rightarrow x° = 180° - 90° + \dfrac{A}{2}$

$\therefore \angle BIC = x° = 90° + \dfrac{A}{2}$

22. (d) Given that, $PQ = 5$ cm, $QR = 12$ cm and QL is a median.

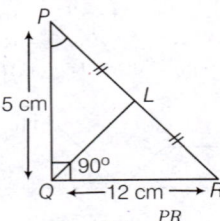

$\therefore \qquad PL = LR = \dfrac{PR}{2}$

In $\triangle PQR$, $(PR)^2 = (PQ)^2 + (QR)^2$

[by Pythagoras theorem]

$= (5)^2 + (12)^2$
$= 25 + 144 = 169 = (13)^2$

$\Rightarrow PR^2 = (13)^2 \Rightarrow PR = 13$ cm

Now, by theorem, if L is the mid-point of the hypotenuse PR of a right angled $\triangle PQR$,

then $QL = \dfrac{1}{2} PR = \dfrac{1}{2}(13) = 6.5$ cm

23. (a) For similar triangles, ratio of areas is equal to the ratio of the squares of corresponding sides.

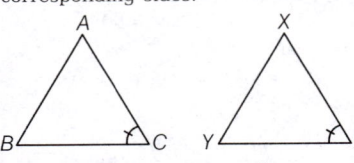

Here, $\dfrac{\text{area of } \triangle ABC}{\text{area of } \triangle XYZ} = \dfrac{AB^2}{XY^2}$

$\Rightarrow \dfrac{32}{60.5} = \dfrac{AB^2}{(7.7)^2}$

$\Rightarrow \dfrac{32 \times 59.29}{60.5} = AB^2$

$\Rightarrow 31.36 = AB^2$

$\therefore AB = \sqrt{31.36} = 5.6$ cm

24. (a) In $\triangle ABD$, $\angle ACB + \angle BCD = 180°$

[linear pair]

$\therefore \qquad \angle BCD = 180° - 70° = 110°$

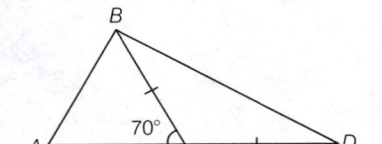

In $\triangle BCD$,

$\Rightarrow \qquad BC = CD$
$\angle CBD = \angle CDB$ ...(i)

[angles opposite to equal sides]

Also, $\angle BCD + \angle CBD + \angle CDB = 180°$
$\Rightarrow \quad 2\angle CDB = 180° - \angle BCD$
$\qquad = 180° - 110° = 70°$
$\therefore \qquad \angle CDB = \angle ADB = \dfrac{70°}{2} = 35°$

25. (*b*) Draw line segment *DG* parallel to *BF*.

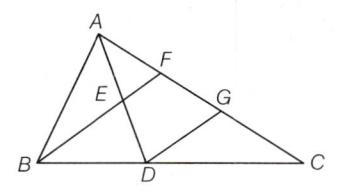

Then, in $\triangle ADG$, $EF \parallel DG$
and $\quad AE = ED$
$\therefore \quad AF = FG$ [by mid-point theorem] ...(i)
Similarly, in $\triangle BCF$, $DG \parallel BF$
and $\quad BD = DC$
$\therefore \quad FG = FG$ [by mid-point theorem] ...(ii)
From Eqs. (i) and (ii),
$\qquad AF = FG = GC$
$\Rightarrow \qquad AC = \dfrac{3}{2} \cdot CF$
$\Rightarrow \qquad CF = \dfrac{2}{3} AC$

26. (*c*) We have,
$\qquad \angle A = 80°$ and $\angle ABC = 60°$

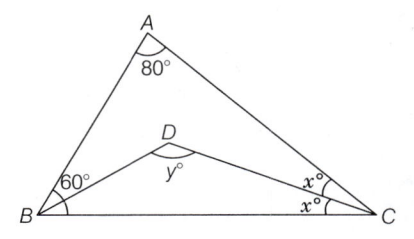

Now, in $\triangle ABC$,
$\qquad \angle A + \angle B + \angle C = 180°$
$\Rightarrow \qquad 80° + 60° + \angle C = 180°$
$\Rightarrow \qquad \angle C = 180° - 140° = 40°$
$\because$ *BD* and *CD* are the bisectors of angles *B* and *C* respectively.
$\therefore \qquad \angle DCB = \dfrac{1}{2} \angle C = \dfrac{1}{2} \times 40° = 20° = x°$
and $\quad \angle DBC = \dfrac{1}{2} \angle B = \dfrac{1}{2} \times 60° = 30°$
Again, in $\triangle BCD$,
$\qquad \angle BDC + \angle DBC + \angle DCB = 180°$
$\Rightarrow \qquad y° + 30° + 20° = 180°$
$\Rightarrow \qquad y° = 180° - 50° = 130°$

27. (*c*) Here in $\triangle ABC$, *AD*, *BE* and *CF* are the altitudes at sides *BC*, *AC* and *AB* respectively and *O* is the orthocentre.

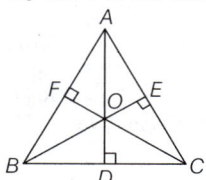

$\therefore \qquad \angle BAC + \angle BOC = 180°$
$\Rightarrow \qquad \angle BAC = 180° - 100°$
$\qquad \qquad [\because \angle BOC = 100°, \text{given}]$
$\therefore \qquad \angle BAC = 80°$

28. (*d*) In $\triangle ABC$, *AD* is the internal angle bisector of $\angle A$.
Using property of internal angle bisector,
$\qquad \dfrac{BD}{CD} = \dfrac{AB}{AC}$

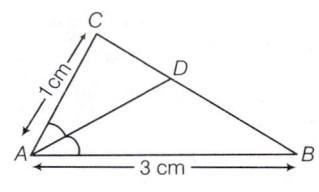

$\Rightarrow \qquad \dfrac{CD}{BD} = \dfrac{AC}{AB}$
On adding 1 to both sides, we get
$\qquad \dfrac{CD}{BD} + 1 = \dfrac{AC}{AB} + 1$
$\Rightarrow \qquad \dfrac{CD + BD}{BD} = \dfrac{AC + AB}{AB}$
$\Rightarrow \qquad \dfrac{BC}{BD} = \dfrac{3 + 1}{3} \Rightarrow \dfrac{BD}{BC} = \dfrac{3}{4}$
$\qquad BD : BC = 3 : 4$

29. (*b*) Since, $QT = PT$ [given]
$\therefore \qquad \angle 1 = \angle 2$ ...(i)
$\qquad [\because$ angles opposite to equal sides are also equal]
Again, $\angle PQR = 90°$
$\therefore \qquad \angle 3 = 90 - \angle 1$ ...(ii)
Now, $\quad \angle 4 = \angle 3 \quad [\because LM \parallel QT]$ ...(iii)

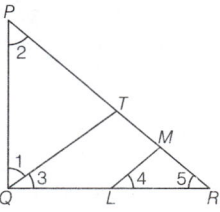

$\angle 4 = 90° - \angle 1$ [from Eq. (ii)] ...(iv)

Again, in $\triangle PQR$,
$$\angle P + \angle Q + \angle R = 180°$$
$$\Rightarrow \angle 2 + 90° + \angle 5 = 180°$$
$$\Rightarrow \angle 5 = 90° - \angle 2$$
$$\Rightarrow \angle 5 = 90 - \angle 1 \quad \text{[from Eq. (i)] ...(iv)}$$
From Eqs. (iv) and (v), we get
$$\angle 4 = \angle 5 \Rightarrow \angle RLM = \angle LRM$$

30. (b) In $\triangle ABC$ and $\triangle DAC$,

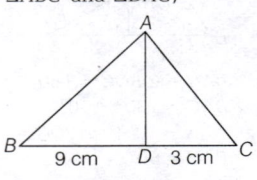

$$\angle BAC = \angle ADC \quad \text{[given]}$$
$$\angle ACB = \angle DCA \quad \text{[common]}$$
$$\therefore \triangle ABC \sim \triangle DAC \quad \text{[by AA similarity]}$$
$$\therefore \frac{AC}{DC} = \frac{BC}{AC} \Rightarrow AC^2 = DC \times BC$$
$$\Rightarrow AC^2 = 3 \times 12 \Rightarrow AC^2 = 36$$
$$\Rightarrow AC = 6 \text{ cm}$$

31. (a) $\because DE \parallel BC$
$$\therefore \frac{AD}{DB} = \frac{AE}{EC}$$

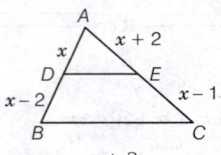

$$\Rightarrow \frac{x}{x-2} = \frac{x+2}{x-1}$$
[by basic proportionality theorem]
$$\Rightarrow x^2 - x = x^2 - 4 \Rightarrow x = 4$$

32. (d) Given that,
$$AB = AC \text{ and } AD = CD = BC$$

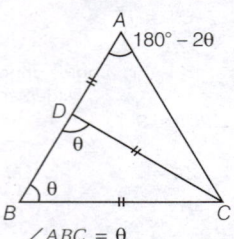

Let $\angle ABC = \theta$
Then, $\angle ACB = \theta \quad [\because AB = AC]$
$$\Rightarrow \angle BAC = 180° - 2\theta$$
$$\Rightarrow \angle ACD = 180° - 2\theta \quad [\because AD = CD]$$
Now, $\angle BCD = \angle ACB - \angle ACD$
$$\Rightarrow \angle BCD = \theta - (180° - 2\theta)$$
$$= 3\theta - 180°$$
and $\angle BDC = \theta \quad [\because CD = BC]$

Now, in $\triangle BCD$,
$$\angle CBD + \angle BDC + \angle BCD = 180°$$
$$\Rightarrow \theta + \theta + 3\theta - 180° = 180°$$
$$\Rightarrow 5\theta = 360° \Rightarrow \theta = 72°$$
$$\therefore \angle BAC = 180° - 2\theta = 180° - 144 = 36°$$

33. (c) Since, X and Y are the mid-points of AB and AC, respectively.
Therefore, $XY = \frac{1}{2} BC$
[by mid-point theorem]

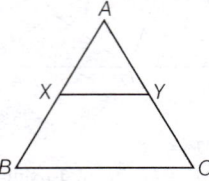

Now, $BC + XY = 12$
or $BC + \frac{1}{2} BC = 12 \quad \left[\because XY = \frac{1}{2} BC\right]$
$$\Rightarrow \frac{3}{2} BC = 12 \Rightarrow BC = 12 \times \frac{2}{3} = 8$$
and $XY = \frac{1}{2} \times 8 = 4 \Rightarrow BC - XY = 8 - 4 = 4$

34. (c) Given, $\angle AED = 90° = \angle ABC$
and $\angle CAB = \angle BAC \quad \text{[common]}$

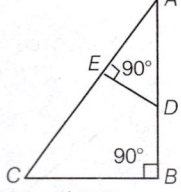

$$\therefore \triangle ABC \sim \triangle AED$$
Now, $EDBC$ is a quadrilateral.
$$\therefore \text{In } EDBC, \angle E + \angle D + \angle B + \angle C = 360°$$
We know that, $\angle E + \angle B = 180°$
$$\Rightarrow \angle C + \angle D = 180°$$
$\therefore \angle C$ and $\angle D$ are supplementry.
So, points may lie on a circle, since $EDBC$ is a cyclic quadrilateral.

35. (c) Given that, $AB = 4.8$ cm, $BC = 7.2$ cm and $BX = 2$ cm

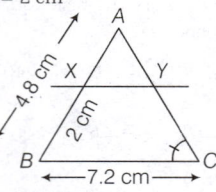

$$\therefore AX = AB - BX$$
$$= 4.8 - 2 = 2.8 \text{ cm}$$
From figure, $\triangle AXY \sim \triangle ABC$
[by AA similarity]

∴ $\dfrac{XY}{BC} = \dfrac{AX}{AB}$

⇒ $XY = \dfrac{AX}{AB} \cdot BC = \dfrac{2.8}{4.8} \times 7.2$

∴ $XY = 4.2$ cm

36. (c) ∵ $\angle PCT + \angle PCB = \pi$ [linear pair]
∴ $\angle PCB = \pi - (\pi - b°) = b°$ ...(i)

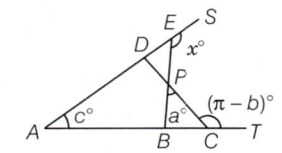

In $\triangle BPC$,
$\angle PCB + \angle BPC + \angle PBC = \pi$
⇒ $\angle PBC = \pi - \angle BPC - \angle PCB$
⇒ $\angle PBC = \pi - a° - b°$ ...(ii)
∵ $\angle ABE + \angle EBC = \pi$ [∵ $\angle PBC = \angle EBC$]
[linear pair]
⇒ $\angle ABE = \pi - (\pi - a° - b°) = a° + b°$ ...(iii)
Now, in $\triangle ABE$,
Sum of two interior opposite angles = Exterior angle
⇒ $\angle EAB + \angle ABE = \angle BES$
⇒ $c° + b° + a° = x°$
∴ $x° = a° + b° + c°$

37. (a) Since, $ZY \| MN$ and $ZX \| LN$
∴ $XNYZ$ is a parallelogram.
⇒ $ZX = YN$ ...(i)
Also, $ZX \| LN$ and $XY \| ML$
∴ $XYLZ$ is a parallelogram.

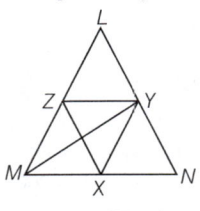

∴ $XZ = YL$ ...(ii)
From Eqs. (i) and (ii), we get $YN = LY$
∴ MY is a median of $\triangle LMN$.

38. (d) In $\triangle PQR$, $\dfrac{PX}{XQ} = \dfrac{1.5}{6} = \dfrac{1}{4}$
and $\dfrac{PY}{YR} = \dfrac{2}{8} = \dfrac{1}{4}$

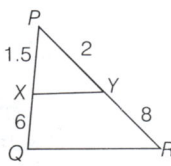

So, $\dfrac{PX}{XQ} = \dfrac{PY}{YR}$
∴ $XY \| QR$
[if a line divides any two sides in the same ratio, then the line is parallel to third side]
I. $\dfrac{PX}{XY} = \dfrac{PQ}{QR} \Rightarrow \dfrac{1.5}{XY} = \dfrac{7.5}{QR}$
⇒ $QR = 5XY$
II. Also, $\dfrac{PX}{PQ} = \dfrac{PY}{PR} = \dfrac{1}{5}$
⇒ QR is parallel of XY.
III. $\triangle PYX$ is similar to $\triangle PRQ$.
[by AA criterion]
∵ $XY \| QR$
⇒ $\angle PXY = \angle PQR$ or $\angle PYX = \angle PRQ$
Hence, all statements are correct.

39. (c) In an equilateral $\triangle ABC$,
$AB = BC = CA$ and $\angle B = \angle C = \angle A$

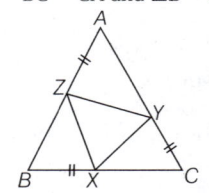

Given that, $BX = CY = AZ$
Now, in $\triangle XYC, \triangle ZYA$ and $\triangle XZB$,
$BX = CY = AZ$ ...(i)
⇒ $(BC - XC) = (AC - AY) = (AB - BZ)$
⇒ $XC = AY = BZ$ ...(ii)
and $\angle B = \angle C = \angle A$ ...(iii)
From Eqs. (i), (ii) and (iii), we get $\triangle XYC$, $\triangle ZYA$ and $\triangle XZB$ are congruent triangles.
∴ $XY = YZ = XZ$
So, $\triangle XYZ$ is an equilateral triangle.
Since, two equilateral triangles are similar.
∴ $\triangle XYZ \sim \triangle ABC$
So, both statements are correct.

40. (c) Given that, $\angle ABC = \angle DEF$,
$\angle ACB = \angle DFE$ and $\angle BAC = \angle EDF$

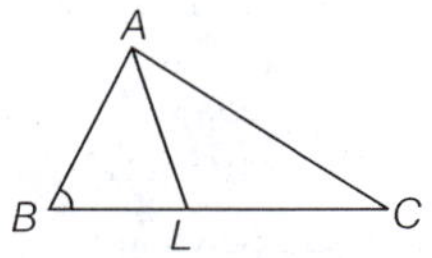

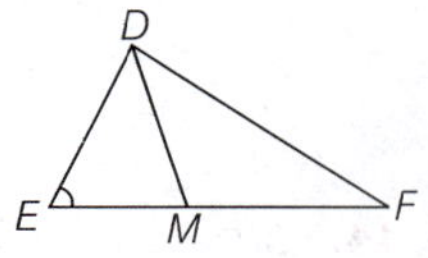

So, $\triangle ABC$ and $\triangle DEF$ are similar.
∴ $\dfrac{AB}{DE} = \dfrac{BC}{EF} = \dfrac{AC}{DF}$

Now, L is mid-point of BC, then $BL = \dfrac{1}{2}BC$
Also, M is the mid-point of EF, then
$EM = \dfrac{1}{2}EF$

$$\Rightarrow \qquad \frac{AB}{DE} = \frac{2BL}{2EM} = \frac{BL}{EM}$$

and $\angle ABL = \angle DEM$

$\therefore \triangle ABL$ is similar to $\triangle DEM$.

[by SAS similarity]

Hence, Statement I is true but Statement II is false.

41. (a) Given, $DE \parallel BC$

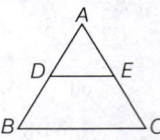

So, $\triangle ADE \sim \triangle ABC$ [by AA similarity]

$$\Rightarrow \frac{\text{Area of } \triangle ADE}{\text{Area of } \triangle ABC} = \frac{AD^2}{AB^2}$$

[by ratio of area of similar triangles]

$$\Rightarrow \frac{1}{2} = \frac{AD^2}{AB^2} \Rightarrow \frac{AB^2}{AD^2} = \frac{2}{1}$$

$$\Rightarrow \frac{AB}{AD} = \frac{\sqrt{2}}{1} \qquad \ldots(i)$$

On subtracting 1 from both sides, we get

$$\frac{AB}{AD} - 1 = \frac{\sqrt{2}}{1} - 1 \Rightarrow \frac{AB - AD}{AD} = \frac{\sqrt{2} - 1}{1}$$

$$\Rightarrow \frac{DB}{AD} = \frac{\sqrt{2} - 1}{1} \qquad \ldots(ii)$$

On dividing Eq. (ii) by Eq. (i), we get

$$\Rightarrow \frac{\frac{DB}{AD}}{\frac{AB}{AD}} = \frac{\sqrt{2} - 1}{\sqrt{2}}$$

$\therefore \qquad DB : AB = (\sqrt{2} - 1) : \sqrt{2}$

42. (b) Let the fourth angle be $x°$.

Then, $80° + 95° + 112° + x° = 360°$

$\Rightarrow 287° + x° = 360°$

$\Rightarrow x° = (360° - 287°) = 73°$

43. (c) Let the smallest angle be $x°$.

Then, its adjacent angle $= (2x - 24)°$

$\therefore \qquad x + 2x - 24 = 180°$

$\Rightarrow \qquad 3x = 204 \Rightarrow x = 68°$

$\therefore$ Largest angle $= (2 \times 68 - 24)°$
$= (136 - 24)° = 112°$

44. (d) External angle of any polygon

$$\frac{360°}{n} = 72° \Rightarrow n = 5$$

$\therefore$ Given, polygon is regular pentagon.

$\therefore$ Every interior angle of it
$= 180° - $ External angle
$= 180° - 72° = 108°$

$\therefore$ Sum of interior angles of it
$= 5 \times 108° = 540°$

45. (c) Given, interior angle of the first polygon
$= 120°$

Let number of sides in first polygon be n_1.

Then, $\frac{n_1 - 2}{n_1} \times 180° = 120°$

$\Rightarrow \qquad 3n_1 - 6 = 2n_1$

$\Rightarrow \qquad n_1 = 6$

$\therefore$ Sides of the second polygon $= 6n$
$= 6 \times 2 = 12$

$\therefore$ Interior angle of the second polygon
$= \frac{12 - 2}{12} \times 180° = 150°$

46. (c) Since, $PQRS$ is a parallelogram.

$\therefore \qquad PQ \parallel SR$

$\therefore \qquad \angle QPS + \angle RSP = 180°$

[$\because$ cointerior angles]

$\Rightarrow \frac{1}{2} \angle QPS + \frac{1}{2} \angle RSP = \frac{1}{2} \times 180°$

$\Rightarrow \angle APS + \angle ASP = 90°$ ...(i)

[$\because PA$ and SA are the angle bisectors of $\angle P$ and $\angle Q$ respectively]

Now, in $\triangle APS$,

$\angle APS + \angle ASP + \angle PAS = 180°$

$\Rightarrow \qquad 90° + \angle PAS = 180°$

$\Rightarrow \qquad \angle PAS = 180° - 90° = 90°$

47. (c) Given, $\angle ABC = 105°$

$\angle DAB = 120°$, $\angle DCX = 120°$

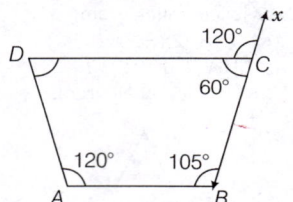

$\Rightarrow \qquad \angle DCB = 180° - 120° = 60°$

We know that, sum of angles of a quadrilateral is equal to $360°$.

$\therefore \qquad \angle ADC = 360° - (120° + 105° + 60°)$
$= 360° - 285° = 75°$

48. (b) We have,

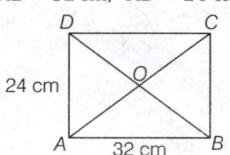

$AB = 32$ cm, $AD = 24$ cm

Since, $\angle A = 90°$

$\therefore \qquad BD^2 = AB^2 + AD^2 = (32)^2 + (24)^2$
$= 1024 + 576 = 1600$

$\therefore \qquad BD = 40$ cm

Since, diagonals of a rectangle bisects each other.

∴ $OD = \dfrac{1}{2}BD = \dfrac{1}{2} \times 40 = 20$ cm

49. (c) Clearly, $AB \parallel DC$ and $AD \parallel BC$. Therefore, $ABCD$ is a parallelogram but it is not necessary that $AB = BC$.
Thus, $ABCD$ is a rectangle.

50. (a) Since, $AB \parallel CD$

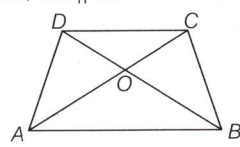

∴ $\angle OAB = \angle OCD$
[Alternate interior angles] …(i)
and $\angle OBA = \angle ODC$
[Alternate interior angles] …(ii)
Now, in $\triangle AOB$ and $\triangle COD$,
$\angle OAB = \angle OCD$ [from Eq. (i)]
$\angle OBA = \angle ODC$ [from Eq. (ii)]
$\angle AOB = \angle COD$
[vertically opposite angle]
Hence, $\triangle AOB \sim \triangle COD$ [by AA similarity]

∴ $\dfrac{OA}{OC} = \dfrac{OB}{OD}$

51. (c) From the figure,

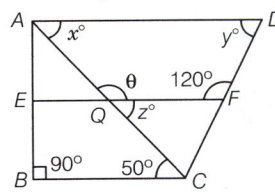

$x° = z° = 50°$ [alternate interior angles and corresponding angles]

∴ $\theta + z° = 180°$ [linear pair]
⇒ $\theta = 180° - 50° = 130°$
Now, in quadrilateral $AQFD$,
$x° + y° + 120° + \theta = 360°$
⇒ $50° + y° + 120° + 130° = 360°$
∴ $y = 360° - 300° = 60°$

52. (c) Draw a perpendicular DE on BC (distance between AD and BC)

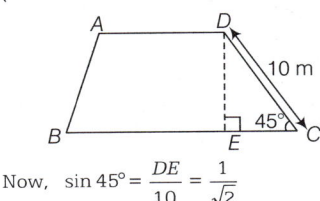

Now, $\sin 45° = \dfrac{DE}{10} = \dfrac{1}{\sqrt{2}}$

⇒ $DE = \dfrac{10}{\sqrt{2}} \times \dfrac{\sqrt{2}}{\sqrt{2}} = 5\sqrt{2}$ m

53. (c) Let length and breadth of rectangle $ABCD$ be $2x$ and $2y$, respectively.

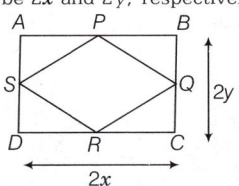

Then, area of $ABCD = 2x \times 2y = 4xy$
Area of all the four triangles
$= 4 \times \dfrac{1}{2} \times x \times y = 2xy$
Area of $PQRS = 4xy - 2xy = 2xy$
$= \dfrac{1}{2}(4xy) = \dfrac{1}{2}$ ar $(ABCD)$

54. (d) In $\triangle SRT$,

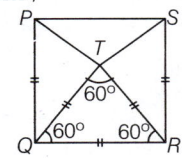

$\angle SRT = 90° - 60° = 30°$
∴ $\angle RTS = \dfrac{1}{2}(180° - 30°) = 75°$
Similarly, $\angle PTQ = 75°$
∵ $\angle PTS + \angle PTQ + \angle QTR + \angle RTS = 360°$
⇒ $\angle PTS + 75° + 60° + 75° = 360°$
⇒ $\angle PTS = 360° - 210° = 150°$

55. (c) Given, $ABCD$ is a rectangle.
∴ $\angle A = \angle B = \angle C = \angle D = 90°$

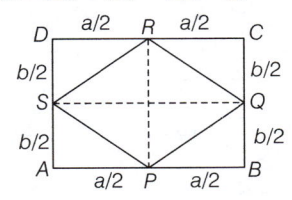

Let $AB = CD = a$
and $BC = AD = b$
Also, given P, Q, R and S are mid-points of AB, BC, CD and AD, respectively.
Now, in $\triangle PBQ$,
$PQ^2 = PB^2 + QB^2$
$PQ^2 = \left(\dfrac{a}{2}\right)^2 + \left(\dfrac{b}{2}\right)^2$
⇒ $PQ = \sqrt{\left(\dfrac{a}{2}\right)^2 + \left(\dfrac{b}{2}\right)^2}$
Similarly, in $\triangle QCR, \triangle RDS$ and $\triangle SAP$,
$PQ = RQ = SR = SP = \sqrt{\left(\dfrac{a}{2}\right)^2 + \left(\dfrac{b}{2}\right)^2}$

∴ Quadrilateral PQRS can be a square or rhombus.
But, the diagonals of PQRS are not equal.
Hence, PQRS is a rhombus.

56. (d) Produce ED and FC to meet at a point M.

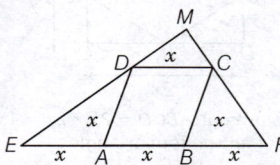

Let each side of the rhombus be x. Then,
AB = BC = CD = AD = BF = AE = x
∵ ∠DAB + ∠CBA = 180°
∴ ∠DAE + ∠CBF = 180°
In △AED,
   ∠EAD + ∠AED + ∠ADE = 180° ...(i)
In △BCF,
   ∠CBF + ∠CFB + ∠BCF = 180° ...(ii)
On adding Eqs. (i) and (ii), we get
(∠EAD + ∠CBF) + (∠AED + ∠ADE)
           + (∠CFB + ∠BCF) = 360°
⇒ 180° + ∠2 AED + 2∠CBF = 360°
           [∵ AE = AD and BC = BF]
⇒ ∠AED + ∠CFB = 90°
⇒ 180° − ∠EMF = 90°
           [angle sum property of a triangle]
⇒ ∠EMF = 90° ⇒ EM ⊥ FM
∴ ED ⊥ CF

57. (c) Join AC.

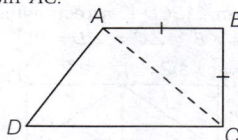

Now, in △ABC,
   AB = BC
∴ ∠BAC = ∠BCA   ...(i)
           [angles opposite to equal sides]
In △ADC,
   CD > AD
∴ ∠DAC > ∠DCA   ...(ii)
           [since, in a triangle, angle opposite to greater side is greater than the angle opposite to smaller side]
On adding Eqs. (i) and (ii), we get
   ∠BAD > ∠BCD

58. (d) In △ADX and △BAY,

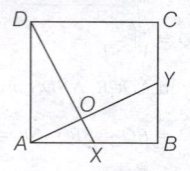

AD = BA   [side of square]
∠DAX = ∠ABY = 90°
AX = BY
$\left[\because AB = BC \Rightarrow \frac{1}{2}AB = \frac{1}{2}BC\right]$
∴ △ADX ≅ △BAY [by SAS congruence rule]
⇒ ∠DXA = ∠AYB   [by CPCT]
Now, △ADX ≅ △BAY
∴ ∠XDA = ∠YAB
Again, ∠OAD = 90° − ∠OAX = 90° − ∠YAB
           = 90° − ∠XDA   ...(i)
In △AOD, ∠ODA + ∠OAD + ∠AOD = 180°
⇒ ∠XDA + (90 − ∠XDA) + ∠AOD = 180°
⇒ ∠AOD = 180° − 90° = 90°   [from Eq. (i)]
Hence, DX is perpendicular to AY.

59. (b) Here, in rectangle PAQR,
   PA = QR = 22 cm
and AR = PQ = 8 cm
Now, in △BAC, M and N are mid-points of BC and BA respectively.
Now, $\frac{BM}{BC} = \frac{MN}{AC}$ or $\frac{BM}{MN} = \frac{BC}{AC}$

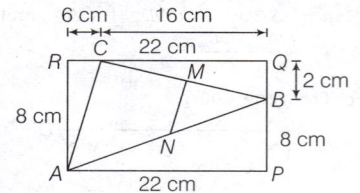

⇒ $\frac{BC}{2 \times MN} = \frac{BC}{10}$
$\left[\because BM = \frac{BC}{2} \text{ and } AC = \sqrt{8^2 + 6^2} = 10 \text{ cm}\right]$
⇒ 2 × MN = 10 cm
⇒ MN = 5 cm

60. (d) Let side of square ABCD and of equilateral △BOC drawn inside ABCD be x.

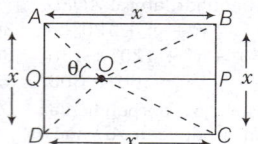

⇒ BO = OC = BC = x
and BP = $\frac{BC}{2} = \frac{x}{2}$
           [∵ BOC is an equilateral triangle]
In right angled △BPO,
$PO = \sqrt{x^2 - \left(\frac{x}{2}\right)^2} = \sqrt{x^2 - \frac{x^2}{4}}$
$= \sqrt{\frac{3x^2}{4}} = \frac{\sqrt{3}}{2}x$

Geometry / 679

and $OQ = AB - PO = PQ - PO$
$= x - \dfrac{\sqrt{3}}{2}x = x\left(\dfrac{2-\sqrt{3}}{2}\right)$

Since, $\angle AOD = 2\theta \Rightarrow \angle AOQ = \theta$
In right angled ΔAOQ,

$\tan\theta = \dfrac{AQ}{OQ} = \dfrac{\dfrac{AD}{2}}{OQ} = \dfrac{\dfrac{x}{2}}{x\left(\dfrac{2-\sqrt{3}}{2}\right)}$

$\left[\tan\theta = \dfrac{\text{perpendicular}}{\text{base}}\right]$

$= \dfrac{x}{2} \times \dfrac{2}{x(2-\sqrt{3})}$

$= \dfrac{1}{2-\sqrt{3}} \times \dfrac{2+\sqrt{3}}{2+\sqrt{3}} = \dfrac{2+\sqrt{3}}{4-3}$

$[\because (a+b)(a-b) = a^2 - b^2]$

$= 2 + \sqrt{3}$

61. (d) I. In regular ploygon,
Sum of exterior angles = 360°
which is always constant.
So, Statement I is incorrect.

II. Sum of interior angles of m sides of regular polygon = $(m-2) \times 180°$
Total sum of interior angles of m sides and n sides of regular polygon
$= (m-2) \times 180° + (n-2) \times 180°$
$= (m+n-4) \times 180°$
So, Statement II is incorrect.
Hence, neither I nor II is correct.

62. (b) Let the common tangent of both circles be PQ.

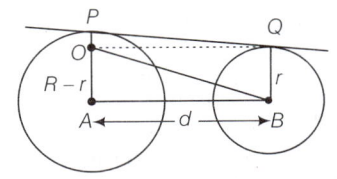

∴ From figure, $PQ = OB = \sqrt{(AB)^2 - (OA)^2}$
$= \sqrt{d^2 - (R-r)^2}$

63. (b) The angle subtended by an arc at the centre of the circle is twice the angle subtended by the arc at any point on the remaining part of the circle.

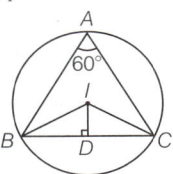

∴ $\angle BIC = 2 \times \angle BAC = 120°$

and $IB = IC$ [radii of circle]
∴ $\angle IBD = \angle ICD = \dfrac{180° - 120°}{2} = 30°$

Now, $\angle BID = 90° - 30° = 60°$

64. (c) The angle subtended by an arc at the centre of the circle is twice the angle subtended by the arc at any point on the remaining part of the circle.

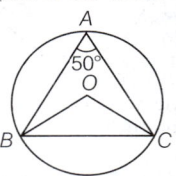

∴ $\angle BOC = 2 \angle BAC = 2 \times 50° = 100°$

Now, in ΔBOC,
$OB = OC$ [radii of circumcircle]
∴ $\angle OBC = \angle OCB = x$ [let]
$\Rightarrow x + x + 100° = 180°$
$\Rightarrow 2x = 80° \Rightarrow x = 40°$

65. (b)

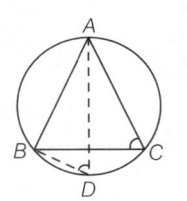

$\angle ADB = \angle ACB = 60°$
[angles in the same segment are equal]

66. (b) Let the radius of circle be r.

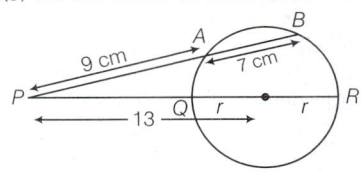

From figure, $PA \times PB = PQ \times PR$
$\Rightarrow 9 \times (9 + 7) = (13 - r)(13 + r)$
$\Rightarrow 169 - r^2 = 144 \Rightarrow r^2 = 25$
$\Rightarrow r = 5$ cm

67. (b) AB and CD are chords when produced meet externally at P.

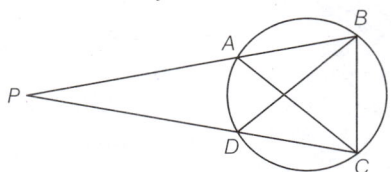

∴ $AP \times BP = DP \times CP$

Now, $\angle P = \angle P$ [common]
and $\angle PBD = \angle PCA$ [angles in the same segment on the arc AD]
$\therefore \triangle PDB \sim \triangle PAC$ [by AA similarity]

68. (a)

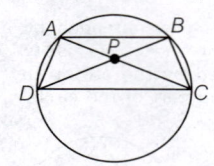

$AP \times PC = BP \times PD$
[when two chords of a circle intersect internally, then they are divided in a proportion]

69. (c) $PA \times PB = PT^2 \Rightarrow 4 \times (4 + x) = 25$
$\Rightarrow 4 + x = \dfrac{25}{4} = 6.25 \Rightarrow x = 2.25$ cm

70. (c) In the given figure,
$\angle DBC = 35°$
and $\angle ACB = 90°$
[$\because$ angle in a semi-circle is a right angle]

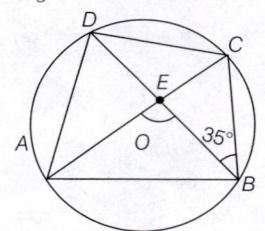

In $\triangle BEC$,
$\angle BEC + \angle ECB + \angle CBE = 180°$
$\Rightarrow \angle BEC = 180° - 90° - 35°$
$\Rightarrow \angle BEC = 55°$
Hence, $\angle AED = \angle BEC = 55°$
[vertically opposite angles]

71. (c) In right angled $\triangle AOB$,
$(AB)^2 = (AO)^2 + (OB)^2$

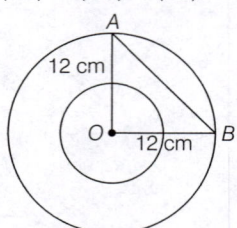

$\Rightarrow (AB)^2 = (12)^2 + (12)^2$
$\Rightarrow (AB)^2 = 144 + 144 = 288$
$\therefore AB = \sqrt{288} = 12\sqrt{2}$ cm
Now, radius $= \dfrac{AB}{2} = \dfrac{12\sqrt{2}}{2} = 6\sqrt{2}$ cm

72. (c) $OA = OB \Rightarrow \angle OBA = \angle OAB = 50°$
In $\triangle OAB$, $\angle OAB + \angle OBA + \angle AOB = 180°$
$\Rightarrow 50° + 50° + \angle AOB = 180°$
$\Rightarrow \angle AOB = 80°$
$\therefore \angle BOD = (180° - 80°) = 100°$

73. (b) Let ABCD be the quadrilateral, which has an inscribed circle.

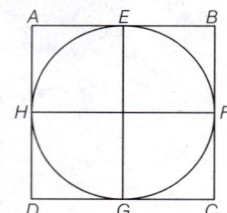

Now, $AE = AH$ ...(i)
$BE = BF$ ...(ii)
$DG = DH$ ...(iii)
and $CG = CF$ ...(iv)
[$\because$ tangents from same point on the circle are equal in lengths]
On adding Eqs. (i), (ii), (iii) and (iv), we get
$AE + BE + DG + CG = AH + BF + DH + CF$
$\Rightarrow AB + DC = AD + BC$
Hence, sum of a pair of opposite sides is equal to sum of the other pair of opposite sides.

74. (a) From the figure,
$AB + CD = BC + AD$
[$\because$ sum of opposite sides of cyclic quadrilateral are equal]

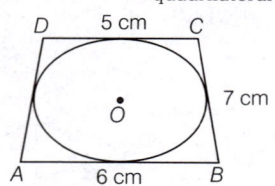

$\Rightarrow 6 + 5 = 7 + AD$
$\Rightarrow AD = 11 - 7 = 4$ cm

75. (d) Given, AB is the diameter of circle.
$\therefore AB = 2r$

Since, CD is parallel and half of AB.
$\therefore CD = \dfrac{1}{2} AB = r$
Let M be the mid-point of CD.

Geometry / 681

Then, $CM = \dfrac{CD}{2} = \dfrac{r}{2}$

By Pythagoras theorem,

$$OM = \sqrt{r^2 - \left(\dfrac{r}{2}\right)^2} = \dfrac{\sqrt{3}}{2}r$$

Hence, the locus of mid-point of CD, i.e. M is a circle of radius $\dfrac{\sqrt{3}}{2}r$.

76. (c) We have, $\angle CED = 180° - 130° = 50°$

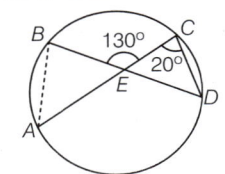

Now, in $\triangle CED$,
$\angle ECD + \angle CED + \angle CDE = 180°$
$\Rightarrow \angle CDE = 180° - 50° - 20° = 110°$
$\therefore \angle BAC = \angle CDE = 110°$
[angles in same segment are equal]

77. (a) $\triangle AOC \cong \triangle AOB$ [by SSS criterion]
$\therefore \angle CAO = \angle BAO = 30°$ [by CPCT]

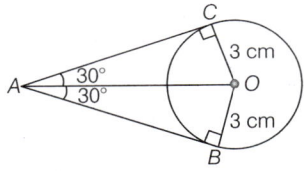

In $\triangle AOC$,

$$\tan 30° = \dfrac{OC}{AC}$$

$\left[\because \tan\theta = \dfrac{\text{perpendicular}}{\text{base}} \text{ and } \tan 30° = \dfrac{1}{\sqrt{3}}\right]$

$\Rightarrow \dfrac{1}{\sqrt{3}} = \dfrac{3}{AC}$

$\therefore AC = 3\sqrt{3}$

Hence, the length of each tangent is $3\sqrt{3}$ cm.

78. (a) I. We know that,

$$\text{Radius} = \dfrac{\text{Arc}}{\text{Angle}}$$

[given arc length is constant]

$\Rightarrow \text{Radius} \propto \dfrac{1}{\text{Angle}}$

$\therefore$ Angular measure (in radian) decreases, if the radius of the arc increases.

II. $1800° \times \dfrac{\pi}{180°} = 10\pi$

$\therefore$ Only Statement I is correct.

79. (c) When two chords of a circle intersects internally, then they are divided in a proportion,
i.e. $PX \cdot PY = PZ \cdot PT$

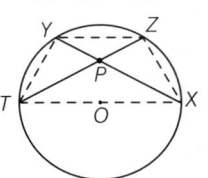

In $\triangle PXZ$ and $\triangle PTY$,
$\angle ZPX = \angle YPT$ [vertically opposite angles]
$\angle PZX = \angle PYT$ [angles in same segment]
$\angle PXZ = \angle PTY$ [angles in same segment]
$\therefore \triangle PXZ \sim \triangle PTY$ [by AA similarity]
Hence, both statements are correct.

80. (d) Here, $CP = CD - PD$
$= 25 - 18 = 7$

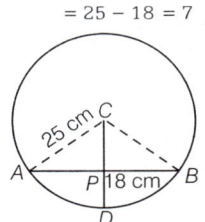

In $\triangle ACP$,
$AC^2 = CP^2 + AP^2$
$\therefore AP = \sqrt{AC^2 - CP^2} = \sqrt{(25)^2 - (7)^2}$
$= \sqrt{625 - 49} = \sqrt{576} = 24$ cm

Similarly, $PB = 24$ cm
$\therefore AB = AP + PB$
$= 24 + 24 = 48$ cm

81. (d) Since, $\triangle ABC$ is right angled at B.

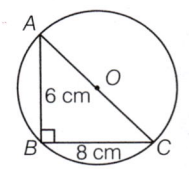

Using Pythagoras theorem,
$AC^2 = AB^2 + BC^2$
$\Rightarrow AC^2 = 6^2 + 8^2$
$\Rightarrow AC^2 = 36 + 64 = 100$
$\therefore AC = 10$ cm

In case of right angled triangle, radius lies on hypotenuse and is the circumcircle of $\triangle ABC$.

$\therefore$ Radius of circumcircle $= \dfrac{10}{2} = 5$ cm

82. (c) From figure,

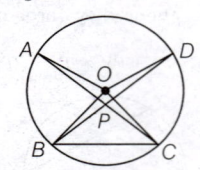

$\angle AOB + \angle COD = 2\angle ACB + 2\angle DBC$
$\Rightarrow 15° + \angle COD = 2(\angle ACB + \angle DBC)$
$\Rightarrow 15° + \angle COD = 2\angle APB$
[exterior angle theorem]
$\Rightarrow \angle COD = 2 \times 30° - 15°$
$\therefore \angle COD = 60° - 15° = 45°$

83. (c) In $\triangle AOX$,
$AO^2 = AX^2 + OX^2$

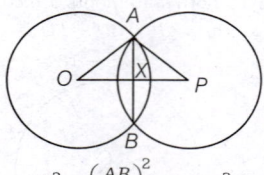

$\Rightarrow (5)^2 = \left(\dfrac{AB}{2}\right)^2 + (OX)^2$

[$\because$ perpendicular from centre to the chord bisects the chord]

$\Rightarrow 25 = \left(\dfrac{8}{2}\right)^2 + (OX)^2 \Rightarrow 25 = (4)^2 + (OX)^2$

$\Rightarrow (OX)^2 = 25 - 16 = 9$
$\therefore OX = \sqrt{9} = 3$ cm

Similarly, in $\triangle APX$, $PX = 3$ cm
$\therefore$ Distance between the centre
$= OX + PX = 3 + 3 = 6$ cm

84. (c) Since, AB and CD are parallel, then OM and ON are perpendicular to AB and CD respectively.

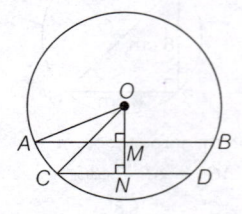

We have, $AB = 12$ cm
$\Rightarrow AM = \dfrac{1}{2} AB = 6$ cm

$CD = 5$ cm $\Rightarrow CN = \dfrac{1}{2} CD = 2.5$ cm

and $OA = OC = \dfrac{1}{2} \times 13 = 6.5$ cm

Now, in $\triangle OAM$,
$OM^2 = OA^2 - AM^2 = 6.5^2 - 6^2 = 6.25$
$\Rightarrow OM = 2.5$ cm

Again, in $\triangle OCN$,
$ON^2 = OC^2 - CN^2 = 6.5^2 - 2.5^2 = 36$
$\Rightarrow ON = 6$ cm

$\therefore$ Distance between AB and $CD = MN$
$= ON - OM = 6 - 2.5 = 3.5$ cm

85. (d) In $\triangle OCM$, $CM^2 = 7^2 - 3^2 = 40$
$\therefore CM = 2\sqrt{10}$ cm

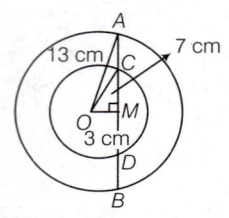

In $\triangle AOM$, $AM = \sqrt{13^2 - 3^2} = 4\sqrt{10}$ cm

Now, $AC = AM - CM = 4\sqrt{10} - 2\sqrt{10}$
$= 2\sqrt{10}$ cm

86. (b) In $\triangle OPC$,
$OP = \sqrt{(3)^2 - (1)^2} = \sqrt{9-1} = \sqrt{8}$

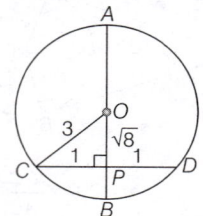

Now, $AP = AO + OP$
$= (3 + \sqrt{8})$ cm   [$\because AO =$ radius]

Also, $BP = BO - OP$
$= (3 - \sqrt{8})$ cm   [$\because BO =$ radius]

$\therefore$ Required ratio $= AP : BP$
$= (3 + \sqrt{8}) : (3 - \sqrt{8})$

87. (c) Since, $AB \parallel PR$ and $QOL \perp AB$
[$\because OQ \perp PR \Rightarrow LOQ \perp PR$]

Since, OL bisects chord AB.
[$\because$ perpendicular from centre to the chord bisects the chord]

So, $AL = LB$
Now, $LQ = LQ$ [common]
and $\angle ALQ = \angle BLQ = 90°$
$\therefore \triangle ALQ \cong \triangle BLQ$ [by SAS criterion]
$\therefore \angle LQA = \angle LQB$ [by CPCT]
But $\angle LQB = (90° - 70°) = 20°$
$\Rightarrow \angle LQA = \angle LQB = 20°$
$\therefore \angle AQB = 40°$

Geometry / 683

88. (b) Join OP. Now, $\angle BPA = 90°$
[angle in semi-circle]
In $\triangle PBA$, $\angle BPA + \angle PBA + \angle BAP = 180°$
$\Rightarrow \quad 90° + 30° + \angle BAP = 180°$
$\therefore \quad \angle BAP = 60°$
But BAT is a straight angle.
$\Rightarrow \quad \angle BAP + \angle PAT = 180°$
$\Rightarrow \quad 60° + \angle PAT = 180°$
$\Rightarrow \quad \angle PAT = 120°$
Now, $OA = OP \Rightarrow \angle OPA = \angle OAP = 60°$
$\Rightarrow \quad \angle OPT = 90°$
$\Rightarrow \quad \angle OPA + \angle APT = 90°$
$\Rightarrow \quad 60° + \angle APT = 90°$
$\therefore \quad \angle APT = 30°$
In $\triangle PAT$, we have
$\angle PAT + \angle APT + \angle PTA = 180°$
$\therefore \quad \angle PTA = 30°$

89. (d) Since, PQL is a tangent and OQ is a radius, so $\angle OQL = 90°$
$\Rightarrow \quad \angle OQS = (90° - 50°) = 40°$
Now, $OQ = OS$
$\Rightarrow \quad \angle OSQ = \angle OQS = 40°$
Similarly, $\angle ORS = (90° - 60°) = 30°$
and $OR = OS$
$\Rightarrow \quad \angle OSR = \angle ORS = 30°$
$\Rightarrow \quad \angle QSR = \angle OSQ + \angle OSR$
$= (40° + 30°) = 70°$

90. (b) $\angle BPQ = 90°$ [angle in a semi-circle]

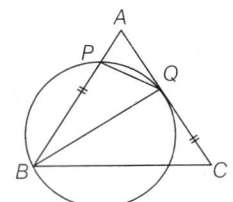

Let $AB = AC = 2x$
Then, $AQ = QC = x$
In $\triangle ABQ$, $BQ^2 = AB^2 - AQ^2$
$= (2x)^2 - x^2 = 3x^2$
[∵ radius is perpendicular to tangent]
$BQ = \sqrt{3}x$
Now, in $\triangle BPQ$, $\quad BQ^2 = BP^2 + PQ^2$ ...(i)
In $\triangle APQ$, $\quad AQ^2 = AP^2 + PQ^2$ ...(ii)
On subtracting Eq. (i) from Eq. (ii), we get
$AQ^2 - BQ^2 = AP^2 - BP^2$
$\Rightarrow \quad BP^2 - AP^2 = BQ^2 - AQ^2$
$\Rightarrow \quad (BP + AP)(BP - AP) = 3x^2 - x^2$
$\Rightarrow \quad AB(BP - AP) = 3x^2 - x^2$
$\Rightarrow \quad 2x(BP - AP) = 2x^2$
$\Rightarrow \quad BP - AP = x$ ...(iii)
and $BP + AP = 2x$ ...(iv)

From Eqs. (iii) and (iv), $AP = \dfrac{x}{2}$
$\therefore$ Required ratio $= \dfrac{AP}{AB} = \dfrac{x/2}{2x} = \dfrac{1}{4}$

91. (b) Let the chord AB of circle C_1 touches the circle C_2 at point M.

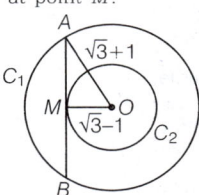

Then, $OA = \sqrt{3} + 1$ and $OM = \sqrt{3} - 1$
Now, in right angled $\triangle OAM$,
$AM^2 = OA^2 - OM^2 = (\sqrt{3} + 1)^2 - (\sqrt{3} - 1)^2$
$= (3 + 1 + 2\sqrt{3}) - (3 + 1 - 2\sqrt{3})$
$\Rightarrow \quad AM^2 = 4\sqrt{3} \Rightarrow AM = 2\sqrt[4]{3}$
$\therefore \quad AB = 2AM = 4\sqrt[4]{3}$
[∵ perpendicular from centre to the chord bisects the chord]

92. (c) Join P and Q with an another point, say T, on the major arc.

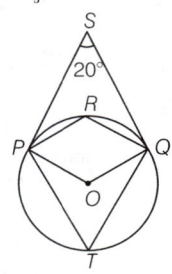

Also, join PO and QO.
In quadrilateral $POQS$,
$\angle PSQ = 20°$
$\angle OPS = \angle OQS = 90°$
[∵ radius is perpendicular to tangent]
$\therefore \quad \angle POQ = 360° - (90° + 90° + 20°)$
$= 160°$
$\therefore \quad \angle PTQ = \dfrac{1}{2} \angle POQ = \dfrac{1}{2} \times 160° = 80°$
Now, $PTQR$ is a cyclic quadrilateral.
$\therefore \quad \angle PRQ = 180° - \angle PTQ$
$= 180° - 80° = 100°$

93. (b) Let side of square be x.

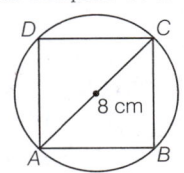

684 / Fast Track Objective Arithmetic

Then, $x^2 + x^2 = 8^2$
$\Rightarrow \quad 2x^2 = 64$
$\Rightarrow \quad x^2 = \dfrac{64}{2} \Rightarrow x = \dfrac{8}{\sqrt{2}}$

∴ Area of square $= \left(\dfrac{8}{\sqrt{2}}\right)^2 = 32$ cm^2

Area of circle $= \dfrac{22}{7} \times \left(\dfrac{8}{2}\right)^2$ cm^2

$= \dfrac{22}{7} \times 16 = \dfrac{352}{7}$

∴ Required difference $= \dfrac{352}{7} - 32$

$= \dfrac{352 - 224}{7} = \dfrac{128}{7} = 18\dfrac{2}{7}$ cm^2

94. (a) Apply Pythagoras theorem in ΔAOP and ΔCOQ.

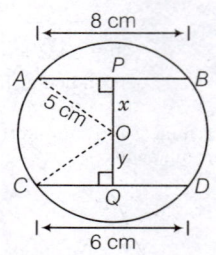

In ΔAOP, $AO^2 = AP^2 + PO^2$
[∵ perpendicular from centre to the chord bisects the chord]

$\Rightarrow \quad AO^2 = \left(\dfrac{AB}{2}\right)^2 + PO^2$

$\Rightarrow \quad (5)^2 = (4)^2 + x^2$

$\Rightarrow \quad 25 - 16 = x^2$

∴ $\quad x = \sqrt{9} = 3$ cm

In ΔCOQ, $CO^2 = OQ^2 + CQ^2$

$\Rightarrow \quad CO^2 = OQ^2 + \left(\dfrac{CD}{2}\right)^2$

$\Rightarrow \quad (5)^2 = y^2 + (3)^2$

$\Rightarrow \quad 25 - 9 = y^2$

$\Rightarrow \quad y = \sqrt{16} = 4$ cm

∴ $PQ = PO + OQ = 3 + 4 = 7$ cm

95. (d) Let $BD = x$

In ΔACD, $AC = \sqrt{16^2 + 12^2}$

$= \sqrt{256 + 144}$

$= \sqrt{400} = 20$ cm

In ΔBCD, $BC = \sqrt{12^2 + x^2}$

$= \sqrt{144 + x^2}$

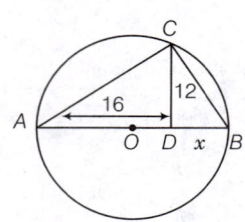

Now, in ΔABC,
$\angle ACB = 90°$
[angle subtended in semi-circle]

∴ $(AB)^2 = (AC)^2 + (BC)^2$

$\Rightarrow (16 + x)^2 = (20)^2 + (\sqrt{144 + x^2})^2$

$\Rightarrow 256 + x^2 + 32x = 400 + 144 + x^2$

$\Rightarrow \quad 32x = 288$

$\Rightarrow \quad x = 9$ cm

96. (b) From the figure,
$AB = 12$ cm $= a$, $BC = 14$ cm $= b$
and $CA = 10$ cm $= c$

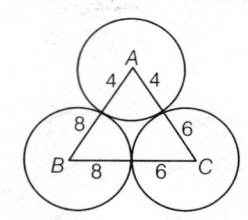

Area of $\Delta ABC = \sqrt{s(s-a)(s-b)(s-c)}$

where, $s = \dfrac{a + b + c}{2} = \dfrac{12 + 14 + 10}{2}$

$= \dfrac{36}{2} = 18$

Now, $s - a = 18 - 12 = 6$
$s - b = 18 - 14 = 4$
and $s - c = 18 - 10 = 8$

∴ Area $= \sqrt{18 \times 6 \times 4 \times 8} = 24\sqrt{6}$ cm^2

97. (c) Here, $\angle ACT = 48°$, $\angle ATC = 36°$

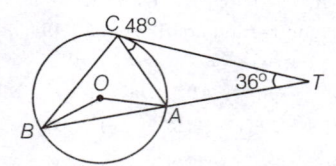

∴ $\angle CAT = 180° - (48° + 36°)$
$= 180° - 84° = 96°$

∴ $\angle CAB = 180° - 96° = 84°$

Geometry / 685

Now, $\angle ABC = \angle ACT = 48°$
[angles made in alternate segments]
$\therefore \quad \angle BCA = 180° - (\angle ABC + \angle CAB)$
$= 180° - (48° + 84°) = 48°$
$\therefore \quad \angle BOA = 2 \times \angle BCA = 2 \times 48° = 96°$

98. (a) Given, $AO = CD$
$\therefore \quad AO = OC = OD = CD$ [radii of circle]

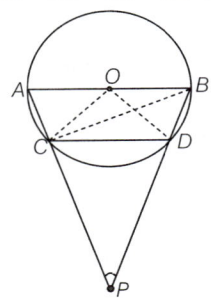

$\therefore \triangle COD$ is an equilateral triangle.
$\Rightarrow \quad \angle COD = 60°$
Now, $\angle CBD = \dfrac{1}{2}\angle COD$
[$\because$ angle subtended at the centre is twice the angle subtended on the circle]
$\Rightarrow \quad \angle CBD = 30°$
Now, $\angle ACB = 90°$ [$\because$ angle in a semi-circle]
$\angle ACB + \angle BCP = 180°$
$\Rightarrow \quad \angle BCP = 180° - 90° = 90°$
Now, in $\triangle CBP$,
$\angle CBP + \angle BCP + \angle BPC = 180°$
$\Rightarrow \quad 30° + 90° + \angle BPC = 180°$
$\therefore \quad \angle BPC = \angle APB = 60°$

99. (a) Let r be the radius of the circle with centre O.

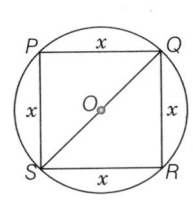

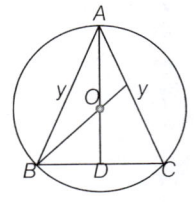

In $\triangle ABC$, AD is the median of equilateral triangle.
$\therefore \quad AD = \dfrac{\sqrt{3}}{2}AB = \dfrac{\sqrt{3}}{2}y$
Centre of circumcircle of equilateral triangle divide the median into the ratio $2:1$.
$\therefore \quad AO = r = \dfrac{2}{3} \times AD = \dfrac{2}{3} \times \dfrac{\sqrt{3}}{2}y$
$\Rightarrow \quad r = \dfrac{y}{\sqrt{3}}$ ...(i)

In the first figure, $PQRS$ is a square inscribed in a circle.
$\therefore \quad \sqrt{2}x = 2r \Rightarrow r = \dfrac{x}{\sqrt{2}}$ ...(ii)
From Eqs. (i) and (ii), we get
$\dfrac{y}{\sqrt{3}} = \dfrac{x}{\sqrt{2}}$
$\Rightarrow \quad \dfrac{x}{y} = \sqrt{\dfrac{2}{3}}$

100. (c) Let the distance from point N to point T be x cm.

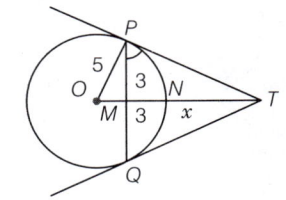

In $\triangle OPM$, $OM = \sqrt{5^2 - 3^2} = 4$
$\therefore \quad MN = ON - OM = 5 - 4 = 1$ cm
In $\triangle POT$, $PT^2 = OT^2 - OP^2 = (5 + x)^2 - 5^2$
$= x^2 + 10x$ ...(i)
In $\triangle PMT$,
$PT^2 = PM^2 + TM^2 = 3^2 + (1 + x)^2$
$= x^2 + 2x + 10$ ...(ii)
From Eqs. (i) and (ii), we get
$x^2 + 2x + 10 = x^2 + 10x \Rightarrow x = \dfrac{10}{8} = \dfrac{5}{4}$
From Eq. (i),
$PT = \sqrt{x^2 + 10x}$
$= \sqrt{\left(\dfrac{5}{4}\right)^2 + 10\left(\dfrac{5}{4}\right)} = 3.75$ cm

101. (b) Let radius of the circle be r.

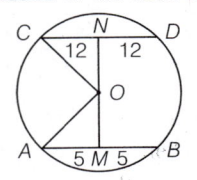

Given, $MN = 17$ cm
Let $ON = x$.
Then, $OM = 17 - x$
In $\triangle AOM$, $OA^2 = OM^2 + AM^2$
$\Rightarrow r^2 = (17 - x)^2 + 5^2 = 289 + x^2 - 34x + 25$
[$\because$ perpendicular from the centre to the chord bisects the chord]
$\Rightarrow \quad r^2 = 314 + x^2 - 34x$ ...(i)

Now, $OC^2 = ON^2 + CN^2$
$\Rightarrow \quad r^2 = x^2 + 12^2 = x^2 + 144 \quad ...(ii)$
From Eqs. (i) and (ii), we get
$314 + x^2 - 34x = x^2 + 144$
$\Rightarrow \quad 34x = 170 \Rightarrow x = 5$
From Eq. (ii), $r^2 = 25 + 144 = 169$
$\Rightarrow \quad r = 13$ cm

102. (c) From the figure, we have

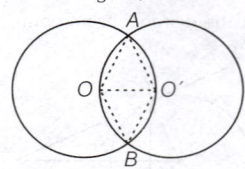

Two circles of same radii passes through the centre of the other.
Here, O and O' are the centre of two circles.
Now, join OO'.
Then, $AO = OB = AO' = BO' = OO' = a$
[radius of circle]
Now, area of quadrilateral $AOBO'$
= Area of equilateral $\triangle AOO'$ + Area of $\triangle BOO'$
$= \dfrac{\sqrt{3}}{4}a^2 + \dfrac{\sqrt{3}}{4}a^2 = \dfrac{2\sqrt{3}a^2}{4} = \dfrac{\sqrt{3}}{2}a^2$

$\left[\because \text{area of equilateral triangle} = \dfrac{\sqrt{3}}{4} \times (\text{side})^2\right]$

103. (c) Since, two circles touch each other externally, if distance between their centres = sum of their radii

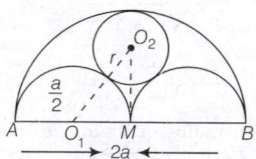

$\therefore O_1O_2 = r + \dfrac{a}{2}$ and $O_2M = a - r$, $O_1M = \dfrac{a}{2}$
Now, $O_2M \perp O_1M$, since O_2M is tangent and O_1M is radius of circle centred at O_1.
In right angled $\triangle O_1MO_2$,
$(O_1O_2)^2 = (O_1M)^2 + (O_2M)^2$
$\Rightarrow \quad \left(r + \dfrac{a}{2}\right)^2 = \dfrac{a^2}{4} + (a - r)^2$
$\Rightarrow \quad r^2 + \dfrac{a^2}{4} + ar = \dfrac{a^2}{4} + a^2 + r^2 - 2ar$
$\Rightarrow \quad 3ar = a^2 \Rightarrow 3r = a$
$\therefore \quad r = \dfrac{a}{3}$

104. (c) In the given figure, OP and OQ are tangents to the circle with centre C.

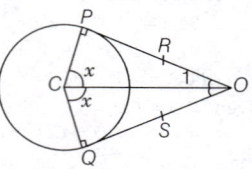

Then, $OP = OQ$ and $\angle CPO = \angle CQO = 90°$
I. We have, $OR \times SQ = OS \times RP$
$\Rightarrow (OP - RP) \times SQ = (OQ - SQ) \times PR$
$\Rightarrow (OP - RP) \times SQ = (OP - SQ) \times PR$
$\Rightarrow OP \times SQ - PR \times SQ$
$\quad = OP \times PR - SQ \times PR$
$\Rightarrow OP \times SQ = OP \times PR$
$\Rightarrow SQ = PR$
$\Rightarrow OQ - OS = OP - OR$
$\Rightarrow OS = OR \quad [\because OQ = OP]$
Radius of circle X = Radius of circle Y
$\Rightarrow X = Y$
Hence, Statement I is correct.
II. $\because \triangle OPC \cong \triangle OQC$ [by SSS criterion]
$\therefore \angle PCO = \angle QCO = x°$ (say)
In $\triangle OPC$, $90° + x + 1 = 180°$
$\Rightarrow \quad x + 1 = 90°$
$\Rightarrow \quad \angle POC + \angle QCO = 1 + x = 90°$
Hence, Statement II is correct.

105. (a) Let P, Q, R and S be the centres of the outer four circles, whose diameter is D. From the figure, it is clear that

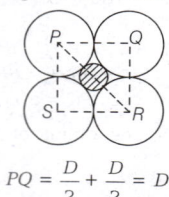

$PQ = \dfrac{D}{2} + \dfrac{D}{2} = D$
Similarly, $QR = D$
Again, let d be the diameter of the shaded circle.
$\therefore PR = \dfrac{D}{2} + d + \dfrac{D}{2} = D + d$
Since, $\triangle PQR$ is a right angled triangle.
$\therefore \quad PR^2 = PQ^2 + QR^2$
$\Rightarrow \quad (D + d)^2 = D^2 + D^2$
$\Rightarrow \quad D^2 + 2Dd + d^2 = 2D^2$
$\Rightarrow \quad d^2 + 2Dd - D^2 = 0$
$\Rightarrow \quad d = \dfrac{-2D \pm \sqrt{4D^2 + 4D^2}}{2} = \dfrac{-2D \pm 2\sqrt{2}D}{2}$
$\Rightarrow \quad d = -D \pm \sqrt{2}D$
$\Rightarrow \quad d = -D + \sqrt{2}D$ or $-D - \sqrt{2}D$
$\Rightarrow \quad d = \sqrt{2}D - D \quad [\because d > 0]$
$\Rightarrow \quad d = D(\sqrt{2} - 1)$

Chapter 37

Coordinate Geometry

It is a system of geometry, where the position of points on the plane is described by using an ordered pair of numbers.

Rectangular Coordinate Axes

The lines XOX' and YOY' are mutually perpendicular to each other and they meet at point O, which is called the origin.

Line XOX' represents X-axis and line YOY' represents Y-axis and together taken, they are called coordinate axes. Any point in coordinate axis can be represented by specifing the position of x and y-**coordinates**.

Quadrants

The X and Y-axes divide the cartesian plane into four regions referred as quadrants.

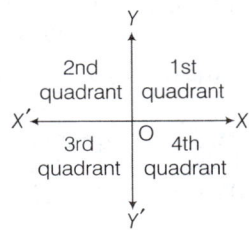

The table of sign conventions of coordinates in various quadrants is given below

Quadrant	Region	Sign of (x, y)	Example
I	XOY	(+, +)	(2, 3)
II	YOX'	(−, +)	(−2, 4)
III	X'OY'	(−, −)	(−1, −2)
IV	Y'OX	(+, −)	(1, −3)

* The coordinates of point O (origin) are taken as $(0, 0)$.
* The coordinates of any point on X-axis are of the form $(x, 0)$.
* The coordinates of any point on Y-axis are of the form $(0, y)$.

Distance Formula

Distance between two points If $A(x_1, y_1)$ and $B(x_2, y_2)$ are two points, then

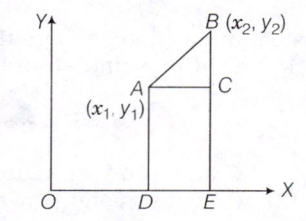

$$AB = \sqrt{(x_2 - x_1)^2 + (y_2 - y_1)^2} = \sqrt{(x_1 - x_2)^2 + (y_1 - y_2)^2}$$

For example The distance between $A(1, 2)$ and $B(5, 6)$ is

$$AB = \sqrt{(5-1)^2 + (6-2)^2} = \sqrt{4^2 + 4^2} = \sqrt{32} = 4\sqrt{2} \text{ units}$$

Distance of a point from the origin The distance of a point $A(x, y)$ from the origin $O(0, 0)$ is given by $OA = \sqrt{x^2 + y^2}$

For example The distance of point $A(5, 3)$ from origin is

$$OA = \sqrt{5^2 + 3^2}$$
$$= \sqrt{25 + 9} = \sqrt{34} \text{ units}$$

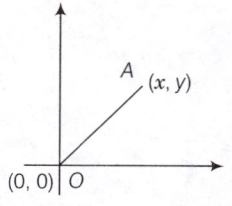

Area of a Triangle

If $A(x_1, y_1)$, $B(x_2, y_2)$ and $C(x_3, y_3)$ are three vertices of a $\triangle ABC$, then its area is given by

$$\text{Area of } \triangle ABC = \frac{1}{2}[x_1(y_2 - y_3) + x_2(y_3 - y_1) + x_3(y_1 - y_2)]$$

For example If we have to find the area of a triangle having the vertices $(0, 0)$, $(5, 0)$ and $(0, 4)$, then

$\therefore$ Area of triangle $= \frac{1}{2}\{0(0-4) + 5(4-0) + 0(0-0)\}$

$= \frac{1}{2} \times 20 = 10$ sq units

$\because (x_1, y_1) = (0, 0)$, $(x_2, y_2) = (5, 0)$ and $(x_3, y_3) = (0, 4)$

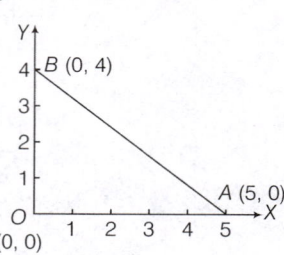

Coordinate Geometry / 689

Collinearity of Three Points
Three points $A(x_1, y_1)$, $B(x_2, y_2)$ and $C(x_3, y_3)$ are collinear, if
  (i) Area of $\triangle ABC$ is 0, i.e. $x_1(y_2 - y_3) + x_2(y_3 - y_1) + x_3(y_1 - y_2) = 0$.
  (ii) Slope of AB = Slope of BC = Slope of AC
  (iii) Distance between A and B + Distance between B and C = Distance between A and C

Centroid of a Triangle
Centroid is the point of intersection of all the three medians of a triangle. If $A(x_1, y_1)$, $B(x_2, y_2)$ and $C(x_3, y_3)$ are the vertices of $\triangle ABC$, then the coordinates of its centroid are

$$\left[\frac{1}{3}(x_1 + x_2 + x_3), \frac{1}{3}(y_1 + y_2 + y_3)\right]$$

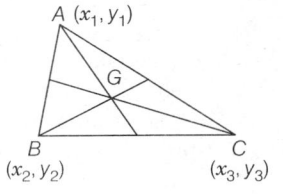

Circumcentre
The circumcentre of a triangle is the point of intersection of the perpendicular bisectors of its sides and is equidistant from all three vertices.
If $A(x_1, y_1)$, $B(x_2, y_2)$ and $C(x_3, y_3)$ are the vertices of triangles and $O(x, y)$ is the circumcentre of $\triangle ABC$, then $OA = OB = OC$
i.e. $\sqrt{(x - x_1)^2 + (y - y_1)^2} = \sqrt{(x - x_2)^2 + (y - y_2)^2} = \sqrt{(x - x_3)^2 + (y - y_3)^2}$

Incentre
The centre of the circle, which touches the sides of a triangle, is called its incentre. Incentre is the point of intersection of internal angle bisectors of a triangle.

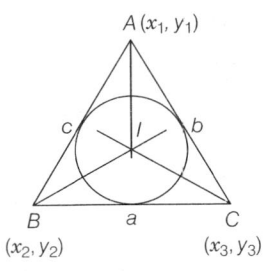

If $A(x_1, y_1)$, $B(x_2, y_2)$ and $C(x_3, y_3)$ are the vertices of a $\triangle ABC$ such that $BC = a$, $CA = b$ and $AB = c$, then coordinates of its incentre I are

$$\left(\frac{ax_1 + bx_2 + cx_3}{a + b + c}, \frac{ay_1 + by_2 + cy_3}{a + b + c}\right)$$

Section formulae
Let $A(x_1, y_1)$ and $B(x_2, y_2)$ be two points on the cartesian plane. Let point $P(x, y)$ divides the line AB in the ratio of $m : n$ internally. Then,

$$x = \frac{mx_2 + nx_1}{m + n}, \quad y = \frac{my_2 + ny_1}{m + n}$$

If P divides AB externally, then
$$x = \frac{mx_2 - nx_1}{m - n}, \quad y = \frac{my_2 - ny_1}{m - n}$$

If P is the mid-point of AB, then
$$x = \frac{x_1 + x_2}{2}, \quad y = \frac{y_1 + y_2}{2}$$

Basic Points Related to Straight Lines

1. General form of equation of straight line is $ax + by + c = 0$, where a, b and c are real constants and x, y are two unknowns.
2. The equation of a line having slope m and intersects at c on X-axis is $y = mx + c$.
   - If the line passes through the origin, i.e. $c = 0$, then the equation of line will become, $y = mx$.
3. **Slope** (Gradient) **of a line $ax + by + c = 0$**

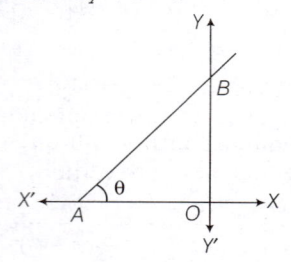

$$\Rightarrow \quad by = -ax - c \quad \Rightarrow \quad y = -\frac{a}{b}x - \frac{c}{b}$$

On comparing with $y = mx + c$, where m is slope

$$\therefore \quad m = \tan\theta = -\frac{a}{b}$$

Slope of the line is always measured in anti-clockwise direction.

4. **Point slope form a line** The equation of a line which passes through a point $p(x_1, y_1)$ and has the slope m is $y - y_1 = m(x - x_1)$

5. **Two points form a line** The equation of a line passing through the points $A(x_1, y_1)$ and $B(x_2, y_2)$ is

$$\frac{y - y_1}{x - x_1} = \frac{y_2 - y_1}{x_2 - x_1} \quad \Rightarrow \quad (y - y_1) = \frac{(y_2 - y_1)}{(x_2 - x_1)}(x - x_1)$$

Slope of such a lines $m = \frac{y_2 - y_1}{x_2 - x_1}$.

6. **Condition of parallel lines** If the slopes m_1 and m_2 of two lines are equal, i.e. $m_1 = m_2$, then lines are parallel.
   - Equation of line parallel to $ax + by + c = 0$ is $ax + by + c_1 = 0$.

7. **Condition of perpendicular lines** If the multiplication of slopes of two lines, i.e. m_1 and m_2 of two lines is equal to -1, i.e. $m_1 \cdot m_2 = -1$, then lines are perpendicular.
   - Equation of line perpendicular to $ax + by + c = 0$ is $bx - ay + c_1 = 0$.

8. **Angle between the two lines is** $\tan\theta = \pm\left(\dfrac{m_2 - m_1}{1 + m_1 m_2}\right)$.

Coordinate Geometry / 691

9. **Intercept form** Equation of line L intersects at a and b on X and Y-axes, respectively is $\dfrac{x}{a} + \dfrac{y}{b} = 1$.

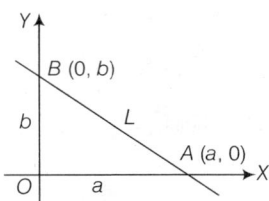

10. **Condition of concurrency of three lines** Let the equation of three lines be $a_1x + b_1y + c_1 = 0$, $a_2x + b_2y + c_2 = 0$ and $a_3x + b_3y + c_3 = 0$. Then, three lines will be concurrent, if $\begin{vmatrix} a_1 & b_1 & c_1 \\ a_2 & b_2 & c_2 \\ a_3 & b_3 & c_3 \end{vmatrix} = 0$.

11. **Distance of a point from the line** Let $ax + by + c = 0$ be any equation of line and $P(x_1, y_1)$ be any point in space. Then, the perpendicular distance (d) of a point P from a line is given by $\left| \dfrac{ax_1 + by_1 + c}{\sqrt{a^2 + b^2}} \right|$.

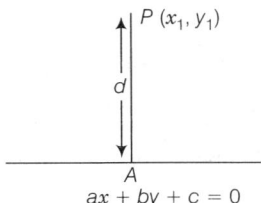

12. The length of perpendicular from the origin to the line $ax + by + c = 0$ is $\dfrac{|c|}{\sqrt{a^2 + b^2}}$.

13. Area of triangle by straight line $ax + by + c = 0$, $a \neq 0$ and $b \neq 0$ with coordinate axes is $\left| \dfrac{c^2}{2ab} \right|$.

14. Distance between parallel lines $ax + by + c = 0$ and $ax + by + d = 0$ is $\left| \dfrac{d - c}{\sqrt{a^2 + b^2}} \right|$.

15. **Area of Trapezium between Two Parallel Lines and Axes**
    Area of trapezium $ABCD$ = Area of $\triangle OCD$ − Area of $\triangle OAB$
    $$= \frac{1}{2} \left| \frac{d^2}{ab} \right| - \frac{1}{2} \left| \frac{c^2}{ab} \right|$$
    $$= \frac{1}{2} \left(\left| \frac{d^2}{ab} \right| - \left| \frac{c^2}{ab} \right| \right)$$

    ◆ Don't write it as $\dfrac{1}{2} \left| \dfrac{d^2 - c^2}{ab} \right|$.

Ex. 1 In which quadrant do the given points lie?
(i) $(-8, -6)$ (ii) $(8, -4)$ (iii) $(-4, 6)$ (iv) $(4, 4)$

Sol. (i) Since, the abscissa (-8) and ordinate (-6) are negative, therefore $(-8, -6)$ lies in 3rd quadrant.

(ii) Since, the abscissa 8 is positive and ordinate (-4) is negative, therefore $(8, -4)$ lies in 4th quadrant.

(iii) Since, the abscissa (-4) is negative and ordinate 6 is positive, therefore $(-4, 6)$ lies in 2nd quadrant.

(iv) Since, the abscissa 4 and ordinate 4 are positive, therefore $(4, 4)$ lies in 1st quadrant.

Ex. 2 Find the distance between the points $A(-6, 8)$ and $B(4, -8)$.

Sol. Here, $A(-6, 8) = A(x_1, y_1)$ and $B(4, -8) = B(x_2, y_2)$

So, $x_1 = -6, y_1 = 8, x_2 = 4$ and $y_2 = -8$

$\therefore$ Required distance, $AB = \sqrt{(x_2 - x_1)^2 + (y_2 - y_1)^2}$

$= \sqrt{\{4 - (-6)\}^2 + (-8 - 8)^2}$

$= \sqrt{(4 + 6)^2 + (-16)^2}$

$= \sqrt{100 + 256} = \sqrt{356} \approx 18.86$ units

Ex. 3 Find the distance of the point $A(8, -6)$ from the origin.

Sol. Distance of a point $A(8, -6)$ from the origin $O(0, 0)$ is

$OA = \sqrt{(8)^2 + (-6)^2} = \sqrt{64 + 36} = \sqrt{100} = 10$ units

Ex. 4 Find the area of $\triangle ABC$, whose vertices are $A(8, -4), B(3, 6)$ and $C(-2, 4)$.

Sol. Here, $A(8, -4)$, then $x_1 = 8, y_1 = -4$
$B(3, 6)$, then $x_2 = 3, y_2 = 6$
$C(-2, 4)$, then $x_3 = -2, y_3 = 4$

$\therefore$ Area of $\triangle ABC = \frac{1}{2} \{x_1(y_2 - y_3) + x_2(y_3 - y_1) + x_3(y_1 - y_2)\}$

$= \frac{1}{2} \{8(6 - 4) + 3(4 - (-4)) + (-2)(-4 - 6)\}$

$= \frac{1}{2} \{16 + 24 + 20\} = \frac{1}{2} \times 60 = 30$ sq units

Ex. 5 Find the value of k for which the points $A(-1, 3), B(2, k)$ and $C(5, -1)$ are collinear.

Sol. Here, $x_1 = -1, x_2 = 2, x_3 = 5, y_1 = 3, y_2 = k$ and $y_3 = -1$

Since, the points are collinear.

$\therefore$ Area of triangle $= 0$

$\Rightarrow$ $x_1(y_2 - y_3) + x_2(y_3 - y_1) + x_3(y_1 - y_2) = 0$

$\Rightarrow$ $-1(k + 1) + 2(-1 - 3) + 5(3 - k) = 0$

$\Rightarrow$ $-k - 1 - 8 + 15 - 5k = 0$

$\Rightarrow$ $6k = 6 \Rightarrow k = 1$

Coordinate Geometry / 693

Ex. 6 What is the slope of the line perpendicular to the line passing through the points $(3, 5)$ and $(-4, 2)$?

Sol. Let m_1 be the slope of the line passing through the points $A(3, 5)$ and $B(-4, 2)$.

$$\therefore \quad m_1 = \frac{y_2 - y_1}{x_2 - x_1} = \frac{2 - 5}{-4 - 3} = \frac{3}{7}$$

Let the line having slope m_2 is perpendicular to the line.

Then, $\quad m_1 \cdot m_2 = -1$

$\Rightarrow \quad \dfrac{3}{7} \times m_2 = -1 \Rightarrow m_2 = \dfrac{-7}{3}$

Hence, the slope of perpendicular line is $\dfrac{-7}{3}$.

Ex. 7 If the coordinates of the mid-points of the sides of a triangle are $(1, 1)$, $(2, -3)$ and $(3, 4)$, then find the coordinates of the centroid.

Sol. Let $P(1, 1)$, $Q(2, -3)$ and $R(3, 4)$ be the mid-points of sides AB, BC and CA, respectively of $\triangle ABC$.

Now, let $A(x_1, y_1)$, $B(x_2, y_2)$ and $C(x_3, y_3)$ be the vertices of $\triangle ABC$.

Since, P is mid-point of AB.

$$\therefore \quad \frac{x_1 + x_2}{2} = 1, \quad \frac{y_1 + y_2}{2} = 1$$

$\Rightarrow \quad x_1 + x_2 = 2, \quad y_1 + y_2 = 2 \qquad \ldots\text{(i)}$

Since, Q is mid-point of BC.

$$\therefore \quad \frac{x_2 + x_3}{2} = 2, \quad \frac{y_2 + y_3}{2} = -3$$

$\Rightarrow \quad x_2 + x_3 = 4, \quad y_2 + y_3 = -6 \qquad \ldots\text{(ii)}$

Since, R is mid-point of AC.

$$\therefore \quad \frac{x_1 + x_3}{2} = 3, \quad \frac{y_1 + y_3}{2} = 4$$

$\Rightarrow \quad x_1 + x_3 = 6, \quad y_1 + y_3 = 8 \qquad \ldots\text{(iii)}$

From Eqs. (i), (ii) and (iii), we get

$(x_1, y_1) = (2, 8), (x_2, y_2) = (0, -6)$ and $(x_3, y_3) = (4, 0)$

Then, coordinates of the centroid

$$= \left(\frac{x_1 + x_2 + x_3}{3}, \frac{y_1 + y_2 + y_3}{3} \right) = \left(\frac{2 + 0 + 4}{3}, \frac{8 - 6 + 0}{3} \right) = \left(2, \frac{2}{3} \right)$$

Ex. 8 If $A(-2, 1)$, $B(2, 3)$ and $C(-2, -4)$ are three points, then find the angle between AB and BC.

Sol. Let m_1 and m_2 be the slopes of line AB and BC, respectively.

$$\therefore \quad m_1 = \frac{3 - 1}{2 - (-2)} = \frac{2}{4} = \frac{1}{2} \text{ and } m_2 = \frac{-4 - 3}{-2 - 2} = \frac{-7}{-4} = \frac{7}{4}$$

Let θ be the angle between AB and BC.

Then, $\quad \tan\theta = \left| \dfrac{m_2 - m_1}{1 + m_1 m_2} \right| = \left| \dfrac{\dfrac{7}{4} - \dfrac{1}{2}}{1 + \dfrac{1}{2} \cdot \dfrac{7}{4}} \right| = \dfrac{2}{3}$

$\therefore \quad \theta = \tan^{-1}\left(\dfrac{2}{3} \right)$

Fast Track Practice

1. On which axis does the point (6, 0) lie?
   (a) X-axis (b) Y-axis
   (c) Either X or Y (d) At origin
   (e) None of these

2. If the distance between the points $(x, 0)$ and $(-7, 0)$ is 10 units, then the possible values of x are [SSC FCI 2012]
   (a) 3 and 17 (b) -3 and 17
   (c) 3 and -17 (d) -3 and -17

3. If the distance of the point $P(x, y)$ from $A(a, 0)$ is $a + x$, then y^2 equals
   (a) $2ax$ (b) $4ax$ (c) $6ax$ (d) $8ax$
   (e) None of these

4. The distance between the points $(4, -8)$ and $(k, 0)$ is 10. Find k. [SSC (10+2) 2017]
   (a) $k = 6$ or -2 (b) $k = 10$ or -2
   (c) $k = 10$ or -4 (d) $k = 6$ or -4

5. Coordinates of a point is (0, 1) and ordinate of an another point is -3. If distance between both the points is 5, then abscissa of second point is
   (a) 3 (b) -3 (c) ± 3 (d) 1
   (e) -1

6. What is the reflection of the point $(6, -3)$ in the line $y = 2$? [SSC CGL (Pre) 2017]
   (a) $(-2, -3)$ (b) $(6, 7)$
   (c) $(-6, 7)$ (d) $(-2, 3)$

7. The points (2, 2), (6, 3) and (4, 11) are the vertices of
   (a) an equilateral
   (b) a right angled triangle
   (c) an isosceles triangle
   (d) a scalene triangle
   (e) Do not form triangle

8. If the point (x, y) is equidistant from points (7, 1) and (3, 5), then find $(x - y)$.
   (a) 2 (b) 4 (c) 6 (d) 8
   (e) None of these

9. If a point (x, y) in a OXY-plane is equidistant from $(-1, 1)$ and $(4, 3)$, then [CLAT 2013]
   (a) $10x + 4y = 23$ (b) $6x + 4y = 23$
   (c) $-x + y = 7$ (d) $4x + 3y = 0$

10. The vertices of a triangle are $A(4, 4)$, $B(3, -2)$ and $C(-3, 16)$. The area of the triangle is
    (a) 30 sq units (b) 36 sq units
    (c) 27 sq units (d) 40 sq units
    (e) None of these

11. Two vertices of an equilateral triangle are origin and (4, 0). What is the area of the triangle?
    (a) 4 sq units (b) $\sqrt{3}$ sq units
    (c) $4\sqrt{3}$ sq units (d) $2\sqrt{3}$ sq units
    (e) 3 sq units

12. If the graph of the equation $2x + 3y = 6$ form a triangle with coordinates axes, then the area of triangle will be [SSC (10+2) 2012]
    (a) 2 sq units (b) 3 sq units
    (c) 6 sq units (d) 1 sq unit

13. Area of the triangle formed by the graph of the straight lines $x - y = 0$, $x + y = 2$ and the X-axis is [SSC CGL 2014]
    (a) 1 sq unit (b) 2 sq units
    (c) 4 sq units (d) None of these

14. The area of the region bounded by $y = |x| - 5$ with the coordinate axis is [SSC CGL 2012]
    (a) 25 sq units (b) 52 sq units
    (c) 50 sq units (d) 20 sq units

15. If the points $A(1, -1)$, $B(5, 2)$ and $C(k, 5)$ are collinear, then k equals
    (a) 2 (b) 4 (c) 6 (d) 9
    (e) None of these

16. If two vertices of a triangle are (5, 4) and $(-2, 4)$ and centroid is (5, 6), then third vertex is
    (a) (12, 10) (b) (10, 12)
    (c) $(-10, 12)$ (d) $(12, -10)$
    (e) $(-10, -12)$

17. A point C divides the line AB, where $A(1, 3)$ and $B(2, 7)$, in the ratio of 3 : 4. The coordinates of C are [RRB 2006]
    (a) $\left(\dfrac{5}{3}, 5\right)$ (b) $(-2, -9)$
    (c) $\left(\dfrac{3}{5}, 5\right)$ (d) $\left(\dfrac{10}{7}, \dfrac{33}{7}\right)$

18. Point $A(4, 2)$ divides segment BC in the ratio 2 : 5. Coordinates of B are (2, 6) and C are $(9, y)$. What is the value of y? [SSC CGL (Pre) 2017]
    (a) 8 (b) -8
    (c) 6 (d) -6

19. In what ratio, the line made by joining the points $A(-4, -3)$ and $B(5, 2)$ intersects X-axis?
    (a) 3 : 2 (b) 2 : 3 (c) -3 : 2 (d) -2 : 3
    (e) None of these

Coordinate Geometry / 695

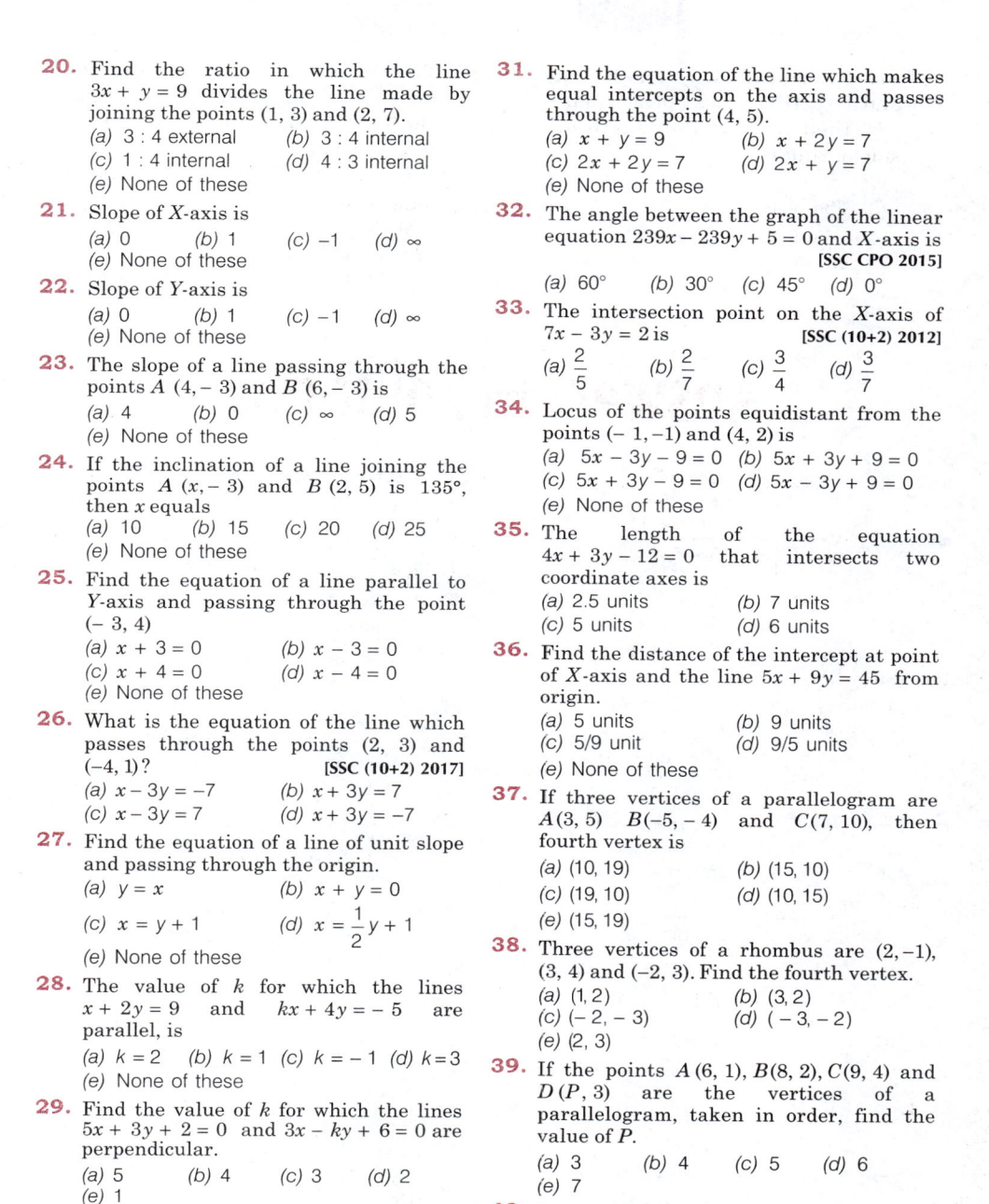

20. Find the ratio in which the line $3x + y = 9$ divides the line made by joining the points (1, 3) and (2, 7).
    (a) 3 : 4 external (b) 3 : 4 internal
    (c) 1 : 4 internal (d) 4 : 3 internal
    (e) None of these

21. Slope of X-axis is
    (a) 0 (b) 1 (c) –1 (d) ∞
    (e) None of these

22. Slope of Y-axis is
    (a) 0 (b) 1 (c) –1 (d) ∞
    (e) None of these

23. The slope of a line passing through the points A (4, – 3) and B (6, – 3) is
    (a) 4 (b) 0 (c) ∞ (d) 5
    (e) None of these

24. If the inclination of a line joining the points A $(x, – 3)$ and B (2, 5) is 135°, then x equals
    (a) 10 (b) 15 (c) 20 (d) 25
    (e) None of these

25. Find the equation of a line parallel to Y-axis and passing through the point (– 3, 4)
    (a) $x + 3 = 0$ (b) $x – 3 = 0$
    (c) $x + 4 = 0$ (d) $x – 4 = 0$
    (e) None of these

26. What is the equation of the line which passes through the points (2, 3) and (–4, 1)? [SSC (10+2) 2017]
    (a) $x – 3y = –7$ (b) $x + 3y = 7$
    (c) $x – 3y = 7$ (d) $x + 3y = –7$

27. Find the equation of a line of unit slope and passing through the origin.
    (a) $y = x$ (b) $x + y = 0$
    (c) $x = y + 1$ (d) $x = \frac{1}{2}y + 1$
    (e) None of these

28. The value of k for which the lines $x + 2y = 9$ and $kx + 4y = – 5$ are parallel, is
    (a) $k = 2$ (b) $k = 1$ (c) $k = – 1$ (d) $k = 3$
    (e) None of these

29. Find the value of k for which the lines $5x + 3y + 2 = 0$ and $3x – ky + 6 = 0$ are perpendicular.
    (a) 5 (b) 4 (c) 3 (d) 2
    (e) 1

30. Find the equation of perpendicular bisector of the line made by joining the points (1, 1) and (3, 5).
    (a) $x + 2y + 8 = 0$ (b) $x – 2y + 8 = 0$
    (c) $x – 2y – 8 = 0$ (d) $x + 2y – 8 = 0$
    (e) None of these

31. Find the equation of the line which makes equal intercepts on the axis and passes through the point (4, 5).
    (a) $x + y = 9$ (b) $x + 2y = 7$
    (c) $2x + 2y = 7$ (d) $2x + y = 7$
    (e) None of these

32. The angle between the graph of the linear equation $239x – 239y + 5 = 0$ and X-axis is [SSC CPO 2015]
    (a) 60° (b) 30° (c) 45° (d) 0°

33. The intersection point on the X-axis of $7x – 3y = 2$ is [SSC (10+2) 2012]
    (a) $\frac{2}{5}$ (b) $\frac{2}{7}$ (c) $\frac{3}{4}$ (d) $\frac{3}{7}$

34. Locus of the points equidistant from the points (– 1, –1) and (4, 2) is
    (a) $5x – 3y – 9 = 0$ (b) $5x + 3y + 9 = 0$
    (c) $5x + 3y – 9 = 0$ (d) $5x – 3y + 9 = 0$
    (e) None of these

35. The length of the equation $4x + 3y – 12 = 0$ that intersects two coordinate axes is
    (a) 2.5 units (b) 7 units
    (c) 5 units (d) 6 units

36. Find the distance of the intercept at point of X-axis and the line $5x + 9y = 45$ from origin.
    (a) 5 units (b) 9 units
    (c) 5/9 unit (d) 9/5 units
    (e) None of these

37. If three vertices of a parallelogram are $A(3, 5)$ $B(–5, – 4)$ and $C(7, 10)$, then fourth vertex is
    (a) (10, 19) (b) (15, 10)
    (c) (19, 10) (d) (10, 15)
    (e) (15, 19)

38. Three vertices of a rhombus are (2, –1), (3, 4) and (–2, 3). Find the fourth vertex.
    (a) (1, 2) (b) (3, 2)
    (c) (– 2, – 3) (d) (– 3, – 2)
    (e) (2, 3)

39. If the points $A(6, 1), B(8, 2), C(9, 4)$ and $D(P, 3)$ are the vertices of a parallelogram, taken in order, find the value of P.
    (a) 3 (b) 4 (c) 5 (d) 6
    (e) 7

40. Do the points (4, 3), (– 4, – 6) and (7, 9) form a triangle? If yes, then find the longest side of the triangle.
    (a) 18.6 units (b) 16.5 units
    (c) 24 units (d) 34 units
    (e) Triangle cannot be formed

696 / Fast Track Objective Arithmetic

41. Are the points (1, 7), (4, 2), (− 1, − 1) and (− 4, 4) vertices of a square? If yes, then what is the length of the side of square?
    (a) 34 units
    (b) $\sqrt{34}$ units
    (c) $\sqrt{68}$ units
    (d) 68 units
    (e) None of the above

42. Two diagonals of a parallelogram intersect each other at coordinates (17.5, 23.5). Two adjacent points of the parallelogram are (5.5, 7.5) and (13.5, 16). Find the lengths of the diagonals. [XAT 2015]
    (a) 15 and 30
    (b) 15 and 40
    (c) 17 and 30
    (d) 17 and 40
    (e) Multiple solutions are possible

Answer with Solutions

1. (a)

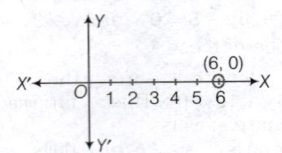

   So, the point (6, 0) lies on X-axis.

2. (c) Given, distance between the points $(x, 0)$ and $(−7, 0) = 10$ units
   Here, $x_1 = x$, $y_1 = 0$, $x_2 = −7$ and $y_2 = 0$
   ∴ Required distance
   $$= \sqrt{(x_2 − x_1)^2 + (y_2 − y_1)^2}$$
   ⇒ $\sqrt{(−7 − x)^2 + (0 − 0)^2} = 10$
   ⇒ $±(x + 7) = 10$
   If $x + 7 = 10$, then $x = 3$
   If $−(x + 7) = 10$, then $x = −17$

3. (b) Given, $AP = a + x$
   ∴ $\sqrt{(x − a)^2 + (y − 0)^2} = a + x$
   ⇒ $(x − a)^2 + y^2 = (a + x)^2$
   ⇒ $x^2 − 2ax + a^2 + y^2 = a^2 + 2ax + x^2$
   ⇒ $x^2 + a^2 + y^2 − a^2 − x^2 = 2ax + 2ax$
   ∴ $y^2 = 4ax$

4. (b) Here, $(k − 4)^2 + (0 + 8)^2 = (10)^2$
   [∵ distance $= \sqrt{(x_2 − x_1)^2 + (y_2 − y_1)^2}$]
   ⇒ $k^2 + 16 − 8k + 64 = 100$
   ⇒ $k^2 − 8k − 20 = 0$
   ⇒ $k^2 − 10k + 2k − 20 = 0$
   ⇒ $k(k − 10) + 2(k − 10) = 0$
   ⇒ $(k + 2)(k − 10) = 0$
   ⇒ $k = −2, k = 10$
   Hence, the value of k is 10 or $−2$.

5. (c) Let the abscissa be x.
   Then, $(x − 0)^2 + (−3 − 1)^2 = 5^2$
   ⇒ $x^2 + 16 = 25$
   ⇒ $x^2 = 9 \Rightarrow x = ±3$

6. (b) Here, the point $(6, −3)$ is five units away from line $y = 2$.

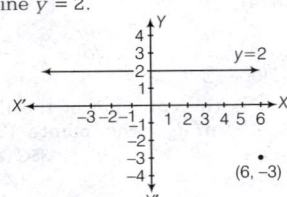

   So, its reflection point will also be 5 units away from the line $y = 2$.
   ∴ Required point $= (6, 7)$

7. (b) Let $A(2, 2)$, $B(6, 3)$ and $C(4, 11)$.
   $AB^2 = (6 − 2)^2 + (3 − 2)^2$
   $= 16 + 1 = 17$
   $BC^2 = (4 − 6)^2 + (11 − 3)^2$
   $= 4 + 64 = 68$
   $CA^2 = (4 − 2)^2 + (11 − 2)^2$
   $= 4 + 81 = 85$
   So, $AB^2 + BC^2 = AC^2$
   Hence, $\triangle ABC$ is right angled triangle.

8. (a) Let $P(x, y)$ be equidistant from points $A(7, 1)$ and $B(3, 5)$.
   Then, $AP = BP$, so $AP^2 = BP^2$
   ∴ $(x − 7)^2 + (y − 1)^2 = (x − 3)^2 + (y − 5)^2$
   ⇒ $x^2 − 14x + 49 + y^2 − 2y + 1$
   $= x^2 − 6x + 9 + y^2 − 10y + 25$
   ⇒ $8x − 8y = 16$
   ∴ After solving, $x − y = 2$

Coordinate Geometry / 697

9. (a) Distance between (x, y) and $(-1, 1)$
$$= \sqrt{(y-1)^2 + (x+1)^2}$$
Distance between (x, y) and $(4, 3)$
$$= \sqrt{(y-3)^2 + (x-4)^2}$$
∵ Points are equidistant,
$$\sqrt{(y-1)^2 + (x+1)^2} = \sqrt{(y-3)^2 + (x-4)^2}$$
On squaring both the sides, we get
$$(y-1)^2 + (x+1)^2 = (y-3)^2 + (x-4)^2$$
$$\Rightarrow y^2 + 1 - 2y + x^2 + 1 + 2x$$
$$= y^2 + 9 - 6y + x^2 + 16 - 8x$$
$$\Rightarrow 2x - 2y + 2 = -6y - 8x + 25$$
$$\Rightarrow 10x + 4y = 23$$

10. (c) Let $x_1 = 4$, $x_2 = 3$, $x_3 = -3$,
$y_1 = 4$, $y_2 = -2$ and $y_3 = 16$
∴ Area of triangle
$$= \frac{1}{2}\{x_1(y_2 - y_3) + x_2(y_3 - y_1) + x_3(y_1 - y_2)\}$$
$$= \frac{1}{2}[4(-2-16) + 3(16-4) + (-3)\{4-(-2)\}]$$
$$= \frac{1}{2}[4 \times (-18) + 3 \times 12 + (-3)(6)]$$
$$= \frac{1}{2}(-72 + 36 - 18)$$
$$= \frac{1}{2} \times (-54) = \frac{1}{2} \times 54 = 27 \text{ sq units}$$
[neglecting negative sign]

11. (c) Since, triangle is equilateral.
∴ $AB = BC = CA = 4 \Rightarrow BD = \frac{4}{2} = 2$

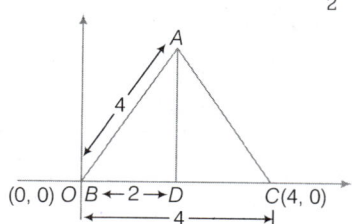

In $\triangle ADC$,
$$AD^2 = 4^2 - 2^2 = 16 - 4 = 12$$
$$\Rightarrow AD = 2\sqrt{3}$$
∴ Area of $\triangle ABC = \frac{1}{2} \times BC \times AD$
$$= \frac{1}{2} \times 4 \times 2\sqrt{3}$$
$$= 4\sqrt{3} \text{ sq units}$$

12. (b) ∵ $2x + 3y = 6$
$$\Rightarrow \frac{2x}{6} + \frac{3y}{6} = 1 \Rightarrow \frac{x}{3} + \frac{y}{2} = 1$$

Comparing with equation of line
$\frac{x}{a} + \frac{y}{b} = 1$, we get
Intercept at X-axis = 3
and intercept at Y-axis = 2

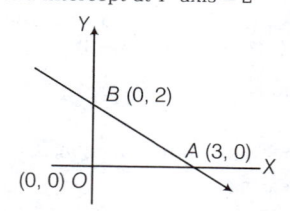

∴ Area of $\triangle OAB = \frac{1}{2} \times 3 \times 2 = 3$ sq units

13. (a) From the figure, $\triangle AOB$ is made by given lines, where $A(1, 1)$, $O(0, 0)$ and $B(2, 0)$ are coordinates.

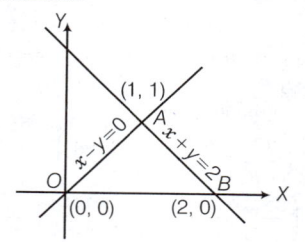

Here, $x_1 = 1$, $y_1 = 1$, $x_2 = 0$,
$y_2 = 0$, $x_3 = 2$ and $y_3 = 0$
∴ Area of $\triangle AOB$
$$= \frac{1}{2}[x_1(y_2 - y_3) + x_2(y_3 - y_1) + x_3(y_1 - y_2)]$$
$$= \frac{1}{2}[1(0-0) + 0(0-1) + 2(1-0)]$$
$$= \frac{1}{2}[1 \times 0 + 0 \times (-1) + 2 \times 1]$$
$$= \frac{1}{2}[0 - 0 + 2] = \frac{1}{2} \times 2 = 1 \text{ sq unit}$$

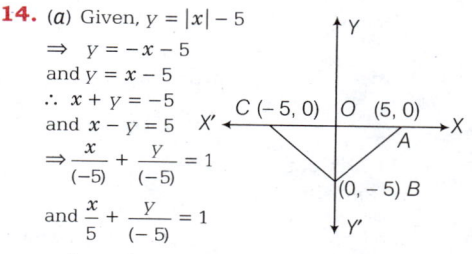

14. (a) Given, $y = |x| - 5$
$\Rightarrow y = -x - 5$
and $y = x - 5$
∴ $x + y = -5$
and $x - y = 5$
$\Rightarrow \frac{x}{(-5)} + \frac{y}{(-5)} = 1$
and $\frac{x}{5} + \frac{y}{(-5)} = 1$

∴ Area of bounded region
$$= \frac{1}{2} \times AC \times OB$$
$$= \frac{1}{2} \times 10 \times 5 = 25 \text{ sq units}$$

15. (d) Given, $x_1 = 1, x_2 = 5, x_3 = k,$
$y_1 = -1, y_2 = 2$ and $y_3 = 5.$
Since, A, B and C are collinear.
$\therefore$ Area of triangle $= 0$
$\Rightarrow \{x_1(y_2 - y_3) + x_2(y_3 - y_1) + x_3(y_1 - y_2)\} = 0$
$\Rightarrow \{1(2-5) + 5(5-(-1)) + k(-1-2)\} = 0$
$\Rightarrow \{-3 + 30 - 3k\} = 0 \Rightarrow 3k = 27$
$\therefore k = 9$

16. (a) Let the third vertex be (x, y).
$\therefore$ Coordinates of centroid
$= \left(\dfrac{x_1 + x_2 + x_3}{3}, \dfrac{y_1 + y_2 + y_3}{3}\right)$
Given, $x_1 = x, x_2 = 5, x_3 = -2, y_1 = y,$
$y_2 = 4, y_3 = 4$ and centroid $= (5, 6)$
$\therefore 5 = \dfrac{x + 5 - 2}{3}$ and $6 = \dfrac{y + 4 + 4}{3}$
$\Rightarrow x = 12$ and $y = 10$

17. (d) Given, $m = 3, n = 4, x_1 = 1, x_2 = 2, y_1 = 3$ and $y_2 = 7$
$\therefore$ Coordinates of C
$= \left(\dfrac{mx_2 + nx_1}{m + n}, \dfrac{my_2 + ny_1}{m + n}\right)$
$= \left(\dfrac{3 \times 2 + 4 \times 1}{3 + 4}, \dfrac{3 \times 7 + 4 \times 3}{3 + 4}\right) = \left(\dfrac{10}{7}, \dfrac{33}{7}\right)$

18. (b) Using section formula, i.e. if a line is divided by a point in certain ratio $(m : n)$, then coordinates of point (x, y)
$x = \dfrac{mx_2 + nx_1}{m + n}, y = \dfrac{my_2 + ny_1}{m + n}$
$\therefore 2 = \dfrac{2y + 5(6)}{2 + 5}$
$\Rightarrow 2y = -30 + 14 \Rightarrow 2y = -16$
$\Rightarrow y = -8$

19. (a) We know that, y-coordinate is zero on X-axis.
Given, $y_1 = -3, y_2 = 2$
$\therefore y = \dfrac{my_2 + ny_1}{m + n}$
$\Rightarrow 0 = \dfrac{m(2) + n(-3)}{m + n}$
$\Rightarrow 2m - 3n = 0 \Rightarrow \dfrac{m}{n} = \dfrac{3}{2}$

20. (b) Let required ratio be $k : 1$. Then, the coordinates of section point are
$\left(\dfrac{2k + 1}{k + 1}, \dfrac{7k + 3}{k + 1}\right)$.
But this point lies on $3x + y - 9 = 0$.

$\therefore 3\left(\dfrac{2k + 1}{k + 1}\right) + \left(\dfrac{7k + 3}{k + 1}\right) - 9 = 0$
$\Rightarrow 6k + 3 + 7k + 3 - 9k - 9 = 0$
$\Rightarrow 4k - 3 = 0 \Rightarrow k = \dfrac{3}{4}$
$\therefore$ Required ratio $= k : 1 = 3/4 : 1$
$= 3 : 4$ internal

21. (a) Slope of X-axis, $m = \tan\theta = \tan 0° = 0$

22. (d) Slope of Y-axis, $m = \tan 90° = \infty$

23. (b) Given, $x_1 = 4, x_2 = 6, y_1 = -3$ and $y_2 = -3$
$\therefore$ Slope $(m) = \dfrac{y_2 - y_1}{x_2 - x_1} = \dfrac{-3 + 3}{6 - 4} = \dfrac{0}{2} = 0$

24. (a) Slope of $AB = \dfrac{5 + 3}{2 - x} = \dfrac{8}{2 - x} = \tan 135°$
$\Rightarrow \tan(180° - 45°) = \dfrac{8}{2 - x}$
$\Rightarrow -\tan 45° = \dfrac{8}{2 - x} \Rightarrow -1 = \dfrac{8}{2 - x}$
$\Rightarrow -2 + x = 8 \Rightarrow x = 10$

25. (a) Required equation of the line is
$\dfrac{y - 4}{x + 3} = \tan 90° = \infty$
$\Rightarrow \dfrac{y - 4}{x + 3} = \dfrac{1}{0} \Rightarrow x + 3 = 0$

26. (a) $x_1 = 2, x_2 = -4, y_1 = 3$ and $y_2 = 1$
Equation of line is
$(y - y_1) = \dfrac{y_2 - y_1}{x_2 - x_1}(x - x_1)$
$\Rightarrow (y - 3) = \dfrac{(1 - 3)}{(-4 - 2)}(x - 2)$
$\Rightarrow y - 3 = \dfrac{-2}{-6}(x - 2)$
$\Rightarrow y - 3 = \dfrac{1}{3}(x - 2)$
$\Rightarrow 3y - 9 = x - 2$
$\Rightarrow 3y - x = 7 \Rightarrow x - 3y = -7$

27. (a) Given, $m = 1$ and $c = 0$ (as it passes through origin)
$\therefore$ Equation of the line is
$y = 1 \times x + 0 \Rightarrow y = x$

28. (a) Given, $x + 2y = 9$
$\Rightarrow y = \dfrac{-1}{2}x + \dfrac{9}{2}$
and $kx + 4y = -5$
$\Rightarrow y = \dfrac{-k}{4}x - \dfrac{5}{4}$
On comparing with $y = mx + c$, we get
$m_1 = \dfrac{-1}{2}$ and $m_2 = \dfrac{-k}{4}$

Coordinate Geometry / 699

Since, the given lines are parallel.
∴ $m_1 = m_2$
$\Rightarrow \dfrac{-1}{2} = \dfrac{-k}{4} \Rightarrow k = 2$

29. (a) Given, $5x + 3y + 2 = 0$
$\Rightarrow y = -\dfrac{5x}{3} - \dfrac{2}{3}$
and $3x - ky + 6 = 0$
$\Rightarrow y = \dfrac{3}{k}x + \dfrac{6}{k}$
On comparing with $y = mx + c$, we get
$m_1 = -\dfrac{5}{3}$ and $m_2 = \dfrac{3}{k}$
Since, both the lines are perpendicular.
∴ $m_1 m_2 = -1$
$\Rightarrow -\dfrac{5}{3} \times \dfrac{3}{k} = -1$
$\Rightarrow k = 5$

30. (d) Slope of the line made by joining the points $A(1,1)$ and $B(3, 5) = \dfrac{5-1}{3-1} = \dfrac{4}{2} = 2$
and mid-point of $AB = \left(\dfrac{1+3}{2}, \dfrac{1+5}{2}\right)$
$\Rightarrow (x_1, y_1) = (2, 3)$
Now, slope of the line perpendicular to the line AB, $m = -\dfrac{1}{2}$
∴ Required equation of perpendicular bisector is
$y - y_1 = m(x - x_1)$
$\Rightarrow y - 3 = -\dfrac{1}{2}(x - 2)$
$\Rightarrow x + 2y - 8 = 0$

31. (a) Let the intercepts made by line $= a$
∴ Equation of the line is
$\dfrac{x}{a} + \dfrac{y}{a} = 1$
$\Rightarrow x + y = a$ ...(i)
But this line passes through (4, 5).
∴ $4 + 5 = a \Rightarrow a = 9$
On putting this value in Eq. (i), we get
$x + y = 9$
which is the required equation.

32. (c) Given, $239x - 239y + 5 = 0$
$\Rightarrow 239y = 239x + 5$
$\Rightarrow y = x + \dfrac{5}{239}$
On comparing $y = mx + c$, we get $m = 1$
and $c = \dfrac{5}{239}$
∴ $m = \tan\theta = 1 \Rightarrow \theta = 45°$

33. (b) Equation of the line is $7x - 3y = 2$
$\Rightarrow \dfrac{7x}{2} - \dfrac{3y}{2} = 1$
$\Rightarrow \dfrac{x}{2/7} - \dfrac{y}{2/3} = 1$ ...(i)
On comparing Eq. (i) with $\dfrac{x}{a} + \dfrac{y}{b} = 1$, we get
Intersection point on X-axis, $a = \dfrac{2}{7}$

34. (c) Let $A(-1, -1)$, $B(4, 2)$ and $P(x, y)$ be the points such that
$AP^2 = BP^2$
∴ $(x+1)^2 + (y+1)^2 = (x-4)^2 + (y-2)^2$
$\Rightarrow (x^2 + 1 + 2x + y^2 + 1 + 2y)$
$= (x^2 + 16 - 8x + y^2 + 4 - 4y)$
$\Rightarrow 2 + 2x + 2y = 20 - 8x - 4y$
$\Rightarrow 10x + 6y - 18 = 0$
$\Rightarrow 5x + 3y - 9 = 0$

35. (c) $4x + 3y - 12 = 0 \Rightarrow \dfrac{x}{3} + \dfrac{y}{4} = 1$

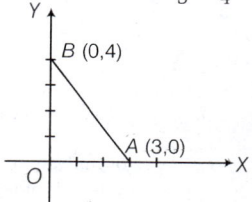

[Graph showing points B(0,4) on Y-axis and A(3,0) on X-axis]

Therefore, the line cuts X-axis at the point (3, 0) and Y-axis at the point (0, 4).
∴ $OA = 3$ units, $OB = 4$ units
∴ $AB = \sqrt{3^2 + 4^2} = 5$ units

36. (b) We know that, $y = 0$ on X-axis.
∴ Putting $y = 0$ in the line $5x + 9y = 45$, we get
$5x = 45 \Rightarrow x = 9$
Now, distance between (9, 0) and (0, 0)
$= \sqrt{(x_2 - x_1)^2 + (y_2 - y_1)^2}$
$= \sqrt{(0-9)^2 + (0-0)^2} = \sqrt{81} = 9$ units

37. (e) Let fourth vertex be $D(x, y)$.
We know that, diagonals of a parallelogram intersect at mid-point. Therefore,
Mid-point of AC = Mid-point of BD
$\left(\dfrac{3+7}{2}, \dfrac{5+10}{2}\right) = \left(\dfrac{-5+x}{2}, \dfrac{-4+y}{2}\right)$
$\Rightarrow \dfrac{-5+x}{2} = 5 \Rightarrow x = 10 + 5 = 15$
and $\dfrac{-4+y}{2} = \dfrac{15}{2}$
∴ $y = 15 + 4 = 19$
Hence, fourth vertex is (15, 19).

38. (d) Let coordinates of D be (x, y) and M be the intersection point of diagonals AC and BD.
Since, M is the mid-point of AC.
∴ Coordinates of M
$$= \left(\frac{-2+2}{2}, \frac{-1+3}{2}\right) = (0, 1)$$

Also, M is the mid-point of BD.
∴ Coordinates of $M = \left(\frac{x+3}{2}, \frac{y+4}{2}\right)$
∴ $\frac{x+3}{2} = 0 \Rightarrow x = -3$
and $\frac{y+4}{2} = 1 \Rightarrow y = -2$
Hence, the fourth vertex is $(-3, -2)$.

39. (e) Diagonals of a parallelogram bisect each other.
So, the coordinates of mid-point of AC
= Coordinates of mid-point of BD
i.e. $\left(\frac{6+9}{2}, \frac{1+4}{2}\right) = \left(\frac{8+P}{2}, \frac{2+3}{2}\right)$
$\Rightarrow \left(\frac{15}{2}, \frac{5}{2}\right) = \left(\frac{8+P}{2}, \frac{5}{2}\right)$
$\Rightarrow \frac{8+P}{2} = \frac{15}{2}$
$\Rightarrow 8 + P = 15$
∴ $P = 7$

40. (a) Let $P(4, 3), Q(-4, -6)$ and $R(7, 9)$ are the given points.
∴ $PQ = \sqrt{(-4-4)^2 + (-6-3)^2}$
$= \sqrt{(-8)^2 + (-9)^2}$
$= \sqrt{64+81} = \sqrt{145} = 12.04$
$QR = \sqrt{[7-(-4)]^2 + [9-(-6)]^2}$
$= \sqrt{11^2 + 15^2}$
$= \sqrt{121+225} = 18.6$
$PR = \sqrt{(7-4)^2 + (9-3)^2}$
$= \sqrt{9+36} = 6.7$
Since, the sum of any two sides is greater than third side, so it will form a triangle, whose longest side is 18.6 units.

41. (b) Let $A(1, 7), B(4, 2), C(-1, -1)$ and $D(-4, 4)$ are the given points.
$AB = \sqrt{(4-1)^2 + (2-7)^2}$
$= \sqrt{9+25} = \sqrt{34}$
$BC = \sqrt{(-1-4)^2 + (-1-2)^2}$
$= \sqrt{25+9} = \sqrt{34}$
$CD = \sqrt{(-4+1)^2 + (4+1)^2}$
$= \sqrt{9+25} = \sqrt{34}$
$DA = \sqrt{(1+4)^2 + (7-4)^2}$
$= \sqrt{25+9} = \sqrt{34}$
$AC = \sqrt{(-1-1)^2 + (-1-7)^2}$
$= \sqrt{4+64} = \sqrt{68}$
$BD = \sqrt{(-4-4)^2 + (4-2)^2}$
$= \sqrt{64+4} = \sqrt{68}$
Now, $AB = BC = CD = DA$ and $AC = BD$, i.e. all the four sides of the quadrilateral $ABCD$ are equal and its diagonals AC and BD are also equal.
Therefore, $ABCD$ is a square.
Hence, length of the side of square is $\sqrt{34}$ units.

42. (d) Since, diagonals of the parallelogram intersect each other at mid-points.

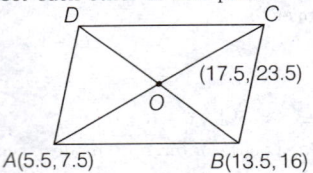

∴ Distance between points $(17.5, 23.5)$ and $(5.5, 7.5)$,
$OA = \sqrt{(17.5-5.5)^2 + (23.5-7.5)^2}$
$\left[\because d = \sqrt{(x_1-x_2)^2 + (y_1-y_2)^2}\right]$
$= \sqrt{(12)^2 + (16)^2} = \sqrt{144+256}$
$= \sqrt{400} = 20$
Now, distance between points $(17.5, 23.5)$ and $(13.5, 16)$,
$OB = \sqrt{(17.5-13.5)^2 + (23.5-16)^2}$
$= \sqrt{(4)^2 + (7.5)^2} = \sqrt{16+56.25}$
$= \sqrt{72.25} = 8.5$
Since, the length of diagonals are double of OA and OB.
∴ Diagonal $AC = 2 \times OA = 2 \times 20 = 40$
and diagonal $BD = 2 \times OB = 2 \times 8.5 = 17$

Chapter 38

Trigonometry

Trigonometry is a branch of Mathematics in which we study right angled triangles and relationship between their sides and angles using trigonometric ratios.

Measurement of Angles

1. **Degree Measure** (Sexagesimal System) In this system, an angle is measured in degrees, minutes and seconds.
   1 right angle = 90° (degree)
   1° = 60′ (60 min); 1′ = 60″ (60 s)

2. **Radian Measure** (Circular System) Angle subtended at the centre by any arc of length 1 unit in a circle of radius 1 unit, is said to have a measure of 1 radian. 1 radian is represented as 1^c.
   2π radian = 360° or π radian = 180°
   1 radian = 57° 16′ 22″

✦ If the angle subtended by an arc of length l to the centre of circle of radius r is θ, then

$$\text{Angle} = \frac{\text{Arc}}{\text{Radius}}, \text{ i.e. } \theta = \frac{l}{r}.$$

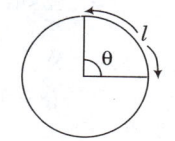

Relation between Radian and Degrees

Radian measure = $\dfrac{\pi}{180} \times$ Degree measure

Degree measure = $\dfrac{180}{\pi} \times$ Radian measure

We can simply put $\pi = 180°$ in radian measure to convert radian measure into degree measure.

Ex. 1 Convert 40° into radian measure.

Sol. $40° = 40 \times \dfrac{\pi}{180}$ radian = $\dfrac{2\pi}{9}$ radian

Ex. 2 Convert $\dfrac{3\pi}{2}$ into degree measure.

Sol. $\dfrac{3\pi}{2}$ radian = $\dfrac{3\pi}{2} \times \dfrac{180}{\pi} = 270°$

Alternate Method

On putting $\pi = 180°$, we get

$$\dfrac{3\pi}{2} = \dfrac{3 \times 180°}{2} = 270°$$

Trigonometric Ratios

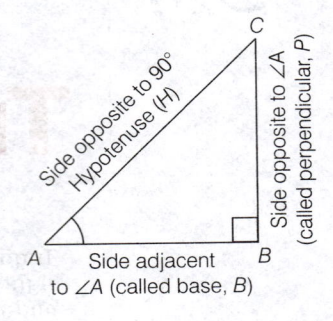

The ratios between different sides of a right angled triangle with respect to its acute angles are called **trigonometric ratios**.

Trigonometric ratios for right angled $\triangle ABC$ with respect to angle A are given below

$\sin A = \dfrac{BC}{AC} = \dfrac{P}{H}$, $\cos A = \dfrac{AB}{AC} = \dfrac{B}{H}$

$\tan A = \dfrac{BC}{AB} = \dfrac{P}{B}$, $\operatorname{cosec} A = \dfrac{AC}{BC} = \dfrac{H}{P}$

$\sec A = \dfrac{AC}{AB} = \dfrac{H}{B}$, $\cot A = \dfrac{AB}{BC} = \dfrac{B}{P}$

By Pythagoras theorem, $H^2 = B^2 + P^2$

Reciprocal Relations

$\sin A = \dfrac{1}{\operatorname{cosec} A}$ or $\operatorname{cosec} A = \dfrac{1}{\sin A}$

$\cos A = \dfrac{1}{\sec A}$ or $\sec A = \dfrac{1}{\cos A}$

$\tan A = \dfrac{\sin A}{\cos A}$ or $\cot A = \dfrac{\cos A}{\sin A}$

Ex. 3 If $\sin A = \dfrac{3}{4}$, then calculate $\cos A$ and $\tan A$.

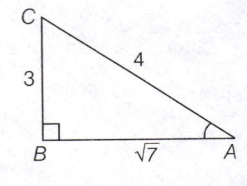

Sol. We have, $\sin A = \dfrac{BC}{AC} = \dfrac{3}{4}$

In right angled $\triangle ABC$, using Pythagoras theorem,

$(AC)^2 = (AB)^2 + (BC)^2$

$\Rightarrow$ $4^2 = (AB)^2 + 3^2 \Rightarrow 16 - 9 = (AB)^2$

$\Rightarrow$ $AB = \sqrt{7}$

$\therefore$ $\cos A = \dfrac{AB}{AC} = \dfrac{\sqrt{7}}{4}$, $\tan A = \dfrac{BC}{AB} = \dfrac{3}{\sqrt{7}}$

Ex. 4 If $\sin A = \dfrac{4}{5}$, then find the value of $(4 + \tan A)(2 + \cos A)$.

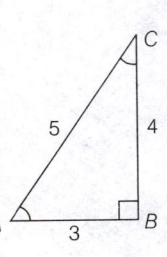

Sol. Given, $\sin A = \dfrac{4}{5} = \dfrac{BC}{AC} = \dfrac{P}{H}$

Now in $\triangle ABC$, using Pythagoras theorem,

$(AC)^2 = (AB)^2 + (BC)^2 \Rightarrow (5)^2 = (AB)^2 + (4)^2$

$\Rightarrow$ $25 = (AB)^2 + 16 \Rightarrow AB = \sqrt{25 - 16} = \sqrt{9} = 3$

So, $\tan A = \dfrac{BC}{AB} = \dfrac{P}{B} = \dfrac{4}{3}$ and $\cos A = \dfrac{AB}{AC} = \dfrac{B}{H} = \dfrac{3}{5}$

Now, $(4 + \tan A)(2 + \cos A)$

$= \left(4 + \dfrac{4}{3}\right)\left(2 + \dfrac{3}{5}\right) = \left(\dfrac{12+4}{3}\right)\left(\dfrac{10+3}{5}\right) = \dfrac{16}{3} \times \dfrac{13}{5} = \dfrac{208}{15}$

Trigonometric Ratios of Some Specific Angles

Trigonometric ratios	0° (0)	30° ($\pi/6$)	45° ($\pi/4$)	60° ($\pi/3$)	90° ($\pi/2$)	180° (π)	270° ($3\pi/2$)	360° (2π)
sin θ	0	$\frac{1}{2}$	$\frac{1}{\sqrt{2}}$	$\frac{\sqrt{3}}{2}$	1	0	−1	0
cos θ	1	$\frac{\sqrt{3}}{2}$	$\frac{1}{\sqrt{2}}$	$\frac{1}{2}$	0	−1	0	1
tan θ	0	$\frac{1}{\sqrt{3}}$	1	$\sqrt{3}$	∞	0	∞	0
cot θ	∞	$\sqrt{3}$	1	$\frac{1}{\sqrt{3}}$	0	∞	0	∞
sec θ	1	$\frac{2}{\sqrt{3}}$	$\sqrt{2}$	2	∞	−1	∞	1
cosec θ	∞	2	$\sqrt{2}$	$\frac{2}{\sqrt{3}}$	1	∞	−1	∞

Ex. 5 Find the value of $2\tan^2 45° + \cos^2 30° - \sin^2 60°$.

Sol. $2\tan^2 45° + \cos^2 30° - \sin^2 60° = 2(1)^2 + \left(\frac{\sqrt{3}}{2}\right)^2 - \left(\frac{\sqrt{3}}{2}\right)^2 = 2$

Ex. 6 Find the value of $\dfrac{2\tan 30°}{1-\tan^2 30°}$.

Sol. $\dfrac{2\tan 30°}{1-\tan^2 30°} = \dfrac{2\times \frac{1}{\sqrt{3}}}{1-\left(\frac{1}{\sqrt{3}}\right)^2} = \dfrac{\frac{2}{\sqrt{3}}}{1-\frac{1}{3}} = \dfrac{\frac{2}{\sqrt{3}}}{\frac{2}{3}} = \dfrac{3}{\sqrt{3}} \times \dfrac{\sqrt{3}}{\sqrt{3}} = \dfrac{3\sqrt{3}}{3} = \sqrt{3}$

Trigonometric Identities

An equation involving trigonometric ratios of an angle is called a trigonometric identity, if it is true for all values of the angles involved.
In a right angled triangle, we have the following identities

1. $\sin^2 A + \cos^2 A = 1$
2. $1 + \tan^2 A = \sec^2 A$
3. $1 + \cot^2 A = \operatorname{cosec}^2 A$

Ex. 7 If $\cot\theta = \dfrac{7}{8}$, then evaluate $\dfrac{(1+\sin\theta)(1-\sin\theta)}{(1+\cos\theta)(1-\cos\theta)}$.

Sol. $\dfrac{(1+\sin\theta)(1-\sin\theta)}{(1+\cos\theta)(1-\cos\theta)} = \dfrac{(1)^2 - (\sin\theta)^2}{(1)^2 - (\cos\theta)^2}$ $\quad [\because (a-b)(a+b) = a^2 - b^2]$

$= \dfrac{1-\sin^2\theta}{1-\cos^2\theta} = \dfrac{\cos^2\theta}{\sin^2\theta} = \left(\dfrac{\cos\theta}{\sin\theta}\right)^2$

$= (\cot\theta)^2 = \left(\dfrac{7}{8}\right)^2 = \dfrac{49}{64}$

Sign of Trigonometric Functions

Trigonometric ratios of complementary and supplementary angles of angle θ are shown below

$\sin\left(\dfrac{\pi}{2}-\theta\right)=\cos\theta$ $\sin\left(\dfrac{\pi}{2}+\theta\right)=\cos\theta$

$\cos\left(\dfrac{\pi}{2}-\theta\right)=\sin\theta$ $\cos\left(\dfrac{\pi}{2}+\theta\right)=-\sin\theta$

$\tan\left(\dfrac{\pi}{2}-\theta\right)=\cot\theta$ $\tan\left(\dfrac{\pi}{2}+\theta\right)=-\cot\theta$

$\cot\left(\dfrac{\pi}{2}-\theta\right)=\tan\theta$ $\cot\left(\dfrac{\pi}{2}+\theta\right)=-\tan\theta$

$\sec\left(\dfrac{\pi}{2}-\theta\right)=\csc\theta$ $\sec\left(\dfrac{\pi}{2}+\theta\right)=-\csc\theta$

$\csc\left(\dfrac{\pi}{2}-\theta\right)=\sec\theta$ $\csc\left(\dfrac{\pi}{2}+\theta\right)=\sec\theta$

$\sin(\pi-\theta)=\sin\theta$ $\sin(\pi+\theta)=-\sin\theta$
$\cos(\pi-\theta)=-\cos\theta$ $\cos(\pi+\theta)=-\cos\theta$
$\tan(\pi-\theta)=-\tan\theta$ $\tan(\pi+\theta)=\tan\theta$
$\cot(\pi-\theta)=-\cot\theta$ $\cot(\pi+\theta)=\cot\theta$
$\sec(\pi-\theta)=-\sec\theta$ $\sec(\pi+\theta)=-\sec\theta$
$\csc(\pi-\theta)=\csc\theta$ $\csc(\pi+\theta)=-\csc\theta$
$\sin(-\theta)=-\sin\theta$ $\csc(-\theta)=-\csc\theta$
$\cos(-\theta)=\cos\theta$ $\sec(-\theta)=\sec\theta$
$\tan(-\theta)=-\tan\theta$ $\cot(-\theta)=-\cot\theta$

II	I
sin, cosec Positive	All Positive
(90° + θ) and (180° − θ)	(90° − θ) and (360° + θ)
III	IV
tan, cot Positive	cos, sec Positive
(180° + θ) and (270° − θ)	(270° + θ) and (360° − θ)

For (90° ± θ) and (270 ± θ), change
$\sin\theta \leftrightarrow \cos\theta$,
$\tan\theta \leftrightarrow \cot\theta$,
$\csc\theta \leftrightarrow \sec\theta$
and for (180 ± θ) and (360 ± θ), trigonometric functions do not change.

✦ Positive or negative sign is used according to the quadrant.

Ex. 8 If $\sin 3A = \cos(A - 26°)$, where $3A$ is an acute angle, find the value of A.

Sol. Given, $\sin 3A = \cos(A - 26°)$
or $\cos(90° - 3A) = \cos(A - 26°)$ [∵ $\cos(90° - \theta) = \sin\theta$]
On equating the angles from both sides, we get
$$90° - 3A = A - 26° \Rightarrow 4A = 116°$$
∴ $A = \dfrac{116°}{4} = 29°$

Ex. 9 What is the value of $\csc^2 68° + \sec^2 56° - \cot^2 34° - \tan^2 22°$? [CDS 2016 (II)]

Sol. We have, $\csc^2 68° + \sec^2 56° - \cot^2 34° - \tan^2 22°$
$= \csc^2 68° - \tan^2 22° + \sec^2 56° - \cot^2 34°$
$= \csc^2 68° - \tan^2(90° - 68°) + \sec^2 56° - \cot^2(90° - 56°)$
$= \csc^2 68° - \cot^2 68° + \sec^2 56° - \tan^2 56° = 1 + 1 = 2$

Ex. 10 Find the value of $\dfrac{\sin^2 63° + \sin^2 27°}{\cos^2 17° + \cos^2 73°}$.

Sol. $\dfrac{\sin^2 63° + \sin^2 27°}{\cos^2 17° + \cos^2 73°} = \dfrac{\sin^2(90° - 27°) + \sin^2 27°}{\cos^2(90° - 73°) + \cos^2 73°} = \dfrac{\cos^2 27° + \sin^2 27°}{\sin^2 73° + \cos^2 73°} = 1$

[∵ $\sin^2\theta + \cos^2\theta = 1$, here θ = 27° and 73°]

Ex. 11 Consider the following statements

I. $\dfrac{\cos 75°}{\sin 15°} + \dfrac{\sin 12°}{\cos 78°} - \dfrac{\cos 18°}{\sin 72°} = 1$

II. $\dfrac{\cos 35°}{\sin 55°} - \dfrac{\sin 11°}{\cos 79°} + \cos 28° \operatorname{cosec} 62° = 1$

III. $\dfrac{\sin 80°}{\cos 10°} - \sin 59° \sec 31° = 0$

Which of the above statements are correct? [CDS 2016 (II)]

Sol. I. LHS $= \dfrac{\cos 75°}{\sin 15°} + \dfrac{\sin 12°}{\cos 78°} - \dfrac{\cos 18°}{\sin 72°}$

$= \dfrac{\cos(90° - 15°)}{\sin 15°} + \dfrac{\sin(90° - 78°)}{\cos 78°} - \dfrac{\cos(90° - 72°)}{\sin 72°}$

$= \dfrac{\sin 15°}{\sin 15°} + \dfrac{\cos 78°}{\cos 78°} - \dfrac{\sin 72°}{\sin 72°} = 1 + 1 - 1 = 1 = $ RHS

II. LHS $= \dfrac{\cos 35°}{\sin 55°} - \dfrac{\sin 11°}{\cos 79°} + \cos 28° \operatorname{cosec} 62°$

$= \dfrac{\cos(90° - 55°)}{\sin 55°} - \dfrac{\sin(90° - 79°)}{\cos 79°} + \cos 28° \cdot \dfrac{1}{\sin 62°}$

$= \dfrac{\sin 55°}{\sin 55°} - \dfrac{\cos 79°}{\cos 79°} + \dfrac{\cos(90° - 62°)}{\sin 62°} = 1 - 1 + 1 = 1 = $ RHS

III. LHS $= \dfrac{\sin 80°}{\cos 10°} - \sin 59° \sec 31° = \dfrac{\sin 80°}{\cos(90° - 80°)} - \sin 59° \sec(90° - 59°)$

$= \dfrac{\sin 80°}{\sin 80°} - \sin 59° \operatorname{cosec} 59° = 1 - \sin 59° \cdot \dfrac{1}{\sin 59°} = 1 - 1 = 0 = $ RHS

Hence, all the statements are correct.

Trigonometric Ratios of Compound Angles

An angle made up of the algebraic sum of two or more angles is called a compound angle.

Sum and Difference Formulae

1. $\sin(x \pm y) = \sin x \cos y \pm \cos x \sin y$
2. $\cos(x \pm y) = \cos x \cos y \mp \sin x \sin y$
3. $\tan(x \pm y) = \dfrac{\tan x \pm \tan y}{1 \mp \tan x \tan y}$
4. $\cot(x + y) = \dfrac{\cot x \cot y - 1}{\cot x + \cot y}$
5. $\cot(x - y) = \dfrac{\cot x \cot y + 1}{\cot y - \cot x}$
6. $\sin(x + y) \cdot \sin(x - y) = (\sin^2 x - \sin^2 y)$ or $(\cos^2 y - \cos^2 x)$
7. $\cos(x + y) \cos(x - y) = (\cos^2 x - \sin^2 y)$ or $(\cos^2 y - \sin^2 x)$
8. $\dfrac{\cos x \pm \sin x}{\cos x \mp \sin x} = \tan(45° \pm x)$

A, B Formulae or Product to Sum Formulae

1. $2 \sin A \cos B = \sin(A + B) + \sin(A - B)$
2. $2 \sin A \sin B = \cos(A - B) - \cos(A + B)$
3. $2 \cos A \sin B = \sin(A + B) - \sin(A - B)$
4. $2 \cos A \cos B = \cos(A + B) + \cos(A - B)$

C, D Formulae or Sum to Product Formulae

1. $\sin C + \sin D = 2 \sin \left(\dfrac{C+D}{2}\right) \cos \left(\dfrac{C-D}{2}\right)$
2. $\sin C - \sin D = 2 \cos \left(\dfrac{C+D}{2}\right) \sin \left(\dfrac{C-D}{2}\right)$
3. $\cos C + \cos D = 2 \cos \left(\dfrac{C+D}{2}\right) \cos \left(\dfrac{C-D}{2}\right)$
4. $\cos C - \cos D = 2 \sin \left(\dfrac{C+D}{2}\right) \sin \left(\dfrac{D-C}{2}\right)$

Trigonometric Ratios of Multiple of an Angle

1. $\sin 2x = 2 \sin x \cos x = \dfrac{2 \tan x}{1 + \tan^2 x}$
2. $\cos 2x = \cos^2 x - \sin^2 x = 2\cos^2 x - 1 = 1 - 2\sin^2 x = \dfrac{1 - \tan^2 x}{1 + \tan^2 x}$
3. $\tan 2x = \dfrac{2 \tan x}{1 - \tan^2 x}$ or $\cot 2x = \dfrac{\cot^2 x - 1}{2 \cot x}$
4. $\sin 3x = 3 \sin x - 4 \sin^3 x$
5. $\cos 3x = 4 \cos^3 x - 3 \cos x$
6. $\tan 3x = \dfrac{3 \tan x - \tan^3 x}{1 - 3 \tan^2 x}$
7. $\cos x \cdot \cos 2x \cdot \cos 4x = \dfrac{1}{4}[\cos(4x - 2x + x)] = \dfrac{1}{4} \cos 3x$
8. $\sin x \cdot \sin 2x \cdot \sin 4x = \dfrac{1}{4}[\sin(4x - 2x + x)] = \dfrac{1}{4} \sin 3x$
9. $\tan x \cdot \tan 2x \cdot \tan 4x = \tan(4x - 2x + x) = \tan 3x$

Trigonometric Ratios of Sub-Multiple Angles

1. $\sin x = 2 \sin \dfrac{x}{2} \cos \dfrac{x}{2} = \dfrac{2 \tan x/2}{1 + \tan^2 (x/2)}$
2. $\cos x = \cos^2 \dfrac{x}{2} - \sin^2 \dfrac{x}{2} = 1 - 2\sin^2 \dfrac{x}{2} = 2\cos^2 \dfrac{x}{2} - 1 = \dfrac{1 - \tan^2 x/2}{1 + \tan^2 x/2}$
3. $1 + \cos x = 2 \cos^2 \dfrac{x}{2}$
4. $1 - \cos x = 2 \sin^2 \dfrac{x}{2}$
5. $\tan x = \dfrac{2 \tan x/2}{1 - \tan^2 x/2}$
6. $\cot x = \dfrac{\cot^2 (x/2) - 1}{2 \cot x/2}$
7. $\sin \dfrac{x}{2} = \sqrt{\dfrac{1 - \cos x}{2}}$
8. $\cos \dfrac{x}{2} = \sqrt{\dfrac{1 + \cos x}{2}}$
9. $\tan \dfrac{x}{2} = \sqrt{\dfrac{1 - \cos x}{1 + \cos x}}$

Sine Rule

In $\triangle ABC$, if $AB = c$, $BC = a$ and $AC = b$.

Then, $\dfrac{\sin A}{a} = \dfrac{\sin B}{b} = \dfrac{\sin C}{c} = K$ (constant)

It can also be written as $\dfrac{a}{\sin A} = \dfrac{b}{\sin B} = \dfrac{c}{\sin C} = K$

Then, $a = K \sin A$, $b = K \sin B$, $c = K \sin C$

✦ Circumradius of a circle $R = \dfrac{a}{2 \sin A} = \dfrac{b}{2 \sin B} = \dfrac{c}{2 \sin C}$, where R is the radius of circumcircle.

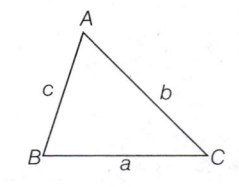

Cosine Rule

In $\triangle ABC$, $\cos A = \dfrac{b^2 + c^2 - a^2}{2bc}$,

$\cos B = \dfrac{c^2 + a^2 - b^2}{2ac}$

and $\cos C = \dfrac{a^2 + b^2 - c^2}{2ab}$

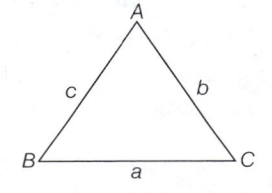

Ex. 12 In a $\triangle ABC$, if $\angle BCA = 60°$ and $AB^2 = BC^2 + CA^2 + X$, then what is the value of X? [CDS 2012]

Sol. By cosine law,

$$\cos 60° = \dfrac{AC^2 + BC^2 - AB^2}{2 \cdot AC \cdot BC} = \dfrac{1}{2}$$

$\Rightarrow AC^2 + BC^2 - AB^2 = AC \cdot BC$

On comparing, we get $X = -(AC)(BC)$

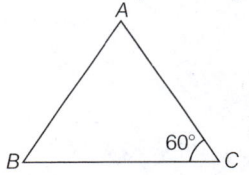

Some Important Results

1. If $\sin x = 0$ or $\tan x = 0$, then $x = n\pi$
2. If $\cos x = 0$ or $\cot x = 0$, then $x = (2n + 1)\dfrac{\pi}{2}$
3. $\sin 22\dfrac{1°}{2} = \dfrac{\sqrt{2 - \sqrt{2}}}{2}$
4. $\cos 22\dfrac{1°}{2} = \dfrac{\sqrt{2 + \sqrt{2}}}{2}$
5. $\tan 22\dfrac{1°}{2} = \sqrt{2} - 1$
6. $\cot 22\dfrac{1°}{2} = \sqrt{2} + 1$
7. $\sin 18° = \cos 72° = \dfrac{\sqrt{5} - 1}{4}$
8. $\cos 18° = \sin 72° = \dfrac{\sqrt{10 + 2\sqrt{5}}}{4}$
9. $\sin 36° = \cos 54° = \dfrac{\sqrt{10 - 2\sqrt{5}}}{4}$
10. $\cos 36° = \sin 54° = \dfrac{\sqrt{5} + 1}{4}$

Ex. 13 Find the value of $\tan 15°$.

Sol. $\tan 15° = \tan(45° - 30°) = \dfrac{\tan 45° - \tan 30°}{1 + \tan 45° \cdot \tan 30°}$

$= \dfrac{1 - \dfrac{1}{\sqrt{3}}}{1 + \dfrac{1}{\sqrt{3}}} = \dfrac{\sqrt{3} - 1}{\sqrt{3} + 1}$ $\left[\because \tan(A - B) = \dfrac{\tan A - \tan B}{1 + \tan A \tan B} \right]$

$$= \frac{\sqrt{3}-1}{\sqrt{3}+1} \times \frac{\sqrt{3}-1}{\sqrt{3}-1} \qquad \text{[on rationalising]}$$

$$= \frac{(\sqrt{3}-1)^2}{(\sqrt{3})^2 - (1)^2} = \frac{3+1-2\sqrt{3}}{3-1} = \frac{4-2\sqrt{3}}{2} = 2 - \sqrt{3}$$

Ex. 14 In $\triangle ABC$, right angled at B, find the value of $\sin A \cos C + \cos A \sin C$.

Sol. Given, $\angle B = 90°$
∴ $\angle A + \angle C = 90°$ [∵ $\angle A + \angle B + \angle C = 180°$]
Now, $\sin A \cos C + \cos A \sin C = \sin(A+C) = \sin 90° = 1$

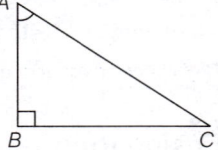

Ex. 15 Find the value of $\cos 20° \cos 40° \cos 60° \cos 80°$.

Sol. $\cos 20° \cos 40° \cos 60° \cos 80° = \cos 60° [\cos 20° \cos 40° \cos 80°]$

$$= \frac{1}{2}\left[\frac{1}{4} \cos(3 \times 20°)\right] \qquad \left[\because \cos\theta \cos 2\theta \cos 4\theta = \frac{1}{4}\cos 3\theta\right]$$

$$= \frac{1}{2}\left[\frac{1}{4} \cos 60°\right] = \frac{1}{2} \times \frac{1}{4} \times \frac{1}{2} = \frac{1}{16}$$

Ex. 16 If $x + y = z$, then find the value of $\cos^2 x + \cos^2 y + \cos^2 z$.

Sol. Given, $x + y = z$
Now, $\cos^2 x + \cos^2 y + \cos^2 z$
$= 1 + (\cos^2 x - \sin^2 y) + \cos^2 z$ [∵ $\cos^2 A = 1 - \sin^2 A$]
$= 1 + \cos(x+y) \cdot \cos(x-y) + \cos^2 z$
$= 1 + \cos z \cdot \cos(x-y) + \cos^2 z$
$= 1 + \cos z [\cos(x-y) + \cos z]$
$= 1 + \cos z [\cos(x-y) + \cos(x+y)]$
$= 1 + \cos z \left[2\cos\frac{(x-y+x+y)}{2} \cdot \cos\frac{(x-y-x-y)}{2}\right]$
$= 1 + 2\cos z \cdot \cos x \cdot \cos y$ $\left[\because \cos A + \cos B = 2\cos\frac{(A+B)}{2} \cdot \cos\frac{(A-B)}{2} \text{ and } \cos(-\theta) = \cos\theta\right]$
$= 1 + 2\cos x \cdot \cos y \cdot \cos z$

Ex. 17 If $(a^2 - b^2)\sin\theta + 2ab\cos\theta = a^2 + b^2$, then find the value of $\tan\theta$.

Sol. Given, $(a^2 - b^2)\sin\theta + 2ab\cos\theta = a^2 + b^2$

$\Rightarrow \quad \dfrac{a^2-b^2}{a^2+b^2} \cdot \sin\theta + \dfrac{2ab}{a^2+b^2} \cdot \cos\theta = 1$

In right angled $\triangle ABC$, $\cos\alpha \cdot \sin\theta + \sin\alpha \cdot \cos\theta = \sin 90°$
$\Rightarrow \quad \sin(\theta + \alpha) = \sin 90°$
   [∵ $\sin A \cos B + \cos A \sin B = \sin(A+B)$]
$\Rightarrow \quad \theta + \alpha = 90° \Rightarrow \theta = 90° - \alpha$
$\Rightarrow \quad \tan\theta = \tan(90° - \alpha) = \cot\alpha$

∴ $\tan\theta = \dfrac{a^2 - b^2}{2ab}$ $\left[\because \sin\alpha = \dfrac{2ab}{a^2+b^2} \text{ and } \cos\alpha = \dfrac{a^2-b^2}{a^2+b^2}\right]$

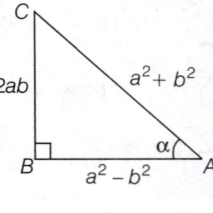

Trigonometry / 709

Ex. 18 If $(1 + \tan A)(1 + \tan B) = 2$, then find the value of $(A + B)$.

Sol. Given, $(1 + \tan A)(1 + \tan B) = 2 \Rightarrow \tan A + \tan B = 1 - \tan A \tan B$

$$\Rightarrow \frac{\tan A + \tan B}{1 - \tan A \tan B} = 1 = \tan 45°$$

$$\Rightarrow \tan(A + B) = \tan 45° \qquad \left[\because \tan(A+B) = \frac{\tan A + \tan B}{1 - \tan A \tan B}\right]$$

$$\Rightarrow A + B = 45° = \frac{\pi}{4}$$

Maximum and Minimum Values of Trigonometric Angles

1. $-1 \leq \sin\theta$ or $\cos\theta \leq 1$
2. $-\infty \leq \tan\theta$ or $\cot\theta \leq \infty$
3. $\sec\theta$ or $\csc\theta \geq 1$
4. $\sec\theta$ or $\csc\theta \leq -1$
5. $-\sqrt{m^2 + n^2} \leq m\sin\theta \pm n\cos\theta \leq \sqrt{m^2 + n^2}$

✦ To obtain the maximum or minimum value of an expression try to convert it into $\sin\theta$ or $\cos\theta$.

Ex. 19 Find the maximum value of $5\tan\theta\cos\theta + 4\cos\theta$.

Sol. $5\tan\theta\cos\theta + 4\cos\theta = \dfrac{5\sin\theta}{\cos\theta} \cdot \cos\theta + 4\cos\theta \qquad \left[\because \tan\theta = \dfrac{\sin\theta}{\cos\theta}\right]$

$$= 5\sin\theta + 4\cos\theta$$

So, $m = 5$ and $n = 4$

∴ Maximum value $= \sqrt{5^2 + 4^2}$

$$= \sqrt{25 + 16} = \sqrt{41}$$

Ex. 20 Find the minimum value of $2\sin^2\theta + 3\cos^2\theta$.

Sol. $2\sin^2\theta + 3\cos^2\theta = 2\sin^2\theta + 2\cos^2\theta + \cos^2\theta = 2(\sin^2\theta + \cos^2\theta) + \cos^2\theta$

$$= 2 + \cos^2\theta \qquad [\because \sin^2\theta + \cos^2\theta = 1]$$

Now, minimum value of $\cos\theta = -1$

∴ Minimum value of expression $= 2 + (-1)^2 = 2 + 1 = 3$

Some Important Results

1. $\tan 1° \cdot \tan 2° \ldots \tan 89° = 1$
2. $\cot 1° \cdot \cot 2° \ldots \cot 89° = 1$
3. $\cos 1° \cdot \cos 2° \cdot \cos 3° \ldots \cos 90° = 0$
4. $\cos 1° \cdot \cos 2° \cdot \cos 3° \ldots$ (above $\cos 90°$) $\ldots = 0$
5. $\sin 1° \cdot \sin 2° \cdot \sin 3° \ldots \sin 180° = 0$
6. $\sin 1° \cdot \sin 2° \cdot \sin 3° \ldots$ (above $\sin 180°$) $= 0$
7. $\dfrac{\sin A}{\cos B} = 1$, when $A + B = 90°$
8. $\tan A \tan B = 1 = \cot A \cot B$, when $A + B = 90°$

Ex. 21 Find the value of $\sin 1° \cdot \sin 2° \cdot \sin 3° \ldots \sin 180° \ldots \sin 196°$.

Sol. We have, $\sin 1° \cdot \sin 2° \cdot \sin 3° \ldots \sin 180° \ldots \sin 196°$.
As, $\sin 180° = 0$
When a value is multiplied with zero (0), then whole expression will result in zero.
$\therefore \quad \sin 1° \cdot \sin 2° \cdot \sin 3° \ldots \sin 180° \ldots \sin 196° = 0$

Ex. 22 If $\sin\theta + \cos\theta = \sqrt{2}$, then find the value of θ.

Sol. Given, $\sin\theta + \cos\theta = \sqrt{2}$
On squaring both sides, we get
$$(\sin\theta + \cos\theta)^2 = (\sqrt{2})^2$$
$\Rightarrow \quad \sin^2\theta + \cos^2\theta + 2\sin\theta\cos\theta = 2$
$\Rightarrow \quad 1 + 2\sin\theta\cos\theta = 2$
$\Rightarrow \quad 2\sin\theta\cos\theta = 1$
$\Rightarrow \quad \sin 2\theta = 1 = \sin 90°$
On equating the angles on both sides, we get $2\theta = 90°$
$\Rightarrow \quad \theta = \dfrac{90°}{2} = 45°$

✦ When $\sin\theta + \cos\theta = x$, then $\sin\theta - \cos\theta = \sqrt{2-x^2}$

Ex. 23 If $\sin\theta + \operatorname{cosec}\theta = 2$, then find the value of $\sin^{36}\theta + \operatorname{cosec}^{36}\theta$.

Sol. Given, $\sin\theta + \operatorname{cosec}\theta = 2$
$\Rightarrow \quad \sin\theta + \dfrac{1}{\sin\theta} - 2 = 0$
$\Rightarrow \quad (\sqrt{\sin\theta})^2 + \dfrac{1}{(\sqrt{\sin\theta})^2} - 2\sqrt{\sin\theta} \times \dfrac{1}{\sqrt{\sin\theta}} = 0$
$\Rightarrow \quad \left(\sqrt{\sin\theta} - \dfrac{1}{\sqrt{\sin\theta}}\right)^2 = 0 \Rightarrow \sqrt{\sin\theta} - \dfrac{1}{\sqrt{\sin\theta}} = 0$
$\Rightarrow \quad \sin\theta = 1 \Rightarrow \theta = 90°$
$\therefore \sin^{36}\theta + \operatorname{cosec}^{36}\theta = \sin^{36}90° + \operatorname{cosec}^{36}90°$
$\qquad = (\sin 90°)^{36} + (\operatorname{cosec} 90°)^{36} = 1 + 1 = 2$

✦ When $\sin\theta + \operatorname{cosec}\theta = 2$, then $\sin^m\theta + \operatorname{cosec}^n\theta = 2$ for all positive values of n.

Ex. 24 Find the value of $\cot\dfrac{\pi}{20} \cdot \cot\dfrac{3\pi}{20} \cdot \cot\dfrac{5\pi}{20} \cdot \cot\dfrac{7\pi}{20} \cdot \cot\dfrac{9\pi}{20}$.

Sol. Given, $\cot\dfrac{\pi}{20} \cdot \cot\dfrac{3\pi}{20} \cdot \cot\dfrac{5\pi}{20} \cdot \cot\dfrac{7\pi}{20} \cdot \cot\dfrac{9\pi}{20}$
Converting them into degree measure, we get
$\cot\dfrac{180°}{20} \cdot \cot\dfrac{3 \times 180°}{20} \cdot \cot\dfrac{5 \times 180°}{20} \cdot \cot\dfrac{7 \times 180°}{20} \cdot \cot\dfrac{9 \times 180°}{20}$
$= \cot 9° \cdot \cot 27° \cdot \cot 45° \cdot \cot 63° \cdot \cot 81°$
Rewriting the above expression, $\cot 9° \cdot \cot 81° \cdot \cot 27° \cdot \cot 63° \cdot \cot 45°$
We know that, $\cot A \cot B = 1$, when $(A + B) = 90°$
So, $\cot 9° \cot 81° = 1$ and $\cot 27° \cot 63° = 1$
$\therefore \quad \cot 9° \cot 81° \cot 27° \cot 63° \cot 45° = 1 \cdot 1 \cdot \cot 45° = 1 \cdot 1 \cdot 1 = 1 \quad [\because \cot 45° = 1]$

Fast Track Practice

Exercise 1 Base Level Questions

1. If $\sin\theta = \dfrac{a^2-1}{a^2+1}$, then the value of $\sec\theta + \tan\theta$ will be [SSC FCI 2012]
   (a) $\dfrac{a}{\sqrt{2}}$ (b) $\dfrac{a}{a^2+1}$
   (c) $\sqrt{2}a$ (d) a

2. If $\dfrac{\sec\theta + \tan\theta}{\sec\theta - \tan\theta} = \dfrac{5}{3}$, then $\sin\theta$ is equal to
   (a) 2/3 (b) 3/4
   (c) 1/4 (d) 1/3
   (e) None of these

3. If α, β and γ are acute angled such that $\sin\alpha = \dfrac{\sqrt{3}}{2}$, $\cos\beta = \dfrac{\sqrt{3}}{2}$ and $\tan\gamma = 1$, then what is $\alpha + \beta + \gamma$ equal to? [CDS 2013]
   (a) 105° (b) 120°
   (c) 135° (d) 150°

4. If $\alpha + \beta = 90°$ and $\alpha : \beta = 2 : 1$, then the value of $\sin\alpha : \sin\beta$ is [SSC FCI 2012]
   (a) $\sqrt{3} : 1$ (b) $2 : 1$
   (c) $1 : 1$ (d) $\sqrt{2} : 1$

5. If $\sin A = \dfrac{5}{13}$, then what is the value of $\tan A + \sec A$? [CMAT 2015]
   (a) 1.8 (b) 1 (c) 1.5 (d) 2

6. Consider the following statements
   I. $\sin 1° > \sin 1^c$ II. $\cos 1° < \cos 1^c$
   III. $\tan 1° > \tan 1^c$
   Which of the above are not correct? [CDS 2016 (II)]
   (a) I and II (b) II and III
   (c) I and III (d) I, II and III

7. If $\tan\theta = \dfrac{4}{3}$, then the value of $\dfrac{3\sin\theta + 2\cos\theta}{3\sin\theta - 2\cos\theta}$ is [SSC (10+2) 2011]
   (a) 0.5 (b) −0.5
   (c) 3 (d) −3.0

8. If $\sec\theta - \csc\theta = 0$, then the value of $\tan\theta + \cot\theta$ is [SSC (10+2) 2013]
   (a) 0 (b) 1
   (c) −1 (d) 2

9. If $\tan^2 x + \dfrac{1}{\tan^2 x} = 2$ and $0° < x < 90°$, then what is the value of x? [CDS 2016 (II)]
   (a) 15° (b) 30° (c) 45° (d) 60°

10. If $\cos\theta \geq 1/2$ in the first quadrant, which one of the following is correct? [CDS 2010]
    (a) $\theta \leq \dfrac{\pi}{3}$ (b) $\theta \geq \dfrac{\pi}{3}$ (c) $\theta \leq \dfrac{\pi}{6}$ (d) $\theta \geq \dfrac{\pi}{6}$

11. If $\sin\theta - \cos\theta = 0$, then what is $\sin^4\theta + \cos^4\theta$ equal to? [CDS 2013]
    (a) 1 (b) $\dfrac{3}{4}$ (c) $\dfrac{1}{2}$ (d) $\dfrac{1}{4}$

12. If $\cos\theta_1 + \cos\theta_2 + \cos\theta_3 = 3$, then what is $\sin\theta_1 + \sin\theta_2 + \sin\theta_3$ equal to? [CDS 2017 (I)]
    (a) 0 (b) 1 (c) 2 (d) 3

13. Consider the following statements
    I. $\dfrac{\cot 30° + 1}{\cot 30° - 1} = 2(\cos 30° + 1)$
    II. $2\sin 45° \cos 45° - \tan 45° \cot 45° = 0$
    Which of the above identities is/are correct? [CDS 2012]
    (a) Only I (b) Only II
    (c) Both I and II (d) Neither I nor II

14. If $0° < \theta < 90°$, then the value of $\sin\theta + \cos\theta$ is [SSC CGL 2012]
    (a) equal to 1 (b) greater than 1
    (c) less than 1 (d) equal to 2

15. If $2\cot\theta = 3$, then what is $\dfrac{2\cos\theta - \sin\theta}{2\cos\theta + \sin\theta}$ equal to? [CDS 2014]
    (a) $\dfrac{2}{3}$ (b) $\dfrac{1}{3}$ (c) $\dfrac{1}{2}$ (d) $\dfrac{3}{4}$

16. If $\dfrac{\sin\theta + \cos\theta}{\sin\theta - \cos\theta} = \dfrac{5}{4}$, then the value of $\dfrac{\tan^2\theta + 1}{\tan^2\theta - 1}$ is [SSC (10+2) 2012]
    (a) $\dfrac{25}{16}$ (b) $\dfrac{41}{9}$ (c) $\dfrac{41}{40}$ (d) $\dfrac{40}{41}$

17. If $\dfrac{\cos^2\theta}{\cot^2\theta - \cos^2\theta} = 3$ and $0° < \theta < 90°$, then the value of θ is [SSC (10+2) 2011]
    (a) 30° (b) 45°
    (c) 60° (d) None of these

18. If $\tan(A+B) = \sqrt{3}$ and $\tan A = 1$, then $\tan(A-B)$ is equal to [SSC (10+2) 2014]
(a) 0 (b) 1 (c) $\dfrac{1}{\sqrt{3}}$ (d) $\sqrt{2}$

19. ABC is a right angled triangle at B and $AB:BC = 3:4$. What is $\sin A + \sin B + \sin C$ equal to? [CDS 2015 (I)]
(a) 2 (b) $\dfrac{11}{5}$ (c) $\dfrac{12}{5}$ (d) 3

20. If $\sin\theta + \cos\theta = 1$, what is the value of $\sin\theta\cos\theta$? [CDS 2009]
(a) 2 (b) 0 (c) 1 (d) 1/2

21. $\dfrac{\cos\theta}{1-\sin\theta}$ is equal to (where, $\theta \neq \dfrac{\pi}{2}$) [CDS 2015 (II)]
(a) $\dfrac{\tan\theta - 1}{\tan\theta + 1}$ (b) $\dfrac{1 + \sin\theta}{\cos\theta}$
(c) $\dfrac{\tan\theta + 1}{\tan\theta - 1}$ (d) $\dfrac{1 + \cos\theta}{\sin\theta}$

22. If $\sec\theta + \tan\theta = 2$, then what is the value of $\sec\theta$? [CDS 2014]
(a) $\dfrac{3}{2}$ (b) $\sqrt{2}$ (c) $\dfrac{5}{2}$ (d) $\dfrac{5}{4}$

23. The numerical value of $\dfrac{5}{\sec^2\theta} + \dfrac{2}{1+\cot^2\theta} + 3\sin^2\theta$ is [SSC (10+2) 2012]
(a) 5 (b) 2
(c) 3 (d) 4

24. $\cot^2 A \cos^2 A$ is equal to [SSC (10+2) 2017]
(a) $\cot^2 A + \cos^2 A$ (b) $\tan^2 A - \cos^2 A$
(c) $\cot^2 A - \cos^2 A$ (d) $\tan^2 A + \cos^2 A$

25. The expression $\sin^2 x + \cos^2 x - 1 = 0$ is satisfied by how many values of x?
(a) Only one value of x [CDS 2012]
(b) Two values of x
(c) Infinite values of x
(d) No value of x

26. If $\sec^2\theta + \tan^2\theta = \dfrac{7}{12}$, then $\sec^4\theta - \tan^4\theta$ is equal to [SSC FCI 2012]
(a) $\dfrac{7}{12}$ (b) $\dfrac{1}{2}$ (c) $\dfrac{5}{12}$ (d) 1

27. The numerical value of $\left(\dfrac{1}{\cos\theta} + \dfrac{1}{\cot\theta}\right)\left(\dfrac{1}{\cos\theta} - \dfrac{1}{\cot\theta}\right)$ is [SSC (10+2) 2012]
(a) 0 (b) −1 (c) 1 (d) 2

28. If $\cos\theta + \sin\theta = \sqrt{2}\cos\theta$, then $\cos\theta - \sin\theta$ is [SSC CGL 2013]
(a) $-\sqrt{2}\cos\theta$ (b) $-\sqrt{2}\sin\theta$
(c) $\sqrt{2}\sin\theta$ (d) $\sqrt{2}\tan\theta$

29. What is $\sin 25° \sin 35° \sec 65° \sec 55°$ equal to? [CDS 2013]
(a) −1 (b) 0 (c) 1/2 (d) 1

30. $\sqrt{\dfrac{1-\cos A}{1+\cos A}}$ is equal to [SSC CGL (Pre) 2016]
(a) $\operatorname{cosec} A - \cot A$ (b) 0
(c) $\sec A - \cot A$ (d) 1

31. The expression $\dfrac{\tan 57° + \cot 37°}{\tan 33° + \cot 53°}$ is equal to [SSC CGL 2012]
(a) $\tan 33° \cot 57°$ (b) $\tan 57° \cot 37°$
(c) $\tan 33° \cot 53°$ (d) $\tan 53° \cot 37°$

32. $\tan 4° \cdot \tan 43° \cdot \tan 47° \cdot \tan 86°$ is equal to [SSC CPO 2011]
(a) 2 (b) 3 (c) 1 (d) 4

33. Consider the following statements
I. $\operatorname{cosec}^2 x + \sec^2 x = \operatorname{cosec}^2 x \sec^2 x$
II. $\sec^2 x + \tan^2 x = \sec^2 x \tan^2 x$
III. $\operatorname{cosec}^2 x + \tan^2 x = \cot^2 x + \sec^2 x$
Which of the above statements are correct? [CDS 2009]
(a) I and II (b) II and III
(c) I and III (d) I, II and III

34. If $\dfrac{\cos x}{1+\operatorname{cosec} x} + \dfrac{\cos x}{\operatorname{cosec} x - 1} = 2$, which one of the following is one of the value of x? [CDS 2009]
(a) $\dfrac{\pi}{2}$ (b) $\dfrac{\pi}{3}$ (c) $\dfrac{\pi}{4}$ (d) $\dfrac{\pi}{6}$

35. The value of $\dfrac{\tan 27° + \cot 63°}{\tan 27°(\sin 25° + \cos 65°)}$ is [SSC CPO 2013]
(a) $\operatorname{cosec} 25°$ (b) $2\tan 27°$
(c) $\sin 25°$ (d) $\tan 65°$

36. $\cot 10° \cdot \cot 20° \cdot \cot 60° \cdot \cot 70° \cdot \cot 80°$ is equal to [SSC (10+2) 2011]
(a) 1 (b) −1 (c) $\sqrt{3}$ (d) $1/\sqrt{3}$

37. If $\cos\theta \operatorname{cosec} 23° = 1$, then the value of θ is [SSC (10+2) 2012]
(a) 23° (b) 37° (c) 63° (d) 67°

38. If $0 < x < \dfrac{\pi}{2}$ and $\sec x = \operatorname{cosec} y$, then the value of $\sin(x+y)$ is [SSC CGL 2013]
(a) 0 (b) 1 (c) $\dfrac{1}{2}$ (d) $\dfrac{1}{\sqrt{3}}$

39. If $\angle A$ and $\angle B$ are complementary to each other, then the value of $\sec^2 A + \sec^2 B - \sec^2 A \sec^2 B$ is [SSC CGL 2012]
(a) 1 (b) −1 (c) 2 (d) 0

40. If $\sin(3x - 20°) = \cos(3y + 20°)$, then the value of $(x + y)$ is [SSC (10+2) 2012]
(a) 20° (b) 30° (c) 40° (d) 45°

41. The numerical value of
$\cot 18° \left[\cot 72° \cos^2 22° + \dfrac{1}{\tan 72° \sec^2 68°} \right]$
is [SSC (10+2) 2011]
(a) 1 (b) $\sqrt{2}$ (c) 3 (d) $\dfrac{1}{\sqrt{3}}$

42. The value of $\text{cosec}^2 67° + \sec^2 57° - \cot^2 33° - \tan^2 23°$ is [CDS 2015 (I)]
(a) $2\sqrt{2}$ (b) 2
(c) $\sqrt{2}$ (d) 0

43. If $\sin x + \sin^2 x = 1$, then what is the value of $\cos^8 x + 2\cos^6 x + \cos^4 x$? [CDS 2016 (II)]
(a) 0 (b) 1 (c) 2 (d) 4

44. If $\sin\theta + \cos\theta = \dfrac{\sqrt{7}}{2}$, then what is $\sin\theta - \cos\theta$ equal to? [CDS 2016 (I)]
(a) 0 (b) 1/2 (c) 1 (d) $\sqrt{2}$

45. If $x = a\cos\theta + b\sin\theta$ and $y = a\sin\theta - b\cos\theta$, then what is $x^2 + y^2$ equal to? [CDS 2017 (I)]
(a) $2ab$ (b) $a + b$
(c) $a^2 + b^2$ (d) $a^2 - b^2$

46. If $7\sin^2 x + 3\cos^2 x = 4$, $0 < x < 90°$, then what is the value of $\tan x$? [CDS 2017 (I)]
(a) $\sqrt{2}$ (b) 1 (c) $\dfrac{\sqrt{3}}{2}$ (d) $\dfrac{1}{\sqrt{3}}$

47. If $x = a\sec\theta\cos\phi$, $y = b\sec\theta\sin\phi$ and $z = c\tan\theta$, then the value of $\dfrac{x^2}{a^2} + \dfrac{y^2}{b^2} - \dfrac{z^2}{c^2}$ is [RRB 2006; SSC CGL 2013]
(a) 9 (b) 0 (c) 1 (d) 4

48. If $\cos^4\theta - \sin^4\theta = 2/3$, then the value of $1 - 2\sin^2\theta$ is [SSC CGL 2013; SSC (10+2) 2011]
(a) 0 (b) $\dfrac{2}{3}$ (c) $\dfrac{1}{3}$ (d) $\dfrac{4}{3}$

49. If $7\sin^2\theta + 3\cos^2\theta = 4$, then the value of $\cos\theta$ $(0° \leq \theta \leq 90°)$ is [SSC CPO 2013]
(a) $\dfrac{\sqrt{6}}{2}$ (b) $\dfrac{\sqrt{3}}{2}$ (c) $\dfrac{\sqrt{2}}{2}$ (d) $\dfrac{\sqrt{5}}{2}$

50. If $x\sin\theta = y\cos\theta = \dfrac{2z\tan\theta}{1-\tan^2\theta}$, then what is $4z^2(x^2 + y^2)$ equal to? [CDS 2017 (I)]
(a) $(x^2 + y^2)^3$ (b) $(x^2 - y^2)^2$
(c) $(x^2 - y^2)^3$ (d) $(x^2 + y^2)^2$

51. If $2(\cos^2\theta - \sin^2\theta) = 1$ (θ is a positive acute angle), then $\cot\theta$ is equal to [SSC (10+2) 2013]
(a) $\dfrac{1}{\sqrt{3}}$ (b) 1
(c) $\sqrt{3}$ (d) $-\sqrt{3}$

52. If $\sec A + \tan A = a$, then the value of $\cos A$ is [SSC CGL (Mains) 2016]
(a) $\dfrac{a^2 + 1}{2a}$ (b) $\dfrac{2a}{a^2 + 1}$
(c) $\dfrac{a^2 - 1}{2a}$ (d) $\dfrac{2a}{a^2 - 1}$

53. If θ is a positive acute angle and $\tan 2\theta \tan 3\theta = 1$, then the value of $\left(2\cos^2\dfrac{5\theta}{2} - 1\right)$ is [SSC CGL 2012]
(a) $-\dfrac{1}{2}$ (b) 1
(c) 0 (d) $\dfrac{1}{2}$

54. If $\dfrac{\sin\theta + \cos\theta}{\sin\theta - \cos\theta} = 3$, then the value of $\sin^4\theta - \cos^4\theta$ is [SSC (10+2) 2011]
(a) $\dfrac{1}{5}$ (b) $\dfrac{2}{5}$
(c) $\dfrac{3}{5}$ (d) $\dfrac{4}{5}$

55. If θ is acute and $\tan\theta + \cot\theta = 2$, then the value of $\tan^5\theta + \cot^{10}\theta$ is [SSC (10+2) 2011]
(a) 1 (b) 2 (c) 3 (d) 4

56. If $0° < \theta < 90°$, then all the trigonometric ratios can be obtained when [CDS 2012]
(a) only $\sin\theta$ is given
(b) only $\cos\theta$ is given
(c) only $\tan\theta$ is given
(d) any one of the six ratios is given

57. If $\sin\theta\cos\theta = \sqrt{3}/4$, then the value of $\sin^4\theta + \cos^4\theta$ is [CDS 2012]
(a) 7/8 (b) 5/8 (c) 3/8 (d) 1/8

58. If $3\sin x + 5\cos x = 5$, then what is the value of $(3\cos x - 5\sin x)$? [CDS 2012]
(a) 0 (b) 2 (c) 3 (d) 5

59. If $\sin 17° = x/y$, then the value of $\sec 17° - \sin 73°$ is [SSC FCI 2012]
(a) $\dfrac{y^2 - x^2}{xy}$ (b) $\dfrac{x^2}{\sqrt{y^2 - x^2}}$
(c) $\dfrac{x^2}{y\sqrt{y^2 + x^2}}$ (d) $\dfrac{x^2}{y\sqrt{y^2 - x^2}}$

714 / Fast Track Objective Arithmetic

60. If $2\sin\left(\dfrac{\pi x}{2}\right) = x^2 + \dfrac{1}{x^2}$, then the value of $\left(x - \dfrac{1}{x}\right)$ is   [SSC CGL 2012]
    (a) -1 (b) 2 (c) 1 (d) 0

61. The value of $\sin^2 1° + \sin^2 3° + \sin^2 5° + \cdots + \sin^2 87° + \sin^2 89°$ is   [SSC (10+2) 2013]
    (a) 22 (b) $22\dfrac{1}{2}$ (c) 23 (d) $22\dfrac{1}{4}$

62. The value of $\sin^2 5° + \sin^2 10° + \sin^2 15° + \cdots + \sin^2 85° + \sin^2 90°$ is   [SSC (10+2) 2012, 2011]
    (a) $7\dfrac{1}{2}$ (b) $8\dfrac{1}{2}$ (c) $10\dfrac{1}{2}$ (d) $9\dfrac{1}{2}$

63. The value of $\dfrac{\sin 39°}{\cos 51°} + 2\tan 11°$ $\tan 31° \tan 45° \tan 59° \tan 79° - 3(\sin^2 21° + \sin^2 69°)$ is   [SSC (10+2) 2011]
    (a) 2 (b) -1 (c) 1 (d) 0

64. The minimum value of $\sin^2\theta + \cos^4\theta$ is   [SSC CPO 2013]
    (a) $\dfrac{1}{\sqrt{2}}$ (b) $\dfrac{3}{5}$ (c) $\dfrac{3}{4}$ (d) $\dfrac{2}{3}$

65. What is the minimum value of $9\tan^2\theta + 4\cot^2\theta$?   [CDS 2017 (I)]
    (a) 6 (b) 9 (c) 12 (d) 13

66. The minimum value of $\cos^2\theta + \sec^2\theta$ is   [SSC (10+2) 2013]
    (a) 0 (b) 1 (c) 2 (d) 3

67. If $\cos x + \cos y = 2$, then the value of $\sin x + \sin y$ is   [SSC FCI 2012]
    (a) 0 (b) 1 (c) 2 (d) -1

68. The equation $\cos^2\theta = \dfrac{(x+y)^2}{4xy}$ is only possible when   [SSC (10+2) 2013]
    (a) $x > y$ (b) $x = y$ (c) $x < y$ (d) $x = -y$

69. What is the value of $\tan 1° \tan 2° \tan 3° \tan 4° \ldots \tan 89°$?   [CDS 2016 (II)]
    (a) 0 (b) 1 (c) 2 (d) $\sqrt{3}$

Exercise 2 Higher Skill Level Questions

1. If $(\sin x + \sin y) = a$ and $(\cos x + \cos y) = b$, what is the value of $\sin x \sin y + \cos x \cos y$?
   (a) $a + b - ab$ (b) $a + b + ab$
   (c) $a^2 + b^2 - 2$ (d) $\dfrac{a^2 + b^2 - 2}{2}$
   (e) None of these

2. Consider the following statements
   I. There exists no value of x such that $\dfrac{1}{1 - \sin x} = 4 + 2\sqrt{3}$, $0 < x < \dfrac{\pi}{2}$
   II. $\sin x = 3^{\sin^2 x}$ does not hold good for any real x.
   Which of the above statement(s) is/are correct?   [CDS 2015 (I)]
   (a) Only I (b) Only II
   (c) Both I and II (d) Neither I nor II

3. If $\sin x + \cos x = c$, then $\sin^6 x + \cos^6 x$ is equal to   [CDS 2015 (I)]
   (a) $\dfrac{1 + 6c^2 - 3c^4}{16}$ (b) $\dfrac{1 + 6c^2 - 3c^4}{4}$
   (c) $\dfrac{1 + 6c^2 + 3c^4}{16}$ (d) $\dfrac{1 + 6c^2 + 3c^4}{4}$

4. If $\sin(10°\ 6'\ 32'') = a$, then the value of $\cos(79°\ 53'\ 28'') + \tan(10°\ 6'\ 32'')$ is   [SSC CGL (Mains) 2012]

   (a) $\dfrac{a(1 + \sqrt{1-a^2})}{\sqrt{1-a^2}}$ (b) $\dfrac{1 + \sqrt{1-a^2}}{\sqrt{1-a^2}}$
   (c) $\dfrac{\sqrt{1-a^2} + a}{\sqrt{1-a^2}}$ (d) $\dfrac{a\sqrt{1-a^2} + 1}{\sqrt{1-a^2}}$

5. If $\cos\theta + \sec\theta = 2$, then the value of $\cos^6\theta + \sec^6\theta$ is   [SSC (10+2) 2012]
   (a) 1 (b) 2 (c) 4 (d) 8

6. The maximum value of $\sin^8\theta + \cos^{14}\theta$, for all real values of θ is   [SSC CGL (Mains) 2012]
   (a) 1 (b) $\sqrt{2}$ (c) $\dfrac{1}{\sqrt{2}}$ (d) 0

7. What is $\csc(75° + \theta) - \sec(15° - \theta) - \tan(55° + \theta) + \cot(35° - \theta)$?   [CDS 2014]
   (a) -1 (b) 0
   (c) 1 (d) 3/2

8. Consider the following statements
   I. $\sqrt{\dfrac{1 - \cos\theta}{1 + \cos\theta}} = \csc\theta - \cot\theta$
   II. $\sqrt{\dfrac{1 + \cos\theta}{1 - \cos\theta}} = \csc\theta + \cot\theta$
   Which of the above identity/identities?   [CDS 2016 (I)]
   (a) Only I (b) Only II
   (c) Both I and II (d) Neither I nor II

Trigonometry / 715

9. What is $\dfrac{(\sin\theta + \cos\theta)(\tan\theta + \cot\theta)}{\sec\theta + \csc\theta}$ equal to? [CDS 2013]
   (a) 1 (b) 2 (c) $\sin\theta$ (d) $\cos\theta$

10. If $5\sin\theta + 12\cos\theta = 13$, then what is $5\cos\theta - 12\sin\theta$ equal to? [CDS 2013]
    (a) -2 (b) -1
    (c) 0 (d) 1

11. If $\sin(A+B) = 1$, where $0° < B < 45°$, what is $\cos(A-B)$ equal to? [CDS 2014]
    (a) $\sin 2B$ (b) $\sin B$
    (c) $\cos 2B$ (d) $\cos B$

12. If α and β are complementary angles, then what is $\sqrt{\cos\alpha \csc\beta - \cos\alpha\sin\beta}$ equal to? [CDS 2014]
    (a) $\sec\beta$ (b) $\cos\alpha$ (c) $\sin\alpha$ (d) $-\tan\beta$

13. $2\csc^2 23° \cot^2 67° - \sin^2 23° - \sin^2 67° - \cot^2 67°$ is equal to [SSC (10+2) 2013]
    (a) $\sec^2 23°$ (b) $\tan^2 23°$
    (c) 0 (d) 1

14. The value of $\cot\theta \cdot \tan(90° - \theta) - \sec(90° - \theta)\csc\theta + (\sin^2 25° + \sin^2 65°) + \sqrt{3}(\tan 5° \cdot \tan 15° \cdot \tan 30° \cdot \tan 75° \cdot \tan 85°)$ is [SSC (10+2) 2012]
    (a) 1 (b) -1 (c) 2 (d) 0

15. The simplified form of the given expression $\sin A \cos A (\tan A - \cot A)$ is (where $0° \leq A \leq 90°$) [SSC CGL (Pre) 2015]
    (a) 1 (b) $1 - \cos^2 A$
    (c) $1 - 2\sin^2 A$ (d) $2\sin^2 A - 1$

16. The minimum value of $\cos^2 x + \cos^2 y - \cos^2 z$ is [CDS 2015 (II)]
    (a) -1 (b) 0 (c) 2 (d) -2

17. The simplified value of $(\sec x \sec y + \tan x \tan y)^2 - (\sec x \tan y + \tan x \sec y)^2$ is [SSC (10+2) 2011]
    (a) -1 (b) 0 (c) $\sec^2 x$ (d) 1

18. If $\tan 15° = 2 - \sqrt{3}$, then the value of $\tan 15° \cot 75° + \tan 75° \cot 15°$ is [SSC (10+2) 2011]
    (a) 14 (b) 12 (c) 10 (d) 8

19. If θ is an acute angle and $\sin\theta \cos\theta = 2\cos^3\theta - 1.5\cos\theta$, then what is $\sin\theta$ equal to? [CDS 2015 (II)]
    (a) $\dfrac{\sqrt{5} - 1}{4}$ (b) $\dfrac{1 - \sqrt{5}}{4}$
    (c) $\dfrac{\sqrt{5} + 1}{4}$ (d) $-\dfrac{\sqrt{5} + 1}{4}$

20. Which of the following is correct in respect of the equation $3 - \tan^2\theta = \alpha(1 - 3\tan^2\theta)$? (Given that α is a real number). [CDS 2016 (I)]
    (a) $\alpha \in \left[\dfrac{1}{3}, 3\right]$
    (b) $\alpha \in \left(-\infty, \dfrac{1}{3}\right] \cup [3, \infty)$
    (c) $\alpha \in \left(\infty, \dfrac{1}{3}\right] \cup [3, \infty)$
    (d) None of the above

21. Consider the following statements
    I. There exists atleast one value of x between 0 and $\pi/2$, which satisfies the equation $\sin^4 x - 2\sin^2 x - 1 = 0$
    II. $\sin 1.5$ is greater than $\cos 1.5$.
    Which of the above statement(s) is/are correct? [CDS 2015 (I)]
    (a) Only I (b) Only II
    (c) Both I and II (d) Neither I nor II

22. What is $\dfrac{(1 + \sec\theta - \tan\theta)\cos\theta}{(1 + \sec\theta + \tan\theta)(1 - \sin\theta)}$ equal to? [CDS 2013]
    (a) 1 (b) 2 (c) $\tan\theta$ (d) $\cot\theta$

23. If $\sin\theta \cos\theta = 1/2$, then what is $\sin^6\theta + \cos^6\theta$ equal to? [CDS 2014]
    (a) 1 (b) 2 (c) 3 (d) 1/4

24. If $\sin\theta + \cos\theta = \sqrt{3}$, then what is $\tan\theta + \cot\theta$ equal to? [CDS 2013]
    (a) 1 (b) $\sqrt{2}$
    (c) 2 (d) $\sqrt{3}$

25. If $\cos x + \sec x = 2$, then what $\cos^n x + \sec^n x$ equal to, where n is a positive integer? [CDS 2014]
    (a) 2 (b) 2^{n-2}
    (c) 2^{n-1} (d) 2^n

26. If $\sin\theta - \cos\theta = 0$, find the value of $\sin\left(\dfrac{\pi}{2} - \theta\right) + \cos\left(\dfrac{\pi}{2} - \theta\right)$ [SSC (10+2) 2012]
    (a) 0 (b) 1
    (c) $\sqrt{2}$ (d) $2\sqrt{2}$

27. If $a\sin\theta + b\cos\theta = c$, then the value of $a\cos\theta - b\sin\theta$ is [SSC CGL 2013]
    (a) $\pm\sqrt{a^2 - b^2 - c^2}$
    (b) $\pm\sqrt{a^2 - b^2 + c^2}$
    (c) $\pm\sqrt{-a^2 + b^2 + c^2}$
    (d) $\pm\sqrt{a^2 + b^2 - c^2}$

28. If $\tan\alpha = n\tan\beta$ and $\sin\alpha = m\sin\beta$, then $\cos^2\alpha$ is [SSC CGL 2013]
   (a) $\dfrac{m^2}{n^2}$ (b) $\dfrac{m^2-1}{n^2-1}$
   (c) $\dfrac{m^2+1}{n^2+1}$ (d) $\dfrac{m^2}{n^2+1}$

29. If $l\cos^2\theta + m\sin^2\theta = \dfrac{\cos^2\theta\,(\csc^2\theta+1)}{\csc^2\theta-1}$, $0° < \theta < 90°$, then $\tan\theta$ is equal to [SSC CGL (Mains) 2012]
   (a) $\sqrt{\dfrac{l-2}{1-m}}$ (b) $\sqrt{\dfrac{2-l}{1-m}}$
   (c) $\sqrt{\dfrac{l-2}{m-1}}$ (d) $\sqrt{\dfrac{l-1}{2-m}}$

30. If $\sin\theta + \cos\theta = \dfrac{1+\sqrt{3}}{2}$, where $0 < \theta < \dfrac{\pi}{2}$, then what is $\tan\theta + \cot\theta$ equal to? [CDS 2016 (II)]
   (a) $\dfrac{\sqrt{3}}{4}$ (b) $\dfrac{1}{\sqrt{3}}$ (c) $\sqrt{3}$ (d) $\dfrac{4}{\sqrt{3}}$

31. If $\cos x + \cos^2 x = 1$, then numerical value of $(\sin^{12}x + 3\sin^{10}x + 3\sin^8 x + \sin^6 x - 1)$ is [SSC CGL 2013]
   (a) 0 (b) 1
   (c) –1 (d) 2

32. If $\tan\theta + \sec\theta = m$, then what is $\sec\theta$ equal to? [CDS 2013]
   (a) $\dfrac{m^2-1}{2m}$ (b) $\dfrac{m^2+1}{2m}$
   (c) $\dfrac{m+1}{m}$ (d) $\dfrac{m^2+1}{m}$

33. If $x\sin\theta - y\cos\theta = \sqrt{x^2+y^2}$ and $\dfrac{\cos^2\theta}{a^2} + \dfrac{\sin^2\theta}{b^2} = \dfrac{1}{x^2+y^2}$, then the correct relation is [SSC (10+2) 2013]
   (a) $\dfrac{x^2}{a^2} + \dfrac{y^2}{b^2} = 1$ (b) $\dfrac{x^2}{b^2} + \dfrac{y^2}{a^2} = 1$
   (c) $\dfrac{x^2}{a^2} - \dfrac{y^2}{b^2} = 1$ (d) $\dfrac{x^2}{b^2} - \dfrac{y^2}{a^2} = 1$

34. If $\sin\theta + 2\cos\theta = -1$, where $0° < \theta < \dfrac{\pi}{2}$, what is $2\sin\theta - \cos\theta$ equal to? [CDS 2014]
   (a) –1 (b) 1/2
   (c) 2 (d) 1

35. In a right angled $\triangle XYZ$, right angled at Y, if $XY = 2\sqrt{6}$ and $XZ - YZ = 2$, then $\sec X + \tan X$ is [SSC CGL 2012]
   (a) $\dfrac{1}{\sqrt{6}}$ (b) $\sqrt{6}$
   (c) $2\sqrt{6}$ (d) $\dfrac{\sqrt{6}}{2}$

36. If $2y\cos\theta = x\sin\theta$ and $2x\sec\theta - y\csc\theta = 3$, then what is $x^2 + 4y^2$ equal to? [CDS 2016 (II)]
   (a) 1 (b) 2
   (c) 4 (d) 8

37. If the angles of a triangle are in the ratio $4:1:1$. Then, the ratio of the largest side to the perimeter is [CDS 2015 (I)]
   (a) $\dfrac{2}{3}$ (b) $\dfrac{1}{2+\sqrt{3}}$
   (c) $\dfrac{\sqrt{3}}{2+\sqrt{3}}$ (d) $\dfrac{2}{1+\sqrt{3}}$

38. A square is inscribed in a right triangle with legs x and y and has common right angle with the triangle. The perimeter of the square is given by [CDS 2015 (II)]
   (a) $\dfrac{2xy}{x+y}$
   (b) $\dfrac{4xy}{x+y}$
   (c) $\dfrac{2xy}{\sqrt{x^2+y^2}}$
   (d) $\dfrac{4xy}{\sqrt{x^2+y^2}}$

39. $\triangle ABC$ is an isosceles right angled triangle having $\angle C = 90°$. If D is any point on AB, then $AD^2 + BD^2$ is equal to [SSC CGL (Mains) 2016]
   (a) CD^2 (b) $2CD^2$
   (c) $3CD^2$ (d) $4CD^2$

Answer with Solutions
Exercise 1 Base Level Questions

1. (d) $\because \sin\theta = \dfrac{a^2-1}{a^2+1}$

In $\triangle ABC$,
$AC^2 = AB^2 + BC^2$
$\Rightarrow BC^2 = AC^2 - AB^2$
$BC = \sqrt{AC^2 - AB^2}$
$= \sqrt{(a^2+1)^2 - (a^2-1)^2}$
$= \sqrt{a^4 + 1 + 2a^2 - a^4 - 1 + 2a^2}$
$= \sqrt{4a^2} = 2a$

$\therefore \sec\theta + \tan\theta = \dfrac{a^2+1}{2a} + \dfrac{a^2-1}{2a} = \dfrac{2a^2}{2a} = a$

2. (c) $\because \dfrac{\sec\theta + \tan\theta}{\sec\theta - \tan\theta} = \dfrac{5}{3}$

$\Rightarrow \dfrac{\dfrac{1}{\cos\theta} + \dfrac{\sin\theta}{\cos\theta}}{\dfrac{1}{\cos\theta} - \dfrac{\sin\theta}{\cos\theta}} = \dfrac{5}{3}$

$\Rightarrow \dfrac{\dfrac{1+\sin\theta}{\cos\theta}}{\dfrac{1-\sin\theta}{\cos\theta}} = \dfrac{5}{3}$

$\Rightarrow \dfrac{1+\sin\theta}{1-\sin\theta} = \dfrac{5}{3}$

$\Rightarrow (1+\sin\theta) \times 3 = 5(1-\sin\theta)$
$\Rightarrow 3 + 3\sin\theta = 5 - 5\sin\theta$
$\Rightarrow 3\sin\theta + 5\sin\theta = 5 - 3$
$\Rightarrow 8\sin\theta = 2$
$\therefore \sin\theta = \dfrac{2}{8} = \dfrac{1}{4}$

3. (c) Given, $\sin\alpha = \dfrac{\sqrt{3}}{2}$
$\therefore \alpha = 60°$ $\left[\because \sin 60 = \dfrac{\sqrt{3}}{2}\right]$

and $\cos\beta = \dfrac{\sqrt{3}}{2}$
$\therefore \beta = 30°$ $\left[\because \cos 30° = \dfrac{\sqrt{3}}{2}\right]$

Also, $\tan\gamma = 1$
$\therefore \gamma = 45°$ $[\because \tan 45° = 1]$

So, $\alpha + \beta + \gamma = 60° + 30° + 45° = 135°$

4. (a) Given, $\alpha + \beta = 90°$ and $\alpha : \beta = 2 : 1$
$\therefore \alpha = \dfrac{2}{2+1} \times 90° = \dfrac{2}{3} \times 90° = 60°$

Similarly, $\beta = \dfrac{1}{3} \times 90° = 30°$

$\therefore \sin\alpha : \sin\beta = \sin 60° : \sin 30°$
$= \dfrac{\sqrt{3}}{2} : \dfrac{1}{2} = \sqrt{3} : 1$

5. (c) Given, $\sin A = \dfrac{5}{13}$

$\therefore \cos A = \sqrt{1 - \sin^2 A} = \sqrt{1 - \dfrac{5^2}{13^2}}$

$= \sqrt{\dfrac{169-25}{169}} = \sqrt{\dfrac{144}{169}} = \dfrac{12}{13}$

$\therefore \tan A = \dfrac{\sin A}{\cos A} = \dfrac{5}{13} \times \dfrac{13}{12} = \dfrac{5}{12}$

and $\sec A = \dfrac{1}{\cos A} = \dfrac{1}{12/13} = \dfrac{13}{12}$

$\therefore \tan A + \sec A = \dfrac{5}{12} + \dfrac{13}{12} = \dfrac{18}{12} = 1.5$

6. (d) We know that,
1 radian $= \dfrac{180}{\pi}$ degree $= 57°17'45''$

Now,
I. $\sin 1° > \sin 1^c \Rightarrow \sin 1° > \sin\dfrac{180°}{\pi}$

False, because $\sin\theta$ is an increasing function for $\theta \in \left[0, \dfrac{\pi}{2}\right]$.

II. $\cos 1° < \cos\dfrac{180°}{\pi}$

False, because $\cos\theta$ is a decreasing function for $\theta \in \left[0, \dfrac{\pi}{2}\right]$.

III. $\tan 1° > \tan\dfrac{180°}{\pi}$

False, since $\tan\theta$ is an increasing function for $\theta \in \left[0, \dfrac{\pi}{2}\right]$.

7. (c) $\because \tan\theta = \dfrac{4}{3}$ [given]

$\therefore \dfrac{3\sin\theta + 2\cos\theta}{3\sin\theta - 2\cos\theta} = \dfrac{3\tan\theta + 2}{3\tan\theta - 2}$

[dividing by $\cos\theta$]

$= \dfrac{3 \times \dfrac{4}{3} + 2}{3 \times \dfrac{4}{3} - 2} = \dfrac{4+2}{4-2} = 3$

8. (d) $\because \sec\theta - \text{cosec}\,\theta = 0 \Rightarrow \sec\theta = \text{cosec}\,\theta$

$\Rightarrow \dfrac{1}{\cos\theta} = \dfrac{1}{\sin\theta} \Rightarrow \dfrac{\sin\theta}{\cos\theta} = 1$

$\Rightarrow \tan\theta = 1$

$\Rightarrow \tan\theta = \tan 45° \Rightarrow \theta = 45°$

$\therefore \tan\theta + \cot\theta = \tan 45° + \cot 45°$
$= 1 + 1 = 2$

9. (c) Given, $\tan^2 x + \dfrac{1}{\tan^2 x} = 2$

$\Rightarrow \tan^4 x + 1 - 2\tan^2 x = 0$

$\Rightarrow (\tan^2 x - 1)^2 = 0$

$\Rightarrow \tan^2 x - 1 = 0$

$\Rightarrow \tan^2 x = 1 \Rightarrow \tan x = \pm 1$

$\therefore \tan x = \tan 45°$

$x = 45°$

10. (a) $\cos\theta \geq \dfrac{1}{2}$ means the value of θ lies between $0°$ and $\dfrac{\pi}{3}$.

$\therefore \theta$ is less than or equal to $\dfrac{\pi}{3}$, i.e. $\theta \leq \dfrac{\pi}{3}$.

[as $\cos\theta$ is a decreasing function]

11. (c) $\because \sin\theta - \cos\theta = 0$

$\Rightarrow \sin\theta = \cos\theta$

Since, $\sin\theta$ and $\cos\theta$ are equal for $\theta = 45°$

So, $\sin^4\theta + \cos^4\theta = (\sin 45°)^4 + (\cos 45°)^4$

$= \left(\dfrac{1}{\sqrt{2}}\right)^4 + \left(\dfrac{1}{\sqrt{2}}\right)^4$

$= \dfrac{1}{4} + \dfrac{1}{4} = \dfrac{1+1}{4} = \dfrac{2}{4} = \dfrac{1}{2}$

12. (a) We have,

$\cos\theta_1 + \cos\theta_2 + \cos\theta_3 = 3$ ...(i)

Since, $0 \leq \cos\theta \leq 1$

$\therefore$ Eq. (i) to be true

$\cos\theta_1 = \cos\theta_2 = \cos\theta_3 = 1$

$\therefore \theta_1 = \theta_2 = \theta_3 = 2n\pi$

Now, $\sin\theta_1 + \sin\theta_2 + \sin\theta_3$

$= \sin 2n\pi + \sin 2n\pi + \sin 2n\pi$

$= 0 + 0 + 0$ $[\because \sin 2n\pi = 0]$

$= 0$

13. (c) Statement I

$\dfrac{\cot 30° + 1}{\cot 30° - 1} = 2(\cos 30° + 1)$

$\Rightarrow \dfrac{\sqrt{3}+1}{\sqrt{3}-1} = 2\left(\dfrac{\sqrt{3}}{2} + 1\right)$

$[\because \cot 30° = \sqrt{3}]$

$\Rightarrow \dfrac{\sqrt{3}+1}{\sqrt{3}-1} \times \dfrac{\sqrt{3}+1}{\sqrt{3}+1} = 2\left(\dfrac{\sqrt{3}+2}{2}\right)$

$\Rightarrow \dfrac{3 + 1 + 2\sqrt{3}}{3 - 1} = \sqrt{3} + 2$

$\Rightarrow \dfrac{4 + 2\sqrt{3}}{2} = \sqrt{3} + 2$

$\Rightarrow \dfrac{2(2+\sqrt{3})}{2} = \sqrt{3} + 2$

$\Rightarrow \sqrt{3} + 2 = \sqrt{3} + 2$

$\therefore$ It is true.

Statement II

$2\sin 45° \cos 45° - \tan 45° \cot 45° = 0$

$\Rightarrow 2 \times \left(\dfrac{1}{\sqrt{2}} \times \dfrac{1}{\sqrt{2}}\right) - 1 \times 1 = 0$

$\Rightarrow 2 \times \dfrac{1}{2} - 1 \times 1 = 0 \Rightarrow 1 - 1 = 0 \Rightarrow 0 = 0$

$\therefore$ Both Statements I and II are true.

14. (c) Let $x = \sin\theta + \cos\theta$

On squaring both sides, we get

$x^2 = \sin^2\theta + \cos^2\theta + 2\sin\theta \cdot \cos\theta$

$= 1 + 2\sin\theta \cdot \cos\theta$

$\because \quad 0 < \theta < 90°$

Value of both $\sin\theta$ and $\cos\theta$ will be less than 1.

$\Rightarrow \sin\theta\cos\theta < 1 \Rightarrow 2\sin\theta \cdot \cos\theta < 2$

$\Rightarrow 1 + 2\sin\theta\cos\theta < 1 + 2$

$\Rightarrow x^2 < 3 \Rightarrow x < \sqrt{3}$

15. (c) In $\triangle ABC$, $\cot\theta = \dfrac{3}{2}$

$\therefore AB = 3$ and $AC = 2$

By Pythagoras theorem,

$BC^2 = (2)^2 + (3)^2$

$BC = \sqrt{13}$

Now, $\cos\theta = \dfrac{3}{\sqrt{13}}$ and $\sin\theta = \dfrac{2}{\sqrt{13}}$

$\therefore \dfrac{2\cos\theta - \sin\theta}{2\cos\theta + \sin\theta} = \dfrac{\dfrac{6}{\sqrt{13}} - \dfrac{2}{\sqrt{13}}}{\dfrac{6}{\sqrt{13}} + \dfrac{2}{\sqrt{13}}} = \dfrac{4}{8} = \dfrac{1}{2}$

Alternate Method

Given, $\dfrac{2\cos\theta - \sin\theta}{2\cos\theta + \sin\theta}$

Dividing numerator and denominator by $\sin\theta$, we get

$\dfrac{2\cot\theta - 1}{2\cot\theta + 1} = \dfrac{2 \times \dfrac{3}{2} - 1}{2 \times \dfrac{3}{2} + 1} = \dfrac{3-1}{3+1} = \dfrac{2}{4} = \dfrac{1}{2}$

16. (c) Given, $\dfrac{\sin\theta + \cos\theta}{\sin\theta - \cos\theta} = \dfrac{5}{4}$

$\Rightarrow \dfrac{\cos\theta\left(\dfrac{\sin\theta}{\cos\theta} + 1\right)}{\cos\theta\left(\dfrac{\sin\theta}{\cos\theta} - 1\right)} = \dfrac{5}{4} \Rightarrow \dfrac{\tan\theta + 1}{\tan\theta - 1} = \dfrac{5}{4}$

$\Rightarrow 4\tan\theta + 4 = 5\tan\theta - 5 \Rightarrow \tan\theta = 9$

Trigonometry / 719

On putting the value of $\tan\theta$, we get

$$\frac{\tan^2\theta + 1}{\tan^2\theta - 1} = \frac{(9)^2 + 1}{(9)^2 - 1} = \frac{81 + 1}{81 - 1} = \frac{82}{80} = \frac{41}{40}$$

17. (c) $\dfrac{\cos^2\theta}{\cot^2\theta - \cos^2\theta} = 3$

$\Rightarrow \cos^2\theta = 3\cot^2\theta - 3\cos^2\theta$

$\Rightarrow 4\cos^2\theta = 3\cot^2\theta$

$\Rightarrow 4\cos^2\theta - \dfrac{3\cos^2\theta}{\sin^2\theta} = 0$

$\Rightarrow \cos^2\theta\left(4 - \dfrac{3}{\sin^2\theta}\right) = 0$

$\therefore 4 - \dfrac{3}{\sin^2\theta} = 0 \Rightarrow 4\sin^2\theta - 3 = 0$

$\Rightarrow \sin^2\theta = \dfrac{3}{4} \Rightarrow \sin\theta = \dfrac{\sqrt{3}}{2} = \sin 60°$

$\Rightarrow \theta = 60°$

18. (c) Given, $\tan(A + B) = \sqrt{3}$

$\Rightarrow \tan(A + B) = \tan 60°$

$\therefore A + B = 60°$ ...(i)

and $\tan A = 1 \Rightarrow \tan A = \tan 45°$

$\therefore A = 45°$

From Eq. (i), $A + B = 60°$

$\Rightarrow 45° + B = 60° \Rightarrow B = 15°$

Now, $\tan(A - B) = \tan(45° - 15°)$

$= \tan 30° = \dfrac{1}{\sqrt{3}}$

Hence, the value of $\tan(A - B)$ is $\dfrac{1}{\sqrt{3}}$.

19. (c) In right angled $\triangle ABC$,

$AB : BC = 3 : 4$ or $\dfrac{AB}{BC} = \dfrac{3}{4}$

Now, in right angled $\triangle ABC$,

$AC^2 = AB^2 + BC^2$

$= 3^2 + 4^2 = 9 + 16$

$\Rightarrow AC^2 = 25 \Rightarrow AC = 5$

$\therefore \sin A = \dfrac{BC}{AC} = \dfrac{4}{5}$

and $\sin B = \sin 90° = 1$, $\sin C = \dfrac{AB}{AC} = \dfrac{3}{5}$

Now, $\sin A + \sin B + \sin C = \dfrac{4}{5} + 1 + \dfrac{3}{5}$

$= \dfrac{4 + 5 + 3}{5} = \dfrac{12}{5}$

20. (b) $\because \sin\theta + \cos\theta = 1$

On squaring both sides, we get

$(\sin\theta + \cos\theta)^2 = 1$

$\Rightarrow \sin^2\theta + \cos^2\theta + 2\sin\theta\cos\theta = 1$

$\Rightarrow 1 + 2\sin\theta\cos\theta = 1$

[$\because \sin^2\theta + \cos^2\theta = 1$]

$\Rightarrow 2\sin\theta\cos\theta = 0$

$\therefore \sin\theta\cos\theta = 0$

21. (b) We have, $\dfrac{\cos\theta}{1 - \sin\theta} \times \dfrac{1 + \sin\theta}{1 + \sin\theta}$

$= \dfrac{\cos\theta(1 + \sin\theta)}{1 - \sin^2\theta}$

$= \dfrac{\cos\theta(1 + \sin\theta)}{\cos^2\theta} = \dfrac{1 + \sin\theta}{\cos\theta}$

22. (d) By trigonometric identity,

$\sec^2\theta - \tan^2\theta = 1$

$\Rightarrow (\sec\theta + \tan\theta)(\sec\theta - \tan\theta) = 1$

Given, $\sec\theta + \tan\theta = 2$ ...(i)

$\therefore \sec\theta - \tan\theta = \dfrac{1}{2}$ ...(ii)

On adding Eqs. (i) and (ii), we get

$2\sec\theta = \dfrac{1}{2} + 2 \Rightarrow \sec\theta = \dfrac{5}{4}$

23. (a) $\dfrac{5}{\sec^2\theta} + \dfrac{2}{1 + \cot^2\theta} + 3\sin^2\theta$

$= 5\cos^2\theta + \dfrac{2}{\csc^2\theta} + 3\sin^2\theta$

[$\because 1 + \cot^2\theta = \csc^2\theta$]

$= 5\cos^2\theta + 2\sin^2\theta + 3\sin^2\theta$

$= 5\cos^2\theta + 5\sin^2\theta$

$= 5(\sin^2\theta + \cos^2\theta)$ [$\because \sin^2\theta + \cos^2\theta = 1$]

$= 5 \times 1 = 5$

24. (c) $\cot^2 A \cos^2 A = \dfrac{\cos^2 A}{\sin^2 A} \times \cos^2 A = \dfrac{\cos^4 A}{\sin^2 A}$

From option (c),

$\cot^2 A - \cos^2 A = \dfrac{\cos^2 A}{\sin^2 A} - \cos^2 A$

$= \dfrac{\cos^2 A - \sin^2 A \cos^2 A}{\sin^2 A}$

$= \dfrac{\cos^2 A(1 - \sin^2 A)}{\sin^2 A}$

$= \dfrac{\cos^2 A \times \cos^2 A}{\sin^2 A}$

$= \dfrac{\cos^4 A}{\sin^2 A}$

Hence, $\cot^2 A \cos^2 A = \cot^2 A - \cos^2 A$.

25. (c) Given that, $\sin^2 x + \cos^2 x - 1 = 0$

$\Rightarrow \sin^2 x + \cos^2 x = 1$

which is an identity of trigonometric ratio and always true for every real value of x. So, the equation have an infinite solutions.

26. (a) $\because \sec^2\theta - \tan^2\theta = 1$ [identity]

Also, $\sec^2\theta + \tan^2\theta = \dfrac{7}{12}$ [given]

$\therefore \sec^4\theta - \tan^4\theta$
$= (\sec^2\theta - \tan^2\theta)(\sec^2\theta + \tan^2\theta)$
$= 1 \times \dfrac{7}{12} = \dfrac{7}{12}$

27. (c) $\left(\dfrac{1}{\cos\theta} + \dfrac{1}{\cot\theta}\right)\left(\dfrac{1}{\cos\theta} - \dfrac{1}{\cot\theta}\right)$

$= \left(\dfrac{1}{\cos\theta} + \dfrac{\sin\theta}{\cos\theta}\right)\left(\dfrac{1}{\cos\theta} - \dfrac{\sin\theta}{\cos\theta}\right)$

$= \left(\dfrac{1+\sin\theta}{\cos\theta}\right)\left(\dfrac{1-\sin\theta}{\cos\theta}\right)$

$= \dfrac{1-\sin^2\theta}{\cos^2\theta} = \dfrac{\cos^2\theta}{\cos^2\theta} = 1$

28. (c) Given, $\cos\theta + \sin\theta = \sqrt{2}\cos\theta$ ...(i)

On squaring both sides, we get
$(\cos\theta + \sin\theta)^2 = (\sqrt{2}\cos\theta)^2$
$\Rightarrow \cos^2\theta + \sin^2\theta + 2\sin\theta\cos\theta = 2\cos^2\theta$
$\Rightarrow 2\sin\theta\cos\theta = \cos^2\theta - \sin^2\theta$
$\Rightarrow 2\sin\theta\cos\theta = (\cos\theta - \sin\theta)(\cos\theta + \sin\theta)$
$\therefore \cos\theta - \sin\theta = \dfrac{2\sin\theta\cos\theta}{(\cos\theta + \sin\theta)}$

$= \dfrac{2\sin\theta\cos\theta}{\sqrt{2}\cos\theta}$ [from Eq. (i)]

$= \sqrt{2}\sin\theta$

29. (d) $\sin 25° \sin 35° \sec 65° \sec 55°$

$= \sin 25° \sin 35° \cdot \dfrac{1}{\cos 65°} \cdot \dfrac{1}{\cos 55°}$

$= \sin 25° \cdot \sin 35° \cdot \dfrac{1}{\cos(90°-25°)}$

$\cdot \dfrac{1}{\cos(90°-35°)}$

$= \sin 25° \cdot \sin 35° \cdot \dfrac{1}{\sin 25°} \cdot \dfrac{1}{\sin 35°} = 1$

30. (a) Given expression

$= \sqrt{\dfrac{1-\cos A}{1+\cos A}} \times \sqrt{\dfrac{1-\cos A}{1-\cos A}}$

$= \dfrac{1-\cos A}{\sqrt{1-\cos^2 A}} = \dfrac{1-\cos A}{\sqrt{\sin^2 A}}$

$= \dfrac{(1-\cos A)}{\sin A} = \dfrac{1}{\sin A} - \dfrac{\cos A}{\sin A}$

$= \mathrm{cosec}\, A - \cot A$

31. (b) $\dfrac{\tan 57° + \cot 37°}{\tan 33° + \cot 53°}$

$= \dfrac{\tan 57° + \cot 37°}{\tan(90°-57°) + \cot(90°-37°)}$

$= \dfrac{\tan 57° + \cot 37°}{\cot 57° + \tan 37°} = \dfrac{\tan 57° + \dfrac{1}{\tan 37°}}{\dfrac{1}{\tan 57°} + \tan 37°}$

$= \dfrac{1+\tan 57°\tan 37°}{1+\tan 37°\tan 57°} \cdot \dfrac{\tan 57°}{\tan 37°}$

$= \dfrac{\tan 57°}{\tan 37°} = \tan 57° \cdot \cot 37°$

32. (c) $\tan 4° \tan 43° \tan 47° \tan 86°$
$= \tan 4° \tan 43° \tan(90°-43°) \tan(90°-4°)$
$= \tan 4° \tan 43° \cot 43° \cot 4°$
$= \tan 4° \tan 43° \cot 43° \cot 4°$
$= 1 \cdot 1 = 1$ [$\because \tan A \cdot \cot A = 1$]

33. (c) I. LHS $= \mathrm{cosec}^2 x + \sec^2 x$

$= \dfrac{1}{\sin^2 x} + \dfrac{1}{\cos^2 x} = \dfrac{\cos^2 x + \sin^2 x}{\sin^2 x \cos^2 x}$

$= \dfrac{1}{\sin^2 x \cos^2 x} = \dfrac{1}{\sin^2 x} \cdot \dfrac{1}{\cos^2 x}$

$= \mathrm{cosec}^2 x \sec^2 x =$ RHS

II. LHS $= \sec^2 x + \tan^2 x$

$= \dfrac{1}{\cos^2 x} + \dfrac{\sin^2 x}{\cos^2 x} = \dfrac{1+\sin^2 x}{\cos^2 x}$

$\neq \sec^2 x \tan^2 x$

III. LHS $= \mathrm{cosec}^2 x + \tan^2 x$

$= 1 + \cot^2 x + \sec^2 x - 1$

$\begin{bmatrix} \because \mathrm{cosec}^2 x = 1 + \cot^2 x \\ \because \sec^2 x = 1 + \tan^2 x \end{bmatrix}$

$= \cot^2 x + \sec^2 x =$ RHS

So, the Statements I and III are correct.

34. (c) $\dfrac{\cos x}{\mathrm{cosec}\, x + 1} + \dfrac{\cos x}{\mathrm{cosec}\, x - 1} = 2$

$\Rightarrow \dfrac{\cos x(\mathrm{cosec}\, x - 1) + \cos x(\mathrm{cosec}\, x + 1)}{(\mathrm{cosec}\, x + 1)(\mathrm{cosec}\, x - 1)} = 2$

$\Rightarrow \dfrac{\cos x\,\mathrm{cosec}\, x - \cos x + \cos x\,\mathrm{cosec}\, x + \cos x}{\mathrm{cosec}^2 x - 1} = 2$

$\Rightarrow \dfrac{2\cos x\,\mathrm{cosec}\, x}{\mathrm{cosec}^2 x - 1} = 2$

$\Rightarrow \dfrac{2\cos x \cdot \dfrac{1}{\sin x}}{\cot^2 x} = 2$

$\therefore \dfrac{2\cot x}{\cot^2 x} = 2$

$\Rightarrow \cot x = 1 = \cot 45°$

$x = 45° = \dfrac{\pi}{4}$

35. (a) $\dfrac{\tan 27° + \cot 63°}{\tan 27° (\sin 25° + \cos 65°)}$

$= \dfrac{\tan 27° + \cot(90° - 27°)}{\tan 27° [\sin 25° + \cos(90° - 25°)]}$

$= \dfrac{\tan 27° + \tan 27°}{\tan 27° [\sin 25° + \sin 25°]}$

$= \dfrac{2}{2\sin 25°} = \cosec 25°$

36. (d) $\cot 10° \cdot \cot 20° \cdot \cot 60° \cdot \cot 70° \cdot \cot 80°$

$= \cot 10° \cot 80° \cdot \cot 20° \cot 70° \cdot \cot 60°$

$= \cot 10° \cdot \tan 10° \cdot \cot 20° \cdot \tan 20° \cdot \cot 60°$

$\qquad [\because \tan(90° - \theta) = \cot \theta]$

$= 1 \cdot 1 \cdot \dfrac{1}{\sqrt{3}} = \dfrac{1}{\sqrt{3}} \quad \left[\because \cot 60° = \dfrac{1}{\sqrt{3}}\right]$

37. (d) $\cos \theta \cdot \cosec 23° = 1$

$\Rightarrow \cosec 23° = \dfrac{1}{\cos \theta} = \sec \theta$

$\Rightarrow \cosec 23° = \cosec(90° - \theta)$

$\Rightarrow 23° = 90° - \theta$

$\Rightarrow \theta = 90° - 23° = 67°$

38. (b) $\sec x = \cosec y \Rightarrow \dfrac{1}{\cos x} = \dfrac{1}{\sin y}$

$\therefore \quad \cos x = \sin y$

$\therefore \quad \sin\left(\dfrac{\pi}{2} - x\right) = \sin y$

$\therefore \dfrac{\pi}{2} - x = y \Rightarrow (x + y) = \dfrac{\pi}{2}$

Now, value of $\sin(x + y) = \sin \dfrac{\pi}{2} = 1$

39. (d) $A + B = 90°$ [$\because$ they are complementary]

$\Rightarrow B = 90° - A$

$\therefore \sec^2 A + \sec^2 B - \sec^2 A \cdot \sec^2 B$

$= \sec^2 A + \sec^2(90° - A) - \sec^2 A$

$\qquad\qquad\qquad\qquad \sec^2(90° - A)$

$= \sec^2 A + \cosec^2 A - \sec^2 A \cdot \cosec^2 A$

$\qquad [\because \sec(90° - \theta) = \cosec \theta]$

$= \dfrac{1}{\cos^2 A} + \dfrac{1}{\sin^2 A} - \dfrac{1}{\sin^2 A \cdot \cos^2 A}$

$= \dfrac{\sin^2 A + \cos^2 A - 1}{\sin^2 A \cdot \cos^2 A} = \dfrac{1 - 1}{\sin^2 A \cdot \cos^2 A} = 0$

40. (b) $\sin(3x - 20°) = \cos(3y + 20°)$

$\Rightarrow \sin(3x - 20°) = \sin[90° - (3y + 20°)]$

$\Rightarrow \sin(3x - 20°) = \sin(70° - 3y)$

$\therefore \quad 3x - 20° = 70° - 3y$

$\Rightarrow \quad 3x + 3y = 90°$

$\Rightarrow \quad 3(x + y) = 90° \Rightarrow x + y = 30°$

41. (a) $\cot 18°$

$\left(\cot 72° \cdot \cos^2 22° + \dfrac{1}{\tan 72° \cdot \sec^2 68°}\right)$

$= \cot 18° \cdot \cot 72° \cdot \cos^2 22° + \dfrac{\cot 18°}{\tan 72° \cdot \sec^2 68°}$

$\qquad [\because \cot(90° - \theta) = \tan \theta]$

$= \cot 18° \cdot \tan 18° \cdot \cos^2 22°$

$\qquad\qquad + \dfrac{\cot 18°}{\cot 18°} \cdot \cos^2 68°$

$= \cos^2 22° + \cos^2 68° = \cos^2 22° + \sin^2 22° = 1$

$\qquad [\because \sin(90° - \theta) = \cos \theta]$

42. (b) $\cosec^2 67° + \sec^2 57°$

$\qquad\qquad\qquad - \cot^2 33° - \tan^2 23°$

$= \cosec^2(90° - 23°) + \sec^2(90° - 33°)$

$\qquad\qquad\qquad - \cot^2 33° - \tan^2 23°$

$= \sec^2 23° + \cosec^2 33° - \cot^2 33° - \tan^2 23°$

$= 1 + \tan^2 23° + 1 + \cot^2 33°$

$\qquad\qquad - \cot^2 33° - \tan^2 23° = 2$

$\qquad [\because 1 + \tan^2 \theta = \sec^2 \theta$ and $1 + \cot^2 \theta = \cosec^2 \theta]$

43. (b) Given, $\sin x + \sin^2 x = 1$

$\Rightarrow \sin x = 1 - \sin^2 x$

$\Rightarrow \sin x = \cos^2 x \qquad\qquad\qquad \ldots(i)$

Now, $\cos^8 x + 2\cos^6 x + \cos^4 x$

$= (\cos^2 x)^4 + 2(\cos^2 x)^3 + (\cos^2 x)^2$

$= \sin^4 x + 2\sin^3 x + \sin^2 x \quad$ [from Eq. (i)]

$= \sin^2 x(\sin^2 x + 2\sin x + 1)$

$= \sin^2 x(\sin x + 1)^2 = [(\sin x)(\sin x + 1)]^2$

$= (\sin^2 x + \sin x)^2$

$= (1)^2 = 1 \qquad [\because \sin^2 x + \sin x = 1]$

44. (b) Given, $\sin \theta + \cos \theta = \dfrac{\sqrt{7}}{2}$

On squaring both sides, we get

$(\sin \theta + \cos \theta)^2 = \left(\dfrac{\sqrt{7}}{2}\right)^2$

$\Rightarrow \sin^2 \theta + \cos^2 \theta + 2\sin \theta \cos \theta = \dfrac{7}{4}$

$\Rightarrow \qquad 1 + 2\sin \theta \cos \theta = \dfrac{7}{4}$

$\Rightarrow \qquad 2\sin \theta \cos \theta = \dfrac{7}{4} - 1 = \dfrac{3}{4}$

Now, $(\sin \theta - \cos \theta)^2 = \sin^2 \theta + \cos^2 \theta$

$\qquad\qquad\qquad\qquad - 2\sin \theta \cos \theta$

$= 1 - \dfrac{3}{4} = \dfrac{4-3}{4}$

$\Rightarrow (\sin \theta - \cos \theta)^2 = \dfrac{1}{4}$

$\therefore \qquad \sin \theta - \cos \theta = \dfrac{1}{2}$

45. (c) We have, $x = a\cos\theta + b\sin\theta$
and $y = a\sin\theta - b\cos\theta$
$\therefore x^2 + y^2 = (a\cos\theta + b\sin\theta)^2 + (a\sin\theta - b\cos\theta)^2$
$= a^2\cos^2\theta + b^2\sin^2\theta + 2ab\cos\theta\sin\theta + a^2\sin^2\theta + b^2\cos^2\theta - 2ab\sin\theta\cos\theta$
$= a^2\cos^2\theta + a^2\sin^2\theta + b^2\sin^2\theta + b^2\cos^2\theta$
$= a^2 + b^2 \quad [\because \sin^2\theta + \cos^2\theta = 1]$

46. (d) We have, $7\sin^2 x + 3\cos^2 x = 4$
$\Rightarrow 7\dfrac{\sin^2 x}{\cos^2 x} + 3\dfrac{\cos^2 x}{\cos^2 x} = \dfrac{4}{\cos^2 x}$
$\Rightarrow 7\tan^2 x + 3 = 4\sec^2 x$
$\Rightarrow 7\tan^2 x + 3 = 4(1 + \tan^2 x)$
$\Rightarrow 7\tan^2 x + 3 = 4 + 4\tan^2 x$
$\Rightarrow 3\tan^2 x = 1$
$\Rightarrow \tan^2 x = \dfrac{1}{3} \quad [\because 0 < x < 90°]$
$\Rightarrow \tan x = \dfrac{1}{\sqrt{3}}$

47. (c) Given, $x = a\sec\theta\cos\phi$, $y = b\sec\theta\cdot\sin\phi$
and $z = c\tan\theta$
Now, $\dfrac{x^2}{a^2} + \dfrac{y^2}{b^2} - \dfrac{z^2}{c^2}$
On putting the values of x, y and z, we get
$\dfrac{a^2\sec^2\theta\cos^2\phi}{a^2} + \dfrac{b^2\sec^2\theta\sin^2\phi}{b^2} - \dfrac{c^2\tan^2\theta}{c^2}$
$= \sec^2\theta(\cos^2\phi + \sin^2\phi) - \tan^2\theta$
$= \sec^2\theta - \tan^2\theta \quad [\because \sec^2\theta - \tan^2\theta = 1]$
$= 1$

48. (b) $\because \cos^4\theta - \sin^4\theta = \dfrac{2}{3}$
$\Rightarrow (\cos^2\theta)^2 - (\sin^2\theta)^2 = \dfrac{2}{3}$
$\Rightarrow (\cos^2\theta - \sin^2\theta)(\cos^2\theta + \sin^2\theta) = \dfrac{2}{3}$
$[\because \cos^2\theta - \sin^2\theta = \cos 2\theta = 1 - 2\sin^2\theta]$
$\Rightarrow \cos^2\theta - \sin^2\theta = \dfrac{2}{3} \Rightarrow \cos 2\theta = \dfrac{2}{3}$
$\Rightarrow 1 - 2\sin^2\theta = \dfrac{2}{3}$

49. (b) $\because 7\sin^2\theta + 3\cos^2\theta = 4$
$\Rightarrow 4\sin^2\theta + 3\sin^2\theta + 3\cos^2\theta = 4$
$\Rightarrow 4\sin^2\theta + 3(\sin^2\theta + \cos^2\theta) = 4$
$\Rightarrow 4\sin^2\theta + 3 = 4 \Rightarrow 4\sin^2\theta = 1$
$\Rightarrow \sin\theta = \dfrac{1}{2} \Rightarrow \theta = 30°$
$\Rightarrow \cos\theta = \cos 30° = \dfrac{\sqrt{3}}{2}$

50. (b) Let $x\sin\theta = y\cos\theta = \dfrac{2z\tan\theta}{1 - \tan^2\theta} = k$
Taking $x\sin\theta = y\cos\theta$
$\Rightarrow \dfrac{y}{x} = \tan\theta \qquad \ldots(i)$
Now, $x = k\csc\theta \qquad \ldots(ii)$
$y = k\sec\theta \qquad \ldots(iii)$
and $z = k\left(\dfrac{1 - \tan^2\theta}{2\tan\theta}\right)$
$= k\left(\dfrac{1 - \dfrac{y^2}{x^2}}{2\cdot\dfrac{y}{x}}\right) = k\left(\dfrac{x^2 - y^2}{x^2} \times \dfrac{x}{2y}\right)$
$= k\left(\dfrac{x^2 - y^2}{2xy}\right) \qquad \ldots(iv)$
From Eqs. (ii) and (iii), we get
$x^2 + y^2 = k^2(\csc^2\theta + \sec^2\theta)$
$= k^2(\tan^2\theta + \cot^2\theta + 2)$
$= k^2(\tan\theta + \cot\theta)^2$
$\Rightarrow x^2 + y^2 = k^2\left(\dfrac{y}{x} + \dfrac{x}{y}\right)^2 = k^2\left(\dfrac{x^2 + y^2}{xy}\right)^2$
$\Rightarrow k^2 = \dfrac{x^2 + y^2}{\left(\dfrac{x^2 + y^2}{xy}\right)^2}$
$= (x^2 + y^2)\cdot\dfrac{x^2 y^2}{(x^2 + y^2)^2}$
$= \dfrac{x^2 y^2}{x^2 + y^2}$
From Eq. (iv), we get
$z^2 = \left(\dfrac{x^2 y^2}{x^2 + y^2}\right)\left(\dfrac{x^2 - y^2}{2xy}\right)^2$
$\therefore 4z^2(x^2 + y^2)$
$= 4\left(\dfrac{x^2 y^2}{x^2 + y^2}\right)\left(\dfrac{x^2 - y^2}{2xy}\right)^2 (x^2 + y^2)$
$= (x^2 - y^2)^2$

51. (c) $2(\cos^2\theta - \sin^2\theta) = 1$
$\Rightarrow \cos^2\theta - \sin^2\theta = \dfrac{1}{2} \qquad \ldots(i)$
We know that,
$\cos^2\theta + \sin^2\theta = 1 \qquad \ldots(ii)$
On solving Eqs. (i) and (ii), we get
$2\cos^2\theta = \dfrac{3}{2} \Rightarrow \cos^2\theta = \dfrac{3}{4}, \sin^2\theta = \dfrac{1}{4}$
$\therefore \cot^2\theta = \dfrac{\cos^2\theta}{\sin^2\theta} = \dfrac{3/4}{1/2} = 3 \Rightarrow \cot\theta = \sqrt{3}$

52. (b) Given, $\sec A + \tan A = a$

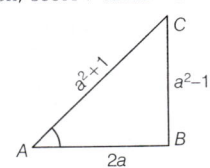

[Triangle ABC with right angle at B, AB = 2a, BC = a^2-1, AC = a^2+1]

$\Rightarrow \sec^2 A + \tan^2 A + 2\sec A \cdot \tan A = a^2$
[on squaring both sides]
$\Rightarrow 1 + \tan^2 A + \tan^2 A + 2\sec A \cdot \tan A = a^2$
$\Rightarrow 1 + 2\tan^2 A + 2\sec A \cdot \tan A = a^2$
$\Rightarrow 2\tan A(\tan A + \sec A) = a^2 - 1$
$\Rightarrow 2\tan A \cdot a = (a^2 - 1)$
$\Rightarrow \tan A = \dfrac{a^2 - 1}{2a}$

Now, in $\triangle ABC$
$(AC)^2 = (AB)^2 + (BC)^2$
$\Rightarrow (AC)^2 = (2a)^2 + (a^2 - 1)^2$
$\Rightarrow (AC)^2 = 4a^2 + (a^2 - 1)^2$
$\Rightarrow (AC)^2 = 4a^2 + a^4 + 1 - 2a^2$
$= a^4 + 1 + 2a^2$
$\Rightarrow (AC)^2 = (a^2 + 1)^2$
$\Rightarrow AC = (a^2 + 1)$
$\therefore \cos A = \dfrac{AB}{AC} \Rightarrow \cos A = \dfrac{2a}{a^2 + 1}$

53. (c) $\because \tan 2\theta \cdot \tan 3\theta = 1$
$\Rightarrow \tan 3\theta = \dfrac{1}{\tan 2\theta} = \cot 2\theta$
$\Rightarrow \tan 3\theta = \tan(90° - 2\theta)$
$\therefore 3\theta = 90° - 2\theta$
$\Rightarrow 5\theta = 90° \Rightarrow \theta = \dfrac{90°}{5} = 18°$
$\therefore 2\cos^2 \dfrac{5\theta}{2} - 1 = 2\cos^2 \dfrac{5 \times 18}{2} - 1$
$= 2\cos^2 45° - 1 = 2 \times \left(\dfrac{1}{\sqrt{2}}\right)^2 - 1$
$= 2 \times \dfrac{1}{2} - 1 = 0$

54. (c) Given, $\dfrac{\sin\theta + \cos\theta}{\sin\theta - \cos\theta} = 3$
$\Rightarrow \sin\theta + \cos\theta = 3\sin\theta - 3\cos\theta$
$\Rightarrow 4\cos\theta = 2\sin\theta \Rightarrow \tan\theta = 2$
$\therefore$ Value of $\sin^4\theta - \cos^4\theta$
$= (\sin^2\theta + \cos^2\theta)(\sin^2\theta - \cos^2\theta)$
$[\because \sin^2\theta + \cos^2\theta = 1]$
$= \sin^2\theta - \cos^2\theta = \cos^2\theta (\tan^2\theta - 1)$
$= \dfrac{\tan^2\theta - 1}{\sec^2\theta} = \dfrac{\tan^2\theta - 1}{1 + \tan^2\theta} = \dfrac{4-1}{1+4} = \dfrac{3}{5}$

55. (b) Given, $\tan\theta + \cot\theta = 2$
$\Rightarrow \tan\theta + \dfrac{1}{\tan\theta} = 2 \Rightarrow \tan^2\theta + 1 = 2\tan\theta$
$\Rightarrow \tan^2\theta - 2\tan\theta + 1 = 0$
$\Rightarrow (\tan\theta - 1)^2 = 0 \Rightarrow \tan\theta = 1$ and $\cot\theta = 1$
$\therefore \tan^5\theta + \cot^{10}\theta = (1)^5 + (1)^{10} = 1 + 1 = 2$

56. (d) If $0° < \theta < 90°$, then all the trigonometric ratios can be obtained when any one of the six ratios is given.
$\because$ We use any of the following identity to get any trigonometric ratios
$\sin^2\theta + \cos^2\theta = 1$, $1 + \tan^2\theta = \sec^2\theta$
and $1 + \cot^2\theta = \csc^2\theta$

57. (b) Given that, $\sin\theta \cdot \cos\theta = \dfrac{\sqrt{3}}{4}$ ...(i)
Now, $\sin^4\theta + \cos^4\theta$
$= (\sin^2\theta + \cos^2\theta)^2 - 2\sin^2\theta \cdot \cos^2\theta$
$= (1)^2 - 2(\sin\theta \cdot \cos\theta)^2$
$= 1 - 2\left(\dfrac{\sqrt{3}}{4}\right)^2 = 1 - 2 \cdot \dfrac{3}{16} = 1 - \dfrac{3}{8} = \dfrac{5}{8}$

58. (c) Given that, $3\sin x + 5\cos x = 5$
On squaring both sides, we get
$9\sin^2 x + 25\cos^2 x + 30\sin x \cos x = 25$
$\Rightarrow 9(1 - \cos^2 x) + 25(1 - \sin^2 x)$
$\qquad + 30\sin x \cos x = 25$
$\Rightarrow 9 - 9\cos^2 x + 25 - 25\sin^2 x$
$\qquad + 30\sin x \cdot \cos x = 25$
$\Rightarrow 9 + 25 - \{9\cos^2 x + 25\sin^2 x$
$\qquad - 30\sin x \cos x\} = 25$
$\Rightarrow 9 = (3\cos x - 5\sin x)^2$
$\Rightarrow 3\cos x - 5\sin x = 3$

59. (d) Given, $\sin 17° = \dfrac{x}{y}$
Now, $\sec 17° - \sin 73°$
$= \sec 17° - \sin(90° - 17°)$
$= \sec 17° - \cos 17°$
$= \dfrac{1}{\cos 17°} - \cos 17°$
$= \dfrac{1 - \cos^2 17°}{\cos 17°} = \dfrac{\sin^2 17°}{\cos 17°}$
$= \dfrac{x^2/y^2}{\sqrt{1 - (x^2/y^2)}}$ $[\because \cos\theta = \sqrt{1-\sin^2\theta}]$
$= \dfrac{x^2/y^2}{\sqrt{\dfrac{y^2 - x^2}{y^2}}} = \dfrac{x^2}{y\sqrt{y^2 - x^2}}$

60. (d) $x^2 + \dfrac{1}{x^2} = 2\sin\left(\dfrac{\pi x}{2}\right)$

$\Rightarrow \left(x - \dfrac{1}{x}\right)^2 + 2 = 2\sin\left(\dfrac{\pi x}{2}\right) = 2$

$[\because a^2 + b^2 = (a-b)^2 + 2ab]$

$\Rightarrow \left(x - \dfrac{1}{x}\right)^2 = 0$

$\therefore \quad x - \dfrac{1}{x} = 0 \quad \left[\begin{array}{l}\because \sin\dfrac{\pi x}{2} = 1 \text{ for all} \\ \text{integer values of } x\end{array}\right]$

61. (b) $\sin^2 1° + \sin^2 3° + \cdots + \sin^2 45°$
$+ \cdots + \sin^2(90°-3°) + \sin^2(90°-1°)$
$= \sin^2 1° + \sin^2 3° + \cdots + \sin^2 45°$
$+ \cdots + \cos^2 3° + \cos^2 1°$
$= (1 \times 22) + \left(\dfrac{1}{\sqrt{2}}\right)^2 = 22 + \dfrac{1}{2} = 22\dfrac{1}{2}$

62. (d) $\sin^2 5° + \sin^2 10° + \cdots + \sin^2 45°$
$+ \cdots + \sin^2 85° + \sin^2 90°$
$= \sin^2 5° + \sin^2 10° + \cdots + \sin^2 45°$
$+ \cdots + \sin^2(90°-5°) + \sin^2 90°$
$= (\sin^2 5° + \cos^2 5°) + (\sin^2 10° + \cos^2 10°)$
$+ \ldots + (\cos^2 40° + \sin^2 40°) + \sin^2 45° + \sin^2 90°$
$= 8 + \left(\dfrac{1}{\sqrt{2}}\right)^2 + 1 = 8 + 1 + \dfrac{1}{2} = 9\dfrac{1}{2}$

63. (d) $\dfrac{\sin 39°}{\cos 51°} + 2\tan 11° \cdot \tan 79° \cdot$

$\quad \tan 31° \cdot \tan 59° \cdot \tan 45° - 3(\sin^2 21 + \sin^2 69°)$

$= \dfrac{\sin 39°}{\cos(90°-39°)} + 2\tan 11° \cdot \tan(90°-11°)$

$\quad \tan 31° \cdot \tan(90°-31°) \cdot 1$
$\quad - 3(\sin^2 21° + \sin^2(90°-21°))$

$[\because \tan(90°-\theta) = \cot\theta, \cos(90°-\theta) = \sin\theta]$

$= \dfrac{\sin 39°}{\sin 39°} + 2\tan 11° \cdot \cot 11° \cdot \tan 31° \cdot \cot 31°$

$\quad - 3(\sin^2 21° + \cos^2 21°)$

$= 1 + 2 - 3 = 0$

$[\because \tan\theta \cdot \cot\theta = 1 \text{ and } \sin^2\theta + \cos^2\theta = 1]$

64. (c) $\therefore$ Minimum value of
$\sin^2\theta + \cos^4\theta$ will be at $\theta = 45°$
$\therefore$ Required minimum value
$= \sin^2 45° + \cos^4 45°$
$= \dfrac{1}{2} + \dfrac{1}{4} = \dfrac{3}{4}$

65. (c) We know that,
$AM \geq GM$
$\therefore \dfrac{9\tan^2\theta + 4\cot^2\theta}{2} \geq [(9\tan^2\theta)(4\cot^2\theta)]^{1/2}$
$\Rightarrow \dfrac{9\tan^2\theta + 4\cot^2\theta}{2} \geq 6 \quad [\because \tan\theta \cdot \cot\theta = 1]$
$\Rightarrow 9\tan^2\theta + 4\cot^2\theta \geq 12$
Hence, the minimum value of
$9\tan^2\theta + 4\cot^2\theta$ is 12.

66. (c) Minimum value of $\cos\theta = -1$
and minimum value of $\sec\theta = -1$
$\therefore \cos^2\theta + \sec^2\theta = (-1)^2 + (-1)^2 = 2$

67. (a) $\cos x + \cos y = 2$
$\because \quad (\cos x)_{\max} \leq 1$
$\Rightarrow \quad \cos x = 1, \cos y = 1 \Rightarrow x = y = 0$
$\therefore \sin x + \sin y = \sin 0° + \sin 0°$
$= 0 + 0 = 0$

68. (b) Given, $\cos^2\theta = \dfrac{(x+y)^2}{4xy}$
We know that, $\cos^2\theta \leq 1$
When $\cos^2\theta < 1$, then $\dfrac{(x+y)^2}{4xy} < 1$
$\Rightarrow \quad x^2 + y^2 + 2xy < 4xy$
$\Rightarrow \quad (x-y)^2 < 0$, which is not possible
$\therefore \quad \cos^2\theta = 1 \Rightarrow (x+y)^2 = 4xy$
$\Rightarrow \quad x^2 + y^2 + 2xy = 4xy$
$\Rightarrow \quad (x-y)^2 = 0 \Rightarrow x = y$

69. (b) We have, $\tan 1° \tan 2° \tan 3° \tan 4° \ldots \tan 89°$
$= \tan 1° \tan 2° \tan 3° \ldots \tan 45° \tan 46° \tan 47°$
$\ldots \tan 89°$
$= (\tan 1° \tan 89°) \cdot (\tan 2° \tan 88°) \ldots$
$(\tan 44° \tan 46°) \cdot \tan 45°$
$= 1 \cdot 1 \ldots 1 = 1 \quad [\because \tan\theta \tan(90°-\theta) = 1]$

Exercise 2 Higher Skill Level Questions

1. (d) $(\sin x + \sin y) = a$ and $(\cos x + \cos y) = b$
On squaring both the equations, we get
$(\sin x + \sin y)^2 = a^2$
$\Rightarrow \sin^2 x + \sin^2 y + 2\sin x \sin y = a^2 \quad \ldots(i)$
and $(\cos x + \cos y)^2 = b^2$
$\Rightarrow \cos^2 x + \cos^2 y + 2\cos x \cos y = b^2 \quad \ldots(ii)$

On adding Eqs. (i) and (ii), we get
$(\sin^2 x + \sin^2 y + 2\sin x \sin y)$
$+ (\cos^2 x + \cos^2 y + 2\cos x \cos y) = a^2 + b^2$
$\Rightarrow \sin^2 x + \cos^2 x + \sin^2 y + \cos^2 y$
$+ 2(\sin x \sin y + \cos x \cos y) = a^2 + b^2$
$\Rightarrow 1 + 1 + 2(\sin x \sin y + \cos x \cos y)$
$= a^2 + b^2$

$\therefore \sin x \sin y + \cos x \cos y = \dfrac{a^2 + b^2 - 2}{2}$

2. (c) I. $\because 1 - \sin x \neq 0 \Rightarrow \sin x \neq 1$
$\Rightarrow x \neq \pi/2$, which does not belong to be given interval. So, there is no value of x exist.

II. $\sin x = 3^{\sin^2 x} \Rightarrow 1 = \sin x \cdot 3^{-\sin^2 x}$
On multiplying 3 both sides, we get
$3 = \sin x \cdot 3^{-\sin^2 x} \cdot 3$
$\Rightarrow \quad 3 = \sin x \cdot 3^{1 - \sin^2 x}$
$\Rightarrow \quad 3 = \sin x \cdot 3^{\cos^2 x}$

The RHS is less than 3 while the LHS is 3. Thus, the equation does not hold for any x.

3. (b) Given, $\sin x + \cos x = c$
On squaring both sides, we get
$(\sin x + \cos x)^2 = c^2$
$\Rightarrow \sin^2 x + \cos^2 x + 2\sin x \cos x = c^2$
$\Rightarrow 1 + 2\sin x \cos x = c^2$
$\Rightarrow \sin x \cos x = \dfrac{c^2 - 1}{2}$ ...(i)

Now, $\sin^6 x + \cos^6 x = (\sin^2 x)^3 + (\cos^2 x)^3$
$= (\sin^2 x + \cos^2 x)[\sin^4 x + \cos^4 x - \sin^2 x \cos^2 x]$
$[\because a^3 + b^3 = (a + b)(a^2 + b^2 - ab)]$
$= 1 [(\sin^2 x + \cos^2 x)^2 - 3\sin^2 x \cos^2 x]$
$= (1 - 3\sin^2 x \cos^2 x)$
$\therefore \sin^6 x + \cos^6 x = 1 - 3\sin^2 x \cos^2 x$
$= 1 - 3\left(\dfrac{c^2 - 1}{2}\right)^2$ [from Eq. (i)]
$= 1 - 3\left(\dfrac{c^4 + 1 - 2c^2}{4}\right)$
$= \dfrac{4 - 3c^4 - 3 + 6c^2}{4} = \dfrac{1 + 6c^2 - 3c^4}{4}$

4. (a) $\sin(10°6'32'') = \dfrac{a}{1} = \dfrac{P}{H}$
$\Rightarrow \sin(90° - 79°53'28'') = a$
$\Rightarrow \cos 79°53'28'' = a$
$\therefore \cos(79°53'28'') + \tan(10°6'32'')$

$= a + \dfrac{a}{\sqrt{1 - a^2}} = \dfrac{a(1 + \sqrt{1 - a^2})}{\sqrt{1 - a^2}}$

5. (b) $\cos\theta + \sec\theta = 2 \Rightarrow \cos\theta + \dfrac{1}{\cos\theta} = 2$
On squaring both sides, we get
$\cos^2\theta + \dfrac{1}{\cos^2\theta} + 2 = 4 \Rightarrow \cos^2\theta + \dfrac{1}{\cos^2\theta} = 2$

On cubing both sides, we get
$\cos^6\theta + \dfrac{1}{\cos^6\theta} + 3\cos^2\theta + \dfrac{3}{\cos^2\theta} = 8$
$\Rightarrow \cos^6\theta + \dfrac{1}{\cos^6\theta} + 3\left(\cos^2\theta + \dfrac{1}{\cos^2\theta}\right) = 8$
$\Rightarrow \cos^6\theta + \dfrac{1}{\cos^6\theta} + 3(2) = 8$
$\left[\because \cos^2\theta + \dfrac{1}{\cos^2\theta} = 2\right]$
$\Rightarrow \cos^6\theta + \dfrac{1}{\cos^6\theta} = 2$
$\Rightarrow \cos^6\theta + \sec^6\theta = 2$

6. (a) Let $f(\theta) = \sin^8\theta + \cos^{14}\theta$
$= \sin^8\theta + (1 - \sin^2\theta)^7$
[$\because$ maximum value of $\sin^2\theta = 1$]
$= 1 + 0 = 1$

7. (b) $\text{cosec}(75° + \theta) - \sec(15° - \theta)$
$\quad - \tan(55° + \theta) + \cot(35° - \theta)$
$\Rightarrow \text{cosec}(75° + \theta) - \text{cosec}[90° - (15° - \theta)]$
$\quad - \tan(55° + \theta) + \tan[90° - (35° - \theta)]$
$\Rightarrow \text{cosec}(75° + \theta) - \text{cosec}(75° + \theta)$
$\quad - \tan(55° + \theta) + \tan(55° + \theta) = 0$

8. (c) I. $\sqrt{\dfrac{1 - \cos\theta}{1 + \cos\theta}} = \sqrt{\dfrac{1 - \cos\theta}{1 + \cos\theta} \times \dfrac{1 - \cos\theta}{1 - \cos\theta}}$
$= \sqrt{\dfrac{(1 - \cos\theta)^2}{1 - \cos^2\theta}} = \sqrt{\dfrac{(1 - \cos\theta)^2}{\sin^2\theta}} = \dfrac{1 - \cos\theta}{\sin\theta}$
$= \text{cosec}\,\theta - \cot\theta$

II. $\sqrt{\dfrac{1 + \cos\theta}{1 - \cos\theta}} = \sqrt{\dfrac{1 + \cos\theta}{1 - \cos\theta} \times \dfrac{1 + \cos\theta}{1 + \cos\theta}}$
$= \sqrt{\dfrac{(1 + \cos\theta)^2}{1 - \cos^2\theta}} = \sqrt{\dfrac{(1 + \cos\theta)^2}{\sin^2\theta}}$
$= \dfrac{1 + \cos\theta}{\sin\theta} = \text{cosec}\,\theta + \cot\theta$

Hence, both Statements I and II are identities.

9. (a) $\dfrac{(\sin\theta + \cos\theta)(\tan\theta + \cot\theta)}{\sec\theta + \text{cosec}\,\theta}$

$= \dfrac{(\sin\theta + \cos\theta)\left(\dfrac{\sin\theta}{\cos\theta} + \dfrac{\cos\theta}{\sin\theta}\right)}{\dfrac{1}{\cos\theta} + \dfrac{1}{\sin\theta}}$

$= \dfrac{(\sin\theta + \cos\theta)\left(\dfrac{\sin^2\theta + \cos^2\theta}{\sin\theta\cos\theta}\right)}{\dfrac{\sin\theta + \cos\theta}{\sin\theta\cos\theta}}$

$$= \frac{(\sin\theta + \cos\theta)\left(\dfrac{1}{\sin\theta\cos\theta}\right)}{\dfrac{\sin\theta + \cos\theta}{\sin\theta\cos\theta}}$$

$$[\because \sin^2\theta + \cos^2\theta = 1]$$

$$= \frac{\dfrac{\sin\theta + \cos\theta}{\sin\theta\cos\theta}}{\dfrac{\sin\theta + \cos\theta}{\sin\theta\cos\theta}} = 1$$

10. (c) $\because 5\sin\theta + 12\cos\theta = 13$
On squaring both sides, we get
$25\sin^2\theta + 144\cos^2\theta + 120\sin\theta\cos\theta = 169$
$\Rightarrow \quad 25(1-\cos^2\theta) + 144(1-\sin^2\theta)$
$\qquad\qquad\qquad + 120\sin\theta\cos\theta = 169$
$\Rightarrow \quad 25 - 25\cos^2\theta + 144 - 144\sin^2\theta$
$\qquad\qquad\qquad + 120\sin\theta\cos\theta = 169$
$\Rightarrow \quad 25\cos^2\theta + 144\sin^2\theta - 120\sin\theta\cos\theta$
$\qquad\qquad\qquad = 169 - 169$
$\Rightarrow \quad (5\cos\theta - 12\sin\theta)^2 = 0$
$\Rightarrow \quad 5\cos\theta - 12\sin\theta = 0$

11. (a) $\because \sin(A+B) = 1$
$\Rightarrow \quad \sin(A+B) = \sin 90°$
$\Rightarrow \quad (A+B) = 90°$
$\therefore \quad B = 90° - A$ or $A = 90° - B$
Now, $\cos(A-B)$
$= \cos A\cos B + \sin A\sin B$
$= \cos(90°-B)\cos B + \sin(90°-B)\sin B$
$= \sin B\cos B + \cos B\sin B$
$= 2\sin B\cos B = \sin 2B$

12. (c) Since, α and β are complementary angles.
$\Rightarrow \alpha = 90° - \beta$
Now, $\sqrt{\cos\alpha\,\text{cosec}\,\beta - \cos\alpha\sin\beta}$
$= \sqrt{\dfrac{\cos\alpha}{\sin\beta} - \cos\alpha\sin\beta}$
$= \sqrt{\dfrac{\cos\alpha}{\cos(90°-\beta)} - \cos\alpha\cos(90°-\beta)}$
$= \sqrt{\dfrac{\cos\alpha}{\cos\alpha} - \cos\alpha\cdot\cos\alpha}$
$= \sqrt{1 - \cos^2\alpha} = \sqrt{\sin^2\alpha} = \sin\alpha$

13. (a) $2\,\text{cosec}^2\,23°\cot^2 67° - \sin^2 23°$
$\qquad\qquad\qquad - \sin^2 67° - \cot^2 67°$
$= \cot^2 67°(2\,\text{cosec}^2\,23°-1)$
$\qquad\qquad\qquad - \sin^2 23° - \cos^2 23°$
$\qquad\qquad\qquad [\because \cos\theta = \sin(90°-\theta)]$
$= \cot^2 67°\,(\text{cosec}^2\,23° + \text{cosec}^2\,23° - 1)$
$\qquad\qquad\qquad - (\sin^2 23° + \cos^2 23°)$
$= \cot^2 67°\,(\text{cosec}^2\,23° + \cot^2 23°) - 1$
$= \cot^2 67°\,\text{cosec}^2\,23° + \cot^2 67°\cot^2 23° - 1$

$= \cot^2 67°\,\text{cosec}^2\,23° + \cot^2 67°\tan^2 67° - 1$
$= \tan^2 23°\,\text{cosec}^2\,23° + 1 - 1$
$= \dfrac{\sin^2 23°}{\cos^2 23°} \times \dfrac{1}{\sin^2 23°} = \dfrac{1}{\cos^2 23°} = \sec^2 23°$

14. (a) $\cot\theta\cdot\tan(90°-\theta) - \sec(90°-\theta)\cdot\text{cosec}\,\theta$
$\qquad\qquad + (\sin^2 25° + \sin^2 65°)$
$\qquad + \sqrt{3}(\tan 5°\cdot\tan 15°\cdot\tan 30°\cdot\tan 75°\cdot\tan 85°)$
$= \cot\theta\cdot\cot\theta - \text{cosec}\,\theta\cdot\text{cosec}\,\theta$
$\qquad\qquad + (\sin^2 25° + \cos^2 25°)$
$\qquad + \sqrt{3}(\tan 5°\cdot\cot 5°\cdot\tan 15°\cdot\cot 15°\cdot\tan 30°)$
$\begin{bmatrix}\because \sec(90°-\theta) = \text{cosec}\,\theta,\\ \sin(90°-\theta) = \cos\theta,\\ \tan(90°-\theta) = \cot\theta\end{bmatrix}$
$= (\cot^2\theta - \text{cosec}^2\theta) + (\sin^2 25°$
$\qquad\qquad + \cos^2 25°) + \sqrt{3}\cdot\dfrac{1}{\sqrt{3}}$
$[\because \sin^2\theta + \cos^2\theta = 1 \text{ and } \cot^2\theta - \text{cosec}^2\theta = -1]$
$= -1 + 1 + 1 = 1$

15. (d) $\sin A\cos A(\tan A - \cot A)$
$= \sin A\cos A\left(\dfrac{\sin A}{\cos A} - \dfrac{\cos A}{\sin A}\right)$
$= \sin A\cos A\left(\dfrac{\sin^2 A - \cos^2 A}{\sin A\cos A}\right)$
$= (\sin^2 A - \cos^2 A)$
$= \sin^2 A - (1 - \sin^2 A)$
$= \sin^2 A - 1 + \sin^2 A = 2\sin^2 A - 1$

16. (a) Since, $0 \leq \cos^2 x \leq 1$,
$0 \leq \cos^2 y \leq 1, 0 \leq \cos^2 z \leq 1$
$\Rightarrow \quad 0 \leq \cos^2 x + \cos^2 y \leq 2$
and $\quad -1 \leq -\cos^2 z \leq 0$
$\therefore \quad -1 \leq \cos^2 x + \cos^2 y - \cos^2 z \leq 2$
$\therefore$ Minimum value of the given expression is -1.

17. (d) $(\sec x\cdot\sec y + \tan x\cdot\tan y)^2$
$\qquad\qquad - (\sec x\cdot\tan y + \tan x\cdot\sec y)^2$
$= (\sec^2 x\cdot\sec^2 y + \tan^2 x\cdot\tan^2 y + 2$
$\qquad \sec x\cdot\sec y\cdot\tan x\cdot\tan y) - (\sec^2 x\cdot\tan^2 y$
$\qquad + \tan^2 x\cdot\sec^2 y + 2\sec x\cdot\sec y\cdot\tan x\cdot\tan y)$
$= \sec^2 x\cdot\sec^2 y + \tan^2 x\cdot\tan^2 y$
$\qquad\qquad - \sec^2 x\cdot\tan^2 y - \tan^2 x\cdot\sec^2 y$
$= \sec^2 x\cdot\sec^2 y - \sec^2 x\cdot\tan^2 y$
$\qquad\qquad - \tan^2 x\cdot\sec^2 y + \tan^2 x\cdot\tan^2 y$
$= \sec^2 x\,(\sec^2 y - \tan^2 y) - \tan^2 x$
$\qquad\qquad\qquad (\sec^2 y - \tan^2 y)$
$= \sec^2 x - \tan^2 x = 1 \quad [\because \sec^2\theta - \tan^2\theta = 1]$

Trigonometry / 727

18. (a) $\tan 15° \cdot \cot 75° + \tan 75° \cdot \cot 15°$
$= \tan 15° \cdot \cot (90° - 15°)$
$\qquad + \tan (90° - 15°) \cdot \cot 15°$
$= \tan^2 15° + \cot^2 15°$ ...(i)
$\begin{bmatrix} \because \tan (90° - \theta) = \cot \theta \\ \text{and } \cot (90° - \theta) = \tan \theta \end{bmatrix}$

Now, $\cot 15° = \dfrac{1}{\tan 15°} = \dfrac{1}{2 - \sqrt{3}}$

$= \dfrac{2 + \sqrt{3}}{(2 - \sqrt{3})(2 + \sqrt{3})} = 2 + \sqrt{3}$

and $\tan 15° = 2 - \sqrt{3}$

$\therefore \tan^2 15° + \cot^2 15° = (2 - \sqrt{3})^2 + (2 + \sqrt{3})^2$
$= 2[(2)^2 + (\sqrt{3})^2] = 2(4 + 3) = 14$

19. (a) Given, $\sin \theta \cos \theta = 2 \cos^3 \theta - \dfrac{3}{2} \cos \theta$

$\Rightarrow 2 \sin \theta \cos \theta = 4 \cos^3 \theta - 3 \cos \theta$
$\Rightarrow \cos \theta \neq 0; 2\sin \theta = 4 \cos^2 \theta - 3$
$\Rightarrow 2 \sin \theta = 4 - 4\sin^2 \theta - 3$
$\Rightarrow 4\sin^2 \theta + 2\sin \theta - 1 = 0$
$\Rightarrow \sin \theta = \dfrac{-2 \pm \sqrt{4 + 16}}{8} = \dfrac{-2 \pm 2\sqrt{5}}{8}$
$\therefore \sin \theta = \dfrac{-1 \pm \sqrt{5}}{4}$

Since, θ is acute angle, $\sin \theta > 0$
$\therefore \sin \theta = \dfrac{\sqrt{5} - 1}{4}$

20. (b) Given, $3 - \tan^2 \theta = \alpha (1 - 3\tan^2 \theta)$
$\Rightarrow (3\alpha - 1) \tan^2 \theta = \alpha - 3$
$\Rightarrow \tan^2 \theta = \dfrac{\alpha - 3}{3\alpha - 1}$

As, $\tan^2 \theta \geq 0$, then $\dfrac{\alpha - 3}{3\alpha - 1} \geq 0$

$\Rightarrow \alpha \geq 3$ or $\alpha \leq \dfrac{1}{3}$

$\therefore \alpha \in \left(-\infty, \dfrac{1}{3}\right] \cup [3, \infty)$

21. (b) I. $\sin^4 x - 2\sin^2 x - 1 = 0$
$\Rightarrow \sin^2 x = \dfrac{2 \pm \sqrt{4 + 4}}{2} = \dfrac{2 \pm \sqrt{8}}{2}$
$\Rightarrow \sin^2 x = \dfrac{2 \pm 2\sqrt{2}}{2} = 1 \pm \sqrt{2}$
$\Rightarrow \sin^2 x = 1 + \sqrt{2} > 1$
or $\sin^2 x = 1 - \sqrt{2} < 0$

Hence, both are not possible.

II. Since, 1.5 radian is in II quadrant.
Therefore, $\sin 1.5 > 0$ and $\cos 1.5 < 0$
$\therefore \sin 1.5 > \cos 1.5$

22. (a) $\dfrac{(1 + \sec \theta - \tan \theta) \cos \theta}{(1 + \sec \theta + \tan \theta)(1 - \sin \theta)}$

$= \dfrac{\left(1 + \dfrac{1}{\cos \theta} - \dfrac{\sin \theta}{\cos \theta}\right) \cos \theta}{\left(1 + \dfrac{1}{\cos \theta} + \dfrac{\sin \theta}{\cos \theta}\right)(1 - \sin \theta)}$

$= \dfrac{\left(\dfrac{\cos \theta + 1 - \sin \theta}{\cos \theta}\right) \cos \theta}{(\cos \theta + 1 + \sin \theta)(1 - \sin \theta)}$

$= \dfrac{\cos \theta + 1 - \sin \theta}{\cos \theta + 1 + \sin \theta - \sin \theta \cos \theta - \sin \theta - \sin^2 \theta}$

$= \dfrac{\cos \theta + 1 - \sin \theta}{\cos \theta + 1 - \sin^2 \theta - \sin \theta \cos \theta}$

$= \dfrac{\cos \theta + 1 - \sin \theta}{\cos \theta + \cos^2 \theta - \sin \theta \cos \theta}$

$[\because 1 - \sin^2 \theta = \cos^2 \theta]$

$= \dfrac{\cos \theta + 1 - \sin \theta}{\cos \theta (\cos \theta + 1 - \sin \theta)}$

$= \dfrac{\cos \theta + 1 - \sin \theta}{\cos \theta + 1 - \sin \theta} = 1$

23. (d) Given, $\sin \theta \cdot \cos \theta = \dfrac{1}{2}$

$\sin^6 \theta + \cos^6 \theta = (\sin^2 \theta)^3 + (\cos^2 \theta)^3$
$= (\sin^2 \theta + \cos^2 \theta)(\sin^4 \theta + \cos^4 \theta$
$\qquad - \sin^2 \theta \cos^2 \theta)$
$[\because \sin^2 \theta + \cos^2 \theta = 1]$
$= (\sin^2 \theta + \cos^2 \theta)^2 - 2\sin^2 \theta \cos^2 \theta$
$\qquad - \sin^2 \theta \cos^2 \theta$
$= (1 - 3\sin^2 \theta \cos^2 \theta)$ $\left[\because \sin \theta \cdot \cos \theta = \dfrac{1}{2}\right]$
$= 1 - 3 \times \dfrac{1}{4} = 1 - \dfrac{3}{4} = \dfrac{1}{4}$

24. (a) $\because \sin \theta + \cos \theta = \sqrt{3}$
On squaring both sides, we get
$(\sin \theta + \cos \theta)^2 = (\sqrt{3})^2$
$\Rightarrow \sin^2 \theta + \cos^2 \theta + 2 \sin \theta \cos \theta = 3$
$\Rightarrow 1 + 2 \sin \theta \cos \theta = 3$
$\Rightarrow \sin \theta \cos \theta = \dfrac{3 - 1}{2} = \dfrac{2}{2} = 1$ ...(i)

Now, $\tan\theta + \cot\theta = \dfrac{\sin\theta}{\cos\theta} + \dfrac{\cos\theta}{\sin\theta}$

$= \dfrac{\sin^2\theta + \cos^2\theta}{\sin\theta\cos\theta} = \dfrac{1}{\sin\theta\cos\theta}$

From Eq. (i), $\tan\theta + \cot\theta = \dfrac{1}{1} = 1$

25. (a) $\cos x + \sec x = 2$ ...(i)

On squaring both sides, we get
$(\cos x + \sec x)^2 = 2^2$
$\Rightarrow \cos^2 x + \sec^2 x + 2\cos x \sec x = 4$
$\Rightarrow \cos^2 x + \sec^2 x + 2 = 4$
$\Rightarrow \cos^2 x + \sec^2 x = 2$ ...(ii)

On cubing both sides of Eq. (i), we get
$\cos^3 x + \sec^3 x + 3(\cos x + \sec x) = 8$
$\Rightarrow \cos^3 x + \sec^3 x + (3 \times 2) = 8$
$\Rightarrow \cos^3 x + \sec^3 x = 2$ ...(iii)

From Eqs. (i), (ii) and (iii), we get
$\cos^n x + \sec^n x = 2$

26. (c) Given, $\sin\theta - \cos\theta = 0$

On squaring both sides, we get
$(\sin\theta - \cos\theta)^2 = 0$
$\Rightarrow \sin^2\theta + \cos^2\theta - 2\sin\theta\cos\theta = 0$

$[\because \sin^2\theta + \cos^2\theta = 1]$

$\Rightarrow 2\sin\theta\cos\theta = 1$ ...(i)

Given, $\sin\left(\dfrac{\pi}{2} - \theta\right) + \cos\left(\dfrac{\pi}{2} - \theta\right)$

$= \cos\theta + \sin\theta = \sqrt{(\sin\theta + \cos\theta)^2}$

$= \sqrt{\sin^2\theta + \cos^2\theta + 2\sin\theta\cos\theta}$

$= \sqrt{1+1} = \sqrt{2}$ [from Eq. (i)]

27. (d) $\because a\sin\theta + b\cos\theta = c$

On squaring both sides, we get
$a^2\sin^2\theta + b^2\cos^2\theta + 2ab\sin\theta\cos\theta = c^2$
$\Rightarrow a^2(1 - \cos^2\theta) + b^2(1 - \sin^2\theta)$
$\quad + 2ab\sin\theta\cos\theta = c^2$
$\Rightarrow a^2 - a^2\cos^2\theta + b^2 - b^2\sin^2\theta$
$\quad + 2ab\sin\theta\cos\theta = c^2$

On rearranging, we get
$a^2 + b^2 - c^2 = a^2\cos^2\theta + b^2\sin^2\theta$
$\quad - 2ab\sin\theta\cos\theta$
$\Rightarrow a^2 + b^2 - c^2 = (a\cos\theta - b\sin\theta)^2$
$\Rightarrow a\cos\theta - b\sin\theta = \pm\sqrt{a^2 + b^2 - c^2}$

28. (b) $\because \tan\alpha = n\tan\beta$

$\Rightarrow \dfrac{\sin\alpha}{\cos\alpha} = n\dfrac{\sin\beta}{\cos\beta}$

$\Rightarrow \dfrac{m\sin\beta}{\cos\alpha} = n\dfrac{\sin\beta}{\cos\beta}$

$[\because \sin\alpha = m\sin\beta, \text{given}]$

$\Rightarrow \cos\alpha = \dfrac{m}{n}\cos\beta$

On squaring both sides, we get
$\cos^2\alpha = \dfrac{m^2}{n^2}\cos^2\beta$ ...(i)

Also, $\sin\alpha = m\sin\beta$

On squaring both sides, we get
$\sin^2\alpha = m^2\sin^2\beta$
$\Rightarrow 1 - \cos^2\alpha = m^2(1 - \cos^2\beta)$
$\Rightarrow 1 - \cos^2\alpha = m^2 - m^2\cos^2\beta$
$\Rightarrow -\dfrac{(1 - \cos^2\alpha - m^2)}{m^2} = \cos^2\beta$
$\Rightarrow \dfrac{(\cos^2\alpha + m^2 - 1)}{m^2} = \cos^2\beta$ ...(ii)

From Eqs. (i) and (ii), we get
$\cos^2\alpha = \dfrac{m^2}{n^2} \times \dfrac{(\cos^2\alpha + m^2 - 1)}{m^2}$
$\Rightarrow n^2\cos^2\alpha = \cos^2\alpha + m^2 - 1$
$\Rightarrow (n^2 - 1)\cos^2\alpha = m^2 - 1$
$\therefore \cos^2\alpha = \dfrac{m^2 - 1}{n^2 - 1}$

29. (d) Given, $l\cos^2\theta + m\sin^2\theta$

$= \dfrac{\cos^2\theta(\csc^2\theta + 1)}{\csc^2\theta - 1}$

$= \dfrac{\cos^2\theta(1 + \sin^2\theta)}{1 - \sin^2\theta} \cdot \dfrac{\sin^2\theta}{\sin^2\theta}$

$= \dfrac{\cos^2\theta(1 + \sin^2\theta)}{\cos^2\theta}$

$= 1 + \sin^2\theta = \cos^2\theta + \sin^2\theta + \sin^2\theta$

$= \cos^2\theta + 2\sin^2\theta$

$\Rightarrow (l-1)\cos^2\theta = (2-m)\sin^2\theta$

$\Rightarrow \tan^2\theta = \dfrac{l-1}{2-m}$

$\Rightarrow \tan\theta = \sqrt{\dfrac{l-1}{2-m}}$

30. (d) Given, $\sin\theta + \cos\theta = \dfrac{1+\sqrt{3}}{2}$

On squaring both sides, we get
$(\sin\theta + \cos\theta)^2 = \left(\dfrac{1+\sqrt{3}}{2}\right)^2$

$\Rightarrow \quad \sin^2\theta + \cos^2\theta + 2\sin\theta\cos\theta$

$$= \frac{1^2 + (\sqrt{3})^2 + 2 \times 1 \times \sqrt{3}}{4}$$

$$[\because (a+b)^2 = a^2 + b^2 + 2ab]$$

$\Rightarrow \quad 1 + 2\sin\theta\cos\theta = \dfrac{1+3+2\sqrt{3}}{4}$

$$[\because \sin^2\theta + \cos^2\theta = 1]$$

$\Rightarrow \quad 2\sin\theta\cos\theta = \dfrac{4+2\sqrt{3}}{4} - 1$

$$= \frac{4+2\sqrt{3}-4}{4} = \frac{2\sqrt{3}}{4}$$

$\Rightarrow \quad 2\sin\theta\cos\theta = \dfrac{\sqrt{3}}{2}$

$\Rightarrow \quad \sin\theta\cos\theta = \dfrac{\sqrt{3}}{4}$ ...(i)

Now, $\tan\theta + \cot\theta = \dfrac{\sin\theta}{\cos\theta} + \dfrac{\cos\theta}{\sin\theta}$

$$= \frac{\sin^2\theta + \cos^2\theta}{\cos\theta\sin\theta} = \frac{1}{\sin\theta\cos\theta}$$

$$= \frac{1}{(\sqrt{3}/4)} = \frac{4}{\sqrt{3}} \quad \text{[from Eq. (i)]}$$

31. (a) $\because \cos x + \cos^2 x = 1$

$\Rightarrow \quad \cos x = 1 - \cos^2 x$

$\Rightarrow \quad \cos x = \sin^2 x$ ...(i)

Again, $\cos x + \cos^2 x = 1$

On cubing both sides, we get

$(\cos x + \cos^2 x)^3 = (1)^3$

$\Rightarrow \cos^3 x + (\cos^2 x)^3 + 3\cos^2 x \cdot \cos^2 x$
$\qquad + 3\cos x \cdot \cos^4 x = 1$

$\Rightarrow \cos^3 x + \cos^6 x + 3\cos^4 x + 3\cos^5 x = 1$

$\Rightarrow \sin^6 x + \sin^{12} x + 3\sin^8 x + 3\sin^{10} x = 1$

[from Eq. (i)]

$\therefore \sin^{12} x + 3\sin^{10} x + 3\sin^8 x + \sin^6 x - 1 = 0$

32. (b) $\because \tan\theta + \sec\theta = m$ ...(i)

$\Rightarrow \quad \sec\theta = m - \tan\theta$

On squaring both sides, we get

$(\sec\theta)^2 = (m - \tan\theta)^2$

$\Rightarrow \quad \sec^2\theta = m^2 + \tan^2\theta - 2m\tan\theta$

$\Rightarrow \quad \sec^2\theta - \tan^2\theta = m^2 - 2m\tan\theta$

$\Rightarrow \quad 1 = m^2 - 2m\tan\theta$

$$[\because \sec^2\theta - \tan^2\theta = 1]$$

$\Rightarrow \quad \tan\theta = \dfrac{m^2-1}{2m}$

On putting the value of $\tan\theta$ in Eq. (i), we get

$\dfrac{m^2-1}{2m} + \sec\theta = m$

$\Rightarrow \quad \sec\theta = m - \left(\dfrac{m^2-1}{2m}\right)$

$\therefore \quad \sec\theta = \dfrac{2m^2 - m^2 + 1}{2m} = \dfrac{m^2+1}{2m}$

33. (b) Given, $\dfrac{\cos^2\theta}{a^2} + \dfrac{\sin^2\theta}{b^2} = \dfrac{1}{x^2+y^2}$

and $x\sin\theta - y\cos\theta = \sqrt{x^2+y^2}$

On squaring both sides, we get

$x^2\sin^2\theta + y^2\cos^2\theta$
$\qquad - 2xy\sin\theta\cos\theta = x^2 + y^2$

$\Rightarrow x^2\sin^2\theta + y^2\cos^2\theta$
$\qquad - 2xy\sin\theta\cos\theta - x^2 - y^2 = 0$

$\Rightarrow x^2(\sin^2\theta - 1) + y^2(\cos^2\theta - 1)$
$\qquad - 2xy\sin\theta\cos\theta = 0$

$\Rightarrow -x^2\cos^2\theta - y^2\sin^2\theta - 2xy\sin\theta\cos\theta = 0$

$\Rightarrow \quad (x\cos\theta + y\sin\theta)^2 = 0$

$\Rightarrow \quad x\cos\theta + y\sin\theta = 0$

$\Rightarrow \quad x\cos\theta = -y\sin\theta$

$\Rightarrow \quad \tan\theta = -\dfrac{x}{y}$

In $\triangle ABC$,

$(AC)^2 = (AB)^2 + (BC)^2$

$\Rightarrow (AC)^2 = (-y)^2 + (x)^2$

$\Rightarrow AC^2 = y^2 + x^2$

$\Rightarrow AC = \sqrt{x^2+y^2}$

$\Rightarrow \sin\theta = \dfrac{x}{\sqrt{x^2+y^2}}$ or $\dfrac{-x}{\sqrt{x^2+y^2}}$

and $\cos\theta = -\dfrac{y}{\sqrt{x^2+y^2}}$ or $\dfrac{y}{\sqrt{x^2+y^2}}$

Now, $\dfrac{\cos^2\theta}{a^2} + \dfrac{\sin^2\theta}{b^2} = \dfrac{1}{x^2+y^2}$

$\Rightarrow \dfrac{y^2}{(x^2+y^2)\cdot a^2} + \dfrac{x^2}{(x^2+y^2)\cdot b^2} = \dfrac{1}{x^2+y^2}$

$\Rightarrow \dfrac{y^2}{a^2} + \dfrac{x^2}{b^2} = 1$

$\Rightarrow \dfrac{x^2}{b^2} + \dfrac{y^2}{a^2} = 1$

34. (c) $\sin\theta + 2\cos\theta = -1$

On squaring both sides, we get

$(\sin\theta + 2\cos\theta)^2 = (-1)^2$

$\Rightarrow \sin^2\theta + 4\cos^2\theta + 4\sin\theta\cos\theta = 1$

$\Rightarrow$ $(1-\cos^2\theta) + 4(1-\sin^2\theta)$
$\qquad + 4\sin\theta\cos\theta = 1$
$\Rightarrow -(\cos^2\theta + 4\sin^2\theta) + 4\sin\theta\cos\theta = 1-5$
$\Rightarrow \cos^2\theta + 4\sin^2\theta - 4\sin\theta\cos\theta = 4$
$\Rightarrow (2\sin\theta - \cos\theta)^2 = 4$
$\Rightarrow 2\sin\theta - \cos\theta = 2$

35. (b) $\because XZ - YZ = 2$ ...(i)
By Pythagoras theorem,
$XY^2 + YZ^2 = XZ^2 \Rightarrow (2\sqrt{6})^2 = XZ^2 - YZ^2$
$\Rightarrow 24 = (XZ - YZ)(XZ + YZ)$
$\qquad [\because XZ - YZ = 2, \text{given}]$

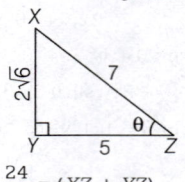

$\Rightarrow \dfrac{24}{2} = (XZ + YZ)$
$\Rightarrow XZ + YZ = 12$ ...(ii)
On adding Eqs. (i) and (ii), we get
$2XZ = 14 \Rightarrow XZ = 7$
$\therefore \quad YZ = 7 - 2 = 5$
Now, $\sec X = \dfrac{7}{2\sqrt{6}}$ and $\tan X = \dfrac{5}{2\sqrt{6}}$
$\therefore \sec X + \tan X = \dfrac{7}{2\sqrt{6}} + \dfrac{5}{2\sqrt{6}}$
$= \dfrac{12}{2\sqrt{6}} = \sqrt{6}$

36. (c) Given, $2y\cos\theta = x\sin\theta$
$\Rightarrow x\sin\theta - 2y\cos\theta = 0$ ...(i)
and $2x\sec\theta - y\csc\theta = 3$
$\Rightarrow \dfrac{2x}{\cos\theta} - \dfrac{y}{\sin\theta} = 3$
$\left[\because \sec\theta = \dfrac{1}{\cos\theta} \text{ and } \csc\theta = \dfrac{1}{\sin\theta}\right]$
$\Rightarrow 2x\sin\theta - y\cos\theta = 3\sin\theta\cos\theta$ ...(ii)
On multiplying both sides of Eq. (ii) by 2, then subtracting the result from Eq. (i), we get
$x\sin\theta - 2y\cos\theta - 4x\sin\theta + 2y\cos\theta$
$\qquad = 0 - 6\sin\theta\cos\theta$
$\Rightarrow -3x\sin\theta = -6\sin\theta\cos\theta$
$\Rightarrow x = \dfrac{-6\sin\theta\cos\theta}{-3\sin\theta}$
$\Rightarrow x = 2\cos\theta$ ...(iii)
On putting $x = 2\cos\theta$ in Eq. (i), we get
$2\cos\theta\sin\theta - 2y\cos\theta = 0$
$\Rightarrow y = \dfrac{2\cos\theta\sin\theta}{2\cos\theta}$
$\Rightarrow y = \sin\theta$ ...(iv)

$\therefore x^2 + 4y^2 = (2\cos\theta)^2 + 4(\sin\theta)^2$
[from Eqs. (iii) and (iv)]
$= 4\cos^2\theta + 4\sin^2\theta$
$= 4(\cos^2\theta + \sin^2\theta)$
$= 4 \qquad [\because \cos^2\theta + \sin^2\theta = 1]$

37. (c) Let the angles of a triangle be $4x$, x and x, respectively.

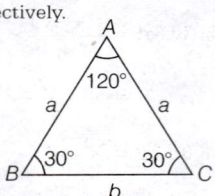

i.e. $\angle A = 4x$, $\angle B = x$ and $\angle C = x$
$\because$ Sum of all angles of a triangle = $180°$
$\therefore \quad \angle A + \angle B + \angle C = 180°$
$\Rightarrow \quad 4x + x + x = 180°$
$\Rightarrow \quad 6x = 180°$
$\Rightarrow \quad x = 30°$
$\therefore \quad \angle A = 4x = 4 \times 30° = 120°$,
$\angle B = x = 30°$
and $\angle C = x = 30°$
So, it is clear that, given triangle is an isosceles triangle.
Let the sides of an isosceles triangle be a, a and b, respectively.
$\therefore$ Perimeter of triangle = $a + a + b = 2a + b$
Using sine rule, $\dfrac{b}{\sin 120°} = \dfrac{a}{\sin 30°}$
$\Rightarrow \dfrac{b}{\sqrt{3}/2} = \dfrac{a}{1/2}$
$\Rightarrow \dfrac{2b}{\sqrt{3}} = 2a \Rightarrow a = \dfrac{b}{\sqrt{3}}$ ...(i)
Now, $\dfrac{b}{2a+b} = \dfrac{b}{2 \times \dfrac{b}{\sqrt{3}} + b}$ [from Eq. (i)]
$= \dfrac{b}{\dfrac{2b+\sqrt{3}b}{\sqrt{3}}} = \dfrac{\sqrt{3}}{2+\sqrt{3}}$

38. (b) Let the length of the side of square be l.

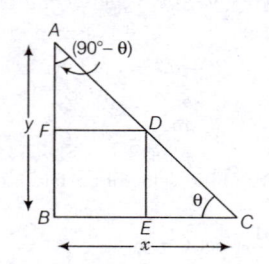

Then, in right angled ΔDEC,

$$\tan\theta = \frac{DE}{EC} = \frac{l}{x-l} \qquad \ldots(i)$$

and in right angled ΔAFD,

$$\tan(90° - \theta) = \frac{FD}{AF}$$

$$\Rightarrow \qquad \cot\theta = \frac{l}{y-l} \qquad \ldots(ii)$$

On multiplying Eqs. (i) and (ii), we get

$$1 = \frac{l^2}{(x-l)(y-l)}$$

$$\Rightarrow \qquad xy - xl - yl + l^2 = l^2$$

Since, $l(x+y) = xy$, then $l = \dfrac{xy}{x+y}$

$\therefore$ Perimeter of square $= 4l = \dfrac{4xy}{x+y}$

39. (b) Using cosine formula,

$$\cos B = \frac{BC^2 + BD^2 - CD^2}{2BC \cdot BD}$$

[here, $\angle B = 45°$ $\because AC = BC$ and $\angle C = 90°$]

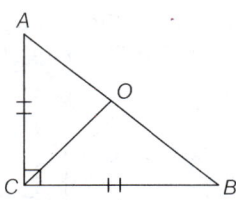

$$\frac{1}{\sqrt{2}} \times 2 \cdot BC \cdot BD = BC^2 + BD^2 - CD^2$$

$$\sqrt{2}BC \cdot BD = BC^2 + BD^2 - CD^2 \qquad \ldots(i)$$

Now, in right angled ΔABC,

$$AC^2 + BC^2 = AB^2$$

$$\Rightarrow \qquad 2BC^2 = AB^2$$

or $$BC^2 = \frac{AB^2}{2} \text{ or } BC = \frac{AB}{\sqrt{2}}$$

On putting the value of BC^2 in Eq. (i), we get

$$\sqrt{2} \cdot \frac{AB}{\sqrt{2}} \cdot BD = \frac{AB^2}{2} + BD^2 - CD^2$$

$$\Rightarrow \qquad AB \cdot BD = \frac{AB^2}{2} + BD^2 - CD^2$$

$$\Rightarrow \qquad AB \cdot BD - BD^2 = \frac{AB^2}{2} - CD^2$$

$$\Rightarrow \qquad BD(AB - BD) = \frac{AB^2}{2} - CD$$

$$[\because AD + BD = AB] \ldots(ii)$$

$$\Rightarrow \qquad BD \cdot AD = \frac{AB^2}{2} - CD^2$$

Now, $AB = AD + BD$

Squaring on both sides, we get

$$AB^2 = AD^2 + BD^2 + 2AD \cdot BD$$

$$\Rightarrow \qquad AD \cdot BD = \frac{AB^2 - AD^2 - BD^2}{2}$$

$$= \frac{AB^2}{2} - \frac{(AD^2 + BD^2)}{2}$$

On putting the value of $AD \cdot BD$ in Eq. (ii), we get

$$\frac{AB^2}{2} - \frac{(AD^2 + BD^2)}{2} = \frac{AB^2}{2} - CD^2$$

$$\Rightarrow \qquad CD^2 = \frac{AD^2 + BD^2}{2}$$

or $$AD^2 + BD^2 = 2CD^2$$

Chapter 39

Height and Distance

It is an important application of trigonometry which helps us to find the height of any object and distance of that object from any point which are not directly measurable, if the angle of elevation/depression from a point is known.

Line of Sight

A line of sight is the line drawn from the eye of an observer to the point, where the object is viewed by the observer.

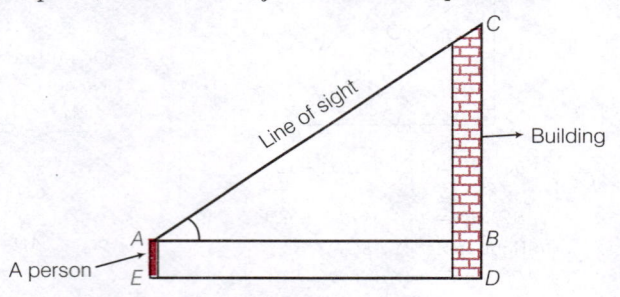

Horizontal Line

The line of sight which is parallel to ground level is known as horizontal line.

Angle of Elevation

The angle of elevation of the point viewed is the angle formed by the line of sight with the horizontal, when the point being viewed is above the horizontal level.

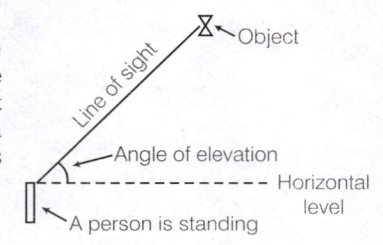

Angle of Depression

When the line of sight is below the horizontal level, the angle so formed by the line of sight with the horizontal is called the angle of depression.

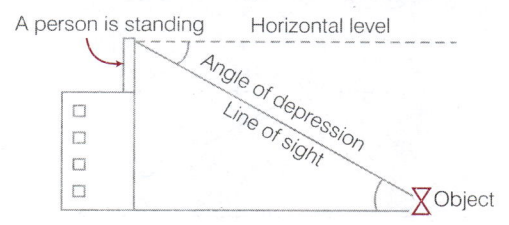

MIND IT!
1. Angle of elevation and depression are always acute angles.
2. Unless stated, it is assumed that the height of the observer is not considered.

Ex. 1 A tower stands vertically on the ground. From a point on the ground which is 30 m away from the foot of a tower, the angle of elevation of the top of the tower is found to be 45. Find the height of the tower.

Sol. Given, angle of elevation is $\angle ACB = 45°$, $BC = 30$ m and $AB = ?$

In $\triangle ABC$, $\tan 45° = \dfrac{AB}{BC}$ $\left[\because \tan\theta = \dfrac{\text{perpendicular}}{\text{base}}\right]$

$\Rightarrow \quad 1 = \dfrac{AB}{30}$

$\therefore \quad AB = 30$ m

Ex. 2 The angle of elevation of the tip of a tower from a point on the ground is 45°. Moving 21 m directly towards the base of the tower, the angle of elevation changes to 60°. What is the height of the tower, to the nearest metre?

Sol. In $\triangle PBC$,

$\tan 60° = \dfrac{h}{x} \Rightarrow \dfrac{h}{x} = \sqrt{3} \Rightarrow x = \dfrac{h}{\sqrt{3}}$ ...(i)

In $\triangle PAC$, $\tan 45° = \dfrac{h}{21+x} = 1 \Rightarrow h = 21 + x$

$\Rightarrow \quad h = 21 + \dfrac{h}{\sqrt{3}}$ [from Eq. (i)]

$\Rightarrow \quad h\left(1 - \dfrac{1}{\sqrt{3}}\right) = 21 \Rightarrow h = \dfrac{21\sqrt{3}}{(\sqrt{3}-1)} \times \dfrac{(\sqrt{3}+1)}{(\sqrt{3}+1)}$

$\Rightarrow \quad h = \dfrac{21\sqrt{3}\,(\sqrt{3}+1)}{2} = 49.68 \approx 50$ m

Alternate Method

$h = \dfrac{21}{\cot 45° - \cot 60°} = \dfrac{21}{1 - \dfrac{1}{\sqrt{3}}} = \dfrac{21\sqrt{3}}{\sqrt{3}-1} \times \dfrac{\sqrt{3}+1}{\sqrt{3}+1} = 49.68 \approx 50$ m

Ex. 3 From a point A on a bridge across a river, the angles of depression of the banks on opposite sides of the river are 30° and 45°, respectively. If the bridge is at a height of 9 m from the surface of river, then find the width of the river.

Sol. Here, width of the river = DC

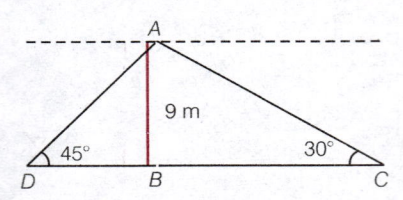

In $\triangle ABC$,

$$\tan 30° = \frac{AB}{BC} \Rightarrow BC = \frac{9}{\tan 30°} = 9\sqrt{3} \text{ m}$$

Now, in $\triangle ABD$,

$$\tan 45° = \frac{AB}{BD}$$

$\Rightarrow \quad AB = BD \Rightarrow BD = 9 \text{ m} \qquad [\because AB = 9 \text{ m}]$

$\therefore \quad DC = DB + BC = 9 + 9\sqrt{3}$

$\qquad = 9(\sqrt{3} + 1) = 24.588 \text{ m} \qquad [\because \sqrt{3} = 1.732]$

Ex. 4 A vertical post 15 ft high is broken at certain height and its upper part, not completely separated, meet the ground at an angle of 30°. Find the height at which the post is broken.

Sol. Given that, height of post = 15 ft
Let the post breaks at point C and the length of lower part is h ft.

or $\qquad BC = h$

So, $\qquad AC = AB - BC = 15 - h = CD$

In $\triangle BCD$,

$$\sin 30° = \frac{BC}{CD} \qquad \left[\because \sin \theta = \frac{\text{perpendicular}}{\text{hypotenuse}}\right]$$

$\Rightarrow \qquad \dfrac{1}{2} = \dfrac{h}{15 - h} \qquad [\because CD = AC]$

$\Rightarrow \qquad 2h = 15 - h$

$\Rightarrow \qquad 2h + h = 15 \Rightarrow 3h = 15$

$\Rightarrow \qquad h = \dfrac{15}{3} = 5 \text{ ft}$

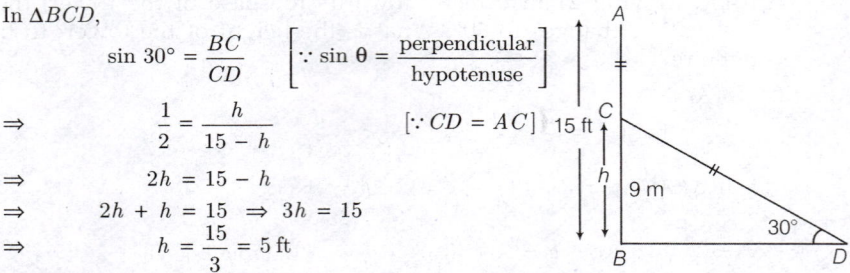

Hence, the height at which the post is broken is 5 ft.

Ex. 5 The angles of depression of two ships from the top of a light-house are 45° and 30°. If the ships are 120 m apart, then find the height of the light-house.

Sol. Let AB, the height of light-house be x m.

Since, $\qquad MN \parallel PQ$

$\therefore \qquad \angle MAP = \angle APB = 30° \qquad$ [alternate angles]

and $\qquad \angle NAQ = \angle AQB = 45°$

Height and Distance / 735

Let the length between P and B be y m. Then, the length between B and Q is $(120 - y)$ m.

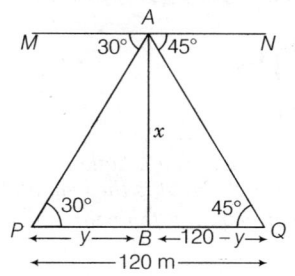

In $\triangle ABP$,
$$\tan 30° = \frac{AB}{BP} \Rightarrow \frac{1}{\sqrt{3}} = \frac{x}{y}$$
$$y = x\sqrt{3} \qquad \ldots(i)$$

In $\triangle ABQ$,
$$\tan 45° = \frac{AB}{BQ} \Rightarrow 1 = \frac{x}{120 - y}$$
$$x = 120 - y \qquad \ldots(ii)$$

From Eqs. (i) and (ii), we get
$$x = 120 - x\sqrt{3}$$
$$\Rightarrow x = \frac{120}{1 + \sqrt{3}} = 43.92 \qquad [\because \sqrt{3} = 1.732]$$
$$\therefore x \approx 44 \text{ m}$$

Alternate Method

When the height of any object from horizontal plane is h. Then, the angle of depression of two consecutive milestone in opposite direction at object is α and β, respectively.

Height of the object,
$$h = \frac{\tan \alpha \cdot \tan \beta}{\tan \alpha + \tan \beta} \times \text{Distance between the objects}$$

Here, $\alpha = 30°$ and $\beta = 45°$

Let height of the light-house be h.

Then, $\quad h = \dfrac{\tan \alpha \cdot \tan \beta}{\tan \alpha + \tan \beta} \times$ Distance between both the ships

$$\Rightarrow h = \frac{\tan 30° \cdot \tan 45°}{\tan 30° + \tan 45°} \times 120$$

$$\Rightarrow h = \frac{\frac{1}{\sqrt{3}} \times 1}{\frac{1}{\sqrt{3}} + 1} \times 120 = \frac{\frac{1}{\sqrt{3}}}{\frac{1 + \sqrt{3}}{\sqrt{3}}} \times 120$$

$$\Rightarrow h = \frac{120}{\sqrt{3} + 1} = 43.92 \qquad [\because \sqrt{3} = 1.732]$$

$$\therefore h \approx 44 \text{ m}$$

Multi Concept Questions

1. A man 2.5 m tall is 32.5 m away from a building. The angle of elevation of the top of the building from his eyes is 60°. What is the height of the building? Also, calculate the distance between the eye of man and top point of the building.
(a) 58.79 m, 65 m
(b) 60 m, 59.5 m
(c) 58 m, 60 m
(d) 59.5 m, 60.5 m

↪ (a) Given, $EC = DB = 2.5$ m, $BC = DE = 32.5$ m,
$\angle AED = 60°$, $AB = ?$ and $AE = ?$

In $\triangle AED$, $\tan 60° = \dfrac{AD}{DE}$

$\Rightarrow \quad \sqrt{3} = \dfrac{AD}{32.5}$ $\quad [\because \tan 60° = \sqrt{3}]$

$\therefore \quad AD = 32.5 \times \sqrt{3} = 32.5 \times 1.732 = 56.29$ m

$\therefore \quad AB = AD + DB = 56.29 + 2.5 = 58.79$ m

Now, $\quad \sin 60° = \dfrac{AD}{AE} = \dfrac{56.29}{AE}$

$\Rightarrow \quad \dfrac{\sqrt{3}}{2} = \dfrac{56.29}{AE}$

$\therefore \quad AE = 65$ m

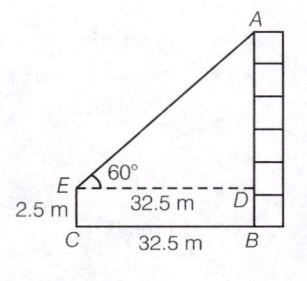

2. From a point A on the ground, the angle of elevation of the top of a 20 m tall building is 45°. A flag is hoisted at the top of the building and the angle of elevation of the top of the flag staff from A is 60°. Find the distance of the building from the point A and length of the flag staff.
(a) 20 m, 15 m
(b) 20 m, 14.64 m
(c) 25 m, 15 m
(d) 20 m, 19 m

↪ (b) Given, $BC = 20$ m, $\angle BAC = 45°$
and $\angle DAC = 60°$

In $\triangle BAC$, $\tan 45° = \dfrac{BC}{AC} = \dfrac{20}{AC}$

$\therefore \quad AC = 20$ m $\quad [\because \tan 45° = 1]$

Let $\quad DB = x$ m

$\therefore \quad DC = DB + BC = (x + 20)$ m

Now, in $\triangle DAC$, $\tan 60° = \dfrac{DC}{AC} = \dfrac{x + 20}{20}$

$\sqrt{3} = \dfrac{x + 20}{20}$

$\therefore \quad x + 20 = 20\sqrt{3} = 34.64$

$\therefore \quad x = 34.64 - 20 = 14.64$ m

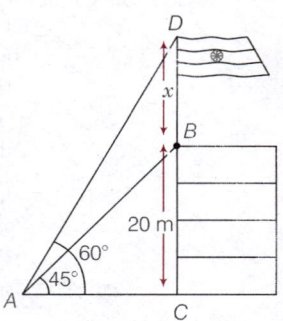

Fast Track Practice

1. If the length of the shadow of a tower is equal to its height, then what is the Sun's altitude at that time? [CDS 2016 (II)]
   (a) 15° (b) 30° (c) 45° (d) 60°

2. A vertical stick 12 m long casts a shadow 8 m long on the ground. At the same time, a tower casts a shadow of 40 m long on the ground. The height of the tower is [SSC CGL 2012]
   (a) 60 m (b) 65 m (c) 70 m (d) 72 m

3. The shadow of a tower standing on a level plane is found to be 50 m longer when the Sun's elevation is 30°, when it is 60°. What is the height of the tower? [CDS 2014]
   (a) 25 m (b) $25\sqrt{3}$ m (c) $\frac{25}{\sqrt{3}}$ m (d) 30 m

4. A man from the top of a 100 m high tower seen a car moving towards the tower at an angle of depression 30°. After some time, the angle of depression becomes 60°. What is the distance travelled by the car during this time? [CDS 2016 (II)]
   (a) $100\sqrt{3}$ m (b) $\frac{200\sqrt{3}}{3}$ m
   (c) $\frac{100\sqrt{3}}{3}$ m (d) $200\sqrt{3}$ m

5. The angle of elevation of the top of an unfinished pillar at a point 150 m from its base is 30°. If the angle of elevation at the same point is to be 45°, then the pillar has to be raised to a height of how many metres? [CDS 2009]
   (a) 59.4 (b) 61.4 (c) 62.4 (d) 63.4

6. The angles of depression of two ships from the top of a light-house are 60° and 45° towards East. If the ships are 300 m apart, the height of the light-house is [SSC CGL (Mains) 2016]
   (a) $200(3+\sqrt{3})$ m (b) $250(3+\sqrt{3})$ m
   (c) $150(3+\sqrt{3})$ m (d) $160(3+\sqrt{3})$ m

7. From the top of a building 60 m high, the angles of depression of the top and bottom of a tower are observed to be 30° and 60°. The height of the tower is [SSC CGL (Pre) 2017]
   (a) 40 m (b) 45 m
   (c) 50 m (d) 55 m

8. The angle of elevation of the top of a tower from the bottom of a building is twice that from its top. What is the height of the building, if the height of the tower is 75 m and the angle of elevation of the top of the tower from the bottom of the building is 60°? [CDS 2011]
   (a) 25 m (b) 37.5 m
   (c) 50 m (d) 60 m

9. The tops of two poles of height 24 m and 36 m are connected by a wire. If the wire makes an angle of 60° with the horizontal, then the length of the wire is [SSC CGL 2012]
   (a) $8\sqrt{3}$ m (b) 8 m (c) $6\sqrt{3}$ m (d) 6 m

10. The angles of elevation of the top of a tower from two points which are at distances of 10 m and 5 m from the base of the tower and in the same straight line with it are complementary. The height of the tower is [CDS 2012]
    (a) 5 m (b) 15 m (c) $\sqrt{50}$ m (d) $\sqrt{75}$ m

11. Two men on either side of a tower 75 m high observed that the angles of elevation of the top of the tower to be 30° and 60°. What is the distance between the two men? [CDS 2016 (II)]
    (a) $100\sqrt{3}$ m (b) $75\sqrt{3}$ m
    (c) $\frac{100\sqrt{3}}{3}$ m (d) $60\sqrt{3}$ m

12. From the top of a tower, the angles of depression of two objects P and Q (situated on the ground on the same side of the tower) separated at a distance of $100(3-\sqrt{3})$ m are 45° and 60°, respectively. The height of the tower is [CDS 2015 (I)]
    (a) 200 m (b) 250 m
    (c) 300 m (d) None of these

13. An aeroplane flying at a height of 300 m above the ground passes vertically above another plane at an instant when the angles of elevation of the two planes from the same point on the ground are 60° and 45°, respectively. What is the height of the lower plane from the ground? [CDS 2017 (I)]
    (a) $100\sqrt{3}$ m (b) $\frac{100}{\sqrt{3}}$ m
    (c) $50\sqrt{3}$ m (d) $150(\sqrt{3}+1)$ m

14. If the angle of elevation of a tower from two distant points a and b ($a > b$) from its foot and in the same straight line and on the same side of it are 30° and 60°, then the height of the tower is [SSC CPO 2013]

    (a) $\sqrt{\dfrac{a}{b}}$　　(b) $\sqrt{a+b}$
    (c) $\sqrt{ab}$　　(d) $\sqrt{a-b}$

15. The angle of elevation of the top of a tower 30 m high from the foot of another tower in the same plane is 60° and the angle of elevation of the top of the second tower from the foot of the first tower is 30°. The distance between the two towers is n times the height of the shorter tower. What is n equal to? [CDS 2014]

    (a) $\sqrt{2}$　　(b) $\sqrt{3}$　　(c) 1/2　　(d) 1/3

16. The angle of elevation of a cloud from a point 200 m above a lake is 30° and the angle of depression of its reflection in the lake is 60°. The height of the cloud is [CDS 2015 (I)]

    (a) 200 m　　(b) 300 m
    (c) 400 m　　(d) 600 m

17. As seen from the top and bottom of a building of height h m, the angles of elevation of the top of a tower of height $\dfrac{(3+\sqrt{3})h}{2}$ m are α and β, respectively.
    If $\beta = 30°$, then what is the value of $\tan\alpha$? [CDS 2013]

    (a) 1/2　　(b) 1/3
    (c) 1/4　　(d) None of these

18. From an aeroplane vertically over a straight horizontal road, the angles of depression of two consecutive kilometre-stones on the opposite sides of the aeroplane are observed to be α and β. The height of the aeroplane above the road is [CDS 2017 (I)]

    (a) $\dfrac{\tan\alpha + \tan\beta}{\tan\alpha \tan\beta}$　　(b) $\dfrac{\tan\alpha \tan\beta}{\tan\alpha + \tan\beta}$
    (c) $\dfrac{\cot\alpha \cot\beta}{\cot\alpha + \cot\beta}$　　(d) $\dfrac{\cot\alpha + \cot\beta}{\cot\alpha \cot\beta}$

19. A person of height 6 ft wants to pluck a fruit which is on a $\dfrac{26}{3}$ ft high tree. If the person is standing $\dfrac{8}{\sqrt{3}}$ ft away from the base of the tree, then at what angle should he throw the stone so that, it hits the fruit? [SSC CGL 2015]

    (a) 30°　　(b) 45°
    (c) 60°　　(d) 75°

20. Peter was standing on the top of a rock cliff facing the sea. He saw a boat coming towards the shore. As he kept seeing time just flew. Ten minutes less than half of an hour, the angle of depression changed from 30° to 60°. How much more time will the boat take to reach the shore? [SNAP 2016]

    (a) 5 min　　(b) 10 min
    (c) 15 min　　(d) 20 min

21. A pole stands vertically inside a triangular park ABC. If the angle of elevation of the top of the pole from each corner of the park is same, then in the ABC, the foot of the pole is at the

    (a) centroid　　[CDS 2016 (II)]
    (b) circumcentre
    (c) incentre
    (d) orthocentre

22. At the foot of a mountain, the elevation of its summit is 45°. After ascending 2 km towards the mountain upon an incline of 30°, the elevation changes to 60°. The height of the mountain is [SSC CGL 2012]

    (a) $(\sqrt{3}-1)$ km　　(b) $(\sqrt{3}+1)$ km
    (c) $(\sqrt{3}-2)$ km　　(d) $(\sqrt{3}-2)$ km

23. A man standing in one corner of a square football field observes that the angle subtended by a pole in the corner just diagonally opposite to this corner is 60°. When he retires 80 m from the corner, along the same straight line, he finds the angle to be 30°. The length of the field is [SSC CGL 2013]

    (a) 20 m　　(b) $40\sqrt{2}$ m
    (c) 40 m　　(d) $20\sqrt{2}$ m

24. A spherical balloon of radius r subtends angle 60° at the eye of an observer. If the angle of elevation of its centre is 60° and h is the height of the centre of the balloon, then which one of the following is correct? [CDS 2013]

    (a) $h = r$
    (b) $h = \sqrt{2}r$
    (c) $h = \sqrt{3}r$
    (d) $h = 2r$

Answer with Solutions

1. **(c)** Let AB and BC be the height of tower and length of the shadow of a tower.

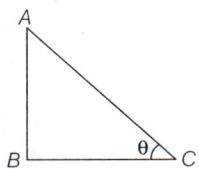

   According to the question,
   $AB = BC$ ...(i)
   Now, in $\triangle ABC$,
   $\tan\theta = \dfrac{AB}{BC} = \dfrac{AB}{AB}$ [from Eq. (i)]
   $\Rightarrow \tan\theta = 1 \Rightarrow \tan\theta = \tan 45° \Rightarrow \theta = 45°$
   Hence, the Sun's altitude is $45°$.

2. **(a)** Given, height of first vertical stick = 12 m and length of its shadow = 8 m
   Let height of tower = x m
   and length of its shadow = 40 m

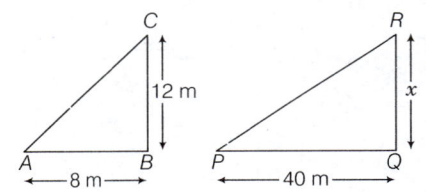

   Now, in $\triangle ABC$ and $\triangle PQR$,
   $\dfrac{12}{8} = \dfrac{x}{40}$ [by proportion]
   $\Rightarrow \dfrac{12 \times 40}{8} = x \Rightarrow x = 60$ m

3. **(b)** Let h be the height of the tower and $BC = x$ m.
   In $\triangle BCA$,
   $\tan 60° = \dfrac{h}{x} \Rightarrow \sqrt{3} = \dfrac{h}{x}$
   $\Rightarrow h = x\sqrt{3}$ ...(i)
   Now, in $\triangle ABD$, $\tan 30° = \dfrac{h}{50 + x}$

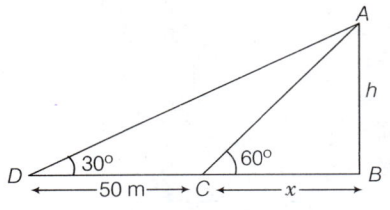

   $\Rightarrow \dfrac{1}{\sqrt{3}} = \dfrac{x\sqrt{3}}{50 + x}$ [$\because h = x\sqrt{3}$]
   $\Rightarrow 50 + x = 3x \Rightarrow x = 25$ m
   $\therefore h = 25\sqrt{3}$ m [from Eq. (i)]

4. **(b)** Let AB be the tower of height 100 m.

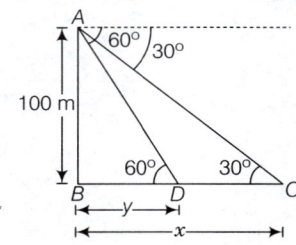

   Now, in $\triangle ABC$,
   $\tan 30° = \dfrac{AB}{BC} = \dfrac{100}{x} \Rightarrow x = 100\sqrt{3}$ m
   Again, in $\triangle ABD$,
   $\tan 60° = \dfrac{AB}{BD} = \dfrac{100}{y} \Rightarrow y = \dfrac{100}{\sqrt{3}}$ m
   $\therefore$ Required distance travelled by car
   $= CD = x - y$
   $= \left(100\sqrt{3} - \dfrac{100}{\sqrt{3}}\right) = \dfrac{300 - 100}{\sqrt{3}}$
   $= \dfrac{200}{\sqrt{3}} = \dfrac{200}{3}\sqrt{3}$ m

5. **(d)** Given, $BC = 150$ m
   $\angle ACB = 30°$ and $\angle DCB = 45°$

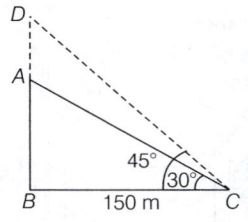

   Then, $AD = ?$
   In $\triangle ABC$, $\tan 30° = \dfrac{AB}{BC} \Rightarrow \dfrac{1}{\sqrt{3}} = \dfrac{AB}{150}$
   $\therefore \quad AB = 86.6$ m
   In $\triangle DBC$, $\tan 45° = \dfrac{DB}{BC} \Rightarrow 1 = \dfrac{AD + AB}{BC}$
   $\Rightarrow \quad BC = AD + 86.6$
   $\therefore \quad AD = 150 - 86.6 = 63.4$ m

6. (c) Let x be the distance BC.

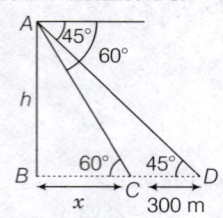

In $\triangle ABC$,
$$\frac{h}{x} = \tan 60° \Rightarrow h = \sqrt{3}x \qquad \ldots(i)$$

In $\triangle ABD$,
$$\frac{h}{x+300} = \tan 45°$$
$$\Rightarrow \quad h = x + 300$$
$$\Rightarrow \quad h = \frac{h}{\sqrt{3}} + 300 \qquad \text{[From Eq. (i)]}$$
$$\Rightarrow \quad h\left(\frac{\sqrt{3}-1}{\sqrt{3}}\right) = 300$$
$$\Rightarrow \quad h = \frac{300\sqrt{3}}{\sqrt{3}-1}$$
$$\Rightarrow \quad h = \frac{300 \times \sqrt{3} \times (\sqrt{3}+1)}{(\sqrt{3}-1)(\sqrt{3}+1)}$$
$$\Rightarrow \quad h = \frac{300 \times (3+\sqrt{3})}{(3-1)} = \frac{300(3+\sqrt{3})}{2}$$
$$\Rightarrow \quad h = 150(3+\sqrt{3})$$

∴ Height of the light house
$= 150(3+\sqrt{3})$ m

7. (a) Let AB be a building and PQ be a tower.
Let height of tower $PQ = h$ m
$AM = (60 - h)$
$BQ = y = PM$,
$AB = 60$, $BM = h$

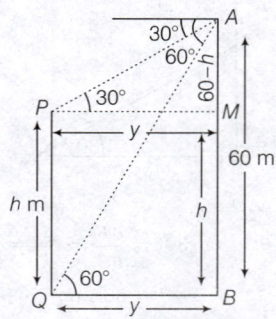

Now, in $\triangle AMP$,
$$\tan 30° = \frac{AM}{PM} = \frac{60-h}{y}$$
$$\Rightarrow \quad \frac{1}{\sqrt{3}} = \frac{60-h}{y}$$

$$\Rightarrow \quad y = (60-h)\sqrt{3} \qquad \ldots(i)$$

Now, in $\triangle AQB$,
$$\tan 60° = \frac{AB}{BQ} = \frac{60}{y}$$
$$\Rightarrow \quad \sqrt{3} = \frac{60}{y} \Rightarrow y = \frac{60}{\sqrt{3}} \qquad \ldots(ii)$$

From Eqs. (i) and (ii), we get
$$(60-h)\sqrt{3} = \frac{60}{\sqrt{3}}$$
$$\Rightarrow \quad 3(60-h) = 60$$
$$\Rightarrow \quad 180 - 3h = 60 \Rightarrow 3h = 180 - 60$$
$$\Rightarrow \quad 3h = 120 \Rightarrow h = 40$$

Hence, height of tower = 40 m

8. (c) We have to find DC.
Given, $2x = 60° \Rightarrow x = 30°$

In $\triangle ABC$, $\tan 60° = \frac{AB}{BC} \Rightarrow \frac{\sqrt{3}}{1} = \frac{75}{BC}$

$$\therefore \quad BC = \frac{75}{\sqrt{3}} = \frac{75 \times \sqrt{3}}{\sqrt{3} \times \sqrt{3}} = 25\sqrt{3} \text{ m}$$

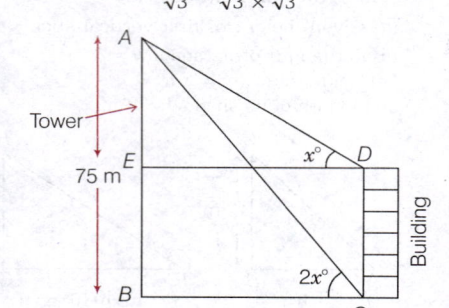

In $\triangle AED$,
$$\tan 30° = \frac{AE}{ED} = \frac{AE}{25\sqrt{3}} \qquad [\because BC = ED]$$
$$\Rightarrow \quad \frac{1}{\sqrt{3}} = \frac{AE}{25\sqrt{3}} \Rightarrow AE = 25 \text{ m}$$
$$\therefore \quad DC = EB = AB - AE$$
$$= 75 - 25 = 50 \text{ m}$$

9. (a) Let AB and DC be two poles and AD be the wire.

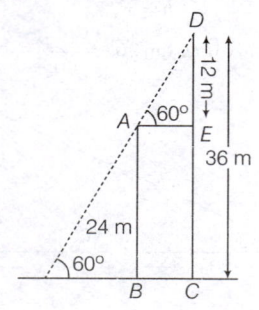

Height and Distance / 741

In $\triangle ADE$,

$\sin 60° = \dfrac{DE}{AD}$ [$\because AD$ = length of wire]

$\Rightarrow AD = \dfrac{DE}{\sin 60°} = \dfrac{12}{\sqrt{3}/2}$

$= \dfrac{12 \times 2}{\sqrt{3}} = \dfrac{24 \times \sqrt{3}}{\sqrt{3} \times \sqrt{3}}$

[on multiplying and dividing by $\sqrt{3}$]

$= \dfrac{24\sqrt{3}}{3} = 8\sqrt{3}$ m

10. (c) Given that, angles are complementary. Let h be the height of the tower.

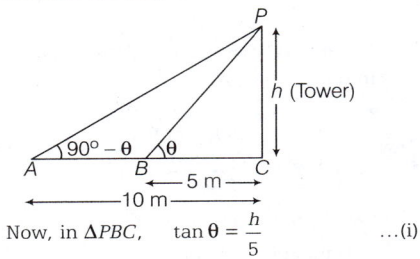

Now, in $\triangle PBC$, $\tan\theta = \dfrac{h}{5}$ ...(i)

and in $\triangle PAC$,

$\tan(90° - \theta) = \dfrac{h}{10}$

$\Rightarrow \cot\theta = \dfrac{h}{10}$ ...(ii)

On multiplying Eqs. (i) and (ii), we get

$\tan\theta \cdot \cot\theta = \dfrac{h}{5} \times \dfrac{h}{10}$

$\Rightarrow \dfrac{h^2}{50} = 1 \Rightarrow h = \sqrt{50}$ m

which is the required height of the tower.

11. (a) Let AB be the tower of height 75 m, C and D denote the positions of two men on either side of a tower.

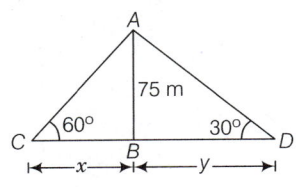

Now, in $\triangle ABC$,

$\tan 60° = \dfrac{AB}{BC} = \dfrac{75}{x}$

$\Rightarrow \sqrt{3} = \dfrac{75}{x} \Rightarrow x = \dfrac{75}{\sqrt{3}}$ m

Again, in $\triangle ABD$,

$\tan 30° = \dfrac{AB}{BD} = \dfrac{75}{y}$

$\Rightarrow \dfrac{1}{\sqrt{3}} = \dfrac{75}{y} \Rightarrow y = 75\sqrt{3}$ m

Hence, distance between two men

$= x + y$

$= \dfrac{75}{\sqrt{3}} + 75\sqrt{3} = \dfrac{75 + 225}{\sqrt{3}}$

$= \dfrac{300}{\sqrt{3}} = 100\sqrt{3}$ m

12. (c) Let $BC = h$ be the height of tower and P and Q be the points, where the angle subtended are 45° and 60°.

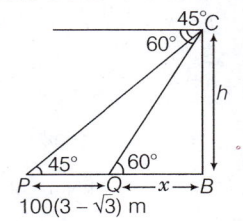

In right angled $\triangle BQC$,

$\tan 60° = \dfrac{BC}{BQ} \Rightarrow \sqrt{3} = \dfrac{h}{x} \Rightarrow x = \dfrac{h}{\sqrt{3}}$ ...(i)

In right angled $\triangle BPC$, $\tan 45° = \dfrac{BC}{PB}$

$= \dfrac{BC}{PQ + QB} \Rightarrow 1 = \dfrac{h}{100(3 - \sqrt{3}) + x}$

$\Rightarrow 100(3 - \sqrt{3}) + x = h$

$\Rightarrow 100(3 - \sqrt{3}) + \dfrac{h}{\sqrt{3}} = h$ [from Eq. (i)]

$\Rightarrow h - \dfrac{h}{\sqrt{3}} = 100(3 - \sqrt{3})$

$\Rightarrow h\left(1 - \dfrac{1}{\sqrt{3}}\right) = 100(3 - \sqrt{3})$

$\Rightarrow \dfrac{h(\sqrt{3} - 1)}{\sqrt{3}} = 100(3 - \sqrt{3})$

$\Rightarrow h = \dfrac{100\sqrt{3}(3 - \sqrt{3})}{(\sqrt{3} - 1)}$

$\therefore h = \dfrac{100\sqrt{3} \times \sqrt{3}(\sqrt{3} - 1)}{(\sqrt{3} - 1)} = 300$ m

13. (a) Let the height of the lower plane from the ground be h m and $PA = x$.

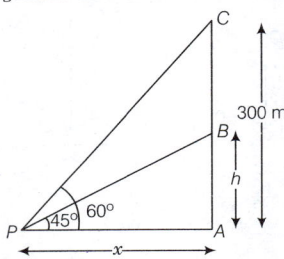

Now, in $\triangle BAP$,

$\tan 45° = \dfrac{AB}{AP}$

$\Rightarrow \qquad 1 = \dfrac{h}{x}$

$\Rightarrow \qquad h = x \qquad \qquad \ldots(i)$

Now, in $\triangle APC$,

$\tan 60° = \dfrac{AC}{AP} = \dfrac{300}{x}$

$\Rightarrow \qquad x = \dfrac{300}{\sqrt{3}} \qquad \qquad \ldots(ii)$

From Eqs. (i) and (ii), we get

$h = x = \dfrac{300}{\sqrt{3}}$

$\Rightarrow \qquad h = \dfrac{300}{\sqrt{3}} \times \dfrac{\sqrt{3}}{\sqrt{3}}$

$= \dfrac{300\sqrt{3}}{3} = 100\sqrt{3}$ m

14. (c) Let $AB = h$ be the height of the tower.

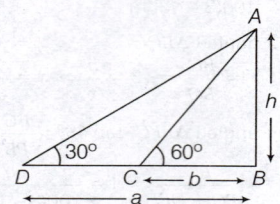

In $\triangle ABC$, $\tan 60° = \dfrac{h}{b}$ ...(i)

and in $\triangle ABD$, $\tan 30° = \dfrac{h}{a}$ ...(ii)

On multiplying Eqs. (i) and (ii), we get

$\tan 60° \tan 30° = \dfrac{h^2}{ab}$

$\Rightarrow \qquad \sqrt{3} \times \dfrac{1}{\sqrt{3}} = \dfrac{h^2}{ab}$

$\Rightarrow \qquad h^2 = ab$

$\Rightarrow \qquad h = \sqrt{ab}$

15. (b) Let h be the height of shorter tower. Then, the distance between the two towers is given by nh m.

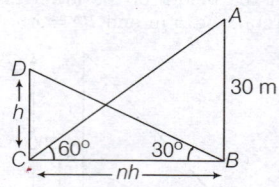

In $\triangle BCD$,

$\tan 30° = \dfrac{h}{nh} \Rightarrow \dfrac{1}{\sqrt{3}} = \dfrac{1}{n}$

$\Rightarrow \qquad n = \sqrt{3}$

16. (c) Let P be the cloud at height H above the level of the water in the lake and Q its image in the water.

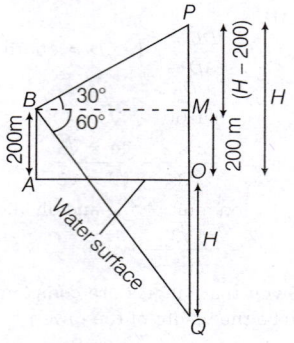

$\therefore \qquad OQ = OP = H$

Given, $\angle PBM = 30°$ and $\angle MBQ = 60°$

In right angled $\triangle PBM$,

$\tan 30° = \dfrac{PM}{BM} = \dfrac{H - 200}{BM}$

$\Rightarrow \qquad \dfrac{1}{\sqrt{3}} = \dfrac{H - 200}{BM}$

$\Rightarrow \qquad BM = \sqrt{3}(H - 200) \qquad \ldots(i)$

In right angled $\triangle QBM$,

$\tan 60° = \dfrac{MQ}{BM} = \dfrac{H + 200}{BM}$

$\Rightarrow \qquad \sqrt{3} = \dfrac{H + 200}{\sqrt{3}(H - 200)}$ [from Eq. (i)]

$\Rightarrow \qquad H + 200 = 3(H - 200)$

$\Rightarrow \qquad H + 200 = 3H - 600$

$\Rightarrow \qquad 2H = 800 \Rightarrow H = 400$ m

Hence, the height of the cloud is 400 m.

17. (b) Let AE and BD be the tower and building, respectively.

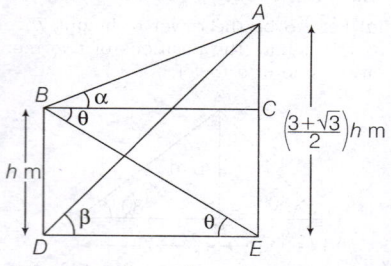

Given that, $\beta = 30°$

In $\triangle ADE$,

$\tan \beta = \tan 30° = \dfrac{AE}{DE} = \dfrac{1}{\sqrt{3}}$

$\Rightarrow \qquad DE = \sqrt{3} \, AE = \sqrt{3}\left(\dfrac{3 + \sqrt{3}}{2}\right)h$

$\Rightarrow \qquad BC = DE = \dfrac{3}{2}(1 + \sqrt{3})h \qquad \ldots(i)$

Now, in $\triangle ABC$, $\tan \alpha = \dfrac{AC}{BC}$

Height and Distance / 743

$\Rightarrow BC \tan\alpha = (AE - CE) = (AE - BD)$
$\qquad\qquad\qquad\qquad\qquad [\because BD = CE]$

$\Rightarrow BC \tan\alpha = \left(\dfrac{3+\sqrt{3}}{2}\right)h - h$

$\Rightarrow \dfrac{3}{2}(1+\sqrt{3})\, h \tan\alpha = \left(\dfrac{1+\sqrt{3}}{2}\right)h$

$\qquad\qquad\qquad\qquad\qquad\text{[from Eq. (i)]}$

$\Rightarrow \tan\alpha = \dfrac{1}{3}$

18. (b) Let the height of aeroplane be h km.

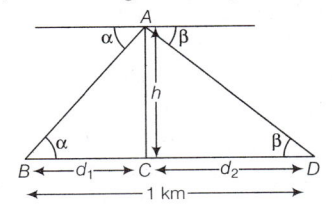

Now, in $\triangle ABC$,

$\tan\alpha = \dfrac{AC}{BC} \Rightarrow \tan\alpha = \dfrac{h}{d_1}$

$\Rightarrow d_1 = h \cot\alpha \qquad \ldots(i)$

Again, in $\triangle ACD$,

$\tan\beta = \dfrac{AC}{CD} \Rightarrow \tan\beta = \dfrac{h}{d_2}$

$\Rightarrow d_2 = h \cot\beta \qquad \ldots(ii)$

On adding Eqs. (i) and (ii), we get

$d_1 + d_2 = h \cot\alpha + h \cot\beta$

$\Rightarrow 1 = h(\cot\alpha + \cot\beta)$

$\Rightarrow h = \dfrac{1}{\cot\alpha + \cot\beta}$

$\qquad = \dfrac{1}{\dfrac{1}{\tan\alpha} + \dfrac{1}{\tan\beta}}$

$\qquad = \dfrac{\tan\alpha \tan\beta}{\tan\alpha + \tan\beta}$

19. (a) Let the required angle be θ.

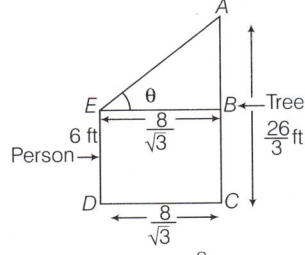

Here, $DC = EB = \dfrac{8}{\sqrt{3}}$ ft

$\therefore AB = AC - BC$

$\qquad = \dfrac{26}{3} - 6 = \dfrac{26-18}{3} = \dfrac{8}{3}$ ft

In right angled $\triangle ABE$,

$\tan\theta = \dfrac{AB}{BE} = \dfrac{8/3}{\dfrac{8}{\sqrt{3}}} = \dfrac{1}{\sqrt{3}} \Rightarrow \tan\theta = \dfrac{1}{\sqrt{3}}$

$\therefore \quad \theta = 30°$

20. (b) Let $DB = x$, $CD = d$ and time taken by boat to cover distance x be 't' min.
Time taken by boat to cover distance CD,
$\qquad d = 20$ min [given]

$\therefore$ Speed of the boat $= \dfrac{d}{20}$

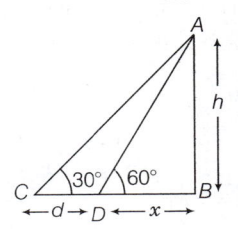

In right angled $\triangle ADB$,

$\tan 60° = \dfrac{AB}{BD} \Rightarrow \sqrt{3} = \dfrac{h}{x}$

$\Rightarrow h = x\sqrt{3} \qquad \ldots(i)$

In right angled $\triangle ABC$,

$\tan 30° = \dfrac{AB}{BC} = \dfrac{AB}{BD + CD} = \dfrac{h}{d+x}$

$\Rightarrow \dfrac{1}{\sqrt{3}} = \dfrac{h}{x+d} \Rightarrow \dfrac{1}{\sqrt{3}} = \dfrac{x\sqrt{3}}{x+d}$ [from Eq. (i)]

$\Rightarrow 3x = x + d \Rightarrow x = \dfrac{d}{2}$

Time taken by boat to cover the distance x

$= \dfrac{x}{\text{Speed of boat}} = \dfrac{d/2}{d/20} = 10$ min

21. (b) Let OP be the pole inside the $\triangle ABC$.

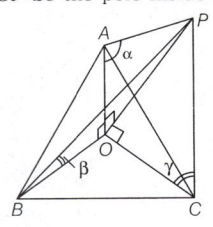

Since, the angle of the elevation of the top of the pole are same from each corner of $\triangle ABC$.

$\therefore$ In $\triangle AOP$,

$\tan\alpha = \dfrac{OP}{OA} \qquad \ldots(i)$

In $\triangle BOP$,

$\tan\beta = \dfrac{OP}{OB} \qquad \ldots(ii)$

In $\triangle COP$,

$\tan\gamma = \dfrac{OP}{OC} \qquad \ldots(iii)$

744 / Fast Track Objective Arithmetic

∵ α = β = γ
∴ OA = OB = OC   [given]
Hence, O is the circumcentre of △ABC.

22. (b)

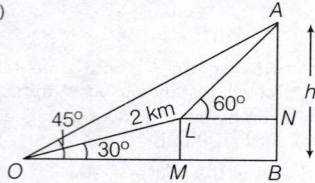

Let AB be the mountain of the height h km.

In △OAB, $\tan 45° = \dfrac{AB}{OB} \Rightarrow OB = h$ km

In △OLM, $\dfrac{OM}{OL} = \cos 30°$

$\Rightarrow \quad OM = 2\cos 30°$ [∵ OL = 2 km]
$\quad = \sqrt{3}$ km

∴ $LN = BM = OB - OM = (h - \sqrt{3})$ km

In △OLM, $\sin 30° = \dfrac{LM}{OL}$

$\Rightarrow \quad LM = 2\sin 30° = 1$ km
∴ $BN = LM = 1$ km

In △ALN,

$\tan 60° = \dfrac{AN}{LN}$

$\Rightarrow \sqrt{3} = \dfrac{AB - BN}{LN} \Rightarrow \sqrt{3} = \dfrac{h - 1}{h - \sqrt{3}}$

$\Rightarrow \sqrt{3}h - 3 = h - 1$

∴ $h = \dfrac{2}{\sqrt{3} - 1} = \dfrac{2}{\sqrt{3} - 1} \times \dfrac{\sqrt{3} + 1}{\sqrt{3} + 1}$
$\quad = (\sqrt{3} + 1)$ km

23. (d) In the figure, let the length of football field be d m and the length of diagonal distance of a field be l m.

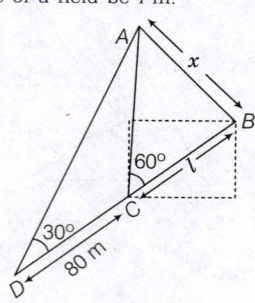

We know that, $\sqrt{2}d = l$

Height of the pole = x m

∴ In △ABC, $\tan 60° = \dfrac{x}{l}$

$\Rightarrow \quad \sqrt{3} = \dfrac{x}{l}$

$\Rightarrow \quad x = \sqrt{3}l$ ...(i)

Now, in △ABD,

$\tan 30° = \dfrac{x}{l + 80}$

$\Rightarrow \quad \dfrac{1}{\sqrt{3}} = \dfrac{x}{l + 80}$

$\Rightarrow \quad l + 80 = \sqrt{3}x$

Now, from Eq. (i), we get

$l + 80 = \sqrt{3}(\sqrt{3}l)$

$\Rightarrow \quad l + 80 = 3l$
$\Rightarrow \quad 80 = 3l - l$

∴ $l = \dfrac{80}{2} = 40$ m

$\Rightarrow \quad d = \dfrac{l}{\sqrt{2}}$

$\quad = \dfrac{40}{\sqrt{2}} = 20\sqrt{2}$ m

24. (c) In △ABO, $\sin 60° = \dfrac{OB}{AO}$

$\Rightarrow \quad AO = \dfrac{OB}{\sin 60°}$ ...(i)

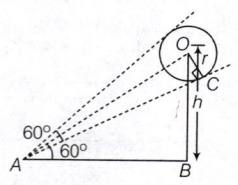

Now, in △AOC,

$\sin \dfrac{60°}{2} = \dfrac{OC}{AO}$

$\Rightarrow \quad AO = \dfrac{OC}{\sin 30°}$ ...(ii)

From Eqs. (i) and (ii), we get

$\dfrac{OB}{\sin 60°} = \dfrac{OC}{\sin 30°}$

$\Rightarrow \quad \dfrac{h}{\sqrt{3}/2} = \dfrac{r}{1/2}$

∴ $h = \sqrt{3}r$

Chapter 40

Set Theory

Sets are used to define the concepts of relations and functions. Sets are the basic tool of Mathematics which are extensively used in developing the foundation of relations and functions, logic and theory, etc.

In this chapter, we will study some basic definitions and operations involving sets. We will also discuss the applications of sets.

Sets

A set is a well-defined collection of objects. When we say well-defined, we mean that the objects follow a given rule or rules. The members of a set are called, elements of the set.

Sets are usually denoted by capital letters of English alphabet, $A, B, C, \ldots$ and elements of sets are denoted by small letters of English alphabet $a, b, c, \ldots$

* If a is an element of a set 'A', then we can write $a \in A$ and read as 'a belongs to A'. The Greek symbol '$\in$' is used to denote the phrase 'belongs to'.
* If b is not an element of the set A, then we can write $b \notin A$ and read as 'b does not belong to A'.

Representation of Sets

Sets are generally represented by following two ways

1. Roster or Tabular or Listing Form

In this form, all the elements of a set are listed, separated by commas and enclosed within curly bracket { }.

For example The set of all even positive integer less than 11 is described in roster form as {2, 4, 6, 8, 10}.

2. Set Builder Form or Rule Form

In this form, instead of listing all elements of a set, we write the set by some special property or properties satisfied by all element of set and write it as $A = \{x : P(x)\}$ or $A = \{x \mid x$ has the property $P(x)\}$ and read it as A is the set of all elements x, such that x has the property P.

The symbol ':' or '|' stands for 'such that'.
For example $P = \{x : x$ is a perfect square and $0 \leq x \leq 100, x \in N\}$
or $P = \{x : x \in N, x = n^2, 0 \leq x \leq 100, n \in Z\}$
Here, Z is a set of integers and N is a set of natural numbers.

- There is no importance of order of elements in a set.
  For example $\{a, e, i, o, u\}$ and $\{e, i, o, u, a\}$ both are same sets.
- There is no effect of repetition of elements in a set.
  For example $\{2, 3, 4\}$ and $\{2, 3, 3, 4\}$ both are same sets.

Types of Sets

Sets are divided in the following type

1. Empty/Null Set

A set which does not contain any element is called an empty set, null set or void set. The empty set is denoted by the symbol ϕ or $\{\}$. Thus, $\phi = \{\}$ as there is no element in the empty set.
For example If $A = \{x : 1 < x < 2, x$ is a natural number$\}$, then A is the empty set, because there is no natural number between 1 and 2.

2. Singleton Set

A set containing only one element is called a singleton set.
For example The set $\{0\}$ is a singleton since, it has only one element 0.

3. Finite and Infinite Sets

A set which is empty or consists of a definite number of elements is called finite set, otherwise the set is called infinite.
For example (i) Let W be the set of the days of the week, then W is the finite set.
(ii) Let S be the set of squares of natural numbers, then S is the infinite set.

- The number of distinct elements in a finite set A is called **cardinal number of set** and it is denoted by $n(A)$.
  For example If $A = \{5, 7, 6, 3\}$, then $n(A) = 4$.

4. Equivalent Sets

Two sets A and B are equivalent, if their cardinal numbers are same.
i.e. $n(A) = n(B)$
For example Let $A = \{a, b, c, d\}$ and $B = \{1, 2, 3, 4\}$, then $n(A) = 4$ and $n(B) = 4$
Therefore, A and B are equivalent sets.

5. Equal Sets

Two sets A and B are equal, if they have same members i.e. if every element of A is an element of B and every element of B is an element of A, then $A = B$.
For example If $A = \{1, 3, 5, 7\}$ and $B = \{7, 3, 1, 5\}$, then $A = B$.

- If two sets are not equal, we write $A \neq B$.
- A set does not change, if its elements are repeated.
- A set does not change even, if the order of its elements is different.

Subset

Let A and B be two non-empty sets. If every element in set A is an element of another set B, then A is called a subset of B. Also, B is said to be super set of A. If A is subset of B, then we write as $A \subseteq B$, which is read as 'A is a subset of B.' If A is not a subset of B, then we write as $A \nsubseteq B$, which is read as 'A is not a subset of B or A is contained in B'.

For example More specifically $A \subseteq B$, if $x \in A \Rightarrow x \in B$ and let $A = \{2, 4, 7\}$, $B = \{1, 2, 3, 4, 7\}$. Then, $A \subseteq B$, since every element of A is in B.

* Every set is a subset of itself.
* Empty set is a subset of every set.
* Total number of subsets of a finite set containing n elements is 2^n.
* If $A \subseteq B$ and $A \neq B$, then A is called a proper subset of B and we write $A \subset B$.

Power Set

The collection of all subsets of a set A is called the power set of A. It is denoted by $P(A)$. For example If $A = (a, b)$, then $P(A) = \{\phi, \{a\}, \{b\}, \{a, b\}\}$

Universal Set

If all the sets under consideration are the subsets of a fixed set U, then U is called the universal set.

For example If $A = \{2, 4, 6\}$, $B = \{1, 3, 5\}$ and $C = \{0, 7\}$, then $U = \{0, 1, 2, 3, 4, 5, 6, 7\}$ is a universal set.

Venn Diagram

In Venn diagram, the universal set is represented by a rectangular region and its subset are represented by circle or a closed geometrical figure inside the universal set.

For example If $U = \{1, 2, 3, \ldots, 10\}$ and $A = \{1, 2, 3\}$, then its Venn diagram is shown in the figure.

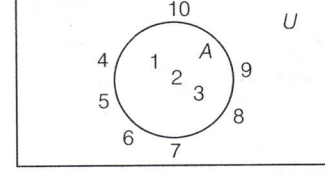

Operations on Sets

Different operations that are performed on sets are given below

1. Union of Sets

Union of two sets A and B is the set of all elements which belongs to A or B.

It is written as $A \cup B$ and is read as 'A union B'.

$\therefore \quad A \cup B = \{x \mid x \in A \text{ or } x \in B\}$

For example If $A = \{1, 3, 5, 7, 9\}$ and $B = \{2, 4, 5, 6, 9\}$.

Then, $\quad A \cup B = \{1, 2, 3, 4, 5, 6, 7, 9\}$.

The union of two sets A and B can be represented by a Venn diagram as shown in figure by shaded portion.

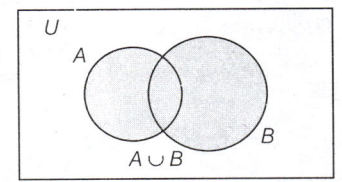

2. Intersection of Sets

The intersection of two sets A and B is the set of all those elements which belongs to both sets A and B.

It is written as $A \cap B$ and is read as 'A intersection B'.

∴ $A \cap B = \{x : x \in A \text{ and } x \in B\}$

For example If $A = \{2, 4, 6, 8\}$ and $B = \{4, 5, 6, 9\}$. Then, $A \cap B = \{4, 6\}$

The shaded portion in figure indicate the intersection of sets A and B.

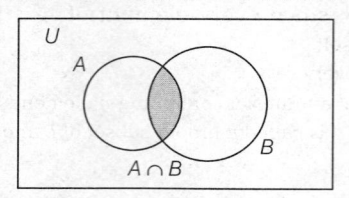

$A \cap B$

3. Difference of Sets

The difference of two non-empty sets A and B is set of elements which belong to A but do not belong to B. This is written as $A - B$ and read as 'A minus B.' If the difference of two non-empty sets is $B - A$, then it is a set of those elements which are in B but not in A.

∴ $(A - B) = \{x \mid x \in A \text{ and } x \notin B\}$ and $B - A = \{x \mid x \in B \text{ and } x \notin A\}$

For example If $A = \{1, 2, 3, 4, 5\}$ and $B = \{4, 5, 6\}$, then

$A - B = \{1, 2, 3, 4, 5\} - \{4, 5, 6\} = \{1, 2, 3\}$ and $B - A = \{4, 5, 6\} - \{1, 2, 3, 4, 5\} = \{6\}$

∴ $A - B \neq B - A$

The difference of two sets A and B can be represented by the following Venn diagrams

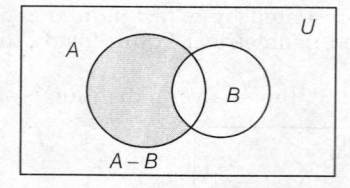

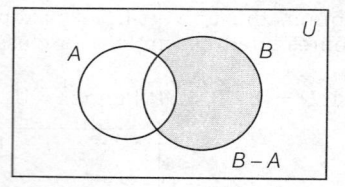

$A - B$ 　　　 $B - A$

* Set $(A - B)$ and B are disjoint, i.e. $(A - B) \cap B = \phi$
* $(A - B) = (A \cup B) - B$
* $(A - B) = A - (A \cap B)$

4. Symmetric Difference of Two Sets

The symmetric difference of A and B is the set $(A - B) \cup (B - A)$.

It is denoted by $A \triangle B$ and read as A symmetric difference B.

∴ $A \triangle B = (A - B) \cup (B - A) = \{x : x \in A \cup B\}$

The symmetric difference of sets A and B is represented by the following Venn diagram

$A - B$ 　　 $B - A$

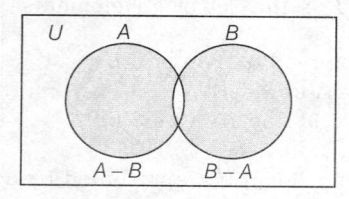

For example If $A = \{1, 2, 3, 4\}$ and $B = \{3, 4, 5, 6\}$, then
$A - B = \{1, 2\}$ and $B - A = \{5, 6\}$
$\therefore$ $A \Delta B = (A - B) \cup (B - A) = \{1, 2, 5, 6\}$

5. Disjoint Sets
If A and B are two sets such that $A \cap B = \phi$, then A and B are called disjoint sets.
For example Let $A = \{2, 4, 6, 8\}$ and $B = \{1, 3, 5, 7\}$. Here, A and B are disjoint sets, because there is no element common to both sets A and B.
In Venn diagram, A and B are disjoint sets.

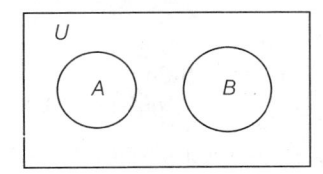

6. Complement of a Set
If A is a subset of universal set U, then the complement of A is denoted by A^c or A' is defined by $\quad A^c = \{x \in U \text{ and } x \notin A\}$
$\therefore \qquad A^c \text{ or } A' = U - A$

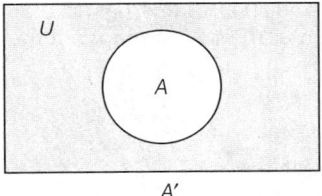

For example If $U = \{1, 2, 3, 4, 5, 6\}$ and $A = \{1, 3, 5\}$, then A' or $A^c = U - A = \{2, 4, 6\}$
Complement of set A i.e. A^c or A' is represented in following Venn diagram by shaded region.

Laws of Sets
If A, B and C are three sets, then

(i) **Idempotent laws**
   (a) $A \cup A = A$
   (b) $A \cap A = A$

(ii) **Complement laws**
   (a) $\phi' = U$ and $U' = \phi$
   (b) $(A')' = A$
   (c) $A \cup A' = U$
   (d) $A \cap A' = \phi$

(iii) **Identity laws**
   (a) $A \cup \phi = A$
   (b) $A \cap U = A$

(iv) **Commutative laws**
   (a) $A \cup B = B \cup A$
   (b) $A \cap B = B \cap A$

(v) **Associative laws**
   (a) $(A \cup B) \cup C = A \cup (B \cup C)$
   (b) $A \cap (B \cap C) = (A \cap B) \cap C$

(vi) **Distributive laws**
   (a) $A \cup (B \cap C) = (A \cup B) \cap (A \cup C)$
   (b) $A \cap (B \cup C) = (A \cap B) \cup (A \cap C)$

(vii) **De-Morgan's laws**
   (a) $(A \cup B)' = A' \cap B'$
   (b) $(A \cap B)' = A' \cup B'$

Important Results Based on Sets

Let A, B and C be finite sets in a finite universal set U. Then,
 (i) $n(A \cup B) = n(A) + n(B) - n(A \cap B)$
 (ii) $n(A \cup B) = n(A) + n(B) \Leftrightarrow A$ and B are disjoint non-void sets.
 (iii) $n(A \cup B \cup C) = n(A) + n(B) + n(C) - n(A \cap B) - n(B \cap C) - n(C \cap A) + n(A \cap B \cap C)$
 (iv) $n(A - B) = n(A) - n(A \cap B)$
 (v) $n(A \Delta B) = n(A) + n(B) - 2n(A \cap B)$
 (vi) $n(A') = n(U) - n(A)$
 (vii) $n(A' \cup B') = n(U) - n(A \cap B)$
 (viii) $n(A' \cap B') = n(U) - n(A \cup B)$
 (ix) If $A_1, A_2, \ldots, A_n$ are n disjoint sets, then
$$n(A_1 \cup A_2 \cup \ldots \cup A_n) = n(A_1) + n(A_2) + \ldots + n(A_n)$$

Ex. 1 Let $S = \{0, 1, 5, 4, 7\}$. Then, the total number of subsets of S is

Sol. Given, number of elements in $S = 5$
$\therefore \quad n = 5$
We know that, number of subsets $= 2^n$
Then, total number of subsets of $S = 2^n = 2^5 = 32$

Ex. 2 If A and B are two sets such that A has 12 elements, B has 17 elements and $A \cup B$ has 21 elements, then how many elements does $A \cap B$ have?

Sol. Given, $n(A) = 12, n(B) = 17$ and $n(A \cup B) = 21$
$\therefore \quad (A \cup B) = n(A) + n(B) - n(A \cap B) \Rightarrow 21 = 12 + 17 - n(A \cap B)$
$\Rightarrow n(A \cap B) = 12 + 17 - 21 = 8$
So, $A \cap B$ has 8 elements.

Ex. 3 If A and B are any two sets, then what is $A \cap (A \cup B)$ equal to?

Sol. $A \cap (A \cup B) = (A \cap A) \cup (A \cap B)$
$= A \cup (A \cap B) \qquad [\because A \cap A = A]$
$= A \qquad$ [by using diagram]
Thus, $A \cap (A \cup B) = A$.

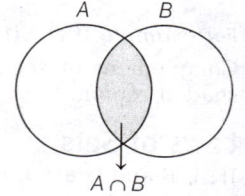

Ex. 4 40% of the people read newspaper X, 50% read newspaper Y and 10% read both the papers. What percentage of the people read neither newspaper?

Sol. Here, $n(A) = 40, n(B) = 50$ and $n(A \cap B) = 10$
We know that, $n(A \cup B) = n(A) + n(B) - n(A \cap B)$
$= 40 + 50 - 10 = 80$
$\therefore$ Percentage reading either or both newspapers $= 80\%$
Hence, percentage reading neither newspaper $= (100 - 80)\% = 20\%$

Ex. 5 If A, B and C are three sets and U is the universal set such that $n(U) = 700$, $n(A) = 200$, $n(B) = 300$ and $n(A \cap B) = 100$, then what is the value of $(A' \cap B')$?

Sol. Given, $n(U) = 700, n(A) = 200, n(B) = 300$ and $n(A \cap B) = 100$
$\therefore \quad n(A \cup B) = n(A) + n(B) - n(A \cap B) = 200 + 300 - 100 = 400$
Now, $n(A' \cap B') = n(U) - n(A \cup B) = 700 - 400 = 300$

Fast Track Practice

1. If $A \equiv \{x \mid x$ is a prime number $\leq 100\}$ and $B \equiv \{x \mid x$ is an odd number $\leq 100\}$. Then, what is the ratio of the number of subsets of set A to set B?
   (a) 2^{25} (b) 2^{-25} (c) 2 (d) $\dfrac{50^2}{25^2}$

2. If N is the set of all positive integers, then $\{n \in N : \mid n - 4 \mid \leq 2\}$ is equal to
   (a) $\{3, 4, 5\}$ (b) $\{2, 3, 4, 5, 6\}$
   (c) $\{2, 3, 4, 5\}$ (d) $\{3, 4, 5, 6\}$

3. How many subsets of $\{1, 2, 3, \ldots, 11\}$ contain at least one even integer?
   (a) 1900 (b) 1964 (c) 1984 (d) 2048

4. Let $A = \{x : x$ is an odd integer$\}$ and $B = \{x : x^2 - 8x + 15 = 0\}$. Then, which one of the following is correct? [CDS 2013 (I)]
   (a) $A = B$ (b) $A \subseteq B$
   (c) $B \subseteq A$ (d) $A \subseteq B^c$

5. Let $A = \{7, 8, 9, 10, 11, 12\}$ and $B = \{7, 10, 14, 15\}$. What is the number of elements in $(A - B)$ and $(B - A)$, respectively? [CDS 2016 (II)]
   (a) 2 and 4 (b) 4 and 2
   (c) 2 and 2 (d) 4 and 4

6. If A and B are any two non-empty subsets of a set E, then what $A \cup (A \cap B)$ is equal to? [CDS 2014 (I)]
   (a) $A \cap B$ (b) $A \cup B$
   (c) A (d) B

7. If A and B are two disjoint sets, then which one of the following is correct?
   (a) $A - B = A - (A \cap B)$
   (b) $B - A' = A \cap B$
   (c) $A \cap B = (A - B) \cap B$
   (d) All of the above

8. If A, B and C are sets, then $A - (B - C)$ equals [UPPSC Pre 2012]
   (a) $(A - B) \cup (A \cap C)$ (b) $(A - B) - C$
   (c) $(A - B) \cap (A - C)$ (d) $(A - B) \cup (A - C)$

9. In an examination out of 100 students, 75 passed in English, 60 passed in Mathematics and 45 passed in both English and Mathematics. What is the number of students passed in exactly one of the two subjects?
   (a) 45 (b) 60 (c) 75 (d) 90

10. In a hotel, 60% had vegetarian lunch while 30% had non-vegetarian lunch and 15% had both types of lunch. If 96 people were present, then how many did not eat either type of lunch?
    (a) 20 (b) 24 (c) 26 (d) 28

11. The given diagram shows the number of students who failed in an examination comprising papers of English, Hindi and Mathematics. The total number of students who took the test is 500. What is the percentage of students who failed in at least two subjects?

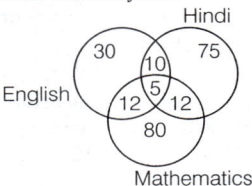

    (a) 8% (b) 8.4%
    (c) 7.8% (d) 7.6%

Directions (Q. Nos. 12-15) *Read the passage below and solve the questions based on it.*

5% of the passengers do not like coffee, tea and lassi and 10% like all the three, 20% like coffee and tea, 25% like lassi and coffee and 25% like lassi and tea. 55% like coffee, 50% like tea, and 50% like lassi.

12. The passengers who like only coffee is greater than the passengers who like only lassi by
    (a) 25% (b) 100%
    (c) 75% (d) 0%

13. The percentage of passengers who like both tea and lassi but not coffee, is
    (a) 15 (b) 25
    (c) 40 (d) 75

14. The percentage of passengers who like atleast 2 of the coffee, tea and lassi, is
    (a) 30 (b) 45
    (c) 50 (d) 60

15. If the number of passengers is 180, then the number of passengers who like lassi only, is
    (a) 10 (b) 18 (c) 27 (d) 36

16. In a class of 50 students, 18 take music, 26 take art and 2 take both art and music. How many students in the class are not enrolled in either music or art?
 [UPPSC 2015]
 (a) 6   (b) 8   (c) 16   (d) 24

17. Out of 130 students appearing in an examination 62 failed in English, 52 failed in Mathematics, whereas 24 failed in both English and Mathematics. The number of students, who passed finally, is [UPSC CSAT 2015]
 (a) 40   (b) 50   (c) 55   (d) 60

18. A group of 50 students appeared for the two examinations one in Physics and the other in Mathematics. 38 students passed in Physics and 37 in Mathematics. If 30 students passed in both subjects, determine how many students failed in both the subjects?
 (a) 2   (b) 3   (c) 4   (d) 5

19. Of the members of three athletic teams in a certain school, 21 are on the basketball team, 26 on the hockey team and 29 on the football team, 14 play hockey and basketball, 15 play hockey and football and 12 play football and basketball, 8 are on all the three teams. How many members are there altogether?
 (a) 38   (b) 47   (c) 51   (d) 43

20. In a survey, it was found that 55% go for jogging, 50% do yoga, 42% do aerobics, 28% do jogging and yoga, 20% do yoga and aerobics, 12% go for jogging and aerobics and 10% do all three. If each one of them go for atleast one of these, then what percentage do exactly one exercise?
 [FCI Grade III 2015]
 (a) 40%   (b) 49%   (c) 57%   (d) 59%

21. In the three intersecting circles given below the numbers in different sections indicate the number of persons speaking different languages. How many persons speak only two languages?
 [UPSC CSAT 2015]

English → / 10 \ / 7 \ 12 ← Hindi
       \ 6 \ 6 / 4 /
             \ 14 / ← Odia

 (a) 13   (b) 17   (c) 11   (d) 23

22. Out of a total of 120 musicians in a club, 5% can play all the three instruments, guitar, violin and flute. It, so happens that the number of musicians who can play any two and only two of the above instruments is 30. The number of musicians who can play the guitar alone is 40. What is the total number of those who can play violin alone or flute alone? [UPSC CSAT 2014]
 (a) 45   (b) 44
 (c) 38   (d) 30

23. There are 50 students admitted to a nursery class. Some students can speak only English and some can speak only Hindi. 10 students can speak both English and Hindi. If the number of students who can speak English is 21, then how many students can speak Hindi, how many can speak only Hindi and how many can speak only English?
 [UPSC CSAT 2014]
 (a) 21, 11 and 29
 (b) 28, 18 and 22
 (c) 37, 27 and 13
 (d) 39, 29 and 11

24. In a gathering of 100 people, 70 of them can speak Hindi, 60 can speak English and 30 can speak French. Further, 30 of them can speak both Hindi and English, 20 can speak both Hindi and French. If x is the number of people who can speak both English and French, then which one of the following is correct? (Assume that everyone can speak at least one of the three languages.) [CDS 2016 (I)]
 (a) $9 < x \le 30$ (b) $0 \le x < 8$
 (c) $x = 9$ (d) $x = 8$

25. In a school, there was a compulsion to learn atleast one Foreign language from the choice given to them, namely German, French and Spanish. Twenty eight students took French, thirty took German and thirty two took Spanish. Six students learnt French and German, eight students learnt German and Spanish, ten students learnt French and Spanish. Fifty four students learnt only one Foreign language while twenty students learnt only German. Find the number of students in the school. [SNAP 2016]
 (a) 60   (b) 62
 (c) 70   (d) None of these

Answer with Solutions

1. **(b)** We know that there are 25 prime number below 100.
   ∴ $n(A) = 25$
   Total number of subsets of set $A = 2^{25}$
   and there are 50 odd numbers below 100.
   Total number of subsets of set $B = 2^{50}$
   ∴ Required ratio $= \dfrac{2^{25}}{2^{50}} = 2^{-25}$

2. **(b)** Given, $A = \{n \in N : |n - 4| \leq 2\}$
   $= \{n \in N : -2 \leq n - 4 \leq 2\}$
   $= \{n \in N : 2 \leq n \leq 6\}$
   $= \{2, 3, 4, 5, 6\}$

3. **(c)** Given, set $= \{1, 2, 3, \ldots, 11\}$
   Here, $n = 11$
   ∴ Total number of subsets $= 2^n = 2^{11}$
   The number of subset which contain odd number $= 2^6$
   The number of subsets which contain atleast one even number
   $= 2^{11} - 2^6 = 2048 - 64 = 1984$

4. **(c)** Given that, $A = \{x : x \text{ is an odd integer}\}$
   $\Rightarrow A = \{1, 3, 5, 7, \ldots\}$
   and $B = \{x : x^2 - 8x + 15 = 0\}$
   $= \{x : x^2 - 5x - 3x + 15 = 0\}$
   $= \{x : x(x - 5) - 3(x - 5) = 0\}$
   $= \{x : (x - 5)(x - 3) = 0\} = \{3, 5\}$
   Since, B has two odd elements.
   ∴ $B \subseteq A$

5. **(b)** Given, $A = \{7, 8, 9, 10, 11, 12\}$
   and $B = \{7, 10, 14, 15\}$
   ∴ $n(A) = 6$ and $n(B) = 4$
   Now, $A \cap B = \{7, 10\}$
   ∴ $n(A \cap B) = n(B \cap A) = 2$
   Hence, $n(A - B) = n(A) - n(A \cap B)$
   $= 6 - 2 = 4$
   and $n(B - A) = n(B) - n(B \cap A)$
   $= 4 - 2 = 2$

6. **(c)** Since, A and B are non-empty subsets of E. 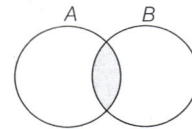
   ∴ $A \cup (A \cap B) = A \cup$ (Shaded portion) $= A$

7. **(a)** Since, A and B are two disjoints, therefore $A \cap B = \phi$
   ∴ $A - B = A - (A \cap B)$

8. **(a)** If A, B and C are sets, then they can be represented as

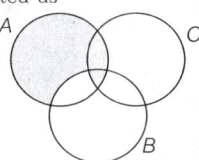

   From the above figure, the only option that can be satisfied is option (a).
   i.e. $(A - B) \cup (A \cap C)$

9. **(a)** Given, total number of students $= 100$
   Let E denotes the students who have passed in English and M denotes the students who have passed in Maths.
   ∴ $n(E) = 75$, $n(M) = 60$
   and $n(E \cap M) = 45$
   We know that,
   $n(E \cup M) = n(E) + n(M) - n(E \cap M)$
   $= 75 + 60 - 45 = 90$
   Required number of students
   $= 90 - 45 = 45$

10. **(b)** Here, $n(A) = 60\%$, $n(B) = 30\%$, $n(A \cap B) = 15\%$
    ∴ $n(A \cup B) = n(A) + n(B) - n(A \cap B)$
    $= 60 + 30 - 15$
    $\Rightarrow n(A \cup B) = 75\%$
    Thus, 25% people who did not eat either type of lunch.
    ∴ Required people $= \dfrac{25}{100} \times 96 = 24$

11. **(c)** Total students who failed in atleast two subjects $= 10 + 12 + 12 + 5 = 39$
    ∴ Required percentage $= \dfrac{39}{500} \times 100 = 7.8\%$

Solutions (Q. Nos. 12-15)

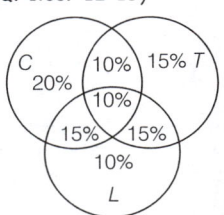

where, C = Coffee, T = Tea and L = Lassi

12. **(b)** The passengers who like only coffee = 20%
    and the passengers who like only lassi = 10%
    ∴ Required passengers $= \dfrac{20 - 10}{10} \times 100\%$
    $= 100\%$

13. (a) It can be seen that the percentage of passengers who like both tea and lassi but not coffee = 15%.
This is the figure representing this area

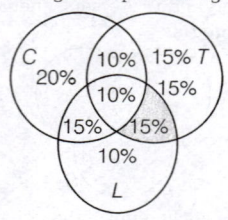

14. (c) The percentage of passengers who like atleast 2 of the coffee, tea and lassi can be seen in the below figure and they are = 50%

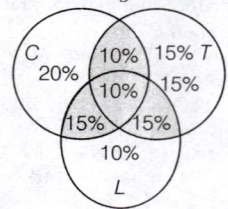

15. (b) 10% of the passengers like only lassi. So, the number of passengers
$= 180 \times 10\% = \dfrac{180 \times 10}{100} = 18$

16. (b) Given, total number of students = 50

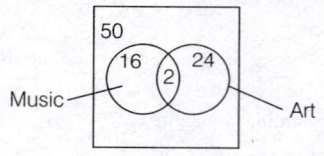

Number of students in the class who are not enrolled in either music or art
$= 50 - (18 + 26 - 2) = 50 - 42 = 8$

17. (a) Here,
Mathematics English

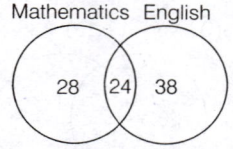

Total number of students who failed in one or both = 28 + 24 + 38 = 90
∴ Number of students who passed finally
$= 130 - 90 = 40$

18. (d) We have, $n(P) = 38$, $n(M) = 37$,
and $n(P \cap M) = 30$
∴ $n(P \cup M) = n(P) + n(M) - n(P \cap M)$
$= 38 + 37 - 30 = 75 - 30 = 45$
Number of students who failed, i.e.
$n(P' \cap M') = n(U) - n(P \cup M)$
$= 50 - 45 = 5$
Hence, the number of students failed in both subjects are 5.

Alternate Method
Let x students failed in both subjects.

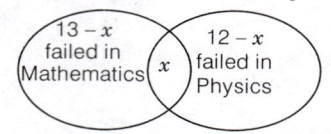

Now, total failed students $= 50 - 30 = 20$
$13 - x + x + 12 - x = 20 \Rightarrow x = 5$

19. (d) Let B, H, F denote the sets of members who are on the basketball team, hockey team and football team respectively.
Then, $n(B) = 21$, $n(H) = 26$,
$n(F) = 29$, $n(H \cap B) = 14$
$n(H \cap F) = 15$, $n(F \cap B) = 12$
and $n(B \cap H \cap F) = 8$
We have to find $n(B \cup H \cup F)$
We know that,
$n(B \cup H \cup F) = n(B) + n(H) + n(F)$
$- n(B \cap H) - n(H \cap F) - n(F \cap B)$
$+ n(B \cap H \cap F)$
∴ $n(B \cup H \cup F)$
$= 21 + 26 + 29 - 14 - 15 - 12 + 8$
$= 84 - 41 = 43$

20. (c) By Venn diagram,

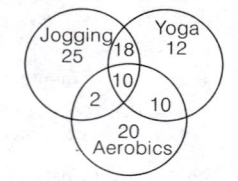

Now, person only doing jogging
$= 55 - 28 - 12 + 10 = 25\%$
Person only doing yoga
$= 50 - 28 - 20 + 10 = 12\%$
Person only doing aerobics
$= 42 - 12 - 20 + 10 = 20\%$
∴ Percentage of person do exactly one exercise $= 25\% + 12\% + 20\% = 57\%$

21. (b) Given, Venn diagram is

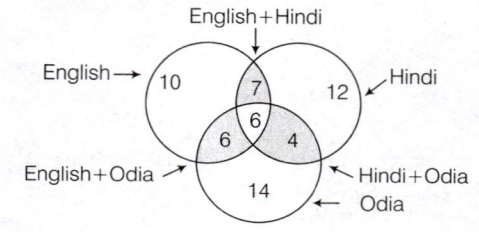

So, according to the given information there are 17 persons speak only two languages.

$$\text{English + Hindi} = 7$$
$$\text{English + Odia} = 6$$
$$\text{Hindi + Odia} = 4$$
$$\text{Total} = 17$$

Hence, option (b) is correct.

22. (b) Let the sets G, V and F represent the persons who can play guitar, violin and flute, respectively.
Then, $n(G \cup V \cup F) = 120$
and $n(G \cap V \cap F) = 5\%$ of $120 = 6$

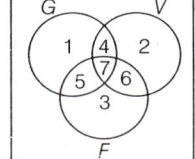

G : Guitar
V : Violin
F : Flute

Let us decompose the sets into various distinct regions and label them using numbers 1 to 7,
i.e., $n(7) = 6$ [as determined above]
Given, $n(4 \cup 5 \cup 6) = 30$ and $n(1) = 40$
We need to determine $n(2 \cup 3)$.
$$n(2 \cup 3) = n(2) + n(3)$$
[since, these regions are disjoint]
Now, $120 = n(1) + n(2) + n(3)$
$\qquad + n(4 \cup 5 \cup 6) + n(7)$
$\Rightarrow \quad 120 = 40 + n(2 \cup 3) + 30 + 6$
$\Rightarrow n(2 \cup 3) = 44$
$\therefore$ Number of musicians who can play violin alone or flute alone = 44

23. (d) Let the sets E and H represent the number of students who can speak English and Hindi, respectively.

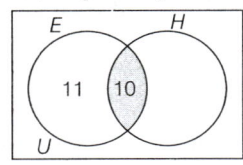

Now, $n(E \cup H) = 50 \Rightarrow n(E \cap H) = 10$
$\therefore \qquad n(E) = 21$
We need to determine $n(H), n(H \cap E^c)$ and $n(E \cap H^c)$
$\qquad n(E \cup H) = n(E) + n(H) - n(E \cap H)$
$\Rightarrow \qquad 50 = 21 + n(H) - 10$
$\Rightarrow \qquad n(H) = 39$ [speak Hindi]

Now, $n(H \cap E^c) = n(H) - n(E \cap H)$
$\qquad = 39 - 10 = 29$ [speak only Hindi]
and $n(E \cap H^c) = n(E) - n(E \cap H)$
$\qquad = 21 - 10 = 11$ [speak only English]

24. (a) Let A, B and C be the number of people who can speak Hindi, English and French.
Then, $n(A) = 70, n(B) = 60, n(C) = 30$
$n(A \cap B) = 30, n(A \cap C) = 20$
and $n(B \cap C) = x$
Venn diagram of the above data is as follows

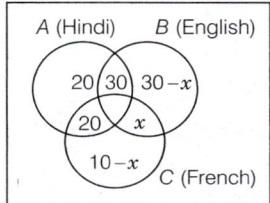

A (Hindi)   B (English)
20 / 30 \ 30 − x
20 x
10 − x
C (French)

It is clear from the above Venn diagram,
$20 + 30 + 20 + 30 - x + x + 10 - x = 100$
$\Rightarrow \qquad 110 - x = 100$
$\Rightarrow \qquad x = 10$
Hence, from the options, the possible value of x is $9 < x \leq 30$.

25. (c) Given, French students = 28 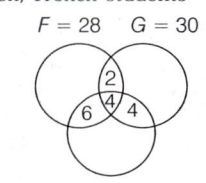
$F = 28 \quad G = 30$

$S = 32$
German students = 30
Spanish students = 32
French and German = 6
German and Spanish = 8
French and Spanish = 10
$\therefore$ Only German and French = $10 - 8 = 2$
Students opting for only one language = 54
Only German = 20
$\therefore$ German students opting for more languages = $30 - 20 = 10$
Students taking only Spanish
$\qquad = 32 - 6 - 4 - 4 = 18$
Students taking only French
$\qquad = 28 - 6 - 4 - 2 = 16$
So, the total number of students
$\qquad = 20 + 18 + 16 + 2 + 4 + 4 + 6$
$\qquad = 70$

Chapter 41

Statistics

Statistics is the branch of Mathematics, which deals with the collection, analysis and interpretation of numerical data. It deals with all the aspects of data, including the planning of data collection in terms of design of surveys and experiments.

In this chapter, we shall extend the study of measures of central tendency, i.e. mean, median and mode from ungrouped data to that of grouped data.

Collection of Data

Collection of data is the first step in statistics towards achieving the goal or conclusion.
On the basis of collection, data are of two types

(i) **Primary data** The data collected actually in the process of investigation by the investigator is called primary data. It is original and first hand information.

(ii) **Secondary data** The data collected by someone and used by any other person known as secondary data.

Presentation of Data

Raw or Ungrouped Data When the data presented is random and is not prepared according to some order, it is known as raw or ungrouped data. It does not give us a clear picture of the class.

Grouped Data When the data is arranged in any manner like ascending or descending order etc., it is called grouped data. It can also be presented in the form of a table called frequency distribution table.

Class Intervals Class intervals are the groups in which all the observations are divided. Each class is bounded by two figures (numbers) which are called class limits. The figure on the left side of a class, is called its lower limit and that on the right side of a class, is called its upper limit.

Class Mark It is the mid-point of the class interval.

i.e. Class mark = $\dfrac{\text{Lower class limit + Upper class limit}}{2}$

Range or a class size Difference between the upper limit and the lower limit of a class is called its class size.
i.e. Range = Upper limit − Lower limit
For example Range of the observations 4, 7, 8, 10, 12 = 12 − 4 = 8
Frequency of an observation The number of times an observation occurs is called its frequency.

Frequency Distribution

The tabular arrangement of data, showing the frequency of each observation is called a frequency distribution. It is a method of presenting the data in a summarised form. Frequency distribution is also known as frequency table.
There are two types of frequency distribution which are as follows

1. Discrete Frequency Distribution

A frequency distribution is called a discrete frequency distribution, if data are presented in a way such that exact measurements of the units are clearly shown.

Marks	Number of students (Frequency)
40	7
60	3
80	3
100	2
Total	15

2. Continuous Frequency Distribution

Exclusive Method A frequency distribution in which upper limit of each class is excluded and lower limit is included, is called an exclusive method.
e.g. In the class 0-10 of marks obtained by students, a student who has obtained 10 marks is not included in this class. It will be counted in the next class, i.e. 10-20.
Inclusive Method In this method, the classes are so formed that the upper limit of a class is included in that class. The following example illustrates the method. In the class 1000-1099, we include workers having wages between ₹ 1000 and ₹ 1099. If the income of a worker is exactly ₹ 1100, then it will be included in the next class 1100-1199.

Exclusive method		Inclusive method	
Wages (in ₹)	Number of workers	Wages (in ₹)	Number of workers
1000-1100	125	1000-1099	125
1100-1200	150	1100-1199	150
1200-1300	200	1200-1299	200
1300-1400	250	1300-1399	250
1400-1500	175	1400-1499	175
1500-1600	100	1500-1599	100
Total	1000	Total	1000

It is clear from the above example that both the inclusive and exclusive methods give us the same class frequency although the class intervals are aparently different in the two cases.

In the above example on inclusive method, the difference between the lower limit of a class and upper limit of the preceding class is 1.

Therefore, we subtract $\frac{1}{2}$ from the lower limit and add $\frac{1}{2}$ to upper limit of each class to make it continuous. *The adjusted classes would be as follows*

Wages (in ₹)	Number of workers
999.5-1099.5	125
1099.5-1199.5	150
1199.5-1299.5	200
1299.5-1399.5	250
1399.5-1499.5	175
1499.5-1599.5	100

Cumulative Frequency

If the frequency of first class interval is added to the frequency of second class and this sum is added to third class and so on, then frequencies so obtained are known as cumulative frequency.

Ex. 1 Consider the table given below

Marks	Number of students (Frequency)	Cumulative frequency
0-10	13	13
10-20	7	20
20-30	5	25
30-40	4	29
40-50	1	30
50-60	7	37
60-70	3	40
70-80	4	44
80-90	5	49
90-100	1	50
Total	50	

Then, find the value of the following.
  (i) Frequency of class 10-20
  (ii) Class size
  (iii) Mid value of 60-70
  (iv) Total frequencies

Sol. (i) Here, frequency of class 10-20 is 7.

(ii) Class size = Upper limit − Lower limit = 30 − 20 = 10

(iii) Mid value = $\dfrac{\text{Upper limit} + \text{Lower limit}}{2}$

$= \dfrac{60 + 70}{2} = 65$

(iv) Total frequencies = 50

Ex. 2 What is the class mark of the interval 12.5-17.5?

Sol. Class mark = $\dfrac{\text{Lower limit} + \text{Upper limit}}{2}$

$= \dfrac{12.5 + 17.5}{2} = \dfrac{30}{2} = 15$

Now, let us study the measures of central tendency.

Mean or Arithmetic Mean (AM)

The sum of all observations divided by the number of observations is called mean and it is denoted by $\bar{x}$.

Mean $(\bar{x}) = \dfrac{\text{Sum of all observations}}{\text{Total number of observations}}$

Mean of Ungrouped Data

If $x_1, x_2, x_3, \ldots, x_n$ are n observations, then mean is given by

(i) **Direct Method** Mean, $\bar{x} = \dfrac{x_1 + x_2 + x_3 + \ldots + x_n}{n}$

$= \dfrac{\sum_{i=1}^{n} x_i}{n} = \dfrac{\Sigma x_i}{n}$

where, the Greek letter 'Σ' (sigma) means 'summation'.

(ii) **Shortcut Method** Mean, $\bar{x} = A + \dfrac{1}{n} \sum_{i=1}^{n} d_i$

where, A = assumed mean and $d_i = x_i - A$.

Ex. 3 Find the mean of 68, 78, 74, 89 and 75.

Sol. Mean = $\dfrac{68 + 78 + 74 + 89 + 75}{5} = \dfrac{384}{5} = 76.8$

Mean of Grouped Data

If $x_1, x_2, x_3, \ldots, x_n$ are n observations with respective frequencies $f_1, f_2, f_3, \ldots, f_n$, then mean is given by

(i) **Direct Method** Mean, $\bar{x} = \dfrac{f_1 x_1 + f_2 x_2 + \ldots + f_n x_n}{f_1 + f_2 + \ldots + f_n} = \dfrac{\sum\limits_{i=1}^{n} f_i x_i}{\sum\limits_{i=1}^{n} f_i}$

where, $i = 1, 2, 3, \ldots, n$.

(ii) **Shortcut Method** Mean, $\bar{x} = A + \dfrac{\sum\limits_{i=1}^{n} f_i d_i}{\sum\limits_{i=1}^{n} f_i}$

where, A = assumed mean and $d_i = x_i - A$ = deviation from assumed mean.

Ex. 4 The following table gives the marks scored by 30 students in a class.

Marks obtained (x_i)	10	20	36	40	50	56	60	70	72	80	88	92	95
Number of students	1	1	3	4	3	2	4	4	1	1	2	3	1

Find the mean of the given data.

Sol. To find the mean marks, we require the product of each x_i with the corresponding frequency f_i, so put them in a column as shown in the following table

Marks obtained (x_i)	Number of students (f_i)	$f_i x_i$
10	1	10
20	1	20
36	3	108
40	4	160
50	3	150
56	2	112
60	4	240
70	4	280
72	1	72
80	1	80
88	2	176
92	3	276
95	1	95
Total	$\Sigma f_i = 30$	$\Sigma f_i x_i = 1779$

$\therefore$ Mean $(\bar{x}) = \dfrac{\Sigma f_i x_i}{\Sigma f_i} = \dfrac{1779}{30} = 59.3$

Ex. 5 The table below shows the number of people within different age group who visited the mall on weekend

Age group (Class interval)	10-25	25-40	40-55	55-70	70-85	85-100
Number of people	3	11	10	8	6	2

Find the mean of the given data.

Sol.

Class interval	Number of people (f_i)	Class mark (x_i)	$f_i x_i$
10-25	3	17.5	52.5
25-40	11	32.5	357.5
40-55	10	47.5	475
55-70	8	62.5	500
70-85	6	77.5	465
85-100	2	92.5	185
Total	$\Sigma f_i = 40$		$\Sigma f_i x_i = 2035$

$$\therefore \text{Mean } (\bar{x}) = \frac{\Sigma f_i x_i}{\Sigma f_i} = \frac{2035}{40} = 50.875$$

Weighted Mean

If $x_1, x_2, \ldots, x_n$ are n observations to which respective weights attached are $w_1, w_2, \ldots, w_n$, then weighted mean is given as

$$\bar{x}_w = \frac{x_1 w_1 + x_2 w_2 + \cdots + x_n w_n}{w_1 + w_2 + \cdots + w_n}$$

Ex. 6 Find the weighted mean of the first n natural numbers, the weight being the corresponding numbers.

Sol. We know that first n natural numbers are $1, 2, 3, \ldots, n$, where the corresponding weights are $1, 2, 3, \ldots$, respectively.

$$\therefore \text{Weighted mean} = \frac{1 \times 1 + 2 \times 2 + 3 \times 3 + \ldots + n \times n}{1 + 2 + 3 + \ldots + n} = \frac{1^2 + 2^2 + \ldots + n^2}{1 + 2 + \ldots + n}$$

$$= \frac{\frac{n(n+1)(2n+1)}{6}}{\frac{n(n+1)}{2}} = \frac{(2n+1)}{3}$$

$$\left[\because \text{ sum of squares of first } n \text{ natural numbers} = \frac{n(n+1)(2n+1)}{6} \right.$$
$$\left. \text{and sum of first } n \text{ natural numbers} = \frac{n(n+1)}{2} \right]$$

Properties of Mean

(i) If mean of observations is $\bar{x}$ and each observation is increased by a, then new mean will be $\bar{x} + a$.

(ii) If mean of observations is $\bar{x}$ and each observation is decreased by a, then new mean will be $\bar{x} - a$.

(iii) If mean of observations is $\bar{x}$ and each observation is multiplied by a, then new mean will be $a\bar{x}$.

(iv) If mean of observations is $\bar{x}$ and each observation is divided by a, then new mean will be $\frac{\bar{x}}{a}$.

(v) If an observation equal to the mean of a series is removed or added to the given series, then the mean of the new series remains the same.

Median

Median is defined as the middle most or the central value of the variable in a set of observations, when the observations are arranged either in ascending or descending order of their magnitudes.

Median of Ungrouped Data

Suppose these are n number of observations. Then, arrange the data in ascending or descending order.

(i) If n is odd, then median = Value of the $\left(\dfrac{n+1}{2}\right)$ th observation

(ii) If n is even, then median = $\dfrac{\text{Value of the } \left[\left(\dfrac{n}{2}\right)\text{th} + \left(\dfrac{n}{2}+1\right)\text{th}\right] \text{observation}}{2}$

Ex. 7 Find the median of the observations 5, 15, 25, 35, 65, 75.

Sol. Here, $n = 6$ (even) and the observations are already arranged in ascending order.

$$\therefore \text{Median} = \dfrac{\left(\dfrac{n}{2}\right)\text{th observation} + \left(\dfrac{n}{2}+1\right)\text{th observation}}{2}$$

$$= \dfrac{\dfrac{6}{2}\text{th observation} + \left(\dfrac{6}{2}+1\right)\text{th observation}}{2}$$

$$= \dfrac{\text{3rd observation} + \text{4th observation}}{2} = \dfrac{25 + 35}{2} = \dfrac{60}{2} = 30$$

Median of a Discrete Frequency Data

Firstly, we arrange the data in the ascending or descending order, then we find the cumulative frequencies of all the classes. Now, find $N/2$ where, $N = \Sigma f_i$. See the cumulative frequency just greater than $N/2$. The corresponding value of x is the median.

Ex. 8 Find out the median of the following frequency distribution.

Variable	11	12	13	14	15	16	17	18
Frequency	5	7	11	9	8	7	3	5

Sol. The frequency distribution table is as follows

Variable (x)	Frequency (f)	Cumulative frequency (cf)
11	5	5
12	7	5 + 7 = 12
13	11	12 + 11 = 23
14	9	23 + 9 = 32
15	8	32 + 8 = 40
16	7	40 + 7 = 47
17	3	47 + 3 = 50
18	5	50 + 5 = 55
Total	$N = 55$	

Here, $N/2 = 55/2 = 27.5$

So, 32 is the cumulative frequency just greater than 27.5. So, median is 14.

Median for Grouped Data

Firstly, we find the cumulative frequencies of all the classes and $\frac{N}{2}$, where $N = \Sigma f_i$. Now, locate the class whose cumulative frequency is just greater than $\frac{N}{2}$ and this class is called median class. After finding the median class, use the following formula

$$\text{Median} = l + \left(\frac{\frac{N}{2} - cf}{f}\right) \times h$$

where, l = Lower limit of median class
cf = Cumulative frequency of class preceding the median class
f = Frequency of median class and h = Class size.

Ex. 9 Find the median of the data given below.

Class boundaries	Frequency
15-25	4
25-35	11
35-45	19
45-55	14
55-65	0
65-75	2

Sol. The frequency distribution table is as follows

Class boundaries	Frequency (f)	Cumulative frequency (cf)
15-25	4	4
25-35	11	4 + 11 = 15
35-45	19	15 + 19 = 34
45-55	14	34 + 14 = 48
55-65	0	48 + 0 = 48
65-75	2	48 + 2 = 50
Total	N = 50	

$\because$ $\frac{N}{2} = \frac{50}{2} = 25$

which is greater than cumulative frequency 15 and less than 34. So, median lies in the class 35-45.

Now, median $= l + \frac{1}{f}\left(\frac{N}{2} - cf\right) \times h$

where, $h = 10, f = 19, cf = 15$ and $l = 35$

$\therefore$ Median $= 35 + \frac{1}{19}(25 - 15) \times 10$

$= 35 + \frac{10}{19} \times 10 = 35 + 5.26 = 40.26$

Mode

Mode is that value of the variable, which occurs most frequently or we can say that the observation which has highest frequency is the mode.

Mode of Ungrouped Data

Mode is that value which is repeated maximum number of times.

Ex. 10 The wickets taken by a bowler in 10 cricket matches are as follows

2 6 4 5 0 2 1 3 2 3

Find the mode of the given data.

Sol. Let us form the frequency distribution table of the given data as follows

Number of wickets	0	1	2	3	4	5	6
Number of matches	1	1	3	2	1	1	1

Clearly, 2 is the number of wickets taken by the bowler in the maximum (i.e. 3) matches. So, the mode of the data is 2.

Mode of Discrete Series

In this case, mode is the value of the variable corresponding of the maximum frequency.

Ex. 11 Find the mode of the following data.

Variables	20	25	30	35	40
Frequency	3	4	6	2	5

Sol. The mode of the distribution is 30 as it has the maximum frequency.

✦ In the cases where two or more items carry the same higher frequency, mode is ill defined or undefined. e.g. 4, 4, 1, 4, 11, 11, 7, 5, 11, 35, 12 etc. or 4, 7, 1, 15, 35, 25, 18 etc. In this case, mode is ill defined.

Mode of Grouped Continuous Data

The class which has maximum frequency is called modal class. The mode is given by the formula,

$$\text{Mode} = l + \left(\frac{f_1 - f_0}{2f_1 - f_0 - f_2}\right) \times h$$

where, l = Lower limit of the modal class

h = Size of the class interval

f_0 = Frequency of the class preceding the modal class

f_1 = Frequency of the modal class

f_2 = Frequency of the class succeeding the modal class

Ex. 12 The table below shows the number of cars (in lakh) on road of 30 different states

Class interval	1-3	3-5	5-7	7-9	9-11
Number of states	10	8	5	6	1

What is the mode of the data shown above?

Sol. Here, modal class = 1-3 [as it has the maximum frequency, i.e. 10]

$\Rightarrow$ $l = 1$, Class size $(h) = 2$,

$f_1 = 10, f_0 = 0, f_2 = 8$

$\therefore$ Mode $= l + \left(\dfrac{f_1 - f_0}{2f_1 - f_0 - f_2}\right) \times h$

$= 1 + \left(\dfrac{10 - 0}{2 \times 10 - 0 - 8}\right) \times 2$

$= 1 + \dfrac{10}{12} \times 2 = 2.66$ (approx.)

Relation between Mean, Median and Mode

For moderately asymmetrical (not symmetrical distribution),

$$\boxed{\text{Mode} = 3 \text{ Median} - 2 \text{ Mean}}$$

For symmetrical distribution,

$$\boxed{\text{Mean} = \text{Median} = \text{Mode}}$$

Ex. 13 For a given data mean is 40 and mode is 25, then find the median.

Sol. Given, mean = 40 and mode = 25

$\therefore$ Mode = 3 Median − 2 Mean

$\Rightarrow$ $25 = 3$ Median $- 2 \times 40$

$\Rightarrow$ 3 Median $= 80 + 25 = 105$

$\therefore$ Median = 35

Fast Track Practice

Mean

1. The sales in rupees of a particular soap from Sunday to Saturday are given. Find the mean of daily sales.
   310, 420, 380, 370, 215, 430, 270
   (a) 342   (b) 342.25
   (c) 342.5   (d) None of these

2. In the following distribution, mean is

x	3	4	5	6	7	8	9	10
f	2	4	2	3	5	4	3	7

   [UPPSC Pre 2014]
   (a) 10   (b) 7   (c) 7.1   (d) 6.5

3. Find out the mean height of plants from the following frequency distribution.

Height (in cm)	61	64	67	70	73
Number of plants	5	18	42	27	8

   [UPPSC Pre 2013]
   (a) 67.00 cm   (b) 66.50 cm
   (c) 68.00 cm   (d) 67.45 cm

4. Consider the following distribution

Class	0-20	20-40	40-60	60-80	80-100
Frequency	17	28	32	f	19

   If the mean of the above distribution is 50, what is the value of f? [CDS 2017 (I)]
   (a) 24   (b) 34   (c) 56   (d) 96

5. If mean of five observations $x+1$, $x+2$, $x+3$, $x+4$ and $x+5$ is 15, then mean of first three observations is [HPPSC Pre 2013]
   (a) 12
   (b) 13
   (c) 14
   (d) 15

6. The mean marks obtained by 300 students in a subject are 60. The mean of top 100 students was found to be 80 and the mean of last 100 students was found to be 50. The mean marks of the remaining 100 students are [CDS 2017 (I)]
   (a) 70   (b) 65
   (c) 60   (d) 50

7. The mean of 5 numbers is 15. If one more number is included, the mean of the 6 numbers becomes 17. What is the included number? [CDS 2017 (I)]
   (a) 24   (b) 25
   (c) 26   (d) 27

8. The mean of 20 observations is 17. On checking it was found that the two observations were wrongly copied as 3 and 6. If wrong observations are replaced by correct values 8 and 9, then what is the correct mean? [CDS 2016 (II)]
   (a) 17.4   (b) 16.6
   (c) 15.8   (d) 14.2

Median

9. Find the median of the observation.
   6, 49, 14, 46, 16, 42, 26, 32, 28
   (a) 26   (b) 28   (c) 30   (d) 32

10. The median of set of 9 distinct observations is 20.5. If each of the largest 4 observations of the set is increased by 2, then the median of the new set [CDS 2016 (II)]
    (a) is increased by 2
    (b) is decreased by 2
    (c) is two times the original median
    (d) remains the same as that of original set

11. For $x > 0$, if a variable takes discrete values $x+4$, $x-3.5$, $x-2.5$, $x-3$, $x-2$, $x+0.5$, $x-0.5$, $x+5$, then what is the median? [CDS 2016 (II)]
    (a) $x - 1.25$ (b) $x - 0.5$
    (c) $x + 0.5$ (d) $x + 1.25$

12. The mean and median of 5 observations are 9 and 8, respectively. If 1 is subtracted from each observation, then the new mean and the new median will respectively be [CDS 2016 (I)]
    (a) 8 and 7
    (b) 9 and 7
    (c) 8 and 9
    (d) Cannot be determined due to insufficient data

Directions (Q. Nos. 13-15) *Consider the following frequency distribution.* [CDS 2015 (I)]

Class	0-10	10-20	20-30	30-40	40-50	50-60	60-70
Frequency	4	5	7	10	12	8	4

Statistics / 767

13. What is the mean of the distribution?
(a) 37.2 (b) 38.1 (c) 39.2 (d) 40.1

14. What is the median class?
(a) 20-30 (b) 30-40 (c) 40-50 (d) 50-60

15. What is the median of the distribution?
(a) 37 (b) 38
(c) 39 (d) 40

Mode

16. Find the mode of the given data.
5, 7, 9, 3, 7, 3, 7, 5, 7 [UPPSC Pre 2014]
(a) 1 (b) 3 (c) 7 (d) 9

17. Find the mode for the given distribution.

Class interval	25-30	30-35	35-40	40-45	45-50	50-55
Frequency	20	53	42	42	41	33

[CDS 2016 (I)]

(a) 31.75 (b) 30.75
(c) 33.75 (d) 35.75

18. The age distribution of 40 children is as follows

Age (in yr)	5-6	6-7	7-8	8-9	9-10	10-11
Number of children	4	7	9	12	6	2

Consider the following statements in respect of the above frequency distribution.

I. The median of the age distribution is 7 yr.

II. 70% of the children are in the age group 6-9 yr.
III. The modal age of the children is 8 yr.

Which of the above statements are correct? [CDS 2016 (I)]
(a) I and II (b) II and III
(c) I and III (d) I, II and III

19. In an asymmetrical distribution, if the mean and median of the distribution are 270 and 220 respectively, then the mode of the data is [CDS 2017(I)]
(a) 120 (b) 220 (c) 280 (d) 370

20. In a moderately asymmetrical distribution, the mode and mean are 32.1 and 35.4, respectively. Calculate the median.
(a) 35 (b) 35.3 (c) 34.3 (d) 34

21. If for a sample data,
Mean < Median < Mode, then the distribution is [UPSC CSAT 2017]
(a) symmetric
(b) skewed to the right
(c) neither symmetric nor skewed
(d) skewed to the left

Answer with Solutions

1. (d) Arithmetic mean,
$$\bar{x} = \frac{310+420+380+370+215+430+270}{7}$$
$$= \frac{2395}{7} = 342.143$$

2. (c) The frequency distribution table is as follows

x	Frequency (f)	fx
3	2	6
4	4	16
5	2	10
6	3	18
7	5	35
8	4	32
9	3	27
10	7	70
Total	$\Sigma f = 30$	$\Sigma fx = 214$

$$\therefore \quad \text{Mean} = \frac{\Sigma fx}{\Sigma f}$$
$$= \frac{214}{30} = 7.1$$

3. (d) We have,

Height (in cm) (x)	Number of plants (f)	fx
61	5	305
64	18	1152
67	42	2814
70	27	1890
73	8	584
Total	$\Sigma f = 100$	$\Sigma fx = 6745$

$$\therefore \quad \text{Mean} = \frac{\Sigma fx}{\Sigma f} = \frac{6745}{100} = 67.45 \text{ cm}$$

4. (a)

Class	Frequency (f_i)	Class Marks (x_i)	$f_i x_i$
0-20	17	10	170
20-40	28	30	840
40-60	32	50	1600
60-80	f	70	70f
80-100	19	90	1710
Total	$\Sigma f_i = 96 + f$		$\Sigma f_i x_i = 4320 + 70f$

Now,

Mean $= \dfrac{\Sigma f_i x_i}{\Sigma f_i} \Rightarrow 50 = \dfrac{4320 + 70f}{96 + f}$

$\Rightarrow 50(96 + f) = 4320 + 70f$
$\Rightarrow 4800 + 50f = 4320 + 70f$
$\Rightarrow 480 = 20f \Rightarrow f = \dfrac{480}{20} = 24$

5. (c) Mean of five observations
$= \dfrac{x + 1 + x + 2 + x + 3 + x + 4 + x + 5}{5}$

$\Rightarrow 15 = \dfrac{5x + 15}{5} \Rightarrow 5x + 15 = 75$

$\Rightarrow x = \dfrac{60}{5} = 12$

∴ First three observations = 13, 14 and 15
Now, mean of first three observations
$= \dfrac{13 + 14 + 15}{3} = \dfrac{42}{3} = 14$

6. (d) Total marks of middle 100 students
$= 300 \times 60 - 100 \times 80 - 100 \times 50$
$= 18000 - 8000 - 5000$
$= 18000 - 13000 = 5000$
∴ Mean marks of remaining 100 students
$= \dfrac{5000}{100} = 50$

7. (d) Since, mean of 5 numbers is 15.
∴ Sum of these 5 numbers
= Mean × Number of data = 15 × 5 = 75
Let the number included be x.

∴ $\dfrac{75 + x}{6} = 17$

$\Rightarrow 75 + x = 17 \times 6 \Rightarrow x = 102 - 75 = 27$

8. (a) Here, mean of 20 observations, $\bar{x} = 17$

∵ $\bar{x} = \dfrac{\sum_{i=1}^{20} x_i}{20}$

$\Rightarrow \Sigma x_i = 17 \times 20 = 340$
∴ $\Sigma x_i = 340 - (3 + 6) + (8 + 9)$

$\Rightarrow \Sigma x_i = 348$

∴ New mean $= \dfrac{348}{20} = 17.4$

9. (b) Arranging the observations in ascending order, we get
6, 14, 16, 26, 28, 32, 42, 46, 49
Here, $n = 9$ (odd)
So, median $= \left(\dfrac{n+1}{2}\right)$th observation

$= \left(\dfrac{9+1}{2}\right)$th observation

= 5th observation = 28

10. (d) Let the 9 observations be
$x_1, x_2, x_3, x_4, x_5, x_6, x_7, x_8, x_9$.
∴ Median of 9 observations is x_5.
∴ $x_5 = 20.5$
If x_6, x_7, x_8, x_9 are increased by 2.
The median of 9 observations is again
$x_5 = 20.5$
∴ Median remains the same as that of original set.

11. (a) We have, $x > 0$, and the observations are,
$x + 4, x - 3.5, x - 2.5, x - 3,$
$x - 2, x + 0.5, x - 0.5, x + 5$
Arrange in ascending order
$x - 3.5, x - 3, x - 2.5, x - 2, x - 0.5, x + 0.5,$
$x + 4, x + 5$
Total number of observations = 8
∴ Median $= \dfrac{(4\text{th} + 5\text{th}) \text{ observations}}{2}$

$= \dfrac{(x - 2) + (x - 0.5)}{2} = \dfrac{2x - 2.5}{2} = x - 1.25$

12. (a) Given, mean of 5 observations is 9.
∴ $\dfrac{\text{Sum of 5 observations}}{\text{Number of observations}} = 9$

$\Rightarrow$ Sum of 5 observations = 9 × 5 = 45
If 1 is subtracted from each observation, then
New mean of 5 observations

$= \dfrac{\text{Sum of 5 observations} - 5}{5}$

$= \dfrac{45 - 5}{5} = \dfrac{40}{5} = 8$

Median of 5 observations $= \left(\dfrac{5+1}{2}\right)$th term

= 3rd term = 8
If 1 is subtracted from each observation, then
New median = 8 − 1 = 7
Hence, the new mean and median are 8 and 7, respectively.

Solutions (Q. Nos. 13-15) We have,

Class	Mid value (x)	Frequency (f)	cf	fx
0-10	5	4	4	20
10-20	15	5	9	75
20-30	25	7	16	175
30-40	35	10	26	350
40-50	45	12	38	540
50-60	55	8	46	440
60-70	65	4	50	260
Total		$\Sigma f = N = 50$		$\Sigma fx = 1860$

13. (a) Mean $= \dfrac{\Sigma fx}{\Sigma f} = \dfrac{1860}{50} = 37.2$

14. (b) Here, $N = 50$
Now, $N/2 = 50/2 = 25$
which lies in the cumulative frequency corresponding to class interval for cf 26, i.e. 30-40.

15. (c) From the table,
$l = 30, h = 10, f = 10$ and $cf = 16$
$\therefore$ Median $= l + \dfrac{1}{f}\left(\dfrac{N}{2} - cf\right) \times h$
$= 30 + \dfrac{1}{10}(25 - 16) \times 10$
$= 30 + \dfrac{10}{10} \times 9 = 30 + 9 = 39$

16. (c) Here, the number 7 occurs maximum number of times, i.e. 4. Hence, mode of the data is 7.

17. (c) The frequency distribution table is as follows

Class interval	Frequency	Cumulative frequency
25-30	20	20
30-35	53	20 + 53 = 73
35-40	42	73 + 42 = 115
40-45	42	115 + 42 = 157
45-50	41	157 + 41 = 198
50-55	33	198 + 33 = 231
Total	$N = 231$	

The greatest frequency 53 lies in the class interval 30-35.
So, 30-35 is the modal class.
Here, $l = 30, h = 5, f_1 = 53, f_0 = 20, f_2 = 42$
$\therefore$ Mode $= l + \left(\dfrac{f_1 - f_0}{2f_1 - f_0 - f_2}\right) \times h$
$= 30 + \left(\dfrac{33}{106 - 62}\right) \times 5 = 30 + \dfrac{165}{44}$
$= 30 + 3.75 = 33.75$

18. (b) The frequency distribution table is as follows

Age (in yr)	Number of children	Cumulative frequency
5-6	4	4
6-7	7	4 + 7 = 11
7-8	9	11 + 9 = 20
8-9	12	20 + 12 = 32
9-10	6	32 + 6 = 38
10-11	2	38 + 2 = 40
Total	$N = 40$	

I. Here, $N = 40 \Rightarrow N/2 = 20$
which lies in the cumulative frequency corresponding to class interval 7-8.
$\therefore l = 7, f = 9, cf = 11, \dfrac{N}{2} = 20$ and $h = 1$
Median $= l + \dfrac{(N/2) - cf}{f} \times h = 7 + \dfrac{20 - 11}{9} \times 1$
$= 7 + 1 = 8$ yr
Hence, Statement I is incorrect.

II. Total number of children $(N) = 40$
Number of children in the age group 6-9
$= 7 + 9 + 12 = 28$
$\therefore$ Required percentage $= \dfrac{28}{40} \times 100\% = 70\%$
Hence, Statement II is correct.

III. $\because$ Modal class of the given data is 8-9 because it has largest frequency among the given classes of the data, i.e. 12.
$\therefore$ Modal age = 8 yr
Since, 9 is not included in this group.
Hence, Statement III is correct.

19. (a) We know that,
3 Median = Mode + 2 Mean
$\Rightarrow$ Mode = 3 Median − 2 Mean
$= 3 \times 220 - 2 \times 270$
$= 660 - 540 = 120$

20. (c) From the relation,
3 Median = 2 Mean + Mode
$\Rightarrow$ 3 Median $= 2 \times 35.4 + 32.1$
$= 70.8 + 32.1 = 102.9$
$\therefore$ Median $= \dfrac{102.9}{3} = 34.3$

21. (d) If mean < median < mode, then the distribution is skewed to the left.

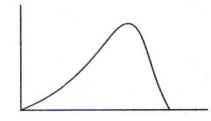

Chapter 42

Data Table

A Data Table is a chart of facts and figures represented in the form of **horizontal rows** and **vertical columns**. These facts and figures can be of imports, exports, income of employees in a factory, students applying for and qualifying a certain field of study, etc.

The amount of data that can be presented on data table is much higher than that which can be presented on any other type of graph or chart.

For example The data table given below shows profit of 3 companies from year 2010 to 2012

Year / Company	2010 (Profit%)	2011 (Profit%)	2012 (Profit%)	Row
SATYAM	30	20	15	1
TCS	15	20	10	2
L&T	15	10	35	3
Column	1	2	3	

Here, to calculate different value related to a data table we must have the knowledge of rows and columns.

(i) **Profit of Satyam in year 2011 can be calculated as** Satyam is in first row and year 2011 is in second column, so the intersection point is at 20.

∴ Profit of Satyam in 2011 is 20%.

(ii) **Sum of profit of TCS and L&T in year 2012 can calculated as**

TCS is in second row and year 2012 is in third column and their intersection point is at 10 in second row. L&T is in third row and year 2012 is in third column and their intersection point is at 35.

So, the required sum of profit = 10 + 35 = 45%

While answering the questions based on tables, read the table title and the column heading carefully. The title of the table will give you a general idea and often the purpose of the information presented. The column headings tell you the specific kind of the information given in that column.

Data Table / 771

Directions (Examples 1-3) *Study the given table carefully and answer the questions that follow.*

Percentage of marks obtained by five students in five different subjects in a school

Subject / Student	English (100)	Science (125)	Mathematics (150)	Social Studies (75)	Hindi (50)
Rahul	67	84	70	64	90
Veena	59	72	74	88	84
Soham	66	90	84	80	76
Shreya	71	66	80	66	86
Varun	63	76	88	68	72

✦ Figures in brackets indicate maximum marks for a particular subject.

Ex. 1 What is Varun's overall percentage in the examination?

Sol. Total marks of Varun

= Sum of marks in [English + Science + Mathematics + Social Studies + Hindi]

= 63% of 100 + 76% of 125 + 88% of 150 + 68% of 75 + 72% of 50

$= 63 + \dfrac{76 \times 125}{100} + \dfrac{88 \times 150}{100} + \dfrac{68 \times 75}{100} + \dfrac{72 \times 50}{100}$

= 63 + 95 + 132 + 51 + 36 = 377

∴ Required percentage = $\dfrac{\text{Total marks of Varun}}{\text{Total marks}} \times 100\% = \dfrac{377}{500} \times 100\% = 75.4\%$

Ex. 2 If in order to pass the exam, a minimum of 95 marks are needed in Science, how many students pass in the exam?

Sol. Let the passing percentage in Science be x, then

$x\%$ of $125 = 95$

$\Rightarrow \quad \dfrac{125 \times x}{100} = 95$

$\Rightarrow \quad x = \dfrac{95 \times 100}{125} = 76$

Here, only three students pass (Rahul, Soham and Varun) in the examination.

Ex. 3 What is the respective ratio of total marks obtained by Veena and Shreya together in Mathematics to the marks obtained by Rahul in the same subject?

Sol. Total marks obtained by Veena and Shreya together in Mathematics

= 74% of 150 + 80% of 150

$= \dfrac{74 \times 150}{100} + \dfrac{80 \times 150}{100} = 111 + 120 = 231$

Marks obtained by Rahul in Mathematics = 70% of $150 = 150 \times \dfrac{70}{100} = 105$

∴ Required ratio $= \dfrac{231}{105} = \dfrac{11}{5} = 11:5$

Directions (Examples 4-7) *Study the given tables carefully and answer the questions that follow.*

Number of candidates (in lakh) appearing in an entrance examination from six different cities and the ratio of candidates passing and failing in the same

City	A	B	C	D	E	F
Number of candidates	1.25	3.14	1.08	2.27	1.85	2.73

Ratio of candidates passing and failing within the city

City	Passing	Failing
A	7	3
B	5	3
C	4	5
D	1	3
E	3	2
F	7	5

Ex. 4 The number of candidates appearing for the exam from city C is what per cent of the number of candidates appearing for the exam from city B? (Rounded off to be nearest integer)

Sol. Required percentage $= \dfrac{1.08 \times 100}{3.14}\% \approx 34\%$

Ex. 5 What is the respective ratio of the numbers of candidates failing in the exam from city D to those failing in the exam from city A?

Sol. Required ratio $= \dfrac{3}{4} \times 2.27 : \dfrac{3}{10} \times 1.25$

$= 1.7025 : 0.375 = 227 : 50$

Ex. 6 Number of candidates passing the exam from city F is what per cent of the total number of candidates appearing from all the cities together? (Rounded off to two digits after the decimal)

Sol. Required percentage $= \dfrac{2.73 \times \dfrac{7}{12}}{1.25 + 3.14 + 1.08 + 2.27 + 1.85 + 2.73} \times 100\%$

$= \dfrac{1.5925}{12.32} \times 100\% = 12.93\%$

Ex. 7 What is the number of unsuccessful candidates in city C?

Sol. Number of unsuccessful candidates in city C

$= \left(1.08 \times \dfrac{5}{9}\right)$ lakh $= 0.6$ lakh

Fast Track Practice

Exercise 1 Base Level Questions

Directions (Q. Nos. 1-4) *Study the table and answer the given questions.*
[IBPS Clerk (Pre) 2016]

Number of muffins sold by five different bakeries in 5 different months

Bakery \ Month	March	April	May	June	July
P	150	109	112	129	138
Q	78	87	58	105	125
R	157	209	211	213	220
S	169	149	122	153	151
T	175	159	141	180	197

1. What is the respective ratio between the total number of muffins sold by bakeries Q and S together in May and the total number of muffins sold by bakeries R and T together in the same month?
   (a) 21 : 23   (b) 45 : 88
   (c) 19 : 23   (d) 20 : 23
   (e) 22 : 29

2. If the number of muffins sold by bakery T in August was 70% more than that sold by the same bakery in June, what was the number of muffins sold by the bakery T in August?
   (a) 255   (b) 221
   (c) 323   (d) 238
   (e) 306

3. The total number of muffins sold by bakeries Q and S together in June is, what per cent more than the number of muffins sold by bakery P in March?
   (a) 70%   (b) 74%
   (c) 72%   (d) 73%
   (e) 75%

4. What is the difference between the total number of muffins sold by bakeries P and Q together in April and the total number of muffins sold by bakeries R and S together in July?
   (a) 175   (b) 152   (c) 165   (d) 143
   (e) 155

Directions (Q. Nos. 5-7) *Study the following table and answer the given questions.* [CLAT 2015]

Number of units manufactured (M) and number of units sold (S) (in hundred) by five different companies over the years

Year	A M	A S	B M	B S	C M	C S	D M	D S	E M	E S
2006	2.8	1.3	3.3	2.2	2.6	1.7	3.0	2.2	1.9	1.4
2007	3.2	2.0	2.4	1.6	2.2	1.5	2.5	1.9	2.0	1.7
2008	1.9	0.9	2.9	1.6	2.1	1.0	2.3	1.5	1.6	1.1
2009	1.0	0.4	2.4	1.3	2.8	1.4	2.1	1.2	3.2	2.5
2010	2.5	1.5	2.3	1.2	2.6	2.1	1.8	1.1	3.1	2.6

5. The number of units sold by company D in the year 2006 is what per cent of the number of units manufactured by it in the year? (Rounded off to two digits after decimal)
   (a) 52.63%   (b) 61.57%
   (c) 85.15   (d) 73.33%

6. What is the respective ratio of total number of units manufactured by companies A and B together in the year 2009 to those sold by them in the same year?
   (a) 2 : 1   (b) 3 : 2
   (c) 5 : 2   (d) None of these

7. What is the average number of units sold by company D over all the years together?
   (a) 166   (b) 158
   (c) 136   (d) 147

8. Refer the below data table and answer the following question.

Year	Ratio Import/Export
2011	0.7
2012	1.1
2013	0.9
2014	1.5
2015	1.2

If the imports in 2012 was ₹ 1000 crores and the total exports in the years 2012 and 2013 together was ₹ 3200 crores, then the imports in 2013 was? [SSC (10+2) 2017]
(a) 2062 (b) 2291
(c) 909   (d) 2545

Directions (Q. Nos. 9-13) *Study the following table to answer the given questions.* [SBI PO 2016]

Data regarding number of candidates appearing for Civil Service (CS) and Engineering Service (ES) Examinations in the years 2007, 2008, 2009, 2010 in the country XYZ

Year	Civil Service - Total number of candidates appeared	Civil Service - Graduates out of the total candidates appeared (in %)	Engineering Service - Total number of candidates appeared	Engineering Service - Graduates out of the total candidates appeared (in %)
2007	58	75	30	52
2008	60	60	36	50
2009	70	65	52	40
2010	76	50	40	60

✦ Figures with regards to total number of candidates appeared are given in thousand.

9. Total number of candidates who appeared for CS and ES together in 2011 was 25% more than the total number of candidates who appeared for the same together in 2010. How many female candidates appeared for both the exams together in 2011 if they formed $\frac{2}{5}$ th of the total number of candidates appearing for both CS and ES that year?
(a) 52000 (b) 58000
(c) 60000 (d) 62000
(e) 64000

10. What is the respective ratio between the number of graduates who had appeared for ES in 2010 and the number of graduates who appeared for CS in 2010?
(a) 13 : 21 (b) 12 : 17
(c) 12 : 19 (d) 11 : 17
(e) 11 : 19

11. Total number of graduates who appeared for ES in 2008 is what per cent of the total number of graduates who appeared for CS in the same year?
(a) 75% (b) 40%
(c) 55% (d) 60%
(e) 50%

12. What is the difference between the average number of candidates who appeared for CS in the years 2007 and 2008 and average number of candidates who appeared for ES in the same years together?
(a) 38400 (b) 24400
(c) 26000 (d) 26400
(e) 24000

13. What is the total number of graduates who appeared for both CS and ES together in the year 2009?
(a) 66300 (b) 64200
(c) 60800 (d) 62800
(e) 66800

Directions (Q. Nos. 14-18) *Study the following table to answer the given questions.*
[NICL AO 2015]

Books sold through online mode and offline mode in a city

Year	Number of books sold (in thousand) (Online mode + Offline mode)	Percentage of books sold through online mode	Respective ratio of number of non-fiction books sold to number of fiction books sold (Online mode + Offline mode)
2010	690	40	2 : 3
2011	720	57.5	4 : 5
2012	945	60	7 : 8
2013	1240	75	1 : 3
2014	1600	79	3 : 5

14. In the years 2010 and 2011 together, the average number of books bought by each buyer through offline mode was 5. If three-fifth of the number of buyers buying books from offline mode, during the years 2010 and 2011 together were of the age group more than or equal to 30 yr, then how many buyers were less than 30 yr of age?
(a) 42000 (b) 41000
(c) 57600 (d) 42500
(e) 45000

15. What is the respective ratio of the total number of books sold through online mode in the years 2010 and 2011 together and the total number of books sold through offline mode in the same years together?
(a) 25 : 32 (b) 24 : 25
(c) 23 : 24 (d) 35 : 36
(e) 25 : 28

16. What is the approximate percentage decrease in the number of books sold through offline mode in the year 2014 from the year 2010?
(a) 15%   (b) 19%   (c) 21%   (d) 23%
(e) 25%

17. In the year 2013, out of the total number of books sold in the fiction category, 4/15th of the books sold were written by Indian authors. Numbers of books written by the Foreign authors in the fiction category form, what per cent of the total number of books sold (non-fiction and fiction category together) in the year 2013?
(a) 62%   (b) 55%   (c) 65%   (d) 60%
(e) 52%

18. What is the difference between the number of fiction books sold in the years 2012 and 2013 together and the number of the non-fiction books sold in the years 2011 and 2012 together?
(a) 873   (b) 876
(c) 573   (d) 673
(e) 676

Directions (Q. Nos. 19-22) *Following table shows the percentage population of six states below poverty line and proportion of male and female.*
[SNAP 2016]

State	Percentage of population below poverty line	Proportion of male and female Below poverty line Male : Female	Above poverty line Male : Female
A	16	2 : 3	3 : 4
B	10	4 : 3	5 : 2
C	14	3 : 4	2 : 3
D	21	5 : 2	4 : 3
E	25	4 : 1	2 : 1
F	20	2 : 3	4 : 2

19. If the total population of state A is 5000, then what is the number of females above poverty line in state A?
(a) 2000   (b) 2400
(c) 2600   (d) Data inadequate

20. If the population of states C and D is 20000 each, then what is the total number of females below poverty line in these states?
(a) 5000   (b) 6000
(c) 7200   (d) None of these

21. If the population of males below poverty line in state C is 6000 and state E is 10000, then what is the ratio of the total population of states C and E?
(a) 2 : 1
(b) 3 : 5
(c) 1 : 5
(d) None of the above

22. If in the state F population of females below poverty line is 16000, then what is the population of males below poverty line in that state?
(a) 8000
(b) 6000
(c) 1200
(d) None of the above

Directions (Q. Nos. 23-26) *Study the table and answer the given questions.*
[SBI PO (Pre) 2017]

Number of voters at 5 different centres A, B, C, D, E

Centre	Total number of registered voters	Percentage of people who voted (Out of the total number of registered voters)
A	2100	80
B	1750	80
C	3000	70
D	2400	76
E	2000	85

23. What per cent of the total number of registered voters cast invalid at centre D, if the number of invalid votes cast at centre D was 10% of the number of votes cast?
(a) 5.5%   (b) 8.5%
(c) 7.6%   (d) 6.5%
(e) Other than those given as options

24. At centre F, the total number of registered voters was 25% less than that at centre C. At centre F, number of people who voted was 450 less than that at centre C and 150 votes cast were declared invalid. What was the respective ratio between the number of valid votes cast and the total number of registered voters at centre F?
(a) 4 : 5   (b) 3 : 4
(c) 2 : 3   (d) 6 : 1
(e) 5 : 8

25. Number of people who did not vote at centre D is what per cent more than that who did not vote at centre A?
(a) $42\frac{6}{7}\%$
(b) $35\frac{2}{3}\%$
(c) $39\frac{4}{7}\%$
(d) $43\frac{5}{8}\%$
(e) Other than those given as options

26. What is the difference between the total number of people who did not vote at centres A and B together and that who did not vote at centres D and E together?
(a) 80
(b) 60
(c) 50
(d) 106
(e) 118

Directions (Q. Nos. 27-29) *The table below shows the number of people who responded to a survey about their favourite style of music. Use this information to answer the following questions to the nearest whole percentage.* [SNAP 2012]

Age	15-20	21-30	31+
Classical	6	4	17
Pop	7	5	5
Rock	6	12	14
Jazz	1	4	11
Blues	2	3	15
Hip-Hop	9	3	4
Ambient	2	2	2
Total	33	33	68

27. What percentage of respondents under 31 indicated that Blues is their favourite style of music?
(a) 7.1%
(b) 7.6%
(c) 8.3%
(d) 14.1%

28. What percentage of respondents aged 21-30 indicated a favourite style other than Rock music?
(a) 64%
(b) 60%
(c) 75%
(d) 36%

29. What percentage of the total sample indicated that Jazz is their favourite style of music?
(a) 6%
(b) 8%
(c) 22%
(d) 12%

Directions (Q. Nos. 30-33) *Study the following table carefully and answer the given questions.* [NICL AO (Pre) 2017]

Number of eggs produced by six farms

Year	Farm					
	A	B	C	D	E	F
2000	420	360	396	528	492	444
2001	564	492	576	612	576	540
2002	588	612	624	648	576	564
2003	600	660	648	636	612	600
2004	648	708	684	672	660	672
2005	732	744	720	756	708	720

30. What is the average number of eggs produced by farm D over the years?
(a) 7703 (b) 7700 (c) 7704 (d) 7604
(e) Other than those given as options

31. The eggs produced by farm B in the year 2003 are approximately what per cent of the eggs produced by farm B over the years?
(a) 18% (b) 16% (c) 19% (d) 20%
(e) Other than those given as options

32. What is the respective ratio of the dozens of eggs produced by farm A to farm E in the year 2005?
(a) 59 : 61
(b) 43 : 59
(c) 61 : 58
(d) 61 : 59
(e) Other than those given as options

33. What is the respective ratio of the dozens of eggs produced by farm A, B and C together in the year 2000, to the dozens of eggs produced by farms D, E and F together in the same year?
(a) 60 : 40
(b) 61 : 49
(c) 49 : 61
(d) 61 : 59
(e) 7 : 61

Directions (Q. Nos. 34-36) *Study the following table and answer the questions that follow.* [SBI Clerk 2015]

Number of pages printed by five printers in five days

Days	Printers				
	M	N	O	P	Q
Monday	196	145	254	169	291
Tuesday	125	172	233	141	297
Wednesday	167	189	243	223	189
Thursday	219	233	158	256	284
Friday	240	289	153	251	271

Data Table / 777

34. If the total number of pages printed by all the given printers together on Saturday was 18% more than the total number of pages printed by all the given printers together on Thursday, what was the total number of pages printed by all the given printers together on Saturday?
(a) 1379 (b) 1299 (c) 1357 (d) 1289
(e) 1331

35. What is the respective ratio between total number of pages printed by printer O on Tuesday and Wednesday together and total number of pages printed by printer Q on Monday and Tuesday together?
(a) 17 : 21 (b) 17 : 19
(c) 19 : 21 (d) 17 : 23
(e) 19 : 23

36. What is the sum of 58% of the total number of pages printed by printers M and O together on Monday and 65% of the total number of pages printed by printers N and Q together on Friday?
(a) 610 (b) 655 (c) 575 (d) 625
(e) 640

Directions (Q. Nos. 37-39) *Study the table and answer the given questions.*
[SBI Clerk (Pre) 2016]

Number of employees in five organisations during five years

Organisation \ Year	2003	2004	2005	2006	2007
A	69	55	56	59	65
B	93	100	79	101	120
C	95	89	85	75	70
D	88	109	100	99	95
E	80	132	147	164	112

37. Number of employees in organisations B and C decreased by 5% and 10%, respectively from 2007 to 2008. What was the total number of employees in organisation B and C together in 2008?
(a) 177 (b) 179 (c) 175 (d) 181
(e) 163

38. What is the respective ratio between total number of employees in organisations B and C together in 2004 and total number of employees in organisations D and E together in 2007?
(a) 21 : 22 (b) 23 : 29
(c) 23 : 27 (d) 21 : 25
(e) 21 : 23

39. What is the average number of employees in organisations A, B and E in 2005?
(a) 98 (b) 96
(c) 92 (d) 88
(e) 94

Directions (Q. Nos. 40-43) *Study the following table carefully and answer the given questions.*
[SBI PO (Pre) 2016]

Coaching institute	Total number of students who have enrolled for the coaching institute	Percentage of students who have enrolled for the given coaching institute from different schools			
		P	Q	R	S
A	80	25	20	15	40
B	100	24	33	21	22
C	200	32	20	17	31
D	250	20	10	20	50

✦ The coaching institutes have students from only the given four schools.

40. In coaching institute B, the total number of students who have enrolled in from schools P and R together is, what per cent less than the total number of students enrolled from schools Q and S together?
(a) $20\frac{1}{13}\%$ (b) $17\frac{3}{11}\%$
(c) $18\frac{2}{11}\%$ (d) $19\frac{1}{5}\%$
(e) $16\frac{2}{5}\%$

41. What is the average number of students who have enrolled in coaching institutes A, C and D from school R?
(a) 32 (b) 33
(c) 34 (d) 31
(e) 30

42. What is the difference between the total number of students who have enrolled in coaching institute A from schools P and S together and the total number of students who have enrolled for coaching institute C from the same school together?
(a) 78 (b) 64
(c) 38 (d) 74
(e) 67

43. In coaching institute D, 42% are females. If 20% of the total females are from school Q, what is the number of male students from school Q who have enrolled from coaching institute D?
(a) 5 (b) 7
(c) 6 (d) 3
(e) 4

Directions (Q. Nos. 44-47) *Study the table and answer the given questions.*
[IBPS SO 2016]

Data regarding population and literacy/illiteracy rate among the population in four different villages

Village	Respective ratio between number of males and females	Percentage of literates (Males and females) out of total population	Number of illiterates (Males and females) out of total population
A	3 : 1	40	360
B	3 : 2	30	350
C	3 : 2	75	50
D	2 : 1	60	120

* Total population = Number of males + Number of females

44. The number of illiterates in village C are what per cent less than that in village D?
(a) $72\frac{1}{3}$% (b) 52%
(c) $60\frac{2}{3}$% (d) $58\frac{1}{3}$%
(e) 60%

45. What is the respective ratio between number of literates in village A and that in village C?
(a) 4 : 3 (b) 5 : 3
(c) 7 : 4 (d) 8 : 5
(e) 9 : 4

46. If in village B, two-third of the literates (males and females) are males, what per cent of total number of females in village B are literate?
(a) 15% (b) 35%
(c) 30% (d) 20%
(e) 25%

47. What is the difference between total number of males in villages C and D together and the total number of females in the same villages together?
(a) 140 (b) 100 (c) 200 (d) 210
(e) 120

Directions (Q. Nos. 48-51) *Based on the following table, answer the given questions.*
[LIC AAO 2016]

Data given in the table for the month of March, 2015

Company	Total number of employees	Number of female employees
A	5550	2410
B	3200	1860
C	2000	1600
D	2500	1220
E	4240	2600
F	3560	1240

48. The number of male employees in company D is what per cent less than the number of female employees in company C?
(a) 12% (b) 15%
(c) 20% (d) 22%
(e) 18%

49. What is the difference between the number of male employees in company A and that in company F?
(a) 840 (b) 810
(c) 820 (d) 740
(e) 790

50. If in April 2015, the number of female employees in company E increased by 10% and the number of total employees in the company remained the same, what was the number of male employees?
(a) 1430 (b) 1420
(c) 1410 (d) 1360
(e) 1380

51. The total number of female employees in companies B and F together is what per cent of the total number of employees in company D?
(a) 124% (b) 122%
(c) 125% (d) 134%
(e) 135%

Data Table / 779

Directions (Q. Nos. 52-54) *Study the table and answer the given questions.*
[SBI Clerk (Mains) 2016]

Percentage of marks obtained by seven students in six different subjects in an examination and maximum marks in each subject is written in parenthesis

Student	Maths (150)	Chemistry (130)	Physics (120)	Geography (100)	History (60)	Computer Science (40)
A	90	50	90	60	70	80
B	100	80	80	40	80	70
C	90	60	70	70	90	70
D	80	65	80	80	60	60
E	80	65	85	95	50	90
F	70	75	65	85	40	60
G	65	35	50	77	80	80

52. What are the average marks obtained by all the seven students in Physics?
(a) 77.26 (b) 89.14
(c) 91.37 (d) 96.11
(e) 103.21

53. What was the aggregate of marks obtained by *B* in all the six subjects?
(a) 409 (b) 419
(c) 429 (d) 439
(e) 466

54. What is the overall percentage secured by *D*?
(a) 52.5%
(b) 55%
(c) 60%
(d) 73.42%
(e) 64.5%

Exercise 2 *Higher Skill Level Questions*

Directions (Q. Nos. 1-4) *Study the following table carefully to answer the questions that follow.* [Bank PO 2010]

Number (N) of six type of electronic products sold by six different stores in a month and the price per product (P) (Price in ₹ 1000) charged by each store

Product	Store A N	A P	B N	B P	C N	C P	D N	D P	E N	E P	F N	F P
L	54	135	48	112	60	104	61	124	40	136	48	126
M	71	4.5	53	3.8	57	5.6	49	4.9	57	5.5	45	4.7
N	48	12	47	18	52	15	54	11.5	62	10.5	56	11
O	52	53	55	48	48	50	54	49	59	47	58	51
P	60	75	61	68	56	92	44	84	46	76	59	78
Q	43	16	44	15	45	14.5	48	15.6	55	18.2	55	14.9

1. What is the total amount earned by store *C* through the sale of *M* and *O* type products together?
(a) ₹ 2719.2 lakh (b) ₹ 271.92 lakh
(c) ₹ 2.7192 lakh (d) ₹ 27.192 lakh
(e) None of these

2. Number of *L* type product sold by store *F* is what per cent of the number same type of products sold by store *E*?
(a) 76.33% (b) 124%
(c) 83.33% (d) 115%
(e) None of these

3. What is the respective ratio of total number of *N* and *L* type products together sold by store *D* and the same products sold by store *A*?
(a) 119 : 104 (b) 102 : 115
(c) 104 : 115 (d) 117 : 103
(e) None of these

4. What is the average price per product charged by all the stores together for product *Q*?
(a) ₹ 14700 (b) ₹ 15700
(c) ₹ 15200 (d) ₹ 14800
(e) None of these

Directions (Q. Nos. 5-8) *Study the following table carefully and answer the questions that follow.* [SBI PO 2016]

University	Total number of faculty members	Percentage of associate professors	Total number of female faculty members	Number of female associate professors
A	100	65	72	52
B	80	55	56	30
C	55	60	40	24
D	90	70	72	48

✦ Faculty members consist of only professors and associate professors in the given universities.

5. The number of male associate professors in university A is what per cent of the number of female associate professors in the same university?
(a) 26%   (b) 28%
(c) 22%   (d) 15%
(e) 25%

6. The total number of professors in universities A and C together is approximately what per cent less than the total number of associate professors in the same universities together?
(a) 37%   (b) 58%
(c) 48%   (d) 42%
(e) 28%

7. What percentage of the total number of faculty members in universities A and B together are male associate professors?
(a) 12%   (b) 14.5%
(c) 15%   (d) 13%
(e) 16.5%

8. What is the difference between the total number of male faculty members in universities B and D together and the total number of female faculty members in the universities A and C together?
(a) 60   (b) 69
(c) 75   (d) 72
(e) 70

Directions (Q. Nos. 9-12) *Study the following table carefully and answer the given questions.* [RBI Officer Grade 2015]

Date related to performance of six batsmen in a tournament

Name of the Batsman	Number of matches played in the tournament	Average runs scored in the tournament	Total balls faced in the tournament	Strike rate in the tournament
M	22	56	—	—
N	18	—	—	153.6
O	—	45	900	110
P	—	36	—	84
Q	—	—	—	140
R	24	51	1368	—

* Strike rate = (Total runs scored/Total balls faced) × 100

* All the given batsmen could bat in all the given matches played by them.
* Few values are missing in the table (indicated by '—'). A candidate is expected to calculate the missing value, if it is required to answer the given question, on the basis of the given data and information.

9. If the runs scored by R in last 3 matches of the tournament are not considered, his average runs scored in the tournament will decrease by 9. If the runs scored by R in the 21st and 22nd match are below 140 and no two scores among these 3 scores are equal, what is the minimum possible runs scored by R in the 24th match?
(a) 65   (b) 85
(c) 70   (d) 60
(e) 75

10. The respective ratio between total number of balls faced by O and that by Q in the tournament is 5 : 3. Total number of runs scored by Q in the tournament is what per cent less than the total number of runs scored by O in the tournament?
(a) $21\frac{3}{11}$%   (b) $25\frac{9}{11}$%
(c) $29\frac{1}{11}$%   (d) $27\frac{5}{11}$%
(e) $23\frac{7}{11}$%

11. Batsman M faced equal number of balls in first 11 matches he played in the tournament and last 11 matches he played in the tournament. If his strike rate in first 11 matches and last 11 matches of the tournament are 83 and 71 respectively, what is the total number of balls faced by him in the tournament?
(a) 1800   (b) 1500
(c) 1700   (d) 1600
(e) 1400

12. In the tournament, the total number of balls faced by batsman N is 402 less than the total number of runs scored by him. What is the average number of runs scored by batsman N in the tournament?
(a) 32   (b) 54
(c) 64   (d) 62
(e) 71

Data Table

Directions (Q. Nos. 13-17) *Study the following table and answer the given questions.* [IBPS PO 2015]

Revenue data of magazine Z during five months

Months	Gross revenue	Amount allocated for commission	Amount allocated for discount and offers	Net revenue
March	₹ 360000	₹ 31200	—	—
April	₹ 320000	₹ 28000	₹ 16000	—
May	—	—	₹ 36000	₹ 336000
June	—	₹ 42000	₹ 30000	₹ 330000
July	—	₹ 40000	₹ 28000	₹ 362000

* I. Net revenue = Gross revenue – Amount allocated for commission – Amount allocated for discount and others
   II. Few values are missing in the table (indicated by '—'). A candidate is expected to calculate the missing value. If it is required to answer the given question, on the basis of the given data and the information.

13. In July, if 40% of the gross revenue of the magazine was collected from advertisement, what was the amount of gross revenue collected from advertisement in that particular month?
(a) ₹ 148000 (b) ₹ 164000
(c) ₹ 144000 (d) ₹ 172000
(e) ₹ 156000

14. In March, if net revenue of the magazine was 85% of its gross revenue, what was the amount allocated for discount and others?
(a) ₹ 23200 (b) ₹ 24200
(c) ₹ 22400 (d) ₹ 22800
(e) ₹ 21600

15. Amount allocated for commission in March is what per cent less than amount allocated for commission in July?
(a) 24% (b) 18% (c) 28% (d) 32%
(e) 22%

16. What is the difference between net revenue of the magazine in April and its gross revenue in June?
(a) ₹ 132000 (b) ₹ 126000
(c) ₹ 118000 (d) ₹ 124000
(e) ₹ 136000

17. In May, the respective ratio of amount allocated for commission and amount allocated for discount and others was 4 : 3. What was the gross revenue of the magazine in May?
(a) ₹ 424000 (b) ₹ 440000
(c) ₹ 380000 (d) ₹ 420000
(e) ₹ 430000

Directions (Q. Nos. 18-29) *Study the table and answer the given questions* [IBPS Clerk (Mains) 2017]

Data related to rating of 5 movies in two cities (X and Y)

Movie	City X Number of raters	City X Average rating obtained	City Y Number of raters	City Y Average rating obtained
A	—	46	—	77
B	120	—	180	—
C	84	72	108	—
D	—	56	—	84
E	80	42.75	90	42

* Few data are missing (indicated by –). You need to calculate the value based on given data, if required, to answer a given question.
* Sum total of ratings obtained = Number of raters × Average rating obtained

18. For movie B, average rating obtained from city Y is 10% less than that obtained from city X. Sum total of ratings obtained from city Y is what per cent more than that obtained from city X?
(a) 35% (b) 32%
(c) 30% (d) 40%
(e) Other than those given as options

19. Combining city X and city Y, the average rating obtained for movie A is 58. Number of raters in city X is what per cent more than the number of raters in city Y?
(a) $27\frac{1}{3}$ (b) $35\frac{2}{3}$
(c) $36\frac{1}{3}$ (d) $33\frac{1}{3}$
(e) Other than those given as options

Answer with Solutions

Exercise 1 Base Level Questions

1. (b) Total number of muffins sold by bakeries Q and S together in May
= 58 + 122 = 180
Total number of muffins sold by bakeries R and T together in May
= 211 + 141 = 352
∴ Required ratio = 180 : 352 = 45 : 88

2. (e) Number of muffins sold by T in June
= 180 muffins
Number of muffins sold by T in August
$= 180 + \left(180 \times \frac{70}{100}\right)$
= 180 + 126 = 306 muffins

3. (c) Total number of muffins sold by bakeries Q and S together in June Month
= 105 + 153 = 258
Number of muffins sold by P in March
= 150
Increase in percentage $= \frac{258 - 150}{150} \times 100$
$= \frac{108}{150} \times 100 = 72\%$

4. (a) Total number of muffins sold by bakeries P and Q together in April
= 109 + 87 = 196
Total number of muffins sold by bakeries R and S together in July
= 220 + 151 = 371
∴ Required difference = 371 − 196 = 175

5. (d) The number of units sold by company D in the year 2006 = 2.2 hundred
The number of units manufactured by company D in the year 2006 = 3 hundred
∴ Required percentage $= \frac{2.2}{3} \times 100\%$
= 73.33%

6. (a) Total number of units manufactured by companies A and B together in the year 2009 = 1.0 + 2.4 = 3.4 hundred
Total number of units sold by company A and B together in the year 2009
= 0.4 + 1.3 = 1.7 hundred
∴ Required ratio $= \frac{3.4}{1.7} = 2 : 1$

7. (b) Total number of units sold by company D over all the years together
= 2.2 + 1.9 + 1.5 + 1.2 + 1.1
= 7.9 hundred = 7.9 × 100 = 790

∴ Average number of units sold by company D over all the years together
$= \frac{790}{5} = 158$

8. (a) Import in 2012 = ₹ 1000 crore
In 2012 = $\frac{\text{Import}}{\text{Export}} = 1.1$
Export = $\frac{\text{Import}}{1.1}$
Export in 2012 = ₹ $\frac{1000}{1.1}$ crore
Total export in year 2013
$= ₹ \left(3200 - \frac{10000}{11}\right)$ crore
$= ₹ \left(\frac{35200 - 10000}{11}\right)$ crore
$= ₹ \frac{25200}{11}$
Import in 2013 = 0.9 × Export
$= 0.9 \times \frac{25200}{11}$
≈ 2062 crore

9. (b) Total number of candidates who appeared for CS and ES together in year 2010 = 76000 + 40000 = 116000
So, total number of candidates who appeared for CS and ES together in year 2011 = $\frac{125}{100} \times 116000 = 145000$
Number of female candidates appeared for both the exams together in year 2011 = $\frac{2}{5}$
Total number of female candidates appeared for both exams together in year 2011
$= \frac{2}{5} \times 145000 = 58000$

10. (c) Number of graduates who had appeared for ES in year 2010 = 60% of 40000
$= \frac{60}{100} \times 40000 = 24000$
Number of graduates who had appeared for CS in year 2010 = 50% of 76000
$= \frac{50}{100} \times 76000 = 38000$
∴ Required ratio $= \frac{24}{38} = \frac{12}{19}$ or 12 : 19

Data Table / 783

11. (e) Total number of graduates who appeared for ES in year 2008 = 50% of 36000
$$= \frac{50}{100} \times 36000 = 18000$$
Total number of graduates who appeared for CS in year 2008
$$= 60\% \text{ of } 60000 = \frac{60}{100} \times 60000 = 36000$$
∴ Required percentage $= \frac{18000}{36000} \times 100 = 50\%$

12. (c) Average number of candidates appeared for CS in the years 2007 and 2008
$$= \frac{58000 + 60000}{2} = \frac{118000}{2} = 59000$$
Average number of candidates appeared for ES in the years 2007 and 2008
$$= \frac{30000 + 36000}{2} = \frac{66000}{2} = 33000$$
∴ Required difference
$$= (59000 - 33000) = 26000$$

13. (a) Total number of graduates who appeared for both CS and ES together in the year 2009
$$= (65\% \text{ of } 70000 + 40\% \text{ of } 52000)$$
$$= \left(\frac{65}{100} \times 70000 + \frac{40}{100} \times 52000\right)$$
$$= (45500 + 20800) = 66300$$

14. (c) Let the number of buyers be x.
According to the question,
Average number of books bought by each buyer through offline mode = 5
$$\therefore \frac{\frac{690 \times 60}{100} + \frac{720 \times 42.5}{100}}{x} = 5$$
$$\Rightarrow \frac{414 + 306}{x} = 5 \Rightarrow x = 144 \text{ thousand}$$
Now, buyers of age group more than or equal to 30 yr $= 144 \times \frac{3}{5}$
$$= 86.4 \text{ thousand}$$
$$= 86400$$
∴ Number of buyers of age group less than 30 yr = 144000 − 86400 = 57600

15. (c) Number of books sold through online mode in the years 2010 and 2011
$$= 690 \times \frac{40}{100} + 720 \times \frac{57.5}{100}$$
$$= 276 + 414 = 690$$
Number of books sold through offline mode in the year 2011
$$= 690 \times \frac{60}{100} + 720 \times \frac{42.5}{100}$$
$$= 414 + 306 = 720$$
∴ Respective ratio = 690 : 720 = 23 : 24

16. (b) Number of books sold through offline mode in year 2014 $= 1600 \times \frac{21}{100} = 336$
Number of books sold through offline mode in year 2010 $= 690 \times \frac{60}{100} = 414$
∴ Percentage decrease $= \frac{414 - 336}{414} \times 100$
$$= \frac{78}{414} \times 100 = 18.84 \approx 19\%$$

17. (b) Total number of books sold in year 2013
$$= 1240$$
Number of fiction books sold
$$= 1240 \times \frac{3}{4} = 930$$
Number of books written by Indian authors
$$= 930 \times \frac{4}{15} = 248$$
Number of books written by Foreign authors
$$= 930 - 248 = 682$$
∴ Required percentage $= \frac{682}{1240} \times 100 = 55\%$

18. (d) Number of fiction books sold in the year 2012 $= 945 \times \frac{8}{15} = 504$
Number of fiction books sold in the year 2013 $= 1240 \times \frac{3}{4} = 310 \times 3 = 930$
Total number of fiction books sold in the years 2012 and 2013 = 504 + 930 = 1434
Number of non-fiction books sold in the year 2011 $= 720 \times \frac{4}{9} = 320$
Number of non-fiction books sold in the year 2012 $= 945 \times \frac{7}{15} = 441$
Total number of non-fiction books sold in the years 2011 and 2012 together
$$= 320 + 441 = 761$$
∴ Required difference = 1434 − 761 = 673

19. (b) Given, total population in state A = 5000
∴ Total females above poverty line
$$= \frac{4}{7} \times 5000 \times \frac{84}{100} = 2400$$

20. (d) Given, population of states C and D
$$= 20000 \text{ each}$$
Number of females below poverty lines in state $C = \frac{4}{7} \times \frac{14}{100} \times 20000 = 1600$
Number of females below poverty line in state $D = \frac{2}{7} \times \frac{21}{100} \times 20000 = 1143$
∴ Total number of females below poverty line in states C and D together
$$= 1600 + 1143 = 2743$$

21. (a) Let the total population of state C be x.
According to the question,
$$14\% \text{ of } x \times \frac{3}{7} = 6000 \Rightarrow \frac{14x \times 3}{100 \times 7} = 6000$$
$$\Rightarrow x = \frac{6000 \times 7 \times 100}{14 \times 3} \Rightarrow x = 100000$$
Again, let total population of state E be y.
According to the question,
$$25\% \text{ of } y \times \frac{4}{5} = 10000 \Rightarrow \frac{25y}{100} \times \frac{4}{5} = 10000$$
$$\Rightarrow y = \frac{10000 \times 5 \times 100}{25 \times 4} \Rightarrow y = 50000$$
$\therefore$ Required ratio $= \frac{100000}{50000} = 2 : 1$

22. (d) Given, females below poverty line in state $F = 16000$.
$\therefore$ Population of males below poverty line
$= \frac{2}{3} \times 16000 = 10666.6 \approx 10667$

23. (c) Number of people voted at centre D
$= 2400 \times \frac{76}{100} = 1824$
Invalid votes $= 1824 \times \frac{10}{100}$
$= 182.4 \approx 182$
$\therefore$ Required percentage
$= \frac{182}{2400} \times 100$
$= 7.583\% \approx 7.6\%$

24. (c) Number of registered voters at centre F
$= 3000 \times \frac{(100-25)}{100}$
$= 3000 \times \frac{75}{100} = 2250$
Number of people voted at centre F
$= 3000 \times \frac{70}{100} - 450$
$= 2100 - 450 = 1650$
Invalid votes at centre $F = 150$
$\therefore$ Valid votes at centre F
$F = 1650 - 150 = 1500$
Now, required ratio = Valid votes
: Registered voters
$= 1500 : 2250$
$= 150 : 225 = 2 : 3$

25. (e) Number of people who did not vote at centre D
$= 2400 \times \frac{24}{100} = 576$
Number of people who did not vote at centre A
$= 2100 \times \frac{20}{100} = 420$

$\therefore$ Required percentage
$= \frac{(576-420)}{420} \times 100$
$= \frac{156}{420} \times 100 = 37\frac{1}{7}\%$

26. (d) Number of people did not vote at centre A
$= 2100 \times \frac{20}{100} = 420$
Number of people did not vote at centre $B = 1750 \times \frac{20}{100} = 350$
$\therefore$ Total number of people did not vote at centre A and B together
$= 420 + 350 = 770$
Now, number of people did not vote at centre D
$= 2400 \times \frac{24}{100} = 576$
Number of people did not vote at centre $E = 2000 \times \frac{15}{100} = 300$
$\therefore$ Total number of people did not vote at centre D and E together
$= 576 + 300 = 876$
$\therefore$ Required difference
$= 876 - 770 = 106$

27. (b) Number of respondents under 31 who's favourite style of music is blues
$= 2 + 3 = 5 \begin{cases} 2 \text{ from age group of } 15\text{-}20 \\ 3 \text{ from age group of } 21\text{-}30 \end{cases}$
Total respondents below $31 = 33 + 33 = 66$
$\therefore$ Percentage of respondents
$= \frac{5}{66} \times 100 = 7.57 = 7.6$

28. (a) Total respondents between 21-30 whose favourite style is other than rock
$= (4 + 5 + 4 + 3 + 3 + 2) = 21$
and total respondent $= 33$
$\therefore$ Per cent of respondents whose favourite style is other than rock $= \frac{21}{33} \times 100 = 64\%$

29. (d) Total respondent whose favourite style of music is Jazz $= (1 + 4 + 11) = 16$
Total respondent $= (33 + 33 + 68) = 134$
$\therefore$ Required percentage of respondent
$= \frac{16}{134} \times 100 \approx 12\%$

30. (e) Average number of eggs produced by farm D over the years
$= \frac{1}{6}(528 + 612 + 648 + 636 + 672 + 756)$
$= \frac{1}{6} \times 3852 = 642$

Data Table / 785

31. (a) Eggs produced by farm B in the year 2003 = 660
Eggs produced by farm B over the years
= 360 + 492 + 612 + 660 + 708 + 744
= 3576
∴ Required percentage
$= \frac{660}{3576} \times 100 = 18.45 \approx 18\%$

32. (d) Eggs produced by farm A in the year
$2005 = \frac{732}{12}$ dozen
and eggs produced by farm E in the year
$2005 = \frac{708}{12}$ dozen
∴ Required ratio $= \frac{732}{12} : \frac{708}{12}$
= 732 : 708 = 61 : 59

33. (c) Eggs produced by farms A, B and C together in the year 2000
$= \frac{420 + 360 + 396}{12}$
$= \frac{1176}{12} = 98$ dozen
Eggs produced by farms D, E and F together in the year 2000
$= \frac{528 + 492 + 444}{12}$
$= \frac{1464}{12} = 122$ dozen
∴ Required ratio = 98 : 122 = 49 : 61

34. (c) Total number of pages printed by all the given printers on Thursday
= 219 + 233 + 158 + 256 + 284 = 1150
∴ Total number of pages printed by all the given printers on Saturday
118% of 1150 = 1357

35. (a) Required ratio = (233 + 243):(291 + 297)
= 476 : 588 = 17 : 21

36. (d) Total number of pages printed by M and O together on Monday = (196 + 254)
Total number of pages printed by N and Q together on Friday = (289 + 271)
∴ Required sum = 58% of (196 + 254)
+ 65% of (289 + 271)
= 58% of 450 + 65% of 560
= 261 + 364 = 625

37. (a) Number of employees in organisation B in 2008 $= \frac{120 \times 95}{100} = 114$
Number of employees in organisation C in 2008 $= \frac{70 \times 90}{100} = 63$
∴ Total number of employees in organisations B and C together in 2008
= 114 + 63 = 177

38. (e) Total number of employees in organisations B and C together in 2004
= 100 + 89 = 189
Total number of employees in organisations D and E together in 2007 = 95 + 112 = 207
∴ Required ratio $= \frac{189}{207} = \frac{63}{69} = \frac{21}{23} = 21 : 23$

39. (e) Average number of employees
$= \frac{56 + 79 + 147}{3} = \frac{282}{3} = 94$

40. (c) Total number of students who have enrolled from schools P and R together in coaching institute B
$= 100 \times \frac{24}{100} + 100 \times \frac{21}{100}$
= 24 + 21 = 45
Total number of students who have enrolled from schools Q and S together in coaching institute B
$= 100 \times \frac{33}{100} + 100 \times \frac{22}{100}$
= 33 + 22 = 55
∴ Required percentage $= \frac{(55 - 45)}{55} \times 100$
$= \frac{10 \times 100}{55} = 18\frac{2}{11}\%$

41. (a) Number of students, enrolled in institute A from school $R = 80 \times \frac{15}{100} = 12$
Number of students, enrolled in institute C from school $R = 200 \times \frac{17}{100} = 34$
Number of students, enrolled in institute D from school $R = 250 \times \frac{20}{100} = 50$
∴ Required average
$= \frac{1}{3}(12 + 34 + 50) = \frac{96}{3} = 32$

42. (d) Total number of students who have enrolled from institute A from schools P and S together $= 80 \times \frac{25}{100} + 80 \times \frac{40}{100}$
= 20 + 32 = 52
Total number of students who have enrolled from institute C from schools P and S together $= 200 \times \frac{32}{100} + 200 \times \frac{31}{100}$
= 64 + 62 = 126
∴ Required difference = 126 − 52 = 74

43. (e) In coaching institute D, females = 42%
∴ Males in coaching institute D
= 100 − 42 = 58%
i.e. total number of females in institute D
$= 250 \times \frac{42}{100} = 105$
∴ Total number of males in institute D
= 250 − 105 = 145

Total number of students from school Q
$$= 250 \times \frac{10}{100} = 25$$
Number of females from school Q
$$= 105 \times \frac{20}{100} = 21$$
∴ Number of male students from school Q who have enrolled from institute D
$$= 25 - 21 = 4$$

44. (d) Number of illiterates in village C = 50
and number of illiterates in village D = 120
∴ Required percentage
$$= \frac{(120 - 50)}{120} \times 100$$
$$= \frac{70 \times 100}{120} = \frac{175}{3} = 58\frac{1}{3}\%$$

45. (d) Let total population of village A = a
Then, percentage of literates = 40%
and percentage of illiterates
$$= (100 - 40) = 60\%$$
$$\therefore a \times \frac{60}{100} = 360$$
$$\Rightarrow a = 600$$
∴ Population of village A is 600.
∴ Number of literates of village A
$$= 600 \times \frac{40}{100} = 240$$
Now, let population of village C = c
Then, percentage of literates = 75%
and percentage of illiterates
$$= (100 - 75) = 25\%$$
$$\Rightarrow c \times \frac{25}{100} = 50 \Rightarrow c = 200$$
Hence, population of village C is 200
∴ Number of literates of village C
$$= 200 \times \frac{75}{100} = 150$$
So, required ratio = 240 : 150 = 8 : 5

46. (e) Let population of village B = b
Then, percentage of literates = 30%
Now, percentage of illiterates
$$= (100 - 30) = 70\%$$
$$\therefore b \times \frac{70}{100} = 350 \Rightarrow b = 500$$
∴ Population of village B = 500
∴ Number of literates in village B
$$= 500 \times \frac{30}{100} = 150$$
Now, number of male literates in village B
$$= 150 \times \frac{2}{3} = 100$$
and number of female literates in village B
$$= 150 - 100 = 50$$

Total number of females in village B
$$= 500 \times \frac{2}{5} = 200$$
∴ Percentage of literate females of village B
$$= \frac{50}{200} \times 100 = 25\%$$

47. (a) Total population of village C
$$= 50 \times \frac{100}{25} = 200$$
Total number of males of village C
$$= 200 \times \frac{3}{5} = 120$$
Total population of village D = $\frac{100}{40} \times 120$
$$= 300$$
Total number of males of village D
$$= 300 \times \frac{2}{3} = 200$$
Total number of females of village C
$$= 200 - 120 = 80$$
Total number of females of village D
$$= 300 - 200 = 100$$
∴ Required difference
$$= (120 + 200) - (80 + 100)$$
$$= 320 - 180 = 140$$

48. (c) Number of male employees in company D = 2500 - 1220 = 1280
Number of female employees in company C = 1600
∴ Required percentage
$$= \frac{1600 - 1280}{1600} \times 100 = \frac{320}{1600} \times 100 = 20\%$$

49. (c) Required difference
$$= [(5550 - 2410) - (3560 - 1240)]$$
$$= 3140 - 2320 = 820$$

50. (e) Number of female employees in March in company E = 2600
Number of female employees in April in company E = $2600 \times \frac{110}{100} = 2860$
∴ Total number of male employees in April in company E = 4240 - 2860 = 1380

51. (a) Required percentage
$$= \frac{(1860 + 1240)}{2500} \times 100 = \frac{3100}{25} = 124\%$$

52. (b) Marks obtained in Physics by
A = 90% of 120 = 108
B = 80% of 120 = 96
C = 70% of 120 = 84
D = 80% of 120 = 96
E = 85% of 120 = 102
F = 65% of 120 = 78
G = 50% of 120 = 60

Data Table / 787

∴ Required average
$$= \frac{(108 + 96 + 84 + 96 + 102 + 78 + 60)}{7}$$
$$= \frac{624}{7} = 89.14$$

53. (e) Marks obtained by B in
Maths = 100% of 150 = 150
Chemistry = 80% of 130 = 104
Physics = 80% of 120 = 96
Geography = 40% of 100 = 40
History = 80% of 60 = 48
Computer Science = 70% of 40 = 28

∴ Required aggregate marks
$$= 150 + 104 + 96 + 40 + 48 + 28$$
$$= 466$$

54. (d) Overall percentage secured by D
$$= \frac{\begin{bmatrix} 80\% \text{ of } 150 + 65\% \text{ of } 130 + 80\% \text{ of } 120 \\ + 80\% \text{ of } 100 + 60\% \text{ of } 60 + 60\% \text{ of } 40 \end{bmatrix}}{150 + 130 + 120 + 100 + 60 + 40} \times 100$$
$$= \frac{120 + 84.5 + 96 + 80 + 36 + 24}{600} \times 100$$
$$= \frac{440.5}{600} \times 100 = 73.42$$

Exercise ❷ Higher Skill Level Questions

1. (d) Total amount earned by store C through the sales of M and O type products together
$$= ₹ (57 \times 5.6 + 48 \times 50) \text{ thousand}$$
$$= ₹ (319.2 + 2400) \text{ thousand}$$
$$= ₹ 27.192 \text{ lakh}$$

2. (e) Number of L type products sold by store F = 48
Number of L type product sold by store E = 40
∴ Required percentage = $\frac{48}{40} \times 100\% = 120\%$

3. (e) Required ratio = (61 + 54) : (54 + 48)
= 115 : 102

4. (b) Required average price per product
$$= ₹ \left(\frac{16 + 15 + 14.5 + 15.6 + 18.2 + 14.9}{6}\right)$$
$$= ₹ \left(\frac{94.2}{6}\right) \text{ thousand} = ₹ 15700$$

5. (e) The number of male associate professors in university A = 65 – 52 = 13
Number of female associate professors in university A = 52
∴ Required percentage = $\frac{13}{52} \times 100\% = 25\%$

6. (d) Number of professors in university A
= 100 – 65 = 35
Number of professors in university C
$$= 55 - \frac{55 \times 60}{100} = 55 - 33 = 22$$
Total number of professors in universities A and C together = 35 + 22 = 57
Total number of associate professors in universities A and C together
= 65 + 33 = 98
∴ Required percentage = $\frac{98 - 57}{98} \times 100\%$
= 41.83% ≈ 42%

7. (c) Total number of faculty members in universities A and B together
= 100 + 80 = 180
Male associate professors in university A
= 65 – 52 = 13
Male associate professors in university B
$$= \frac{80 \times 55}{100} - 30$$
$$= 44 - 30 = 14$$
Total number of associate professors in universities A and B = 13 + 14 = 27
∴ Required percentage = $\frac{27}{180} \times 100 = 15\%$

8. (e) Total number of male faculty members in universities B and D
= (80 – 56) + (90 – 72)
= 24 + 18 = 42
Total number of female faculty members in universities A and C = 72 + 40 = 112
∴ Required difference = 112 – 42 = 70

9. (a) Total run scored by R in 24 matches
= 24 × 51 = 1224
Total run scored by R in 21 matches
= 21 × (51 – 9) = 21 × 42 = 882
∴ Runs scored by R in last 3 matches
= 1224 – 882 = 342
So, maximum runs scored by R in 22nd and 23rd matches = 139 + 138 = 277
∴ Minimum possible runs scored by R in 24th match = 342 – 277 = 65

10. (e) Total balls faced by O = 900
∴ Total balls faced by Q = $\frac{3}{5} \times 900 = 540$
∴ Total runs of O
$$= \frac{\text{Strike rate} \times \text{Total balls faced}}{100}$$
$$= \frac{110 \times 900}{100} = 990$$

Total run of Q
$$= \frac{\text{Strike rate} \times \text{Total balls faced}}{100}$$
$$= \frac{140 \times 540}{100} = 756$$
∴ Required percentage
$$= \frac{990 - 756}{990} \times 100 = 23\frac{7}{11}\%$$

11. (d) Total runs scored by batsman M in 22 matches $= 22 \times 56 = 1232$
Now, total number of balls faced by batsman M in tournament
$$= \frac{2 \times \text{Total runs} \times 100}{\text{Strike rate of (first 11 mathces} + \text{last 11 matches)}}$$
$$= \frac{2 \times 1232 \times 100}{(83 + 71)} = \frac{2 \times 1232 \times 100}{154}$$
$$= 2 \times 8 \times 100 = 1600$$

12. (c) Let the average runs scored by N be x.
∴ $153.6 = \frac{18x}{18x - 402} \times 100$
⇒ $2764.8x - 61747.2 = 1800x$
⇒ $964.8x = 61747.2$
⇒ $x = 64$

13. (d) Gross revenue of July
$$= 362000 + 28000 + 40000$$
$$= 430000$$
Gross revenue collected from advertisement $= 430000 \times \frac{40}{100} = 4300 \times 40$
$$= ₹ 172000$$

14. (d) In March,
Net revenue $= 360000 \times \frac{85}{100} = 306000$
∴ Amount allocated for discount and others
$$= 360000 - 306000 - 31200$$
$$= 54000 - 31200 = ₹ 22800$$

15. (e) Required percentage
$$= \frac{40000 - 31200}{40000} \times 100\%$$
$$= \frac{8800}{40000} \times 100\% = 22\%$$

16. (b) Net revenue in April
$$= 320000 - 28000 - 16000$$
$$= 320000 - 44000 = ₹ 276000$$
Gross revenue in June
$$= 330000 + 30000 + 42000$$
$$= 330000 + 72000 = ₹ 402000$$
∴ Required difference
$$= 402000 - 276000 = ₹ 126000$$

17. (d) In May, ratio of amount allocated for commission and amount allocated for discount $= 4 : 3$
∴ Amount allocated for commission
$$= 36000 \times \frac{4}{3} = ₹ 48000$$
∴ Gross revenue in May
$$= 336000 + 48000 + 36000 = ₹ 420000$$

18. (a) Let the average rating for movie B from city X be x.
∴ Average rating for movie B from city Y
$$= \frac{90}{100} \times x$$
Sum total of ratings for movie B from city X
$$= 120x$$
Similarly, sum total of ratings for movie B from city $Y = 180 \times \frac{90}{100} \times x = 162x$
Required percentage
$$= \frac{162x - 120x}{120x} \times 100$$
$$= \frac{42}{120} \times 100 = 35\%$$

19. (e) Let the number of rates of city X and city Y be x and y, respectively.
Then, $58(x + y) = 46x + 77y$
⇒ $58x + 58y = 46x + 77y$
⇒ $12x = 19y$
∴ Required percentage
$$= \frac{x - y}{y} \times 100 = \left(\frac{x}{y} - 1\right) \times 100$$
$$= \left(\frac{19}{12} - 1\right) \times 100$$
$$= \frac{7}{12} \times 100 = 58\frac{1}{3}\%$$

Chapter 43

Pie Chart

Pie chart is a circular chart divided into sectors in which the arc length, its central angle and area are proportional to the quantities that it represents.

- Circular chart is called pie chart because of its shape. Each sector of chart is allotted to each category and shows the portion of the entire chart.
- In the questions of pie chart, the total quantity is distributed over a total angle of 360° or 100%. Here, the data can be plotted with respect to only one parameter.
- Uses of pie charts are restricted to represent limited type of information.
- Pie chart is also useful for representing proportions or percentages of various elements with respect to the total quantity.

Some Important Formulae

- If pie chart is specialised in degree, then

$$\text{Value of any sector} = \frac{\text{Angle of the sector}}{360°} \times \text{Total value}$$

- If pie chart is specialised in percentage, then

$$\text{Value of any sector} = \frac{\text{Per cent of the sector}}{100} \times \text{Total value}$$

The following pie chart gives the distribution of the population in different geographical zones.

Distribution of population in geographical zones

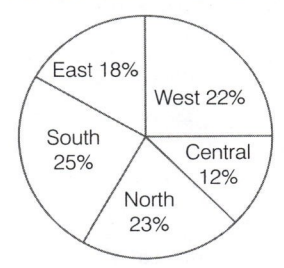

- East 18%
- West 22%
- Central 12%
- North 23%
- South 25%

790 / *Fast Track* Objective Arithmetic

From the given pie chart, we can calculate the following
+ Population of any zone when the total population is given.
+ Population of any zone as a percentage of another zone.
+ Percentage increase in the total population, given that percentage increase in the population of one or more zones.

Directions (Examples 1-2) *Study the following pie chart and answer the questions that follow.*

National budget expenditure in the year 2012
(Percentage distribution)

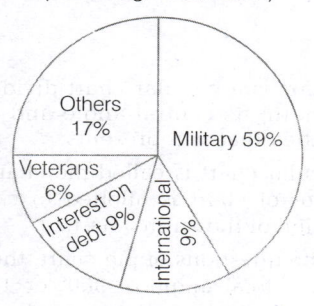

Others 17%
Military 59%
Veterans 6%
Interest on debt 9%
International 9%

Ex. 1 In year 2012, if India had a total expenditure of ₹ 120 billion, then how many billions did it spend on interest on debt?

Sol. Total expenditure = ₹ 120 billion

∴ Expenditure of interest on debt = 9% of 120 = $\frac{9}{100} \times 120$ = ₹ 10.8 billion

Ex. 2 If ₹ 9 billion were spent in year 2012 for veterans, then what would have been the total expenditure for that year (in billions)?

Sol. ₹ 9 billion were spent for veterans. 6% of the total expenditure was spent on veterans in the year 2012.

Hence, the total expenditure = $\frac{9}{6} \times 100$ = ₹ 150 billion

Directions (Examples 3-5) *Study the following pie charts and answer the questions that follow.*

Sales by location of company
(in million pounds)

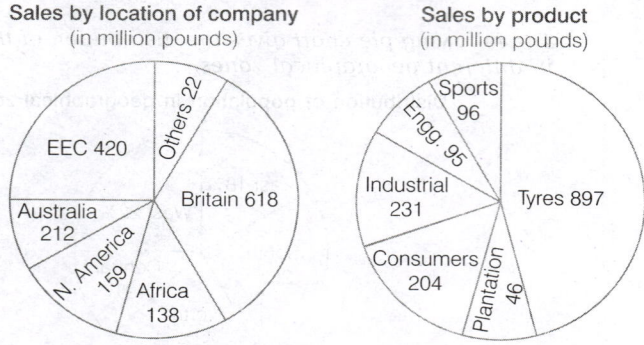

EEC 420
Others 22
Britain 618
Australia 212
N. America 159
Africa 138

Sales by product
(in million pounds)

Sports 96
Engg. 95
Industrial 231
Tyres 897
Consumers 204
Plantation 46

Total sales = 1569 million pounds

Pie Chart / 791

Ex. 3 If in the next year, the sales of sports goods were expected to double, then assuming that the total sales do not change, what would be the approximate percentage share of sports goods in the total sales?

Sol. Total sales = 1569 million pounds
Sports goods sales next year = 2 × 96 = 192 million pounds
Therefore, percentage share of sports goods sales = $\dfrac{192}{1569} \times 100\% \approx 12\%$

Ex. 4 If in the subsequent year, consumers are to increase their shares by 7%, then assuming that the total sales remain constant, the consumer sales would have to increase by how many million of pounds?

Sol. Currently, share of consumer products = $\dfrac{204}{1569} \times 100\% = 13\%$

Let increase in consumer products sales be x million.
Therefore, if the share of consumer products increases by 7%, then
$\dfrac{204 + x}{1569} \times 100 = 20 \Rightarrow 20400 + 100x = 31380$
$\Rightarrow \qquad 100x = 31380 - 20400$
$\Rightarrow \qquad 100x = 10980 \Rightarrow x = 110$ million pounds (approx.)

Ex. 5 If 20% of the tyre sales were in the EEC countries, then what was the approximate value of sales of other products in the EEC countries in million of pounds?

Sol. 20% of tyre sales = $\dfrac{20}{100} \times 897 = 179.4$ million

∴ EEC sales = 420 million = Sales of tyre + Sales of other products
∴ 420 million = 179.4 + Sales of other products in EEC
∴ Sales of other products in EEC = 420 − 179.4 = 240.6 million pounds (approx.)

Directions (Examples 6-10) *Study the following pie chart and answer the questions that follow.*

Countrywise global exports presentation

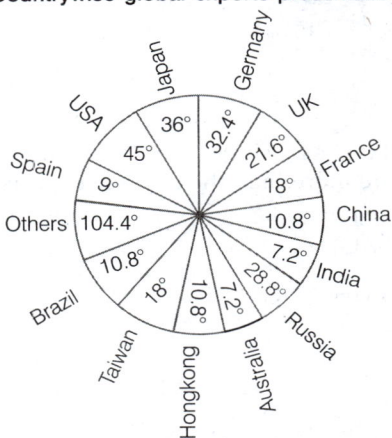

USA 45°, Japan 36°, Germany 32.4°, UK 21.6°, France 18°, China 10.8°, India 7.2°, Russia 28.8°, Australia 7.2°, Hongkong 10.8°, Taiwan 18°, Brazil 10.8°, Others 104.4°, Spain 9°

Total = 72000 billion

Ex. 6 By how much does the value of the exports of USA exceed that of Germany?

Sol. The difference in the angles subtended by USA and Germany = $45° - 32.4° = 12.6°$

∴ Difference in the exports of USA and Germany (in billion)
$$= 72000 \times \frac{12.6°}{360°} = 2520 \text{ billion}$$

Ex. 7 The difference in the values of the exports of Japan and France is how many times that of UK and Taiwan?

Sol. The difference in the angles subtended by Japan and France
$$= 36° - 18° = 18° \quad \ldots\text{(i)}$$
The difference in the angles subtended by UK and Taiwan
$$= 21.6° - 18° = 3.6° \quad \ldots\text{(ii)}$$
Clearly, Eq. (i) is 5 times of Eq. (ii).
So, the difference in the values of exports of Japan and France (i.e. 18) is 5 times that of UK and Taiwan.

Ex. 8 The value of the exports of the OPEC countries is how much more than the value of the exports of India and Australia put together, given that OPEC has a 20% share in the value of the exports of others?

Sol. Value of the exports of India and Australia
$$= 7.2° + 7.2° = 14.4° \quad \ldots\text{(i)}$$
Value of exports of OPEC countries
$$= 104.4° \times \frac{20}{100} = 20.88° \quad \ldots\text{(ii)}$$
Difference $= 20.88° - 14.4° = 6.48°$
∴ Required value $= 72000 \times \frac{6.48°}{360°} = 1296$ billion

Ex. 9 If exports of developing countries accounted for 36% of the total worldwide exports, then what is the value of the exports of Japan as a percentage of the exports of the developing countries?

Sol. Exports of developing countries = 36% of total exports
Exports of Japan $= \frac{36}{360} \times 100\% = 10\%$ of total exports
∴ Required percentage $= \frac{10}{36} \times 100\% = 0.2777 \times 100\% = 27.77\%$

Ex. 10 Considering 'others' as a single country, what is the number of countries whose exports are more than the average exports per country?

Sol. Total number of countries = 14
Average angle subtended by each country $= \frac{360°}{14} \approx 25.7°$
Only USA, Japan, Germany, Russia and others have exports greater than 25.7°.

Fast Track Practice

Exercise 1 Base Level Questions

Directions (Q. Nos. 1-2) *Given below is the pie chart which shows the percentage distribution of a book 'XYZ' published in 5 different stores.* [RRB PO (Pre) 2017]

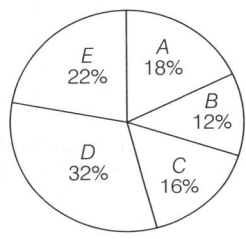

E 22%, A 18%, B 12%, C 16%, D 32%

Total number of books = 550

1. Find the central angle for the book D.
   (a) 117.5°
   (b) 115.2°
   (c) 112.8°
   (d) 108.5°
   (e) 118.8°

2. What is the difference between average of book sold by store A and E together and average books sold by store C and D together?
   (a) 33   (b) 11
   (c) 22   (d) 44
   (e) 20

Directions (Q. Nos. 3-5) *The following pie chart shows the performance in an examination in a particular year for 360 students. Study the pie chart and answer the questions given below.* [SSC (10+2) 2012]

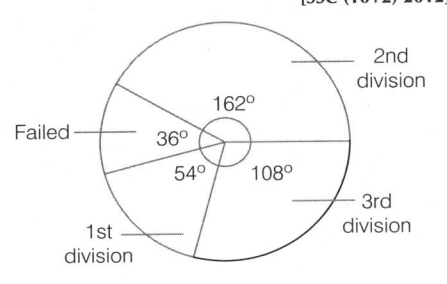

Failed 36°, 1st division 54°, 3rd division 108°, 2nd division 162°

3. The number of students who passed in 2nd division is more than those in Ist division by
   (a) 111   (b) 112   (c) 109   (d) 108

4. The ratio of successful students to the failed students is
   (a) 9 : 1   (b) 5 : 1
   (c) 1 : 9   (d) 2 : 7

5. The percentage of students who have failed in the examination is
   (a) 20%   (b) 36%   (c) 10%   (d) 30%

Directions (Q. Nos. 6-9) *The following pie chart shows the analysis of the result of an examination in which 5 candidates have failed. Study the chart and answer the questions given below.* [SSC CGL 2012]

210°, 15°, 135° — Failed candidates, Passed male candidates, Passed female candidates

6. The total number of examinees was
   (a) 100   (b) 120   (c) 135   (d) 150

7. Percentage of passed female candidates with respect to total examinees is
   (a) 30%   (b) 37.5% (c) 40%   (d) 45%

8. Percentage of passed male candidates with respect to total passed candidates is
   (a) 60.8%   (b) 56%
   (c) 71%   (d) 58%

9. Ratio of passed male candidates to the successful female candidates is
   (a) 9 : 1   (b) 1 : 14
   (c) 14 : 9   (d) 9 : 14

Directions (Q. Nos. 10-11) *In the following pie chart, percentage expenses on various items during the production of a book are given. Based upon the information given in the pie chart, answer the questions given below.* [CGPSC 2013]

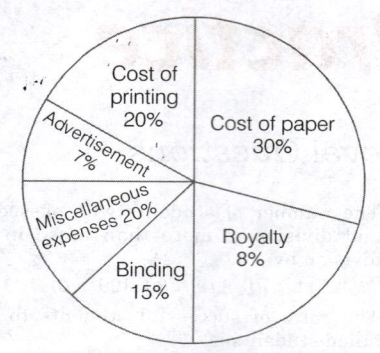

794 / Fast Track Objective Arithmetic

Directions (Q. Nos. 15-17) *In the following questions, the expenses of a country for a particular year is given in pie chart. Read the pie chart and answer the questions that follow.* [SSC CPO 2015]

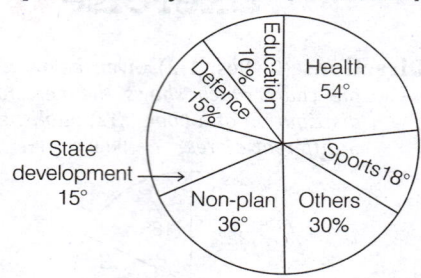

10. If the cost of paper is ₹ 150000, then the expense on advertisement is
 (a) ₹ 35000 (b) ₹ 3500
 (c) ₹ 40000 (d) ₹ 25000
 (e) None of these

11. The central angle corresponding to the cost of printing is
 (a) 60° (b) 72°
 (c) 45° (d) 102°
 (e) None of these

Directions (Q. Nos. 12-14) *The following pie chart represents a total expenditure of ₹ 540000 on different items in constructing a flat in a town. Study the pie chart and answer the questions that follow.* [SSC CGL 2013]

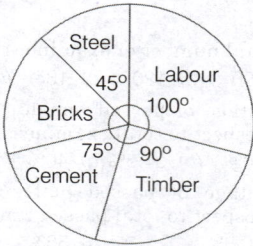

12. The expenditure (in ₹) on bricks is
 (a) 75000 (b) 67500
 (c) 150000 (d) 70000

13. The percentage of the total expenditure spent on steel and cement is
 (a) 33.23% (b) 25%
 (c) $33\frac{1}{3}$% (d) 30%

14. The expenditure (in ₹) on cement is
 (a) 75000 (b) 90000
 (c) 135000 (d) 112500

15. The per cent of less money spent on non-plan than that on defence is
 (a) 15% (b) 12% (c) 5% (d) 10%

16. The per cent of excess money spent on others than that on sports, is
 (a) 26% (b) 25% (c) 27% (d) 28%

17. If the total amount spent by the government during the year was ₹ 100000 crore, then the amount spent on defence and education together was
 (a) ₹ 30000 crore (b) ₹ 15000 crore
 (c) ₹ 25000 crore (d) ₹ 20000 crore

Directions (Q. Nos. 18-20) *Study the following pie chart carefully to answer the questions that follow.* [Bank PO 2013]

Expenditure on different sectors by university for various purposes

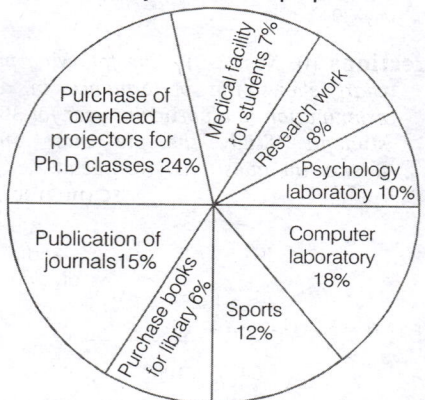

Total expenditure = ₹ 60 lakh

Pie Chart / 795

18. What is total sum of expenditures on research work, purchase of overhead projectors for Ph.D classes and purchase of books for library together?
    (a) ₹ 22.6 lakh   (b) ₹ 22.8 lakh
    (c) ₹ 23.4 lakh   (d) ₹ 20.8 lakh
    (e) None of these

19. If the expenditure on purchase of overhead projectors for Ph.D students is decreased by 7%, then what will be the expenditure on the same after the decrease?
    (a) ₹ 133920   (b) ₹ 1339200
    (c) ₹ 102000   (d) ₹ 108000
    (e) None of these

20. Which of the following is definitely true?
    (a) Ratio between expenditure of university on purchase of library books and expenditure on computer laboratory is 3 : 1 respectively
    (b) Expenditure on medical facilities for students is ₹ 4.6 lakh
    (c) Difference between the expenditure on research and medical facilities for students is ₹ 60000
    (d) All are true
    (e) None of the above

Directions (Q. Nos. 21-22) *The following pie chart represents the proposed outlay of the fifth five year plan of ₹ 40000 crore. Examine the chart and answer the questions that follow.*

[SSC Multitasking 2011]

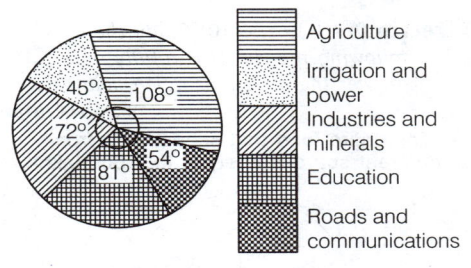

21. The amount proposed on agriculture is more than that on industries and minerals by
    (a) 7.5%   (b) 50%
    (c) 12%   (d) 12.5%

22. The amount (in ₹ crore) proposed on irrigation and power is less than that on industries and minerals by
    (a) 3000   (b) 3500
    (c) 2000   (d) 2500

Exercise 2 Higher Skill Level Questions

Directions (Q. Nos. 1-5) *Refer to the pie charts and answer the given questions.*
[IBPS Clerk (Mains) 2017]

Number of students (boys + girls) studying in standard Xth of various schools (A, B, C, D and E in the year 2010)

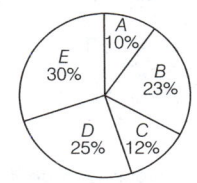

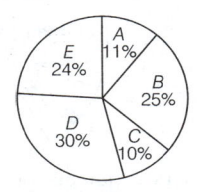

Total number of students (Boys + Girls) = 3000

Total number of boys = 1600

1. What is the difference between the total number of students (boys + girls) in school A and B together and the total number of students (boy + girls) in schools D and E together?
    (a) 640   (b) 600   (c) 620   (d) 680
    (e) 660

2. Number of girls studying in school C, are what per cent of total number of students (boys + girls) studying in the same school?
    (a) 48   (b) $60\frac{1}{3}$
    (c) 50   (d) $55\frac{5}{9}$
    (e) $42\frac{5}{8}$

3. What is the average number of boys studying in school A and B?
    (a) 280   (b) 288
    (c) 272   (d) 248
    (e) 278

4. Number of girls studying in school E are approximately what per cent more than the number of boys studying in the same school?
    (a) 45%   (b) 30%
    (c) 50%   (d) 40%
    (e) 35%

5. What is the respective ratio between the number of boys and the number of girls studying in school *D*?
 (a) 12 : 7 (b) 16 : 9
 (c) 16 : 7 (d) 18 : 7
 (e) 14 : 9

Directions (Q. Nos. 6-10) *Study the following pie charts carefully and answer the questions given below it.*
 [Bank PO 2010]

The entire fund that school gets from different sources is equal to ₹ 500 lakh

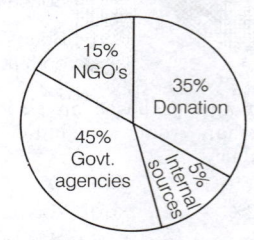

Sources of funds in school
- 15% NGO's
- 35% Donation
- 45% Govt. agencies
- 5% Internal sources

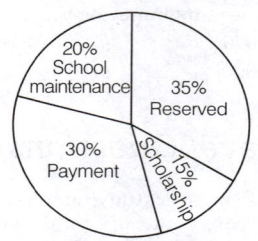

Uses of funds by school
- 20% School maintenance
- 35% Reserved
- 30% Payment
- 15% Scholarship

6. What is the difference between the funds acquired by school from NGO's and internal sources?
 (a) ₹ 50 lakh (b) ₹ 45 lakh
 (c) ₹ 75 lakh (d) ₹ 25 lakh
 (e) None of these

7. If the school managed school maintenance from the government agencies fund only, how much fund from government agencies would still left for other use?
 (a) ₹ 120 lakh (b) ₹ 150 lakh
 (c) ₹ 110 lakh (d) ₹ 95 lakh
 (e) None of these

8. If scholarship has to be paid out of the donation fund, then what is the approximate per cent of donation fund used for this purpose?
 (a) 43% (b) 53%
 (c) 37% (d) 45%
 (e) 32%

9. What is the total amount used by the school for payment?
 (a) ₹ 100 lakh (b) ₹ 110 lakh
 (c) ₹ 150 lakh (d) ₹ 140 lakh
 (e) None of these

10. What amount of the fund is acquired by the school from government agencies?
 (a) ₹ 220 lakh (b) ₹ 310 lakh
 (c) ₹ 255 lakh (d) ₹ 225 lakh
 (e) None of these

Directions (Q. Nos. 11-14) *Study the given pie charts carefully to answer the questions that follow.* [Bank PO 2010]

Break-up of number of employees working in different departments of an organisation, the number of males and the number of employees who recently got promoted in each department

Break-up of employees working in different departments
- Accounts 20%
- Production 35%
- Marketing 18%
- HR 12%
- IT 15%

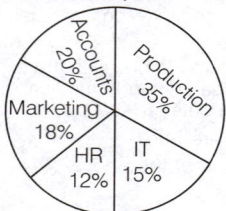

Total number of employees = 3600
Break-up of number of males working in each department

Break-up of number of employees who recently got promoted in each department
- Accounts 8%
- Production 33%
- Marketing 22%
- HR 11%
- IT 26%

Total number of employees who got promoted = 1200

11. The number of employees who got promoted from the HR department was what per cent of the total number of employees working in that department? (rounded off to two digits after decimal)
 (a) 36.18% (b) 30.56%
 (c) 47.22% (d) 28.16%
 (e) None of these

12. The total number of employees who got promoted from all the departments together was what per cent of the total number of employees working in all the departments together? (rounded off the nearest integer)
    (a) 56% (b) 21%
    (c) 45% (d) 33%
    (e) 51%

13. What is the total number of females working in the production and marketing departments together?
    (a) 468 (b) 812
    (c) 582 (d) 972
    (e) None of these

14. If half of the number of employees who promoted from the IT department were males, then what was the approximate percentage of males who got promoted from the number of working males in IT department?
    (a) 61% (b) 29%
    (c) 54% (d) 42%
    (e) 38%

Directions (Q. Nos. 15-17) *Refer to the pie chart and answer the given questions.*
[IBPS SO 2016]

Percentage of employees in different departments of branch 'XYZ' in the year 2014

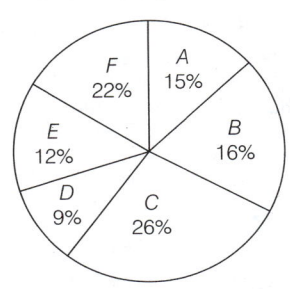

F 22%, A 15%, B 16%, C 26%, D 9%, E 12%

Total number of employees = 450

15. In 2014, the number of female employees in department C was 5/13 of the total number of employees in same department. If the number of female employees in department F was 4 less than that in department C, what is the number of male employees department F?
    (a) 41 (b) 42 (c) 58 (d) 54
    (e) 48

16. In department E, the respective ratio between the number of female employees and male employees was 5 : 4. There were equal number of unmarried males and unmarried females in department E. If the respective ratio between married females and married males was 3 : 2, what is the number of unmarried females?
    (a) 6 (b) 15 (c) 12 (d) 4
    (e) 8

17. The number of employees in department E is what per cent less than the number of employees in departments A and C together?
    (a) 72% (b) 60%
    (c) 65% (d) 70%
    (e) 68%

Directions (Q. Nos. 18-20) *Refer to the pie chart and answer the given questions.*
[IBPS SO 2016]

Percentage of Associate Professors in different universities in January 2013

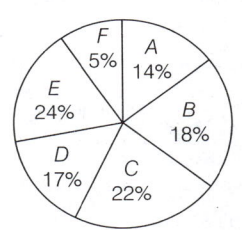

F 5%, A 14%, B 18%, C 22%, D 17%, E 24%

Total number = 300

18. In 2013, the number of female Associate Professors in university B was double the number of male Associate Professors in the same university. If in university B, the number of female Associate Professors is same as that in university D, what is the number of male associate professors in university D?
    (a) 24 (b) 26 (c) 15 (d) 25
    (e) 20

19. In 2014, equal number of Associate Professors merged from universities B and D. If the resultant respective ratio between the number of Associate Professors in university B and that in university D is 8 : 7, what is the number of Associate Professors who resigned from university B in 2014?
    (a) 10 (b) 24
    (c) 12 (d) 5
    (e) 16

798 / Fast Track Objective Arithmetic

20. In January 2013, 24% of the Assistant Professors in university E were promoted to Associate Professors. If university E had 54 associate professors in December 2012, what was the number of Assistant Professors in December 2012?

(Note No Associate Professor was recruited and no Assistant Professor left university E in the same time.)
(a) 150 (b) 50
(c) 126 (d) 100
(e) 75

Answer with Solutions

Exercise 1 Base Level Questions

1. (b) Central angle for book $D = \dfrac{32}{100} \times 360°$
   $= 115.2°$

2. (c) Required difference
   $= \dfrac{1}{2}[(32\% + 16\%) - (18\% + 22\%)]550$
   $= \dfrac{1}{2} \times 8\%$ of $550 = 4\%$ of $550 = 22$

3. (d) The number of students passed in 1st division $= \dfrac{54°}{360°} \times 360 = 54$
   The number of students passed in 2nd division $= \dfrac{162°}{360°} \times 360 = 162$
   ∴ Required difference $= 162 - 54 = 108$

4. (a) Required ratio $= \dfrac{54° + 108° + 162°}{36°}$
   $= \dfrac{324}{36} = \dfrac{9}{1} = 9 : 1$

5. (c) The percentage of failed students in the examination $= \dfrac{36°}{360°} \times 100 = 10\%$

6. (b) Let total number of candidates be x.
   By pie chart, $\dfrac{15°}{360°} \times x = 5$
   [failed students = 5, given]
   ∴ $x = 24 \times 5 = 120$

7. (b) Number of passed female candidates
   $= \dfrac{135°}{360°} \times 100\% = 37.5\%$

8. (a) Percentage of passed male candidates
   $= \dfrac{210°}{360° - 15°} \times 100\%$
   $= \dfrac{210°}{345°} \times 100\% = 60.8\%$

9. (c) Required ratio $= \dfrac{210°}{135°} = 14 : 9$

10. (a) Let the expense on book = ₹ x
    Then, $\dfrac{x \times 30}{100} = 150000$
    $\Rightarrow x = \dfrac{150000 \times 10}{3} = ₹ 500000$
    ∴ Expense on advertisement
    $= \dfrac{500000 \times 7}{100} = ₹ 35000$

11. (b) The central angle corresponding to the cost of printing $= \dfrac{20}{100} \times 360° = 72°$

12. (a) Angle made by the expenditure on bricks $= 360° - (45° + 100° + 90° + 75°)$
    $= 360° - 310° = 50°$
    Thus, expenditure on bricks
    $= \dfrac{50°}{360°} \times 540000$
    $= ₹ 75000$

13. (c) Expenditure on steel and cement
    $= 75° + 45° = 120°$
    ∴ Required percentage
    $= \dfrac{120°}{360°} \times 100\%$
    $= \dfrac{100}{3}\% = 33\dfrac{1}{3}\%$

14. (d) Expenditure on cement
    $= \dfrac{75°}{360°} \times 540000 = ₹ 112500$

15. (c) Money spent on non-plan
    $= \dfrac{36°}{360°} \times 100 = 10\%$
    Money spent on defence = 15%
    ∴ Required difference = 15% − 10% = 5%

16. (b) Money spent on sports
    $= \dfrac{18°}{360°} \times 100 = 5\%$
    Money spent on others = 30%
    ∴ Required difference = 30% − 5% = 25%

Pie Chart / 799

17. (c) Money spent on defence and education together

$= 100000 \times \dfrac{10 + 15}{100}$

$= 100000 \times \dfrac{25}{100}$

$= ₹ 25000$ crore

18. (b) Total sum of expenditures

$= \dfrac{(8 + 24 + 6)}{100} \times 60$

$= \dfrac{38}{100} \times 60 = ₹ 22.8$ lakh

19. (e) Required expenditures after decrease

$= \dfrac{(24 - 7)}{100} \times 60$ lakh

$= \dfrac{17 \times 60}{100}$ lakh $= ₹ 1020000$

20. (c) From option (c),
Required difference

$=$ Expenditure on research work
$-$ Expenditure on medical facilities

$= \dfrac{8 - 7}{100} \times 60$ lakh

$= ₹ 60000$

21. (b) Amount spend on agriculture

$= \dfrac{108°}{360°} \times 40000 = ₹ 12000$ crore

Amount spend on industries and minerals

$= \dfrac{72°}{360°} \times 40000 = ₹ 8000$ crore

∴ Required percentage

$= \dfrac{12000 - 8000}{8000} \times 100 = 50\%$

22. (a) Required amount

$= \dfrac{72° - 45°}{360°} \times 40000$

$= \dfrac{27°}{360°} \times 40000$

$= ₹ 3000$ crore

Exercise 2 Higher Skill Level Questions

1. (e) Total number of students in school
$A + B = (10\% + 23\%) \times 3000$
$= 33\% \times 3000$

Total number of students in school
$D + E = (25\% + 30\%) \times 3000$
$= 55\%$ of 3000

∴ Required difference $= \left(\dfrac{55 - 33}{100}\right) \times 3000$

$= 22 \times 30 = 660$

2. (d) Number of boys in school C

$= 1600 \times \dfrac{10}{100} = 160$

∴ Number of girls in school C

$= \left(3000 \times \dfrac{12}{100} - 160\right)$

$= 360 - 160$
$= 200$

Total number of students in school C

$= 3000 \times \dfrac{12}{100} = 360$

∴ Required percentage

$= \dfrac{200}{360} \times 100\% = \dfrac{500}{9}\%$

$= 55\dfrac{5}{9}\%$

3. (b) Number of boys in school A

$= 1600 \times \dfrac{11}{100} = 176$

Number of boys in school B

$= 1600 \times \dfrac{25}{100} = 400$

∴ Average number of boys in schools A and B

$= \dfrac{176 + 400}{2} = \dfrac{576}{2}$

$= 288$

4. (e) Number of boys in school E

$= 1600 \times \dfrac{24}{100} = 384$

Number of girls in school E

$= \left(3000 \times \dfrac{30}{100} - 384\right)$

$= 900 - 384 = 516$

∴ Required percentage

$= \left(\dfrac{516}{384} \times 100 - 100\right)\%$

$= (134.375 - 100)\%$

$= 34.375\% \approx 35\%$

5. (b) Number of boys in school D

$= 1600 \times \dfrac{30}{100} = 480$

and number of girls in school D

$= \left(3000 \times \dfrac{25}{100} - 480\right)$

$= 750 - 480 = 270$

∴ Required ratio $= 480 : 270 = 16 : 9$

6. (a) Required difference = (Percentage of funds acquired from NGO − Percentage of funds acquired from internal sources) of 500 lakh = (15 − 5)% of 500 lakh
$$= \frac{500 \times 10}{100} \text{ lakh} = ₹ 50 \text{ lakh}$$

7. (e) Fund from government agencies
$$= \frac{500 \times 45}{100} = ₹ 225 \text{ lakh}$$
Expenses in school maintenance
$$= \frac{500 \times 20}{100} = ₹ 100 \text{ lakh}$$
∴ Remaining fund = (225 − 100) lakh
= ₹ 125 lakh

8. (a) Fund from donation $= \frac{500 \times 35}{100}$
= ₹ 175 lakh
Scholarship amount $= \frac{15 \times 500}{100} = ₹ 75$ lakh
∴ Required percentage
$$= \frac{75}{175} \times 100 = 42.85\% = 43\% \text{ (approx.)}$$

9. (c) Total amount used by the school for payment
$$= \frac{500 \times 30}{100} = ₹ 150 \text{ lakh}$$

10. (d) Fund acquired from government agencies
$$= \frac{500 \times 45}{100} = ₹ 225 \text{ lakh}$$

11. (b) Number of promoted employees in HR department = 1200 × 0.11 = 132
Number of employees working in HR department = 3600 × 0.12 = 432
∴ Required percentage
$$= \frac{132}{432} \times 100\% = 30.56\%$$

12. (d) Number of promoted employees in all departments = 1200
Number of working employees in all departments = 3600
∴ Required percentage
$$= \frac{1200}{3600} \times 100\% \approx 33\%$$

13. (c) Number of employees working in production and marketing
= 3600 × (0.35 + 0.18) = 1908
Number of male employees in production and marketing
= 2040 × (0.50 + 0.15) = 1326
∴ Number of female employees in production and marketing
= 1908 − 1326 = 582

14. (e) Number of promoted employees in IT department = 1200 × 0.26 = 312
Number of promoted male employees in IT department = 156
Number of male employees working in IT department = 2040 × 0.20 = 408
∴ Required percentage
$$= \frac{156}{408} \times 100\%$$
= 38.23% ≈ 38%

15. (c) Total number of employees in department $C = 450 \times \frac{26}{100} = 117$
Female employees in department C
$$= 117 \times \frac{5}{13} = 45$$
Now, total number of employees in department F
$$= 450 \times \frac{22}{100} = 99$$
Female employees in department $F = 45 − 4 = 41$
∴ Male employees in department F
= 99 − 41 = 58

16. (c) Total number of employees in department $E = \frac{450 \times 12}{100} = 54$
∴ Female employees in department E
$$= \frac{5}{5+4} \times 54$$
$$= \frac{5}{9} \times 54 = 5 \times 6 = 30$$
∴ Male employees in department E
= 54 − 30 = 24
Let unmarried males and females in department $E = x$
∴ Married females in department E
= (30 − x)
and married males in department E
= (30 − x)
Now, according to the question,
$$\frac{30-x}{24-x} = \frac{3}{2}$$
⇒ 72 − 3x = 60 − 2x
⇒ −3x + 2x = 60 − 72
⇒ −x = −12 ⇒ x = 12
Hence, number of unmarried females is 12.

17. (d) Total number of employees in department $A = \frac{135}{2}$
Total number of employees in department C
= 117

Pie Chart / 801

∴ Total number of employees in department

$(A + C) = \dfrac{135}{2} + 117 = \dfrac{369}{2}$

and total number of employees in department $E = \dfrac{12}{100} \times 450 = 54$

∴ Required percentage $= \dfrac{\dfrac{369}{2} - 54}{369/2} \times 100$

$= \dfrac{261/2}{369/2} \times 100 = \dfrac{261}{369} \times 100$

$= \dfrac{29}{41} \times 100 = 70.73 \approx 70\%$

18. (c) Total number of Associate Professors in university B in the year 2013

$= 300 \times \dfrac{18}{100} = 54$

According to the question,
  Female associate professors
  $= 2 \times$ Male Associate Professors
$\Rightarrow 36 = 2 \times 18$
and $36 + 18 = 54$

Now, total number of Associate Professors in university $D = 300 \times \dfrac{17}{100} = 51$

Total number of female Associate Professors in university $D = 36$
∴ Number of male Associate Professors in university $D = 51 - 36 = 15$
Hence, number of male Associate Professors in university D is 15.

19. (b) Let number of merged Professors from universities B and $D = x$
Then, according to the question,

$\dfrac{54 - x}{51 - x} = \dfrac{8}{7}$

$\Rightarrow 378 - 7x = 408 - 8x$
$\Rightarrow 8x - 7x = 408 - 378 \Rightarrow x = 30$

∴ Number of Associate Professors resigned from university B in 2014 $= 54 - 30 = 24$

20. (e) Total number of Associate Professors in university E in the year 2013

$= 300 \times \dfrac{24}{100} = 72$

Let Associate Professors in December 2012 be x.

Then, $\dfrac{24}{100} \times x + 54 = 72 \Rightarrow \dfrac{24x}{100} = 18$

$\Rightarrow \quad x = \dfrac{1800}{24} = 75$

Chapter 44

Bar Chart

A bar chart is a chart with **rectangular bars** with lengths proportional to the values that they represent. Bar charts are diagramatic representation of discrete data.

A bar is a thick line whose width is shown merely for attention. In this method of data representation, the data is plotted on the X and Y-axes as bars.

One of the axes (normally the X-axis) of the bar diagram represents a discrete variable, while the other axis represents the scale for continuous variable.

Each bar diagram has a title indicating the subject matter represented in the diagram.

Types of Bar Charts

Following are the main bar charts

1. **Vertical/Simple bar chart** A vertical bar chart is really a simple bar chart which has one independent (discrete) variable and one dependent (continuous) variable.

    For example Following chart shows the percentage distribution of total expenditures of a company

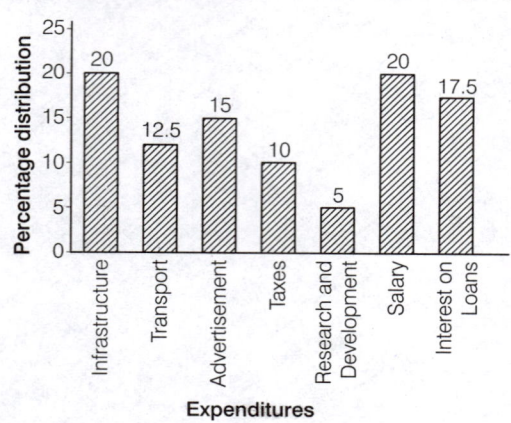

Bar Chart / 803

Directions (Examples 1-2) *Study the following bar chart and answer the questions that follow.*

Sales of cellular phones over the years

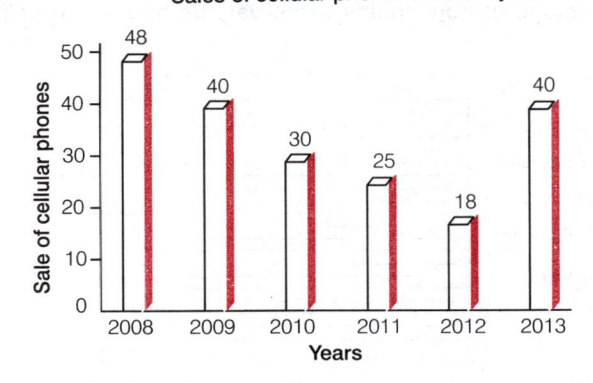

Ex. 1 Find the percentage increase in sales from 2012 to 2013.

Sol. Percentage increase in sales = $\dfrac{40 - 18}{18} \times 100\% = \dfrac{22}{18} \times 100\% \approx 122\%$

Ex. 2 Find the two years between which the rate of change of cellular phones is minimum.

Sol. Rate of change during years

$2008\text{-}09 = \dfrac{8}{48} \times 100\% = 16.6\%$, $2009\text{-}10 = \dfrac{10}{40} \times 100\% = 25\%$

$2010\text{-}11 = \dfrac{5}{30} \times 100\% = 16.6\%$, $2011\text{-}12 = \dfrac{7}{25} \times 100\% = 28\%$

$2012\text{-}13 = \dfrac{22}{18} \times 100\% \approx 122\%$

Hence, it is minimum for years 2008-09 and 2010-11.

2. **Horizontal bar chart** A bar chart which is shown with the dependent variable on the horizontal scale is typically referred to as a horizontal bar charts. Otherwise, the layout is similar to the vertical bar chart.

For example Following chart shows the distribution of ozone layer in different locations.

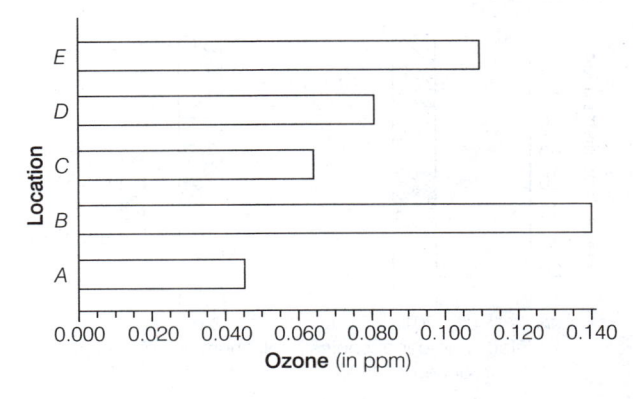

Directions (Examples 3-4) *Read the following bar chart carefully and answer the questions that follow.*

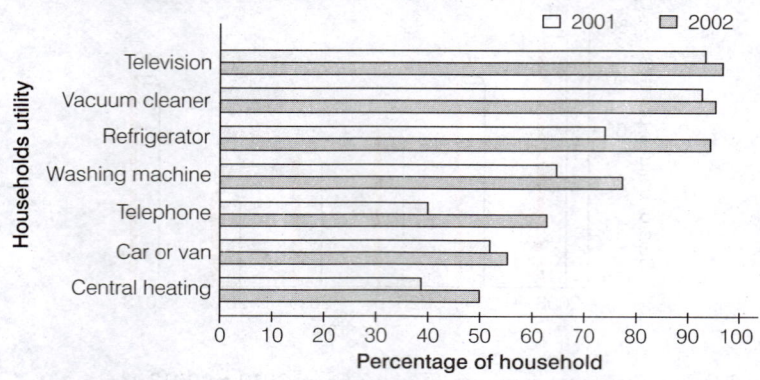

Percentage of households using various household utilities

□ 2001   □ 2002

(Bar chart showing Households utility on y-axis: Television, Vacuum cleaner, Refrigerator, Washing machine, Telephone, Car or van, Central heating; Percentage of household on x-axis: 0 to 100)

Ex. 3 In 2001, if the total number of households was 403 million, then what would be the number of households having washing machine?

Sol. Percentage of households having washing machines in 2001 = 65%
Hence, number of households having washing machines in 2001
= Total number of households × 65% = 403 million × 65% = 261.95 million ≈ 262 million

Ex. 4 If total number of households in years 2001 and 2002 were 360 and 403 million respectively, then find the ratio of households using refrigerator in year 2001 to that of using car in year 2002.

Sol. Number of households using refrigerators in year 2001 = 75% of 360 = 270 million
Number of households using cars in year 2002 = 55% of 403 = 221.65 million
∴ Required ratio = $\dfrac{270}{221.65}$ = 1.2

3. **Sub-divided bar chart** A sub-divided bar chart is used to represent various parts of total magnitude of a given variable.

For example Following chart shows percentage distribution of sales of four books in two different years.

(Sub-divided bar chart showing Percentage distribution (0-100) for years 2012 and 2013)

Book A
Book B
Book C
Book D

2012
Total number of copies produced = 25000

2013
Total number of copies produced = 35000

Years

Bar Chart / 805

Directions (Examples 5-6) *The following bar chart represents the number of students who passed the CAT exam or the XAT exam or the CET exam or none of these exams. (Assume that there is no student who passed more than one exam)*

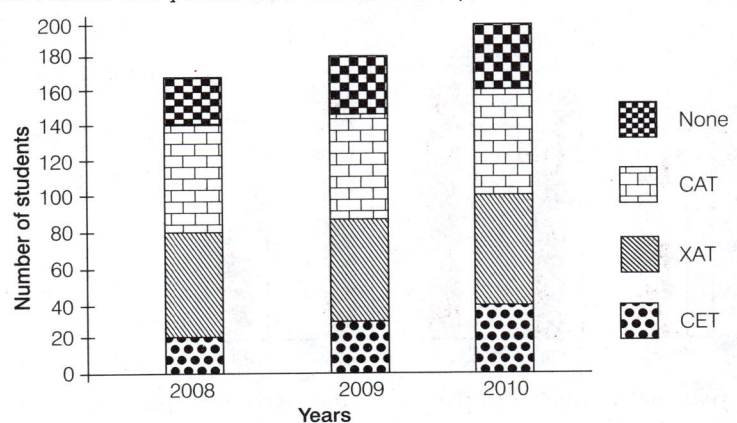

Ex. 5 Which year showed the best result in MBA entrance exams (in terms of percentage of students who cleared)?

Sol. Result of MBA = $\dfrac{\text{Number of passed students}}{\text{Total number of students}} \times 100\%$

In 2008 = $\dfrac{140}{170} \times 100\% = 82.35\%$; In 2009 = $\dfrac{150}{180} \times 100\% = 83.33\%$

In 2010 = $\dfrac{160}{200} \times 100\% = 80\%$

Thus, in the year 2009, the result is best.

Ex. 6 What is the percentage increase in the number of students in 2010 over 2008?

Sol. Number of students in 2008 = 170 and in 2010 = 200

∴ Required percentage increase = $\dfrac{200 - 170}{170} \times 100\% = \dfrac{30}{170} \times 100\% = 17.64\%$

4. **Grouped/Multiple bar chart** In this type, two or more bars are constructed adjoining each other to represent either different components of a complete data or to show multiple variables.

For example Following below chart shows the production of 3 types of car in three different years.

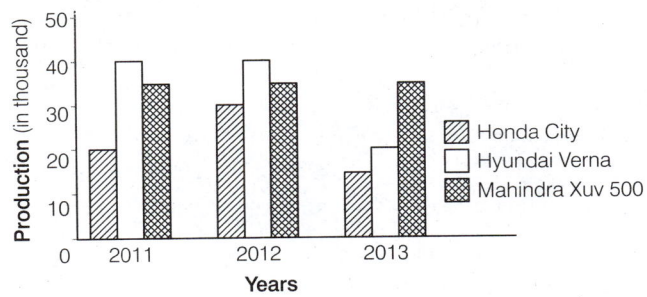

Directions (Examples 7-9) *Read the following bar chart carefully and answer the questions that follows.*

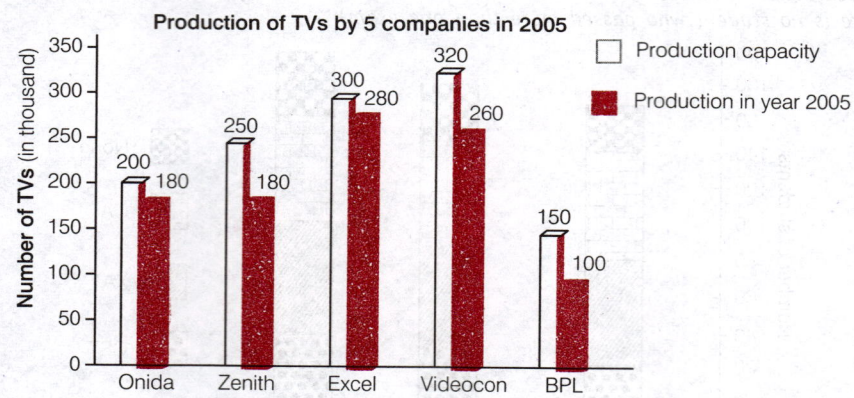

Production of TVs by 5 companies in 2005

[Bar chart showing Number of TVs (in thousand) for each company — Production capacity and Production in year 2005:
- Onida: 200, 180
- Zenith: 250, 180
- Excel: 300, 280
- Videocon: 320, 260
- BPL: 150, 100]

Ex. 7 In year 2005, which company had the maximum percentage of unutilised capacity?

Sol. Percentage of unutilised capacity = $\dfrac{\text{Production capacity} - \text{Production}}{\text{Production capacity}} \times 100\%$

$$\text{Onida} = \frac{20}{200} \times 100\% = 10\%$$

$$\text{Zenith} = \frac{70}{250} \times 100\% = 28\%$$

$$\text{Excel} = \frac{20}{300} \times 100\% = 6.67\%$$

$$\text{Videocon} = \frac{60}{320} \times 100\% = 18.75\%$$

$$\text{BPL} = \frac{50}{150} \times 100\% = 33.3\%$$

Hence, BPL had maximum unutilised capacity.

Ex. 8 The TVs produced by Excel form what percentage of the total production?

Sol. Total production = 180 + 180 + 280 + 260 + 100 = 1000

∴ Required percentage = $\dfrac{280}{1000} \times 100\% = 28\%$

Ex. 9 A new company CASINO was set up in 2006 and sold 122000 pieces in that year. Due to this, the other five given companies together reduced their production by the same number of sets sold by CASINO in the ratio of their production capacities. What is the production of Excel (in thousand sets) in 2006?

Sol. Total number of TVs sold by CASINO = 122000

Total production capacity of all the companies put together

= 200 + 250 + 300 + 320 + 150 = 1220

Decrease in production of Excel = $\dfrac{300}{1220} \times 122000 = 30000$

∴ Actual production of Excel in year 2006 = 280000 − 30000 = 250000

Fast Track Practice

Exercise 1 Base Level Questions

Directions (Q. Nos. 1-2) *The following two questions are based on the given histogram that shows the percentage of villages in the states, which are not electrified.* [SSC CPO 2012]

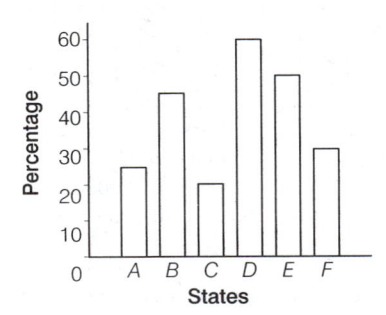

1. Which of the following states has twice the percentage of villages electrified in comparison to state *D*?
   (a) A (b) C (c) E (d) F

2. How many states have atleast 50% electrified villages?
   (a) 1 (b) 2 (c) 3 (d) 5

Directions (Q. Nos. 3-5) *Study the following bar chart and answer the questions that follow.* [SSC CPO 2009]

Demand and production of colour TVs of five companies for Jan 2006

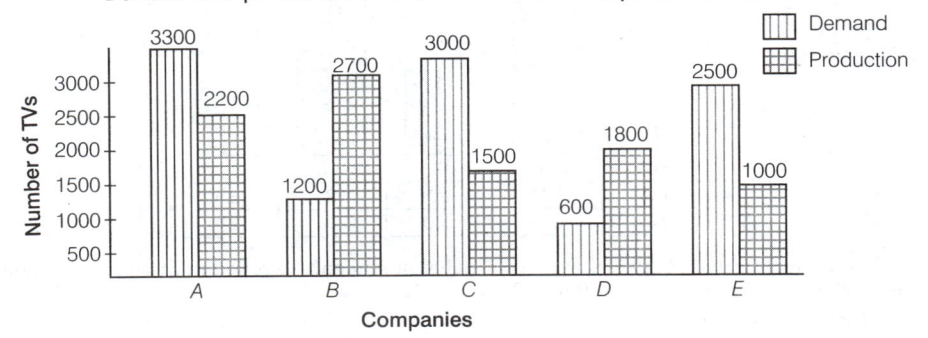

3. What is the ratio of the number of companies having more demand than production to the number of companies having more production than demand?
   (a) 2 : 3 (b) 4 : 1 (c) 2 : 2 (d) 3 : 2

4. What is the difference between average demand and average production of the five companies taken together?
   (a) 1400 (b) 400 (c) 280 (d) 138

5. Demand of company *D* is approximately what per cent of demand of company *E*?
   (a) 12% (b) 20% (c) 24% (d) 30%

Directions (Q. Nos. 6-8) *Study the following bar chart and answer the questions that follow.*

[SSC CGL 2013]

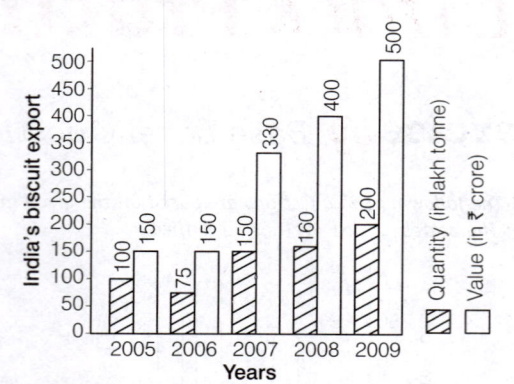

Bar chart: India's biscuit export — Quantity (in lakh tonne) and Value (in ₹ crore)
- 2005: 100, 150
- 2006: 75, 150
- 2007: 150, 330
- 2008: 160, 400
- 2009: 200, 500

6. In which two years, was the value per tonne equal?
   (a) 2006 and 2007   (b) 2005 and 2006   (c) 2008 and 2009   (d) 2007 and 2008

7. The year, in which the percentage increase in export was maximum from its preceding year, is
   (a) 2009   (b) 2007   (c) 2008   (d) None of these

8. What was the percentage drop in export quantity from 2005 to 2006?
   (a) 75%   (b) 0%   (c) 25%   (d) 50%

Directions (Q. Nos. 9-12) *Refer to the bar graph and answer the given questions.*

[SBI PO (Pre) 2016]

Data related to number of students enrolled for a vocational course in two institutes (A and B) during five years

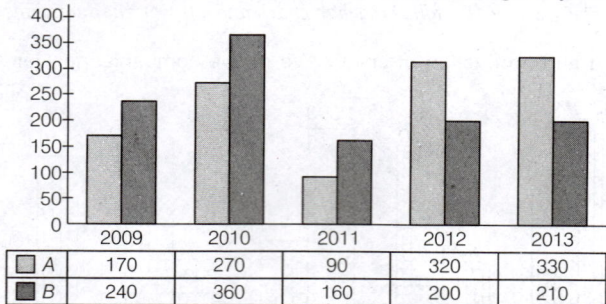

	2009	2010	2011	2012	2013
A	170	270	90	320	330
B	240	360	160	200	210

9. What is the difference between average number of students enrolled in institute A in 2009 and 2010 and that in institute B in 2011 and 2012?
   (a) 20   (b) 40   (c) 30   (d) 10   (e) 60

10. If the number of enrolled students in institute A in 2013 is 25% less than that in 2014, how many students were enrolled in institute A in 2014?
   (a) 450   (b) 480   (c) 440   (d) 430   (e) 415

11. In 2009, 65% of students enrolled in institute B were male. If the respective ratio between number of male students enrolled in institutes A and B in 2009 was 3 : 4, what was the number of male students enrolled in institute A in the same year?
   (a) 111   (b) 117   (c) 123   (d) 114   (e) 105

12. Number of students enrolled in institute B decreased by what per cent from 2010 to 2013?
   (a) $41\frac{2}{3}\%$ (b) $36\frac{2}{3}\%$ (c) $43\frac{1}{3}\%$ (d) $29\frac{1}{3}\%$ (e) $20\frac{1}{3}\%$

Bar Chart / 809

Directions (Q. Nos. 13-17) *Study the following bar chart and answer the given questions.*
[SBI PO 2016]

Data regarding number of boys and girls studying in Stand. X of five different schools

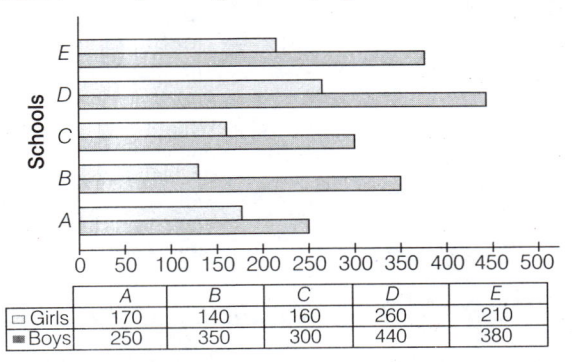

	A	B	C	D	E
Girls	170	140	160	260	210
Boys	250	350	300	440	380

13. What is the difference between total number of boys studying in schools B and C together and the total number of girls in the same schools together?
(a) 350 (b) 200 (c) 400 (d) 309
(e) 300

14. The number of students (boys and girls together) studying in school B are what per cent less than that in school D?
(a) 20% (b) 15% (c) 40% (d) 25%
(e) 30%

15. Number of students (boys and girls together) in Std. V of school E is 20% less than those in Std. X of the same school. How many students study in Std. V in the same school?
(a) 506 (b) 472 (c) 420 (d) 464
(e) 524

16. Number of boys studying in school E is what per cent more than the number of boys studying in school A?
(a) 48% (b) 52% (c) 32% (d) 45%
(e) 64%

17. What is the respective ratio between the number of students (boys and girls together) studying in school A and that in school C?
(a) 7 : 11 (b) 25 : 27 (c) 21 : 23 (d) 21 : 22
(e) 8 : 9

Directions (Q. Nos 18-21) *Study the following graph carefully and answer the questions.*
[NICL AO (Pre) 2017]

Water level of four major rivers (in metre) in four different months

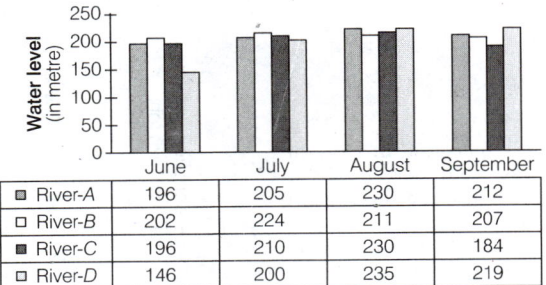

	June	July	August	September
River-A	196	205	230	212
River-B	202	224	211	207
River-C	196	210	230	184
River-D	146	200	235	219

810 / Fast Track Objective Arithmetic

18. If the water level of River-A in July is decreased by 20%, then what will be the water level of River-A in July?
(a) 156 m (b) 162 m (c) 164 m (d) 152 m
(e) Other than those given as options

19. In which river and in which month respectively the water level is the highest?
(a) River-C in August
(b) River-D in September
(c) River-A in July
(d) River-B in August
(e) River-D in July

20. What is the average water level of River-A in all the four months together?
(a) 224.50 m
(b) 212.25 m
(c) 210.75 m
(d) 222.25 m
(e) Other than those given as options

21. What was the respective ratio between the level of River-C in September and the water level of River-B in June?
(a) 91 : 101 (b) 94 : 101
(c) 51 : 103 (d) 31 : 101
(e) 92 : 101

Exercise 2 — Higher Skill Level Questions

Directions (Q. Nos. 1-5) *Study the following graph carefully to answer the questions that follow.* [Bank PO 2011]

Number of soldiers recruited (in thousand) **in three different forces in six different years**

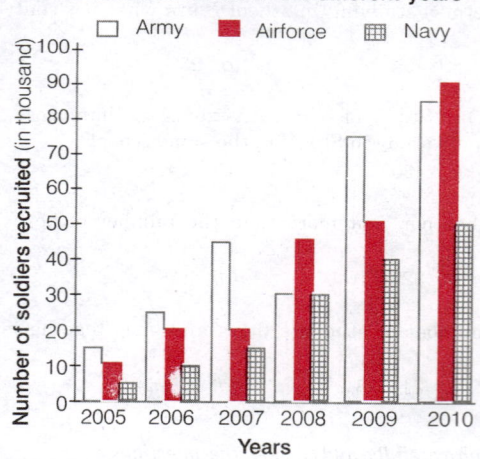

☐ Army ■ Airforce ▦ Navy

(Bar graph: Number of soldiers recruited (in thousand) vs Years 2005–2010)

1. What was the average number of soldiers recruited in the navy over all the years together?
(a) 25000
(b) 24000
(c) 2400
(d) 28000
(e) None of the above

2. Number of soldiers recruited in navy in the year 2009 was what percentage of number of soldiers recruited in army in the year 2006?
(a) 140   (b) 150
(c) 160   (d) 180
(e) None of these

3. If 30% of soldiers recruited in airforce in the year 2010 was female, what is the number of males recruited in airforce in that year?
(a) 63000   (b) 6300
(c) 61000   (d) 6100
(e) None of these

4. What was the respective ratio between the number of soldiers recruited for airforce in the year 2005 and the number of soldiers recruited in army in the year 2009?
(a) 2 : 15   (b) 5 : 13
(c) 2 : 17   (d) 15 : 4
(e) None of these

5. What was the approximate percentage decrease in number of soldiers recruited in army in the year 2008 as compared to the previous year?
(a) 20   (b) 23
(c) 38   (d) 30
(e) 33

Directions (Q. Nos. 6-10) *The following bar chart shows the production of cement (in lakh tonne) of four factories A, B, C and D over the years.* [SSC CPO 2013]

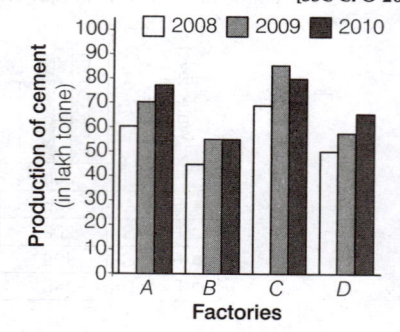

(Bar chart: Production of cement (in lakh tonne) for factories A, B, C, D in years 2008, 2009, 2010)

Bar Chart / 811

6. The percentage increase in production of cement by factory B from 2008 to 2009 is about
   (a) 30.8 (b) 16.7 (c) 18.2 (d) 22.2
7. The production of cement by factory B in 2009 and production of cement by factory D in 2010 together is what per cent of production by factory A in 2008?
   (a) 200% (b) 50% (c) 100% (d) 150%
8. Which of the four factories has recorded the maximum percentage growth in production of cement from 2008 to 2009?
   (a) D (b) A (c) B (d) C
9. Which of the given factories has recorded the maximum percentage growth in production of cement from 2009 to 2010?
   (a) D (b) A (c) B (d) C
10. The difference (in lakh tonne) between the average production of cement by four factories in 2009 and average production by the same factories in 2008 is
    (a) 12.75 (b) 11.25 (c) 11.75 (d) 12.50

Directions (Q. Nos. 11-15) *Study the following bar graph carefully and answer the questions that follow.* [MAT 2015]

11. In which year the annual growth rate of total production (of all products) is highest?
    (a) 2009 (b) 2010 (c) 2011 (d) 2013
12. If the stability of the production during 2008 to 2013 is defined as
    $$\frac{\text{Average production}}{\text{(Maximum production – Minimum production)}}$$, then which product is most stable?
    (a) Product P (b) Product Q (c) Product R (d) Product S
13. If four products P, Q, R and S shown in the graph are sold at price of ₹ 9, ₹ 4, ₹ 13 and ₹ 3, respectively during 2008-13, then the total revenue of all the products is lowest in which year?
    (a) 2009 (b) 2010 (c) 2011 (d) None of these
14. Individual revenue of P, Q, R and S for the entire period (2008-11) is calculated based on the price of ₹ 9, ₹ 4, ₹ 13 and ₹ 3, respectively. Which product fetches the lowest revenue?
    (a) Product P (b) Product Q (c) Product R (d) Product S

15. Four products P, Q, R and S shown in the graph are sold at price of ₹ 9, ₹ 4, ₹ 13 and ₹ 3 respectively during 2008-13. Which of the following statements is true?
(a) Product R fetches second highest revenue across products in 2009
(b) Sum of revenue of P, Q and S is more than the revenue of R in 2012
(c) Cumulative revenue of P and Q is more than the cumulative revenue of S in 2011
(d) None of the above

Directions (Q. Nos. 16-20) *Read the following bar chart carefully and answer the questions that follow.*

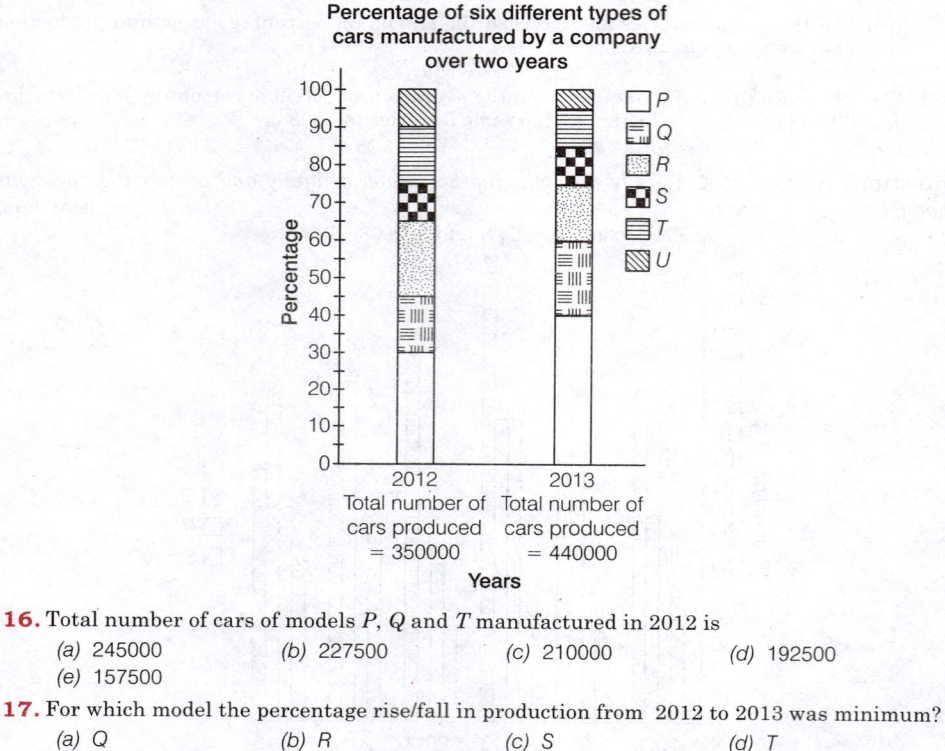

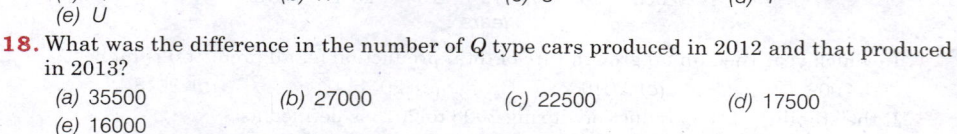

Percentage of six different types of cars manufactured by a company over two years

2012 Total number of cars produced = 350000
2013 Total number of cars produced = 440000
Years

16. Total number of cars of models P, Q and T manufactured in 2012 is
(a) 245000   (b) 227500   (c) 210000   (d) 192500
(e) 157500

17. For which model the percentage rise/fall in production from 2012 to 2013 was minimum?
(a) Q   (b) R   (c) S   (d) T
(e) U

18. What was the difference in the number of Q type cars produced in 2012 and that produced in 2013?
(a) 35500   (b) 27000   (c) 22500   (d) 17500
(e) 16000

19. If the production percentage of P type cars in 2013 was the same as that in 2012, then the number of P type cars produced in 2013 would have been
(a) 140000   (b) 132000   (c) 117000   (d) 105000
(e) 97000

20. If 85% of the S type cars produced in each year were sold by the company, how many S type cars remained unsold?
(a) 7650   (b) 9350   (c) 11850   (d) 12250
(e) 13350

Answer with Solutions

Exercise 1 Base Level Questions

1. **(b)** State D has 40% villages electrified and State C has 80% villages electrified. Thus, correct option is (b).

2. **(d)** From the question, we have 6 states and only one State D doesn't meet this condition, because it has less than 50% villages electrified, it has 40% electrification. Thus, 6 − 1 = 5 states have atleast 50% villages electrified.

3. **(d)** Companies which have more demand than production = A, C and E, i.e. 3 companies.
   Companies which have more production than demand = B, D, i.e. 2 companies
   ∴ Required ratio = 3 : 2

4. **(c)** Average demand
   $= \dfrac{\text{Sum of all demands}}{\text{Number of all demands}}$
   $= \dfrac{3300 + 1200 + 3000 + 600 + 2500}{5}$
   $= \dfrac{10600}{5} = 2120$
   Similarly, average production
   $= \dfrac{2200 + 2700 + 1500 + 1800 + 1000}{5}$
   $= \dfrac{9200}{5} = 1840$
   ∴ Required difference = 2120 − 1840 = 280

5. **(c)** Required percentage $= \dfrac{600}{2500} \times 100\%$
   $= 24\%$

6. **(c)** Value of biscuit per tonne
   in 2005 $= \dfrac{150 \times 10000000}{100 \times 100000} = ₹\ 150$
   in 2006 $= \dfrac{150 \times 10000000}{75 \times 100000} = ₹\ 200$
   in 2007 $= \dfrac{330 \times 10000000}{150 \times 100000} = ₹\ 220$
   in 2008 $= \dfrac{400 \times 10000000}{160 \times 100000} = ₹\ 250$
   in 2009 $= \dfrac{500 \times 10000000}{200 \times 100000} = ₹\ 250$
   Thus, the value per tonne is equal in the years 2008 and 2009.

7. **(b)** Percentage increase in export from preceding year
   in 2007 $= \dfrac{150 - 75}{75} \times 100\% = 100\%$
   in 2008 $= \dfrac{160 - 150}{150} \times 100\% = 6.66\%$
   in 2009 $= \dfrac{200 - 160}{160} \times 100\% = 25\%$
   Thus, required year is 2007.

8. **(c)** The percentage drop in export quantity from 2005 to 2006
   $= \dfrac{100 - 75}{100} \times 100\% = 25\%$

9. **(b)** Average number of students enrolled in institute A in years 2009 and 2010
   $= \dfrac{170 + 270}{2}$
   $= \dfrac{440}{2} = 220$
   and average number of students enrolled in institute B in years 2011 and 2012
   $= \dfrac{160 + 200}{2} = \dfrac{360}{2} = 180$
   ∴ Required difference = 220 − 180 = 40

10. **(c)** Let number of students in institute A in 2014 = x
    Now, according to the question,
    $x - x \times \dfrac{25}{100} = 330 \Rightarrow x - \dfrac{x}{4} = 330$
    $\Rightarrow \dfrac{3x}{4} = 330$
    $\Rightarrow x = \dfrac{330 \times 4}{4} = 110 \times 4 = 440$

11. **(b)** In 2009, male students enrolled in institute $B = \dfrac{240 \times 65}{100} = 156$
    Given that in 2009, ratio of male students of institute A and B was 3 : 4.
    So, we can say that number of male students in institutes A and B are $3x$ and $4x$, respectively.
    Now, $4x = 156$
    $\Rightarrow x = \dfrac{156}{4} = 39$
    ∴ Male students enrolled in institute A in year 2009 = $3x = 3 \times 39 = 117$

12. (a) Required percentage
$$= \frac{(360 - 210)}{360} \times 100$$
$$= \frac{150}{360} \times 100$$
$$= \frac{15}{36} \times 100$$
$$= \frac{125}{3} = 41\frac{2}{3}\%$$

13. (a) Total number of boys in schools B and C together = 350 + 300 = 650
Total number of girls in schools B and C together = 140 + 160 = 300
$\therefore$ Required difference = 650 − 300 = 350

14. (e) Total number of students in school B
= 140 + 350 = 490
Total number of students in school D
= 260 + 440 = 700
$\therefore$ Required percentage
$$= \frac{700 - 490}{700} \times 100\% = 30\%$$

15. (b) Total number of students in Std. X of school E = 210 + 380 = 590
According to the question,
Number of students in Std. V of the school $E = \frac{590 \times 80}{100} = 472$

16. (b) $\because$ Number of boys in school E = 380 and number of boys in school A = 250
$\therefore$ Required percentage
$$= \frac{380 - 250}{250} \times 100\%$$
$$= \frac{130}{250} \times 100\% = 52\%$$

17. (c) Required ratio
= (170 + 250) : (160 + 300)
= 420 : 460 = 21 : 23

18. (c) Water level of River-A in July
$$= 205 \times \frac{(100 - 20)}{100}$$
$$= 205 \times \frac{80}{100} \text{ m} = 164 \text{ m}$$

19. (a) From given table, it is clear that the water level of River-C was the highest in the month of August.

20. (c) Average water level of River-A in all the months together
$$= \frac{1}{4}(196 + 205 + 230 + 212)$$
$$= \frac{843}{4} \text{ m} = 210.75 \text{ m}$$

21. (e) Water level of River-C in September
= 184 m
and water level of River-B in June = 202 m
$\therefore$ Required ratio = 184 : 202 = 92 : 101

Exercise 2 *Higher Skill Level Questions*

1. (a) Required average
$$= \frac{\begin{bmatrix} 5000 + 10000 + 15000 \\ + 30000 + 40000 + 50000 \end{bmatrix}}{6}$$
$$= \frac{150000}{6} = 25000$$

2. (c) Required percentage
$$= \frac{40000}{25000} \times 100\% = 160\%$$

3. (a) Required number = $90000 \times \frac{70}{100}$
= 63000

4. (a) Required ratio = $\frac{10000}{75000} = 2 : 15$

5. (e) Required percentage
$$= \frac{45000 - 30000}{45000} \times 100\% \approx 33\%$$

6. (d) Required percentage increase
$$= \frac{55 - 45}{45} \times 100\% = 22.2\%$$

7. (a) Production of cement by factory B in 2009 and by factory D in 2010 together
= 55 + 65 = 120 lakh tonne
Production of cement by factory A in 2008
= 60 lakh tonne
$\therefore$ Required percentage
$$= \frac{120}{60} \times 100\% = 200\%$$

8. (c) Percentage increase in the production of cement from the year 2008 to the year 2009
for factory $A = \frac{70 - 60}{60} \times 100\% = 16.66\%$
for factory $B = \frac{55 - 45}{45} \times 100\% = 22.2\%$
for factory $C = \frac{85 - 70}{70} \times 100\% = 21.42\%$
for factory $D = \frac{60 - 50}{50} \times 100\% = 20\%$
Thus, the percentage increase was maximum for factory B.

Bar Chart / 815

9. (a) Percentage increase in the production of cement from 2009 to 2010

for factory $A = \dfrac{75-70}{70} \times 100\% = 7.14\%$

for factory B = No growth
for factory C = Negative
for factory $D = \dfrac{65-60}{60} \times 100\% = 8.33\%$

Thus, percentage increase was maximum for factory D.

10. (b) Required difference

$= \left(\dfrac{70+55+85+60}{4}\right) - \left(\dfrac{60+45+70+50}{4}\right)$

$= \dfrac{270}{4} - \dfrac{225}{4} = \dfrac{45}{4}$

= 11.25 lakh tonne

11. (b) It is clear from the bar graph that, in 2010, the annual growth rate of total production is highest.

12. (c) For product P,
Average production of P during 2008-13

$= \dfrac{275}{6} = 45.83$

Maximum production of P = 75
Minimum production of P = 25

∴ Stability of $P = \dfrac{45.83}{75-25} = 0.9166$

For product Q,
Average production of Q during 2008-13

$= \dfrac{420}{6} = 70$

Maximum production = 115
Minimum production = 40

∴ Stability of $Q = \dfrac{70}{115-40} = \dfrac{70}{75} = 0.9333$

For product R,
Average production of R during 2008-13

$= \dfrac{595}{6} = 99.166$

Maximum production = 125
Minimum production = 65

∴ Stability of $R = \dfrac{99.16}{125-65} = 1.652$

For product S,
Average production of S during 2008-13

$= \dfrac{755}{6} = 125.83$

Maximum production = 165
Minimum production = 85

∴ Stability of $S = \dfrac{125.83}{165-85} = 1.572$

So, R is the most stable product.

13. (c) In 2008,
Total revenue
$= 45 \times 9 + 95 \times 4 + 75 \times 13 + 115 \times 3$
$= 405 + 380 + 975 + 345 = ₹ 2105$

In 2009,
Total revenue
$= 25 \times 9 + 40 \times 4 + 95 \times 13 + 155 \times 3$
$= 225 + 160 + 1235 + 465 = ₹ 2085$

In 2010,
Total revenue
$= 40 \times 9 + 115 \times 4 + 115 \times 13 + 165 \times 3$
$= 360 + 460 + 1495 + 495 = ₹ 2810$

In 2011,
Total revenue
$= 35 \times 9 + 60 \times 4 + 65 \times 13 + 140 \times 3$
$= 315 + 240 + 845 + 420 = ₹1820$

In 2012,
Total revenue
$= 75 \times 9 + 40 \times 4 + 125 \times 13 + 85 \times 3$
$= 675 + 160 + 1625 + 255 = ₹2715$

In 2013,
Total revenue
$= 55 \times 9 + 70 \times 4 + 120 \times 13 + 95 \times 3$
$= 495 + 280 + 1560 + 285 = ₹ 2620$

So, total revenue of all products is lowest in 2011.

14. (b) For product P,
Total revenue for the period 2008-11
$= 45 \times 9 + 25 \times 9 + 40 \times 9 + 35 \times 9$
$= 405 + 225 + 360 + 315 = 1305$

For product Q,
Revenue
$= 95 \times 4 + 40 \times 4 + 115 \times 4 + 60 \times 4$
$= 380 + 160 + 460 + 240 = 1240$

For product R,
Revenue
$= 75 \times 13 + 95 \times 13 + 115 \times 13 + 65 \times 13$
$= 975 + 1235 + 1495 + 845 = 4550$

For product S,
Revenue
$= 115 \times 3 + 155 \times 3 + 165 \times 3 + 140 \times 3$
$= 345 + 465 + 495 + 420 = 1725$

So, product Q fetches the lowest revenue.

15. (c) By option (c),
In 2011,
Cumulative revenue of P and Q
$= 35 \times 9 + 60 \times 4 = 555$
Cumulative revenue of $S = 140 \times 3$
$= 420$

So, we can see that cumulative revenue of P and Q is more than cumulative revenue of S in 2011.

Solutions (Q. Nos. 16-20) We shall first determine the number of cars for each model produced by the company during the two years.

In 2012,
Total number of cars produced = 350000
$P = (30 - 0)\%$ of 350000
$= 30\%$ of $350000 = 105000$
$Q = (45 - 30)\%$ of 350000
$= 15\%$ of $350000 = 52500$
$R = (65 - 45)\%$ of 350000
$= 20\%$ of $350000 = 70000$
$S = (75 - 65)\%$ of 350000
$= 10\%$ of $350000 = 35000$
$T = (90 - 75)\%$ of 350000
$= 15\%$ of $350000 = 52500$
$U = (100 - 90)\%$ of 350000
$= 10\%$ of $350000 = 35000$

In 2013,
Total number of cars produced = 440000
$P = (40 - 0)\%$ of 440000
$= 40\%$ of $440000 = 176000$
$Q = (60 - 40)\%$ of 440000
$= 20\%$ of $440000 = 88000$
$R = (75 - 60)\%$ of 440000
$= 15\%$ of $440000 = 66000$
$S = (85 - 75)\%$ of 440000
$= 10\%$ of $440000 = 44000$
$T = (95 - 85)\%$ of 440000
$= 10\%$ of $440000 = 44000$
$U = (100 - 95)\%$ of 440000
$= 5\%$ of $440000 = 22000$

16. (c) Total number of cars of models P, Q and T manufactured in 2012
$= 105000 + 52500 + 52500$
$= 210000$

17. (b) Using the above calcualtion, the percentage change (rise/fall) in production from 2012 to 2013

for $Q = \left[\dfrac{(88000 - 52500)}{52500}\right] \times 100 \%$
$= 67.62\%$, rise

for $R = \left[\dfrac{(70000 - 66000)}{70000}\right] \times 100 \%$
$= 5.71\%$, fall

for $S = \left[\dfrac{(44000 - 35000)}{35000}\right] \times 100 \%$
$= 25.71\%$, rise

for $T = \left[\dfrac{(52500 - 44000)}{52500}\right] \times 100 \%$
$= 16.19\%$, fall

for $U = \left[\dfrac{(35000 - 22000)}{35000}\right] \times 100 \%$
$= 37.14\%$, fall

$\therefore$ Minimum percentage rise/fall in production of model R.

18. (a) Required difference
$= 88000 - 52500 = 35500$
(using calculations from above)

19. (b) If the production percentage of P type cars in 2013 = Production percentage of P type cars in 2012 = 30%, then
Number of P type cars produced in 2013
$= 30\%$ of 440000
$= 132000$

20. (c) Number of S type cars which remained unsold in 2012 = 15% of 35000
and number of S type cars which remained unsold in 2013 = 15% of 44000
$\therefore$ Total number of S type cars which remains unsold
$= 15\%$ of $(35000 + 44000)$
$= 15\%$ of 79000
$= 11850$

Chapter 45

Line Graph

A line graph (cartesian graph) indicates the variation of a quantity with respect to the two parameters caliberated (plotted) on X and Y-axes, respectively.

A line graph shows the quantitative information or a relationship between two changing quantities (variables) with a **line or curve** that connects a series of successive data points.

Types of Line Graph

Different types of line graph are discussed below

1. **Single Line Graph** Used for single variable representation.

   For example The following single line graph represents the yearly sales figure of a company in the years 2001-2010.

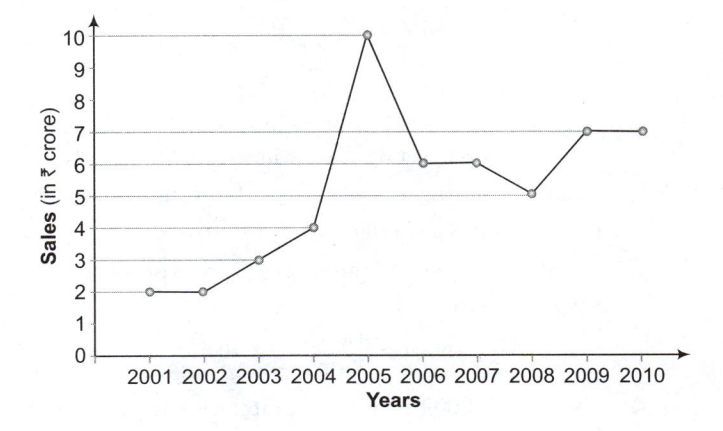

Directions (Examples 1-5) *Study the following graph carefully and answer the given questions.*

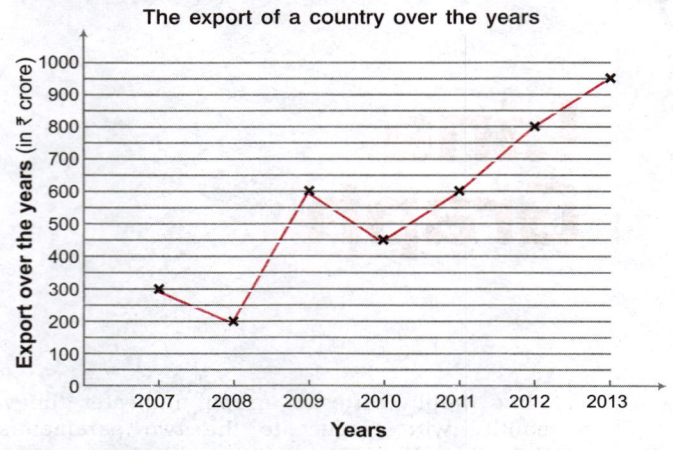

The export of a country over the years

Export over the years (in ₹ crore) vs *Years*

Data points: 2007: 300, 2008: 200, 2009: 600, 2010: 450, 2011: 600, 2012: 800, 2013: 950

Ex. 1 Which year has the highest per cent increase/decrease in exports as compared to the preceding year?

Sol. The per cent increase/decrease in exports as compared to the preceding year for the year

$$2008 = \frac{300-200}{300} \times 100\% = 33.33\% \text{ decrease}$$

$$2009 = \frac{600-200}{200} \times 100\% = 200\% \text{ increase}$$

$$2010 = \frac{600-450}{600} \times 100\% = 25\% \text{ decrease}$$

$$2011 = \frac{600-450}{450} \times 100\% = 33.33\% \text{ increase}$$

$$2012 = \frac{800-600}{600} \times 100\% = 33.33\% \text{ increase}$$

$$2013 = \frac{950-800}{800} \times 100\% = 18.75\% \text{ increase}$$

Thus, the calculations show that

(i) Highest per cent increase is in year 2009, i.e. 200%.

(ii) Highest per cent decrease is in year 2008, i.e. 33.33%.

Ex. 2 What is the difference in exports in the years 2009 and 2010?

Sol. Required difference = 600 − 450 = ₹ 150 crore

Ex. 3 What is the percentage increase in exports from the lowest to the highest for the given years?

Sol. Required percentage = $\frac{950-200}{200} \times 100\% = 375\%$

Ex. 4 Exports in 2009 is approximately what per cent of that of year 2010?

Sol. Required percentage = $\frac{600}{450} \times 100\% \approx 134\%$

Ex. 5 What is the total exports for the given years?

Sol. Total exports = 300 + 200 + 600 + 450 + 600 + 800 + 950 = ₹ 3900 crore

Line Graph / 819

2. **Multiple Line Graph** Used for more than one variable representation.
   For example The following multiple line graph represents the maximum and minimum temperature recorded everyday in a certain week.

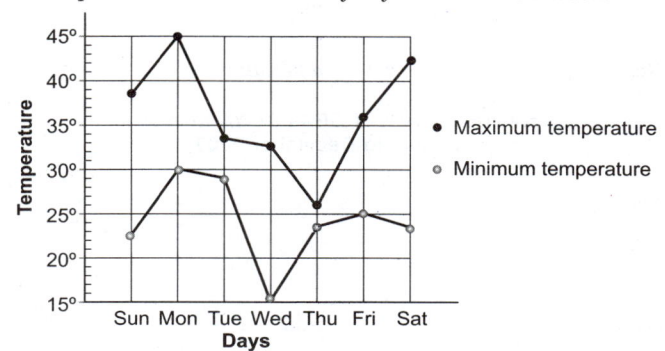

Directions (Examples 6-8) *Refer to the graph and answer the given questions.* [IBPS SO 2016]

Number of tourists (in hundred) in two cities A and B in five different months in the year 2005

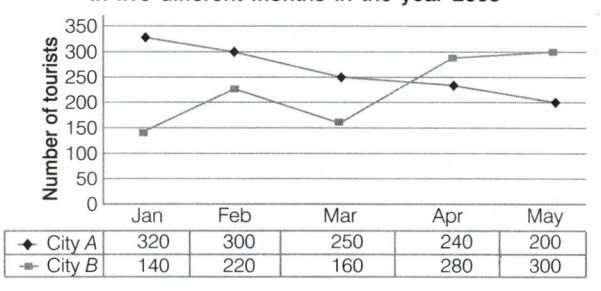

	Jan	Feb	Mar	Apr	May
City A	320	300	250	240	200
City B	140	220	160	280	300

Ex. 6 Number of tourists in city B in April are what per cent more than that in city A in March?

Sol. ∵ Number of tourist in city B in April = 280
and number of tourist in city A in March = 250
∴ Required percentage = $\left(\dfrac{280-250}{250}\right) \times 100 = \dfrac{30}{250} \times 100 = 12\%$

Ex. 7 What is the respective ratio between the total number of tourists in city A in January and February together and that in city B in the same months together?

Sol. Total number of tourist in city A in the month of January and February
= 320 + 300 = 620
and total number of tourist in city B in the month of January and February.
= 140 + 220 = 360
∴ Required ratio = 620 : 360 = 31 : 18

Ex. 8 The number of tourists in city A in April is what per cent less than that in the same city in January?

Sol. Number of tourist in city A in the month of April = 240
and number of tourist in city A in the month of January = 320
∴ Required percentage = $\dfrac{(320-240)}{320} \times 100 = \dfrac{80 \times 100}{320} = 25\%$

Fast Track Practice

Directions (Q. Nos. 1-4) *Study the following graph carefully and answer the questions that follow.*

Circulation growth of GRAMSEWA magazine from July to December 2003

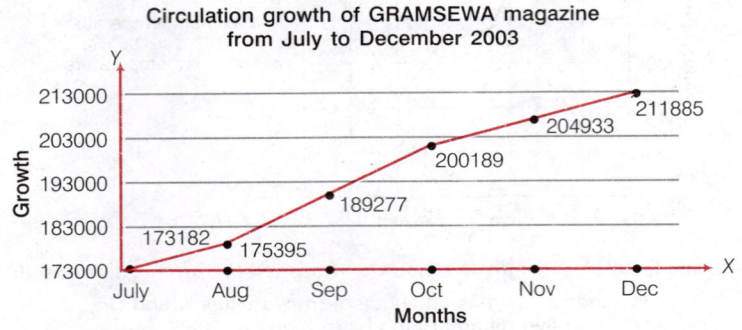

[Graph showing monthly circulation: July 173182, Aug 175395, Sep 189277, Oct 200189, Nov 204933, Dec 211885]

1. During November and December, there is an even growth rate, the average of which is
   (a) 2.36%   (b) 2%
   (c) 2.88%   (d) 3.36%
   (e) None of these

2. The circulation in October is .... times than that of July.
   (a) 1.5   (b) 2
   (c) 1   (d) 1.15
   (e) None of these

3. The growth rate is very marginal during the month of
   (a) August   (b) October
   (c) November   (d) December
   (e) None of these

4. What is the total circulation of magazine from July to December?
   (a) 1154681   (b) 1154861
   (c) 1145861   (d) 1150862
   (e) None of these

Directions (Q. Nos. 5-8) *Study the following graph carefully and answer the questions that follow.* [Bank PO 2007]

Per cent rise in production from the years 1999 to 2006

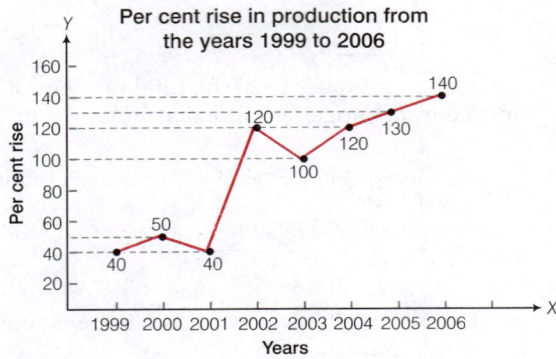

[Graph: 1999: 40, 2000: 50, 2001: 40, 2002: 120, 2003: 100, 2004: 120, 2005: 130, 2006: 140]

5. For how many years, the per cent rise was more than 100%?
   (a) One   (b) Two   (c) Three   (d) Five
   (e) None of these

6. For how many years, the per cent rise was lower than the average of the per cent rise over the given years?
   (a) Two (b) One (c) Five (d) Three
   (e) None of these

7. For which of the given years, the per cent rise (from the previous year) was the least? [Years 2001 and 2003 are not to be considered]
   (a) 2000 (b) 2004 (c) 2006
   (d) Cannot be determined
   (e) None of these

8. If the production in year 1998 was 1000 units, how much was the production in year 2002?
   (a) 35280 units
   (b) 64680 units
   (c) 46200 units
   (d) Cannot be determined
   (e) None of the above

Directions (Q. Nos. 9-12) *Refer to the graph and answer the given questions.*
[SBI PO (Pre) 2017]

Number of students who opted for 2 courses A, B from 2010 to 2014

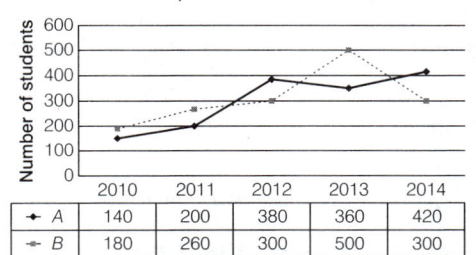

	2010	2011	2012	2013	2014
A	140	200	380	360	420
B	180	260	300	500	300

9. Number of students who opted for course B in 2013 was what per cent more than those who opted for course A in 2013?
   (a) $45\frac{1}{6}$%   (b) $42\frac{4}{9}$%
   (c) $38\frac{8}{9}$%   (d) $56\frac{7}{9}$%
   (e) $62\frac{1}{9}$%

10. In 2014, if 'X' students passed in courses A and B each and the ratio of number of students that failed in courses A and B respectively was 5 : 2, what is the value of 'X'?
    (a) 190 (b) 220 (c) 160 (d) 150
    (e) 180

11. What is the average number of students who opted for course A in 2010, 2011 and 2012?
    (a) 225 (b) 250 (c) 230 (d) 240
    (e) 260

12. The number of students who opted for courses A and B in 2011 was respectively 25% more and 35% less than that in 2009. What was the total number of students who opted for courses A and B together in 2009?
    (a) 600 (b) 540 (c) 575 (d) 560
    (e) 584

Directions (Q. Nos. 13-16) *Study the following graph carefully and answer the questions that follow.*
[IDBI SO 2012]

Runs scored by three different teams in five different cricket matches

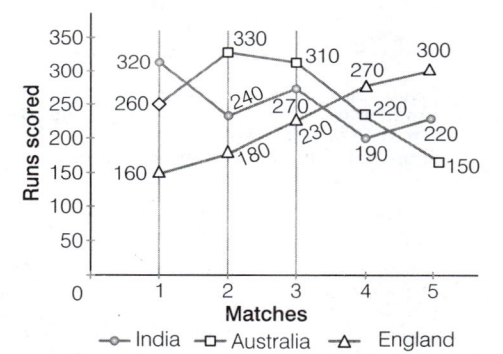

13. Total runs scored by India and Australia in Match 4 together is approximately what percentage of the total runs scored by England in all the five matches together?
    (a) 42% (b) 18% (c) 36% (d) 24%
    (e) 28%

14. In which match, is the difference between the runs scored by Australia and England the second lowest?
    (a) 1 (b) 2 (c) 3 (d) 4
    (e) 5

15. In which match the total runs scored by India and England together is the third highest/lowest?
    (a) 1 (b) 2
    (c) 3 (d) 4
    (e) 5

16. What is the respective ratio between the runs scored by India in Match 5, Australia in Match 1 and England in Match 2?
    (a) 11 : 13 : 7 (b) 11 : 7 : 13
    (c) 11 : 3 : 9   (d) 11 : 13 : 9
    (e) None of these

Directions (Q. Nos. 17-20) *Study the following graph carefully and answer the given questions.*
[IBPS Clerk 2015]

Data related to number of calories burned by two individuals (A and B) on treadmill during five days

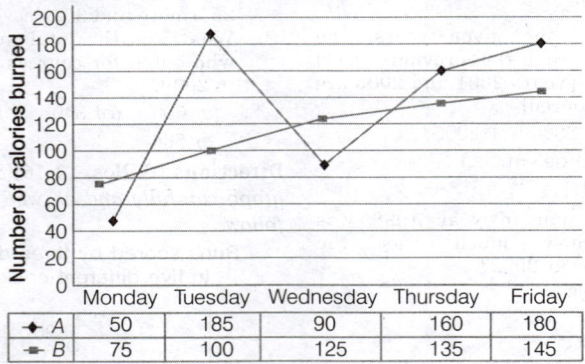

	Monday	Tuesday	Wednesday	Thursday	Friday
A	50	185	90	160	180
B	75	100	125	135	145

17. What is the respective ratio of total number of calories burned by A and B together on Wednesday and that by the same individuals together on Tuesday?
(a) 45 : 59 (b) 43 : 57
(c) 41 : 57 (d) 43 : 61
(e) 47 : 61

18. If the number of calories burned by A and B increased by 10% and 20% respectively from Friday to Saturday, what was the total number of calories burned by them together on Saturday?
(a) 378 (b) 372 (c) 368 (d) 384
(e) 364

19. If the average number of calories burned by B on Thursday, Friday and Saturday together is 125, what was the number of calories burned by B on Saturday?
(a) 110 (b) 95
(c) 115 (d) 90
(e) 105

20. Number of calories burned by B increased by what per cent from Monday to Thursday?
(a) 80% (b) 60%
(c) 70% (d) 75%
(e) 65%

Directions (Q. Nos. 21-24) *Study the following graph carefully and answer the questions that follow.*
[Bank Clerk 2010]

Number of students in three schools A, B and C over the years from 2002-07

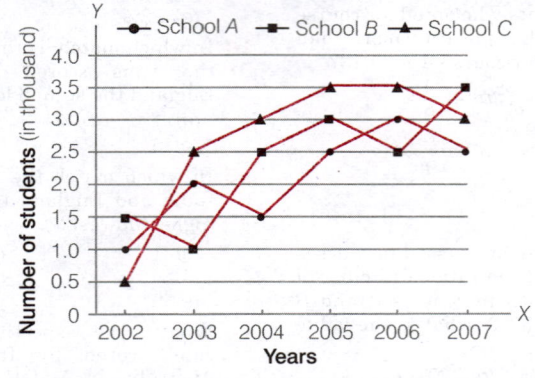

21. What was the average number of students in all the schools together in the year 2006?
(a) 30000 (b) 9000 (c) 3000
(d) 6000 (e) None of these

22. How many times the total number of students in all the three schools A, B and C together was exactly equal among the given years?
(a) 2 (b) 5 (c) 4 (d) 3
(e) None of these

Line Graph / 823

23. Total number of students in school B and school C together in the year 2004 was approximately what per cent of the total number of students in school B and school C together in the year 2007?
(a) 85% (b) 80% (c) 75% (d) 184%
(e) 131%

24. What was the approximate average number of students in school A over all the years together?
(a) 1990   (b) 2090
(c) 2300   (d) 1800
(e) 2700

Directions (Q. Nos. 25-27) *Study the following graph carefully and answer the questions that follow.*   [Bank PO 2010]

Per cent rise in profit of two companies L and M from the year 2004 to 2009

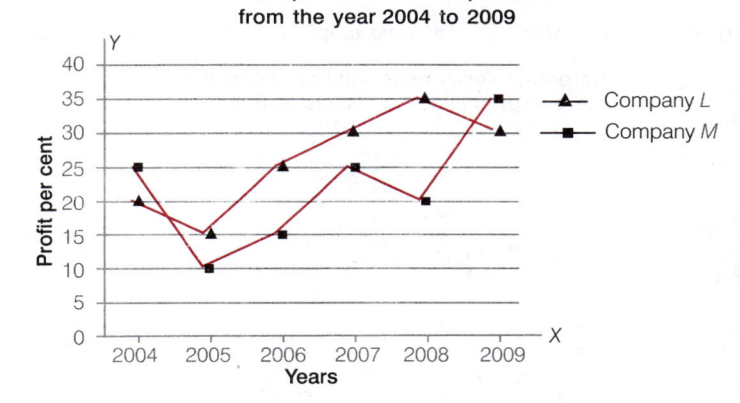

25. If the profit earned by company L in the year 2005 was ₹ 1.84 lakh, what was the profit earned by the company in the year 2006?
(a) ₹ 2.12 lakh   (b) ₹ 2.3 lakh
(c) ₹ 2.04 lakh   (d) ₹ 3.4 lakh
(e) None of these

26. Which of the following statements is true with respect to the above graph?
(a) Company M made the highest profit in the year 2009
(b) Company L made the least profit in the year 2005
(c) The ratio of the profits earned by company L in the years 2007 and 2006 was 6 : 5, if the income was same in both the years
(d) Company L made the highest profit in the year 2008
(e) All of the above

27. If the profit earned by company M in the year 2008 was ₹ 3.63 lakh, what was the amount of profit earned by it in the year 2006?
(a) ₹ 2.16 lakh
(b) ₹ 1.98 lakh
(c) ₹ 2.42 lakh
(d) ₹ 4.02 lakh
(e) None of the above

Directions (Q. Nos. 28-29) *Study the following line graph carefully and answer the following questions.*   [RRB PO (Pre) 2017]

Number of males and number of females are given and they are visiting a place from Monday to Friday

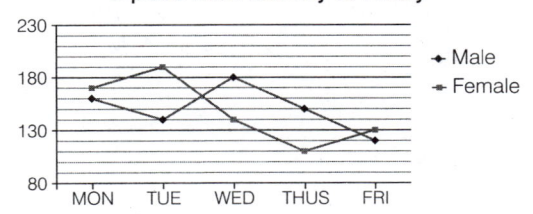

28. Total number of males and females together visited the place on Tuesday are what per cent more/less than the total number of male and females together visited the place on Thursday?
(a) $26\frac{12}{13}$%   (b) $25\frac{3}{13}$%
(c) $26\frac{3}{13}$%   (d) $25\frac{7}{13}$%
(e) None of these

29. If on Saturday, the number of males and number of females increased by 25% and 20% respectively as compared to that on Friday, then find the total number of males and females together visited the place on Saturday.
(a) 196   (b) 306
(c) 316   (d) 206
(e) 216

Directions (Q. Nos. 30-34) *Study the following graph to answer the given questions.*
[MAT 2015]

Corporate social responsibility spending for the arts promotion by sectors in three years

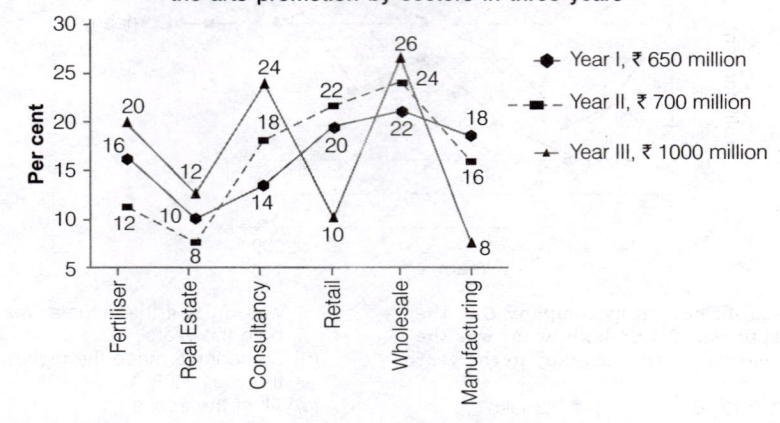

— Year I, ₹ 650 million
-- Year II, ₹ 700 million
— Year III, ₹ 1000 million

30. The corporate sectors that continuously increased their support for the arts from Year III made a total contribution of approximately how many million rupees in Year III?
(a) 100   (b) 200
(c) 300   (d) 500

31. How many of the given corporate sectors contributed more than 126 million in Year I?
(a) Three   (b) One
(c) Two   (d) Four

32. From Year II to Year III, the per cent increase is minimum for which sector?
(a) Consultancy
(b) Wholesale
(c) Manufacturing
(d) Fertiliser

33. The contribution of retail sector in Year II is what per cent more than contribution consultancy sector in Year I?
(a) 69.23%   (b) 59%
(c) 49%   (d) 79%

34. Of the total consultancy sector's contribution in the Year II and Year III together, 1/3 of the contribution went to village orchestras and 1/4 of the remainder went to rural television. Approximate contribute village orchestras by the consultancy sector is what per cent greater than contribute to television?
(a) 40%
(b) 70%
(c) 50%
(d) 100%

Line Graph / 825

Directions (Q. Nos. 35-38) *Refer to the graph and answer the given questions.* [LIC AAO 2016]

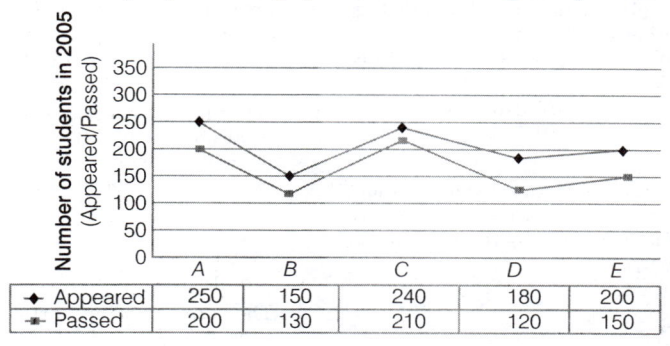

Number of students in 2005 (Appeared/Passed)

	A	B	C	D	E
◆ Appeared	250	150	240	180	200
■ Passed	200	130	210	120	150

35. In 2006, the number of students who appeared the exam from school A was 12% more than that who appeared the exam from same school in previous year, how many students appeared the exam from school A in 2006?
(a) 272　　(b) 280
(c) 254　　(d) 260
(e) 285

36. What is the respective ratio between the total number of students who appeared the exam from schools C and E together and that who passed the exam from the same schools together?
(a) 11 : 7　　(b) 13 : 9
(c) 11 : 8　　(d) 11 : 9
(e) 4 : 3

37. What is the average number of students who passed the exam from schools A, B and C?
(a) 190
(b) 160
(c) 180
(d) 150
(e) 140

38. The number of students who appeared the exam from school D was what per cent more than that who appeared from school B?
(a) 50%
(b) 30%
(c) 20%
(d) 10%
(e) 25%

Directions (Q. Nos. 39-43) *Refer to the graph and answer the given questions.* [IBPS SO 2016]

Data regarding number of gift articles sold by two shops A and B in five different days in December in a particular year

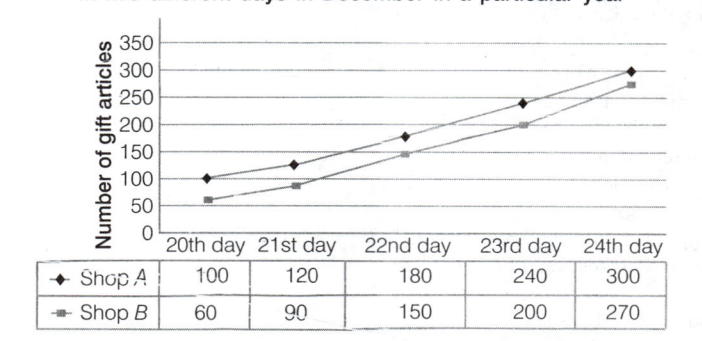

	20th day	21st day	22nd day	23rd day	24th day
◆ Shop A	100	120	180	240	300
■ Shop B	60	90	150	200	270

826 / **Fast Track** Objective Arithmetic

39. What is the difference between total number of gift articles sold by both the shops together on December 21 and that sold by both the shops together on December 23?
(a) 230 (b) 240
(c) 170 (d) 180
(e) 270

40. The number of gift articles sold by shop B on December 20 is what per cent of number of articles sold by the same shop on December 24?
(a) 25%
(b) $35\frac{1}{4}$%
(c) $30\frac{1}{4}$%
(d) $22\frac{1}{4}$%
(e) 14%

41. What is the average number of gift articles sold by shop B on December 20 and 23?
(a) 70 (b) 65 (c) 80 (d) 60
(e) 130

42. What is the respective ratio between the total number of gift articles sold by shop A on December 23 and 24 together and by the same shop on December 21 and 22 together?
(a) 9 : 4 (b) 9 : 5
(c) 11 : 7 (d) 11 : 9
(e) 9 : 7

43. The number of gift articles sold by the shop A on December 22 is what per cent more than that sold by shop B on the same day?
(a) 20% (b) 15%
(c) 50% (d) 30%
(e) 25%

Answer with Solutions

1. (c) Growth in November
$= \frac{204933 - 200189}{200189} \times 100\% = 2.36\%$
Growth in December
$= \frac{211885 - 204933}{204933} \times 100\% = 3.39\%$
∴ Average $= \frac{2.36 + 3.39}{2}\% = \frac{5.75}{2}\%$
$= 2.875\% \approx 2.88\%$

2. (d) Let the circulation in October be x times to that of July.
Then, $200189 = x \times 173182$
∴ $x = \frac{200189}{173182} = 1.15$

3. (a) In the month of August, the growth rate is least, as is obvious from the graph.

4. (b) Total circulation $= 173182 + 175395$
$+ 189277 + 200189 + 204933 + 211885$
$= 1154861$

5. (e) For four years, i.e. 2002, 2004, 2005 and 2006, the percentage rise was more than 100%.

6. (d) Average per cent rise
$= \frac{40 + 50 + 40 + 120 + 100 + 120 + 130 + 140}{8}$
$= 92.50\%$
The relevant years were 1999, 2000 and 2001.

7. (c) In the year 2000 $= \frac{50 - 40}{40} \times 100\%$
$= 25\%$
In the year 2004 $= \frac{120 - 100}{100} \times 100\%$
$= 20\%$
In the year 2005 $= \frac{130 - 120}{120} \times 100\%$
$= 8.33\%$
In the year 2006 $= \frac{140 - 130}{130} \times 100\%$
$= 7.692\%$
Clearly, per cent rise was the least for the year 2006.

8. (e) Production in year 2002
$= 1000 \times \frac{140}{100} \times \frac{150}{100} \times \frac{140}{100} \times \frac{220}{100}$
$= 6468$

9. (c) Required percentage
$= \left(\frac{500 - 360}{360}\right) \times 100$
$= \frac{140}{360} \times 100$
$= \frac{7}{18} \times 100$
$= \frac{7 \times 50}{9} = \frac{350}{9}$
$= 38\frac{8}{9}\%$

10. (b) Number of passed student in each course = X
Ratio of failed students in each course = 5 : 2
Now, according to the question,
$$\frac{420 - X}{300 - X} = \frac{5}{2}$$
$\Rightarrow \quad 1500 - 5X = 840 - 2X$
$\Rightarrow \quad 5X - 2X = 1500 - 840$
$\Rightarrow \quad 3X = 660$
$\Rightarrow \quad X = 220$

11. (d) Average number of students who opted for course A in 2010, 2011 and 2012
$$= \frac{140 + 200 + 380}{3}$$
$$= \frac{720}{3} = 240$$

12. (d) Let number of students in course A in 2009 = x
Then, according to the question,
$$x \times \frac{(100 + 25)}{100} = 200$$
$\Rightarrow \quad \dfrac{x \times 125}{100} = 200$
$\Rightarrow \quad x = \dfrac{200 \times 100}{125} = 160$

Now, let number of students in course B in 2009 = y
Then, according to the question,
$$y \times \frac{(100 - 35)}{100} = 260$$
$\Rightarrow \quad y \times \dfrac{65}{100} = 260$
$\Rightarrow \quad y = \dfrac{260 \times 100}{65} = 400$

∴ Total number of students in courses A and B together in 2009 = 160 + 400 = 560

Solutions (Q. Nos. 13-16)

Match	India	Australia	England
1	320	260	160
2	240	330	180
3	270	310	230
4	190	220	270
5	220	150	300

13. (c) Required percentage
$$= \frac{190 + 220}{160 + 180 + 230 + 270 + 300} \times 100$$
$$= \frac{410}{1140} \times 100 = 36\%$$

14. (c) Difference between the runs scored by Australia and England in
match 1 = 260 − 160 = 100
match 2 = 330 − 180 = 150
match 3 = 310 − 230 = 80
match 4 = 270 − 220 = 50
match 5 = 300 − 150 = 150
Hence, the second lowest difference is in match 3.

15. (a) Total runs scored by India and England in
match 1 = 320 + 160 = 480
match 2 = 240 + 180 = 420
match 3 = 270 + 230 = 500
match 4 = 190 + 270 = 460
match 5 = 220 + 300 = 520
Hence, the third highest/lowest score is in match 1.

16. (d) Required ratio = 220 : 260 : 180
= 11 : 13 : 9

17. (b) Total number of calories burned by A and B together on Wednesday
= 90 + 125 = 215
Total number of calories burned by A and B together on Tuesday
= 185 + 100 = 285
∴ Required ratio = 215 : 285 = 43 : 57

18. (b) Total number of calories burnt on Saturday
$$= 180 \times \frac{110}{100} + 145 \times \frac{120}{100}$$
= 198 + 174 = 372

19. (b) Average calories = 125
Total calories = 3 × 125 = 375
Number of calories burnt on Thursday and Friday by B = 135 + 145 = 280
Number of calories burnt on Saturday
= 375 − 280 = 95

20. (a) Required percentage
$$= \frac{135 - 75}{75} \times 100\%$$
$$= \frac{60}{75} \times 100\% = 80\%$$

21. (c) Required average number of students in 2006 = $\dfrac{(2.5 + 3 + 3.5) \times 1000}{3} = 3000$

22. (d) Total number of students in the year 2002 = (0.5 + 1 + 1.5) × 1000
= 3 × 1000 = 3000
Total number of students in the year 2003
= (1 + 2 + 2.5) × 1000
= 5.5 × 1000
= 5500
Total number of students in the year 2004
= (1.5 + 2.5 + 3) × 1000
= 7 × 1000 = 7000
Total number of students in the year 2005
= (2.5 + 3 + 3.5) × 1000
= 9 × 1000 = 9000

Total number of students in the year 2006
= (2.5 + 3 + 3.5) × 1000
= 9 × 1000 = 9000
Total number of students in the year 2007
= (2.5 + 3 + 3.5) × 1000
= 9 × 1000 = 9000
Hence, the total number of students is same in the years 2005, 2006 and 2007.

23. (a) In the year 2004,
School B + School C = (2.5 + 3)
= 5.5 × 1000 = 5500
In the year 2007,
School B + School C = (3.5 + 3)
= 6.5 × 1000 = 6500
∴ Required percentage = $\frac{5500}{6500} \times 100$
= 84.61% ≈ 85%

24. (b) Required average
= $\frac{(1 + 2 + 1.5 + 2.5 + 3 + 2.5)}{6} \times 1000$
= $\frac{12.5 \times 1000}{6}$
= 2083 ≈ 2090

25. (b) Profit of the company L in the year 2005
= 1.84 lakh
In the year 2006, 25% rise in the profit of the company L.
So, the profit of the company L in the year 2006 = 1.84 × $\frac{(100 + 25)}{100}$
= 1.84 × 1.25 = ₹ 2.3 lakh

26. (e) Company L made maximum and minimum profit in years 2008 and 2005, respectively. Similarly, company M made maximum and minimum profits in years 2009 and 2005, respectively.

27. (c) Profit of company M in the year 2008
= ₹ 3.63 lakh
∴ Profit of company M in the year 2006
= 3.63 × $\frac{100}{120} \times \frac{100}{125}$
= ₹ 2.42 lakh

28. (a) Total number of males and females together on Tuesday = 140 + 190 = 330
Total number of males and females together on Thursday = 150 + 110 = 260
∴ Required percentage = $\frac{330 - 260}{260} \times 100\%$
= $26\frac{12}{13}\%$

29. (b) On Saturday, total number of males visited the place
= $\frac{125}{100} \times 120 = 150$

and total number of females visited the place
= $\frac{120}{100} \times 130 = 156$
∴ Required males and females = 150 + 156
= 306

30. (d) Two corporate sectors namely, consultancy and wholesale experience a continuous increase from Year I to Year III.
∴ The contribution of Year III
= (24% + 26%) of 1000 million
= 50% of 1000 million
= ₹ 500 million

31. (c) ₹ 126 million is approx, 19.38% of ₹ 650 million.
So, we have to consider the sectors which contribute more than 19.38%.
It is clear form the graph that, retail sector and wholesale sector contributed more than ₹ 126 million in Year I.

32. (b) From Year II to Year III, the per cent increase is minimum for wholesale sector
= $\frac{26 - 24}{24} \times 100 = \frac{2}{24} \times 100 = 8.33\%$

33. (a) The contribution of retail sector in Year II
= $\frac{22 \times 700}{100}$ = ₹ 154 million
Contribution of consultancy sector in Year I
= $\frac{14 \times 650}{100}$ = ₹ 91 million
∴ Required percentage
= $\frac{154 - 91}{91} \times 100$
= $\frac{63}{91} \times 100 = 69.23\%$
i.e. 69.23% more than the contribution of consultancy in Year I.

34. (d) Total contribution of consultancy sector in Year II and Year III together
= 126 + 240 = 366 million
Now, 1/3 of 366 went to village orchestra, then, village orchestra
= $366 \times \frac{1}{3}$ = 122 million
and $\frac{1}{4}$ of (366 − 122), went to rural television, then rural television
= $\frac{1}{4} \times 244 = 61$ million
∴ Required percentage
= $\frac{122 - 61}{61} \times 100\% = 100\%$

35. (b) Number of students in school A who appeared for exam in 2005 = 250

Line Graph / 829

∴ Number of students who appeared in 2006
$= 250 + \dfrac{12}{100} \times 250 = 250 + 30 = 280$

36. (d) Required ratio $= \dfrac{240 + 200}{210 + 150} = \dfrac{440}{360}$

$= \dfrac{22}{18} = \dfrac{11}{9} = 11 : 9$

37. (c) Average number of students who passed the exam from schools A, B and C
$= \dfrac{200 + 130 + 210}{3} = \dfrac{540}{3} = 180$

38. (c) Number of students who appeared from school $D = 180$
Number of students who appeared from school $B = 150$
∴ Required percentage
$= \dfrac{180 - 150}{150} \times 100$
$= \dfrac{30}{150} \times 100 = 20\%$

39. (a) Total number of gift articles sold by shops A and B on December 21
$= 120 + 90 = 210$
and total number of gift articles sold by shops A and B on December 23
$= 240 + 200 = 440$

∴ Required difference
$= 440 - 210 = 230$

40. (d) Required percentage
$= \dfrac{60}{270} \times 100 = \dfrac{600}{27}$
$= 22.2222 \approx 22.25 = 22\dfrac{1}{4}\%$

41. (e) Average number of gift articles sold by shop B on December 20 and 23
$= \dfrac{1}{2}(60 + 200) = \dfrac{260}{2} = 130$

42. (b) Total number of gift articles sold by shop A on December 23 and 24
$= 240 + 300 = 540$
and total number of gift articles sold by shop A on December 21 and 22
$= 120 + 180 = 300$

∴ Required ratio $= \dfrac{540}{300} = \dfrac{54}{30}$

$= \dfrac{18}{10} = \dfrac{9}{5} = 9 : 5$

43. (a) Required percentage
$= \dfrac{(180 - 150)}{150} \times 100$
$= \dfrac{30 \times 100}{150} = 20\%$

Chapter 46

Mixed Graph

Mixed graph is used when the desired parameter is a function of two or three variables and the data are to be presented as a combination of two or more forms of data presentation.

In mixed graph, we study various types of graph based questions, i.e. based on data tables, pie chart, bar chart, line graph etc. when the data are represented by any two of them, it is called **mixed graph**.

It may be the combination of data table and pie chart or data table and line graph or data table and bar chart or pie chart and line graph or bar chart and line graph.

Such type of questions requires a very careful analysis of data because the result is dependent on two or more graphs.

Examples given below will give a clear idea about the types of questions.

Directions (Examples 1-3) *Study the graph carefully and answer the questions that follow.*

Degreewise distribution of employees working in various departments of an organisation and the ratio of number of men to number of women

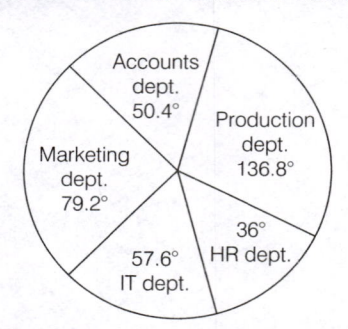

Accounts dept. 50.4°
Production dept. 136.8°
Marketing dept. 79.2°
IT dept. 57.6°
HR dept. 36°

Total number of employees = 3250

Mixed Graph / 831

Respective ratio of number of men to number of women in each department

Department	Men	Women
Production	4	1
HR	12	13
IT	7	3
Marketing	3	2
Accounts	6	7

+ Above mixed graph is combination of pie chart and data table.

Ex. 1 What is the number of men working in the marketing department?

Sol. Number of men working in the marketing department
$$= 3250 \times \frac{79.2°}{360°} \times \frac{3}{3+2} = 715 \times \frac{3}{5} = 143 \times 3 = 429$$

Ex. 2 What is the ratio of the number of women working in the HR department to the number of men working in the IT department?

Sol. Number of women working in the HR department $= 3250 \times \frac{36°}{360°} \times \frac{13}{13+12} = 169$

Number of men working in the IT department $= 3250 \times \frac{57.6°}{360°} \times \frac{7}{7+3} = 364$

∴ Required ratio $= \frac{169}{364} = \frac{13}{28} = 13 : 28$

Ex. 3 The number of men working in the production department of the organisation is what per cent of the total number of employees working in that department?

Sol. Total number of employees in production department $= 3250 \times \frac{136.8}{360} = 1235$

Number of men working in the production department $= 1235 \times \frac{4}{4+1} = 988$

∴ Required percentage $= \frac{988}{1235} \times 100\% = 80\%$

Radar Chart/Graph

Radar chart, sometimes known as spider chart, is a two-dimensional chart/graph designed to plot one or more series of values over multiple common quantitative variables by providing an axis for each variable, arranged radially as equiangular spokes around a central point. In this diagram, every value is represented with respect to a central point. All the changes in the values are expressed in the form of distance from the centre point.

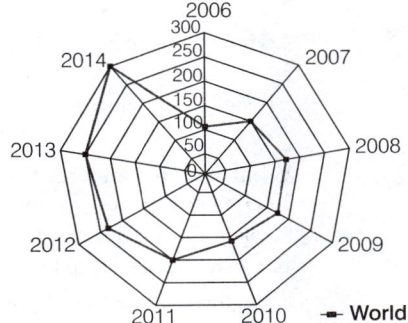

The given radar graph shows the trade growth (in $ billion) of world in comparison to the previous year. There is a centre in each of the radar graphs.

The magnitude of the centre of the radar graph is zero like in the origin of bar graph or line graph. Each polygonal ring denotes the magnitude of data in each of the lines/sectors like in the bar graphs or notations of line chart.

For example In the above radar graph, world trade growth from 2005 to 2006 is $ 100 billion. Similarly, world trade growth from 2013 to 2014 is $ 50 billion.

Directions (Examples 4-6) *Study the radar graph carefully and answer the questions that follows.*

Monthly salary (in thousand) of five different persons in three different years

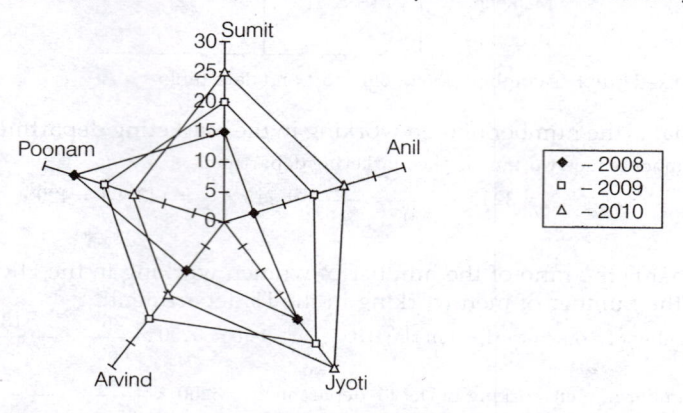

◆ – 2008
□ – 2009
△ – 2010

Ex. 4 What was the average monthly salary of Sumit in the year 2008, Anil in the year 2009 and Jyoti in the year 2010 together?

Sol. Monthly salary of Sumit in the year 2008 = ₹ 15000, monthly salary of Anil in the year 2009 = ₹ 15000
and monthly salary of Jyoti in the year 2010 = ₹ 30000

∴ Required average monthly salary = $\dfrac{15000 + 15000 + 30000}{3}$ = ₹ 20000

Ex. 5 Total monthly salary of Arvind in all the years together was what per cent of the total monthly salary of all the five persons together in the year 2008?

Sol. Total monthly salary of all the persons in 2008
= 15000 + 5000 + 20000 + 10000 + 25000
= ₹ 75000

Arvind's total monthly salary in all the years together = 10000 + 15000 + 20000
= ₹ 45000

∴ Required percentage = $\dfrac{45000}{75000}$ × 100% = 60%

Ex. 6 What was the per cent decrease in monthly salary of Poonam in the year 2009 as compared to her monthly salary in previous year?

Sol. Poonam's monthly salary in the year 2008 = ₹ 25000
In the year 2009 = ₹ 20000

∴ Required percentage decrease = $\dfrac{25000 - 20000}{25000}$ × 100% = 20%

Fast Track Practice

Exercise 1 Base Level Questions

Directions (Q. Nos. 1-4) *Refer to pie chart and answer the given questions.*
[SBI Clerk (Mains) 2016]

Five different companies A, B, C, D and E make two items I and II. The total number of items produced by these five companies is 80000. The cost of production of each item is ₹ 5000. The distribution of the total production by these companies is given in the following pie chart and the table shows the ratio of production of item I to that of item II and the percentage profit earned by these companies on each of these items.

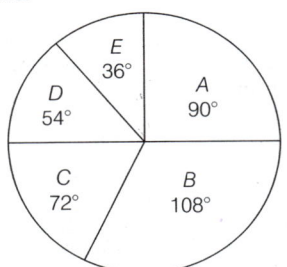

[Pie chart: A 90°, B 108°, C 72°, D 54°, E 36°]

Company	Ratio of production Item I	Ratio of production Item II	Profit earned (in %) Item I	Profit earned (in %) Item II
A	2	3	20	15
B	1	2	25	30
C	2	3	10	12
D	3	2	15	25
E	4	1	30	24

1. What is the profit earned by company C on item II?
   (a) ₹ 57.6 lakh   (b) ₹ 55.4 lakh
   (c) ₹ 56.8 lakh   (d) ₹ 54 lakh
   (e) None of these

2. What is the total cost of production of item I by companies A and B together?
   (a) ₹ 5 crore   (b) ₹ 6 crore
   (c) ₹ 8 crore   (d) ₹ 9 crore
   (e) None of these

3. What is the total of the profit earned by company E on production of item I and the profit of company D on production of item II?
   (a) ₹ 1.56 crore   (b) ₹ 2.2 crore
   (c) ₹ 1.3 crore   (d) ₹ 2.6 crore
   (e) None of these

4. What is the ratio of the cost of production of item II by company A to the cost of production of item I by company E?
   (a) 17 : 12   (b) 4 : 5
   (c) 7 : 4   (d) 15 : 8
   (e) 1 : 2

Directions (Q. Nos. 5-9) *Study the pie chart and the table and answer the given questions.*
[IBPS PO 2015]

Distribution of total number of shirts (linen and cotton) **sold by six different stores in 2003**

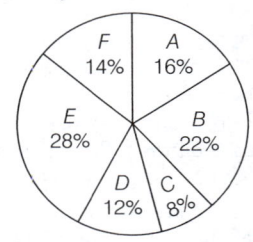

[Pie chart: A 16%, B 22%, C 8%, D 12%, E 28%, F 14%]

Total number = 84000

Store	Respective ratio of number of linen shirts to cotton shirts sold
A	7 : 5
B	5 : 6
C	3 : 2
D	5 : 3
E	4 : 3
F	7 : 3

5. What is the difference between average number of linen shirts sold by stores D and E together and average number of cotton shirts sold by the same stores together?
   (a) 2920   (b) 2880
   (c) 2940   (d) 3140
   (e) 3060

6. What is the respective ratio between number of shirts (linen and cotton) sold by store C and number of linen shirts sold by store F?
   (a) 22 : 31
   (b) 30 : 41
   (c) 40 : 49
   (d) 20 : 29
   (e) 44 : 57

7. Total number of cotton shirts sold by stores A and B together is what per cent of the number of shirts (linen and cotton) sold by store E?
   (a) $62\frac{1}{3}\%$
   (b) $64\frac{1}{3}\%$
   (c) $61\frac{2}{3}\%$
   (d) $68\frac{1}{2}\%$
   (e) $66\frac{2}{3}\%$

8. What is the central angle corresponding to the number of shirts (linen and cotton) sold by store E?
   (a) 100.8°
   (b) 96.4°
   (c) 104.2°
   (d) 98.8°
   (e) 102.6°

9. Number of shirts (linen and cotton) sold by store D is what per cent more than the number of linen shirts sold by store B?
   (a) 18%
   (b) 22%
   (c) 16%
   (d) 24%
   (e) 20%

Directions (Q. Nos. 10-14) *Answer the questions based on the following information.*

The first table gives the percentage of students in MBA class, who sought employment in the areas of finance, marketing and software. The second table gives the average starting salaries of the students per month (in ₹) in these areas. The third table gives the number of students who passed out in each year.

Percentage of students

Years	Finance	Marketing	Software	Others
1992	22	36	19	23
1993	17	48	23	12
1994	23	43	21	13
1995	19	37	16	28
1996	32	32	20	16

Average of starting salary (in ₹)

Year	Finance	Marketing	Software
1992	5450	5170	5290
1993	6380	6390	6440
1994	7550	7630	7050
1995	8920	8960	8160
1996	9810	10220	8640

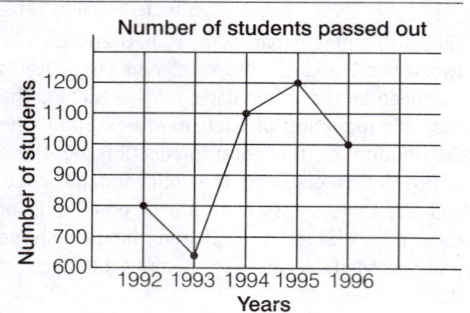

Number of students passed out

10. The number of students, who get jobs in finance is less than the students getting marketing jobs, in the 5 yr by
    (a) 826
    (b) 650
    (c) 750
    (d) 548
    (e) None of these

11. What is the percentage increase in the average salary of finance from 1992-96?
    (a) 60%
    (b) 32%
    (c) 96%
    (d) 80%
    (e) None of these

12. The average annual rate at which the initial salary offered in software increase, is
    (a) 21%
    (b) 33%
    (c) 15.9%
    (d) 65%
    (e) None of these

13. What is the percentage increase in the average monthly salary offered to a marketing student over the given 5 yr?
    (a) 98%
    (b) 117%
    (c) 56%
    (d) 80%
    (e) None of these

14. In 1994, students seeking jobs in finance earned more than those opting for software (per annum) by
    (a) ₹ 43 lakh
    (b) ₹ 33.8 lakh
    (c) ₹ 28.4 lakh
    (d) ₹ 38.8 lakh
    (e) None of these

Mixed Graph / 835

Directions (Q. Nos. 15-19) *Study the given graphs to answer these questions.* **[MAT 2015]**

The pie chart shows sources of income for an NGO. The total income is ₹ 40 crore. The bar chart gives the expenditure incurred on various items A-Food for poor, B-Education to illiterate, C-Mid-day meal programme, D-General expenses, E-Eye-camp expenses, F-Integrated street children programme. Total expenditure = ₹ 39 crore.

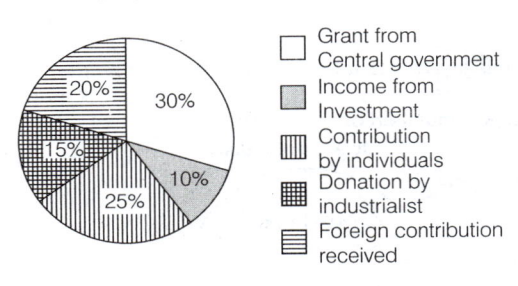

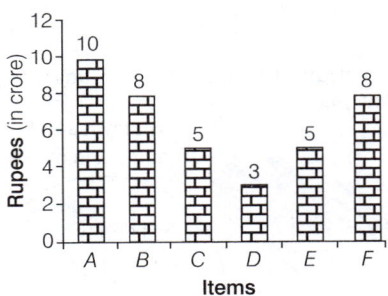

15. What percentage of money is saved?
    (a) 3.5%   (b) 3.0%
    (c) 2.5%   (d) 4.0%

16. If the industrialist stops donation and the expenditure pattern remains the same, then what will be the decrease in money spent for mid-day meal programme?
    (a) ₹ 1.55 crore   (b) ₹ 1.95 crore
    (c) ₹ 0.50 crore   (d) ₹ 0.77 crore

17. What is the ratio of expenditure on food for poor and mid-day meal programmes together to that of grant from central government?
    (a) 7 : 6   (b) 6 : 7
    (c) 5 : 4   (d) 4 : 5

18. The general expenses is how many times 'Income from investment'?
    (a) 0.75   (b) 0.57   (c) 0.65   (d) 0.58

19. Suppose in the next year, grant from central government increases by 10%, foreign contribution decreases by 10% and other income amounts remain same. If the expense pattern, remains same, what is the per cent increase in 'Food for poor' sector?
    (a) 5%   (b) 1%
    (c) 4%   (d) 2%

Exercise 2 — Higher Skill Level Questions

Directions (Q. Nos. 1-5) *Study the radar graph carefully and answer the questions that follow.* **[SBI PO 2013]**

Number of students (in thousand) **in two different universities in six different years**

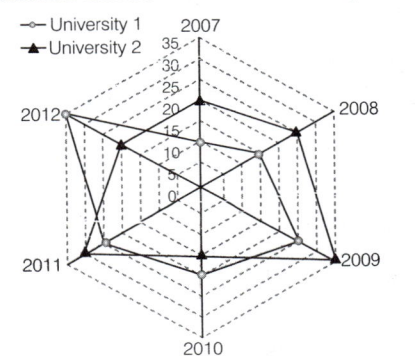

1. What was the difference between the number of students in University 1 in the year 2010 and the number of students in University 2 in the year 2012?
   (a) 0   (b) 5000   (c) 15000   (d) 10000
   (e) 1000

2. What is the sum of the number of students in University 1 in the year 2007 and the number of students in University 2 in the year 2011 together?
   (a) 50000   (b) 55000
   (c) 45000   (d) 57000
   (e) 40000

3. If 25% of the students in University 2 in the year 2010 were females, what was the number of male students in the University 2 in the same year?
   (a) 11250   (b) 12350   (c) 12500   (d) 11500
   (e) 11750

836 / Fast Track Objective Arithmetic

4. What was the per cent increase in the number of students in University 1 in the year 2011 as compared to the previous year?
   (a) 135   (b) 15   (c) 115   (d) 25
   (e) 35

5. In which year was the difference between the number of students in University 1 and the number of students in University 2 highest?
   (a) 2008   (b) 2009   (c) 2010   (d) 2011
   (e) 2012

Directions (Q. Nos. 6-10) *Study the following pie charts and table to answer the questions that follow.* [RBI 2009]

Statewise details of adult population of a country

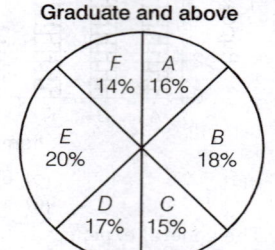

Graduate and above
- A 16%
- B 18%
- C 15%
- D 17%
- E 20%
- F 14%

Total number = 24 lakh

Upto XII std. pass
- A 15%
- B 16%
- C 18%
- D 12%
- E 19%
- F 20%

Total number = 32 lakh

M : F Ratio [M = Male and F = Female]

State	Graduate and above M : F	Upto XII Std. pass M : F
A	7 : 5	7 : 9
B	5 : 3	3 : 5
C	5 : 4	4 : 5
D	9 : 8	5 : 7
E	9 : 7	9 : 10
F	4 : 3	3 : 2

6. What is the difference between graduate male population and XII Std. male population from State A?
   (a) 24000   (b) 14000
   (c) 28000   (d) 36000
   (e) None of these

7. What is the ratio of graduate female population of State E to XII Std. female population of State D, respectively?
   (a) 7 : 5   (b) 5 : 7
   (c) 16 : 15   (d) 15 : 16
   (e) None of these

8. Graduate female population of State C is what per cent of the XII Std. female population of that state?
   (a) 40%   (b) 62.5%   (c) 50%   (d) 52.5%
   (e) None of these

9. Class XII pass male population of State C is what per cent of the total XII Std. population of all the states together?
   (a) 8%   (b) 12%   (c) 11%   (d) 9%
   (e) None of these

10. What is the ratio of graduate male population of State E to XII Std. female population of that state?
    (a) 28 : 35   (b) 35 : 28
    (c) 32 : 45   (d) 45 : 32
    (e) None of these

Directions (Q. Nos. 11-15) *Read the following graph and answer the questions below.* [NICL AO (Pre) 2017]

Percentage expenditure of various company
- A 30%
- B 18%
- C 9%
- D 11%
- E 17%
- F 15%

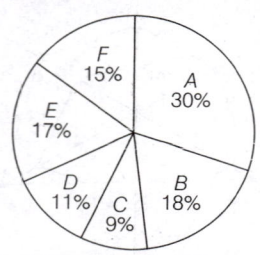

Total expenditure = ₹ 40 crore

Mixed Graph

Company	Profit
A	40
B	25
C	50
D	60
E	70
F	80

✦ Profit = $\dfrac{\text{Income} - \text{Expenditure}}{\text{Expenditure}} \times 100$

11. What is the income of company A?
(a) ₹ 21.6 crore
(b) ₹ 16.8 crore
(c) ₹ 24.4 crore
(d) ₹ 27.5 crore
(e) Other than those given as options

12. What is the sum of net profit of companies C, D and F?
(a) ₹ 12 crore   (b) ₹ 10.48 crore
(c) ₹ 12.46 crore   (d) ₹ 9.24 crore
(e) ₹ 8.64 crore

13. Find the difference between the total income and the total expenditure of all the six companies?
(a) ₹ 29.5 crore
(b) ₹ 25.8 crore
(c) ₹ 27.2 crore
(d) ₹ 20.6 crore
(e) Other than those given as options

14. Profit of company E is what percentage of the income of company C?
(a) 120%
(b) 88.14%
(c) 900%
(d) 40%
(e) Other than those given as options

15. Income of company B is how much percentage more than the expenditure of company C?
(a) 125%   (b) 150%
(c) 30%   (d) 14%
(e) Other than those given as options

Answer with Solutions

Exercise ❶ Base Level Questions

1. (a) Number of items produced by company
$C = \left[\dfrac{72}{360}\right] \times 80000 = X$

Cost of production $= X \times 5000 = Y$

Cost of production of item II $= \left[\dfrac{3}{5}\right] \times Y = Z$

Per cent profit earned on item II

$= 12\%$ of $Z = \left[\dfrac{12}{100}\right] \times Z$

$= \left[\dfrac{3}{5}\right] \times \left[\dfrac{12}{100}\right] \times Y$

$= \dfrac{3}{5} \times \dfrac{12}{100} \times X \times 5000$

$= \dfrac{3}{5} \times \dfrac{12}{100} \times \dfrac{72}{360} \times 80000 \times 5000$

$= 5760000 = 57.6$ lakh

2. (c) Cost of production of item I by company A
$= \left[\dfrac{90}{360}\right] \times 80000 \times 5000 \times \dfrac{2}{5}$

Cost of production of item I by company B
$= \left[\dfrac{108}{360}\right] \times 80000 \times \left[\dfrac{1}{3}\right] \times 5000$

∴ Total cost $= 80000 \times 5000$
$\times \left[\left(\dfrac{90}{360}\right) \times \left(\dfrac{2}{5}\right) + \left(\dfrac{108}{360}\right) \times \left(\dfrac{1}{3}\right)\right]$

$= 80000 \times 5000 \times \left[\dfrac{1}{5}\right] = 8$ crore

3. (a) Total cost

$= 80000 \times 5000 \times \left[\left(\dfrac{36}{360}\right) \times \left(\dfrac{4}{5}\right)\right.$

$\left. \times \left(\dfrac{30}{100}\right) + \left(\dfrac{54}{360}\right) \times \left(\dfrac{2}{5}\right) \times \left(\dfrac{25}{100}\right)\right]$

$= 1.56$ crore

4. (d) Required ratio

$$= \left[\frac{90}{360}\right] \times \left[\frac{3}{5}\right] : \left[\frac{36}{360}\right] \times \left[\frac{4}{5}\right]$$

$= 90 \times 3 : 36 \times 4$

$= 30 : 16 = 15 : 8$

5. (c) Total shirts sold by store D

$$= \frac{12 \times 84000}{100} = 10080$$

Number of linen shirts sold by store D

$$= \frac{5}{8} \times 10080 = 6300$$

Number of cotton shirts sold by store D

$= 10080 - 6300 = 3780$

Total shirts sold by store E

$$= \frac{28 \times 84000}{100} = 23520$$

Number of linen shirts sold by store E

$$= \frac{4}{7} \times 23520 = 13440$$

Number of cotton shirts sold by store E

$= 23520 - 13440 = 10080$

Now, average number of linen shirts sold by stores D and E

$$= \frac{6300 + 13440}{2} = 9870$$

and average number of cotton shirts sold by stores D and E

$$= \frac{3780 + 10080}{2} = 6930$$

∴ Required difference

$= 9870 - 6930 = 2940$

6. (c) Total number of shirts sold by store C

$= 8\%$ of 84000

$$= \frac{8}{100} \times 84000 = 6720$$

Number of linen shirts sold by store F

$$= \frac{7}{10} \text{ of } (14\% \text{ of } 84000)$$

$$= \frac{7}{10} \times 14 \times 840$$

$= 8232$

∴ Required ratio $= 6720 : 8232 = 40 : 49$

7. (e) Total number of cotton shirts sold by store A

$$= \frac{5}{12} \text{ of } (16\% \text{ of } 84000)$$

$$= \frac{5}{12} \times \frac{16}{100} \times 84000 = 5600$$

Total number of cotton shirts sold by store B

$$= \frac{6}{11} \text{ of } (22\% \text{ of } 84000)$$

$$= \frac{6}{11} \times \frac{22}{100} \times 84000 = 10080$$

Total number of cotton shirts sold by stores A and $B = 5600 + 10080 = 15680$

Total number of shirts (linen and cotton) sold by store $E = 28\%$ of $84000 = 23520$

∴ Required percentage

$$= \frac{15680}{23520} \times 100 = 66\frac{2}{3}\%$$

8. (a) Central angle corresponding to the number of shirts sold by store E

$$= 28\% \text{ of } 360° = \frac{28}{100} \times 360°$$

$= 100.8°$

9. (e) Total number of shirts sold by store D

$= 12\%$ of $84000 = 10080$

Number of linen shirts sold by store B

$$= \frac{5}{11} \text{ of } (22\% \text{ of } 84000)$$

$$= \frac{5}{11} \times \frac{22}{100} \times 84000 = 8400$$

∴ Required percentage

$$= \frac{10080 - 8400}{8400} \times 100\% = 20\%$$

10. (c)

Years	Number of students passed	Number of students employed from finance	Number of students employed from marketing
1992	800	$0.22 \times 800 = 176$	$0.36 \times 800 = 288$
1993	650	$0.17 \times 650 = 110.5$	$0.48 \times 650 = 312$
1994	1100	$0.23 \times 1100 = 253$	$0.43 \times 1100 = 473$
1995	1200	$0.19 \times 1200 = 228$	$0.37 \times 1200 = 444$
1996	1000	$0.32 \times 1000 = 320$	$0.32 \times 1000 = 320$
Total		1087.50	1837

∴ Required difference $= 1837 - 1087.5$
$= 749.5 \approx 750$

Mixed Graph / 839

11. (d) Average salary of finance in 1992
= ₹ 5450
Average salary of finance in 1996 = ₹9810
∴ Required percentage increase
$= \frac{9810 - 5450}{5450} \times 100\%$
$= \frac{4360}{5450} \times 100\% = 80\%$

12. (c) Salary offered in software
in 1992 = ₹5290 and in 1996 = ₹8640
∴ Percentage increase
$= \frac{8640 - 5290}{5290} \times 100\%$
$= \frac{3350}{5290} \times 100 = 63.32\%$
Thus, required average annual increase
rate $= \frac{1}{4} \times 63.32 = 15.83\% \approx 15.9\%$

13. (a) Average monthly salary to a marketing student
in 1992 = ₹ 5170 and in 1996 = ₹ 10220
∴ Required percentage increase
$= \frac{10220 - 5170}{5170} \times 100\%$
$= \frac{5050}{5170} \times 100\% \approx 98\%$

14. (b) In 1994, students seeking jobs in finance earned
= 23% of 1100 × 7550
= ₹1910150
Students seeking jobs in software earned
= 21% of 1100 × 7050
= ₹1628550
∴ Difference in the amount earned
= 1910150 − 1628550
= ₹281600
≈ ₹ 2.81 lakh per month
= ₹2.81 × 12 lakh per annum
≈ ₹33.8 lakh per annum

15. (c) We know that, total income = ₹40 crore
Total expenditure = ₹39 crore
Then,
Saving = Total income − Expenditure
= 40 − 39 = ₹1 crore
Percentage of saving $= \frac{1}{40} \times 100\% = 2.5\%$

16. (d) Decrease in income = 15% of ₹ 40 crore
∴ New income $= \frac{85}{100} \times 40 = ₹34$ crore
Percentage of expenditure on mid-day meal
$= \frac{5}{39} \times 100\% = 12.8\%$

Now, new expenditure after gradual decrease in expenditure with respect to
income $= \frac{39}{40} \times 34 = ₹33.15$ crore
∴ Expenditure on mid-day meal
$= \frac{12.8}{100} \times 33.15 = ₹ 4.24$ crore
∴ Decrease in expenditure on mid-day meal
= 5 − 4.24
= ₹ 0.76 crore
≈ ₹ 0.77 crore

17. (c) Expenditure on food for poor
= ₹10 crore
Expenditure for mid-day meal programme
= ₹5 crore
Total expense = (10 + 5) crore = ₹15 crore
Grant from central government
$= \frac{30 \times 40}{100} = ₹ 12$ crore
∴ Required ratio = 15 : 12 = 5 : 4

18. (a) General expenses = ₹ 3 crore
Income from investment $= \frac{10 \times 40}{100}$
= ₹ 4 crore
Suppose, general expenses is x times of income from investment, then
$3 = 4 \times x \Rightarrow x = \frac{3}{4}$
⇒ $x = 0.75$ times

19. (b) Initially, grant from central government
= 30% of 40 crore $= \frac{30 \times 40}{100} = ₹ 12$ crore
After 10% increase, it will become
$= \frac{12 \times 110}{100} = ₹ 13.2$ crore
Similarly, foreign contribution decreases 10%, then it will become
$= \frac{8 \times 90}{100} = ₹ 7.2$ crore
Total increase in donation
= (13.2 + 7.2) − (12 + 8)
= 20.4 − 20 = ₹ 0.4 crore
∴ Gradual increase in expenditure
$= \frac{39}{40} \times 40.4 = ₹ 39.39$ crore
≈ ₹ 39.40 crore (approx.)
Gradual increase in A
$= \frac{10}{39} \times 39.4 = ₹ 10.10$ crore
∴ Per cent increase $= \frac{10.10 - 10}{10} \times 100\%$
$= \frac{0.10}{10} \times 100\% = 1\%$

840 / Fast Track Objective Arithmetic

Exercise 2 Higher Skill Level Questions

1. (a) Number of students in University 1 in
2010 = 20000
Number of students in University 2 in
2012 = 20000
∴ Required difference = 20000 − 20000 = 0

2. (e) Number of students in University 1 in
2007 = 10000
Number of students in University 2 in 2011
= 30000
∴ Required sum
= 10000 + 30000 = 40000

3. (a) Total students in University 2 in 2010
= 15000
∴ Number of females
$= 15000 \times \dfrac{25}{100} = 3750$
and number of males in university in 2010
= 15000 − 3750 = 11250

4. (d) Number of students in University 1 in
2011 = 25000 and number of students in
University 1 in 2010 = 20000
∴ Required percentage
$= \dfrac{25000 - 20000}{20000} \times 100$
$= \dfrac{5000}{20000} \times 100 = 25\%$

5. (e) Difference in number of students of
University 1 and University 2 in 2007
= 20000 − 10000 = 10000
Difference in number of students of
University 1 and University 2 in 2008
= 25000 − 15000 = 10000
Difference in number of students of
University 1 and University 2 in 2009
= 35000 − 25000 = 10000
Difference in number of students of
University 1 and University 2 in 2010
= 20000 − 15000 = 5000
Difference in number of students of
University 1 and University 2 in 2011
= 30000 − 25000 = 5000
Difference in number of students of
University 1 and University 2 in 2012
= 35000 − 20000 = 15000
It is clear from above that the difference between the number of students in University 1 and the number of students in University 2 is the highest in the year 2012.

6. (b) Graduate male population of State A
$= \left(24 \times \dfrac{16}{100} \times \dfrac{7}{12}\right)$ lakh = 2.24 lakh

Now, XII Std. male population of State A
$= \left(32 \times \dfrac{15}{100} \times \dfrac{7}{16}\right)$ lakh = 2.1 lakh
∴ Required difference
= (2.24 − 2.1) lakh = 14000

7. (d) Graduate female population of State E
$= 24 \times \dfrac{20}{100} \times \dfrac{7}{16} = 2.1$ lakh
XII Std. female population of State D
$= 32 \times \dfrac{12}{100} \times \dfrac{7}{12} = 2.24$ lakh
∴ Required ratio = 2.1 : 2.24
= 210 : 224 = 15 : 16

8. (c) Graduate female population of State C
$= 24 \times \dfrac{15}{100} \times \dfrac{4}{9} = 1.6$ lakh
XII Std. female population of State C
$= 32 \times \dfrac{18}{100} \times \dfrac{5}{9} = 3.2$ lakh
∴ Required percentage $= \dfrac{1.6}{3.2} \times 100\% = 50\%$

9. (a) XII Std. pass male population of State C
$= 32 \times \dfrac{18}{100} \times \dfrac{4}{9} = 2.56$ lakh
∴ Required percentage $= \dfrac{2.56}{32} \times 100 = 8\%$

10. (e) Graduate male population of State E
$= 24 \times \dfrac{20}{100} \times \dfrac{9}{16} = 2.7$ lakh
XII Std. female population of State E
$= 32 \times \dfrac{19}{100} \times \dfrac{10}{19} = 3.2$ lakh
∴ Required ratio = 27 : 32

11. (b) Expenditure of company A
$= \dfrac{40 \times 30}{100} = ₹ 12$ crore
∴ $P\% = \dfrac{\text{Income} - \text{Expenditure}}{\text{Expenditure}} \times 100$
⇒ $40 = \dfrac{\text{Income} - 12}{12} \times 100$
⇒ 100 × Income − 1200 = 480
⇒ 100 × Income = 480 + 1200 = 1680
⇒ Income $= \dfrac{1680}{100}$
= ₹ 16.8 crore

12. (d) Expenditure of company C
$= 40 \times \dfrac{9}{100} = ₹ 3.6$ crore
Expenditure of company D
$= 40 \times \dfrac{11}{100} = ₹ 4.4$ crore

Expenditure of company F
$$= 40 \times \frac{15}{100} = ₹\ 6 \text{ crore}$$
Now, income of company C,
$$50 = \left(\frac{\text{Income} - 3.6}{3.6}\right) \times 100$$
$\Rightarrow \quad 180 = \text{Income} \times 100 - 360$
$\Rightarrow \quad \text{Income} = \dfrac{540}{100} = ₹\ 5.4 \text{ crore}$
$\Rightarrow \quad C\text{'s Income} = ₹\ 5.4 \text{ crore}$
Now, income of company D,
$$60 = \left(\frac{\text{Income} - 4.4}{4.4}\right) \times 100$$
$\Rightarrow \quad 264 = \text{Income} \times 100 - 440$
$\Rightarrow \quad \text{Income} = \dfrac{704}{100} = ₹\ 7.04 \text{ crore}$
$\Rightarrow \quad D\text{'s Income} = ₹\ 7.04 \text{ crore}$
Now, income of company F,
$$80 = \left(\frac{\text{Income} - 6}{6}\right) \times 100$$
$\Rightarrow \quad 480 = \text{Income} \times 100 - 600$
$\Rightarrow \quad \text{Income} = \dfrac{1080}{100} = ₹\ 10.8 \text{ crore}$
$\Rightarrow \quad F\text{'s Income} = ₹\ 10.8 \text{ crore}$
$\therefore$ Net profit of company C
  = Income − Expenditure
  = 5.4 − 3.6 = ₹ 1.8 crore
Net profit of company D
  = Income − Expenditure
  = 7.04 − 4.4 = ₹ 2.64 crore

Net profit of company F
  = Income − Expenditure
  = 10.8 − 6 = ₹ 4.8 crore
$\therefore$ Sum of net profit of company C, D and F
  = ₹ (1.8 + 2.64 + 4.8) = ₹ 9.24 crore

13. (d) $\because$ Incomes made by company $A, B, C, D,$ E and F are 16.8 crore, 9 crore, 5.4 crore, 7.04 crore, 11.56 crore and 10.8 crore, respectively.
$\therefore$ Total income by these 6 companies
  = ₹ 60.6 crore
and total expenditure = ₹ 40 crore
$\therefore$ Required difference = ₹ (60.6 − 40) crore
  = ₹ 20.6 crore

14. (b) Income of company E = ₹ 11.56 crore
and expenditure of company E = ₹ 6.8 crore
$\therefore$ Profit of company E
  = 11.56 − 6.8 = ₹ 4.76 crore
and income of company C = ₹ 5.4 crore
$\therefore$ Required percentage = $\dfrac{4.76}{5.4} \times 100$
  $= \dfrac{476}{540} \times 100$
  $= \dfrac{4760}{54} = 88.14\%$

15. (b) Income of company B = ₹ 9 crore
and Expenditure of company C = ₹ 3.6 crore
$\therefore$ Required percentage
  $= \dfrac{9 - 3.6}{3.6} \times 100 = 150\%$

Chapter 47

Data Sufficiency

Analysis of given data to reach conclusion is known as **Data Sufficiency**. Given data may or may not be sufficient for a definite conclusion. You have to decide whether the problem can be solved by using the information from the given statements combined or individually or it cannot be answered using these statements. These type of questions basically check the analytical ability of a candidate and his basic knowledge in different topics.

Data sufficiency questions are not new topics in themselves. Hence, to solve the questions on Data Sufficiency, basic knowledge of algebra, geometry, arithmetic and statistics is prerequisite. The problems based on this topic consist of a mathematical or logical problem followed by two or more than two statements containing the information related to it.

Directions (Examples 1-3) *Each of the question below consists of a question and two statements number I and II given below it. You have to decide whether the data provided in the statements are sufficient to answer the question. Read both the statements and give answer.* **[IBPS SO 2016]**

(a) If the data in Statement I alone is sufficient to answer the question, while the data in Statement II alone is not sufficient to answer the question

(b) If the data in Statement II alone is sufficient to answer the question, while the data in Statement I alone is not sufficient to answer the question

(c) If the data in Statement I alone or in Statement II alone is sufficient to answer the question

(d) If the data in both the Statements I and II are not sufficient to answer the question

(e) If the data in both the Statements I and II together are necessary to answer the question

Data Sufficiency

Ex. 1 What is Rasika's present age?
  I. Rasika's age four years hence will be three times Manisha's age that time.
  II. Rasika's age two years ago was five times Manisha's age that time.

Sol. (e) Let present age of Rasika = R and present age of Manisha = M

From both statements, $(R + 4) = 3(M + 4) \Rightarrow R + 4 = 3M + 12$
$$\Rightarrow R - 3M = 8 \quad \ldots(i)$$
and $(R - 2) = 5(M - 2) \Rightarrow R - 2 = 5M - 10 \Rightarrow R - 5M = -8 \quad \ldots(ii)$

On solving Eqs. (i) and (ii), we get, $R = 32$ and $M = 8$
$\therefore$ Present age of Rasika = 32 yr
Hence, the data in both the Statements I and II together are necessary to answer the question.

Ex. 2 What is the area of the square?
  I. Area of the largest circle that can be inscribed in the given square is 616 cm^2.
  II. Area of the smallest circle in which the given square can be inscribed is 1212 cm^2.

Sol. (c) Let side of square = a cm

We know that, area of the largest circle that can be inscribed in the given square is $\dfrac{\pi a^2}{4}$.

From Statement I,

Area of the largest circle inscribed = $\dfrac{\pi a^2}{4} \Rightarrow 616 = \dfrac{a^2 \times \pi}{4} \Rightarrow \dfrac{a^2 \times 22}{4 \times 7} = 616$

$\Rightarrow a^2 = \dfrac{616 \times 4 \times 7}{22} \Rightarrow a^2 = 56 \times 7 \times 2 \Rightarrow a^2 = 784$

Now, area of square = $a^2 = 784$ cm^2

From Statement II,

Area of smallest circle = $\pi \left(\dfrac{\sqrt{2}a}{2}\right)^2 \Rightarrow 1212 = \dfrac{2a^2}{4} \times \dfrac{22}{7} \Rightarrow a^2 = \dfrac{1212 \times 2 \times 7}{22}$

$\Rightarrow a^2 = 771.27$

$\therefore$ Area of square = $a^2 = 771.27$ cm^2

Hence, the data given in Statement I alone or data in Statement II alone are sufficient to answer to the question.

Ex. 3 What is the cost of painting four walls of the rectangular hall wall at ₹ 135 per m^2, the hall has a door measuring 3.5 m × 1.5 m and no windows?
  I. Perimeter of the floor of the hall is equal to the perimeter of a square field having side 12 m. Length and breadth of the hall are in the ratio of 5 : 1, respectively.
  II. Perimeter of a smaller wall is 15 m.

Sol. (e) From Statement I, let the length of hall = $5x$ m and breadth of hall = x m
$\therefore$ Perimeter of hall = 2 (Length + Breadth)
$\Rightarrow 12 \times 4 = 2(5x + x) \Rightarrow 6x = 24 \Rightarrow x = 4$
$\therefore$ Length = $5x = 5 \times 4 = 20$ m and breadth = $x = 4$ m

From Statement II, perimeter of smaller wall = 15 m
$\Rightarrow 2(h + b) = 15 \Rightarrow 2h + 8 = 15 \Rightarrow h = 3.5$ m

Now, area of four walls of the hall = 2 (Length + Breadth) × Height
$= 2(20 + 4) \times 3.5 = 168$ m^2

∴ Required area of four walls = Total area − Area of door = 168 − (3.5 × 1.5)
= 168 − 5.25 = 162.75 m²

∴ Cost of painting = 135 × 162.75 = ₹ 21971.25

Hence, data given in both the Statements I and II together are necessary to answer the question.

Directions (Examples 4-6) *Each of these questions is followed by information in Statements I, II and III. You have to study the questions and statements and decide which of the statement(s) is/are necessary to answer the questions.*

Ex. 4 What is the capacity of the cylindrical tank?
I. Radius of the base is half of its height.
II. Area of the base is 616 sq m.
III. Height of the cylinder is 28 m.
(a) I and II   (b) II and III   (c) I and III   (d) All I, II and III
(e) Any two of the three

Sol. (e) To know the capacity, we have to find the volume of the cylinder, i.e. $\pi r^2 h$.
For this, any two of the three are enough.

e.g. take Statements I and III, $h = 28$ m, then $r = \dfrac{28}{2} = 14$ m

Then, capacity of cylindrical tank $= \pi r^2 h = \dfrac{22}{7} \times 14 \times 14 \times 28 = (22 \times 28 \times 28)$ m³

From Statements II and III, area of base $= \pi r^2 = 616$ sq m, $h = 28$
∵ Capacity = Area of base × h = $\pi r^2 h = (616 \times 28)$ m³

From Statements I and II, capacity $= \left(616 \times 2 \times \sqrt{\dfrac{616}{\pi}}\right)$ m³

Ex. 5 What is staff strength of company X?
I. Male and female employees are in the ratio of 2 : 3, respectively.
II. Of the officer employees, 80% are males.
III. Total number of officers is 132.
(a) I and III   (b) II and either III or I
(c) All I, II and III   (d) Any two of the three
(e) Data is insufficient

Sol. (e) Data is insufficient because combinely all statements fail to provide sufficient information required to answer.

Ex. 6 What is the two-digit number?
I. Number obtained by interchanging the digits is more than the original number by 9.
II. Sum of the digits is 7.
III. Difference between the digits is 1.
(a) I and III   (b) I and II   (c) II and III   (d) II and either I or III
(e) Any two of I, II and III

Sol. (d) Let unit's and ten's digits be x and y, respectively.
Then, original number $= 10y + x$
From Statement I, $(10x + y) - (10y + x) = 9 \Rightarrow x - y = 1$
From Statement II, $x + y = 7$
From Statement III, $x - y = 1$

∴ Number can be found using Statements II and either I or III. As two number of variables require two equation for knowing unknowns.

Fast Track Practice

Directions (Q. Nos. 1-4) *Each of the questions below consists of two statements numbered I and II given below it. You have to decide whether the data provided in the statements are sufficient to answer the questions. Read both the statements and give the answer.* **[OBC PO 2010]**

(a) If the data in Statement I alone is sufficient to answer the question while the data in Statement II alone is not sufficient to answer the question

(b) If the data in Statement II alone is sufficient to answer the question while the data in Statement I alone is not sufficient to answer the question

(c) If the data in Statement I alone or in Statement II alone is sufficient to answer the question

(d) If the data in both the Statements I and II are not sufficient to answer the question

(e) If the data in both the Statements I and II together are necessary to answer the question

1. What is the exact average of n, 35, 39, 42, p and w?
   I. n is six more than w.
   II. w is four less than p.

2. What was the profit/loss per cent earned/incurred by selling an article for ₹ 24000?
   I. The ratio of the selling price to the cost price of the article is 5 : 3.
   II. The difference between the cost price and the selling price is ₹ 9600.

3. What will be the difference between two numbers?
   I. The square of the first number is 9 times the second number.
   II. The ratio of the first number to the second number is 3 : 4.

4. What is the ratio of two numbers x and y?
   I. 40% of x is 20% of 50.
   II. 30% of y is 25% of 72.

Directions (Q. Nos. 5-9) *Each of the following questions is followed by information in three statements. You have to find out which statement(s) is/are sufficient to answer the question and mark your answer accordingly.*

5. What is the average age of the six members A, B, C, D, E and F in the family?
   I. Average age of D and E is 14 yr.
   II. Average age of A, B, C and F is 50 yr.
   III. Average age of A, B, D and E is 40 yr.
   (a) I and II   (b) I and III
   (c) II and III   (d) I, II and III
   (e) None of these

6. What is the area of the right angled triangle?
   I. Base of the triangle is X cm.
   II. Height of the triangle is Y cm.
   III. Hypotenuse of the triangle is Z cm.
   (a) I and II   (b) Only II
   (c) II and III   (d) Any two of three
   (e) None of these

7. In how many days will B alone complete the work?
   I. A and B together can complete the work in 8 days.
   II. B and C together can complete the work in 10 days.
   III. A and C together can complete the work in 12 days.
   (a) I and II
   (b) II and III
   (c) I, II and III
   (d) Data is insufficient
   (e) None of the above

8. What is the rate of interest percentage per annum?
   I. An amount doubles itself at simple interest in 10 yr.
   II. Difference between the compound interest and simple interest on an amount of ₹ 15000 in 2 yr is ₹ 150.
   III. The compound interest calculated after 8 yr is more than the amount principle.
   (a) Only I   (b) Only II
   (c) II and III   (d) I and III
   (e) Either I or II

9. What are the marks scored by Abhijit in English?
   I. Marks scored by Abhijit in Mathematics are more than his marks in Science by 20.
   II. Total marks scored by Abhijit in Mathematics, Science and English are 197.

III. Marks scored by Abhijit in Science are more than his marks in English by 12.
(a) Any two of the three
(b) II and III
(c) I, II and III
(d) Data is insufficient
(e) None of the above

Directions (Q. Nos. 10-14) *In each of the given questions, one question and two statements numbered* I *and* II *are given. You have to decide whether the data given in both the statements are sufficient to answer the question or not. Read both the statements and give answer.* [NICL AO 2015]

(a) If the data in Statement I alone is sufficient to answer the question while the data in Statement II alone is not sufficient to answer the question
(b) If the data in Statement II alone is sufficient to answer the question while the data in Statement I alone is not sufficient to answer the question
(c) If the data either in Statement I or in Statement II alone is sufficient to answer the question
(d) If the data in both the statements together are not sufficient to answer the question
(e) If the data in both the Statements I and II together are sufficient to answer the question

10. How much time will train A take to completely cross train B running in opposite direction (towards each other)?
    I. The total length of train A and train B together is 777 m.
    II. Train B can cross an electric pole in 6 s. The respective ratio of speed of train A and train B is 7 : 8.

11. What is the volume (in m^3) of the cylinder?
    I. The radius of the cylinder is 21 m.
    II. The sum of the height and radius of the cylinder is 29 m. The curved surface area of the cylinder is 1056 m^2.

12. What was the percentage of discount given while selling table?
    I. The cost price of the table is ₹ 8000. 12% profit was earned by selling the table.
    II. If there were no discount, then earned profit would have been 18%.

13. In how many days 14 men can complete a piece of work?
    I. If 18 women can complete the same piece of work in 24 days.
    II. If 28 children can complete the same piece of work in 56 days.

14. What is the number?
    I. 25% of that number is one-fourth of that number.
    II. Two-third of that number is 226 less than that number.

Directions (Q. Nos. 15-17) *Each question below consists of statements numbered* I *and* II *given below it. You have to decide, whether the data provided in the statement are sufficient to answer the questions. Read both the statements and give answer.*
[SBI Clerk (Mains) 2016]

(a) If the data in Statement I alone is sufficient to answer the question, while the data in Statement II alone is not sufficient to answer the question
(b) If the data in Statement II alone is sufficient to answer the question, while the data in Statement I alone is not sufficient to answer the question
(c) If the data in Statement I alone or in Statement II alone is sufficient to answer the question
(d) If the data in both the Statements I and II are not sufficient to answer the question
(e) If the data in both the Statements I and II together are necessary to answer the question

15. How many workers are required for completing the construction work in 10 days?
    I. 20% of the work can be completed by 8 workers in 8 days.
    II. 20 workers can complete the work in 16 days.

16. What is the monthly salary of Praveen?
    I. Praveen gets 15% more than Sumit while Sumit gets 10% less than Lokesh.
    II. Lokesh's monthly salary is ₹ 2500.

17. What is the distance between city P and city Q?
    I. Two persons started simultaneously from P to Q, with their speeds in the ratio 4 : 5.
    II. B reaches P one hour earlier than A to Q. The difference between speeds of A and B is 20 km/h.

Data Sufficiency / 847

Directions (Q. Nos. 18-22) *Each of these questions consists of a question and two statements numbered I and II. You have to decide whether the data provided in the statements are sufficient to answer the question. Read both the statements and give answer.* **[MAT 2015]**

(a) If the data in Statement I alone is sufficient to answer the question while the data in Statement II alone is not sufficient to answer the question

(b) If the data in Statement II alone is sufficient to answer the question while the data in Statement I alone is not sufficient to answer the question

(c) If the data in both Statements I and II are not sufficient to answer the question

(d) If the data in both Statements I and II together are not sufficient to answer the question

18. There were 90 students to be interviewed for selection to a computer course. To finish the interview in one day, three panels X, Y and Z were formed. How many students are interviewed by Panel X?
  I. The three panels on an average interviewed 30 students.
  II. The number of students interviewed by Panel X is more by 4 than the students interviewed by Panel Z. The number of students interviewed by Panel Y is more by 2 than the students interviewed by Panel Z.

19. A, B, C and D are four numbers. What is the standard deviation of these four numbers?
  I. The sum of squares of A, B, C and D is 360.
  II. The sum of A, B, C and D is 36.

20. Mukesh wants to go to the roof top of a building using a ladder. The roof level is 120 m above ground level. How many steps are there in the ladder that extends from ground level to roof level?
  I. Each step is 8 inch high.
  II. Each step is 2.5 inch wide.

21. What is the length of the side of a rectangle formed by a rope?
  I. The total length of the rope is a multiple of 4.
  II. The area of triangle formed by the same rope is twice the area of the circle formed by the rope.

22. Akbar spends 20% of his pocket money on purchase of toffees, 10% on ice-cream and 5% on cold drinks. He saves the balance pocket money in a bank.
The bank pays him interest at the rate of 8% per annum, Rahim spends 30% of his pocket money on purchase of toffees, 12% on ice-cream and 6% on cold drinks. Rahim saves the balance pocket money for purchase of story books.
Who spends more on purchase of toffees?
  I. Akbar spends more on ice-cream than Rahim.
  II. Rahim spends more on cold drinks than Akbar.

Directions (Q. Nos. 23-27) *Each of the questions below consists of a question and two statements numbered I and II given below it. You have to decide whether the data provided in the statements are sufficient to answer the question. Read the question and both the statements and give answer.* **[IBPS SO 2016]**

(a) If the data in Statement I alone is sufficient to answer the question while the data in Statement II alone is not sufficient to answer the question

(b) If the data in Statement II alone is sufficient to answer the question while the data in Statement I alone is not sufficient to answer the question

(c) If the data either in Statement I alone or in Statement II alone is sufficient to answer the question

(d) If the data in both Statements I and II together are not sufficient to answer the question

(e) If the data in both the Statements I and II together are necessary to answer the question

23. What is the area of a circular field?
  I. The area of the largest square that can be inscribed in the given circular field is 392 cm^2.
  II. The area of the smallest square in which the given circular field can be inscribed is 784 cm^2.

24. What was the initial quantity of a mixture of juice and water?
  I. Juice and water were in the ratio of 6 : 1 in the mixture initially.
  II. When 7 L of the mixture is taken out and 5 L of water is added, the ratio of juice to water becomes 8 : 3.

25. What is the curved surface area of a right circular cylinder?
   I. The area of the base of the cylinder is 616 cm^2.
   II. The volume of the cylinder is 9240 cm^3.

26. In how many days can 'B' alone complete the work?
   I. A, B and C together can complete the work in $4\frac{32}{37}$ days.
   II. A and B together can complete the work in $6\frac{2}{3}$ days, B and C together can complete the work together in $8\frac{2}{11}$ days and A and C together can complete the work in $7\frac{1}{3}$ days.

27. How much money did Ms. Malini receive as retirement fund?
   I. Out of the total money received, Ms. Malini gave 25% to her husband and 10% to her daughter. Out of the remaining, she invested 30% in Mutual Funds, 60% in Pension fund scheme and the remaining ₹ 260000 she spent on miscellaneous items.
   II. Out of the total money received, Ms. Malini invested 58.5% in various schemes, gave 35% of the total money received to her husband and daughter and the remaining money she spent on miscellaneous items.

Directions (Q. Nos. 28-30) *Each question below has a question and two/three statements. You have to decide whether the data provided in the statements are sufficient to answer the question. Choose appropriate option in each case.* [IBPS Clerk (Mains) 2017]

28. There are certain number of cards in a bag which are numbered serially (1, 2, 3, 4, and so on). How many cards are there in the bag?
   I. If two cards are drawn at random, one after another without replacement, the probability that both the cards are even numbered, is 7/31.
   II. The respective ratio between number of odd numbered cards and number cards in the bag is 16 :15.
   (a) Only Statement I is sufficient to answer the question
   (b) Only Statement II is sufficient to answer the question
   (c) Any of the statement is sufficient to answer the question
   (d) Both the statements together are required to answer the question
   (e) Both the statements together are not sufficient to answer the question

29. What is the rate of interest per annum?
   I. The amount becomes ₹ 9331.20 in 2 yr at compound interest.
   II. The difference between CI and SI at the same rate of interest in 2 yr is ₹ 51.20.
   III. The amount invested is ₹ 8000.
   (a) Statements I and III are sufficient to answer the question
   (b) Statements I and II are sufficient to answer the question
   (c) Statement II and either Statement I or III are sufficient to answer the question
   (d) Statement III and either Statement I or II are sufficient to answer the question
   (e) Any two of the three statements is sufficient to answer the question

30. A and B are two positive integers. What is the sum of the cubes of A and B?
   I. The sum of A and B is 14.
   II. The sum of the squares of A and B is 106.
   III. The value of A is 4 less than the value of B.
   (a) Statements I and III are sufficient to answer the question
   (b) Statements I and II are sufficient to answer the question
   (c) Statement II and either Statement I or Statement III are sufficient to answer the question
   (d) Statement III and either Statement I or Statement II are sufficient to answer the question
   (e) Any two of the three statements is sufficient to answer the question

Answer with Solutions

1. (d) From Statement I,
$$n = w + 6$$
From Statement II,
$$p = w + 4$$
The value of w is not given.
So, exact average cannot be determined.

2. (a) Selling price of an article = ₹ 24000
From Statement I,
$$\text{Cost price} = \frac{3}{5} \times 24000 = ₹ 14400$$
$$\therefore \text{Profit} = \frac{24000 - 14400}{14400} \times 100\%$$
$$= 66.66\%$$
From Statement II,
It is not clear whether cost price is more than selling price or vice-versa.

3. (e) Suppose numbers be x and y.
From Statement I, $x^2 = 9y$
From Statement II, $\dfrac{x}{y} = \dfrac{3}{4}$
From both statements, we get
$$x = 12 \text{ and } y = 16$$
$\therefore$ Difference $= 16 - 12 = 4$

4. (e) From Statement I,
$$\frac{40}{100} \times x = \frac{20}{100} \times 50 \Rightarrow x = 25$$
From Statement II, $\dfrac{30}{100} \times y = \dfrac{25}{100} \times 72$
$$\Rightarrow y = 60$$
From both statements,
$$x : y = 25 : 60 = 5 : 12$$

5. (a) From Statement I, $D + E = 14 \times 2 = 28$
From Statement II,
$$A + B + C + F = 4 \times 50 = 200$$
$$\therefore \frac{A + B + C + D + E + F}{6} = \frac{28 + 200}{6}$$
$$= \frac{228}{6} = 38$$
$\therefore$ Statements I and II are sufficient to answer the question.

6. (d) Area of triangle $= \dfrac{1}{2} \times \text{Base} \times \text{Height}$
$$= \frac{1}{2} \times X \times Y = \frac{1}{2} XY$$
So, to find the area, the measure of height and base is required which can be obtained by any of the two statements.

7. (c) I. $(A + B)$'s 1 day work $= \dfrac{1}{8}$ ...(i)
II. $(B + C)$'s 1 day work $= \dfrac{1}{10}$ ...(ii)
III. $(C + A)$'s 1 day work $= \dfrac{1}{12}$ ...(iii)
According to the questions,
$2(A + B + C)$'s 1 day work
$$= \frac{1}{8} + \frac{1}{10} + \frac{1}{12}$$
$\Rightarrow (A + B + C)$'s 1 day work
$$= \frac{1}{2}\left[\frac{1}{8} + \frac{1}{10} + \frac{1}{12}\right] \quad ...(iv)$$
On subtracting Eq. (iii) from Eq. (iv), we get B's 1 day work.
Then, required number of days
$$= \frac{1}{B\text{'s 1 day work}}$$

8. (e) From Statement I,
$$SI = \frac{P \times R \times T}{100} \Rightarrow P = \frac{P \times R \times 10}{100}$$
$\Rightarrow R = 10\%$
From Statement II,
$$\text{Difference }(D) = \frac{PR^2}{(100)^2}$$
$$\Rightarrow 150 = \frac{15000 \times R^2}{10000}$$
$\Rightarrow R = 10\%$
Thus, either Statement I or II is sufficient.

9. (c) From Statements I, II and III.
Let marks in English be x.
Marks in Science $= x + 12$
and marks in Mathematics $= x + 32$
From Statement II,
$$E + S + M = 197$$
$$x + x + 12 + x + 32 = 197$$
$$\Rightarrow 3x + 44 = 197$$
$$\Rightarrow 3x = 197 - 44 = 153 \Rightarrow x = 51$$
$\therefore$ Marks in English $= 51$

10. (d) From both statements,
Total length of both trains A and $B = 777$ m
Time taken to cross the electric pole by train $B = 6$ s.
Speed of train A : Speed of train $B = 7 : 8$
From above data, we are unable to find the time taken to cross both the trains to each other.

11. (e) From Statement I,
Radius of cylinder $(r) = 21$ m
From Statement II,
Let radius and height of cylinder be r and h, respectively.
According to the Statement II,
$$r + h = 29 \quad ...(i)$$
and $2\pi rh = 1056$

$\Rightarrow \quad 2\pi(29-h)h = 1056$ [from Eq. (i)]
$\Rightarrow \quad 2 \times \dfrac{22}{7}(29h - h^2) = 1056$
$\Rightarrow \quad 29h - h^2 = \dfrac{1056 \times 7}{2 \times 22}$
$\Rightarrow \quad h^2 - 29h + 168 = 0$
$\Rightarrow \quad h^2 - 21h - 8h + 168 = 0$
$\Rightarrow \quad h(h-21) - 8(h-21) = 0$
$\Rightarrow \quad (h-8)(h-21) = 0$
$\Rightarrow \quad h = 8, 21$
$\therefore \quad r = 21$ m, when $h = 8$ m
and $r = 8$ m, when $h = 21$ m
∵ We can't select one value of radius or height.
∴ We select the value of $r = 21$ m with the help of Statement I.
∴ Both Statement are necessary.

12. (e) From Statement I,
CP of table = ₹ 8000
and profit percentage = 12%
∴ SP of table = $\dfrac{8000 \times 112}{100}$ = ₹ 8960
From Statement II,
MP of table = $\dfrac{8000 \times 118}{100}$ = ₹ 9440
Now, discount = 9440 − 8960 = ₹ 480
∴ Discount percentage = $\dfrac{480}{9440} \times 100\%$
= $\dfrac{300}{59}\% = 5\dfrac{5}{59}\%$

13. (d) ∵ It is not given in question that the man is equal to woman or child or not.
So, we are unable to find the days required by men to complete the work.

14. (b) Let the number be x.
Statement I data is insufficient.
From Statement II, $x \times \dfrac{2}{3} = x - 226$
$\Rightarrow \quad 2x = 3x - 678$
$\Rightarrow \quad 3x - 2x = 678 \Rightarrow x = 678$
Hence, the number is 678.

15. (c) I. Work done by 8 workers in 8 days = $\dfrac{1}{5}$
∴ $\dfrac{8 \times 8}{1/5} = \dfrac{M \times 10}{1}$
$\Rightarrow \quad M = \dfrac{64 \times 5}{10} = 32$
So, 32 workers are required to complete the work in 10 days.

II. Men required to complete the work in 16 days = 20
∴ Men required to complete the work in 10 days = $\dfrac{20 \times 16}{10} = 32$

So, either Statement I alone or Statement II alone is sufficient to answer the question.

16. (e) I. Salary of Sumit = 90% of salary of Lokesh.
and salary of Praveen = 115% of salary of Sumit.
II. Salary of Lokesh = ₹ 2500
∴ Salary of Praveen
= $\dfrac{115}{100}\left(\dfrac{90}{100} \times 2500\right)$ = ₹ 2587.5
So, both Statements I and II are required to answer the question.

17. (e) I. Ratio of speeds = 4 : 5
Let speed be $4x$ and $5x$.
II. Difference between speeds = 20 km/h
$\Rightarrow 5x - 4x = 20 \Rightarrow x = 20$
∴ Speeds are $5 \times 20 = 100$ km/h
and $4 \times 20 = 80$ km/h
Let the distance be d between P and Q.
According to the statement,
$\dfrac{d}{80} - \dfrac{d}{100} = 1$ h
$\Rightarrow \dfrac{100d - 80d}{8000} = 1$
$\Rightarrow 20d = 8000 \Rightarrow d = 400$ km
Hence, data is Statements I and II both are necessary to answer the question.

18. (b) From Statement I,
$\dfrac{X + Y + Z}{3} = 30$ ...(a)
From Statement II,
Let interviewed by
Panel $Z = m$ ...(i)
Panel $Y = m + 2$ ...(ii)
Panel $X = m + 4$ ...(iii)
Then, $m + 4 + m + 2 + m = 90$
[∵ total students = 90]
$\Rightarrow \quad 3m + 6 = 90 \Rightarrow m = 28$
Then, interviewed by $X = m + 4 = 28 + 4$
= 32 students
So only Statement II is required to answer the question.

19. (d) Both Statements I and II are not sufficient to answer the question.

20. (a) From Statement I, height of each step is 8 inch. Let the ladder has 'n' steps.

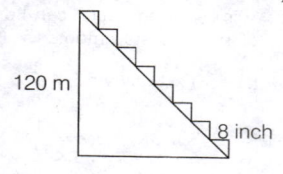

Data Sufficiency / 851

Now, height of wall = 'n' × Height of step.
$\Rightarrow \quad 120 = n(0.2032)$ [∵ 1 inch = 0.0254m]
$\therefore \quad n \approx 590$
But, we cannot find the number of steps using width only.
So, data in Statement I alone is sufficient to answer the question.

21. (d) Both Statements I and II together are not sufficient.

22. (d) Let the pocket money of Akbar be 100%.
Balance pocket money
= 100% − (20% + 10% + 5%) = 65%
Let the pocket money of Rahim be 100%.
Balance pocket money
= 100% − (30% + 12% + 6%) = 52%
As we have not been provided base values of their pocket money, so the percentage values of expenses made on ice-cream and cold drinks cannot be compared.

23. (c) From Statement I,
Area of largest square inscribed in a circle of radius r is $2r^2$.
Then, $2r^2 = 392 \Rightarrow r = \sqrt{196} = 14$ cm
Thus, we can find the area of the circular field.
From Statement II,
Area of the smallest square in which a circle can be inscribed = $4r^2$
Then, $4r^2 = 784 \Rightarrow r = \sqrt{196} = 14$ cm
Thus, we can find the area of the circular field.
Hence, Statement I alone or Statement II alone is sufficient to answer the question.

24. (e) From Statements I and II,

	Juice	Water
Initially	$6x$	x
When 7 L is taken out	$6x − 6$	$x − 1$
When 5 L water is added	$6x − 6$	$x + 4$

Now, $\dfrac{6x-6}{x+4} = \dfrac{8}{3}$
$\Rightarrow \quad 18x - 18 = 8x + 32$
$\Rightarrow \quad 10x = 50 \Rightarrow x = 5$
Hence, intially the quantity of mixture
= $7 \times 5 = 35$ L

25. (e) Curved surface area of the cylinder
$= 2\pi rh$
From Statement I,
Area of the base = 616 cm^2
$\Rightarrow \quad \pi r^2 = 616 \Rightarrow r = 14$ cm
From Statement II,
Volume of the cylinder = $\pi r^2 h = 9240$ cm^3
From Statements I and II,
Volume of the cylinder = $\pi r^2 h$
$= \dfrac{22}{7} \times 14 \times 14 \times h = 9240$
$\Rightarrow \quad h = 15$ cm

∴ Curved surface area of the cylinder = $2\pi rh$
$= 2 \times \dfrac{22}{7} \times 14 \times 15 = 1320$ cm^2

26. (b) From Statement II,
(A and B)'s 1 day work
$\dfrac{1}{A} + \dfrac{1}{B} = \dfrac{3}{20}$ ...(i)
Similarly, $\dfrac{1}{B} + \dfrac{1}{C} = \dfrac{11}{90}$ ...(ii)
and $\dfrac{1}{C} + \dfrac{1}{A} = \dfrac{3}{22}$ ...(iii)
Now, $2\left(\dfrac{1}{A} + \dfrac{1}{B} + \dfrac{1}{C}\right) = \dfrac{3}{20} + \dfrac{11}{90} + \dfrac{3}{22}$
$= \dfrac{297 + 242 + 270}{1980}$
$= \dfrac{809}{1980}$
or $\dfrac{1}{A} + \dfrac{1}{B} + \dfrac{1}{C} = \dfrac{809}{1980 \times 2}$ ...(iv)
On subtracting Eq. (iii) from Eq. (iv), we get
$\dfrac{1}{B} = \dfrac{809 - 540}{1980 \times 2} = \dfrac{269}{3960}$
Hence, B alone can complete the work in $\dfrac{3960}{269}$ days.

27. (a) From Statement I,
Suppose Ms. Malini receives ₹ $100x$ as retirement fund.
Then, $100x - (25 + 10)x = 65x$
Again, $65x - \dfrac{65x \times 90}{100} = 6.5x$
So, $6.5x = 260000$
$\therefore \quad x = \dfrac{260000 \times 10}{65} = ₹\ 40000$
Malini's retirement fund = $100x$
$= 100 \times 40000 = ₹\ 4000000$
Hence, Statement I alone is sufficient.
From Statement II,
Amount is not given, so we cannot find Malini's retirement fund.

28. (e) Let the total number of cards be n.
From Statement II,
Even numbered cards = $\dfrac{15}{31}n$
From Statement I, $\dfrac{^{\frac{15}{31}n}C_2}{^nC_2} = \dfrac{7}{31}$
$\Rightarrow \quad \dfrac{\dfrac{\frac{15}{31}n\left(\frac{15n}{31} - 1\right)}{2}}{\dfrac{n(n-1)}{2}} = \dfrac{7}{31}$
$\Rightarrow \quad \dfrac{15}{31}n\left(\dfrac{15n}{31} - 1\right) \times 31 = 7[n(n-1)]$

On solving the above equation, we can get the total number of cards.
Hence, the data given in both the statements are required to answer the question.

29. (e) From Statements I and III,
Here, $A = ₹ 9331.20, P = ₹ 8000$
$n = 2, r = ?$

From $\qquad A = P\left(1 + \dfrac{r}{100}\right)^n$

$\Rightarrow \qquad 9331.20 = 8000\left(1 + \dfrac{r}{100}\right)^2$

$\Rightarrow \qquad \dfrac{9331.20}{8000} = \left(1 + \dfrac{r}{100}\right)^2$

$\Rightarrow \qquad \left(1 + \dfrac{r}{100}\right)^2 = \dfrac{93312}{80000} = \dfrac{46656}{40000}$

$\Rightarrow \qquad 1 + \dfrac{r}{100} = \dfrac{216}{200}$

$\Rightarrow \qquad 1 + \dfrac{r}{100} = \dfrac{108}{100}$

$\Rightarrow \qquad \dfrac{r}{100} = \dfrac{108}{100} - 1 = \dfrac{8}{100}$

$\Rightarrow \qquad r = 8\%$

Now from Statements II, and III,

$\qquad \text{Difference} = P\left(\dfrac{R}{100}\right)^2$

$\Rightarrow \qquad 51.20 = 8000\left(\dfrac{R}{100}\right)^2$

$\Rightarrow \qquad R = 8\%$

Now, from Statements I and II,

$\qquad 9331.20 = P\left(1 + \dfrac{R}{100}\right)^2$

and $\qquad 51.20 = P\left(\dfrac{R}{100}\right)^2$

On dividing both the equations, we get

$\left(\dfrac{100 + R}{R}\right)^2 = 182.25$

$\Rightarrow \qquad \dfrac{100 + R}{R} = 13.5$

$\Rightarrow \qquad 12.5R = 100$

$\therefore \qquad R = 8\%$

Hence, any two of the three statements is sufficient to answer the question.

30. (e) From Statements I and II,
$\qquad A + B = 14 \qquad \qquad ...(i)$
and $\qquad A^2 + B^2 = 106 \qquad ...(ii)$
On solving Eqs. (i) and (ii), we get
$\qquad A = 9 \text{ or } 5$
and $\qquad B = 5 \text{ or } 9$
But, it does not matter which of them is bigger as we have to find the sum of their cubes.
$\qquad A^3 + B^3 = 9^3 + 5^3$
$\qquad \qquad = 729 + 125$
$\qquad \qquad = 854$

From Statements II and III,
$\qquad A^2 + B^2 = 106 \qquad ...(iii)$
$\qquad A = B - 4 \qquad \qquad ...(iv)$
On solving Eqs. (iii) and (iv), we get
$\qquad B = 9, A = 5$
So, $\qquad A^3 + B^3 = 5^3 + 9^3$
$\qquad \qquad = 125 + 729$
$\qquad \qquad = 854$

From Statements I and III,
$\qquad A + B = 14 \qquad \qquad ...(v)$
$\Rightarrow \qquad A = B - 4 \qquad \qquad ...(vi)$
On solving Eqs. (v) and (vi), we get
$\qquad A = 5, B = 9$
So, $\qquad A^3 + B^3 = 125 + 729 = 854$

So, any two of the three statements is sufficient to answer the question.

Practice Set 1

1. The value of $\sin^2\theta \cos^2\theta (\sec^2\theta + \csc^2\theta)$ is
   (a) 2   (b) 4
   (c) 1   (d) 3

2. If $(\csc\theta - \cot\theta) = 2$, then $(\csc\theta + \cot\theta)$ is equal to
   (a) 2   (b) $\dfrac{1}{2}$
   (c) 1   (d) $\dfrac{3}{2}$

3. If $4x = \sec\theta$ and $\dfrac{4}{x} = \tan\theta$, then $8\left(x^2 - \dfrac{1}{x^2}\right)$ is
   (a) $\dfrac{1}{2}$ (b) $\dfrac{1}{4}$ (c) $\dfrac{1}{16}$ (d) $\dfrac{1}{8}$

4. If $\sec\theta + \tan\theta = P$, then $\cos\theta$ is equal to
   (a) $\dfrac{P^2+1}{P^2-1}$ (b) $\dfrac{P^2-1}{(P^2+1)^2}$
   (c) $\dfrac{2P}{P^2+1}$ (d) $\dfrac{4P^2}{(P^2+1)^2}$

5. If $x^4 + \dfrac{1}{x^4} = 119$, then value of $x^3 - \dfrac{1}{x^3}$ is
   (a) 27   (b) 36   (c) 45   (d) 54

6. The simplified value of
   $\left(1-\dfrac{1}{3}\right)\left(1-\dfrac{1}{4}\right)\left(1-\dfrac{1}{5}\right)\ldots\left(1-\dfrac{1}{99}\right)\left(1-\dfrac{1}{100}\right)$ is
   (a) $\dfrac{2}{99}$ (b) $\dfrac{1}{25}$ (c) $\dfrac{1}{50}$ (d) $\dfrac{1}{100}$

7. If $1^3 + 2^3 + 3^3 + \ldots + 10^3 = 3025$, then find the value of $2^3 + 4^3 + 6^3 + \ldots + 20^3$.
   (a) 6050   (b) 9075   (c) 12100   (d) 24200

8. The total number of prime factors in $6^{10} \times 7^{17} \times 11^{27}$ is
   (a) 50   (b) 51   (c) 52   (d) 64

9. The simplified value of
   $\dfrac{\sqrt{8+\sqrt{28}} - \sqrt{8-\sqrt{28}}}{\sqrt{8+\sqrt{28}} + \sqrt{8-\sqrt{28}}}$ is
   (a) $\dfrac{1}{\sqrt{7}}$ (b) $\sqrt{7}$ (c) $-2\sqrt{7}$ (d) $\dfrac{1}{-2\sqrt{7}}$

10. Product of three natural numbers is 24000 and their HCF is 10. How many such triplets of numbers are there?
    (a) 5   (b) 4
    (c) 7   (d) 6

11. The average temperature of Delhi for Monday, Tuesday and Wednesday was 25°C, while for Thursday, Wednesday and Tuesday, it was 26°C. If the temperature on Thursday was 27°C, then what was the temperature on Monday?
    (a) 21°C   (b) 24°C
    (c) 27°C   (d) 30°C

12. What will be the ratio of petrol and kerosene in the final solution formed by mixing petrol and kerosene that are present in three vessels in the ratio 4 : 1, 5 : 2 and 6 : 1, respectively?
    (a) 166 : 22   (b) 83 : 22
    (c) 83 : 44   (d) None of these

13. In an examination, there were 1000 boys and 800 girls. 60% of the boys and 50% of the girls passed. Find the per cent of candidates failed.
    (a) 46.4%   (b) 48.4%   (c) 44.4%   (d) 49.6%

14. If the price of sugar rises from ₹ 6 per kg to ₹ 7.50 per kg, a person having no increase in his expenditure on sugar, will have to reduce his consumption of sugar by
    (a) 15%   (b) 20%   (c) 25%   (d) 30%

15. The monthly incomes of two persons are in the ratio 4 : 7 and their expenses are in the ratio 11 : 20. If each saves ₹ 400 per month, then their monthly incomes must be, respectively
    (a) ₹ 3600 and ₹ 4200
    (b) ₹ 4000 and ₹ 7000
    (c) ₹ 4200 and ₹ 7350
    (d) ₹ 4800 and ₹ 8400

16. The present ages of A and B are in the ratio 4 : 5 and after 5 yr, they will be in the ratio 5 : 6. The present age of A is
    (a) 10 yr   (b) 20 yr
    (c) 25 yr   (d) 40 yr

17. A, B and C are partners of a company. During a particular year A received one-third of the profit, B received one-fourth of the profit and C received the remaining ₹ 5000. How much did A receive?
    (a) ₹ 5000   (b) ₹ 4000
    (c) ₹ 3000   (d) ₹ 1000

18. The sum of money, that will give ₹ 1 as interest per day at the rate of 5% per annum simple interest is
    (a) ₹ 3650   (b) ₹ 36500
    (c) ₹ 730   (d) ₹ 7300

19. By selling an article for ₹ 72, there is a loss of 10%. In order to gain 5%, its selling price should be
    (a) ₹ 87   (b) ₹ 85   (c) ₹ 80   (d) ₹ 84

20. Two trains, one 160 m and the other 140 m long, are running in opposite directions on parallel trails, the first at 77 km an hour and other 67 km an hour. How long will the trains take to cross each other?
    (a) 7 s   (b) $7\frac{1}{2}$ s
    (c) 6 s   (d) 10 s

21. A starts from a place P to go to a place Q. At the same time, B starts from Q to P. If after meeting each other A and B took 4 and 9 h more respectively, to reach their destinations, then ratio of their speeds is
    (a) 3 : 2   (b) 5 : 2
    (c) 9 : 4   (d) 9 : 13

22. A person buys some pencils at 5 for a rupee and sells them at 3 for a rupee. This gain per cent will be
    (a) $66\frac{2}{3}$%   (b) $76\frac{2}{3}$%
    (c) $56\frac{2}{3}$%   (d) $46\frac{2}{3}$%

23. Pipe A alone can fill a tank in 8 h. Pipe B alone can fill it in 6 h. If both the pipes are opened and after 2 h pipe A is closed, then the other pipe will fill the tank in
    (a) 4 h   (b) $2\frac{1}{2}$ h
    (c) 6 h   (d) $3\frac{1}{2}$ h

24. What is difference between compound interest on ₹ 5000 for $1\frac{1}{2}$ yr at 4% per annum when the interest is compounded yearly and half-yearly?
    (a) ₹ 2.04   (b) ₹ 3.06
    (c) ₹ 8.30   (d) ₹ 4.80

25. A can do a job in 15 days, B in 10 days and C in 30 days. If A is helped by B and C on every third day, then the job will be completed in
    (a) $6\frac{1}{3}$ days   (b) $8\frac{1}{3}$ days
    (c) 8 days   (d) 9 days

26. The river flows at 4 km/h. A boat can go downstream thrice as fast as upstream. The speed of boat in still water is
    (a) 12 km/h   (b) 16 km/h
    (c) 8 km/h   (d) 10 km/h

27. Let BE and CF be the two medians of a $\triangle ABC$ and G be the intersection. Also, let EF cut AG at O, then AG : AO is
    (a) 1 : 1   (b) 1 : 2
    (c) 2 : 1   (d) 3 : 1

28. A chord of length 8 cm is at a distance 3 cm from the centre of the circle. The length of the radius of the circle is
    (a) $\sqrt{73}$ cm   (b) $\sqrt{55}$ cm
    (c) 5 cm   (d) 10 cm

29. Volume of two cones are in the ratio 1 : 4 and their diameter are in the ratio 4 : 5. The ratio of their height is
    (a) 1 : 5   (b) 5 : 4
    (c) 5 : 16   (d) 25 : 64

30. A sum of money becomes ₹ 4500 after 2 yr and ₹ 6750 after 4 yr on compound interest. The sum is
    (a) ₹ 4000   (b) ₹ 2500
    (c) ₹ 3000   (d) ₹ 3050

31. Three-fifth of the square of a certain number is 126.15. What is the number?
    (a) 210.25   (b) 75.69
    (c) 14.5   (d) 145

32. If I is the incentre of $\triangle ABC$ and $\angle A = 60°$, then the value of $\angle BIC$ is
    (a) 100°   (b) 120°
    (c) 150°   (d) 110°

33. The perimeter of square and a circular field are the same. If the area of the circular field is 3850 m^2, then what is the area (in m^2) of the square?
    (a) 4225   (b) 3025
    (c) 2500   (d) 2025

Practice Set 1 / 855

34. If the length of a rectangle is increased by 25% and the width is decreased by 20%, then the area of rectangle will be
  (a) increased by 5%
  (b) decreased by 5%
  (c) remains unchanged
  (d) increased by 10%

35. If base of a prism is a square of side 4 cm. If the height of prism is 10 cm, then what will be total surface area of that prism?
  (a) 192 cm²
  (b) 212 cm²
  (c) 214 cm²
  (d) None of the above

36. A conical vessel whose internal radius is 12 cm and height 50 cm, is full of liquid. The content are emptied into a cylindrical vessel with radius (internal) 10 cm. The height of which the liquid rises in the cylinder vessel is
  (a) 25 cm   (b) 20 cm
  (c) 24 cm   (d) 22 cm

37. A tree breaks due to storm and the broken part bends, so to that the top of the tree touches the ground making an angle 30° with it. The distance between the feet of the tree to the point, where the top touches the ground, is 8 m. Find the height of the tree.
  (a) 8 m   (b) $8\sqrt{3}$ m
  (c) 24 m   (d) 12 m

38. Two points A (−3, b) and B (1, b + 4) and the coordinates of the middle point of AB are (−1, 1). The value of b is
  (a) 1   (b) −1
  (c) 2   (d) 0

39. If $x^3 - \dfrac{1}{x^3} = 14$, then the value of $x - \dfrac{1}{x}$ will be
  (a) 2   (b) 3
  (c) 4   (d) 5

40. If tan (x + y) tan (x − y) = 1, then the value of $\tan\left(\dfrac{2x}{3}\right)$ is
  (a) $\dfrac{1}{\sqrt{3}}$ (b) $\dfrac{2}{\sqrt{3}}$
  (c) $\sqrt{3}$ (d) 1

Directions (Q. Nos. 41-45) *Study the following graph carefully and answer the questions that follow.*

Per cent rise in profit of two companies L and M from the year 2004-09

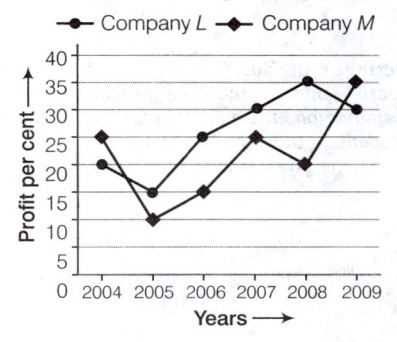

41. If the profit earned by company L in the year 2005 was ₹ 1.84 lakh, what was the profit earned by the company in the year 2006?
  (a) ₹ 2.12 lakh
  (b) ₹ 2.3 lakh
  (c) ₹ 2.04 lakh
  (d) Cannot be determined

42. Which of the following statements is true with respect to the given graph?
  (a) Company M made the highest profit in the year 2009
  (b) Company L made the least profit in the year 2005
  (c) The ratio of the profits earned by company L and company M in the year 2006 was 6 : 5
  (d) Company L made the highest profit in the year 2008

43. What was the percentage increase in per cent rise in profit of company M in the year 2009 from the previous year?
  (a) 25   (b) 15
  (c) 50   (d) 75

44. If the profit earned by company M in the year 2008 was ₹ 3.63 lakh, what was the amount of profit earned by it in the year 2006?
  (a) ₹ 2.16 lakh
  (b) ₹ 1.98 lakh
  (c) ₹ 2.42 lakh
  (d) Cannot be determined

856 / *Fast Track* Objective Arithmetic

45. What was the average per cent rise in profit of company L over all the years together?

(a) $15\frac{1}{3}$ (b) $25\frac{1}{3}$

(c) $18\frac{5}{6}$ (d) None of these

Directions (Q. Nos. 46-50) *The following pie chart shows the performance in an examination in a particular year for 360 students. Study the pie chart and answer the questions given below.*

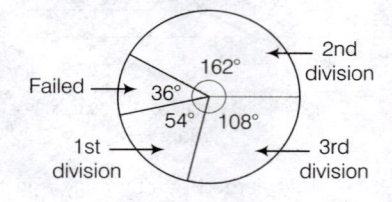

46. The number of students who passed in 1st division is
(a) 45   (b) 54   (c) 64   (d) 74

47. The number of students who passed in 2nd division is more than those in 1st division, is
(a) 111   (b) 112   (c) 109   (d) 108

48. The ratio of successful students to the failed students is
(a) 9 : 1   (b) 5 : 1   (c) 1 : 9   (d) 2 : 7

49. The percentage of students who have failed in the examination, is
(a) 20%   (b) 36%   (c) 10%   (d) 30%

50. The total number of students who have passed in 2nd or 3rd division, is
(a) 162   (b) 270
(c) 108   (d) None of these

Answers

1. (c)	2. (b)	3. (a)	4. (c)	5. (b)	6. (c)	7. (d)	8. (d)	9. (a)	10. (a)
11. (b)	12. (b)	13. (c)	14. (b)	15. (d)	16. (b)	17. (b)	18. (d)	19. (d)	20. (b)
21. (a)	22. (a)	23. (b)	24. (a)	25. (d)	26. (c)	27. (d)	28. (c)	29. (d)	30. (c)
31. (c)	32. (b)	33. (b)	34. (c)	35. (a)	36. (c)	37. (b)	38. (b)	39. (a)	40. (a)
41. (b)	42. (a)	43. (d)	44. (c)	45. (d)	46. (b)	47. (d)	48. (a)	49. (c)	50. (b)

Practice Set 2

1. A rectangular sump of dimensions 6 m × 5 m × 4 m is to be built by using bricks to make the outer dimensions 6.2 m × 5.2 m × 4.2 m. Approximately, how many bricks of size 20 cm × 10 cm × 5 cm are required to build the sump for storing water?
   (a) 15408 (b) 3000 (c) 15000 (d) 14508

2. In a survey, it was found that 80% of those surveyed owned a car while 60% of those surveyed owned a mobile phone. If 55% owned both a car and mobile phone, then what per cent of those surveyed owned a car or a mobile phone or both?
   (a) 65% (b) 80% (c) 85% (d) 97.5%

3. If $\frac{1}{8}$ of $\frac{2}{3}$ of $\frac{4}{5}$ of a number is 12, then 30% of the number will be
   (a) 42 (b) 48 (c) 54 (d) 64

4. Three bells ring at intervals of 9, 12 and 15 min. All the three began to ring at 8:00 am. At what time will they ring together again?
   (a) 8 : 45 am (b) 10 : 30 am
   (c) 11 : 00 am (d) 1 : 30 pm

5. The average of x_1, x_2 and x_3 is 14. Twice the sum of x_2 and x_3 is 30. What is the value of x_1?
   (a) 20 (b) 27 (c) 16 (d) 2

6. A circle of 1 m radius is drawn inside a square as shows in the figure given below. What is the area (in m^2) of the shaded portion?

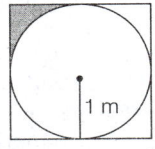

   (a) $(4 - \pi)$ (b) $\left(1 - \frac{\pi}{2}\right)$
   (c) $\left(\frac{1}{4} - \frac{\pi}{2}\right)$ (d) $\left(1 - \frac{\pi}{4}\right)$

7. The average speed of a train in the onward journey is 25% more than that of the return journey. The train halts for one hour on reaching the destination. The total time taken for the complete journey is 17 h covering a distance of 800 km. The speed of the train in the onward journey is
   (a) 45 km/h (b) 47.06 km/h
   (c) 50.00 km/h (d) 56.25 km/h

8. If $2x + y = 6$ and $x = 2$ are two linear equations, then graph of two equations meet at a point
   (a) (2, 0) (b) (0, 2) (c) (2, 2) (d) (1, 2)

9. In the given figure, ABCD is a cyclic quadrilateral, $AB = BC$ and $\angle BAC = 70°$, then $\angle ADC$ is equal to

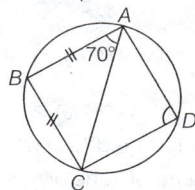

   (a) 40° (b) 80° (c) 110° (d) 140°

10. ABC is a right angled triangle with $AB = 12$ cm and $AC = 13$ cm. A circle, with centre O, has been inscribed inside the triangle. Calculate the value of x, the radius of the inscribed circle.

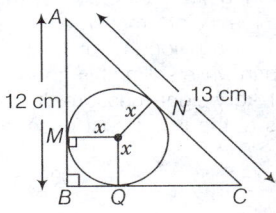

    (a) 5 cm (b) 2 cm (c) 4 cm (d) 6 cm

11. If $\sin\alpha \sec(30° + \alpha) = 1 \, (0 < \alpha < 60°)$, then the value of $\sin\alpha + \cos 2\alpha$ is
    (a) 1 (b) $\frac{2 + \sqrt{3}}{2\sqrt{3}}$ (c) 0 (d) 2

12. If $2\cos\theta - \sin\theta = \frac{1}{\sqrt{2}}$ $(0° < \theta < 90°)$, then the value of $2\sin\theta + \cos\theta$ is
    (a) $\frac{1}{\sqrt{2}}$ (b) $\sqrt{2}$ (c) $\frac{3}{\sqrt{2}}$ (d) $\frac{\sqrt{2}}{3}$

13. If $\sin\theta + \csc\theta = 2$, then the value of $\sin^{100}\theta + \csc^{100}\theta$ is equal to
    (a) 1 (b) 2 (c) 3 (d) 100

14. A kite flying at a height of 45 m from the level ground is attached to a string inclined at 60° to the horizontal. The length of the string is
    (a) $\frac{30}{\sqrt{2}}$ m (b) $30\sqrt{5}$ m
    (c) $30\sqrt{3}$ m (d) $30\sqrt{2}$ m

15. Find the value of
    $\frac{\sin 80°}{\cos 10°} + \sin 59° \sec 31° + \tan 49° \cot 49°$.
    (a) 3 (b) 1 (c) 2 (d) 0

16. There are four consecutive prime numbers. The product of first three is 385 and the product of last three is 1001. The first and last numbers are respectively
    (a) 5, 11 (b) 5, 13 (c) 7, 11 (d) 7, 13

17. If $1x5x01$ is divisible by 11, then the value of x is
    (a) 6 (b) 4 (c) 8 (d) 9

18. A student is asked to multiply a number by $\frac{8}{17}$ instead of it, he divides the number by $\frac{8}{17}$ and get 225 more than the original result. The given number is
    (a) 8 (b) 17 (c) 64 (d) 136

19. Left pan of a faulty balance weighs 100 g more than its right pan. A shopkeeper keeps the weight measure in the left pan while buying goods but keeps it in the right pan while selling his goods. He uses only 1 kg weight measure. If he sells his goods at the listed cost price, then what is his gain?
    (a) $\frac{200}{11}$% (b) $\frac{100}{11}$% (c) $\frac{1000}{9}$% (d) $\frac{200}{9}$%

20. In three vessels, the ratio of water and milk is 6 : 7, 5 : 9 and 8 : 7, respectively. If the mixture of the three vessels is mixed, then what will be the ratio of water and milk?
    (a) 2431 : 3781 (b) 3691 : 4499
    (c) 4381 : 5469 (d) None of these

21. If $4b^2 + \frac{1}{b^2} = 2$, then the value of $8b^3 + \frac{1}{b^3}$ is
    (a) 0 (b) 1 (c) 2 (d) 5

22. $\left(x + \frac{1}{x}\right)\left(x - \frac{1}{x}\right)\left(x^2 + \frac{1}{x^2} - 1\right)\left(x^2 + \frac{1}{x^2} + 1\right)$ is equal to
    (a) $x^6 + \frac{1}{x^6}$ (b) $x^8 + \frac{1}{x^8}$
    (c) $x^8 - \frac{1}{x^8}$ (d) $x^6 - \frac{1}{x^6}$

23. If a right circular cone of height 24 cm has a volume of 1232 cm^3, then the area of its curved surface is $\left(\text{take } \pi = \frac{22}{7}\right)$
    (a) 1254 cm^2 (b) 704 cm^2
    (c) 550 cm^2 (d) 154 cm^2

24. The area of a regular hexagon of side $2\sqrt{3}$ cm is
    (a) $18\sqrt{3}$ cm^2 (b) $12\sqrt{3}$ cm^2
    (c) $36\sqrt{3}$ cm^2 (d) $27\sqrt{3}$ cm^2

25. A pipe can fill a tank with water in 3 h. Due to leakage in bottom, it takes $3\frac{1}{2}$ h to fill it. In what time, the leak will empty the fully tank?
    (a) 12 h (b) 21 h (c) $6\frac{1}{2}$ h (d) $10\frac{1}{2}$ h

26. The largest three-digit number which is a perfect cube, is
    (a) 986 (b) 729 (c) 981 (d) 864

27. The LCM of two numbers is 280 and the ratio of the numbers is 7 : 8. Find the numbers.
    (a) 70 and 48 (b) 42 and 48
    (c) 35 and 40 (d) 28 and 32

28. A certain number of men complete a work in 45 days. If there were 5 men more, the work could be finished in 9 days less. How many men were originally there?
    (a) 30 (b) 15 (c) 25 (d) 20

Practice Set 2 / 859

29. The average age of a group of four men whose ages are in the ratio of 2 : 3 : 4 : 5 is 42 yr, what is the age of the eldest person in this group?
    (a) 60 yr (b) 48 yr (c) 36 yr (d) 24 yr

30. The price of an article decreased by 20% as a result of which the sale increased by 10%. What will be the effect on the total revenue of the shop?
    (a) 10% increase (b) 12% increase
    (c) 12% decrease (d) 10% decrease

31. A mixture of 20 L of milk and water contains 20% water. How much water should be added to this mixture so that the new mixture contains 25% water?
    (a) $1\frac{1}{3}$ L (b) $1\frac{1}{2}$ L (c) $1\frac{1}{4}$ L (d) $1\frac{1}{5}$ L

32. In a zoo, there are some pigeons and some rabbits. If their heads are counted, these are 100 and if their legs are counted, these are 320. How many pigeons are there?
    (a) 66 (b) 60 (c) 40 (d) 45

33. Working together, P and Q can do a job in 6 days, Q and R can do the same job in 10 days while P and R can do it in 5 days. How long will it take, if all of them work together to complete the job?
    (a) $4\frac{2}{7}$ days (b) $4\frac{3}{7}$ days
    (c) $4\frac{4}{7}$ days (d) $4\frac{5}{7}$ days

34. A sum becomes 10/9 times itself in 1 yr. Find the rate of simple interest.
    (a) $11\frac{1}{2}$% (b) $11\frac{1}{9}$%
    (c) $12\frac{1}{2}$% (d) $12\frac{1}{9}$%

35. The ratio of the areas of a square to that of the square drawn on its diagonal is
    (a) 1 : 1 (b) 1 : 2 (c) 1 : 3 (d) 1 : 4

36. The length of longest pole that can be placed in the floor of a room is 12 m and the length of longest pole that can be placed in the room is 15 m. The height of the room is
    (a) 3 m (b) 6 m (c) 9 m (d) 12 m

37. The ratio of heights of two cylinders is 3 : 2 and the ratio of their radii is 6 : 7. What is the ratio of their curved surface areas?
    (a) 9 : 7 (b) 1 : 1 (c) 7 : 9 (d) 7 : 4

38. Three unbiased coins are tossed. Find the probability of getting atleast two heads.
    (a) $\frac{2}{3}$ (b) $\frac{2}{5}$ (c) $\frac{1}{4}$ (d) $\frac{1}{2}$

39. $9 - 1\frac{2}{9}$ of $3\frac{3}{11} \div 5\frac{1}{7}$ of $\frac{7}{9}$ is equal to
    (a) 8 (b) 9
    (c) $8\frac{32}{81}$ (d) $\frac{3}{4}$

40. A shopkeeper earns a profit of 10% after allowing a discount of 20% on the marked price. The cost price of the article whose marked price is ₹ 880, is
    (a) ₹ 704 (b) ₹ 640
    (c) ₹ 774 (d) ₹ 680

Directions (Q. Nos. 41-45) *Each question below is followed by two Statements I and II. You are to determine whether the given data in the statement is sufficient for answering the question. You should use the data and your knowledge of Mathematics to choose between the possible answers.*
Give answer
   (a) if the Statement I alone is sufficient to answer the question but the Statement II alone is not sufficient
   (b) if the Statement II alone is sufficient to answer the question but the Statement I alone is not sufficient
   (c) if both Statements I and II together are needed to answer the question
   (d) if cannot get the answer from the Statements I and II together and need even more data

41. Find x.
    I. 20% of x is equal to the $\frac{1}{5}$th part of an another number.
    II. $\frac{7}{20}$th part of x is equal to the 35% of an another number.

42. Find the ratio of men, women and children of the city.
    I. The population of the city is 93280 in which 56100 are men.
    II. Ratio of the numbers of men and children is 5 : 2 and the number of women is double the number of children.

43. Find the SP of rice.
  I. 50 kg rice are bought for ₹ 3350 and ₹ 150 are spent in travelling.
  II. Profit was 5%.

44. What is the speed of train?
  I. Length of train is 120 m.
  II. It crosses an another train of length 180 m, in 4 s.

45. What is the CI after 3 yr?
  I. Rate is 5%.
  II. After two years, difference between CI and SI is ₹ 20.

Directions (Q. Nos. 46-50) *Study the following table carefully and answer the questions given below.*

Percentage of marks obtained by different students in different subjects

Students	Hindi (150)	English (150)	Math (150)	S.St. (150)	Science (75)	Marathi (50)	Physical Education (75)
A	88	85	86	74	78	80	85
B	92	80	79	82	70	70	97
C	75	89	85	90	83	91	85
D	63	66	69	71	85	64	62
E	80	76	89	95	79	70	73
F	69	81	86	76	69	85	76

46. How many marks did E get in all the subjects together?
  (a) 659
  (b) 600
  (c) 625
  (d) 708

47. What are the average marks obtained by all the students together in English?
  (a) 110.27
  (b) 113.76
  (c) 121.52
  (d) 119.25

48. How many students have scored the highest marks in more than one subject?
  (a) One
  (b) Two
  (c) Three
  (d) Four

49. Who has scored the highest marks in all the subjects together?
  (a) E
  (b) C
  (c) F
  (d) A

50. Marks obtained by F in S.St. are what per cent of marks obtained by E in the same subject?
  (a) 74
  (b) 85
  (c) 76
  (d) 80

Answers

1. (a)	2. (c)	3. (c)	4. (c)	5. (b)	6. (d)	7. (d)	8. (c)	9. (d)	10. (b)
11. (a)	12. (c)	13. (b)	14. (c)	15. (a)	16. (b)	17. (c)	18. (d)	19. (d)	20. (b)
21. (a)	22. (d)	23. (c)	24. (a)	25. (b)	26. (b)	27. (c)	28. (d)	29. (a)	30. (c)
31. (a)	32. (c)	33. (a)	34. (b)	35. (b)	36. (c)	37. (a)	38. (d)	39. (a)	40. (b)
41. (d)	42. (b)	43. (c)	44. (d)	45. (c)	46. (a)	47. (d)	48. (c)	49. (b)	50. (d)

Practice Set 3

Directions (Q. Nos. 1-6) *What will come in place of question mark (?) in the following number series?*

1. 311, 300, 278, 245, 201, 146, ?
   (a) 70   (b) 90   (c) 80   (d) 110
2. 17, 22, 32 47, 67, 92, ?
   (a) 112   (b) 132   (c) 111   (d) 122
3. 123, 183, 213, 228, 235.5, ?
   (a) 238.25   (b) 239.25
   (c) 275.50   (d) 238.50
4. 9, 20, 42, 75, 119, ?
   (a) 174   (b) 170   (c) 168   (d) 180
5. 23, 32, 45, 62, 83, ?
   (a) 116   (b) 108
   (c) 102   (d) 118
6. 17, 23, 35, 59, 107, ?
   (a) 217   (b) 223
   (c) 203   (d) 227
7. The average speed of a tractor is two-fifth the average speed of a car. The car covers 450 km in 6 h. How much distance will the tractor cover in 8 h?
   (a) 210 km/h
   (b) 240 km/h
   (c) 420 km/h
   (d) 480 km/h
8. The marks of six boys in a group are 48, 59, 87, 37, 78 and 57. What are the average marks of all six boys?
   (a) 62   (b) 64
   (c) 61   (d) 63
9. What is 74% of five-eighth of 1200?
   (a) 555   (b) 565
   (c) 445   (d) 455
10. Sumit purchased an item for ₹ 6500 and sold it at the gain of 24%. From that amount, he purchased another item and sold it at the loss of 20%. What is his over all gain/loss?
    (a) Loss of ₹ 42
    (b) Gain of ₹ 42
    (c) Loss of ₹ 52
    (d) Neither gain nor loss
11. How many sacks are required for filling 1026 kg of rice, if each sack is filled with 114 kg of rice?
    (a) 19   (b) 15   (c) 7   (d) 9
12. Mani's monthly income is three-fourth Rakhi's monthly income. Rakhi's monthly income is ₹ 38000. What is Mani's annual income?
    (a) ₹ 4.32 lakh   (b) ₹ 3.42 lakh
    (c) ₹ 3.22 lakh   (d) ₹ 4.22 lakh
13. What value will be obtained, if the square of 22 is subtracted from the cube of 12?
    (a) 1244   (b) 1344   (c) 1454   (d) 1354
14. The ratio between the present ages of Tarun and Varun is 3 : 7, respectively. After 4 yr, Varun's age will be 39 yr. What was Tarun's age 4 yr ago?
    (a) 12 yr   (b) 13 yr   (c) 19 yr   (d) 11 yr
15. The simple interest accrued in 3 yr on a principal of ₹ 25000 is three-twentieth the principal. What is the rate of simple interest per cent per annum?
    (a) 5   (b) 4   (c) 6   (d) 3
16. The sum of five consecutive even numbers is equal to 170. What is the sum of the second largest number amongs them and the square of the smallest number amongs them together?
    (a) 940   (b) 932   (c) 938   (d) 936
17. The area of a square is four times the area of a rectangle. The length of the rectangle is 25 cm and its breadth is 1 cm less than one-fifth its length. What is the perimeter of the square?
    (a) 40 cm   (b) 60 cm   (c) 80 cm
    (d) Couldn't be determined
18. Sohan got 54 marks in Hindi, 65 marks in Science, 89 marks in Maths, 69 marks in Social Science and 68 marks in English. The maximum marks of each subject are 100. How much over all percentage of marks did he get?
    (a) 74   (b) 69   (c) 68   (d) 72

19. 8 women can complete a work in 15 h. In how many hours will 12 women complete the same work?
    (a) 12 (b) 6
    (c) 8 (d) 10

20. 25% of Reena's yearly income is equal to 75% of Anubhav's monthly income. If Anubhav's yearly income is ₹ 240000, then what is Reena's monthly income?
    (a) ₹ 60000 (b) ₹ 12000 (c) ₹ 5000
    (d) Couldn't be determined

21. What will be the compound interest obtained on a principal amount of ₹ 5500 at the rate of 3% per annum after 2 yr?
    (a) ₹ 343.95 (b) ₹ 324.95
    (c) ₹ 354.95 (d) ₹ 334.95

22. The average of five numbers is 371.8. The average of the first and second numbers is 256.5 and the average of the fourth and fifth numbers is 508. Which of the following is the third number?
    (a) 360 (b) 310
    (c) 330 (d) 380

23. Find the sum of the squares of first 50 natural numbers.
    (a) 42925 (b) 42900
    (c) 42860 (d) 42875

24. The ratio of two numbers is 3 : 4 and their HCF is 4. Their LCM is
    (a) 12 (b) 16 (c) 24 (d) 48

25. Which is the least number which when doubled will be exactly divisible by 12, 18, 21 and 30?
    (a) 2520 (b) 1260 (c) 630 (d) 196

26. If $m : n = 3 : 2$, then $(4m + 5n) : (4m - 5n)$ is equal to
    (a) 4 : 9 (b) 9 : 4
    (c) 11 : 1 (d) 9 : 1

27. Two numbers are in the ratio 3 : 5. If 9 is subtracted from each, they are in the ratio 12 : 23. Find the smaller number.
    (a) 27 (b) 33
    (c) 49 (d) 55

28. A, B and C centered into a partnership. A invested ₹ 2560 and B ₹ 2000. At the end of the year, they gained ₹ 1105 out of which A got ₹ 320. C's capital was
    (a) ₹ 4280 (b) ₹ 2840
    (c) ₹ 4820 (d) ₹ 4028

29. A mixture contains spirit and water in the ratio 3 : 2. If it contains 32 L more than water, the quantity of spirit in the mixture is
    (a) 100 L (b) 120 L
    (c) 80 L (d) 96 L

30. On selling an article for ₹ 651, there is a loss of 7%. The cost price of that article is
    (a) ₹ 744 (b) ₹ 751
    (c) ₹ 793 (d) ₹ 700

31. The selling price of 5 articles is same as the cost price of 3 articles. The gain or loss is
    (a) 20% gain (b) 25% gain
    (c) 33.33% loss (d) 40% loss

32. A can do a work in 6 days and B in 9 days. How many days will both take together to complete the work?
    (a) 7.5 (b) 5.4
    (c) 3.6 (d) 3

33. Some carpenters promised to do a job in 9 days but 5 of them were absent and remaining men did the job in 12 days. The original number of carpenters was
    (a) 24 (b) 20
    (c) 16 (d) 18

34. A cistern has two pipes. One can fill it water in 8 h and other can empty it in 5 h. In how many hours will the cistern be emptied, if both the pipes are opened together when $\frac{3}{4}$ of the cistern is already full of the water?
    (a) $13\frac{1}{3}$ h (b) 10 h
    (c) 6 h (d) $3\frac{1}{3}$ h

35. A man walking at the rate of 5 km/h crosses a bridge in 15 min. The length of the bridge (in m) is
    (a) 600 (b) 750
    (c) 1000 (d) 1250

36. A track covers a distance of 550 m in 1 min, whereas a bus covers a distance of 33 km in 45 min. The ratio of their speeds is
    (a) 4 : 3 (b) 3 : 5
    (c) 3 : 4 (d) 50 : 3

Practice Set 3 / 863

37. A motorboat in still water travels at a speed of 36 km/h. It goes 56 km upstream in 1 h 45 min. The time taken by it to cover the same distance down the stream will be
    (a) 2 h 25 min   (b) 3 h
    (c) 1 h 24 min   (d) 2 h 21 min

38. Two trains are running in opposite directions with the same speed. If the length of each train is 120 m and they cross each other in 12 s, the speed (in km/h) of each train is
    (a) 72   (b) 10   (c) 36   (d) 18

39. The length of a plot is five times is breadth. A playground measuring 245 m² occupies half of the total area of the plot. What is the length of the plot?
    (a) $35\sqrt{2}$ m   (b) $175\sqrt{2}$ m
    (c) 490 m   (d) $5\sqrt{2}$ m

40. The area of a circle of a radius 5 is numerically, what per cent of its circumference?
    (a) 200   (b) 225   (c) 240   (d) 250

41. With a given data rate of simple interest, the ratio of principal and amount for a certain period of time is 4 : 5. After 3 yr with the same rate of interest, the ratio of principal and amount becomes 5 : 7. The rate of interest is
    (a) 4%   (b) 6%   (c) 5%   (d) 7%

42. The compound interest on a certain sum of money at 5% for 2 yr is ₹ 328. The simple interest on that sum at the same rate and for the same period of time will be
    (a) ₹ 320   (b) ₹ 322   (c) ₹ 325   (d) ₹ 326

Directions (Q. Nos. 43-45) *What should come in place of the question mark (?) in the following questions?*

43. $(6)^2 + (8)^2 \times (2)^2 - (9)^2 = ?$
    (a) 215   (b) 209   (c) 221   (d) 211

44. $7008 \div 24 + 6208 \div 16 = ?$
    (a) 640   (b) 720   (c) 700   (d) 680

45. $\frac{3}{4}$th of $\frac{3}{5}$th of $\frac{2}{3}$rd of ? = 3174
    (a) 10550   (b) 10540   (c) 10580   (d) 10500

Directions (Q. Nos. 46-50) *The circle graph (pie chart) given here shows the spendings of a country on various sports during a particular year. Study the graph carefully and answer the questions given below it.*

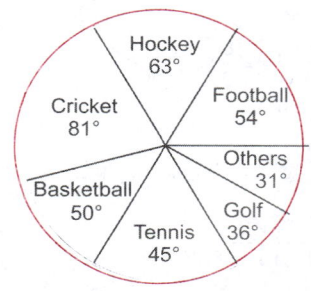

Hockey 63°
Football 54°
Others 31°
Golf 36°
Tennis 45°
Basketball 50°
Cricket 81°

46. What per cent of the total spendings is spent on tennis?
    (a) $12\frac{1}{2}$%   (b) $22\frac{1}{2}$%   (c) 25%   (d) 45%

47. How much per cent more is spent on hockey than that on golf?
    (a) 27%   (b) 35%   (c) 37.5%   (d) 75%

48. How much per cent less is spent on football than that on cricket?
    (a) $22\frac{2}{9}$%   (b) 27%   (c) $33\frac{1}{3}$%   (d) $37\frac{1}{2}$%

49. If the total amount spent on sports during the year was ₹ 2 crore, then amount spent on cricket and hockey together was
    (a) ₹ 800000   (b) ₹ 8000000
    (c) ₹ 12000000   (d) ₹ 16000000

50. If the total amount spent on sports during the year be ₹ 18000000, then amount spent on basketball exceeds that on tennis by
    (a) ₹ 250000   (b) ₹ 360000
    (c) ₹ 375000   (d) ₹ 410000

Answers

1. (c)	2. (d)	3. (b)	4. (a)	5. (b)	6. (c)	7. (b)	8. (c)	9. (a)	10. (c)
11. (d)	12. (b)	13. (a)	14. (d)	15. (a)	16. (d)	17. (c)	18. (b)	19. (d)	20. (c)
21. (d)	22. (c)	23. (a)	24. (d)	25. (c)	26. (c)	27. (b)	28. (a)	29. (d)	30. (d)
31. (d)	32. (c)	33. (b)	34. (b)	35. (d)	36. (c)	37. (c)	38. (c)	39. (a)	40. (d)
41. (c)	42. (a)	43. (d)	44. (d)	45. (c)	46. (a)	47. (d)	48. (c)	49. (b)	50. (a)

Practice Set 4

Directions (Q. Nos. 1-14) *What should come in place of the question mark (?) in the following questions?*

1. $4\frac{1}{2} + \left(1 \div 2\frac{8}{9}\right) - 3\frac{1}{13} = ?$
   (a) $1\frac{9}{26}$ (b) $2\frac{7}{13}$
   (c) $1\frac{11}{26}$ (d) $1\frac{10}{13}$

2. $\dfrac{6 \times 136 \div 8 + 132}{628 \div 16 - 26.25} = ?$
   (a) 15   (b) 24   (c) 18   (d) 12
   (e) 28

3. $\{(441)^{1/2} \times 207 \times (343)^{1/3}\} \div \{(14)^2 \times (529)^{1/2}\} = ?$
   (a) $6\frac{1}{2}$ (b) $5\frac{1}{2}$ (c) $5\frac{3}{4}$ (d) $6\frac{3}{4}$

4. $\{\sqrt{7744} \times (11)^2\} \div (2)^3 = (?)^3$
   (a) 7   (b) 9   (c) 11   (d) 13

5. $(4356)^{1/2} \div \dfrac{11}{4} = \sqrt{?} \times 6$
   (a) 2   (b) 4   (c) 8   (d) 16

6. $\dfrac{3}{8}$ of $\{4624 \div (564 - 428)\} = ?$
   (a) $13\frac{1}{4}$ (b) $14\frac{1}{2}$ (c) $11\frac{5}{6}$ (d) $12\frac{3}{4}$

7. $456 \div 24 \times 38 - 958 + 364 = ?$
   (a) 112   (b) 154   (c) 128   (d) 136

8. $(43)^2 + 841 = (?)^2 + 1465$
   (a) 41   (b) 35   (c) 38   (d) 33

9. $3\frac{3}{8} \times 6\frac{5}{12} - 2\frac{3}{16} \times 3\frac{1}{2} = ?$
   (a) 21   (b) 18
   (c) 14   (d) 15

10. $(34.5 \times 14 \times 42) \div 2.8 = ?$
    (a) 7150   (b) 7365   (c) 7245   (d) 7575

11. $(216)^4 \div (36)^4 \times (6)^5 = (6)^?$
    (a) 13   (b) 11   (c) 7   (d) 9

12. $\dfrac{\sqrt{4356} \times \sqrt{?}}{\sqrt{6084}} = 11$
    (a) 144   (b) 196   (c) 169   (d) 81

13. $\left(3\dfrac{6}{17} \div 2\dfrac{7}{34} - 1\dfrac{9}{25}\right) = (?)^2$
    (a) 2/5   (b) 1/3   (c) 4/5   (d) 1/5

14. $(1097.63 + 2197.36 - 2607.24) \div 3.5 = ?$
    (a) 211.5   (b) 196.5
    (c) 209.5   (d) 192.5

15. The LCM of two numbers is 20 times their HCF. The sum of HCF and LCM is 2520. If one of the numbers is 480, the other number is
    (a) 400   (b) 480   (c) 520   (d) 600

16. Three numbers which are coprimes to one another are such that the product of the first two is 551 and that of the last two is 1073. The sum of the three numbers is
    (a) 75   (b) 81   (c) 85   (d) 89

17. Number of students studying in colleges A and B are in the ratio of 3 : 4 respectively. If 50 more students join college A and there is no change in the number of students in college B, the respective ratio becomes 5 : 6. What is the number of students in college B?
    (a) 450   (b) 500   (c) 400   (d) 600

18. Cost of 12 belts and 30 wallets is ₹ 8940. What is the cost of 4 belts and 10 wallets?
    (a) ₹ 2890   (b) ₹ 2980
    (c) ₹ 2780   (d) ₹ 2870

19. 75% of a number is equal to three-seventh of another number. What is the ratio between the first number and the second number respectively?
    (a) 4 : 7   (b) 7 : 4
    (c) 12 : 7   (d) 7 : 12

20. A 275 m long train crosses a platform of equal length in 33 s. What is the speed (in km/h) of the train?
    (a) 66   (b) 60   (c) 64   (d) 72

Practice Set 4 / 865

21. Present ages of father and the son are in the ratio of 6 : 1 and 4 yr hence, the ratio of their ages will become 4 : 1 respectively. What is the son's present age?
    (a) 10 yr (b) 6 yr (c) 4 yr (d) 8 yr

22. Ghanshyam purchased an article for ₹ 1850. At what price should he sell it so that 30% profit is earned?
    (a) ₹ 2450 (b) ₹ 2245
    (c) ₹ 2405 (d) ₹ 2425

23. Cost of 6 dozen apples and 8 dozen bananas is ₹ 1400. What will be the cost of 15 dozen apples and 20 dozen bananas?
    (a) ₹ 3200 (b) ₹ 3500
    (c) ₹ 3600 (d) ₹ 4200

24. Beena and Meena started a boutique investing amounts of ₹ 35000 and ₹ 56000, respectively. If Beena's share in the profit earned by them is ₹ 45000, what is the total profit earned?
    (a) ₹ 81000 (b) ₹ 127000
    (c) ₹ 72000 (d) ₹ 117000

25. Latika spends 45% of her monthly income on food and 30% of the monthly income on transport. Remaining amount of ₹ 4500 she saves. What is her monthly income?
    (a) ₹ 16000 (b) ₹ 18000
    (c) ₹ 16500 (d) ₹ 18500

26. The amount of simple interest accrued on an amount of ₹ 28500 in seven years is ₹ 23940, what is the rate of interest percentage per annum?
    (a) 10.5 (b) 12.5 (c) 11 (d) 12

27. Mr. Sharma invested an amount of ₹ 25000 in fixed deposit at the rate of compound interest 8% per annum for two years. What amount Mr. Sharma will get on maturity?
    (a) ₹ 28540 (b) ₹ 29160
    (c) ₹ 29240 (d) ₹ 28240

28. Four-seventh of a number is equal to 40% of another number. What is the ratio between the first number and the second number respectively?
    (a) 5 : 4 (b) 4 : 5
    (c) 10 : 7 (d) 7 : 10

29. Which of the following has the fractions in ascending order?
    (a) $\frac{5}{11}, \frac{3}{8}, \frac{4}{9}, \frac{2}{7}$ (b) $\frac{5}{11}, \frac{4}{9}, \frac{3}{8}, \frac{2}{7}$
    (c) $\frac{2}{7}, \frac{3}{8}, \frac{4}{9}, \frac{5}{11}$ (d) $\frac{2}{7}, \frac{4}{9}, \frac{3}{8}, \frac{5}{11}$

30. Sum of the digits of a two-digit number is 8 and the digit in the ten's place is three times the digit in the unit's place. What is the number?
    (a) 26 (b) 36
    (c) 71 (d) 62

31. A started a business with a capital of ₹ 100000. 1 yr later, B joined him with a capital of ₹ 200000. At the end of 3 yr from the start of the business, the profit earned was ₹ 84000. The share of B in the profit exceeded the share of A, is
    (a) ₹ 10000 (b) ₹ 12000
    (c) ₹ 14000 (d) ₹ 15000

32. The wheel of a motorcar makes 1000 revolutions in moving 440 m. Then, the diameter (in m) of the wheel is
    (a) 0.44 (b) 0.14 (c) 0.24 (d) 0.34

33. A container contains 60 kg of milk from this container 6 kg of milk was taken out and replaced by water. This process was repeated further two times. The amount of milk left in the container is
    (a) 34.24 kg (b) 39.64 kg
    (c) 43.74 kg (d) 47.6 kg

34. In what time will ₹ 1000 amount to ₹ 1331 at 20% per annum, compounded half yearly?
    (a) $1\frac{1}{2}$ yr (b) 2 yr
    (c) 1 yr (d) $2\frac{1}{2}$ yr

35. If 6 persons working 8 h a day earn ₹ 8400 per week, then 9 persons working 6 h a day will earn per week
    (a) ₹ 8400 (b) ₹ 16800
    (c) ₹ 9450 (d) ₹ 16200

36. Two pipes A and B can separately fill a cistern in 60 min and 75 min, respectively. There is a third pipe in the bottom of the cistern to empty it. If all the three pipes are simultaneously opened, then the cistern is full in 50 min. In how much time the third pipe alone can empty the cistern?
    (a) 110 min (b) 100 min
    (c) 120 min (d) 90 min

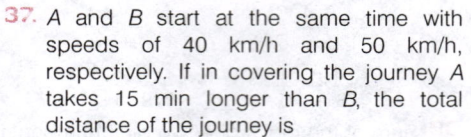

37. A and B start at the same time with speeds of 40 km/h and 50 km/h, respectively. If in covering the journey A takes 15 min longer than B, the total distance of the journey is
   (a) 45 km (b) 48 km
   (c) 50 km (d) 52 km

Directions (Q. Nos. 38-41) *What should come in place of the question mark (?) in the following questions?*

38. 37, ?, 103, 169, 257, 367
   (a) 61 (b) 59
   (c) 67 (d) 55

39. 3, ?, 14, 55, 274, 1643
   (a) 11 (b) 5
   (c) 6 (d) 8

40. 960, 839, 758, 709, ?, 675
   (a) 696 (b) 700
   (c) 688 (d) 684

41. 61, 72, ?, 73, 59, 74, 58
   (a) 70 (b) 60 (c) 71 (d) 62

42. A boat running downstream covers a distance of 20 km in 2 h while it covers the same distance upstream in 5 h. Then, the speed of the boat in still water is
   (a) 7 km/h (b) 8 km/h
   (c) 9 km/h (d) 10 km/h

43. The difference between the length and breadth of a rectangle is 23 m. If its perimeter is 206 m, then its area is
   (a) 1520 m^2 (b) 2420 m^2
   (c) 2480 m^2 (d) 2520 m^2

44. If the diagonals of two squares are in the ratio 2 : 5, their areas will be in the ratio of
   (a) $\sqrt{2} : \sqrt{5}$ (b) 2 : 5
   (c) 4 : 25 (d) 4 : 5

45. If $A = \frac{4}{5}$ of B and $B = \frac{5}{2}$ of C. Then, the ratio of A : C is
   (a) 1 : 2 (b) 2 : 1
   (c) 2 : 3 (d) 1 : 3

Directions (Q. Nos. 46-50) *Study the following graph carefully and answer the questions given below.*

Production of three types of tyres by a company over the years (in lakh)

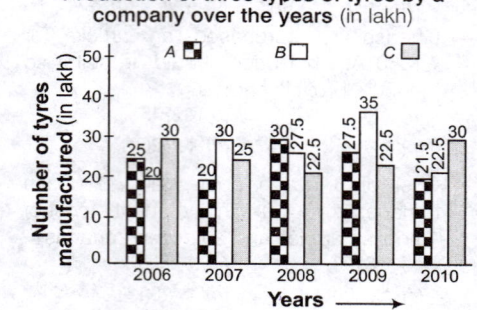

46. What was the percentage drop in the number of C type tyres manufactured from 2007 to 2008?
   (a) 25% (b) 10%
   (c) 15% (d) 2.5%

47. What was the difference between the number of B type tyres manufactured in 2008 and 2009?
   (a) 100000 (b) 2000000
   (c) 1000000 (d) None of these

48. The total number of all the three types of tyres manufactured was the least in which of the following years?
   (a) 2009 (b) 2010
   (c) 2006 (d) 2008

49. In which of the following years, was the percentage production of B type to C type the maximum?
   (a) 2008 (b) 2009 (c) 2010 (d) 2007

50. The total production of C type tyres in 2006 and 2007 together was what percentage of production of B type tyres in 2008?
   (a) 50% (b) 100%
   (c) 150% (d) 200%

Answers

1. (d)	2. (c)	3. (d)	4. (c)	5. (d)	6. (d)	7. (c)	8. (b)	9. (c)	10. (c)
11. (d)	12. (c)	13. (a)	14. (b)	15. (d)	16. (c)	17. (d)	18. (b)	19. (a)	20. (b)
21. (b)	22. (c)	23. (b)	24. (d)	25. (b)	26. (d)	27. (b)	28. (d)	29. (c)	30. (d)
31. (b)	32. (b)	33. (c)	34. (a)	35. (c)	36. (b)	37. (c)	38. (b)	39. (b)	40. (d)
41. (b)	42. (a)	43. (d)	44. (c)	45. (b)	46. (b)	47. (d)	48. (b)	49. (b)	50. (d)

Practice Set 5

1. Three bells ring at intervals of 9, 12 and 15 min. All the three began to ring at 8 : 00 am. At what time will they ring together again?
   (a) 8 : 45 am (b) 10 : 30 am
   (c) 11 : 00 am (d) 1 : 30 am

2. Groups, each containing 3 boys are to be formed out of 5 boys, A, B, C, D and E such that no group can contain both C and D together. What is the maximum number of such different groups?
   (a) 5 (b) 6
   (c) 7 (d) 8

3. Which one of the following is rational number?
   (a) 1.010010001.....
   (b) 2.371371.....
   (c) $\dfrac{\sqrt{2}-1}{\sqrt{2}+1}$
   (d) π

4. Which one of the following is the correct sequence in respect of the roman numerals C, D, L and M?
   (a) C > D > L > M
   (b) M > L > D > C
   (c) M > D > C > L
   (d) L > C > D > M

5. In the given figure, $\angle OQP = 30°$ and $\angle ORP = 20°$, then $\angle QOR$ is equal to

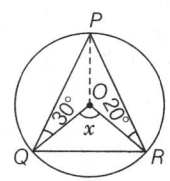

   (a) 100° (b) 120°
   (c) 130° (d) 140°

6. PQ is a chord of length 8 cm of a circle of radius 5 cm. The tangents at P and Q intersect at a point T, find the length TP.

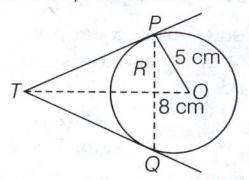

   (a) $\dfrac{30}{4}$ cm (b) $\dfrac{20}{3}$ cm
   (c) $\dfrac{18}{4}$ cm (d) $\dfrac{22}{3}$ cm

7. In the following distribution, mean is

x	3	4	5	6	7	8	9	10
f	2	4	2	3	5	4	3	7

   (a) 10 (b) 7
   (c) 7.1 (d) 6.5

8. The mode of the following data is

Class interval	Frequency
10-25	2
25-40	3
40-55	7
55-70	6
70-85	6
85-100	6

   (a) 41.5 (b) 52.0
   (c) 54.0 (d) 50.5

9. The denominator of fraction is one more than twice the numerator. If the sum of fraction and their reciprocal is $2\dfrac{16}{21}$, then the value of fraction is
   (a) $\dfrac{7}{3}$ (b) $\dfrac{3}{7}$
   (c) $\dfrac{4}{9}$ (d) $\dfrac{5}{11}$

10. If the zeroes of the polynomials $x^3 - 3x^2 + x + 1 = 0$ are $a - b, a, a + b$, then b is equal to
    (a) $\sqrt{2}$ (b) 1
    (c) -1 (d) $\sqrt{2} - 1$

11. Find the value of $\sqrt{42 + \sqrt{42 + \sqrt{42 + ...}}}$ upto ∞.
    (a) 7 (b) -7
    (c) -6 (d) None of these

12. The average of x_1, x_2 and x_3 is 14. Twice the sum of x_2 and x_3 is 30. What is the value of x_1?
    (a) 20 (b) 27
    (c) 16 (d) 2

13. A city has a population of 300000 out of which 180000 are males. 50% of the population is literate. If 70% of the males are literate, then the number of literate females are
    (a) 24000 (b) 30000
    (c) 54000 (d) 60000

14. If $\frac{5}{12}$ of $\frac{2}{5}$ of 20% of a number is 252, then the number is
    (a) 5040 (b) 2520
    (c) 7560 (d) 7650

15. The value of $\sqrt{48} \times \sqrt{192} \times \sqrt{225}$ is
    (a) 1320 (b) 1440 (c) 1560 (d) 1680

16. The value of $\left[\left\{\left(\frac{-1}{3}\right)^2\right\}^{-2}\right]^{-1}$ is
    (a) $\frac{-1}{81}$ (b) $\frac{1}{81}$
    (c) -81 (d) 81

17. If $9^{3n+2} = 27^{5n+1}$, then the value of n is
    (a) $\frac{1}{2}$ (b) $\frac{2}{9}$ (c) $\frac{1}{6}$ (d) $\frac{1}{9}$

18. A person purchases 100 pens at a discount of 10%. Then, net amount of money spent by the person to purchase the pens is ₹ 600. The selling expenses incurred by the person are 15% on the net cost price. What should be the selling price for 100 pens in order to earn a profit of 25%?
    (a) ₹ 802.50 (b) ₹ 811.25
    (c) ₹ 862.50 (d) ₹ 875

19. If Sohan, while selling two goats at the same price makes a profit of 10% on one goat and suffers a loan of 10% on the other
    (a) he makes no profit and no loss
    (b) he makes a profit of 1%
    (c) he suffers a loss of 1%
    (d) he suffers a loss of 2%

20. If $X : Y = 4 : 7$ and $Y : Z = 5 : 11$, then $X : Y : Z$ is equal to
    (a) 4 : 35 : 55 (b) 20 : 35 : 77
    (c) 4 : 35 : 77 (d) 35 : 20 : 77

21. Amit started a business by investing ₹ 30000. Rahul joined the business after sometime and invested ₹ 20000. At the end of the year, profit was divided in the ratio of 2 : 1. After how many months did Rahul join the business?
    (a) 2 months (b) 3 months
    (c) 4 months (d) 5 months

22. A and B can complete work together in 5 days. If A work twice his speed and B at half of his speed, this work can be finished in 4 days. How many days would it take for A done to complete the job?
    (a) 10 days (b) 12 days
    (c) 15 days (d) 18 days

23. If it takes 15 women to sow an acre in two days how many women does it take to sow three acres in a day?
    (a) 60 (b) 90
    (c) 120 (d) 180

24. A person travels from X to Y at a speed of 40 km/h and returns by increasing his speed of 50%. What is his average speed of both the trips?
    (a) 36 km/h (b) 45 km/h
    (c) 48 km/h (d) 50 km/h

25. If $\left(a + \frac{1}{a}\right)^2 = 3$, then what is the value of $a^3 + \frac{1}{a^3}$?
    (a) $3\sqrt{3}$ (b) 3 (c) 0 (d) 9

26. If $x^2 - 3x + 1 = 0$, then find the value of $x + \frac{1}{x}$.
    (a) 0 (b) 2
    (c) 3 (d) None of these

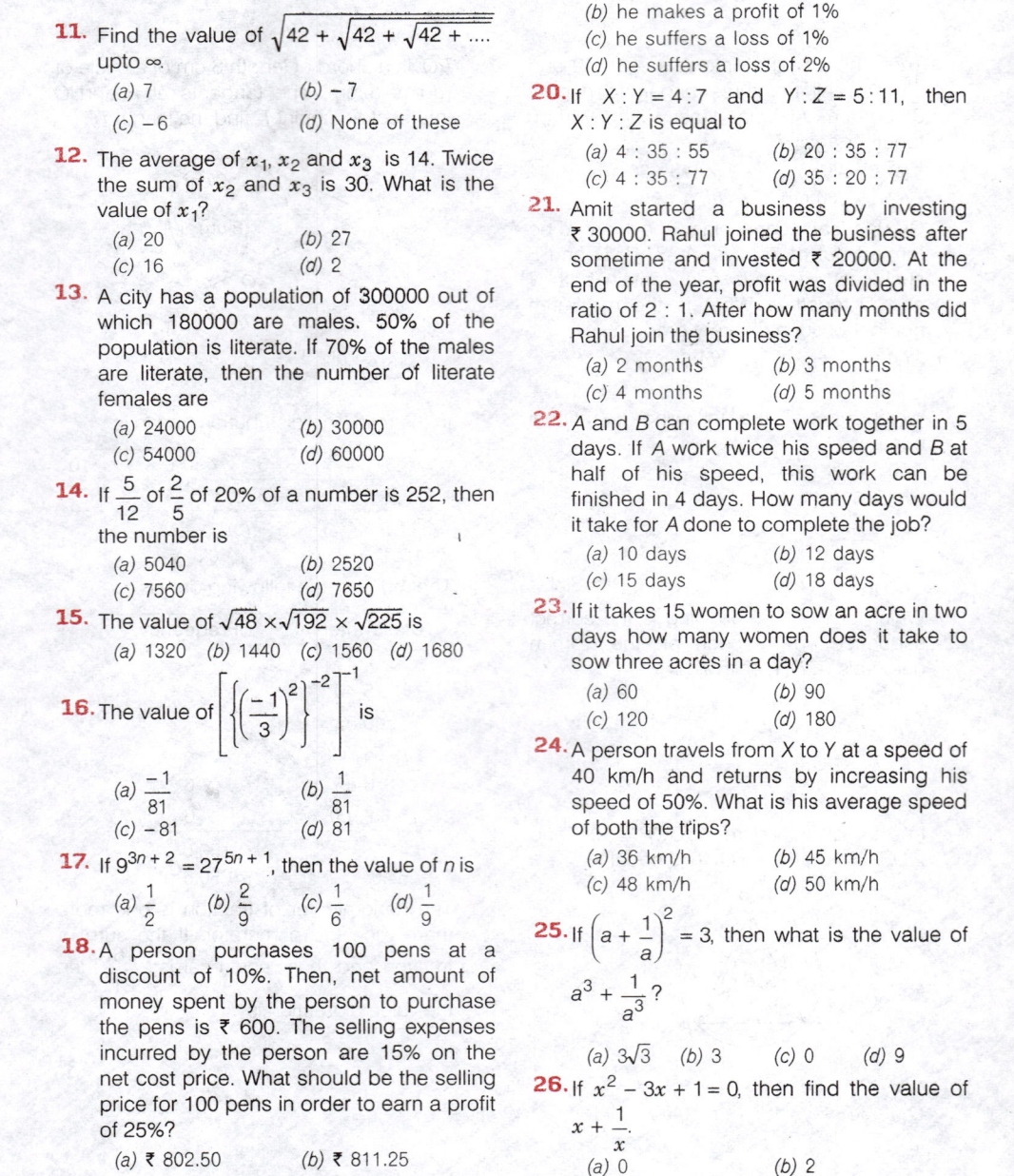

27. If $x - y = -1$, then find the value of $x^3 - y^3 + 3xy$.

    (a) 1   (b) –1   (c) 3   (d) –3

28. A bus is moving at a speed of 30 km/h ahead of a car with speed of 50 km/h. How many kilometres apart are they, if it takes 15 min for the car to catch up the bus?

    (a) 5 km   (b) 7.5 km
    (c) 12.5 km   (d) 15 km

29. A train travels at a certain average speed for a distance of 63 km and then travels a distance of 72 km at an average speed of 6 km/h more than its original speed. If it takes 3 h to complete the total journey, then what is the original speed of the train?

    (a) 24 km/h   (b) 33 km/h
    (c) 42 km/h   (d) 66 km/h

30. The area of a trapezium is 384 cm². If its parallel sides are in the ratio 3 : 5 and the perpendicular distance between them is 12 cm, the smaller of the parallel sides is

    (a) 20 cm   (b) 24 cm
    (c) 30 cm   (d) 36 cm

31. The perimeter of the triangular base of a right prism is 60 cm and the sides of the base are in the ratio 5 : 12 : 13. Then, its volume will be (height of the prism being 50 cm)

    (a) 6000 cm³   (b) 6600 cm³
    (c) 5400 cm³   (d) 9600 cm³

32. If the mean of 4, 5 a, 6, b, 9 and 11 is 10, then find the value of $(a + b)$.

    (a) 25   (b) 15   (c) 35   (d) 45

33. Find the value of k, if the mean of the following distribution is 20.

x	15	17	19	20 + k	23
f	2	3	4	5k	6

    (a) 5   (b) 4   (c) 2   (d) 1

34. Water flows out through a circular pipe whose internal diameter is 2 cm, at the rate of 6 m/s into a cylindrical tank, the radius of whose base is 60 cm. By how much will the level of water rise in 30 min?

    (a) 3 m   (b) 3.5 m   (c) 4 m   (d) 4.5 m

35. A hollow garden roller 63 cm wide with a girth of 440 cm is made of iron 4 cm thick. The volume of iron used is

    (a) 107712 cm³   (b) 170112 cm³
    (c) 102217 cm³   (d) 107212 cm³

36. A village having the population of 4000, requires 150 L of water per head per day. It has a tank measuring $20 \times 15 \times 6$ m³. For how many days will the water of this tank last?

    (a) 2 days (b) 3 days (c) 1 day (d) 4 days

37. The given venn diagram shows the number of students, who failed in an examination comprising papers of English, Hindi and Mathematics. The total number of students who took the test is 500. What is the percentage of students, who failed in atleast two subjects?

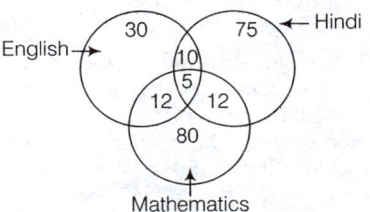

    (a) 8%   (b) 8.4%   (c) 7.8%   (d) 7.6%

38. In an examination, 70% of the students passed in the paper I and 60% of the students passed in the paper II. 15% of the students failed in both the papers, while 270 students passed in both the papers. What is the total number of students?

    (a) 600   (b) 580   (c) 560   (d) 540

39. How many terms of the GP $3, \frac{3}{2}, \frac{3}{4}, ...$ are needed to give the sum $\frac{3069}{512}$?

    (a) 20   (b) 30   (c) 25   (d) 10

40. A person has 2 parents, 4 grandparents, 8 great grandparents and so on. Find the number of his ancestors during the ten generations preceding his own.

    (a) 2046   (b) 1846   (c) 2056   (d) 2066

41. If AM and GM of two positive numbers a and b are 10 and 8 respectively, then find the numbers.

    (a) 4, 16   (b) 8, 32   (c) 2, 8   (d) 8, 2

870 / Fast Track Objective Arithmetic

42. A square is divided into 9 identical smaller squares. Six identical balls are to be placed in these smaller squares such that each of the three rows gets atleast one ball (one ball in one square only). In how many different ways can this be done?
    (a) 27   (b) 36   (c) 54   (d) 81

43. In how many different ways can six players be arranged in a line such that two of them, Ajeet and Mukherjee are never together?
    (a) 120   (b) 240   (c) 360   (d) 480

44. Numbers 1 to 25 are marked on tokens of equal size one on each. If a token is drawn at random, then find the probability of getting a number divisible by both 2 and 3.
    (a) $\frac{4}{25}$ (b) $\frac{7}{25}$ (c) $\frac{2}{25}$ (d) $\frac{9}{25}$

45. Three students are picked at random from a school having a total of 1000 students. The probability that these students will have identical date and month of their birth, is
    (a) $\frac{3}{1000}$ (b) $\frac{3}{365}$
    (c) $\frac{1}{(365)^2}$ (d) None of these

46. A person has only ₹ 1 and ₹ 2 coins with her. If the total number of coins that she has, is 50 and the amount of money with her is ₹ 75, then the number of ₹ 1 and ₹ 2 coins are, respectively
    (a) 15 and 35   (b) 35 and 15
    (c) 30 and 20   (d) 25 and 25

47. A two-digit number is such that the product of the digits is 8. When 18 is added to the number, then the digits are reversed. The number is
    (a) 18   (b) 24
    (c) 42   (d) 81

48. Consider the following statements
    If p is a prime such that $p + 2$ is also a prime, then
    I. $p(p + 2) + 1$ is a perfect square.
    II. 12 is a divisor of $p + (p + 2)$, if $p > 3$.
    Which of the above statements is/are correct?
    (a) Only I
    (b) Only II
    (c) Both I and II
    (d) Neither I nor II

49. By adding x to 1254934, the resulting number becomes divisible by 11, while adding y to 1254934 makes the resulting number divisible by 3. Which one of the following is the set of values for x and y?
    (a) $x = 1, y = 1$
    (b) $x = 1, y = -1$
    (c) $x = -1, y = 1$
    (d) $x = -1, y = -1$

50. In an examination, every candidate took Physics or Mathematics or both, 65.8% took Physics and 59.2% took mathematics. The total number of candidates was 2000. How many candidates took both Physics and Mathematics?
    (a) 750   (b) 500
    (c) 250   (d) 125

Answers

1. (c)	2. (c)	3. (b)	4. (c)	5. (a)	6. (b)	7. (c)	8. (b)	9. (b)	10. (a)
11. (a)	12. (b)	13. (a)	14. (c)	15. (b)	16. (b)	17. (d)	18. (c)	19. (c)	20. (b)
21. (b)	22. (a)	23. (b)	24. (c)	25. (c)	26. (c)	27. (b)	28. (a)	29. (c)	30. (b)
31. (a)	32. (c)	33. (d)	34. (a)	35. (a)	36. (b)	37. (c)	38. (a)	39. (d)	40. (a)
41. (a)	42. (d)	43. (d)	44. (a)	45. (c)	46. (d)	47. (b)	48. (c)	49. (b)	50. (b)